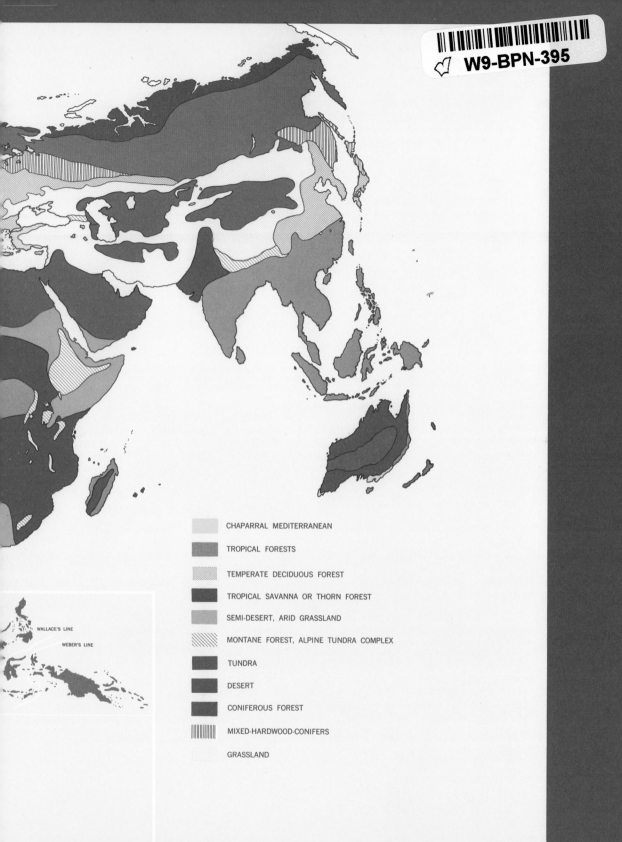

WALLACE'S LINE

WEBER'S LINE

CHAPARRAL MEDITERRANEAN

TROPICAL FORESTS

TEMPERATE DECIDUOUS FOREST

TROPICAL SAVANNA OR THORN FOREST

SEMI-DESERT, ARID GRASSLAND

MONTANE FOREST, ALPINE TUNDRA COMPLEX

TUNDRA

DESERT

CONIFEROUS FOREST

MIXED-HARDWOOD-CONIFERS

GRASSLAND

ECOLOGY AND FIELD BIOLOGY

FOURTH EDITION

ECOLOGY AND FIELD BIOLOGY

Robert Leo Smith
West Virginia University

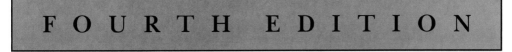 HarperCollins*Publishers*

To the F$_2$ generation
Patricia, Gretchen, Erin, and Alexandra

Sponsoring Editor: Glyn Davies
Project Editor: Paula Cousin
Art Direction: Kathie Vaccaro
Text Design: Caliber Design Planning, Inc.
Cover Coordinator: Mary Archondes
Cover Design: Heather A. Ziegler
Cover Illustration: Robert Leo Smith, Jr.
Production: Kewal Sharma

Ecology and Field Biology, Fourth Edition

Library of Congress Cataloging-in-Publication Data

Smith, Robert Leo.
   Ecology and field biology.

   Includes bibliographical references.
   1. Ecology.   2. Biology—Field work.   I. Title.
QH541.S6   1990   574.5   89-26801
ISBN 0-06-046331-7 (student ed.)
ISBN 0-06-046332-5 (teacher ed.)

93   94   95   96   97   CWI   10   9   8   7

BRIEF CONTENTS

DETAILED CONTENTS

Preface

Ten years have passed since the third edition of *Ecology and Field Biology* appeared. Over that decade, ecology experienced an exponential expansion in research, published papers, new books, and new journals. Old concepts faded away and new concepts emerged. For most of the decade, population ecology dominated the field; but now, as the 1980s have faded into the 1990s, the growing problems of environment are pushing ecosystem ecology to the forefront again, but with a difference. There is a growing fusion between theoretical and population ecology and ecosystem ecology as major environmental problems—including global warming, fragmentation of habitats, and insularization of populations—and decline in biodiversity call for an integrated approach to find ecological solutions acceptable to a nonecological society.

The great expansion in ecology and the age of the book demanded a thorough revision and rewrite of *Ecology and Field Biology,* a task that has taken me three years. Although at a cursory glance the book may appear the same, I have made major organizational and content changes with the addition of new chapters, and a major expansion of the material in population ecology.

Writing an ecology text has become a challenging task. The fragmentation of ecology into specialized subdisciplines with their own journals and books requires many hours devoted to reading and becoming acquainted with new concepts and advances in many areas. Keeping up with the literature is more difficult and demanding than ever. While most ecologists become more specialized and restrict their reading and research to a narrow area of interest, a textbook author is forced to become more and more a generalist, ranging over the broad landscape of ecology, attempting to make something of developing patterns, discovering interrelations among the various fields, and synthesizing that information into some kind of coherent whole.

You have two choices in writing an ecology text. The first and the easiest is to write about what you know the best and ignore or skim over the rest of the field. The result is an idiosyncratic book reflecting the author's own, often narrow, view of ecology. Such a book is inflexible and forces the user to accept that author's concept of the subject. The second, more difficult approach is to write a general text reflecting the broad diversity that makes ecology such an exciting field. A text in general ecology should provide an overview of the field just as texts in general biology, general chemistry, and general physics do for those fields. Such texts provide the basic ideas and concepts with which students should be familiar. A general ecology text should do no less for ecology.

My objectives in writing *Ecology and Field Biology* are to provide a balanced overview of the various areas of ecology—population, evolutionary, and ecosystems ecology, theoretical and applied, plant and animal—to explore the various areas of ecology in some detail, to provide a basic reference source in ecology, and to produce a readable, flexible text around which instructors can develop their own courses according to the time available, and to aid in that flexibility by providing numerous chapter cross-references throughout the book.

Unfortunately, many curricula restrict ecology to a one-quarter or one-semester course. It is no longer possible to cover ecology in one semester, let alone one quarter. Under such restrictions all one can do is present selected topics in ecology. My own required course in general ecology five years ago expanded from a one-semester, 4-hour course with laboratory to a two-semester course of 3 hours, each with laboratory. Such a broad background is necessary before students move on to more specialized advanced courses in ecology.

In revising a book in a field that is growing as rapidly as ecology, the author faces the task of deciding what to eliminate, what to keep, and what to add. The tendency is to add and as a result revisions tend to become thicker. Population ecology now makes up a significant portion of this fourth edition, and the treatment is more extensive and rigorous than in the previous edition. I eliminated the material on species and speciation to make room for new material on population genetics most applicable to the current concerns of conservation biology and biodiversity—genetic drift, inbreeding, and minimum viable populations. I eliminated also the chapter on behavioral ecology, not because behavioral ecology is unimportant, but because many of its aspects, such as optimal foraging, spacing and distribution, mating systems, and parental care, have become so closely linked with population ecology that it makes no sense to separate those topics from population

ecology. I incorporated much of the material from the behavioral ecology chapter in the chapters on life history patterns, predation, and mutualism. I have added new chapters on herbivory, parasitism and disease, and mutualism.

Other sections also were thoroughly rewritten and updated. I broke up the material on community ecology into several chapters, adding one new one on disturbance and patch dynamics, an important new area of study that is beginning to link ecosystem and population ecology. I expanded the material on paleoecology to include recent studies on vegetation development as influenced by the Pleistocene glaciation.

Part V on comparative ecosystem ecology was moved from the middle to the back of the book, because it is based on many of the ecological concepts considered throughout the rest of the text.

As the title of Part V, "Comparative Ecosystem Ecology," implies, the chapters provide a comparative study of the major ecosystems relative to structure and function, including differences in physical structure, microclimates, energy flow, nutrient cycling, and adaptation of organisms to those environments. The common thread through all these chapters is the nitrogen cycle that serves to emphasize the functional differences among the various major ecosystems.

The material in Part V not only introduces ecological concepts unique to certain ecosystems, such as nutrient spiraling in streams and microbial loops in marine ecosystems, but it also provides a basic understanding of major problems receiving so much attention in ecology and environmental conservation—the destruction of old growth forest, tropical deforestation, wetland drainage, habitat fragmentation, water pollution, acidification, and the like.

I added considerable new material to the text, some of it found in no other ecology texts. These topics include experimental approaches to ecology, effects of flooding on plants, heterothermy, habitat selection, dispersal, nutrient spiraling, microbial loops, hydrothermal vents, food web theory, cascading in food webs, population and evolutionary responses of parasites, dispersal, and current successional theory.

Ecology and Field Biology is two books in one: a textbook covering the broad range of ecology and a manual on methodology. As a textbook it provides certain features not found in most other texts: chapter outlines, summaries, and a glossary of well over 600 terms. As a sourcebook this text provides an entrance to ecological literature including a list of field identification manuals, a survey of ecological journals, and an extensive bibliography. The appendix offers an introductory manual of ecological methodology, useful for conducting field studies and analyzing data. The appendix eliminates the need for a supplementary laboratory manual.

Like past editions the text is divided into five parts. Part I consists of two introductory chapters. Chapter 1 provides a brief overview of the nature of ecology with a strong emphasis on its history. Chapter 2, "Ecological Experimentation and Models," is new to this edition. Students are told that ecology is an empirical science and involves the testing of hypotheses, experiments, development of mathematical models to help explain and understand how populations, communities, and systems function.

Part II considers the structure and function of ecosystems. Chapter 3 introduces the concept of a system and examines the two major processes in the functioning of an ecosystem: energy fixation by photosynthesis and recycling of nutrients through decomposition. Chapter 4 provides an overview of climate and its major influence on ecosystems and the environment in which organisms live. Chapters 5, 6, and 7 cover the three major abiotic influences and limitations on organisms: temperature, moisture, and light. The approach is mostly from an organismal and ecophysiological point of view. This subject matter has been reorganized relative to the third edition and separated into individual chapters with new material added. Chapter 8 is new, consolidating material otherwise treated in several places in the third edition, including nutrient budgets and models of nutrient flow. Chapter 9 is a revision of the soil material from the third edition, and new Chapter 10 considers the physical aspects of the aquatic environment, to balance the soil chapter, which relates to the terrestrial environment.

Chapters 11 and 12 are two keystone chapters on ecosystem functioning, considering energy flow and nutrient cycling. They are referenced extensively throughout the text. New to the energy chapter is a discussion of current food web theory current to 1989. The material on biogeochemical cycling, Chapter 12, has been extensively revised and updated, including a discussion on acid rain.

Part III, "Population Ecology," has been rewritten and expanded to reflect the advances in population ecology since the third edition. Chapter 13 considers population genetics with an emphasis on the genetics of small and fragmented populations. Chapter 14 deals largely with demography and considers both unitary and modular populations. Chapter 15 considers population growth and density-dependent and density-independent influences, with examples from both plant and animal populations. Chapter 16 expands on intraspecific competition introduced in

Chapter 15. New material added is an expanded discussion of dispersal from the standpoint of both genetic and somatic fitness. Following in Chapter 17 is a discussion of life history patterns, replacing in part the old chapter on behavioral ecology. This chapter considers reproductive effort, including energy costs, reproductive values and costs, mating patterns, r-selection, K-selection, and habitat selection. Chapter 18 looks at interspecific competition, including examples of experimental approaches to the study of interspecific competition. Chapter 19 is an overview of the theory of predation, including optimal foraging theory. Chapter 20 discusses plants and their herbivorous predators. Chapter 21 looks at carnivorous predators and their prey, including intraspecific predation or cannibalism and human exploitation of wild populations. Chapter 22, new to this edition, discusses parasites and diseases, including their population dynamics and evolutionary responses. Chapter 23, the concluding chapter of Part III, considers the coevolution of various facets of mutualism, its possible origins and effects on the interacting populations.

Part IV, "The Community," covers much of what is considered community ecology, although no distinct line can be drawn between population ecology and community ecology. Chapter 24, on community basics, contains much new material. It considers physical and biological structure of communities, island biogeography, and the roles of predation, competition, parasites and diseases, and mutualism on community structure. This chapter discusses horizontal heterogeneity, edges, ecotones, and patch dynamics. These topics are basic to landscape ecology and to an understanding of island biogeography as applied to fragmented habitats.

Succession has been divided into three chapters to emphasize the advances made in the development of successional theory and the study of the roles of disturbance and patch dynamics in community structure. Chapter 25 deals with succession and provides the most current thinking, concepts, and models of succession. Chapter 26 examines scales and frequency of disturbance, its role in nutrient cycling and ecosystem stability. Chapter 27 looks at the influence of the past on present-day communities.

Part V is concerned with comparative ecosystem ecology. It deals with the major terrestrial, freshwater, and marine ecosystems. All of these chapters contain a good deal of ecophysiology and evolutionary adaptation, as well as comparative study of ecosystem function with emphasis on the nitrogen cycle. Chapter 28 introduces the various classifications of ecosystems, including the more recent ecoregion approach. Chapter 29 looks at the soil as a living system. Chapter 30 reviews the structure and function of grasslands, tropical savannas, shrubland, and deserts. Chapter 31 covers various types of forests and has a greater emphasis on heathlands and tropical forests than in the third edition. Chapter 32 deals with freshwater ecosystems—lakes, running water, and wetlands. Freshwater ecology has expanded greatly during the past 10 years and this chapter reflects that growth. Discussed are the concepts of the river continuum, nutrient spiraling in streams, and role of macrophytes in lake ecosystems. Chapter 33 discusses marine ecosystems from salt marshes to open sea. New to this discussion are hydrothermal vents and microbial loops in open sea food webs.

This edition is accompanied by a student resource manual containing three appendixes on methodology. These appendixes provide both a techniques manual for conducting laboratory exercises and field studies and an expansion of the mathematical analysis introduced in the main body of the text. I have rewritten all of these, replacing some old examples with new, reworked and rechecked manually and on computer all of the calculations, added new material on line transects and the calculation of reproductive values, and completely revised the section on estimating animal numbers.

In addition, a glossary of more than 600 terms (including most of the ones italicized in the text), an extensive bibliography, and an index are included in the main text.

An annotated list of journals, selected and general references, and abstracts are also included in the student resource manual.

Another feature is the illustration program. Some of the line drawings date to the first edition and were done by the late Ned Smith. New line drawings (excluding all graphs) and spots in this fourth edition and the cover were done by my son Robert Leo Smith, Jr., a medical illustrator with the University of Virginia and a freelance natural history artist.

In keeping with the spirit of ecology, the design of the cover provides a visual synthesis of some ecological concepts. A careful study of the cover both front and back will reveal them. The cover design features population and community ecology. Illustrated are two biomes, the North American plains and the African savanna; most of the organisms included are ecological equivalents. The bison represents the dominant grazing herbivore on the plains; the wildebeest occupies the same position in the African savanna. Predators are represented by the African lion, meer-

kats, and Bateleer eagle on the African savanna, and the badger and golden eagle on the plains. The herds of buffalo and wildebeest suggest population size and growth and intraspecific competition and social interactions. The prairie dog and meerkat represent colonial social organization and the concept of genetic interchange between colonies. The prairie dog calling an alarm at the sight of predators suggests predator defense, altrustic behavior, and the concept of kin selection and predator defense. The growth of conifers on the hill dropping onto the plain represents an ecotone between short-grass plains and the western coniferous forest. The kopje, a rocky island in a sea of grass, with its own unique and restricted forms of life, represents the concept of island biogeography. There are still other concepts hidden within the cover which I will leave to you to find.

Finally, a textbook improves only with input from its readers. Although the manuscript has been reviewed, copyedited, checked and rechecked for typographical and conceptual errors, some will slip through. If you find errors, please write them down and call them to my attention. I would also appreciate your comments on how to improve the book. I try to keep on the "cutting edge" of ecology, but I am sure I have overlooked significant papers. Please call them to my attention. Send your comments, list of errors, and suggestions to Glyn Davies, Biology Editor, Harper & Row Publishers, Inc., 10 East 53rd Street, New York, NY 10022. He will send them on to me. Because of time constraints, I cannot answer all suggestions personally, but all comments are deeply appreciated.

Robert Leo Smith

Acknowledgments

No textbook is the product of the author alone. Although the author writes the text, the material of which it is composed represents the work of hundreds of ecological researchers who have spent lifetimes in the field and laboratory. Their published works on experimental results, observation, and thinking provide the raw material out of which a textbook is fashioned.

Revisions of a textbook depend heavily on the input of users. I am very grateful for the comments offered by a number of instructors and students. The following reviewers provided detailed critiques of the entire book: Robert C. Clover, Cal State University; Dr. George Dalrymple, Flordia International University; Hal DeSelm, University of Tennessee; Dr. Richard T. Hartman, University of Pittsburgh; Dr. Thomas Kunz, Boston University; Dr. Leslie Real, North Carolina State University; Dr. James Runkle, Wright State University; Dr. Harold Underwood, Texas A&M University; Dr. Frederick M. Williams, Penn State University. I took all of their suggestions seriously and accepted most of their recommendations. Steven Jenkins of the University of Nevada at Reno reviewed the material on population demography and cleared up inconsistencies. George Hall, professor of physical chemistry at West Virginia, as well as an ornithologist and ecologist, helped revise the section on the physical structure of water. Rich Greer and Bob Whitmore, of West Virginia University, both ecological statisticians, Charles Hall of Syracuse University, and Tom Smith of the University of Virginia, both ecological modelers, provided input and incisive criticism of Chapter 2. Stan Tajchman, a forest microclimatologist at WVU, provided help in that area. Dean Urban and Hank Shugart of the University of Virginia and Jim Mc-Graw of West Virginia University came through with last-minute aid in locating and providing needed information.

I am indebted to four of my former students, Mohd. Nawayai Yasak, Mohd. Tajaddin Abdullah, Zabba Zabidin, and Burhanuddin Modh. Nor, of the Department of Wildlife and National Parks, Peninsular Malaysia, and Mohd. Khan bin Momim Khan, director of the Department of Wildlife and National Parks, for providing me the opportunity to visit and travel in Peninsular Malaysia on two occasions and introducing me to the southeastern Asia rain forest, its wildlife, plants, and conservation problems. Another graduate student, Alexine Keuroghlian, of São Paulo, Brazil, who is working on the black lion tamarin project, provided photos of the Amazonian rain forest and insights into the rain forest deforestation from a Brazilian point of view.

The book could not have arrived at its present stage without the help and encouragement of the staff at Harper & Row. Former biology editors Claudia Wilson and Sally Cheney were both patient and prodding in getting this edition into shape, and Glyn Davies gave the project its last big shove. Kristin Zimet, free-lance copy editor, was the finest with whom I have worked. In addition to her outstanding ability as a copy editor, she has an exceptional knowledge of ecology as well. She picked up inconsistencies and errors, eliminated redundancies, and suggested organizational changes within and between chapters that greatly improved the book. All the hard work in getting out a book falls to the project editor, whose job it is to pull together manuscript, art, captions, galleys, pages, and to prod author, designers, illustrators, and compositors. Paula Cousin was more than equal to the task. It was a real pleasure to work with her.

As usual this book is sort of a family affair. My wife Alice took care of all the problems of living, as well as working on the bibliography, while I spent my time, with little to spare, working on this book. My son Robert Leo, Jr., did all of the new drawings and cover for the text. My second son, Thomas Michael, an ecologist and ecological modeler, provided insights into successional theory and critiques, and helped me expand my ecological knowledge in numerous ways including guidance through Africa and Australia. My daughter, Pauline, provided the opportunity to visit and coaxed me to go to Maui, the results of which are evident at places in the text, and gave last minute help on the bibliography. And my second daughter, Maureen, handled personal and business matter affairs while Alice and I were off to other places gathering ecological data.

Many ecologists' comments, suggestions, and ideas, offered at various times and places, have influenced to varying degrees the further refinement of this book. To them go my thanks.

Finally, West Virginia University has provided both support and tolerance over the years.

RLS

PART I

INTRODUCTION

C H A P T E R 1

Ecology: Its Meaning and Scope

- **Ecology Defined**

- **The Development of Ecology**

  Plant Studies and Energy Flow
  Population Studies
  Animal Studies
  Cooperative Studies

- **Disputes**

- **Applied Ecology**

  The Public Awakens
  The Future of Life

- **Summary**

■ Ecology Defined

What is ecology? Ask nonecologists and they will probably answer that ecology has something to do with the environment or with saving it. Before the 1960s, few of them could have given you any answer. If you had asked a biologist in the same time period, you would probably have gotten some vague answer that ecology was "quantified natural history." Ecology only became a household word in the 1960s, through the environmental movement.

The origin of the word *ecology* is the Greek *oikos*, meaning "household," "home," "place to live." Clearly, ecology deals with the organism in its place to live,

its environment. Ecologists still are working toward a deeper definition. Here is a sampling of their attempts:

"the study of structure and function of nature" (Odum 1971:3).

"the scientific study of the distribution and abundance of animals" (Andrewartha 1961:10).

"the scientific study of the interactions that determine the distribution and abundance of organisms" (Krebs 1985:4).

"the scientific study of the relationships between organisms and their environments" (McNaughton and Wolfe 1979:1).

"the study of the relationships between organisms and the totality of the physical and biological factors affecting them or influenced by them" (Pianka 1988:4).

"the study of the adaptation of organisms to their environment" (Emlen 1973:1).

"the study of the principles which govern temporal and spatial patterns for assemblages of organisms" (Fenchel 1987:12).

"the study of the patterns of nature and how those patterns came to be, and how they change in space and time" (Kingsland 1985:1).

"the study of organisms and their environment—and the interrelationships between the two" (Putman and Wratten 1984:13).

"the study of the relationship between organisms and their physical and biological environments" (Ehrlich and Roughgarden 1987:3).

The author of the term, Ernst Haeckel (1869), defined ecology as "the body of knowledge concerning the economy of nature—the investigation of the total relationships of the animal both to its inorganic and its organic environment; including, above all, its friendly and inimical relations with those animals and plants with which it directly or indirectly comes into contact—in a word, ecology is the study of all those complex interrelations referred to by Darwin as the conditions for the struggle for existence."

None of these definitions is satisfactory. They are either too restrictive or too vague. The original definition applies only to animals; most others focus on populations and overlook systems. The subject has outgrown them.

For now, let us use a wider working definition. Ecology is the study of the structure and function of nature. Structure includes the distribution and abundance of organisms as influenced by the biotic and abiotic elements of the environment; and function includes how populations grow and interact, including competition, predation, parasitism, mutualisms, and transfers of nutrients and energy.

The term *ecology* is derived from the same root word as *economics*, "management of the household." In effect, ecology could be considered the economics of nature. Wells, Huxley, and Wells (1939) commented that "Ecology is really an extension of economics to the whole world of life." Some economic concepts, such as resource allocation, cost-benefit ratios, and optimization theory, have found a place in ecology.

■ The Development of Ecology

Just as there is no consensus on the definition of ecology, so there is no agreement on its beginnings. It is more like a multistemmed bush than a tree with a single trunk. Some historians trace the beginnings of ecology to Darwin, Thoreau, and Haeckel; others to the Greek scholar Theophrastus, a friend and associate of Aristotle, who wrote about the interrelationships between organisms and the environment.

Plant Studies and Energy Flow

The modern impetus to ecology came from the plant geographers. They discovered that, although plants differed in various parts of the world, certain similarities and differences demanded explanation. One of the early influential plant geographers, Carl Ludwig Willdenow (1765–1812), pointed out that similar climates support similar vegetation. His ideas stimulated the thinking of a wealthy young Prussian, Friedrich Heinrich Alexander von Humboldt (1769–1859). He traveled in Mexico, Cuba, Venezuela, and Peru, explored the Orinoco and Amazon rivers, and described his travels in a 30-volume work, *Voyage to the Equatorial Regions*. In these books Humboldt described vegetation by physical type in terms of physiognomy, correlated vegetation types with environmental characteristics, and coined the term *association*.

Other plant geographers stayed closer to home and still made major contributions. J. F. Schouw (1789–1852) studied the effects of temperature on plant distribution and introduced the idea of naming plant associations after dominant species. Anton Kerner (1831–1898), commissioned by the Hungarian government to describe the vegetation of eastern Hungary and Transylvania, advanced the concept of *succession*—vegetation change through time—in his book *Plant Life of the Danube Basin*. He pioneered the use of experimental transplant gardens to study the behavior of plants taken from different elevations.

The work of these botanists influenced another generation of plant geographers. One was Johannes Warming (1841–1924). He also studied the tropical vegetation of Brazil, advanced the idea of life forms and dominant plants to describe vegetational changes, and noted the effects of fire and time. His major contribution was his book *Plantesamfund: Grundtrak af den okologiske Plantegeografi* in 1895, in which he unified plant morphology, physiology, taxonomy, and biogeography into a coherent whole. The book fixed the modern concept of ecology and greatly influenced its development.

Andreas Schimper (1856–1901) also traveled extensively in the tropics and in his book *Plant Geography on a Physiological Basis* tried to explain regional differences in vegetation. Meanwhile, Jozef Paczoski (1864–1941) in Poland studied vegetation–environment interactions on a more local scale. He wrote a text in which he described how plants modify their environment by creating microenvironments and introduced such concepts as shade tolerance, competition, plant succession, and the role of fire. Because his book was published in Slavic, it was belatedly discovered by other ecologists.

Out of the roots of plant geography grew the study of plant communities. It developed separately in Europe and the United States. In Europe, two pioneering plant ecologists were Christen Raunkaier of Denmark and A. E. Tansley of Great Britain. Raunkaier (1860–1938) contributed a scheme of life form classification and quantitative methods of sampling vegetation, the data from which could be treated statistically. Tansley (1935), one of the giants of modern ecology, introduced both the term and concept *ecosystem* and urged a more experimental approach to plant ecology. He founded the British Ecological Society and headed a conservation movement in England. His views on ecology and ecological research anticipated by years the ecological studies typical of the 1970s.

Prominent among the early plant ecologists in the United States were J. M. Coulter of the University of Chicago and C. E. Bessey of the University of Nebraska. Coulter was the major professor of H. C. Cowles, who received his doctoral degree in 1897 for his study "Ecological Relations of the Sand Dune Flora of Northern Indiana." This work marked the beginning of pioneering studies of plant succession, one of the central concepts of modern ecology.

At the same time another graduate student, F. E. Clements, working under Bessey, studied the plant geography of Nebraska. Dogmatic and convincing, Clements quickly became the major theorist of plant ecology in the United States. He gave ecology a hierarchal framework, introduced innumerable terms (no longer used) and the idea of environmental indicators, and developed a theory of succession that still colors ecology today.

Early plant ecologists were concerned mostly with terrestrial vegetation; but in Europe a group of biologists was interested in the natural history of fresh waters. Prominent among these biologists were A. Thienemann (1931) and F. A. Forel (1901). Thienemann developed an ecological approach to freshwater biology. He introduced the ideas of organic nutrient cycling and trophic feeding levels, using the terms *producers* and *consumers.* Forel was more interested in the physical parameters of freshwater habitats, particularly lakes. He described thermal stratification and wave motion within lakes and in his monograph on Lake Leman introduced the term *limnology* for the study of freshwater life.

In the United States S. A. Forbes, an entomologist at the University of Illinois and the Illinois State Laboratory of Natural History (Illinois Natural History Survey), wrote in 1887 a classic of ecology, "The Lake as a Microcosm," which concerned the interrelations of life in a lake, particularly through food chains, and the role of natural selection in the regulation of numbers of predators and prey.

Unrelated to limnology, but nevertheless destined to have an important influence on its future development and that of ecology, was the work of Edgar Transeau in an Illinois cornfield. Transeau was not an ecologist, much less a limnologist. He was interested in improving the production of agriculture by understanding the photosynthetic efficiency of corn plants. His landmark paper "The Accumulation of Energy in Plants," which appeared in the *Ohio Journal of Science* in 1926, marked the beginning of the study of primary production and energy budgets.

Thienemann's concept of trophic levels and producers and consumers and Transeau's concept of energy budgets and primary production stimulated the study of lakes by E. A. Birge and especially by C. Juday of the Wisconsin Natural History Survey. In a classic paper "The Annual Energy Budget of an Inland Lake," Juday (1940) summarized not only the accumulation of energy by aquatic plants over a year but also its movement through various feeding groups, including the decomposers.

The work of Juday and Birge influenced the research of a young limnologist at the University of Minnesota, R. A. Lindeman. In his 1942 paper "The Trophic-Dynamic Aspect of Ecology" Lindeman described energy-available relationships within a community. The paper marked the beginning of ecosystem ecology, a significant advance.

Lindeman's study stimulated further pioneering work on energy flow and nutrient budgets by G. E. Hutchinson (1957, 1969) of Yale University and H. T. Odum and E. P. Odum in the 1950s. J. Ovington (1962) in England and R. E. Rodin and N. I. Bazilevic (1967) in the Soviet Union investigated nutrient cycling in forests. The increased ability to measure energy flows and nutrient cycling by means of radioactive tracers and to analyze large amounts of data with computers permitted the development of *system ecology,* the application of general systems theory and methods to ecology.

Population Studies

As plant ecology was evolving out of plant geography, other developments were assuming an important role in the evolution of ecology. One was the voyage of Charles Darwin on the *Beagle,* during which he collected numerous biological specimens, made detailed notes, and mentally framed his view of life on Earth. Darwin observed the relationships between organisms and environment, the similarities and dissimilarities of organisms within continental land masses and among continents, which he attributed to geographical barriers separating the inhabitants. He noted from his collection of fossils how successive groups of plants and animals, distinct yet obviously related, replaced one another over geological time.

Darwin's theory of evolution and the origin of species was influenced by the writings of Thomas Malthus (1798). An economist, Malthus advanced the principle that populations grew in geometric fashion, doubling after some period of time. Experiencing such rapid growth, a population would outstrip its food supply. Ultimately the population would be restrained by a "strong, constantly operating force—among plants and animals the waste of seeds, sickness, and premature death. Among mankind, misery and vice." From this concept Darwin developed the idea of "the survival of the fittest" as a mechanism of natural selection and evolution.

Meanwhile, unknown to Darwin, an Austrian monk, Gregor Mendel, was studying in his garden the transmission of inheritable characters from one generation of pea plants to another. The work of Mendel would have answered a number of Darwin's questions on the mechanisms of inheritance and provided for his theory of natural selection the firm base it needed. Belatedly, Darwin's theory of evolution and Mendelian genetics were combined to form the study of evolution and adaptation, two central themes in ecology. The theoretical basis of the role of inheritance in evolution was advanced by Sewell Wright (1931), R. A. Fisher (1930), and J. Haldane (1932, 1954) who developed the field of *population genetics.*

The Malthusian concept of population growth and limitations stimulated the study of population dynamics. P. F. Ver-hulst (1838) of Italy formulated the mathematical basis of the nature of population growth under limiting conditions. Verhulst's work, later expanded by R. Pearl and L. J. Reed (1929), was the basis for the contributions of A. Lotka and V. Volterra (1926) on population growth, predation, and interspecific competition. Their work established the foundations of *population ecology,* concerned with population growth, regulation, and intraspecific and interspecific competition. The mathematical models of Lotka and Volterra were tested experimentally in the Soviet Union by G. F. Gause (1934) with laboratory populations of protozoans and in the United States by Thomas Park (1954) with flour beetles. Many of the concepts of population genetics have been combined with ideas from population ecology to make up the field of *population biology,* or more precisely *evolutionary ecology,* concerned with the interactions of population dynamics, genetics, natural selection, and evolution.

Population ecology, because of its quantitative approach, led to *theoretical mathematical ecology.* First concerned with sampling, statistical analysis, and distribution of organisms, theoretical population ecology quickly moved into quantitative studies of competition, predation, community and population stability, cycles, community structure, community association, and species diversity.

Animal Studies

The concepts of natural selection, evolution, and population dynamics arose not out of plant ecology (in which the concepts of population ecology have only recently been applied) but out of the areas of natural history related to animals. Early animal ecology developed later than plant ecology and along lines divorced from it. The beginnings of animal ecology, which developed later than plant ecology, can be traced to two Europeans, R. Hesse of Germany and Charles Elton of England. Elton's *Animal Ecology* (1927) and Hesse's *Tiergeographie auf oekologischer Grundlage* (1924), later translated into English as *Ecological Animal Geography*, strongly influenced the development of animal ecology in the United States.

Two pioneering animal ecologists there were Charles Adams and Victor Shelford. Adams published the first text on animal ecology, *A Guide to the Study of Animal Ecology* (1913). Shelford's *Animal Communities in Temperate America* (1913) was a landmark work because he stressed the relationship of plants and animals and emphasized the concept of ecology as a science of communities. The community concept had been developed much earlier by Humboldt and by the marine biologist Karl Mobius. In his essay "An Oyster Bank Is a Biocenose" (1877) Mobius explained that the oyster bank, although dominated by one animal, was really a complex community of many interdependent organisms. He proposed the word *biocenose* for such a community. The word comes from the Greek meaning "life having something in common."

In 1949 the encyclopedic *Principles of Animal Ecology* by five second-generation ecologists from the University of Chicago, W. C. Allee, A. E. Emerson, Thomas Park, Orlando Park, and K. P. Schmidt, pointed out the direction modern ecology was to take, with its emphasis on trophic structure and energy budgets, population dynamics, and natural selection and evolution.

Still another area of biology, animal behavior, grafted its roots onto ecology. Although Darwin, Wallace, and others described behavior of animals, formal study began with George John Romanes (1848–1894), who introduced the comparative method of studying nonhuman animals to gain insights into human behavior. His approach depended largely on inferences, but C. Lloyd Morgan (1852–1936), an English behaviorist, emphasized the use of direct observation and experiment. Since the early 1900s, animal behavior studies developed along four major lines. One was the study of behavioral mechanisms, perceptual and physiological. A second, more relevant to ecology, included comparative physiology and *ethology*, the systematic study of the function and evolution of behavioral patterns. The three major founders of ethology were Konrad Lorenz, noted for his studies of genetically programmed behavior; Niko Tinbergen, who developed the scheme of four areas of inquiry: causation, development, evolution, and function; and Karl von Frisch, who pioneered studies of bee communication and behavior. After World War II a third, new field of animal behavior, wedded to ecology, appeared. It was *behavioral ecology*, which investigates the way animals interact with their living and nonliving environments, with a special emphasis on how that behavior is influenced by natural selection and ecological conditions. Behavioral ecology in 1975 begot a controversial offspring, *sociobiology*, pioneered by E. O. Wilson (1975) and summarized in his book *Sociobiology: The New Synthesis*. Sociobiology, concentrating on field observations of social groups of animals, applies the principles of evolution to social behavior in animals. It became controversial when some scientists applied it to human social interactions.

Developing simultaneously was *physiological ecology*, concerned with the responses of individual organisms to temperature, moisture, light, nutrients, and other such stresses. Physiological ecology dates to Justus Leibig (1840), who studied the role of limited supplies of nutrients in the growth and development of plants. The idea of limiting factors was extended—too much of a good thing—by F. F. Blackman (1905). He also advanced the idea of factor interaction and studied the relationship among light, carbon dioxide, temperature, and the rate of assimilation in plants. V. E. Shelford applied the concept of limiting factors to animals in a law of tolerance. It incorporated both the physiology of an organism and the environment. He suggested that organisms had both a negative and an optimal response to environmental conditions, and that these responses had a role in the distribution of organisms. L. J. Henderson (1913) in a classic book, *The Fitness of the Environment,* explored the biological significance of the properties of matter.

Observations and later experimental work on such phenomena as the effects of plant exudates on the growth of associated species and the use of chemical defensive mechanisms by animals led to investigations of chemical substances in the natural world. Scientists studied the nature and use of chemicals in animal recognition, trail making, and courtship, and defense in both plants and animals. Such work has grown into the specialized field of *chemical ecology.*

Cooperative Studies

Ecology has developed from so many roots and has grown so many stems that it probably will always remain, as Robert McIntosh (1980) calls it, "a polymorphic discipline." It now ranges over many diverse areas—marine, freshwater, and terrestrial. It involves all taxonomic groups, from bacteria and protozoa to mammals and forest trees, and it deals with them at different levels: individuals, populations, ecosystems. Any of these levels and groups may be studied from various points of view: behavioral, physiological, mathematical, chemical. As a result ecol-

ogy, by necessity, involves specialists with minimal common ground.

Pulling some of these groups together in the 1960s was the International Biological Program, known as IBP. The organization of the program in 1960 was stimulated by a growing concern over environmental problems facing the world. The United States' IBP, initiated in 1967, focused on a cooperative study and analysis of ecosystems, including the tundra, the coniferous forest, the eastern deciduous forest, the desert, and Mediterranean types. The goals, as summarized by McIntosh (1976), included (1) understanding the interactions of the many components of complex ecological systems; (2) exploiting this understanding to increase biological productivity; (3) increasing the capacity to predict the effects of environmental impacts; (4) enhancing the capacity to manage natural resources; and (5) advancing the knowledge of human genetic, physiological, and behavioral adaptations.

IBP's greatest contribution was to increase our understanding of processes in ecosystems, particularly photosynthesis and productivity, water and mineral cycling, decomposition, and the role of detritus. Although it lacked strong organization and coordinated direction, it did advance modern ecosystem ecology. Summaries of IBP research, published and being published in numerous volumes, will provide a base for future ecological research.

■ Disputes

Ecology's complex past has encouraged controversy. Ecologists have often found themselves in warring camps.

The first major split in ecology was

the failure of plant ecology and animal ecology to meet on common ground. In England plant ecology was influenced strongly by A. E. Tansley and animal ecology by Charles Elton. At that time the field of ecology was covered by one journal, *The Journal of Ecology*, sponsored by the British Ecological Society. In a few years Elton started *The Journal of Animal Ecology*. The two, plant and animal ecology, went their separate ways.

In the United States the split was less amicable. Early on, a controversy developed over the term *ecology*. Botanists decided at the Madison (Wisconsin) Botanical Congress in 1893 to drop the *o* from *oecology* and adopt an anglicized spelling. Zoologists refused to recognize the term. The entomologist William Morton Wheeler complained that the botanists had usurped the word and had distorted the science. He urged zoologists to drop the term and adopt the word *ethology*. The schism was widened by a fundamental difference in approach. Plant ecologists ignored any interaction between plants and animals. In effect, they viewed plants as growing in a world without parasitic insects and grazing herbivores. For years plant and animal ecologists went their separate ways. At last F. E. Clements and V. E. Shelford attempted to bring the two sides together with *Bio-Ecology* (1939), in which they suggested that plants and animals be considered as interacting components of broad biotic communities or biomes.

Although the gap between plant and animal ecology narrowed, a new division was to plague ecology. It had its roots in the influential ideas of Clements. Clements viewed the plant community as an organism. Like an individual organism, vegetation moved through several stages of development, from colonizing bare ground to mature, self-reproducing climax in balance with its climatically determined environment. The climax was the end, or goal, toward which all vegetation progressed. If disturbed, vegetation responded by retracing its developmental stages to the climax again.

Clements' organismal approach was not lost on animal ecologists. The American zoologist and animal behaviorist William Morton Wheeler, an international authority on ants and termites, advanced the idea that ant colonies too behaved as organisms. They carried out such functions as food gathering, nutrition, self-defense, and reproduction. Drawing on the ideas of Lloyd Morgan, a biological philosopher, Wheeler applied emergence theory to ecology. He proposed that natural associations had certain emergent properties as a whole—predators and prey, parasites and hosts—that arose from lower levels of organization. All levels occurred together in an ecological community or biocenosis. The biocenosis modified its component species through behavioral changes and new levels of integration. Everything in the biocenosis was related to everything else. His view of a tight, orderly nature contrasted with the chaos imposed on nature by humans.

This organismic, levels-of-hierarchy view of nature advanced by Clements and Wheeler captured the thinking of that influential group of ecologists at the University of Chicago—the authors of *The Principles of Animal Ecology*, Allee, Emerson, Park, Park, and Schmidt. In that book they stated that the organismic concept of ecology was "one of the fruitful ideas contributed by biological science to modern civilization."

Although the organismal concept dominated ecology until the 1960s, not all ecologists accepted it. Two sharp critics were H. A. Gleason and A. E. Tansley. In 1926 Gleason published a paper in the *Bulletin of the Torrey Botanical Club* titled

"The Individualistic Concept of the Plant Association." In it he challenged the organismic concept, pointing out that the plant association was hardly an organism capable of self-reproduction. Instead, he argued, each community is unique. It arises randomly through environmental selection of seeds, spores, and other reproductive parts of plants that enter a particular area. The English ecologist A. E. Tansley, once enamored with the organismic concept, ultimately rejected it too. Although he admitted to certain striking similarities between the development of vegetation and the development of individual organisms, vegetational development was quite different from the ontogeny of plants and animals. Vegetation, he allowed, might be called a quasi-organism, but certainly not an organism or a complex organism. In fact, Tansley rejected the whole idea of a biotic community as anthropomorphic. No social relationship exists among plants or between plants and animals as the term connotes, he argued. In its place Tansley substituted the term *ecosystem:* Plants and animals were components of a system that also included the physical factors shaping it. The individualistic concept of Gleason supplanted the organismic concept—almost. It lived on in the "new ecology" of the 1960s. The new ecology, as defined by E. P. Odum (1964, 1971), is a *systems ecology,* an "integrative discipline that deals with supraindividual levels of organization."

According to this concept, ecosystems develop from youth to maturity. Each stage of development exhibits unique characteristics. Interactions among populations and between plants and animals result in a hierarchal organization. Interacting components produce large functional wholes. The new system has properties that are not evident at the level below

(Odum 1971, 1982). These emergent properties account for most of the changes in species and growth over time. Study must be holistic because systems are too complex to study in bits. Because the whole is greater than the sum of its parts, ecosystems can be studied only as functional units.

This holistic approach has its critics, reductionists who consider the ecosystem as the sum of its parts. By understanding how each part—the species, its numbers and characteristics—functions, they seek to discover how the whole system operates. Rather than guiding the evolution of species, ecosystems derive their character from the evolution of species.

Fenchel (1987:17) puts the reductionists' point of view well:

> I find the entire argument as nonsensical as stating that an alarm clock is qualitatively different from its constituent wheel, bolts, and springs. A holist approach to an alarm clock . . . is to observe that when wound it will run. To arrive at a real understanding of the device one must take it apart in order to see how it works . . . to take a reductionist's approach.

The holist would counter that studying the wheels, bolts, and springs tells nothing about the way the whole system functions, what the clock really does. They are outside the context of the whole clock. Only when all parts of the system are functioning as a unit can the clock function. Then its emergent property, telling time, becomes apparent. Is the clock greater than the sum of its parts or not? Allen and Starr (1982) in their book *Hierarchy* argue that the whole problem of emergent properties is a matter of scale. They agree that some properties of the whole are emergent and cannot be derived from the behavior of the parts alone. They also point out that the ecosystem models of holists are simply large-scale reductionism. Ecosystem ecologists cannot possibly study an entire ecosystem. They can only study pieces of it. The major difference between a reductionist and a holist is that the holist studies larger pieces, made up parts studied by the reductionist.

The reductionists belong to another new ecology that emerged in the 1960s, *theoretical ecology*, led by the late Robert MacArthur. At first theoretical ecology was largely animal ecology, but in the 1970s theoretical ecology was adopted by plant ecologists, especially in the field of population biology, led by J. L. Harper (1977).

The bases of theoretical ecology are the mathematical models of Lotka and Volterra, especially their models of interspecific competition and predation, and the experimental work of G. Gause and T. Park. From that foundation theoretical ecologists have developed a body of theory relating to competition, population growth, life-history strategies, resource utilization, niche, coevolution, community structure, food webs, and the like. Theoretical ecologists apply theories and equations developed in pure mathematics, physics, and even economics to ecological questions. One of the strengths of theoretical ecology is its attempt to provide a substantive mathematical foundation on which predictions can be based. Some of the theories and field research stimulated by theoretical ecologists have provided new insights into the relationships among species, utilization of resources, and life-history patterns. At the same time critics argue that too many hypotheses are untested or untestable in the field.

What keeps ecosystem ecologists and population and evolutionary ecologists apart? Population ecologists focus on species interactions with their environment. They are interested in the historical reasons why natural selection has favored particular adaptive responses. Ecosystem ecologists are more interested in how interactions in populations, communities, and ecosystems affect the present and future. These differences may not be as great as they appear.

What can bring the two groups together? Loehle and Peckmann (1988) explore this problem, especially how systems ecologists might use evolutionary theory. Population ecologists could approach population growth and interactions such as mutualism, parasitism, predation, and competition as interacting systems (Berry 1981) and as components of a hierarchy of systems. Systems ecologists could integrate some evolutionary theory into system models, particularly in the area of ecosystem development and organization. Food web theory and analysis, for example, cross over the line into both evolutionary and systems ecology, involving as they do both species interactions and the transfer of energy and nutrients through a hierarchical structure. Ecosystem functioning ultimately depends upon species adaptations, which are the outcomes of evolution. For example, efficiency of water use by certain ecosystems such as grasslands and deserts (see Chapter 30) results from the water use efficiency of the individual plants. The natural assemblage of plants and animals that comprise the living component of an ecosystem is not a random collection of species but rather one that has been determined by the competitive abilities and other attributes of the component species (H. Odum 1983). Open for study are how evolution has affected ecosystem dynamics and conversely, how ecosystem dynamics can influence natural selection of species.

■ Applied Ecology

Ecological theories and models are little more than academically significant if they cannot contribute to an understanding of human-impacted environments and provide a basis for ecosystem and natural resource management, preservation, and restoration, all making up *applied ecology*. Applied ecology is the application of ecological insights and principles to the management of forest, range, wilderness, wildlife, and fisheries, all of which involve predictive modeling at population and ecosystem levels.

Applied ecology began to take shape in the 1930s. In 1932 Herbert Stoddard introduced the idea of the role of fire in the control of plant succession in his book *The Bobwhite Quail*. Aldo Leopold expounded the application of ecological principles in the management of wildlife in his classic *Game Management* (1933). In *Forest Soils* (1954), H. L. Lutz and R. F. Chandler discussed nutrient cycles and their role in the forest ecosystem; and J. Kittredge pointed out the impact of forests on the environment in *Forest Influences* (1948).

We have begun to apply ecosystem and theoretical ecology to resource management in the past decade or so, even though economics often takes precedence over ecology. Forestry, once concerned with the raising of trees for harvest, now emphasizes biomass accumulation, nutrient cycling, the effects of timber harvesting on nutrient budgets, and the role of fire in forest ecosystems. Range management now is interested in the functioning of grassland ecosystems, the effects of grazing intensities on aboveground and belowground production by plants, and

species structure. Wildlife management, once emphasizing only game species, now considers the entire wildlife spectrum, including species not hunted. The range of interest covers both population ecology of wildlife and the maintenance and management of plant communities as wildlife habitat. Wildlife management has developed an interest in population genetics, with an emphasis on the effects of hunting on game species.

A related developing field comprising both applied and theoretical ecology is *conservation biology*. It is best defined as "the science of scarcity and diversity" (Soulé 1986). Conservation biology addresses the problems of gross habitat destruction and a great reduction in population size of species.

A second developing field tied to applied ecology is *landscape ecology,* which is concerned with spatial patterns in landscape and how they develop, with emphasis on the role of disturbance, including human impacts on the landscape. It explores "how a heterogenous combination of ecosystems is structured, functions, and changes" (Forman and Godron 1986).

A third new field is *restoration ecology,* which applies experimental research to the reproduction of ecosystems on highly disturbed lands (Jordan et al. 1987).

The Public Awakens

Applied ecology has been around at least since the early 1930s, but it did not gain visibility until the 1970s, when ecology was at the heart of social, political, and economic issues. The public grew aware of the problems of pollution, toxic wastes, overpopulation, vanishing wildlife, and a degraded environment.

Although the public treated these issues in the 1970s as if they were new, the ecological or environmental movement, called *ecologism* by Anna Bramwell (1989), goes back to the late 1800s. The ecological movement, a linkage of ecological ideas and energy economics, has it roots in Europe, especially Germany. In fact, an early founder of political ecology was Ernst Haeckel, who coined the term ecology. Two of his books, *The Riddle of the Universe* (1900) and *The Wonders of Life* (1905), were translated into English. From Germany, where historically there has been a great interest in a back-to-the-land movement, organic farming, and soil and forest conservation, the ecological or "green" movement moved to northern Europe, Great Britain, and America. In Britain and northern Europe environmentalists used a literary approach to create an ecological ideology. Notable are the novels of the Norwegian Kurt Hamsun (1859–1952) *(Growth of the Soil)* and of the English authors Robert Jefferies (1848–1887) *(The Life of the Field)* and Henry Williamson (1896–1977) (*The Story of a Norfolk Farm* and many others). The English ecologist, Charles Elton, called attention to the close relationship between animals and their environment in his classic book *Animal Ecology,* and was instrumental in the founding of The Nature Conservancy Council. One of the most influential recent books was E. F. Schumacher's *Small Is Beautiful* (1973), with its emphasis on biologically sound agriculture and small-scale technology.

In America the approach was more direct and less literary. George Perkin Marsh in 1865 called attention to the effects of poor land use on the human environment in his dramatic book *Man and Nature.* In the 1930s F. E. Clements urged that the Great Plains be managed as grazing land and not be broken by the plow, which they ultimately were, resulting in

the dust bowl. Paul Sears wrote *Deserts on the March* (1935) in response to the dust bowl of the 1930s, and William Vogt's *Road to Survival* (1948) and Fairfield Osborn's *Our Plundered Planet* called attention to the growing population-resource problem. Aldo Leopold's *A Sand County Almanac* (1949), which called for an ecological land ethic, was read largely by those interested in wildlife management until the 1970s, when it became the bible of the environmental movement.

Rachael Carson probably did more than anyone else to bring environmental problems to the attention of the public. Since the publication of her book *Silent Spring* (1962), people have become more aware that chemical poisons and other pollutants are recycled through the environment. Once castigated as more fiction than fact, Carson's predictions came only too true as carnivorous birds fell victim to chemical pesticides. With a ban on DDT, hawks and fish-eating birds began a gradual comeback. Carson made people quick to recognize other continuing chemical dangers such as dioxin and PCBs.

The Future of Life

In the past quarter century since people became concerned about growing environmental problems that reduce the quality of life, how has the situation changed? We started off well enough with environmental legislation and a National Environmental Protection Act, designed to protect the environment from overzealous development and to mitigate losses; but government is less sensitive now to environmental issues and funding for environmental protection and research, especially at the federal level, has shrunk. During the 1980s there was even an environmental backlash at the federal level, as the administration attempted to undo all the progress of the previous two decades. In academics ecology has again become the poor cousin of molecular biology.

There has, of course, been some progress; but we have discovered that our environmental problems are not only more difficult to solve than once believed: Many are growing worse. Toxic wastes pollute groundwater and land. Air is becoming more polluted continent-wide. Regional haze has cut visibility in the eastern United States by more than 50 percent in the past 40 years. Acid rain is killing lakes and streams and possibly destroying spruce forests in North America and Europe. Increased concentrations of carbon dioxide and ozone threaten climatic stability. Roads cut into open country and suburban expansion eat away at the hinterlands and farmlands. Continued deforestation in both temperate and tropical regions is fragmenting wildlife habitat, threatening many species with extinction. A rapidly growing urban and suburban population with increasing interest in outdoor recreation is placing pressures on state and national parks that threaten their ecological integrity. Even the oceans have not escaped, as human debris and chemicals have been deadening the seas and destroying marine life. In spite of surplus agricultural production, wetlands are still being drained at an alarming rate, threatening already dangerously declining wetland wildlife.

The future of life on Earth depends upon far more ecological knowledge than we now possess, and upon applying all we know. For the first time in the history of Earth *Homo sapiens* has become the dominant organism, changing Earth at will with little regard for the consequences. It is little wonder, then, that some of the most challenging problems of our time lie in that ecotone between theoretical and applied ecology.

■ Summary

Ecology, difficult to define precisely, is the study of the interrelations of organisms with their total environment, physical and biological. Its origins are diverse, but a main root goes back to early natural history and plant geography, which evolved into the study of plant communities, ecosystems, trophic levels, and energy budgets. Another major root that developed out of natural history of animals was the study of natural selection and evolution, beginning with the major contributions of Darwin and Wallace. It branched into evolutionary ecology, population genetics, population ecology, and theoretical ecology. Another root gave rise to behavioral ecology, concerned with the way animals interact with their living and nonliving environment as influenced by natural selection and environmental conditions, and physiological ecology, concerned with the responses of individual organisms to temperature, moisture, and the like. Studies of certain chemical reactions of organisms to their environment stimulated the development of chemical ecology with emphasis on uses of chemicals by plants and animals as attractants, repellents, and defensive mechanisms, their evolution and chemical structure. In fact, ecology has so many roots that it has become a polymorphic discipline.

As its various disciplines expand, ecology is becoming fragmented into isolated specialties. Two major divisions are holistic ecosystem ecology and reductionist evolutionary and population ecology, although differences between the two are not so great as they sometimes appear.

Ecological theory matters only if it helps us solve major environmental and resource management problems. Traditionally, applied ecology meant forest, range, and wildlife management. Recently, applied ecology has spawned the new fields of conservation biology, restoration ecology, and landscape ecology. The future quality of life on Earth in all its aspects depends upon our ability to apply ecological principles to its management.

Ecological Experimentation and Models

- **Theories and Hypotheses**

- **Experimentation**

- **Inductive and Deductive Methods**

- **Models**

    Analytical and Simulation Models
    Validation and Verification

- **Summary**

Until the 1960s, ecological studies were largely descriptive, with numerous papers on the nature of vegetational communities, animal communities, and succession. The descriptive approach has been replaced with an experimental, empirical approach emphasizing hypothesis testing, sampling design, statistical analysis, interpretation, and the development of models that mimic real-world phenomena to provide insights and develop new hypotheses.

Theories and Hypotheses

Theory is the basis of all science. A *theory* is a statement of cause that seems plausible but cannot be directly confirmed or disproved by experimentation. Theories, however, can beget testable hypotheses. A *hypothesis* is a statement about causative agents that can be tested experimentally. A testable hypothesis is, in fact, both a research hypothesis and a statistical hypothesis (Romesburg 1981). The research hypothesis states the part of the theory intended for testing; it is a generalized statement about causal processes. The statistical hypothesis develops explicit statements of relationships.

Testing the hypothesis and reaching a conclusion requires a null hypothesis, usually designated as H_0. A null hypothesis must be falsifiable. We can never prove it correct, but we reject or fail to reject it with known risks. Rejection of a null hypothesis involves the acceptance of a second or alternative hypothesis, H_1. The risks involve the probability of the chance p (α) that the null hypothesis H_0 will be rejected (and H_1 accepted) when in fact H_0 is true (Type I error), and the probability of the chance p (β) that H_1 will be rejected (and H_0 accepted) when H_1 is correct (Type II error). The largest acceptable risk of error is conventionally set at alpha = 0.05, the 95 percent level of significance. However, depending on the study, the alpha levels may be increased or decreased.

■ Experimentation

Testing hypotheses entails first the collection of data by direct observation or experimentation. Direct observation and comparison is usually employed in natural experiments (Diamond 1986)—observations of systems as they existed and now exist, or reconstruction of changes following some sort of disturbance over time. Observed similarities between phenomena in such studies yield correlation coefficients, but provide no evidence of cause and effect, only associations.

An experimental study usually involves simplification. By manipulating one or a few variables while holding others constant, the investigator can relate observed responses to the manipulated system. A laboratory experiment involves perturbations produced by the experimenter, who controls the biotic and abiotic environment by the regulation of dependent and independent variables. A *dependent variable* is the one whose value depends upon the value assigned to another variable, the one taking on the range of values. The latter is the *indepen-*

dent variable. These experiments usually involve synthetic communities of a few species. Based on simple starting conditions and assumptions, such experiments can reveal a range of possible outcomes to evaluate in the field. Because of the difficulties of maintaining and breeding laboratory populations, these experiments generally use small, short-lived organisms, such as beetles and Paramecia, as models of larger ones.

Field experiments involve manipulations of one or a few independent variables in natural communities. This purpose is often accomplished by adding or removing a species, by erecting exclosures that prevent access to a space or resource, or by adding or withholding nutrients or water. To reduce the risk that a difference in the dependent variables in the manipulated and unmanipulated plots is due to natural variations among plots or experimental units and not to treatment, the experimenter must replicate manipulated and unmanipulated plots in space and time.

Field experiments gain in realism, spatial scale, and scope over laboratory experiments, and in control over natural experiments, but they are subject to certain limitations. One is the danger of pseudo-replication of sites (Hurlbert 1984), which occurs when the experimenter fails to define the experimental unit precisely. Experiments require replication of experimental units, some receiving treatments, others remaining as controls. These units must be interspersed in space and time so that one experimental unit has no effect on the data collected from another. In other words, the experimental units must be independent. If there is only one replicate per treatment, if replicates are really subsamples within a larger unit, if data from replicates are pooled prior to analysis, or if successive samples taken over time from a single experimental unit are treated as separate samples rather than repeated samples from the same unit, the experimental units are not independent.

■ Inductive and Deductive Methods

There are two approaches to testing hypotheses: the inductive method, which goes from the specific to the general; and the deductive method, more specifically called the hypothetico-deductive (H-D) method, which goes from the general to the specific. Both have their advantages, but modern ecologists lean strongly toward the hypothetico-deductive approach.

The *inductive method* is useful for investigating correlation or association between classes of facts. It infers generalities from a particular collection of observations or facts. Examples of an inductive approach are regression and correlation analyses. An experimenter accumulates facts or observations, plots the data, fits a regression equation to them, and depending upon the functional relationship of the data, uses them to arrive at some conclusions about the hypothesis. The weaknesses of the inductive approach are that it is time-intensive, it minimally addresses functional relationships, and it may lead to extrapolation beyond the range of data; but it can yield information upon which to base experimental studies.

An example of an inductive approach is a study of the influence of understory density on snowshoe hare habitat use (Litviatis, Sherburne, and Bissonette 1985). A long-standing theory is that a close association exists between forest understory characteristics and density of snowshoe hares. The investigators established two

49 ha study sites in a regenerating spruce-fir and hardwood stand. They determined snowshoe hare densities in spring and fall by mark-recapture methods (see Appendix A), quantified habitat use with fecal pellet counts, measured intensity of twig-clipping, and sampled understory vegetation. After data were collected, the investigators compared the estimated snowshoe hare density with stem cover density and compared overwinter survival with understory stem density. They found that both overwinter survival and spring population densities were associated with dense understory (Figures 2.1 and 2.2), which possibly provided escape and thermal cover. The investigators concluded that patterns of habitat use by hares were influenced primarily by understory density and not by plant species composition. However, they found that plant species composition did influence hare population density. Dense spruce-fir understories supported greater densities of hares than hardwood stands, presumably because conifers provided the hares with superior cover. This inductive study suggests an experimental study involving habitat manipulations to test the hypothesis that dense understory conifer cover supports higher populations of snowshoe hares than hardwood cover.

Figure 2.1 Relationship between stem cover unit density and overwinter survival of snowshoe hares in Maine, 1981–1983. (Litvaitis et al. 1985:871.)

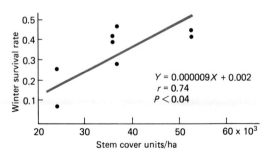

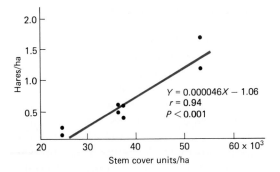

Figure 2.2 Relationship between stem cover unit density and estimated snowshoe density in Maine, 1981–1983. (Litvaitis et al. 1985:872.)

In the *hypothetico-deductive method,* the investigator develops a general statement or research hypothesis first, then collects data to support or refute the statement, develops a mathematical model, and attempts to fit the model to the data. Depending on the observed fit, the investigator may modify the model. The deductive method also has its weaknesses. If the study involves laboratory experimental procedures under controlled conditions, the results might not reflect field conditions. The results of such experiments, to be of any predictive value and application, must be verified by field experiment and observation. In the field the experimenter faces the risk that differences observed in the dependent variables in the manipulated and unmanipulated plots is due to natural differences rather than to the manipulation.

Another snowshoe hare study is a good example. Observational field studies and conceptual models suggest that initial decline from peak densities in the ten-year cycle of snowshoe hare populations is triggered by a shortage of winter food brought about by high density of hares, which is reflected in decreased survival and fecundity. Vaughn and Keith (1981) tested this hypothesis by studying the de-

mographic responses of experimental snowshoe hare populations of high and low densities to available food supplies.

Vaughn and Keith (1981) established eight exclosures of 2.9 to 5.8 ha of natural habitat during the fall near Rochester, Alberta. Four of the exclosures were stocked with high density populations, an average of 13.2 hares per ha, and four with low density populations, an average of 3.2 hares per ha at the start of the experiment. One-half of high density and low density populations were provided with commercial rabbit food ad libitum; the other half had to survive on available browse in the exclosures, which was not sufficient to support the initial fall populations over winter. This setup gave the experimenters four treatments: high and low density × abundant and scarce food, replicated 3 to 7 times. The experimenters monitored the change in numbers, weight, and survival from December through May over a three-year period.

Vaughn and Keith found that the mean winter weights in all the populations were lower, but that the overwinter weight losses were greater among hares in the scarce food treatment plots. Adult survival in the scarce food plots was only slightly reduced by food shortages, but juvenile survival in those plots was much lower. All major components of reproduction were affected by food shortages. Hares living under scarce food conditions experienced a shorter breeding season, and a reduction in mean natality from 18.4 to 8.5 young per female surviving. Juveniles born to parents from food-scarce treatments averaged a weight gain of 9.8 g/day compared to a weight gain of 12.1g/day for juveniles born to parents in the food-abundant treatments. The results of this experiment support the hypothesis that declines in snowshoe hare populations are initiated by a winter food shortage.

■ Models

The results of ecological studies may generate new hypotheses of cause and effect and stimulate the development of a model. Modeling is not a necessary part of ecological study, but if experimenters want to extrapolate patterns from processes, they need to develop a model. A *model* is an abstraction and simplification of a natural phenomenon developed to predict a new phenomenon or to provide insights into existing ones.

A verbal or word model is a written statement about a phenomenon. Hypotheses, in effect, are word models. Other models are graphic, providing a visual assessment of a phenomenon. Graphic models, of which there are many examples in this book, can serve as a basis for mathematical formulation. The experimenter may wish to formalize the process by constructing a mathematical model.

Analytical and Simulation Models

A number of modeling approaches is used in ecological studies. It is convenient to divide the approaches into analytical and simulation models.

Analytical models are mathematical formulations that can be solved in closed form. They include approaches such as differential equations and Markov models. A familiar and simple analytical model used in population ecology is the exponential growth equation (see Chapter 15). Population growth is modeled as the outcome of two life history features, birth and death. Simply stated, the rate of population growth is equal to births − deaths: $r = b - d$. The parameter r is the difference between the instantaneous birthrate and the instantaneous death rate. Popula-

tion growth may then be expressed as a differential equation $dN/dt = rN$, where dN/dt is the change in population size N with change in time t. This equation can be solved analytically and expressed as the integrated form $N_t = e^{rt}$, where N_t is the population at time t, N_0 is the initial population, e is the base of natural logarithms, r is the instantaneous coefficient of population growth, and t is the time interval. Verbally, the model states that a population size is the function of the initial size of the population, N_0, the rate of increase r, and time. If the predicted changing population size is plotted arithmetically against time, a J-shaped curve results. If the model is stated logarithmically, $\log_e N_t = \log_e N_0 rt$, a plot of the result gives a straight line with a slope equal to r (Figure 2.3).

The assumptions of the exponential growth model are that the environment is

Figure 2.3 Exponential growth curve plotted arithmetically and logarithmically. (See also Figure 15.1.)

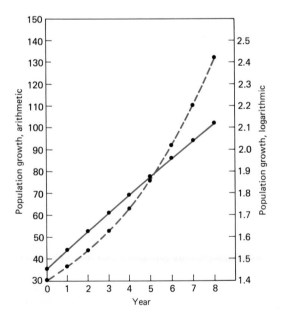

unlimited (there is nothing to inhibit growth), conditions for growth are always favorable, and age distribution and reproductive rate are constant. These fairly restrictive assumptions do not often hold for natural populations in changing environments. The more we relax these assumptions, the more complex the equations necessary to describe the population's behavior become. In certain cases the assumptions result in a model that cannot be solved analytically, and a simulation approach is required.

Simulation models take a variety of forms, including differential equations such as the exponential growth model discussed above. However, because they cannot be solved analytically, they require the use of a computer to arrive at a solution. Although simulation models sacrifice mathematical tractability, they allow the model structure to incorporate much richer biological detail. One example of an area of ecological research where simulation models have been widely used is the study of population dynamics in a spatially heterogeneous environment. These applications include examining the consequences of spatial heterogeneity of resources and patterns of seed dispersal and species migration.

Fahrig and Merriam (1985) used a simulation model approach to examine the consequences of habitat patchiness on the population dynamics of white-footed mice inhabiting forest patches in an agricultural landscape. The question that the ecologists wished to address was whether population dynamics within a patch is influenced by the degree to which it is isolated from other patches.

The experimenters developed a patch dynamics model to simulate changing sizes of resident mice populations in a series of habitat patches (forest woodlots) located in a landscape of otherwise unsuitable habitat (agricultural fields). These habitat patches were either isolated or connected to other patches (Figure 2.4). Connected patches allow for movement of individuals between patches, whereas isolated patches do not. The model has two components of population dynamics: (1) within-patch dynamics, and (2) between-patch dynamics. Within-patch dynamics included recruitment into the population by reproduction, movement of individuals from one age class to the next (aging), and age-specific mortality. Between-patch dynamics was simulated as rates of immigration and emigration, which were dependent on the state of the population in each patch and whether the patch was connected to or isolated from other patches. The basic operator of the model was a series of transition matrices, one for each time period, whose elements represented the population processes discussed above.

In addition to model simulations, field data on population dynamics of white-footed mice inhabiting seven woodlots were collected for comparison. A number of these woodlots were connected by fence-rows, which functioned as corridors between forest patches.

Figure 2.4 Diagrammatic model of isolated and interconnected woodland islands in an agricultural landscape.

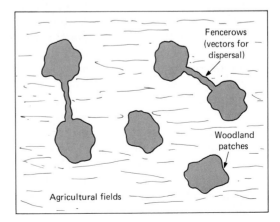

The model results suggested that population growth is lower in isolated than in connected woodlots and that population growth rates within a patch type (isolated vs. connected) do not differ. The field study of the populations in the seven woodlots showed similar results, with population growth lower in the isolated patches.

Simulation models are useful in studies of energy flow and nutrient cycling in ecosystems, of succession, and of impacts of human disturbances and insect outbreaks on forests and grasslands. They are useful in the development of production and management models in forestry, exploitation models in fisheries, and integrated pest management in entomology, among others. One of the values of simulator models is their ability to convert quantitative details of natural history and physiology into parameters for analytical models and for simulations at the community and landscape levels (see Shugart 1984).

Validation and Verification

Once developed, the model should be validated and verified with experimental and field data. *Verification* is the process of testing whether or not the model is a reasonable representation of the real-life system being investigated and whether its parts or components agree or coincide with known mechanisms of the system. A model is verified if its output can be matched to a given set of data by using those data to estimate its parameters. *Validation* is the explicit and objective test of the basic hypothesis. Validation measures quantitatively the extent to which the output of the model agrees with the behavior of the real-life system (Jeffers 1988). The process compares the model's results with the data upon which it is based as well as new

sets of observations that are independent of the data used to frame the model. Validation indicates (1) how well the model predicts both the data used in its construction and other independent data, and (2) how consistently the model predicts other occurrences.

Validation is a most important step. Ecology has many models, but few trustworthy, validated ones. Some validated simulator models involve incremental forest growth, responses of forests to fire, and effects of hurricanes on tropical forest ecosystems (see Horn, Shugart, and Urban 1989; Shugart 1984).

Although usually regarded as predictors, models, even unvalidated ones, have other uses. They serve as a means of arriving at different forms of possible solutions to a problem and yield insights that might otherwise go undiscovered. They can also be used together with empirical methods to generate new hypotheses and to design more powerful research programs.

■ Summary

Ecology has evolved from a descriptive approach to an empirical and experimental approach emphasizing hypothesis testing, statistical analysis, and development of models to provide new insights and develop new hypotheses. A hypothesis is a statement about causative agents that can be tested experimentally. Testing hypotheses entails the collection of data by direct observation or by experimentation, which involves simplification by manipulating one or a few variables while holding others constant. Field experiments involve manipulations of one or a few independent variables in natural communities.

Two approaches to testing hypotheses are the inductive method, which goes from the specific to the general, and the deductive method, more specifically the hypothetico-deductive method, which goes from the general to the specific. The inductive approach is useful for investigating correlation between classes of facts. In the hypothetico-deductive method the investigator develops a research hypothesis, collects data to support or refute it, develops a mathematical model, attempts to fit the model to the data, and then, if necessary, modifies the model.

A model is an abstraction and simplification of natural phenomena developed to predict new phenomena or provide insights into existing ones. A verbal or graphic model may serve as the basis for a more formal mathematical model. A mathematical model may be analytical or simulation. Analytical models are mathematical formulations that can be solved in closed form. They include such approaches as differential equations. Simulation models may take on a variety of forms, including differential equations. Because they cannot be solved analytically, simulation models require the use of a computer to arrive at a solution.

Once developed, the model should be validated and verified. Verification is the process of testing whether or not the model functions as a reasonable representation of a real-life system being investigated. Validation is the explicit and objective test of the basic hypothesis. It measures quantitatively the extent to which the output of the model agrees with the behavior of the real-life system.

PART II

THE ECOSYSTEM

Ecosystem Basics

- **A System Defined**

- **Components of Ecosystems**

- **Essential Processes**

   Photosynthesis
   Decomposition

- **Summary**

Imagine yourself in a spacecraft far from Earth. Through the window, the planet appears as a bluish ball suspended in the black void of space. Its surface is kaleidoscopic. Patterns of whites, blues, reds, browns, and greens are constantly changing upon its surface as it turns and as clouds swirl across it. It hangs alone, self-contained, dependent on the outer reaches of space only for the energy of sunlight. It is close enough to the sun to be warmed by radiant energy; yet it does not become overheated or overcooled. It is protected from damaging radiation by a layer of gases, an *atmosphere*, unlike that of any of its sister planets.

As the spacecraft approaches Earth, the patterns sharpen into broad outlines of mountains, deserts, plains, and seas. In the vast area from outer space to the core of Earth, it is only at the narrow interface of land, air, and water that life exists. This thin blanket of life surrounding Earth,

from mountaintops to sea bottom, is called the *biosphere* (Figure 3.1). Rooted plants cease above the snowline, and only specialized heterotrophic communities exist above it and thrive on the deep ocean floor.

Supporting the biosphere is the *lithosphere*, the rocky material of Earth's outer shell from the surface to about 100 km deep. The thin, uppermost layer of the crust is the substrate of life—the foundation on which it rests, the material of which it is made, and its primary source of nutrients.

The body of liquid on or near the surface of Earth is the *hydrosphere*. Most of the water is in the oceans, some is in the lithosphere as groundwater, and a small fraction is in the atmosphere.

As the spaceship comes even closer to Earth, the mountains, deserts, plains, and seas focus into expanses of grasslands, forests, croplands, rivers, lakes, oceans. Each

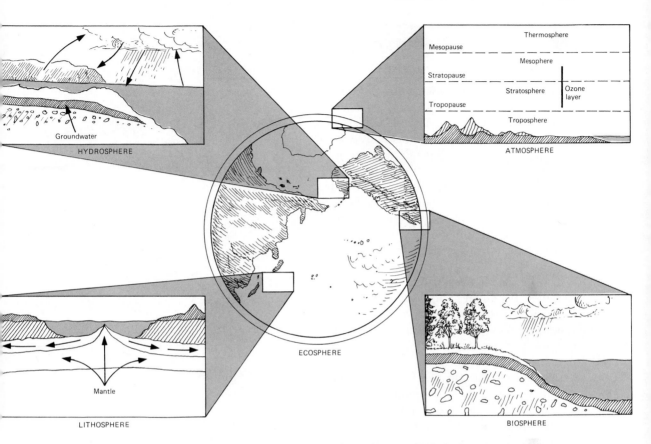

Figure 3.1 Schematic diagram of the structural features of the outer part of Earth. The atmosphere, biosphere, lithosphere, and hydrosphere are often collectively called the *ecosphere*.

region is physically different, and each is inhabited by different organisms that are well adapted to their environment. In spite of their differences, each functions in the same manner. Energy is fixed by plants and transferred to animal components. Nutrients are withdrawn from the substrate, deposited in the tissues of plants and animals, cycled from one feeding group to another, released by decomposition to the soil, water, and air, and then recycled. Each of these regions constitutes an ecosystem. These ecosystems are not independent. Energy and nu-

trients in one find their way to another, so that ultimately all parts of Earth are interrelated in one whole.

A. E. Tansley (1935) coined the term *ecosystem*, borrowing the idea of a system from physics.

> The more fundamental conception is . . . the whole *system* (in the sense of physics) including not only the organism-complex, but also the whole complex of physical factors forming what we call the environment. . . .
>
> We cannot separate (the organisms) from their special environment with which they form one physical system. . . . It is the system so formed which (provides) the basic units of nature on the face of the earth. . . . These *ecosystems,* as we may call them, are of the most various kinds and sizes.

■ A System Defined

If the various biotic units such as forests and lakes in the landscape are ecological systems, then we should have some idea of what a system is. A *system* consists of a set or collection of interdependent parts or subsystems (the inner boxes of the diagram in Figure 3.2) enclosed in a defined boundary (the dotted line). Outside is an environment, which provides the inputs necessary for its functioning. An *input* or forcing function is any resource from outside to which the system responds. The system's *output* is any attribute transmitted to the environment. The system processes inputs in a certain manner, with each subsystem—separate mechanisms within the system—performing set functions. Each part has a specific role, but its expression depends upon the proper functioning of all parts. The output from the system is directly related to the input. If input ceases, the system no longer functions.

Many systems have a feedback mechanism, which provides a degree of control or homeostasis. Some of the output is fed back into the system to influence future output. A feedback system involves an ideal state or set point toward which the system adjusts or away from which the system moves. If the feedback accelerates a deviation away from the set point, it is called *positive*. In this case, output is continually reinvested into input. Exponential growth of population and compound interest on principal are examples of positive feedback. Positive feedback is necessary in the biological world; but unless controlled, it can destroy the system. Counteracting positive feedback is *negative* feedback. It halts or reverses a movement away from the set point by controlling the behavior of the input.

A room dehumidifier is an example of

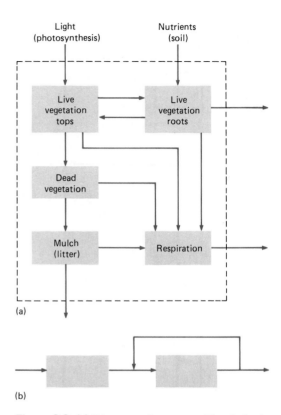

Figure 3.2 (a) Diagram of a system. The dashed line represents the boundary with the system's environment. The boxes inside represent the various subsystems, and the arrows show flows between components. The lines crossing the boundary are input from and output to the environment. (b) A system with a feedback loop.

a purely mechanical system in which a specific set point can be fixed. Suppose the desired humidity of a room is 50 percent. The dehumidifier set point is placed at 50 percent. When the humidity of the air exceeds that point, the moisture-sensitive device in the dehumidifier trips the switch to allow an inflow of electricity to turn the fan that pulls the air over the refrigerated coils on which the water condenses and drips off to be carried away. When the humidity falls below the set point, the moisture-sensitive device responds by shutting off the input of electricity to stop the fan.

Temperature regulation in birds and mammals is an example of the functioning of a biological system. The body temperature of humans is 98.6° F(37° C). When the temperature of the environment rises, sensory mechanisms, primarily in the skin, detect the change and transmit the information to the brain. Acting on the information, the brain sends a message to the effector mechanisms that increase blood flow to the skin and induce sweating. Water excreted through the skin evaporates, cooling the body. If the environmental temperature becomes much warmer than the normal body temperature, the body is unable to lose heat fast enough to hold the temperature at normal. Positive feedback takes over and body metabolism increases, further raising body temperature. The homeostatic mechanisms break down and heat stroke or death results. If the environmental temperature falls below a certain point,

the temperature-regulating system responds by reducing blood flow to the skin and causing shivering, an involuntary muscular exercise producing more heat. If the environmental temperature becomes too cold, metabolic processes slow down, further decreasing body temperature and eventually causing death by freezing.

We shall explore systems at the organismal, population, and ecosystem levels throughout this book.

■ Components of Ecosystems

An ecosystem is basically an energy-processing and nutrient-regenerating system whose components have evolved over a long period of time (Figure 3.3). The

Figure 3.3 Schematic diagram of an ecosystem. (Adapted in part from O'Neill 1976.)

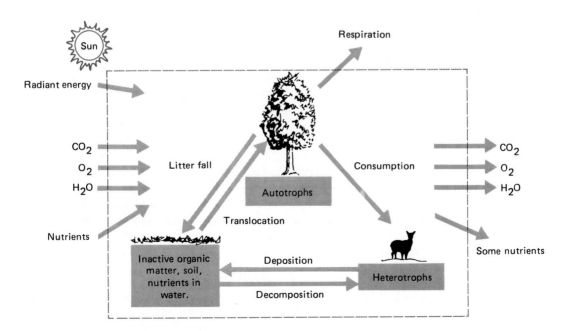

boundaries of the system are determined by the environment; only certain forms of life can be sustained by given conditions. Plant and animal populations within the system represent the subsystems through which it functions.

Inputs into the system are both biotic and abiotic. The abiotic inputs are energy and inorganic matter. Radiant energy, both heat and light, imposes restraints on the system by influencing temperature, moisture, seasonality, and photosynthesis. These factors affect what organisms survive in a system and its productive capability. Inorganic matter consists of all nutrients—water, carbon dioxide, oxygen, and so forth—that affect the growth, reproduction, and replacement of biotic material and the maintenance of energy flow. Some of the materials in this chemical environment are necessary for the maintenance and functioning of the system, whereas others are toxic or detrimental.

The biotic inputs include other organisms that move into the ecosystem as well as other ecosystems in the landscape. No ecosystem stands alone. One is affected by the other. A stream ecosystem, for example, is strongly influenced by the terrestrial ecosystem through which it flows.

In simplest terms all ecosystems, aquatic and terrestrial, consist of three basic components: the producers, the consumers, and abiotic matter. The *producers* or *autotrophs*, the energy-capturing base of the system, are largely green plants. They fix the energy of the sun and manufacture food from simple inorganic and organic substances. Autotrophic metabolism is greatest in the upper layers of the ecosystem—the canopy of the forest and the surface water of lakes and oceans.

The *consumers* or *heterotrophs* utilize the food stored by the autotrophs, rearrange it, and finally decompose the complex materials into simple, inorganic substances. In this role they influence the rate of energy flow and nutrient cycling.

The heterotrophic component is often subdivided into two subsystems, consumers and decomposers. The consumers feed largely on living tissue, and the decomposers break down dead matter or *detritus* into inorganic substances. However, all heterotrophic organisms are consumers, and all in some way act directly or indirectly as decomposers. Heterotrophic activity in the ecosystem is most intense where organic matter accumulates—in the upper layer of the soil, in the litter of terrestrial ecosystems, and in the sediments of aquatic ecosystems.

The third or *abiotic* component consists of inactive or dead organic matter, dissolved organic matter and nutrients in aquatic systems, and the soil matrix. Inactive organic matter comes from plant and consumer remains and is acted upon by the decomposer subsystem of the heterotrophs. Such inorganic matter is the basis of the internal cycling of nutrients in the ecosystem.

Traditionally other abiotic materials— CO_2, O_2, nutrients derived from the weathering of materials and from precipitation, and so on—are considered as abiotic components of the ecosystem. We shall consider them as inputs instead.

The driving force of the system is the energy of the sun, which causes all other inputs to circulate through the system (Hydrothermal vent is an exception). The various outputs, or more correctly, outflows, from one subsystem become inflows to another. While energy is utilized and dissipated as the heat of respiration, the chemical elements from the environment are recycled by organisms within the system. How fast nutrients turn over in the system is influenced by these consumers.

■ Essential Processes

The inputs of energy and nutrients into the ecosystem are handled by photosynthesis, herbivory, carnivory, and decomposition. Herbivory and carnivory process nutrients and energy, move them along the routes, and retain them in the system as long as possible. But the two basic processes are photosynthesis—the fixing of energy and the incorporation of nutrients into active plant tissue—and decomposition—the final dissipation of energy and the reduction of organic matter into inorganic substances.

Photosynthesis

The ecosystem operates on the fixation of energy and the production of organic compounds through the photosynthetic activity of autotrophs. Energy enters the ecosystem as visible light and is stored as reduced carbon compounds by plants in photosynthesis. From that point biochemical changes involve the oxidation of reduced carbon compounds into ones of less chemical energy. These chemical rearrangements are accompanied by the degradation of energy, production of respiratory heat, and outputs of water, carbon dioxide, and nitrogenous compounds, which are recycled through the system.

The formula for photosynthesis is

$$6CO_2 + 6H_2O + energy \rightarrow C_6H_{12}O_6 + 6H_2O + 6O_2$$

Intermediate compounds of photosynthesis include glucose and the by-products of water and oxygen, but further syntheses produce free amino acids, proteins, fatty acids and fats, vitamins, pigments, and coenzymes. The synthesis of various products may take place in different parts of the plant. Mature leaves of certain species of plants may produce only simple sugars and young shoots and rapidly developing leaves may produce fats, proteins, and other constituents.

Chlorophyll-bearing vascular plants and algae, both terrestrial and aquatic, account for most photosynthesis. Photosynthetic bacteria that use hydrogen, hydrogen sulfide, and various organic compounds instead of water as electron donors make minor contributions.

Photosynthesis is a complicated process. Let us outline the essential features. When light energy strikes a leaf, a portion of it is absorbed by the chlorophyll pigments and moved to reaction centers, where the energy is transferred to an electron of the chlorophyll molecule. The energy of the molecule is now raised to a higher state, and the electron is passed along through a series of chemical reactions. These result in the synthesis of adenosine triphosphate (ATP) from adenosine diphosphate (ADP), in the production of a strong reductant NADPH from nicotinamide adenine dinucleotide phosphate, and in the oxidation of water to H^+ and OH^- ions. The H^+ ions go to the reduction of NADP and the OH^- ions form water, release some molecular oxygen, and supply the electron to replace the one lost to the chlorophyll molecule. Further reactions involve the synthesis of carbohydrate from CO_2 and a series of reactions that incorporates energy as well as hydrogen and CO_2 into carbohydrates.

In most plants photosynthesis involves the immediate reaction of CO_2 from the atmosphere with ribulose bisphosphate (RuP_2), a phosphophorylated sugar with five carbon atoms, catalyzed by a single enzyme, ribulose bisphosphate carboxylase-oxygenase. The reaction of carbon dioxide molecules with RuP_2 forms two molecules of a three-carbon product, 3-phosphoglycolate (PGA). The plant converts this product into six-carbon sugars, starches, and other products, but uses some of the PGA to reform RuP_2 molecules that act as CO_2 acceptors again. This three-carbon photosynthetic pathway is called the Calvin-Benson cycle or C_3 cycle, and plants employing it are known as C_3 plants (Figure 3.4).

The C_3 pathway has one drawback. RuP_2 not only has an affinity for CO_2; it will combine with oxygen as well. When RuP_2 combines with O_2 instead of CO_2, it forms a two-carbon molecule, phosphoglycolate, and PGA, in contrast to the two PGAs formed in the reaction with CO_2. This pathway, called *photorespiration,* uses more energy than it produces and reduces the efficiency of C_3 photosynthesis (Ogren and Challet 1982).

Some plants have another pathway, facilitated by an internal leaf anatomy different from that of C_3 plants. C_3 plants have vascular bundles surrounded by a layer or sheath of large colorless cells; cells containing chlorophyll are irregularly distributed throughout the mesophyll of the leaf. Plants exhibiting the C_4 cycle have vascular bundles surrounded by a sheath or wreath of cells rich in chlorophyll, mitochondria, and starch. Mesophyll cells are arranged laterally around the bundle sheaths and have few chloroplasts.

C_4 plants have acceptor molecules of phospho-enol-pyruvate (PEP), a three-carbon compound, within the chlorophyll. CO_2 entering the leaf reacts with PEP to form malic acid or aspartic acid, each having four-carbon molecules (Figure 3.4). The plant breaks down malic and aspartic acids by enzymatic action to release the fixed CO_2 and form PGA with three-carbon molecules, as in the Calvin cycle. At the same time the two acids also produce pyruvic acid to form more PEP. Plants us-

ing the four-carbon method are called C_4 plants. This extra step does not replace the Calvin cycle; rather, it concentrates CO_2 for the Calvin cycle.

The major difference between the two types of plants is in the initial fixation of CO_2. C_4 plants have that extra step in CO_2 fixation, which gives them a certain advantage. PEP is more efficient than RuP_2 when the concentration of CO_2 is low and the concentration of O_2 is high. Mesophyll cells capture CO_2 in the outer leaf tissue and release CO_2 to the bundle sheath cells in the inner tissues. This process draws down the CO_2 in the leaf to near zero and increases the CO_2 pressure in the chloroplast, allowing CO_2 to compete effectively with oxygen for binding with RuP_2. This ability to fix CO_2 at very low levels allows C_4 plants to carry on photosynthesis when the temperature is high or moisture is low, conditions that close down stomata in the leaf to reduce loss of moisture. By creating a CO_2 sink in the mesophyll of the leaf, C_4 plants increase the steepness of the gradient of CO_2 concentration from the leaf to the atmosphere. Carbon dioxide can then diffuse into the leaf when stomatal resistance is high.

The C_4 pathway appears to be an adaptation to environments with high light intensity, high water stress, and high leaf temperatures. Such an environment places an extra premium on efficient CO_2 fixation while reducing water loss through the stomata. C_4 plants have a high water use efficiency—the number of grams of dry weight gained per kilogram of water transpired during the growing season. Compared to C_4 plants, C_3 plants have a lower light saturation threshold (Figure 3.5). They also have a lower optimum temperature range (16° to 25° C) for photosynthesis than C_4 plants (30° to 45° C) (Figure 3.6, on page 35).

The C_4 method is not found in algae,

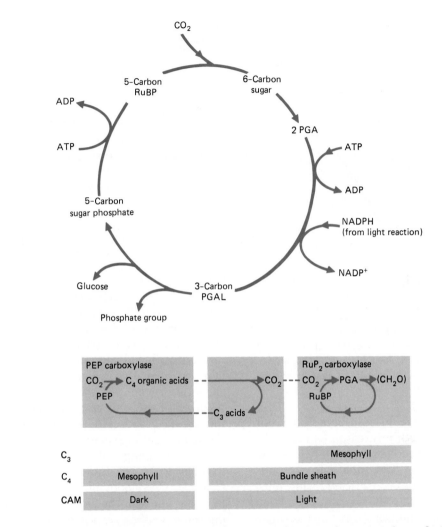

Figure 3.4 (a) The Calvin Benson or C_3 cycle. (b) Basic features of C_3, C_4, and CAM photosynthetic pathways compared. Carboxylations in C_4 plants take place in different cells; in CAM plants they take place in the same cell but at different times of day. (Adapted from Jones 1983.)

bryophytes, ferns, gymnosperms, or the more primitive angiosperms. C_4 species are mostly herbs associated with tropical and subtropical grasslands, and some shrubs in desert and saline regions, all environments where high water use efficiency is important. Grasses make up about half of the known C_4 species. Most

North American species of C_4 grasses have a subtropical distribution. From Florida to Texas 65 to 82 percent of the grass species are C_4 species, in the Central Plains 31 to 61 percent, and in the northern part of the continent 0 to 23 percent (Terri 1979; Terri and Stowe 1976). No known C_4 grasses grow on the tundra.

The ability of C_4 plants to carry on photosynthesis in full sun, at high temperatures, and in low water regimes gives them a competitive edge over C_3 plants in those conditions. But C_3 plants, in spite of their photorespiratory limitations, are

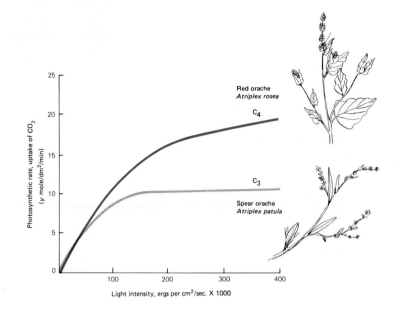

Figure 3.5 Effect of changes in light intensity on the photosynthetic rates of C_3 and C_4 plants grown under identical conditions of a 16-hour day, 25° C at day, 20° C at night with ample water and nutrients. The C_3 species spear orache *(Atriplex patula)* exhibits a decline in the rate of photosynthesis, as measured by CO_2 uptake, as light intensity increases. The C_4 species red orache *(Atriplex rosea)* shows no such inhibition. (After Bjorkman 1973:53.)

more productive in low light and cool temperatures. For this reason C_4 plants do not dominate vegetation. They grow poorly in shade and their advance into temperate regions is restricted by low temperatures.

Some agricultural crops, such as sugar cane and corn, are C_4 plants, but many others are C_3 species with a low photosynthetic rate (Bassham 1977; Bjorkman and Berry 1973). Agricultural plants escape competition only because humans eliminate weeds, many of which are C_4 plants, including crabgrass *(Digitaria)*, pigweed *(Setaria)*, and barnyard grass *(Echinochloa)*.

There is yet another photosynthetic process, known as the Crassulacean Acid Metabolism (CAM). It is found among a number of succulent semidesert plants in some 15 families, including Cactaceae, Euphorbiaceae, and Crassulaceae, from

which the method of carbon fixation received its name. CAM plants do not have specialized bundle sheath cells, and except during periods of adequate moisture they open their stomata during the cool nights rather than by day to minimize water loss. During the night PEP carboxylase fixes CO_2 and stores it as malic acid as C_4 in the vacuoles of the cells (Figure 3.4). During the day the plants close their stomata, and the malic acid enzymatically gives up the CO_2 which, along with respiratory CO_2, is fixed by RuP_2 and incorporated into the C_3 cycle. CAM does not avoid photorespiration, but with the stomata closed during the day, photorespiratory CO_2 cannot escape except by loss through diffusion. Oxygen does not affect CO_2 uptake in the dark, because PEP is insensitive to O_2. Although appearing similar in function to C_4 plants, CAM plants differ; they separate the initial fixation of CO_2 from the Calvin cycle in different times rather than in different locations. The CAM pathway is slow and inefficient, but well adapted to a rigorous arid environment.

Leaf Area Index and Photosynthesis.
Photosynthesis is carried on by individual leaves, but the net photosynthetic produc-

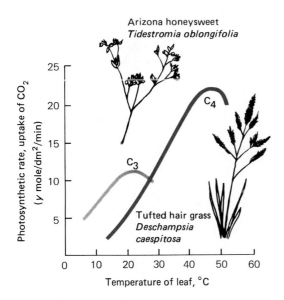

Figure 3.6 Effect of changes in leaf temperature on the photosynthetic rates of C_3 and C_4 plants. C_3 plants, represented by tufted hair grass *(Deschampsia caespitosa)* of the arctic tundra, exhibit a decline in the rate of photosynthesis as the temperature of the leaf increases. They also reach maximum photosynthetic output at a lower temperature than C_4 species, represented by Arizona honeysweet *(Tidestromia oblongifolia).* This desert species increases its rate of photosynthesis as the temperature of the leaf increases, up to about 50° C. (After Bjorkman and Berry 1973.)

the beginning of the growing season and increases with full leaf and plant maturity. A low LAI indicates less leaf area than ground area and represents wasted sunlight on the ground; a high LAI indicates energy wasted by respiration of shaded leaves. Most plants have an optimal LAI, the point at which there is minimal shading of one leaf by another and net photosynthesis is maximum.

The optimal LAI of a plant or plant community depends on the intensity of light radiation, the shape and arrangement of leaves in the canopy, and the angle of the sun. It changes seasonally, daily, and even hourly. The optimal LAI increases with leaf area up to a point, as light intensity increases and as leaves are more perpendicular to the ground. Leaves that are perpendicular to incoming light

Figure 3.7 Relation of photosynthesis to leaf orientation, leaf area, and canopy structure. Although horizontal leaves capture the most sunlight, the upper layers shade the lower, reducing the overall interception of light. An upright orientation of leaves is more efficient in capturing the sun's energy. That type of leaf arrangement is typical of communities in which the individual plants are growing closely, as among the grasses.

tion of a plant or community of plants depends upon the contribution of all leaves. That contribution depends upon their position in the vegetative profile and the degree to which the leaves above deplete the light resource for the leaves below.

The intensity of light reaching a plant is influenced by the local light regime, the modification of that light by the structure of vegetation, and the position of the leaf (Figure 3.7). The relationship of the leaf-to-light interception can be described by the *leaf area index* (LAI) (Figure 3.8). This number is the ratio of leaf area per unit of ground area. The LAI is usually low at

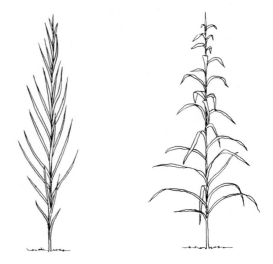

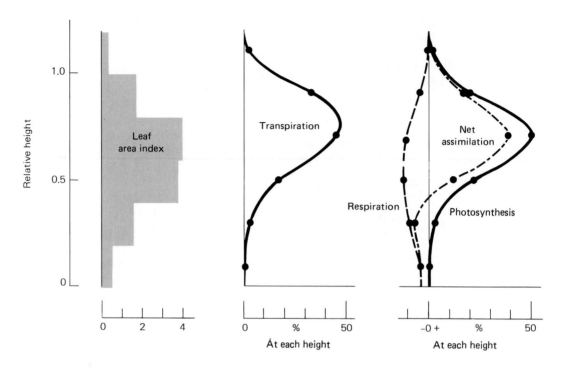

Figure 3.8 The amount of photosynthetic area is measured in terms of leaf area index (LAI). LAI is the ratio of the surface area of a leaf (upper or lower surface only) to a given surface area of land as measured in hectares or acres. An LAI value of 1 means one hectare of leaf surface to 1 hectare of ground surface; a value of 2 means 2 hectares of leaf surface to 1 hectare of ground surface. The LAI varies with the type of plant, leaf orientation, type of leaf, and canopy structure. Note that the LAI index varies with the vertical structure of the canopy. The greatest amount of transpiration and photosynthesis takes place where the LAI is the highest. (After Baumgartner 1968).

(horizontal leaves) intercept the most light, but a number of layers of such leaves reduces the amount of light reaching the lower ones. Leaves growing at an angle to the ground require a much higher LAI to intercept the same quantity of light as horizontal leaves; but as their LAI increases above a certain value, angled leaves carry on photosynthesis at a faster rate. Such an adaptation in leaf po-

sition, resulting in maximum photosynthesis despite a high LAI, is characteristic of corn, grasses, beets, turnips, and other row crops (Loomis et al. 1967).

Efficiency of Photosynthesis. The photosynthetic efficiency of converting energy of the sun to organic matter can be assessed in two ways: the amount of energy required for the evolution of a molecule of oxygen, or by the ratio of calories per unit area of harvested vegetation to solar radiation intercepted. In terms of energy input photosynthesis is an efficient process. To release 1 mole of oxygen and to fix 1 mole of carbon dioxide a green plant needs an estimated 320 kcal of light energy. For each mole of oxygen evolved, approximately 120 kcal of energy are fixed. This corresponds to a gross efficiency of approximately 38 percent. Efficiencies calculated for isolated chloroplasts and for some algae amount to 21 to 33 percent (Bassam 1965; Kok 1965: Wassink 1968).

From the standpoint of calories stored in relation to light energy available, the efficiency is considerably less. The usable spectrum, 0.4 to 0.7μ wavelengths, is only about one-half the total energy incident upon vegetation. Highest short-term net efficiency measured over periods of weeks of active growth may amount to 12 to 19 percent (Wassink 1968). In most instances, however, photosynthetic efficiency is computed either for year or growing season. Efficiencies for converting radiation to biomass in coniferous forests of the northern temperate region range from 0.1 to 3 percent, for deciduous forests from 0.5 to 1 percent, and for deserts from 0.01 to 0.2 percent (Webb et al. 1983). A Puerto Rican tropical rain forest has an estimated net photosynthetic efficiency of 7 percent (H. T. Odum 1970). Croplands range from 3 to 10 percent; corn, a C_4 plant, has an efficiency of about 9 percent, and sugar cane, also a C_4 plant, about 8 percent (Cooper 1975). The net photosynthetic efficiency of land areas is about 0.3 percent, and of the ocean about 0.13 percent. Total yields of solar energy on Earth amount to about 0.15 to 0.18 percent (Wassink 1968).

Decomposition

Decomposition is the reduction of energy-rich organic matter by consumers (largely decomposers and detritivores) to CO_2, H_2O, and inorganic nutrients. Whereas photosynthesis involves incorporation of solar energy and of organic matter into biomass, decomposition involves the loss of heat energy and the conversion of organic nutrients into inorganic ones. Decomposition is a complex of many processes, including fragmentation, mixing, change in physical structure, ingestion, egestion, and concentration, accomplished by diverse organisms linked together in highly tangled food webs. It involves all consumers to some degree, because the passage of organic matter through the digestive system results in the degradation of organic matter and the contribution of changed and partially decomposed material in the form of feces as a substrate for other feeding groups. However, true decomposition is accomplished largely by decomposers, the bacteria and fungi, and by detritivores, which feed on dead matter of all kinds.

The Decomposers. The innumerable organisms involved in decomposition fall into several major functional groups. Organisms most commonly associated with decomposition are the *microflora,* the bacteria and fungi. Bacteria may be aerobic, requiring oxygen as the electron acceptor, or they may be anaerobic, able to carry on their metabolic functions without oxygen by using some inorganic compound as the oxidant. This type of respiration by anaerobic bacteria, common in aquatic muds and sediments and in the rumen of ungulate herbivores, is fermentation. Fermentation, which converts sugars to organic acids and alcohols, is less efficient in the breakdown of organic matter, lowers the pH of the substrate, and favors fungal activity. Many decomposer bacteria are facultative anaerobes. They use oxygen when it is present, but in its absence they can use inorganic compounds as their energy source.

Bacteria are the dominant microfloral decomposers of animal matter. Fungi are the major decomposers of plant material. They extend their hyphae over and into the detrital material to withdraw nutrients. Fungi range from "sugar fungi" that feed on highly soluble materials to more complex hyphal fungi that invade tissues.

Bacteria and fungi secrete enzymes into plant and animal material. Some of the products are absorbed as food, and the remainder is left for other organisms to utilize. Once one group has exploited the material to its capabilities, another group of bacteria and fungi able to utilize the remaining material more resistant to decomposition (such as cellulose and lignin) move in. Thus a succession of microflora occurs in the detritus until the material is finally reduced to inorganic nutrients.

Decomposition is aided by the fragmentation of detritus by litter-feeding invertebrates, the *detritivores.* They fall into four major groups: (1) microfauna, represented by protozoans; (2) mesofauna, whose body length falls between 100 μ and 2 mm, represented by mites, springtails, and potworms; (3) macrofauna, between 2mm and 20mm, represented by nematodes, caddisfly larvae, and mayfly and stonefly nymphs; and (4) megafauna, longer than 20 mm, represented by snails, earthworms, and millipedes. These detritivores feed on plant and animal remains and on fecal material. In aquatic and semi-aquatic ecosystems this group includes mollusks and crabs.

Energy and nutrients incorporated in bacterial and fungal biomass do not go unexploited in the decomposer world. Feeding on bacteria and fungi are the *microbivores.* Making up this group are protozoans such as amoebas, springtails (Collembola), nematodes, larval forms of beetles (Coleoptera), flies (Diptera), and mites (Acarina). Smaller forms feed only on bacteria and fungal hyphae. Because larger forms may feed on both microflora and detritus, members of this group are often difficult to separate from detritivores.

Microbivores in a way act as regulators of decomposition. They may so reduce microbial populations that they delay ordinary decomposition, or they may promote microbial activity by preventing the microflora from overpopulation and by maintaining it at a level of maximum productivity or rate of division. Thus microbivores may prevent aging and senescence of bacterial and fungal populations. If microbivores reduce fungal populations, they also stimulate their growth by dispersing fungal spores.

Stages of Decomposition. Decomposition moves through several stages, from the deposition of dead organic matter to its final breakdown into inorganic nutrients (Figure 3.9). Early stages of decomposition involve *leaching,* the loss of soluble sugars and other compounds carried away by water. An abiotic process, it results in weight loss and changes in chemical composition of the organic substance. Early stages also involve *fragmentation,* the reduction of leaves and other organic matter into smaller particles. It may be accomplished physically by wind and trampling or by the action of detritus-consuming invertebrates. Consumption of organic matter by decomposers results in the oxidation of organic compounds accompanied by the release of energy, and their subsequent degradation into smaller and simpler products, processes collectively called *catabolism.* This conversion of materials from organic to inorganic form is *mineralization.* It results in the gradual disintegration of dead organic matter into nutrients available to primary producers and microbes.

The same decomposer organisms may resynthesize these compounds into decomposer tissue in processes called *anabolism* and for a time tie up these nutrients in microbial tissue. As long as these nu-

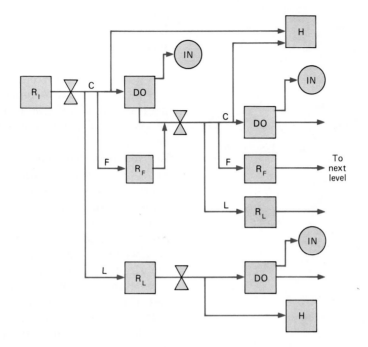

Figure 3.9 A model of the decomposition of a resource (R) over time. The decomposition process involves catabolism (C); fragmentation (F), the reduction of particle size of chemical unchanged litter; and leaching (L) of soluble materials in an unchanged chemical form to another site. All these processes at some time result in chemical changes, including mineralization to inorganic forms (IN), resynthesis (anabolism) into decomposer tissue (DO), and formation of humus (H). Each of these products undergoes further decomposition, until the primary resource is completely mineralized. (Adapted from Swift, Heal, and Anderson 1979: 51, 53, 54.)

trients are a part of decomposer biomass, they are unavailable for recycling. This situation is known as *nutrient immobilization*. The amount of mineral matter that can be tied up by microbes and detritivores varies greatly (Figure 3.10). Many microbes exhibit luxury consumption, consuming more than they need for maintenance and growth. Such consumption of nutrients, particularly potassium, calcium, and nitrogen, can affect primary production. The amount of nutrients available for primary producers depends in part on the magnitude of uptake and subsequent release through death by microbial decomposers.

Decomposition in Action. Microbial decomposition of plant leaves can begin while the leaves are still on the plant. Living plant leaves (called the *phylloplane*) produce varying quantities of exudates that support an abundance of surface microflora. These organisms feed on the exudates and on any cellular material that sloughs off. The same exudates account for the nutrients leached from the leaves during a rain. In tropical rain forests leaves are heavily colonized by bacteria, actinomycetes, and fungi (Ruinen 1962).

While microbes are utilizing exudates

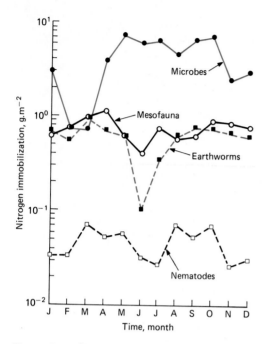

Figure 3.10 Seasonal immobilizaiton of nitrogen in forest litter by mesofauna and microbes. Microbes immobilize much greater quantities of nitrogen than mesofaunal arthropods, earthworms, and nematodes. Mesofaunal immobilization is much more constant throughout the year. Earthworm immobilization is the lowest during the summer, when microbial immobilization is the highest. However, in the summer minimal litter is available to the earthworms. During the fall and winter earthworms immobilize as much nitrogen as arthropods and microbes, which are at their lowest level of activity. (After Ausmus, Edwards, and Witkamp 1976: 411.)

cellaneous compounds have been identified in the rhizosphere. Obviously, not all such exudates occur in the rhizosphere of all plants. The absence of certain exudates can influence the quantity and quality in the microflora of rhizospheres of different plant species (Clark 1969a, b).

When the plant body becomes senescent, decomposition accelerates. The plant is invaded by both bacteria and fungi. If moisture is sufficient, fungi colonize dead culms of grass plants. A favorite point of invasion is the internode where the leaf is attached to the stem. The species of fungal flora involved are influenced by the distance of the internodes and leaves from the ground (the closer to the ground, the more humid the habitat) and by the species of grass (Hudson and Webster 1958; J. Webster 1956–1957). Fungi infect pine needles five to six months before the needles fall (Burges 1963). Destruction of the palisade layers of deciduous leaves by leaf miners opens up the affected leaves to microbial attack while they are still hanging on the tree. Still, the bulk of decomposition does not take place until the dead vegetation comes in contact with the soil.

Once on the ground, plant debris is subject to attack by other bacteria and fungi, mostly the latter. Among the first to invade the materials are the sugar-consuming fungi, such as Penicillium, which utilize the readily decomposable organic compounds (Steward et al. 1966). When the glucose is gone, the debris is invaded by other bacteria that feed on more complex carbohydrates, hemicellulose, and cellulose. The rate at which these organisms feed on the debris depends upon moisture and temperature. Higher temperature favors more rapid decomposition, and continuous moisture is more favorable than alternate wetting and drying.

The accessibility of detritus to the microflora is aided by detritivores that open

of leaves, other organisms are utilizing organic material from the roots of living plants. The soil region immediately surrounding the roots (known as the *rhizosphere*) and the root surface itself (the *rhizoplane*) support a host of microbial feeders on root litter and root exudates. Root exudates may consist of simple sugars, fatty acids, and amino acids. In fact some 10 sugars, 21 amino acids, 10 vitamins, 11 organic acids, 4 nucleotides, and 11 misup entries and break the litter into smaller parts. The action of such litter feeders as millipedes and earthworms may increase exposed leaf area up to 15 times its original size (Ghilarov 1970). Because the net assimilation of plant detritus by litter-feeders is on the average less than 10 percent—they utilize only the easily digested proteins and carbohydrates—a great deal of the material with concentrated mineral matter passes through the gut of these organisms. The fecal material is readily attacked by other microbes. Some litter-feeders, such as earthworms, enrich the soil with vitamin B_{12}. In addition they mix organic matter with the soil, bringing the material in contact with other microbes. Although the mineral pool in the contained biomass and the contribution of energy flow by the detritivores is relatively small (about 4 to 8 percent), they make a major indirect contribution to decomposition (van der Drift 1971).

The detritivores have an essential role in the decomposition of woody material—standing dead trees, fallen logs, and the like. Such material is nutritionally poor, dominated by lignin, hard to digest, and protected initially by water-repellent bark. The entry of the early microbial invaders, the Basidomycetes (white rot), is made possible by bark-feeding (cambium) beetles, wood-boring beetles, carpenter ants, and termites. Their tunnels and drillings provide entry pathways; their frass and the debris of pulverized wood provide a soft substrate for initial fungal growth; and their bodies carry fungal spores into the wood.

Many of the invading fungi are able to colonize woody debris when the chemical composition is most diverse. The initial decomposition by fungi prepares the way for an invasion of a host of organisms, many of them feeding specialists: ambrosial beetles that encourage and then consume fungal mycelia, collembola, mites, nematodes, slugs, snails, and predators, among them pseudoscorpions, spiders, centipedes, salamanders, and shrews. Mosses, lichens, ferns, and woody plants take root, aided by the mycorrhizal fungi (see Chapter 23). As the nutrients are depleted, the decaying wood diminishes in value and supports fewer species, until finally the material is reduced to humus (for detailed treatment, see Maser and Trappe 1984).

The importance of detritivores has been demonstrated in several experiments involving the use of nylon litter bags to exclude mesofauna and macrofauna from the litter sample or the use of naphthalene, which drives away arthropods but does not inhibit the activity of the microflora. Whitcamp and Olsen (1963) placed white oak leaves, some confined in litter bags and some unconfined, in pine, oak, and maple stands in November. By the following June both the unconfined and confined leaves showed a similar loss in weight. After June the unconfined leaves showed a sudden increase in microbial activity. Before June both types of leaves lost weight by the breakdown of easily decomposable substances through the action of microflora and microfauna. Because the unconfined leaves were broken into fragments by the detritivores as well as by birds, mice, wind, and rain, they were more available to microorganisms for further decay. Experiments by Kurcheva (1964), Edwards and Heath (1963), and Whitcamp and Crossley (1966) show that the suppression of activities of detritivores results in a marked slowdown in the rate of microbial decomposition. In the absence of these invertebrates, the decomposition of wood is slowed by half. Not only do the detritivores physically fragment the substrate, they also inoculate it with fungi and bacteria (Ghilarov 1970).

Without the activity of the detritivores, nutrient elements would stagnate in the litter, bound energy in the ecosystem would increase, and both primary and secondary productivity would decrease.

Decomposition of animal matter is more straightforward than decomposition of plant matter. The chemical breakdown of flesh does not require all of the specialized enzymes needed to digest plant matter. Decomposition is accomplished largely by bacteria rather than fungi and by certain arthropods such as blowflies—if scavengers and large detritivores such as crows, vultures, and foxes leave any carcass behind for decomposers.

In summer and fall, when temperatures are high, microbial activity and colonization of dead flesh by blowflies (Calliphoridae) is intense. Blowfly maggots emerging from eggs laid in the carcass can consume a small mammal carcass in seven to eight days (Putnam 1978a, b). Between bacteria and maggots, 70 percent of the organic material of a small mammal carcass is released, leaving only hair and bones. Because of low temperatures, decomposition in winter and spring is restricted to the relatively slow activity of microorganisms. The carcass, nearly 85 percent remaining, gradually becomes mummified through reduced pH and chemical changes in the tissues and is eventually fragmented and scattered.

Most fecal material has little left to decompose. However, the dung of large grazing herbivores still contains an abundance of partially digested organic matter that provides a rich resource for specialized detritivores, in addition to bacteria, fungi, and earthworms. Among them are many species of flies that lay their eggs in the dung, upon which the larvae will feed. The most notable are the dung beetles (Scarabaeinae, Aphodiinae, and Geotrupinae). Some species of the Scarabaeinid dung beetles live beneath the dung upon which their larvae feed. Others, the tumblebugs, form a mass of dung into a ball in which they lay their eggs, roll it a distance, dig a hole, and bury it as a food supply for the larvae. Aphodiinae dung beetles tunnel and form a dung ball underground; and the earth-boring dung beetles, Geotrupinae, spend most of their lives in deep burrows, usually beneath carrion or dung. The female lays her eggs in a plug of dung at the end of the burrow.

Humus Formation: The End Point. As the amount of energy decreases with time, the least decomposable material, largely derived from lignin, is left behind as humus. *Humus* is a structureless, noncellular, dark-colored, chemically complex organic material whose characteristic constituents are humin, a group of unchanged plant chemicals, and other organic compounds such as fulvic acid and humic acid. The latter is derived from lignin and plant flavonoids, which undergo degradation and conjugation with amino compounds, carbohydrates, and silicate materials. Because of the chemical complexity of the material, subsequent decomposition proceeds so slowly that the amount of organic material in the soil changes little each year. Annual loss by decomposition is balanced by the formation of new humic material. Carbon dating indicates that humus in podzol soils (see Chapter 8) has a mean residence time of 250 ± 60 years and that in chernozems 870 ± 50 years (Campbell et al. 1967).

Influences on Decomposition. The rate of decomposition is influenced by a number of environmental and biotic variables. Among them are moisture, temperature, exposure, altitude, type of vegetation, and the ratio of carbon to other elements.

Both temperature and moisture greatly influence microbial activity by affecting metabolic rates. Alternate wetting and drying and continuous dry spells tend to reduce both the activity and populations of microflora. Slope exposure, especially as it relates to temperature and moisture, and type of vegetation can increase or decrease decomposition. Whitcamp (1963) found that bacterial counts on north-facing slopes were nine times higher in hardwoods than in coniferous stands, but counts from hardwood and coniferous stands on south slopes did not differ. This finding undoubtedly was due to drier conditions on south slopes.

The species composition of leaves in the litter has the greatest influence. Easily decomposable and highly palatable leaves from such species as redbud, mulberry, and aspen support higher populations of decomposers than litter from oak and from pine, which is high in lignin. Earthworms have a pronounced preference for such species as aspen, white ash, and basswood, take with less relish and do not entirely consume sugar maple and red maple leaves, and do not eat red oak leaves at all (Johnson 1936). In a European study (Lindquist 1942) earthworms preferred the dead leaves of elm, ash, and birch, ate sparingly of oak and beech, and did not touch pine or spruce needles. Millipedes likewise show a species preference (van der Drift 1951). Therefore litter from certain species decomposes more slowly than litter from others. On easily decomposable material initially high populations and diversity of microflora decline with time as energy is depleted, but on more resistant oak and pine litter, initially low population densities increase as decomposition proceeds (Witcamp 1963).

Mineralization can take place only when the nutrient itself in nonlimiting. The availability of the nutrient in question depends upon the energy-nutrient ratio, which expresses the limitation to accessibility of energy imposed by the concentration of the nutrient. This ratio is usually expressed as a carbon-nutrient or C:X ratio. If the C:X ratio is high, then the supply of that nutrient relative to the energy source is low and activity of the microflora and decomposition are inhibited. The rate of mineralization of any element is related to the availability of that element to the decomposers. As long as the supply of energy or carbon is high and the demand for X is greater than that available, the limited nutrient is conserved (immobilized) in the biomass of the decomposers. As C is utilized and reduced to CO_2, the ratio of C to X declines. When the ratio of the detrital material drops to the ratio of the organisms themselves, usually 20 or below, the nutrient is no longer in demand (Table 3.1).

Thus, if decomposer organisms are in a situation in which the nutrients they require are limited (they usually are), the microorganisms will incorporate these nutrients into their own biomass, removing it from circulation. If the organism has a high demand for nitrogen and the detritus it colonizes has a high C:N ratio, then N is limited relative to the energy available. As N is utilized, no more is immediately available, restricting further decomposition, unless the organism can draw some nitrogen from another source, for example the soil. In this case the decomposers may deplete soil nitrogen to a point where plants experience nitrogen deficiency.

Aquatic Decomposition. In terrestrial ecosystems bacteria and fungi play the major role in the mineralization of organic matter. The same cannot be said for aquatic ecosystems. Although heterotrophic bacteria obtain both energy and carbon from the decomposition of organic matter and break down the material into smaller organic molecules, evidence suggests that bacteria act more as converters than as regenerators of nutrients. Phytoplankton

TABLE 3.1 Carbon-nutrient ratios for resource and decomposer organisms, and the approximate extent of concentration for the nutrients during uptake from one level to the next. Carbon contents assumed for each level are wood, 47 percent, fungus, 45 percent, insect, 46 percent.

	C:N	C:P	C:K	C:Ca	C:Mg
Resource (wood)	157	1424	224	147	1022
Fungal decomposer	26	94	136	11	346
Concentration	6	15	2	13	3
Animal decomposer	6	51	66	157	235
Concentration (to wood)	26	28	3	1	4
(to fungus)	4	2	2	xs	2

and zooplankton take the major part in the cycling of nutrients; but only the bacteria are able to utilize the array of organic molecules that occur in low concentrations in the water, convert them into biomass, and make them available for further transfer.

Phytoplankton, macroalgae, and zooplankton furnish dissolved organic matter, algae being the main contributors. Phytoplankton and other algae excrete quantities of organic matter at certain stages of their life cycle, particularly during rapid growth and reproduction. During photosynthesis the marine alga *Fucus vesticulosus* produces as exudate on the average 42 mg C/100 g dry weight of algae/hr. Total exudate accounts for nearly 40 percent of the net carbon fixed (Sieburth and Jensen 1970). Johannes (1968) points out that 25 to 75 percent of the regeneration of nitrogen and phosphorus takes place in the presence of microorganisms by autolysis and solution rather than by bacterial decomposition. In fact, 30 percent of the nitrogen contained in the bodies of zooplankton is lost by autolysis within 15 to 30 minutes after death, too rapidly for any bacterial action to occur.

Important in the concentration of organic nutrients are the bacteria. Only they are able to assimilate the many kinds of dissolved organic molecules that occur in low concentrations in the water. Dissolved organic matter then becomes a substrate for the growth of bacteria. Both dissolved and colloidal matter condense on the surface of air bubbles in the water, forming organic particles on which bacteria flourish (Riley 1963; R. T. Wright 1970). Fragments of cellulose supply another substrate for bacteria. Bits of plant detritus, bacteria, and phytoplankton are consumed both by bacteria and by planktonic animals (Strickland 1965). As in terrestrial ecosystems the utilization of these organic nutrients by bacteria results in both increase and immobilization of nutrients. Bacteria and algae can use nutrients such as phosphorus in excess of their need. Such luxury consumption can reduce the supply of available nutrients for each other, reducing or enhancing algal blooms.

Ciliates and zooplankton consume bacteria and in turn excrete nutrients in the form of exudates and fecal pellets in the water. Zooplankton, too, in the presence of an abundance of food, consumes more than it needs and will excrete half or more of the ingested material as fecal pellets. These pellets make up a significant fraction of the suspended material. They are attacked by bacteria, which utilize the nutrients and growth substances they contain. Thus the cycle starts over again.

Aquatic muds are largely anaerobic habitats. Fungi are absent except in plant litter lying on the bottom (Willoughby 1974), and the decomposer bacteria are largely facultative anaerobes. Incomplete decomposition of organic matter often results in the accumulation of peat and organic muck; but the particulate matter from dead plants and animals nevertheless supports a rich bacterial flora. Plant fragments are colonized by bacteria and both become food for snails and mollusks. Newell (1965) points out that the snail *Hydrobai* feeds on detritus found in mud flats. Its fecal pellets are devoid of nitrogenous compounds but are rich in carbohydrates, suggesting that the snail, whose diet is low in nitrogen, cannot digest cellulose and other complex carbohydrates. If the pellets are held in filtered seawater, their nitrogen content quickly rises, because marine bacteria colonize and rapidly grow on the fecal matter and utilize nitrogenous compounds dissolved in seawater. The fecal pellets, now enriched in nitrogen in the form of bacterial protein, are reingested by the snail. The snail then digests the bacteria, the resultant fecal pellets are again devoid of nitrogen, and are recolonized by bacteria. Thus a major function of bacteria in aquatic environments is to concentrate nutrients, as well as to release them to the environment by mineralization.

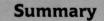

Summary

The various biotic units that make up the biosphere, the thin layer of life on Earth, are known as ecosystems. They may be as large as vast, unbroken tracts of forest and grasslands or smaller than a pond. The ecosystem is an energy-processing system, receiving abiotic and biotic inputs. The driving force is the energy of the sun. Abiotic inputs include oxygen, carbon dioxide, and nutrients from weathering of Earth's crust and from precipitation. Biotic inputs include organic materials from surrounding ecosystems.

The ecosystem itself consists of three components: (1) the autotrophs, producers that fix energy of the sun; (2) the heterotrophs, consumers and decomposers that utilize the energy and nutrients fixed by the producers and return nutrients to the system; and (3) dead organic material and inorganic substrate that act as short-term nutrient pools and maintain cycling of nutrients within the system.

The most basic processes in the functioning of the ecosystem are photosynthesis and decomposition. Photosynthesis is the process by which green plants utilize the energy of the sun to convert carbon dioxide and water into carbohydrates. Most plants utilize the Calvin cycle alone, which involves the formation of a three-carbon phosphoglyceric acid used in subsequent reactions. Other plants utilize a four-carbon process in which carbon dioxide taken into the leaf reacts to form malic or aspartic acid, stored in mesophyll cells. The CO_2 fixed in these compounds is then released to the Calvin cycle in the bundle sheath cells. These two groups of plants, C_3 and C_4 respectively, possess structural and physiological differences that are important ecologically. C_4 plants can carry on photosynthesis at higher leaf temperatures, at higher light intensities, and at lower CO_2 concentrations than C_3 plants. Succulent plants of the semiarid deserts utilize the Crassulacean Acid Metabolism (CAM). These plants open their stomata and fix CO_2 as malic acid by night. By day they close their stomata and utilize both fixed and respiratory photosynthesis in the Calvin cycle.

Involved in the return of nutrients to the ecosystem and the final dissipation of energy are the decomposer organisms. True decomposer organisms are bacteria and fungi, but decomposition depends upon other associated organisms as well. The detritivores, an array of organisms from microfauna such as protozoans to earthworms and caddisflies fragment and digest large detrital material into smaller pieces more easily attacked by microflora. Feeding on the microflora are the microbivores that act as regulators of microbial populations, preventing senescence and stimulating bacterial and fungal growth. Decomposers delay the loss of and regulate the movements of nutrients by incorporating or immobolizing them as part of decomposer tissue and by releasing nutrients through mineralization. Initial decomposition rates are high because of the leaching of soluble compounds and consumption of highly palatable tissue by microbes. Later decomposition slows because less decomposable material remains. In terrestrial ecosystems bacteria and fungi take the major role in decomposition; in aquatic ecosystems bacteria and fungi act more as converters, whereas phytoplankton and zooplankton take a major part in the cycling of nutrients.

CHAPTER 4

Climate

The abiotic environment of an ecosystem ultimately is controlled by climate, the prevailing weather conditions. Climate determines the availability of water (Chapter 5) and the degree of heat (Chapter 6); thus it influences the development of soil, the nature of vegetation, and the type of biological community. Because climate determines moisture and temperature, it influences nutrient cycling, photosynthesis, and decomposition. The major structural and functional aspects of an ecosystem relate directly and indirectly to climate.

Throughout Earth's history climate has changed. Today's climate is not the same as that of thousands and millions of years ago. It has gone through periods of warming and cooling (see Chapter 27), affecting the nature and distribution of life and the well-being and extinction of species. Currently we may be entering a warming period, induced, as some evidence suggests, by increased concentration of CO_2 in the atmosphere (see Chapter 12), brought about by the burning of fossils fuels, release of methane, and destruction and burning of tropical forests. Such warming could have major ecological effects, such as a rising sea level and changing patterns of precipitation, soil moisture, plant distribution, and agricultural crop production (Schneider 1989). To appreciate current ecological problems and the distribution of life, one needs some understanding of Earth's climate.

■ Global Patterns

Solar Radiation: Determiner of Climate

The major determiner of climate is solar radiation. Variations in the heat budgets of Earth along with Earth's daily rotation and path around the sun produce the prevailing winds and move ocean currents. Wind and ocean currents in turn influence rainfall patterns over Earth.

The amount of solar radiation that reaches the outer limit of the atmosphere as measured on a surface held perpendicular to the sun's rays is 2.0 cal/cm^2/min. This value, called the *solar constant,* fluctuates slightly, about 15 percent, through the year. The amount of solar radiation that reaches Earth's surface is one-half of the solar constant or 1.0 cal/cm^2/min.

Considering the amount of solar radiation that penetrates Earth's atmosphere as 100 percent, 21 percent is reflected from the clouds back to space, 5 percent is reflected by dust and aerosols, and 6 percent is reflected by Earth. The total reflective loss to space by Earth and atmosphere is 32 percent. Another portion of solar radiation, 3 percent, is absorbed by clouds, and 15 percent is absorbed by dust, water vapor, and carbon dioxide. The absorptive loss is 18 percent. Thus reflection and absorption account for 50 percent loss of solar radiation as it passes through Earth's atmosphere to its surface.

In passing through the atmosphere, the energy of certain wavelengths is absorbed, so the spectrum of energy that reaches Earth's surface is different from that of incident solar radiation (Figure 4.1). Ultraviolet wavelengths are nearly all removed by the atmosphere. (Humans are

Figure 4.1 Energy in the solar spectrum before and after depletion by the atmosphere from a solar altitude of 30°. The figures above the bars indicate 1, near infrared with wavelength over 1 micron; 2, near infrared, 0.7–1.0 microns; 3–5, visible light; 3, red; 4, green, yellow, and orange; 5, violet and blue; 6–7, ultraviolet. Note the strong reduction in ultraviolet. Nearly all wavelengths are absorbed by ozone at high levels. The region of peak energy is shifted toward the red end of the spectrum. Visible light in blue wavelengths is scattered rather than absorbed, producing the blue light of the sky. (From Reifsnyder and Lull 1965:21.)

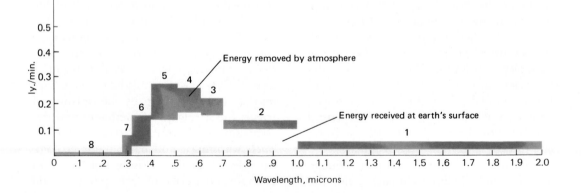

changing this protective situation, as discussed in Chapter 12.) Molecules of atmospheric gas scatter the shorter wavelengths, giving a bluish color to the sky and causing Earth to shine out in space. Water vapor scatters radiation of all wavelengths, so an atmosphere with much water vapor is whitish—think of the grayish appearance of a cloudy day. Dust scatters long wavelengths to produce reds and yellows in the atmosphere. Because of the scattering of solar radiation by dust and water vapor, part of it reaches Earth as diffuse light from the sky, called *skylight.* Infrared radiation that reaches Earth and is felt as heat (sensible heat) is absorbed, and a portion is reradiated as far infrared (4 to 100 μ). What we see as light is visible radiation, which can be broken down into the spectrum that ranges from violet to red.

Solar radiation absorbed as short-wave energy by Earth and reradiated as long-wave radiation is largely responsible for heating Earth's atmosphere. Long-wave radiation, prevented from escaping into space by the ozone layer of the stratosphere, warms the upper atmosphere. Thus the atmosphere receives most of its heat directly from Earth and only indirectly from the sun (see Chapter 12).

Albedo

Earth does not absorb all the solar radiation impinging on it. A percentage of solar radiation, called *albedo,* is reflected by Earth's surface. The amount reflected determines how fast and to what degree the surface is heated. Water's surface has a low albedo for direct rays, approximately 2 percent, and a very high albedo for low angle rays, a fact experienced on the water in late afternoons of summer. As you might expect, the albedo for snow and ice is high, 40 to 90 percent; hence the problem of snow blindness on sunny days in a snowy landscape. For forests and grasslands, the albedo ranges between 5 and 30 percent; and for clouds overhead it is about 90 percent.

Because of the reflectance of the surface and the angles of the sun's rays, Earth's albedo varies from region to region. Interestingly, the albedos for land do not vary much during the summer. From tundra to desert in North America, summer albedos are about 16 percent. Greens and browns of the summer landscape absorb far more solar radiation than they reflect. During the winter, the story is different. The treeless snow-covered tundra has a winter albedo of about 85 percent, compared to 46 to 50 percent for forest and grassland at 45° to 55°N latitudes and 18 to 29 percent for western desert and shrubland and eastern croplands and woodlands at 25° to 35°N latitude.

Global albedos measured from outer space beyond the atmosphere range from high values of 50 to 60 percent at the polar regions to 20 to 30 percent in tropical and equatorial latitudes. Although the Southern Hemisphere has less land mass and more ocean than the Northern Hemisphere, their albedos are similar. This fact suggests that global albedos are influenced more by cloud cover than by surfaces of land and sea. However, the albedo over the Northern Hemisphere has an irregular pattern compared to a more uniform distribution over the Southern Hemisphere.

Creation of Air and Ocean Currents

Sunlight does not strike Earth uniformly (Figure 4.2). Because of Earth's shape, the sun's rays strike more directly on the equator than on the polar regions. The

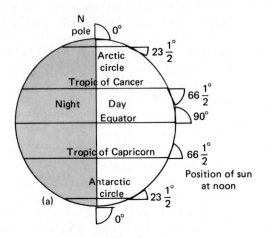
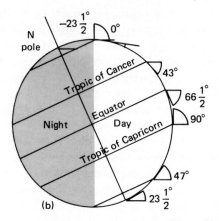

Figure 4.2 Altitude of the sun at the equinoxes and solstices. Latitude and inclination of Earth's axis at an angle of 66½° and the rotation of Earth about the sun determine the amount of solar radiation reaching any point on Earth at any time. Earth's surface at all times lies half in the sun's rays and half in shadow, marked by a dividing line, the circle of illumination. The circle of illumination is bisected by the equator and always lies at right angles to the sun's rays. Two times a year, at vernal and autumnal equinoxes (March 20 or 21, September 22 or 23), the circle of illumination passes through the poles. At the time of summer and winter solstices (June 21 or 22, December 22 or 23) the circle of illumination is tangent to the Arctic and Antarctic Circles (66½° N and S latitudes). Thus at the time of fall and spring equinoxes *(left)* the sun's rays fall directly on the equator, so at noon the altitude of the sun at the equator is 90°. At the North Pole the sun is at the horizon and keeps that position as Earth rotates. At the time of the winter solstice *(right)* the noon altitude of the sun is 66½° at the equator and 90° at the Tropic of Capricorn. The sun remains below the horizon at the North Pole and above the horizon at an altitude of 23½° at the South Pole. At the summer solstice the position of the sun is 90° at the Tropic of Cancer. To visualize this situation, turn the diagram upside down and mentally change North to read South. At this time of year the sun remains above the horizon at the North Pole for the entire 24 hours. (From A. Strahler 1971:52, 54.)

lower latitudes are heated more than the polar ones. Air heated at the equato-

rial regions rises until it eventually reaches the stratosphere, where the temperature no longer decreases with altitude. There the air whose temperature is the same or lower than that of the stratosphere is blocked from any further upward movement. With more air rising, the air mass is forced to spread out north and south toward the poles. As the air masses approach the poles, they cool, become heavier, and sink over the arctic regions. This heavier cold air then flows toward the equator, displacing the warm air rising over the tropics (Figure 4.3).

If Earth were stationary and without any irregular land masses and oceans, the atmosphere would flow in this unmodifed circulatory pattern. Earth, however, spins on its axis, rotating from west to east. An object or parcel of air moving over Earth (or any other rotating body) tends to move in a straight line and will travel with Earth at the same speed. At the equator, Earth, far from its axis, moves most rapidly at 425 m/sec. (24,000 miles a day). At 30°N and S latitudes an object on the surface moves at a reduced speed of 403 m/sec., and at 60° latitude at 232 m/sec. By the law of motion a moving object will maintain its momentum unless acted upon by some outside force such as friction. If an

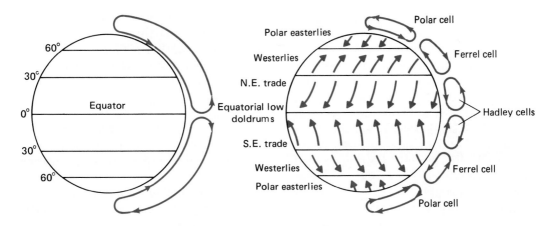

Figure 4.3 Circulation of air cells and prevailing winds. On an imaginary, nonrotating Earth *(left)* air heated at the equator would rise, spread out north and south, and upon cooling at the two poles, would descend and move back to the equator. On a rotating Earth *(right)* the air current in the Northern Hemisphere starts as a south wind that is moving north, but is deflected to the right by Earth's spin and becomes a southwest or west wind. A north wind is deflected to the right and becomes a northeast or east wind. Because the northward air flow aloft just north of the equatorial regions becomes a nearly true westerly flow, northward movement is slowed, air piles up at about 30° north latitude and loses considerable heat by radiation. Because of the piling up and heat loss, some of the air descends, producing a surface high-pressure belt. Air that has descended flows both northward to the pole and southward to the equator at the surface. The northward-flowing air current turns right to become the prevailing westerlies. The southward-flowing current, also deflected to the right, becomes the northeast trades of the low latitudes. The air aloft gradually moves northward, continues to lose heat, descends at the polar region, gives up additional heat at the surface, and flows southward. This flow of air is deflected to the right to become the polar easterlies. Similar flows take place in the Southern Hemisphere, but the air flow is deflected to the left by the rotating Earth. The pattern of rising and descending air forms tubes about Earth. The direct circulation cells near the equator are called Hadley cells and the indirect midlatitude cells are called Ferrel cells, after the meteorologists who described them. The polar cells are referred to simply as polar cells.

object or parcel moves north from the equator, where it is traveling most rapidly, it will speed up relative to Earth's rotation, causing it to veer in the direction of Earth's rotation; its path will appear to curve or deflect to the right (east). If the object moves southward from the equator, it will also curve to the right, but the deflection will appear to be left or west relative to the Northern Hemisphere. This deflection is explained as an imaginary force, the *Coriolis force.* The force depends upon the change in distance from Earth's axis; at the equator the Coriolis force is 0, and at the poles it is maximum. This explains the series of belts of prevailing east winds, known in the polar regions as polar easterlies and above the equator as the easterly trade winds. In the middle latitudes is a region of west winds known as the westerlies (Figure 4.3).

These belts break the simple flow of air toward the equator and the flow aloft toward the poles into a series of cells. The flow is divided into three cells in each hemisphere. The air that flows up from the equator forms the equatorial zone of low pressure. The equatorial air cools, loses its moisture, and descends. By the time the air has reached 30° latitude north or south, it has lost enough heat to sink, forming a cell of semipermanent high pressure encircling the earth and a region of light winds known as the horse latitudes. The air, warmed again, picks up moisture and rises once more at 60° latitude. Some of it flows toward the poles, some toward the equator. Meanwhile at each pole, another cell builds up and flows outward in a southerly and northerly direction to meet the rising warm air at approximately 60° latitude flowing toward the poles. This convergence produces a semipermanent low-pressure area at about 60° north and south latitude.

Although tropical regions about the equator are always exposed to warm temperatures, the sun is directly over the equator only two times a year, March 21 and September 21. On June 21, summer in the Northern Hemisphere, the sun is directly over the Tropic of Cancer; on December 22, summer in the Southern Hemisphere, it is directly over the Tropic of Capricorn. Aside from causing seasonal changes, the sun also affects the heating of land, water, and air masses as its vertical orientation changes north and south. This results in the movement of the junction of rising and falling air masses over the equatorial regions, the *intertropical convergence* (Figure 4.4). Because air and land heat slowly, a time lag exists between the

change in the vertical orientation of the sun and the intertropical convergence by about one month and does not move as far north and south as does the sun (about 15° north and south instead of 30°). The rainy season in the tropics moves with the intertropical convergence.

Because the masses of land and water are not uniform and because land heats and cools more rapidly than the oceans, the surface of Earth experiences uneven heating and cooling. At any given time the temperature changes are much greater over continental areas than over oceans. Oceans act as heat reservoirs; continents affect the circulation. In winter, for example, the west coast of a continent is warmer than the east coast because the air reaching the west coast has traveled over warmer ocean areas. Also produced are monsoon winds, dry winds that blow from

Figure 4.4 Oscillations in precipitation over tropical regions result in intertropical convergences, producing rainy seasons and dry seasons. As the distance from the equator increases, the longer the dry season and the less rainfall. These oscillations result from changes in the altitude of the sun between the equinoxes and the summer and winter solstices, diagramed in Figure 4.2. (From H. Walter 1971.)

Summer Autumn and Spring Winter

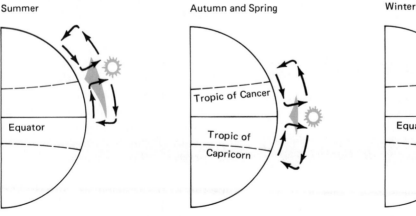
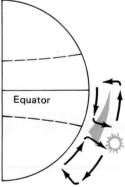

continental interiors to the oceans in summer and winds heavy with moisture that blow from the oceans to the interior in winter, bringing with them torrential rains. Last are moving air masses with cyclonic and anticyclonic frontal systems.

These circulatory patterns and the moisture regimes they influence are mostly but not entirely responsible for forests, grasslands, and deserts, which in turn reflect the climatic patterns of the planet (Figure 4.5).

The turning of Earth, solar energy, and winds produce ocean currents, the horizontal movements of water around the planet. In the absence of any land masses oceanic waters could circulate unimpeded around the globe, as does the flow of water around the Antarctic continent. But land masses divide the ocean into two main bodies, the Atlantic and Pacific. Both oceans are unbroken from high latitudes, north and south, to the equator; and both are bounded by land masses on either side (Figure 4.6).

Each ocean is dominated by two great circular water motions or *gyres*, each centered on a subtropical high pressure area. Within each gyre the current moves clockwise in the Northern Hemisphere and counterclockwise in the Southern Hemisphere. The movements of the currents are caused partly by the prevailing winds, the trades or tropical easterlies on the equator side and prevailing westerlies on the pole side. The two gyres, north and south, in both oceans are separated by an equatorial countercurrent that flows eastward. This current results from the return of lighter (less dense) surface water piled up on the western side of the ocean basin by the equatorial currents.

As the currents flow westward they become narrower and increase their speed. Deflected by the continental basin,

they turn poleward, carrying warm water with them. Two major currents in the Northern Hemisphere are the Gulf Stream in the Atlantic and the Kuroshio Current in the Pacific. Their counterparts in the Southern Hemisphere are the Australian Current in the Pacific and the Brazil Current in the Atlantic. As an example, the Gulf Stream flows north from the Caribbean, presses close to Florida, swings along the southeast Atlantic coast of North America, flows northeastward across the North Atlantic, and divides. One part becomes the Norway Current (or North Atlantic Drift), carrying warm water past Scotland and along the Norwegian coast. The other part swings south to become the Canary Current, completing the gyre (Figure 4.6). The counterpart of the Canary Current in the Southern Atlantic is the Benguela Current, which flows north along the African coast. In the Pacific the two currents are the California and the Peruvian or Humboldt. These cold currents coming from the Arctic and Antarctic regions result in upwellings that recharge surface waters with nutrients from the deep and in fogs that move over coastal areas, providing an abundance of moisture and cooling the air at latitudes where the world's hottest deserts exist inland.

These ocean currents influence ecosystems, especially in coastal regions. The Gulf Stream, for example, brings milder temperatures and more moisture to Great Britain and Norway than would normally occur at their latitudes. The cold California Current is responsible for the mild, wet climate of the northwestern Pacific coast of North America that holds the tall, dense coniferous rain forests, dominated by Douglas-fir and Sitka spruce. The fogs and low clouds of summer reduce moisture stress during that dry season.

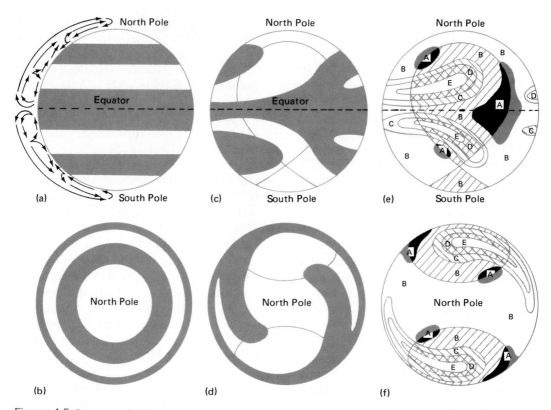

Figure 4.5 Patterns of air currents and rainfall and associated patterns of vegetation. If Earth were uniform, without irregular masses of land and oceans, the general circulation of the atmosphere and rainfall belts would appear as in (a) looking down on the Equator and (b) looking down on the pole. Because of the differential heating of Earth and its rotation, air flow is broken into three cells in each hemisphere, described in Figure 4.3. This air flow influences rainfall. Shaded portions are areas of maximum rainfall; unshaded portions are dry areas.

Unequal masses of land and water prevent the development of rain belts that correspond to the belts of ascending air. Land heats and cools more rapidly than water, the oceans act as heat reservoirs, and the continents affect the pattern of circulation of air. The generalized rainfall pattern as modified by oceans and continents is shown in (c) and (d). Again the shaded portions are areas of maximum rainfall; unshaded areas are dry. The loop formed by the dark line in (c) is a generalized continental area. It roughly rep-

resents North and South America in the Western Hemisphere and Europe, Asia, and Africa in the Eastern Hemisphere. In (d) the continents are represented by an egg-shaped loop in each hemisphere. The dry area forms an S in the Northern Hemisphere, the center being the pole.

The circulation pattern can be carried one step further. Rainfall varies greatly over Earth, from less than 25 mm to over 225 cm. This variation is reflected in the climatic regions and vegetation types. The letters and shadings in (e) and (f) represent the climatic conditions of Earth according to Thornwaite's classification:

Climatic region	Vegetation type	Code
superhumid	rainforest	A
humid	forest	B
subhumid	grassland	C
semiarid	steppe	D
arid	desert	E

(U.S.D.A. Yearbook of Agriculture 1941:100)

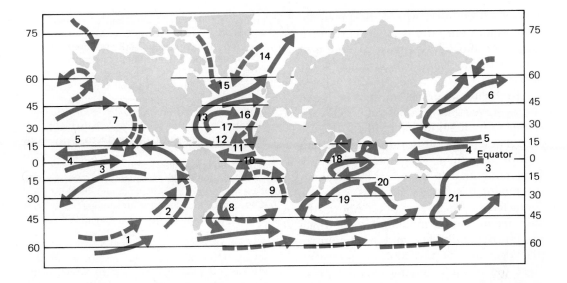

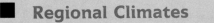

Figure 4.6 Ocean currents of the world. Note how the circulation is influenced by continental land masses and how oceans are interconnected by the currents. 1, Antarctic West Wind Drift; 2, Peru Current (Humboldt); 3, South Equatorial Current; 4, Counter Equatorial Current; 5, North Equatorial Current; 6, Kuroshio Current; 7, California Current; 8, Brazil Current; 9, Benguela Current; 10, South Equatorial Current; 11, Guinea Current; 12, North Equatorial Current; 13, Gulf Stream; 15, North Atlantic Current; 16, Canaries Drift; 17, Sargasso Sea; 18, Monsoon Drift (summer east; winter west); 19, Mozambique Current; 20, West Australian Current; 21, East Australian Current. (Dashed arrows represent cold water.) (From Coker 1947:121.)

■ Regional Climates

Interaction of Temperature and Moisture

The climate of any given region results from a combination of patterns of temperature and moisture. These patterns are determined not only by latitude, but also by the location of the region within the continental land mass. Nearby bodies of water and geographical features such as

mountains further affect these patterns. In turn, the interaction of temperature and moisture is reflected in the continental distribution of vegetation.

Zonation of vegetation in North America usually follows the pattern of moisture rather than that of temperature. In Europe the zonation of natural vegetation more nearly follows that of temperature, resulting in broad belts of vegetation running east and west. Only in the far north in North America do the vegetation zones (tundra and boreal coniferous forest) stretch in these directions. Below, the vegetation is mostly controlled by precipitation and evaporation, the latter influenced considerably by temperature. Because the available moisture becomes less from east to west, vegetation follows a similar pattern with belts running north and south. Humid regions along the coast support natural forest vegetation. This zone is broadest in the east. West of the eastern forest region is a subhumid zone where precipitation is less than evaporation. Here the ratio of precipitation to evaporation is about 60:80 percent, and the land supports a tall-grass prairie. Beyond is semiarid country, where the precipitation-evaporation ratio is 20:40 percent. It supports a short-grass prairie. To the west and on the lee of the mountains is the desert.

In the mountainous country, both east and west, vegetation zones reflect climatic changes on an altitudinal gradient (Figure 4.7). These belts often duplicate the pattern of latitudinal vegetation distribution. In general the belts include the land about the mountain's base, which has a climate characteristic of the region. Above the base region is the montane level, which has greater humidity and temperatures that decrease as the altitude increases. Here the forest vegetation changes from deciduous to coniferous. Beyond is the subalpine zone, which includes coniferous trees adapted to a more rigorous climate than the montane species. Above the subalpine lies the krumholtz, a land of stunted trees. Above that is the alpine or tundra zone, where the climate is cold. Here trees are replaced by grasses, sedges, and small tufted plants. On the very top of the highest mountains is a land of perpetual ice and snow.

Mountains influence regional climates by modifying the patterns of precipitation. Mountain ranges intercept air flows. As an air mass reaches the mountains it ascends, cools, becomes saturated (because cool air holds less moisture than warm air), and releases much of its moisture on the windward side. As the cool and dry air descends on the leeward side, it warms and picks up moisture (Figure 4.8). As a result the windward side of a mountain range supports more lush vegetation than the leeward side, where dry, often desertlike conditions exist. Thus in North America the westerly winds that flow over the Sierra Nevada and the Rockies drop their moisture on the west-facing slopes, which support excellent forest growth, while the leeward sides are desert or semidesert.

Figure 4.7 Altitudinal zonation of vegetation in mountains. (a) On Mount Marcy in the Adirondack Mountains of New York State, the forest on the lower slope is Transition, consisting chiefly of northern hardwoods. The forest on the middle slope is Canadian, with paper birch, red spruce, and balsam fir. The upper slope is Hudsonian and Alpine, characterized by dwarf spruce, willows, and heaths. In the southern Appalachians oaks and hickories replace the northern hardwoods, northern hardwoods replace the spruce, and spruce replaces willows and stunted spruce.

(b) In the Rocky Mountains the Sonoran zone is characterized by grassland and shrubby vegetation, such as chaparral and juniper; the Transition by oaks and, higher up, lodgepole pine. The Canadian zone contains lodgepole pine, Engelmann spruce, Douglas-fir, and silver fir; the Hudsonian, mountain hemlock, western white pine; the Arctic-Alpine, willows, and so on. Note that the life zones extend higher on the southwest slope than on the northeast slope. Elevations are approximate only.

(c) Zonation in the Andes Mountains of South America is more complex. This cross-section of the Andes is at the latitude of Lima, Peru, from the coast to the high Andean region. Moving up the mountains one passes from the fog-green slopes or lomas of the coastal zone (1) through deserts in the lowlands (2), stands of rootless *Tillandsias* in the desert (3), a zone of pillar cacti (4), stands of *Carica* and *Jatropha* (5), scrub steppe (6), open evergreen forest (7), grass steppe with scattered shrubs (8), valley slopes with summer-green herbaceous cover (9), rocky slopes with bromeliads (10), the puna grasslands (11), subalpine cushion plants (12), and finally glaciers and snowfields (13). (After Walter 1971.)

Similar conditions exist in the Himalayas and the mountain ranges in Europe.

A picture of regional climates and the progression of temperature and precipitation can be obtained from a climograph. A *climograph*, a composite picture of the climate of an area, is a plot of the mean monthly temperatures against the mean monthly relative humidities or precipitation (Figure 4.9 on page 59). The points for each month connected together form an irregular polygon, which can be compared with the polygon for another area.

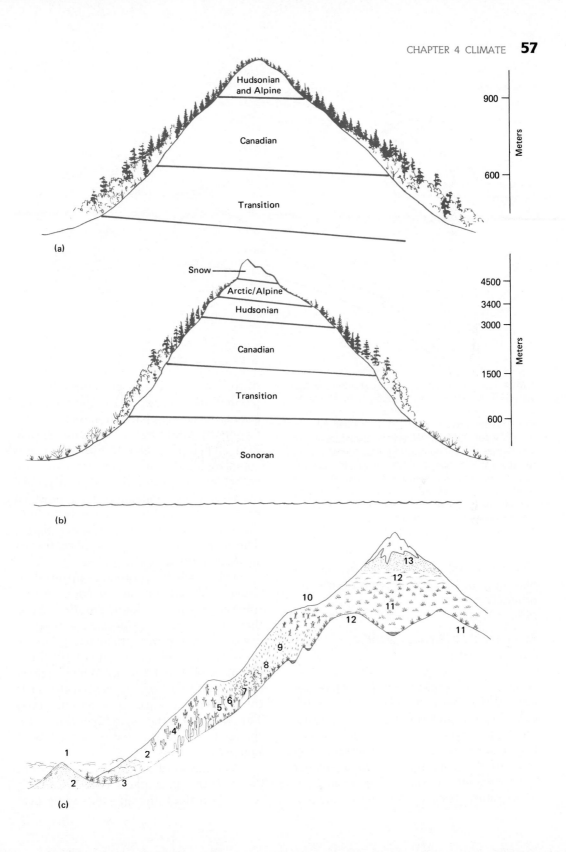

(a)

Hudsonian and Alpine

Canadian

Transition

900

600

Meters

Snow

Arctic/Alpine

Hudsonian

Canadian

Transition

Sonoran

4500
3400
3000

1500

600

Meters

(b)

(c)

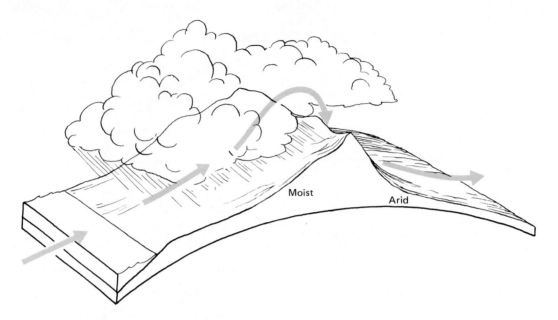

Figure 4.8 Formation of a rain shadow. When an air mass encounters a topographical barrier such as a mountain range, it is forced to go over it. As it rises, the air cools at a dry adiabatic rate of 10° C per 1000 m. When the moisture in the air reaches the condensation point, it cools at the moist adiabatic rate of 6° and drops its moisture as rain on the windward side. This abundant moisture stimulates heavy vegetational growth. The air, no longer saturated, descends on the leeward slope, warms, and picks up moisture, creating arid to semiarid conditions.

The climograph is simpler to use than tables in comparing one region with another.

Patterns of Atmospheric Movements

Adiabatic Processes. An important aspect of regional and local climates is the daily heating and cooling of air masses. Earth's atmosphere is a gas. A gas allowed to expand becomes cooler; the same volume of gas compressed into a smaller space becomes warmer. The change in temperature results from the change in degree of crowding of gas molecules. Expanded into a larger space, the molecules are more widely separated and collide less frequently, resulting in a drop in sensible heat. When the same volume of gas is compressed, the molecules are closer together and collide more frequently, raising sensible heat. Such a process, in which heat is neither lost to nor brought in from the outside, is termed an *adiabatic process.* The temperature change is called the *dry adiabatic temperature change.*

The adiabatic process is involved in rising and sinking air masses. Atmospheric pressure at high altitudes is less than near Earth's surface. As a mass of dry air rises, it moves into an area of lower pressure and cools at the rate of 10° C per kilometer (1° C per 100 m). When a parcel of air sinks to a lower elevation, the air is compressed and warmed at the same rate. This rate of change in temperature with change in height is called the *dry adiabatic lapse rate.*

As a parcel of air moves up or down, the moisture it contains condenses as it cools. As it does, the air gains some heat

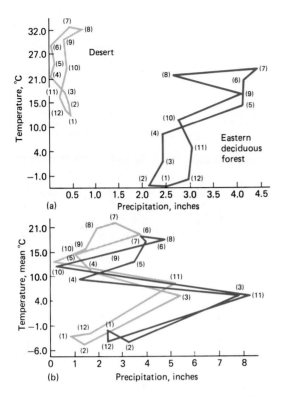

Figure 4.9 Temperature-moisture climographs. (a) The hot, dry desert climate differs graphically from the cool, temperate, moist climate of the East. Data for the graph are mean temperature and precipitation from 1941 to 1950 for Yuma, Arizona, and Albany, New York. (b) Comparison of conditions on the rain shadow side and the high rainfall side in the Appalachian Mountains in West Virginia. (Numbers in parentheses indicate month.)

water areas and lower regions of the atmosphere creates instability in the atmosphere. By day Earth is heated by short-wave solar radiation (Figure 4.10), which is absorbed in different amounts over the land depending on vegetation, slope, soil, season, and so on. Lower layers of the atmosphere, heated by the Earth's surface by radiation, conduction, and convection, rise in small volumes; colder air falls.

Consider a parcel of warm air near the surface lifted by turbulence to a higher elevation and cooling adiabatically. As it rises it is surrounded by cooler and therefore heavier air. Because the parcel of air is lighter, it continues to rise and cool at a rate less than surrounding air. As it rises it pulls in more air behind it, creating eddies of turbulent adiabatic air. This turbulence creates an *unstable* air mass. Thus an air mass cooling more rapidly than adiabatic air favors vertical motion and instability (Figure 4.11).

As air rises and adiabatic cooling continues, the air temperature approaches dew point (the temperature at which water content reaches saturation and condenses). The dew point temperature decreases at the rate of 0.2° C per 100 m. Dew point and air temperature approach each other at the rate of about 0.9° C per 100 m. As the air mass reaches dew point, the water it contains condenses to form clouds of fog. This condensation of moisture in rising turbulent air masses gives substance to unstable air masses and developing thunderheads on a hot summer day.

Unstable air conditions result when rising air masses cool at a lapse rate faster than adiabatic. If a rising air mass cools at a rate slower than adiabatic, it is cooler and denser than the surrounding air and sinks lower. Such an air mass is not subject to eddies and turbulence. When such conditions prevail, an air mass is considered

of condensation, generating sensible heat that counteracts the cooling. The rate of cooling takes on a reduced value known as the *moist adiabatic lapse rate*. The moist adiabatic lapse rate has no single value because the rate varies with the temperature of and the moisture in the air; but the rate averages about 6° C per kilometer.

Under ideal conditions air rises and cools and sinks and heats at adiabatic rates, but differential heating of land and

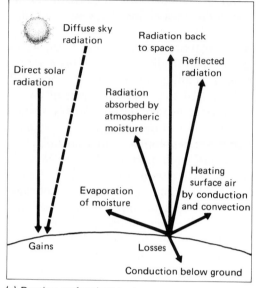

(a) Daytime surface heat exchange

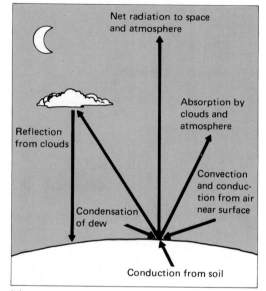

(b) Nighttime surface heat exchange

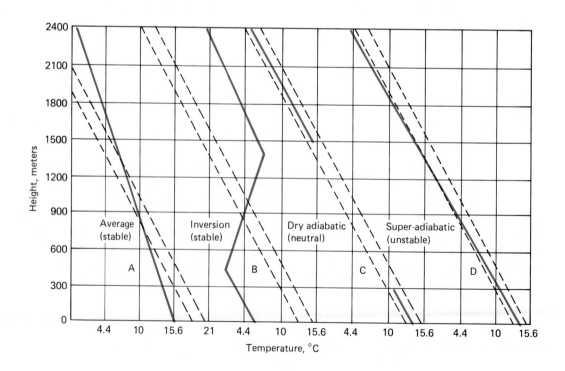

Figure 4.10 Radiant heating of Earth. (a) Solar radiation that reaches Earth's surface in the daytime is dissipated in several ways, but heat gains exceed heat losses. (b) At night there is a net cooling of the surface, although some heat is returned by various processes. (Adapted from Schroeder and Buck 1970:14.)

Figure 4.11 Adiabatic processes plotted for hypothetical stable, neutral, and unstable air masses. The dashed lines represent the dry-adiabatic lapse rate; the solid line the measured lapse rate of the hypothetical air mass and the arrow line the temperature of a rising or lowering parcel of unsaturated air in which the surrounding air mass is 10° C at 914 m (3000 ft). In A, the air has a lapse rate of 2° C per 300 m, but the parcel of air is lifted and cooled at a rate of 3° C per 300 m and becomes progressively colder and more dense than the surrounding air. At 914 m the temperature of the parcel of air would be 4° C, while that of the surrounding air would be 6° C. The parcel of air would then drop, warm at the dry adiabatic rate, and become warmer than its environment, whereupon the parcel of air would move up again. In B, the parcel of air is initially in an inversion layer where the temperature increases at the rate of 1.7° C per 300 m altitude. As the parcel of air lifts, it cools while the surrounding air remains warm. The parcel of air will then drop back to its original level as soon as the lifting force is removed. In C the parcel of air has a measured lapse rate the same as that of surrounding air. If moved up or down, the parcel of air will change at the same rate as that of its environment. In D, the plotted temperature lapse rate is 3.4° C per 300 m, greater than the dry adiabatic lapse rate. If this parcel of air is lifted, it will cool at the dry adiabatic rate, or 0.3° C less per 300 m than its surroundings. Thus the parcel of air is warmer and less dense than its surroundings. Its buoyancy will cause it to move upward rapidly as long as it remains warmer than the surrounding air. If the parcel of air moves downward, it will cool more rapidly than the surrounding air. Thus a parcel of air having a lapse rate greater than the dry adiabatic rate exhibits vertical motion and overturning. (After Schroeder and Buck 1970:52.)

stable. Therefore a temperature lapse rate less than the dry adiabatic rate is stable. If the lapse rate is the same as dry adiabatic, then the air mass is considered *neutral.*

Inversions. Gradual radiational cooling by night cools the surface air above Earth. This layer of air gradually deepens as the night progresses and forms a nighttime surface temperature inversion in which the temperature increases with height (Figure 4.12a). This night inversion is gradually eliminated by morning surface heating (Figure 4.12b).

Such inversions are particularly pronounced when the air mass is stable and the weather is calm and clear. Consider a typical summer day. At night the layer of surface air is cooled while the air aloft remains near the warmer daytime temperature. In mountainous or hilly country, cold dense air flows down slopes and gathers in the valleys. The cold air then is trapped beneath a layer of warm air (Figure 4.13). Such radiational inversions trap impurities and other air pollutants. Smoke from industry and other heated pollutants rise until their temperature matches the surrounding air. Then they flatten out and spread horizontally. As pollutants continue to accumulate, they may fill the entire area with smog. These inversions break up when surface air is heated during the day to create vertical convections and turbulence or when a new air mass moves in.

Similar but more widespread inversions occur when a high-pressure area stagnates over a region. In a high-pressure area the air flow is clockwise and spreads outward. The air flowing away from the high must be replaced, and the only source for replacement air is from above. Thus surface high-pressure areas are regions of sinking air movements from aloft, called *subsidence.*

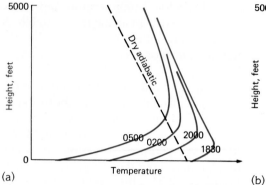

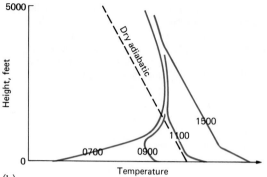

Figure 4.12 Formation and elimination of a nighttime surface inversion typical on clear, cool nights. (a) As the ground cools rapidly after sundown, a shallow surface inversion forms (1830). As cooling continues during the night, the inversion deepens from the surface upward, reaching its maximum depth just before dawn (0500). (b) After sunrise, the surface begins to warm (0700) and the night surface inversion is gradually eliminated during the forenoon of a clear summer day. A surface superadiabatic layer and a dry adiabatic layer above deepen until they reach their maximum depth about midafternoon (1500). (After Schroeder and Buck 1970:27, 28.)

When high-level winds slow down, heavy cold air at high levels in the atmosphere tends to sink. As the parcel of air sinks, it compresses, heats, and becomes drier. A layer of warm air then develops at a higher level in the atmosphere with no chance to descend. It hangs several hundred to several thousand feet above the earth, forming a *subsidence inversion* (Figure 4.14). Such inversions tend to prolong the period of stagnation and increase

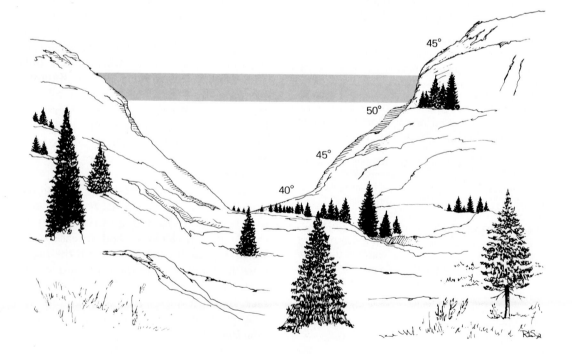

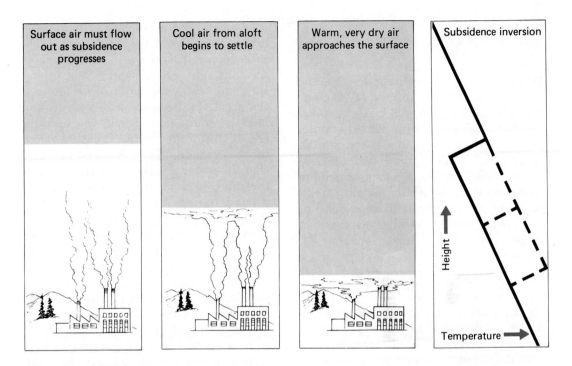

the intensity of pollution. Subsidence inversions that bring about our highest concentrations of pollution are often accompanied by lower-level radiation inversions.

Along the west coast of the United States, and occasionally along the east coast, the warm seasons often produce a

Figure 4.13 Topography plays an important role in the formation and intensity of nighttime inversions. At night air cools next to the ground, forming a weak surface inversion in which temperature increases with height. As cooling continues during the night, the layer of cool air gradually deepens. At the same time cool air descends downslope. Both cause the inversion to become deeper and stronger. In mountain areas the top of the night inversion is usually below the main ridge. If air is sufficiently cool and moist, fog may form in the valley. Smoke released in such an inversion will rise only until its temperature equals that of the surrounding air. Then smoke flattens out and spreads horizontally just below the thermal belt. (Adapted from Schroeder and Buck 1970:29.)

Figure 4.14 Descent of a subsidence inversion. The movement of the inversion is traced by successive temperature measurements, indicated by the dashed lines. The nearly horizontal dashed lines indicate the descending base of the inversion. The solid line indicates temperature. The temperature lapse rate in the descending layer is nearly dry adiabatic. The bottom surface is marked by a temperature inversion. Two features, temperature inversion and a marked decrease in moisture, identify the base of the subsiding layer. Below the inversion is an abrupt rise in the moisture content of the air. (After Schroeder and Buck 1970:62.)

coastal or *marine inversion*. In this case cool, moist air from the ocean spreads over low land. This layer of cool air, which may vary in depth from a few hundred to several hundred thousand feet, is topped by warmer, drier air, which also traps pollutants in the lower layers. Such inversions intensify the effects of air pollution on plant and animal life and human health (see Chapter 12).

■ Microclimates

When the weather report states that the temperature is 24° C and the sky is clear, the information may reflect the general weather conditions for the day. But on the surface of the ground, in and beneath the vegetation, on slopes and cliff tops, in crannies and pockets, the climate is quite different. Heat, moisture, air movement, and light all vary greatly from one part of the community to another to create a range of "little" or "micro" climates.

Climate near the Ground

On a summer afternoon the temperature under a calm, clear sky may be 28° C at 1.83 m, the standard level of temperature recording. But on or near the ground—at the 5 cm level—the temperature may be 5° C higher; and at sunrise, when the temperature for the 24-hour period is the lowest, the temperature may be 3° C lower at ground level (Biel 1961). Thus in the middle eastern part of the United States, the afternoon temperature near the ground may correspond to the temperature at 1.83 m in Florida, 700 miles to the south; and at sunrise the temperature may correspond to the 1.83 m temperature in southern Canada. Even greater extremes occur above and below the ground surface. In New Jersey, March temperatures about the stolons of clover plants 1.5 cm above the surface of the ground may be 21° C, while 7.5 cm below the surface the temperature about the roots is − 1° C (Biel, 1961). The temperature range for a vertical distance of 9 cm is 2° C. Under such climatic extremes most organisms exist.

The chief reason for the great differences between temperature at ground level and at 1.83 m is solar radiation. During the day the soil, the active surface, absorbs solar radiation, which comes in short waves as light, and radiates it back as long waves to heat a thin layer of air above it (Figure 4.15). Because air flow at ground level is almost nonexistent, the heat radiated from the surface remains close to the ground. Temperatures decrease sharply in the air above this layer and in the soil below. The heat absorbed by the ground during the day is reradiated by the ground at night. This heat is partly absorbed by the water vapor in the air above. The drier the air, the greater is the outgoing heat and the stronger is the cooling of the surface of the ground and the vegetation. Eventually the ground and the vegetation are cooled to the dew point, and water vapor in the air may condense as dew. After a heavy dew a thin layer of chilled air lies over the surface, the result of rapid absorption of heat in the evaporation of dew.

Figure 4.15 Idealized temperature profiles in the ground and air for various times of day. C = heat transport by convection; L = heat transport by latent heat of evaporation or condensation. (From Gates 1962).

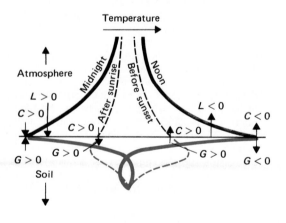

Influences of Vegetation and Soil

By alternating wind movement, evaporation, moisture, and soil temperatures, vegetation influences or moderates the microclimate of an area, especially near the ground. Temperatures at ground level under the shade are lower than those in places exposed to the sun and wind. Average maximum soil temperatures at 2.5 cm below the surface in a northern hardwoods forest and an aspen-birch forest in New York State were 15.5° C from mid-May to late June and 20° C (absolute maximum 25° C) from early July to mid-August (Spaeth and Diebold, 1938). Mean differences between maximum temperatures for the forest and adjacent fields ranged from 5° to 12° C. Maximum soil temperatures at 2.5 cm in open chestnut-oak forests were 19° C from mid-May to late June and 24° C (absolute maximum,

34° C) from July to mid-August. On fair summer days a dense forest cover reduces the daily range of temperatures at 2.5 cm by 10° to 18° C, compared with the temperature in the soils of bare fields.

Vegetation also reduces the steepness of the temperature gradient and influences the height of the active surface, the area that intercepts the maximum quantity of solar insolation. In the total absence of or in the presence of very thin vegetation, temperature increases sharply near the soil; but as plant cover increases in height and density, the leaves of the plants intercept more solar radiation (Figure 4.16). Plant crowns then become the active surface and raise it above the

Figure 4.16 Vertical temperature gradients at midday in a cornfield from seedling stages to harvest. Note the increasing height of the active surface. (Adapted from Wolfe et al. 1949.)

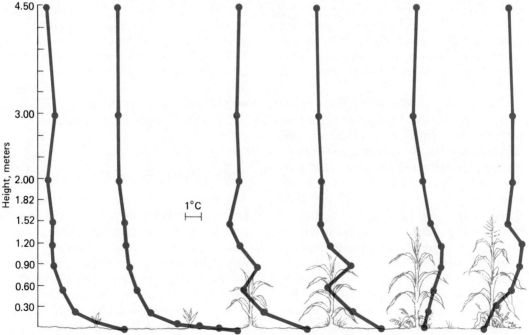

ground. As a result temperatures are highest just above the dense crown surface and lowest at the surface of the ground. Maximum absorption of solar radiation in tall grass occurs just below the upper surface of the vegetation, whereas in short grass maximum temperatures are at the ground level (Waterhouse 1955). As the grasses grow taller the level of maximum temperature falls into and rises with the upper level of the grass stalks until the temperature eventually reaches an approximate equilibrium with the air above. (Among broad-leafed plants daily maximums occur on the upper leaf surfaces.) At night minimum temperatures are some distance above the ground because the air is cooled above the tops of plants and the dense stalks prevent the chilled air from settling to the ground.

Within dense vegetation air movements are reduced to convection and diffusion (Figure 4.17). In dense grass and low plant cover, complete calm exists at ground level. This calm is an outstanding feature of the microclimate near the ground because it influences both temperature and humidity and creates a favorable environment for insects and other animals.

Vegetation deflects wind flow up and over its top. If the vegetation is narrow, such as a windbreak or a hedgerow, the microclimate on the leeward side may be greatly affected. Deflection of wind produces an area of eddies immediately behind the vegetation, in which the wind speed is low and small particles such as seeds are deposited. Beyond is an area of turbulence, in which the climate tends to be colder and drier than normal. If some wind passes through the barrier and some goes over it, no turbulence develops, but the mean temperature behind the barrier is high in the morning and lower in the afternoon.

Humidity changes greatly from the ground up. Because evaporation takes place at the surface of the soil or at the active surface of plant cover, the vapor content (absolute humidity) decreases rapidly from a maximum at the bottom to atmospheric equilibrium above. Relative humidity increases above the surface, because actual vapor content increases only slowly during the day, whereas the capacity of the heated air over the surface to hold moisture increases rather rapidly. During the night little difference exists above and on the ground. Within growing vegetation, however, relative humidity is much higher than above the plant cover. In fact, near-saturation conditions may exist.

Soil properties, too, enter the microclimatic picture. In a soil that conducts heat well, considerable heat energy will be transferred to the substratum, from which it radiates to the surface at night. On such soils surface temperatures are lower by day and higher by night than the surface temperatures of poorly conducting soils. This influences the occurrence of frost. Moist soils are better conductors of heat than dry soils. Light-colored sandy soils increase reflection and reduce the rate at which heat energy is absorbed. Dark soils, on the other hand, absorb more heat.

Valleys and Frost Pockets

The widest climatic extremes occur in valleys and pockets, areas of convex slopes, and low concave surfaces. These places have much lower temperatures at night, especially in winter, much higher temperatures during the day, especially in summer, and a higher relative humidity. Protected from the circulating influences of the wind, the air becomes stagnant. It is heated by insolation and cooled by terrestrial radiation, in sharp contrast to the wind-exposed, well-mixed air layers of the upper slopes. In the evening cool air from the uplands flows down the slope into the pockets and valleys to form a lake of cool air. Often when the warm air in the valley comes into contact with the inflowing cold air, the moisture in the warm air may condense as valley fog.

A similar phenomenon takes place on small concave surfaces. Like the larger valley, these concave surfaces radiate heat

rapidly on still, cold nights, and cold air flows in from surrounding higher levels. On such sites the air temperature near the ground may be 8° C lower than the surrounding terrain, causing a temperature inversion. Because low ground temperatures in these areas tend to result in late spring frosts, early fall frosts, and a sub-

Figure 4.17 Comparative wind velocities. (a) The distribution of wind velocities with height as affected by the timber canopy of coniferous forests for wind velocities of 8, 16, and 24 km/hr at 43 m above the ground. (b) The average wind velocity during a June day (based on 1938, 1939, and 1940 data) inside a coniferous forest with a cedar understory in northern Idaho. Note the decrease in velocity near the ground. (From Gisborne 1941.)

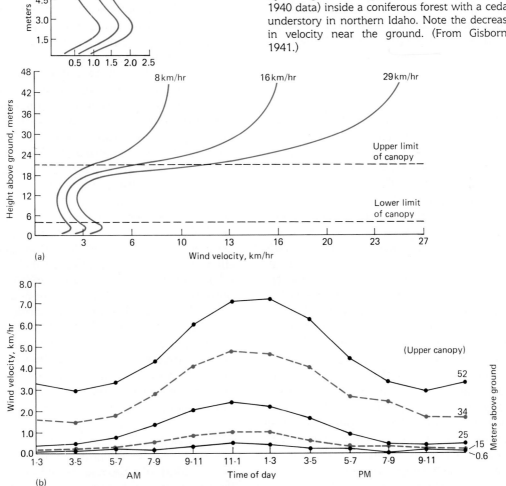

sequent short growing season, these depressions are called *frost pockets*. The pockets need not be deep. Minimum temperatures in small depressions only 1 to 1.2 m deep were equivalent to those of a nearby valley 60 m below the general level of the land (Spurr 1957). Such variations in temperature due to local microrelief can strongly influence the distribution and growth of plants. Tree growth is inhibited; and because the low surfaces more often than not accumulate water as well as cold air, such sites may contain plants of a more northern distribution. Frost pockets may also develop in small forest clearings. The surface of the tree crowns channels cold air into the clearings as terrestrial radiation cools the layer of air just above.

North-Facing and South-Facing Slopes

The greatest microclimatic differences exist between north-facing and south-facing slopes (Figure 4.18). South-facing slopes in the Northern Hemisphere receive the most solar energy. The amount of energy is maximal when the slope grade equals the sun's angle from the zenith point. North-facing slopes receive the least energy, especially when the slope grade equals or exceeds the angles of sun-ray inflection. At latitude 41° north (about central New Jersey and southern Pennsylvania) midday insolation on a 20° slope is, on the average, 40 percent greater on the south-facing slopes than on the north-facing slopes during all seasons. This difference has a marked effect on the moisture and heat budget of the two sites. High temperatures and associated low vapor pressures induce evapotranspiration of moisture from soil and plants. The evaporation rate often is 50 percent higher, the average temperature higher, the soil moisture lower, and the extremes

more variable on south-facing slopes. Thus the microclimate ranges from warm and dry (xeric) conditions with wide extremes on south-facing slopes to cool and moist (mesic), less variable conditions on north-facing slopes (Figure 4.19). Xeric conditions are most highly developed on the top of south-facing slopes, where air movement is greatest, whereas the most mesic conditions are at the bottom of the north-facing slopes.

The whole north-facing and south-facing slope complex is the result of a long chain of interactions: Solar radiation influences moisture regimes; the moisture regime influences the species of trees and other plants occupying the slopes; the species of trees influence mineral recycling, which is reflected in the nature and chemistry of the surface soil and the nature of the herbaceous ground cover.

Being mobile, few if any animals are typical of only north-facing or south-facing slopes, as far as we now know. However, their movements may be limited to some extent by the differences in conditions and food supplies on the slopes. Deer tend to use south-facing slopes more heavily in winter and early spring and north-facing slopes in summer (Taber and Dasmann 1958). In the central Appalachians the red-backed vole, normally an inhabitant of cool fir, spruce, aspen, and northern hardwood forests throughout its range, is restricted in its local distribution to forested, mesic, north-facing slopes. Those species of soil invertebrates intolerant of humidity, such as some mites, can exist only in a dry habitat and therefore are confined to the xeric conditions of the south-facing slope.

Urban Microclimate

The building of cities has altered not only the slope, soil, and vegetation of the land

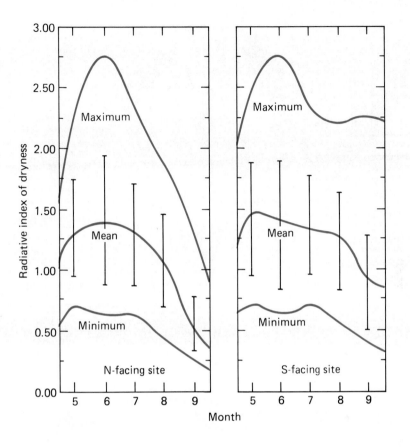

Figure 4.18 Microclimates of north-facing and south-facing slopes as described by the seasonal course of the radiative index of dryness during 1948–1982 on the Little Laurel Run watershed in the Appalachian mountains of northern West Virginia (39°41' N, 79°45' W). The radiative index of dryness is the ratio of seasonal sums of net radiation to those of latent heat of precipitation. When over a period of time net radiation is greater than the latent heat of precipitation, there is an excess of energy and the site experiences a dry period. On both the north-facing and south-facing slopes, the excess of energy supply over water supply follows a seasonal course. In June, for example, net radiation may exceed the latent heat of precipitation by a factor of 2.75. Note, however, that the excess of energy supply over water supply on the south-facing slope lasts longer than that on the north-facing slope. Thus north-facing slopes tend to be more mesic and cooler than south-facing slopes. (After Tajchman and Harris 1987.)

they occupy, but also has altered the atmosphere, creating a distinctive urban microclimate.

The urban microclimate is a product of the morphology of the city and the density and activity of its occupants. In the urban complex, stone, asphalt, and concrete pavement and buildings with a high capacity for absorbing and reradiating heat replace natural vegetation with low conductivity of heat. Rainfall on impervious surfaces is drained away as fast as possible, reducing evaporation. Metabolic heat from masses of people and waste heat from buildings, industrial combustion, and vehicles raise the temperature of the surrounding air. Industrial ac-

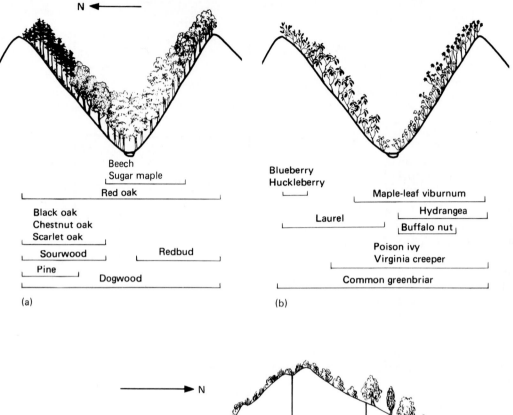

(a)

(b)

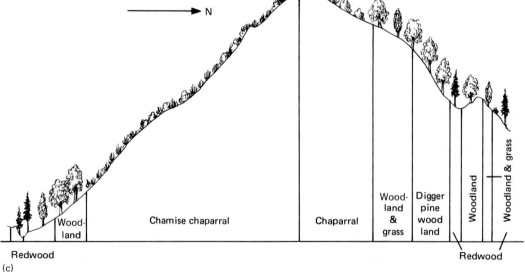

(c)

Figure 4.19 Influence of microclimate on the distribution of vegetation on north-facing and south-facing slopes. (a, b) The distribution of trees and shrubs in the hill country of southwestern West Virginia. (c) Vegetation in the Point Sur area in California. In each case note the similarity of vegetation on the lower parts of north-facing and south-facing slopes. (West Virginia, original data; California, adapted from Griffin and Critchfield, 1972.)

tivities, power production, and vehicles pour water vapor, gases, and particulate matter into the atmosphere in great quantities. The effect of this storage and reradiation of heat is the formation of a heat island about cities (Figure 4.20) in which the temperature may be 6° to 8° C higher than the surrounding countryside (Landsberg 1970; SMIC 1971).

Heat islands are characterized by high temperature gradients about the city. Highest temperatures are associated with areas of highest density and activity; temperatures decline markedly toward the periphery of the city (Figure 4.21). Although detectable throughout the year, heat islands are most pronounced during summer and early winter and are most noticeable at night when heat stored by pavements and buildings is reradiated to the air. The magnitude of the heat island is influenced strongly by local climatic conditions such as wind and cloud cover. If the wind speed, for example, is above some varying critical value, a heat island cannot be detected.

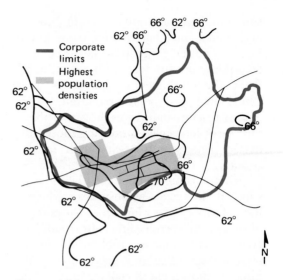

Figure 4.21 Thermal pattern of night air in a small city, Chapel Hill, North Carolina. The highest temperatures are inside the corporate limits, where the population and activity are the greatest. (From Kopec 1970.)

During the summer the buildings and pavement of the inner city absorb and store considerably more heat than does the vegetation of the countryside. In cities with narrow streets and tall buildings the walls radiate heat toward each other instead of toward the sky. At night these structures slowly give off heat stored during the day. Although daytime differences in temperature between the city and the country may not differ noticeably, nighttime differences become pronounced shortly after sunset and persist through the night. The nighttime heating of the air from below counteracts radiative cooling and produces a positive temperature lapse rate while an inversion is forming over the countryside. This, along with the surface temperature gradient, sets the air in motion, producing "country breezes" to flow into the city.

In winter solar radiation is considerably less because of the low angle of the

Figure 4.20 Idealized scheme of nighttime circulation above a city in clear, calm weather. A heat island develops over the city. At the same time a surface inversion develops in the country. As a result cool air flows toward the city, producing a country breeze in the city at night. Lines are temperature isotherms; arrows represent wind. (Adapted from H. Landsberg 1970.)

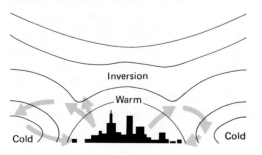

sun, but heat accumulates from human and animal metabolism and from home heating, power generation, industry, and transportation. In fact, heat contributed from these sources is 2.5 times that contributed by solar radiation. This energy reaches and warms the atmosphere directly or indirectly, producing more moderate winters in the city than in the country.

Urban centers influence the flow of wind. Buildings act as obstacles, reducing the velocity of the wind up to 20 percent of that of the surrounding countryside, increasing its turbulence, robbing the urban area of the ventilation it needs, and inhibiting the movement of cool air in from the outside. Strong regional winds, however, can produce thermal and pollution plumes, transporting both heat and particulate matter out of the city and modifying the rural radiation balance a few miles downwind (Clarke 1969; Oke and East 1971).

Throughout the year urban areas are blanketed with particulate matter, carbon dioxide, and water vapor. The haze reduces solar radiation reaching the city by as much as 10 to 20 percent. At the same time, the blanket of haze absorbs part of the heat radiating upward and reflects it back, warming both the air and the ground. The higher the concentration of pollutants, the more intense is the heat island.

The particulate matter has other microclimatic effects. Because of the low evaporation rate and the lack of vegetation, relative humidity is lower in the city than in surrounding rural areas, but the particulate matter acts as condensation nuclei for water vapor in the air, producing fog and haze. Fogs are much more frequent in urban areas than in the country, especially in winter (Table 4.1).

TABLE 4.1 Climate of the City Compared to the Country

Elements	Comparison with Rural Environment
Condensation nuclei and particles	10 times more
Gaseous admixtures	5–25 times more
Cloud cover	5–10 percent more
Winter fog	100 percent more
Summer fog	30 percent more
Total precipitation	5–10 percent more
Relative humidity, winter	2 percent less
Relative humidity, summer	8 percent less
Radiation, global	15–20 percent less
Duration of sunshine	5–15 percent less
Annual mean temperature	0.5°–1.0° C more
Annual mean wind speed	20–30 percent less
Calms	5–20 percent more

Source: Adapted from H. E. Landsberg 1970.

Another consequence of the heat island is increased convection over the city. Updrafts, together with particulate matter and large amounts of water vapor from combustion processes and steam power, lead to increased cloudiness over cities and increased local rainfall both over cities and over regions downwind. An evidence of weather modification by pollution is the increase in precipitation and stormy weather about La Porte, Indiana, downwind from the heavily polluted areas of Chicago, Illinois, and Gary, Indiana, and close to moisture-laden air over Lake Michigan. Since 1925 there has been a 31 percent increase in precipitation, a 35 percent increase in thunderstorms, and a 240 percent increase in the occurrence of hail (Changnon 1968).

■ Summary

Solar radiation, the major source of thermal energy for Earth, is the major determinant of climate. Variations in heat budgets influenced by the position of incoming rays on a spherical Earth, the distribution of land and water masses, and the albedo (the percentage of solar radiation reflected back toward space), and Earth's daily rotation produce the prevailing winds, move ocean currents, and affect rainfall patterns over the planet. The climate of any given region is a combination of patterns of temperature and moisture, which are influenced by latitude and location in the continental land mass. In mountains climatic changes on an altitudinal gradient are reflected in zonation of vegetation. At the same time mountain ranges influence regional climates by intercepting air flow, causing its moisture to drop out on the windward side and creating a rain shadow on the leeward side.

Daily heating and cooling causes air masses to sink and rise. Under certain conditions the temperature of air masses increases with height rather than decreases. Such an air mass is very stable, creating a temperature inversion that can trap atmospheric pollutants and hold them close to the ground. Inversions break up when air close to the ground heats, causing it to circulate and rise through the inversion, or when a new air mass moves into the area.

Although regional climates determine conditions over an area, the actual climatic conditions under which organisms live vary considerably from one area within a region to another. These variations, or microclimates, are influenced by topographic differences, height above the ground, exposure, and other factors. Most pronounced are environmental differences between the ground level and upper strata of vegetation and between north-facing and south-facing slopes. Other microclimates exist over urban areas. Urban areas are characterized by the presence of a heat island. Compared to surrounding rural areas, a city has a higher average temperature, particularly at night, more cloudy days, more fog, more precipitation, a lower rate of evaporation, and lower humidity.

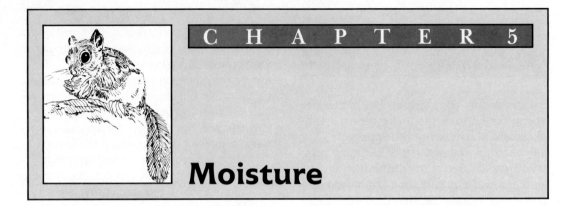

Moisture

Water is essential to all life. Means of obtaining and conserving it have shaped the nature of terrestrial communities. Means of living within it have been the pervading influence on aquatic life. The seasonal, regional, and global distribution of precipitation influences the distribution of plants and animals. How evenly precipitation is distributed throughout the year is more important to life than the average annual rainfall. Major differences exist between regions receiving 130 cm of precipitation well distributed throughout the year and a region in which nearly all of the 130 cm of precipitation falls during a few months. In the latter situation, typical of regions with wet and dry seasons, organisms must face a long period of drought. Thus the nature of the hydrologic cycle and its spatial and temporal variations, along with temperature, strongly influence the nature of ecosystems and the distribution of organisms.

■ The Water Cycle

Leonardo da Vinci wrote, "Water is the driver of nature." Perceptive as he was, even he could not have appreciated the full meaning of his statement based on the scientific knowledge of his time. Without the cycling of water, biogeochemical cycles could not exist, ecosystems could not function, and life could not be maintained. Water is the medium by which materials make their never-ending odyssey through the ecosystem.

Distribution of Water

Although water such as a stream or autumn rains looks like a local phenomenon,

it forms a single worldwide resource distributed in land, sea, and atmosphere and unified by the hydrological cycle. It is influenced by solar energy, by the currents of the air and oceans, by heat budgets, and by water balances of land and sea. Through historical time the balance of free water has remained relatively stable, although the balances between land and sea have fluctuated. According to the Russian hydrologist Shinitnikov (Kalinin and Bykov 1969), at present we are passing from a humid period in earth's history to a dry one, in which the land areas are losing water at the rate of 105 mi 3/yr, and the oceans are gaining that amount, rising on the average of 0.05 in./yr.

Oceans cover 71 percent of the earth's surface (Table 5.1). With a mean depth of

TABLE 5.1 World's Water Resources

Resource	Volume (W) (10^3 km^3)	Annual Rate of Removal (Q)		Renewal Period (T = W/Q)
		Volume (10^3 kg^3)	Process	
Total water on earth	1,460,000.0	520.0	evaporation	2,800 years
Total water in the oceans	1,370,000.0	449.0	evaporation	3,100 years
		37.0	difference between precipitation and evaporation	37,000 years
Free gravitational waters in the earth's crust to a depth of 5 km in the zone of active water exchange	60,000.0	13.0	underground runoff	4,600 years
	4,000.0	13.0	underground runoff	300 years
Lakes	750.0	—	—	—
Glaciers and permanent snow	29,000.0	1.8	runoff	16,000 years
Soil and subsoil moisture	65.0	85.0	evaporation and	280 days
Atmospheric moisture	14.0	520.0	underground runoff precipitation	9 days
River waters	1.2	36.3*	runoff	12 (20) days

Note: Average error is probably 10–15 percent.
*Not counting the melting of Antarctic and Arctic glaciers.
Source: Kalinin and Bykov, 1969.

3.8 km (2.36 mi), they hold 93 to 97 percent of all the earth's waters (depending on the estimate used). Fresh water represents only 3 percent of the planet's water supply. Of the total fresh water on Earth, 75 percent is locked up in glaciers and ice sheets, enough to maintain all the rivers of the world at their present rate of flow for the next 900 years. Thus about 2 percent of the world's water is tied up in ice. This leaves less than 1 percent of the world's water available as fresh. Freshwater lakes contain 0.3 percent of the freshwater supply, and at any one time rivers and streams contain only 0.005 percent of it. Soil moisture accounts for approximately 0.3 percent. Another small portion of the earth's water is tied up in living material.

Groundwater accounts for 25 percent of our fresh water. Groundwater fills the pores and hollows within the earth just as water fills pockets and depressions on the surface. Estimates, necessarily rough and inaccurate, place renewable and cyclic groundwater at 7×10^6 km^3 (Nace 1969), or approximately 11 percent of the freshwater supply. Some of the groundwater is "inherited," as in aquifers in desert regions, where the water is thousands of years old. Because inherited water is not rechargeable, heavy use of these aquifers for irrigation and other purposes is depleting the supply. In the foreseeable future the supply could be exhausted. A portion of the groundwater, approximately 14 percent, lies below 1000 m. Known as fossil water, it is often saline, and does not participate in the hydrological cycle.

The atmosphere, for all its clouds and obvious close association with the water cycle, contains only 0.035 percent fresh water; yet it is the atmosphere and its relation to land and oceans that keeps the water circulating over Earth.

The Local Water Cycle

Outside it is raining. The rain strikes the house and runs down the windows and walls into the ground. It disappears into the grass, drips from the leaves of trees and shrubs, and trickles down the trunks. When the rain stops, the windows and walls dry. The entire episode—the spring shower, the infiltration into the ground, the throughfall in the trees and bushes, the runoff from the walks, the evaporation—epitomizes the water cycle on a local scale (Figure 5.1).

Precipitation is the driving force of the water cycle. Whatever its form, precipitation begins as water vapor in the atmosphere. When air rises it is cooled adiabatically, and when it rises beyond the temperature level at which condensation takes place, clouds form. The condensing moisture coalesces into droplets 1 to 100 μ in diameter and then into rain droplets with a diameter of approximately 1000 μ (1 mm). Where temperatures are cold enough, ice crystals may form instead. Particulate matter smaller than 10 mμ in the atmosphere may act as nuclei on which water vapor condenses. At some point the droplets or ice crystals fall as some form of precipitation. As the precipitation reaches the earth, some of the water reaches the ground directly, and some is intercepted by vegetation, litter on the ground, and urban structures and streets. It may be stored, run off, or in time infiltrate the soil.

Because of *interception*, various amounts of water evaporate into the atmosphere without ever reaching the ground. Grass in the Great Plains may intercept 5 to 13 percent of the annual precipitation (average loss, 7.9 percent), and litter beneath a stand of grass may intercept 2.8 to 8.0 percent of annual precipitation (average loss, 4.3 percent) (Corbett

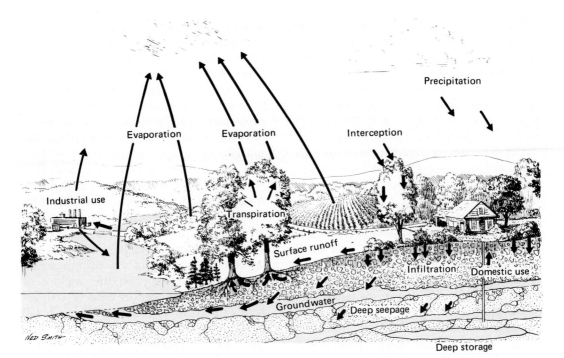

Figure 5.1 The water cycle, showing major pathways through the ecosystem.

and Crouse 1968). Precipitation striking forest trees or forest canopy must penetrate the crowns before it reaches the ground. A forest in full leaf in the summer can intercept a significant portion of a light summer rain.

The amount of rainfall intercepted depends upon the type and the age of the forest. In general, conifers intercept more rainfall on an annual basis than hardwoods. Mature pine stands, for example, will intercept 20 percent of a summer rainfall and 14 to 18 percent of the annual precipitation, whereas an oak forest will intercept 24 percent of a summer rain, but only 11 percent of the total annual precipitation (Lull 1967).

In a deciduous forest in summer a relatively greater proportion of rainfall is intercepted during a light shower than during a heavy rain. During a light shower the rain does not exceed the storage capacity of the canopy, and the water held by the leaves subsequently evaporates. Water exceeding the storage capacity of the canopy either drips off the leaves as *throughfall* or runs down the stem, twigs, and trunk as *stemflow*. Water then enters the soil in a relatively narrow band around the base of a tree. In winter deciduous trees intercept very little precipitation.

In urban areas a great portion of the rain falls on roofs and sidewalks, which are impervious to water. The water runs down gutters and drains to be hurried off to rivers. The percentage of rainfall striking impervious surfaces varies with the nature of the urban area. In downtown sections 100 percent of the surface may be impervious to rain. In residential areas imperviousness varies with the size of the lot. For lots of about 6000 ft^2 (557 m^2) about 80 percent of the area is impervious; for lots 6000 to 10,000 ft^2 (557 to

929 m^2), 40 percent; and for those 15,000 ft^2 (1392 m^2) and over, 25 percent (Antoine 1964). In suburbia, with its expanse of lawns, impervious areas are smaller. On a 1.8-acre (0.73 ha) lot about 92 percent is lawn, 8 percent pavement and roof. A typical 0.4-acre (0.16 ha) lot is 24 percent roof and pavement (Felton and Lull 1963). The largest paved areas in the suburbs are shopping centers, which require three to four times as much area for parking as separate stores.

City streets may intercept, store, and eventually lose by evaporation 0.04 to 0.10 in (1 to 2.5 mm) of water, and buildings 0.04 in (1 mm) (Viessman 1966). Estimates place interception by urban areas at about 16 percent, approximately the same amount intercepted by forest trees in summer leaf. Residential areas, surprisingly, intercept less; lawn grass 2 in. (5 cm) high intercepts 0.01 in (0.245 mm) of water, and storage in leaf depression is about 0.04 in. (1 mm) (Felton and Lull 1963). Total annual interception by residential areas comes to about 13 percent of a 45-in. (114 cm) rainfall (Lull and Sopper 1969).

The precipitation that reaches the soil moves into the ground by *infiltration*. The rate of infiltration is governed by soil, slope, type of vegetation, and the characteristics of the precipitation itself. In general, the more intense the rains, the greater the rate of infiltration, until the infiltration capacity of the soil, determined by soil porosity, is reached. Because water moves through the soil by the action of two forces, capillary attraction and gravity, soils with considerable capillary pore space have initial rapid infiltration rates, but the pores rapidly fill with water. Water will infiltrate longer into soils with a high proportion of noncapillary pore space. Vegetation that tends to roughen the surface retards surface flow and allows the water to move into the soil. Slope, impeding layers of stone or frozen soil, and other conditions influence the rate at which water moves into the soil.

Long wet spells and heavy storms may saturate the soil and intense rainfall or rapid melting of snow can exceed the infiltration capacity of the soil. When the soil can no longer absorb the precipitation, the water becomes *overland flow*. Sheet flow over the surface is changed to channelized flow as the water becomes concentrated in depressions and rills. This process can be observed even on city streets as water moves in sheets over the pavement and becomes concentrated in streetside gutters. Again the amount of runoff and associated erosion of soil depends upon slope, texture of the soil, soil moisture conditions, and the type and condition of vegetation.

In the undisturbed forest infiltration rates usually are greater than intensity of rainfall, and surface runoff does not occur. In urban areas infiltration rates may range from zero to a value exceeding the intensity of rainfall on certain areas where soil is open and uncompacted. Because of low infiltration, runoff from urban areas might be as much as 85 percent of the precipitation (Lull and Sopper 1969). Because they are so compacted by frequent tramping and mowing, lawns have a low infiltration rate. Felton and Lull (1963) have demonstrated that water infiltrates in lawns at an average rate of 0.01 in. (0.25 mm)/min, compared to 0.58 in. (0.142 mm)/min in forest soil.

Water entering the soil will *percolate* or seep down to an impervious layer of clay or rock to collect as groundwater. From here the water finds its way into springs, streams, and eventually to rivers and seas. A great portion of this water is utilized by humans for domestic and industrial purposes, after which it reenters the water cycle by discharge into streams or into the atmosphere.

A part of the water is retained in the soil. The portion held by capillary forces between the soil particles is called *capillary water*. Another portion adheres as a thin film to soil particles. This *hygroscopic water* is unavailable to plants. The maximum amount of water that a soil can hold at one-third atmosphere of pressure after gravitational water is drained away is *field capacity*. Highly porous sandy soils have a low field capacity, whereas fine-textured and humic soils have a high field capacity. Humus may retain 100 to 200 percent of its own weight in water. For each 2.5 cm of humus about 2 cm of water is stored for as long as two days after a rain and is slowly discharged into streams (Lull 1967). Storage in urban areas obviously is considerably less, almost nothing for paved areas.

Water remaining on the surface of the ground, in the upper layers of the soil, and collected on the surface of vegetation and water in the surface layers of streams, lakes, and oceans returns to the atmosphere by *evaporation*. Evaporation is the movement of water molecules from the surface into the atmosphere at a rate governed by how much moisture the air contains (vapor-pressure deficit).

As the surface layers of the soil dry out, a dry barrier through which little soil water moves develops and evaporation ceases. Further water losses from the soil take place through the leaves of plants. Plants take in water through the roots and lose it through the leaves in a process called *transpiration*. In the presence of sufficient moisture leaves remain turgid with the stomata fully open, which permits an easy inflow of carbon dioxide, but at the same time permits a large leakage of water. This leakage continues as long as moisture is available for roots in the soil, as long as the roots are capable of removing water from the soil, and as long as the amount of energy striking the leaf is enough to supply the necessary latent heat of evaporation. Thus plants can continue to remove water from the soil until the capillary water is exhausted.

The temperate deciduous forest and the urbanized areas of the northeastern United States represent two environmental extremes in water cycling. In comparison to the forest, urbanized areas are characterized by reduced interception, less infiltration, much less soil moisture storage, less evapotranspiration, and reduced water quality. Urban areas also exhibit increased overland or surface flow, increased runoff, and increased peak flows of streams and rivers.

The Global Water Cycle

Once evaporated into the atmosphere or carried away by surface runoff, the water involved in the local hydrological cycle enters the global water cycle. The molecules of water that fell in the spring shower might well have been a part of the Gulf Stream a few weeks before and perhaps spent some time in the Amazon tropical rain forest before that. The local storm is simply a part of the mass movement and circulation of water about the earth, a movement suggested by the changing cloud patterns over the face of the earth. The atmosphere, oceans, and land masses form a single gigantic water system that is driven by solar energy. The presence and movement of water in any one part of the system affects the presence and movements in all other parts.

The atmosphere is one key element in the world's water system. At any one time the atmosphere holds no more than a 10- to 11-day supply of rainfall in the form of vapor, clouds, and ice crystals. The turnover rate of water molecules is rapid. The source of water in the atmosphere is evaporation from land and sea, and therefore the amount of moisture in the atmosphere at any given point around the globe depends upon global variations in the rate of evaporation. Evaporation is considerably greater at lower latitudes than at higher latitudes, reflecting the greater heat budgets produced by the direct rays of the sun. The amount of moisture in the atmosphere over oceans is greater than over land, not only because there is more free water to evaporate, but also because oceans contain well over 90 percent of the world's water. Oceans provide 84 percent of total annual evaporation, considerably more than they receive in return from precipitation. Land areas contribute 16 percent, intercepting more water than evaporates from them.

Moisture in the atmosphere moves with the general circulation of the air. Air currents, hundreds of kilometers wide, are in fact giant unseen rivers moving above Earth. Only a part of this moisture falls as precipitation in any one place. For example, in a year's time the United States receives an unequally distributed 6000 km^3 or 75 cm in depth of precipitation, but the liquid equivalent of water vapor passing over the country is 10 times that much. Atmospheric precipitation for Earth as a whole is approximately 100 cm in depth, and the average resident time for a water molecule in the atmosphere is approximately 10 days.

Variations in evaporation and precipitation follow the pattern of the global air currents. The trade winds move moisture-laden air toward the equator, where it is warmed, rises, cools, and drops its moisture as rain. Thus the equatorial regions are areas of maximum precipitation. The air that rises over the equator descends earthward in two subtropical zones around 30° N and 30° S latitude. As the

air descends, it warms, and picks up moisture from land and sea. The highest annual losses to evaporation occur in the subtropics of the western North Atlantic (the Gulf Stream) and the Northern Pacific (the Kuroshio Current). North of them are two more zones of ascending air and low pressure, which produce the west-coast areas of maximum rainfall. In high latitudes the air descends again in the polar regions, where it remains dry.

Global detention of precipitation varies with region and season. Maximum detention occurs in the polar regions of the Northern and Southern hemispheres. In tropical regions the period of maximum detention time is the early part of the rainy season. On the average the residence time of water on land is 10 to 120 days.

Precipitation on land in excess of evaporation is eventually transported to the sea by rivers. Rivers are the prime movers of water over the globe and carry many more times the amount of water their channels hold. By returning water to the sea they tend to balance the evaporation deficit of the oceans. Sixteen major rivers discharge 13,600 cm annually, 45 percent of all water carried by rivers. Adding the next 50 largest rivers brings the total to 17,600 cm, 60 percent of all water discharged to the sea.

Evaporation, precipitation, detention, and transportation maintain a stable water balance on Earth. Consider the amount of water that falls on Earth in terms of 100 units (Figure 5.2). On the average 84 units are lost from the ocean by evaporation, while 77 units are gained from precipitation. Land areas lose 16 units by evaporation and gain 23 units by precipitation. Runoff from land to oceans makes up 7 units, which balances the evaporative deficit of the ocean. The remaining 7 units circulate as atmospheric moisture.

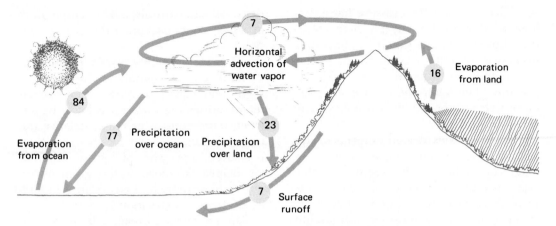

Figure 5.2 Global water budget. The mean annual global precipitation of 83.6 cm has been converted into 100 units.

In its global circulation the water also influences the heat budgets of Earth. The highest heat budgets are in the low latitudes, the lowest are in the polar regions, and a balance between incoming and outgoing cold and hot is achieved at 38° to 39° latitude. Excessive cooling of higher latitudes is prevented by the north and south transfer of heat by the atmosphere in the form of sensible and latent heat in water vapor and by warm ocean currents.

Examined from a global point of view, the water cycle emphasizes the close interaction between the physical environment of and geographical locations on Earth. Thus the water problem, often considered in local terms, is actually a global problem, and local water management schemes can affect the planet as a whole. Problems result not because an inadequate amount of water reaches Earth, but because it is unevenly distributed, especially relative to human population centers. Because humanity has strongly interjected itself into the water cycle, the natural usable water resources have decreased, and water quality has declined. The natural water cycle

has not been able to compensate for the deteriorating effects on water resources of human actions.

■ Humidity

The moisture of water vapor content of the air is expressed as humidity. Water vapor gets into the air by evaporation from moist surfaces and bodies of water and from transpiration by plants. In the air water vapor acts as an independent gas. Like air, it has weight and exerts pressure. The amount of pressure water vapor exerts independent of dry air is *vapor pressure.*

In the presence of dry air, water vapor molecules will leave a moist surface and mix with the dry air above it. Rapid at first, the process slows until the amount of water molecules entering the air balances the number leaving it. When a volume of air holds all the water vapor it can, it is *saturated.* The pressure that water vapor exerts when the air is saturated is *saturation vapor pressure.*

The amount of water vapor a volume of air can hold varies with its temperature. The warmer the air, the more water vapor it can hold, so the greater its vapor pressure. Saturation vapor pressure has a definite fixed value for any given temperature; however, the amount of water vapor needed to saturate a given weight of air (in contrast to volume) varies with air pressure. The lower the air pressure, the more water vapor the air will hold at any given temperature.

Humidity is expressed in terms of the amount of water vapor a given volume or weight of air holds. The weight of water vapor per unit volume of air, its *absolute humidity,* is measured as grams per cubic meter. However, the measure of humidity of greatest interest to ecologists is relative humidity. *Relative humidity* is the measure of the amount of water vapor actually in the air relative to the amount of water vapor the air could hold if saturated. Thus if the actual amount of water vapor in the air is only 20 percent of the amount it can hold at a given temperature, the relative humidity is 20 percent. If the air holds all the water vapor it can, it is saturated and its relative humidity is 100 percent.

A reduction in temperature reduces the air's capacity to hold water. For example, if air with 20 percent humidity is cooled until its saturation vapor pressure is reduced by one-half, the relative humidity increases to 40 percent. Thus relative humidity moves in reverse of temperature. Assuming the amount of water vapor remains constant, relative humidity decreases as the temperature goes up, because warm air can hold more water vapor. Conversely, it increases as the temperature goes down.

Another useful measure of air moisture is *dew point.* If air containing a given amount of water vapor, measured as relative humidity, is cooled to the point at which actual vapor pressure equals saturation vapor pressure, the relative humidity is 100 percent. Any further cooling causes some of the water vapor to condense as water droplets. The temperature at which condensation begins is the dew point.

Relative humidity varies during the 24-hour day. Generally it is higher at night and early morning, when the air temperatures are lower; it is lower by day when temperatures increase. Relative humidity is also affected by changes in air pressure. As pressure of a given weight of air decreases, more water vapor is required to saturate it. Thus relative humidity decreases as elevation increases.

In any one area relative humidity varies widely from one place to another, depending upon the terrain. Variations in humidity are most pronounced in mountainous country. Low elevations warm up and dry out earlier in the spring than high elevations, and soil moisture becomes depleted later in the summer. The daily range of humidity is greatest in the valleys and decreases at higher elevations. Because daytime temperatures decrease with altitude, as does the dew point, relative humidities are greater on tops of mountains than in the valleys.

As nighttime cooling begins, the temperature change with elevation reverses. Cold air rushes downslope and accumulates at the bottom. Through the night, if additional cooling occurs, the air becomes saturated with moisture and fog or dew forms by morning. These differences in humidity can produce vegetative differences on mountain slopes. They are most pronounced on slopes of mountains along the Pacific coast.

Temperature and wind together exert a considerable influence on evaporation and relative humidity. An increase in air temperature causes convection currents. This air turbulence mixes surface layers with drier air above. Wind movements associated with cyclonic disturbances also mix moisture-laden air with drier air above. As a result, the vapor pressure of the air is lowered and evaporation from the surface increases.

■ Plant Responses to Moisture

Plants obtain the two major ingredients they need for photosynthesis, water and carbon dioxide, through roots and leaves. Roots withdraw water from the soil; the stomata on the leaves facilitate the diffusion of carbon dioxide from the atmosphere to the cells of the leaf. But as CO_2 moves into the leaf through the stomatal openings, water vapor goes out. In fact, plants lose most of the water taken up by their roots through transpiration. When plants face a moisture stress during a period of short or protracted drought, they are faced with the prospect of either closing the stomata to retain water, thus reducing photosynthesis, or opening the stomata to take in CO_2, losing needed water. How plants respond to water stress affects not only their own fitness but ecological relationships.

Lack of water is a major selective force in the evolution of the plant's ability to cope with moisture stress. A large group—algae, fungi, lichens, and most mosses—has no protective mechanism against water loss. The internal water status of these plants tends to match atmospheric moisture conditions. As water becomes less available, the plant dries out. Its cells shrink without disturbing the fine protoplasmic structures within them, and their vital processes gradually become suppressed. When moisture conditions improve, the plant imbibes water and the cells fill and resume normal functioning. Such plants (called *poikilohydric*) restrict their growth to moist periods, and their biomass is always small.

Other plants (termed *homoiohydric*), mostly ferns and seed plants, are able to maintain within limits a stable water balance independent of fluctuations of atmospheric moisture conditions. That is made possible by water stored in a vacuole within the cell, a protective cuticle that slows down evaporation, stomata that regulate transpiration, and an extensive root system that draws water from the soil.

Homoiohydric plants lose water to the atmosphere by transpiration. Water evaporates from the internal surfaces of the plant and subsequently diffuses from it. As long as vapor pressure inside the leaf is greater than the vapor pressure of the atmosphere, a plant loses water to the environment; but homoiohydric plants do not give up water easily. They hold it back by a combination of osmotic pressure and turgor pressure of water within the cells pushing outward against the cell walls. These two forces make up *water potential*, a measure of energy in water. Water tends to flow in the direction of lower water potential. Plants maintain a water balance by drawing water from the soil, a process made possible by a gradient of negative water potential from soil, where under normal moisture conditions water potential is the highest, to the atmosphere, where on relatively dry days water potential is the lowest.

Responses to Drought

Leaves reflect water stress in plants. Some, such as evergreen rhododendrons, show water stress by an inward curling of the leaves, which reduces surface areas exposed to the air. Others show it in a wilted appearance, caused by a lack of turgor in the leaves. A less obvious, but more significant response is the closure of the stomata. This response reduces transpirational loss of water, but it also raises the internal temperature of the leaf, causing additional heat stress, which can affect protein synthesis (Walbot and Cullis 1985). Stomatal closure reduces CO_2 diffusion into the leaf. It ultimately results in reduced growth and smaller plants, leaves, buds, and other parts (Kramer 1983). Severe drought inhibits the production of chlorophyll, causing the leaves to turn yellow or later in the summer to exhibit premature autumn coloration. As conditions worsen, deciduous trees may prematurely shed their leaves, the oldest ones dying first. Such premature shedding can result in dieback of twigs and branches.

Certain other physiological changes take place during drought. Water-stressed plants reduce their osmotic potential by accumulating such inorganic ions as calcium, magnesium, potassium, and sodium (Mattson and Haack 1987), amino acids, sugars, and sugar alcohols (Kramer 1983). A decrease in soil water, increase in soil temperature, and reduced root growth alter mineral uptake from the soil. These changes may make water-stressed plants more vulnerable to insect attack (Mattson and Haack 1987).

Conifers and other evergreens experience winter drought. It comes about when winter temperatures are warm enough to thaw the ice in the water ducts of woody plants. The trees lose this water by transpiration through the leaves; but because the soil is frozen, making water unavailable, the plants cannot replace it. The result is browning of needles and dying back of twigs from dehydration.

Adaptations to Drought

Plants in semiarid and desert regions have evolved adaptive mechanisms to tolerate drought. One method is to reduce the leaf area of the canopy by dropping the leaves during periods of extreme water stress and developing new leaves at the onset of rain. Although such a response usually results in decreased production, some plants maintain some of their photosynthetic capacity by increasing photosynthetic activity in green stems.

When water is limiting, some species of plants reorient the angle of their leaves so that their surfaces are parallel to the sun's rays (Ehleringer and Forseth 1980; Schultz et al. 1987). This shift in position reduces leaf temperatures and transpiration rates. With adequate water, these plants orient their leaves perpendicular to the sun's rays.

Another means of conserving water is succulence, or storage of water within the plant's tissues. Notable are the cacti. Because their root systems are shallow, cacti can draw up considerable amounts of water during periods of rain. The plants swell rapidly as they store water in enlarged vacuoles of the cells. During periods of moisture stress the cacti draw on this store of water. Many plants close their stomata, especially during the day, and open them at night to take in and fix CO_2 for use during the day. Some employ the CAM type of photosynthesis (see Chapter 3).

Plants may meet moisture stress by means of various leaf adaptations. In some the leaves are small, the cell walls thickened, the stomata smaller but more numerous per unit of surface area, the vascular system denser, and the palisade tissue more highly developed than the mesophyll. These adaptations increase the ratio of the internal exposed cellular surface of the leaf to the external surface. Many have leaves coated with hairs that scatter incoming solar radiation; others have leaves coated with waxes and resins that reflect light and reduce their absorptivity (Mulroy 1979).

Root adaptations provide still other means of maintaining water balance. Some woody plants, such as cottonwood and willows, have deep roots that reach an underground water supply. Growing in arid and semiarid regions, these plants, known as *phreatophytes,* have roots in constant contact with a fringe of capillary water above the water table. Other plants have spreading root systems covering a large shallow surface area. This arrangement enables them to secure the maximum amount of water when moisture is available.

Some plants adopt a life cycle in which the population survives the dry period as dormant seeds ready to sprout quickly when the rains come. They complete the active stages of their life cycle within a short period before moisture is completely gone and then return to seed dormancy. The cycle may be based on annual wet and dry seasons; or the plant seeds may remain dormant for years, waiting for adequate moisture.

Plants of the salt marsh grow in a physiologically dry substrate because salinity limits the amount of water they can absorb by osmosis; and what they do take in contains a high level of solutes. These plants, known as *halophytes,* compensate for an increase in sodium and chlorine uptake by diluting internal solutions with water stored in the tissues. Some plants exhibit high internal osmotic pressure many times that of freshwater and terrestrial plants, possess salt-secreting glands, secrete heavy cutin on their leaves, and are succulent.

Responses to Flooding

Too much water can create as much stress on plants as too little water. Symptoms of such stress are similar to those under drought conditions—stomatal closure, yellowing and premature loss of leaves, wilting, and rapid reduction in photosynthesis (Jackson and Drew 1984; Kozlowski and Pallardy 1984)—but the causes are different.

Growing plants need both sufficient water and a rapid gas exchange with their environment. Much of this exchange takes place between the roots and soil atmosphere. When soil pores are filled with water, no gas exchange can take place. The plants, experiencing depressed CO_2 levels, in effect, are asphyxiated. Anaerobic conditions in the soil inhibit uptake and transport of ions and depress the concentration of nitrogen, phosphorus, and potassium in the shoot. These conditions also result in an accumulation in the plant of ethylene. Ethylene gas, highly insoluble in water, is produced normally in small amounts in the roots. Under flood conditions ethylene diffusion from the roots is slowed, and oxygen diffusion into the roots is inhibited. Ethylene, therefore, accumulates to high levels in the root. It stimulates adjacent cortical cells in the cortex of the root to lyse and separate to form interconnected gas-filled chambers (Drew et al. 1979; Jackson 1982) called *aerenchyma.* These chambers, typical of wetland plants, allow some exchange of gases between submerged and better aerated roots.

Excess water around the roots also stimulates the plant to resist the movement of water through the roots to the shoots, resulting in wilting of leaves, and it brings about certain morphological changes in the plant. In response to the lack of oxygen and the accumulation of trapped gas, adventitious roots grow horizontally along the oxygenated zone of the soil surface, where anaerobically generated toxins are absent. These roots emerge from the submerged part of the stem when original roots succumb, to re-place the functions of the original roots. Perennially high water tables force non-wetland plants to develop shallow, horizontal root systems, which make such plants highly susceptible to drought conditions.

Prolonged flooding often results in dieback or death of plants, particularly woody species. The response of trees to flooding depends upon duration, movement of water, and the species of trees. Trees are injured more by standing in stagnant water than in oxygenated flowing water. If a tree's roots are flooded for more than half the growing season, death results (Kozlowski 1984).

Adaptations to Flooding

Wetland plants have evolved certain adaptations to cope with a waterlogged environment. Most have gas-filled chambers through which oxygen diffuses from the shoots to the roots. These interconnected gas spaces throughout the leaves and roots occupy nearly 50 percent of the plant tissue. Some plants of cold, wet, alpine tundras have similar spaces in leaf, stem, and root to carry oxygen to the roots (Keeley et al. 1984). In contrast to nonwetland species, wetland plants lose less oxygen from the roots, maintaining their oxygen concentrations (Jackson and Drew 1984).

Few woody species can grow on permanently flooded sites. Among the few are bald cypress, mangroves, willows, and swamp tupelo. Cypress growing over fluctuating water tables develop knees or pneumatophores, which may be beneficial but not necessary for survival. Pneumatophores on mangroves provide oxygen to roots during tidal cycles. These plants may also modify their metabolism to adapt to short-term hypoxia and reduce ethanol toxicity (Hook 1984).

Plant Distribution

The availability or lack of moisture determines the worldwide distribution of vegetation: deserts, tundra, grasslands, deciduous, coniferous, and tropical forests. Of more immediate interest is the influence of moisture on smaller regional and local scales.

Regional vegetation patterns are influenced by altitude, topography, aspect, and soils. Altitudinal distribution and growth forms of coniferous trees in high mountains are influenced by limited water uptake because of frozen soil, high atmospheric desiccation, and loss of water through the cuticle of terminal shoots. Tree forms of lower elevations are replaced by elfinwood of the krumholtz or tree line, which in turn gives way to tundra vegetation.

High mountains intercept moisture on the windward side to create a dry rainshadow on the leeward side (see Figure 4.8). The windward side of the Coastal Range and the Cascade Range in the Pacific Northwest intercepts the maritime air masses from the Pacific. As a result precipitation is very high, exceeding 250 cm a year on the western slopes, whereas in the rainshadow of the eastern slopes precipitation may be as low as 30 to 40 cm a year (Figure 5.3). The west-facing slopes are heavily forested with sitka spruce (*Picea sitchensis*) on the Coastal Range and Douglas-fir (*Pseudotsuga menziesii*) and western hemlock (*Tsuga heterophylla*) on the Cascades. On the rainshadow side the slopes are dominated by deciduous oaks (*Quercus*) and the evergreen Pacific madrone (*Arbutus menziesii*). (See Franklin and Dyrness 1973.)

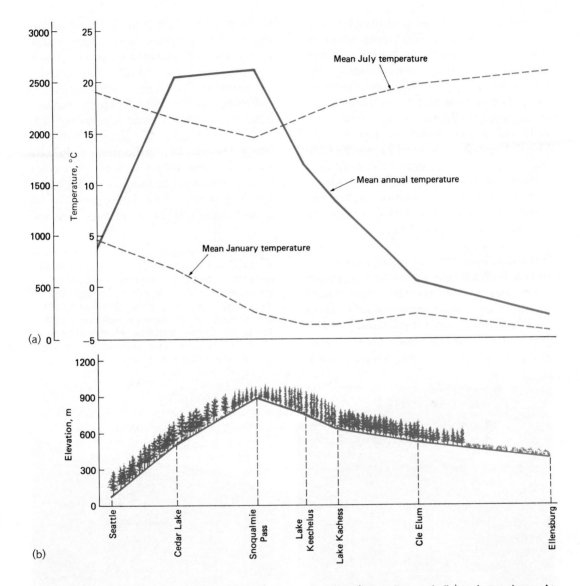

Figure 5.3 A climatic cross-section of the Cascade Range in the vicinity of Snoqualmie Pass, Washington (47° 25′ N lat.) covering a distance of approximately 152 km. (a) The strong rainshadow effects and the relationship between annual precipitation and high mean July temperature on the leeward side. Low annual precipitation is accompanied by high summer temperatures, intensifying xeric conditions. (b) The effect of the rainshadow and temperature on vegetational distribution. The coniferous rain forest of the west-facing slopes is dominated by western hemlock *(Tsuga heterophylla)* on lower slopes. At high elevations this forest gives way to subalpine forest dominated by Pacific silver fir *(Abies amabilis)*, subalpine fir *(Abies lasiocarpa)*, and mountain hemlock *(Tsuga mertensia)*. The high east-facing slopes are dominated by grand fir *(Abies grandis)* and Douglas-fir *(Pseudostuga menziesii)*. Lower, more xeric elevations are dominated by ponderosa pine *(Pinus ponderosa)*, giving way to xeric shrub steppe dominated by sage *(Artemisia tridentata)* and grasses. (After Franklin and Dyrness 1973:43–45.)

Some of the most pronounced rain-shadow effects in a small area occur in some of the islands in the Hawaiian chain. Maui, for example, is small, only 892 square kilometers; yet climatic conditions range from desert dryness to tropical rain forest wetness. Rainfall ranges from over 1016 cm at Puu Kukui on top of West Maui Mountain to 38 to 51 cm in the Central Valley. This difference in rainfall is reflected in the vegetation, which ranges from scrubby xeric vegetation on the leeward side to moist forested slopes on the windward side (Figure 5.4).

On an even more local scale moisture influences vegetational composition on slopes. For example, within several square kilometers in the central Appalachians one may move from mesic streambottoms and streamside slopes dominated by oaks, yellow-poplar, beech, black cherry, flowering dogwood, and many other species,

to moist coves and mesic north-facing slopes, dry ridgetops, and upper south-facing slopes dominated by scarlet, black, and chestnut oaks, and mountain laurel.

Even in flat country microrelief can influence distribution of plant communities. In a study of prairie, meadow, and marsh vegetation of Nelson Country, North Dakota, Dix and Smeins (1967) divided the soils into ten drainage classes, from excessively drained to permanently standing water. They determined the indicator species for each drainage class and

Figure 5.4 Rainshadow effects on the mountains of Maui, Hawaiian Islands. (a) The windward, east-facing slopes, intercepting the trade winds, are cloaked with a wet forest. The light-colored canopy trees are *kukui* or candlenut trees *(Aleurites moluccana)*. (b) Low-growing xeric vegetation of the dry western side. Much of this vegetation is disturbed and dominated by exotic species. (Photos by R. L. Smith.)

(a)   (b)

then divided the vegetational display into six units corresponding to the drainage patterns (Figure 5.5). The uplands fell into high, mid, and low prairie, and the lowlands into meadow, marsh, and cultivated depressions. The excessively drained high prairie supported stands of needle-and-thread grass, western wheatgrass, and prairie sandweed. The mid prairie, considered to be climax or true prairie, was dominated by big bluestem, little bluestem, and prairie dropseed. Low prairie on soils of moderate moisture supported big bluestem, little bluestem, and yellow Indian grass. Lowlands and areas of sluggish drainage where the water table was within the rooting zone of most plants supported canarygrass, sedge, and rivergrass. Meadows on even wetter soil were dominated by northern reed grass, woolly sedge, and spike rush. Marshes with standing water supported stands of reeds, cattails, and tule bulrush. Cultivated depressions were colonized by spikerush and water plaintain.

Although each drainage class supported a characteristic stand of vegetation, no one species was associated solely with another species. At any particular site the drainage conditions influenced the combination of dominant plants and each community blended into the other. The only sharp breaks came where drainage conditions were sharp and severe.

Aridity, even seasonal in nature, can influence growth, survival, and distribution of plants. Seeds arriving on dry sites are subject to dehydration. To germinate successfully, they need to absorb quantities of water. If they do germinate, the seedlings often fail to survive after early stages of development. Trees that become established on perennially dry sites rarely grow tall. They grow slowly, with just enough photosynthetic production to achieve microscopic annual growth. By producing only a small amount of living tissue, they are able to reduce water stress and avoid senescence and death. Such trees tend to be very long-lived. Extreme examples are the bristlecone pines of the Rocky Mountains and Great Basin of the United States, *Pinus aristata* and *P. aristata* var. *longaeva*.

■ Animal Responses to Moisture

Animals' responses to moisture stresses are more complex than those of plants. Unlike plants, animals possess a more or less universal mechanism to rid the body of excess water and solutes or to conserve them. These mechanisms range from the contractile vacuoles of protozoans to the complex kidney and urinary systems of birds and mammals.

If an organism is not to dehydrate, the input of water must equal losses. For the most part moisture becomes a general problem in only two environments, exclusive of periodic droughts in other areas— the saltwater environment and the desert. Interestingly, many organisms of these two environments use the same strategy to overcome the problem.

Adaptations to Saltwater

Animals of the marine and brackish environments have to inhibit the loss of water by osmosis through the body wall to prevent an accumulation of salts in the system. Invertebrates get around the problem by possessing body fluids with the same osmotic pressure as seawater. Marine teleost fish absorb salty water into the gut. They secrete magnesium and calcium through the kidneys and pass these ions off as a partially crystalline

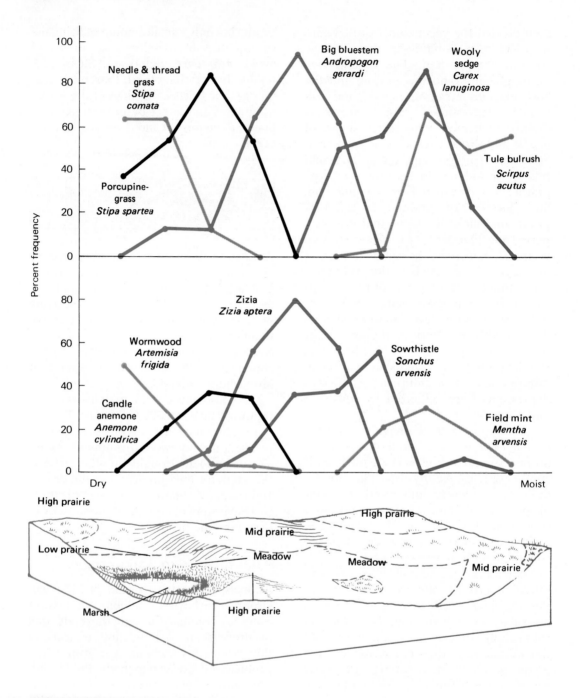

Figure 5.5 The influence of topography and drainage regimes on prairie vegetation. The block diagram of a hypothetical North Dakota landscape shows the relative positions of vegetational units. The distributional curves of selected species are based on a drainage gradient. The high prairie is the vegetational unit of excessive drainage. (From Dix and Smeins 1967:33, 43.)

paste. The fish excrete sodium and chlorine by pumping the ions across membranes of special cells in the gills. This pumping process is called *active transport.* It involves the movement of salts against a concentration gradient at the cost of metabolic energy. Sharks and rays retain urea to maintain a slightly higher concentration of salt in the body than in surrounding seawater. Birds of the open sea can utilize seawater because they possess a special salt-secreting gland, located on the surface of the cranium (Schmidt-Nielsen 1960). Gulls, petrels, and other seabirds excrete from these glands fluids in excess of 5 percent salt. Petrels and other tube-nosed swimmers forcibly eject the fluid through the nostrils; others drip the fluid out of the internal or external nares.

The kidney is the main avenue for the elimination of salt among marine mammals. Porpoises have highly developed renal capacities to eliminate salt loads rapidly (Malvin and Rayner 1968). The urine of marine mammals has a greater osmotic pressure than blood and seawater; it is hyperosmotic.

Among the reptiles the diamondbacked terrapin *(Macaclemys terrapin),* an inhabitant of the salt marsh, lives in a variably saline environment. The terrapin retains its osmotic pressure when the water is dilute; yet it possesses the ability to accumulate substantial amounts of urea in the blood through the functioning of salt glands when it finds itself in water more concentrated than 50 percent seawater.

Adaptations to Aridity

Animals in arid environments possess adaptations suggestive of those of arid land plants. One is to become dormant during periods of environmental stress. The animal may adopt an annual cycle of active and dormant stages. For example, for eight or nine months, the spadefoot toad estivates in an underground cell lined with a gelatinous substance that reduces evaporative losses through the skin. It emerges when rainfall saturates the ground, moves to the nearest puddle, mates, and lays eggs. Young tadpoles hatch in a day or two, mature rapidly, and metamorphose into functioning adults. They dig their own retreats in which to estivate until the next rainy period.

Many desert animals are active throughout the year. They have evolved ways of circumventing the lack of water by physiological and behavioral adaptations. The kangaroo rat of the southwestern North American desert and its ecological counterparts, the jerboas and gerbils of Africa and the Middle East, and the marsupial kangaroo mice and pitchi-pitchi of Australia, feed on dry seeds and dry plant material even when succulent green plants are available. These mammals obtain water from their own metabolic processes. To conserve water these rodents remain by day in sealed burrows; they possess no sweat glands, their urine is highly concentrated (25 percent urea and salt, twice the concentration of seawater), and their feces are dry. In addition, some desert mammals can tolerate a certain degree of dehydration. Desert rabbits may withstand water losses up to 50 percent of their body weight; camels can tolerate a 27 percent loss.

Extreme adaptations to aridity exist among some of the African antelopes. Outstanding is the oryx *(Oryx leucoryx)* (Figure 5.6) Many African ungulates migrate to escape the heat and dryness, but the oryx remains. During the day it stores heat in its body, causing *hyperthermia,* a substantial rise in body temperature (Taylor 1969). Because evaporative loss is due to the difference between the body and air

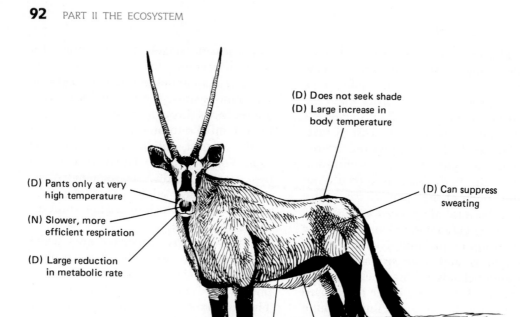

(D) Does not seek shade
(D) Large increase in
    body temperature

(D) Pants only at very
    high temperature

(N) Slower, more
    efficient respiration

(D) Large reduction
    in metabolic rate

(D) Can suppress
    sweating

-R.L.SMITH jR.

(N) Low body temperature

(N) Large reduction in
    cutaneous evaporation

Figure 5.6 The physiological adaptations to aridity and heat of the African ungulate, the oryx. D = day; N = night. (Adapted from M. S. Gordon 1972; based on data from C. R. Taylor 1969.)

temperature, a high body temperature might mean loss of moisture. The oryx reduces daytime evaporative losses by suppressing sweating and panting only at very high temperatures. It reduces its metabolic rate, lowering the rate of internal production of calories and thus the amount of needed evaporative cooling. By night the oryx reduces its nonsweating evaporation across the skin by reducing its metabolic rate below that of daytime. The oryx's respiratory rates are proportional to its respiratory efficiency and inversely proportional to body temperature. A cool animal breathes more slowly than a warm one, using a greater proportion of inspired oxygen. With a lowered nighttime body temperature, the saturation level for water vapor in the exhaled air is lower. The oryx normally does not drink water. It depends upon metabolic water and grasses and succulent shrubs. In fact, the oryx can obtain all the water it needs by eating food containing an average of 30 percent water.

Some desert birds, like marine birds, utilize a salt gland to maintain a water balance. Otherwise marine, desert, and salt marsh birds all have the same basic adaptations for the conservation of water and the elimination of sodium and chlorine ions. Because birds normally secrete semi-solid uric acid as an adaptation related to weight reduction for flying, they are well fitted for water conservation in arid environments.

Responses to Drought and Flooding

Plants are fixed to their site; but animals can move to more favorable environments when moisture conditions become critical. Of course, under extreme conditions, animals die from dehydration or drown. Drying of potholes, for example, causes nesting losses among waterfowl and muskrats and makes them highly vulnerable to predation. Flooding and heavy rains can drown young birds and mammals in the nest.

Drought, however, can have more pervasive effects on animal abundance and distribution. During the wet season on the African savanna, the African buffalo finds food abundant; but during the dry season, the quality of food declines as the grasses dry. Buffalo become more selective, seeking green leaves, moving to the moist riverine habitat, breaking into smaller units, and utilizing different areas. As the dry season progresses, buffalo become less selective, consuming dry stems and leaves they would have rejected earlier. As food quantity and quality decline, food availability for the buffalo declines (Sinclair 1977). The more buffalo present, the less food there is available for each individual.

Eventually, available protein drops below maintenance level, and the animals use up their fat reserves. Undernourished and lacking the protein intake necessary to maintain the immune system, old animals become most vulnerable to the diseases and parasites they normally harbor. The number dying depends upon the rapidity with which the adults use up their energy reserves before the coming of the rainy season. If the next season sees more rainfall and that rainfall extends sporadically into the dry season, the mortality of adults the following year is reduced. Thus rainfall indirectly regulates the density of buffalo populations (Sinclair 1977).

Drought stress promotes outbreaks of leaf-eating insects by influencing thermal and nutritional conditions favoring insect growth (Mattson and Haack 1987). High temperatures raise the rate and efficiency of enzymatic reactions in insects, enhance the insects' detoxification systems allowing them to override plant defenses, and may promote genetic changes favoring population growth. The increased content of nitrogen, minerals, and sugars in the leaves of drought-stressed plants provide a nutritionally rich food source for the leaf-eating insects. These conditions favor outbreaks of some plant-eating insects, such as pine sawflies and forest tent caterpillars, during periods of drought.

Moisture influences the speed of development and fecundity of some insects. If the air is too dry, the eggs of some locusts and other insects become quiescent. There is an optimum humidity at which the nymphs of some insects develop the fastest. Some species of insects lay more eggs at certain relative humidities than above or below those points. Excessive moisture and cloudy weather kill insect nymphs, inhibit insect pollination of plants, and spread parasitic fungi, bacteria, and viruses among both plants and animals.

■ Summary

Most of Earth's water is in the oceans. Less than 1 percent is available as free fresh water. This water moves through the water cycle, involving precipitation, interception, infiltration, surface flow, and evaporation. The key to the water cycle is the atmosphere. Evaporation into and precipitation from the atmosphere, detentions in oceans and land, and atmospheric transport maintain the global water balance. Oceans lose more water by evaporation and gain less from precipitation than land areas. The loss is made up by runoff from land.

Atmospheric moisture is expressed in terms of relative humidity, the measure of the amount of water vapor actually in the air relative to the amount of water vapor the air could hold if saturated. The amount of water vapor a parcel of air can hold increases with its temperature. Relative humidity varies during the 24-hour day. It is higher by night than by day and is greater at high elevations than at low, but it varies widely with topography, vegetation, wind, and temperature. Another useful measure of air moisture is dew point, the temperature at which water vapor in the atmosphere condenses as water droplets.

Water stress is a major selective force in the evolution of plants and animals. Fungi, mosses, lichens, and algae have no means of regulating internal moisture and only grow in moist conditions. Most plants maintain a stable water balance independent of environmental fluctuations by a combination of osmotic pressure and turgor pressure known as water potential. Under normal moisture conditions decreasing water potential pulls water from the soil through the plant to the leaf.

Plants respond to moisture deficits first by closing the stomata, reducing transpiration. Severe drought decreases photosynthesis, causes leaves to turn yellow, and can result in premature shedding of leaves and even death.

To maintain a water balance during long periods of dry weather, plants depend upon drought resistance—a combination of drought avoidance and drought tolerance. To avoid drought desert annuals germinate and bloom only when rains are sufficient. Some plants reduce transpirational water loss by closing their stomata, increasing leaf thickness, possessing a waxy cuticle, storing water in cells (succulence), or shedding leaves during the dry season. Other plants extend roots into areas of soil where the water has not been tapped. Some species in riverine habitats have deep roots that reach a permanent water supply. Plants in saline environments may accumulate and excrete salt through the leaves or store water in them.

Too much water can also create stress. In extreme conditions anaerobic respiration in the soil asphyxiates plants. Flooding increases toxic material in the plant, depresses its concentration of ions and oxygen, and may stimulate the growth of adventitious roots along the oxygenated soil surface. External symptoms suggest those of drought stress. Prolonged flooding can result in death.

Wetland plants have developed certain adaptations to withstand flooding. Notable are aerenchyma tissues, which form interconnected gas spaces through the leaves to the roots, carrying oxygen to the waterlogged roots.

Moisture or the lack of it has a major influence on the distribution of plants, both geographically and locally. Some are restricted to moist or mesic sites, others to dry or xeric sites.

Animals' responses to moisture stresses are more complex than those of plants. Most animals maintain a water balance by some sort of excretory system. Many animals inhabiting saline environments have salt-excreting glands. Some animals of arid regions reduce water loss by becoming nocturnal, avoiding the heat of day; by producing highly concentrated urine; by using only metabolic water; and by tolerating a certain degree of dehydration. Drought stress can lower food quality and in extreme cases cause death. Drought also promotes outbreaks of leaf-eating insects by influencing thermal and nutritional conditions in plants in favor of insect growth.

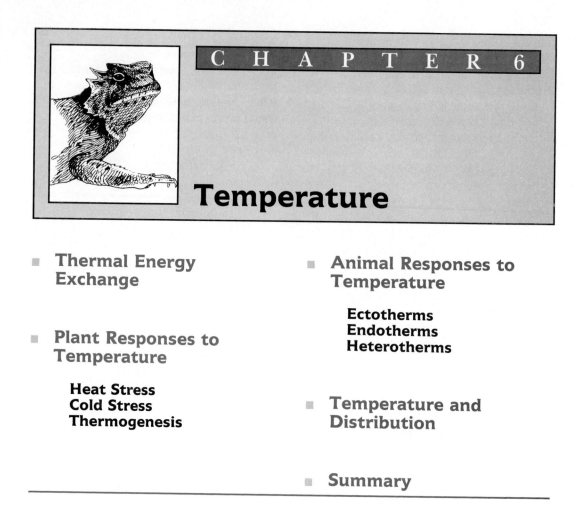

Temperature

The temperatures in which plants and animals live depend on solar radiation (Chapter 4). The amount of solar radiation reaching any point on Earth varies with the time of year, slope, aspect, cloud cover, time of day, and other factors. Seasonal fluctuations can be extreme. In North Dakota, for example, where the annual mean temperature is between 3° and 9° C, temperatures fluctuate from a low of −43° C in winter to 49° in summer. In West Virginia, where the mean annual temperature is 12° C, temperatures range from −37° to 44° C. Temperatures fluctuate from sunlight to shade, from day-

light to dark. Surface temperatures of soil may be 30° higher in the sunlight than in the shade. Daytime temperatures may be 17° C higher than nighttime temperatures; on deserts this spread may be as high as 40° C. Temperatures can rise to 38° C on tidal flats exposed to direct sunlight and sink to 10° C within a few hours when the flats are covered by water.

The ability to withstand extremes in temperatures varies widely among plants and animals, but there are temperatures above and below which no life can exist. A temperature of 52° C is about the highest at which any animal can still grow and multiply. Observed thermal limits for the survival of metabolically active vascular plants range from about $+60°$ C to $-60°$ C in different species. Some hot-spring algae can live in water as warm as 73° C under favorable conditions (Brock 1966); and some arctic algae can complete their life cycles in places where temperatures barely rise above 0° C. Nonphotosynthetic bacteria inhabiting hot springs can actively grow at temperatures greater than 90° C (Bott and Brock 1969; see also Brock 1979).

■ Thermal Energy Exchange

Living organisms must exchange energy with their environment and maintain a balance between heat energy gained and heat energy lost. Thermal energy absorbed from the environment plus metabolic energy produced must equal thermal energy lost from the body and energy stored.

A major source of heat absorbed is solar radiation. Direct sunlight can be intense. At noon in summer on a clear day it can amount to 1.2 to 1.4 cal/cm^2/min (calories per square centimeter per minute). Another source is skylight or diffuse radiation, sunlight scattered by moisture and dust in the atmosphere. This heat may amount to 0.2 cal/cm^2/min. A third source is reflected sunlight, bounced off objects. It can range from 0.1 to 0.3 cal/cm^2/min., depending upon the reflecting surface. In addition, thermal radiation, longwave infrared radiation from 4 to 30 u and beyond, is emitted from the surfaces of soil, rocks, organisms, and all other objects in the environment at their ambient temperatures (Figures 6.1, 6.2). The rate at which a surface emits radiation depends upon its surface temperature and its ability to give off thermal radiation. A final source is the heat of metabolism—basal heat production necessary to sustain life plus heat added by the physical acts of digestion and activity.

Just as the organism gains heat from the environment, so it loses heat to the environment. One source of loss is infrared radiation. The organism continually is emitting radiation in the form of electromagnetic waves. The difference between radiant energy lost from an organism and that absorbed from the environment represents the net energy exchange.

Another source of heat transfer is *conduction*, the direct transfer of heat from one substance to another. The amount of heat lost (or gained) by conduction varies with the surface area exposed and the distance and temperature difference of the two surfaces. An animal's heat loss varies with its body position (sitting, standing, curled) and its movement from one place to another.

Convection is the transfer of heat by the circulation of fluid (liquid or gas). Convection may occur naturally in the fluid surrounding an object; or it may be forced, with pressures from fluids passing by or over an object. The amount of heat transferred by convection depends upon the shape and area of an organism, the velocity of the fluid, and the physical properties of the fluid and the organism.

Forced convection can speed up another process, evaporation. *Evaporation* depends upon the vapor pressure gradient between air and the object and upon boundary (surface) resistance. Every organism is coupled to the air temperature by a boundary layer of air adhering to its surface (Gates 1972). The resistance this layer offers to absorption and loss of heat depends upon the size, shape, texture, and orientation of the organism, wind

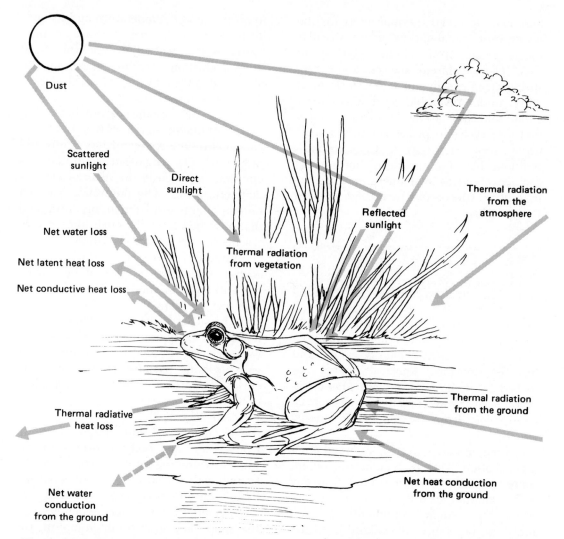

Figure 6.1 Exchange of energy and water between a frog and its environment. The flow of energy to a frog's upper surface includes (1) short-wave radiation directly from the sun and scattered from clouds and other parts of the environment; and (2) long-wave thermal radiation emitted from substrate, vegetation, and atmosphere. The frog loses energy by (1) emitting long-wave infrared radiation; (2) exchanging energy with ambient air; and (3) convection, evaporation, and condensation. It will also lose or gain energy by conduction of heat to or from its body core and by conduction and thermal radiation to the ground. (After Tracy 1976: 295.)

speed, and temperature differences between the surface of the organism and the air. With plants we must add the size and density of stomata and the resistance of the cuticle to loss of water.

The relationship of all these components in heat exchange can be summarized: Net heat gain by solar radiation, infrared radiation, and metabolism plus gain in energy storage must equal the total heat lost by radiation, conduction, convection, evaporation, and losses in energy

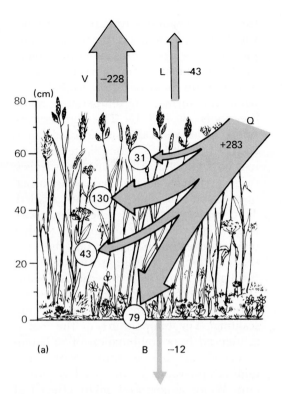

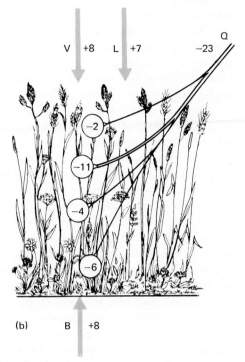

Figure 6.2 (a) Energy exchange—absorption and emission—in a meadow on a sunny day. Q = net radiation; V = evaporation; L = sensible heat convection; B = soil heat flux; figures are cal/cm². The active layer by day lies between 30 and 55 cm; it absorbs 45 percent of net radiation. The second most active layer, the lowermost, absorbs 28 percent. During input 80 percent of the radiant energy is used for evaporation of water, 15 percent for sensible heat convection, 5 percent for raising soil temperature. (b) At night net radiation as well as heat exchange is reversed. (After Cernusca 1976.)

storage. Animals have evolved adaptations to exploit or decrease the effects of one or more of the components in this heat equilibrium equation.

One aspect is important: Heat produced continuously by organisms is lost passively to the environment only when the ambient temperature is lower than the core body temperature. When ambient temperature equals core temperature, the route for passing heat off to the environment is lost. When ambient temperature exceeds body temperature, the flow is reversed and heat moves from the environment to the organism.

Thermal balance in the core of an animal is influenced by the heat of metabolism, heat stored, heat flow to skin (affected by the thickness and conductivity of fat, fur or hair, and feathers), heat flow to the ground, and heat lost by evaporation (Bakken 1976). A general formula for heat balance is

$$M^* = K_o(T_b - T_e)$$

where M^* = effective net metabolic heat production, K_o = overall thermal conductance of organism, T_b = body core temperature, T_e = operative environmental temperature.

Depending upon the values of these several components, organisms are continually either emitting or absorbing radiation. Within a few minutes or even seconds an organism may undergo great temperature changes on the surface. Energy gained at one moment is lost at another. For example, Moen (1968) measured the fluctuations on the surface of a ring-necked pheasant. Under a bright sun and free convection conditions, the surface temperature of the bird was 45° C, 4° to 5° higher than core temperature. Heat flowed from the outside surface to the interior of the animal. With a cloud cover and a slight breeze the bird's surface temperature decreased nearly to air temperature (21° C), and heat flow was from body core to the environment. These variations took place in a manner of seconds, while the air temperature remained the same. Extreme temperatures in the bird were buffered by the insulation of feathers.

■ Plant Responses to Temperature

Although both plants and animals live in the same thermal environment, there are fundamental differences in their responses to the temperature regime. Plants cannot move away to escape adverse effects of heat and cold, and they derive their heat for metabolic processes largely from the external environment.

The thermal environment in which a plant lives at any one time can vary immensely across its living structure. Roots are buffered from temperature extremes by the protection and relatively stable seasonal temperatures of the soil, while the above ground structures experience widely varying temperatures. For example, March temperatures about the stolons of clover plants 1.5 cm above the surface of the ground may be 21° C, while 7.5 cm below the surface the temperature about the roots is −1° C. The temperature range for a vertical distance of 9 cm is 22° C (Biel 1961). To complicate the thermal environment of a plant, some leaves, twigs, and buds are exposed to the insolation of the sun while others are shaded. Tree trunks, cacti, and other plants with large stem diameters experience higher temperatures on the sunny side than the shaded side, a condition that changes through a sunny day for cactus (Nobel 1976). Smooth-barked trees exposed to intense solar radiation may undergo sun scalding. The temperature of the leaf is influenced by a combination of radiation absorbed, air temperature, relative humidity, and wind. Wind speed is important. Winds as low as 1 mi/hr affect leaf temperature and increase moisture loss.

Plants balance the input of heat by radiation, convection, and transpiration. The role of evapotranspiration is modified by the opening and closing of stomata and by changes in the shape and position of the leaf. Stomatal closing to reduce loss of water under drought conditions increases the internal temperature of leaves. Leaves exposed to the sun may be smaller and thinner than those in the shade. Sun leaves of oaks have deeper lobes, increasing the surface area exposed to the air for cooling. Desert plants may have very small leaves, or dispense with leaves and carry on photosynthesis through green stems. Plants of desert and semiarid regions may change the orientation of their leaves to the sun's rays. During times of heat stress the plants hold their leaves parallel rather than horizontal to the sun's rays, reducing surface area exposed to solar radiation

(Ehleringer 1980; Ehleringer and Forseth 1980; Mooney et al. 1977).

Plants do not follow ambient temperatures exactly. During warm summer days and beneath the desert sun, plant tissue temperatures can be above ambient. For example, desert succulent cacti have surface temperatures 10° to 15° C above ambient (Nobel 1978); winter annuals may be warmer than the surrounding air (Regehr and Bazzaz 1976), and some desert ephemerals may have temperatures below ambient (Mulroy and Rundel 1977).

Plants living in permanently high or low temperature regimes have evolved physiological properties that enable them

Figure 6.3 The effect of temperature on net photosynthesis. Graphs show incident solar radiation, air temperature in relation to leaf temperature, transpiration rate, and relative metabolic rates for an exposed leaf and a shaded leaf. The exposed leaf is inhibited photosynthetically when its temperature is high. (From Gates 1968b:6.)

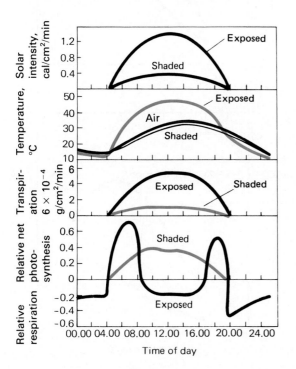

to endure their environment, but plants living in moderate environments must adjust physiologically to a rapidly changing thermal environment. Their survival depends upon their ability to avoid or tolerate thermal stress. Avoidance usually involves some morphological adaptation, such as dormant bulbs and tubers, or leaf fall and associated dormancy during winter or dry seasons. Tolerance involves the ability of the plant to meet thermal stress, either heat or cold, at cellular or subcellular levels.

Optimal temperature varies with the stages of the plant's life cycle. Optimal temperatures for photosynthesis differ among species, particularly between C_4 and C_3 plants (see Chapter 3). To germinate, seeds of many plants require chilling under moist conditions after a period of maturation. Some plants require low temperatures during or shortly after germination. The temperature necessary to stimulate flowering may be lower than the temperature that favors flower development.

Heat Stress

Heat affects the physiological processes of plants. A rate process such as photosynthesis is quite sensitive to heat stress (Figure 6.3). On a hot summer day a sunlit leaf can become too warm for photosynthesis. Following an early morning burst of activity, net photosynthesis ceases and respiration becomes dominant. How different species or even races within species respond photosynthetically to high temperatures appears to be genetically determined. For example, Pearcy (1976, 1977) studied the populations of the perennial *Atriplex lentiformis*, which grows both in southern coastal California and in Death Valley. Cloned plants from separate populations of this plant showed similar growth rates in a controlled environment (Figure 6.4). Leaves of the two clones had similar photosynthetic responses in a whole range of moderate temperatures (23° C day/18° C night), but in a hot temperature regime (43°/30° C), the desert clones grew better than the coastal clones. The desert population shifted its temperature responses, so photosynthesis improved in the heat. The coastal clone, however, was adversely affected by high temperatures; the rate of photosynthesis fell. The coastal clone obviously lacked the ability to acclimate to high growth temperatures.

Other responses of plants, particularly those growing at moderate temperatures, to heat stress depend on the intensity and nature of the stress. Most plants die at temperatures between 44° and 50° C. If high temperature comes on rapidly, the heat coagulates the proteins and disrupts the plasmic structure of the plant. If it comes on gradually, the plant experiences a breakdown in proteins and the release of toxic ammonia. Plants highly tolerant of heat generally have high levels of bound water and high plasmic viscosity. They are able to carry on protein synthesis at a sufficiently high rate in the face of rising temperatures to equal the protein breakdown rate and thus avoid ammonia poisoning (for details see Steponkus

1981). Heat tolerance of plants varies seasonally and with developmental stages and aging. Germinating and young plants and growing organs are more sensitive to heat than adult organs.

Cold Stress

When temperatures drop below the minimum for growth, a plant becomes dormant, even though respiration and photosynthesis may continue slowly. At low

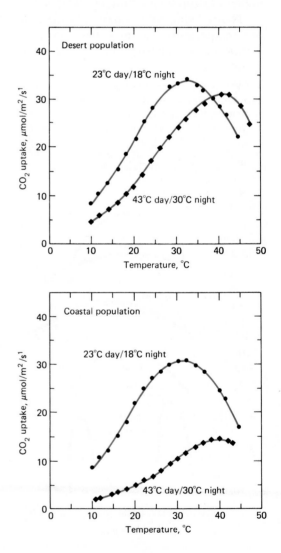

Figure 6.4 The temperature dependence of photosynthetic uptake by leaves of cloned plants of *Atriplex lentiformis,* big saltbush, from a coastal and a desert population. Both were grown under high light (1700–1800 mol quanta $m^{-2}s^{-1}$) and normal concentrations of CO_2 and O_2. Both clones were grown at moderate temperatures. When placed under a high temperature regime, only the desert population was able to carry on photosynthesis efficiently. These two populations represent ecotypes of the same species, adapted to different environmental conditions. (After Pearcy 1977: 485.)

temperatures photosynthesis is restricted in all plants. In plants sensitive to *chilling,* the action of low temperatures above freezing, photosynthesis ceases at $+5°$ to $+10°$ C. *Chilling tolerant* plants will carry on photosynthesis within that range of temperatures.

The ability of the plant to survive depends upon the specific lower limits at which the metabolic processes can continue to function under the stress of cold. Survival is influenced by *cold resistance* (the ability to resist low temperature stress without injury), the magnitude of the drop in temperature, and how long the cold persists. An outburst of cold air from an invading polar front, especially if accompanied by radiational cooling at night, may be more than the metabolically active plant can accommodate. With too little resistance to cold, the plant succumbs, but if the onset of cold is periodic, the plant may be able to adapt gradually, as plants do with the coming of winter.

If freezing occurs slowly enough, ice crystals form outside the cells. This intercellular ice draws additional water out of the cells, dehydrating them. The degree of dehydration depends upon a thermodynamic balance established between inside and outside the cell and the concentration of solutes within the cell. If the temperature falls too rapidly for this dehydration to take place, ice crystals form within the cell and damage the plasma membrane. When the plant tissues thaw, the cellular contents spill out, producing the watery appearance so characteristic of frozen plants.

The ability of plants to tolerate cold temperatures is genetic. Plants of tropical and subtropical regions and the leaves of evergreen shrubs and trees in warm-temperate coastal regions have little resistance to low temperature and may suffer lethal damage at temperatures just above freezing (Larcher and Bauer 1981). Plants in seasonally cold climates build up a tolerance to frost coupled with winter dormancy.

These plants develop frost tolerance or hardening gradually in the fall of the year. The first level of hardening develops at $+5°$ to $0°$ C. At this stage the plants can survive moderate frost. If frost is progressive and persistent, plants reach their limit of potential freezing tolerance. They form or add protective antifreeze compounds in the cells, such as certain sugars, amino acids, and nontoxic substances. If frost is less severe, resistance to freezing drops to a lower level, but it is quickly restored by a severe frost. Plants retain their tolerance to cold until growth starts in spring. Then a sudden drop in temperature may kill tissues that were able to survive far lower temperatures during the winter.

Tolerance to freezing is not uniformly distributed throughout the plant. Roots, bulbs, and rhizomes are most sensitive to freezing, succumbing to temperatures between $-10°$ to $-30°$ C; they are protected underground. Buds of woody plants, too, vary in their tolerance. Terminal buds are less resistant to cold than lateral buds; most resistant are the basal reserve buds, the ones that would replace new spring growth killed by a late frost. Woody stems are more cold resistant than leaves and buds. In fully frost-hardened twigs and trunks, the cambium is most resistant and the xylem is the most sensitive.

During the growing season some plants avoid frost damage by increasing their sugars and sugar alcohols to lower the freezing point of cell fluids. This change allows *supercooling* for short periods of time: Cell sap is lowered to a temperature somewhat below freezing without freezing immediately. Further resistance to chilling is obtained by insulation. Some species of arctic and alpine and early spring flowers of temperate regions possess hairs that act as heat traps and prevent cold injury. The interior temperature of cushion-type and rosette plants may be 20° C higher than the surrounding air. Thus some winter annuals can carry on photosynthesis at low midwinter temperatures (Regehr and Bazzaz 1976).

Thermogenesis

A very few plants are thermogenic, mostly members of the largely tropical arum lily family, which includes philodendrons, calla lily, jack-in-the-pulpits, and the skunk cabbage (*Symplocarpus foetidus*) of North America. The arum family is characterized by a fleshy spike, the *spadix*, which carries dense clusters of small, inconspicuous flowers, enclosed by a leafy hood, the *spathe*. Some tropical species, notably the philodendrons, can temporarily maintain a core temperature in the spadix up to 22° C higher than the environment. Sterile male flowers produce this heat by a regulated rate of oxidative metabolism (for biochemistry see Meeuse 1975). The flowers consume oxygen at a rate approaching that of the sphinx moth and flying hummingbirds. The heat serves to volatilize odoriferous compounds—often amines or indole—that mimic decomposing flesh and attract certain pollinating insects.

The most thermogenic is skunk cabbage, which pushes its green and purple spathe and spadix up through the cold mud and snow of early spring. R. M. Knutson (1974) measured the temperatures of skunk cabbage spadices over a three-year period. He found that the spadix maintained an internal temperature 15° to 30° C above ambient air temperatures of −15° to +15° C. In doing so skunk cabbage consumed oxygen at a rate comparable to that of a homoiothermic animal the size of a shrew or a hummingbird. As long as temperatures remained above freezing, spadix temperatures remained nearly constant at 21° to 22° C. On colder days and during nights when the temperature dropped, skunk cabbage generated more heat. In such a manner the skunk cabbage remained endothermic over the 14-day flowering period. Skunk cabbage obtains its energy from the large quantity of respiratory tissue in the spadix as well as from the deep, fleshy root. Attracted to the warmth of the spadix are the heterothermic insects of early spring.

■ Animal Responses to Temperature

How animals confront thermal stress is influenced heavily by their physiology and the nature of their environment. Because air has a lower specific heat than water and absorbs less solar radiation, terrestrial animals are subject to more radical and potentially dangerous changes in their thermal environment than aquatic organisms. Incoming solar radiation can produce lethally high temperatures, and radiational loss of heat to space can result in lethally low temperatures. Aquatic animals live in a more stable energy environment, but they have a lower tolerance for temperature change.

Physiologically animals can be split into three groups: (1) those that utilize sources of heat energy such as solar radiation and reradiation rather than metabolism; (2) those that generate their heat energy metabolically; (3) those that use both internally and externally derived energy. Animals belonging to the first group are called *ectotherms* because their principal source of heat is external. If the environmental temperature varies appreciably during the 24-hour day, the body temperature of these animals also varies. Because of this variation in body temperature, such animals are also called *poikilotherms*. Animals of the second group are called *endotherms* because their body temperature is a product of their own oxidative metabolism. Because the body temperature of endothermic animals remains relatively constant regardless of environmental temperatures, they are called *homoiotherms*. The third group is animals that cross over between groups at some stage. These animals are called *heterotherms*.

Ectotherms

Ectotherms have a high thermal conductance between the body and the environment and a low metabolic rate. They gain body heat easily from the environment and lose it just as fast. Within the range of temperatures that poikilothermic animals can tolerate, the rate of metabolism and therefore oxygen consumption increase according to Van Hoff's rule: For every temperature rise of 10° C, the rate of oxygen consumption doubles. Rising temperatures increase the rate of enzymatic activity, which controls metabolism and oxidation of carbohydrates. Lacking any homeostatic devices, terrestrial poikilotherms must depend upon some behavioral control over body temperatures. Lizards vary their body temperatures no more than 4° to 5° C when active and amphibians 10° C. The range of body temperatures over which ectotherms carry out their daily activities is called the *active temperature range (ACT)*. By limiting their ACT, poikilotherms can adapt their physiological and developmental processes to a limited range at a low metabolic cost.

Most aquatic poikilotherms generally encounter temperature fluctuations of lesser magnitude and usually have more poorly developed behavioral and physiological thermoregulatory capabilities than terrestrial forms. Immersed in a watery environment, most of these animals do not maintain any appreciable difference between their body temperature and that of the environment. Any heat produced in the muscles is transferred to blood flowing through them, and carried to the gills and skin, where it is lost to the water by convection and conduction. Thus fish are ideal poikilotherms. Because of the close relationship between body temperature and environmental temperature, fish are readily victimized by any rapid change in temperature.

Being ectothermic has its advantages and disadvantages. Prisoners of environmental temperatures, poikilotherms of temperate regions can become highly active only during the warmer parts of the day. Because metabolic activity declines with temperature, these animals become sluggish in the cool of morning and evening. Similarly, they have to restrict their active life to the late spring, summer, and early fall. Energy production in ectotherms is largely anaerobic—50 to 98 percent of it—which depletes cellular energy stores and accumulates lactic acid in the tissues. This fact severely limits bursts of poikilothermic activity to a few minutes; then exhaustion sets in. This is one reason why so many predatory terrestrial poikilotherms secure prey by ambush rather than by chase.

Ectotherms can allocate more energy to biomass production than to metabolism. Because they do not depend upon internally generated body heat, they can reduce metabolic activity during periods of temperature extremes and of food or water shortage. Low energy demands enable ectotherms to colonize areas of limited food and water, such as deserts. They are not obliged to exceed a minimum body size by metabolic heat losses, a point to be discussed later. That fact enables ectotherms to exploit resources and habitats unavailable to endotherms. On the other hand, the same metabolic restrictions impose an upper size limit: Ectotherms would not be able to absorb enough heat to warm a very large body. For this reason some paleontologists argue that the large dinosaurs had to be endothermic. A counterargument is that large ectotherms could develop and maintain body temperatures above air temperatures in a tropical environment because their low surface-to-volume ratio would limit cooling. (For both arguments see papers in Thomas and Olson, eds. 1980.)

Aquatic poikilotherms adapt to their thermal environment largely through acclimation and terrestrial forms by behavioral responses. We can view the total range of the thermal environment of an aquatic poikilotherm as a group of zones (Figure 6.5). Within the central zone of *thermal tolerance* an aquatic poikilotherm, particularly a fish, is most at home. Within this range fish seek certain *preferred temperatures*. The zone of thermal tolerance is bounded by an upper and lower zone of *thermal resistance*, temperature ranges within which the organism can survive for an indefinite period. The upper and lower bounds of thermal resistance are marked by the upper and lower *incipient lethal temperature*. At this temperature a stated fraction of a fish population (generally 50 percent), will die with prolonged exposure when brought to it rapidly from a different temperature.

The incipient lethal temperature is not fixed. Its value is affected by the previous thermal history of the organism. Within limits, poikilotherms can adjust or *acclimatize* to higher or lower temperatures. At this point it is best to define two similar terms with different meanings, ac-

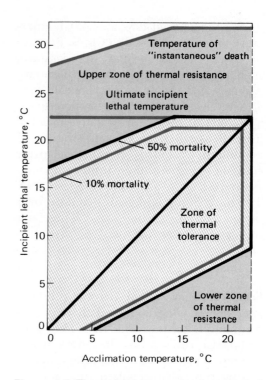

Figure 6.5 Thermal tolerance of a hypothetical fish in relation to thermal acclimation. (From C. C. Coutant 1970: 350.)

climation and acclimatization. *Acclimation* refers to differences in the physiology of an animal after exposure to environments differing in one or two well-defined variables, such as temperature and oxygen level, usually under controlled conditions. *Acclimatization* refers to differences after exposure to natural environments that vary in a number of parameters simultaneously. If the organism lives at the higher end of the tolerable range, it acclimatizes so that both the lethal temperatures and the lower limits are higher than if it were living within the cooler end of the range. Aquatic poikilotherms can adjust slowly to seasonal temperature changes.

Once acclimatized to a given temperature, fish adapt more readily to an increase than to a decrease in temperature. Acclimatization both increases the length of time an organism can survive at an elevated or lower temperature and raises the maximal or minimal temperature at which it can survive for a given length of time. However, acclimatization has upper and lower limits, and ultimately a temperature will be reached that will be lethal. This temperature is the *ultimate incipient lethal temperature* (Coutant 1970).

Amphibians present a somewhat different situation. Permanently aquatic forms, such as many salamanders, maintain body temperature in the same manner as fish: They seek preferred temperatures within their habitat. For semiterrestrial and terrestrial frogs and salamanders the problem of temperature is more complex. Salamanders exhibit seasonal variations in temperature and possess little behavioral thermoregulation. Generally they are restricted to moist, shaded environments, which seasonally are thermally homogeneous.

Semiterrestrial frogs, such as bullfrogs and green frogs, exert considerable control over their body temperature, which does not, as is often assumed, simply follow air temperature. By basking in the sun (*heliothermism*), frogs can raise their body temperature as much as 10° C above ambient temperature. Because of associated evaporative water losses, such amphibians must either be near water or sit partially submerged in it (Figure 6.6). Forms that live near water also use evaporative cooling through the skin to reduce body heat loads. By changing position or location or by seeking a warmer or cooler substrate, amphibians can maintain body temperatures within a narrow range of variation (Lillywhite 1970).

Relatively few reptiles are aquatic. Most are terrestrial and lack the buffering effects of water. Exposed to widely fluctuating temperatures of the terrestrial environment, reptiles must possess more refined means of temperature regulation.

Although poikilothermic, reptiles exhibit little relationship between their core body temperature (T_b) and ambient temperature. Evaporative cooling by panting and by water loss through the skin keep the body temperature from reaching *critical thermal maximum* (CTM), the temperature at which the animal's capacity to move is so reduced that it cannot escape from thermal conditions that will lead to its death (Cowles and Bogert 1944). Thus

Figure 6.6 The body temperature of a bullfrog measured telemetrically. Dips in the black bulb temperature indicate effects of cloud cover, convection, or both. Water temperature around the pond's edge varies from one location to another by as much as 2° to 3° C. Thus a frog in shallow water may show a temperature higher than the edge water. Note the relative uniformity of temperature the bullfrog maintains by moving in and out of the water. (From Lillywhite 1970: 164.)

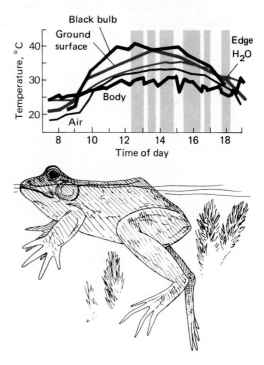

reptiles possess some of the basic physiological mechanisms characteristic of and highly developed in endotherms.

The simplest behavioral way for a reptile to regulate body temperature is heliothermism. Basking in the sun raises the core body temperature. When it reaches a preferred level, the animal moves to the shade and remains there until the body temperature drops below the preferred range. The reptile then moves out into the sun again.

More elaborate behavior common to many lizards is *proportional control*. If ambient temperature is lower than preferred, the lizard can spread its ribs, flatten its body, and orient itself so that its body is at right angles to the sun to gain the maximum amount of heat. If the temperature is too high, the lizard can appress its ribs and orient its body parallel to the sun, decreasing the surface area exposed. Other behavioral means include burrowing into the soil, panting, and possibly changing color. Thus reptiles have a variety of behavioral mechanisms for temperature regulation (Figure 6.7).

Some reptiles possess physiological mechanisms allowing them some control over maintenance of a preferred body

Figure 6.7 Behavioral mechanisms for the regulation of body temperature in the horned lizard (*Phrynosoma coronatum*). (From Heath 1965.)

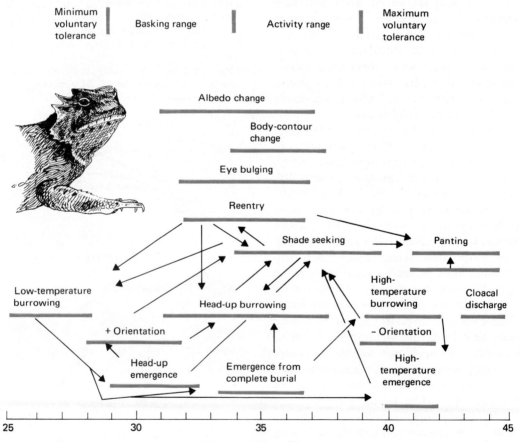

temperature. At least four families of lizards (Iguanidae, Gekkonidae, Varanidae, and Agamidae) can control the rate of change in body temperature by changing the rate of heartbeat and varying the rate of metabolism.

Long periods of below-freezing temperatures in winter face many poikilothermic animals of temperate and arctic regions. They escape the cold through supercooling and resistance to freezing.

Supercooling takes place when the body temperature falls below freezing without freezing body fluids. The amount of supercooling that can take place is influenced by the presence of certain solutes in the body. Some arctic fish, certain insects of temperate and cold climates, and reptiles exposed to occasional cold nights employ supercooling.

Some intertidal invertebrates of high latitudes and certain aquatic insects survive the cold by actually freezing and then thawing when the temperature moderates. In some, more than 90 percent of the body fluids may become frozen, and the remaining fluids contain highly concentrated solutes. Ice forms outside shrunken cells, and muscles and organs are distorted. After thawing, they quickly resume normal shape. Other animals, particularly arctic and antarctic fish and many insects, resist freezing because of glycerol in body fluids. Glycerol protects against freezing damage and lowers the freezing point, increasing the degree of supercooling. Wood frogs (*Rana sylvatica*), spring peepers (*Hyla crucifer*), and gray treefrogs (*H. versicolor*) can overwinter just beneath the leaf litter because they accumulate glycerol in their body fluids (Schmid 1982). Dormant poikilotherms also experience such physiological changes as lower blood sugar, increased liver glycogen, altered concentrations of blood hemoglobin, altered carbon dioxide and oxygen con-

tent in the blood, changed muscle tone, and darkened skin.

In addition to supercooling, many insects exhibiting frost hardiness enter a resting stage called *diapause,* characterized by a cessation of feeding, growth, mobility, and reproduction. Among many insects diapause is a genetically determined, obligatory resting stage before development can proceed. It is timed mostly through photoperiod (see Chapter 7) and associated with falling temperatures. Diapause prevents the appearance of a sensitive stage of development at a time when low temperatures would kill individuals. Diapause ends with the lengthening of photoperiod and the return of warm temperatures.

Endotherms

Birds and mammals escape the thermal constraints of the environment by being homoiothermic. Instead of depending upon the environment for heat, they produce it by metabolic oxidation. They regulate the gradient between body and air temperatures by changes in insulation (the type and thickness of fur, structure of feathers, and layer of fat), by evaporative cooling, and by increasing or decreasing metabolic heat production. Homoiothermy allows animals to remain active regardless of environmental temperatures, although at high energy costs.

High body temperatures may have evolved from an inability of large animals to dissipate rapidly the heat produced during periods of high activity (Heinrich 1976, 1979). That situation would favor enzyme systems that function at a high temperature with a set point around 40° C. Otherwise the animal would have to expend enormous amounts of water for evaporative cooling to ambient temperatures. The ability to operate at high temperatures provides homoiotherms with greater endurance and the means of remaining active at low temperatures.

However, homoiothermy has placed a size restraint on animals possessing it. A close relationship exists between body size and basal metabolic rate of a resting, post-absorptive animal in a range of temperature at which metabolic rate does not vary with temperature, the *thermoneutral zone* (Figure 6.8). Basal metabolic rate is proportional to body mass raised to the 3/4 power. As body weight increases, the weight-specific metabolic rate decreases. A doubling of body mass, for example, would decrease the basal metabolic rate by 75 percent. Conversely, as body mass decreases, basal metabolism increases, exponentially with very small body size (Figure 6.9). Within any taxonomic group of endotherms, small animals have a higher metabolism and require more food per unit body weight than large ones. Part of the reason lies in the ratio of surface area to body mass. An animal loses heat to the

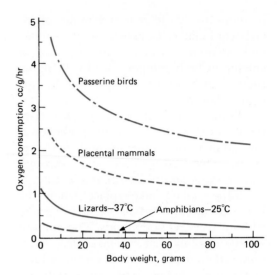

Figure 6.9 Weight-specific or standard metabolic rate as a function of body weight in four groups of vertebrates according to the following equations, where M/W is weight-specific metabolic rate expressed in cc O_2/g/hr and W is body weight expressed in grams. In placental mammals: $M/W = 3.8W^{-0.27}$. In passerine birds: $M/W = 7.54W^{-0.276}$. In lizards at a body temperature of 37° C: $M/W = 1.33W^{-0.35}$. In temperature zone amphibians at a body temperature of 25° C: $M/W = 0.335W^{-0.33}$. Note how basal metabolism rises with a decrease in body weight. (From Hill 1976: 22.)

Figure 6.8 General resting metabolic response of homoiotherms to changes in ambient temperature, indicating the thermoneutral zone and upper and lower critical temperatures.

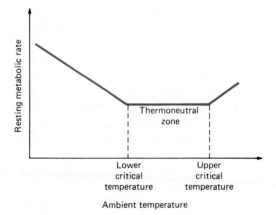

environment in proportion to the surface area exposed. All things being equal, a large animal loses less heat to the environment than a small one. To maintain homeostasis, small endothermic animals have to burn energy rapidly. In fact, weight-specific rates of small endotherms rise so rapidly that below a certain size they could not meet energy demands. Five grams is about as small as an endotherm can be and still maintain a metabolic heat balance. Few endotherms ever get that small. Because of the problem of heat balance, most young birds and mammals are born in altricial state and begin life as ectotherms. They depend upon the body heat of the parents to maintain their body temperatures. That allows the young animals to allocate most of their energy to growth rather than to heat production.

To regulate heat exchange with the environment homoiotherms utilize a number of physiological and morphological mechanisms. One is a countercurrent heat exchanger that can heat or cool vital body parts and maintain normal body core temperature (T_b). Some mammals, particularly those of the arctic such as porpoises and whales, have extensive areas in the extremities where the veins are closely juxtaposed to the arteries. Much of the heat lost by outgoing arterial blood is picked up or exchanged to the returning venous blood. Thus the venous blood is warmed on its return and reenters the body core only slightly cooler than the outgoing arterial blood. Such vascular arrangements are common in the legs of mammals and birds, and the tails of rodents, especially the beaver.

Many animals have arteries and veins divided into a large number of small, parallel, intermingling vessels that form a discrete vascular bundle or net known as the *rete*. Within the rete the blood flows in two directions and a heat exchange takes place. Such a rete in the head cools the highly heat-sensitive brain of the oryx, an African antelope exposed to high daytime temperatures (Figure 5.6), and African gazelles fleeing from predators (Figure 6.10). The external carotid artery passes through a cavernous sinus filled with venous blood cooled by evaporation from the moist mucous membranes of the nasal passages. Arterial blood passing through the sinus cavernosus is cooled on the way to the brain, reducing its temperature 2° to 3° C lower than the body core.

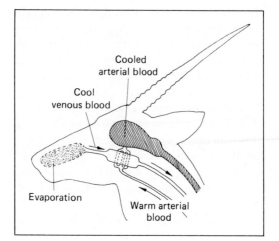

Figure 6.10 The desert gazelle can keep a cool head in spite of a high body core temperature. Arterial blood passes in small arteries through a pool of venous blood cooled by an evaporative process as it drains from the nasal region. (After Taylor 1972.)

Among some arctic mammals countercurrent circulation prevents excessive heat loss through the extremities. In addition, fat of low melting point is selectively deposited in those parts of the extremities subject to excessive cooling. Fat in the footpads of the arctic fox and the caribou has a melting point of 0°, about 30° C lower than the fat of the body core; yet the fat remains soft and flexible (Hill 1976; Irving 1972).

Physiologically it is more difficult for a homoiotherm to adapt to high temperatures than to cold. Endogenously produced heat must be transferred to the atmosphere by evaporation, largely through sweating and panting. Because birds possess a body temperature some 4° to 5° C higher than mammals, they are more tolerant of intense heat. Birds do not sweat, and water loss through the skin is inhibited by the insulating covering of feathers. Body heat is lost largely by radiation, conduction, and convection. But when conditions demand it, birds can decrease their heat load by evaporative cooling through panting. However, panting requires work, and work only adds more metabolic heat. Some groups of birds, particularly the goatsuckers, owls, pelicans, boobies, doves, and gallinaceous birds get around this dilemma by *gular fluttering,* movements of parts of the gullet. Evaporative cooling by gular fluttering uses less energy than the heavy breathing of panting.

Another approach to the heat problem, especially in birds, is *hyperthermia.* A rise in body temperature adjusts the difference between the body and the environment, and thermal homeostasis is reestablished at a higher temperature. (Hyperthermia among mammals was considered in Chapter 5.)

Parts of the body may serve as thermal windows in the conservation or radiation of heat. Horns and antlers of Artyodactyla, such as deer and goats, are vascular and only incompletely shielded by hair. Dilation of the blood vessels in response to heat stress allows increased radiation of heat, and constriction in the cold reduces radiation. The large ears of such desert mammals as the kit fox and jackrabbit function as efficient radiators to the cooler desert sky, which on clear days may have a radiation temperature of 25° C below that of the animal (Schmidt-Nielsen 1964). By seeking shade, where the ground temperatures are low and solar radiation is screened out, or by sitting in depressions, where radiation from hot ground surface is obstructed, the jackrabbit could radiate 5 kcal/day through its two large ears (400 cm^2). This loss is equal to one-third of the metabolic heat produced in a 3-kg rabbit. Such a radiation loss alone may be sufficient to take care of

the necessary heat loss without much loss of water.

Some animals of arid regions simply avoid heat by adopting nocturnal habits and remaining underground or in the shade during the day. Some desert rodents that are active by day periodically seek burrows and passively lose heat through conduction by pressing their bodies against burrow walls. Some birds, such as the whippoorwill and certain hummingbirds, and some bats go into a daily torpor.

Within limits warm-blooded animals can maintain their basal heat production by changing insulation; but with declining temperatures there is a point beyond which insulation is no longer effective and body heat must be maintained by increased metabolism (Figure 6.11). This *critical temperature* (see Figure 6.8) varies greatly between tropical and arctic animals (Scholander et al. 1950). Tropical birds and mammals exposed to temperatures below 23.5° to 29.5° C increase their heat production. If the air temperature is lowered to 10° C, the tropical animal must triple its heat production; and if lowered to freezing, the animal is no longer able to produce heat as rapidly as it is being lost. Arctic small mammals, on the other hand, do not increase their heat production until the air temperature has fallen to −29° C. Large arctic mammals can sustain the coldest weather without heat beyond that produced by normal basal metabolism. Eskimo dogs and arctic foxes can sleep outdoors at temperatures of −40° C without stress. This ability is not due to any difference in metabolism itself but to effective insulation and cold acclimation. A number of mammalian species significantly increase their basal metabolism during cold acclimation without the intervention of shivering.

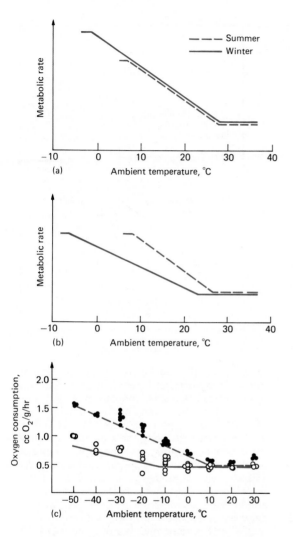

Figure 6.11 (a) Simple metabolic acclimatization. Note that in winter the metabolic rate increases as the ambient temperature declines. The plateau on the left indicates maximal metabolic rates. (b) Simple insulatory acclimatization. Insulation reduces the metabolic rate in winter and permits tolerance of much lower temperatures. (c) A real-life example of a single arctic fox (*Vulpes vulpes*) during summer and winter. Compare these curves with the hypothetical curves in (a) and (b). (From Hart 1957: 134.)

ized around the head, neck, thorax, and major blood vessels. It is found in the young of most species and in animals that hibernate, and it increases in mass when animals are chronically exposed to low temperatures (for detailed reviews, see Chaffee and Roberts 1971; Smith and Horwitz 1969). Heat generated from the metabolism of this fat is transported to the heart and brain. Nonshivering thermogenesis allows animals to be active at lower temperatures than if they had to depend on shivering alone. This form of heat production and exercise are additive; the two together can be important in maintaining body temperatures.

Heterotherms

Among the poikilotherms are a number of species that assume a degree of endothermism, and among the homoiotherms are species that assume some degree of ectothermism. Because these animals at some stage of their daily or seasonal cycle or environmental situation take on the characteristics of both groups, they might be considered *heterotherms,* animals that are endothermic part of the time and ectothermic part of the time. They are characterized by rapid, drastic, repeated changes in body temperature.

When body temperature falls below the critical level metabolism may be further increased by shivering. Among birds exposed to cold, shivering and muscular activity are primary sources of extra heat.

Mammals acclimated to cold temperatures can increase their heat production by nonshivering thermogenesis. Nonshivering thermogenesis is the generation of heat from the metabolism of brown fat, a highly vascular brown adipose tissue local-

Insects are ectothermic and poikilothermic; yet in the adult stage most species of flying insects are heterothermic. When flying they have high rates of metabolism with heat production as great as or greater than homoiotherms, and show the same general relationship between body mass and energy metabolism (Figure 6.12). They reach this high metabolic state in a simpler fashion than homoiotherms because they are not energetically constrained by pulmonary and cardiovascular pumping systems. Insects take in oxygen by demand through openings or spiracles on the body wall and transport it throughout the body by a tracheal system. Nor do they have to maintain equal temperatures through the body.

Temperature is critical to the flight of insects. Most cannot fly if the temperature of the thoracic muscles is below 30° C; nor can they fly if the muscle temperature is over 44° C. This constraint means the insect has to warm up before it takes off and it has to get rid of excess heat in flight. With wings beating up to 200 times a second, flying insects can produce a prodigious amount of heat.

Some insects, such as butterflies and dragonflies, can warm up by orienting their bodies and spreading their wings to the sun. Most warm up by shivering the flight muscles in the thorax. Moths and butterflies may vibrate their wings to raise thoracic temperatures above ambient (Pivnick and McNeil 1986; Rawlins 1980). Bumblebees do the same through abdominal pumping without any external wing movements, utilizing flight muscles uncoupled from the wings (Heinrich 1975, 1976). The problem now becomes maintaining a constant thoracic temperature during flight and not losing too much heat through the abdomen. That depends upon maximizing the rate of heat loss at a

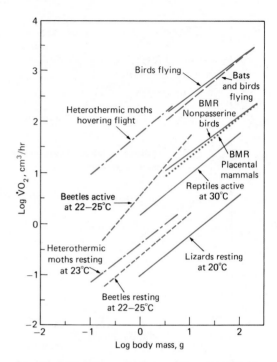

Figure 6.12 The regressions of energy metabolism on body mass of heterothermic insects and some selected terrestrial vertebrates at rest, during terrestrial activity, and during flight. The rate of oxygen consumption by heterothermic moths at rest scales with body mass at approximately the 0.78 power. During hovering flight the rate of oxygen consumption by moths increases nearly 150-fold with little change in the slope of the regression line. At rest the moths are ectothermic and in flight heterothermic. Note the relationship between metabolism of resting heterothermic moths and ectothermic lizards, and between flying heterothermic moths and homoiothermic birds and bats. (From Bartholomew 1981: 60.)

high ambient temperature (T_a) and minimizing heat loss at a low T_a.

One heat retention mechanism is insulation. The bumblebee, the sphinx and other moths, and butterflies have the thorax and dorsal surface of the abdomen insulated with pile (which appears as fine hair). The ventral surface of the abdomen, free of pile, acts as a thermal window. The bumblebee has abdominal air sacs, which act as another insulating device to retard heat flow from the thorax to the abdomen. The blood flow between the cooler abdomen and the warm thorax is heated and cooled by a countercurrent mechanism. Cool blood pumped anteriorly by the heart is warmed in the aorta, passing through the flight muscles, and if necessary can be cooled on the way back to the abdomen (Figure 6.13). Without

Figure 6.13 (a) Temperature regulation in the bumblebee involves a heavy insulation of pile on the thorax and the dorsal side of the abdomen and a lack of pile on the ventral side of the abdomen, which acts as a thermal window. A narrow petiole between the thorax and abdomen and air sacs in the anterior part of the abdomen retard heat flow from the thorax to the abdomen.

The heart pumps cool blood (hemolymph) anterior to the thorax. As the blood passes in the aorta through the petiole, it is heated by warm hemolymph flowing back from the thorax. In the thorax the aorta loops between the right and left dorsal and longitudinal muscles and then enters the head. (After Heinrich 1976: 564.) (b) The sphinx moth produces heat in its massive thoracic flight muscles. The thorax is insulated by a thick coat of scales and partially insulated from the abdomen by an air space. Hemolymph flows ventrally from the thorax to the abdomen and is pumped anteriorly by a dorsally located heart. As in the bumblebee the aorta loops between the right and left dorsal longitudinal muscles. Heart rate increases as thoracic temperature rises and the rate of heat transfer from the thorax to the poorly insulated abdomen increases. Listed are typical temperatures during flight at two different ambient temperatures, 15° C and 35° C. Body temperatures for T_a = 35° C are given in brackets. The sphinx moth regulates its thoracic temperature by increasing rates of heat loss at higher T_a rather than increasing rate of heat production at lower T_a. (After Heinrich 1971: 155.)

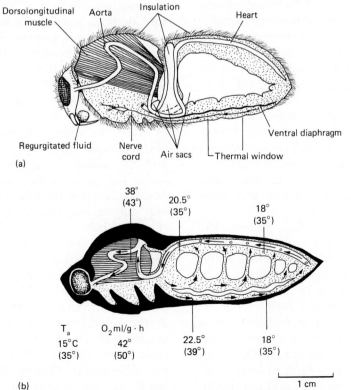

(a)

(b)

such heat retention mechanisms cool blood circulating into the thorax would cool the muscles, intefering with their functioning. In flight on a hot summer day, the insect may need to reduce thoracic temperatures to prevent overheating. By increasing blood flow to the ventral abdomen, bumblebees in particular lose excess heat to the air. In very warm situations a bumblebee will resort to evaporative cooling by regurgitating fluid, wetting its proboscis, and moving it about in the air.

When crawling about flowers securing nectar, flying insects suddenly become ectothermic, and thoracic temperatures may drop to ambient, except in bumblebees. They maintain a relatively high thoracic temperature, permitting immediate flight without another warmup.

Certain vertebrate poikilotherms that exhibit some degree of thermoregulation employ countercurrent heat exchangers. The swift, highly predaceous tuna and mackerel sharks (Lamnidae) possess a rete in the band of dark muscle tissue used for sustained swimming effort. Metabolic heat produced in the muscle warms up the venous blood, which gives up heat to the adjoining, newly oxygenated blood returning from the gills. Such a countercurrent heat exchange increases the temperature and power of the muscles, because warm muscles are able to contract and relax more rapidly.

The swordfish *Xiphias gladius,* like most fish, maintains a body temperature essentially that of the surrounding water, but it protects its eyes and brain from rapid cooling during its excursions into deep cold water with a heat exchanger (Carey 1982). The heat exchanger is a rete, associated with the eye muscle, that arises from the carotid artery. The brain heater, rich in mitochrondia and cytochrome *c,* keeps the brain and eye significantly warmer than the water. White marlin, sailfish, and other bill fish possess similar brain heaters.

A few homoiotherms allow their core body temperatures to fall to near ambient temperature as a means of circumventing thermally unfavorable environmental conditions. The simplest response is daily torpor, experienced by a number of birds, such as hummingbirds and poorwills, and small mammals such as bats, pocket mice, kangaroo mice, and even white-footed mice. Such daily torpor is not necessarily associated with any scarcity of food or water. Rather it seems to have evolved as a means of reducing energy demands over that part of the day in which the animals are inactive. Nocturnal mammals, such as bats, go into torpor by day, and hummingbirds enter torpor by night. As the animal goes into torpor, its homoiothermic response relaxes and its temperature declines to within a few degrees of ambient. Arousal returns body temperature rapidly to normal as the animal renews its metabolic heat.

A deeper state of torpor is *hibernation,* characterized by a cessation of coordinated locomotory movements; a reduction of heart rate, respiration, and total metabolism; a body temperature below 10° C; and the ability, even the necessity, to warm up spontaneously and to emerge periodically from the hibernating state using only endogenously generated heat (French 1988).

Entrance into hibernation is a controlled physiological process akin to sleep (Walker et al. 1979), difficult to explain and difficult to generalize from one species to another. Some hibernators, such as the woodchuck, feed heavily in late summer to lay on large fat reserves, from which they will draw energy during hibernation. Others, like the chipmunk, lay up a store of food instead.

All hibernators, however, have to acquire a metabolic regulatory mechanism different from that of the active state. Tissues of deeply hibernating mammals must be able to function adequately at both of the temperatures at which they exist during the year.

During hibernation the lowered heartbeat may be even, or a period of relatively rapid beats may be followed by a period of up to a minute in which the heart does not beat. Greatly reduced respiration may take place at evenly spaced intervals, or long periods of no breathing may be followed by several deep respirations. Some hibernators respond to ambient temperature below 0° C by increasing both heart and respiratory rates to speed the metabolic rate. All need to awaken spontaneously from time to time to maintain their physiological integrity. Those hibernators that do not build up fat reserves feed on stored food and, if the weather is mild, emerge briefly above ground.

Associated with entrance into hibernation are a number of other physiological changes (see Lyman et al. 1982). One of the most important involves high CO_2 levels in the body and a change in blood acid levels. As the animal enters hibernation, it accumulates metabolically produced CO_2 by reducing CO_2 output in breathing (ventilation) below that produced by metabolism. The increase in CO_2 builds up respiratory acidosis. Acidosis affects cellular processes, inhibits glycolysis, lowers the threshold for shivering, and reduces the metabolic rate.

As the animal arouses from hibernation, it hyperventilates without any change in breathing intervals, CO_2 in the body decreases, and blood pH rises. These changes are followed by shivering in the muscles (Figure 6.14), resulting in high lactic acid production and a rise in metabolism.

Some hibernators, such as woodchucks and marmots, prepare for entrance into hibernation by changing the burrow atmosphere. They plug up the entrance to the burrow with soil and nesting material, which greatly reduces air circulation and allows the CO_2 to build up in the burrow. That facilitates an increase of CO_2 in the body. Once in hibernation, the animal has a metabolic rate low enough to permit the O_2 level to rise and the CO_2 to

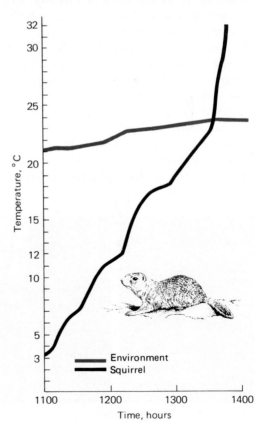

Figure 6.14 Rise in body temperature of the arctic ground squirrel upon arousal from hibernation. With a second rapid rise at 24° C, the animal opens its eyes and sits up. (After Mayer 1960.)

drop by diffusion through the ground and burrow wall. When the animal arouses, the O_2 limits in the burrow are near normal, providing favorable conditions for a rapid return to normal metabolic rates. The high metabolic rate of arousal then increases the CO_2 in the burrow atmosphere, allowing the animal to enter hibernation again.

Hibernation in bears is unique, if in fact bears can be considered true hibernators or even heterotherms. Bears do not undergo profound hypothermia. It would take them much too long to warm up a large body. Their large size allows them to build sufficient fat reserves to last them through the period of winter dormancy even though metabolism is near normal. Bears move into the hibernating state gradually in the fall when acorns and berries are abundant. The bear gorges itself, consuming up to 20,000 calories a day. By the time it enters hibernation, November for the female black bear in eastern North America and January for the male, the bear will have a layer of fat some four inches thick. It undergoes other physiological changes. In late August and September the animal attains the biochemical state of hibernation characterized by a decrease in urea in blood plasma and an increase in creatinine. It seeks a den, which may be a shallow cave, a shelter beneath overhanging rocks or low branches of a conifer, upturned roots of a fallen tree, a large brush pile, or a large hollow in a tree (Figure 6.15).

When the bear enters hibernation the body temperature drops only slightly from 37° C to 35° C, within normal limits for an active bear. Its heartbeat decreases from a summer sleeping rate of 40 to 50 beats per minute to 8 to 10 beats per minute. Its metabolism drops to 50 to 60 percent of normal, requiring the burning of some 4000 calories of energy a day; the

Figure 6.15 (a) The den of a hibernating female black bear was located in the base of this large hollow tree, which she entered from the top. (b) The square-cut area near the bottom of the trunk is the "door" scientists cut into the tree to remove the tranquilized bear for study. (c) The tranquilized 18-year-old sow has been refitted with a new radio collar and weighed. Blood samples and milk samples were collected for analysis. The bear was then returned to the den and the "door" replaced and nailed fast. This bear had four cubs weighing four pounds each; when returned to the den she was given a fifth orphaned cub for adoption. Camphor was smeared on the sow's nose to mask the scent of the introduced cub. Such adoptions are highly successful. (Photos by R. L. Smith.)

(a)　　　　　(b)　　　　　(c)

bear will lose 20 to 25 percent of body weight. It does not arouse periodically during hibernation, like other hibernators; but the bear's nearly normal body temperature permits it to awaken suddenly and move about immediately if disturbed. Bears do not drink, eat, defecate, or urinate, although urine is formed and enters the urinary bladder (Nelson 1980). Even after the bear emerges from hibernation five to six months later, it may continue in the physiological state of hibernation and not eat or drink for several weeks.

The bear is able to starve itself successfully because it can control urea metabolism, so uremia and dehydration do not occur. Urea, water, and other constituents in urine are absorbed by the bladder wall and enter the bloodstream, and urea is degraded by two mechanisms. One is the diffusion of urea into the lumen of the intestine, where it is hydrolyzed to produce ammonia and CO_2. After reabsorption, the ammonia reacts with glycerol released from the breakdown of fat to form amino acids, which are reincorporated into plasma proteins. A second mechanism involves a direct pathway of urea nitrogen into amino acids. Thus during the period of starvation in the hibernating state, the bear has an anabolic pathway in which urea nitrogen is converted into plasma proteins (Nelson and Beck 1984; Nelson et al. 1983). Research on the physiology of hibernating black bears is of great interest to medicine because it may offer clues to such human health problems as comas and kidney functioning.

For the female bear hibernation is also the time of giving birth to two to four young. The blind, almost naked cubs are only eight inches long and weigh 220 to 280 grams. By the time the female comes out of hibernation, the young suckling on

the sleeping sow will weigh 1.8 to 2.5 kilograms.

A dormancy similar to hibernation is employed by some desert mammals and birds during the hottest and driest parts of the year. Such summer dormancy is called *estivation*. Because little evidence exists to show that estivation is triggered by either heat or drought and the physiological processes are similar, Hudson (1973) considers it synonymous with hibernation. Walker et al. (1979) suggest that estivation be termed shallow torpor instead.

■ Temperature and Distribution

Because the optimum temperature for the completion of the several stages of the life cycle of many organisms varies, temperature imposes a restriction on the distribution of species. Generally, many species are restricted in their range by the lowest critical temperature of their life cycle, usually the reproductive stage. For example, the green frog (*Rana clamitans*) does not breed until the water in breeding ponds is about 25° C, and the eggs develop in water at 33° C, a temperature lethal for some more northern frogs (Moore 1949a). Its eggs, however, will not develop at all until the temperature exceeds 11° C. Therefore the range of the green frog extends only slightly above the northern boundary of the United States.

Further south the vampire bat (*Desmodus rotundus*) ranges from central Mexico to northern Argentina. Both the northern and southern limits of its range parallel the 10° C minimal isotherm of January (McNab 1973) (Figure 6.16). Its distributional limits are associated with the bat's poor capacity for temperature

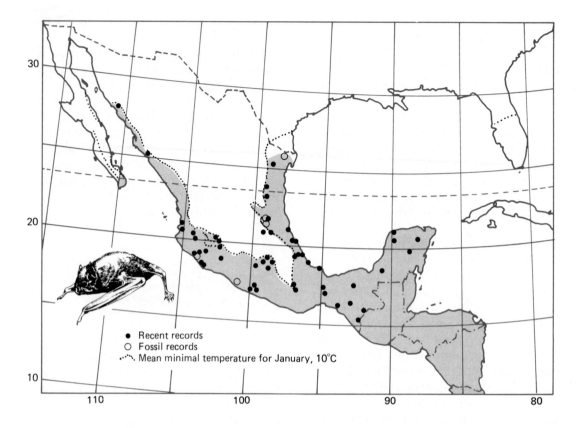

Figure 6.16 The northward distribution of the vampire bat (*Desmodus rotundus*) parallels closely the 10° C minimal isotherm for January. Its southward distribution in Argentina and Chile is also limited by the same isotherm. (After McNab 1973: 140.)

regulation at ambient temperatures below 10° C.

Temperature is a major influence on the natural distribution of plants on a continental, regional, and local scale. Their ranges often coincide with a certain minimal or maximal isotherm; but suitable temperature regimes allow some plants such as red spruce and balsam fir to occupy local pockets of habitat outside their major range. The northward distribution of C_4 plants (see Chapter 3), especially the grasses, appears to be limited by cool temperatures (Stowe and Terri 1978). The role of temperature may be more evident in the distribution of C_3 and C_4 grasses on an elevational gradient along tropical mountain slopes on Hawaiian volcanos (Rundel 1980) and Mt. Kenya (Tieszen et al. 1979). On the Hawaiian volcanoes C_4 species dominate the low, hot elevations and C_3 species dominate the cool, high elevations (Figure 6.17). The vegetational change with elevation and temperature suggests that C_3 plants outcompete C_4 plants at cool temperatures and vice versa. The mean maximum daily temperature at the midpoint of replacement both in Hawaii and Africa is about 22° C.

On a continental scale, the ranges of forest trees reflect temperature regimes.

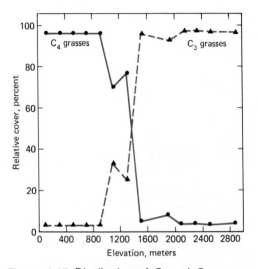

Figure 6.17 Distribution of C₃ and C₄ grasses, measured by relative ground cover along an elevational gradient in the Hawaii Volcanoes National Park. C₄ grasses dominate the lower elevations with higher temperatures. At cool higher elevations, C₃ grasses dominate. Midpoint of replacement seems to be a mean maximum daily temperature of 22° C. (After Rundel 1980: 355.)

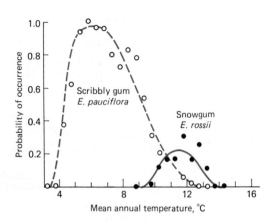

Figure 6.18 The distribution of trees is influenced by maximum and minimum mean annual temperatures. *E. pauciflora* has a much wider distribution than *E. rossii*, · which is restricted to a narrow high temperature range, just at the edge of the range of *E. pauciflora*. (After Austin et al. 1984.)

In southern Australia the distribution of snow gum (*Eucalyptus pauciflora*) differs from that of another species (*E. rossii*) on the basis of temperature alone (Austin et al. 1984) (Figure 6.18). Snow gum can occupy a much wider range of temperature regimes than *E. rossii* because its seeds do not germinate until they are least likely to be affected by snow, drought, or high temperature (Abrecht 1985), and it has the ability to maintain high photosynthetic activity over a range of temperature optimums, closely tracking the long-term mean monthly temperatures (Slayter 1977). The northernmost limit of the range of the evergreen live oak (*Quercus virginiana*) is stopped by average annual minimum temperature between −3.9° C and −6.7° C. The northern limit of the

sugar maple (*Acer saccharum*) parallels the 2° C mean annual isotherm. Paper birch (*Betula papyrifera*), a cold climate species, is found as far north as the 12° C July isotherm and seldom grows naturally where the average July temperature exceeds 21° C. The distribution of black spruce (*Picea mariana*) follows a similar pattern.

Some plants, such as blueberries, grow beyond their range but will not flower or fruit successfully unless chilled. Other plants grow in a region during the summer, but are unable to reproduce or grow to normal size because the twigs freeze back in winter or are killed by late spring frosts. Only within their natural ranges are temperatures favorable for growth and reproduction.

It is often difficult to separate the influence of temperature from that of moisture in terrestrial environments. Together, the two determine in large measure the climate of a region and the distribution of vegetation (Figure 6.19).

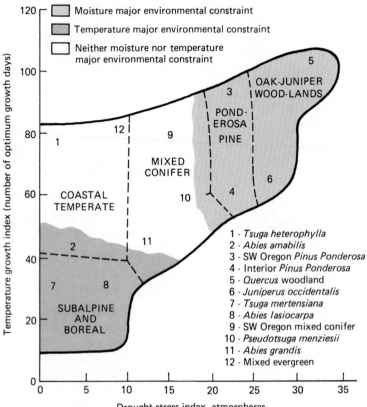

Figure 6.19 Distribution of some of the major forest zones in the Pacific Northwest, based on moisture (maximum plant moisture stress during the dry season) and temperature (optimum growth days). Temperature is the major influence separating subalpine types from temperate. In the xerophytic zone moisture is limiting. Within the temperate zones neither moisture nor temperature conditions are severely limiting for forest tree species. (After Franklin and Dyrness 1973: 50.)

■ Summary

The ultimate heat source for life is solar radiation. Life exists within a restricted range of temperatures, from about 52° C for animals and 60° C for plants to about −60° C, depending upon the species.

Organisms maintain a balance between heat energy gained from and heat energy lost to the environment. Heat energy gains come from reflected sunlight, diffuse radiation, long-range infrared radiation, and heat of metabolism. Heat losses involve infrared radiation. Heat transfer between organism and environment involves conduction, convection, and evaporation. Heat loss to the environment can occur only when the ambient temperature is less than body core temperature.

Response to the thermal energy environment differs between plant and animals. Plants receive their energy from the environment. Heat affects the physiological processes of plants, especially photosynthesis; and various phases of the life cycle have their optimum range of temperature. Plant metabolism contributes little to internal plant temperature, but plants have evolved certain mechanisms to resist extremes of heat and cold. These include reflectivity of leaves and bark, leaf size and shape, orientation of leaves toward the sun, and frost hardening. The last involves the synthesis of protective antifreeze substances in cells. Plants are ectothermic, but a few members of the arum family are seasonally endothermic, a feature that appears to attract pollinating insects.

Animal responses are more complex. Animals depending upon the environment as a source of heat are ectothermic; those depending upon internally produced metabolic heat to maintain body temperature are endothermic. Animals that depend upon environmental heat are generally poikilothermic: Their body temperatures fluctuate to a degree with ambient temperatures. They regulate body temperatures behaviorally by moving in and out of warmer and cooler areas. Animals that are heated metabolically are homoiothermic. They maintain their body temperature by means of a closely regulated feedback system. Some homoiothermic and poikilothermic animals are actually heterothermic. Depending upon the environmental and physiological situation, they function either as endotherms or ectotherms. They may regulate body temperature metabolically or allow body temperature to drop to ambient, a mechanism that allows heterotherms either to increase or to conserve energy.

Animals respond to temperature changes by acclimatization. Within limits poikilotherms acclimatize to higher or lower temperatures by avoiding temperature extremes, by adjusting tolerances to given temperature ranges, by heliothermism, and in some instances by evaporative cooling. Some reptiles change their body temperatures by changing the rate of heartbeat and metabolism. Some cold-tolerant poikilotherms utilize supercooling, the synthesis of glycerol in body fluids to resist freezing in winter. Homoiotherms and heterotherms acclimatize by changes in body insulation, vascularization, evaporative heat loss, and nonshivering thermogenesis. Many also employ a well-developed countercurrent circulation, which involves the exchange of body heat between arterial and venous blood or their equivalent in insects. This allows the animals to retain body heat by reducing heat loss through body parts, or to cool blood flowing to such vital organs as the brain.

Some animals enter a state of dormancy during environmental extremes to reduce the high energy costs of staying warm or cool. Such a dormancy or torpor involves a slowdown in metabolism, including heartbeat, respiration, and body temperature. Some birds, such as hummingbirds, and mammals, such as bats, undergo daily torpor, the equivalent of deep sleep. Such torpor does not involve the extensive metabolic changes characteristic of seasonal torpor. A number of mammals undergo hibernation, a whole rearrangement of metabolic activity to a low level. What separates hibernation from winter dormancy is the ability of these mammals to arouse periodically during hibernation.

Temperature regimes set upper and lower limits within which organisms can grow and reproduce, influencing the spatial distribution of plants and animals.

Light

The role of light in ecosystems extends beyond its role in photosynthesis. It affects the local distribution of plants in both terrestrial and aquatic environments, and thus the structure of ecosystems. It also influences the daily and seasonal activities of plants and animals.

■ The Nature of Light

Light is that part of solar radiation in the visible range, embracing wavelengths of 0.40 microns to 0.70 microns. These wavelengths collectively are known as *pho-*

tosynthetically active radiation (PAR), because it is utilized in photosynthesis (see Chapter 3).

Earth's atmosphere absorbs much arriving light (see Figure 4.1). Light that reaches the surface of Earth is either reflected, absorbed, or transmitted through objects. About 6 to 12 percent of photosynthetically active light striking a leaf is reflected. The degree of reflection varies with the nature of the leaf surface. Because they reflect green light most strongly, leaves appear green. Most of the red light is absorbed by chlorophyll in the mesophyll of the leaf and used in photosynthesis. A remaining fraction of light is transmitted through the leaf, the amount depending upon thickness and structure. Transmitted light is mostly green and far red, a point to be considered shortly.

Light that strikes the surface of water is further attenuated, producing a wider range of light environments than on land (Figure 7.1). Depending upon the angle at which light strikes the water, a greater or lesser amount is reflected from the surface. For example, at an angle of inci-

dence 60° from perpendicular, only 6 percent of light is reflected; at 80°, 35 percent is reflected. The nearer the angle is to perpendicular, the more light will enter the water.

Light that enters the water is absorbed rapidly. Only about 40 percent reaches a meter deep in clear lake water. In the process of depletion water absorbs wave lengths differentially. First to go are all infrared and visible red light in wavelengths greater than 750 nm, reducing solar energy by one-half. In clear water yellow goes next, followed by green and violet, leaving only blue wavelengths to penetrate deeper. As with the wavelengths behind it, a fraction of the remaining intensity is lost with increasing depth. In the

Figure 7.1 The spectral distribution of solar energy at Earth's surface and after it has been modified by passage through varying depths of pure water. Note how rapidly red wavelengths are attenuated. At approximately 10 meters red light is depleted; but at 100 meters, blue wavelengths still retain nearly one-half their relative intensity. (From G. Clarke 1939.)

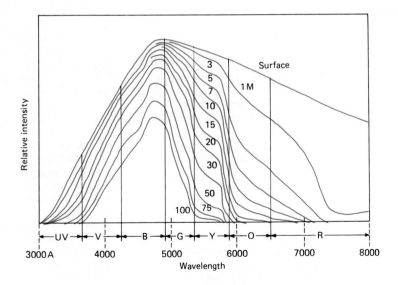

clearest of seawater, only about 10 percent of the blue wavelengths reach more than 100 m.

Of course, all water is not clear. Freshwater lakes, streams, and ponds, and coastal waters have a certain degree of turbidity, which greatly affects light penetration. Because of the high level of yellow substances, such as clays, colloids, and fine detrital material washed into the water from terrestrial ecosystems, the blue wave band is most strongly attenuated and is removed at very shallow depths. Green is the most penetrating wavelength; and where concentrations of yellow materials are high, red wavelengths may penetrate as far as the green. In very yellow water

red is the last wavelength to be extinguished. Heavy growth of phytoplankton and blue-green algae can shut out light to deeper waters just as effectively, if not more so, than the dense canopy of a deciduous or tropical rain forest.

The point at which the intensity of light reaching a certain depth is insufficient for photosynthesis is the *extinction coefficient*. It is the ratio between the intensity of light at a given depth with intensity at the surface. The light intensity below which plants can no longer carry on sufficient photosynthesis to maintain a balance between energy demands and respiration is called *compensation intensity*. Few plants have a compensation intensity of less than

TABLE 7.1 Adaptations of Seedlings in Five Different Light and Habitat Situations

Type of habitat	Illumination 0–40 cm above ground surface	Adaptive features
Dry, severe	Continuously high through growing season	Annual life cycle; small seeds efficiently dispersed; creeping shoots; compact rosettes; vertical leaves; high resistance to wilting
Recently cleared, nutrient-rich, moist	Initially high	Prolific seed production; efficient dispersal; tall stature; high growth rate; competitive; rapid extension of stem on shading; high leaf area index; minimal mutual shading
Grassland	Vertical variation, small increase in intensity with increase in height above ground	Tall stature; rapid extension of growth on shading
Open woodland	High intensity early in growing season before canopy closes (spring); moderately high late in season after canopy thins (fall)	Growing and flowering before leaf canopy closes; flowering after canopy thins; tall stature; low resistance to wilting; large leaf area with minimal mutual shading; horizontal leaves; rapid extension of growth on shading; potential for high rates of photosynthesis
Dense woodland	Low intensity over most of growing season; small increase in intensity with increase in height above ground	Low respiration; horizontal leaves; limited extension of growth on shading; high resistance to fungal attack

Source: After J. P. Grime 1966.

about 100 footcandles or 1 percent of sunlight. In aquatic environments phytoplankton can carry on photosynthesis only to the depth at which photosynthetically active radiation falls to 1 percent of that just below the surface.

■ Plant Adaptations to Light

In Chapter 3 we saw how intensity of light relates to photosynthesis. Some plants, both aquatic and terrestrial, are able to maintain a better balance between energy production and respiration than others. The relative ability of plants to carry on photosynthesis at low light intensities is termed their *shade tolerance*. This concept is somewhat oversimplified. There is no clear-cut distinction between shade-tolerant and shade-intolerant species based solely on photosynthetic efficiencies in low light (Loach 1969). Other influences are involved in the maintenance of energy balance in shade-tolerant plants in low light environments, such as root competition, clipping by herbivores, and moisture. Table 7.1 shows how seedlings adapt to the combined influences of light and habitat type.

Terrestrial Environments

In both grasslands and forest most of the sunlight that floods open spaces is intercepted by a leafy canopy (Figure 7.2). Only about 1 to 5 percent of the light striking the canopy of a typical temperate hardwood forest reaches the forest floor. In a tropical rain forest only 0.25 to 2 percent gets through. More light travels through pine stands—about 0 to 15 percent—but densely crowned spruce allows only 2.5 percent of open sunlight to reach the forest floor. Woodlands with trees pos-

sessing relatively open crowns, such as oaks and birches, allow light to filter through. There light dims gradually, as it does in grasslands. In grasslands most of the light is intercepted by the middle and lower layers.

Light conditions within the forest vary from unfiltered light coming through the gaps to filtered light in which the spec-

Figure 7.2 (a) Attenuation of radiation in a boreal mixed forest. Ten percent of the incident photosynthetically active radiation (PAR) is reflected in the upper crown, and the greatest absorption occurs in the crown. (b) In the meadow 20 percent of the photosynthetically active radiation is reflected from the upper surface, whereas the greatest absorption occurs in the middle and lower regions, where the leaves are most dense. Only 2 to 5 percent of the incident PAR reaches the ground. (Adapted from Larcher 1975: 15.)

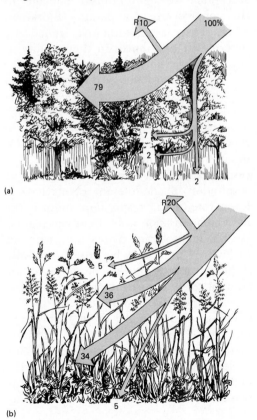

(a)

(b)

trum is altered. There are large differences in attenuation between far red and visible radiation. Troughs of blue and red, the most highly attenuated wavelengths, vary with peaks of green and far red wavelengths. Far red wavelengths are less strongly attenuated than visible light, enriching the shaded environment. This large change in its spectral quality makes the light in the subcanopy less active photosynthetically.

Leaves of shade-tolerant plants have photosynthetic properties different from those of leaves grown in high light. Adaptations to low light in shade-tolerant plants involve some adjustments in light gathering, energy fixation, and energy conservation in photosynthesis. In general shade plants have a poor ability to adapt to high light intensities. They are highly susceptible to light-induced damage to their photosynthetic apparatus (Bjorkman 1981). There are quantitative, genetically determined differences in their ability to adjust to different light regimes.

The relative ability of different species of plants to tolerate shade involves differences in the efficiency with which they lower respiration rates or increase photosynthetic rates at low light intensity. Of greatest importance in determining success or failure in shade may be the rate of respiration. Shade-tolerant plants have a lower rate of leaf respiration and a lower light compensation point than sun plants.

Shade-intolerant plants may respond to lowered intensities by growing rapidly in height, increasing their leaf area in an attempt to emerge from the shade. This initial rapid growth rate and their relatively thin leaves make them highly susceptible to drought. Thin cell walls of their stems reduce supporting tissue and subject them to fungal infections. Under high light intensities of the open sun plants have both a high rate of photosyn-

thesis and a high rate of respiration and rapidly convert photosynthetate into biomass.

By contrast, shade-tolerant plants are characterized by increased chlorophyll per unit of leaf weight and increased leaf area per unit weight invested in shoot biomass. They have lower photosynthetic, respiration, metabolic, and growth rates than sun plants, so they conserve nutrient and energy reserves. The photosynthetic system becomes light-saturated at lower intensities than in sun plants, but maintains maximum photosynthesis for a longer period of time. Such plants cannot compete with sun plants under high light intensities. However, because of their low metabolism and slow growth in low light, shade plants can sustain themselves under poor light conditions for a long period of time. This ability is further strengthened by the shade plant's inherent resistance to fungal infection (Grime 1966).

Typical among shade species are sugar maple, white cedar, and hemlock (Table 7.2). These species exist successfully under a dense forest canopy at low light intensities but reach maximum growth rates only when the canopy is opened sufficiently to allow considerably more light to enter. Thus sun and shade plants have evolved characteristics that enable them to colonize certain habitats successfully and inhibit them from occupying others. These differential abilities influence community composition and succession (see Chapter 25).

Aquatic Environments

Because of the rapid attenuation of light with depth, most aquatic plants live in the equivalent of a shaded environment, but with a major difference. Far red wave-

TABLE 7.2 Relative Shade Tolerance of Some North American Trees

	Very Intolerant	
Jack pine	Whitebark pine	Aspens
Virginia pine	Pin cherry	Paper birch
Longleaf pine	Tamarack	Black locust
Lodgepole pine	Western larch	Cottonwoods
	Intolerant	
Red pine	Junipers	Black cherry
Shortleaf pine	Sycamore	Sweet gum
Loblolly pine	Sassafras	Black walnut
Ponderosa pine	Scarlet oak	Red alder
Eastern red cedar	Yellow-poplar	Madrone
	Intermediate	
Eastern white pine	Douglas-fir	Silver maple
Western white pine	Silver maple	Oaks
Sugar pine	Hickories	White ash
Slash pine	Yellow birch	Elms
	Tolerant	
Spruces	Grand fir	Basswood
Redwood	White fir	Red maple
Western red cedar	Alpine fir	
	Very tolerant	
Eastern hemlock	Balsam fir	Sugar maple
Western hemlock	Pacific silver fir	American beech
Mountain hemlock	Pacific yew	Flowering dogwood
Red spruce	White cedar	Hop hornbeam

lengths are strongly absorbed by water, but not by terrestrial foliage. The red to far red ratio in aquatic environments is 3.6 to 4.7 at one meter, compared to 1.15 for forest interiors.

Aquatic plants can so modify the light environment for plants lower in the water column that they greatly reduce PAR. Because phytoplankton species can become photoinhibited at high light intensities at the surface, especially in sunny weather, they carry on high photosynthetic rates at some depth below the surface (Figure 7.3). Some species of phytoplankton can move up and down through the water col-umn to escape inhibitory effects of high light and to reach a depth at which light intensity is most favorable. Photosynthetic ability is further influenced by seasonal variations in water temperature and light, and by the chromatic adaptations of phytoplankton and macrophytes. Most macrophytes grow at a depth at which light intensity is most favorable.

In marine environments the zonation of red, green, and brown algae typically has been explained by the different absorptive spectra of their pigments, affecting their ability to carry out photosynthesis. Because they contain certain

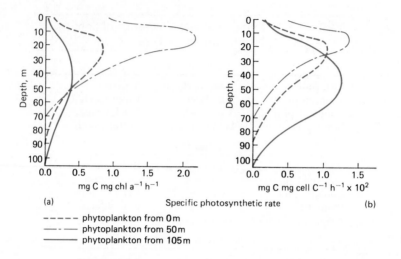

Figure 7.3 Vertical profiles of (a) specific photosynthetic rate per unit of chlorophyll and (b) specific rate per unit cellular biomass, expressed as carbon of phytoplankton collected at three depths, Om, 50m, and 105m, and incubated between the lake surface and 105m depth in late summer (9 September 1974), at Lake Tahoe, Nevada, and California. The depth at which phytoplankton occurred was influenced in part by thermal stratification. In early September thermal stratification (see Chapter 10) was at its maximum and phytoplankton at shallow depths received considerably more light than that in deeper water. Phytoplankton collected at shallow depths was light-adapted and deep populations were dark or shade-adapted. (From Tilzer and Goldman 1978: 816.)

pigments accessory to chlorophyll *a* that absorb green light, red algae have been considered typical of the deepest water. Green algae with chlorophyll *a* and *b*, similar to terrestrial vegetation, is considered a shallow water inhabitant. Brown algae, with chlorophyll *a* and *c* and a special carotenoid pigment, have been considered the seaweeds of intermediate depths.

Many recent studies, however, point out that zonation of seaweeds is not that simple. No significant difference exists in either the minimum or maximum depths inhabited by red, green, and brown algae.

Red algae do not have an exclusive claim to the deepest waters; and the green algae, supposedly restricted to shallow water, may grow in greater abundance at depths of 90 to 100 m than red algae (Figure 7.4). In Hawaiian waters green algae may penetrate as deep as red algae and more deeply than the brown. Seaweeds appear to respond physiologically more to irradiance than to the spectral quality of light (Ramus 1983). Other important influences on zonation include structural anatomy (such as the thickness of blade), type of substrate, salinity, temperature, predation, and competition for settling space (Saffo 1987).

Ultraviolet Radiation

Perhaps of some ecological significance is solar ultraviolet radiation. Most of the ultraviolet radiation is blocked from reaching Earth by the ozone layer. What does get through is further scattered and absorbed by atmospheric particles and molecules. Receiving most exposure are arctic and alpine plants, growing at high altitudes and latitudes where the atmosphere is thin. These plants have the ability to absorb ultraviolet radiation in their epider-

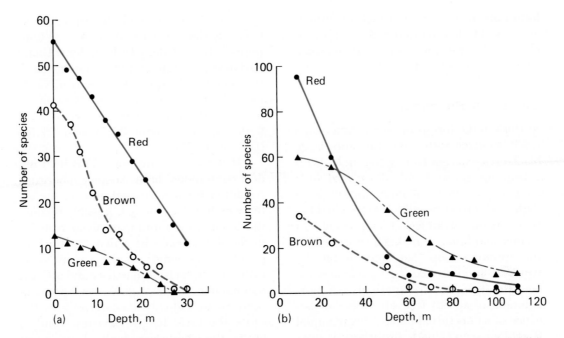

Figure 7.4 Variations in the distribution of red, green, and brown algae with depth. (a) In New England waters red algae show a gradual decline in the number of species with depth. Brown algae are more abundant at shallow depth and show a marked decline with depth, with a few species still prevalent at 30 meters. The less abundant green algae have relatively more species adapted to light conditions at deeper waters. (From Kirk 1983: 294, based on data from Mathieson 1979.) (b) In the clear tropical waters of the Caribbean, all three types of marine benthic algae reach much greater depths. Typically, the red algae decline sharply in abundance below 50 meters; less abundant brown algae disappear more rapidly. Below 50 meters green algae dominate at all depths down to 110 meters. (From Kirk 1983:296, based on data of Taylor 1959.)

mal tissues, protecting the internal tissues and the photosynthetic apparatus.

Of some concern is the diminution of the ozone layer by atmospheric pollution and its future effect on life on Earth. An increase in ultraviolet radiation over that to which plants are adapted can adversely affect nucleic acids, cause genetic altera-

tion, increase mutation rates, and inhibit photosynthesis, reducing growth and biomass production (Caldwell 1981). Increased solar ultraviolet radiation can change the competitive balance among plant species by favoring increased growth and productivity of UV-adapted species, if they can utilize more of the available resources (Fox and Caldwell 1978).

■ Photoperiodism

No one can ignore the rhythmicity of Earth, its recurring daily and seasonal changes. Life of the daylight hours retreats before the night, night to day. As the year progresses, the seasons change. The progression of lengthening and shortening days are marked by the flush of new growth and the senescence of the old, by the arrival and departure of migrating birds. These daily and seasonal rhythms are driven by the daily rotation of Earth on its axis and its 365-day revolution about the sun. Through evolutionary time life became attuned to and synchronized with the daily and seasonal changes in the environment.

Circadian Rhythms

At dusk in the forests of North America, a small squirrel with silky fur and large black eyes emerges from a tree hole. With a leap the squirrel sails downward in a long sloping glide, maintaining itself in flight with broad membranes stretched between its outspread legs. Using its tail as a rudder and brake, it makes a short, graceful upward swoop that lands it on the trunk of another tree. It is *Glaucomys volans*, the flying squirrel, perhaps the most common of all our tree squirrels, but because of its nocturnal habits, this mammal is seldom seen. Unless disturbed, it does not come out by day. It emerges into the forest world with the arrival of darkness; it retires to its nest before the first light of dawn.

The squirrel's activity forms a 24-hour cycle. The correlation of the onset of activity of the flying squirrel with the time of sunset suggests that light has some regulatory effect on it. If the flying squirrel is brought indoors and confined under artificial conditions of night and day, the animal will restrict its periods of activity to darkness, and its periods of inactivity to light. Whether the conditions under which the animal lives are 12 hours of darkness and 12 hours of light or 8 hours of darkness and 16 hours of light, the onset of activity begins shortly after dark.

Such behavior in itself does not mean that the animal has any special time-keeping mechanisms. It could be responding behaviorally to nightfall and daybreak. However, if the same squirrel is kept in constant darkness, it still maintains that constant activity pattern from day to day in the absence of all time cues. Under these conditions, though, the squirrel's activity rhythm deviates from the 24-hour periodicity. The daily cycle under constant darkness (the light condition under which the nocturnal squirrel is active) varies from 22 hours 58 minutes to 24 hours 21 minutes, the average being less than 24 hours (most frequent 23:50 and 23:59) (DeCoursey 1961).

If the same flying squirrel is held under continuous light, a very abnormal condition for a nocturnal animal, the activity cycle is lengthened, possibly because the animal, attempting to avoid running in the light, delays its activity as much as possible. The length of the activity cycle, including periods of inactivity and activity, maintained under a given set of conditions is an individual characteristic. Because the cycle length deviates from 24 hours, the squirrel gradually drifts out of phase with the day-night changes of the external world (Figure 7.5).

This innate rhythm of activity and inactivity covering approximately 24 hours is characteristic of most living organisms except bacteria. Because these rhythms approximate, but seldom match, the periods of Earth's rotation, they are called *circadian* (from the Latin *circa*, "about," and *dies*, "day"). The period of the circadian rhythm, the number of hours from the beginning of a period of activity one day to the beginning of activity on the next, is called the *free-running cycle*. In other words, the rhythm of activity exhibits a self-sustained oscillation under constant conditions. The length of the free-running cycle usually is a function of the intensity of light provided under constant conditions. Increasing the light intensity lengthens the free-running cycle in organisms active by night and shortens the cycle of organisms active by day (Hoffman 1965).

Circadian rhythms are genetic, incorporated in the geome, transmitted from

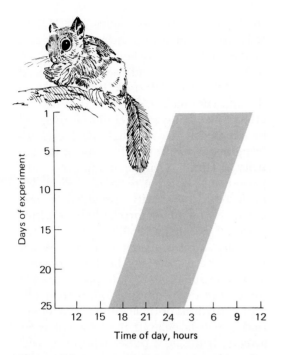

Days of experiment

Time of day, hours

Figure 7.5 Drift in the phase of activity of a flying squirrel held in continuous darkness at 20° C for 25 days. Note that the onset of activity gradually became later each day. (After De-Coursey 1960: 51.)

one generation to another. They function endogenously by some internal cellular mechanism. They are affected little by temperature changes, are insensitive to a great variety of chemical inhibitors, and are not learned from or imprinted upon the organism by the environment. They influence not only activity but also physiological processes and metabolic rates. Nor have they any special adaptation to specific local or regional environmental conditions. What circadian rhythms do is provide a mechanism by which organisms can maintain synchrony with their environment (for excellent examples in bats see Erkert 1984).

Thus plants and animals are influenced by two periodicities, the external rhythm of 24 hours and the internal circadian rhythm of approximately 24 hours.

If the two rhythms are to be in phase, some external environmental "timesetter" must adjust the endogenous rhythm to the exogenous. The most obvious time-keepers, cues, synchronizers, or *Zeitgebers* (Aschoff 1958) are temperature and light. Of the two the master Zeitgeber is light. It brings the circadian rhythm of organisms into phase with the 24-hour photoperiod of the external environment.

We might have difficulty in the field demonstrating the role or even need of light to synchronize the circadian rhythm with the environment. The activity pattern of organisms could just as well be explained by a behavioral response to changing light conditions (Enright 1970). However, it can be demonstrated in the laboratory by holding the organism under constant conditions of continuous darkness or continuous light to allow the circadian rhythm to drift out of phase with the natural environment or even fade away.

The length of time before this change depends upon the organism and the conditions of light and dark (Kramm 1975). The activity rhythms of rodents and bats may continue for several months in constant darkness. Other rhythms, such as the leaf movements of plants, fade much more quickly. Once a rhythm has faded, a new one can be started by some exposure to light or dark. It may be the interruption of continuous darkness by a short flash of light or the interruption of continuous light by darkness, the change from continuous darkness to continuous light, or vice versa. With some organisms a change in temperature may start a new rhythm.

The activity period of organisms shows an entrainment to light-dark cycles. The flying squirrel, both in its natural environment and in artificial day-night schedules, synchronizes its daily cycle of activity to specific phases of the light-dark cycle, as do bats. DeCoursey (1960a, 1961) demonstrated this experimentally in the flying squirrel. She held her experimental animals in constant darkness until their circadian rhythms no longer were in phase with the natural environment. Then she subjected them to a light-dark cycle out of phase with their free-running cycle. If the light period fell into the animals' subjective night (natural period of activity), they delayed their onset of activity. Synchronization, or *entrainment,* took place in a series of stepwise delays until the animals' rhythms were stabilized with

the light-dark change (Figure 7.6). If light fell into the subjective dawn or at the end of the dark period (when the animals' activity was about to end), activity advanced toward the period of dusk. If light fell into the animals' inactive day period, it had no effect. The flying squirrel did not need to be exposed to a whole light-dark cycle to bring about a shift in the phase of the activity rhythm. A single ten-minute light period was sufficient to cause a *phase shift* in locomotary activity, provided it was given during the squirrels' light-sensitive period (DeCoursey 1960b).

Bats shows a similar response. When several species of bats held in constant dark were subjected to light pulses toward the end of the resting period of their circadian rhythms or during their activity period, the bats delayed the onset of their activity. If the light pulses fell in the bats' inactivity or rest period they gave rise to only slight delays or an advance reaction (Erkert 1984).

The degree of rigidity of circadian rhythms varies among bats. Circadian rhythms are fairly rigid among fruit-eating and nectar-feeding bats of the neotropical regions, where environmental and Zeitgeber conditions are fairly constant. Among insectivorous bats circadian rhythms are relatively susceptible to exogenous influences such as changes in light conditions (bright moonlight, for example), differences in the activity rhythms of changing prey species, and their own vulnerability to late-flying, day-active birds of prey (for detailed review see Erkert 1984).

The antelope ground squirrel *(Ammospermophilus leucurus)* and the red squirrel *(Tamiasciurus hudsonicus),* two diurnal mammals, show a similar response. When Kramm (1975, 1976) exposed experimental animals held under constant darkness to light during their subjective night (inactivity period), they shortened the activ-

Figure 7.6 Synchronization of flying squirrels possessing a circadian rhythm of less than 24 hours in constant darkness to a cycle of 10 hours of light, 14 hours of darkness. For squirrel A rephasing light fell during the squirrel's subjective night; synchronization was accomplished by a stepwise delay, and the onset of activity was stabilized shortly after light-dark change. For squirrel B light fell in the subjective day, and the free-running period continued unchanged until the onset drifted up against "dusk" light change. Then a delaying action of light prevented it from drifting forward. When the squirrel was returned to constant darkness, its onset of activity continued a forward drift. (After DeCoursey 1960: 52.)

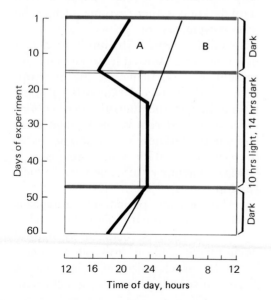

ity period of the cycle. When exposed to light in their subjective day (activity period), the animals lengthened their activity period. In effect, if the animal's activity was too late in the day, the activity could phase shift toward dawn; and if the activity period was too early in the day, the phase shift would be delayed.

Biological Clocks

Involved in the maintenance of a circadian rhythm is some biological clock. Basically the biological clock is cellular. In one-celled organisms and plants, the clock appears to be located in individual cells, but in multicellular animals the clock is associated with the brain.

Skillful surgical procedures have allowed circadian physiologists to discover the location of clocks in some mammals, birds, and insects. In most insects studied, the clock, including the photoreceptors, is located either in the optic lobes or in the tissue between the optic lobes and the brain. In the cockroach and cricket the receptors for the entrainment of circadian rhythm or locomotion and stridulation are located in the compound eye, but the controlling clock is in the brain (Beck 1980; Saunders 1982). In birds the clock evidently is located in the pineal gland, deep in the lower central part of the brain (Farner and Lewis 1974; Gwinner 1978). In mammals it appears to be located in a number of specialized cells (suprachiasmatic nuclei) just above the optic chiasm, the place where the optic nerves from the eyes intersect.

To function as a timekeeper, the clock has to have an internal mechanism with a natural rhythm of approximately 24 hours, which can be reset by recurring environmental signals, such as the changes in the time of dawn and dusk. The clock has to be able to run continuously in the absence of any environmental timesetter.

It also has to be able to run the same at all temperatures. Cold temperatures must not slow it down, nor warm temperatures speed it up.

Two basic models of biological clocks have been proposed. One is an oscillating circadian rhythm sensitive to light (Bünning 1960) (Figure 7.7). The cycle or time measuring process begins with the onset of light or dawn. The first half is light-sensitive and the second half requires dark-

Figure 7.7 The Bünning model of the entrainment of circadian rhythms to daylength. Oscillations of the clock cause an alternation of half-cycles with quantitatively different sensitivities to light (white vs. solid). The free-running clock in continuous light or continuous darkness tends to drift out of phase with the 24-hour photoperiod. Short-day conditions allow the dark to fall into the white half-cycle; in the long day the light falls into the black half-cycle. (From Bünning 1960: 253.)

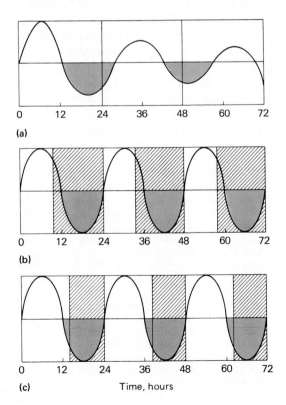

ness. Short-day effects are produced when light does not extend into the dark period. Long-day effects are produced when it does. Because this simple model does not explain all photoperiodic responses, a variation is a two-oscillator model in which one oscillation is regulated by dawn and the other by dusk.

A second model is the hourglass (Saunders 1982). It does not involve circadian components, nor is it free-running. It is started by a light-on, light-off stimulus. The timing process begins with darkness, and it shuts off in prolonged darkness or at the beginning of a light period. A light period then prepares it for resetting by darkness. Although such a model can account for the activity periods of many organisms, it cannot account for the biomodal period of activity that persists in many organisms such as birds in the absence of time cues (Aschoff 1966); nor can it account for other ecologically important time-related activities of organisms.

A third model combines the hourglass and the single oscillator model (Pittendrigh 1966). The light-sensitive phase of the end of the prolonged light period is a resetting mechanism. The oscillation is reset to a particular phase at the end of a prolonged light period, producing an

hourglass effect (Figure 7.8). There are other more complex variations on these models described by Beck (1980) and Saunders (1982).

How and why circadian rhythms and biological clocks function are the domain of the physiologist. Of much greater interest to the ecologist is the adaptive value of circadian rhythms. For one, a circadian rhythm provides the organism a time-dependent mechanism, enabling it to respond consistently. In other words, it enables organisms to time their activities to a daily cycle in a manner appropriate to their ecology.

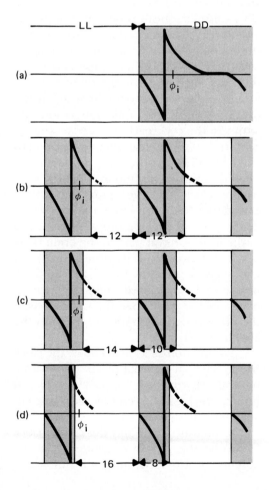

Figure 7.8 Hourglass oscillation in pupal eclosion (emergence) in the fruit fly *Drosophila pseudoobscura*. It involves a circadian oscillator (O) or pacemaker as part of the total circadian system of the fly. (a) Constant light (LL) suppresses the oscillation; on transfer to constant darkness (DD), it resumes its motion, starting from the circadian time scale (Ct) of 12 hours. (b, c, d) Photoperiods of 12 hours or more also dampen the oscillation, which starts up again from Ct12 with the onset of darkness each night. As the photoperiod increases, light at dawn falls back into late subjective night. The photoinducible phase at the end of the critical night is not illuminated by photoperiods of 12 and 14 hours, but it is illuminated by a 16-hour photoperiod. (After Pittendrigh 1966: 297.)

Certain circadian activities, for example, relate to aspects of the environment ecologically more important than light or dark alone. The transition from night to day is accompanied by such environmental changes as a rise in humidity and a drop in temperature. Woodlice, centipedes, and millipedes, which lose water rapidly in dry air, spend the day in the relatively constant environment of darkness and dampness under stones, logs, and leaves. At dusk they emerge, when the humidity of the air is more favorable. These animals show an increased tendency to escape from light as the length of time they spend in darkness increases. On the other hand, their intensity of response to low humidity decreases with darkness. Thus these invertebrates come out at night into places too dry for them during the day; and they quickly retreat to their dark hiding places as light comes (Cloudsley-Thompson 1956, 1960).

The circadian rhythms of many organisms relate more to the biotic than the physical aspects of their environment. Predators such as insectivorous bats must relate their feeding activity to the activity rhythm of their prey. Moths and bees must visit flowers when they are open, and the flowers must have a rhythm of opening and closing that coincides with the time when insects that pollinate them are flying. The circadian clock is intimately involved in sun orientation for insects, reptiles, and birds. The best way for organisms to economize their energies is to adapt to the periodicities of their environment.

Critical Daylengths

In the northern and southern latitudes the daily periods of light and dark lengthen and shorten with the seasons. The activities of plants and animals are geared to the changing seasonal rhythms of night and day. The flying squirrel, for example, starts its daily activity with nightfall and maintains this relation through the year. As the short days of winter turn to the longer days of spring, the squirrel begins its activity a little later each day (Figure 7.9).

Most animals and plants of temperate regions have reproductive periods that closely follow changing daylengths of the seasons. For most birds the height of the breeding season is the lengthening days of spring; for deer the mating season is the shortening days of fall. Trilliums and violets bloom in the lengthening days of spring before the forest leaves are out, while an abundance of sunlight reaches the forest floor. Asters and goldenrods flower in the shortening days of fall.

The signal for these responses is critical daylength. When a period of light (or dark) reaches a certain portion of the 24-hour day, it inhibits or promotes a photoperiodic response (Figure 7.10). Critical daylength varies among organisms, but it

Figure 7.9 Onset of running wheel activity for one flying squirrel in natural light conditions throughout the year. The graph is the time of local sunset through the year. Observe how the onset of activity of the flying squirrel grows later each day as the days lengthen and the nights grow shorter. In the long nights of winter, the squirrel becomes active much earlier in the afternoon. (From DeCoursey 1960: 50.)

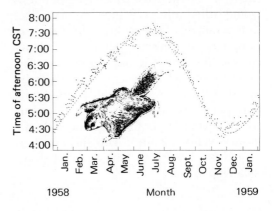

1958   Month   1959

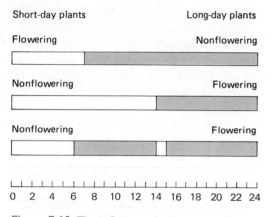

Figure 7.10 The influence of photoperiod on the time of flowering in long-day and short-day plants. If exposure to light is experimentally controlled, short-day plants are stimulated to flower under short-day conditions, inhibited from flowering under long-day conditions, and respond to an interruption of a long dark period as though they had been exposed to long-day conditions. Long-day plants do not flower under short-day conditions, only under long-day conditions and under interrupted short-day conditions. Animals show a similar response, particularly in reproductive cycles, to long-day and short-day conditions.

usually falls somewhere between 10 and 14 hours. Through the year plants and animals compare that time scale with the actual length of day or night. As soon as the actual daylength or nightlength is greater or less than the critical daylength, the organism responds appropriately. Some organisms can be classed as day-neutral—not affected by daylength, but by some other influence such as rainfall or temperature. Others are short-day or long-day organisms. Short-day plants, whose flowering is stimulated by daylengths shorter than critical daylengths, include fall asters and goldenrods. Long-day plants like trilliums, whose flowering is stimulated by daylengths longer than a critical value, bloom in late spring. Many organisms have both long-day and short-day responses.

Because the same period of dark and light occurs two times a year as the days lengthen in spring or shorten in fall, the organism could get its signals mixed. For some insects that would be impossible, because the sensitive period occurs only once in their development as larva or pupa. Other organisms depend on the direction from which the critical daylength is approached. In one situation the critical daylength is reached as long days move into short, and in the other as short days move into long.

There are numerous examples of photoperiodic responses in plants and animals. Horticulturists use them to bring plants into flowering off season for the floral trade. Diapause, a stage of arrested growth in insects during winter in the temperate regions, is controlled by photoperiod (Chapter 6). The time measurement in such insects is precise, usually falling in a light phase somewhere between 12 and 13 hours. A quarter-hour difference in the light period can determine whether an insect goes into diapause or not (see Saunders 1982). Adkisson (1966) found that the pink cotton bollworm, for example, failed to enter diapause if the larvae were exposed to a light period of 13.25 hours; its critical day length is 13 hours. Experimentally, the bollworm terminated diapause most rapidly under photoperiods of 14 hours, less rapidly at 16 and 12 hours. Thus to the pink cotton bollworm the shortening days of late summer and fall forecast the coming of winter and call for diapause; and the lengthening days of late winter and early spring are the signals for the insect to resume development, pupate, emerge as an adult, and reproduce.

That increasing daylength increases gonadal development and spring migratory behavior in birds was experimentally demonstrated over 60 years ago when Rowan (1925) forced juncos into the reproductive stage out of season by artificial increases in daylength. Since then numerous experimental studies involving many species have demonstrated that the reproductive cycle in birds is controlled by the seasonal rhythm of changing daylengths.

After the breeding season the gonads of birds regress spontaneously. They enter a *refractory period* when light cannot induce gonadal activity, the duration of which is regulated by daylength (see Farner 1959, 1964a, 1964b; Wolfson 1959, 1960). In general, short days hasten the refractory period; long days prolong it. However, birds may possess a refractory period with two distinct parts (Hamner 1968). One is an absolute refractory period, whose length is independent of photoperiod. The other is a relative refractory period, when photosensitive birds will not respond to daylengths equal to or shorter than the one to which they were previously exposed.

As winter approaches and the natural day shortens, the bird's timing mechanism continually readjusts, so the progressively shorter days become photoperiodically stimulatory. After the refractory period, the *progressive phase* begins in the late fall and winter. During this period the birds fatten, they migrate, and their reproductive organs increase in size. This process can be speeded by exposing the birds to a long-day photoperiod. Completion of the progressive phase brings the bird into the *reproductive stage*.

Photoperiods influence the breeding cycles and such activity cycles as food storage (Muul 1969) in many mammals. Antler growth of male deer is also a photoperiodic response (Figure 7.11). Growth is triggered by the lengthening days of spring and velvet is shed in the shortening days of fall (Goss et al. 1974). Normally the deer is in velvet about one-third of the

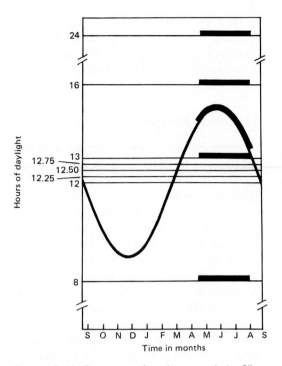

Figure 7.11 Response of antler growth in Sika deer to constant daylengths of different durations. The curve represents the natural course of daylength during a year at 42° north latitude. Under natural conditions, antlers are in velvet from early May to early September. When yearling Sika deer were held in light/dark conditions of 8L/16D, 13L/11D, 16L/8D, and 24L/0D starting with the autumnal equinox, all shed and regrew antlers in the spring in synchrony with the outdoors. Deer exposed to light periods of 12, 12.25, 12.50, and 12.75 hours failed to replace their antlers the following year. The critical daylength was just over 12.75 hours, approaching 13 hours. (After Goss et al. 1974: 407.)

year. When the duration of that year is changed by altering the frequency of daylength, antlers may be replaced as often as two, three, or four times a year, or only once every other year, depending on how much the light cycle has been increased or decreased. If deer are held under constant photoperiod, they will still develop antlers, but only if the daylength ratio of light to dark exceeds 12.75L: 11.25D. Under constant daylength of 12L:12D, the annual cycle of antler growth will not be expressed, and antlers may not be replaced for several years.

Seasonal Morphs

Photoperiod can also result in seasonal morphs among a number of insects. One example is the Pierid butterfly, the veined, white *Pieris napa*, in western North America. Populations of this butterfly along the California coast have two generations a year and occur in two phenotypes. One is a dark spring form with heavy black scaling on the veins of the hind wings, and the other a lighter summer form almost devoid of black scales. Inland populations have only one generation a year, and all adults are the dark form. Both populations, however, are capable of producing both phenotypes. The dark forms are induced by the short days of autumn and emerge from overwinter diapause pupae. The light summer forms are induced by the long days of summer and emerge from nondiapause pupae (Shapiro 1977). Such photoperiodic responses may be important in the reproductive isolation of closely related populations and in the evolutionary process of speciation (see Beck 1980; Tauber and Tauber 1976a, b.)

Circannual Clocks

All of the activities discussed above exhibit a yearly periodicity, suggesting the existence of a circannual clock. To prove its existence, we need to demonstrate conditions similar to those required of a circadian clock: (1) a free-running rhythm in the absence of environmental cues, (2) a period that approximates but deviates from a 365-day cycle, and (3) temperature independence.

Considerable evidence exists that a circannual clock controls body weight and hibernation in the ground squirrel (*Citellus lateralis*) of western North America. When held for several years under constant temperature regimes and constant artificial days of 12L/12D, the squirrels exhibited circannual rhythms of about a year (Pengelley and Asmundson 1974). Such a rhythm has advantages. It enables the animals to prepare well in advance for winter and to arouse in spring without the need of environmental cues, because the animals are shut from the outside world in a deep burrow. It also brings the animal into breeding condition almost immediately after hibernation, so reproduction can take place in spring.

Most birds show strong circannual rhythms in molt, body weight (fat deposition), testis growth, and nocturnal restlessness (Berthold 1974); but whether we need to invoke a circannual clock is questionable. The cycles can be explained by seasonal photoperiod responses and a circadian clock.

■ Seasonality

Seasonality is "the occurrence of certain obvious biotic and abiotic events within a definite limited period or periods of the astronomic year" (Leith 1974). The unfolding of leaves in spring and the dropping of leaves in fall, the blooming of flowers and the ripening of seeds, the migration of birds, and other biological events recurring with the passage of seasons are influenced by the interaction of light with temperature and moisture. The study of their timing, of biotic and abiotic forces affecting it, and of the relations among phases of the same or different species is called *phenology* (Leith 1974).

Seasonality in temperate regions results largely from changes in light and temperature. Seasonality in tropical regions is keyed to rainfall. In a broad way seasonal changes in temperature and light regimes result in alternate warm and cold periods. However, the progression is gradual in temperate zones; seasons can be identified as early or late spring, early or late fall, and so on. In the tropics the seasons are alternately wet and dry, and their onset is abrupt. The beginning of the rainy season is a dependable environmental cue by which plants and animals of the tropics can become synchronized to seasonal changes. The dry season in the wetter tropics is a period with less than 100 cm rainfall per month. In extreme conditions, of course, no rain falls. The wet season is the period with more than 100 cm rainfall per month. The onset of the rainy season, which may last up to six months, varies with the movement of the intertropical convergence (see Figure 4.4).

For this reason the wet season and the dry season are more or less predictable.

Plants

Phenological responses reflect altitudinal and latitudinal changes in light, temperature, and moisture. The advance of spring in the temperate regions is marked by progressively later timing of flowering of the same species of trees and herbs across a region. Although these progressive changes are most pronounced across broad geographical areas (Figure 7.12), distinct variations exist within a given region. These variations reflect local microclimates, which act as selection pressures on local populations, resulting in ecotypic variations in environmental response. In

Figure 7.12 The arrival of spring 1970 across North Carolina, as indicated by the opening of the flowers of dogwood *(Cornus florida)* and redbud *(Cercis canadensis).* (From Leith 1974.)

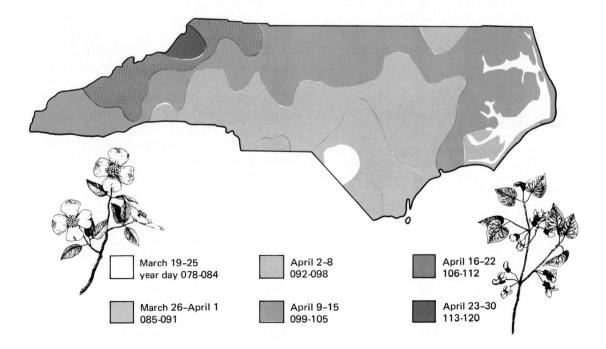

| | March 19–25 year day 078-084 | | April 2–8 092-098 | | April 16–22 106-112 |
| | March 26–April 1 085-091 | | April 9–15 099-105 | | April 23–30 113-120 |

tropical regions wet and dry seasons show a similar advance. For example, seedlings within tree species of northern origin stop growth sooner than seedlings of the same species with a southern origin; and seedlings of the same species from the highest altitudes stop growing before those from lower altitudes (Flint 1974).

Some seasonal changes relate to single abiotic factors. Some responses result from photoperiod, others from temperature, and still others from both. Plants' response to temperature is related to *temperature sums,* the summation of mean daily temperatures. In general, the breaking of seed dormancy depends upon chilling by an accumulation of low temperatures. Induction of dormancy and seedling growth are photoperiodic responses, as is the cessation of shoot growth in fall. Depending upon the species, seed germination requires either chilling or long days accompanied by high temperatures.

These seasonal responses combine to give a seasonal aspect to flowering and vegetative growth in the temperate regions (Figure 7.13). The timing of flowering and fruiting for the spring and fall has certain ecological and evolutionary advan-

tages. Plants flower without experiencing a heavy demand for energy resources for vegetative growth. Flowering and fruiting in spring is over or nearly so before much vegetative growth takes place, and the plants draw on energy and nutrient reserves in the roots. Seeds get into the ground early, allowing seedlings to develop a strong root system before winter. Flowering and fruiting in the fall takes place after vegetative growth is over, so energy can be diverted to reproduction. Seeds produced then usually require chilling before they can germinate.

Seasonality in plant activity influences biomass accumulation and energy fixation. Studies on such seasonality are few, but one example is the phenological development of the mayapple (*Podophyllum peltatum*) (Figure 7.14), studied by Taylor (1974) at Oak Ridge in the Cumberland Mountains of Tennessee. Mayapple grows in colonies in eastern North American woodlands and develops rapidly in spring before the forest canopy closes. Shoot development and flowering are spring events; rhizome growth is a summer activity. During the period when light is most available in the understory, biomass of leaves, stems, rhizomes, roots, flowers, and fruits increases. Biomass production reaches its maximum in early summer during the phase of unripe seeds and fruits, just prior to the onset of senescence. During the one-year study the rhizome-root compartment increased 21 percent in dry weight above the standing crop estimate at the time of leaf emergence. Leaf and aboveground stems accounted for another 57 percent of the total dry weight biomass; unripe fruits and seeds accounted for 17 percent; and an increase in underground shoots prior to senescence added an additional 4 percent. At the onset of senescence leaves change color; then the leaves and fruits drop. At the end of the growth period, fruits are

Figure 7.13 Phenogram of the oak-hickory forest association in eastern Tennessee, depicting flowering seasons distinguished by spring and early summer flora. Data include flowering of 36 trees and shrubs and 97 herbaceous species and represent the mean date of first flowering between 1963 and 1970. (After Taylor 1974: 239.)

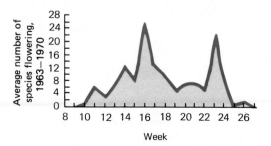

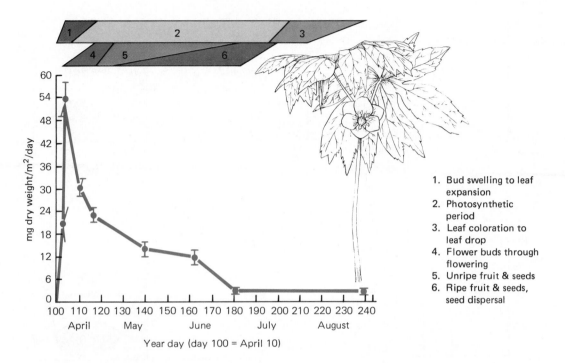

1. Bud swelling to leaf expansion
2. Photosynthetic period
3. Leaf coloration to leaf drop
4. Flower buds through flowering
5. Unripe fruit & seeds
6. Ripe fruit & seeds, seed dispersal

Figure 7.14 Phenological spectrum and seasonality in the productivity of mayapple *(Podophyllum peltatum)* in deciduous forests of eastern Tennessee. Productivity is maximal during the reproductive phase. It is reduced when small mammals and insects consume the fruit, which coincides with the period of ripe fruits and seeds and seed dispersal. (After Taylor 1974: 246.)

eaten by insects and small mammals and the seeds distributed.

In mediterranean-type climates with dry summers and wet winters, characteristics of southern California, the Mediterranean regions, and parts of Chile and Australia (see Chapter 30), plants are most active metabolically in late spring, when temperatures become warm and the stress of summer drought has not yet developed (Figure 7.15). Summer brings a sharp drop in flowering and plant activity and in the presence of pollinating insects, which are active when a food base is available (Mooney and Parsons 1973).

The evergreen shrub *Heteromeles arbutifolia* provides an example of how plants in mediterranean climate ecosystems allocate energy and accumulate biomass. Although the plant fixes carbon throughout the year, peak activity takes place during the winter and spring, when moisture is not limiting (Mooney, Parsons, and Kummerow 1974). The plant uses energy fixed early for the growth of new leaves during the late spring and early summer, when temperatures are more favorable. Stem growth ceases when energy demands for reproductive functions are high.

In tropical regions flowering and fruiting and leafy growth of plants reflect the alternation of wet and dry seasons. The coming of the rainy season is marked by a flush of vegetative growth, just as warming spring temperatures trigger leafy growth in temperate regions. Over much of the tropics flowering and fruiting coincide with the dry season (Janzen 1967). Some species flower at the end of the rainy season, when soil moisture is still high. Other species flower at the end of

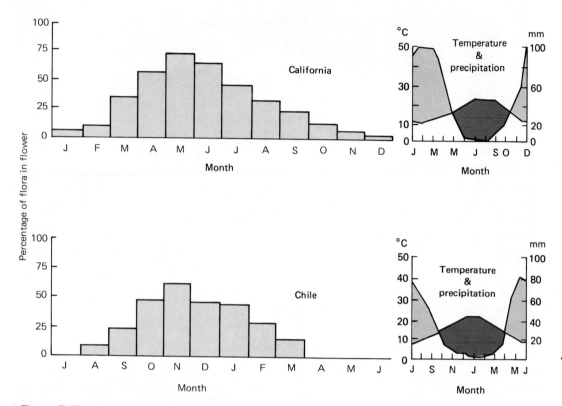

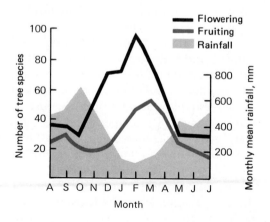

Figure 7.15 Seasonal flowering activity of plants growing in the Mediterranean climatic regions of California and Chile. The California data are for the flora of San Dimas Experimental Forest, southern California. Chilean data are for shrubs only in the Santiago region. (After Mooney, Parsons, and Kummerow 1974: 260.)

the dry season. Although there may be definite peaks, flowering and fruiting continue through the dry season (Figure 7.16). In a Costa Rican tropical forest, for example, some 60 trees that flower during the dry season spread their flowering over 3.5 months, minimizing overlap in flowering. This difference in flowering times may be the outcome of two selective forces: competition for pollinators and interspecific pollen transfer. Because the trees are self-incompatible, they require the assistance of animals for the transfer of pollen. Many trees blooming at the

Figure 7.16 Synchronization of flowering and fruiting in rain forest tree species in Golfito, Costa Rica, with mean monthly rainfall. Note that flowering and fruiting reach their highest during the dry season, the months of January, February, and March. (Adapted from D. H. Janzen 1967.)

same time could compete for the attention of pollinators; flowering at different times could reduce that competition. Because the pollinators visit a number of different plant species they could transfer incompatible pollen from one plant to another and reduce seed set. A sequence in blooming time would reduce improper pollen transfer (Rathcke 1983).

There are other ecological and evolutionary advantages to plant reproduction in the dry season. It reduces or eliminates conflict in the plant between energy demands for leaf growth and for reproduction. Plants can draw on nutrient and energy reserves stored in the roots during the rainy season. Flowers blooming in leafless trees are available to pollinating insects, birds, and mammals, particularly bats, and fruits are conspicuous and accessible to fruit-eating animals, aiding in the dispersal of seeds.

Not all plants, of course, flower and fruit in the dry season. Shrubs in Costa Rican forests bloom shortly before and just after the maximum flowering of trees. Vines and herbaceous plants flower during the wet season. Plants with large-seeded fruits, whose seeds may be damaged by fruit-eating animals, exhibit synchronized fruiting, so some seeds escape damage. Plants with smaller-seeded fruits ripen throughout the year. These plants are able to utilize animal agents of dispersal with minimal competition for their services (Snow 1966, 1976).

Animals

Seasonal activities of animals center about reproduction and changing food resources. Appearance and disappearance of many species of insects is usually associated with diapause. In temperate regions diapause is usually controlled by photoperiod. Insects show up when food is abundant and go into diapause during the time of year when it is scarce. Growth after diapause is influenced by temperature and moisture.

In a similar manner photoperiod influences the seasonality of birds in temperate regions (Figure 7.17). Seasonal phases such as migration, reproduction, and molt usually do not overlap. In winter birds channel energy demands into survival and thermoregulation. Within this pattern some seasonal environmental mechanisms may modify bird activity. Migration, triggered by photoperiod and physiological processes, is highly correlated with weather patterns. Spring migration occurs with the onset of a warm front with an air flow to the north or northeast. The birds move on the wind and have a good chance of arriving in favorable weather at the end of the flight. When a cold front arrives, migratory movements stop, not to resume again until high pressure areas have passed. In fact, radar studies show that a cold front in spring may immediately start a reverse southward movement (Drury et al. 1961; Eastwood 1971). Fall migrations to the south appear to be timed to start after the pas-

Figure 7.17 Events in the annual life cycle of a migratory songbird. Note the strong seasonal periodicity, related to the months and thus to daylength. (From C. M. Weise 1974: 140.)

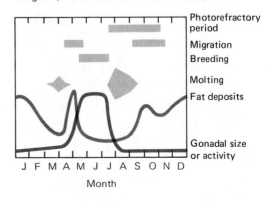

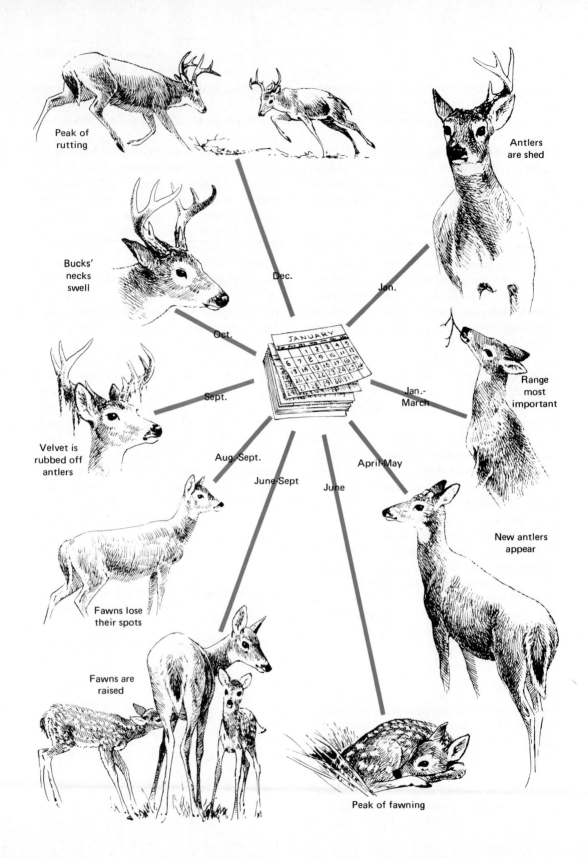

Peak of
rutting

Antlers
are shed

Bucks'
necks
swell

Dec.

Jan.

Oct.

Sept.

Jan.-
March

Range
most
important

Velvet is
rubbed off
antlers

Aug.-Sept.

April-May

June-Sept

June

New antlers
appear

Fawns lose
their spots

Fawns are
raised

Peak of fawning

sage of a cold front, with its flow of continental polar air.

Although photoperiod controls the timing of nesting in birds, the initiation of nesting across a geographic range also is influenced by climatic gradients, as well as seasonal development, availability of nesting material, and food for the young. In a study of the nesting phenology of the robin (*Turdus migratorius*) James and Shugart (1974) used a stepwise multiple regression analysis involving seven climatic variables. They found that a combination of April wet-bulb and dry-bulb temperatures was the best predictor of the beginning of the nesting period: "If the mean noon relative humidity is near 50 percent in April, the beginning of the nesting season will be in late April or early May, regardless of the dry-bulb temperature; if the mean noon relative humidity is either higher or lower than 50 percent in April, the beginning of the nesting period will be later." Over the eastern United States, robins experience a three-day retardation of the breeding season per degree of increasing latitude.

Among mammals photoperiod influences reproductive and related behavior and physiological responses. The reproductive cycle of the white-tailed deer, for example, is initiated in the fall, and the young are born in spring when the highest quality food for the lactating mother and her young is available (Figure 7.18). Antler growth, a photoperiod response discussed earlier, is seasonal in male deer. Antler growth, development, and loss is controlled by the hormones testosterone and prolactin, the production of which is mediated by the photoperiodic responses

Figure 7.18 Seasonal cycle of the white-tailed deer. The annual cycle is attuned to the decreasing daylength of fall, during which the breeding season begins, and to the lengthening days of spring, when antler growth begins.

of the hypothalamus. These hormones control the reproductive cycle in the buck (Figure 7.19).

A close interrelationship exists between flowering and fruiting phenology and animal reproductive activities in the tropics. The breeding season and seasonal changes in bird populations are related to the availability of food (Karr 1976). In some tropical regions nesting is greatest during the rainy season, but in the area of pronounced wet and dry seasons, breeding is more evenly distributed through the year. In some tropical regions of South America influx of northern migrants during the temperate winter depresses the local breeding bird populations and shifts most nesting to the dry season. Reproduction reaches a maximum at the beginning and end of the dry season.

In tropical Central America, the home of numerous species of frugivorous and insectivorous bats, the reproductive periods track with the seasonal production of food. Frugivorous bats have two peaks of birth, one in the last half of the dry season and the other in the middle of the wet season. The females are polyestrous and typically produce two young annually. The birth periods coincide with just after one fruiting peak and just before another. Thus young are born when both females and young will have adequate food. Insectivorous bats, however, are monoestrous and produce young only during the wet season, when insect life is abundant. Insect numbers are depressed during the dry season. By contrast, the vampire bat is polyestrous and breeds throughout the year. Among the tropical bats, it alone has a constant supply of food—animal blood—throughout the year.

Some rodents, the meadow vole *Microtus pennsylvanicus* for example, have seasonal periods of growth. Experimentally, the adult voles gain weight under a long photoperiod (18L:6D), and young voles

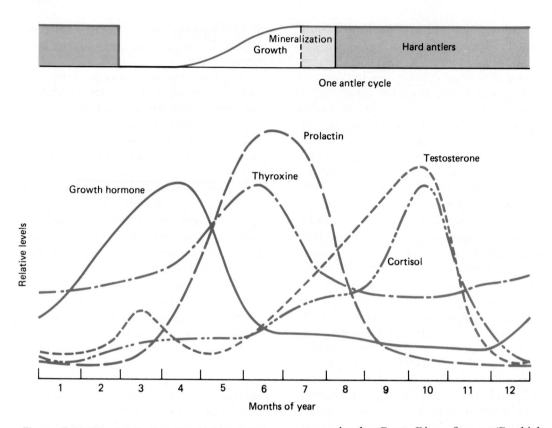

Figure 7.19 The seasonal course of hormonal levels in the male white-tailed deer and its relationship to antler growth. Note the response of growth hormones and prolactin to lengthening days and the decline in prolactin and the increase in the production of testosterone during the shortening days of fall. Prolactin stimulates the growth of antlers, but testosterone is necessary to harden the antlers and bring the buck into reproductive condition. The decline in testosterone in winter results in the shedding of antlers. (After Goss et al. 1976.)

season in the Costa Rican forests (Buskirk and Buskirk 1976). Spiders are most abundant in the early rainy season, decline shortly before the end of the rains, and continue to decline steadily until the end of the dry season (Robinson and Robinson 1970). The tropical dry season initiates other seasonal activities as well: estivation, diapause, and local and regional migrations to moist areas, exemplified by the large-scale migrations of African ungulates.

Decomposers

Decomposer activity is influenced by temperature, moisture, carbon-nitrogen ratio of leaves, and litter fall. In tropical rain forests litter fall continues through the year, but in seasonal tropical forests and temperate forests litter fall is seasonal. In

have a rapid growth rate. Under a short photoperiod (6L:18D) adults lost weight and the young grew more slowly. Under natural conditions of temperature and light adult meadow voles lost weight until the winter solstice, and the growth rate of subadults declined during November and December (Pistole and Cranford 1982).

Insects and other arthropods reach their greatest biomass early in the rainy

temperate forests this seasonal litter fall is mostly in the fall, with some additions from seasonal herbaceous growth. In evergreen forests, both broadleaf and needle-leaf, litter fall is less seasonal. With a heavy deposition of litter in the fall in temperate forests, microbial growth and the amount of carbon immobilized in bacterial and fungal tissue increases rapidly, then declines when cold weather comes (Figure 7.20). At that time CO_2 evolution from the litter declines and fungal propagules and accumulation of mycelium reach a maximum. As the temperature rises in spring, growth of bacteria and

fungi resumes, and soil animals increase their activity with a corresponding rise in CO_2 evolution and microbial immobilization of carbon in soil and litter. Although total availability of carbon is not limiting to decomposers during any period of the year, the biochemical form of carbon limits the rate of microbial and faunal catabolism. Similarly microbial activity results in immobilization of nitrogen. In fact, nearly all of the nitrogen in the O_1 litter layer of the eastern deciduous forest in the late spring may be in microbial cells (Ausmus et al. 1975).

Figure 7.20 Seasonal microbial activity in the eastern deciduous forest. The graph shows microbial immobilization of carbon in the soil in upper O_1 and lower O_2 litter layers, calculated monthly through the year. Note the sharp rise in the carbon pool of O_1 litter in the fall. Periods of maximum litterfall and root sloughing are indicated by vertical arrows, and maximum and minimum CO_2 efflux by horizontal arrows. (After Burgess and O'Neill 1976.)

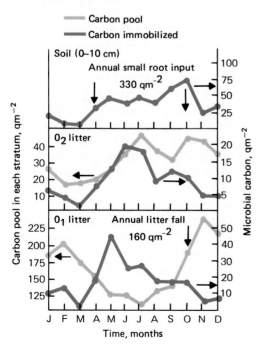

Tidal and Lunar Cycles

For some organisms environmental time-setters associated with tidal and lunar rhythms are of greater ecological importance than light-dark cycles. Animals that inhabit the intertidal zones of the sea show rhythms in their behavior that coincide with cycles of high and low tides. These endogenous timing processes also show persistent internal rhythms that are comparable to circadian rhythms of animals from nonintertidal environments. Among a number of animals showing activity rhythms entrained to tidal cycles are the European shore crabs (*Carcinus maenas*) and fiddler crabs (*Uca minax* and *U. crenulata*) (for reviews see Enright 1975; Palmer 1974).

Reproduction in some marine organisms is restricted to a period that bears some relationship to tides. These rhythmic phenomena occur every lunar cycle of 28 days or in some instances every semilunar cycle of 14 to 15 days. Among some species, such as the grunion (*Leusesthes tenuis*), a small California fish that swarms in from the sea to lay eggs on sandy beaches, and the intertidal midge *(Clunio marinus)* the periodicities are so exact the activities of these animals can be predicted ahead of time. Laboratory studies confirm the entrainment of activity cycles to moonlight (see Enright 1975).

■ Summary

The influence of light on individuals and communities goes beyond photosynthesis. First, it affects the vertical distribution of plants within communities, aquatic and terrestrial, by its intensity and spectral qualities. Plants that function best under different light intensities may be classified as sun plants (shade-intolerant) or shade plants (shade-tolerant). Each group is characterized by adaptations to certain light regimes. Shade-adapted plants have low photosynthetic, respiratory, metabolic, and growth rates and are resistant to fungal infections. Sun-tolerant plants have a high rate of respiration, are adapted to high light intensity, rarely reach light saturation, and are highly susceptible to fungal infections under low light conditions.

Daily periodicities of plants and animals are under the influence of day-night cycles. The timing of daily activities is controlled by an internal physiological biological clock, whose basic structure is probably chemical. It is free-running under constant conditions with an oscillation that has its own inherent frequency. For most organisms the inherent clock deviates more or less from 24 hours. Under natural conditions this clock is set by external time cues, which synchronize the activity of plants and animals with the environment. Because the most dependable timesetter is light (day and night), most of the species studied so far are entrained to a 24-hour period. The onset and cessation are usually synchronized with dusk and dawn, the response depending upon whether the organisms are diurnal (light-active) or nocturnal (dark-active).

The biological clock is useful not only to synchronize the daily activities of plants and animals with night and day, but also to time the activities with the seasons of the year. The possession of a self-sustained rhythm with approximately the same frequency as that of environmental rhythms enables organisms to "predict" such advance situations as the coming of spring. It brings plants and animals into a reproductive state at a time of year when the probability for survival of offspring is the highest; it synchronizes within a population such activities as mating and migration, dormancy and flowering. The acquisition and refinement of a physiological timekeeper, geared to cues that provide organisms with a distinct and species-specific synchronization with the environment, are the results of natural selection.

Seasonal changes in light, along with accompanying changes in temperature and precipitation, bring about seasonal periodicities in recurring biological events, the study of which is phenology. In temperate regions seasonality is influenced by changes in light and temperature and in tropical regions by rainfall. Such periodicities are reflected in leaf growth, flowering, and fruiting of plants and in the breeding cycles, feeding activities, and migration of animals.

Other periodicities, particularly among marine organisms, are strongly influenced by lunar cycles and by tidal cycles associated with them.

Nutrients

In 1840 Justus von Leibig, a German organic chemist, set forth a simple statement based on his analyses of the fertility of surface soil and plant growth: "The crops on a field diminish or increase in the exact proportion to the diminuation or increase in the mineral substances conveyed to it in manure." What he implied was that each plant requires certain kinds and quantities of nutrients. If one of them is absent, the plant dies. If it is present in minimal quantity only, the growth of the plant will be minimal. This concept became known as the *law of the minimum*.

Continued investigations through the years disclosed that not only nutrients but other environmental conditions, such as moisture and temperature, affected the growth and distribution of plants. Eventually the law of the minimum was extended to cover all environmental requirements of plants and animals.

In 1905 a British plant physiologist, F. E. Blackman, pointed out that too much as well as too little of a substance could limit the presence, growth, or success of an organism. Organisms, then, live within a range of too much or too little, the limits of tolerance. This concept of maximum substances or conditions limiting the presence or success of organisms was incorporated by V. E. Shelford in 1913 in the *law of tolerance.*

Modern ecologists, however, recognize that plants and animals are not necessarily limited by one particular basic requirement, but rather by an interaction of a number of them. The low level of one resource may be partially compensated for by a certain level of another. For example, some plants respond to sodium when potassium is inadequate (Reitemeier 1957). Also, the influence of one resource item may increase as others reach maximum or minimum levels.

■ Important Nutrients

Among the 30 to 40 elements plants and animals require for growth and development, some are needed in relatively large amounts. These nutrients, known as *macronutrients,* include, among others, oxygen, nitrogen, carbon, calcium, phosphorus, potassium, magnesium, and sulfur. Others, needed in only minute quantities, are *micronutrients* or trace elements. They include copper, zinc, boron, manganese, molybdenum, cobalt, vanadium, and iron. Some are essential to all organisms; others appear to be essential for some species, but not for others. If micronutrients are lacking, plants and animals fail as completely as if they lacked nitrogen, calcium, or any other major element.

Molecular *oxygen,* O_2, which makes up 21 percent of the earth's atmosphere, is a by-product of photosynthesis. Three major pools of oxygen are carbon dioxide, water, and molecular oxygen, all of which interchange atoms with one another. Other sources include nitrate and sulfate ions, which release oxygen upon their decomposition and supply oxygen for a

number of living organisms. Molecular oxygen is a building block of protoplasm and is necessary for biological oxidation, in which it serves as a hydrogen acceptor, producing water. Plants and animals use large amounts of oxygen in respiration. In terrestrial habitats the supply of oxygen is rarely inadequate for life, except at high altitudes, deep in the soil, and in soils saturated with water. In aquatic communities, however, oxygen may be limited. Here the supply comes from photosynthesis and from the diffusion of atmospheric oxygen into the water, the rate of which is proportional to the surface area of water exposed to the air. The total quantity of oxygen that water can hold at saturation varies with temperature, salinity, and depth. It is depleted by the respiration of aquatic plants and animals. Oxygen content in aquatic environments fluctuates daily from a low at night, when respiration is greatest, to a high at midday, when photosynthesis is maximal.

Carbon is the basic constituent of all organic compounds. In the ecosystem it occurs as carbon dioxide, carbonates, fossil fuel, and as part of living tissue; it makes up only 0.03 percent of the atmosphere. The amount of carbon dioxide in natural waters is highly variable, for it occurs in free and combined states. The pH of aquatic media and of the soil has a pronounced influence on the proportions of the two types. Carbon dioxide combines with water to form a weak carbonic acid, H_2CO_3, which dissociates:

$$CO_2 + H_2O \rightleftharpoons H_2CO_3 \rightleftharpoons H^+ + HCO_3^- \rightleftharpoons H^+ + CO_3^{2+}$$

Carbon dioxide in solution and carbonic acid make up free carbon dioxide; the bicarbonate (HCO_3^-) and the carbonate (CO_3^{2+}) ions are the combined forms.

The presence of bicarbonate and carbonate ions in soil or water helps to buffer or maintain a certain pH of that medium. This pH is significant in ecological systems because an increase in soil pH lowers the availability of most nutrients to plants. The amount of carbon dioxide fluctuates during the day, but in a manner opposite to daily oxygen fluctuations. The carbon dioxide concentration is lowest at midday and highest at night, when only respiration is taking place.

Nitrogen makes up 78 percent of the atmosphere as molecular nitrogen, N_2, but most plants can utilize it only in a fixed form, such as in nitrites and nitrates (the exceptions are nitrogen-fixing bacteria and blue-green algae). Most of the nitrogen in the soil is found in organic matter. Nitrates leached from the soil and transported by drainage water are an important source of nitrogen for aquatic communities. During the summer the nitrogen supply in them may be utilized completely by phytoplankton and nitrates may disappear from the surface water. As a result phytoplankton growth, or "bloom," is reduced greatly in late summer. Nitrates build up again in winter. On the other hand, nitrogenous wastes draining from agricultural fields, sewage disposal plants, and other sources often overload aquatic ecosystems with nitrogen. This situation results in massive plankton growth and other undesirable changes in community structure.

All life processes depend upon nitrogen. Even chlorophyll is a nitrogenous compound. Nitrogen is a building block of protein and a part of enzymes. It is needed in an abundant supply for reproduction, growth, and respiration.

Two elements needed in appreciable quantities are *calcium* and *phosphorus*, which are closely associated in the metabolism of animals. Together the two make up 70 percent of the total ash present in the animal body and nearly 1 percent of the wet body mass. In animals calcium is necessary for proper acid-base relationships, for the clotting of blood, for contraction and relaxation of the heart muscles, and for the control of fluid passage through the cells. It gives rigidity to the skeleton of vertebrates and is the principal component in the exoskeleton of insects and the shells of mollusks and arthropods. A number of mollusks and bivalves are restricted to hard water because there is insufficient calcium in soft water to harden the shells. In plants calcium is especially important in combining with pectin to form calcium pectate, a cementing material between cells. Plant roots need a supply of this element at the growing tips to develop normally.

Phosphorus not only is involved in photosynthesis, but also plays a major role in energy transfer within the plant and animal. It is a major component of the nuclear material of the cell, where it is involved in cellular organization (DNA) and in the transfer of genetic material. Animals require an adequate supply of calcium and phosphorus in the proper ratio, preferably 2:1 in the presence of vitamin D. An inadequate supply of either may limit the nutritive value of other elements. The lack of either in animals is associated with rickets, a condition involving the improper calcification of bones. When the supply of phosphorus in plants is low, growth is arrested, maturity delayed, and roots stunted. An excessive intake of calcium can inhibit the assimilation of other mineral elements.

Accompanying calcium and phosphorus is *magnesium*. In animals it is concentrated in the bones, and in plants it is an integral part of chlorophyll. It is essential for the maximum rates of enzymatic reactions involving the transfer of phosphates from ATP to ADP and is crucial to protein synthesis in plants. Low intake of magnesium by ruminants causes a serious disease, grass tetany, which may result in death.

Potassium is utilized in large quantities by plants. If the element is readily available and growing conditions are favorable, the uptake (in crop plants at least) may be above average total requirements (Reitemeier 1957). In plants potassium is involved in the formation of sugars and starches, in protein synthesis, in maintaining turgor and activating enzymes, and in movements of stomata. It is the primary charge-balancing cation during the transport of anions from one plant part to another. In animals the synthesis of proteins, normal cell division and growth, and carbohydrate metabolism all depend upon potassium. Unlike calcium, phosphorus, and magnesium, potassium occurs in body fluids and soft tissues. In animals it is readily absorbed metabolically and may interfere with sodium availability; excess over needs is immediately excreted. Because plants usually contain an adequate supply, deficiencies rarely occur among grazing animals.

Iron and *manganese* are involved in the production of chlorophyll. Iron is part of the complex protein compounds that serve as activators and carriers of oxygen and as transporters of electrons. Manganese enhances electron transfer from H_2O to chlorophyll and activates certain enzymes involved in fatty acid synthesis. Over half the iron present in the animal body is in the hemoglobin of blood. Lack of iron results in anemia. A low level of manganese in an animal can result in malformation of bones, in delayed sexual maturity, and in impaired reproduction.

Sodium and *chlorine* are required in minute quantities by plants. Chlorine enhances the electron transfer from H_2O to chlorophyll. Animals require the two in much greater quantities. Obtained from common salt, these elements, particularly sodium, are important for the maintenance of the acid-base balance of the body and osmotic homeostasis, for the formation of and flow of gastric and intestinal secretions, for nerve transmission, lactation, growth, and maintenance of body weight and appetite.

Sulfur, like nitrogen, is a basic constituent of protein, and many plants utilize as much of this element as they do phosphorus. Sulfur supplied by rainwater and by organic matter in the soil is sufficient to meet plant needs. Considerable quantities are released into the atmosphere in industrial areas and carried to the soil by rainfall and dry fallout (see Chapter 12). Excessive sulfur can be toxic to plants. Exposure for only an hour to air containing 1 ppm (part per million) of sulfur dioxide is sufficient to kill vegetation. Plants, especially in arid and semiarid country, are affected by high concentrations of soluble sulfates in the soil, which limit the uptake of calcium.

Boron and *cobalt* are two micronutrients whose deficiency effects, one in plants and the other in animals, are notable. Some 15 functions have been ascribed to boron, including cell division, pollen germination, carbohydrate metabolism, water metabolism, maintenance of conductive tissue, and translocation of sugar in plants. Plants with boron deficiency are stunted both in leaves and roots. This condition is most common in the croplands of eastern and central North America where vegetation is removed annually.

Without cobalt, an element not required by plants, animals become anemic and waste away. Cobalt deficiency is most pronounced in ruminants such as deer, cattle, and sheep, which require the element for the synthesis of Vitamin B_{12} by bacteria in the rumen. The quantity of cobalt required is very small. An 0.4 ha of grassland supporting seven sheep needs to supply only 0.03 gm/yr.

In addition to producing deficiency symptoms when undersupplied, some elements can be toxic in excess. *Copper* and *molybdenum* are two of them. Molybdenum acts as a catalyst in the conversion of gaseous nitrogen into a usable form by free-living, nitrogen-fixing bacteria and blue-green algae; but high concentrations of molybdenum cause "teart" disease in ruminants. This disease in cattle and deer is characterized by diarrhea, debilitation, and permanent fading of hair color. In plants copper is concentrated in chloroplasts, where it influences photosynthetic rate, is involved in oxidation-reduction reactions, and acts as an enzyme activator. When present in excess, copper interferes with phosphorus uptake in plants, depresses iron concentration in the leaves, and reduces growth. Copper can interact with molybdenum, which ties it up. Copper deficiency in animals may cause poor utilization of iron. As a result, iron concentrates in the liver, anemia develops, and calcification of bones decreases.

Zinc usually is abundant in the soil, but it may be unavailable to plants. Zinc may exist as insoluble compounds in the soil when the pH is around 7. Zinc is needed in the formation of auxins in plant growth substances, is associated with water relations in plants, and is a component of several plant enzyme systems. In animals, zinc functions in several enzyme systems, especially the respiratory enzyme carbonic anhydrase in red blood cells, where it plays an essential role in the elimination of carbon dioxide. Insufficient zinc in the diet of mammals can cause a dermatitis known as parakeratosis.

Iodine and *selenium* are two other micronutrients of note. Iodine, deficient in the soils of northeastern North America, the Andes Mountains of South America, and the mountainous parts of Europe,

Asia, and Africa, is involved in thyroid metabolism. In animals the lack of iodine results in goiter, hairlessness, and poor reproduction. Selenium, closely related to vitamin E in its function, prevents white muscle disease in newborn ruminants. The amount necessary is on the order of 0.1 mg/kg of ration. The borderline between the requirement level and toxicity level is narrow. Too much selenium results in the loss of hair, sloughing of hooves, liver injury, and death.

■ Nutrients and Plant Life

Moisture is the most pervasive influence on the distribution of plants, but nutrient availability has pronounced local or regional effects. Plants require some 16 essential elements. Not all plants require them in the same quantities or in the same ratios, but all do require certain minimal amounts of essential ions for growth, and the requirements are specific.

Each species of plant has a specific ability to exploit the nutrient supply that may not be duplicated by other species. Thus plants growing in the same environment exploit slightly different nutrient sources. Shallow-rooted plants, for example, utilize the nutrient supply in the upper surface soil, whereas plants with taproots draw on deeper supplies of nutrients. Some species growing on soils poor in nutrients have become adapted to low nutrient levels and show little response to increased availability, whereas species growing on highly fertile soils have become adapted to higher levels of nutrition. In general, the effects of nutrient availability, especially nitrogen and phosphorus, on plant abundance and distribution are reflected in the competitive abilities of plants.

Nowhere has this fact been better demonstrated than on the long-term grassland fertilizer trial plots at Rothamstead Experimental Farm in England. Among the plots, established in 1856, are some containing natural vegetation that probably resembles the original. The unfertilized, unmanaged plots contain some 60 species of higher plants, representing not only every other plant found on all other plots, but also some restricted to natural plots (Thurston 1969). Species diversity on these plots was high, and no species was clearly dominant. The vegetation was short and the yield of hay low. On plots that received applications of phosphorus, potassium, sodium, and magnesium, but no nitrogen, legumes became dominant at the expense of other species. The addition of nitrogen discouraged the legumes, reduced their growth, and encouraged the grasses. In general nutrient addition to grassland enhances the growth of a few highly competitive and lush-growing grasses at the expense of creeping and rosette-type species (Willis 1963).

Shifts and changes in species composition reflect the competitive abilities of plants under nutrient regimes, as well as their ability to exploit a rich source of nutrients or to get along on a poor source. Nitrogen-fixing legumes outcompete grasses only when the nitrogen content of the soil is low (Wolf and Smith 1964). When nitrogen is added to the system, grasses dominate.

Calcicoles and Calcifuges

Soil acidity can affect the nutrient uptake of plants and indirectly the distribution of plants and the nature of the plant community. Because of the close relationship between soil pH and available calcium, plants have been broadly classified as *calcicole* (lime-loving), *calcifuge* (lime-hating), and *neutrophilus*. The relationship is not so simple, however, because calcium availability is only one of a number of pH-related soil conditions, including the toxicity of iron and aluminum ions and other heavy metals, and deficiencies of nitrogen, phosphorus, and magnesium. Calcium deficiency and low pH are closely associated only because calcium is the most abundant cation in the soil (see Chapter 9).

True calcifuges (Figure 8.1a), those that have a low lime requirement and can live in soils with a pH of 4.0 or less, have certain characteristics, the opposites of which would describe a calcicole. Highly acidic soils associated with calcifuges invariably have a high concentration of aluminum ions, toxic to most plants. Calcifuge plants are tolerant of aluminum. They either possess specific sites within the cytoplasm where aluminum may accumulate harmlessly or have the ability to chelate aluminum or precipitate it at the cell surface. Calcifuge plants are especially sensitive to calcium and have a high demand for iron. If grown on calcareous soils, they suffer from lime chlorosis, a disease in which roots and leaves become stunted and the leaves yellowed. This sensitivity may be due in part to the plant's mechanism for chelating aluminum, which has an affinity for iron at higher pH. This deficiency can be corrected by foliar application of ferrous iron salts. Thus it is the ability to grow in the presence of toxic ions, especially aluminum, that sets calcifuge plants apart from calcicoles.

True calcicole plants (Figure 8.1b) are restricted to soils of high pH, not because they have any particular demand for calcium but because they are susceptible to aluminum toxicity, acidity, and other factors influenced by calcium. Aluminum toxicity begins to appear at a pH of 4.5. Free aluminum accumulates in the surface of the root and in the root cortex, and it interacts with phosphorus to form highly

(a)

(b)

Figure 8.1 (a) The eastern red cedar *Juniperus virginiana* is typically associated with thin soil with underlying limestone and dolomite outcrops. Because of the tree's ability to return calcium to the soil, it tends to make the soil more alkaline. (b) Rhododendron (*Rhododendron maximum*), one of the heaths, is a typical calcifuge. In central Appalachians it is associated with emergent boulders, on which the plant roots in a mossy seed bed. The sites are strongly acidic and high in aluminum. (Photos by R. L. Smith.)

insoluble compounds. Calcareous soils tend to be porous and well-drained, with relatively low soil moisture and higher soil temperatures than noncalcareous soils. These conditions may restrict certain plants, less competitive on more mesic sites, to calcium-rich soils.

Plants of Serpentine and Toxic Soils

Heavy metals can affect plants by direct toxicity, resulting in chlorosis and stunted growth, by interfering with the uptake of other nutrients, and by inhibiting root growth and penetration. For these reasons

high concentrations of heavy metals such as iron, nickel, chromium, zinc, and cobalt can be a powerful force of natural selection on plants, resulting in flora specialized to endure their toxic effects. Nowhere is this effect more evident than on the chemically distinct *serpentine soils*, derived from certain ultrabasic rocks, usually greenish in color. These soils are very high in iron and magnesium, high in nickel, chromium, and cobalt, and low in calcium, phosphorus, sodium, and aluminum. Serpentine soils possess a visually distinctive flora, including a number of rare and endemic species (Kruckeberg 1954). The transition is often sharp. In Oregon, for example, open, stunted stands of pine replace Douglas-fir; in California chaparral vegetation replaces oak woodland; and in New Zealand tussock grassland replaces southern beech (Whittaker 1954). Vegetation on serpentine soils looks xeric because of reduced vegetative structure and stunted appearance. Vegetation is restricted largely because of calcium deficiency and the toxicity of heavy metals (Walker 1954). Plants of serpentine soils appear to substitute magnesium for other bases and to accumulate nickel and chromium into their tissues at levels highly toxic to other plants.

The evolution of tolerance to heavy metals has produced certain species that are able to contain the toxic effects. Modification of the root's uptake mechanism may exclude heavy metals from the plant. The plant may carry the metals unchanged in its tissues or degrade them to some innoxious form. A few species are site-specific for heavy metals (Antonovic et al. 1971), indicating their presence. For example, in England and Europe the pansy *(Viola calaminaria)* and the pennycress *(Thlaspi calaminare)* grow on zinc carbonate and silicate (calamine) soils. In central Africa a group of copper-tolerant plants, especially *Becium homblei*, indicate the possible location of commercial ore deposits (Howard-Williams 1970).

Figure 8.2 Old, unreclaimed strip mine overburdens, as well as mine tailings, are high in toxic heavy metals. They act as a strong selective force on vegetation. (Photo by R. L. Smith.)

The industrial age has produced large acreages of heavy metal soils, including mine tailings, coal spoils, and strip-mined lands (Figure 8.2). These areas have favored the rapid evolution of *ecotypes* (within-species matches between organisms and environment) of certain plants, particularly grasses, tolerant of heavy metals within 30 to 100 years. The evolution of such polymorphic races depends upon three factors: (1) the selection of tolerant seedlings from a normal surrounding population; (2) continued natural selection for metal tolerance and against susceptible genotypes, despite gene flow from the surrounding population; and (3) selection for the ability to survive a physically harsh and nutrient-poor environment. Such selection usually favors a tolerance for a specific heavy metal such as zinc or nickel, not a broad range of them. These ecotypes generally are poor competitors in adjacent normal habitats. Races of tolerant grasses are now being utilized to revegetate toxic soils (Gemmel and Goodman 1980; Smith and Bradshaw 1979).

Halophytes

Soils high in sodium chloride, such as the solonchaks or alkaline soils of semiarid regions, tidal salt marshes and mangrove swamps, support their own distinctive vegetation. It is difficult to generalize whether such plants are adapted to circumvent the osmotic problems created by saline conditions—in which case the response relates more to moisture than to salts (see Chapter 5)—or whether they are plants that require high salinity. Examples are *Salicornia* (Figure 8.3) and sea blite *(Suaeda maritima)* of the coastal salt marshes.

Salt-tolerant plants react in one of a number of ways. They may restrict uptake of sodium and chlorine or dilute the ions through succulence. Some excrete excess salt intake through salt glands in the leaves, or accumulate ions in tissues away from metabolic sites and then shed the leaves and their accumulated salts (Greenway and Murrs 1980).

Figure 8.3 *Salicornia,* a succulent annual or perennial, depending upon the species, is a true halophyte, able to exist only in saline situations. Here the plant is growing on a salt pan in a Virginia tidal marsh. (Photo by R. L. Smith.)

Atmospheric Pollutants

Atmospheric pollutants, particularly sulfur dioxide, ozone, oxides of nitrogen, and hydrogen fluoride can have a pronounced effect on the growth of plants and the composition of plant communities. The gases, products of industrial processes and energy production, are carried by winds and air currents to regions distant from their sources. Under low dosage the vegetation becomes a sink for the contaminants. At intermediate doses individual tree species or individuals of a species may suffer impaired metabolism and nutrient stress. This problem reduces their competitive ability and subjects them to insect attack and diseases. High dosages may kill one or many species of plants (Smith 1981).

The San Bernardino National Forest northeast of Los Angeles has been in the path of air masses heavily contaminated with photochemical oxidants for 30 years. These pollutants have caused the decline and death of ponderosa pine, largely through bark beetle infestation of pollution-stressed trees. As the pines die, the forest composition is shifting to more pol-

lution-tolerant white fir and incense cedar (see Smith 1981).

Sulfur dioxide in high dosages can destroy vegetation. There are several well-known examples, but the best documented is the destruction of a mixed boreal forest in the vicinity of an iron smelter in Wana, northern Ontario. Gordon and Gorham (1963) established a series of vegetation sampling transects running 56 km out from the smelter. Based on their studies, they recognized four zones of damage. Within 8 km of the source all woody vegetation with the exception of some damaged elder was destroyed, and no continuous plant cover existed. In the zone extending 16 to 19 km, the forest canopy was eliminated, but slightly injured elder and some damaged mountain maple survived. Shrubs dominated the zone from 19 to 27 km, interspersed with some scattered white spruce and white birch. From 27 km the forest was intact, but many trees exhibited stress.

Pollution stress can have much the same effect on forest trees as drought (see Chapter 5), altering nitrogen metabolism and improving the food quality of the plant for defoliating, wood-boring, and other insects. For example, pollution-damaged ponderosa pine is subject to infestations of bark beetles and pine loopers (see Chapter 12).

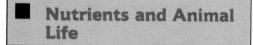

Nutrients and Animal Life

Because all animals depend directly or indirectly on plants for food, plant quantity and quality affect the well-being of animals. In a severe shortage of food, animals suffer from acute malnutrition, leave the area, or starve. In other situations the quantity of food available may be sufficient to allay hunger, but the low quality affects reproduction, health, and longevity.

Some correlation seems to exist between the presence and absence of certain nutrients and the abundance and fitness of certain animals. One essential nutrient that has received attention is sodium, the least stable nutrient in forest and arctic ecosystems (Jordan et al. 1972). In areas of sodium-deficient soil, herbivorous animals face an inadequate supply of sodium in their diets. The problem has been noted in Australian herbivores (Blair-West et al. 1986), in African elephants (Wier 1972), in rodents (Aumann and Emlen 1965; Weeks and Kirkpatrick 1978), in white-tailed deer (Weeks and Kirkpatrick 1976), and in moose (Belovsky and Jordan 1981).

Deficiency can influence distribution and behavioral and physiological adaptations. The spatial distribution of elephants in Wankie National Park in central Africa appears to be closely correlated with the sodium content of drinking water, with the highest number of elephants found at waterholes with the highest sodium content. Three herbivorous mammals, the European rabbit and the white-tailed deer (in parts of their range) and moose experience sodium deficiencies. In sodium-deficient areas in southwestern Australia the European rabbit builds up sodium reserves during the nonbreeding season. These reserves appear to be exhausted near the end of the breeding season, which ends abruptly. The sodium reserves appear to be drained by nematode infections. During the breeding season, the rabbits selectively graze on sodium-rich plants to the point of depletion, exhausting sodium availability. In sodium-deficient areas of southern Indiana, white-tailed deer experience a rapid loss in sodium ions in spring because of an excessive intake of potassium and water in the resurgence of succulent green foods. The deer respond with an intense sodium drive, seeking natural mineral licks and increasing production of mineralocorticoids, indicated by a significant enlargement of the zona glomerulosa (Weeks and Kirkpatrick 1976).

Moose of the northern forest ecosystem face a particular problem obtaining sufficient sodium (Belovsky and Jordan 1981). Northern terrestrial vegetation contains only 3 to 28 ppm Na, far below the nutritional requirements of ruminants. To circumvent sodium stress, moose feed in summer on submerged aquatic plants, which contain between 2 and 9400 ppm Na. Moose may consume from 63 to 95 percent of the production of submerged aquatics, particularly pond lilies. Such heavy consumption appears to trigger declines in aquatic vegetation. The resulting decreased availability of sodium may be one reason for the periodic decline of moose populations in parts of their range (Belovsky 1981).

Other studies show a more general relationship between the level of soil fertility and the health of mammals. Studies of the relationship between soil and cottontail rabbits *(Sylvilagus floridanus)* in Missouri showed no significant difference in body weight of rabbits collected from areas with soils of contrasting fertility (Williams 1965); but there was a positive relationship between soil fertility and fecundity both in Missouri (Williams and Caskey 1965) and in Alabama (Hill 1972) (Figure 8.4).

The size of deer, their antler development, and reproductive success all relate to nutrition (Moen and Severinghaus 1983). Other factors being equal, only deer obtaining high quality foods grow large antlers. Deer on diets low in calcium, phosphorus, and protein show stunted

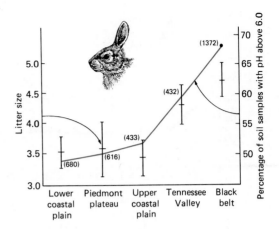

Figure 8.4 The mean size of second cottontail litters from five Alabama soil regions plotted against the percentage of soil samples above 6.0 pH, indicative of high fertility soils. Vertical lines give the range ± two standard deviations. The numbers in parentheses are the numbers of soil samples analyzed for each soil region. Note the strong relationship between soil fertility and litter size, which shows a marked increase in regions with a high percentage of soils above 6.0 pH. (From Hill 1972: 1201.)

growth, and the bucks develop only thin spike antlers (French et al. 1955). Reproductive success of does is highest where food is abundant and nutritious. On the best range in New York State, 1.71 fawns on the average were born for each reproductive doe (Figure 8.5). On a poor range, however, average fawn production per doe was only 1.06 (Cheatum and Severinghaus 1950).

The need for quality foods differs among herbivores. Ruminant animals, for example, can subsist on rougher or lower quality forage because the rumen can synthesize such requirements as vitamin B_1 and certain amino acids from simple nitrogen compounds. Neither the caloric content nor the nutrient status of a food item limits its nutritive value for the ruminant. Nonruminant herbivores require more complex proteins and may carry

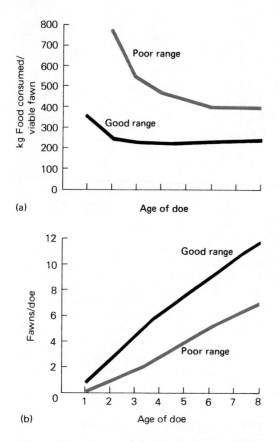

(a)

(b)

Figure 8.5 Differences in the reproductive success of female white-tailed deer on good and poor range in New York State. (a) Food consumed per viable fawn on poor range (Adirondack Mountains) was much greater than food consumed on good range (western New York). (b) Reproductive success was considerably greater on good range than poor range. (Data from Cheatum and Severinghaus 1950.)

bacterial symbionts in the caecum. Seed-eating herbivores exploit the concentration of nutrient in the seeds. Such animals are not likely to have dietary problems.

Among the carnivores, quantity of food is more important than quality. Carnivores rarely have a dietary problem because they consume other animals that have resynthesized and stored protein and other nutrients from plants in their tissues. What they fail to obtain from an animal source, carnivores can obtain by eating plant concentrates such as fruits.

■ Nutrient Budgets

Nutrients upon which organisms depend are constantly being added to, stored, and removed from ecosystems by artificial or natural processes (Figure 8.6). In woodland, shrub, and grassland ecosystems nutrients are imported by wind, rain, dust, and animal life and are returned annually to the soil by leaves, litter, roots, animal excreta, and the bodies of the dead. Released to the soil by decomposition, these nutrients again are taken up first by plants and then by animals. Some nutrients are retained in plant and animal biomass, some are stored in the soil, some are lost to ecosystems by leaching, erosion, and harvesting of plants and animals. In freshwater and marine ecosystems nutrients are imported by precipitation and runoff, taken up by phytoplankton, and consumed by animal life. The remains of plants and animals drift to the bottom, where decomposition takes place. The nutrients again are recirculated to the upper layers by annual overturn and by upwellings from the deep. The inputs and outflows balanced one against the other make up the nutrient budget.

A nutrient budget is the measure of the input and outflow of elements through the various components of an

Figure 8.6 Generalized nutrient cycle in a forest ecosystem. Input of nutrients comes from precipitation, windblown dust (dry deposition), litterfall, and release by weathering and root decomposition. Outgo is through wood harvest, hunting, runoff, erosion, and leaching.

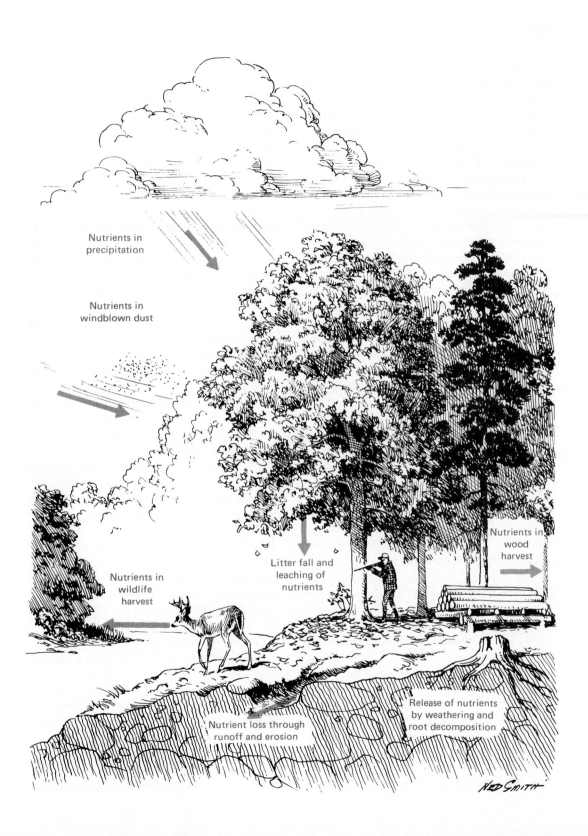

Nutrients in
precipitation

Nutrients in
windblown dust

Nutrients in
wildlife
harvest

Litter fall and
leaching of
nutrients

Nutrients in
wood
harvest

Nutrient loss through
runoff and erosion

Release of nutrients
by weathering and
root decomposition

NED SMITH

ecosystem. It involves tracing the pathways of each element through individuals, through parts of the ecosystem, and into and out of the system and measuring amounts of the elements so circulated.

Obtaining data for such a budget is difficult and time-consuming. Inputs can be determined by measuring the quantities of nutrients carried in by precipitation and aerosols collected over the system. Estimates of the standing crop of nutrients can be obtained by sampling a number of trees of the various species, by determining the nutrient distribution within the biomass, and by measuring mineral content of leaf wash and stemflow. Transfers from the forest soil to the roots can be measured by collecting samples of soil solution. The uptake by vegetation can be estimated by sampling the mineral content of current years' growth, including foliage, branches, and bole. Pathways of nutrients within the system can be followed by means of radioactive tracers. Outflow from terrestrial systems can be determined by analyzing the nutrient content of streamflow from the system.

Because nutrient budgets of temperate forests have been studied more intensively than any other type of ecosystem, data from them can serve to illustrate nutrient budgets of ecosystems.

In the temperate forest ecosystem appreciable quantities of plant nutrients are carried in by rain and snow (Eaton et al. 1973; Emanuelsson, Eriksson, and Egner 1954; Likens et al. 1985; Patterson 1975) and by aerosols (White and Turner 1970; Elwood and Henderson 1975). One study of forested areas in western Europe showed the weight of nutrients carried in by precipitation to be roughly equivalent to the quantity removed by timber harvest (Neuwirth 1957). For some elements the amount carried in by aerosols, known as dryfall, may exceed that carried in by pre-

cipitation. Estimates of annual income to an English mixed deciduous woodland by dryfall were 125.2 kg/ha of sodium, 6.3 kg/ha of potassium, 4.2 kg/ha of calcium, 16.2 kg/ha of magnesium, and 0.34 kg/ha of phosphorus. In an eastern United States deciduous forest ecosystem over 50 percent of the annual magnesium and potassium input, over 40 percent of the calcium input, and one-third of the sulfur input were dryfall (Eaton et al. 1978; Elwood and Henderson 1975). The income from aerosols is greater for elements known to occur as droplets, such as sodium, potassium, and magnesium. The income of calcium, associated with terrestrial sources, and phosphorus, associated with biological activity, is greater in rainfall (White and Turner 1970).

Seventy to 90 percent of gross rainfall reaches the forest floor, mostly as throughfall and stemflow. On its way through the canopy and down the stems rainwater carries with it nutrients deposited as dust on the leaves and stems or leached from them. Such throughfall is richer in calcium, sodium, potassium, phosphorus, iron, magnesium, and silica than rainwater collected in the open at the same time (Eaton, Likens, and Bormann 1975; Madgwick and Ovington 1959; Patterson 1975; Tamm 1951). Throughfall of rain in an English oak woodland accounted for 17 percent of the nitrogen, 37 percent of the phosphorus, 72 percent of the potassium, and 97 percent of the sodium added by the canopy to the soil; the remainder was added by fallen leaves (Carlisle et al. 1967).

Although stemflow amounts to only 5 percent of the total rainfall reaching the forest floor, it is so concentrated about the trunk that with some species the moisture it supplies is five to ten times as great as measured rainfall reaching the forest floor. How much stemflow reaches the

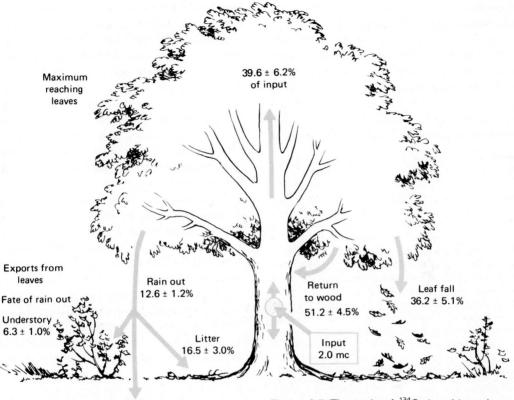

Maximum
reaching
leaves

39.6 ± 6.2%
of input

Exports from
leaves

Fate of rain out

Understory
6.3 ± 1.0%

Rain out
12.6 ± 1.2%

Return
to wood
51.2 ± 4.5%

Leaf fall
36.2 ± 5.1%

Litter
16.5 ± 3.0%

Input
2.0 mc

Soil (0-4 in.)
77.2 ± 2.9%

Figure 8.7 The cycle of ^{134}Cs in white oak, an example of nutrient cycling through plants. The figures are for an average of 12 trees at the end of the 1960 growing season. (After Witherspoon et al. 1960; courtesy Oak Ridge National Laboratory.)

ground varies with the species. Smooth-barked beech has considerably more stem-flow than oaks, whose bark absorbs water (Patterson 1975). Although throughfall provides more nutrients to the soil because of its large volume of flow, stemflow provides a more concentrated nutrient solution. Again the concentration of nutrients depends upon the species. Beech and hickories return considerably more calcium and potassium than oaks, and pines return smaller amounts of calcium, magnesium, potassium, and manganese than hardwoods. In northern North American forests of balsam fir, epiphytic lichens may remove nitrogen and add calcium and magnesium to stemflow (Lang et al. 1976).

The nutrients carried to the forest floor by throughfall, stemflow, and other sources are taken up in time by the surface roots of trees and translocated to the canopy. Such short-term internal cycling has been investigated by means of radioactive tracers. By inoculating white oak trees with 20 microcuries of ^{134}Cs (cesium-134), Witherspoon and others (1962) were able to follow the gains, losses, and transfers of this radioisotope. About 40 percent of the ^{134}Cs inoculated into the oaks in April moved into the leaves by early June (Figure 8.7). When the first rains fell after inoculation, leaching of radiocesium from the leaves began. By September this loss amounted to 15 percent of the maximum concentration in the leaves. Seventy percent of this rainwater loss reached mineral soil; the remaining 30 percent found its way into the litter and understory. When the leaves fell in autumn, they carried with them twice as much radiocesium as

was leached from the crown by rain. Over the winter, half was leached out to mineral soil. Of the radiocesium in the soil, 92 percent still remained in the upper 10 cm nearly two years after the inoculation. Eighty percent of the cesium was confined to an area within the crown perimeter and 19 percent was located in a small area around the trunk. This finding suggests that cesium distribution in the soil was greatly influenced by leaching from rainfall and stemflow.

The use of ^{137}Cs to trace the pathway of mineral cycling in the forest was carried one step further in a study of cesium cycling in a tulip poplar (*Liriodendron tulipifera*) stand (Olsen et al., 1970). Dominant trees were tagged with ^{137}Cs. The fate of the tracer was continuously monitored for five years, and its pathway through the forest ecosystem was traced as illustrated in Figure 8.8. The leaves were richest in cesium in the spring and poorest in the

Figure 8.8 The biogeochemical cycle of ^{137}Cs in a tulip poplar (*Liriodendron tulipifera*) forest ecosystem. The trees of this forest were tagged originally with a total of 467 microcuries of ^{137}Cs in May 1971. Continuous inventory followed the seasonal and annual distribution and fluxes of ^{137}Cs among the abiotic and biotic components of the forest ecosystem. Data are for the 1965 season. Numbers in the boxes are microcuries of ^{137}Cs per square meter of ground surface. Arrows indicate the pathways of ^{137}Cs transfer between compartments. Numbers by the arrows are estimates of annual fluxes; numbers in parentheses are the averaged transfer coefficients (days) of ^{137}Cs for the seasons during which the process occurs. (Courtesy Oak Ridge National Laboratory.)

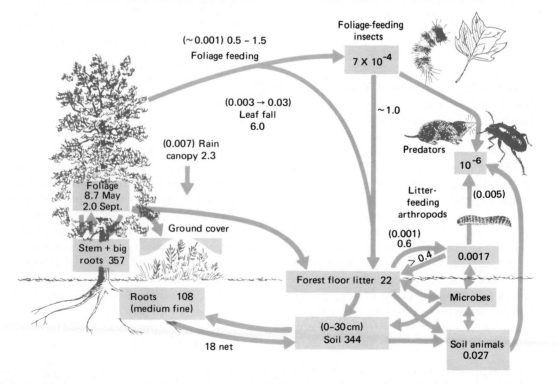

fall. The reason for the decline in the leaves was the movement of ^{137}Cs from the foliage to the woody tissue when the leaves were mature and senescent. There was a cumulative loss of the tracer from the foliage both from rainfall and litterfall, but the greatest transfer to the soil was through the roots. The ^{137}Cs accumulated in the litter and organic layers of the forest floor and in the upper 10 cm of the mineral soil. From here cesium was picked up by decomposers, soil organisms, and litter-feeding arthropods and their predators.

Much of the nutrient pool in the forest ecosystem is involved in short-term cycling. Nutrients taken up by trees are returned to the forest floor by litterfall,

throughfall, and stemflow (Table 8.1). However, a portion of the nutrient uptake is stored in tree limbs, trunk, bark, and roots as accumulated biomass and additional quantities are stored in the living biomass in the forest floor and consumer organisms. This portion is effectively removed from short-term cycling. Additional nutrients are leached from the soil, transported to streamflow, and carried out of the ecosystem by underground waterflow. The nutrient cycle in a forest ecosystem undisturbed by human activity can function and its nutrient budget can be balanced only by withdrawing nutrients from the soil.

Considerable quantities are withdrawn permanently from ecosystems by harvest-

---

TABLE 8.1 Nutrient Budget of a Yellow-Poplar *Liriodendron* Forest, Oak Ridge, Tennessee, USA

	Org. Mat.	N	K	Elements Ca	P	Mg
Amounts (kg/ha)						
Overstory foliage total	3200	53.9	52.3	35.2	6.3	
Overstory branches	27100	78.6	46.1	143.7	30.6	
Overstory boles	94400	172.0	74.8	276.8	9.7	
Overstory roots	36000	122.7	111.5	150.8	18.5	
Understory vegetation	8800	23.8	32.4	115.2	5.7	
Forest litter layer	6000	77.9	9.2	99.6	5.3	
Soil-rooting zone	159000	7650	38960	8130	2840	
Increments (kg/ha/yr)						
Overstory foliage	3200	53.9	52.3	35.2	6.3	
Overstory branches	575	1.7	1.0	3.1	0.7	
Overstory boles	2254	5.4	2.3	6.3	0.3	
Overstory roots	513	1.8	1.2	2.2	0.2	
Understory vegetation	81	0.3	0.2	0.6	0.1	
Fluxes (kg/ha/yr)						
Atmosphere precipitation		7.7	0.73	6.1	0.06	
Atmosphere particulates			2.56	4.5	0.7	
Overstory litterfall	4290	31.3	19.0	77.7	2.8	
Leaf wash		9.4	31.7	17.8	0.4	
Stem flow		0.1	1.0	0.4	0.002	
Leaching—forest floor						
Leaching—rooting depth (60 cm)		3.5	8.9	44.5	0.05	
Leaching—watershed						

Reichle 1981: 393

ing (Table 8.2), especially in agricultural ecosystems and in forest ecosystems harvested in short-term rotations. In such ecosystems these losses must be replaced by artificial fertilization; otherwise the system becomes impoverished.

Nutrient budgets for a number of aquatic and terrestrial ecosystems are discussed and compared in Chapters 29 through 33. The budgets developed, especially for forest ecosystems, suggest several characteristics of nutrient cycling. The accumulation of nutrients in the rooting zone of the soil and in the tree biomass removes them from short-term circulation and makes them unavailable for varying periods of time. Thus the rate of nutrient cycling can be maintained only by the pumping of nutrients from deep soil reserves and by the weathering of parent rock. Nutrient cycling is influenced by climate, the nature of the soil, and the nature of the plants occupying the area, including their ability to pump, transport, store, and return certain mineral elements. Seasonal variations in utilization, storage, and retention complicate the cycling. Except for a sharp peak in mineral uptake, which occurs when the productivity of the forest seems to be at its peak, age seems to have little influence on mineral cycling.

The gain to the ecosystem from precipitation, extraneous material, and mineral weathering is offset by losses. Water draining away removes more mineral matter than is supplied by precipitation (Likens et al. 1967, 1985; Viro 1953; Whittaker and Woodwell 1969).

■ A Model of Nutrient Flow

A general model of nutrient flow, Figure 8.9, describes the relationship among the various components of the ecosystem. The food base depends on (1) the available nutrients, which in turn depend upon inputs into the detrital pool; (2) the rate at which detritus is decomposed; (3) the amounts of detritus and available nutrients that go into the storage in soil, humus, detrital sediment, and the like; and (4) the release of nutrients from the reserve.

How nutrient utilization takes place in an ecosystem depends upon the size of the abiotic nutrient reserve, the proportions of nutrients stored in the biotic component and the abiotic environment, and the rate of turnover of the recycling pool of nutrients, that is, how fast nutrients are passed from the detrital pool to available

TABLE 8.2 Comparison of the Average Annual Yield and Nutrient Removal by a 16-Year-Old Loblolly Pine Plantation with That of Agricultural Crops

Crop	Yield	Removal			
		N	P	K	Ca
	Tons/ha	Kilograms/ha			
Loblolly pine, whole tree	14.5	17.5	2.4	12.6	12.8
Loblolly pine, pulpwood	8.75	6.5	0.9	5.1	6.4
Corn (grain)	11.75	130.5	29.7	37.3	—
Soybeans (beans)	3.0	145.0	14.85	46.7	—
Alfalfa (forage)	10.0	212.5	23.25	185.6	75.4

Data from Jorgensen and Wells 1986

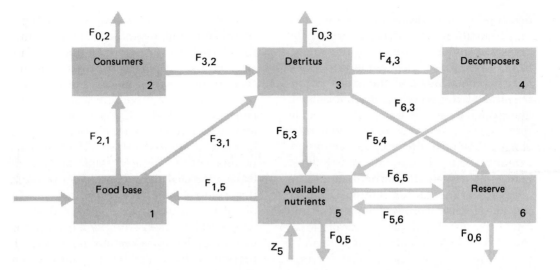

Figure 8.9 Model of nutrient flow of an ecosystem. Arrows indicate the direction of flow (F) of nutrients between compartments represented by numbered blocks. Subscripts represent compartment numbers. Thus $F_{3,1}$ indicates the flow of nutrients from compartment 1 to compartment 3; F_0 indicates outflow to the environment; and $F_{0,3}$ is outflow to the environment from compartment 3. (From Webster, Waide, and Patten 1975: 13.)

form through leaching and decomposition and their subsequent uptake. Because of nutrient limitation and the necessity to control the rate of nutrient flow to prevent overutilization, two mechanisms relate nutrient utilization to energy flow (O'Neill et al. 1975).

One mechanism is to put energy conversion in a number of plant populations of low biomass with different responses to environmental conditions. These plant populations are capable of rapid reproduction when conditions are optimal. One or more populations can expand rapidly and take up nutrients quickly. A succession of plant populations through the season ensures continuous energy fixation and utilization of nutrients.

A second mechanism places energy conversion in individuals of great bulk

and with slow reproductive rates. Large individuals are able to survive unfavorable conditions, and large quantities of nutrients are stored in biomass.

Although the first mechanism is characteristic of aquatic ecosystems and the second of terrestrial ecosystems, especially forests, neither mechanism operates exclusively in only one type of ecosystem. Lakes have rooted aquatics with biomass accumulation, and forests have a seasonal parade of herbaceous understory plants.

Each mechanism has provisions for the conservation of nutrients. One provision is a large pool of organic matter. Litter and soil organic matter and the structural mass of autotrophs form such a pool in terrestrial ecosystems. In aquatic ecosystems the pool is particulate and dissolved organic matter. Turnover of the organic pool is relatively slow, roughly one magnitude lower than the turnover of the vegetative component (O'Neill et al. 1975). Organic matter has a key role in recycling nutrients because it prevents rapid losses from the system. Large quantities of nutrients are bound tightly in organic matter structure and are not readily available, but their release can be activated by decomposer organisms. It takes place mostly in those ecosystems in which the persistent standing crop of organic matter in the presence of adequate moisture and energy is determined by the supply of nutrients and the recycling mechanisms. If energy and especially water are limited, as they are in desert ecosystems, then recycling is minimal because the ecosystem cannot attain a standing crop large enough to deplete the nutrient supply.

Another mechanism is to partition the nutrient reserve between short-term and long-term nutrient pools. For example, of the structural components of individual plants—wood, bark, twigs, and leaves—leaves are recycled the fastest and wood the slowest. The leaves represent a short-term nutrient pool and wood a long-term reservoir. Day and McGinty (1975) studied nutrient cycling by several tree species on the Coweeta Watershed at Franklin, North Carolina. Among the forest tree species studied were chestnut oak *(Quercus prinus)*, rhododendron *(Rhododendron maxima)*, and flowering dogwood *(Cornus florida)*. Although chestnut oak had the largest total standing crop biomass and the largest total standing crop of nutrients, nutrients were apportioned among the various structural components of the different species in a manner that influenced nutrient cycling. The largest standing crop of potassium was in the wood of chestnut oak, and the largest standing crop of calcium and magnesium was in the bark of chestnut oak. The evergreen rhododendron had the highest concentration and the highest nutrient standing crop of nitrogen, calcium, magnesium, and potassium in the twig compartment, which was slowly recycled. Rhododendron also had the highest standing crop of leaf biomass. Chestnut oak, however, had the highest standing crop of nitrogen and potassium in that compartment. Rhododen-

dron leaves had the highest standing crop of calcium and magnesium, but because the leaves are evergreen, the nutrients are recycled over a period of seven years instead of one year. The leaves of dogwood concentrate over three times as much calcium per unit leaf biomass as oak and one and a half times as much per unit leaf biomass as rhododendron. Chestnut oak, because of its greater total biomass of leaves, recycled the most calcium through that compartment. However, the amount of calcium recycled by dogwood through its low leaf biomass—175 kg/ha compared to 845 kg/ha for chestnut oak—was 66 percent of that cycled by chestnut oak and 150 times more than that cycled by rhododendron. Because of its high wood-to-leaf ratio and its high concentration of calcium, the understory dogwood tree is important in short-term cycling. Thus in the mature forest nutrients stored in vegetation are recycled at various time intervals from 1 to 100 years. This partitioning prevents excessive losses of nutrients and releases nutrients slowly to biogeochemical cycles.

■ Summary

The major source of nutrients is the weathering of soil minerals, which releases nutrients to terrestrial and aquatic ecosystems through leaching. The kind and quantity of elements and nutrients available for circulation in the ecosystem affect the growth, reproduction, and distribution of plants. Species of both vary in their requirements and tolerances for different elements. Some elements, the macronutrients, including nitrogen, calcium, phosphorus, and potassium, are required in relatively large quantities by all living organisms. Others, the trace elements, including copper, zinc, and iron, are needed in lesser, often minute quantities; without them plants and animals fail.

Each species has specific nutrient requirements or exploits a nutrient supply in a manner somewhat different from other species. This fact is often reflected in the competitive abilities of plants. Some plants grow well on acid soils, are tolerant of aluminum, and are sensitive to calcium. These plants are known as calcifuges. Other plants are susceptible to aluminum toxicity and do best in soils high in calcium. They are calcicoles. A few plants have evolved tolerances to toxic heavy metals such as copper, nickel, and zinc; and a special group of plants, the halophytes, are tolerant of high salinities. Having adverse effects on plants are atmospheric pollutants such as sulfur dioxide.

The source of nutrients for animals, directly or indirectly, is plants. A low level of nutrients in plants can have adverse effects on the growth, development, and fitness of herbivores. One essential nutrient that influences the distribution, behavior, and fitness of grazing herbivores is sodium. Among herbivores the quality of food is of critical importance. Among carnivores quantity is more important than quality, because they consume animals that have stored up and resynthesized proteins and other nutrients from plants.

Nutrients are constantly being added, stored, and removed from ecosystems. A measure of this inflow and outflow makes up a nutrient budget. The rate at which nutrients are added, retained, and recycled over short-term and long-term periods influences the nature of the ecosystem.

Soils

■ **The Soil Profile**

■ **Soil Properties**

    Color
    Texture
    Moisture
    Chemistry

■ **Soil Formation**

    Weathering
    Biotic Influences
    The Organic Horizon

■ **Soil Development**

    Podzolization
    Laterization
    Calcification
    Gleyization
    Toposequences
    Soils and Time

■ **Soil Classification**

■ **Summary**

Soil is the foundation of terrestrial communities. It is the site of decomposition of organic matter and the return of mineral elements to the nutrient cycle (see Chapter 12). Roots occupy a considerable portion of the soil, to which they tie the vegetation and from which they pump water and minerals in solution needed by plants for photosynthesis. Vegetation, in turn, influences the development of soil, its chemical and physical properties, and its organic matter content. Thus soil acts as a pathway between the organic and mineral worlds.

■ **Definition of Soil**

As familiar as it is, soil is difficult to define. One definition has soil as a natural product formed from weathered rock by the action of climate and living organisms. Another states that soil is a collection of natural bodies of Earth that is composed of mineral and organic matter and is capable of supporting plant growth. Such definitions seem inadequate or stilted. Indeed, one eminent soil scientist, a pioneer of modern soil studies, Hans Jenny, will not give an exact definition of soil. In his book *The Soil Resource* (1980: 364), he writes:

> Popularly, soil is the stratum below the vegetation and above hard rock, but questions come quickly to mind. Many soils are bare of plants, temporarily or permanently; or they may be at the bottom of a pond growing cattails. Soil may be shallow or deep, but how deep? Soil may be stony, but surveyors (soil) exclude the larger stones. Most analyses pertain to fine earth only. Some pretend that soil in a flower pot is not soil, but soil material. It is embarrassing not to be able to agree on what soil is. In this pedologists are not alone. Biologists cannot agree on the definition of life and philosophers on philosophy.

Of one fact we are sure: Soil is not just an abiotic environment of plants. It is teeming with life—billions of minute animals, bacteria, and fungi. The interaction between the abiotic and biotic make soil a living system.

■ The Soil Profile

Soil, Hans Jenny (1980: 6) wrote, "is a body of nature that has its own internal organization and history of genesis." Before proceeding we should have some appreciation of its internal organization; its genesis we can consider later.

A fresh cut along a roadbank or an excavation tells something about a soil. A close-up and even cursory look reveals bands and blotches of color from the surface downward. An even closer examination, involving some handling of the material, reveals changes in texture and structure. This vertical cut through a body of soil is the *soil profile*. The apparent layers are called the *horizons*, the product of local chemical and physical processes in the soil. Within a horizon, a particular property, such as color or mineral matter, reaches its maximum intensity. Away from this level, this property decreases gradually in both directions. Each horizon varies in thickness, color, texture, structure, consistency, porosity, acidity, and composition.

In general soils have four major horizons: *O*, an organic layer, and *A*, *B*, and *C*, the mineral layers. Below the four may lie the *R* or nonsoil horizon. Because the soil profile is essentially a continuum, often there is no clear-cut distinction between one horizon and another, and the content of each horizon varies considerably (Figure 9.1).

The *O* horizon is the surface layer, formed or forming above the mineral layer and composed of fresh or partially decomposed organic material. It is usually absent in cultivated soils. This layer and the upper part of the *A* horizon constitute the zone of maximum biological activity. They are subject to greatest changes in soil temperatures and moisture conditions, contain the most organic carbon, and are the sites where most or all decomposition takes place.

The *A* horizon is characterized by an accumulation of organic matter, by the loss of clay, iron, and aluminum, and by the development of a granular, platy, or crumbly structure. The *B* horizon is characterized by a concentration of all or any of the silicates, clay, iron, aluminum, and humus, alone or in combination, and by the development of blocky, prismatic, or columnar structure. The *C* horizon con-

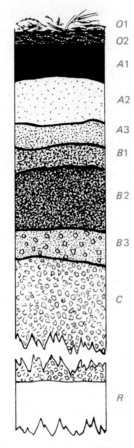

Figure 9.1 A generalized profile of the soil. Rarely does any one soil possess all of the horizons shown. O_1: loose leaves and organic debris. O_2: Organic debris partially decomposed or matted. A_1: A dark-colored horizon with high content of organic matter mixed with mineral matter. The A horizon, together with the O horizon, is the zone of maximum biological activity. A_2: A light-colored horizon of maximum leaching. Prominent in spodosols, it may be faintly developed in other soils. A_3: Transitional to B, but more like A than B; sometimes absent. B_1: Transitional to B, but more like B than A, sometimes absent. B_2: A deeper-colored horizon of maximum accumulation of clay minerals or of iron and organic matter; maximum development of blocky or prismatic structure or both. B_3: Transitional to C. C: The weathered material, either like or unlike the material from which the soil presumably formed. A gley layer may occur, as well as layers of calcium carbonate, especially in grassland. R: Consolidated bedrock.

tains weathered material, either like or unlike the material from which the soil is presumed to have developed.

■ Soil Properties

Horizons are distinguished by the properties of soil, some of which are color, texture, structure, and moisture. All are highly variable from one soil to another.

Color

Color has little direct influence on the function of a soil, but considered with other properties, it can tell a good deal about the soil. In fact, it is one of the most useful and important characteristics for the identification of soil (see page 185). In temperate regions dark-colored soils generally are higher in organic matter than light-colored ones. Well-drained soil may range anywhere from pale brown to dark black, depending upon the organic matter content. However, it does not always follow that dark-colored soils are high in organic matter. Soils of volcanic origin, for example, are dark in color. In warm temperate and tropical regions, dark clays may have less than 3 percent organic matter. Red and yellow soils show the presence of iron oxides, the bright colors indicating good drainage and good aeration. Other red soils obtain their color from parent material such as red lava rock and not from soil-forming processes. Well-drained yellowish sands are white sands containing a small amount of organic matter and such coloring material as iron oxide. Red and yellow colors increase from cool regions to the equator. Quartz, kaolin, carbonates of calcium and magnesium, gypsum, and various compounds of ferrous iron give whitish and grayish colors to the soils. The grayish are permanently saturated soils in which iron is in the ferrous form. Poorly drained soils are mottled with various shades of yellow-brown and gray. The colors of soils are determined by the use of standardized color charts.

Texture

The texture of a soil is determined by the proportion of different-sized soil particles (Figure 9.2). Texture is partly inherited from the parent material from which the soil was derived and partly from the soil-forming process.

Particles are classified on the basis of size into gravel, sand, silt, and clay. Gravel consists of particles larger than 2.0 mm. Sand ranges from 0.05 to 2.0 mm, is easily

seen, and feels gritty. Silt consists of particles from 0.002 to 0.05 mm in diameter, which can scarcely be seen by the naked eye, and feels and looks like flour. Clay particles, less than 0.002 mm, too small to be seen under an ordinary microscope, are colloidal. Clay controls the most important properties of soils, including plasticity and the exchange of ions between soil particles and soil solution. The nature of a soil's texture is the percentage (by weight) of sand, silt, and clay. Based on these proportions soils are divided into textural classes (Table 9.1).

Figure 9.2 A soil texture chart, showing the percentages of clay (below 0.002 mm), silt (0.002 to 0.05 mm), and sand (0.05 to 2.0 mm) in the basic soil textural classes.

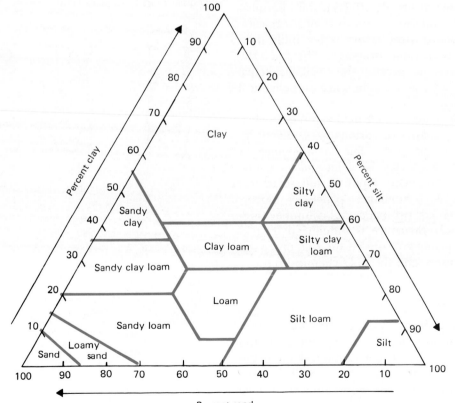

Percent sand

TABLE 9.1 Texture Classes of Soils

Common Names	Texture	Class Names
Sandy soils	Coarse	Sandy Loamy sands
Loamy soils	Moderately coarse	Sandy loam Fine sandy loam Very fine sandy loam
	Medium	Loam Silt loam Silt
	Moderately fine	Clay loam Sandy clay loam Silty clay loam
Clayey soils	Fine	Sandy clay Silty clay Clay

Texture plays a major role in the movement of air and water and in penetration by roots. It greatly affects a soil's permeability and its water storage capacity. Coarser texture favors water infiltration and more rapid drainage. The finer the texture, the greater the available active surface for water adherence and chemical activity.

Soil particles are held together in clusters or shapes of various sizes, called aggregates or peds. The arrangement of these aggregates is called *soil structure*. There are many types of soil structure: granular, crumblike, platelike, blocky, subangular, prismatic, and columnar (Figure 9.3). Structure is influenced by texture, plants growing on the soil, other soil organisms, and the soil's chemical status.

Moisture

Dig into the surface layer of a soil about a day after a soaking rain and note the depth of water penetration. Unless the soil was a heavy clay, you should discover that the transition between wet surface soil and dry soil is sharp. The water that

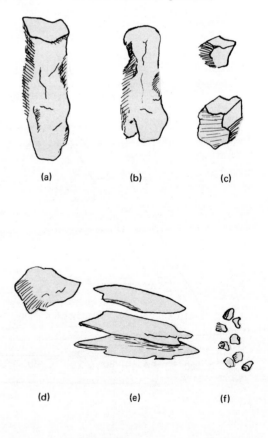

Figure 9.3 Some types of soil structure: (a) prismatic, (b) columnar, (c) angular blocky, (d) subangular blocky, (e) platelike, (f) granular.

fell on the ground infiltrated the soil, filling the pore spaces and draining into the dry soil below. Depending upon the amount of water falling on the soil surface, the downward flow halts within two or three days and the water hangs in the soil capillaries.

The maximum amount of water the soil will hold following free drainage is called field capacity (FC). It represents the moisture conditions in each individual soil after the large pore spaces have drained fully. A more precise definition is the amount of water under negative (suction) pressure of approximately 1/3 bars (atmospheres of tension).

Field capacity is influenced heavily by soil texture, and to a lesser extent by clay minerals, stoniness, and soil structure. Sand has 30 to 40 percent of its volume in pore space and clays and loams 40 to 60 percent.

Field capacity, however, does not represent the water available to plants. That is *available water capacity* (AWC), the supply of water available to plants in a well-drained soil. It is the water retained between FC and the *permanent wilting point* (PWP) or between 1/3 and 15 atmospheres of tension. It represents the soil's renewable storage volume.

The permanent wilting point occurs when the soil dries out to a point where the water potential and conductivity assume such low values that the plant is unable to extract sufficient water to meet its demands. The plant wilts permanently and will not revive when placed in a saturated atmosphere.

The ability of a soil to retain moisture against drainage determines soil moisture regimes. Texture plays an important role. Clays and clay loams retain more moisture than sandy soils. AWC is lowest in coarse-textured soils and highest in medium-textured soils. In fine-textured soils films of water are held on soil particles at increasingly high tensions and become less available to plants. The amount of organic matter in the soil adds to a soil's water holding capacity, as does stoniness. Although stoniness reduces the amount of space for water storage, stones do impede runoff and favor infiltration.

The topographic position of a soil affects the movement of water both on and in the soil. In general, water tends to drain downslope, leaving soils on higher slopes and ridgetops relatively dry, and creating a moisture gradient from ridgetops to streams. However, after a dry period more moisture may be stored from rain on the upper slope than on the more moist lower slope.

Chemistry

Chemical elements in the soil are adsorbed on soil particles and dissolved in soil solution and are a constituent of soil organic matter. These ions move from soil to plant, from plant to animals, and into the biogeochemical cycle (see Chapter 12). In aquatic systems ions are dissolved and obey the laws of diffusion and dilute solutions. In soils ions are limited in their mobility because they are closely held to solid particles of clay and humus.

The key to the availability of nutrients in the soil is the nature of the clay-humus complex. It is made up of platelike particles in the soil, called *micelles* (Figure 9.4). Micelles consist of sheets of tetrahedron and octahedron aluminosilicates (silicate, aluminum, iron combined with oxygen, and hydroxyl ions). The interior of the plates is electrically balanced, but the edges and sides are negatively charged. They attract positive ions, water molecules, and organic substances. The number of negatively charged sites on soil particles that can attract positively charged

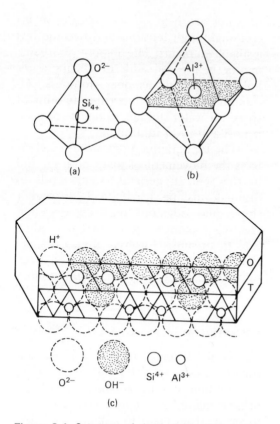

Figure 9.4 Structure of clay colloids. Basic building units of aluminosilicate crystals are (a) the four-sided tetrahedron with small Si^{4+} in the central cavity formed by $4O^{2-}$; and (b) the eight-sided octahedron with Al^{3+} or Mg^{2+} in the interstices of six touching O^{2-} and OH^{-} ions. (Spheres not drawn to size.) (c) Front view of a kaolinite-type clay platlet, showing a layer of octahedrons (O) and a layer of tetrahedrons (T). The view shows an exposed horizontal sheet of hydroxyl (OH^{-}) ions. The OH^{-} ion at the upper left corner is dissociated into an oxygen ion (O^{2-}) and an exchangeable hydrogen ion (H^{+}) ion. (After Jenny 1980: 51, 61.)

cations is called the *cation exchange capacity*. These positively charged ions can be replaced by still other ions in the soil solution.

Cation exchange capacity varies among soils, depending upon the structure of the clays. Some are made up of octahedron and tetrahedron sheets arranged in large lattices that swell and expand when moist. They possess an extensive surface area, externally and internally, which allows ion exchanges not only between micelles but within the micelle as well. Others have octahedrons and tetrahedrons arranged randomly. They are amorphous clays that do not expand when moist, and cation exchange is of lesser magnitude.

Negatively charged ions attract cations of Ca^{+}, Na^{+}, Mg^{+}, K^{+}, and H^{+}, among others. H^{+} ions are especially tenacious. Hydrogen ions added by rainwater, cationic acids from organic matter, and metabolic acids from roots displace other cations, such as Ca^{+}. Ions are available to plants only when dissolved in soil solution. Ions in soil solution maintain an equilibrium with absorbed ions in micelles. As plants remove ions from soil solution in the vicinity of roots, other ions diffuse to the region. That, in turn, enhances release of ions from the micelles. If not taken up by plants, these cations are carried deeper into the soil or are removed altogether through the groundwater and frequently move into aquatic systems. The cation exchange capacity has a pronounced effect on soil fertility and availability of nutrients to plants.

Acidity is one of the most familiar of all chemical conditions in the soil. Typically, soils range from a very acid pH of 3 to a pH of 8, strongly alkaline. Soils just over a pH of 7 (neutral) are considered basic, and those of 6.6 or less acid. Soil acidification results from the removal of bases by the leaching effects of water moving through the soil profile, the withdrawal of exchangeable ions by plants, the release of organic acids by roots and microorganisms, and the dissociation of $CaCO_3$. If the soil is poorly buffered against acidic inputs, then soil acidity increases.

Soil acidity has a pronounced effect on nutrient availability. As soil acidity increases, the proportion of exchangeable Al^+ increases and Ca^+, Ca^+, Na^+, and others decrease. Such changes bring about not only nutrient deprivation but also aluminum toxicity (see Chapter 8). Harmful effects of low pH in both soil and aquatic environments are due not so much to the acid as to the toxic Al^+ and Fe^+ ions released. (We will discuss these effects in Chapter 12.)

■ Soil Formation

Weathering

The formation and development of soil begin with the weathering of rocks and their minerals. Exposed to the combined action of water, wind, and temperature, rock surfaces flake and peel away. Water seeps into crevices, freezes, expands, and cracks the rock into smaller pieces. Accompanying this physical weathering and continuing long afterward is chemical weathering, resulting in the decomposition of minerals, dominated by the primary silicates olivine, augite, hornblende, and quartz and by the aluminosilicates, the feldspars and micas. Easily weathered, these minerals, particularly the aluminosilicates, produce most of the clays and other fine mineral matter of soils and provide mineral nutrition of plants.

Water and carbon dioxide combine to form carbonic acid, which reacts with hydroxides of potassium, sodium, calcium, and magnesium to form carbonates and bicarbonates. They either accumulate deeper in the material or are carried away,

depending upon the amount of water passing through. The aluminosilicates are converted to secondary minerals, particularly clays such as kalonite and montmorillonite. As iron is especially reactive with water and oxygen, iron-bearing minerals are prone to rapid decomposition. Iron remains oxidized in the red ferric state or is reduced to the gray ferrous state. Fine particles, especially clays, are shifted and rearranged within the mass by percolating water and on the surface by runoff, wind, or ice.

Eventually the rock is decomposed into loose material. This material may remain in place, but more often than not, much of it is lifted, sorted, and carried away. Some moves downslope by force of gravity and surface runoff. Material transported from one area to another by wind is known as *loess;* that transported by water as *alluvial, lacustrine* (or lake), and *marine deposits;* and that moved by glacial ice as *till.* In a few places soil material comes from accumulated organic matter, such as peat. Materials remaining in place are called *residual.*

The mantle of unconsolidated material is called the *regolith.* It may consist of slightly weathered material with fresh primary minerals; or it may be intensely weathered and consist of highly resistant minerals such as quartz. Because of variations in slope, climate, and native vegetation, many different soils can develop on the same regolith. The thickness of the regolith, the kind of rock from which it was formed, and the degree of weathering affect fertility and moisture.

Biotic Influences

Over time plants and animals have a pronounced influence on soil development. Plants colonize the weathered material. More often than not, intense weathering, mostly chemical, goes on under some plant cover. Plant roots penetrate and further break down the regolith. They pump nutrients up from its depths and add them to the surface. In doing so plants recapture minerals carried deep into the material by weathering processes. Through photosynthesis, plants capture the sun's energy and add a portion of it in the form of organic carbon—approximately 18 billion metric tons, or 1.7×10^{17} kilocalories annually—to the soil. This energy source of plant debris enables bacteria, fungi, earthworms, and other soil organisms to colonize the area.

The breakdown of organic debris into humus is accomplished by decomposition and, finally, mineralization. Higher organisms in the soil—millipedes, centipedes, earthworms, mites, springtails, grasshoppers, and others—consume fresh material and leave partially decomposed products in their excreta. This material is further decomposed by microorganisms into various carbohydrates, proteins, lignins, fats, waxes, resins, and ash. These compounds are then broken down into simpler products by mineralization.

The fraction of organic matter that remains is called humus, dark-colored organic tissue too small to be seen by the naked eye. It is not stable, as it represents a stage in the decomposition of organic matter. New humus is being formed as old humus is being destroyed by mineralization. The equilibrium set up between the formation of new humus and the destruction of old determines the amount of humus in the soil.

The activities of soil organisms, the acids produced by them, and the continual addition of organic matter to mineral matter produce profound changes in the weathered material. Rain falling upon and filtering through the accumulating organic matter picks up acids and minerals in solution, reaches the mineral soil, and sets up a chain of complex chemical reactions that continue in the regolith. Calcium, potassium, sodium, and other mineral elements, soluble salts, and carbonates are carried in solution by percolating water deeper into the soil or are washed into streams, rivers, and eventually the sea.

The greater the rainfall, the more water moves down through the soil and the less moves upward. Thus high precipitation results in heavy leaching and chemical weathering, particularly in regions of high temperatures. These chemical reactions tend to be localized within the regolith. Organic carbon, for instance, is oxidized near the surface, whereas free carbonates precipitate deeper into the rock material. Fine particles, especially the clays, move downward. These localized chemical and physical processes in the parent material result in the development of soil horizons, which impart to the soil a distinctive profile.

The Organic Horizon

Of all the horizons of the soil, none is more important or ecologically more interesting than the organic horizon of the forest floor. A close relationship exists among litter, humus, and the environmental conditions in the forest community—the internal microclimate of the soil, the moisture regime, its chemical composition, and its biological activity. The forest organic layer plays a dominant role in the life and distribution of many forest plants and animals, in the maintenance of soil fertility, and in many soil-forming processes.

The nature and quality of the forest organic layer depend in part on the kind and quality of forest litter. The fate of that litter and the development of a horizon are conditioned by the activity of micro-flora and soil animals. In fact, many forms of humus undergo initial breakdown in the bodies of animal organisms. To complete the circle, the composition and density of the soil fauna are influenced by the litter.

The importance of the organic layer was stressed early in the history of ecology. Darwin, in his famous work "The Formation of Vegetable Mould through the Action of Worms, with Observations on Their Habits" (1881), pointed out the influence of these animals on the soil. At about the same time, in 1879 and 1884, the Danish forester P. E. Muller described the existence of two types of humus formation in the temperate forest soil; these he called mull and mor. Not only did he observe differences in vegetation, soil structure, and chemical composition, but he discovered differences in their fauna also. Muller considered mull and mor as biological rather than purely physiochemical systems, and he regarded the fauna present as aiding in their formation. Others have regarded mull and mor from the physical and chemical point of view, with little regard for biological mechanisms. Actually mull and mor result from an interaction of all three (Figure 9.5).

Mor. Mor, characteristic of dry or moist acid habitats, especially heathland and coniferous forest, has a well-defined, unincorporated, and matted or compacted organic deposit resting on mineral soil. It results from an accumulation of litter that is slowly mineralized and remains unmixed with mineral soil. Thus a sharp distinction or break exists between the *O* and *A* horizons.

Slow though mineralization may be, it is the manner in which the process proceeds that distinguishes mor from other humus types. The main decomposing agents are fungi, both free-living and mycorrhizal, which tend to depress soil animal activity and produce acids; nitrifying bacteria may be absent. The vascular cells of leaves disappear first, leaving behind a residue of mesophyll tissue. Proteins within the leaf litter are stabilized by protein-precipitating material, making them, in some cases, resistant to decomposition. Because of limited volume, pore space, acidity, and type of litter involved and the

Figure 9.5 The sequence of humus types and related processes. Note the inverse relationship between bacteria and fungi as the humus sequence goes from mor to mull, as well as the pronounced changes in invertebrate life. (From Wallwork, 1973: 53.)

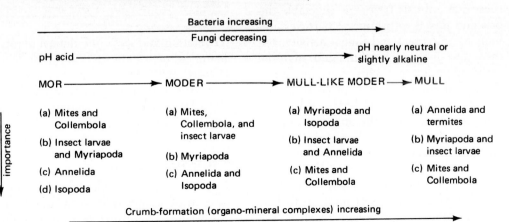

nature of its breakdown, mor is inhabited by a small biomass of soil animals. They have little mechanical influence on the soil. Instead, they live in an environment of organic material cut off from mineral soil beneath.

Mull. Mull, on the other hand, results from a different process. Characteristic of mixed and deciduous woods on fresh and moist soils with a reasonable supply of calcium, mull possesses only a thin scattering of litter on the surface, and the mineral soil is high in organic matter. All organic materials convert to true humic substances. Because of animal activity, they are inseparably bound to the mineral fraction, which absorbs them like a dye. There is no sharp break between the *O* and *A* horizons.

Because of less acidity and a more equitable base status, bacteria tend to replace fungi as the chief decomposers, and nitrification is rapid. Soil animals are more diverse and possess a greater biomass, reflecting a more equitable distribution of living space, oxygen, food, and moisture, and a smaller fungal component. This faunal diversity is one of mull's greatest assets, because the humification process flows through a wide variety of organisms with differing metabolisms. Not only do these soil animals fragment vegetable debris and mix it with mineral particles, enhancing microbial and fungal activity, but they also incorporate humified material with mineral soil. This constant interchange of material takes place from the surface to soil and back again. Plants extract nutrients from the soil and deposit them on the surface. Then the soil flora and fauna reverse the process.

Moder. On the continuum from mull to mor lies *moder*, the insect mull of Muller. In this humus type, plant residues are transformed into the droppings of small arthropods, particularly Collembola and mites. Residues not consumed by the fauna are reduced to small fragments, little humified and still showing cell structure. The droppings, plant fragments, and mineral particles all form a loose, netlike structure held together by chains of small droppings. In acid mor the shape of the droppings is destroyed by the washing action of rainwater. Under more extreme conditions humus leached from the droppings acts as a binding substance to form a dense, matted litter approaching a mor. On the continuum between moder and mull, the droppings of large arthropods, which are capable of taking in considerable quantities of mineral matter with food, are common. However, moder differs from mull in its higher organic content, restricted nitrification, and a more or less mechanical mixture of the organic components with the mineral, the two being held together by humic substances, yet separable. In other words, the organic crumbs are deficient in mineral matter, in contrast to mull, in which mineral and organic parts are inseparably bound together.

■ Soil Development

Vegetation and its prime determinant, climate, influence the development of soils (Figure 9.6). Four major processes involved in soil development are podzolization, laterization, calcification, and gleyization.

Podzolization

In general, podzolization involves the migration of sesquioxides from the *A* to the *B* horizon. The insoluble iron and alumi-

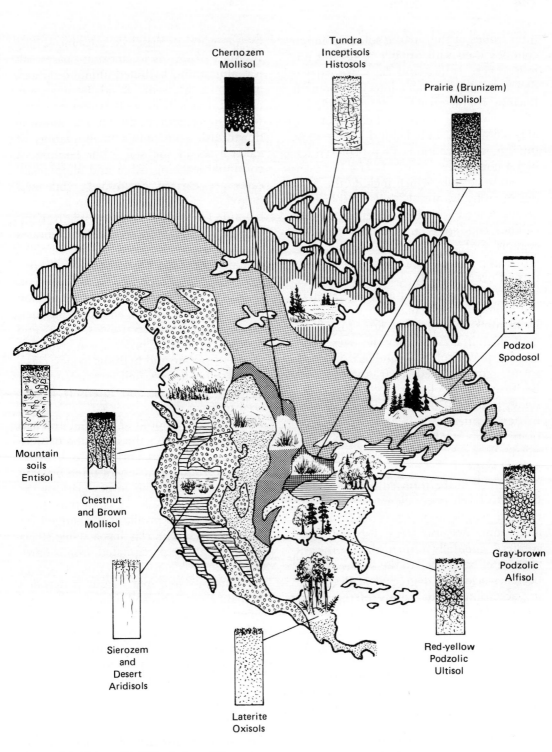

Figure 9.6 Major soils of North America.

num oxides of the surface soil combine in complex ways with organic acids and humus. These oxides and clays are carried downward by rainwater into the spodic *B* horizon, characterized by an accumulation of active organic matter and oxides of iron and aluminum. Left behind under most intense podzolization is a light-colored (albic) *A* horizon.

Podzolic soils, which include the Spodosols, Alfisols, and Ultisols (Table 9.2), are associated with humid temperate deciduous coniferous and deciduous forest regions. Only a part of the organic matter—leaves, trees, and some trunks—is turned over annually. Leaves, the largest source of organic matter and vegetation of the ground layer, remain on the surface. Dead roots add relatively little to soil organic matter because they die over an irregular period of time and are not uniformly concentrated near the surface. Because only the leaves are returned annually to the soil and much of the mineral matter is tied up in the trunk and branches, most of the currently available nutrients turned back to soil come from annual leaf fall.

Rainfall in the forested regions is sufficient to leach many elements, especially calcium, magnesium, potassium, iron, and aluminum. Because trees, especially the conifers, generally return an insufficient amount of bases back to the surface soil, it becomes acid, the degree of acidity varying according to the nature of the forest and its site.

Laterization

In humid subtropical and tropical forested regions of the world, where rainfall is heavy and temperatures high, the soil development processes are much more intense. Because temperatures are uniformly high, weathering in these regions is almost entirely chemical, brought about by water and its dissolved substances. The residues from this weathering—bases, silica, aluminum, hydrated aluminosilicates, and iron oxides—are freed. Because precipitation usually exceeds evaporation, the water movement is almost continuously downward. With only a small quantity of electrolytes in the soil water because of continual leaching, silica and aluminosilicates are carried downward, while sesquioxides of aluminum and iron remain behind. The sesquioxides are relatively insoluble in pure rainwater, but the silicates tend to precipitate as a gel in solutions containing humic substances and electrolytes. If humic substances are present, they act as protective colloids about iron and aluminum oxides and prevent their precipitation by electrolytes. The end product of such a process is a soil composed of silicate and hydrous oxides, clays, and residual quartz, deficient in bases, low in plant nutrients, and intensely weathered to great depths.

The large amount of residual iron and aluminum left after the depletion of silica and bases becomes enriched as hydrous oxides, forming a variety of often brilliant reddish colors in the upper part of the soil, which generally lacks distinct horizons. Below, the profile is unchanged for many meters. The clay has a stable structure, and unless precipitated, iron is hardened into a cemented laterite. It is very pervious to water and is easily penetrated by plant roots. This soil-forming process is called *laterization*. True lateritic soils are called *oxisols*, but the process of laterization is involved in the development of Ultisols, soils formerly classified as Red-Yellow Podzolic and Red Earths.

Calcification

The subhumid-to-arid and temperate-to-tropical regions of the world—the plains and prairies of North America, the

TABLE 9.2 New Soil Orders

Order	Derivation and Meaning	Description	Approximate Equivalents
Entisol	Coined from *recent*	Dominance of mineral soil materials; absence of distinct horizons; found on floodplains	Alluvial soils, azonal soils, regosol, lithosol
Vertisol	L. *verto,* "inverted"	Dark clay soils that exhibit wide, deep cracks when dry	Grumusols
Inceptisol	L. *inceptum,* "beginning"	Texture finer than loamy sand; little translocation of clay; often shallow; moderate development of horizons	Brown forest soil, sol brun acide, acide, humic gley, weak podzols
Aridisol	L. *aridus,* "arid"	Dry for extended periods; low in humus, high in base content; may have carbonate, gypsum, and clay horizons	Sierozems, red desert soils, solonchak
Mollisol	L. *mollis,* "soft"	Surface horizons dark brown to black with soft consistency; rich in bases; soils of semihumid regions	Chestnut, chernozem, prairie; some brown and brown forest and associated humic gleys
Spodosol	Gr. *spodos,* "ashy"	Light gray, whitish A_2 horizon on top of a black and reddish B horizon high in extractable iron and aluminum	Podzol, brown podzolic soils
Alfisol	Coined from *Al* and *Fe*	Shallow penetration of humus, translocation of clay; well-developed horizons	Gray-brown podzolic, gray wooded soils, noncalcic brown soils, some planisols
Ultisol	L. *ultimus,* "last"	Intensely leached; strong clay translocation, low base content: humid, warm climate	Red-yellow podzolic, red-brown laterite, some latisols
Oxisol	Fr. *oxide,* "oxidized"	Highly weathered soils; red, yellow, or gray; rich in kalolinite, iron oxides, and often humus; in tropics and subtropics	Laterites, latosols
Histosol	Gr. *histos,* "organic"	High content of organic matter	Bog soils, muck

steppes of Russia, the veldts and savannas of Africa, and the pampas of South America—support grassland vegetation. Dense root systems may extend many feet below the surface. Each year nearly all of the vegetative material above ground and a part of the root system are turned back to the soil as organic matter. Although the material decomposes rapidly the following spring, it is not completely gone before the next cycle of death and decay begins. Soil inhabitants mix the humus with mineral soil, developing a soil high in organic matter.

Because the amount of rainfall in grassland regions generally is insufficient to remove calcium and magnesium carbonates, they are carried down only to the average depth that percolating waters reach. Grass maintains a high calcium content in the surface soil by absorbing large quantities from the lower horizons and redepositing them on the surface. Little clay is lost from the surface. This process of soil development is called *calcification*. The soils so formed are called *mollisols*.

Soils developed by calcification have a distinct A horizon of great thickness and an indistinct B horizon, characterized by an accumulation of calcium carbonate. The A horizon is high in organic matter and in nitrogen, even in tropical and subtropical regions.

Arid and semiarid regions have relatively sparse vegetation. Because plant growth is limited, little organic matter and nitrogen accumulate in the soil. Scant precipitation results in slightly weathered and slightly leached soils high in plant nutrients. The horizons are usually faint and thin. Within these regions are areas where soils contain excessive amounts of soluble salts, either from the parent material or from the evaporation of water draining in from adjoining land. Infrequent rains penetrate the soil, but soon afterward surface evaporation draws the salt-laden water upward. The water evaporates, leaving saline and alkaline salts at or near the surface to form a crust or caliche. Soils of arid and semiarid regions are *aridosols,* soils low in humus, high in base content, and often having gypsum, clay, or carbonate horizons.

Gleyization

Calcification, laterization, and podzolization are processes that take place on well-drained soil. Under poorer drainage conditions a different soil development process is at work. Slope of the land determines to a considerable extent the amount of rainfall that will enter and pass through the soil, the concentration of erosion materials, the amount of soil moisture, and the height at which water will stand in the soil. The amount of water that passes through or remains in the soil determines the degree of oxidation and breakdown of soil minerals. Iron in soils where water stays near or at the surface of the ground most of the time is reduced to ferrous compounds. They give a dull gray or bluish color to the horizons. This process, called *gleyization,* may result in compact, structureless horizons. Gley soils are high in organic matter because more organic matter is produced by vegetation than can be broken down by humification, which is greatly reduced by an absence of soil microorganisms. On gentle to moderate slopes, where drainage conditions are improved, gleyization is reduced and takes place deeper in the profile. As a result the subsoil shows varying degrees of mottling of grays and browns. On hilltops, ridges, and steep slopes, where the water table is deep and the soil well drained, the subsoil is reddish to yellowish-brown from oxidized iron compounds.

In all, five drainage classes are recognized (Figure 9.7). (1) Well-drained soils are those in which plant roots can grow to a depth of 90 cm without restriction due to excess water. (2) On moderately well-drained soils plant roots can grow to a depth of 50 cm without restriction. (3) Somewhat poorly drained soils restrict the growth of plant roots beyond a depth of 36 cm. (4) Poorly drained soils are wet most of the time and are usually charac-terized by the growth of alders, willows, and sedges. (5) On very poorly drained soils water stands on or near the surface most of the year.

Toposequences

The major soil development processes define broad soil patterns, but local soil patterns are controlled by topography. As the topography of an area develops over time through the action of physical and chemical weathering, a group of soils develops along with it. This group of soils is known as a *toposequence*. If the parent material is the same throughout all or part of the toposequence, the group of related soils is called a *catena*, from the Latin meaning "chain." In Figure 9.8, the group of soils from A through G represent a toposequence ranging from well-drained ridge-top to very poorly drained soils and allu-

Figure 9.7 Effect of drainage on the development of an alfisol. Wetness increases from left to right. The diagram represents the topographic position the profiles might occupy. Note that the strongest soil development takes place on well-drained sites where weathering is maximum. The least amount of weathering takes place on the very poorly drained soils, where the wet season water table lies above the surface of the soil. G or g indicates mottling; t indicates translocated silicate clays. (Adapted in part from Knox 1952.)

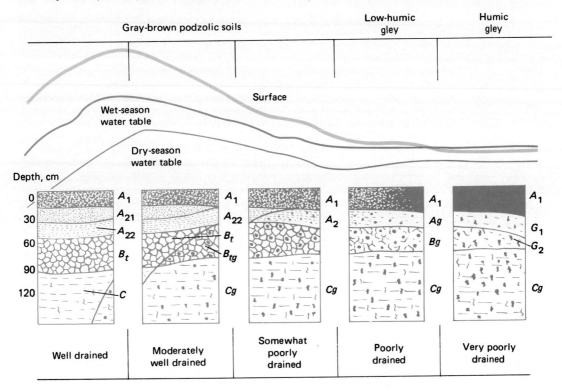

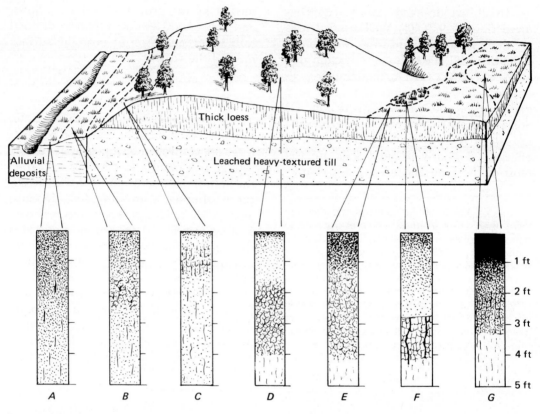

Figure 9.8 Topography and vegetation acting together produce a toposequence of soil types. This diagram shows the normal sequence of eight soil types from the Mississippi to the uplands in Illinois. It also illustrates how bodies of soil types fit together in the landscape. Boundaries between adjacent bodies are gradations or continuums, rather than sharp lines.

The lower part of the diagram pictures the profiles of seven of the soils, showing the color and thickness of the surface horizon and the structure of the subsoil. Note how the natural vegetation that once covered the land (trees for forested areas, grass clumps for grass) influenced surface color. The diagram also shows how topographic position and distance from the bluff influence subsoil development.

Profile A (Sawmill) is an entisol, a bottom-land soil formed from recent sediments and not subject to much weathering. Profile B (Worthen) at the foot of the slope is also an entisol. Developed from recent alluvial material, it shows little structure. Profile C (Hooper) on the slope developed from a thick loess on top of leached till, whereas the soil on the bottom of the slope de-veloped directly from the till. Profile D (Seaton) is an upland soil formerly covered with timber. It possesses a light surface color and lacks structure, the result of rapid deposition of loess during early soil formation, holding soil weathering to a mini-mum. These two soils probably should be consid-ered entisols because they lack well-formed hori-zons, which are the product of strong soil weathering. Profile E (Joy) represents upland soil developed under grass. Note the dark surface and the lack of structure, again the result of a rapid deposition of loess. Profile F (Edgington) is a depressional wet spot. Extra water flowing in from adjacent fields increased the rate of weath-ering, resulting in a light grayish surface and subsurface and a blocky structure to the subsoil, which indicates strongly developed gley soil. The depth of subsoil suggests that consider-able sediment has been washed in from the surrounding area. Profile G (Sable) represents a depressional upland prairie soil. The deep, dark surface and the coarse blocky structure mark it as a mollisol. Abundant grass growth produced the dark color. (After Veale and Wascher 1956.)

vial deposits. However, only soils C through G, those soils derived from the same parent material (in this case, loess), make up a catena.

Soils and Time

The weathering of rock material, the accumulation, decomposition, and mineralization of organic material, the loss of minerals from the upper surface and gains in minerals and clay in the lower horizons, and horizon differentiation, all require considerable time. Well-developed soils in equilibrium with weathering, erosion, and biotic influences may require 2000 to 20,000 years for their formation; but soil differentiation from parent material may take place in as short a time as 30 years. Certain acid soils in humid regions develop in 2100 years because the leaching process is speeded by acidic materials. Parent materials heavy in texture require a much longer time to develop into "climax" soils, because of an impeded downward flow of water. Soils develop more slowly in dry regions than in humid ones. Soils on steep slopes often remain young regardless of geological age, because rapid erosion removes soil nearly as fast as it is formed. Floodplain soils age little through time, because of the continuous accumulation of new materials. Young soils are not as deeply weathered as and are more fertile than old soils, because they have not been exposed to the leaching process as long. The latter tend to be infertile because of longtime leaching of nutrients without replacement from fresh material.

Groups of soils of different ages derived from the same parent materials but subject to soil-forming processes over different lengths of time make up *chronosequences*. Best examples of chronosequences occur in areas of retreating glaciers, volcanic flows, and sand dune systems.

■ Soil Classification

Each combination of climate, vegetation, soil material, slope, and time results in a unique soil, the smallest repetitive unit of which is a *pedon*. Soils may vary considerably even within a small area (Figure 9.9). Changes in slope, drainage, and soil material account for local differences between individual soils. These individuals are roughly equivalent to the lowest category in the soil classification system—the soil series.

The present soil taxonomic system consists of orders (see Table 9.2), suborders, great groups, families, and series. In the new classification system for the United States (USDA 1982) the names of the lower classification units always end on the formative syllable of their respective order, preceded by other syllables connotative of various soil properties. For example, a suborder of Mollisol is Aquoll, with *aqu* meaning water. There are other classification systems, including the FAO-UNESCO system (see Landon 1984, *Booker Tropical Soil Manual*), which is somewhat more practical and easier to use in the field.

At the highest taxonomic level in any system, emphasis is placed on the presence or absence of certain diagnostic soil horizons that result from the interaction of soil-forming factors, primarily climate and vegetation. In the U.S. system, soil series are named after the locality in which they were first described. For example, the Ovid Series in New York was named after the town of Ovid and the Miami after the Miami River in western Ohio. Like species among plants and animals, soil series are defined in terms of the largest number of differentiating characteristics and occur in fairly limited areas. Higher categories of taxonomy combine series

Figure 9.9 A local soil map, indicating soil associations and soil series. The symbol H indicates the Hagerstown series and C the Chilhowie series. These soils are deep, well-drained, and on fairly level land. The symbols indicate the characteristics of the soils. For example, HbB is Hagerstown silt loam, 2 to 6 percent slope; HgB is very rocky silt loam, 2 to 6 percent slope; HgC is very rocky silt loam, 6 to 12 percent slope; and CdB is Chilhowie silty clay, 2 to 6 percent slope. (Courtesy of Soil Conservation Service.)

into larger groupings distinguished by even fewer differentiating properties. At the highest level of classification, classes correspond roughly with broad climatic zones.

Every soil series has its neighboring soil series with unlike properties, into which it grades abruptly or gradually. If these several soils found side by side have developed from the same soil material but differ mainly in natural drainage and slope, they form a catena (see Toposequences). When soils are mapped, unlike soils may be grouped together into associations for reasons of scale or practical use. When occurring in inseparable patterns, soils are mapped as complexes. In detailed mapping, a soil series can be subdivided into types, phases, and variants.

Summary

Soils are the base for terrestrial ecosystems. Soil is the site of decomposition of organic matter and of the return of mineral elements to the nutrient cycle. It is the habitat of animal life, the anchoring medium for plants, and their source of water and nutrients. Soil begins with the weathering of rock and minerals, which involves the leaching out and carrying away of mineral matter. Its development is guided by climate, vegetation, original material, and topography. Plants rooted in weathering material further break down the substratum, pump up nutrients from its depths, and add all-important organic material. Through decomposition and mineralization this material is converted into humus, an unstable product that is continuously being formed and destroyed by mineralization.

As a result of the weathering process, accumulation and breakdown of organic matter, and the leaching of mineral matter, horizons, or layers, are formed in the soil. There are four major horizons: the *O,* or organic, layer; the *A* horizon, characterized by accumulation of organic matter and loss of clay and mineral matter; the *B* horizon, in which mineral matter accumulates; and *C,* the underlying material. These horizons may divide into subhorizons.

Of all the horizons, none is more important than the humus layer, which plays a dominant role in the life and distribution of plants and animals, in the maintenance of soil fertility, and in much of the soil-forming process. Humus is usually grouped into three types: mor, characteristic of acid habitats, whose chief decomposing agents are fungi; mull, characteristic of deciduous and mixed woodlands, whose chief decomposing agents are bacteria; and finally, moder, which is highly modified by the action of soil animals.

Profile development is influenced over large areas by vegetation and climate. In grassland regions the chief soil-forming process is calcification, in which calcium accumulates at the average depth reached by percolating water. In forest regions, podzolization—involving the leaching of calcium, magnesium, iron, and aluminum from the upper horizon, and the retention of silica—takes place. In tropical regions laterization, in which silica is leached and iron and aluminum oxides are retained in the upper horizon, is the major soil-forming process. Gleyization takes place in poorly drained soils. Organic matter decomposes slowly, and iron is reduced to the ferrous state.

Differences between soils and between horizons within soils are reflected by variations in texture, structure, and color. Texture of the soil is determined by the proportion of soil particles of different sizes, broadly sand, silt, and clay. It is important in the movement and retention of water in the soil. Soil particles, particularly the clay-humus complex, are the key to nutrient availability and the cation exchange capacity of the soil—the number of negatively charged sites on soil particles that can attract positively charged ions.

Each combination of climate, vegetation, soil material, topography, and time results in a unique soil, of which the smallest repetitive unit is the pedon. Soil individuals, the lowest category in the soil classification system, are the soil series. They may be further categorized into families, great groups, suborders, and orders.

The Aquatic Environment

■ Properties of Water

Structure

Because of the physical arrangement of its hydrogen atoms and hydrogen bonds, liquid water consists of branching chains of oxygen tetrahedra. The physical state of water, whether liquid, gas, or solid, depends upon the kinetic energy compared to the attractive energy of hydrogen bonds. Kinetic energy depends upon temperature alone, so the higher the temperature, the more H bonds are broken. Now, the kinetic energy is reflected in the speed of molecules, but this speed does not depend upon the rate at which bonds are being formed. In water in a liquid state hydrogen bonds are being broken as fast as they form. At low temperatures the tetrahedral arrangement is almost perfect; but when water freezes, the arrangement becomes a perfect lattice. This open, less dense structure is the reason why ice floats. As the temperature of frozen water increases, the molecular arrangement becomes looser and more diffuse. Hydrogen bonds continuously break and reform, resulting in random packing and contracting the distance between molecules. As the temperature becomes even higher, the more freely the aggregrates of H_2O molecules, which are $(H_2O)_x$, move around. When ice melts completely, the volume of water contracts, and its density increases up to a temperature of $3.98°$ C. Beyond this point the loose arrangement of molecules results in a reduction in density again. The fact that H_2O attains maximum density at approximately 4° C is of fundamental importance to aquatic life.

Seawater behaves somewhat differently. Seawater is defined as water with a minimum salinity of 27.7 o/oo (o/oo = parts per thousand). Its density, or rather its specific gravity relative to that of an equal volume of pure water (sp. gra. = 1) at atmospheric pressure, is correlated with salinity. At 0° C the density of seawater with a salinity of 35 o/oo is 1.028. The lower its temperature, the greater becomes the density of seawater; the higher its temperature, the lower becomes its density. No definite freezing point exists for seawater. Ice crystals begin to form at a temperature that varies with salinity. As pure water freezes out, the remaining unfrozen water is even saltier and has an even lower freezing point. If the temperature decreases enough, a solid block of ice crystals and salt is ultimately formed.

Physical Properties

Water is capable of storing tremendous quantities of heat with a relatively small rise in temperature. It is exceeded in this capacity only by liquid ammonia, liquid hydrogen, and liquid lithium. Therefore water is described as having a high *specific heat,* the number of calories necessary to raise 1 g of a substance 1° C. The specific heat of water is given a value of 1.

Not only does water have a high specific heat; it also possesses the highest heat of fusion and heat of evaporation of all known substances that are liquid at ordinary temperatures. Large quantities of heat must be removed before water can change from a liquid to its solid form, ice; and conversely, ice must absorb considerable heat before it can be converted to a liquid. It takes approximately 80 calories

of heat to convert 1 g of ice at 0° C to a liquid state at 0° C. This amount of heat would raise the same quantity of water from 0° to 80° C.

Evaporation occurs at the interface between air and water at all ranges of temperature. Here again, considerable amounts of heat are involved; 536 cal are needed to overcome the attraction between molecules and convert 1 g of water at 100° C into vapor. This amount of heat would raise 536 g of water 1° C. When evaporation occurs, the source of thermal energy may come from the sun, from the water itself, or from objects in or around it. Heat involved at the point of evaporation is returned to actual heat at the point of condensation (the conversion from vapor to liquid). Such phenomena play a major role in meteorological cycles.

The property of *viscosity* also has biological importance. The viscosity of water is high because water molecules interact with neighboring molecules by forming hydrogen bonds. Viscosity can be visualized best if we imagine or observe a liquid flowing through a clear glass tube. The liquid moving through the tube behaves as if it consisted of parallel concentric layers flowing over one another. The rate of flow is greatest at the center; because of friction between layers, the flow decreases toward the sides of the tubes. This type of resistance between layers is called *lateral* or *laminar viscosity.* This phenomenon can be observed along the side of any stream or river with uniform banks. The water at streamside is nearly still, whereas the current in the center may be swift.

Viscosity of flowing water is complicated by another type of resistance, *eddy viscosity,* in which the water masses pass from one layer to another, creating turbulence both horizontally and vertically. Biologically important, eddy viscosity is many times greater than laminar viscosity.

Viscosity is the source of frictional resistance to objects moving through the water. Because this resistance is 100 times that of air, animals must expend considerable muscular energy to move through the water. This resistance has been an important selection pressure in the evolution of a streamlined fusiform body shape in fish and such marine mammals as whales and porpoises, enabling them to move with the greatest economy in their watery environment.

Another property of water significant for biological processes is surface tension. Within all substances particles of the same matter are attracted to one another. Molecules of water below the surface are symmetrically surrounded by other molecules, so the forces of attraction are the same on all sides. At the water's surface, the molecules exist under a different set of conditions. Below is a hemisphere of strongly attractive similar water molecules; above is the much smaller attractive force of the air. The molecules on the surface, then, are drawn into the liquid; the liquid surface tends to contract and become taut. This property is *surface tension.*

In aquatic ecosystems surface tension is a barrier to some organisms and a support for others. It is the force that draws liquids through the pores of the soil and conducting networks of plants. Aquatic insects and plants have evolved structural adaptations that prevent the penetration of water into the tracheal systems of the former and the stomata and internal air spaces of the latter.

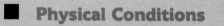

■ Physical Conditions

Temperature

The thermal properties of water give the aquatic environment a more stable temperature regime than the terrestrial environment. It is not subject to sharp daily fluctuation in temperature, and seasonal changes come on gradually. Slow warming and cooling of surface waters—bodies of water gain heat by solar radiation and lose heat by evaporation, radiation, and melting of ice only at the top—results in seasonal temperature stratification and overturn in many lakes, in ponds, and to a certain degree in the seas.

As the ice melts in early spring, the surface water is heated by the sun. When it reaches 4° C and achieves its greatest density, a slight temporary stratification develops, which sets up convection currents. Aided by the strong winds of spring, these currents mix the water throughout the basin until the water is uniformly 4° C. At a uniform temperature even the slightest winds can cause a complete circulation of the water between the surface and the bottom. During this spring overturn, the nutrients on the bottom, the oxygen on the top, and the plankton within are mixed.

With the coming of summer the sun's intensity increases, and the temperature of the surface water rises. The higher the temperature of the surface water, the greater is the difference in the density between the surface and deeper layers. The thermal density gradient opposes the energy of the wind, and it becomes more difficult for the waters to mix. As a result, a mixing barrier is established. The freely circulating surface water, with a small but variable temperature gradient, is the *epilimnion.* Below is the barrier, the *metalimnion,* a zone characterized by a steep and rapid decline in temperature. Within the metalimnion is the *thermocline,* the plane at which the temperature drops most rapidly—1° C for each meter of depth. Below these two layers is the *hypolimnion,* a deep, cold layer, in which the temperature drop is gentle (Figure 10.1).

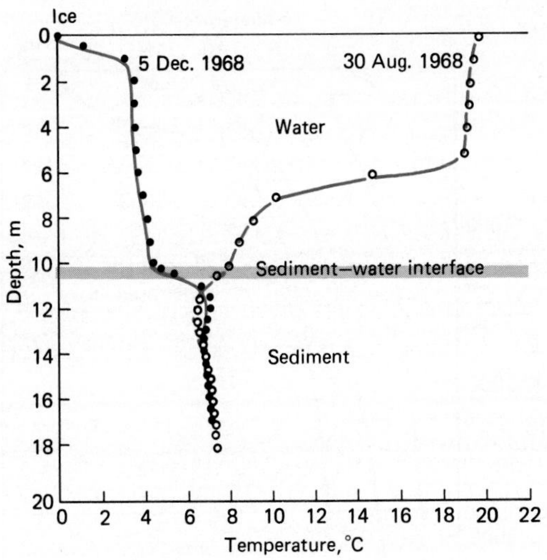

Figure 10.1 Temperature stratification during summer (August) and winter (December) in MIrror Lake, New Hampshire. Note the relatively uniform temperature in December through the water column below 2 m and the increase in temperature in the bottom sediments. In August the water is uniformly warm to the depth of 6 m (the epilimnion). The sharp decline in temperature between 6 and 8 m represents the thermocline. Note that the coolest and densest bottom water is 8° rather than 4° C. The bottom sediment, except for a narrow surface layer, maintains a fairly constant temperature. (From Likens 1985: 116.)

With the coming of autumn, the air temperature falls. The surface water loses heat to the atmosphere through evaporation, convection, and conduction. The temperature of the surface water drops and the thermocline sinks; the epilimnion increases until it includes the entire lake. The temperature once again is uniform from top to bottom, the lake waters circulate, and oxygen and nutrients are recharged throughout the lake. This fall overturn, through stirring caused by the slightest wind, may last until ice forms.

As the surface water cools below 4° C, it becomes lighter and remains on the surface. If the climate is cold enough, it freezes; otherwise, it remains close to 0° C. A slight inverse temperature stratification may develop, in which the water becomes warmer up to 4° C with depth. The water immediately beneath the ice may be warmed by solar radiation through the ice. Because this warming increases its density, this water subsequently flows to the bottom, where it mixes with water warmed by heat conducted from bottom mud. The result is a higher temperature at the bottom, although the overall stability of the water is undisturbed. As the ice melts in spring, the surface water again becomes warm, currents pass unhindered, and the spring overturn takes place.

This general picture of seasonal changes in temperature stratification must not be considered a uniform condition in all bodies of water. In shallow lakes and ponds temporary stratification of short duration may occur; in others stratification may exist, but without a thermocline. In some very deep lakes the thermocline simply descends during periods of overturn, but does not disappear. In such lakes the bottom water never becomes mixed with the top layers.

Temperature regimes in the seas are somewhat different, reflecting the greater size, depth, and currents. Temperatures range from −3° C in arctic waters to 27° C in tropical waters; and currents may be warmer or colder than the waters through which they flow. Daily and seasonal temperature changes are greater in coastal waters than in open seas. In general, seawater is never more than 2° to 3° C below the freezing point of fresh water or higher than 27° C. At any given place the temperature of deep water is almost constant and cold, below the freezing point of fresh water. Seawater has no definite freezing point, although there is a temperature for seawater of any given salinity at which ice crystals form.

Unlike fresh water, seawater (with a salinity of 27.7 o/oo or higher) becomes heavier as it cools and does not reach its greatest density at 4° C. The limitation of 4° C as the temperature of bottom water does not apply to the sea. Nevertheless, bottom temperatures of the sea rarely go below the freezing point of fresh water, generally averaging around 2° C, even in the tropics if the water is deep enough. The temperature of the ocean floor over 1 mile deep is 3° C.

The upper layers of the ocean exhibit a stratification of temperature superimposed on the deep layers (Figure 10.2). Depths below 300 m are usually thermally stable. At high and low latitudes temperatures remain fairly constant through the year. At middle latitudes temperatures vary with the season. In summer the surface waters become warmer and lighter, forming a temporary seasonal thermocline. In subtropical regions the surface waters are constantly heated, developing a marked permanent thermocline.

The temperature of flowing waters of streams and rivers is not constant. In general, small, shallow streams tend to follow, but lag behind, air temperatures, warming and cooling with the seasons but never falling below freezing in winter. Streams with large areas exposed to direct sunlight are warmer than those shaded by trees, shrubs, and high steep banks (Figure 10.3). This fact is ecologically important because temperature affects the composition of flowing water communities.

Light

The quality of light in the aquatic environment is not the same as that in the terrestrial environment. Some of the differences have already been discussed in Chapter 7. Light is rapidly attenuated as it passes through water. Limnologists use

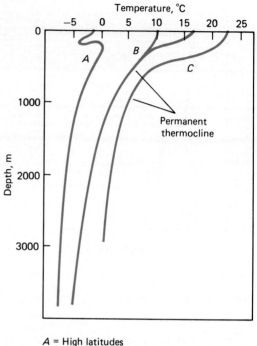

A = High latitudes
B = Middle latitudes
C = Low latitudes

Figure 10.2 A generalized graph of temperature stratification in the oceans at high, middle, and low latitudes. At high latitudes, colder, less saline surface waters overlie warmer, more saline waters. At middle latitudes, the ocean exhibits seasonally varying surface temperatures and a seasonal thermocline. Waters of the lower latitudes possess stable surface temperatures and a permanent thermocline. (After R. V. Tait 1968.)

different methods, physical and mathematical, to measure the diminution or vertical absorption of light, which they describe by different terms, such as *coefficient of extinction*, *total coefficient of absorption*, and *coefficient of attenuation*. What is important ecologically is not so much the amount of light absorbed but the amount and quality of light transmitted. That will vary from one body to another, because light will be

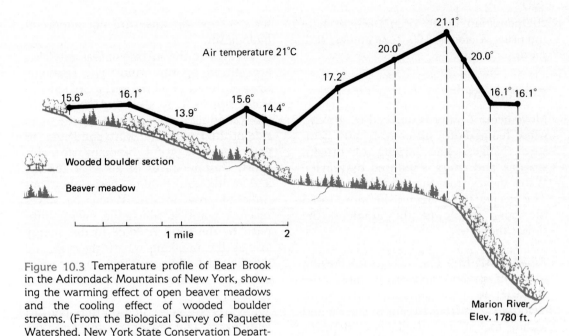

Figure 10.3 Temperature profile of Bear Brook in the Adirondack Mountains of New York, showing the warming effect of open beaver meadows and the cooling effect of wooded boulder streams. (From the Biological Survey of Raquette Watershed, New York State Conservation Department 1934.)

altered by dissolved and suspended material in the water and by phytoplankton.

For example, in the clear waters of the Gulf Stream, most light below 15 m is confined to the blue-green wavelengths, which gives the deep blue color to the water. In inland waters, with their higher concentrations of yellow substances, blue light is removed completely at very shallow depth, which is the reason why such waters rarely appear blue. In turbid waters with high levels of humic material both in solution and in the particulate state, green light is rapidly attenuated and may appear yellowish. Waters with high humic concentration appear dark, and nutrient-poor lakes and ponds, low in humic material and phytoplankton, are clear and bluish, with much of the color coming from molecular scattering of light. Highly productive waters supporting a dense growth of phytoplankton take on a decidedly greenish appearance.

The color of water, then, depends upon the concentration of particulate matter in the water, which influences the intensity of light scattering. The concentration of particulate matter depends a great deal on the nature of the watershed, its climate, topography, soil type, vegetation, and land use. Land disturbance may increase soil erosion, increasing the amount of particulate and colloidal matter in the water.

The depth to which light can penetrate is limited by the turbidity of water and the absorption of light rays. On this basis standing bodies of water can be divided into two basic layers—the *trophogenic zone*, in which photosynthesis dominates, and the *tropholytic zone*, where decomposition is most active. The boundary between the two zones is the *compensation depth*, at which photosynthesis balances respiration and beyond which light penetration is so low that it is no longer effective. Generally compensation depth occurs where light intensity is about 100 footcandles (Edmondson 1956).

Movement

Movement of water is induced by gravity, wind, temperature differences, and pressure changes. The flowing of brooks, streams, and rivers is induced by gravity. Water moves from higher to lower levels. The steeper the gradient, the more rapid the flow; the lower the gradient, the slower the flow.

Flowing Water. The velocity of flowing water of streams and rivers molds the character of their aquatic environment. Velocity varies from stream to stream and within the stream itself, and depends upon the size, shape, and steepness of the stream channel, the roughness of the bottom, the depth, and rainfall.

The velocity of flow also influences the degree of silt deposition and the nature of the bottom. The current in shallow, fast-flowing sections does not allow deposition of silt, although coarser particles will drop out in smooth or quiet sections of the stream. When streams and rivers slow their flow, the load of sediment is deposited on the bottom or along the edges of the channel where the current is slowest. High water increases the velocity; it moves bottom stones, scours the streambed, and cuts new banks and channels. In very steep streambeds, the current may remove all but the very large rocks and leave a boulder-strewn stream.

Waves. The most conspicuous water movements in ponds, lakes, and seas are traveling surface waves, caused by pressure of the wind on their surfaces (Figure 10.4). Except for the effect they have on shoreline and shore organisms (see Chap-

ter 33), surface waves are not important biologically.

Waves on the surface of seas and lakes are stirred by the wind. The frictional drag of the wind on the surface of smooth water ripples it. As the wind continues to blow, it applies more pressure to the steep side of the ripple, and wave size begins to grow. As the wind becomes stronger, short, choppy waves of all sizes appear; and as they absorb more energy, they continue to grow in size. When the waves reach a point at which the energy supplied by the wind is equal to the energy lost by the breaking waves, they become the familiar whitecaps. Up to a certain point, the stronger the wind, the higher the waves.

Waves are generated on the open sea and lakes. The stronger the wind, the longer the fetch (the distance the waves can run without obstruction under the drive of the wind blowing in a constant direction) and the higher the waves. As the waves travel out of the fetch, or if the winds die down, sharp-crested waves change into smooth long-crested waves or swells that can travel great distances because they lose little energy as they travel. Swells are characterized by troughs and ridges, the height of which is measured from the bottom of the trough to the crest. The length of the wave is measured from the crest to the following wave and its period by the time required for successive crests to pass a fixed point. None of these measures is static, for all depend upon the wind and the depth of the wave.

Figure 10.4 (a) The nature of waves and the change of their form as they enter shallow water. (b) The cork float in the water demonstrates that although the water in a wave seems to move, it remains in place while the wave form itself passes along. (From Nybakken 1982: 14.)

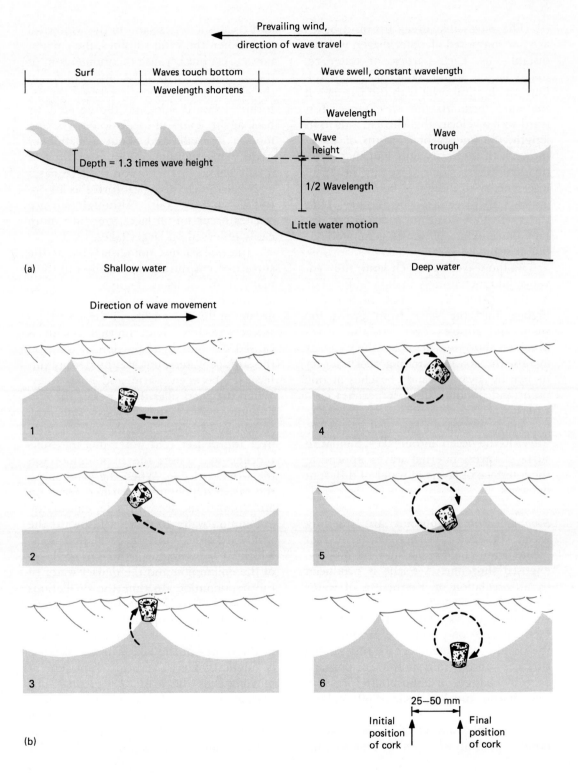

Prevailing wind, direction of wave travel

Surf

Waves touch bottom
Wavelength shortens

Wave swell, constant wavelength

Wavelength

Wave height

Wave trough

Depth = 1.3 times wave height

1/2 Wavelength

Little water motion

(a)

Shallow water

Deep water

Direction of wave movement

1

2

3

4

5

6

(b)

25–50 mm

Initial position of cork

Final position of cork

The waves that break up on a beach are not composed of water driven in from distant seas. Each particle of water remains largely in the same place and follows an elliptical orbit with the passage of the wave form. As the wave moves forward with a velocity that corresponds to its length, the energy of a group of waves moves with a velocity only half that of individual waves. The wave at the front loses energy to the waves behind and disappears, its place taken by another. Thus the swells that break on a beach are distant descendants of waves generated far out at sea.

As the waves approach land, they advance into increasingly shallow water. The height of the waves rises higher and higher until the wave front grows too steep and topples over. As the waves break on shore they dissipate their energy against the coast, pounding rocky shores or tearing away at sandy beaches at one point and building up new beaches elsewhere.

There are internal as well as surface waves in the ocean and in lakes. Similar to surface waves, internal waves appear at the interface of layers of water of different densities.

Seiches. More important are standing waves, or *seiches* (sāshes), a term that comes from the French and means dry, exposed shoreline. A seiche is produced by an oscillation of a structure of water about a point or node.

There are two kinds of seiches, surface and internal. Both are produced by the wind's movement across the water's surface, by heavy rain showers, or perhaps even by changes in atmospheric pressure (see Bryson and Ragotzkie 1960; Vallentyne 1957).

When the wind blows across a lake or pond, it piles up water on the leeward end

and creates a depression on the windward end. When the wind subsides, the current flows back, but because the momentum of the returning currents is not broken on the shore, a depression is created on the former leeward side, and the water flows back again. Thus an oscillation or rocking motion is established about a stationary node. This motion continues until it is finally halted by friction on the lake basin or by such meteorological forces as an opposite wind or rain. Although surface seiches occur on all lakes, they are more easily observed on larger lakes.

Internal seiches, not observable on the surface, occur during the summer in thermally stratified lakes (Figure 10.5). They are much more pronounced and exert a greater influence on life in the lake than surface seiches. The internal seiche is caused not only by the action of wind on the surface waters, but also by density differences between warm and cold water. When the wind piles the water of the epilimnion up on the leeward side, the weight and circulation of the lighter water over the denser cold water tilts the thermocline and raises the hypolimnion on the windward side. When the wind stops, the raised hypolimnion, pushed down on and toward the leeward side, causes the epilimnion water to flow back toward the windward side. Thus an oscillation is established between the lighter water layer of the epilimnion and the denser water of the hypolimnion. In time the oscillations are slowed by friction on the lake basin or by wind from an opposite direction.

These oscillations move about a point in the center of the lake where resistance is highest. Here vertical displacement movement is the least; the greatest displacement is on the windward and leeward ends. The position of an internal seiche can best be determined by charting the variations in the depth of a particular

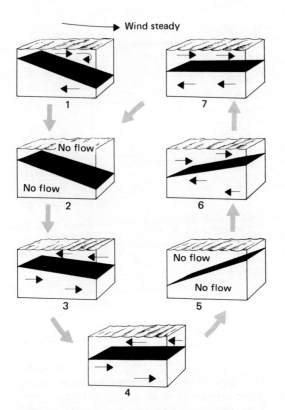

Figure 10.5 The thermocline slope and an internal oscillating wave or seiche as affected by winds across the surface.

temperature over a period of hours. This movement makes difficult the determination of the true position of the thermocline.

Internal seiches are important ecologically, because they distribute heat and nutrients vertically in the lake and transport them into shallow water. The seiche sets up two rhythmic but opposite current systems above and below the thermocline, whose speeds are greatest when the thermocline is level. A turbulence develops in the hypolimnion; without this turbulence the hypolimnion would become stagnant. In addition, plankton is moved up or down with the water mass. Even fish and other organisms are influenced by internal seiches.

Eddy Currents. The mixing of water and the transfer of nutrients are also carried on by *eddy currents* (Figure 10.6), small, turbulent currents whose energy is dissipated at right angles to the major currents. Because the major currents are largely horizontal and the turbulence is vertical, an interchange of adjacent water masses takes place. Oxygen may be transferred downward and heat upward; nutrients and even plankton are intermixed. The degree of intermixing depends upon the intensity of the turbulence, which changes gradually from one depth to another.

Langmuir Cells. Wind blowing across the surface may also generate circulating horizontal spirals of water spinning just below the surface. They are called *Langmuir cells* after the physical chemist I. Langmuir (1938) who studied them in Lake George, New York. These spirals of water, which may range from a few centimeters to many meters in diameter, are parallel to the water's surface and the direction of the wind (Figure 10.7). Each circulates in a direction opposite to the adjacent one. One circulates clockwise, the adjacent one

Figure 10.6 Vertical transfer of water by eddies across current boundaries. Generated by differential movements of currents, particularly along density layers, eddies exchange characteristics of adjacent water masses. (After Clarke 1954: 247.)

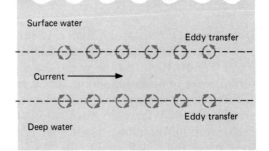

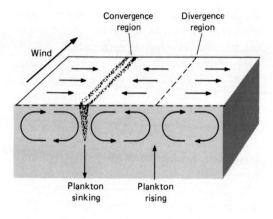

Figure 10.7 Wind-induced circulatory currents, Langmuir cells in a body of water. These cells are marked by accumulated materials on the surface where the adjacent cells downwell.

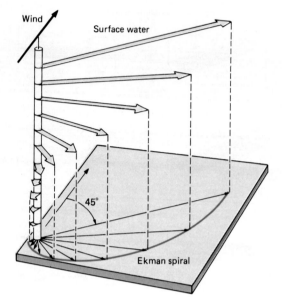

Figure 10.8 The Ekman spiral. The wind-driven surface current moves at an angle of 45° to the direction of the wind, to the right in the Northern Hemisphere, to the left in the Southern. Successively deeper water layers are deflected further than those immediately above them and move at slower speeds. Net water movement is at 90° to the wind. (Adapted from Strahler 1971: 258.)

counterclockwise, so that between one pair there are upwelling or divergent currents and between the next two, downwelling or convergent currents. The position of the downwelling currents is marked on the surface by streaks of accumulated materials, such as leaves and strands of submerged aquatic vegetation lifted to the surface. Where upwelling with its divergent currents occurs, no such streaks occur. These internal currents mix heat and nutrients in the surface water during the summer and circulate phytoplankton, accounting for its often patchy and changing distribution in lakes and ponds.

The Ekman Spiral. Also influencing subsurface currents in oceans is the Ekman spiral (Figure 10.8). As the wind sets the surface layer of water in motion, that layer in turn sets in motion a layer of water beneath, and that layer in turn another. Each layer moves more slowly than the layer above and is deflected to the right of it because of the Coriolis effect. At the base of the spiral the movement of water is counter to the flow on the surface, although the average flow of the spiral is at right angles to the wind.

In coastal regions the Ekman transport of surface water can bring deep waters up to the surface, a process called upwelling. Wind blowing parallel to a coast causes surface water to be blown offshore. This water is replaced by water moving upward from the deep. Although cold and containing less dissolved oxygen, upwelling water is full of nutrients that support a rich growth of phytoplankton. For this reason regions of upwelling are highly productive.

■ Chemical Conditions

Oxygen

Oxygen is rarely limiting in the terrestrial environment; but even at saturation levels, oxygen in the aquatic environment is relatively meager and problematic. Oxygen enters water by absorption from the atmosphere—molecular diffusion across the surface boundary layer—and by photosynthesis. The amount of oxygen and other gases water can hold depends upon temperature, pressure, and salinity. The solubility of a gas in water decreases as the temperature rises; for this reason cold water holds more oxygen than warm water. The solubility decreases as salinity increases and increases as pressure increases. The relative saturation of oxygen in water is a measure of the amount of gas present under existing conditions compared with the equilibrium content expected at the same temperature and partial pressure. (Partial pressure of a gas in a liquid is the amount of gas absorbed at a constant temperature by a given volume of liquid relative to the pressure in atmospheres that the gas exerts. Partial pressures and thus relative saturation change with altitude.) When oxygen in the water is in equilibrium with oxygen in the atmosphere, the pressure of oxygen in both media is the same, although the amounts of the gas are quite different. Difference in partial pressure between air and water is the driving force accounting for the flux of oxygen from air across the surface of the water.

Oxygen absorbed by surface water is mixed with deeper water by turbulence and internal currents. In shallow, rapidly flowing water and in wind-driven sprays, oxygen may reach and maintain saturation and even supersaturated levels, because of the increase of absorptive surfaces at the air-water interface. Water loses its oxygen through increased temperatures, which decrease holding capacity, through the respiration of plants and animals, and through aerobic decomposers. Further losses take place at the interface between bottom sediments and overlying water and during chemical oxidation in bottom sediments. Aerobic decomposition and plant respiration at night can reduce oxygen levels in water to zero.

During the summer oxygen, like temperature, may become stratified in lakes and ponds (Figure 10.9). In general, the

Figure 10.9 Oxygen stratification in Mirror Lake, New Hampshire, in winter (January), summer (August), and late fall (November), parallels that of temperature. During late fall overturn results in both uniform temperature and uniform distribution of oxygen throughout the lake basin. In summer there is a pronounced stratification of oxygen, which declines sharply in the thermocline and is nonexistent on the bottom, because of decomposition taking place in bottom sediments. In winter oxygen is also stratified, but it is present at low concentration in deep water. (From Likens 1985: 98.)

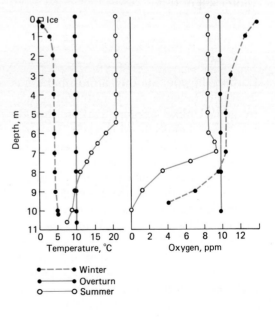

amount of oxygen is greatest near the surface, where there is an interchange between the water and the atmosphere and some stirring by the wind. The quantity decreases with depth, a decrease caused in part by the respiration of decomposer organisms working on organic matter sinking down from the surface. In some lakes oxygen varies little from top to bottom; every layer is saturated for its temperature and pressure. Water in some other lakes is so clear that light penetrates below the depth of the thermocline and permits the development of phytoplankton. Because of photosynthesis, the oxygen content may be greater in deep water than on the surface.

During spring and fall overturn, when water recirculates through the lake, oxygen is replenished in deep water and nutrients are returned to the top. In winter the reduction of oxygen in unfrozen water is slight, because bacterial decomposition is reduced by the cold and water at low temperatures holds the maximum amount of oxygen. Under ice, however, oxygen depletion may be serious, causing a heavy winter kill of fish.

Carbon Dioxide, Alkalinity, and pH

Carbon dioxide, another atmospheric gas in water, behaves much differently than oxygen. Unlike oxygen, carbon dioxide combines chemically with water, so it occurs both in free and bound states. Like oxygen, free carbon dioxide in fast-flowing water is in equilibrium with the atmosphere, according to the laws of the behavior of gases. Because the concentration of CO_2 in the atmosphere is low, the concentration of CO_2 is also low in water, about 0.5 cc/1 at 10° C and 0.2 cc/1 at 24° C. Considerably more CO_2 is held as carbonate and bicarbonate ions.

CO_2 dissolved in water combines with water to form carbonic acid:

$$CO_2 + H_2O \rightleftharpoons H_2CO_3 \rightleftharpoons H^+$$
$$+ HCO_3 \rightleftharpoons H^+ + CO_3^=$$

Carbon dioxide in simple solution and as $H_2CO_3^-$ is free carbon dioxide. Carbon dioxide in bicarbonate ions HCO_3^- and in the carbonate ion $CO_3^=$ is called combined or bound CO_2. In the presence of an acid combined carbon dioxide is converted to free CO_2.

The amount of acid needed to free CO_2 is the measure of the water's *alkalinity*, the amount of anions of weak acid in water and of the cations balanced against them. H_2CO_3 dissociates, releasing the hydrogen ion and thus affecting the *pH*, the measure of the degree to which water is acid. At neutral (pH 7) most of the CO_2 is present as HCO_3^-. At a high pH more CO_2 is present as $CO_3^=$ than at a low pH, where more carbon dioxide is present in the free condition. Addition or removal of CO_2 affects pH, and a change in pH affects CO_2.

Bicarbonate and carbonate ions (and other anions of weak acids) form a *buffer system*, which resists changes in pH. The buffer capacity of solutions is determined by the abundance of their anions, which is directly related to alkalinity.

Seawater and hard fresh water are highly buffered. Surface water of the oceans has a pH of 8.0 to 8.4, deep water 7.4 to 7.9, and in areas of heavy seaweed growth, 9.0. The pH of fresh water varies from 3 to 10; the usual range in streams and lakes is 6.5 to 8.5, but many lakes and streams may have a pH as low as 3.5 seasonally, depending upon the nature of the watershed. Waters draining from watersheds dominated geologically by limestone will have a much higher pH and be well buffered compared to waters from watersheds dominated by acid sandstones and granite. The pH of soft waters can fluctuate widely, especially when receiving acid rain (Chapter 12).

In soft waters carbon dioxide may be reduced by photosynthesis of aquatic plants. If photosynthesis is high, corresponding respiration of organic matter, absorption from the atmosphere, and absorption from groundwater may not be enough to balance the losses. This fact can affect the physiology of aquatic invertebrates, especially in the oxygen affinity and alkalinity of the blood and the development of the exoskeleton.

A close relationship exists between the formation and dissolution of calcium carbonate and the photosynthetic and respiratory activity of aquatic plants:

$$CaCO_3 \rightarrow Ca(HCO_3) \qquad (1)$$

$$CO_3^- + Ca^{++} \leftrightarrows CaCO_3 \qquad (2)$$

A loss of CO_2 causes a shift to the left (1) and absorption of CO_2 a shift to the right (2).

In hard water (high pH), plant activity removes carbon dioxide and causes the precipitation of calcium carbonate. In soft water, Ca^{++} and $CO^=$ tend to remain in solution. The limited supplies of Ca^{++} and $CO_3^=$ in soft waters restrict the distribution of aquatic invertebrates, whose exoskeletons are composed largely of calcium carbonate.

Salinity and Dissolved Substances

The aquatic environment contains numerous dissolved substances introduced into the natural waters from soil, surface flows, geological sources, precipitation, and the atmosphere. The amounts of dissolved materials in water have a pronounced effect on the osmotic regulation of aquatic organisms (see Chapter 5). Common ions of dissolved solids found in natural waters include sodium, potassium, calcium, magnesium, iron, and chlorine, among others. The amounts and relative abundances of these ions determine whether the waters are soft, hard, or saline. In soft fresh water the calcium and carbonate ions are less concentrated than sodium and chlorine. Hard fresh water contains more dissolved salts than soft fresh water, especially calcium and carbonate ions. In seawater chlorine and sodium are the first and second most abundant ions. Seawater is fairly constant in the relative abundance of its constituents. Hard and fresh water may be highly variable, depending upon input from surrounding terrestrial environments.

Two elements, sodium and chlorine, make up some 86 percent of sea salts. Along with other such major elements as sulfur, magnesium, potassium, and calcium they comprise 99 percent of sea salts. Seawater, however, differs from a simple sodium chloride solution in that the equivalent amounts of cations and anions are not balanced against each other. The cations exceed the anions by 2.38 milliequivalents. As a result, seawater is weakly alkaline (pH 8.0 to 8.3) and strongly buffered, a condition that is biologically important.

The amount of dissolved salt in seawater is usually expressed as chlorinity or salinity. Because oceans are usually well mixed, sea salt has a constant composition; that is, the relative proportions of major elements change little. Thus the determination of the most abundant element, chlorine (Table 10.1), can be used as an index of the amount of salt present in a given volume of seawater. Chlorine expressed in o/oo is the amount of chlo-

TABLE 10.1 Composition of Seawater of 35‰ Salinity

Elements	g/kg	Millimole/kg	Milliequiva-lent/kg
CATIONS			
Sodium	10.752	467.56	467.56
Potassium	0.395	10.10	10.10
Magnesium	1.295	53.25	106.50
Calcium	0.416	10.38	20.76
Strontium	0.008	0.09	0.18
Total			605.10
ANIONS			
Chlorine	19.345	545.59	545.59
Bromine	0.066	0.83	0.83
Fluorine	0.0013	0.07	0.07
Sulphate	2.701	28.12	56.23
Bicarbon-ate	0.145	2.38	—
Boric acid	0.027	0.44	—
Total			602.72

Note: Surplus of cations over strong anions (alkalinity): 2.38.

Source: Kalle 1971.

rine in grams in a kilogram of seawater. Chlorinity can be converted to salinity, the total amount of solid matter in grams per kilogram of seawater. The relationship of salinity to chlorinity is expressed by

$$S (0/00) = 1.80655 \times chlorinity$$

The salinity of parts of the ocean is variable because physical processes change the amount of water in the seas. Salinity is affected by evaporation, precipitation, movement of water masses, mixing of water masses of different salinities, formation of insoluble precipitates that sink to the ocean floor, and diffusion of one water mass to another. Salinities are most variable near the interface of sea and air.

Salinity, like temperature, exhibits a vertical gradient in the sea. This gradient, called a *halocline*, is especially pronounced

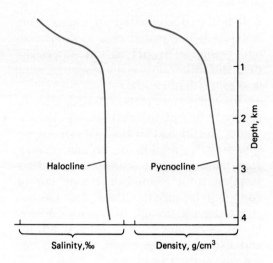

Figure 10.10 The halocline, the stratification of salinity in the ocean, and the pycnocline, stable stratification of density. (After M. Gross 1972.)

at higher latitudes (Figure 10.10). There abundant precipitation causes a marked change in salinity with depth. In the middle latitudes, salinity together with temperature produces a marked gradient in density. Because density of seawater increases with depth but does not reach its greatest density at 4° C as with fresh water, there is no seasonal overturn. This fact results in a normally stable stratification of density, known as the *pycnocline*.

■ Summary

The nature of the aquatic environment is shaped by the physical and chemical properties of water. Because of its molecular structure, water is the most dense at 4° C, which is the underlying cause of summer stratification of lake and ocean water and is the reason why ice floats. Water has the ability to absorb considerable quantities of heat with a small rise in temperature and gives up heat just as slowly. It has high viscosity, which affects its flow, and it exhibits a high surface tension, caused by the stronger attraction of water molecules for each other than for air above the surface. These properties have a pronounced influence on aquatic life.

Because of the thermal properties of water, aquatic environments exhibit a more stable temperature regime than terrestrial environments. Because water reaches its greatest density at 4° C, a temperature stratification develops in lakes and seas in summer: a circulating surface layer of warmer, lighter water, the epilimnion; a middle zone, the metalimnion, in which the temperature rapidly drops; and a bottom layer of dense water at approximately 4° C, the hypolimnion. When surface water cools in the fall and warms in spring, the difference in density between the layers decreases and the water circulates throughout the lake. A similar stratification, paralleling that of temperature, takes place with dissolved gases. Like lakes, the marine environment also experiences stratification. Because there is no true seasonal overturn, a normally stable stratification occurs, known as the pycnocline.

Light is rapidly attenuated in the aquatic environment. The wavelengths absorbed affect the apparent color of the water. The depth to which light can penetrate is limited by the turbidity of the water. The point at which light penetration is insufficient to carry on photosynthesis is the compensation depth.

Another feature of the aquatic environment is the movement of water. Movements include traveling surface waves of lakes and oceans, caused by pressure of wind on the surface; seiches or internal standing waves, caused by an oscillation of a structure of water about a point or node;

eddy currents, small internal turbulent currents whose energy is dissipated at right angles to major currents; Langmuir cells, circulating horizontal spirals of water parallel to the surface of the water and direction of the wind that spin just below the surface, each circulating in a direction opposite to the other; and in the sea, Ekman spirals, caused by layers of water moving more slowly at right angles to the wind than the layers above it.

Oxygen, rarely limiting in terrestrial environments, is limiting in aquatic environments. Oxygen achieves its greatest saturation in aquatic environments in cold, fast-moving water. In any aquatic environment oxygen in the water is in equilibrium with oxygen in the atmosphere, with the amount of oxygen varying with temperature and atmospheric pressure. Carbon dioxide combines with water, so it may exist in a free or bound state. Carbon dioxide in simple solution and in the form of carbonic acid is free, and carbon dioxide in carbonate and bicarbonate ions is combined or bound. Bicarbonate and carbonate ions act as a buffer system, resisting changes in pH.

Among the dissolved substances in aquatic environments are the ions of such substances as sodium, potassium, calcium, chlorine, iron, and others. Fresh water contains minimal amounts of sodium and chlorine ions, the ones associated with salinity. These two ions make up 86 percent of sea salt. Although sea salt has a constant composition, salinity varies throughout the ocean. It is affected by evaporation, precipitation, movement of water masses, and mixing of water masses of different salinities. Because of its salinity, seawater does not reach its greatest density at 4° C but becomes heavier as it cools.

Energy Flow

Sunlight floods Earth with two forms of energy that keep the planet functioning (Figure 11.1). One is heat energy, which warms Earth, heats the atmosphere (Chapter 6), drives the water cycle (Chapter 5), and moves currents of air and water (Chapter 4). The other is photochemical energy, which plants use in photosynthesis (Chapter 3), fix in carbohydrates and other compounds, and supply to other living organisms.

■ The Nature of Energy

Energy is measured in several units. For ecologists the *gram calorie* (g cal), the amount of heat necessary to raise 1 gram of water 1° C at 15° C, is the most convenient unit of energy. When large quantities of energy are involved, the *kilogram calorie* (kcal) is more appropriate. A kcal is the amount of heat required to raise 1 kil-

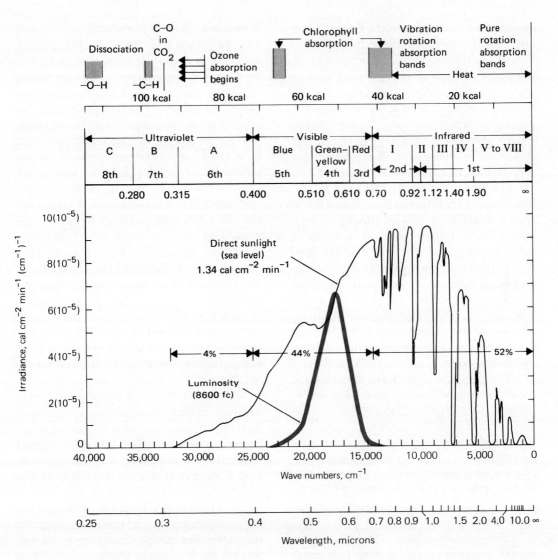

Figure 11.1 Spectral distribution of solar radiation, showing the division of the spectrum involved in plant processes. Shown at the top are the basic photochemical and radiation processes that may occur with the absorption of radiation by plants and other objects. The quantum content, expressed on a scale as kcal, is indicated for each frequency of incident radiation. Compare this figure with Figure 4.1. (From Gates 1965: 9.)

ogram of water 1° C at 15° C. Because ecologists are concerned with energy flow for a given area, they measure energy per unit area, say a square centimeter or square meter. Other often used units of measure are the *langley*, 1 g cal/cm²/min; the *joule*, 0.24 g cal; the *watt*, 1 joule/sec or 14.3 g cal/m²; and the *British thermal unit* (BTU), the amount of heat necessary to raise 1 pound of water 1° F or 252 g cal.

The transfer of radiation involves the

movement or flow of units of energy from one point to another, known as flux. This flux of energy requires an energy source, the point from which energy flows, and an energy sink or receiver, the point to which it flows. Without a sink for thermal energy, the sun could not be an energy source. Earth receives energy from the sun, absorbs a part of it, and gives up energy as heat to a sink, outer space.

Energy flow is mediated at the molecular level. Characteristically, thermal energy is distributed rapidly among all molecules in a system without necessarily causing a chemical reaction. The effect of thermal energy is to set the molecules into a state of random motion and vibration. The hotter the object, the more the molecules are moving, vibrating, and rotating. These motions spread from a hot body to a cooler one, transferring energy from one to the other.

The energy of light waves, on the one hand, especially the red and blue wavelengths, causes electronic transitions within atoms and molecules, called *excitations*. These excitations can lead to photochemical reactions. The energy of light sends one electron of a pair of bonding electrons to a higher state or orbit. Uncoupled from its partner, it is free to be involved in photochemical reactions.

Energy is either *potential* or *kinetic*. Potential energy is energy at rest. It is capable of and available for work (defined as a force that causes a particle or other body to be moved or displaced). Kinetic energy is due to motion and results in work. Work that results from the expenditure of energy can either store energy (as potential energy) or arrange or order matter without storing energy.

The expenditure and storage of energy is described by two laws of thermodynamics. The first law states that energy is neither created nor destroyed. It may change forms, pass from one place to another, or act upon matter, to transform it to energy in various ways; but regardless of what transfers and transformations take place, no gain or loss in total energy occurs. Energy is simply transferred from one form or place to another. When wood is burned, the potential present in the molecules of wood equals the kinetic energy released, and heat is evolved to the surroundings. This reaction is *exothermic*. On the other hand, energy from the surroundings may be paid into a reaction. Here, too, the first law holds true. In photosynthesis, for example, the molecules of the products store more energy than the reactants. The extra energy is acquired from the sunlight, but again, there is no gain or loss in total energy. When energy from outside is put into a system to raise it to a higher energy state, the reaction is *endothermic*.

Although the total amount of energy involved in any chemical reaction, such as burning wood, does not increase or decrease, much of the potential energy stored in the substance undergoing reaction is degraded during the reaction into a form incapable of doing any further work. This energy ends up as heat, serving to disorganize or randomly disperse the molecules involved, making them useless for further work. The measure of this relative disorder is named *entropy*.

The second law of thermodynamics makes an important generalization about energy transfer. It states that when energy is transferred or transformed, part of the energy assumes a form that cannot be passed on any further. When coal is burned in a boiler to produce steam, some of the energy creates steam that performs work, but part of the energy is dispersed as heat to the surrounding air. The same thing happens to energy in the ecosystem. As energy is transferred from one organism to another in the form of food, a large part of that energy is degraded as heat and as a net increase in the disorder of energy. The remainder is stored as living tissue. But biological systems seemingly do not conform to the second law, for the tendency of life is to produce order out of disorder, to decrease rather than increase entropy.

The second law, theoretically, applies to the isolated, closed system in which there is no exchange of energy or matter between the system and its surroundings. An isolated system approaches thermodynamic equilibrium, that is, a point at which all the energy has assumed a form that cannot do work. A closed system tends toward a state of minimum free energy (energy available to do work) and maximum entropy, whereas an open system maintains a state of higher free energy and lower entropy. In other words, the closed system tends to run down; the open one does not. As long as there is a constant input of matter and free energy to the system and a constant outflow of entropy (in the form of heat and waste), the system maintains a steady state. Life is an open system maintained in a steady state.

Energy enters the biosphere as visible light that is stored in energetic covalent bonds during photosynthesis (Chapter 3). From that point biochemical changes involve a series of rearrangements of matter into compounds of less potential chemical energy. These chemical rearrangements are accompanied by the production of heat that eventually goes into the energy sink. This loss of heat is accompanied by a loss of carbon dioxide, water, and nitrogenous compounds, which are recycled through the biosphere. Although some

energy is irrevocably lost from the biosphere, some of it is stored in the system. The more organized the system, the longer energy is stored.

Primary Production

The flow of energy through the ecosystem starts with its fixation by plants through photosynthesis, a process that in itself demands expenditure of energy (Chapter 3). Plants depend upon energy stored in seeds until their own production machinery is working. Then green plants begin to accumulate energy. Energy accumulated by plants is called *production*, more specifically *primary production* because it is the first and basic form of energy storage in an ecosystem. The rate at which energy is stored by photosynthetic activity is known as *primary productivity*. All of the sun's energy that is assimilated, that is, total photosynthesis, is *gross primary production*. Like other organisms, plants require energy for reproduction and maintenance. The energy for these needs is provided by a reverse of the photosynthetic process, *respiration*. Energy remaining after respiration and stored as organic matter is *net primary production*.

Production is usually expressed in kilocalories per square meter per year (kcal/m^2/yr^1). However, production may be expressed as dry organic matter in grams per square meter per year (g/m^2/yr^2). If either of the two measures is employed to estimate efficiencies and other ratios, the same unit must be used for both the numerator and denominator of the ratio. Only calories can be compared with calories, dry weight with dry weight.

Biomass

Net primary production accumulates over time as plant biomass. Part of this accumulation is turned over seasonally through decomposition (Chapter 3) and part is retained over a longer period as living material. The accumulated organic matter found in a given area at a given time is the *standing crop biomass*. Biomass is usually expressed as grams dry weight of organic matter per unit area. Thus biomass differs from productivity, which is the rate at which organic matter is created by photosynthesis. Biomass present at any given time is not the same as total production, and high biomass does not necessarily imply high productivity.

Plants allocate net production to leaves, twigs, stem, bark, roots, flowers, and seed. How much is allocated reflects the life history strategies of various plants (see Chapter 17). Determination of such allocation is difficult and tedious, particularly for woody plants. Involved are cutting sample trees, weighing the various components, and determining both the energy content and nutrient content. Such estimates have been made for a number of forest stands (for a summary of some forest types of eastern United States, see Whittaker et al. 1974). For example, trees of a young oak-pine forest on Long Island, New York, allocated 25 percent of net primary production to stem wood and bark, 40 percent to roots, 33 percent to twigs and leaves, and 2 percent to flowers and seeds. Shrubs allocated 54 percent of net primary production to roots, 21 percent to stems, and 25 percent to leaves (Whittaker and Woodwell 1968).

The proportionate allocation of net primary production to belowground and aboveground biomass tells much about different components within the ecosystem. A high root-to-shoot ratio (R/S) indicates that most of the production goes into the supportive function of plants. Plants with a large root biomass are more effective competitors for water and nutrients and can survive more successfully in harsh environments, because most of their active biomass is below ground. Plants with a low R/S ratio have most of their biomass above ground and assimilate more light energy, resulting in higher productivity. Tundra sedge and grass meadows, characteristic of an environment with a long cold winter and a short growing season, have R/S ratios ranging from 5 to 11. Tundra shrubs may range from 4 to 10. Further south, midwest prairie grasses have an R/S ratio of about 3, indicative of cold winters and low moisture supply. Forest ecosystems, with their high aboveground biomass, have a low R/S ratio. For the Hubbard Brook forest in New Hampshire the R/S ratio (based on data of Gosz et al. 1976) for trees is 0.213, for shrubs 0.5, and for herbs 1.0. As we could predict, the R/S ratio decreases through the vertical strata from the herbaceous layer to the canopy.

The aboveground biomass is distributed vertically in the ecosystem. The vertical distribution of leaf biomass or in aquatic systems the concentration of plankton influences the penetration of light, which in turn influences the distribution of production in the ecosystem. The region of maximum productivity in the aquatic ecosystem is not the upper sunlit surface (strong sunlight inhibits photosynthesis), but some depth below, depending upon the clarity of the water and the density of plankton growth. As depth increases, light intensity decreases until it reaches a point at which the light received by the plankton is just sufficient to meet respiratory needs and production equals respiration (Figure 11.2). This is known as the *compensation intensity*. In the forest ecosystem a similar situ-

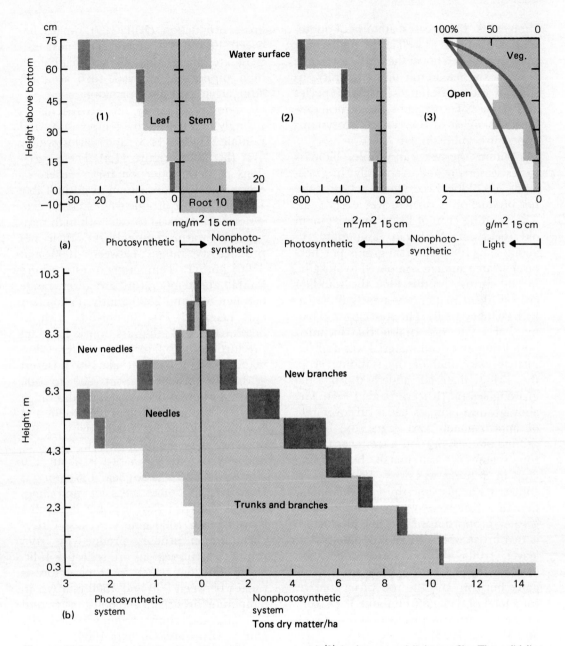

Figure 11.2 Vertical distribution of production and biomass in aquatic and terrestrial communities. (a) Three graphs for a pondweed community: (1) division of biomass into leaf, stem, and root (solid areas represent winter buds); (2) concentration of chlorophyll in the plant community; and (3) leaf area and light profile. The solid line is light in the community; the broken light represents light in open water. (After Ikusima 1965.) (b) Structure and productive systems of a pine-spruce-fir forest in Japan. (After Monsi 1968.)

ation exists. The greatest amount of photosynthetic biomass as well as the highest net photosynthesis is not at the top, but at some point below maximum light intensity. In spite of wide differences in plant species and in types of ecosystems, the vertical profiles of biomass of the various ecosystems appear to be quite similar.

Within the vertical profile, biomass varies seasonally and even daily. In grasslands and oldfield ecosystems much of the net production is turned over every year. The standing crop of living material in an old field in Michigan amounted to about 4×10^3 kg/ha (kilograms per hectare) in late summer, compared to 80 kg/ha in late spring. At this time the standing crop in dead matter was nearly 3×10^3 kg/ha (Golley 1960). The aboveground biomass of a tall-grass prairie that included both living and dead material was approximately twice that of the standing crop, the living material added during the growing season (Kucera et al. 1967). The aboveground biomass has a turnover rate of approximately two years, the belowground biomass of roots a turnover rate of four years. In a salt marsh the standing crop in autumn was 9×10^3 kg/ha; in winter it was just one-third of this quantity. In a forest ecosystem a considerably greater proportion of the net production is tied up as wood. In an oak-pine forest, leaves, fruits, flowers, dead wood, and bark contributed 342 g/m²/yr to the organic horizon, and the roots 311 g/m²/yr, for a total of 653 g/m² or about 58 percent of the net primary production (Woodwell and Marpels 1968).

Net Productivity

Plants fix energy continually during the daylight hours of the growing season, and spend a minimum of it on respiration. The ratio of net primary production to gross production (NPP/GPP) ranges between 40 and 80 percent. The most efficient are plants that do not maintain a high supporting biomass, such as grass, large algae, and phytoplankton.

The productivity of ecosystems is strongly influenced by temperature and rainfall (Figure 11.3) and varies widely over the globe (Figure 11.4). These variations in net production for a variety of ecosystem types are summarized in Table 11.1. The most productive terrestrial ecosystems are tropical forests with high rainfall and warm temperatures; their net productivity ranges between 1000 and 3500 g/m²/yr. Temperate forests, where rainfall and temperature are lower, range between 600 and 2500 g/m²/yr (Whittaker and Likens 1975). Shrublands such as heath balds and tall-grass prairie have net productions of 700 to 1500 g/m²/yr (Whittaker 1963; Kucera et al. 1967) Desert grasslands produce about 200 to 300 g/m²/yr, whereas deserts and tundra range between 100 and 250 g/m²/yr (Rodin and Bazilevic 1967). Net productivity of the open sea is generally quite low. The productivity of the North Sea is about 170 g/m²/yr, the Sargasso Sea 180 g/m²/yr. However, in some areas of upwelling, such as the Peru Currents, net productivity can reach 1000 g/m²/yr. These differences in net primary productivity from tropic to arctic regions are reflected in litter production, which in tropical forests ranges between 900 and 1500 g/m²/yr, in temperate forests 200 and 600 g/m²/yr, and in arctic and alpine regions 0 and 200 g/m²/yr (Bray and Gorham 1964).

Some ecosystems have consistently high production. Such high productivity usually results from an additional energy subsidy to the system. This subsidy may be a warmer temperature, greater rainfall, circulating or moving water that carries food or additional nutrients into the com-

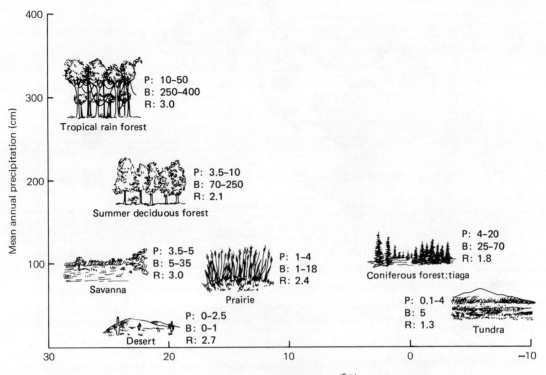

Figure 11.3 Climatic distribution of primary production, biomass, and radiation input. P = primary production (tn/ha); B = biomass (tn/ha); R = solar radiant input (kcal/m^{-2}/yr^{-1} 0.3–3.0 microns). (From Etherington 1976: 355.)

munity, or in the case of agricultural crops, the use of fossil fuel for cultivation and irrigation, the application of fertilizer, and the control of pests. Swamps and marshes, ecosystems at the interface of land and water, may have a net productivity of 3300 g/m^2/yr. Estuaries, because of input of nutrients from rivers and tides, and coral reefs, because of input from changing tides, may have a net productivity between 1000 and 2500 g/m^2/yr. Among agricultural ecosystems sugarcane has a net productivity of 1700 to 1800 g/m^2/yr, hybrid corn 1000 g/m^2/yr, and some tropical crops 3000 g/m^2/yr.

In any ecosystem annual net production changes with time and age. For example, mean annual net primary production of a Scots pine plantation achieved maximum productivity of 22 × 10^3 kg/ha at the age of 20; it then declined to 12 × 10^3 kg/ha at 30 years of age (Ovington 1961). Woodlands apparently achieve their maximum annual production in the pole stage, when the dominance of the trees is the greatest and the understory is at a minimum (Figure 11.5). The understory makes its greatest contribution during the juvenile and mature stages of the forest. As age increases, more and more of the gross production is needed for maintenance, and less remains for net production.

Net productivity also declines from a young ecosystem such as a weedy field or an agricultural crop to a mature plant

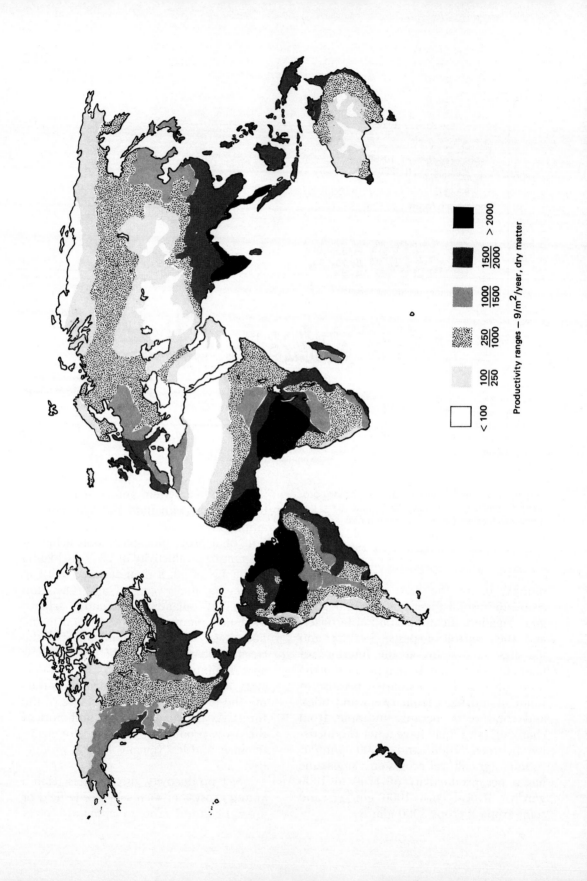

Productivity ranges — 9/m²/year, dry matter

> 2000

$\dfrac{1500}{2000}$

$\dfrac{1000}{1500}$

$\dfrac{250}{1000}$

$\dfrac{100}{250}$

< 100

Figure 11.4 A map of world terrestrial primary production. Note the high productivity of tropical regions. (Based on Golley and Leith 1972.)

Figure 11.5 Relation of aboveground biomass and production to size of tree for 63 sample trees of three major species—sugar maple, yellow birch, and beech. Note that as trees increase in size, the ratio of branches to stems increases. In larger trees branches account for a greater proportion of net primary production than stems. In smaller trees current leaves and twigs account for the greater percentage of primary production. This percentage declines rather rapidly as trees approach pole stage, then increases as trees mature. (From Whittaker et al. 1974: 239.)

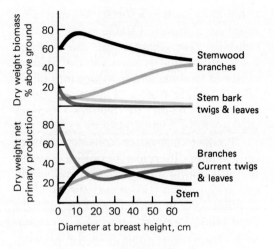

TABLE 11.1 Net Primary Production and Plant Biomass of World Ecosystems

Ecosystems (in Order of Productivity)	Area $(10^6\ km^2)$	Mean Net Primary Production per Unit Area $(g/m^2/yr)$	World Net Primary Production $(10^9\ mtr/yr)$	Mean Biomass per Unit Area (kg/m^2)
CONTINENTAL				
Tropical rain forest	17.0	2000.0	34.00	44.00
Tropical seasonal forest	7.5	1500.0	11.30	36.00
Temperate evergreen forest	5.0	1300.0	6.40	36.00
Temperate deciduous forest	7.0	1200.0	8.40	30.00
Boreal forest	12.0	800.0	9.50	20.00
Savanna	15.0	700.0	10.40	4.00
Cultivated land	14.0	644.0	9.10	1.10
Woodland and shrubland	8.0	600.0	4.90	6.80
Temperate grassland	9.0	500.0	4.40	1.60
Tundra and alpine meadow	8.0	144.0	1.10	0.67
Desert shrub	18.0	71.0	1.30	0.67
Rock, ice, sand	24.0	3.3	0.09	0.02
Swamp and marsh	2.0	2500.0	4.90	15.00
Lake and stream	2.5	500.0	1.30	0.02
Total continental	149.0	720.0	107.09	12.30
MARINE				
Algal beds and reefs	0.6	2000.0	1.10	2.00
Estuaries	1.4	1800.0	2.40	1.00
Upwelling zones	0.4	500.0	0.22	0.02
Continental shelf	26.6	360.0	9.60	0.01
Open ocean	332.0	127.0	42.00	1.00
Total marine	361.0	153.0	55.32	0.01
World total	510.0	320.0	162.41	3.62

Source: Adapted from Whittaker and Likens 1973.

community such as a forest. As plant communities approach a stable or steady-state condition (see Chapter 25), more of the gross production is used for the maintenance of biomass and less goes into newly added organic matter. Thus the ratio of gross production to biomass declines through time (Figure 11.6).

Productivity varies considerably not only among different types of ecosystems, but also among similar systems and within one system from year to year. Productivity is influenced by such factors as nutrient availability, moisture, especially precipitation, temperature, length of the growing season, animal utilization, and fire. For example, the herbage yields of a grassland may vary by a factor of eight between wet and dry years (Weaver and Albertson 1956). Overgrazing of grasslands by cattle and sheep or defoliation of forests by such insects as the saddled prominent and gypsy moth can seriously reduce net production. Fire in grasslands may result in increased productivity if moisture is normal, but in reduced productivity if precip-

itation is low (Kucera et al. 1967). An insufficient supply of nutrients, especially nitrogen and phosphorus, can limit net productivity, as can the mechanical injury of plants, atmospheric pollution, and the like.

Although the size of the standing crop is not synonymous with high productivity, the size of the standing crop does influence the capacity to produce. A pond with too few fish or a forest with too few trees does not have the capacity to utilize the energy available. On the other hand, too many fish or too many trees means less energy available to each individual. This crowding lowers the efficiency of use and influences the storage and transfer of energy through the ecosystem. For example, Hayne and Ball (1956) found that the production of bottom-dwelling invertebrates in a pond amounted to nearly 17 times standing crop when predatory fish were present. In the absence of fish the production rate of bottom fauna declined and finally stopped with a large standing crop. Thus, in the presence of predatory fish the size of the standing crop of bottom fauna was depressed, but production increased. In ponds with fish the annual production amounted to 811 lb/acre of bottom fauna and 181 lb of fish. This quantity represents the efficiency of 18 percent in energy conversion from bottom fauna to fish.

■ Secondary Production

Net production is the energy available to the heterotrophic components of the ecosystem. Theoretically, at least, all of it is available to the grazers or even to the decomposers, but rarely is it all utilized in this manner. The net production of any given ecosystem may be dispersed to an-

Figure 11.6 Change through time in ratios between gross community photosynthesis and biomass or production efficiency, and between biomass and gross community photosynthesis, or maintenance efficiency. Note the early high production efficiency and the later accumulation of biomass. (After G. D. Cooke 1967: 71.)

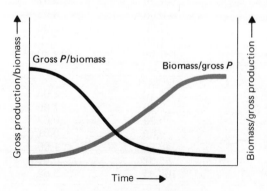

other food chain outside of the ecosystem by humans, wind, or water. Much of the living material is physically unavailable to the grazers—they cannot reach many plants. The organic matter of live organisms is unavailable to decomposers and detritus-feeders, and that of dead plants may not be relished by grazers. The amount of net production available to herbivores may vary from year to year and from place to place. The quantity consumed varies with the type of herbivore and the density of the population. Once consumed, a considerable portion of the plant material, again depending upon the kind of plant involved and the digestive efficiency of the herbivore, may pass through the animal's body undigested. A grasshopper assimilates only about 30 percent of the grass it consumes, leaving 70 percent available to the detrital food chain (Smalley 1960). Mice, on the other hand, assimilate about 85 to 90 percent of what they consume (Golley 1960; R. L. Smith 1962).

Energy, once consumed, either is diverted to maintenance, growth, and reproduction or is passed from the body as feces and urine (Figure 11.7). The energy content of feces is transferred to the detritus food chain. The energy lost through urine is high (Coo et al. 1952). Another portion is lost as fermentation gases. Of the energy left after losses through feces, urine, and gases, part is utilized as *heat increment,* which is heat required for metabolism above that required for basal or resting metabolism. The remainder of the energy is *net energy,* available for maintenance and production. It includes energy involved in capturing or harvesting food, muscular work expended in the animal's daily routine, and energy needed to keep up with the wear and tear on the animal's body. The energy used for maintenance is lost as heat. Maintenance costs, highest in

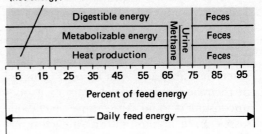

Figure 11.7 Relative values of the end products of energy metabolism in white-tailed deer. Note the small amount of net energy gained (body weight) in relation to that lost as heat, gas, urine, and feces. The deer is a herbivore, a first-level consumer. (After Cowan 1962: 5.)

small, active, warm-blooded animals, are fixed or irreducible. In small invertebrates energy costs vary with temperature, and a positive energy balance exists only within a fairly narrow range of temperatures. Below 5° C spiders become sluggish, cease feeding, and have to utilize stored energy to meet metabolic needs. At approximately 5° C assimilated energy approaches energy lost through respiration. From 5° to 20.5° C spiders assimilate more energy than they respire. Above 25° C, their ability to maintain a positive energy balance declines rapidly (Van Hooke 1971).

Energy left over from maintenance and respiration goes into the production of new tissue, fat tissue, growth, and new individuals. This net energy of production is *secondary production* or *consumer production.* Within secondary production there is no portion known as gross production. What is analogous to gross production is actually assimilation. Secondary production is greatest when the birthrate of the population and the growth rates of the individuals are the highest. This peak usually coincides, for evolutionary reasons, with the time when net primary production is also the highest.

This scheme is summarized in Figure 11.8. It is applicable to any consumer organism, herbivore or carnivore, poikilotherm or homoiotherm. Herbivores are the energy source for carnivores, and when they are eaten, not all of the energy in their bodies is utilized. Part of it goes unconsumed, and the same metabolic losses can be accounted for. At each transfer considerably less energy is available for the next consumer level.

The energy budget of a consumer population can be summarized by

$$C = A + F + U$$

where C is the energy ingested or consumed, A is the energy assimilated, and F and U are the energy lost through feces and nitrogenous wastes.

The term A can be refined further:

$$A = P + R$$

where P is secondary production and R is energy lost through respiration. U, representing nitrogenous wastes, should be included as part of A (A = P + R + U), because they are involved in the homeostasis of organisms. Because of the difficulty of separating nitrogenous from fecal wastes, U typically is not included in A. Thus

$$C = P + R + F + U$$

Figure 11.8 (a) General model of components of energy metabolism in secondary production. (b) A field example for a ground-dwelling spider population. (Data from Moulder and Reichle 1972: 496.)

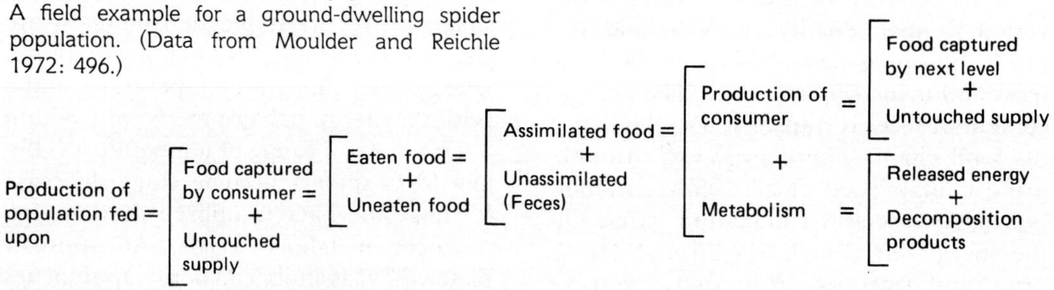

(a)

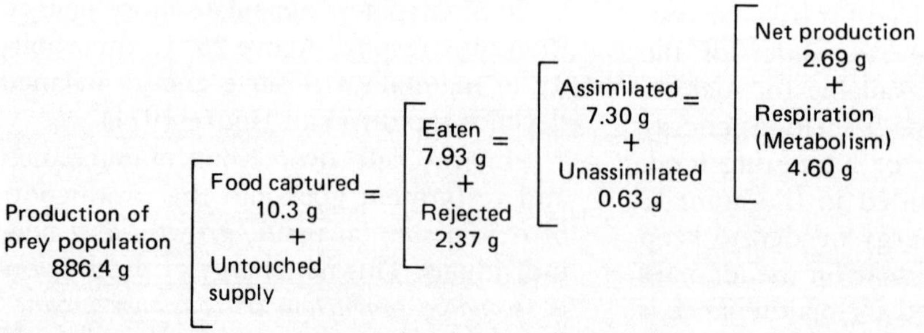

(b)

or secondary production is

$$P = C - R - F - U$$

Just as net primary production is limited by a number of variables, so is secondary production. The quantity, quality (including nutrient status and digestibility), and availability of net production are three limitations. So is the degree to which primary and available secondary production are utilized.

Secondary production can be examined from the viewpoint of three different ratios. One is the ratio of assimilation to consumption, A/C, the measure of the efficiency of the consumer at extracting energy from the food it consumes. It relates to food quality and effectiveness of digestion. The second is ratio of production to assimilation, P/A. It is a measure of the efficiency of a consumer in incorporating assimilated energy into new tissue, or secondary production. These two ratios underlie the magnitude of a third ratio of production to consumption, P/C. It indicates how much energy consumed by the animal is available to the next group of consumers.

The ability of the consumer population to convert the energy it ingests varies with the type of consumer and the species (Table 11.2). Homoiotherms use about 98 percent of their assimilated energy in me-tabolism and only about 2 percent in secondary production. Poikilotherms, on the average, convert about 44 percent of their assimilated energy in secondary production. They turn a greater proportion of their assimilated energy (A) into biomass (P). However, there is a major difference in assimilation efficiency between poikilotherms and homoiotherms. Poikilotherms have an efficiency of about 42 percent, whereas homoiotherms have an average efficiency of over 70 percent. Therefore the poikilotherm has to consume more calories than a homoiotherm to obtain sufficient energy for maintenance, growth, and reproduction.

■ Food Chains

Net production theoretically represents the energy available either directly or indirectly to consumer organisms; it is the base upon which the rest of life on Earth depends. This energy stored by plants is passed along through the ecosystem in a series of steps of eating and being eaten known as a *food chain*.

Food chains are descriptive. When worked out diagrammatically, they consist of a series of arrows, each pointing from one species to another, for which it is a source of food. In Figure 11.9, for exam-

---

TABLE 11.2 Assimilation Efficiency and Production Efficiencies for Homoiotherms and Poikilotherms

Efficiency	All Homoiotherms	Grazing Arthropods	Sap-feeding Herbivores	Lepidoptera	All Poikilotherms
Assimilation					
A/C	77.5 ± 6.4	37.7 ± 3.5	48.9 ± 4.5	46.2 ± 4.0	41.9 ± 2.3
Production					
P/C	2.0 ± 0.46	16.6 ± 1.2	13.5 ± 1.8	22.8 ± 1.4	17.7 ± 1
P/A	2.46 ± 0.46	45.0 ± 1.9	29.2 ± 4.8	50.0 ± 3.9	44.6 ± 2.1

Source: Based on data from Andrzejewska and Gyllenberg 1980.

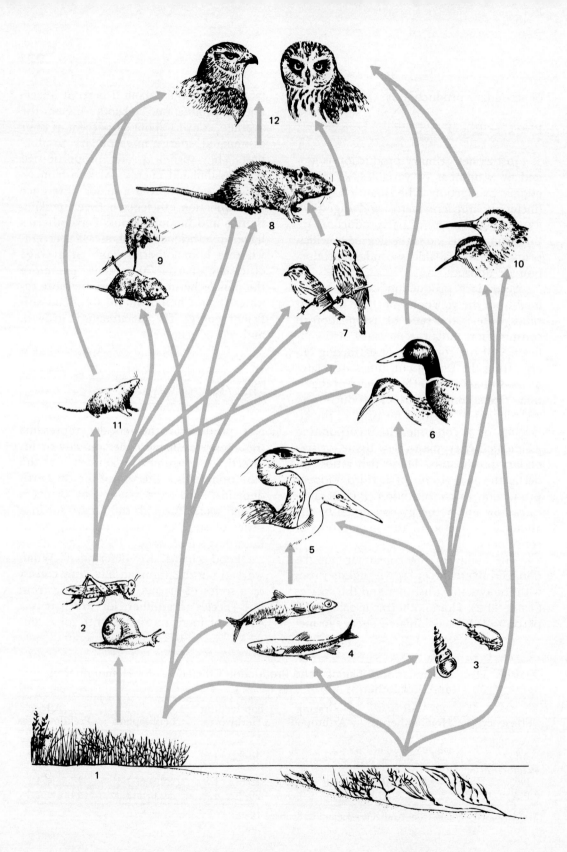

Figure 11.9 A midwinter food web in a *Salicornia* salt marsh (San Francisco Bay area). Producer organisms are terrestrial and aquatic plants (1). The plants are consumed by terrestrial herbivorous invertebrates, represented by grasshopper and snail (2) and by herbivorous marine and intertidal invertebrates (3). Fish, represented by smelt and anchovy (4), feed on vegetable matter from both ecosystems. The fish in turn are eaten by first-level carnivores, represented by the great blue heron and common egret (5). Continuing through the web are the following omnivores: clapper rail and mallard ducks (6), savanna and song sparrows (7), Norway rat (8), California vole and salt marsh harvest mouse (9), and the least and western sandpipers (10). The vagrant shrew (11) is a first-level carnivore. Top carnivores are the marsh hawk and short-eared owl (12). (Adapted from R. F. Johnson 1956: 99.)

ple, the marsh vegetation is eaten by the grasshopper, the grasshopper is consumed by the shrew, the shrew by the marsh hawk or the owl.

No one organism lives wholly on another; the resources are shared, especially at the beginning of the chain. The marsh plants are eaten by a variety of invertebrates, birds, mammals, and fish; and some of the animals are consumed by several predators. Thus food chains become interlinked to form a *food web*, the complexity of which varies within and between ecosystems.

Components

Herbivores. Feeding on plant tissues is a whole host of plant consumers, the herbivores, which are capable of converting energy stored in plant tissue into animal tissue. Their role in the community is essential, for without them the higher trophic levels could not exist. The English ecologist Charles Elton (1927) in his classic book *Animal Ecology,* suggested that the term *key industry* be used to denote animals that feed on plants and are so abundant

that many other animals depend upon them for food.

Only herbivores are adapted to live on a diet high in cellulose. The structure of the teeth, complicated stomachs, long intestines, a well-developed cecum, and symbiotic flora and fauna enable these animals to use plant tissues. For example, ruminants, such as deer, have a four-compartment stomach. As they graze, these animals chew their food hurriedly. The material consumed descends to the first and second stomachs (the rumen and reticulum), where it is softened to a pulp by the addition of water, kneaded by muscular action, and fermented by bacteria. At leisure the animals regurgitate the food, chew it more thoroughly, and swallow it again. The mass again enters the rumen, where the coarse particles remain behind for further bacterial digestion. The finer material is pulled into the reticulum and from there forced by contraction into the third compartment, or omasum. There the material is further digested and finally forced into the abomasum, or true glandular stomach.

The digestive process in ruminants relies heavily on bacterial fermentation in the rumen, reticulum, and omasum. Millions of microorganisms attack various digestive materials such as cellulose, starch, pectin, and hemicellulose sugars and convert part of them to short-chain volatile fatty acids. These acids are rapidly absorbed through the wall of the rumen into the bloodstream and are oxidized to form the animal's chief source of energy. Part of the material is converted to methane and lost to the animal and part remains as fermentation products. Many of the microbial cells involved are digested in the abomasum to recapture still more of the energy and nutrients. In addition to fermentation, the bacteria also synthesize B-complex vitamins and essential amino acids.

Another outstanding group of herbivores is the lagomorphs—the rabbits, hares, and pikas. In contrast to ruminants, these herbivores have a simple stomach and a large cecum. During digestion part of the ingested plant material enters the cecum and part enters the intestine to form dry pellets. In the cecum the ingested material is attacked by microorganisms and is expelled into the large intestine as moist soft pellets surrounded by a proteinaceous membrane. The soft pellets, much higher in protein and lower in crude fiber than the hard fecal pellets, are reingested (*coprophagy*) by the lagomorphs. The amount of feces recycled by coprophagy ranges from 50 to 80 percent. The reingestion is important because it provides bacterially synthesized B vitamins and ensures a more complete digestion of dry material and a better utilization of protein (see McBee 1971).

Carnivores. Herbivores are the energy source for carnivores, the flesh-eaters. Organisms that feed directly upon the herbivores are termed first-level carnivores or second-level consumers. First-level carnivores represent an energy source for the second-level carnivores. Still higher categories of carnivorous animals feeding on secondary carnivores exist in some communities. As the feeding level of carnivores increases, their numbers decrease and their fierceness, agility, and size increase. Finally, the energy stored at the top carnivore level is utilized by the decomposers.

Omnivores. Not all consumers confine their feeding to one level. The omnivores consume both plants and animals. The red fox feeds on berries, small rodents, and even dead animals. Thus it occupies herbivorous and carnivorous levels, as well as acting as a scavenger. Some fish feed on both plant and animal matter. The basically herbivorous white-footed mouse also feeds on insects, small birds, and bird eggs. The food habits of many animals vary with the seasons, with stages in the life cycle, and with their size and growth (Figure 11.10).

Decomposers. Conventionally, decomposers make up the final consumer group. Their basic function is to release nutrients contained in plant and animal biomass back into nutrient cycles. The work of the decomposers, then, is opposite to that of producers, which convert nutrients and energy into plant biomass. The process of decomposition (see Chapter 3) involves a complex series of food chains in which organisms of decay use the energy and nutrients of dead plant and animal matter.

There are two groups of decomposer organisms, macroorganisms and microorganisms. The macroorganisms or detritivores include such small detritus-feeding animals as mites, earthworms, millipedes, and snails in terrestrial ecosystems and crabs, mollusks, and worms in aquatic ecosystems. Acting as reducer-decomposers, they ingest organic matter, break it down into smaller pieces, mix it with the substrate, excrete it, spread fungal spores, break down microbial antagonisms, and even add substances to the material that stimulate microbial growth. They consume bacteria and fungi associated with detritus, as well as protozoans and small invertebrates that cling to the material. True decomposers are the microorganisms—bacteria and fungi. A succession of these organisms absorbs a portion of the detrital material as food, leaving unconsumed material as food for still other microorganisms.

Although frequently considered unclassifiable in the general scheme of food chains, decomposers function as herbivores or carnivores, depending upon the nature of their food, plant or animal.

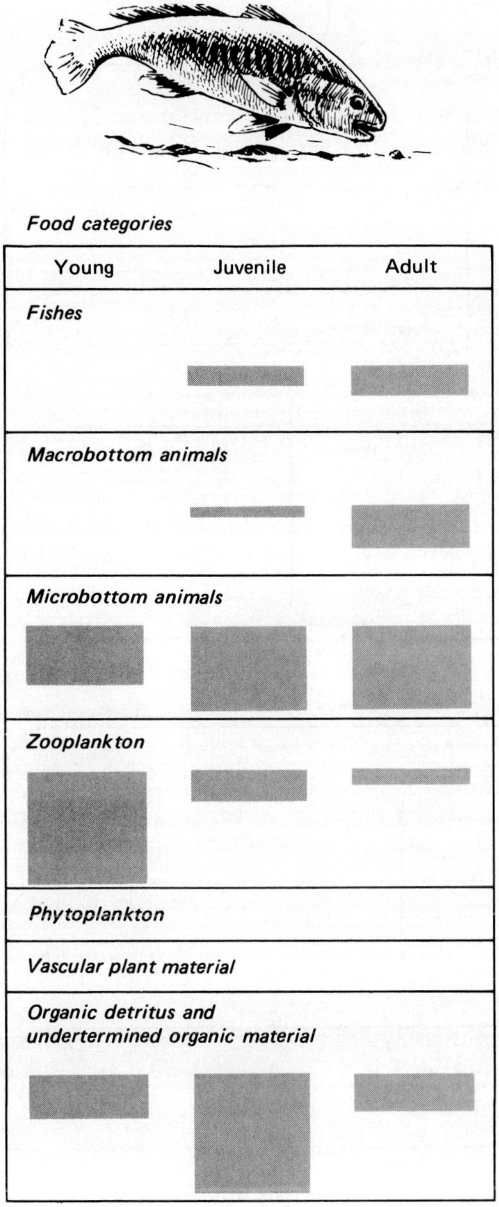
Food categories
Young
Juvenile
Adult
Fishes
Macrobottom animals
Microbottom animals
Zooplankton
Phytoplankton
Vascular plant material
Organic detritus and
undertermined organic material

Figure 11.10 Trophic spectra for young, juvenile, and adult stages of the Atlantic croaker from Lake Pontchartrain, Louisiana. The young fish subsist largely on zooplankton, the juveniles on organic matter, the adults on bottom animals and fish. (After Darnell 1961: 566.)

Other Feeding Groups. Several other feeding groups are also involved in energy transfer. *Parasites* spend a considerable part of their life cycle living on or in and drawing their nourishment from their hosts. Most do not kill their hosts, but some insect larvae, *the parasitoids,* slowly consume their host and then transform into another stage of their life cycle. Functionally, parasites are specialized carnivores and herbivores.

Scavengers are animals that eat dead plant and animal material. Among them are termites and various beetles that feed in dead and decaying wood, and crabs and other marine invertebrates that feed on plant particles in water. Botflies, dermestid beetles, vultures, and gulls are among the many animals that feed on animal remains. Scavengers may be either herbivores or carnivores.

Saprophytes are plant counterparts of scavengers. They draw their nourishment from dead plant and animal material, chiefly the former. Because they do not require sunlight as an energy source, they can live in deep shade or dark caves. Examples of saprophytes are fungi, Indian pipe, and beechdrops. The majority are herbivores, but others feed on animal matter.

Major Food Chains

Within any ecosystem there are two major food chains, the grazing food chain and the detrital food chain (Figure 11.11). In most terrestrial and shallow-water ecosystems, with their high standing crop and relatively low harvest of primary production, the detrital food chain is dominant. In deep-water aquatic ecosystems, with their low biomass, rapid turnover of organisms, and high rate of harvest, the grazing food chain may be dominant.

The amount of energy shunted down the two routes varies among communities.

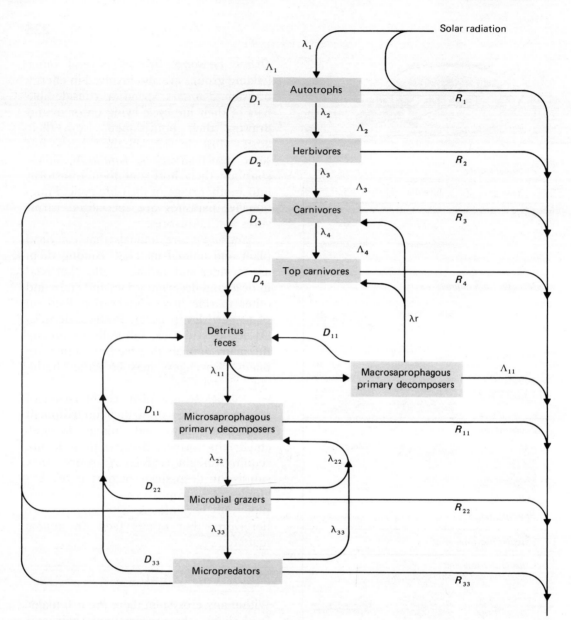

Figure 11.11 Model of a detrital and a grazing food chain. Two pathways lead from the autotrophs, one to the grazing herbivores and the other to detrital feeders. Note the connections between the two food chains. (After Paris 1968.)

In an intertidal salt marsh less than 10 percent of living plant material is consumed by herbivores and 90 percent goes the way of the detritus-feeders and decomposers (Teal 1962). In fact, most of the organisms of the intertidal salt marsh obtain the bulk of their energy from dead plant material. In a Scots pine plantation 50 percent of the energy fixed annually is utilized by decomposers (Figure 11.12). The remainder is removed as yield or is stored in tree trunks (Ovington 1961). In some communities, particularly undergrazed, unconsumed organic matter may accumulate and remain out of circulation for some time, especially when conditions are not favorable for microbial action. The decomposer or detritus food chain receives additional materials from the waste products and dead bodies of both the herbivores and carnivores.

Detrital Food Chains. The detrital food chain is common to all ecosystems, but in terrestrial and littoral ecosystems it is the major pathway of energy flow. In a yellow-poplar forest *(Liriodendron)*, 50 percent of gross primary production goes into maintenance and respiration, 13 percent is accumulated as new tissue, 2 per-

cent is consumed by herbivores, and 35 percent goes to the detrital food chain (Edwards, unpublished, cited by O'Neill 1975). Two-thirds to three-fourths of the energy stored in a grassland ecosystem ungrazed by domestic animals is returned to the soil as dead plant material, and less than one-fourth is consumed by herbivores (Hyder 1969). Of the quantity consumed by herbivores, about one-half is returned to the soil as feces. In the salt marsh ecosystem the dominant grazing herbivore, the grasshopper, consumes just 2 percent of the net production available to it (Smalley 1960).

Enough has been said about the functional role of decomposers (see Chapter 3) in the detrital food chain. A quantified budget of energy flow through a detrital food chain is difficult to measure, although the use of radioactive tracers gives

Figure 11.12 The fate of the 1.3 percent of solar energy assimilated as net production by a 23-year-old Scots pine plantation. (Data from J. D. Ovington 1961.)

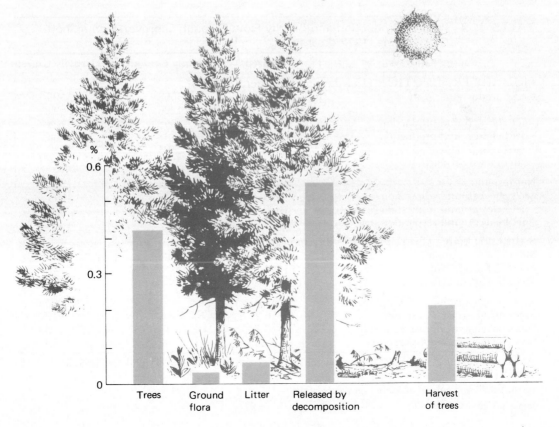

some idea of energy flow through selected food webs. Andrews et al. (1975), basing their estimates on a number of studies, determined that microbial activity accounted for 99 percent of total saprophytic assimilation in a short-grass prairie ecosystem. Saprophagic grazers accounted for the remaining 1 percent. A summary of energy flow in ungrazed and heavily grazed ecosystems is given in Table 11.3.

Gist and Crossley's (1975) study of a selected invertebrate population living in forest litter provides an example of a detrital food chain (Figure 11.13). Although no data for energy flow were collected, the flux of radioactive calcium and phosphorus permitted the construction of a food web. The quantity of elements involved in the flux provided some idea of the energy flow involved. The litter is utilized by five

groups: millipedes (Diploda), orbatid mites (Cryptostigmata), springtails (Collembola), cave crickets (Orthoptera), and pulmonate snails (Pulmonata). The mites and springtails are the most important of these litter feeders. These herbivores are preyed upon by small spiders (Araneidae) and predatory mites (Mesostigmata). The predatory mites feed on annelids, mollusks, insects, and other arthropods. Spiders feed on predatory mites; springtails, pulmonate snails, small spiders, and cave crickets are preyed upon by carabid beetles; whereas medium-sized spiders feed on cave crickets and other insects. The medium-sized spiders, in turn, become additional items in the diet of the beetles. Beetles, spiders, and snails are eaten by birds and small mammals, members of the grazing food chain not shown on the orig-

TABLE 11.3 Primary Production and Energy Flow Through Saprovores on a Short-Grass Prairie, 1972 (kcal/m^2)

Energy Pathways	Ungrazed	Lightly Grazed	Heavily Grazed
PRIMARY PRODUCTION (154-day measurement period)*			
Gross production	5838	6508	5596
Net production	3852	4296	3694
net aboveground production	562	958	728
net belowground production	3290	3338	2966
net root production	2256	2594	2539
net crown production	1034	744	427
Net respiration	1986	2212	1902
net aboveground respiration	936	1134	944
net belowground respiration (model)	1050	1078	958
net belowground respiration (lab)	598	—†	461
SAPROPHAGIC ACTIVITY (200-day measurement period)			
Microbial respiration	2990	—	2304
Microbial production	1594	—	—
Microbial assimilatin	4584	—	—
Saprophagic-grazing consumption	45.0	—	35.0
Saprophagic-grazing respiration	16.2	—	12.5
Saprophagic-grazing production	3.7	—	3.0
Saprophagic-grazing assimilation	19.9	—	15.5

*Solar radiation for 154-day plant growing season: received = 8.28×10^5 kcal/m^2, photosynthetically active = 3.89×10^5 kcal/m^2.

†Dashes indicate item not measured on this treatment.

Source: Andrews et al. 1974.

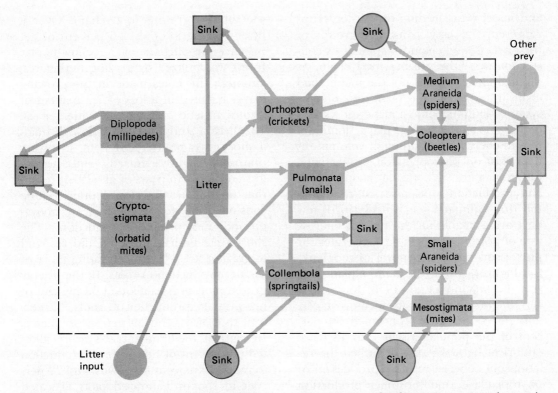

Figure 11.13 Model of a detrital food web involving litter-dwelling invertebrates. The dashed line represents the boundaries of the system. Note that the detrital food chain involves a herbivorous component—millipedes and mites to the left and crickets and springtails to the right. The herbivores support a carnivorous component. The detrital food web, like the grazing food web, can become complex. (After Gist and Crossley, Jr. 1975: 86.)

inal web. Through predation detrital food webs feed into grazing food chains at higher consumer levels. Detrital and grazing food webs are separate compartments in an ecosystem only at the detritus and primary consumer levels.

Food chains involving saprophages may take two directions (as shown in Figure 11.11), toward the carnivores or toward microorganisms. The role of these feeding groups in the final dissipation of energy has already been mentioned; but

they also are food to numerous other animals. Slugs eat the larvae of certain flies and beetles, which live in the heads of fungi and feed on the soft material. Mammals, particularly the red squirrel and chipmunks, eat woodland fungi. Dead plant remains are food sources for springtails and mites, which in turn are eaten by carnivorous insects and spiders. They in turn are energy sources for insectivorous birds and small mammals. Blowflies lay their eggs in dead animals, and within 24 hours the maggot larvae hatch. Unable to eat solid tissue, they reduce the flesh to a fetid mass by enzymatic action in which they feed on the proteinaceous material. These insects are food for other organisms.

Grazing Food Chains. The grazing food chain is the one most obvious to us. Cattle grazing on pastureland, deer browsing in the forest, rabbits feeding in old fields, and insect pests feeding on garden crops represent the basic consumer groups of the grazing food chain. In spite of its conspicuousness, the grazing food chain, as pointed out earlier, is not the major food chain in terrestrial and many aquatic ecosystems. Only in some aquatic ecosystems do the grazing herbivores play a dominant role in energy flow. Although voluminous data exist on phytoplankton productivity, filtration rates by grazing zooplankton, and production efficiencies of zooplankton (for summary, see Wetzel 1975), few data are available on the flow of energy, rate of grazing, biomass turnover rates for phytoplankton, and turnover of zooplankton biomass within the same aquatic system. Carter and Lund (1968) found that certain grazing protozoans feeding on certain planktonic algae consumed 99 percent of the populations in 7 to 14 days. Hillbricht-Ilkowska (1975) studied the relationship between primary production of phytoplankton and consumer production in freshwater lakes. He found that the direct utilization of primary production by filter-feeding zooplankton was intense and its efficiency of energy transfer was high when the size and structure of phytoplankton were favorable for filter feeding. In lakes where the net form of phytoplankton was dominant, direct utilization was low and energy transfer inefficient. As an example, in lakes in which the production of phytoplankton ranged from 200 to 1200 $kcal/m^2/yr$, production of filter-feeding zooplankton amounted to 10 to 250 $kcal/m^2/yr$ with a transfer efficiency of 2 to 30 percent. In lakes in which filter-feeding zooplankton production ranged from 50 to 150 $kcal/m^2/yr$, the production of predacious zooplankton ranged from 2 to 50 $kcal/m^2/yr$ with an energy transfer efficiency of 5 to 40 percent.

In terrestrial systems a relatively small proportion of primary production goes by way of the grazing food chain. Over a three-year period only 2.6 percent of net primary production of a yellow-poplar forest was utilized by grazing herbivores, although the holes made in the growing leaves resulted in a loss of 7.2 percent of photosynthetic surface (Reichle et al. 1973). In a study of energy flow through a short-grass prairie ecosystem, involving ungrazed, lightly grazed, and heavily grazed plots, Andrews et al. (1974) found that on the heavily grazed prairie cattle consumed 30 to 50 percent of aboveground net primary production. The short-grass prairie, however, did respond to grazing stress by concentrating more of the net production in roots. In the heavily grazed plots grasses allotted 69 percent of net primary production to roots, 12 percent to crowns, and 19 percent to shoots. In contrast, the lightly grazed prairie allotted 60 percent of net primary production to roots, 18 percent to crowns, and 22 percent to shoots. Ungrazed plots allocated only 14 percent of production to shoots. These figures suggest that light grazing stimulates primary production above ground. In both lightly and heavily grazed plots about 40 to 50 percent of energy consumed by the cattle is returned to the ecosystem and the detrital food chain as feces (Dean et al. 1975).

Although the aboveground herbivores are the conspicuous grazers, belowground herbivores can have a pronounced impact on primary production and the grazing food chain. Andrews et al. (1975) found that belowground herbivores consisting mainly of nematodes (Nematoda), scarab beetles (Scarabaeidae), and adult ground beetles (Carabidae) accounted for 81.7 percent of total herbivore assimilation on the ungrazed short-grass prairie, 49.5 percent on the lightly grazed prairie, and 29.1 percent on the heavily grazed prairie. Ninety percent of the invertebrate herbivore consumption took place below ground, and 50 percent of the total energy was processed by nematodes. On the lightly grazed prairie cattle consumed 46 kcal/m^2 during the grazing season and the belowground invertebrates consumed 43 kcal/m^2. Thus belowground herbivorous consumption can impose a greater stress on the ecosystem than the aboveground herbivores. When a nematicide was added to a mid-grass prairie, aboveground net production increased 30 to 60 percent.

Size has considerable influence on the direction a food chain takes, because there are upper and lower limits to the size of food an animal can capture. Some animals are large enough to defend themselves successfully. Some foods are too small to be collected economically; it takes too long to secure enough to meet the animal's metabolic needs. Thus the upper limit to the size of an organism's food is determined by its ability to handle and process the item, and the lower limit by the animal's ability to secure enough.

There are exceptions, of course. By injecting poisons, spiders and snakes kill prey much larger than themselves; wolves hunting in packs can kill an elk or a caribou. The idea that food chains involve animals of progressively larger sizes is true only in a general way. In the parasitic chain the opposite situation exists. The larger animals are at the base, and as the number of links increases, the size of the parasites become smaller.

A grazing food chain that has been carefully worked out (Golley 1960) involves old field vegetation, meadow mice, and weasels (Figure 11.14). The mice are almost exclusively herbivorous, and the weasels live mainly on mice. The vegetation converts abut 1 percent of the solar energy into net production, or plant tissue. The mice consume about 2 percent of the plant food available to them, and the

weasels about 31 percent of the mice. Of the energy assimilated, the plants lose about 15 percent through respiration, the mice 68 percent, and the weasels 93 percent. The weasels use so much of their assimilated energy in maintenance that a carnivore preying on weasels could not exist.

In a very general way, energy transformed through the ecosystem by way of the grazing chain is reduced in magnitude by 10 from one level to another. Thus if an average of 1000 kcal of plant energy is consumed by herbivores, about 100 kcal is converted to herbivore tissue, 10 kcal to first-level carnivore production, and 1 kcal to second-level carnivores. The amount of energy available to second- and third-level carnivores is so small that few organisms could be supported if they depended on that source alone. For all practical purposes, each food chain has from three to four links, rarely five (Pimm 1982). The fifth link is distinctly a luxury item in the ecosystem.

Supplementary Food Chains. Other feeding groups, such as the parasites and scavengers, form supplementary food chains in the community. Parasitic food chains are highly complicated because of the life cycle of the parasites. Some parasites are passed from one host to another by predators in the food chain. External parasites (ectoparasites) may transfer from one host to another. Other parasites are transmitted by insects from one host to another through the bloodstream or plant fluids.

Food chains also exist among parasites themselves. Fleas that parasitize mammals and birds are in turn parasitized by a protozoan, *Leptomonas*. Chalcid wasps lay eggs in the ichneumon or tachinid fly grub, which in turn is parasitic on other insect larvae.

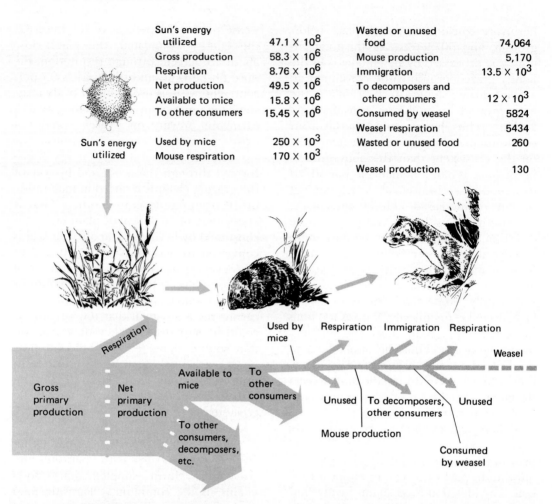

Sun's energy utilized	47.1×10^8	Wasted or unused food	74,064
Gross production	58.3×10^6	Mouse production	5,170
Respiration	8.76×10^6	Immigration	13.5×10^3
Net production	49.5×10^6	To decomposers and other consumers	12×10^3
Available to mice	15.8×10^6	Consumed by weasel	5824
To other consumers	15.45×10^6	Weasel respiration	5434
Used by mice	250×10^3	Wasted or unused food	260
Mouse respiration	170×10^3	Weasel production	130

Figure 11.14 Energy flow through a food chain in an old field community in southern Michigan. The relative size of the blocks suggests the quantity of energy flowing through each channel. Values are cal/ha^{-2}/yr^{-1}. (Based on data from Golley 1960.)

Models of Energy Flow

The concept of energy flow in ecological systems is one of the cornerstones of ecology. The model was first developed by Raymond Lindeman in 1942 in a study of the trophic dynamic structure of Lake Mendota in Wisconsin. The trophic dynamic concept is based on the assumptions that the laws of thermodynamics hold for plants and animals, that plants and animals can be arranged into feeding groups or trophic levels, that at least three trophic levels—producer, herbivore, and carnivore—exist, that the net energy content of one trophic level is passed on to the next trophic level above, and that the system is in equilibrium.

Whether at the producer or consumer level, energy flow through the ecosystem is mediated by the individual (Figure 11.15a). A quantity of energy in food is

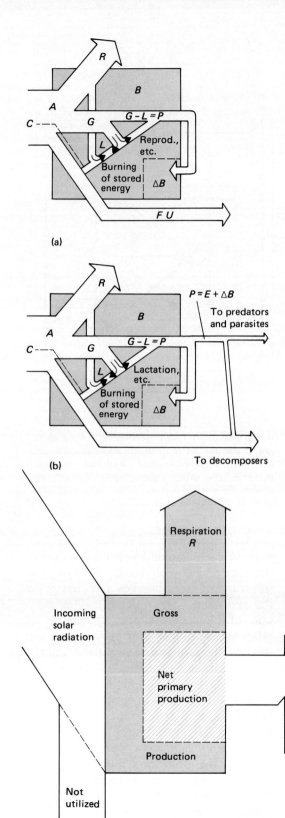

(a)

(b)

(c)

consumed. Part of it is assimilated and part is lost as feces, urine, and gas. Part of the assimilated energy is used for respiration, and part is stored as new tissue which can be used to some extent as an energy reserve. Some is used for growth and the production of new individuals.

A model of energy flow through a population (Figure 11.15b) shows some additions. Growth in individual biomass becomes changes in size of standing crop. Part of the biomass goes to predators and parasites, and there are gains and losses of energy and biomass from and to other ecosystems.

The population boxes can be fitted into several trophic levels and linked to form a model of energy flow through the ecosystem (Figure 11.15c). In this model

Figure 11.15 Models of energy flow (a) through the individual organism, (b) through the population, and (c) through the ecosystem. Note the losses and the portion of energy accumulated as growth in the organism. The ecosystem model considers the decomposers as occupying one of several trophic levels rather than as a separate trophic pathway. A = assimilation, B = biomass, B = changes in biomass, C = consumption, E = material removed from population to ecosystem, FU = rejecta, G = total biomass growth, L = decrement of biomass through weight loss and death, R = respiration.

there is no attempt to separate out the decomposers and detritus feeders, because they must fall into one of the several trophic levels when the food web is collapsed.

To provide for the grazing and detritus paths of energy transfer, Wiegert and Owens (1971) propose a somewhat different model of energy flow, with transfers through two channels beyond the autotrophs (Figure 11.16). They define all organisms utilizing living material as biophages, and all organisms utilizing nonliving matter as saprophages. First-order biophages, who eat living plants, are the traditional herbivores; whereas first-order saprophages feed on dead plant material as well as organic material egested by first-order biophages. First-order biophages in turn are utilized at death by second-order saprophages, which are functional carnivores. In turn, first-order saprophages may be preyed upon by second-order biophages. Although this model, like others, cannot separate out organisms that occupy several trophic levels, it does modify the Lindeman model so that the decomposers and detritivores are assigned to functional components.

Trophic Levels and Ecological Pyramids

Feeding relationships in an ecosystem may be considered in terms of trophic or feeding levels. From a functional rather than a species point of view, all organisms that obtain their energy in the same number of steps from the producers belong to the same trophic level in the ecosystem. The first trophic level belongs to the producers, the second level to the herbivores, and higher levels to the carnivores. Some animals occupy a single trophic level, but many others occupy more than one trophic level.

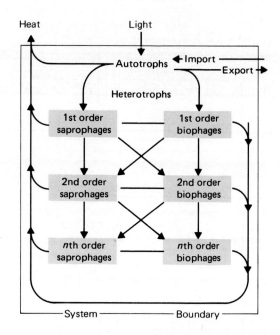

Figure 11.16 Model of two-channel energy flow. The model separates energy flow into two pathways, one utilized by organisms that feed on living material, the biophages, and the other utilized by organisms that feed on nonliving material, the saprophages. (After Weigert and Owen 1971: 71.)

By considering all of the living biomass in each trophic level and the energy transferred between levels, we can construct pyramids of biomass and energy for the ecosystem (Figure 11.17). The pyramid of biomass indicates by weight or other measurement of living material the total bulk of organisms or fixed energy present at any one time—the standing crop (Figure 11.17b). Because some energy or material is lost in each successive link, the total mass supported at each level is limited by the rate at which energy is being stored below. In general, the biomass of the producers must be greater than the biomass of the herbivores they support, and the biomass of the herbivores must be greater than that of the carnivores. This fact usually results in a gradually sloping pyramid, particularly for terrestrial and shallow-water communities where the producers are large, accumulation of organic matter is large, life cycles are long, and the rate of harvesting is low.

However, for some ecosystems the pyramid of biomass is inverted. Primary production in aquatic ecosystems such as lakes and open seas is concentrated in the microscopic algae. The algae have a short life cycle, multiply rapidly, accumulate little organic matter, and are heavily exploited by herbivorous zooplankton. At any point the standing crop is low. As a result, the base of the pyramid is much smaller than the structure it supports.

When production is considered in terms of energy, the pyramid indicates not only the amount of energy flow at each level, but, more important, the role the various organisms play in the transfer of energy. The base upon which the pyramid of energy is constructed is the quantity of organisms produced per unit or, stated differently, the rate at which food material passes through the food chain (Figure 11.17c). Some organisms have a small biomass, but they assimilate and pass on total energy considerably greater than that of organisms with a much larger biomass. On a pyramid of biomass these smaller organisms would appear much less important in the community than they really are.

Another pyramid commonly found in ecological literature is the pyramid of numbers (Figure 11.17a). This pyramid was advanced by Charles Elton (1927), who pointed out the great difference in the numbers of organisms involved in each step of the food chain. The animals at the lower end of the chain are the most abundant. Successive links of carnivores decrease rapidly in number until there are very few carnivores at the top. The pyramid of numbers often is confused with a similar one in which organisms are grouped into size categories and then arranged in order of abundance. Here the smaller organisms again are the most abundant, but such a pyramid does not indicate the relationship of one group to another.

The pyramid of numbers ignores the biomass of organisms. Although the numbers of a certain organism may be greater, their total weight may not be equal to that of the larger organisms. Neither does the pyramid of numbers indicate the energy transferred or the use of energy by the groups involved. And because the abundance of members varies so widely, it is difficult to show the whole community on the same numerical scale.

Ecological Efficiencies

Ecological efficiency is the ratio of any of the various parameters of energy flow within or between trophic levels, populations, and individual organisms (Table 11.4) (Kozlovsky 1968). Efficiencies of individuals, however, are more physiological than ecological.

An ecological rule of thumb allows a magnitude of 10 reduction in energy as it passes from one trophic level to another (Slobodkin 1962). If 1000 kilocalories were consumed by herbivores, about 100 kilocalories would be converted into herbivore tissue, 10 kilocalories into first-level carnivore production, and 1 kilocalorie into the second-level carnivore. However, based on data available, a 90 percent loss of energy from one trophic position to another may be too high. Certainly a wide range in the efficiency of conversion exists among various feeding groups. Production efficiency in plants (net production/light absorbed) is low, ranging from 0.34 percent in some phytoplankton to 0.8 to

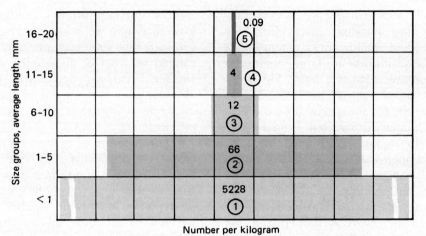

(a)

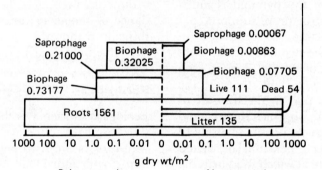

(b)

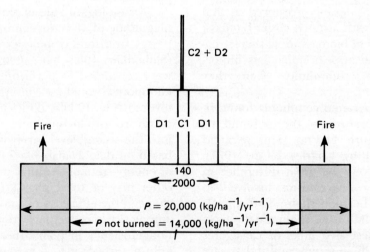

(c)

Figure 11.17 Examples of ecological pyramids. (a) A pyramid of numbers among the metazoans of the forest floor in a deciduous forest. (After Park, Allee, and Shelford 1939.) (b) A pyramid of biomass for a northern short-grass prairie for July. The base of the pyramid represents biomass (g dry weight/m^2) of producers; the second (middle) level, primary consumers; and the top secondary consumers. Aboveground biomass (right) and belowground biomass (left) are separated by a dashed vertical line. The trophic level magnitudes are plotted on a horizontal logarithmic scale. The compartments are divided on a vertical linear scale according to live, standing dead, and litter biomass, or biophagic and saprophagic consumer biomass. Unlike the conventional pyramids, this one recognizes the detrital and grazing food chain components on the same trophic levels. (After French, Steinhorst, and Swift 1979.) (c) A pyramid of energy for the Lamto Savanna, Ivory Coast. P is primary production; C1, primary consumers; C2, secondary consumers; D1, decomposers of vegetable matter; D2, decomposers of animal matter. Again, the detrital and grazing food chains have been collapsed into the same trophic levels. (After Lamotte 1975: 216.)

0.9 percent in grassland vegetation. Plant production consumed by herbivores is utilized with a varying degree of efficiency. Herbivores consuming green plants are wasteful feeders, but not nearly as wasteful as those that feed on plant sap (Table 11.5). More energy loss occurs in assimilation. Assimilation efficiencies vary widely among poikilotherms and homoiotherms (Table 11.2). Homoiotherms are much more efficient in assimilation than poikilotherms (see Chapter 6); but all carnivores, even poikilotherms, have a high assimilation efficiency. Predaceous spiders feeding on invertebrates have assimilation efficiencies of over 90 percent (Moulder, Reichle, and Auerbach 1970; Van Hooke 1971). About 2 to 10 percent of the energy consumed by herbivorous homoiotherms goes into biomass production, less than the suggested 10 percent average; however, poikilotherms convert 17 percent of their consumption to herbivorous biomass. On midwestern grasslands, average herbivore production efficiency, involving mostly poikilotherms, ranges from

TABLE 11.4 Definitions of Selected Energy Efficiencies

BETWEEN TROPHIC LEVELS

Energy intake(Lindeman) efficiency	$\dfrac{P}{L}$	$\dfrac{\text{gross production}}{\text{light absorbed}}$
Assimilation efficiency	$\dfrac{A_t}{A_{t-1}}$	$\dfrac{\text{assimilation } t}{\text{assimilation } t-1}$
Production efficiency	$\dfrac{P_t}{P_{t-1}}$	$\dfrac{\text{production of biomass } t}{\text{production of biomass } t-1}$

WITHIN TROPHIC LEVELS

Production efficiency	$\dfrac{P_t}{A_t}$	$\dfrac{\text{production at } t}{\text{assimilation at } t}$
Growth efficiency	$\dfrac{P_t}{I_t}$	$\dfrac{\text{production at } t}{\text{energy ingested at } t}$
Assimilation efficiency	$\dfrac{A_t}{I_t}$	$\dfrac{\text{assimilation at } t}{\text{energy intake at } t}$

TABLE 11.5 Consumer Waste (kcal/m²/season) in a Mixed Prairie Grassland

Consumers	Consumption	Wastage	Production
ABOVEGROUND			
Primary consumers			
Plant-tissue–feeders			
Mammals	1.60	.53	.018
Arthropods	6.33	4.35	1.13
Plant-sap–feeding arthropods	24.72	49.44	5.16
Pollen-nectar–feeding arthropods	.98	.20	.26
Seed-feeders			
Birds	.16	.00	$.505 \times 10^{-8}$
Mammals	.20	.00	.002
Arthropods	.17	.03	.04
Dead plant-litter–feeding arthropods	9.88	4.26	1.48
Secondary consumers			
Predators			
Birds	.88	.00	5.66×10^{-8}
Mammals	.20	.00	.003
Arthropods	3.25	.65	.94
Scavenger arthropods	.34	.17	.04
BELOWGROUND			
Primary consumers			
Plant-tissue–feeders	244.26	107.05	34.39
Plant-sap–feeders	382.38	240.10	64.74
Fungal-feeders	33.66	18.55	5.51
Bacteria-feeders	66.53	13.31	14.63
Secondary consumers			
Predators	117.83	23.57	27.63
Protozoa-feeders	31.65	6.33	7.92

Source: Scott, French, and Leetham 1979: 97, 98, 99.

5.3 to 16.5 percent (Table 11.6). Production efficiency on the secondary consumer level ranges from 13 to 24 percent.

Transfer of energy from one trophic level to another tells the real story, but such data are hard to collect. The ratio of phytoplankton to secondary zooplankton production in freshwater ecosystems is 1:7.1, according to Brylinsky (1980), and the ratio of herbivore zooplankton production to carnivorous zooplankton production is 1:2.1. Efficiencies are lower in the benthic community—2.2 for herbivores and 0.3 for carnivores.

Energy transfer efficiency (consumption at trophic level t/secondary production $t - 1$) among invertebrate consumers on a short-grass plain is 9 percent for herbivores (Gyllenberg 1980), 28 percent for saprophages (Kajah et al. 1980), 38 percent for aboveground predators, and 56 percent for belowground predators. Reichle (1971) calculated the trophic level production efficiency (production at trophic level n/ production at level $n - 1$) for soil and litter invertebrates in a deciduous forest ecosystem: saprophages, 0.11 to 0.17; phytophages, 0.02 to 0.07; and predators, 0.02.

TABLE 11.6 Consumer Efficiency (Secondary Production/Secondary Consumption)

Habitat	Growing Season, Days	Producers		Herbivores		Carnivores	
		Production	% Efficiency	Production	% Efficiency	Production	% Efficiency
Short-grass plains	206	3.767	0.8	53	11.9	6	13.2
Mid-grass prairie	200	3.591	0.9	127	16.5	37	23.7
Tall-grass prairie	275	5.022	0.9	162	5.3	15	13.9

Source: Based on data from Andrzejewska and Gyllenberg 1980.

Community Energy Budgets

Because energy flow involves both inputs and outputs, the efficiency and production of an ecosystem can be estimated by measuring the quantity of energy entering the community through the various trophic levels and the amount leaving it. Such information can give *net ecosystem production* (or net community production, as it is often called), which can be expressed as:

net ecosystem production (biomass accumulation) = gross primary production − plant respiration − animal respiration − decomposer respiration

A balance sheet for energy flow and production with debit and credit sides can be drawn up for a community, provided we can calculate energy flow through populations within the ecosystem and then relate this information to the flow of energy from one trophic level to the other. Herein lies the problem with community energy budgets. Those drawn to date are based in part on assumed rather than known values of energy flow. Such budgets assume that energy flow through a population is constant, when in reality it is fluctuating. For example, animal populations may have a pronounced influence on the rate of energy fixation of plants. Nutrient inputs can limit energy fixation and storage in both primary and secondary production.

Few communities have been studied intensively enough to present such a broad picture, but some studies are available. One is a salt marsh, an autotrophic community (Teal 1962), and another is a cold spring, a largely heterotrophic community (Teal 1957) whose major energy source is plant material fallen into the water. Table 11.7 shows the data.

■ **Food Web Theory**

Trophic relationships in nature are not simple, straight-line food chains. Numerous food chains mesh in a complex food web with links leading from producers through an array of primary and secondary consumers. Producers support numerous primary consumers, who support many higher consumers.

It is hard to apply the concept of trophic levels to food webs, because some organisms feed on more than one trophic level. Just what is the position of an organism such as the sparrows in Figure 11.9? They feed both on primary producers

---

TABLE 11.7 Community Energy Balance Sheets

Autotrophic Community: Salt Marsh

Input as light	6000,000 kcal/m²/yr
Loss in photosynthesis	563,620, or 93.9%
Gross production	36,380, or 6.1% of light
Producer respiration	28,175, or 77% of gross production
Net production	8,205 kcal/m²/year
Bacterial respiration	3,890, or 47% of net production
First-level consumer respiration	596, or 7% of net production
Second-level consumer respiration	48, or 0.6% of net production
Total energy dissipation by consumers	4,534, or 55% of net production
Export	3,671, or 45% of net production

Heterotrophic Community: Temperate Cold Spring

Organic debris	2,350 kcal/m²/year, or 76.1% of available energy
Gross photosynthetic production	710 kcal/m²/year, or 23.0% of available energy
Immigration of caddis larvae	18 kcal/m²/year, or 0.6% of available energy
Decrease in standing crop	8 kcal/m²/year, or 0.3% of available energy
Total energy dissipation to heat	2,185 kcal/m²/year, or 71% of available energy
Deposition	868 kcal/m²/year, or 28% of available energy
Emigration of adult insects	33 kcal/m²/year, or 1% of available energy

Source: For the salt marsh, Teal 1962; for the cold spring, Teal 1957.

(seeds), putting them on level 2, and on herbivorous insects, placing them on level 3. How do we measure the amount of energy that the species contributes to each of the trophic levels it occupies? Some species have different food habits during different seasons or stages of their life cycle. How do such species figure in determining the maximum number of trophic levels?

Recent interest in the nature and role of food webs in community structure and energy flow makes such questions important. Ecologists working with food web theory have arrived at a somewhat different view of trophic levels (Yodzis 1988, Cohen 1989, Pimm 1982). They call species in a food web with the same diets and same predators *trophic species,* and classify them as basal, intermediate, and top species. Basal species, which include both primary producers and detritus, are at the bottom of the food web and feed on no

other species. The top species, occupying the apex of the food web, are ones on which no other species feed. Intermediate species are neither basal nor top; they may feed on more than one trophic level. Predators are species that feed on other species in the web, and prey are species that are fed on by some other species.

Instead of attempting to define trophic levels, Yodzis (1988) suggests the idea of *trophic height,* which he defines as 1 plus the length of the shortest food chain that links a particular species in the food web to the basal species. Species in a food web are linked together by several food chains of differing energy transfer efficiencies. At any given trophic height, an invertebrate ectotherm should produce more energy than a vertebrate ectotherm, and the latter should produce more energy than an endotherm (see Chapter 6). Because such efficiencies are difficult to quantify, Yodzis assigns basal species and invertebrate ectotherms a length of 1, vertebrate ecto-therms a length of 1.5, and endotherms a length of 2.

Figure 11.18, a section of the food web in Figure 11.9, is a simple example of trophic heights and food chain links. In the conventional approach the sparrows occupy trophic levels 2 and 3. If we consider the shortest food chain that links the sparrows to basal species, they occupy trophic height 2 (1 + length 1). The shrew occupies trophic height 3 (1 + length 1 + length 1). The top species, the marsh hawk, also occupies trophic height 3. In the maximal food chain, the trophic height of the hawk would be 4.5. The utility of such an approach is obvious.

Current interest in food webs has arisen out of two long-standing generalizations. One is that the maximum number of trophic levels in a food web is controlled by the second law of thermodynamics, that energy transfer from one trophic level to another is always inefficient. The other is that complex food webs

are more stable than simple ones. In the 1970s these two generalizations were questioned (May 1973; Pimm and Lawton 1977; Pimm 1982). Scientists began to ask new questions: What determines the size of food webs and the number of trophic levels? How are food webs organized and structured? How are food webs affected by the invasion of new species and the deletion of current ones? What makes food webs stable? What processes, including population dynamics and energy flow, affect food webs?

J. E. Cohen (1978), F. Briand (1983a, 1983b), and Briand and Cohen (1984) assembled published food webs—terrestrial, marine, and freshwater—in an effort to detect structure or nonrandomness in them. These food webs became the data base for much of the development of food web theory. Such data have their shortcomings. None of the food webs collected were obtained originally with a food web study in mind, but rather to characterize certain environments and relationships among a few species. Published food webs tend to be rather general at the basal or producer level and most detailed at the upper trophic levels. Often these food relations are based on general observations and not on detailed food studies. Further, most published food webs are grazing ones that ignore the associated detrital food webs. In spite of deficiencies, the analysis of the food webs has provided some interesting insights into food webs and community structure.

Briand subjected 40 food webs to multivariate analysis. He was able to separate food webs into two large groups: those found in fluctuating environments characterized by variations in temperature, salinity, pH, moisture, and other conditions, and those found in constant environments. Food webs in constant environments are characterized by greater species

Figure 11.18 A portion of the food web from Figure 11.9. Each link in the web is marked with its length, and each species is numbered with its trophic height. (Figure suggested by Yodzis 1988: 241.)

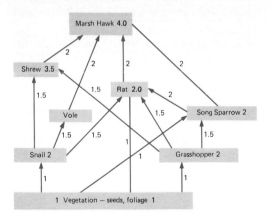

richness and more trophic links (connectance) than food chains in fluctuating environments. The latter tended to be shorter, with fewer trophic links, possibly because environmental conditions imposed greater rigidity on the shape of the webs. Organisms in fluctuating environments may be forced to optimize their foraging (see Chapter 19), restricting their choice of prey. The widest food chains, those with the greatest number of herbivorous species, were the shortest. Narrow food webs, by contrast, had the greatest fraction of top carnivores. In general, Briand found that distinctive habitats or ecosystems possessed their own distinctive food web structure. Food webs in benthic regions, for example, are shorter than pelagic food webs, probably because of their detrital base.

Food chains tend to be short, rarely exceeding four links (Pimm 1982). No matter how productive an ecosystem, the number of links remains essentially the same. Increased productivity of an ecosystem does not, as often assumed, support longer food chains. It may support more species and thus more complex food webs, but as the number of species in the web increases, the connectance decreases. In general the number of links in a food web is about twice the number of species (Cohen 1989).

In any food web there are four possible connectances: (1) top and intermediate, (2) different intermediates, (3) intermediate and basal, and (4) top and basal. The proportion in each is about the same for every habitat: 35 percent in the first category, 30 percent in the second, 27 percent in the third, and 8 percent in the fourth (Cohen 1989). Rarely does the top link with the basal. The proportion of species within each web that are basal, intermediate, and top appears to be independent of species richness (Briand and Cohen 1984).

How food chains might be assembled is another question. Are they a product of random assemblage or not? To seek some answer to that question Yodzis (1981, 1982, 1983) simulated the development of a food web based on natural parameters. Starting with a number, N, of basal species, Yodzis added additional species, each of which had a certain ecological efficiency, making a fraction of its own consumption available to the next trophic level. Each new species added had to obtain its energy from other species. Each species had to choose a food source already utilized by another. At the same time, a newly arriving species had a total production, a fraction of which had to be made available to a subsequently arriving species. There is, however, an upper and lower limit to the number of species a predator can exploit and a point at which total production is too low to allow a new species to enter. Thus the limitation of energy transformation (the second law) forces a pattern to a food web. Yodzis found a certain degree of similarity between his simulated food webs and real food webs.

The kind of species that can enter a food web are also a subject for simulation studies. It turns out that generalist species most easily invade simple food webs, and specialists, capable of exploiting a restricted source of energy, are best able to invade complex webs (Drake 1983). These studies suggest that food webs development is not random but regulated by properties of existing food webs and invading species.

What happens if a species is removed from or reduced in a food web? Theoretically, there is little effect if the species removed is a generalized prey species or a generalist predator species, or a member of a simple, straight-line food chain (Pimm 1980). If the species removed is a top predator, then the effect moves from the top species down through the intermediate to the basal. Some of the best examples come from experiments in fisheries and lake productivity (Carpenter, Kitchell, and Hodgson 1985). Consider a top piscivore, a predatory fish such as a bass. It reduces the biomass of vertebrate planktivores, allowing invertebrate planktivore biomass and large zooplankton herbivores to increase, and causing small herbivore zooplankton and primary production to decrease. If this top predator is reduced, vertebrate planktivores increase, invertebrate planktivores and large herbivorous zooplankton decrease, and small herbivore zooplankton and primary production increase.

J. Cohen (1989, Cohen and Newman 1986) has proposed a quantitative cascade model for food webs. A community may contain any number of species from l to S. Their numbering specifies a pecking order of feeding. Any species j in a hierarchy of feeding can, but does not necessarily, feed on any species i with a lower number $i < j$. In turn j cannot feed on any species with a number k at least as large as $k > j$. The model assumes that each species actually eats any species below it with a probability d/S independently of what else is going on. The probability that the species does not eat species $i < j$ is $1 - d/S$. The two assumptions of an ordering and a probability of feeding proportional to $1/S$ is the cascade model. Models of food webs may provide a mechanism to study the accumulation of environmental toxins along food chains and to derive quantitative predictions of the effects of introducing or eliminating species in a community.

This brief introduction to developing food web theory passes over some of the more theoretical aspects (see Pimm 1982; Yodzis 1988). Ecologists now are beginning to test hypotheses in the field (Pimm and Kitching 1987).

■ Summary

A basic functional characteristic of the ecosystem is the flow of energy. The energy of sunlight is fixed by the autotrophic component of the ecosystem, green plants, as primary production. This energy is then available to the heterotrophic component of the ecosystem, of which the herbivores are the primary consumers. The herbivores in turn are a source of food for the carnivores. At each step or transfer of energy in the food chain a considerable amount of potential energy is lost as heat, until ultimately the amount of available energy is so small that few organisms can be supported. The animals above the herbivore level on the chain often utilize several sources of energy, including plants, and thus become omnivores. All food chains eventually end with the decomposers, mainly bacteria and fungi, that reduce the remains of plants and animals into simple substances. Energy flow in the ecosystem may take two routes. One goes through the grazing food chain, the other through the detritus food chain, in which the bulk of the production is utilized as dead organic matter by the decomposers.

The loss of energy at each transfer limits the number of trophic levels or steps in the food chain from three to five. At each level the biomass declines; if the total weight of individuals at each successive tropic level is plotted, a sloping pyramid is formed. In certain aquatic situations, however, where there is a rapid turnover of small aquatic consumers, the pyramid of biomass may be inverted. Energy, however, always decreases from one trophic level to another and is pyramidal.

The ratio of energy flow in or between trophic levels of natural communities or in or between individual organisms is ecological efficiency. Because efficiencies are dimensionless, any number of ratios can be determined. Among some of the most useful are assimilation efficiencies, growth efficiencies, and utilization efficiencies.

Energy flow involves complex trophic relationships known as food webs, about which food web theory has grown. This theory considers the size, organization, and structure of food webs, as influenced by environment, numbers of, invasions by, and loss of species, and the relationships of one trophic level to another.

Although knowledge of energy flow in ecosystems is fragmentary and difficult to come by, the concept of energy flow is a valuable guide for the study and understanding of both ecosystem functioning and the relationship of humans to their environment.

Biogeochemical Cycling

The living world depends upon the flow of energy and the circulation of materials through the ecosystem. Both influence the abundance of organisms, the metabolic rate at which they live, and the complexity and structure of the ecosystem. Energy and materials flow through the community as organic matter; one cannot be separated from the other. Certain elements are essential for energy fixation and flow; mineral cycling cannot function without the input of energy. One feeds upon the other in an apparent positive feedback loop: The availability of nutrients limits primary productivity; the more nutrients available, the more energy flows; increased energy flow produces greater cycling of nutrients. What prevents the sys-

tem from running away is a negative feedback loop of limits on available nutrients, the energy requirements of resource utilization, and limits on the ability of the system to regenerate nutrients. In essence two forces are at work. One is the tendency within ecosystems for living organisms to store resources in the form of growth. The other is the tendency for organic matter to degrade into inorganic compounds.

■ Gaseous Cycles

There are two types of biogeochemical cycles, the *gaseous* and the *sedimentary*. The main reservoirs of nutrients possessing a gaseous cycle are the atmosphere and the ocean, so those cycles are pronouncedly global. In sedimentary cycles the main reservoir is the soil and rocks of Earth's crust. Both types involve biological and nonbiological agents, both are driven by the flow of energy (Chapter 11), and both are tied to the water cycle (Chapter 5). Gaseous cycles basic to biological life are those of oxygen, carbon dioxide, nitrogen, and hydrogen.

The Oxygen Cycle

Oxygen, the by-product of photosynthesis, is very active chemically. It can combine with a wide range of chemicals in Earth's crust and react spontaneously with organic compounds and reduced substances. It is involved in the oxidation of carbohydrates with the release of energy, carbon dioxide, and water. Its primary role in biological oxidation is that of a hydrogen acceptor. The breakdown and decomposition of organic molecules proceeds primarily by dehydrogenation. Hydrogen is removed by enzymatic action from organic molecules in a series of reactions and is finally accepted by oxygen, forming water.

Oxygen, although necessary for life, can be toxic, as it is to anaerobic bacteria. Higher organisms have evolved means to protect themselves from the toxic effects of molecular oxygen, including organelles called peroxisomes within the cells. The oxidative reactions they mediate produce hydrogen peroxide. Peroxide in the presence of certain enzymes becomes an acceptor in the oxidation of other compounds, but the energy involved is not utilized by cells.

At the same time higher organisms have also evolved elaborate mechanisms to ensure a supply of oxygen. Plants have stomata to regulate the movement of gases into and out of the leaf. Some animals obtain oxygen from air or water by diffusion through the skin; others have gills or lungs. Organisms have an array of catalysts such as iron-containing molecules, copper-containing enzymes, cytochromes, and cytochrome oxidases to mediate the transfer of hydrogen to oxygen molecules. In higher organisms the mitochrondia house this oxidative system, which acts as a low temperature furnace in which organic molecules are slowly burned. The energy evolved is used to form the energy-rich bonds of ATP, stable enough to be transported but unstable enough to break down when energy is needed.

The major supply of free oxygen that supports life is provided by two significant sources in the atmosphere. One source is the photodissociation of water vapor, in which most of the hydrogen released escapes into outer space. The other source is photosynthesis, active only since life began on Earth. Because photosynthesis and respiration are cyclic, involving both the release and utilization of oxygen, one would seem to balance the other, so no significant quantity of oxygen would accumulate in the atmosphere. At some time in Earth's history the amount of oxygen introduced into the atmosphere had to exceed the amount involved in the decay of organic matter and oxidation of sedimentary rocks. Part of the atmospheric oxygen today would have been used to decompose organic matter buried as the unoxidized reserves of photosynthesis—coal, oil, gas, and organic carbon in sedimentary rocks. The amount of stored carbon in Earth suggests that 150×10^{20} g oxygen was once available to the atmosphere, over ten times as much as now is present, 10×10^{20} g (F. S. Johnson 1970).

The major nonliving oxygen pool consists of molecular oxygen, water, and carbon dioxide, all intimately linked in photosynthesis and other oxidation-reduction reactions. Oxygen is biologically exchangeable in such compounds as nitrates and sulfates, utilized by organisms that reduce them to ammonia and hydrogen sulfide.

Because oxygen is so reactive, its cycling is complex. It reacts rapidly with reduced organic matter in the soil and reduced mineral constituents of Earth's crust. Buried organic matter that escapes oxidation acts to conserve atmospheric oxygen by contributing its portion to replace the oxygen consumed by weathering, stabilizing the atmospheric reservoir (Figure 12.1). The amount of oxygen remaining in the atmosphere after oxidation is on the order of 1 part per 10,000, but the amount in the atmosphere remains resis-

tant to short-term changes of 100 to 1000 years. The residence time of atmospheric oxygen before its removal by respiration and decay and replacement by photosynthesis is on the order of 4500 years (Walker 1984).

As a constituent of carbon dioxide, oxygen circulates freely throughout the bio-

Figure 12.1 The oxygen cycle. Bold letters denote reservoirs of oxygen or reduced matter with which oxygen can react. The units are 10^{12} moles of either oxygen or the capacity of the reservoir to combine with oxygen. Processes are indicated by lowercase letters. The rates are expressed in 10^{12} moles per year of oxygen or equivalent capacity to combine with oxygen. (Adapted from Walker 1980.)

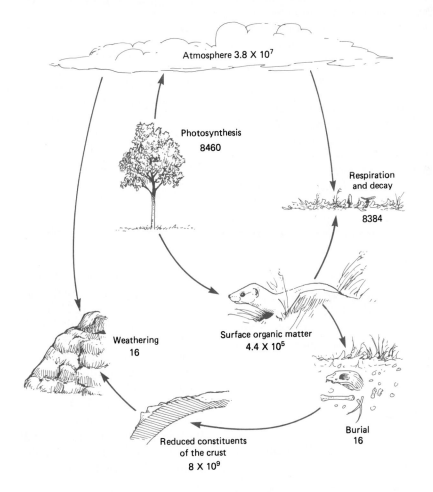

Atmosphere 3.8×10^7

Photosynthesis
8460

Respiration
and decay
8384

Weathering
16

Surface organic matter
4.4×10^5

Burial
16

Reduced constituents
of the crust
8×10^9

sphere. Some carbon dioxide combines with calcium to form carbonates. Oxygen combines with nitrogen to form nitrates, with iron to form ferric oxides, and with many other minerals to form various other oxides. Part of the atmospheric oxygen that reaches the higher levels of the troposphere reduces to ozone (O_3) in the presence of high-energy ultraviolet radiation.

The Ozone Layer

Most of the ozone resides in the stratosphere, the region characterized by increasing temperature with altitude 10 to 50 kilometers above Earth. The temperature increases because the upper relatively thin layer of ozone absorbs most of the ultraviolet radiation. A downward intrusion of stratospheric air supplies the troposphere with the O_3 necessary to initiate photochemical processes in the lower atmosphere. Most importantly, the ozone layer shields Earth against biologically harmful solar ultraviolet radiation.

Ozone in the outer atmosphere is maintained by a cyclic photolytic reaction. Production of ozone requires the breakage of the O–O bond in O_2. Once freed, the O atoms rapidly combine with O_2 to form O_3:

$$O + O_2 + M \rightarrow O_3 + M$$

where M is any third-body molecule such as N_2 or O_2 that aids in the reaction but is not changed by it. In the stratosphere the O_2 bond is broken (dissociated) by solar radiation:

$$O_2 + hv \rightarrow O + O$$

where h is Planck's constant (cal/sec) and v is frequency of the light. This stratospheric reaction consumes a large amount of solar energy.

At the same time a reverse reaction consumes O_3:

$$O_3 + hv \rightarrow O + O_2$$

$$O + O_3 \rightarrow 2O_2$$

Net reaction: $2O_3 + hv \rightarrow 3O_2$

There are a number of other catalytic reactions that tend to destroy O_3 and maintain a balance between O_2 and O_3, especially the gaseous oxides of nitrogen and hydrogen. For example, NO consumes O_3 in the following catalytic cycle:

$$NO + O_3 \rightarrow NO_2 + O_2$$

$$O_3 + hv \rightarrow O + O_2$$

$$NO_2 + O \rightarrow NO + O_2$$

Net reaction: $2O_3 + hv \rightarrow 3O_2$

Thus the NO consumed in the first reaction is replaced by another molecule available for further destruction of ozone in the final reaction, and O_3 is consumed.

Under natural conditions in the atmosphere a balance exists between the rates of ozone formation and destruction. However, in recent times a number of anthropogenic and even biological catalysts injected into the stratosphere are reactive enough to cause a decrease in stratospheric ozone (see Cicerone 1987). Among them are chlorofluorocarbons (CCl_2F_2 and CCl_3F), methane (CH_4), both natural and anthropogenic, and nitrous oxide (NO_2) from denitrification and synthetic nitrogen fertilizer. Of particular concern is chlorine monoxide (ClO) derived from chlorofluorocarbons in aerosol spray propellants (banned in the United States), refrigerants, solvents, and other sources. Its chlorine can be be tied up in harmless chlorine nitrates ($ClNO_3$), or it can be involved in the catalytic destruction of ozone (Figure 12.2):

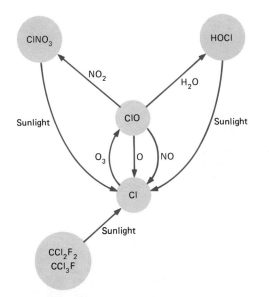

Figure 12.2 The cycle of ozone destruction induced by chlorine monoxide (ClO) derived from human use of chlorofluorocarbons. The chlorine can either be tied up in harmless chlorine nitrates or be involved in the photolytic dissociation of ozone. (After Kerr 1987: 1182.)

$$Cl + O_3 \rightarrow ClO + O_2$$

$$O_3 + h\nu \rightarrow O + O_2$$

$$ClO + O \rightarrow Cl + O_2$$

Net reaction: $2O_3 + h\nu \rightarrow 3O_2$

In 1985 atmospheric scientists discovered a pronounced springtime thinning in the ozone layer over Antarctica. The question is whether the thinning results from chemical destruction by chlorofluorocarbons or from a natural effect of upper-level winds. Detection of chlorine monoxide in the hole hints that ClO is the problem. Models of the behavior of chlorofluorocarbons in the stratosphere suggest that total ozone could be diminished as much as 16 percent in the lower stratosphere, where aerosols recycle the chlorine from inactive reservoirs (Solomon et al. 1984; Kerr 1987).

Reduced amounts of ozone allow disproportionate amounts of ultraviolet radiation to penetrate Earth's atmosphere. For example, under the conditions of an overhead sun and average amounts of O_3, a 1 percent decrease in ozone would result in a 20 percent increase in ultraviolet penetration at wavelengths of 250 nm, a 250 percent increase at 290 nm, and a 500 percent increase at 287 nm (Walker 1977). Although solar ultraviolet wavelengths of less than 240 nm are absorbed by both atmospheric O_2 and O_3, only O_3 is effective for wavelengths between 240 and 320 mn (Cicerone 1987). Such reductions in the ozone layer can have adverse ecological effects on Earth. In addition to altering DNA and increasing skin cancer, changes in the ozone layer could increase the temperature of the lower atmosphere, change air circulation patterns, and contribute to the greenhouse effect (see page 257).

The Carbon Cycle

Because it is a basic constituent of all organic compounds and a major element in the fixation of energy by photosynthesis, carbon is so closely tied to energy flow that the two are inseparable. In fact the measurement of productivity (see Appendix C) is commonly expressed in terms of grams of carbon fixed per square meter per year. The source of all fixed carbon, both in living organisms and fossil deposits, is carbon dioxide, CO_2, found in the atmosphere and dissolved in the waters of Earth. To trace its cycling through the ecosystem is to redescribe photosynthesis and energy flow (Figure 12.3).

The cycling of carbon dioxide involves assimilation by plants and conversion to glucose. From glucose plants synthesize polysaccharides and fat and store them in the form of plant tissue. In digesting plant tissue herbivores synthesize these compounds into other carbon compounds.

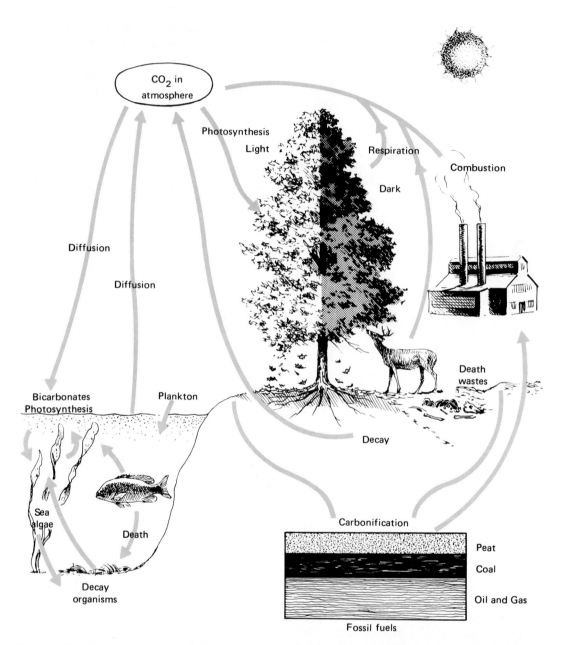

Figure 12.3 The carbon cycle. Although the main reservoir is the gas CO_2, considerable quantities are tied up in organic and inorganic compounds of carbon in the biosphere.

Meat-eating animals feed on herbivores and the carbon compounds are redigested and resynthesized into other forms.

Some of the carbon is returned directly by both plants and animals in the form of CO_2 as a by-product of respiration. The remainder for a time becomes incorporated in the living biomass.

Carbon contained in animal wastes and in the protoplasm of plants and animals is released eventually by assorted decomposer organisms. The rate of release depends upon environmental conditions such as soil moisture, temperature, and precipitation. In tropical forests most of the carbon in plant detritus is quickly recycled, for there is little accumulation in the soil. The turnover rate of atmospheric carbon over a tropical forest is about 0.8 year (Leith 1963). In drier regions, such as grasslands, considerable quantities of carbon are stored as humus. In swamps and marshes, where dead material falls into the water, organic carbon is not completely mineralized; it is stored as raw humus or peat and is circulated only slowly. The turnover rate of atmospheric carbon over peat bogs is somewhere on the order of 3 to 5 years (Leith 1963).

Similar cycling takes place in the freshwater and marine environments. Phytoplankton utilizes the carbon dioxide that has diffused into the upper layers of water or is present as carbonates and converts it into carbohydrates. The carbohydrates so produced pass through the aquatic food chains. The carbon dioxide produced by respiration is reutilized by the phytoplankton in the production of more carbohydrates. Under proper conditions a portion is reintroduced into the atmosphere. Significant portions of carbon, bound as carbonates in the shells of mollusks and foraminifers become buried in the bottom mud at varying depths when the organisms die. Isolated from biotic activity, that carbon is removed from cycling and becomes incorporated into bottom sediments, which through geological time may appear on the surface as limestone rocks or as coral reefs. Other organic carbon is slowly deposited as gas, petroleum, and coal at an estimated global rate of 10 to 13 $g/m^2/yr$.

Local Patterns. The concentration of carbon dioxide in the atmosphere around plant life fluctuates throughout the day

(Figure 12.4a). At daylight, when photosynthesis begins, plants start to withdraw carbon dioxide from the air and the concentration declines sharply. By afternoon, when the temperature is increasing and the humidity is decreasing, the respiration rate of plants increases, the assimilation rate of carbon dioxide in the atmosphere increases. By sunset, when the light phase of photosynthesis ceases, carbon dioxide is no longer being withdrawn from the atmosphere, and its concentration in the atmosphere increases sharply. A similar diurnal fluctuation takes place in aquatic ecosystems.

There is also an annual course in the production and utilization of carbon dioxide (Figure 12.4b). This seasonal change relates both to temperature and to growing seasons. In the spring, when land is greening and phytoplankton is actively growing, the daily production of carbon dioxide is high. Measured by nocturnal accumulation in spring and summer, the rate of carbon dioxide production by respiration may be two to three times higher than winter rates at the same temperature. The rate increases dramatically about the time of the opening of buds and falls off just as rapidly about the time the leaves of deciduous trees start to drop in the fall. This increased production of CO_2 does not accumulate in the atmosphere, however, because of the high rate of CO_2 fixation by photosynthesis.

The Global Pattern. Like water, the carbon budget of the earth is closely linked to the atmosphere, land, and oceans and to the mass movements of air around the planet. The carbon pool involved in the global carbon cycle amounts to an estimated 55,000 Gt (Gt is a gigaton, equal to 1 billion or 10^9 metric tons). Fossil fuels and shale account for an estimated 12,000 Gt. The oceans contain 93 percent of the active carbon pool, that is, over 39,000 Gt,

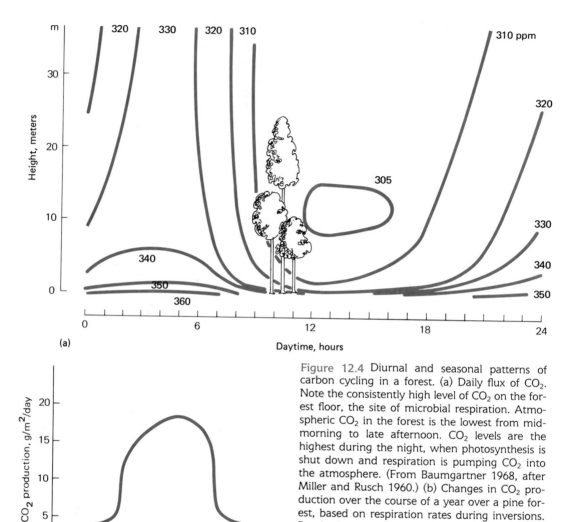

Figure 12.4 Diurnal and seasonal patterns of carbon cycling in a forest. (a) Daily flux of CO_2. Note the consistently high level of CO_2 on the forest floor, the site of microbial respiration. Atmospheric CO_2 in the forest is the lowest from midmorning to late afternoon. CO_2 levels are the highest during the night, when photosynthesis is shut down and respiration is pumping CO_2 into the atmosphere. (From Baumgartner 1968, after Miller and Rusch 1960.) (b) Changes in CO_2 production over the course of a year over a pine forest, based on respiration rates during inversions. Respiration rates are the highest during the summer, when photosynthesis and decomposition are the greatest. (Woodwell and Dykeman 1966.)

mostly as bicarbonate ions (HCO_3^-) and carbonate ions (CO_3^-). (Figure 12.5). Dead organic matter in the oceans accounts for 1650 Gt of carbon, and living matter, mostly phytoplankton, 1 Gt. The terrestrial biosphere contains an estimated 1456 Gt as dead organic matter and 826 Gt as living matter. The atmosphere, the major coupling mechanism in the cycling

of CO_2, holds about 702 Gt of carbon (Baes et al. 1977).

Carbon cycling in the sea is nearly a closed system. The surface water acts as the site of main exchange of carbon between atmosphere and ocean. The ability of the surface waters to take up CO_2 is governed largely by the reaction of CO_2 with the carbonate ion to form bicarbonates:

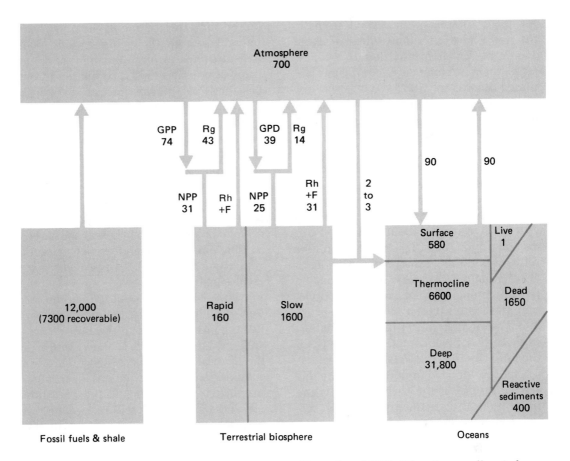

Figure 12.5 Global compartments of CO_2. The diagram shows three interconnected subsystems. Arrows indicate the major fluxes (in Gt/yr) and pool sizes (in Gt) of carbon. Fluxes include gross primary production (GPP), green plant respiration (R_g), net primary production (NPP = GPP/R_g), respiration by heterotrophs (R_h), and fires (F). (From Baes et al. 1977: 314.)

$$CO_2 + CO_3^- + H_2O \rightleftharpoons 2HCO_3^-$$

In the surface water carbon circulates physically by means of currents and biologically through assimilation by phytoplankton and movement through the food chain. Only about 10 percent of the carbon that comes up from the deep in upwellings is used in organic production; the other 90 percent goes back to the deep

(Broecker 1973). The thermocline (a layer between the warmer upper zones and colder zones; see Chapter 10) separates the surface pool of carbon from deep waters. However, some mix between the surface and the deep takes place by eddy diffusion, and 10 to 20 percent of the particulate matter sinks to the bottom. Because 80 percent of the ocean floor is bathed in water high in CO_2 and unsaturated in carbonates, about 85 percent of the calcium carbonate produced in the sea is destroyed by dissolution:

$$CaCO_3 + CO_2 + H_2O \rightleftharpoons Ca^{2+} + 2HCO_3^-$$

About 15 percent of the calcium carbonate becomes incorporated in deep sediments. Carbon in the sediments is trapped for about 10^8 years, but carbon atoms in the water have a residence time of about 10^5 years, which means that in 100,000 years carbon atoms are completely replaced (Baes et al. 1977).

Most of the carbon in the land mass is in slowly exchanging matter, about 1456 Gt in humus and recent peat and about 600 Gt in large stems and roots. A small amount, 160 Gt, is in rapidly exchanging material. Exchanges between the land mass and the atmosphere are nearly in equilibrium. Photosynthesis by terrestrial vegetation removes about 113 Gt. Plant decomposition, plant respiration, and fires return about the same amount. Forests are the main consumers, fixing about 36 Gt of an estimated total world net production of 65 Gt/yr (Olsen 1970). Forests are also a major reservoir of the terrestrial organic carbon pool, containing 1485 Gt of the estimated total 2216 Gt.

Of considerable importance in the terrestrial system of the carbon cycle is the proportion of detrital carbon, that is, carbon contained in the dead organic matter on the soil's surface and in the underlying mineral soil, to the living organic pool. Recent estimates place the amount of detrital carbon at 1456 Gt compared to 826 Gt for total world terrestrial biomass (Schlesinger 1977). This estimate of the total carbon pool is larger than previous ones, which failed to include the carbon in the lower soil profile. On a worldwide basis this pool exceeds carbon in the surface soil by a factor of 25. The lower soil layers of tropical forests and both tropical and temperate grasslands contain a large percentage of soil carbon, whereas soil carbon in temperate forest and arctic regions is confined largely to the surface soil layers (Schlesinger 1977).

The average amount of carbon per unit of soil profile increases from the tropical regions poleward to the boreal forest and tundra (Table 12.1). Low values for the tropical forest reflect high rates of decomposition, which compensate for high productivity and litter fall. Frozen tundra soil and waterlogged soils of swamps and marshes have the greatest accumulation of

TABLE 12.1 Distribution of Detritus and Biomass by Ecosystem Types

Ecosystem Type	Mean Total Profile Detritus (kg C/m^{-2})	CV* (%)	World Area (10^9 ha)	Total World Detritus (10^9 mtn C)	Total World Biomass (10^9 mtn C)
Woodland and shrubland	6.9	59	8.5	59	22.0
Tropical savanna	3.7	87	15.0	56	27.0
Tropical forest	10.4	44	24.5	255	460.0
Temperate forest	11.8	35	12.0	142	175.0
Boreal forest	14.9	53	12.0	179	108.0
Temperate grassland	19.2	25	9.0	173	6.3
Tundra and alpine	21.6	68	8.0	173	2.4
Desert scrub	5.6	38	18.0	101	5.4
Extreme desert	0.1	—	24.0	3	0.2
Cultivated	12.7	—	14.0	178	7.0
Swamp and marsh	68.6	63	2.0	137	13.6
Total			147.0	1456	826.9

*CV = coefficient of variation = standard deviation/mean × 100.

Source: Adapted from W. H. Schlesinger 1977.

detritus, because decay is inhibited by moisture, low temperature, or both. Detritus greatly exceeds biomass in boreal forests and tundra. Detritus likewise exceeds biomass in temperate grasslands, but in tropical savannas the reverse situation exists, probably because of recurring fires. In forest ecosystems carbon loss in soil respiration averages two times the total annual detrital input of carbon from aboveground and belowground sources. The world output of carbon from soil respiration in terrestrial ecosystems amounts to an estimated 75 Gt per year.

The global cycle of carbon dioxide, like local cycles, exhibits a marked annual variation, particularly in the terrestrial-dominated Northern Hemisphere (Figure 12.6). Carbon dioxide content of the atmosphere up to the lower stratosphere declines markedly during early spring north of 30° North Latitude, when greening vegetation and increasing biomass storage

Figure 12.6 This section from the long-range monthly average values of the concentration of CO_2 in the atmosphere at Mauna Loa Observatory, Hawaii, points out the seasonal effects of photosynthesis and respiration on the annual cycle of carbon. Atmospheric concentrations of CO_2 in the Northern Hemisphere are the highest in winter and lowest in summer.

withdraws more carbon dioxide from the atmosphere than is replaced by respiration. In July, as spring progresses from 20° North Latitude to the North Pole, the concentration of carbon dioxide declines sharply until it reaches a low in August. By October the level increases again. The decline of CO_2 during the arctic summer suggests a rapid and heavy removal of carbon dioxide by the photosynthetic activity of arctic plant life. The scrubbing effect is strongest in August, when arctic ecosystems are most active and polar ice is at a minimum.

Intrusions. The CO_2 flux among land, sea, and atmosphere has been disturbed by a rapid injection of carbon dioxide into the atmosphere from the burning of fossil fuels and from the clearing of forests. Clearing increases the input of CO_2 from burning and decomposing trees, and it reduces the forest sink for the storage of CO_2 scrubbed from the atmosphere by photosynthesis and biomass accumulation.

Adding to the problem of increasing carbon dioxide are increases in other atmospheric gases. One of them is methane gas, CH_4, major sources of which are ruminant animals, microbial decomposition in swamps, marshes, and tundra, and industrial gases released to the atmosphere.

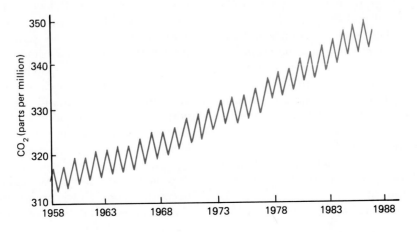

Over time (resident time of 3.2 years) methane oxidizes to H_2O, especially in the stratosphere. Atmospheric methane has approximately doubled over the past 200 years (Rasmussen and Khalil 1986), and the mixing ratio of methane in the trophosphere increased by 11 percent between 1978 and 1987 alone (Blake and Rowland 1988). This increase is linked to population growth and agricultural development associated with increased worldwide cattle and rice paddy production. Another increasing atmospheric gas is carbon monoxide (Khalil and Rasmussen 1984), the major source of which is the gasoline engine. The annual worldwide input amounts to 27×10^6 tons, 95 percent of which is produced in the Northern Hemisphere.

Input of CO_2 has been increasing at an exponential rate in the 200 years since the beginning of the Industrial Revolution. During that time the CO_2 concentration in the atmosphere has risen from an estimated 260 ppm (assuming the atmosphere had a CO_2 steady state, which it probably did not) to about 360 ppm in 1986. Up to 1950 two-thirds of the carbon dioxide injected into the atmosphere came from biospheric sources and one-third from anthropogenic sources. Since 1950 most has come from the burning of fossil fuels (Broecker et al. 1979), although the rate of increase declined from 4.5 percent in 1973 to 2.25 percent in 1983. The destruction of forests and changes in land use add another 5 Gt/yr according to some estimates (Woodwell 1978, Houghton et. al. 1983).

About 40 to 50 percent of the total CO_2 injected into the atmosphere remains there; the rest must be stored in two possible sinks, the terrestrial biomass and the ocean. In view of increasing dependence on fossil carbon, the ability of the sinks to take care of excess carbon becomes increasingly important.

The ocean has been credited as the major carbon sink (Stuiver 1978). The storage of excess CO_2 by oceans depends heavily upon the eddy current circulation between the surface and deep water (see Chapter 10). Evidence based on oceanic studies suggests that the major CO_2 sink is the thermocline region of large ocean gyres. Thirty-four percent of excess CO_2 generated is stored in the surface and thermocline gyres, and 13 percent is carried to deep seas, leaving 53 percent in the atmosphere; so the oceans cannot absorb all the carbon unaccounted for. These models fail to take into account the significant uptake and storage of carbon by marine biota (Smith 1981; Betzer et al. 1984).

The role of terrestrial biota in absorbing a significant fraction is subject to debate (Hobbie et al. 1984). Some ecologists (Botkin 1977; Houghton et al. 1983) argue that terrestrial vegetation is a net source, not a sink. Field evidence does show that increased atmospheric CO_2 increases the growth of subalpine conifers in the western United States (LaMarch, Jr. et al. 1984). Experimental studies show that lower CO_2 concentrations in high altitude environments limit the growth of subalpine trees. Moreover, most models assume that mature terrestrial vegetation is in a carbon-steady state with the atmosphere, returning as much as it consumes. However, mature terrestrial vegetation is not necessarily either in a vegetational steady state or in a CO_2-steady state (Lugo and Brown 1986) over short time intervals. All mature systems have patches of natural and human disturbances and developing vegetation, and even old growth forests add some growth. In addition to their own carbon uptake, these terrestrial systems export carbon to marine and freshwater ecosystems; they also store it at a rate of deposition and burial exceeding rate of decomposition. However, in spite of all uncertainties and debates over CO_2 inputs and outputs, there is no disagreement that atmospheric CO_2 has increased by 25 percent over the past century. If it continues to increase at its present rate, the CO_2 concentration in the atmosphere will reach 400 ppm in the next decade or so.

The Greenhouse Effect. Concern over increasing atmospheric CO_2 and CH_4 has to do with the "greenhouse effect" (Figure 12.7). Acting as a shield over Earth, carbon dioxide allows incoming short-wave radiation to penetrate the atmosphere but absorbs long-wave radiation and redirects some of it toward Earth. As the concentration of CO_2 in the atmosphere increases, it traps more heat. Each doubling of CO_2 concentration results in a doubling of the average temperature of the troposphere, with a greater increase at the higher latitudes and a smaller increase at the lower latitudes. Molecule for molecule, methane is still more effective than CO_2 in contributing to the greenhouse effect, because its

Figure 12.7 The greenhouse effect. Solar short-wave radiation easily penetrates Earth's atmosphere, although some radiation is blocked by atmospheric dust particles and pollutants. Carbon dioxide and other atmospheric gases (ozone, methane) inhibit the escape of long-wave radiation back to outer space, heating Earth. If the retention of more long-range radiation than normal is not balanced by some blockage of incoming short-wave radiation, Earth will continue to warm.

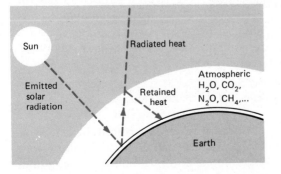

infrared absorptions fall into wavelength regions not strongly absorbed by existing concentration of H_2, CO_2, and O_3 and because it contributes to an increase in heat-absorbing stratospheric water vapor (see Blake and Rowland 1988).

Concern over the rapid rise in carbon dioxide in the atmosphere has stimulated intensive studies and modeling of atmospheric changes in CO_2, past and present (DOE 1986; Kasting and Ackerman 1986; COHMAP 1988). Current models predict a global warming of 1° C from the base year 1850 by the year 2000. Such a rise would lengthen the growing season in the northern United States and southern Canada by ten days. It could also reduce precipitation in the central plains, bring an earlier summer, and increase evaporation from the soil (Manake and Wetherald 1986). At the current rate of increase in atmospheric CO_2, temperatures would rise 2° to 3° C by the year 2100.

Such rises in temperature would have pronounced ecological effects, let alone social and economic impacts. One effect would be a short-term rise in plant productivity and related phenomena. C_3 plants (see Chapter 3) would have the most to gain (Kramer 1981; Lemon 1983; Waggoner 1984). They would increase their net productivity because CO_2 would no longer be limiting; they would increase their water-use efficiency because the stomata would not need to open fully to gain sufficient CO_2, reducing transpiration; and photorespiration would cease to be a problem. Provided moisture did not become limiting nor the temperatures too warm, C_3 plants would become competitive with C_4 plants, which would not respond to increased CO_2. However, increased net photosynthesis carries with it certain demands. With more carbohydrates available for growth, plants would need more water and more nutrients.

Without that, increased CO_2 would be to no avail.

Increases in plant productivity could have other outcomes. Increased growth could result in changes in allocation of photosynthesis to various parts of the plant and in changes in leaf area and leaf structure. These changes and increased branching in herbaceous and woody plants and in root/shoot ratios could alter the competitive position of plants, influencing ecosystem composition.

A general warming undoubtedly would result in a northward shift of ecosystems toward the poles. Temperature changes (and associated hydrological balances) have had a profound influence on vegetational distribution through Earth's history (see Woodward 1987). Simulation models of global warming (Emanuel, Shugart, and Stevenson 1985; Solomon 1986; Woodward 1987; COHMAN 1988) involving changes in temperature and precipitation predict major vegetational changes. Tundra, boreal, and deciduous forests would shift northward, and deciduous forests would expand as coniferous forests decreased or replaced tundra. Verification of these models by comparing paleoecological data with model simulations shows them to be highly accurate.

Other outcomes have been hypothesized for global warming (Peters and Darling 1985). Melting permafrost would release methane and carbon dioxide locked up in poorly decomposed detritus. Southern animals would extend their ranges northward; and confronted with new competitors and a loss of habitat, arctic animals would face extinction. Small relict populations, lacking genetic diversity to adapt to changing climates and habitats, and left behind in limited refuges in a changing environment, would perish. Along continental coastlines rising sea levels would gradually submerge estuaries, tidal marshes, and human settlements.

The Nitrogen Cycle

Processes. Nitrogen is an essential constitutent of protein, a building block of all living material. It is also the major constituent, about 79 percent, of the atmosphere. The paradox is that in its gaseous state, abundant though it is, nitrogen is unavailable to most life. It must first be converted to some chemically usable form, and getting it into that form comprises a major part of the nitrogen cycle. The processes involved are fixation, ammonification, nitrification, and denitrification.

Fixation is the conversion of nitrogen in its gaseous state to ammonia or nitrate. Ammonia is the product of biological fixation; nitrate is the product of high-energy fixation by lightning or occasionally cosmic radiation or meteorite trails. In high-energy fixation nitrogen and oxygen in the atmosphere combine into nitrates, which are carried to the earth in rainwater as nitric acid, H_2NO_3. Estimates suggest that less than 8.9 kg N/ha is brought to the earth by high-energy fixation.

Biological fixation, the more important method, makes available 100 to 200 kg N/ha, roughly 90 percent of the fixed nitrogen contributed to Earth each year. In biological fixation molecular (or gaseous) nitrogen, N_2, is split into two atoms:

$$N_2 \rightarrow 2N$$

This step requires an input of 160 kcal for each mole (28 g) of nitrogen. The free N atoms can combine with hydrogen to form ammonia, with the release of about 13 kcal of energy:

$$2N + 3H_2 \rightarrow 2NH_3$$

This fixation is accomplished by symbiotic bacteria living in association with leguminous and root-noduled nonleguminous plants, by free-living aerobic bacteria, and by blue-green algae. In agricultural ecosystems approximately 200 species of nodulated legumes are the preeminent nitrogen fixers. In nonagricultural systems some 12,000 species of plants, from free-living bacteria and blue-green algae to nodule-bearing plants, are responsible for nitrogen fixation.

Legumes, the most conspicuous of the nitrogen-fixing plants, have a symbiotic relationship with members of the bacterial genus *Rhizobium*. Rhizobia are aerobic, non-spore-forming rod-shaped bacteria. They exist in the immediate surroundings of the plant roots, called the rhizosphere, where, stimulated by secretions from the legumes, they multiply. The secretions, together with enzymes secreted by the legumes in response to the exudates of the bacteria, loosen the fibrils of the root hair walls. Swarming rhizobia enter the root hair tips and penetrate the inner corticular cells, where they multiply and increase in size, resulting in swollen infected cells in which hemoglobin develops. These cells make up the central tissues of the nodules. Inside the nodules the bacteria change from rod-shaped to a nonmobile form that carry on nitrogen fixation.

A large number of nonleguminous nodule-bearing plants, most of them associated with early pioneering species on sites where the soil was low in nitrogen, make significant contributions of nitrogen to wildlands. Among such plants are *Alnus, Ceanothus,* and *Elaeagnus.*

Also contributing to the fixation of nitrogen are free-living soil bacteria. The most prominent of the 15 known genera are the aerobic *Azotobacter* and the anaerobic *Clostridium* (see Nishustin and Shilnikova 1969). *Azotobacter* prefers soils with a pH of 6 to 7 that are rich in mineral salts and low in nitrogen. Less efficient at fixing nitrogen, *Clostridium* is ubiquitous, found in nearly all soils. Although it prefers a neutral pH, it is more tolerant of

acidic conditions than *Azotobacter.* Both genera produce ammonia as the first stable end product. The free-living and symbiotic bacteria both require molybdenum as an activator and are inhibited by an accumulation of nitrates and ammonia in the soil.

Blue-green algae are another important group of largely nonsymbiotic nitrogen fixers. Of the some 40 known species the most common are in the genera *Nostoc* and *Calothrix,* found both in soil and aquatic habitats. Blue-green algae are well adapted to exist on the barest requirements for living. They are often pioneers on bare mineral soil. Especially successful in waterlogged soil, they appear to be nitrogen fixers in rice paddies of Asia (Singh 1961), where studies indicate that they annually fix 30 to 50 kg N/ha. Blue-greens are perhaps the only fixers of nitrogen over a wide range of temperatures in aquatic habitats from arctic and antarctic seas to freshwater ponds and hot springs. In the hot springs of Yellowstone the blue-greens, which are responsible for the bluish-green color, fix nitrogen at a temperature of 55° C (W. D. P. Stewart 1967). Like bacteria, blue-greens require molybdenum for nitrogen fixation.

Other plants may be involved in nitrogen fixation. In humid tropical forests epiphytes growing on tree branches and bacteria and algae growing on leaves may fix appreciable amounts of nitrogen. Certain lichens (*Collema tunaeforme* and *Peltigera rufescens*) have been implicated in nitrogen fixation (Henriksson 1971). Lichens with nitrogen-fixing ability possess nitrogen-fixing blue-green species as their algal component.

Nitrogen is also made available through the breakdown of organic matter by ammonification, nitrification, and denitrification, three other processes in the nitrogen cycle. In *ammonification* amino acids are broken down by decomposer organisms to produce ammonia, with a yield of energy. It is a one-way reaction. Ammonium, or the ammonia ion, is absorbed directly by plant roots and incorporated into amino acids which are subsequently passed along through the food chain. Wastes and dead animal and plant tissues are broken down to amino acids by heterotrophic bacteria and fungi in soil and water. Amino acids are oxidized to carbon dioxide, water, and ammonia, with a yield of energy:

$$CH_2NH_2COOH + 1\frac{1}{2} O_2 \rightarrow 2CO_2$$
$$+ H_2O + NH_3 + 176 \text{ kcal}$$

Part of the ammonia is dissolved in water, part is trapped in the soil, and some is trapped and fixed in both acid clay and certain base-saturated clay minerals near the point where first broken down or introduced into the soil (Nommik 1965).

Nitrification is a biological process in which ammonia is oxidized to nitrate and nitrite, yielding energy. Two groups of microorganisms are involved. *Nitrosomonas* utilize the ammonia in the soil as their sole source of energy because they can promote its oxidation to nitrite ions and water:

$$NH_3 + 1\frac{1}{2} O_2 \rightarrow HNO_2 + H_2O$$
$$+ 165 \text{ kcal } HNO_2 \rightarrow H^+ + NO_2^-$$

Nitrite ions can be oxidized further to nitrate ions in an energy-releasing reaction. This energy left in the nitrite ion is exploited by another group of bacteria, the *Nitrobacter*, which oxidizes the nitrite ion to nitrate with a release of 18 kcal of energy (M. Alexander 1965):

$$NO_2 + \frac{1}{2} O_2 \rightarrow NO_3^-$$

Thus nitrification is a process in which the oxidation state (or valence) of nitrogen is increased. *Nitrosomonas* oxidizes 35 mols of nitrogen for each mol of CO_2 assimilated; *Nitrobacter* oxidizes 100 mols.

Although nitrification is generally considered a beneficial process, it may in some situations have deleterious effects. Nitrification involves the conversion of slowly leached forms of nitrogen into readily leached nitrates. If quantities of nitrates are large enough and sufficient water percolates through the soil, the nitrates can be removed faster than they can be taken up by the roots, a situation that results in pollution of water. An abundance of nitrates also leads to increased losses of gaseous nitrogen.

Nitrates are a necessary substrate for *denitrification,* in which the nitrates are reduced to gaseous nitrogen by certain organisms to obtain oxygen. The denitrifiers, represented by fungi and the bacteria *Pseudomonas*, are facultative anaerobes. They prefer an oxygenated environment, but if oxygen is limited, they can use NO_3^- instead of O_2 as the hydrogen acceptor and can release N_2 in the gaseous state as a by-product:

$$C_6H_{12}O_6 + 4NO_3^- \rightarrow 6CO_2$$
$$+ H_2O + 2N_2$$

Nitrification and denitrification both require certain conditions: a sufficient supply of organic matter, a limited supply of molecular oxygen, a pH range of 6 to 7, and an optimum temperature of 15° C.

Cycling of Nitrogen. Let us follow the nitrogen cycle briefly (Figure 12.8, Table 12.2). The sources of nitrogen under natural conditions are the fixation of atmospheric nitrogen, additions of inorganic nitrogen in rain from such sources as lightning fixation and fixed "juvenile" ni-

Figure 12.8 The nitrogen cycle. Although the atmosphere is the major reservoir, nitrogen becomes available to life largely through biological processes.

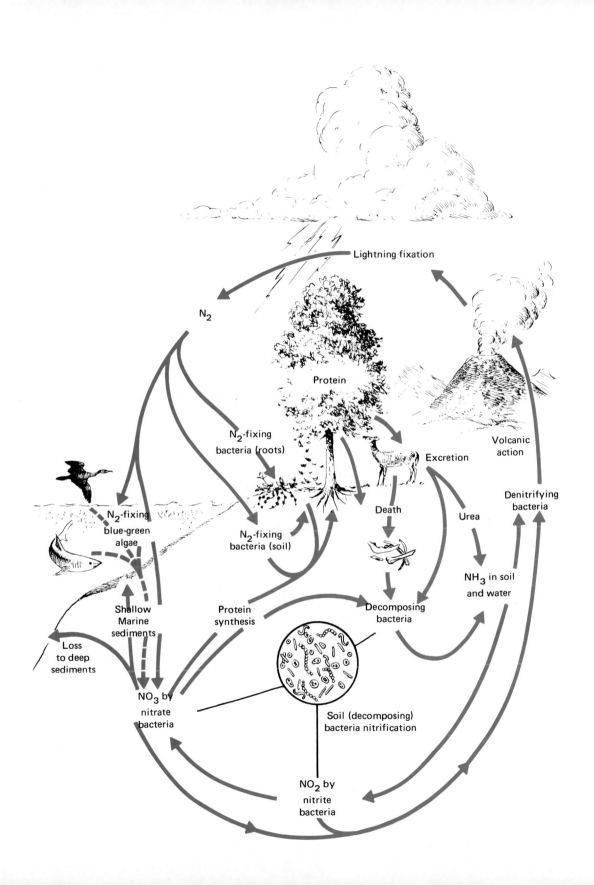

Lightning fixation

N_2

Protein

N_2-fixing
bacteria (roots)

Volcanic
action

N_2-fixing
blue-green
algae

Excretion

Denitrifying
bacteria

N_2-fixing
bacteria (soil)

Death

Urea

NH_3 in soil
and water

Shallow
Marine
sediments

Protein
synthesis

Decomposing
bacteria

Loss
to deep
sediments

NO_3 by
nitrate
bacteria

Soil (decomposing)
bacteria nitrification

NO_2 by
nitrite
bacteria

TABLE 12.2 Budget for the Nitrogen Cycle

	LAND		SEA		ATMOSPHERE	
	Rate (10^6 mtn/yr)	Error (%)	Rate (10^6 mtn/yr)	Error (%)	Rate (10^6 mtn/yr)	Error (%)
INPUT						
Biological nitrogen fixation	—	—	10	50	—	—
Symbiotic	14	25	—	—	—	—
Nonsymbiotic	30	50	—	—	—	—
Atmospheric nitrogen fixation	4	100	4	100	—	—
Industrially fixed nitrogen fertilizer	30	5	—	—	—	—
N-oxides from combustion	14	25	6	25	20	25
Return of volatile nitrogen compounds in rain	?	—	?	—	—	—
River influx	—	—	30	50	—	—
N_2 from biological dentrification	—	—	—	—	83	100
Natural NO_2	—	—	—	—	?	—
Volatilization (HN_3)	—	—	—	—	?	—
Total input	92+		50		103+	
STORAGE						
Plants	12,000	30	800	50	—	—
Animals	200	30	170	50	—	—
Dead organic matter	760,000	50	900,000	100	—	—
Inorganic nitrogen	140,000	50	100,000	50	—	—
Dissolved nitrogen	—	—	20,000,000	10	—	—
Nitrogen gas	—	—	—	—	3,800,000,000	3
$NO + NH_4$	—	—	—	—	Less than 1	50
$NH_3 + NH_4$	—	—	—	—	12	50
N_2O	—	—	—	—	1,000	50
Total storage	912,200		21,000,970		3,800,001,013	
LOSS						
Denitrification	43	—	40	100	—	—
Volatilization	?	—	?	—	—	—
River runoff (includes enrichment from fertilizers)	30	50	—	—	—	—
Sedimentation	—	—	0.2	50	—	—
N_2 in all fixation processes	—	—	—	—	92	50
NH_3 in rain	—	—	—	—	Less than 40	50
NO_2 in rain	—	—	—	—	?	—
N_2O in rain	—	—	—	—	?	—
Total loss	73		40.2		132+	

*The error columns list plus-or-minus probable errors as a percentage of the estimate.

Source: Inger et al. 1972. © 1972 by the Board of Regents of the University of Wisconsin System.

trogen from volcanic activity, ammonia absorption from the atmosphere by plants and soil, and nitrogen accretion from windblown aerosols, which contain both organic and inorganic forms of nitrogen. In terrestrial ecosystems nitrogen, largely in the form of ammonia or nitrates, depending upon a number of variable conditions, is taken up by plants, which convert it into amino acids. The amino acids are transferred to consumers, which convert them to different types of amino acids. Eventually their wastes (urea and excreta) and the decay of dead plant and animal tissue are broken down by bacteria and fungi into ammonia. Ammonia may be volatilized (lost as gas to the atmosphere), may be acted upon by nitrifying bacteria, or may be taken up directly by plants. Nitrates may be taken up directly by plants. Nitrates may be utilized by plants, immobilized by microbes, stored in decomposing humus, or leached away. This material is carried to streams, lakes, and eventually the sea, where it is available for use in aquatic ecosystems. There nitrogen is cycled in a similar manner, except that the large reserves contained in the soil humus are largely lacking. Life in the water contributes organic matter and dead organisms that undergo decomposition and subsequent release of ammonia and ultimately nitrates.

Tracer studies with ^{15}N, a short-lived nonradioactive isotope, show that in marine ecosystems, ammonia is recycled rapidly and preferentially by phytoplankton (Dugdale and Goering 1967). As a result there is little ammonia in most natural waters, and nitrate is utilized only in the virtual absence of ammonia. In addition to biological cycling there are small but steady losses from the biosphere to the deep sediments of the ocean and to sedimentary rocks. In return there is a small addition of "new" nitrogen from the weathering of igneous rocks and juvenile nitrogen from volcanic activity.

Under natural conditions nitrogen lost from ecosystems by denitrification, volatilization, leaching, erosion, windblown aerosols, and transportation out of the system is balanced by biological fixation and other sources. Both chemically and biologically, terrestrial and aquatic ecosystems constitute a dynamic equilibrium system in which a change in one phase affects the other.

Intrusions. Human intrusion into the nitrogen cycle often results either in a reduction in the nitrogen cycle or an overload of the system.

Cultivation of grasslands, for example, has resulted in a steady decline in the nitrogen content of the soil (Jenny 1933). Mixing and breaking up of the soil exposes more organic matter to decomposition and decreases the amount of root material, thus decreasing new additions of organic matter. The removal of nitrogen through harvested crops or grazing causes additional losses. Harvest of timber may result in a heavy outflow of nitrogen from the forest ecosystem, not only in timber removed, but also in short-term nitrate losses from the soil. In such situations outputs exceed inputs, impoverishing the system.

On the other hand, excessive amounts of nitrogen may be added to the system, creating various problems. Heavy addition of commercial fertilizer, especially in the form of anhydrous ammonia, increases the amount of nitrogen in cropland ecosystems. However, if fertilizer is not properly applied, a considerable portion of the added nitrogen may be lost as nitrate nitrogen to the groundwater. The amount of nitrates lost from the soil by percolating water varies with the type of crop, rate and time of fertilizer application, soil permeability, the ratio of precipitation or irrigation to evaporation, hydrology of the area, the portion of watershed in crops, and general climate.

Another source of agricultural nitrate pollution is animal waste. Only recently has this waste become an environmental problem. Prior to the use of concentrated livestock feeding yards, animal wastes were recycled to croplands, and nutrients were eventually returned to the animals as grain and hay. Inputs of feed and fertilizer were converted into outputs of meat and milk. Increasing costs of chemical fertilizers and a renewed interest in organic farming may result in the return of feedlot wastes to the land.

A third source of nitrate pollution is human waste, particularly sewage. In spite of the magnitude of water pollution from agricultural sources, human effluents contribute even heavier loads. Municipal sewage treatment facilities contribute 25 percent of the nitrogen and 50 percent of the phosphorus found in the surface waters of Wisconsin (Corey et al. 1967, cited by Frink 1971). Groundwater from rural areas contributes 50 percent of the nitrogen and only 2 percent of the phosphorus. In the Potomac River estuary waste water accounts for 51 percent of the nitrogen, compared to 31 percent from agricultural runoff (Jaworski and Helling 1970). As with agricultural wastes, the potential solution is to return human wastes to the land. This solution is ecologically sounder than diluting sewage effluents in streams. The major drawback in such use of sewage effluents and sludge is the frequent high concentration of such heavy metals as cadmium.

Automobiles and power plants are the major sources of other nitrogenous pollutants, nitrogen oxides. The major nitrogenous air pollutant is nitrogen dioxide, NO_2. In the atmosphere nitrogen dioxide is reduced by ultraviolet light to nitrogen monoxide and atomic oxygen:

$$NO_2 \rightarrow NO + O$$

Atomic oxygen reacts with oxygen to form ozone:

$$O_2 + O \rightarrow O_3$$

Ozone, in a never-ending cycle, reacts with nitrogen monoxide to form nitrogen dioxide and oxygen:

$$NO + O_3 \rightarrow NO_2 + O_2$$

This cycle illustrates only a few of the reactions that nitrogen oxides undergo or trigger. In the presence of sunlight atomic oxygen from nitrogen dioxide also reacts with a number of reactive hydrocarbons to form radicals. These radicals then take part in a series of reactions to form still more radicals that combine with oxygen, hydrocarbons, and nitrogen dioxide. As a result, nitrogen dioxide is regenerated, nitrogen monoxide disappears, ozone accumulates, and there form a number of secondary pollutants, such as formaldehydes, aldehydes, and peroxyacytnitrates, known as PAN (see Am. Chem. Soc. 1969). All of these substances collectively form photochemical smog.

Nitrogen dioxide and the secondary pollutants are harmful to both humans and plants. Nitrogen dioxide, a pungent gas that produces a brownish haze, causes nose and eye irritations at 13 ppm and pulmonary discomfort at 25 ppm. Ozone irritates the nose and throat at 0.05 ppm and causes dryness of the throat, headaches, and difficulty in breathing at 0.1 ppm. All of these contaminants, along with others such as sulfur dioxide and carbon monoxide, usually act synergistically. The total effect is much greater than individual effects alone. Air pollution has been positively correlated with increase in asthma, bronchitis, emphysema, and lung cancer (Lave and Seskin 1970).

Ozone, PAN, and nitrogen dioxide severely injure many forms of plant life. PAN is extremely toxic. By destroying some of the lower epidermal cells of the leaves and by damaging the chloroplasts, it interferes with the plant's metabolic processes. Although the mechanism has not been defined, PAN also interferes with enzymes important in providing the energy necessary to split the water molecules in photosynthesis (Treshow 1970).

A number of important leafy vegetable plants are extremely sensitive to PAN. Among them are spinach, endive, and tobacco. Other sensitive crops are oats, alfalfa, beets, beans, corn, celery, and peppers. Affected plants usually show some form of glazing, silvering, or bronzing in irregular patches on the undersides of the leaves (see Hill and Heggestad 1970).

Whereas ozone in the stratosphere is important in reducing ultraviolet radiation, ozone accumulating in the surface atmosphere is harmful. Reactions to it vary among plants. Highly sensitive tobacco is flecked with white lesions, bean leaves show stippling and bleached areas, and leaves of woody plants may have reddish-brown lesions. Plants especially sensitive to ozone include white and ponderosa pines, alfalfa, oats, spinach, tobacco, and tomato. Nitrogen dioxide too causes direct injury to plants. Symptoms of damage from nitrogen dioxide are white or brown collapsed lesions or tissues between the veins and near the margins of the leaves. These symptoms closely resemble reactions to both ozone and sulfur dioxide.

■ Sedimentary Cycles

Mineral elements which are required by living organisms are obtained initially from inorganic sources. Available forms occur as salts dissolved in soil water or in lakes, streams, and seas. The mineral cycle varies from one element to another, but essentially it consists of two phases: the salt-solution phase and the rock phase. Mineral salts come directly from Earth's crust by weathering. Soluble salts then enter the water cycle. With water they move through the soil to streams and lakes and eventually reach the seas, where they remain indefinitely. Other salts are returned to Earth's crust through sedimentation. They become incorporated into salt beds, silts, and limestones; after weathering they again enter the cycle.

Plants and many animals fulfill their mineral requirements from mineral solutions in their environments. Other animals acquire the bulk of their minerals from plants and animals they consume. After the death of living organisms the minerals are returned to soil and water through the action of organisms and processes of decay.

Most elements enter the sedimentary cycle. Two will serve as examples. The sulfur cycle is a hybrid of the gaseous and sedimentary processes. Phosphorus, on the other hand, is wholly sedimentary.

The Sulfur Cycle

Sulfur is one of those elements whose cycle is both sedimentary and gaseous. It involves a long-term sedimentary phase in which sulfur is tied up in organic (coal, oil, and peat) and inorganic (pyritic rocks and sulfur) deposits. It is released by weathering of rocks, erosional runoff, decomposition of organic matter, and industrial production and is carried to terrestrial and aquatic ecosystems in a salt solution. But the bulk of sulfur first appears in the gaseous phase as a volatile gas, hydrogen sulfide (H_2S), in the atmosphere. It comes from several sources: the combustion of fossil fuels, volcanic eruptions, the surface of oceans, and gases released in terrestrial and aquatic decomposition. Hydrogen sulfide quickly oxidizes into another volatile form, sulfur dioxide (SO_2). The concentration of sulfur as H_2S in the unpolluted atmosphere is estimated at $6g/m^3$, as SO_2 at 1 g/m^3. Atmospheric sulfur dioxide, soluble in water, is carried back to Earth in rainwater as weak sulfuric acid, H_2SO_4. Whatever the source, sulfur in soluble form is taken up by plants. Starting with photosynthesis, it is incorporated through a series of metabolic pro-

cesses into such sulfur-bearing amino acids as cystine. From the producers the sulfur in amino acids is transferred to consumers (Figure 12.9).

Excretions and death carry sulfur in living material back to the soil and to the bottoms of ponds, lakes, and seas, where microorganisms release it as hydrogen sulfide or as a sulfate. Other microorganisms in the forest soil convert inorganic sulfates to organic forms and incorporate it into forest ecosystems (Swank et al. 1983). Colorless sulfur bacteria both reduce hydrogen sulfide to elemental sulfur and oxidize it to sulfuric acid. Green and purple bacteria utilize hydrogen sulfide in

Figure 12.9 The sulfur cycle. Note the two phases, sedimentary and gaseous. Major sources from human activity are burning of fossil fuels and acidic drainage from coal mines. Major natural sources are volcanic eruption and decomposition of organic matter.

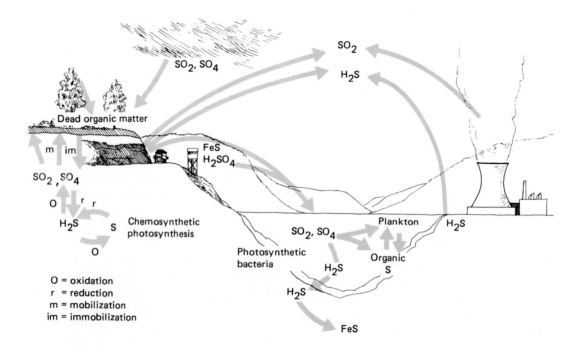

the presence of light as an oxygen acceptor in the photosynthetic reduction of carbon dioxide. Best known are the purple bacteria found in salt marshes and in mud flats of estuaries. These organisms are able to carry the oxidation of hydrogen sulfide as far as sulfate, which may be recirculated and taken up by the producers or may be used by sulfate-reducing bacteria. The green sulfur bacteria can carry the reduction of hydrogen sulfide to elemental sulfur.

Sulfur, in the presence of iron and under anaerobic conditions, will precipitate as ferrous sulfide, FeS_2. This compound is highly insoluble under neutral and alkaline conditions and is firmly held in mud and wet soil. Sedimentary rocks containing ferrous sulfide (called pyritic rocks) may overlie coal deposits. Exposed to the air in deep and surface mining, the ferrous sulfide oxidizes and in the presence of water produces ferrous sulfate ($FeSO_4$) and sulfuric acid:

$$2FeS_2 + 7O_2 + 2H_2O \rightarrow 2FeSO_4 + H_2SO_4$$

In other reactions ferric sulfate (Fe_2SO_4) and ferrous hydroxide ($FeOH_3$) are produced:

$$12FeSO_4 + 3O_2 + 6H_2O \rightarrow 4Fe_2(SO_4)_3 + 4Fe(OH_3) \downarrow$$

In this manner sulfur in pyritic rocks, suddenly exposed to weathering by human activities, discharges heavy slugs of sulfuric acid, ferric sulfate, and ferrous hydroxide into aquatic ecosystems. These compounds destroy aquatic life and have converted hundreds of miles of streams and rivers in the eastern United States to highly acidic water.

The Global Sulfur Cycle. The gaseous phase of the sulfur cycle permits it circulation on a global scale. The atmosphere contains not only sulfur dioxide and hydrogen sulfide but sulfate particles as well. The sulfate particles become part of the dry deposition; the gaseous forms combine with moisture to be recirculated to land and sea by precipitation. The concentration of sulfur dioxide in rainwater falling over land, according to estimates, is about 0.6 mg/liter and over oceans is 0.2 mg/liter, excluding sea spray. It is almost impossible to estimate the biological turnover of sulfur dioxide because of the complicated cycling within the biosphere. Erikisson (1963) estimates that the net annual assimilation of sulfur by marine plants is on the order of 130 million tons. Adding the anaerobic oxidation of organic matter brings the total to an estimated 200 million tons. Both industrially emitted sulfur and fertilizer sulfur are eventually carried to the sea; these two sources probably account for the 50-million-ton annual increase of sulfur in the ocean. The balance sheet of sulfur in the global cycle appears in Table 12.3.

Intrusions. Annually we pour into the atmosphere some 147 million tons of sulfur dioxide (Erikisson 1962). Seventy percent of it comes from the burning of coal. Once in the atmosphere gaseous sulfur dioxide reacts with moisture to form sulfuric acid.

Sulfuric acid in the atmosphere has a number of effects. It is irritating to the respiratory tract in concentrations of a few parts per million. In a fine mist or absorbed in small particles, it can be carried deep into the lungs to attack sensitive tissue. High concentrations of sulfur dioxide (over 1000 micromilligrams/m^3) have been implicated as a prime cause in many air

TABLE 12.3 Budget of Sulfur in Nature (10^6 tn/yr)

Item	ATMOSPHERE To	ATMOSPHERE From	LITHOSPHERE To	LITHOSPHERE From	PEDOSPHERE To	PEDOSPHERE From	HYDROSPHERE To	HYDROSPHERE From
River discharge						80	80	
Weathering				15	15			
Fertilizers				10	10			
Precipitation		165			65		100	
Sea spray	45							45
Dry deposition		200			100		100	
Sedimentation			15					15
Industrial	40			40				50
Increase in sea			50*					50
Balance								
soils—atmosphere	110					110		
oceans—atmosphere	170							170
Total	365	365	65	65	190	190	280	280
Specification								
as SO_2 sulfur	45	165			90	80	180	95
as SO_2 sulfur	40	200			100		100	
as H_2S sulfur	280					110		170
as other forms of sulfur			65	65				15

* For the balance this has to be treated as an item borrowed by the ocean from the lithosphere.

Source: E. Erikisson 1963. Copyright by American Geophysical Union.

pollution disasters characterized by higher than expected death rates and increased incidence of bronchial asthma. Among such disasters are the Meuse Valley in Belgium in 1930, Donora, Pennsylvania, in 1938, London in 1952, and New York and Tokyo in the 1960s.

Plants exposed to atmospheric sulfur are injured or killed outright. Injury to plants is caused largely by acidic aerosols during periods of foggy weather, during light rains, or during periods of high relative humidity and moderate temperatures. Pines, more susceptible than broadleaf trees, react by partial defoliation and reduced growth. Exposure of plants to sulfur dioxide with as low a concentration as 0.3 ppm for 8 hours can produce both acute and chronic injury. Acute injury is characterized by dead tissue between the veins and along the margins of plant leaves; chronic injury is marked by brownish-red or blackish areas in the blade of the leaf.

Acid Rain

The amounts of sulfur dioxide and nitrogen oxides injected into the atmosphere have increased substantially during the past 100 years. The input of sulfur from human activities equals the amount normally injected into the atmosphere from natural sources, about 75 to 100 million tons per year. Most of it comes from the Northern Hemisphere. Motor vehicles and internal combustion engines are the major sources of nitrogen oxides (NO_x);

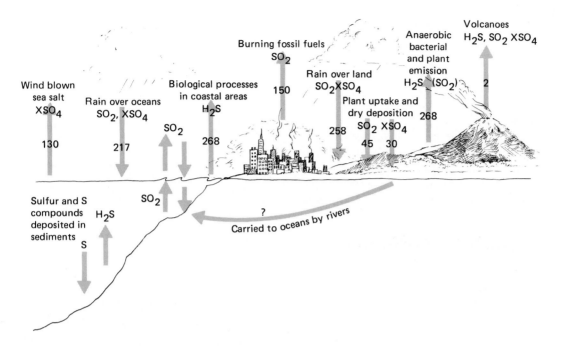

Figure 12.10 The pathways for the formation of atmospheric sulfates and nitrates, constituents of acid rain, and their transport, leading to acid deposition.

in North America and West Germany they account for one-half of the total NO_x emissions.

Sulfur dioxide and nitrogen oxides emissions are transported and mixed in the atmosphere, and the pathways they take are strongly influenced by the general atmospheric circulation (National Research Council 1983). Some of the pollutants return to Earth relatively soon as dry deposition; a major portion is transported for a long time, during which the SO_2 and NO_x and their oxidative products become involved in complex chemical reactions (Figure 12.10) that transform these pollutants into sulfates and nitrates. Ultimately they reach Earth in the form of weak sulfuric and nitric acid. As a result the acidity of precipitation downwind from major industrial centers in eastern North America and northern and central Europe has increased at least 2 to 16 times over that of precipitation in geographical areas remote from industrial pollution (Galloway et al. 1984). The annual pH of rain and snow in those continental regions averages between 4 and 4.5 (Figure 12.11) and extends from 2.3 to 4.6, compared to a more natural pH of 5.6 (Likens and Bormann 1974). Added to precipitation is acidic fog, whose pH may range from 2.2 to 4.0 (Waldman et al. 1982).

Acid deposition, wet or dry, can have a pronounced effect on aquatic and terrestrial ecosystems. Those that are not well buffered are highly sensitive to acid inputs. Acid precipitation has been implicated in the acidification of mountain streams and lakes in the northern United States, especially the Adirondacks, eastern Canada, and Scandinavia (National Research Council of Canada 1981; Swedish Ministry of Agriculture 1982), and is suspected as one of the agents in the decline and death of spruce and fir in Germany, Canada, and the high Appalachians.

Acidic inputs into aquatic ecosystems comes from rain, snow, and groundwater leaching from adjacent watersheds. Of particular importance is snowfall, which accumulates over winter and melts quickly

in the spring, discharging much of the winter precipitation in a slug of acidic water into streams and lakes (Hornbeck et al.

Figure 12.11 Zones of mean annual deposition of (a) sulfate (m moles/m²), (b) nitrate (m moles/m²), and (c) mean annual wet deposition of hydrogen ions (m moles/m²) of precipitation weighted by precipitation amount in United States and Canada in 1980. (From Impact Assessment, Work Group 1, United States-Canada Memorandum of Intent on Transboundary Air Pollution, final report, January 1983.)

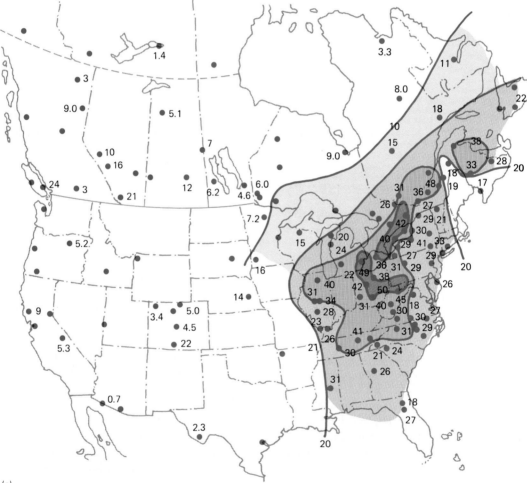

(a)

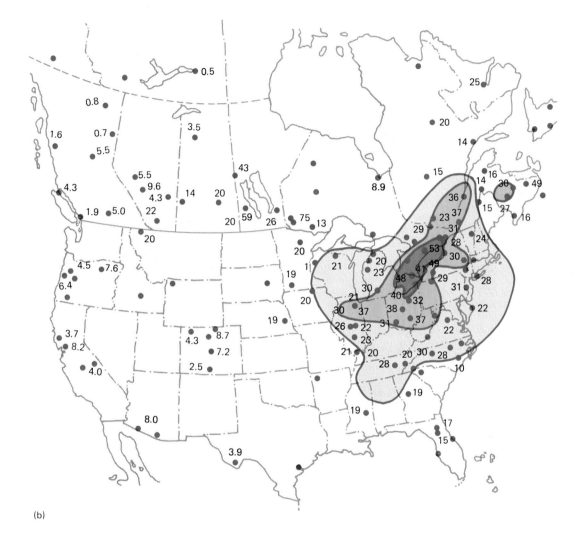

(b)

1976). This sudden, heavy input can drop the pH levels of aquatic ecosystems quickly and release aluminum ions, which at the level of 0.1 to 0.3 mg/l retards growth, gonadal development, and egg production of fish and increases mortality. The aluminum, rather than acidity (which some species of fish can tolerate), may be the most significant pollutant (Schofield and Trojnar 1980), although recent studies suggest that acidity alone is sufficient to cause mortality in developing brook trout (Hunn et al. 1987).

Experimental acidification with H_2SO_4 of a small headwater stream at Hubbard Brook from a normal pH of 5.4 to a pH of 4.0 points out the effects acidification can have on stream life. Following acidification, stream-water concentrations of Al, Ca, Mg, and K increased, as did the downward drift of immature insects. Emergence of adult mayflies and stoneflies de-

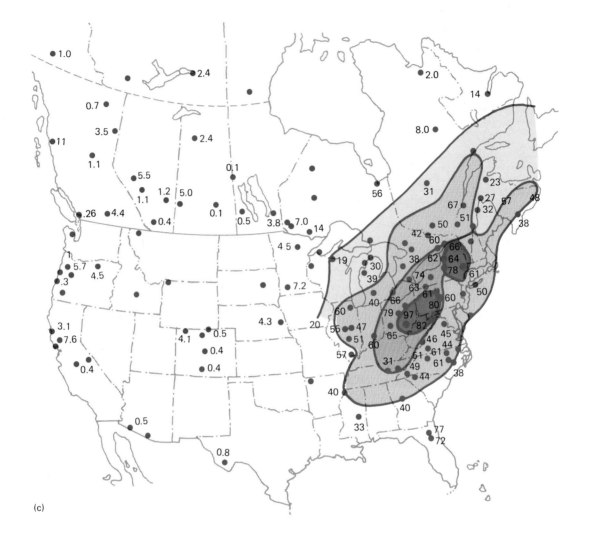

(c)

creased, periphyton biomass increased, and most trout moved downstream to areas of higher pH. Acidified streams in Norway and in Nova Scotia have lost their trout. Low pH inhibits gonadal development and egg production.

Acidification of lakes reduces bacterial activity in sediments (Rao et al. 1984), inhibiting decomposition and nutrient regeneration. Reduced nutrient regenera-

tion can impact aquatic food webs by reducing phytoplankton and invertebrates upon which fish depend for food. Loss of fish reduces the survival and reproductive success of such fish-eating birds as great blue herons and loons.

Some amphibians are highly vulnerable to acidic waters, and runoff from acidic snow into temporary ponds may destroy reproduction. Most vulnerable to acidity is the fertilization stage, because sperm apparently disintegrate at low pHs (Schlichter 1981). Embryonic stages of most species of frogs studied appear to be tolerant of low pHs (3.9–4.5), but the embryonic stages of the leopard frog, (*Rana pipens*) and the spotted salamander (*Ambystoma maculatum*) are particularly sensitive to acidity, with 100 percent mortality at pHs of 4 to 5 (Pierce 1985). Tolerance to acidity among amphibians tends to increase through the larval to adult stages.

The effects of acid precipitation on terrestrial ecosystems are more difficult to document or demonstrate, except in localized vicinities of smelters or acid mine drainage. Except for the strongly podzolic ones, soils throughout much of eastern North America are sufficiently buffered to neutralize acidic precipitation. In some cases acidic precipitation can improve soil fertility by addition of sulfur (Krug and Fring 1983); but in time the deposition may exceed the forest soil's capacity to utilize it. Acidic precipitation can have adverse effects on plant life directly. Substantial leaching of nutrients in acidic soils can reduce nutrient availability and increase the solubility of aluminum. Free aluminum attacks trees' root systems, reducing their ability to take up nutrients and moisture. Wet surfaces of vegetation permit the uptake of pollutants ten-fold, and when bathed in fog the entire plant takes them in (Smith 1981).

Atmospheric pollution is suspected as one of the causes of yellowing and early loss of needles of spruces as well as dieoff of these trees (Figure 12.12) in the Adirondack Mountains of New York, White Mountains of New Hampshire, Green Mountains of Vermont, Laurentian

Mountains of Quebec, and high Appalachians of West Virginia (Johnson 1983; Johnson and Siccama 1983; McLaughlin et al. 1983; Johnson et al. 1983; Raynal 1983; Puckett 1982) and a similar phenomenon in the mountains of West Germany (Postel 1984). These high elevation forests as well as pine forests in the mountains of southern California are bathed in fogs, which bring both beneficial substances such as inorganic nitrogen and detrimental ones such as sulfur dioxide, nitrous oxides, and heavy metals (Waldman et al. 1982; Lovett et al. 1982). Cloud droplets generally are more acidic and hold higher concentrations of other pollutants than rainfall. When immersed in fog, needle-leaved evergreens more effectively comb moisture out of the air. Dry deposition dusting moist foliage leaches nutrients from the leaves. In addition, the trees are subjected to ozone and other photochemical oxidants. Weakened by air pollution, these trees are subject to further damage by insects, fungi, and drought.

More important, the deposition of sulfur, nitrates, and ammonia significantly influences plant nutrition and soil chemistry. Recent studies (Schulze 1989) in the spruce forests of Germany show that spruce roots take up ammonium rather than nitrate. Nitrate left in the soil solution together with sulfate leaches out to the groundwater, accelerating soil acidification. Soil acidification reduces the amount of available magnesium, calcium, and potassium in the soil solution and affects nutrient uptake and root growth. More roots in declining stands grow in the litter layer than in the lower soil, which affects the pattern of water uptake and the susceptibility of the trees to short-term summer drought. Spruce trees also take

Figure 12.12 Death and decline of spruce and fir in the high Appalachians has been attributed to air pollutants which stress the trees. This stress makes them vulnerable to attacks by such insects as balsam woolly adelgid (*Adelges piceae*) which has killed stands of Frasier fir. (Photo by Hugh Morton.)

up atmospheric nitrogen pollution through the needles, which stimulates canopy growth while depleting other nutrient reserves in the acidic soils. This nitrogen to cation imbalance results in some of the symptoms associated with spruce decline.

Although no definitive statement can be made, evidence suggests that wet and dry deposition of sulfur dioxide, nitrous oxides, ozone, and other air pollutants are having an adverse effect on ecosystem function, plant community composition, and species survival in both plants and animals.

The Phosphorus Cycle

The phosphorus cycle differs from the sulfur cycle in that the element is unknown in the atmosphere and none of its known compounds have any appreciable vapor pressure. Thus the phosphorus cycle can follow the hydrological cycle only partway from land to sea (Figure 12.13).

Under undisturbed natural conditions phosphorus is in short supply. It is freely soluble only in acid solutions and under reducing conditions. In the soil it becomes immobilized as phosphates of either calcium or iron. Even superphosphate (a soluble mixture of phosphates) applied to croplands may be converted rapidly to unavailable inorganic compounds. Its natural limitations in aquatic ecosystems are emphasized by the explosive growth of algae in water receiving heavy discharges of

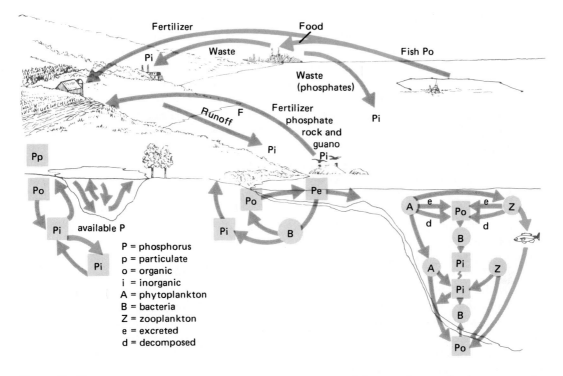

Figure 12.13 The phosphorus cycle in terrestrial and aquatic ecosystems.

phosphorus of terrestrial ecosystems esphosphorus-rich wastes (Stevenson and Stoermer 1982).

The main reservoirs of phosphorus are rock and natural phosphate deposits, from which the element is released by weathering, by leaching, by erosion, and by mining for agricultural use. Some of it passes through terrestrial and aquatic ecosystems by way of plants, grazers, predators, and parasites; and it is returned to the ecosystem by excretion, death, and decay. In terrestrial ecosystems organic phosphates are reduced by bacteria to inorganic phosphates. Some are recycled to plants, some become immobilized as unavailable chemical compounds, and some are immobilized by incorporation into bodies of microorganisms. Some of the

capes to lakes and seas, both as organic phosphates and as particulate organic matter.

In marine and freshwater systems the phosphorus cycle involves four fractions: particulate phosphorus, the largest reservoir involving both living and dead particulate matter (including phytoplankton); inorganic phosphates, mostly soluble orthophosphate (PO_4^{3-}) with an extremely short turnover time; an organic phosphorus compound, XP, with low molecular weight (about 250), excreted by organisms; and a soluble macromolecular colloidal phosphorus. The major exchange is between the inorganic phosphate and the particulate fractions. The organic compound fraction converts to the colloidal compound fraction, and both, especially the colloid, release phosphate to the soluble inorganic fraction. This is rapidly recycled through the plankton (Lean 1973a, 1973b).

Phosphorus in phytoplankton may be ingested by zooplankton or detritus-feeding organisms. Zooplankton in turn may excrete as much phosphorus daily as is stored in its biomass (Pomeroy et al. 1963). By excreting phosphorus zooplankton is instrumental in keeping the aquatic cycle going. More than half of the phosphorus zooplankton excretes is inorganic phosphate, which is taken up by phytoplankton. In some instances 80 percent of this excreted phosphorus is sufficient to meet the needs of phytoplankton. The remainder of the phosphorus in aquatic ecosystems is in organic forms. They are utilized by bacteria, which fail to regenerate much dissolved inorganic phosphate themselves (Johannes 1968).

Part of the phosphorus in aquatic ecosystems is deposited in deep and shallow sediments. Precipitated largely as calcium compounds, much of it becomes immobilized for long periods of time in bottom sediments. Seasonal overturns (Chapter 10) and ocean upwellings return some of it to photosynthetic zones, where the phosphorus becomes available to phytoplankton. The phytoplankton then maintains the phosphorus in the lighted zone during the period of active growth (bloom). Ketchum and Corwin (1965) and Ketchum (1967) found that prior to a phytoplankton bloom 28 percent of the needed phosphorus was supplied by the inorganic fraction in solution in the upper waters and 72 percent was supplied by vertical transport and upwelling from the deep. During the period of bloom, however, phytoplankton obtained 86 percent of its phosphorus from the inorganic phosphorus dissolved in solution, 12 percent from vertical mixing, and 2 percent from regeneration by biological cycling. Thus during the bloom stage much of the phosphorus was tied up in organic matter, and only by a rapid turnover in its populations could phytoplankton meet its phosphorus requirements.

The role of organisms in the cycling of phosphorus in aquatic ecosystems is further illustrated by processes in tidal marshes. In the intertidal marshes of the southern United States marsh grass (*Spartina alternifolia*) withdraws phosphorus from subsurface sediments. Half of the phosphorus withdrawn is fixed in plant tissue. The other half is leached from leaves by rain and tides and carried out by tidal exchange (Reimold 1972). When the grass dies, animal life in the marsh and adjacent waters and tidal creeks use the detritus as food.

One of the detritus feeders is the ribbed mussel (*Modiolus dimissus*). It plays a major role in the turnover of phosphorus through the tidal marsh ecosystem (Kuenzler 1961). To obtain its food, which consists of small organisms as well as particles rich in phosphorus suspended in the tidal waters, the mussel must filter great quantities of seawater. It ingests some of the particles, but rejects most which settle on the intertidal mud. These particles, rich in phosphorus, stay in the marsh instead of being carried out to sea. Each day the mussel removes from the water one-third of the phosphorus found in suspended matter. The particulate matter deposited on the mud is food for the deposit feeders, which release the phosphate back to the ecosystem.

Human activities have altered the phosphorus cycle. As the natural supply of phosphorus in the soil is depleted, farmers apply fertilizer. The source of that fertilizer is phosphate rock. Because of the abundance of calcium, iron, and ammonium in the soil, most of the phosphate applied as fertilizer becomes immobilized as insoluble salts, and little escapes in runoff.

Part of the phosphorus used as fertilizer is removed in crops when harvested. Transported far from the point of fixation, this phosphorus eventually is released as waste in the processing and consumption of food. Concentration of phosphorus in wastes of food-processing plants and of feedlots adds a quantity of phosphates to natural waters. Greater quantities are supplied by urban areas, where phosphates are concentrated in sewage systems.

Most of the phosphorus enrichment of aquatic ecosystems comes from sewage disposal plants. Primary sewage treatment removes only 10 percent of the total phosphorus. Secondary treatment removes only 30 percent at best, and even some of this phosphorus finds its way into streams. Feedlots contribute runoff, but phosphorus has such a strong affinity to the soil particles that the problem is not as severe. Sewage, with its heavy load of phosphate detergents, contributes nearly all of the phosphorus reaching lakes and rivers.

Phosphorus is more intimately involved in the nutrient enrichment (eutrophication) of fresh water than nitrogen (see Chapter 32). Of the three nutrients required for aquatic plant growth, potassium is usually present in excess, nitrogen is supplemented by fixation, and phosphorus tends to be precipitated in the sediments and cannot be supplemented naturally (Am. Chem. Soc. 1969). Thus phosphorus is usually limiting, and in the presence of a luxury supply, algae respond with luxurious growth.

Cycling of Heavy Metals, Hydrocarbons, and PCBs

Bringing biogeochemical cycles forcibly to the attention of the public has been the accumulation of toxic materials in the biosphere. Heavy metals such as lead, mercury, and cadmium, for example, have always been present in the environment and have entered into natural cycles of the ecosystem. However, human activities have markedly increased their concentration in the environment, and the passage of these metals through the food chain has markedly increased their concentration in the upper trophic levels.

Lead. An example of a heavy metal injected into ecosystems by human acitivity is lead. Ninety-eight percent of the lead in the biosphere comes from automobile emissions, the burning of lead alkyl additives to gasoline. In 1968 more than 2.3×10^8 kg of lead was emitted to the atmosphere, which roughly amounts to 1 kg per person (Chow and Earl 1970; Commoner 1971). This quantity has been greatly reduced in the United States, Japan, Brazil, and the European Common Market countries by the use of unleaded gasoline.

Minor sources are metal smelting plants and agricultural areas where lead arsenate is used as an orchard spray. Introduced as a fine aerosol, lead eventually falls out either in precipitation or in dust to contaminate the soil. Once on or in plants, lead enters the food chain (Chow 1970). Plant roots take up lead from the soil and leaves take it from contaminated air or from particulate matter on the leaf surface. From there lead enters the food chain, picked up by herbivorous insects and rodents and passed to frogs, birds, bats, and other high-level consumers. This uptake through the food chain is most pronounced along highway roadsides (Smith 1976; Bull et al. 1983; Birdsall et al. 1983; Grue et al. 1986). There is evidence that microbial systems are also capable of taking up and immobilizing substantial quantities of lead (Tornabene and Edwards 1972).

A third source of lead pollution is lead

shot. Ingested by waterfowl mistaking shot for seeds of aquatic plants, lead is a major cause of mortality in these birds.

Atmospheric lead pollution poses a serious health problem in industrialized areas of Earth (Goldsmith and Hexter 1967; Wessel and Dominski 1977). The long-term increase in concentrations of atmospheric lead has resulted in a significant increase in the concentrations of lead in humans. The average body burden of lead among adults and children in the United States is 100 times greater than the natural burden, and existing rates of lead absorption are 30 times the levels in preindustrial society. An intake of lead can cause palsy, partial paralysis, loss of hearing, and death.

DDT and Other Chlorinated Hydrocarbons. Of all human intrusions into biogeochemical cycles, none has done more to call attention to nutrient cycling than the widespread application of DDT. During World War II this pesticide began to be used in huge quantities to control disease-carrying and crop-destroying insects. As early as 1946 Clarence Cottam called attention to its damaging effects on ecosystems and nontarget species, but the impact of pesticides on ecosystems remained obscure until Rachel Carson's (1962) *Silent Spring* exposed the dangers of hydrocarbons. The detection of DDT in the tissues of animals in the Antarctic (Risebrough et al. 1976), far removed from any applied source of the insecticide, emphasized the fact that DDT does indeed enter the global biogeochemical cycle and become dispersed around the earth.

DDT (along with other chlorinated hydrocarbons) has certain characteristics that enable it to enter global circulation. Because it is highly soluble in lipids or fats and not very soluble in water, it tends to accumulate in the lipids of plants and an-

imals. It is persistent and stable, undergoes little degradation (largely from DDT to DDE), and has a half-life of approximately 20 years. It has a vapor pressure high enough to ensure direct losses from plants. It can become adsorbed by particles or remain as a vapor; in either state it can be transported by atmospheric circulation and then return to land and sea with rainwater.

Most DDT was applied to croplands, forests, and marshes for insect control (Figure 12.14). Back in 1963 the maximum production of DDT was 8.13×10^{10}g. Production declined sharply in 1970s, and its use was banned in the United States in 1972. Production continues, however, and because DDT is widely used in third world countries, great quantities are still being injected into the biosphere, and imported back to other countries in foodstuffs.

Insecticides are applied on a large scale by aerial spraying. Half or more of a toxicant applied in this manner is dispersed to the atmosphere and never reaches the ground. For example, in a massive spraying of DDT over 66,000 acres of forest in eastern Oregon, only about 26 percent reached the forest floor (Tarrant 1971).

On the ground or on the water's surface the pesticide is subject to further dispersion. There is apparently little movement of DDT into the soil, where its mean life is about 4.5 years. Pesticides reaching the soil are lost through volatization, chemical degradation, bacterial decomposition, runoff, and the harvest of organic matter, which can amount to about 1 percent of the total DDT used on the crop.

In flowing water pesticides are subject to further distribution and dilution as they move downstream. Insecticides released in oil solution penetrate to the bottom and cause mortality of fish and aquatic inver-

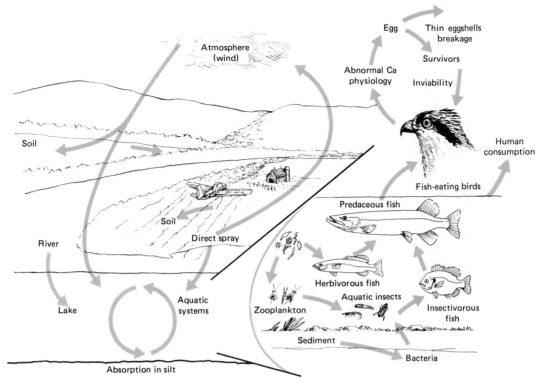

Figure 12.14 The movement of chlorinated hydrocarbons in terrestrial and aquatic ecosystems. The initial input comes from spraying on vegetation. A relatively large portion fails to reach the ground and is carried on water droplets and particulate matter through the atmosphere.

tebrates (see reviews in Cope 1971; Pimentel 1971a). Trapped in the bottom rubble and mud, the insecticide may continue to circulate locally and kill for some days.

In lakes and ponds emulsifiable forms of DDT tend to disperse through the water, but not necessarily in a uniform way. DDT in oil solutions tends to float on the surface and move about in response to the wind.

Eventually pesticides reach the ocean. In surface slicks their concentration may be 10,000 times greater than in lower waters. These slicks, which attract plankton,

are carried by ocean currents. In the oceans part of the DDT residues may circulate in the mixed layer, some may be transferred below the thermocline to the abyss, and more may be lost through sedimentation of organic matter.

Not only does the atmosphere receive the bulk of pesticidal sprays (well over 50 percent of that applied), but it also picks up that fraction volatilized from soils, vegetation, and water. The adsorption of residues to particulate matter increases the capacity of the atmosphere to hold DDT. Thus the atmosphere becomes a large circulating reservoir of DDT and other chlorinated hydrocarbons. Chemical degradation, diffusion across air-sea interfaces, rainfall, and dry fallout remove residues from the atmosphere (SCEP 1970). In turn pesticides carried to land and sea return to the atmosphere by volatilization.

Although the quantity of residues of DDT and other chlorinated hydrocarbons may be relatively small, the concentrations are sufficient to have a deleterious effect on marine, terrestrial, and freshwater ecosystems (Beyer and Gish 1980; Matthiessen 1985). DDT and related compounds tend to concentrate in the lipids of living organisms, where they undergo little degradation (see Menzie 1969; Bitman 1970; Peakall 1970). As long as the rate of accumulation of DDT is not too great, birds tend to eliminate some of the DDT, which reduces the body burden. Adeline penguins in the Antarctic, for example, accumulate some 45 μg DDE, the degradation product of DDT, in their first year and 89 μg every year thereafter for a burden of 400 μg DDE by their fifth year; but the actual accumulation is much less, indicating a daily loss rate of 0.12 percent, giving DDE a half-life of 580 days. In contrast the half-life of PCB accumulated by the penguins was 270 days, with a daily excretion rate of 0.26 percent (Subramanian et al. 1987).

The high solubility of DDT in lipids allows the magnification of its concentration through the food chain. Most of the DDT contained in the food is retained in the fatty tissues of the consumer organism. Because it breaks down slowly, DDT accumulates to high and even toxic levels. The concentrated DDT is passed on to the consumer at the next trophic level, where again it is retained and accumulated. The carnivores on the top level of the food chain receive massive amounts of pesticides.

There are a number of examples of species concentrating this pollutant. Eastern oysters held for 40 days in flowing seawater containing only 0.1 ppb of DDT concentrated the pesticide some 70,000 times that contained in the water (Butler 1964). Four species of algae in water containing 1.0 ppm for 7 days concentrated the pesticide 227-fold (Vance and Drummond 1969). *Daphnia,* a genus of zooplankton, concentrated DDT 100,000-fold during a 14-day exposure in water containing 0.5 ppb of DDT (Preuster 1965). Slugs and worms in a cotton field concentrated DDT 18 and 11 times that of the level in the soil. Pesticides concentrated by first-level consumers are then passed on to second-level consumers, carnivores, who in turn magnify the pesticide even more.

The concentration of DDT in a Long Island salt marsh sprayed for mosquito control over a period of years was 13 lb/acre. The actual concentration of DDT in the water was 0.00005 ppm. The residue in consumers showed the increase in concentration along the food chain: 0.04 ppm in plankton, 0.16 ppm in shrimp, 0.28 ppm in eels, 2.07 ppm in predacious fish, and 75.0 ppm in ringbilled gulls. Thus the ringbilled gull, a predacious bird near the top of the food chain, contained a level of DDT a million times greater than the water (Woodwell et al. 1967).

High concentrations of DDT in the tissues often result in death or impaired reproduction and genetic constitution of organisms. Laboratory populations of zooplankton, shrimps, and crabs are killed outright by exposure to DDT in parts per billion. The continuous exposure of shrimp to DDT in 0.2 ppb, a concentration that has been detected in waters flowing into shrimp nursery areas, killed the entire population in less than 20 days (SCEP 1972). A residue level of 5.0 ppm in the lipid tissues of the ovaries of freshwater trout causes 100 percent die-off of the fry, which pick up lethal doses as they utilize the yolk sac. High levels of DDT are correlated with the decline of such fish as sea trout and California mackerel.

DDT and its degradation product DDE interfere with calcium metabolism in such birds as the peregrine falcon, osprey, and bald eagle. Chlorinated hydrocarbons block ion transport by inhibiting the enzyme ATPase, which makes available the needed energy. The reduced transport of ionic calcium across membranes can cause death. DDE also inhibits the enzyme carbonic anhydrase (Bitman 1970; Peakall 1970), essential for the deposition of calcium carbonate in the eggshell and for the maintenance of a pH gradient across the membranes of the shell gland.

Since the prohibition of DDT in the United States birds of prey are recovering from low reproductive rates induced by DDT. Grier (1982) reports that reproduction in the bald eagles of northwest Ontario declined from 1.26 young per breeding area in 1966 to 0.46 in 1974. Although the eagles nested in an area with little or no DDT, the birds accumulated the pesticide by feeding on contaminated fish in their wintering areas in the United States. Since 1974 reproduction has increased steadily to 1.21 per breeding area in 1981. The recovery of bald eagles and other birds occupying high trophic levels, particularly the osprey and brown pelican, attests to the importance of controlling the substances we add to ecosystems.

The amount of pesticides we add to the environment is enormous and increasing. The United States manufactures 50,000 individual pesticide products, and in 1985 used 500,000 metric tons—two kilograms for each American. Of this amount, 92 percent was used by households and not by agriculture.

Polychlorinated Biphenyls. Another contaminant widespread in the environment is polychlorinated biphenyl (PCB), a ma-

jor toxic waste problem. PCB is a generic name used for a number of synthetic organic compounds characterized by biphenyl molecules containing chlorine atoms. They are widely used as dielectric fluids in capacitors and transformers, in plastics, solvents, and printing inks. The production of PCBs increased steadily from their introduction in 1930 to a high of 34,000 tons in 1970. Production and use are now declining since the discovery that PCBs are an environmental contaminant. (For a good review of PCBs, see Ahmed 1976.)

A Swedish scientist, S. Jensen (1966), first recognized the threat of PCBs to the environment when he discovered them in fish and birds he was examining for the presence of DDT. Since then biologists have found PCBs in human tissue, foodstuffs, and many species of fish and birds. Like DDT, PCBs have an affinity for fatty tissue, degrade slowly, and accumulate in the food chain.

Except for the source, the pathway of PCBs is similar to that of DDT. PCBs are discharged into the environment through sewage outfalls and industrial discharges. Rivers are a means of transport, but the atmosphere appears to be the major mode (see A. L. Hammond 1972). PCBs accumulate in bottom sediments, adsorbed in silt and fine particles, which hold them in the aquatic environment. Fish take up PCBs and concentrate them in their tissues (Stein et al. 1987), where the chemical residue is often higher than DDT. Typical concentrations of PCBs in fish, both ocean and freshwater, range from 0.01 to 1.0 ppm away from heavily polluted industrial areas. In polluted waters, such as parts of the Great Lakes and the Hudson River in New York State, PCBs range from 10 to 85 ppm, with individual fish carrying as much as 400 ppm. Federal

Drug Administration acceptable tolerance levels of PCBs in food range from 0.5 ppm to 5.0 ppm, depending upon the food item. Predatory fish-eating birds, particularly ospreys and cormorants, have concentrations that range from 300 to 1000 ppm. Traces have been found in human tissue. Like DDT, PCBs appear to cause thinning of eggshells, deformities in newly hatched birds (Gilbertson et al. 1976), and reduction in growth rates of certain marine diatoms (Mosser et al. 1972).

■ Radionuclides

Ever since the atomic bomb ushered in the atomic age, the impact of nuclear radiation on life on earth has been a major concern. Involved are high-energy, short wavelength radiations, known as ionizing radiations. They are so called because they are able to remove electrons from some atoms and attract them to other atoms, producing positive and negative ion pairs. Of greatest interest is ionizing electromagnetic or gamma radiation, which has a short wavelength, travels a great distance, and penetrates matter easily.

Sources of gamma radiation are atomic blasts from weapons testing, nuclear reactors, and radioactive wastes. By-products of both weapons testing and nuclear reactors such as zinc-65 (^{65}Zn), strontium-90 (^{90}Sr), cesium-137 (^{137}Cs), iodine-131 (^{131}I), and phosphorus-32 (^{32}P) are radioactive. When the uranium atom is split or fissioned into smaller parts, it produces, in addition to tremendous quantities of energy, a number of new elements or fission products, including strontium, cesium, barium, and iodine. Some of these fission products last only a few seconds; others can remain active for several thousand years. These radioactive elements enter the food chain and become incorporated in living organisms in which they can cause cancer and genetic defects.

In the same atomic reaction some particles with no electrical charges, called neutrons, get in the way of high-energy particles. Nonfission products are the result. They include the radioisotopes of such biologically important elements as carbon, zinc, iron, and phosphorus, which are useful in tracer studies.

Both fission and nonfission products are released to the atmosphere by nuclear testing and by wastes from nuclear reactors unless carefully handled. Later they return to earth along with rain, dust, and other material as radioactive fallout. In the case of weapons testing in the atmosphere, the fallout is worldwide. Once the isotopes reach the earth, they enter the food chain and become concentrated in organisms in amounts that exceed by many times the quantities in the surrounding environment. In effect, local radiation fields develop in the tissues of plants and animals.

Of particular concern is the radioactive output of nuclear power plants, especially since the Three Mile Island and Chernobyl accidents. Pressurized water reactors, commonly used in nuclear power plants, release low levels of radioactivity to the air and condenser water. When water passes through the intense neutron flux of the reactor, it is contaminated by radioactivity. Trace elements in the water are activated, producing radioisotopes. Added to them are radioactive corrosive products from the surface of metal cooling tubes. Except for tritium, most of the radioisotopes are removed in a radioactive waste removal process, which creates additional problems of nuclear waste disposal. Those left have a short half-life and rapidly decay below detection levels.

Terrestrial Ecosystems

Dispersion of radionuclides in terrestrial ecosystems comes from gaseous, particulate, and aerosol deposition and from liquid and solid wastes. Plants both intercept particulate contaminants and absorb radionuclides through the foliage. Additional input into vegetation may come by uptake of radionuclides from soil and litter. From the plants radionuclides move through the ecosystem along the food chain.

Strontium-90 and cesium-137 are two of the most destructive radioactive materials released into the biogeochemical cycle (Figure 12.15). Ecologically they be-

Figure 12.15 Radionuclide cycling through the food chain. Strontium and cesium are transferred through fallout to the grazing food chain and on to humans through meat and milk. This cycling occurred during the early days of nuclear weapon testing and more recently following the nuclear power plant explosion at Chernobyl. (Courtesy Oak Ridge National Laboratory.)

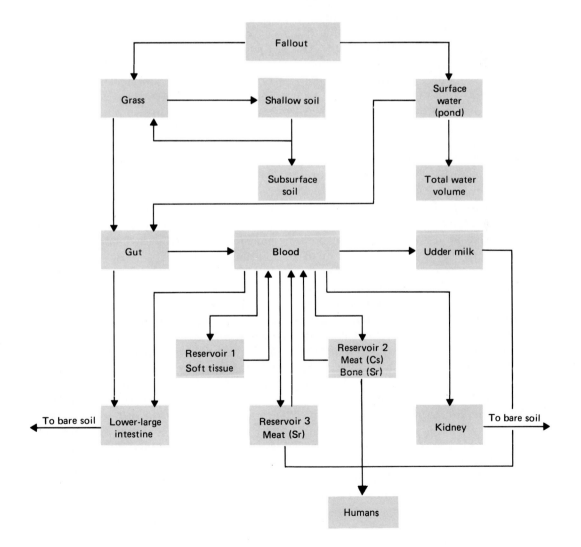

have like calcium and follow it in the cycling of nutrients. Both easily enter the grazing food chain, especially in regions with high rainfall or abundant soil moisture and with low levels of calcium and other mineral nutrients in the soil. One such region, the arctic tundra, has been subject to heavy nuclear fallout from weapons testing in the past. It received another input as fallout from the Chernobyl nuclear power plant explosion in 1986.

Lichens, the dominant plants of the tundra, absorb virtually 100 percent of the radioactive particles and gases drifting on them. From lichens the contaminants ^{90}Sr and ^{137}Cs travel up the food chain, from caribou and reindeer to wild carnivores and humans.

Early studies on the effects of nuclear fallout on human food chains involved Eskimos and northern Alaskan Indians (Palmer et al. 1963; W. C. Hanson 1971). Caribou are a major food source for northern Alaskan natives, who kill the animals during the northward migration in spring and stockpile the meat for late spring and early summer food. Because the caribou feed all winter on lichens, their flesh in spring during the study contained three to six times as much ^{137}Cs as it did in the fall. In spring the Indians showed a corresponding rise in ^{137}Cs level, often 50 percent (Figure 12.16). The level decreased when the people changed to a diet of fish in the summer.

The accident at Chernobyl produced a nuclear cloud that drifted northward over the Arctic, depositing much of its contamination over arctic Eurasia (Davidson et al. 1987) and causing a repeat of the story of the 1960s. Radioactive-contaminated lichens passed strontium and cesium on to the reindeer, making both meat and milk unsuitable for human consumption, se-

verely impacting the Laplanders' economy and their future.

In vertebrate food chains strontium and cesium usually accumulate throughout the body at higher trophic levels, but cobalt, ruthenium, iodine, and some other radionuclides do not. Some, such as iodine, concentrate in certain tissues. In arthropod food chains potassium, sodium, and phosphorus accumulate, whereas strontium and cesium do not (see Table 12.4).

Aquatic Ecosystems

Radionuclides contaminate aquatic ecosystems largely through waste from nuclear power plants and from nuclear processing industry. Radioactive materials that enter the water become incorporated in bottom sediments and circulate between mud and water. Some become absorbed by bottom-dwelling insects and fish downstream from the source. In fact, they may be exposed to chronic low-level radiation. Under such conditions a sort of equilibrium is established between retention in the organisms, the bottom sediments, daily input, and decay.

Concentrations of radionuclides in aquatic food chains vary with the system and the species. ^{32}P concentrations increase at higher trophic levels; ^{90}Sr becomes concentrated in the bones of fish and in invertebrates possessing calcareous exoskeletons, yet the concentration in fish flesh is less than that in plants and invertebrates. For example, cobalt-60 concentrations in algae (plants which absorb more nutrients physically through cell membranes than they take in and concentrate biologically) are 2500 to 6200 times that of the surrounding water; concentrations in fish are only 25 to 50 times that of the water. Thus the concentration of

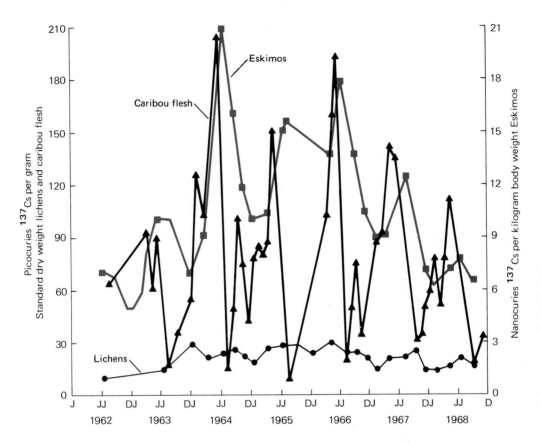

Figure 12.16 Cesium-137 concentrations in lichens, caribou flesh, and Eskimos at Anaktuvuk Pass, Alaska, during the period 1962–1968. Note the relationship between the concentration of ^{137}Cs in caribou flesh and the amount in humans. As the concentration in caribou declined seasonally, so did the concentration in humans. (From W. C. Hanson 1971.)

radionuclides does not necessarily increase consistently through the food chain as does the concentration of chlorinated hydrocarbons. In many situations the concentration of radionuclides decreases at higher trophic levels. Aquatic organisms tend to concentrate radionuclides the same as they do the stable elements.

One isotope, ^{137}Cs, related biogeochemically to potassium, does not appear to increase at higher trophic levels in aquatic ecosystems, but its accumulation in the bones of fish and a similar accumulation of ^{90}Sr in the shells of clams serve as an index of low-level radiocontamination of the environment. Because they continuously feed on particulate matter, clams are excellent indicator species. There is no turnover of radionuclides deposited in the growth rings of the shell, so their shells are a record of the radionuclide contamination in their environment (D. J. Nelson 1962).

In spite of considerable study, we still know little about uptake, assimilation, distribution in tissues, turnover rates, and equilibrium levels of radionuclides for many taxonomic groups. We know even

TABLE 12.4 Aquatic and Terrestrial Food-Chain Concentration of Elements

| Trophic Level | ELEMENT CONCENTRATION FACTORS* | | | | | | | | | | | | |
	Ca	Sr	K	Cs	Na	Co	Zn	Mn	Ru	Fe	P	Ra	I
AQUATIC													
Water	1.0	1.0	1.0	1.0	1.0	1.0	1.0	1.0	1.0	1.0	1.0	1.0	1.0
Algae and higher plants	1–400	10–3,000		50–25,000		2,500–6,200	140–33,500	700–35,000	80–2,000	2,400–200,000	36,000–50,000	0.5	60–200
Invertebrates													
saprovores	16	10–4,000		60–11,000					130		2–100,000	0.5	20–1,000
herbivores		1		600		325	150	6,000–140,000		125	2,000		
carnivores				800									
Fish													
omnivores		1	300–2,500	125–6,000			4–40			10,000	3,000–100,000	0.5	25–50
carnivores	0.5–300	1–150	400–2,700	640–9,500								1.5	
TERRESTRIAL													
Plants	1.0	1.0	1.0	1.0	1.0	1.0	1.0	1.0	1.0	1.0	1.0	1.0	1.0
Invertebrates													
saprovores	0.1–18	0.1	3.5	0.2	17						11		
herbivores	0.1	0.1	3.0	0.3–0.5	21	0.4			0.4		17		
carnivores	0.1		2.0	0.1–0.5	27	0.5			1.2		18		
Mammals													
herbivores		0.5–4.5		0.3–2.0	0.3	0.3			0.4	0.8		0.01	0.5
omnivores				1.2–2.0					0.2				0.2
carnivores				3.8–7.0									0.1

*Ratio of element level in consumer to element level in food-chain base, with base value normalized at 1.0.

Source: D. E. Reichle et al. 1970.

less about the role of the environment in the cycling of radionuclides through various ecosystems. As the number of nuclear power plants throughout the world and associated nuclear wastes increase, our knowledge of the behavior of radionuclides needs to be more sophisticated. The knowledge of hazards must extend not only to humans, but also to the biota upon which they depend.

■ Summary

Nutrients flow from the living to the nonliving components of the ecosystem and back in a perpetual cycle. By means of these cycles plants and animals obtain nutrients necessary for their well-being.

There are two basic types of biogeochemical cycles: the gaseous, represented by oxygen, carbon, and nitrogen cycles; and the sedimentary, represented by the phosphorus cycle. Because it is so reactive chemically, oxygen, the main life support on Earth, has a complex cycle involving oxidative reactions with organic matter and reduced mineral components of Earth's crust. An important constituent of the atmospheric reservoir of oxygen is ozone (O_3). The ozone layer blocks out much of the solar ultraviolet radiation. Injection of chlorofluorocarbons and nitrogen and other oxides into the atmosphere appears to be causing some destruction of stratospheric ozone, with serious ecological implications.

The carbon cycle is so closely tied to energy flow that the two are inseparable. It involves the assimilation of and respiration of carbon dioxide by plants, its consumption in the form of carbohydrates in plant and animal tissue, its release through respiration, the mineralization of litter and wood, soil respiration, accumulation of carbon in standing crop biomass, and withdrawal into long-term reserves. The carbon dioxide cycle exhibits both diurnal and annual curves. The equilibrium of carbon dioxide exchange among land, sea, and air has been disturbed by rapid injection of CO_2 into the atmosphere by burning fossil fuels; but over one-half is removed from the atmosphere by oceans and terrestrial and marine vegetation. Increased CO_2 in the atmosphere has the potential of raising the average temperature of Earth by several degrees, with serious ecological implications.

The nitrogen cycle is characterized by fixation of atmospheric nitrogen by nitrogen-fixing plants, largely legumes and blue-green algae. Involved in the nitrogen cycle are the processes of ammonification, nitrification, and denitrification.

Human intrusion into the nitrogen cycle involves inputs of nitrogen dioxide into the atmosphere and nitrates into the aquatic ecosystems. The major sources of nitrogen dioxide are automobiles and burning of fossil fuels. Nitrogen dioxide is reduced by ultraviolet light to nitrogen monoxide and atomic oxygen. These substances react with hydrocarbons in the atmosphere to produce a number of pollutants, including ozone and PAN, which make up photochemical smog, a pollutant harmful to plants and animals. Excessive quantities of nitrates are added to aquatic ecosystems by improper use of nitrogen fertilizer on agricultural crops, by animal wastes, and by sewage effluents (the largest source). More closely involved with the pollution of aquatic systems is phosphorus from sewage effluents.

The sedimentary cycle involves two phases, salt solution and rock. Minerals become available through the weathering of the earth's crust, enter the water cycle as salt solution, take diverse pathways through the ecosystem, and ultimately return to the sea or back to the earth's crust through sedimentation.

The phosphorus cycle is wholly sedimentary, with reserves coming largely from phosphate rock. Much of the phosphate used as fertilizer becomes immobilized in the soil, but great quantities are lost in detergents and other wastes carried by sewage effluents.

The sulfur cycle is a combination of the gaseous and sedimentary cycles because it has reservoirs in both the earth's crust and the atmosphere. It involves a long-term sedimentary phase in which sulfur is tied up in organic and inorganic deposits, is released by weathering and decomposition, and is carried to terrestrial and aquatic ecosystems in salt solution. A considerable portion of sulfur is cycled in the gaseous state, which permits its circulation on a global scale. Sulfur enters the atmosphere from the combustion of fossil fuel, volcanic eruptions, the surface of the ocean, and gases released by decomposition. Entering the gaseous cycle initially as hydrogen sulfide, sulfur quickly oxidizes to sulfur dioxide. Sulfur dioxide, soluble in water, is carried to Earth as weak sulfuric acid. Whatever the source, sulfur is taken up by plants and incorporated into sulfur-bearing amino acids, later to be released by decomposition. Injected into the atmosphere by industrial consumption of fossil fuels, sulfur dioxide has become a major pollutant, affecting and even killing plant growth, causing respiratory afflictions in humans and animals, and producing acid rain. Acid deposition, either in cloud droplets, precipitation, or dry particles, is implicated in the acidification of lakes and streams in northeastern United States, Canada, and Scandinavia where the soils are poorly buffered, and in the decline of spruce forests in Europe and eastern North America.

Industrial use of such heavy metals as lead and mercury, always present at low levels in the biosphere, has significantly increased their occurrence. Both pose potential and actual health problems as they enter the food chain.

Of still more serious consequence globally are the chlorinated hydrocarbons. Used in insect control, these pesticides have contaminated the global ecosystem and entered food chains. Because they become concentrated at higher trophic levels, chlorinated hydrocarbons affect predaceous animals most adversely. Fish-eating birds are endangered because chlorinated hydrocarbons interfere with their reproductive capability.

Radionuclides from nuclear weapons testing and from nuclear power plants can enter and become concentrated in food chains, particularly grazing ones, and transfer to higher trophic levels. In some situations concentrations of radionuclides decrease at higher trophic levels, especially in aquatic ecosystems.

Further intensive study of the cycling of nutrients and toxic materials is necessary to maintain the integrity of exploited ecosystems and the safe handling and disposal of toxic wastes.

PART III

POPULATION ECOLOGY

C H A P T E R 1 3

Population Genetics

Populations of different species of plants and animals make up the structural components of the ecosystem through which energy and nutrients flow. Considered ecologically, a *population* is a group of interbreeding organisms of the same kind occupying a particular space. It is characterized by density, the number of organisms occupying a definite unit of space. It has an age structure, the ratio of one age class to another. It acquires new members through birth and immigration and loses members through death and emigration. The differences between gains and losses determine the rate of population growth.

Because it is composed of interbreeding organisms, a population is also a genetic unit. Each individual carries a certain combination of genes, a sample of the population's total genetic information. The sum of all genetic information carried by all individuals of an interbreeding population is the *gene pool*. Gene flow, the exchange of genetic information between populations, comes about through immigration and emigration.

Populations consist of numerous subpopulations or *demes* of varying densities. These subpopulations are both ecological and genetic, but an ecological population is not necessarily the same as the genetic population. To define this incongruity, Antonovics and Levin (1980) have proposed the neighborhood concept of population density and gene flow. To visualize the concept consider a large number of even-aged plants distributed across the landscape. The distance between the plants beyond which any density effects are absent is the ecological effective distance. Draw a circle about each plant outside of which density effects are absent; the area within the circle represents the plant's ecological neighborhood. The number of plants in each such neighborhood is 1, and all plants within the idealized population have the same neighborhood area (Figure 13.1). If we increase the density of plants by some magnitude, we will arrive at a point where ecological neighborhoods will be congruent, but density effects are still absent. The density of such a population as a whole is the ecological effective density, the highest density the population can attain before density-dependent interactions become important (see Chapter 16). If we increase the density further, ecological neighborhoods will overlap. Although the ecological neighborhood of each plant remains the same, it has to share its neighborhood with other plants. When plants are less than a neighborhood apart, density-dependent interactions set in (see Chapter 16). Thus the ecological neighborhood is defined in terms of distances over which density-dependent effects operate.

The deme also represents a genetic population, but it is defined in terms of the distances genes travel, which may have little relationship to the ecological neighborhood. For example, wind-pollinated plants whose density is regulated by intraspecific competition may spread their pollen over a much larger distance. Moths that release sex-attractant pheromones to the air have an ecological neighborhood smaller than their genetic neighborhood. Self-incompatible plants that compete for pollinators may have similar genetic and ecological neighborhoods. Animals whose foraging distances are about the same as their search distances for mates have ecological and genetic neighborhoods of similar sizes. Plants regulated by herbivores may have ecological neighborhoods larger than their genetic ones. Thus ecological effective distances may be smaller, larger, or about the same as the genetic neighborhoods. This point is important to remember when considering the interrelations of individuals and demes of a species at both ecological and genetic levels.

Populations may also be considered as evolutionary units. Evolution involves changes in gene frequency through time in the gene pool, with consequent changes in physical expression. These irreversible changes in genetic constitution result from selective pressures brought to bear by the environment on individuals of the population. If the population contains sufficient variation in both its gene pool and forms of genetic expression, then selection (anything that produces a systematic heritable change in a population) favors certain types of individuals over others. These individuals have more off-

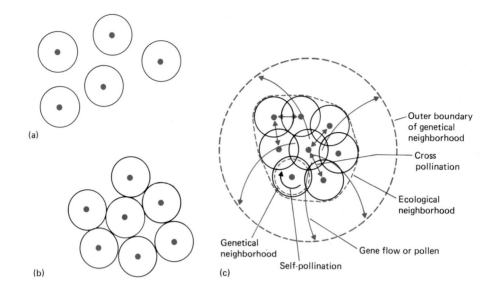

Figure 13.1 Ecological and genetic neighborhoods. (a) Individuals are so widely spaced that effects of density are absent. The circle about each plant indicates its ecological neighborhood. (b) The highest density possible before density-dependent effects take place, forming a neighborhood continuum. (c) Plants are more crowded and the number of plants per ecological neighborhood is increased, indicated by overlapping circles. The dotted line encircling the ecological neighborhoods indicates the genetic neighborhood of the population and arrows the gene flow. If plants are self-pollinated, the genetic neighborhood is smaller than the ecological neighborhood; if gene flows remains within the total ecological neighborhood, both the genetic and ecological neighborhood are the same; and if gene flow occurs over much larger distances, then the genetic neighborhood is larger than the ecological neighborhood.

spring than the less fit. As a result the less adapted types in the population decrease and the better adapted ones increase. In other words, certain variations within the gene pool increase at the expense of other variations to produce a population better adapted to the environment. The environ-ment is always changing; but the popula-tion through survival of certain individu-als is able to track slow environmental change and survive with a modest amount of evolution.

■ Genetic Variation

Wherever you go along the seashore, whether it be long stretches of beaches or harbors and docks, you see and hear the gulls, especially the ubiquitous herring gull. Even a moderately alert observer will detect obvious differences among the her-ring gulls. Most conspicuous are the adults with their bluish-gray back, their white head, neck, underparts and tail, their black-tipped primary wing feathers, and their yellow bill with a bright red spot near the tip of the lower mandible. Among the adult gulls are younger birds with a different pattern. Some are darkish brown-gray, mottled and barred on the back with white and grayish buff. Others are lighter in tone with some gray on the back. Still others are similar to adults but with some dusky spotting on the tail and wings. Their bills may have only a suggestion of the red spot.

If you examine the gulls more closely, you will detect other differences. The size, the shades of gray on the back, the length of the bill, the shape of the red spot, the length of the wing, and other characteristics vary among the birds. In fact, so widespread are these smaller differences that a person who looks carefully at the birds and becomes acquainted with a colony can distinguish one bird from another.

Types of Variation

The most obvious type of variation among members of a population is discontinuous, that is, a variation in a specific character or set of characters that separates individuals into discrete categories. Thus the gulls can be divided by their plumage into first-year birds, second-year birds, third-year birds, and mature adults. Another type of discontinuous variation is morphological, such as male and female or the red and gray phase of the screech owl. Other discontinuous differences are biochemical, such as blood groups in humans, or even behavioral, such as song dialects in birds.

A second type of variation, the one commonly used in taxonomic and evolutionary studies, is the continuous variable, a variation in a character that can be placed along a range of values. Characters subject to continuous variation can be measured—for example, tail length of a species of mouse, number of scales on the belly of a snake, rows of kernels on ears of corn, and shape and size of sepals and petals. The measure-

ments of a character or set of characters for several individuals in a population can be tabulated as a frequency distribution and arranged graphically as a histogram (Figure 13.2). The *mean* of the character for the population is the sum of all the values divided by the number of specimens. Many specimens will have approximately the same numerical value. The most frequent numerical value is the *mode* or modal class. Measurements of other specimens will vary above and below the mode. The *frequency* of occurrence of measured values will fall away from the mode, with fewer and fewer individuals in the more distant classes. The *frequency distribution* of these variable characters tends to follow a bell-shaped curve, the normal frequency distribution.

Figure 13.2 Histogram showing the frequency distribution of the hind tibia lengths of nymphal exuvia (shed skin) of the periodical cicada *(Magicicada septendecim)*. (After Dybas and Lloyd 1962.)

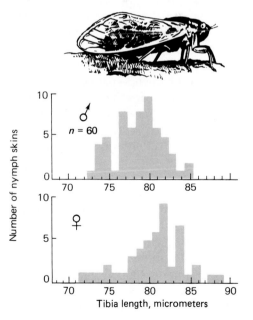

Variations may also be typed according to whether they are genetic or nongenetic. The characteristics of a species and variations in individuals are transmitted from parent to offspring. The sum of the hereditary information carried by the individual is the *genotype*. The genotype directs the development of the individual and produces the characters that make up the morphological, physiological, and behavioral characteristics of the individual. The external or observable expression of the genotype is the *phenotype*. The expression of some genotypic characters may be influenced by external and internal conditions. For example, a seedling with the gene for the formation of chlorophyll will develop the normal green color if germinated in the light, but it will be white, that is, it will have a different appearance, if germinated in the dark. The gene directs the character of green color, but its expression is affected by environmental conditions. Thus, the phenotype of an individual (P) is determined by the genetic endowment (G) modified by the environment (E) and a factor of interaction between the genes and the environment or selective pressure.

The ability of a genotype to give rise to a range of phenotypic expressions under different environmental situations is known as *phenotypic plasticity*. Some genotypes have a narrow range of reaction to environmental conditions and therefore give rise to a fairly constant phenotypic expression. But many plants and animals that can survive under a wide range of environmental conditions possess variable and diverse phenotypic responses. Some of the best examples of such phenotypic plasticity are found among plants. The size of plants, the ratio of reproductive tissue to vegetative tissue, and even the shape of the leaf may vary widely at different levels of nutrition, light, and moisture (Figure 13.3). Lacking the mobility of animals, plants must possess more flexibility in their response to environmental conditions in order to survive. Phenotypic plasticity represents nongenetic variation. An environmentally induced modification of a character is not inherited. What is inherited is the ability of the organism to modify such a character.

Sources of Variation

The primary genetic control mechanism, found within the nucleus of every cell in the organism, is deoxyribonucleic acid, DNA. DNA, the information template from which all cells in the organism are

Figure 13.3 Plasticity of response to light by leaves of dandelion. A leaf growing in the shade exhibits minimal lobing. Leaves growing in the sun are deeply lobed. Increased lobing of leaves may relate to increased heating of the leaf in full sunlight. Lobed leaves present less surface area for absorption of heat and more edge per surface area, which increases the dissipation of heat.

copied, is a complex molecule in the shape of a double helix, resembling a twisted ladder in construction. The long strands, comparable to the uprights of the ladder, are formed by an alternating sequence of deoxyribose dsugar and phosphate groups. The connections between the strands, or the rungs, consist of pairs of the nitrogen bases adenine, guanine, cytosine, and thymine. In the formation of the rungs adenine is always paired with thymine and cytosine is always paired with guanine. The DNA molecule is divided into smaller units, called nucleotides, consisting of three elements: phosphate, deoxyribose, and one of the nitrogen bases bonded to the strand at the deoxyribose. The information of heredity is coded in the sequential pattern in which the base pairs occur. According to current theory, each species is unique in that its base pairs are arranged in a different order and probably in different proportions from every other species.

DNA is present in larger units called chromosomes, which are found in most living organisms. Each species has a characteristic number of chromosomes in every cell, and the chromosomes occur in pairs. When cells reproduce (a process of division called *mitosis*), each resulting cell nucleus receives the full complement of chromosomes, or the *diploid* number (for example, 46 in humans). In organisms that reproduce sexually the germ cells or gametes (sperm and egg) result from a process of cell division, *meiosis*, in which the pairs of chromosomes are split so that each resulting cell nucleus receives only one-half the full complement, or the *haploid* number (23 in humans). When egg and sperm unite to form a new individual, the diploid number is restored. The chromosomes recombine in a great array of combinations. This segregation and recombination of chromosomes and the he-

reditary information they carry are the primary sources of variation.

Each chromosome carries units of heredity called *genes,* the informational units of the DNA molecule. Because chromosomes are paired, genes are also paired in the body cells. The position a gene occupies on a chromosome is known as a *locus.* Genes occupying the same locus on a pair of chromosomes are termed *alleles.* If each member of the pair of alleles affects a given trait in the same manner, the two alleles are called *homozygous.* If each affects a given trait in a different manner, the pair is called *heterozygous.* During meiosis the alleles are separated as the chromosomes separate. At the time of fertilization the alleles, one from the sperm and one from the egg, recombine as the chromosomes recombine.

Although variation in individuals comes from differences in genetic materials, in macro- and microenvironments, and in gene-environment interaction, major interest lies in the sources of new genetic variation. One source of differences in genetic material is the reassortment and recombination of existing genes both at the level of the gene and at the level of the chromosome. The other source is mutation, or a change in the genetic material in the gene or chromosome.

Recombination of Genetic Material. When two gametes combine to form a zygote, the gene contents of the chromosomes of the parents are mixed in the offspring. Because the number of possible recombinations is extremely large, recombination rather than mutation is the immediate and major source of variation. Recombination does not result in any change in genetic information, as mutation does, but it does provide different combinations of genes upon which selection can act. Because some combinations of interacting genes are more adaptive than others, selection determines the variations or new types that will survive in the population. The poorer combinations are eliminated by selection and the better ones retained.

The amount or degree of recombination influencing the amount of variability in a population is limited by a number of characteristics of the species. One limitation is the number of chromosomes and the number of genes involved. Another is the frequency of crossing-over, the exchange of corresponding segments of homologous chromosomes during meiosis. Others include gene flow between populations, the length of generation time, and the type of breeding, for example, single versus multiple broods in a season in animals and self-pollination versus cross-pollination in plants.

Mutation. A *mutation* is an inheritable change of genetic material in the gene or chromosome. Organisms that possess such changes are called mutants.

Micromutations, or gene mutations, are alterations in the DNA sequence of one or a few nucleotides. During meiosis the gene at a given locus usually is copied exactly and eventually becomes part of the egg or sperm. On occasion the precision of this duplication process breaks down and the offspring DNA is not an exact replication of the parent DNA. The alteration may be a change in the order of nucleotide pairs, a substitution of one nucleotide pair for another, a deletion of a pair, or various kinds of transpositions.

The rate of mutation in general is low. Most common mutations involve the change of one allele into another. Even in a population homozygous for *A,* for example, *A* eventually will mutate to *a* in some of the gametes; and in a population

having both genes, mutations may be forward to *a* or backward to *A*. If *A* mutates to *a* faster than *a* to *A*, the frequency of allele *A* decreases. Rarely is one of the alleles lost to the population, for reversibility prevents a long-term or permanent loss. Eventually such mutations arrive at an equilibrium. Even if one allele is lost from the population, it will usually reappear by mutation.

Macromutations, or chromosomal mutations, result from a change in the number of chromosomes or a change in the structure of the chromosome.

A change in chromosomal number can arise in two ways: the complete or partial duplication of the diploid number rather than the transmission of the haploid number, or the deletion of some of the chromosomes.

Polyploidy is the duplication of entire sets of chromosomes. It can arise from an irregularity in meiosis or from the failure of the whole cell to divide at the end of the meiotic division of the nucleus. The individual body cell is ordinarily diploid ($2n$ or twice the haploid number). Forms of polyploidy are triploid ($3n$ or three haploid sets), tetraploid ($4n$), and so on.

Polyploidy exists mostly in plants. The condition is rare in animals because an increase in sex chromosomes would interfere with the mechanisms of sex determination and the animal would be sterile. Polyploid plants differ from normal diploid individuals of the same species in appearance and are usually larger, more vigorous, and occasionally more productive.

Another form of macromutation is the duplication or deletion of a part of a normal complement of chromosomes. Such deletions or duplications result in abnormal phenotypic conditions. (One such condition is Down's syndrome or mongolism in humans.)

A change in the physical structure of

a chromosome may occur in the form of deletion, duplication, inversion, or translocation of segments (Figure 13.4).

Deletion is the loss of a part of a chromosome; a definite segment and the genes thereon are missing in the offspring cell.

Occasionally the functions of the genes at the missing loci will be assumed by genes in some other part of the chromosome, but if the segment lost is large, the individual dies. Also lethal is a deletion in individuals that are homozygous for the character involved. In heterozygous individuals deletion may permit the manifestation of characters determined by recessive genes. The loss of a short segment has a marked effect on development.

Duplication involves an addition to a chromosome. In general duplication is

Figure 13.4 Types of chromosomal aberration. When these altered chromosomes join with normal homologs during the first meiotic division, they assume characteristic configurations that allow locus-by-locus matching. Synaptic configurations of deletion and inversion are shown. (After Wilson and Bossert 1971.)

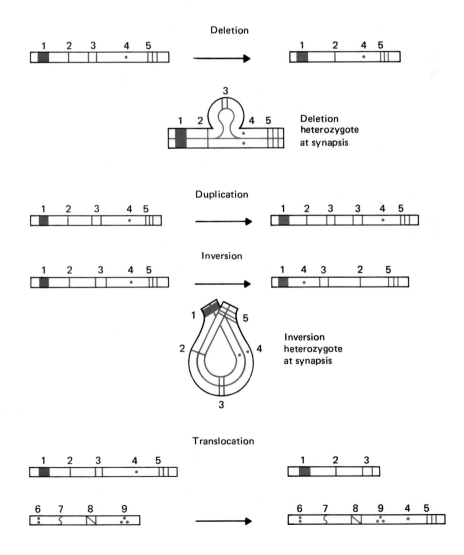

less harmful than deletion, and in some cases it has little or no effect on the phenotype. Duplication may increase both the genetic material and the effect of certain genes on development, or it may cause an imbalance of gene activity, reducing the viability of an organism.

An *inversion* is an alteration of the sequence of genes in a chromosome. It may occur when a chromosome breaks in two places and the segment between the breaks becomes turned around, reversing the order of genes in respect to an unbroken chromosome. When the altered chromosome is paired with a normal chromosome in a heterozygous individual, the alteration interferes with pairings. The chromosomes must bend, loop, or in some way cross over so that each gene aligns itself with its homolog. This crossing-over between inverted and normal chromosomes usually produces abnormal, nonviable gametes.

Translocation is the exchange of segments between two nonpaired (nonhomologous) chromosomes. The genes in the translocated segment become linked to those of the recipient chromosome. If the translocation is reciprocal, all of the chromosome material is present in the individual even though it is rearranged. Individuals carrying such translocations are usually normal. If the translocation is not reciprocal, some genes will be transferred to completely different chromosomes and the linkage relationship becomes altered drastically. The effect of the translocation becomes evident during meiosis and the formation of gametes. During the segregation of chromosomes, the two within the translocated segments will produce balanced and unbalanced gametes. Some will have excessive deletions and others will have a duplication of material. If an unbalanced gamete fertilizes a balanced gamete, the fertilized egg is not viable.

Hardy-Weinberg Equilibrium

If genes occur in two forms, A and a, then any individual carrying them can fall into three possible diploid classes: *AA, aa,* and *Aa.* Individuals in which the alleles are the same, *AA* or *aa,* are called homozygous; and those in which the alleles are different, *Aa,* are heterozygous. The haploid gametes produced by the homozygous individuals are either all A or all a; those by the heterozygous, half A and half a. They can recombine in three possible ways: *AA, aa, Aa.* Thus the proportion of gametes carrying A and a is determined by the individual genotypes, the genes received from the parents. Eggs and sperm unite at random, enabling the prediction of the proportion of offspring of different genotypes based on parental genotypes.

Assume that a population homozygous for the dominant *AA* is mixed with an equal number from a population homozygous for the recessive *aa.* Their offspring, the F_1 generation, then will consist of 0.25 *AA,* 0.50 *Aa,* and 0.25 *aa* (Figure 13.5). These proportions are called genotypic frequencies. The gene frequencies, of course, are 0.50 of A and 0.50 of a. This proportion will be maintained through successive generations of a bisex-

Figure 13.5 Mixing two homozygous populations.

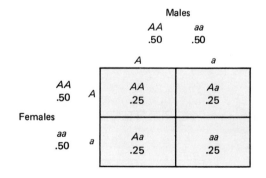

ual population (Figure 13.6) if at least five conditions exist: (1) reproduction is random; (2) mutations either do not occur or they occur in equilibrium, that is, the rate of mutation from A to a is the same as a to A; (3) the population is large enough that changes by chance in the frequency of genes are insignificant; (4) the population is closed, with no immigration or emigration; and (5) no natural selection is occurring.

The equilibrium of these three genotypes can be expressed as a general statistical law, known as the Hardy-Weinberg law, which can be stated simply as follows:

p = frequency of allele A

q = frequency of allele a

$p + q = 1.0$

Then

$p^2 + 2pq + q^2 = 1.0$

in which

p^2 = frequency of homozygous individuals, AA

q^2 = frequency of homozygous individuals, aa

$2pq$ = frequency of heterozygous individuals, Aa

In the hypothetical population above, the proportion of the genotypes in the F_1 generation will be $(0.5)^2 + 2(0.5 \times 0.5) + (0.5)^2$; and the same tendency can be demonstrated even if the ratio is not the classical Mendelian 1:2:1. Imagine a population in which the ratio of A alleles (p) to the a alleles (q) is 0.6 to 0.4 (Figure 13.7). The frequency of the genotypes in the F_1 generation will be 0.36 AA, 0.48 Aa, and 0.16 aa, and the gene frequency will be

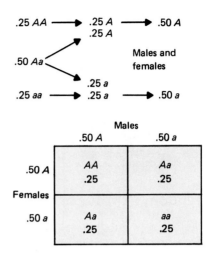

Figure 13.6 Proportions in the F_2 generation.

$(0.6)^2 + 2(0.6 \times 0.4) + (0.4)^2$. We can conclude that all succeeding generations will carry the same proportions of the three genotypes, provided that the assumptions mentioned earlier are fulfilled.

The stated assumptions are never perfectly fulfilled in any real population, so the Hardy-Weinberg law must be consid-

Figure 13.7 The Hardy-Weinberg law applied to a hypothetical case.

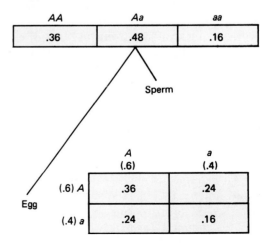

ered wholly theoretical, a distribution against which actual observations can be compared with limitations (Wallace 1968). Nevertheless, the Hardy-Weinberg law is of fundamental importance in theoretical population genetics.

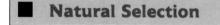

■ Natural Selection

Nonrandom Reproduction

Variation in a population seldom is constant from generation to generation. One reason is gene mutation, the ultimate source of genetic variation. Of more immediate consequence is nonrandomness of reproduction within a population. Not every individual is able to contribute its genetic characteristics to the next generation or to leave surviving offspring. It is this selectivity that is natural selection.

Before a given individual in a population can contribute to the succeeding generation, it must first survive to reproduce. Survival begins from the time of fertilization through the periods of development, growth, and sexual maturation. Fertilized eggs may fail to develop fully and die from physiological or environmental causes. Disease, predation, and accidents eliminate those young not quite as swift, as quick, or as strong as their siblings. In such survival genetic variation plays a key role, for natural selection influences the frequency of alleles in a population. If a mutation arises that places its carrier at a disadvantage, selective pressures eliminate the individual; on the other hand, an advantageous mutation is retained.

An example of such selection can be found among the flies. When DDT was first used as an insecticide against houseflies, the chemical was highly effective, destroying the bulk of local populations.

Among the flies were a few that did not die, that carried a mutation or a certain combination of genes that made them resistant to the spray. Resistance in one strain of flies was due to a recessive gene. Flies homozygous for this gene tolerated a high concentration of DDT, while homozygous dominants and heterozygotes were killed. These flies survived to multiply. Many of their offspring were as resistant to the sprays as the parents; some were even more resistant. The least resistant were selected against; the most highly resistant were retained in the reproductive population. Later applications of DDT continually selected for a combination of genes most resistant to the insecticide. As a result DDT became ineffective in fly control, and newer, stronger sprays are required. Eventually these sprays will select resistant strains of flies, which will become adapted to the new environmental conditions. However, to acquire this resistance the flies pay a price. In the absence of DDT the resistant flies are inferior competitors to the nonresistant flies, which have a shorter development time (Pimentel et al. 1951). If the spraying is stopped, evolution will be reversed and the resistance will largely disappear from the fly population.

Once they reach reproductive age, more individuals are eliminated from the parental population. The maintenance of genetic equilibrium is based on random mating, but mating is not random. Many species of animals, particularly among birds, fish, and some insects, have elaborate courtship and mating rituals. Any courtship pattern that deviates from the commonly accepted pattern is selected against, and the deviating individual and its genes are eliminated from the reproductive population. On the other hand, animals possessing a color pattern or movement that accents the typical pattern are selected for. Any new mutations that improve on courtship, mating signals, and ritual would possess a favored position in subsequent generations. Among polygamous species, in particular, the majority of males go mateless, for the females mate with dominant males that tolerate no interference from younger or less aggressive males. States of psychological and physiological readiness also are involved in mate selection. Unless both male and female are of the same state of sexual readiness, mating will not occur.

Neither is fecundity random. Some families or lines increase in number through time; others fade away. Obviously those who produce more offspring increase the frequency of their genes in a population and affect natural selection. For example, if individuals with allele *A* produce ten offspring to every one produced by those with allele *a*, the proportion of *A* in the population will increase. There is a limit, however, for natural selection does not always favor fecundity. If an increased number of young per female results in reduced maternal care, survival of offspring may be reduced, particularly among those animals whose chances of individual survival are high. Those organisms whose chances of individual survival are low—for example, ground-nesting game birds and oceanic fish—have become very fecund.

Fitness and Modes of Selection

If an organism can tolerate a given set of conditions so that it can leave fertile progeny, thus contributing its genetic traits to the population gene pool, it can be said to be adapted to its environment. If an organism survives only as an individual and leaves few or no mature, reproducing progeny, thus contributing little or nothing to the gene pool of the population, it is poorly adapted. Those individuals that contribute the most to the gene pool are

said to be the most fit, and those that contribute little or nothing are the least fit. The fitness of the individual is measured by its reproducing offspring. Natural selection is not a measure of individual survival, but of differential reproduction, the ability to leave the most offspring capable of further reproduction.

In simplest terms fitness is measured by comparing the number of offspring produced by one genotype to the number produced by another. Suppose genotype *AA* produces 250 offspring and genotype *BB* produces 200. The reproductive success of genotype *BB* compared to *AA* is reduced by 50, or expressed in fractional terms $50/250 = 0.20$. Obviously *AA* is the more successful genotype. In measuring selection or the adaptative value of a genotype, fitness is frequently designated as *W*, the value of which ranges from 1.00 for the most productive genotype to 0 for no reproduction (lethal genes). In our simple example the value of *W* (for *AA*) would be designated as 1.00; the value of *W* for *BB* would be $1.00 - 0.20$, or 0.80. The selective pressure acting on a genotype is designated as a *selection coefficient, s*. It can be stated as the difference between 1.00 and the fitness value. In the example the selection coefficient for *AA* is 0; for *BB* it is 0.20. Thus, *W* (fitness) $= 1 - s$; similarly, $s = 1 - W$. For *BB*, $W = 1 - 0.20 = 0.80$; $s = 1 - 0.80 = 0.20$. This extremely simple example and these figures are given here only to provide some appreciation of the term fitness. The calculations used to determine fitness values and selection coefficients are more complex, but not difficult. Good discussions are given in Haldane (1954), Wallace (1968), and Ricklefs (1979).

Within a population selection may act in three ways. Given an optimum intermediate genotype, *stabilizing selection* favors the average expression of the phenotype at the expense of both extremes.

Directional selection favors one extreme phenotype at the expense of all others. The mean phenotype is shifted toward the extreme, provided that heritable variations of an effective kind are present. *Disruptive selection* favors both extremes, although not necessarily to the same extent, at the expense of the average (Figure 13.8).

Disruptive selection is most apt to occur in a population living in a heterogeneous environment in which there is a strong selection for adaptability or phenotypic flexibility. Increased competition within the population may select for a closer adaptation to habitat, with the result that the population may subdivide. This division would give rise either to a polymorphic situation or to separation into different populations with different characteristics. The latter is most likely to take place in areas where selection is intense and where optimum habitat adjoins or is penetrated by less than optimum habitat. Organisms settling in these habitats will adapt to the local environment. If disruptive selection is strong enough, it will lead to positive assortative mating and eventually to genetic divergence of two or more groups.

An example of the development of a polymorphic species through selection is the peppered moth *(Biston betularia)* in England. Before the middle of the nineteenth century the moth, as far as is known, was always white with black speckling in the wings and body (Figure 13.9). In 1850 near the manufacturing center of Manchester, a melanistic form of the species was caught for the first time. The black form, *carbonaria,* increased steadily through the years until it became extremely common, often reaching a frequency of 95 percent or more in Manchester and other industrial areas. From these places *carbonaria* spread mostly westward into rural areas far from the industrial cities. The black form came about by

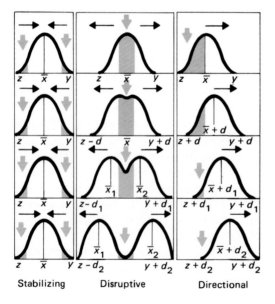

Figure 13.8 The three main kinds of selection. Directional selection accounts for most of the change observed in evolution. The curves represent the frequency of organisms within a range of values. The shaded areas are the phenotypes being eliminated by selection. The arrows indicate the amount and direction of evolutionary change.

Stabilizing   Disruptive   Directional

Figure 13.9 Normal and melanistic forms of the polymorphic peppered moth *Biston betularia* at rest on a lichen-covered tree.

the spread of dominant and semidominant mutant genes, none of which are recessive. This increased frequency and spread was brought about by natural selection. The normal form of the peppered moth has a color pattern that renders it inconspicuous when it rests on lichen-covered tree trunks; but the grime and soot of industrial areas carried great distances over the English countryside by prevailing westerly winds killed or reduced the lichen on trees and turned the bark of the trees a nearly uniform black. The dark form is conspicuous against the lichen-covered trunk, but inconspicuous against the black.

A British biologist, H. R. D. Kettlewell (1961; see also Kettlewell 1965) experimentally demonstrated the role of natural selection in the spread of the dark form. He reared, marked, and released melanistic and light forms in polluted woods. The melanistic form had a better survival rate than the light form and therefore left more offspring. To confirm the role of natural selection, Kettlewell released light and dark forms in unpolluted woods. There the light form survived better. The reason was selective predation. In woods with lichen-covered trees the melanistic form was more easily seen by several species of insect-feeding birds and was therefore subject to heavier predation. In polluted woods the light form bore the brunt of predation. For this reason the normal form has virtually disappeared from polluted country but it is still common in the unpolluted areas in western and northern Great Britain.

Although selective predation is considered to be the major influence in maintaining melanic polymorphism in the peppered moth, the polymorphism may also be maintained by a heterozygous advantage of the dark-colored forms independent of selection by predation (Lees and Creed 1975). The dark-colored individuals (those possessing the *carbonaria* allele) apparently have a physiological advantage over the nonmelanistic form in withstanding the effects of air pollution. Predation aside, the dark form will increase in those areas where it has the physiological advantage. In some cases the visual advantage is less than the physiological advantage.

Group and Kin Selection

Selection pressures, according to evolution theory, impinge only on individuals (phenotypes) of a species. Nevertheless, traits such as warning calls, warning coloration, and helping behaviors often benefit a group at the expense of the individual. Such acts and traits are called *altruistic.* Altruism, strictly defined, is the sacrifice of one's own well-being in the service of another. In genetic terms, the altruist contributes to the genetic fitness of another while decreasing its own fitness (Michod 1982). How are such traits maintained in a population when the altruist and thus the altruistic gene is selected against and the nonaltruistic or selfish gene is at a selective advantage?

To explain such traits, some population and behavioral ecologists have suggested the concept of *group selection,* which operates on the differential productivity of local populations (Wilson 1980). The characteristic selected improves the fitness of the group, though it may decrease the fitness of any individual in the group. Any genetic differences among local populations that decrease the likelihood of local extinction or increase the likelihood that the local population will produce emigrants or colonists who will affect the genetic composition of other local populations will favor that group over another.

The idea of group selection was suggested by Darwin in *The Origin of Species* and by Sewall Wright (1931, 1935) as *interdemic selection*. It began to receive serious attention and stir considerable controversy when a Scottish ecologist, V. C. Wynne-Edwards, published in 1962 a book *Animal Dispersion in Relation to Social Behavior*. In it Wynne-Edwards advanced the idea that animals tend to avoid overexploitation of their habitat, especially the food supply, by altruistic restraints in population growth, either by reducing or refraining from reproduction. Restraint was achieved through the mechanism of social behavior, in which displays provided information about local population. Local populations that restrained reproduction were more likely to survive than populations that grew beyond the ability of the resource to support them. Such populations would decline or go extinct, leaving empty habitats. These habitats would then be colonized by dispersers from altruistic populations.

The idea of group selection was vigorously challenged by many evolutionary ecologists (see Williams 1966; Wiens 1966; Lack 1966; Ghiselin 1974; Maynard-Smith 1976) and accepted in modified forms by others (Lewontin 1965; J. Emlen 1973; Gilpin 1975; Wilson 1975; Alexander and Borgia 1978). D. W. Wilson (1975, 1977, 1979, 1980, 1983) has framed it in a somewhat different context. He views group selection as a component of natural selection that operates on the differential productivity of local populations within a global system. Local populations or demes are groups of individuals that interact with one another sometime in their life cycle, giving any two of them a equal opportunity of becoming neighbors (Mayr 1963; Wilson 1977). Within demes are subgroups that interact with their neighbors

in a number of ways, such as mating, competition, defense against predation, and the like. These ecological groups can be defined more precisely as "the smallest collection of individuals within a population defined such that genotypic fitness calculated within each group is not a (frequency-dependent) function of the composition of any other group" (Uyenoyama and Feldman 1980: 395). Wilson (1979, 1980) calls them trait groups. Thus we can consider demes not only a population of individuals but also a population of groups isolated by certain traits. The population of a given organism consists of the total population over a large given area, the global population, which in turn consists of numerous local populations, or demes, isolated in varying degrees from one another; and the deme consists of numbers of trait groups. Within these trait groups intrademic group selection as well as individual selection takes place.

Individual selection is that component of natural selection that operates on the differential fitness of individuals within local populations. Traits promoted by individual selection are considered selfish, and those promoted by group selection altruistic. Altruistic or group-benefit traits are those that increase the relative productivity of local populations within a global system. Altruistic traits may be weak, that is, not strongly sacrificial, or strong. Selection for an altruistic trait might take place in species with small freely interbreeding local populations sufficiently isolated to allow some differentiation in gene frequency, and with little gene flow between other local groups.

A model of intrademic group selection illustrates how an altruistic trait, although selected against in an individual, may be selected for in a group (Wilson 1983). Assume that an altruistic allele *A* benefiting the group at the expense of itself exists in a global population with a frequency of $p = 0.5$. Individuals, however, are distributed in local populations in which p varies from 0.3 to 0.7, and selection operates within each group (Figure 13.10). Note that in each local population, the A allele declines in frequency because of individual selection, but the populations N' increase relative to p. When we weigh the new global frequency of A by group size, the frequency of p actually increases, suggesting that an altruistic allele can evolve, even though it is selected against within each group. As with individual selection, the model suggests that group fitness involves some heritable phenotypic variation that influences group productivity or persistence. Natural selection will favor that characteristic, just as genetic variation influences individual selection. Only within groups can an allele have low relative individual fitness and still be selected.

Consider a local population or deme possessing a high frequency of a gene that decreases mortality or increases reproduction, surrounded by demes possessing the gene at lower frequencies. The deme possessing a high frequency of the adaptive gene will develop some selective advantage over neighboring demes. It will produce a greater surplus population that will emigrate to surrounding demes or into empty habitat patches (created perhaps by local extinctions of populations possessing the adaptive gene in low frequencies). In fact, group selection may be particularly important for traits affecting dispersal (see Slatkin 1987). Group selection can then be defined as changes in gene frequencies resulting from differential extinction or productivity of groups. Because the groups themselves are short-lived, extinction of groups is not as relevant as differential

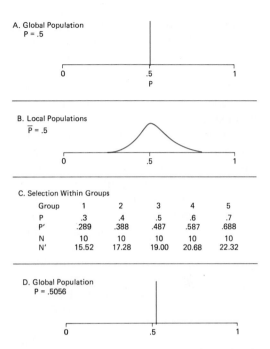

Figure 13.10 A numerical example of intrademic selection. The x axis of the graphs shows the given frequencies (p) of the A allele and the y axis the relative abundances of groups with a given frequency p. (a) Global population represented by a single frequency of A_1. (b) Local groups that vary in p values. (c) Table showing selection within five representative groups, using Wright's model with $b = 2$ and $s = 0.05$. Wright's (1945) model is

Genotype	Frequency	Selective value
A_1A_1	p^2	$(1 + bp)(1 - 2s)$
A_1A_2	$2p(1 - p)$	$(1 + bp)(1 - s)$
A_2A_2	$(1 - p)$	$(1 + bp)$

productivity of groups or the contribution of groups to the mating pool (Michod 1982).

Evolution of an altruistic trait may take place more easily when groups involved are made up of closely related individuals sharing genes by common descent. In such a setting an individual may increase its own fitness in the long run and the fitness of close relatives by decreasing its fitness in the short run. This idea was advanced by W. H. Hamilton in 1964 as *inclusive fitness,* subsequently called *kin selection* by Maynard Smith (1964).

The cost and benefit of an altruistic trait depends upon the closeness of the relationship. Genes for altruism can be selected only when the relative benefit to the recipient is greater than the reciprocal of the coefficient of relationship between the altruist and the recipient. This relation was defined by Hamilton as $k > 1/r$, where r is the proportion of the genes of two individuals identical because of common descent and k is the relationship of the recipient's benefit (b) to the altruist's cost $c (k = b/c)$.

The value of r varies from 1 for identical twins to 0 for no relationship. For parent-offspring the value is 0.5, because the individual offspring receives half of its genes from a parent; conversely, a parent contributes half of its genes to an individual offspring. The value for a full sibling would be 0.5; for a half-sib, 0.25; for an uncle, aunt, niece or nephew, 0.25; and for a first cousin 0.125. The closer the relationship, the higher the value of r. If the value of r is large, that is, the relationship is close, the value of $1/r$ is small. If k, the ratio of fitness to the loss of fitness, exceeds $1/r$, then the altruistic gene would be selected for. If the value of $1/r$ is small, then the ratio k of benefits b to cost c can be relatively large and still favor the altruistic gene. In other words, an altruist can afford greater risk to help a close relative than a distant one. Even in that situation the benefit/cost ratio k would have to exceed a value of 2 (1/.5).

Consider an individual A who, facing a 100 percent chance of death, gives a warning call to a full sibling. If this alarm warns only a sibling then the benefit, survival of 1, relative to the cost, death of 1, would be 1/1, less than 1/.5 or 2, and al-truism would be selected against. If the warning saved four full sibs, then 4/1 > 1/.5 and the trait would be selected for. If the act resulted in only 50 percent chance of death, then the benefit, 1/.5, would be the same as the 1/r, 1/.5, and there would be no benefit from the act. By the same token, the altruist in the initial situation would have to save over eight first cousins before the trait would be selected for.

An altruistic act does not mean sacrifice of life. Dispersal from a kin group, which improves the reproductive opportunities for the remaining sibs or parents, as in black bear (see Rogers 1987) and ground squirrel (Holekamp and Sherman 1989), and helping parents raise full or half sibs are examples of altruistic acts. Both increase the fitness of the recipients and decrease the individual fitness of the altruist. The altruist may actually be improving its own fitness by improving the fitness of its closely related recipients.

Kin selection, then, involves the evolution of a genetic trait expressed by one individual that affects the genotypic fitness of one or more directly related individuals (Michod 1982). It is favored when an increase in fitness of closely related individuals is great enough to compensate for the loss in fitness of the altruistic individual. If the dispersers and helpers are very close kin, as they usually are, those individuals serve to increase the fitness of the genetic traits they hold in common. This effect is particularly pronounced in eusocial insects such as bees and termites in which the workers are nonreproducing females waiting on a sister, the reproductive queen. Although they do not leave behind offspring of their own (at least for one breeding season), altruists improve their own fitness and that of their kin because they have a very close genetic relationship with the offspring.

Two aspects of fitness are involved in

kin selection. One is personal fitness, passing on genes to an individual's own offspring (classic selection). The other is the additional fitness acquired by improving the fitness of very close relatives, especially parents and sibs (kin selection), who share many genes. Personal fitness added to fitness acquired indirectly by improving the fitness of close relatives is *inclusive fitness* (Hamilton 1964). This concept is used to explain many aspects of social behavior (Hamilton 1964).

An example of altruistic behavior and kin selection may be found in the Florida scrub jay *(Aphelocoma coerulescens)*. Nonbreeding helpers aid their parents or close kin in rearing their young (Woolfenden 1975). Florida scrub jays rarely breed before two years of age and may not breed for several years after that. Many unmated males and females remain in the local area on familiar ground among familiar birds. Female helpers rarely remain for more than two years, because they join the breeding population. Unmated males may be around longer. Because of the male's assistance, the family size of the pairs they help increases. As family groups expand, their territories expand. Eventually a dominant male helper may claim a part of the enlarged territory as his own or inherit it and become a breeder (Woolfenden and Fitzpatrick 1977, 1984). By helping, the male has come out ahead. He has increased his inclusive fitness and improved his own direct fitness by increasing his opportunity to become a breeder.

The male may not be altruistic at all. He may be practicing a form of selfish behavior, increasing his own opportunities by cooperating with breeding pairs (Woolfenden 1971; Ligon 1981; for a worldwide survey of helpers at bird nests see Skutch 1986). Nevertheless his actions still fall under the concept of kin selection, as defined.

Kin selection is open to some questions. Is it an integral part of the evolution of sibling cooperation and helping at the nest? Is it an important part and cause of group living? Or is kin selection a form of selfish individual selection and pseudoaltruistic behavior under the guise of cooperation? These and many other questions need to be answered before kin and group selection can be understood (see Alexander and Tinkle 1981).

■ Inbreeding

Although a local population of a species consists of all individuals occupying a given area, it really is made up of a number of semi-isolated subpopulations, often with minimal interchange (see Ralls, Harvey, and Lyles 1986). Isolation becomes most pronounced when subpopulations occupy patches or islands of habitats, such as house mice restricted to barns or white-footed mice confined to islands of forest growth scattered through agricultural or suburban landscapes. Such small populations may be subject to inbreeding.

Inbreeding, simply defined, is breeding between relatives. With inbreeding, mates on the average are more closely related than they would be if they had been chosen at random from the population. Some reasons for inbreeding are small isolated populations, close proximity of potential mates, ecological preferences, morphological resemblances among individuals, and the like. The principal effect of inbreeding is an increased frequency of homozygous genotypes.

The extreme in inbreeding is self-fertilization, as in some plants. That extreme provides a measure of comparison for degrees of inbreeding. Consider the population described by the Hardy-Weinberg equilibrium earlier, in which $p(A) = q(a)$ and both are equal to 0.5. Then $p^2 = .25$, $2pq = .50$, and $q^2 = .25$. In our hypothetical population, all breeding involves selfing from the start. Within the population all homozygotes AA and aa breed true. Offspring from the heterozygotes will be one-half heterozygotes and one-half homozygotes each generation. The homozygotes produced will be added to the pool of homozygotes in the population; the remaining heterozygotes in the next generation again will produce offspring one-half homozygotes and one-half heterozygotes. Eventually the self-fertilizing population will become exclusively homozygous.

In the second generation the gene frequencies will be [1/4 (1) + (1/2)(1/4)], [1/2 (1/2)] [1/4 (1) + 1/2 (1/4)] or 3/8 AA, 2/8 Aa, 3/8 aa (Table 13.1). Within one generation the degree of heterozygosity has declined from 1/2 to 1/4. The expected frequency of heterozygosity in a self-fertilizing population is

$$2pq \times (1/2)^n$$

where $2pq$ is the frequency of heterozygotes in the first generation and n is the number of consecutive generations. Note that with inbreeding, the frequency of alleles does not change, as predicted by the Hardy-Weinberg equilibrium, but homozygosity increases at the expense of heterozygosity.

The Inbreeding Coefficient

Exclusively self-fertilizing populations, the extreme of inbreeding, are uncommon. Even populations of self-fertilizing plants experience some periods of cross-fertilization or outbreeding. Close inbreeding usually involves mating between brother and sister, parent and offspring, or more frequently cousins, individuals who share a number of like genes (Figure 13.11). However, possession of two like alleles by different individuals does not imply inbreeding. The two alleles in a homozygote may be identical because they are the result of independent mutations. Such individuals are *allozygous* for the alleles involved. Inbred individuals possess alleles identical by descent; they can be traced back to a common ancestor (Figure 13.12). Such homozygotes are *autozygous*. One outcome of inbreeding is an increase in the frequency of autozygous individuals.

The degree of inbreeding is measured

TABLE 13.1 Decrease in Heterozygosity Under Systematic Self-fertilization Starting with an Equilibrium Population ($p = q = \frac{1}{2}$)

Generations	GENOTYPIC FREQUENCIES			F	q
	A/A	A/a	a/a		
0	$\frac{1}{4}$	$\frac{1}{2}$	$\frac{1}{4}$	0	$\frac{1}{2}$
1	$\frac{3}{8}$	$\frac{1}{4}$	$\frac{3}{8}$	$\frac{1}{2}$	$\frac{1}{2}$
2	$\frac{7}{16}$	$\frac{1}{8}$	$\frac{7}{16}$	$\frac{3}{4}$	$\frac{1}{2}$
3	$\frac{15}{32}$	$\frac{1}{16}$	$\frac{15}{32}$	$\frac{7}{8}$	$\frac{1}{2}$
4	$\frac{31}{64}$	$\frac{1}{32}$	$\frac{31}{64}$	$\frac{15}{16}$	$\frac{1}{2}$
n	$\dfrac{1 - (\frac{1}{2})^n}{2}$	$(\frac{1}{2})^n$	$\dfrac{1 - (\frac{1}{2})^n}{2}$	$1 - (\frac{1}{2})^n$	$\frac{1}{2}$
∞	$\frac{1}{2}$	0	$\frac{1}{2}$	1	$\frac{1}{2}$

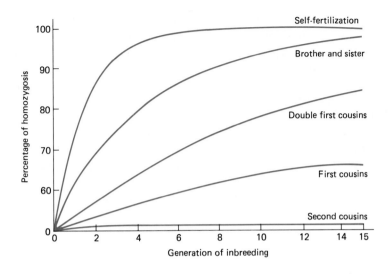

Figure 13.11 The percentage of homozygous offspring from systematic matings with different degrees of inbreeding. Note how rapidly homozygosity declines as the relationship of offspring becomes further removed from original parental stock.

by the *coefficient of inbreeding, F. F* is the probability that an individual receives at a given locus two genes that are identical by descent; stated differently, it is the amount of heterozygosity that has been lost. For a self-compatible population the inbreeding coefficient is

$$F = \frac{H_0 - H}{H_0}$$

where $H_o = 2pq$. F is a measure of the fractional reduction of the loss of heterozygosity in an inbreeding population relative to reduction in a randomly mating population with the same frequency of alleles.

Consider a heterozygous population possessing two kinds of gametes in which no two alleles are alike. The total number of kinds of gametes produced in the population is $2N_0$. When the gametes unite to form the zygotes of the next generation, a probability of $1/2N_0$ exists that two identical gametes will unite to form a homozygote. Thus within an inbreeding population the probability an individual in the first filial or F_1 generation will have two alleles identical by descent is the same as the probability of having two identical alleles, $1/2N_0$. The inbreeding coefficient of the first generation is $1/2N_0$.

Figure 13.12 Pedigrees charting inbreeding due to (a) sibling mating and (b) parent-offspring mating illustrate autozygosity, possession of alleles identical by descent. The inbreeding coefficient is 1/4 in both cases.

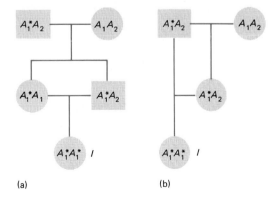

(a)

(b)

The second inbreeding generation will not consist only of homozygotes produced by union of two gametes with alleles identical by descent from the first generation. The probability also exists that two gametes from different homozygous individuals in the F_1 generation but descending from the same ancestors in generation 0, F_0, will also be present. Therefore the population consists not only of homozygotes produced by new breeding but also homozygotes attributed to previous inbreeding (Table 13.2):

$$F_1 = \frac{1}{2N}$$

$$F_2 = \frac{1}{2N_1} + \left(1 - \frac{1}{2N}\right) F_1$$

$$F_3 = \frac{1}{2N_2} + \left(1 - \frac{1}{2N_2}\right) F_2$$

$$F_{n+1} = \frac{1}{2N_{n+1}} + \left(1 - \frac{1}{2N_{n-1}}\right) F_{n-1}$$

The inbreeding coefficient consists of two parts. The first part $\frac{1}{2}N_{n-1}$ is derived from new inbreeding. The second part $(1 - \frac{1}{2}N_{n-1})F_{n-1}$ is attributed to previous inbreeding.

The above coefficients relate to an idealized population. This constraint can be removed if the term $1/2N_{n-1}$ is replaced by F, the rate at which heterozygosity is lost and the rate at which genes become fixed:

$$F = \frac{F_n - F_{n-1}}{1 - F_{n-1}}$$

When $F = 0$, no inbreeding occurs, and when $F = 1$, complete inbreeding occurs.

In summary, upon inbreeding the Hardy-Weinberg frequencies change to

$$[p^2(1 - F) + pF] + [(2pq - 2pqF)] + [q^2 + 2pqF)] = 1$$

Consequences of Inbreeding

In normally outcrossing populations, close inbreeding is detrimental. It increases autosomal homozygosity with all its attendant problems. Rare recessive deleterious genes become expressed. They can result in death, decreased fertility, smaller body size, loss of vigor, reduced fitness, reduced pollen and seed fertility in plants, and various meiotic abnormalities such as poor chromosomal pairing. These consequences are termed *inbreeding depression*.

Not all inbreeding is bad for a population. Occasional inbreeding will main-

TABLE 13.2 Genotypic Frequencies with Inbreeding

	FREQUENCY IN POPULATION			
GENOTYPE	With Inbreeding Coefficient F		With $F = 0$ (Random Mating)	With $F = 1$ (Complete Inbreeding)
	Allozygous genes	Autozygous genes		
AA	$p^2(1 - F)$ +	pF	p^2	p
Aa	$2pq(1 - F)$		$2pq$	0
aa	$q^2(1 - F)$ +	qF	q^2	q

tain rare alleles that might be lost with continual outbreeding. Close inbreeding under artificial situations is used in the breeding of domestic plants and animals to fix certain desirable genes that will breed true, in spite of the expression of deleterious ones. These inbred lines are then outcrossed to produce "hybrid" vigor. However, among normally outbred individuals there is no safe amount of inbreeding. Heterozygous individuals are much better off than homozygous ones.

Close inbreeding in nature is rare, less than 2 percent in natural populations of vertebrates for which there are data (Ralls, Harvey, and Lyles 1986). In fact, natural mechanisms exist to reduce inbreeding. One is spatial separation, or differences between the sexes in dispersal of young (Greenwood 1980; Holmes and Sherman 1983). One sex stays behind; the other leaves. Among birds, juvenile females most frequently leave the home area; the males tend to return to the vicinity of their birth. Separation is further enhanced in many species by monogamous mating habits and the frequent loss of a mate during the nesting season or between seasons. Among mammals young females stay close to the home place, while young males, often driven away by the females, seek new places to live. Adult male lions defend groups of females from competitors, but in a few years they are forced to relinquish the pride to younger males (Pusey and Packer 1986). Hoogland (1982) has pointed out how sex-biased dispersal among black-tailed prairie dogs *(Cynomys ludovicianus)* reduces breeding between first-degree relatives by 90 percent. Dispersal involves virtually all males. Young males leave the family group or coterie before breeding and young females remain behind. Adult male prairie dogs are more likely to move to new breeding groups if adult daughters are in the home

colony. Although dispersal does reduce inbreeding, sex-biased dispersal may have evolved for other reasons, such as enhanced reproductive success, increased access to food resources and space, and other amenities (Greenwood 1980).

A second mechanism reducing close inbreeding is kin recognition (Holmes and Sherman 1982, 1983). Because of their close association during early life, siblings recognize one another over time. Females of both the Beldings ground squirrel and Arctic ground squirrel can distinguish between full and half sisters, although this discrimination fades over a long absence. Females mate with unrelated males, leave the group if a related male returns, or fail to come into estrus, especially if their father is in the group (Hoogland 1982).

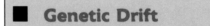

■ Genetic Drift

In sexual reproduction only a few of the gametes produced actually form a new generation. In general, all an individual's genes will be represented somewhere among its gametes, but not in any two of them; yet under conditions of stable population size, two gametes are about all that an individual can leave behind. For a heterozygote *Aa*, there is a 50:50 chance that the two gametes will either be *A, A,* or *a, a,* assuming no natural selection. Thus a 50:50 chance exists that a heterozygote will fail to pass on one of its genes. In a whole population these losses tend to balance each other, so that the gene frequencies of the filial generation are a replica, but never an exact one, of the parents' gene frequencies. The familiar law of averages is at work here. The larger the population, the more closely the gene frequencies of each generation will resemble those of the previous generation (Table 13.3).

However, large populations consist of numerous subpopulations, often more or less isolated, like the imaginary population of white-footed mice in the small woodlot. Because of habitat fragmentation, more and more populations of species are being isolated to varying degrees. In these cases, the subpopulations represent random sampling among gametes not representative of the gene pool of the larger population. Each subpopulation has its own distinctive sample. Chance fluctuation in allele frequencies in these small populations as a result of random sampling is *random genetic drift*. Over time some genes will continue to segregate while others will become fixed. After some time the population will become fixed or homozygous for one allele or another and other alleles will be lost. Allele frequencies spread out progressively as the proportion of fixed genes steadily increases, ultimately resulting in homozygous populations (Figure 13.13).

How rapidly this outcome occurs depends upon the size of the population. Consider the data in Table 13.3, which compares the variances over generations for small and large populations when $p = q = 0.5$, the number of individuals is constant, and the generations are discrete. The variance at any given time depends upon the size of the population, the number of generations elapsed, and initial gene frequencies. Note that a population of 500 shows little variance over 50 generations and minimal after 100 generations; but allele frequencies in natural populations behave so erratically that we cannot predict the probability of ultimate fixation.

Genetic drift is akin to inbreeding. The major difference is that inbreeding involves nonrandom mating, whereas genetic drift involves random mating; and inbreeding comes about because the population is small. Both result in an increase in homozygosity and a loss of heterozygosity. The measure of genetic drift is given by the *fixation index*, the reduction of heterozygosity of a subpopulation due to random genetic drift and thus the amount of inbreeding due solely to population subdivision:

$$F_{\text{ST}} = \frac{H_{\text{T}} - H_{\text{S}}}{H_{\text{T}}}$$

TABLE 13.3 Comparison of Variances over Generations for Populations of Different Sizes with Discrete Generations and a Constant Number of Individuals, all with $p_0 = q_0 = 0.5$

NUMBER OF GENERATIONS	SIZE OF POPULATION					
	6	10	50	100	500	1,000
1	0.02	0.01				
2	0.04	0.02	0.01			
3	0.06	0.04	0.01	0.01		
4	0.07	0.05	0.01	0.01		
5	0.09	0.06	0.01	0.01		
10	0.15	0.10	0.02	0.01		
50	0.25	0.23	0.10	0.06	0.01	
100	0.25	0.25	0.16	0.10	0.05	0.01

Source: Mettler and Gregg 1969:51.

Figure 13.13 A computer simulation (Monte Carlo) of the dispersal of gene frequencies among 400 hypothetical populations over 32 generations. The genetic model had the following conditions: (1) each population of eight diploid individuals—four randomly formed pairs—is constant from generation to generation; (2) each individual mates but once and the number of offspring produced by each mating varies—a Poisson distribution with a mean of two; (3) selection and mutation are absent; and (4) each population started with 2 *AA*, 4 *Aa*, and 2 *aa*, so the initial gene frequency was 0.5 for each allele at a single autosomal locus. The figure shows the populations classed according to gene frequencies in generations 1, 2, 4, 8, 16, and 32. Because of chance variations—random genetic drift—in such small populations, gene frequencies show an increasing spread toward fixation of one allele or the other in most of the populations. This simulation demonstrates why small populations exhibit fixed genes and a lack of genetic diversity. (From Mettler and Gregg 1969: 52.)

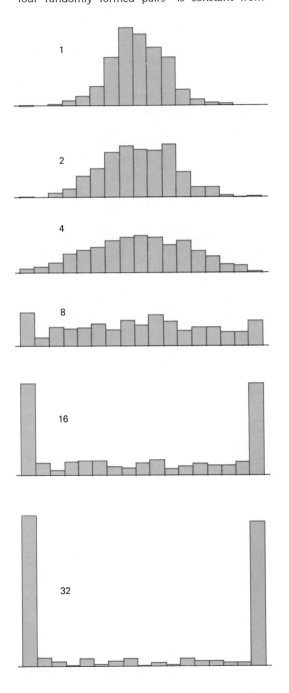

where H_S represents the heterozygosity of a random-mating subpopulation and H_T represents the heterozygosity in an equivalent random-mating total population.

Thus the fixation index F_{ST} is the probability that two alleles chosen at random in the same subpopulation are identical by descent. (For a full discussion of the derivation of the fixation index see D. Hartyl 1981.)

As in inbreeding, the value of F will change as genetic drift continues generation after generation. This changing value of F (dropping the subscript ST), identified as F_t to represent the average value of the fixation index in subpopulations in generation t, is

$$F_t = 1 - \left(1 - \frac{1}{2N}\right)^t$$

The value of F_t will range from 0 for no homozygosity to 1, complete homozygosity.

Effective Population Size

The values of F_t are based on an "ideal" population, characterized by a constant population size, equal sex ratios, equal probability of mating among all individuals, and a constant dispersal rate. Most populations are not ideal. There are age-related differences in reproduction, and particularly in polygamous populations the ratio of breeding males to females is unequal. In such populations the number of males is more important than the number of females in determining the amount of random drift. For these reasons the actual size of a small or subpopulation is of little meaning. Of greatest importance is the genetically *effective population size*, N_e. N_e is not the same as the actual number of breeding individuals. Unless the sexes are equal, N_e is less than N. N_e is defined as the size of an ideal population subject to the same degree of genetic drift as a particular real population. The ideal population is a randomly breeding one with a 1:1 sex ratio and with the number of progeny per family randomly distributed (Poisson distribution).

Unequal Sex Ratios. In a monogamous population in which one male mates with one female, all offspring are less closely related than in a polygamous population with, say, a ratio of one breeding male to five females. In the latter situation the offspring would be half or full sibs. The chance of an allele being lost or fixed and thus the amount of genetic drift is much greater in such a population.

The effective population size is given by

$$N_e = \frac{4N_m N_f}{N_m + N_f}$$

where N_m and N_f are the numbers of breeding males and females, respectively. As the disparity in the ratio of males to females widens, the effective population diminishes (Figure 13.14).

Consider a population of white-tailed deer consisting of 100 adult does and 50 adult bucks. The actual size of the population is 150; but because of the unequal sex ratio the effective population size is $4(50 \times 100)/150 = 133$. However, the white-tailed deer is a polygamous species with a dominance hierarchy among the males. Only the dominant males breed, and subdominant, potentially breeding males cannot contribute their genes to the next generation during any particular breeding season. Assume that 25 dominant males mate with 100 does, an average of four does per breeding male, not an unrealistic breeding ratio. Under these conditions the actual breeding population consists of 100 does and 25 bucks. The effective breeding population, N_e, now is $4(25 \times 100)/150 = 66$. What N_e tells us is that the sampling error or genetic drift in our deer population of 150 animals with a sex ratio of 1:4 is equal to that of a population of 66 with an equal number of males and females.

Effective population size becomes more significant in exploited populations, especially those in which trophy animals,

Figure 13.14 The depression of effective population N_e due to the disparity in the sex ratio of reproducing animals. As the ratio widens with fewer males to females, as it may in polygamous species, especially where males are hunted, the effective population size drops dramatically. (From Foose and Foose 1982.)

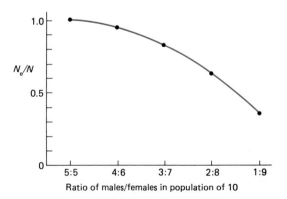

Ratio of males/females in population of 10

usually dominant males, are selected by hunters and removed from the population. For example, antler size in the white-tailed deer is genetic (Harmel 1983; Smith et al. 1983), controlled by a single set of dominant-recessive alleles (Templeton et al. 1983). A single dominant allele has a major effect on the phenotypic expression of five to ten points and a single recessive determines the phenotypic expression of two to four points. If these males are selected against by sport hunting (as they are in North America), then subdominant males become the dominants and genes for smaller antlers ultimately may become fixed in the population.

Individual Distance. The concept of effective population size also applies to a population whose individuals are more or less continuously distributed over a large area, defined as being larger than the greatest dispersal distance of the species. This area is known as the neighborhood area for the species. An estimate of the neighborhood area is the root mean square of dispersal distance of individuals about their natal origin (Dobzhansky and Wright 1947). The effective breeding size of such populations (often called neighborhood size) depends upon the number of breeding individuals per unit area and the amount of dispersion between an individual's birthplace and that of its offspring. The latter is denoted by σ^2, the one-way variance (1/2 the standard deviation squared) of the distance between birth and breeding site.

If dispersion follows a normal curve, then 39 percent of all individuals will have their offspring within a radius of σ centered at their own birthplace; 87 percent will have their offspring within a radius of 2σ; and 99 percent will have their offspring within a radius of 3σ (Wright 1969, 1978). In terms of dispersal distance, σ, and the number of breeding individuals

per unit area, δ, the effective or neighborhood size of the population is given by

$$N_e = 4\pi\sigma^2\delta$$

where π is 3.1416.

The equation can be applied to data on bannertail kangaroo rat *(Dipodomys spectabilis)* obtained by W. T. Jones (1987) who studied the dispersal of this rodent on a 36-ha study site in southeastern Arizona. The kangaroo rat lives in mounds that persist for and are occupied over many generations. Dispersing juveniles settle in vacant mounds. Jones determined that the adult breeding population consisted of 70 males and 72 females for a density of 2.27/ha. Two hundred and eighteen juveniles (107 males, 111 females) moved a mean distance of 50 m from natal to breeding site. The estimated mean dispersal distance d for the juvenile population,

$$\sum_{i=1}^{n} d_i f(d_i)$$

(where d_i = dispersal distance and $f(d_i)$ = the fraction of individuals dispersing distance d_i) (see Moore and Dolbeer 1989) is 140.25 m and the mean square dispersal distance, σ^2,

$$\sum_{i=1}^{n} d^2 f(d_i)/2,$$

is 10943 m^2/2 = 5472 m^2 or .5472 ha. The estimated effective population size is N_e = 4(3.1416)(.5476)(2.27) = 15.75. The neighborhood size is small because of the very limited dispersal of the kangaroo rat.

Therefore even within large areas of habitat, the overall population is a mosaic of subpopulations restricted by dispersal distance, which may promote a degree of inbreeding. Even northern white-tailed deer subpopulations occupying winter deer yards inhabit largely exclusive summer ranges that rarely overlap with other subpopulations (Nelson and Mech 1987).

Population Fluctuations. Actual populations are dynamic; they fluctuate through time. In fact, numbers may change dramatically. Under adverse environmental conditions or sudden loss of habitat, the population may decline sharply or "crash" (Figure 13.15). The survivors of the crash, the progenitors of future populations, possess only a sample of the original gene pool. Suppose that the effective size of a particular population for generation 1 is N_1, for generation 2, N_2, and so on. The overall increase in F_{st} is given as

$$1 - F_{st} = \left(1 - \frac{1}{2N_e}\right)^t$$

where

$$\frac{1}{N_e} = \left(\frac{1}{t}\right)\left(\frac{1}{N_1} + N_2 + \ldots \frac{1}{N_t}\right).$$

The effective size of a population fluctuating over time is the harmonic mean or reciprocals of the effective number of each generation. This harmonic mean is strongly influenced by the smaller values (Crow and Kimura 1970). Thus a population crash would tend to reduce the average effective size over time. The sharp reduction in numbers creates a severe population *bottleneck*, which can severely reduce genetic diversity in the remaining population and future generations.

There are several examples from natural populations. One is the cheetah (*Acinonyx jubatus*), the populations of which are sparse and isolated. S. T. O'Brien and associates (1983) sampled the blood of 55 cheetahs from two geographically isolated populations in South Africa and found them genetically the same (monomorphic) at each of 44 allozyme loci. Analysis of 155 abundant soluble proteins from cheetah fibroblasts (cells found in vertebrate connective tissue that form and maintain collagen) revealed a low frequency of polymorphism. The average heterogeneity was only 0.013, compared with an average heterogeneity of 0.036 for mammal populations studied so far. The homozygosity of the cheetah populations suggests that the species experienced a severe bottleneck perhaps 100 generations ago, perhaps as a result of intensive poaching and

Figure 13.15 A population faced with an environmental catastrophe or overexploitation enters a "bottleneck," in which the surviving population consists of only a sample of the total gene pool. If the population makes a complete recovery, it may lack the genetic diversity found in the original population. If the population remains small, as in captive or insular (fragmented, isolated) situations, it is subject to random genetic drift. (From Frankel and Soule 1981: 32.)

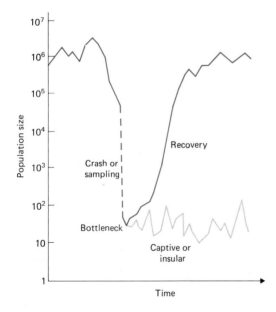

decimation by farmers, resulting in a severe range contraction. A similar situation exists with the elephant seal (Bonnell and Selander 1974). A bottleneck also results when a small group of emigrants from one subpopulation founds a new subpopulation and when small populations of animals, such as wild turkey or otter, are introduced into new or empty habitats. The emigrants or introduced population carries only a sample of genes from the parent population, so it is subject to random genetic drift. This drift is known as a *founder effect* (Figure 13.16).

Dispersal and Mating Strategies

Unless separated by a wide expanse of inhospitable habitat, some interchange takes place among subpopulations of organisms. A few white-footed mice move from one woodlot to another, and home ranges of subpopulations of deer overlap at the peripheries. If these immigrants or dispersers become part of the breeding population, they introduce a different genetic sample that tends to reduce or slow random genetic drift and helps maintain genetic diversity.

The degree to which dispersal influences genetic drift depends upon effective population size, amount of dispersal or immigration, and the degree to which individuals in the population are monogamous or polygynous. In small populations of a constant size, monogamous species should experience inbreeding depression more slowly than polygynous species, because of their maximum population size. The greater the sex disparity in polygynous species and the larger the harem, the greater the rate of inbreeding. Where immigration or dispersal involves genetic interchange, loss in homozygosity declines (Figure 13.17). Little genetic interchange is needed to prevent significant random genetic drift among subpopulations and to maintain genetic diversity. Change in homozygosity in subpopulations experiencing some immigration is given by

$$\hat{F} = \frac{1}{4N_m + 1}$$

where N_m = the actual number of migrants per generation.

F_{st} decreases as the number of immigrants increases. For example, one immigrant per generation would give a value of $F = 1/4(1) + 1 = 1/5 = 0.2$. Five immigrants per generation would greatly reduce F: $F = 1/4(5) + 1 = 1/21 = .05$. Conservation geneticists believe that one reproductively successful immigrant per generation is the minimal number needed

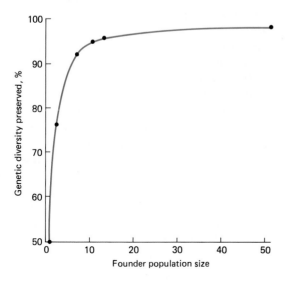

Figure 13.16 The genetic diversity in founder populations of various sizes. Founder populations of 10 may hold 90 percent of the genetic diversity in the parent population. Populations of 50 may contain a nearly complete sample. That genetic diversity can be maintained only if the population expands. The concept of a founder population is important in the introduction of a species into vacant or new habitat. (From Foose 1983: 388.)

Figure 13.17 The alleviation of inbreeding over time in monogamous and polygamous populations with either sex dispersing. In this model population size is excluded, using instead the proportion of the population replaced by immigrants. Immigrant males were always considered successful in polygamous matings. Numbers on the curves represent dispersal rates. Note that when dispersal rates in monogamous populations are low relative to population size, inbreeding continues to increase rapidly because of the slow diffusion of new alleles into the population. Contrast this case with the rapid alleviation of inbreeding depression in polygamous matings with male dispersal and a harem of five. Even in po-lygamous populations that reach significantly large breeding coefficients, only a few dispersing males are necessary to reduce genetic drift. (From Chesser 1983: 74, 75.)

to slow genetic drift and five immigrants the maximum (Frankel and Soulé 1981: 129); the percentage of immigrants needed depends upon whether the population is monogamous or polygynous. A large number of immigrants could swamp the genetic character of the subpopulation.

Information from the field on genetic effects of immigration is lacking but R. K. Chesser (1983) provides a simulation model. It suggests that in small monagamous subpopulations a slow immigration rate does not alleviate inbreeding depression, because of the slow diffusion of unrelated alleles into the gene pool and the equally slow decline of inbreeding among kin. In polygynous populations, however, inbreeding is quickly alleviated when the harem size is large and immigrant males are involved in breeding. Dominant immigrant males with a harem of five (Figure 13.17) can rapidly reduce inbreeding coefficients.

Remember that certain advantages and disadvantages derive from monogamy and polygyny. Although inbreeding may be more severe in small monogamous populations, the effective size is maximum (assuming equal numbers of both sexes). Monogamous populations do retain rare alleles and great qualitative genetic variation. Polygynous populations retain greater quantitative genetic variation, but they are subject to the loss of rare alleles and the potential genetic variation carried by nonbreeding males. Given sufficient numbers and some immigration, cycles of inbreeding and outbreeding will retain genetic variation.

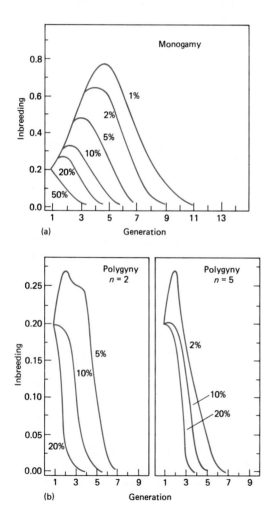

■ Minimum Viable Populations

Effective population size tells us something about how the breeding population relates to the total population, but we need to know more. Breeding populations experience mortality and dispersal, and these individuals must be replaced by younger animals. For this reason the total population must have an adequate age structure (Chapter 14) to possess a maximum effective population. The population must also be above a critical size and age structure. Below that size inbreeding and the loss of selectable genetic variation become a problem for continued survival.

The threshold number of individuals that will ensure the persistence of subpopulation in a viable state for a given interval of time, say several hundred years, is the *minimum viable population*, MVP (Gilpin and Soulé 1986; Shaffer 1981). The minimum viable population has to be large enough to cope with chance variations in individual births and deaths, random series of environmental changes, random changes in allele frequency or genetic drift, and catastrophes. Conservation geneticists consider a minimum viable population to be 500 individuals with an absolute minimal N_e of 50 (see Table 13.3, Figure 13.18). Any smaller population is subject to serious genetic drift, to loss of the genetic variability necessary to track environmental changes, and to high probability of stochastic extinction. Such estimates of MVP assume the absence of management; and they may be so large that no large animal population of that size could be maintained in a reserve or sanctuary.

Sizes of MVP vary among species. Those species in which reproduction is highly density dependent and which live in a more or less constant environment may persist for a long time in spite of a decline in genetic diversity. Species with highly variable population sizes, such as

Figure 13.18 Loss of genetic diversity as measured by heterozygosity due to random genetic drift for various effective population sizes, based on a rate of decline in heterozygosity of $\frac{1}{2}N_e \times 100$ percent per generation. Note how rapidly genetic diversity declines when N_e is small. Even with $N_e = 500$ some loss of genetic diversity takes place. (From Foose 1983: 376.)

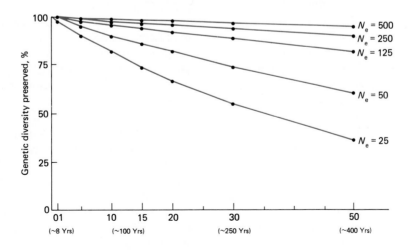

lagomorphs, may need large populations to counteract environmental stochastic effects on population growth. MVP should be a guide to the size at which populations should be maintained, but it should never be a rule of thumb deciding a species fate; otherwise we would have given up on the whooping crane when it was down to a remnant population of 20.

■ Summary

The raw material of evolution and adaptation to local environments is the genetic variability of individuals making up local populations. Natural selection acts upon this variability, reducing the influence of the less fit and favoring the more fit, with fitness measured by the number of reproducing offspring contributed to the next generation. Natural selection works in three ways: nonrandom mating, nonrandom fecundity, and nonrandom survival. Certain genetic combinations are more fit than others; they transmit more genetic information to future generations than less fit combinations.

Two sources of genetic variation acted upon by natural selection are mutations and recombination of genes provided by parents in bisexual populations. Theoretically, variations in biparental populations, reflected in gene frequencies and genotypic ratios, remain in Hardy-Weinberg equilibrium, $p^2 + 2pq + q^2 = 1$, which gives the expected genotypic frequencies of AA, Aa, and aa, respectively, if the conditions of random reproduction, equilibrium in mutation, and a relatively large, closed population exist. In nature such conditions do not occur, and there is a departure from genetic equilibrium. The direction evolution takes depends upon the genetic characteristics of those individuals in the population that survive and leave behind viable progeny. Natural selection may be stabilizing, maintaining the current genetic equilibrium and favoring intermediate phenotypes; directional, favoring phenotypes at one extreme of the range; or disruptive, favoring genotypes at both extremes of the range at the same time.

Although natural selection seems to impinge only on individuals, the evolution of certain altruistic traits seems to require some form of group and kin selection. Group selection, which operates on the differential production of local populations, favors characteristics which improve the fitness of the group but may decrease the fitness of any individual within the group. Kin selection involves the evolution of altruistic traits in a group of closely related individuals possessing genes by common descent. It involves the evolution of a genetic trait expressed by one individual that affects the genotypic fitness of one or more directly related individuals. Individual fitness plus any additional fitness acquired by improving the fitness of very close relatives is inclusive fitness.

Because of habitat fragmentation and human exploitation of the landscape, populations of many species of plants and animals are being reduced to isolated or semi-isolated small populations. These small populations carry only a sample of the genetic variability of the total population. Two potential outcomes are inbreeding and genetic drift.

Inbreeding, mating between relatives, brings out hidden genetic variation, increases homozygous genotypes at the expense of heterozygous ones, and reveals the effects of homozygous rare alleles that often result in reduced fecundity, viability, and even death. The degree of inbreeding is given by the inbreeding coefficient, F, which measures the probability that two alleles at a locus in an individual are identical by descent, that is, derived by replication of a single allele in an ancestral population.

Small populations are subject to random genetic drift, chance fluctuations in allele frequency as a result of random sampling among gametes. Drift results in the accumulation of fixed populations lacking in genetic variability. Genetic drift mimics inbreeding by increasing homozygosity, but it is the result of random mating and the alleles involved are not necessarily identical by descent. The effects of random drift are measured in terms of heterozygosity of individuals, subpopulations, and the total population, and the associated F statistics. The value $F_{st} = (H_t - H_s)/H_t$ measures the effects of random genetic drift among subpopulations.

Influencing the degree of random genetic drift is the effective population size, N_e, the size of an ideal population having the same rate of increase in F_{st} as the population in question. In monogamous populations with an equal number of breeding males and females, the actual population size and the effective population size are the same. In polygamous populations a wide disparity in sex ratios can strongly reduce the effective population size and increase genetic drift. The effective population size is given by $4N_m N_f/(N_m + N_f)$ where N_m and N_f are the number of males and females, respectively. If the population is spread out more or less uniformly across the landscape, the effective size becomes $4\pi\sigma^2 d$, where d is the density of breeding individuals per unit area and σ^2 is the root mean square of dispersal distance. If the population fluctuates over time, the effective population is given by the harmonic mean of the various population values.

Exchange of individuals among populations (emigration and immigration) can reduce genetic drift and inhibit genetic divergence among subpopulations. The equilibrium value of F_{st} for immigration or dispersal is given by $1/(4N_m + 1)$ where N_m is the number of migrants per generation. Only a few per generation are needed to reduce random genetic drift and keep F_{st} below 0.10.

If a subpopulation is to persist over time, it has a threshold number below which the population must not fall to maintain genetic diversity. This number is called the minimum viable population. Although a figure of 500 is generally given, this value may be much too low; we must consider age structure and sex ratios of the population and chance variations in the environment. Because of the adverse effects of population fragmentation, population genetics is becoming important in the preservation and management of wild species in the face of human development of planet Earth.

Properties of Populations

■ Unitary and Modular Populations

Populations are groups of individuals of the same kind living in the same space at the same time. This definition presents two problems. One is defining the boundaries of a population. Most populations know no boundaries other than those drawn by the ecologist studying them. The second problem is defining the individuals. For most of the animal kingdom there is no major problem. You can recognize a deer as an individual. It consists of morphological parts: head, four legs, a tail, fur, and other sharply defined characteristics. Other deer, regardless of age, have a similar appearance. The growth form is highly determinate.

What about the trees among which the deer are standing (Figure 14.1)? Each stem appears as an individual tree, but is it? If you were to do some digging, you would discover that many of the trees are connected by the same root structure. Further, each tree is not quite the same as others in shape, number of leaves, size of

crown, and amount of branching. In fact, if you give the tree some careful thought you will appreciate the fact that it is made up of populations of smaller units: leaves, buds, twigs, branches, and seasonally, flowers and fruits. Its growth is indeterminate. The deer obviously is a real individual, a unit. Many of the trees probably are not; each stem is a part of a larger unit, the clump of closely associated trees. Each is a module, a part of a whole.

An individual, strictly speaking, is derived from a zygote. The deer is such an individual, and many trees also go back to sexual reproduction, produced when wind-carried pollen fertilized egg cells in the ovules of the female catkins. Once established, though, these sexually produced trees gave rise to modules or new "individuals" from buds on shallow horizontal roots, a form of asexual reproduction. The individual trees produced by sexual reproduction and established by

Figure 14.1 Populations consist of either unitary or modular organisms. The white-tailed deer is a unitary organism with a discrete growth form. The surrounding trees appear as individuals, but many of them are probably modules of the original parent tree, arising from buds on the horizontal roots to form clones. In addition, the tree is made up of iterated growth forms of buds, leaves, stems, and twigs.

seeds are termed *genets*. Genets are the genetic, evolutionary units. The clones, modular units arising from the genets, are of different ages and different sizes, but all genetically the same. They are *ramets*. Such modular growth is not a feature restricted to plants. A number of animal groups are also modular, including sponges, corals, and hydroids, all sessile organisms.

There is more to modular growth in plants than ramets. Both ramets and genets grow by accumulating modules of leaf and axillary buds and the section of stem on which they grow. Buds give rise to leaves, new buds, and stems. Stems eventually develop into branches, limbs, and vertical extensions of trunks. Some modules develop into flowers, producing potential new genets.

Modular growth in itself is a demographic process (see Harper 1977; White 1979; Harper and Bell 1979). The leaves, buds, stems, and roots are also populations in their own right. They compete with neighboring modules for moisture, light, and nutrients, and respond in their own way to environmental conditions. They experience their own birthrates, death rates, and changes in age structure. The distribution of birthrates and death rates of the modules influences the growth form the plant will take, including the order of branching in trees.

Modular growth may be vertical or horizontal. Among herbaceous plants modular growth is horizontal. Stolons or runners, characteristic of strawberries, or rhizomes as in mayapple *(Podophyllum peltatum)*, Indian cucumber root *(Medeola virginiana)*, and quackgrass *(Agropyron repens)*, produce new plants laterally away from the parent plant or genet. Usually the connections between the original zygote and its clones die, resulting in separate ramets that go on to produce their own lateral extensions (Figure 14.2). Woody plants exhibit vertical growth of modules, giving rise to new leaves and other structures one above the other while maintaining strong woody connections to older modules, rather than rotting away.

A number of species of perennial plants and trees exhibit horizontal growth as well—for example, aspen *(Populus)*, sumac *(Rhus)*, and locust *(Robinia)*, trees that invade old fields. Once established, the genets extend their roots horizontally away from the parent plant, giving rise to dense stands of ramets (Figure 14.3). These clones may cover considerable area and be of different ages. Because they are a part of the original growth iterations from the parent plant, the genet may be very old, although the "individual" trees or plants may be very young. As a result the life spans of genets are indeterminate for many plants. Such characteristics of plants complicate studies of their demography.

Populations of unitary or modular organisms can be described by such measurable properties as density, dispersion, age ratios, mortality, and natality. The study of such vital statistics is known as *demography*.

■ Density and Dispersion

Crude Versus Ecological Density

The size of a population in relation to a definite unit of space is its density. Every ten years the census bureau counts the number of people living in the United States; wildlife biologists determine the number of game in a particular area; a forester determines the number and volume of trees in a timber stand. The mea-

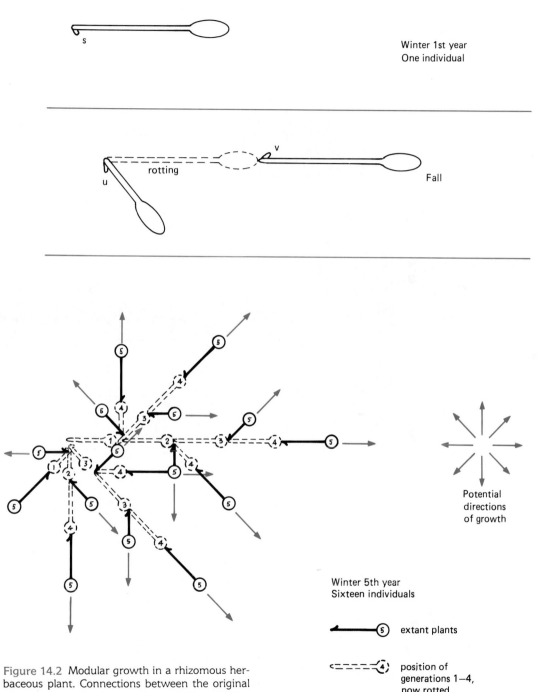

Winter 1st year
One individual

rotting

Fall

Potential
directions
of growth

Winter 5th year
Sixteen individuals

extant plants

position of
generations 1–4,
now rotted

Figure 14.2 Modular growth in a rhizomous herbaceous plant. Connections between the original
genet and subsequent ramets are lost as old rhizomes die, so the genet grows as a series of physiologically independent units. (Adapted from Bell
1974.)

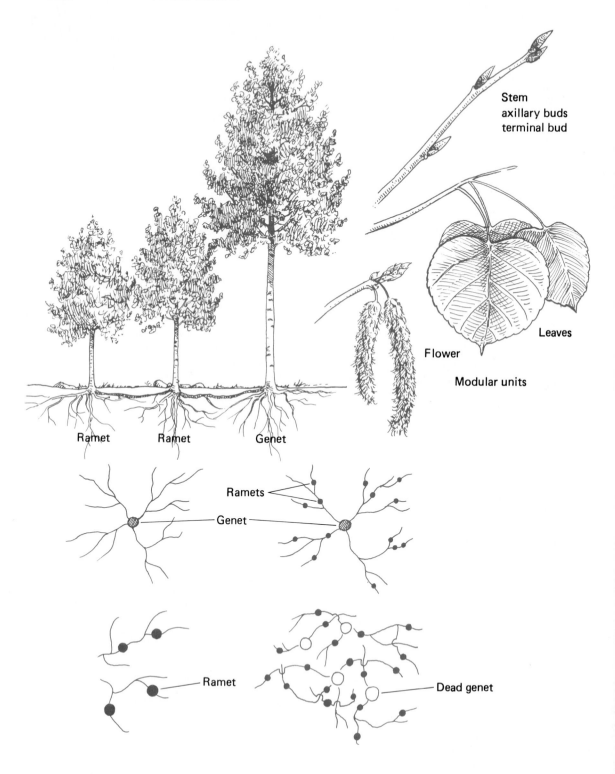

Stem
axillary buds
terminal bud

Flower

Leaves

Modular units

Ramet   Ramet   Genet

Ramets

Genet

Ramet

Dead genet

Figure 14.3 Horizontal and vertical modular growth in an aspen. Vertical modular growth involves iterations of units consisting of leaf, axillary bud, and associate stem (internode). Horizontal growth involves iterations of roots and root buds, which give rise to clones of the genet. The clones are various ages, with the youngest individuals forming the leading edge. The clones and vertical modular units remain in close connection through maintenance of woody tissue.

sure of number of individuals per unit area is called *crude density*.

Populations do not occupy all the space within a unit because it is not all a suitable habitat. A biologist might estimate the number of deer per square mile, but the deer might not utilize the entire area because of such factors as human habitation and land use practices, lack of cover, or lack of food. A soil sample may contain 2 million arthropods per square meter, but these arthropods inhabit only the pore spaces in the soil, not the entire substrate. Goldenrods inhabiting old fields grow in scattered groups or clumps because of soil conditions and competition from other old field plants. No matter how uniform a habitat may appear, it is not uniformly habitable because of microdifferences in light, moisture, temperature, or exposure, to mention a few conditions. Each organism occupies only areas that can adequately meet its requirements, resulting in patchy distribution. Density measured in terms of the amount of area available as living space is *ecological density*.

Attempts have been made to make such ecologically realistic measurements. For example, one study in Wisconsin expressed the density of bobwhite quail as the number of birds per mile of hedgerow rather than per acre (Kabat and Thompson 1963). However, ecological densities are rarely estimated because it is difficult to determine what portion of a habitat represents living space.

The density of organisms in any one area varies with the seasons, weather conditions, food supply, and many other factors. However, an upper limit to the density is imposed by the size of the organism and its trophic level. Generally the smaller the organism, the greater its abundance per unit area. A 40 ha forest will support more woodland mice than deer and more trees 5.1 to 7.6 cm inches dbh (diameter breast height) than trees 30 to 35 cm inches dbh. The lower the trophic level, the greater the density of the organism. The same 40 ha forest will support more plans than herbivores, more herbivores than carnivores.

From a practical point of view, density is one of the more important parameters of populations. It determines in part energy flow, resource availability and utilization, physiological stress, dispersal, and productivity of a population. The density of human populations relates to economic growth and the expansion and management of towns, cities, regions, states, and nations. Increasing or decreasing populations can place strains on economic and social institutions. The distribution of humans in a given region affects certain land use and pollution problems. Wildlife biologists need to know about densities of game populations to regulate hunting and manage habitats. Foresters base timber management and evaluation of site quality in part on the density of trees.

Patterns of Dispersion

Crude density also tells us nothing about how evenly individuals within the population are distributed over space and time. Determining this dispersion is a major field problem.

Spatial Dispersion. Individuals may be distributed randomly, uniformly, or in clumps (Figure 14.4). Distribution is considered random if the position of each individual is independent of the others. Random distribution is relatively rare (depending upon the scale of the landscape considered), for it can occur only where the environment is uniform, resources are equally available throughout the year, and interaction among members of the population produces no patterns of attraction or avoidance. Some invertebrates of the forest floor, particularly spiders (Cole 1946; Kuenzler 1958), the clam *Mulinia lateralis* of the intertidal mud flats of the northeastern coast of North America, and certain forest trees (Pielou 1974) appear to be randomly distributed.

Uniform or regular distribution is the more even spacing of individuals than would occur by chance. Regular patterns of distribution result from intraspecific competition among members of a population. For example, territoriality under relatively homogeneous environmental conditions can produce uniform distribution (Figure 14.5). In plants it may result from severe competition for crown and root space among forest trees (see Gill 1975) and for moisture among desert plants (Beals 1968) and savanna trees

(Smith and Goodman 1986). Autotoxicity (the production of exudates toxic to seedlings of the same species), a characteristic common among plants of arid country, is another mechanism for achieving uniform distribution.

The most common type of distribution is clumped, also called clustered, contagious, and aggregated. This pattern of dispersion results from responses by plants and animals to habitat differences, daily and seasonal weather and environmental changes, reproductive patterns, and social behavior. The distribution of human beings is clumped because of social behavior, economics, and geography.

There are various degrees and types of aggregated distribution. Groups of varying sizes and densities may be randomly or nonrandomly distributed over an area; individuals within the clumps may be distributed randomly or uniformly. Aggregations may be small or large. Population clusters may tend to concentrate around a geographical feature that provides nutrients or shelter (see Figure 14.6).

Aggregations among plants are often influenced by the nature of propagation and specific environmental requirements. Nonmobile seeds, such as those of oaks and cedar, are clumped near the parent plant or where they are placed by animals.

Figure 14.4 Patterns of distribution: (a) uniform, (b) random, and (c) clumped.

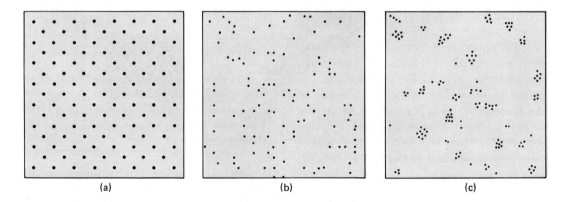

(a)   (b)   (c)

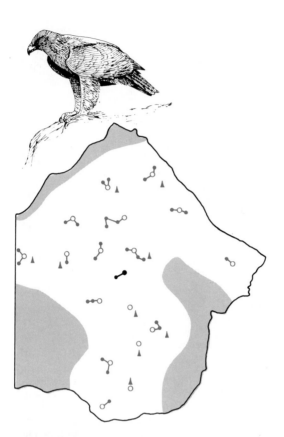

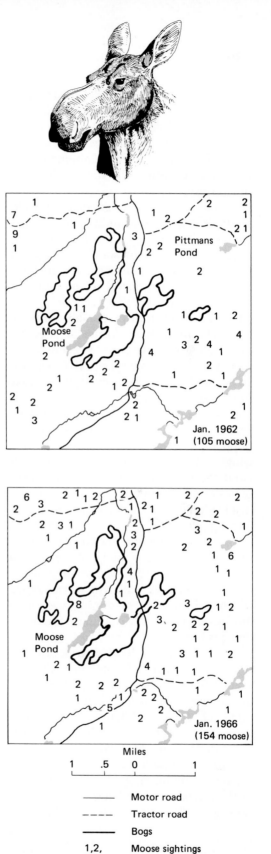

Jan. 1962
(105 moose)

Jan. 1966
(154 moose)

Miles
1   .5   0   1

	Motor road
- - - -	Tractor road
▬▬▬	Bogs
1,2,	Moose sightings

Group of sites belonging to one pair

o   Single site

●—● Marginal site not regularly occupied

▲ ▲ Breeding, year of survey 1967

Low ground unsuited to breeding eagles

Figure 14.5 Golden eagle territories in Scotland, showing the even dispersion of breeding sites. Territoriality in birds and other animals, a function of intraspecific competition, usually results in a relatively uniform distribution over an area of suitable habitat. (From Brown 1976: 109.)

Figure 14.6 Randomly distributed aggregate distribution of moose in the Lower Noel Paul River, Newfoundland. The small aggregates reflect the low sociality of moose. Individuals and groups of moose are randomly distributed over a rather homogeneous habitat. (From Bergerud and Manuel 1969: 912.)

Mobile seeds are more widely distributed, but even they tend to concentrate near the parent plant. Vegetative propagation produces clumping. Seed germination, survival of seedlings, and competitive relationships also affect the degree and type of aggregation.

Some animal aggregations represent separate responses of individuals to environmental conditions. They may be drawn together by a common source of food, water, or shelter. Moths attracted to light, earthworms congregated in a moist pasture field, barnacles clustered on a rock, all have low levels of or no social interaction. The individuals do not aid one another and only passively prevent other members of the same species from sharing the condition that brought each of them to the same location.

Aggregations on a higher social level reflect some degree of interaction among population members. Prairie chickens congregate for communal courtship; elk band together in herds with some social organization, usually with a cow as the head (Altmann 1952); birds congregate on feeding grounds away from territorial sites, yet show intolerance for each other near the nest. Aggregations of the highest social structure are found among insect societies, such as ants and termites, in which individual members are organized into social castes according to the work they perform.

The pattern of local distribution has two other features of note: intensity and grain (Pielou 1974). The *intensity* of population dispersion is high if there is a wide range of densities (Figure 14.7), low if there is little variation in density. A pattern is *fine-grained* if each individual has an equal likelihood that its neighbors will belong to any other species in proportion to their relative abundance. A pattern is *coarse-grained* when each individual is

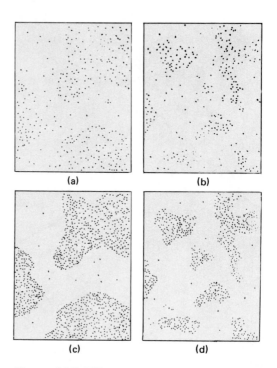

Figure 14.7 Different grains and intensities of population distribution. The intensity of a pattern is high if a wide range of densities is present and low if density contrasts are slight. The grain of a pattern is coarse if both the clumps and the gaps between them are large. (a) Low intensity and coarse grain; (b) low intensity and fine grain; (c) high intensity and coarse grain; (d) high intensity and fine grain. (From Pielou 1974: 149.)

likely to have as neighbors members of its own species. Coarse-grained distribution is clumped with relatively large spaces between groups; in other words, the distribution is patchy.

A change in density in the local population can affect both intensity and grain of population dispersion. As density increases, fine-grained species spread randomly (Figure 14.8). Coarse-grained species, preferring certain patches of habitat, become more aggregated as density increases; but as density becomes very high, individuals may be forced to occupy marginal patches in a fine-grained manner. As

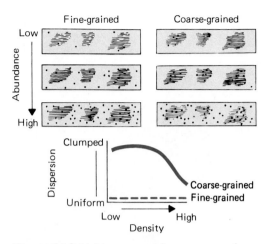

Figure 14.8 Habitat or patch occupancy for a fine-grained and a coarse-grained population as density increases from low to high levels. (From Wiens 1976: 94.)

density increases, dispersion may become more uniform.

On a regional scale, individual populations of a species are not distributed continuously over the landscape but in clusters or patches and exhibit some degree of variability in the nature of the habitats they occupy. Regional distributions of populations make up the *range* of the species (Figure 14.9). The boundaries of a range or biogeographical distribution are not fixed; they may fluctuate greatly. Habitat changes, competition, predation, and climatic changes can influence the extent of a species range, expanding it one year and contracting it another.

Temporal Dispersion. Organisms in populations are distributed not only in space, but also in time. Temporal distribution can be circadian, relating to daily changes in light and dark. The environmental rhythm of daylight and dark (see Chapter 7) is responsible for the daily movement of some animal populations, such as nectar-feeding insects seeking patches of open flowers, the daily movement of plankton from deeper to upper layers of water, and the withdrawal and emergence of nocturnal and diurnal animals. Other distributions relate to changes in humidity and temperature, seasons, lunar cycles, and tidal cycles. Seasonal changes result in the appearance and disappearance of populations of wildflowers in forest and field and in the return and departure of migrant animals. The populations of forests and fields are quite different in spring, summer, fall, and winter. Distributions may also be related to longer periods of time, which encompass annual cycles, successional stages, and evolutionary changes.

Movements. Most organisms disperse at some stage of their life cycle. They leave their immediate environments either permanently or seasonally for more favorable habitats. Such movements are essential for individual survival, especially of the young, the group most prone to disperse, for there is no room for all in any one immediate environment. Insufficient resources, deteriorating habitats, and alleviation of inbreeding are major impetuses for dispersal. Whatever the reasons, the dispersers improve their potential fitness by moving to a new territory, in spite of the risks of travel. Dispersal leads to colonization of suitable new areas, expansion of a species range, and the spread of genes. Isolated pockets of colonizers, subject to genetic drift, establish locally adapted populations that serve as a focal point for further colonization.

Dispersal movements may be one-way out of one habitat and one-way into another with no return trip. The former is termed *emigration* and the latter *immigration*. Both involve the same movement; the distinction is one of viewpoint. Emigrants from one area become the immi-

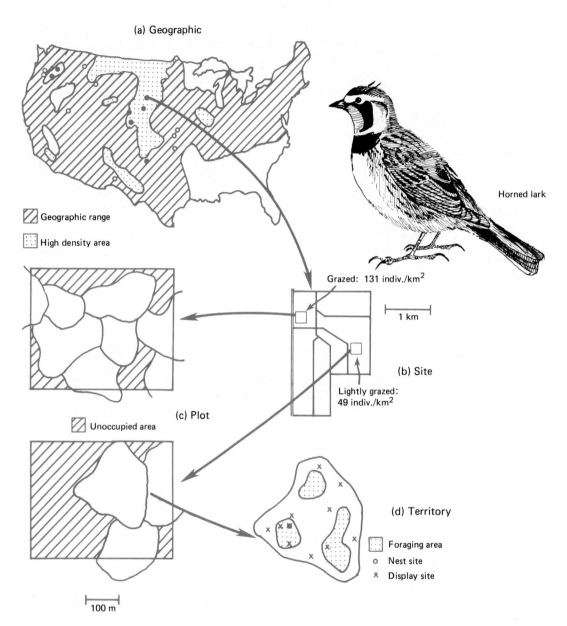

Figure 14.9 Population distribution on different scales for the horned lark *Eromophila alpestris*. (a) Although the horned lark is distributed over a wide biogeographical breeding range, its regional distribution ranges from areas of low to high densities. (b) On a local scale the distribution of the bird is influenced by the availability of habitat. (c) Within a plot of given habitat, the bird's distribution is influenced by territorial behavior. (d) Within each territory the bird allocates space to different activities. (From Wiens 1973: 237.)

grants to another. Dispersal with a return to the place of origin is termed *migration*.

For mobile animals dispersal is active; but many sessile organisms, particularly plants, depend on passive means of dispersal, involving gravity, wind, water, the coats of mammals, the feathers of birds, and the guts of both. The distance these organisms travel depends upon the quality of their dispersal agent. Seeds of most plants fall near the parent, and their density falls off quickly with distance (Figure 14.10). Heavier seeds, such as those of oaks, have a much shorter dispersal range than the lighter wind-carried seeds of maples, birch, milkweed, and dandelion. Wind also is the means of dispersal for many animals, such as the young of some species of spider, larval gypsy moths, and cysts of brine shrimp. In streams the larval forms of some invertebrates disperse downstream in the current to suitable microhabitats; but larval offspring of sessile marine organisms such as barnacles become active planktonic swimmers in the marine environment and determine where they will settle. Many plants depend upon some active carrier, such as birds and mammals, to disperse their seeds, which in some cases has evolved to mutual advantage (see Chapter 23). Seeds of some plants are armed with spines and hooks that catch on the fur of mammals, feathers of birds, and clothing of humans. Other plants, such as the cherries and viburnums, depend upon birds and mammals to consume their fruits and carry the seeds to some distant point in their guts. The seeds of such passive dispersers usually are relatively heavy and few in number, but they have higher chances of being deposited in suitable places for germination than wind-dispersed seeds.

Migratory movements fall into three categories, the most familiar of which is

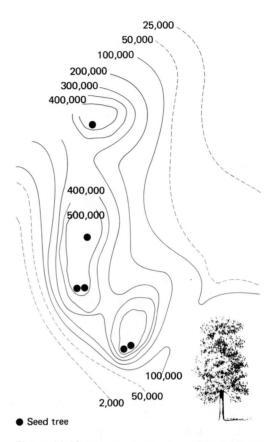

● Seed tree

Figure 14.10 Pattern of annual seedfall of yellow-poplar *Liriodendron tulipfera*. Lines show equal seeding density. Typically with this wind-dispersed species, seedfall drops off rapidly away from the parent tree. (After Engle 1960.)

the repeated return trip made by individuals. Such round trip migrations may be daily or annual, short-range or long-range. Zooplankton in oceans move down to lower depths by day and move up to the surface water at night (Figure 14.11). The movement appears to be a response to light intensity. Bats leave their roosting places in the evening hours to travel to their feeding ground and return by daybreak. Annually earthworms make a vertical migration deeper into the soil to

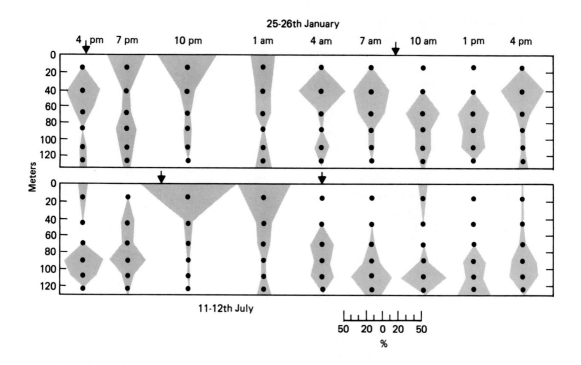

Figure 14.11 The daily upward migration of the copepod *Calanus finmarchicus* in the Clyde sea area during 24-hour periods in January and July. The time of ascent in the evening and descent in the morning, as well as the depth to which the copepod migrates, is correlated with differences in time of sunrise and sunset. The journey represents a distance of 15,000 times the animals' own length. (After Nichols 1933.)

spend the winter and move back to the upper soil in spring and summer. Mule deer in the western mountains move from their summer ranges on north-facing slopes to wintering grounds on south-facing slopes (Figure 14.12). Similarly, elk move down from their high mountain summer ranges to lowland winter ranges. On a larger scale caribou move from summer calving ranges in the taiga to the arctic tundra for the winter, where lichens comprise their major food source (Figure 14.13). Gray whales move down from Arc-

tic waters in summer to their wintering waters off the California coast, and humpbacked whales migrate from northern oceans to the central Pacific off the Hawaiian Islands. The most familiar of all migrations is the annual spring return and fall departure of birds.

A second type of migration involves only one return trip. Such migrations are common to some species of Pacific salmon. Young hatch and grow in headwaters of coastal streams and rivers. The young move downstream and out to the open sea, where they reach sexual maturity. At this stage they return to their home streams to spawn and die.

A third type, characterized by the monarch butterfly, is unusual because the fall migrants do not return north but their offspring do. About 70 percent of the last generation of monarch butterflies in summer move south in noticeable flights to their wintering grounds in the highlands

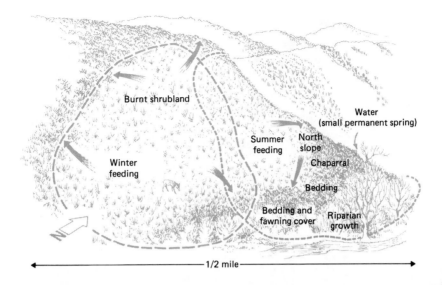

Figure 14.12 The seasonal short-range migration of mule deer in California. In summer the deer keep to the cooler, moister north-facing slopes. In winter they move to the south-facing slopes. (From Taber and Dasmann 1958: 34.)

of Mexico, a trip that covers about 14,000 kilometers. From the wintering grounds monarchs undertake a northward movement in January and arrive in the deep south of the United States in early spring, where they start a new generation. The first spring generation continues the northward trek, following the milkweed north with the spring. One generation succeeds another until the monarchs that finally arrive on the northern breeding ground are several generations removed from the ancestors that migrated south the previous fall. Such migrations are not confined to the monarch butterfly. Similar but less extensive migrations are undertaken by other insects, including two leafhoppers *(Macrosteles fascifrons* and *Empoasca fabae)*, the harlequin bug *(Murgantia histrionica)*, and the milkweed bug *(Oncopeltus fasciatus)*.

The proximal reasons for migratory dispersal vary with the species. They include extreme weather conditions, changes in thermal and photoperiodic environments, interspecific and intraspecific competition for resources, predation and parasitism, and changes in availability of food, cover, and other resources. The ultimate reason is that gains in fitness exceed the costs of migration. For example, whales fatten in the food-rich arctic waters during the summer but move down to tropical waters to calve. Because baleen whales do not feed during their southern stay in spite of lactation, living in tropical waters greatly reduces their metabolic energy demands. George Cox (1985) has hypothesized that migration in birds evolved from different selection pressures, including increased seasonality of climate, speciation of geographically isolated segments of migrant populations, and elimination of permanent resident population segments by interspecific competition (Chapter 18). These factors lead to disjunct breeding and nonbreeding ranges, between which birds move on an annual basis.

Figure 14.13 Migratory pathways of four vertebrates. Ring-necked ducks *(Aythya collaris)* breeding in the northeast migrate in a corridor along the coast to wintering grounds in South Carolina and Florida. Not shown are the migratory pathways of the prairie populations. The canvasback duck *(Aythya valisinera),* whose major breeding areas are in the prairie pothole region, has a number of corridors; but most canvasbacks diverge either to the Atlantic or Pacific Coast. The major corridor (the broad path) leads to the Chesapeake Bay and down the Mississippi to Louisiana. The barren-ground caribou winters in the taiga; with the coming of spring, bands of caribou coalesce and move northward in great numbers to the calving grounds on the arctic tundra, where they spend the summer. Gray whales summer in the arctic waters and Bering Sea, where they store enough fat to sustain them without feeding the rest of the year. In winter they move south along the Pacific Coast to their winter breeding ground in the Gulf of California and the waters off Baja California.

■ Mortality and Survival

A major influence on population size is mortality. Mortality, which begins even in the egg, can be expressed either as the probability of dying (mortality rate) or as a death rate. The *death rate* is the number of deaths during a given time interval divided by the average population; it is an instantaneous rate. The *probability of dying* is the number that died during a given time interval divided by the number alive at the beginning of the period. The complement of the mortality rate is the *probability of living*, the number of survivors divided by the number alive at the beginning of the period. Because the number of survivors is more important to the population than the number dying, mortality is better expressed in terms of survival or in terms of *life expectancy*, the average number of years to be lived in the future by members of a population.

The Life Table

A clear and systematic picture of mortality and survival is best provided by a *life table*, a device first developed by students of human populations and widely used by actuaries of life insurance companies. The life table consists of a series of columns, each of which describes an aspect of mortality statistics for members of a population according to age. Figures are presented in terms of a standard number of individuals, or cohort, usually, but not always, 1000 at birth or hatching. The columns include x, the units of age or age level; l_x, the number of individuals in a cohort that survive to the listed age level; d_x, the number of a cohort that die in an age interval, x to $x + 1$ (from one age level listed to the next); and, q_x, the probability of dying or age-specific mortality rate, determined by dividing the number of individuals that died during the age interval by the number of individuals alive at the beginning of the age interval. Another column, s or the survival rate, may be added. It can be calculated from $1 - q$. In order to calculate life expectancy, given in column e_x, two additional statistics are needed, L_x and T_x. L_x, the average years lived by all individuals in each category in the population, is obtained by summing the number alive at the age interval x and the number at the age $x + 1$ and dividing the sum by 2. T_x, the number of time units left for all individuals to live from age x onward, is calculated by summing all the values of L_x from the bottom of the table up to the age interval of interest. Life expectancy, e_x, is obtained by dividing T_x for the particular age class x by the survivors for that age as given in the l_x column. (Construction of a life table is given in Appendix B of the *Student Resource Manual*).

Data for life tables are relatively easy to obtain for laboratory animals and for human beings. Data on mortality, survivorship, and age for organisms in the wild are much more difficult to obtain. Mortality (d_x) can be estimated by determining the ages at death of a large number of animals born at the same time, if they were marked or banded. We can record the ages at death of animals marked at birth, but not necessarily of those born during the same season of the year. Data from several years and several cohorts may be pooled to provide the information for the d_x column. Another approach to gathering data is to determine the age of death of a representative sample of carcasses of the species concerned. Age can be determined by examining wear and replacement of teeth in deer, growth rings in the cementum of teeth of ungulates and carnivores, annual rings in the horns of mountain sheep, and weight of the lens of the eye. This information also goes into the d_x column. Recording the ages at death of a sample of a population wiped out by some catastrophe could provide data for the l_x series. Determining the age of animals taken during a hunting season provides information for the l_x column, because the sample is from a living population; but the data are often biased in favor of older age classes, especially if they are collected between breeding seasons.

There are three basic types of life tables. One is the *cohort, horizontal,* or *dynamic life table.* Its construction involves following a cohort of individuals, a group all born within a single short span of time, from the birth of the first to the death of the last. Construction of such life tables is most easily accomplished when the species are annuals, so the generations are discrete. Growth of annuals is determinate, and seeds germinate the following spring. If seeds lie dormant in the ground (seed bank) for a longer period of time, then the situation is more complicated.

McGraw (1989) constructed a life table for shoots of *Rhododendron maximum* (Table 14.1). A shoot is an apical meristem and the live leaves produced by that meristem. McGraw was interested in whether shoot life histories were patterned more clearly by age or size and whether interactions between age and size determined shoot fate (death, growth, flowering, or branching). He was able to use a life table approach to his study because the shoots are analogs of individuals in traditional populations: They grow, reproduce (branch), and die. They are easily aged by counting annual bud scars along the stem; new leaves and stems are produced annually. McGraw sampled the entire shoot population on large rhododendron, judged median for the rhodo-

TABLE 14.1 Life Table for Shoots of *Rhododendron maximum*

Age	l_x	d_x	q_x	e_x
0	1.000	0	0	5.60
1	1.000	0.016	0.016	4.60
2	0.984	0	0	3.67
3	0.984	0.075	0.077	2.67
4	0.909	0.185	0.024	1.85
5	0.724	0.346	0.477	1.19
6	0.378	0.270	0.714	0.82
7	0.108	0.095	0.882	0.62
8	0.013	0.013	1.000	0.50

Source: Adapted from McGraw 1989.

dendron population based on previous studies. McGraw divided the shoots into age classes, noted shoot death by counting those that no longer supported any live green leaves, and determined fertility by the number of daughter branches produced over a period of one year.

A second example of a cohort life table is one for an insect, the gypsy moth (Table 14.2). With insects the life table is best divided into developmental stages rather than discrete time intervals. The l_x values are obtained by observing a natural population over the annual season and by estimating the size of the surviving population at each stage of development: eggs, larvae, pupae, and adults. If we keep records of weather conditions, abundance of predators, parasites, and diseases, we can also estimate how many die from various causes (d_{xf}).

Developing a cohort life table for organisms with overlapping generations is much more difficult. We have to mark or in some manner follow all the individuals born in one year through to the death of the last survivor. Because many animals are long-lived and get thoroughly mixed with individuals of other age classes, tracking becomes a real problem. Table 14.3 is

TABLE 14.2 Life Table Typical of Sparse Gypsy Moth Populations in Northeastern Connecticut

x	l_x	d_{xf}	d_x	$100\ q_x$
Eggs	550.0	Parasites	82.5	15
		Other	82.5	15
		Total	165.0	30
Instars I–III	385.0	Dispersion, etc.	142.4	37
Instars IV–VI	242.5	Deer mice	48.5	20
		Parasites and disease	12.1	5
		Other	167.3	69
		Total	227.9	94
Prepupae	14.6	Predators, etc.	2.9	20
Pupae	11.7	Vertebrate predators	9.8	84
		Other	0.5	4
		Total	10.3	88
Adults	1.4	Sex (SR = 30:70)	1.0	70
Adult, female	0.4	—	—	—
Generation	—	—	549.6	99.93

an example of a cohort life table for such a vertebrate, the red deer. Lowe (1969) followed the 1957 cohort of red deer on the Isle of Rhum by making a total count of calves in 1957. From that year on Lowe aged all deer shot under rigorously controlled conditions and by that means was able to sort out members of the 1957 cohort from all the rest. By 1966, 92 percent of the cohort was dead.

The second type of life table is the *time-specific, static*, or *vertical life table* (Table 14.4). It is constructed by sampling the population in some manner (such as hunting take) and aging the animals to obtain a distribution of age classes during a specific period of time. It involves the assumptions that each age class is sampled in proportion to its numbers in the population and age at death, that the birthrates and death rates are constant, and that the population is stationary. It assumes, for example, that survivors of one-year classes were survivors from the year

TABLE 14.3 Cohort Life Table for Red Deer on Isle of Rhum, 1957

x	l_x	d_x	$100\ q_x$	ϵ_x
STAGS				
1	1000	84	84.0	4.76
2	916	19	20.7	4.15
3	897	0	0	3.25
4	897	150	167.2	2.23
5	747	321	430.0	1.58
6	426	218	512.0	1.39
7	208	58	278.8	1.31
8	150	130	866.5	0.63
9	20	20	1000.0	0.50
HINDS				
1	1000	0	0	4.35
2	1000	61	61.0	3.35
3	939	185	197.0	2.53
4	754	249	330.2	2.03
5	505	200	396.0	1.79
6	305	119	390.1	1.63
7	186	54	290.3	1.35
8	132	107	810.5	0.70
9	25	25	1000.0	0.50

Source: V. P. W. Lowe 1969.

before and so on, as they would have been if they were a single cohort. Such assumptions, of course, are false, because when data are so collected the number of survivors in one year may be greater than in the previous year. In this case the data have to be smoothed (see Caughley 1977).

The third type is the *dynamic-composite life table*. It records the same information as the dynamic life table, but it takes as the cohort a composite of a number of animals marked over a period of years rather than at just one birth period (Table 14.5). For example, wildlife biologists may mark or tag a number of newly hatched young birds or newly born young mammals each year over a period of several years. After following the fate of each year's group, they pool the data and treat all of the marked animals as one cohort (see Barkalow et al. 1970). In another approach, biologists record the ages at death of animals found over a series of years and pool those data to construct a life table. This method was employed for the life table of Yellowstone elk in Table 14.5.

Both the static and the dynamic-composite life tables are inaccurate. Mor-

TABLE 14.4 Time-Specific Life Table for Red Deer on Isle of Rhum, 1957

x	l_x	d_x	1000 q_x	e_x
STAGS				
1	1000	282	282.0	5.81
2	718	7	9.8	6.89
3	711	7	9.8	5.95
4	704	7	9.9	5.01
5	697	7	10.0	4.05
6	690	7	10.1	3.09
7	684	182	266.0	2.11
8	502	253	504.0	1.70
9	249	157	630.6	1.91
10	92	14	152.1	3.31
11	78	14	179.4	2.81
12	64	14	218.7	2.31
13	50	14	279.9	1.82
14	36	14	388.9	1.33
15	22	14	636.3	0.86
16	8	8	1000.0	0.50
HINDS				
1	1000	137	137.0	5.19
2	863	85	97.3	4.94
3	778	84	107.8	4.42
4	694	84	120.8	3.89
5	610	84	137.4	3.36
6	526	84	159.3	2.82
7	442	85	189.5	2.26
8	357	176	501.6	1.67
9	181	122	672.7	1.82
10	59	8	141.2	3.54
11	51	9	164.6	3.00
12	42	8	197.5	2.55
13	34	9	246.8	2.03
14	25	8	328.8	1.56
15	17	8	492.4	1.06
16	9	9	1000.0	0.50

TABLE 14.5 Composite-Dynamic Life Table for Northern Yellowstone Female Elk

x	l_x	d_x	q_x	e_x
0	1000	323	.323	11.8
1	677	13	.019	16.2
2	664	2	.003	15.5
3	662	2	.003	14.6
4	660	4	.006	13.6
5	656	4	.006	12.7
6	652	9	.014	11.7
7	643	3	.005	11.0
8	640	3	.005	10.0
9	637	9	.014	9.0
10	628	7	.001	8.1
11	621	12	.019	7.3
12	609	13	.021	6.4
13	596	41	.069	5.5
14	555	34	.061	4.9
15	521	20	.038	4.2
16	501	59	.118	3.3
17	442	75	.170	2.7
18	367	93	.253	2.1
19	274	82	.299	1.7
20	192	57	.297	1.2
21 +	135	135	1.000	0.5

Source: Adapted from Houston 1982.

tality and reproduction vary from year to year over which the data are collected. The data reflect standing age distributions; yet the life table is based on stable age distributions. If age distributions are unstable, populations are changing continuously and the data from which life tables are constructed do not reflect the true nature of the population. In spite of these shortcomings, such life tables may present a reasonable assessment of average conditions in the population.

Mortality and Survivorship Curves

From life tables two kinds of curves can be plotted—mortality curves based on the q_x column and survivorship curves based on the l_x column.

Mortality Curves. A mortality curve is derived by plotting data in the q_x or mortality rate column of the life table against age. It consists of two parts: (1) the juvenile phase, in which the rate of mortality is high, and (2) the post-juvenile phase, in which the rate first decreases as age increases and then increases with age after a low point in mortality is reached (Figure 14.14). For most populations, a roughly J-shaped curve results.

Being a ratio of the number dying during an age interval to the number alive at the beginning of the period (or surviving the previous age period), q_x is independent of the frequency of the younger age classes. Therefore it is free of the biases inherent in the l_x column and survivorship curves. Most life tables of wild populations are subject to bias because the one-year age class is not adequately represented. This error distorts each succeeding l_x and d_x value. If the first values are inaccurate, all succeeding ones are inaccurate. But if the first values of q_x are

wrong, the error does not affect the other values. For this reason mortality curves, which indicate the rate of mortality indirectly by the slope of the line, are more informative than survivorship.

Survivorship Curves. The survivorship curve depicts age-specific mortality through survivorship. It is obtained by plotting the number of individuals of a particular age cohort against time. It may be constructed in two ways. One is to plot logarithms (so that a strong curved line represents constant fractions) of the number of survivors, or the l_x column, against time.

Survivorship curves may be classified into at least three hypothetical types (Figure 14.15) (Deevey 1947). They are conceptual models only, not ones to which survivorships must conform; but they do serve as models to which real-life survivorship curves can be compared. The Type I curve is convex. It is typical of populations whose individuals tend to live out their physiological life span; they exhibit a high degree of survival at all ages and experience heavy mortality at the end of the life span. Such a curve is typical of some plants, such as *Phlox drummondi* (Figure 14.16 on page 344), and many mammals, such as the red deer (Figure 14.17 on page 345). The Type II curve is linear and typical of organisms with constant mortality rates. Such a curve is characteristic of the adult stages of many birds (Figure 14.18 on page 345) and rodents and some plants (Figure 14.19 on page 346). Type III is concave and is typical of organisms with extremely high mortality rates in early life, such as many species of invertebrates (Figure 14.20 on page 346), fish, and some plants.

Among plants life table attributes and survivorship apply not only to individuals but also to the modular units: the popu-

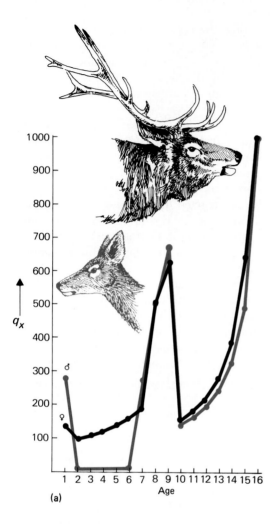

(a)

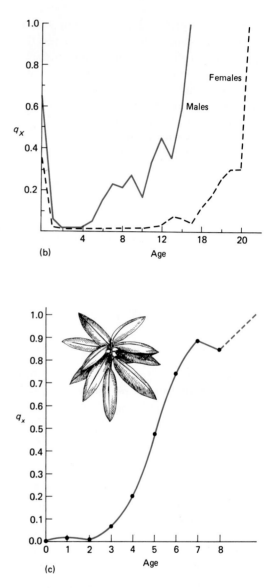

(b)

(c)

lations of leaves, buds, flowers, fruits, and seeds, as well as clones (Figure 14.21 on page 346). Mortality and survivorship of these subpopulations influence the individual growth form of the plant, its ability to compete, and its photosynthetic and reproductive performance.

The validity of the survivorship curves depends upon the validity of the life table and the l_x column. Life tables and therefore survivorship curves are not typical of some standard population of a species but rather reflect specific places, times, and conditions. They assume stable age distri-

Figure 14.14 Mortality curves. (a) The red deer *(Cervus elapus)*, males and females. (Data from Lowe 1969.) (b) Elk or wapiti *(Cervus canadensis)* of Yellowstone. (From Houston 1982: 57.) The mortality curves of the males of these two related species are similar, with a sharp peak in mortality in the middle years. The female red deer show a similar peak, but female elk do not (see Chapter 15). The mortality curves, however, assume a J-shape. (c) The mortality curve of *Rhododendron* also assumes a strong J-shape, with maximum mortality late in life. (Data from McGraw 1989.)

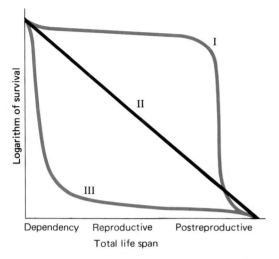

Figure 14.15 Three basic types of survivorship curves. The vertical axis may be scaled arithmetically or logarithmically. If it is logarithmic, the slope of the lines will show the following rates of change: Type I, curve for organisms living out their full physiological lifetime; Type II, curve for organisms in which the rate of change in mortality is fairly constant at all age levels, so there is more or less a uniform percentage decrease in survivorship over time; Type III, organisms experiencing high mortality and low survivorship early in life.

bution (proportions of the age classes do not change with an increasing population). Their greatest usefulness is for the comparison of populations of one area, time (Figure 14.22 on page 347), sex, or species (Figure 14.23) with populations of another.

■ Natality

The greatest influence on population increase is natality, the production of new individuals in the population. Natality may be described as maximum or physiological natality or as realized natality. *Physiological natality* is the maximum possible number of births under ideal environmental conditions, the biological limit. Because this maximum is rarely achieved in wild populations, its measure is of little value to a field biologist, except as a yardstick against which to compare realized natality. *Realized natality* is the amount of successful reproduction that actually occurs over a period of time. It reflects the type of breeding season (continuous, discontinuous, or strongly seasonal), the number of litters or broods per year, the length of gestation or incubation, and so on. It is influenced by environmental conditions, nutrition, and density of the population.

Animal Natality

Natality, measured as a rate, may be expressed either as crude birthrate or specific birthrate. *Crude birthrate* is expressed in terms of population size, for example, 50 births per 1000 population. *Specific birthrate*, a more accurate measure, is expressed relative to a specific criterion such as age. The most usual form is an age-specific schedule of births, the number of offspring produced per unit time by females in different age classes.

Because population increase depends upon the number of females in the population, the age-specific birth schedule counts only females giving rise to females. The age-specific schedule is obtained by determining the mean number of females born in each age group of females, the statistic m_x. Given the l_x or survivorship from the life table and the m_x schedule, we can obtain the *net reproductive rate, R_o,* the number of females left during a lifetime by a newborn female. In human demography the net reproductive rate is usually modified into a *fertility rate,* the number of births per 1000 women 15 to 40 years of age.

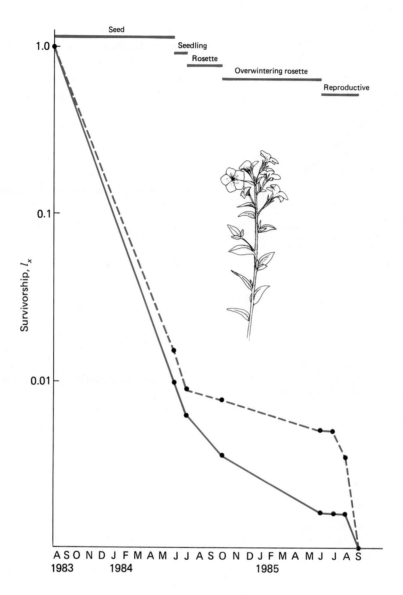

Figure 14.16 Type I survivorship curve for the annual *Phlox drummondi* at Nixon, Texas, 1974–1975. Age is indicated in both days (day 0 = May 29, 1974) and percent deviation from mean life span of 122 days. The survivorship curve in effect is in two parts, one for the seeds and the other for vegetative growth. (From Leverich and Levin 1979: 885.)

Net reproductive rates may be determined from a fecundity schedule based on the time-specific life table for red deer (Table 14.6). The fecundity table uses the survivorship column l_x from the life table (Table 14.4) and an m_x column, the mean number of females born to females in each age group. For calculation of the net reproductive rate, the l_x column is converted to a proportionality, in this case by dividing each age value by 1000. Female

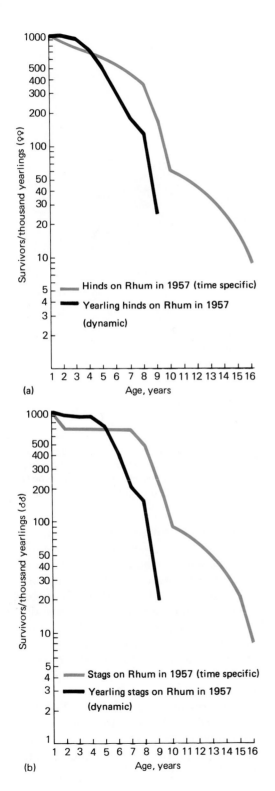

(a)

(b)

Figure 14.17 Type I survivorship curves for hinds (females) and stags (males) of red deer on Isle of Rhum. The green curve is derived from the time-specific life table; the black curve from the 1957 cohort life table. (From Lowe 1969: 436, 437.)

red deer aged 1 and 2 years produced no young; therefore, their m_x value is 0. The m_x values for females 3 years old and above increase with age until 9 years, when a decline in mean number of births begins. If the m_x values for each age group are summed, the sum 4.569 is the mean number of females that would be produced by each doe in a lifetime of 16 years if survival were complete. Obviously, each doe does not live a full 16 years. To adjust for mortality in each age group the m_x value for each age class is multiplied by the corresponding l_x or survivorship value. The resulting value $l_x m_x$ gives the mean number of females born in each age

Figure 14.18 Survivorship curve for the song sparrow *(Melospiza melodia)*. The curve is typical of birds. After a period of high juvenile mortality, characterized by a concave or Type III curve, the survivorship curve becomes linear, suggestive of Type II. (From Johnson 1956.)

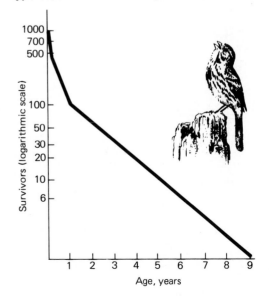

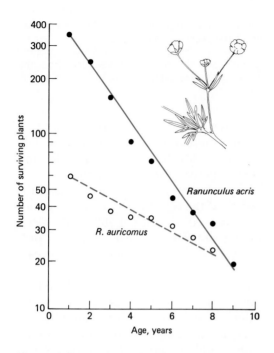

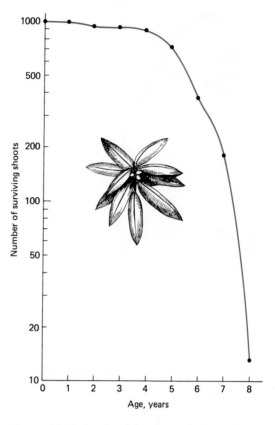

Figure 14.19 Survivorship curves of the buttercups *Ranunculus acris* and *Ranunculus auricomus.* These curves are linear or Type II. (From Sarukhan and Harper 1974.)

Figure 14.21 Survivorship curve of plant modules: shoots of rhododendron *(Rhododendron maximum).* The survivorship curve is Type I. (Based on data from McGraw 1989.)

Figure 14.20 A Type III survivorship curve exhibited by the oystershell scale *(Lepidosaphes ulma).* (Adapted from Price 1975.)

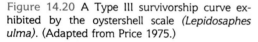

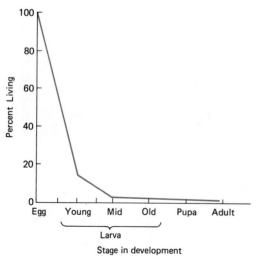

group adjusted for survivorship; thus for age class 3, the m_x value is 0.311, but adjusted for survivorship, the value drops to 0.242. When these adjusted values are summoned over all ages at which reproduction occurs, the sum represents the number of females that will be left during a lifetime by a newborn female or R_o, the net reproductive rate. For a red deer population with the survivorship indicated, the reproductive rate is 1.316.

Reproductive rate is measured on the basis of mature females and omits males and immature females. Useful for comparative purposes, the reproductive rate

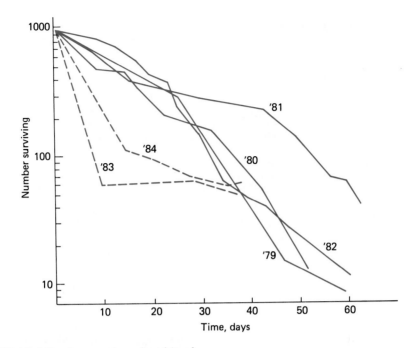

Figure 14.22 Variation in annual survivorship of small-mouthed salamander *(Ambostyma tex-anum)* larvae. Low survivorship in 1983 and 1984 resulted from severe flooding of their stream habitat. Note the variations in the survivorship. The curve for 1981 is Type I and that of 1979 Type II. (From Pefranka and Sih 1986: 731.)

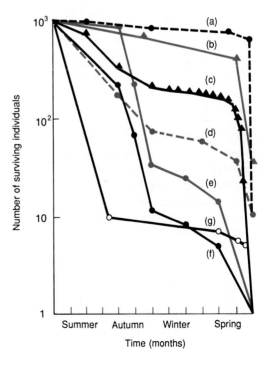

Figure 14.23 Survivorship curves for natural populations of seven winter annuals from seed production to maturity, reflecting the influence of mean fecundity (average number of seeds, following the species name). The higher the fecundity, the lower the survivorship. Possessing Type I style survivorship curves are (a) *Viola fascicu-lata*, 2, and (b) *Avena barbata*, 4. (c) *Phlox drum-mondi*, 23, and (d) *Bromus mollis*, 47, have characteristics of Type I and Type II. (e) *Sedum smallii*, 114, and (f) *Minuartia uniflora*, 305, have survivorship curves of Type II and Type III. Basically a Type III is (g) *Spergula vernalis*, 100–414. (From Watkinson 1986: 149.)

TABLE 14.6 Fecundity Table for Red Deer

x	l_x	m_x	l_xm_x	xl_{xmx}
1	1.000	0	0	0
2	.863	0	0	0
3	.778	.311	.242	0.726
4	.694	.278	.193	0.772
5	.610	.308	.134	0.920
6	.526	.400	.210	1.260
7	.442	.476	.210	1.470
8	.357	.358	.128	1.024
9	.181	.447	.081	0.729
10	.059	.289	.017	0.170
11	.051	.283	.014	0.154
12	.042	.285	.012	0.144
13	.034	.283	.010	0.130
14	.025	.282	.007	0.098
15	.017	.285	.005	0.075
16	.009	.284	.003	0.048
			$R_0 = \Sigma l_xm_x = 1.316$	$\Sigma xl_xm_x = 7.72$

summarizes information on the frequency of pregnancy, the number of females born, and the length of breeding season. Because the simplest methods of obtaining reproductive rates involve counting embryos, placental scars, number of eggs, and unfledged young in birds, the reproductive rate often incorporates a measure of mortality for the original group of ova.

Plant Natality

The distinct demographic differences between plants and most animals create problems in determining natality in plants. Plants accumulate structural units: new buds, leaves, flowers, fruits, new roots below ground, and new clones or ramets, extensions of the parent plants. Each of these modular populations has its own natality rate. At the same time plants produce new individuals through sexual reproduction. Sexual reproduction in plants involves the production of seeds and the germination of seeds. Except for annuals and biennials it is difficult to estimate seed

production by individual plants and by populations. Perennial and woody plants, even within a population, vary in longevity and in seed production, which is not necessarily an annual event. Individual plants vary widely in seed production from year to year and from age class to age class, and the amount of seed produced over a lifetime is largely unknown. Seeds usually undergo a period of dormancy, which in some species last for years until conditions are right for germination (Figure 14.24). Germination is the formal equivalent of birth in plants.

The simplest situation exists with annual and biennial plants. Both produce seeds in one reproductive effort at the end of their individual life spans. Censusing the number of seeds produced by individuals, we can arrive at some idea of fecundity. Leverich and Levin (1979) did so for the annual Drummond's phlox, arriving at a fecundity schedule for both ovules and seeds (Table 14.7). From this fecundity schedule they were able to determine R, the net rate of increase, in a manner sim-

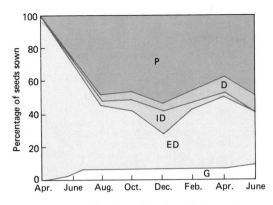

Figure 14.24 The fate of seeds of *Ranunculus repens* through 14 months. P = seeds lost to predation; D = seeds that died without germinating; ID = seeds in innate or induced dormancy; ED = seeds in enforced dormancy; G = seeds observed to germinate. Note that seeds in enforced dormancy will remain in that state in the soil until they are exposed to conditions suitable for germination. Only a fraction of the seeds present germinate in any one year, influencing the survivorship curves of seeds. (From Sarukhan 1974.)

ilar to that used for the red deer; but in this case, involving an annual species with no overlapping generations, R_0 not only gives us the average number of offspring produced per individual over a lifetime, but it provides the factor needed to determine the size of the next generation, a point to be considered later. Fecundity schedules can also be obtained for plant modules such as new shoots, as McGraw has done for rhododendron (Table 14.8).

Fecundity Curves

The m_x column of the fecundity table provides the data needed to plot a fecundity curve (Figure 14.25), which depicts the number of offspring per individual per age class. The fecundity curve is useful to compare births with mortality and survivorship (see Chapter 17). It suggests that individuals of certain age classes contribute more to population growth than others. In other words, certain age classes

TABLE 14.7 Fecundity Schedule for *Phlox drummondi* at Nixon, Texas, Based on Seed Production

$x - x'$	B_x^{seed}	N_x	b_x^{seed}	l_x	$l_x b_x$
0–299	.000	996	.0000	1.0000	.0000
299–306	52.954	158	.3394	.1586	.0532
306–313	122.630	154	.7963	.1546	.1231
313–320	362.317	151	2.3995	.1516	.3638
320–327	457.077	147	3.1904	.1476	.4589
327–334	345.594	136	2.5411	.1365	.3470
334–341	331.659	105	3.1589	.1054	.3330
341–348	641.023	74	8.6625	.0743	.6436
348–355	94.760	22	4.3072	.0221	.0951
355–362	.000	0	.0000	.0000	.0000
					$\Sigma = 2.4177$

$R_0 = \Sigma l_x b_x = 2.42$ (per capita)

$R = \dfrac{\ln R_0}{365} = 0.0024$ (per capita per day)

Note: $x - x'$ = age interval; B_x^{seed} = total no. of seeds produced during interval; N_x = no. surviving to day x; b_x^{seed} = average no. of seeds per individual during interval; l_x = survivorship; $l_x b_x$ = contribution to net reproductive rate during interval.
Source: From Leverich and Levin 1979.

TABLE 14.8 Fecundity Table of Shoots of *Rhododendron maximum* in Northern West Virginia

x	l_x	m_x	$l_x m_x$
0	1.000	0.012	0.012
1	1.000	0.032	0.032
2	0.984	0.000	0.000
3	0.984	0.115	0.113
4	0.909	0.037	0.034
5	0.724	0.046	0.033
6	0.378	0.000	0.000
7	0.108	0.000	0.000
8	0.013	0.077	0.001
			$R = 0.225$

Source: Adapted from McGraw 1989.

have a higher reproductive value than others, a point that will be considered in Chapter 15.

■ Life Equations

A picture of the limitations of the growth of a population, seasonal gains and losses, and other important events occurring throughout the year can be summarized in a life equation table (Table 14.9). Because slight changes in reproduction, survival, or sex ratios can influence the rate of increase considerably from year to year, wildlife biologists, especially, find the information summarized in the life equation highly useful.

The life equation is a modification of the life table. The life table is a mathematical expression of the vital statistics of a population. The life equation illustrates changes within the population. It involves a census or inventory of an identifiable population. Age in the life equation is referred to as stages in the life history of a population within one breeding cycle, instead of within a day, a month, or a year.

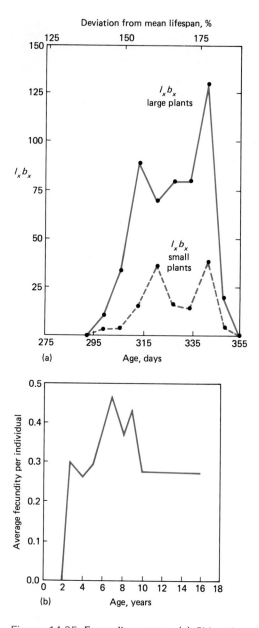

Figure 14.25 Fecundity curves. (a) *Phlox drummondi* at Nixon, Texas. (After Leverich and Levin 1979: 888.) (b) Red deer. (Data from Lowe 1974.) In *Phlox* fecundity is highest in the old age classes, which we would expect in an annual plant with one reproductive bout. In red deer the highest fecundity is in the middle age groups of females.

TABLE 14.9 Life Equation of a Black-Tailed Deer Population on a 36,000-Acre Area, 1949 to 1954

Year	Type of Gain or Loss	MALES			FEMALES			Total
		Adults	Yearlings	Fawns	Adults	Yearlings	Fawns	
1949	Prehunting population	312	140	456	1003	274	475	2690
	Legal hunting kill	204	3	—	—	—	—	−207
	Crippling loss and illegal kill	10	11	8	38	8	8	− 83
	Winter losses	13	19	304	135	40	229	−740
1950	Prefawning population	85	107	144	860	226	238	1,660
	Fawning season gain	192	144	+707	1098	238	+589	2968
	Summer mortality	2	1	221	52	12	83	−371
	Prehunting population	190	143	486	1046	226	506	2597
	Legal hunting kill	125	25	71	160	43	86	−510
	Crippling loss	6	12	21	50	13	26	−128
	Winter losses	2	6	80	37	10	61	−196
1951	Prefawning population	57	100	314	799	160	333	1763
	Fawning season gain	157	314	+617	959	333	+515	2895
	Summer mortality	2	3	311	48	17	195	−576
	Prehunting population	155	311	306	911	316	320	2319
	Legal hunting kill	96	9	—	—	—	—	−105
	Crippling loss	5	10	3	15	6	3	− 42
	Winter loss	2	8	89	42	9	67	−217
1952	Prefawning population	52	284	214	854	301	250	1955
	Fawning season gain	336	214	+762	1155	250	+601	3318
	Summer mortality	3	2	436	58	12	260	−771
	Prehunting population	333	212	326	1097	238	341	2547
	Legal hunting kill	178	80	62	205	55	64	−644
	Crippling loss	9	20	19	70	17	20	−161
	Winter loss	5	6	71	30	8	54	−174
1953	Prefawning population	141	106	174	786	158	203	1568
	Fawning season gain	247	174	+608	944	203	+506	2672
	Summer mortality	2	2	338	47	10	225	−624
	Prehunting population	245	172	270	897	193	281	2058
	Legal hunting kill	85	26	38	70	18	36	−273
	Crippling loss	4	9	8	27	7	8	− 63
	Winter loss	5	7	71	29	7	53	−172
1954	Prefawning population	151	130	153	771	161	184	1568

Source: Adapted from E. R. Brown 1961.

Life equations, like life tables, begin with a given population, usually 1000, broken down into sex and age categories. If a game animal is involved, the table begins with a prehunting population. Hunting losses then are subtracted, according to sex and age, leaving a posthunting population. The number left after winter losses comprises the prebreeding season population. To it is added the breeding season gains. Finally, a new prehunting population estimate is obtained for one year later.

These tabular data of gains and losses can be presented as a sort of survival curve (Figure 14.26). Curves of several

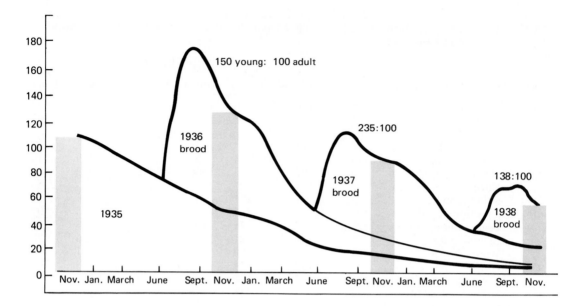

Figure 14.26 Diagrammatic life equation of a population of the California quail *(Lophortyx californica)*. The graphs follow the population trends of each year class. The bars indicate the ratios of immature to mature birds in November. Note the decline in the age ratio from 1936 to 1938. In spite of a high ratio of young to adults in 1937, which should indicate an increasing population, the population actually was declining sharply. (After J. T. Emlen 1940: 96.)

years can be joined to illustrate numerical changes over a period of years. In addition, changes in age structure can be indicated by showing the proportions of several age classes.

Life equations are not precise. Some categories in the equation cannot be measured accurately and must be estimated. However, the information derived from these estimates may be quite accurate. Properly constructed, the life equation shows the magnitude of population losses due to several causes and where and when the heaviest losses occur. The life equation also indicates the extent and importance of production of young to the future of the population, gaps in knowledge of population behavior of the species involved, and the most important research problems for future study.

Age, Stage, and Size Structure

Individuals in a population lend themselves to classification for various purposes into age classes of various refinements, into life history stages (preproductive, reproductive, and postreproductive in birds and mammals; and eggs, pupae, larvae, or instars in insects), and into size classes (heights of herbaceous plants or seedlings and diameters of trees). Age structure or some aspect of it influences population growth, gene frequency changes (see Chapter 13), and genetic equilibrium.

Age Structure in Animals

Theoretically, all continuously breeding populations tend toward a *stable age distri-*

bution; that is, the ratio of each age group in a growing population remains the same if the age-specific birthrate and the age-specific death rate do not change. If the stable age distribution is disrupted by any cause, such as natural catastrophe, disease, starvation, or emigration, the age composition will tend to restore itself upon return to the previous conditions, provided, of course, the rates of birth and death are still the same.

When mortality balances natality and the population is closed, experiencing neither movements in nor out, then the population has reached a constant size. It assumes a fixed age or life table distribution, the *stationary age distribution.*

Basically a population can do three things—increase, decrease, or remain stable. The ratio of young to adults in a relatively stable population of most mammals and birds is approximately 2:1 (Figure 14.27). A normally increasing population should have an increasing number of young; a decreasing population a decreasing number of young. Within this framework are a number of variations. A population, for example, may be decreasing yet show an increasing percentage of young. Whatever the relationships, the number in one age class that enters the next age class influences the age structure of a population from year to year. A deficiency of young can lead to an aging population, but a high proportion of old individuals may not be a cause of a deficiency of young. By combining information on population density, age ratios, and reproduction, a biologist

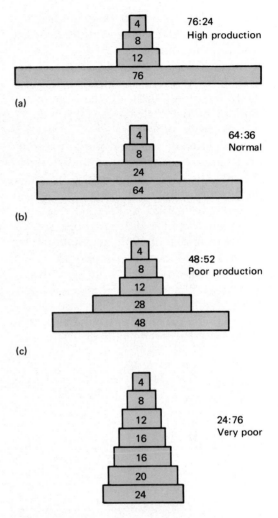

(a)

(b)

(c)

(d)

Figure 14.27 Theoretical age pyramids, especially applicable to mammals and birds. Growing populations in general are characterized by a large number of young. The age pyramid of a growing population has a broad base and a narrow, pinched top, indicating that increasing numbers may enter the reproductive age and increase all age classes (a). If the population is neither growing nor declining, then the number of individuals in each age class tends to remain the same (stationary age distribution). The pyramid of such a population is a narrow one with a narrow base of young (d). If the population is declining, few individuals are being added to the reproductive and other age classes in the population and a high proportion may be in the older age classes. In some populations the age pyramid has a marked indentation in the reproductive age classes and a relatively large percentage of the population in the pre- and postreproductive age classes. (After M. Alexander 1958.)

can correlate changes in population structure with habitat changes and ecological and human influences.

The history of a population can be detected in a series of age distributions. D. B. Houston (1982) compiled age distributions of the Yellowstone elk herd obtained from winter range information over a period of years and from them constructed a series of age pyramids (Figure 14.28). There were significant differences in age distributions from one period to the next from 1951 to 1964. The age distribution for 1964 did not differ significantly from that of 1965, nor that of 1965 from 1966. The age pyramid for 1951 with its proportionately large numbers of 6- to 9-year olds reflects changes in mortality, natality, and rate of increase following the removal of a large number of animals in a herd reduction in 1943. The

Figure 14.28 Age pyramids for the northern Yellowstone elk herd, based on standing age distribution. Note in the series of female age pyramids how lower age classes move into and influence the older age classes. In 1951 the female segment of the population had a disproportionately large number of 6- to 9-year olds, the result of an earlier culling program. Note how that age class was responsible for a relatively high proportion of 16- to 20-year-old animals 11 years later. These age pyramids suggest an unstable population, but there is little significant difference between 1964, 1965, and 1966. In 1951 a few old males existed in the population, but few males live beyond 8 years of age. (From Houston 1982: 52, 54.)

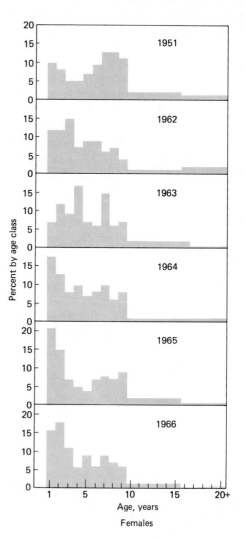

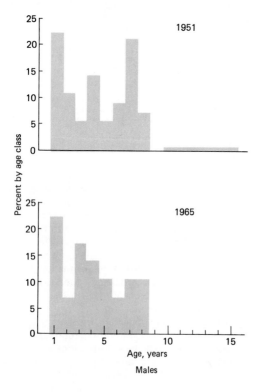

1951 pyramid with its proportionately few yearlings and two-year-olds and relatively high proportion of older females hints at a stationary distribution, which would develop in the absence of human exploitation. From 1951 to 1966 the proportion of yearlings and two-year-olds increased significantly, while the proportion of females 10 years and older declined. This change indicates an unstable age distribution, which could have been caused by intense human predation on the herd, influencing changes in fecundity, mortality, and rates of increase. The age pyramids for males show that few males lived beyond 10 years. The high proportion of yearlings may reflect greater vulnerability to hunting, distorting their actual proportion in the population.

Although useful in looking at a population's history and trends, age pyramids, especially for wild populations subject to exploitation, rarely indicate whether the rate of increase is positive, negative, or zero. When age distributions are not stable, any change in age distribution does not imply changes in survivorship or fecundity (see Caughley 1977) and cannot be safely used to predict future population trends.

Age Structure in Plants

Only recently has age structure of plant populations been employed in demographic studies. The modular structure of plants imposes certain difficulties not common in animal populations: a combination of asexual and sexual reproduction. Determining age in plants, as in animals, involves following a cohort of marked individuals over time (Figure 14.29), destroying a sample of the population, and determining the ages of individuals by growth rings, bud scars, or other indicators (Figure 14.30).

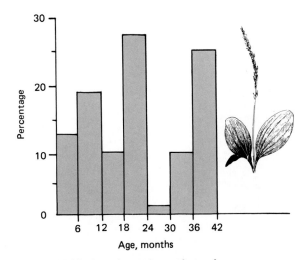

Figure 14.29 Age classes in a colony of common plantain *(Plantago rugelii)*. The age structure is based on marked individuals of known ages. (From Hawthorne and Caver 1976: 516.)

Among many plant species, especially woody ones, age may not tell enough. In even-aged stands of trees, the bulk of the individuals will fall into a very few age classes because they dominate the site, competitively excluding young age classes. Because of competition among individuals, some trees will achieve greater size than others, giving the impression of age differences, when in fact the trees are the same age. Age structure in such stands is replaced by some size criterion such as diameter classes (Figure 14.31). A similar approach is necessary for uneven-aged stands, because older individuals are not necessarily the larger, dominant ones (Figure 14.32). What appear to be younger, smaller trees are in fact often the same age as the dominant individuals. One or two age classes dominate the site until they die or are removed, allowing young age classes to develop. Size classes and numbers within each class change as the population ages and mortality increases (Table 14.10). In such populations age

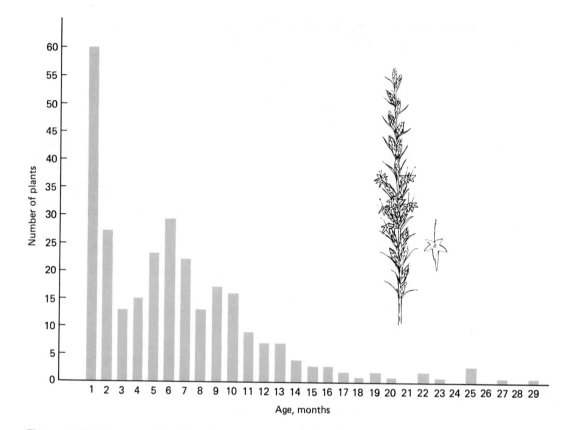

Figure 14.30 Age pyramid of a population of sharp gayfeather *(Liatris acidota)* on the Texas Gulf Coast. Ages were estimated by ring counts in the corms. The maximum number of mature individuals is in age class 6. (From Schall 1978: 95.)

alone is a poor criterion of population dynamics.

Plant ecologists have also applied analysis of age and size to modular units of plants, particular leaves and twigs, and ramets. Age distributions, influenced by growth and senescence of leaves, have an important effect on the structure, growth form, and productivity of plants population. For example, Noble et al. (1979) have shown how the application of fertilizer to a dune population of sand sedge increased the birth and death rates of shoots; age structure was dominated by younger shoots, in contrast to older shoots of unfertilized stand. Turkington (1983) showed how neighboring plant species can affect leaf birth and death rates, influencing the age structure of leaf populations. Dickerman and Wetzel (1985) studied the clonal growth and ramet demography of cattails *(Typha latifolia)* in south-central Michigan, providing a detailed picture of shoot emergence, mortality, density, height, and competition. They followed the fate of marked individual shoots over a two-year period and were able to develop both age and height classification of the population (Figure 14.33). Shoots emerged in three main pulses each year, resulting in three major cohorts. The first cohort emerged in early spring and was gone by late autumn. The second cohort

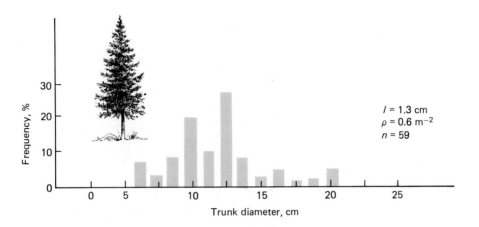

l = 1.3 cm
ρ = 0.6 m^{-2}
n = 59

Trunk diameter, cm

Figure 14.31 Size classes in an even-aged 54-year-old stand of balsam fir *(Abies balsamea)*. Although all trees are the same age, note the difference in the size classes, as measured by diameter breast height. The smaller trees give the impression of being younger, when instead they are suppressed. l = interval of trunk diameter between successive bars, determined by dividing the range of observed diameters into 12 equal intervals; p = stand density; n = number of trees in sample. (From Mohler et al. 1978.)

appeared in midsummer, and about 80 percent of these shoots died by autumn. The remainder resumed growth the following spring. The third cohort emerged in late summer, and about 90 percent of its shoots resumed growth the following spring.

Sex Ratios

An almost universal characteristic of plants and animals is sexual reproduction, although some organisms reproduce asexually at times. Even these species have some provision for sexual reproduction in their life cycle, for only by mixing and recombining genes can a population maintain genetic variability.

Populations of most organisms tend toward a 1:1 sex ratio (the proportion of males to females). The primary sex ratio (the ratio at conception) tends to be 1:1. (This may not be universally true and is, of course, difficult to determine.)

The secondary sex ratio (the ratio at birth) among mammals is often weighted toward males, but it shifts toward females in older age groups. For example, the ratio among fetuses of elk in western Canadian national parks was 113 males to 100 females (Flook 1970). Between ages 1½ and 2½ the ratio of males to females dropped abruptly until it remained at about 85:100, although in certain areas it dropped as low as 37:100. The greatest decline in the number of males occurred between the ages of 7 and 14. The decline of females was much less rapid. The difference in the rate of decline has an additional effect on the higher ratio of females to males. The loss of males allows more food and space for the females and young, thus further increasing the female rate of survival.

Among humans, too, males exceed females at birth, but as age increases the ratio swings in favor of females. In 1965 the ratio of males to females based on a stable age distribution in the United States was age 0 to 4, 104:100; age 40 to 44,

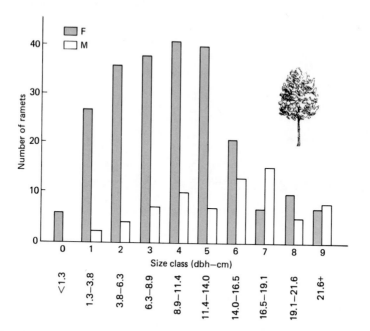

Figure 14.32 Size classes of a population of ramets of quaking aspen *(Populus tremuloides)* by sex in lower northern Michigan. Female ramets have greater numbers and greater basal areas than male ramets. (From Sakai and Burris 1985: 1925.)

Figure 14.33 Age classes of shoots of cattail *(Typha latifolia)* at Lawrence Lake, Michigan, over two years. As an overstory develops, smaller shoot size classes make up a proportionately small segment of the population; but by summer's end the overstory shoots have died and smaller shoots, the young segment of the population, become proportionately large. Note the advancement and fate of the age classes over time. (From Dickerman and Wetzel 1985: 542.)

TABLE 14.10 Yield Table for Douglas-fir on Fully Stocked Hectare

Age (Years)	Trees	Av. DBH (cm)	Basal Area (m²)
20	1427	14.5	9.4
30	875	29.9	14.3
40	600	31.0	18.1
50	440	38.8	20.8
60	345	46.2	23.1
70	283	53.0	24.9
80	243	59.1	26.5
90	210	65.0	27.8
100	188	70.1	29.0
110	172	74.7	30.7
120	158	79.0	30.9
130	148	83.0	31.7
140	138	87.1	32.6
150	128	90.9	33.2
160	120	94.5	33.8

*Derived from McArdle et al. 1949.

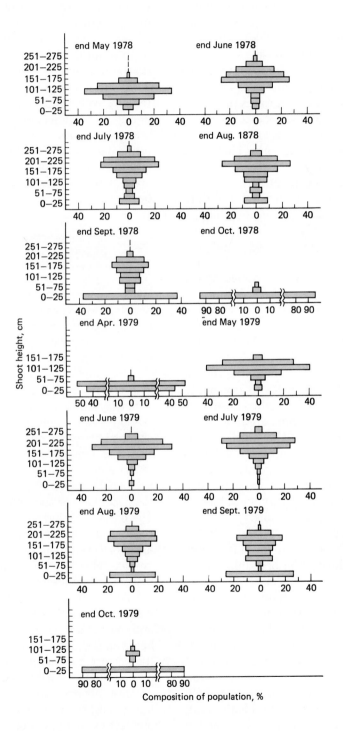

Composition of population, %

100:100; age 60 to 64, 88:100; age 80 to 84, 54:100. Lower mortality among females is characteristic of both advanced and underdeveloped countries.

Among birds the sex ratios tend to remain weighted toward the males (Bellrose 1961). Fall and winter sex ratios of prairie chickens show a preponderance of males in adult groups; similar ratios are characteristic of other gallinaceous birds and some passerine birds.

It is not easy to explain why sex ratios should shift from an equal ratio at birth to an unequal one later in life. Perhaps a partial answer lies in factors related to the genetic determination of sex and the physiology and behavior of the two sexes.

Sex in organisms is determined by the X and Y chromosomes. The XY combination of chromosomes produces males in mammals and females in birds and some insects, notably butterflies. Perhaps the chromosome combination itself may be partially responsible for biased losses of females in birds and males in mammals. Each gene on the X and Y chromosome is expressed in the XY combination, whereas in the XX combination the heterozygous combination of alleles can mask the harmful effects of single recessive genes. Therefore, XY adults may be more susceptible to disease, physiological stress, and aging than XX organisms.

Physiological and behavioral patterns affect mortality of the sexes differently. For example, during the breeding season the male elk battles other males for dominance of a harem, defends his harem from rivals, and mates with the females. These activities not only consume considerable energy, but leave little time for feeding, and the male often ends the breeding season in poor physical condition. Among birds the female may help the male defend territory, builds the nest, lays and incubates the eggs, broods the young, and, often with the help of the male, feeds the young. While incubating the eggs and brooding the young, the female is much more vulnerable to predation and other dangers than is the male. Thus adult males in mammals and adult females in birds apparently expend more energy, are subject to greater stress, and are more vulnerable to predation and other dangers. Their higher vulnerability and mortality is consistent with the imbalance in the sex ratio in the older age cohorts.

■ Summary

Living organisms exist in groups of the same species. These groups, or populations, occupy a particular space; the size of the population in relation to a definite unit of space is its density.

Populations are distributed in some kind of pattern over an area. Some are uniformly distributed, a very few are randomly distributed, and most exhibit a clumped or contagious distribution, which results in aggregations. Some aggregations reflect a degree of sociality on the part of the population members, which may lead to cooperative or competitive situations. The pattern of clumped distribution is called coarse-grained when the clumps and areas between them are large and fine-grained when the clumps and areas between them are small.

Population density depends upon the number of individuals added to the group and the number leaving—the difference between the birthrate and death rate and the balance between emigration and immigration. Birthrate usually has the greatest influence on the addition of new individuals. Mortality is a reducer and is the greatest in young and old. Mortality and survivorship are best analyzed by means of a life table, which is an age-specific summary of mortality operating on a population. From the life table we can derive both mortality curves and survivorship curves that are useful in comparing demographic trends within a population and between populations living under different environmental conditions. Reproductive rates of populations can be best analyzed by means of a life table that relates reproduction with age and survivorship. Age and sex ratios influence natality and mortality rates of populations. Reproduction is limited to certain age classes; mortality is most prominent in others. Changes in age class distribution bring about changes in production of young and mortality. The sex ratio tends to be balanced between males and females at birth, but the ratio changes to favor females in mammals and males in birds as cohorts age.

Population Growth and Regulation

Populations are dynamic. Depending upon the organism, populations may change from hour to hour, day to day, season to season, year to year. In some years certain organisms are abundant; in other years they are scarce. Local populations appear, expand, decline, and occasionally become extinct. Eventually the area may be recolonized by individuals moving in from other populations. Changes come about as the interaction of organisms and their environment affects birthrates, death rates, and the movement of individuals in and out of populations.

■ Rate of Increase

Possessing sufficient data to construct a life table and knowing the age-specific fecundity (discussed in Chapter 14), we can determine some characteristics of population growth. By multiplying age-specific survivorship, l_x, by age-specific fecundity, m_x, and summing all the $l_x m_x$ values for the entire lifetime, we obtain the net reproductive rate R_0, defined as

$$\sum_{x=0}^{\infty} l_x m_x$$

If R_0 equals 1, the birthrate equals the death rate; individuals are replacing themselves and the population is remaining stable. If the value is greater than 1, the population is increasing; if it is less than 1, it is decreasing.

With the survivorship schedule (l_x) and the fecundity schedule (m_x), and an additional parameter, p_x, the proportion of animals surviving in each age class, we can chart the growth of a population by constructing a *population projection table*. The parameter p_x is the complement of q_x (from the life table), expressed as $1 - q_x$.

The construction of a population projection table can be illustrated using a hypothetical squirrel population. The survivorship schedule, fecundity schedule, and p_x are given in Table 15.1. For ease in calculation the l_x column is expressed as a decimal fraction of 1 rather than in terms of 1000. Let the initial population (Table 15.2) be ten females aged one year, who

TABLE 15.1 Life Table for an Artificial Squirrel Population

x	l_x	q_x	p_x	e_x	m_x	$l_x m_x$	$x l_x m_x$
0	1.0	0.7	0.3	1.09	0	0	
1	0.3	0.5	0.5	1.47	2.0	0.60	0.60
2	0.15	0.4	0.6	1.43	3.0	0.45	0.90
3	0.09	0.55	0.45	1.05	3.0	0.27	0.81
4	0.04	0.75	0.25	0.75	2.0	0.08	0.32
5	0.01	1.0	0.00	0.5	0.00	0.00	0.00
Σ						1.40	2.63

will produce a litter the year of introduction, year 0. During year 0 the initial population of 10 one-year-old females gives birth to 20 females, which are added to age class 0 of that year. The survival of these two age groups is obtained by multiplying the number of each by the p_x value. Because the p_x of the females in age 1 is 0.5, 5 individuals ($10 \times 0.5 = 5$) survive to year 1 (age 2). The p_x value of age 0 is 0.3, so only 6 of the 20 in this age class in year 0 survive ($20 \times 0.3 = 6$) to year 1 (age 1). Survivorship is tabulated year by year diagonally down the table to the right through the various age classes. In year 1 the 6 one-year-olds and the 5 two-year-olds together contribute 27 young to age class 0. The m_x value of the 6 one-year-olds is 2.0, so they produce 12 offspring. The 5 two-year-olds (m_x value of 3.0) produce 15 offspring. The steps for determin-

TABLE 15.2 Population Prediction Table for an Artificial Squirrel Population

	Year										
Age	0	1	2	3	4	5	6	7	8	9	10
0	20	27	34.1	40.71	48.21	58.37	70.31	84.8	101.86	122.88	148.06
1	10	6	8.1	10.23	12.05	14.46	17.51	21.0	25.44	30.56	36.86
2	0	5	3.0	4.05	5.1	6.03	7.23	8.7	10.50	12.72	15.28
3	0	0	3.0	1.8	2.43	3.06	3.62	4.4	5.22	6.30	7.63
4	0	0	0	1.35	0.81	1.09	1.38	1.6	1.94	2.35	2.83
5	0	0	0	0	0.33	0.20	0.27	0.35	0.40	0.49	0.59
Total	30	38	48.2	58.14	68.93	83.21	100.32	120.85	145.36	175.30	211.25
Lambda	λ	1.27	1.27	1.21	1.19	1.21	1.20	1.20	1.20	1.20	1.20

ing the number of offspring in year t, N_{t0} are given by the equation

$$N_{t0} = \sum_{x=1}^{\infty} Nt_x m_x$$

where N_{t0} = the number in age class 0 at the given year t and N_{tx} = the number of age x in year t. For year 0 the calculation is

$$N_0 = (10)(2) = 20$$

and for succeeding years

$$N_1 = (6)(2) + (5)(3) = 27$$

$$N_2 = (8.1)(2) + (3)(3) + (3)(3) = 34.1$$

and so on. The number of offspring is obtained by multiplying the number in each age group by the m_x value for that age and summing these values over all ages.

From such a population projection table we can calculate *age distribution,* the relative proportion of individuals in the various age classes, for any one year by dividing the number in each age group by the total population size for that year. The general equation is

$$C_{tx} = \frac{N_{tx}}{\sum_{y=0}^{\infty} N_{ty}}$$

where c_{tx} = the proportion of age group x in year t, N_{tx} = the number in each age

group x at year t, and N_{ty} = the number in age group y at year t.

Comparing the age distribution of the artificial squirrel population in year 3 with that of the population in year 7, we observe that the population attained a stable or unchanging age distribution by year 7 (Table 15.3). From that year on, the proportions of each age group in the population and the rate of growth remain the same year after year, even though the population is steadily increasing. A special case of a stable age distribution is a *stationary age distribution* (Chapter 14), in which both the population size and age structure are unchanging.

The construction of the population projection table demonstrates some concepts of population growth. The constant rate of increase of the population from year to year and stable age distribution depend upon survivorship, l_x, from each age class to the next and upon fecundities of each age class, m_x. Both factors were used in the development of the population projection table. By dividing the total number of individuals in year $x + 1$ by total number of individuals in previous year x, we arrive at the finite multiplication rate, λ (lambda), for each time period.

The method of determining λ from changes in the size of a natural population from one year to another is useful. Obtaining reasonably accurate counts of a population from one year to another, we can divide the current population size by

---

TABLE 15.3 Approximation of Stable Age Distribution, Artificial Squirrel Population

	Proportion in Each Age Class Time										
AGE	0	1	2	3	4	5	6	7	8	9	10
0	.67	.71	.71	.71	.69	.70	.70	.70	.70	.70	.70
1	.33	.16	.17	.17	.20	.17	.18	.18	.18	.18	.18
2		.13	.06	.07	.06	.07	.07	.07	.07	.07	.07
3			.06	.03	.03	.04	.04	.03	.03	.03	.03
4				.02	.01	.01	.01	.01	.01	.01	.01
5					.01	.01	.01	.01	.01	.01	.01

the population size of the year previous and gain some rough idea of the current rate of population increase.

Lambda can be used as a multiplier to predict population size some time in the future:

$$N_t = N_0 \lambda^T$$

For our imaginary squirrel population, we can multiply the population size, 30, at time 0 by 1.20 to obtain a population size of 36 for year 1. If we multiply 36 again by 1.20, or the initial population size 30 by λ^2 (1.20^2), we get a population size of 43 for year 2; and if we multiply the population at N_0, 30, by λ^{10} we arrive at a projected population size of 186. These population sizes do not correspond to the population sizes early in the population projection table, because in the lower years λ is higher and the population has not reached a stable age distribution. Only after the population achieves a stable age distribution does the λ value of 1.20 predict future population size. For example, if population size N in year 7 is used as N_0, then 142×1.20^3 predicts a population size of 245 in the year 10. This fact emphasizes that population projections using λ assume a stable age distribution.

Generations in the squirrel population are not discrete, as the population projection table makes evident; they overlap. The female squirrels in the initial population at time 0 continue to contribute to population growth, although at a reduced rate, at the same time their offspring are adding their own reproductive efforts. In Table 15.1, the net reproductive rate, R_0, assumes a discrete generation time, T. If generations are discrete, as they are in annual plants and many insects, then the unit of time, t, and generation time, T, are the same. In populations with overlapping generations the value of T becomes T_c, mean cohort generation time, which is the

mean period of time elapsing between the birth of the parents and the birth of the offspring. Mean cohort generation time is computed from the fecundity schedule by multiplying each $l_x m_x$ by its appropriate age x; all values are summed and divided by the sum of the $l_x m_x$ or R_0 to obtain T_c:

$$T_c = x l_x m_x / R_0$$

The net reproductive rate R_0, provided it is near 1, can be used to estimate the intrinsic rate of natural increase, r, a measure of the instantaneous rate of change of a population size per individual, by the formula

$$r \approx \ln R_0 / T_c$$

For the squirrel population described in Table 15.1 the cohort generation time is $\Sigma x l_x m_x / R_0 = 2.63/1.40 = 1.87$. The intrinsic rate of increase can be approximated by

$$\ln 1.40/1.87 = .336/1.87 = .1797$$

The intrinsic rate of increase can be converted to the finite rate of increase by the formula $\lambda = e^r$. Determining e^r with an r value of 0.1797 or 0.180, we obtain $\lambda = 1.198$ or 1.20. This value agrees with the value for λ obtained from the population projection after stable age distribution was achieved.

The intrinsic rate of increase can be obtained more precisely from the Euler equation:

$$\sum_{\lambda}^{\infty} e^{-rx} l_x m_x = 1$$

which can be solved only by iteration (see Appendix B). The iterated value of r for the squirrel population is .186, which gives an λ value of 1.203.

Because r is derived from R, which involves schedules of survivorship, l_x, and fecundity, m_x, r includes both births and deaths. In a closed population, one in which no individual enters or leaves, the intrinsic rate of increase is the instantaneous birthrate minus the instantaneous death rate or $r = b - d$. In an open population r also must include immigration (i) and emigration (e), or $r = (b + i) - (d + e)$.

Deriving any sort of value for r for natural populations can be difficult, yet in the management of wild populations, whether game, endangered, or pest species, some estimate of r is needed.

A known value of r for a population allows the calculation of doubling time of population growth. The doubling time of a population is when $N_t/N_0 = 2$. Therefore $N_0 e^{rt}/N_t = 2$. Thus $e^{rt} = 2$ and $rt = ln\ 2 = .693$; then $t = .693/r$. The doubling time of the artificial squirrel population is $.693/.186 = 3.72$ years.

However useful r may be, the parameter has certain limitations. It refers to a particular time interval; it is free from the effects of the population's own density and its own age structure. It is influenced by life history events, such as frequency of reproduction, and by local environmental conditions; and it expresses the rate of increase under particular conditions before population density begins to affect birthrates and death rates. In effect, values of r relate to specific populations under specific conditions.

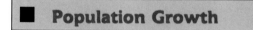

Population Growth

Exponential Growth

If a population were suddenly presented with an unlimited environment, as can happen when a small number of animals is introduced into a suitable but unoccupied habitat, it would tend to expand geometrically. Assuming there was no movement in or out of the population and no mortality, then birthrate alone would account for population changes. Under this condition population growth would simulate compound interest. However, growth of populations is tempered by death, so a death rate must be factored in with the birthrate. This factoring has already been done in the determination of the rate of increase, r. Exponential growth of a population can be expressed in a general growth equation:

$$dn/dt = rN$$
$$= (b_0 - d_0)N$$

where b_0 = individual birthrate, measured when the population is very small, and d_0 = individual death rate, measured at an early stage of population growth.

This differential equation, $dN/dt = rN$, can be solved to obtain a more useful equation for calculating exponential growth:

$$N_t = N_0 e^{rt}$$

where e = base of natural logarithms $2.71828 . . .$, r = rate of increase, and t = unit of time.

Given the starting population of 30 from the life history table, the exponential growth of the hypothetical squirrel population with an r of 0.186 would go as follows:

Year	Population Growth	
0		30
1	$30\ (e^{.186}) =$	36
2	$30\ (e^{.372}) =$	43
3	$30\ (e^{.555}) =$	52
4	$30(e^{.774c}) =$	65
5	$30\ (e^{.930}) =$	76
10	$30\ (e^{1.86}) =$	192

This growth is exponential for that value of r and initial population size. The exponential growth of any population depends upon and varies with the value of r.

The rate of growth at first is influenced by heredity or life history features, such as the age at beginning of reproduction, the number of young produced, survival of young, and length of the reproductive period. Regardless of the initial age of the colonizers, the number of animals in the prereproductive age class would increase because of births, whereas those in the older age categories for a time would remain the same. As the young mature, more would enter the reproductive stage and more young would be produced. If the number of animals is plotted against time, the points will fall into an exponential growth curve (Figure 15.1) defined by the foregoing formula. If the logarithms of the numbers of animals are plotted against time, the points will fall into a straight line.

A population can increase exponentially until it overshoots the ability of the environment to support it. Then the population declines sharply or "crashes" through disease, starvation, or emigration. From the low point the population may recover to undergo another phase of exponential growth, it may decline to extinction, or it may recover and fluctuate about some lower level. A J-shaped curve is characteristic of many organisms introduced into a new and unfilled environment.

An example of an exponential growth curve is the rise and fall of a reindeer herd on St. Paul, one of the Pribilof Islands, Alaska (Figure 15.2). Introduced on St. Paul in 1910, reindeer expanded rapidly from 4 males and 22 females to a herd of 2000 in only 30 years. Then the

Figure 15.1 Exponential growth curve plotted arithmetically and logarithmically. This growth curve is for the hypothetical squirrel population with an r of .131. The initial slope of the exponential curve is influenced by the value of r. A low value gives a flatter curve in the early stages of growth; a high r gives a much steeper curve.

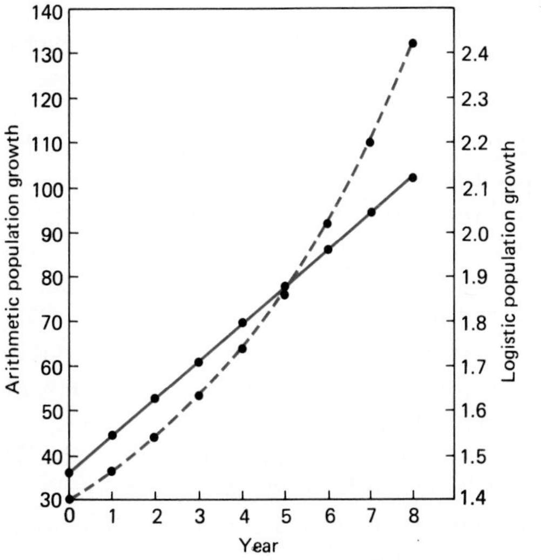

Figure 15.2 Exponential growth of the St. Paul reindeer herd and its subsequent decline. The early population growth was relatively slow, but within 20 years the population grew rapidly, until the herd exceeded the carrying capacity of the range and declined more rapidly than it grew. (From V. C. Scheffer 1951.)

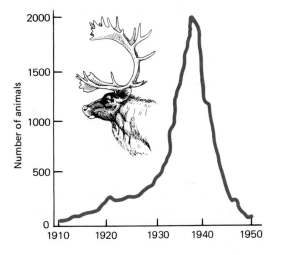

population decreased dramatically. The decline produced a curve typical of a population exceeding the ability of the resource base to support it. Growth stops abruptly and the population declines sharply in the face of environmental deterioration. So severely did the reindeer overgraze their range that the herd plummeted to 8 animals in 1950.

Logistic Growth

For populations in the real world, the environment is not constant, and resources are not unlimited. As population density increases, competition for available resources among its members also increases. Eventually the detrimental effects of increased density—increased mortality, decreased fecundity, or both—slow population growth until it ceases. This level, called *carrying capacity*, is expressed as *K*. At this level the population theoretically is in equilibrium with its resources or environment.

Inhibition of the growth of populations can be described mathematically by adding to the exponential equation

$$N_t = rN$$

a variable to describe the effects of density:

$$\frac{(K - N)}{K}$$

Then

$$\frac{dN}{dt} = rN\left(\frac{K - N}{K}\right)$$

where dN/dt = instantaneous rate of change, K = carrying capacity, and $(K - N)/K$ = unutilized opportunity for population growth. The equation says that the rate

of increase of a population over a unit of time is equal to the potential increase of the population times the unutilized portion of the resources.

The logistic growth rate of the hypothetical squirrel population serves as an example. The population will have a starting size of 30 with a rate of increase of 0.186, inhabiting an imaginary woodlot with a carrying capacity of 50.

Year	Growth	Size
0		30
1	.186(30)(0.4) = 2.2	32
2	.186(32)(0.36) = 2.14	34
3	.186(34)(0.28) = 1.87	36
6	.186(36)(0.22) = 1.59	38
10	.186(44)(0.12) = 0.98	45
15	.186(47)(0.06) = 0.52	47.5
20	.186(49)(0.02) = 0.18	49.18

Note how the rate of increase is slow at first, accelerates, and then slows. The point in the logistic growth curve where population growth is maximal is $K/2$, known as the *inflection point*. From this point on population growth slows (Figure 15.3). As population density approaches carrying capacity, N approaches K and the rate declines.

The logistic equation involves several assumptions. Age distribution is stable, initially at least. No immigration or emigration takes place. Increasing density depresses the rate of growth instantaneously without any time lags. The relationship between density and the rate of growth is linear. We also have to assume a predetermined level for K.

We can view the logistic equation in two ways. One is to consider it as a law of population growth. That was the original proposition elaborated by Pearl (1927) (see Kingsland 1985). A second, preferable view is to regard the logistic equation as a mathematical description of how populations might grow under favorable conditions. Although natural populations appear to grow logistically, rarely, if ever, do they do so.

For example, consider the growth of human population in Monroe County, West Virginia, whose economic base has always been agriculture and small industry. It was settled by Europeans in the early 1700s, and the population was well established in 1800, the year for which U.S. Census data are first available. The population reached 13,200 in 1900 and has fluctuated about that number since then. The population grew most rapidly from 1800 to 1850, so growth during that period provided the data to estimate r, the rate of increase, as 0.074. K, based on the

population levels since 1900, was set at 13,200. With these values of r and K the logistic equation predicted that the population would reach the asymptote around 1870, 30 years before it actually did. The theoretical growth curve rises much more steeply than the actual growth curve (Figure 15.4); in effect, the human population growth did not conform to the logistic. The value for Chi-square goodness-of-fit (830, 9 df) well exceeds the table value (16.9, 9 df 95 percent), so we reject the hypothesis that actual population growth of humans in Monroe County conformed to the logistic. The reasons for nonconformity are obvious. The age structure was not stable, birthrates and deaths varied from census period to census period; immigration and emigration were common to the population. The most surprising feature of the population is its relative stability after reaching K.

The logistic equation suggests that populations function as systems, regulated by positive and negative feedback (Chapter 3). Growth results from positive feedback (illustrated by the exponential growth curve), ultimately slowed by the negative feedback of competition and resource availability. As the population reaches the upper limit of the environment, it theoretically responds instantaneously as density-dependent reactions set in. Rarely does such feedback work

Figure 15.3 (a) The logistic growth curve is sigmoidal or S-shaped. Its upper limit is termed the carrying capacity, K. The inflection point, at which density begins to slow population growth, is located at $K/2$. When populations exceed K, they decline exponentially until they reach K. (b) Three kinds of range of fluctuations can take place when the population overshoots K. One is chaotic; the population fluctuates wildly with no regulation. Such a situation can lead to sudden extinction. Another is stable limit cycles, in which the population fluctuates about some equilibrium level. Oscillations also may be damped. After overshooting or rising to K, the population levels off and maintains itself at K through compensating birth and death rates.

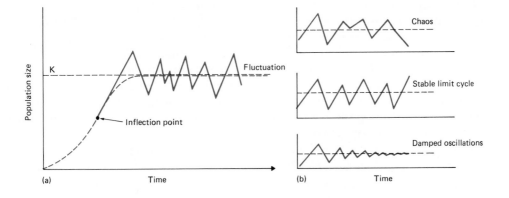

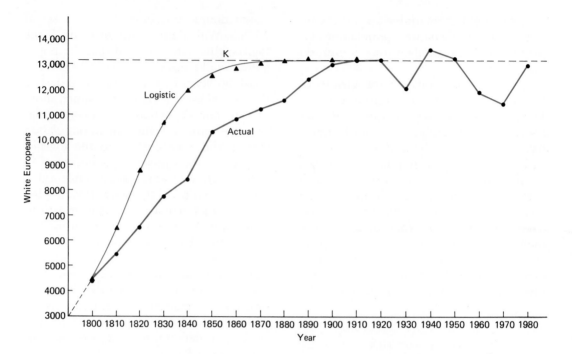

Figure 15.4 Actual and logistically predicted population growth of white European population in Monroe County, West Virginia. (Data from U.S. Census Bureau.)

$$\frac{dN}{dt} = rN\left(\frac{K - N_{t-w}}{K}\right)$$

Another lag is a *reproductive time lag* (*g*), influenced by the length of gestation or its equivalent:

$$\frac{dN}{dt} = rN_{t-g}\left(\frac{K - N_{t-w}}{K}\right)$$

Means of incorporating time lags into the logistic are detailed by Krebs (1985) and Berryman (1981).

Time lags result in fluctuations in populations (Figure 15.3). The population may fluctuate widely without any reference to equilibrium size. Such populations may be influenced by some powerful outside force such as weather or by some chaotic changes inherent in the population. (For an introduction to chaos in populations, in which simple process can lead to complex behaviors, see May 1976; Schaf-

as smoothly in real life. Often adjustments lag and available resources may be sufficient to allow the population to overshoot equilibrium. Unable to sustain itself on the available resources, the population declines to some point below carrying capacity, but not before it has altered resource availability for future generations. Its recovery as determined by reproductive rates is influenced by the density of the previous generation and the recovery of the resources, especially food supply. These factors build a *time lag* into population recovery, which can be incorporated into the logistic equation as a *reaction time lag* (*W*), a lag between environmental change and a corresponding change in the rate of population growth:

fer 1985; Schaffer and Kot 1985; Yodzis 1989). A population may fluctuate instead about the equilibrium level, *K,* rising and falling between some upper and lower limits. Such fluctuations are called *stable limit cycles.*

■ Extinction

When deaths exceed births and emigration exceeds immigration, then populations decline. R_0 becomes less than 1; r becomes negative. Unless the population can reverse the trend, it at worst faces extinction or at best increases its probability of becoming extinct.

There are several causes for the decline of sparse populations. When only a few individuals are present, females of reproductive age may have only a small chance of meeting a male in the same reproductive condition. Many females remain unfertilized, reducing average fecundity. A small population faces the prospect of an increased death rate from predation and sudden environmental changes, because there are fewer individuals to survive. Losses feed upon losses in positive feedback until the population disappears.

Extinction is a natural process. Through millions of years of Earth's history, species have appeared and disappeared, leaving a record of their existence in fossil bones and footprints in sedimentary rock. Some species could not adapt to geological and climatic changes. Others diverged into new species while the parent stock disappeared. Still others could not withstand the predatory pressures of the relatively new species *Homo sapiens,* who appeared to have the first great impact in the Pleistocene. Massive extinctions oc-

curred at several points in Earth's history: the late Ordovician; the late Devonian; the Late Permian, which witnessed the extinction of up to 96 percent of its species; and the Cretaceous-Tertiary, which saw the end of the dinosaurs.

Mass extinction is happening today at an accelerated pace. The greatest number of extinctions has taken place since A.D. 1600. Well over 75 percent of modern-day extinctions have been caused by humans through the alteration and destruction of habitat, introduced predators and parasites, predator and pest control, competition for resources, and hunting of various types.

Despite the popular impression, extinction does not take place simultaneously over the full range of a species. It begins with isolated local extinctions, when environmental conditions deteriorate or the population is unable to replace itself. Local extinctions often begin when habitats are destroyed and the dispossessed find remaining habitats filled. Restricted to marginal habitats, the individuals may persist for a while as nonreproducing members of a population or succumb to predation and starvation. As the habitat becomes more and more fragmented, the species is broken down into small isolated or "island" populations, out of contact with other populations of its species. As a result the population is subject to inbreeding and genetic drift, reducing the ability of the small population to withstand environmental changes (see Chapter 13).

The maintenance of local populations often depends heavily on the immigration of new individuals. As the distance between local populations or islands increases and as the size of the local population declines, their continued existence becomes more precarious. As the number falls below some minimum level (see Chapter 13), the local population may become extinct simply through random fluctuations.

Although we equate such situations mostly with rarer species, even common species experience local extinctions. They often go unnoticed because the loss is masked by the influx of immigrants from surrounding areas. One study of suburban population of robins showed that because of the losses of nests and young through predation by cats and interference by humans, the robins were not replacing themselves (Howard 1974). Robins sang each spring only because new birds moved into the area. Thus suburbia bacame a population sink rather than a population source.

In fact, most local populations do not thrive for long. They are revived fast enough by new immigrants that replace the losses and keep the population going. As one local population slides down the slope toward extinction, a local population somewhere else is experiencing overcrowding and supplies new recruits for depleted habitats.

Extinctions are of two sorts: deterministic and stochastic. Deterministic extinction comes about through some force or change from which there is no escape. The Cretaceous-Tertiary extinctions are an example (see Stanley 1989). So is destruction of habitat on a local or regional scale. Habitat destruction alone causes the extinction of a species only when it is a local endemic or it is already on the verge of extinction, such as the dusky seaside sparrow (*Ammordramus maritimus subsp*) and potentially the spotted owl (*Stix occidentalis*).

Usually what local deterministic extinction brings about is the second type, stochastic. Stochastic extinctions come about from normal random changes within the population or environment. Such changes normally do not destroy a population, but rather thin it out; but this smaller population faces an increased risk of extinction from some decimating event.

Stochastic events may be demographic or environmental. Demographic stochasticity is chance variation in individual births and deaths. The road to demographic stochasticity is habitat deterioration or loss, reducing population size. In a small population high death rate or low birthrate can lead to a random or accidental extinction. When a population falls below minimum viable size (see Chapter 13) it faces great risk of going extinct. Environmental stochasticity is a random series of adverse environmental changes, mostly through deterioration in environmental quality. If all members of a local population are affected equally by an adverse environmental change, the population may be reduced to a level at which demographic stochasticity takes over. It can be accompanied by a reduced effective population size, N_e and an increase in genetic drift and inbreeding (Chapter 13). How long a population can exist at a lower level depends upon the size of individuals, longevity, and mode of reproduction (see Chapter 17).

A classic example of a recent extinction is that of the heath hen (*Tympanuchus cupido cupido*). Formerly abundant in New England the heath hen, an eastern form of the prairie chicken, was driven by excessive hunting and habitat destruction to the island of Martha's Vineyard off the Massachusetts coast and to the pine barrens of New Jersey. By 1880 it was restricted to Martha's Vineyard. At this point the population was subjected to deterministic extinction over most of its range. The small population, confined to a small island, was highly vulnerable to stochastic events. At first the population prospered, growing from a population of 200 birds in 1890 to over 2000 in 1920. Then a major environmental stochastic event, a combination of fire, winter gales, and cold weather, reduced the population to 50. The heath hen never recovered, and the last bird died in 1932.

■ Density-Dependent Regulation

No population grows indefinitely. Sooner or later it has to come into equilibrium with its environment and resources. Equilibrium comes about through changes in the birthrate, death rate, or some combination of the two (Figure 15.5). As a population increases, competition among its members and scarcity of resources result in increased mortality, decreased natality, or both. If a population drops to some lower level and resources become more abundant, it increases through some combination of decreased mortality and increased natality.

There is positive feedback of population growth when conditions are favorable and negative feedback of population decline when conditions are unfavorable.

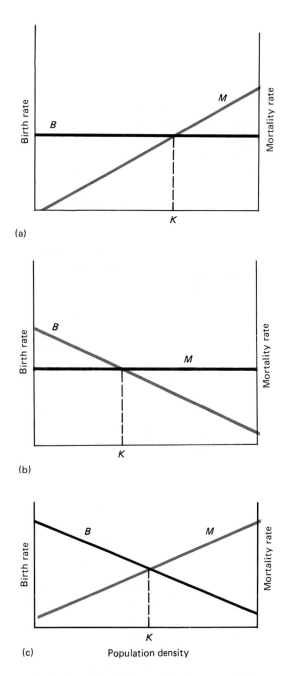

(a)

(b)

(c) Population density

Figure 15.5 For population regulation, both birthrate and death rate must be density-dependent. In (a) the birth rate is independent of population density, as indicated by the horizontal line. It remains unchanged as the population increases, but the death rate increases. As long as the birthrate exceeds the death rate, the population increases toward K. At K the population reaches equilibrium, maintained by increasing mortality. In (b) the situation is reversed. The mortality rate is independent of population density, but the birthrate declines as density increases until the population reaches K. At that point equilibrium is maintained by a decreased birthrate. In (c) both birthrates and death rates are density-dependent, and the population reaches equilibrium when the birth rate exceeds the death rate. Fluctuations in each will tend to hold the population at or near equilibrium point and influence population density. If the birthrate increases, then mortality increases.

Throughout population growth, both types of feedback operate, with a change in their relative importance. In the early stages of population growth, positive feedback dominates. In the later stages, negative feedback brakes positive feedback when the population approaches carrying capacity. How rapidly these two responses function relates to the population's impact on resources and future growth. If individuals remove a resource faster than it is replaced, then the present population impoverishes the environment for the next generation, slowing population growth. Through such positive and negative feedbacks a population arrives at some level of balance with its environment.

Implicit in such a concept of population regulation is density dependence. Density-dependent effects influence a population in proportion to its size. At some low density there is no interaction and mortality is independent of population size. As population density increases, density-dependent effects become apparent as mortality increases, but the number dying is less than the number surviving. Density-dependent effects *undercompensate* for increased density, so the population increases. Eventually there comes a point when mortality exceeds survival. Then density-dependent mechanisms *overcompensate* for the increased density and the population declines (Bellows 1981). Density-dependent mechanisms affecting natality and mortality act largely through environmental shortages and competitive interactions among members of the population to obtain those resources.

Intraspecific Competition

Population regulation involves intraspecific competition. *Competition* results only when a needed resource is in short supply relative to the number seeking it. As long as resources are abundant enough to al-

low each individual a sufficient amount for survival and reproduction, no competition exists. When resources are insufficient to satisfy adequately the needs of all individuals, the means by which they are allocated has a marked influence on the welfare of the population.

In a long-term experiment involving sheep blowflies (*Lucilia cuprina*) Nicholson (1954) demonstrated the influence of intraspecific competition in a population. Although the experimental population lacked all the complex interactions we would expect to find in nature, the work does show what might happen.

In one experiment Nicholson fed to a culture of blowflies containing both adults and larvae a daily quantity of beef liver for the larvae and an ample supply of dry sugar and water for the adults. The number of adults in the cages varied, with pronounced oscillations (Figure 15.6). When

Figure 15.6 Oscillations in the numbers of adult sheep blowflies (*Lucilia cuprina*) and in the daily rate of egg production in a laboratory population. Larvae received unlimited food; adults received a limited supply (0.5 g of liver daily). Peaks in the adult cycle alternate with peaks in the egg cycle. (After Nicholson 1954.)

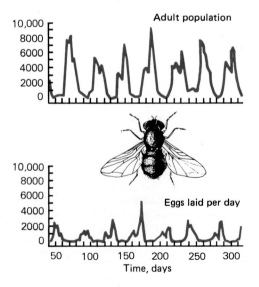

the population of adults was high, the flies laid such a great number of eggs that resulting larvae consumed all the food before they were large enough to pupate. As a result no adult offspring came from the eggs laid during that period. Through natural mortality the number of adults progressively declined, and few eggs were laid. Eventually a point was reached where the intensity of larval competition was so reduced that some of the larvae secured sufficient food to grow to a size large enough to pupate. These larvae in turn gave rise to egg-laying adults. Because of the developmental time lag between the survival of larvae and an increase in egg-laying adults, the population continued to decline, further reducing the intensity of larval competition and permitting an increasing number of larvae to survive. Eventually the adult population again rose to a very high level and the whole process started again.

Competition for limited food held this blowfly population in a stage of stability and prevented any continuing increase or decrease. The time lag involved in the addition of egg-laying adults to the declining population resulted in an alternate over- and undershooting of the equilibrium position, causing an oscillating population density.

In a second experiment Nicholson supplied the adults with a surplus of suitable food, which was unavailable to the larvae. As a result of the enormous quantity of eggs laid by the adults, larval competition intensified and eventually the density of adults decreased in a manner comparable to the other experiment.

In another variation the larvae were supplied with a surplus of food, and the adults were given a constant daily quota of protein food (Figure 15.7). Again the adult population oscillated. The adults produced a high number of eggs that, because of the lack of larval competition, nearly all developed into adults. The adults competed intensely for a limited amount of food. Lacking sufficient protein for the production of eggs, the adults laid fewer eggs; and for the lack of replacements the adult population declined. Competition was gradually relaxed to a point where some of the flies obtained enough protein to produce eggs. After a two-week lag the adult population began to build up again.

From these results Nicholson hypothesized that the magnitude of the oscillations would be reduced if the larvae and the adults each competed for a limited quantity of food not available to the other. This assumption was confirmed experimentally. Under the conditions described not only were the fluctuations slight, but they lost their periodicity, and the mean population level was nearly quadrupled.

In these competitive situations the larvae and the adults were seeking food, the rate of supply of which was not influenced by the activity of the flies. In effect the resource, or food available, could be subdivided into many small parts to which the competitors, the larvae and the adult flies, had general access. The individuals "scrambled" for their food, which under gross crowding resulted in wastage because each competitor got such a small fraction of the food that it was unable to survive.

Nicholson called this type of competition *scramble*. In its purest form, all competing individuals garner such a small share of the resources that none survive. Among some populations outcomes are less severe; competition is scramble-like rather than pure scramble. Such competition, in which each individual is affected by the amount of shared resource remaining, might also be called *exploitative*. Competing individuals do not necessarily react to each other, only to the level of resources. Scramble competition tends to produce sharp fluctuations in the popula-

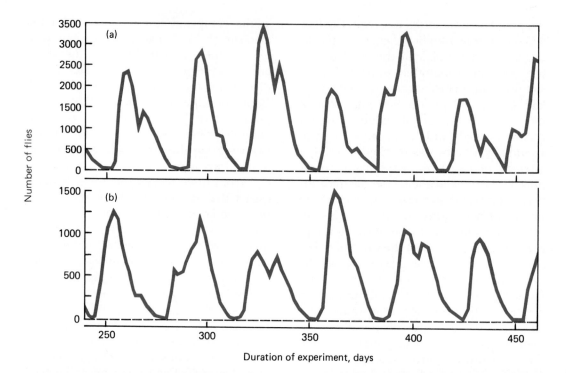

Duration of experiment, days

Figure 15.7 Fluctuations in the number of adult blowflies in two cultures subjected to the same constant conditions, but restricted to different daily quotas of food: (a) 50 gm; (b) 25 gm. Although a greater food supply (a) permitted a greater increase in density, the final outcome in both situations was similar. The adults, experiencing scramble competition, increased rapidly, then declined sharply as they reached the limits that the food could support. (From Nicholson 1957.)

tion over time, and it limits the average density of the population below that which the resources could support if an adequate amount of resources were supplied to only part of the population.

That is exactly what takes place with *contest competition*. The deleterious effects of limited resources are confined to a fraction of the population, and members of the population interact directly. For that reason contest competition can also be called *interference competition*. Once a pop-

ulation characterized by contest competition passes the point where resources become limiting, a fraction of the individuals obtain all the resources they need. The remaining individuals get less and produce no offspring or die.

Intraspecific competition influences births, deaths, and growth of individuals in a density-dependent manner. Its effects come slowly, involving at first the quality of life rather than survival of individuals. Later, as its impacts become accentuated, intraspecific competition affects individual fitness.

Growth and Fecundity

As population density increases to a point where resources are insufficient to meet their needs, individuals in populations characterized by scramble-like competition reduce their intake of food. That slows the rate of growth and inhibits reproduction.

Examples of this inverse relationship between density and rate of body growth may be found among populations of poikilothermic vertebrates. Dash and Hota (1980) discovered that frog larvae reared experimentally at high densities failed to grow normally (Figure 15.8). The tadpoles experienced slower growth, required a longer time to reach the size at which transformation from the tadpole stage takes place, and had a lower probability of reaching metamorphosis. Those that did reach threshold size were smaller than those living in less dense populations.

These relationships among density, growth, and fecundity extend to other vertebrate groups. Harp seals *(Phoca groenlandica)* become sexually mature when they reach 87 percent of their mature body weight. At low population densities young animals attain this weight at a much faster rate than when population densities are high (Figure 15.9). Fertility in harp seals, too, is density-dependent (Figure 15.10a) (Lett et al. 1981). Likewise, recruitment of young into the grizzly bear *(Ursus arctos)* declines as the number

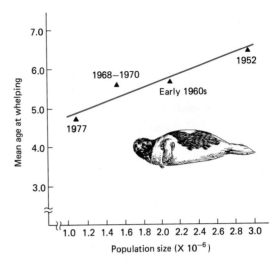

Figure 15.9 The mean age of maturity of harp seals (and other marine and terrestrial mammals) is related not so much to age as to weight. Seals arrive at sexual maturity when they reach 87 percent of adult body weight. Seals attain this weight at an earlier age when population density is low. (Note that the harp seal population in the Northwest Atlantic from which this data comes declined from 1952 to 1977.) (From Lett et al. 1981: 144.)

Figure 15.8 Influence of density on the growth rate of the tadpole *Rana tigrina*. Note how rapidly the growth rate declines as density increases from 5 to 160 individuals confined in the same space. (After Dash and Hota 1980: 1027.)

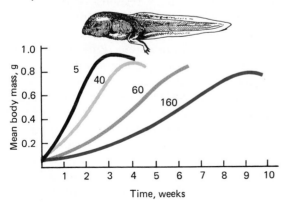

of adult bears increases (Figure 15.10b) (McCullough 1981). In fact, 71 percent of the variation in recruitment in the grizzly bear population in Yellowstone National Park can be explained by the number of adults in the population. In the red deer on the Isle of Rhum, calf mortality in winter, fecundity, and calf/hind ratios declined as population increased, and conception dates became later. Among the stags antler growth declined as density increased, and antler cleaning and casting dates became later (Clutton-Brock et al. 1982). Some birds also exhibit a similar density-dependent relationship to fecundity. Roseberry and Klimstra (1984) found that the higher the breeding population relative to carrying capacity on their

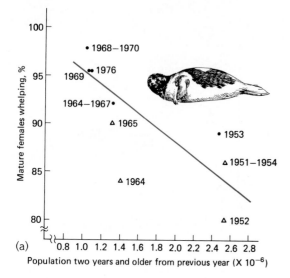

(a)

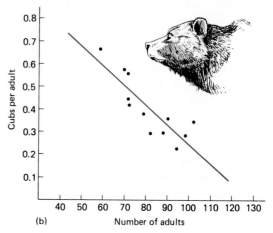

(b)

Figure 15.10 (a) Fertility is density-dependent in harp seals. As the population of seals (measured by including only animals two years and older from the previous year) increases, the percentage of females giving birth to young decreases markedly. (From Lett et al. 1982: 146.) (b) The grizzly bear shows a similar decrease in number of young produced per adult female as the population increases. (From McCullough 1982: 177.)

southern Illinois study area, the lower was the rate of summer gain (Figure 15.11).

These examples suggest that vertebrate populations do respond to increasing numbers in a density-dependent fashion through intraspecific competition. The timing of the response depends upon the nature of the population itself. Fowler (1981) has hypothesized that among large mammals with a long life span and relatively low reproduction, regulating mechanisms do not function until the population approaches carrying capacity. In other words, mortality and natality are compensatory until the population is close to K, at which point density-dependent mechanisms set in and tend to overcompensate. The birthrate of bison as related

to population density shows such a response (Figure 15.12), but the birthrates or juvenile survival of elk (Figure 15.13a), red deer (Figure 15.13b), and harp seal (Figure 15.13c) appear to be linear. Organisms characterized by high reproduction, short life spans, and highly variable populations show a curvilinear response to density at much lower population levels.

Many plants respond to density in a scramble-like fashion through reduction in individual growth. Individuals adjust their growth form, size, shape, number of leaves, flowers, and the production of seeds in a scramble fashion to the limited resources available, a response known as *phenotypic plasticity*. In some plants reproduction by seed at high densities is extremely low or nonexistent (Putwain et al. 1968); instead the plant reproduces by vegetative means. As the density of plants increases, the number of vegetative offspring also declines. For example, at high densities the genets of perennial rye grass (*Lolium perenne*) produce few tillers, and their weight is low. At low densities the genets produce many tillers, and their weight is high (Kays and Harper 1974).

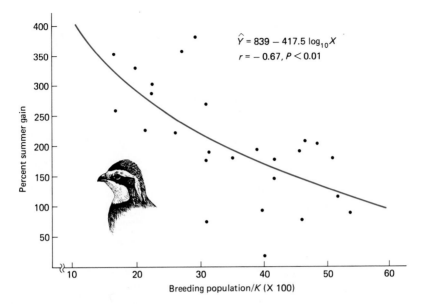

$$\hat{Y} = 839 - 417.5 \log_{10} X$$
$$r = -0.67, P < 0.01$$

Figure 15.11 The bobwhite quail shows a curvilinear density-dependent response to natality. Natality declines at a slower rate in early stages of population growth and continues the decline much more rapidly as density increases. The breeding population of bobwhite quail on the Southern Illinois Carbondale Research area, 1954–1979, was adjusted to account for changes in land use and therefore carrying capacity of the area. (From Roseberry and Klimstra 1984: 96.)

Plant Biomass

If you walk into pine woods of the same age but at two contrasting densities—one thinned, the other unthinned—you will be aware of some obvious differences (Figure 15.14). The trees of the artifically thinned stand are fewer, larger in diameter, and taller. Those in the unthinned stand are thicker on the ground, smaller in diameter, and shorter in height. If you were to cut and weigh uniform samples of the stands, you would find that the total biomass on both plots was similar, but mean weight of the plant was greatest on the thinned plots with the lower density. These observations suggest a close rela-

tionship between plant density and the growth of individual plants as measured by biomass accumulation.

Assume a group of seedlings occupying the same habitat but starting out at

Figure 15.12 An example of nonlinear density-dependent change in a large mammal population, the American bison. The birthrate of the bison, expressed as young per female, is independent of density until the population reaches a certain size; then birthrates overcompensate for increased density. (From Fowler 1981: 607, from data of J. E. Gross et al.)

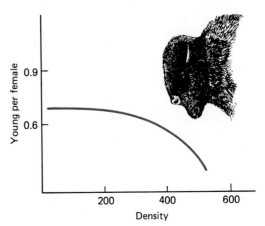

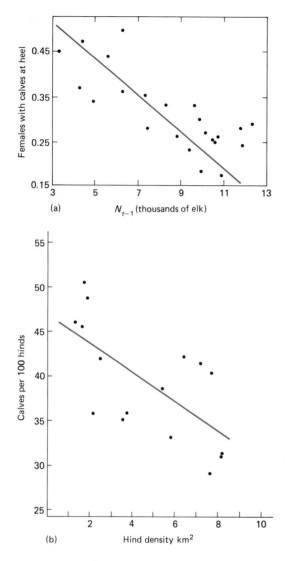

(a)

(b)

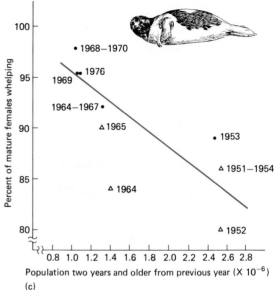

(c)

Figure 15.13 The recruitment of young in two closely related large mammals, (a) the North American elk (Houston 1982: 45) and (b) European red deer (Clutton-Brock et al. 1982: 268), shows a similar linear density-dependent response to population size. (c) Density-dependent recruitment of young in two populations of the harp seal, one (△) in the Gulf of St. Lawrence, and the other (●) off southern Laborador. (From Lett et al. 1981: 46.)

different densities. The pattern of growth among all will follow a similar pattern. At first all individuals will grow at about the same rate, accumulating biomass in a density-independent fashion. Eventually all populations, regardless of initial density, will reach a point at which individuals begin to compete with one another for light, nutrients, moisture, and space. Individuals respond in a plastic manner. They it-

erate fewer modules (leaves, stems, buds), they reduce individual plant size and fecundity, and they alter the distribution of weight and height. A negative density-dependent relationship between mean plant size and density becomes evident: The denser the plants, the smaller their size. As plant size increases, contest competition develops, resulting in the unequal distribution of resources, especially light and space. At this point some individuals lack the ability to respond in a plastic manner and they succumb. The result is a developing hierarchy of size, with few large and many small individuals. The total yield per unit area will increase only as

(a)

(b)

Figure 15.14 Two same-age pine stands, one unthinned and the other artificially thinned, illustrate the relation between density and mean weight or size. The unthinned stand has a greater density of smaller individuals. In the thinned stand biomass accumulates on fewer individuals. Overall the total biomass is probably similar. (Photos by R. L. Smith.)

the number of individuals declines. Ultimately, further growth of survivors is possible only if some of the remaining individuals die and if growth exactly compensates losses from mortality.

The rate at which plant populations experience competition and the continued depressing effect on rate of increase on the mean weight of plants depends upon the initial density. If the density of seedlings is high, that population will experi-

ence the effects of competition, plastic response, and mortality much sooner than less dense populations. Eventually, though, populations starting growth with widely differing densities converge upon a common density, which will decrease through time.

This relationship between mean plant weight and density was demonstrated by Yoda and others in 1963 in experimental monocultures of buckwheat *(Erigeron canadensis)* with a fixed sowing density and wide range of fertilities. Prompted by the work of Yoda, others have carried out similar experiments, including Lonsdale and Watkinson (1982). They sowed ryegrass *(Lolium perenne)* at densities of 1000, 5000, 10,000, 50,000, and 100,000 per square meter at two different light intensities, high and low (17 percent of natural light). They harvested samples of the population at 14, 35, 76, 104, and 146 days. Plant weight at first increased independent of density, but at some point plant weight increased as density decreased because of increasing mortality (Figure 15.15). The populations with highest density experienced increased mortality much sooner than the low density populations. In time all populations, regardless of initial density, reached the same position where density declined as plant weight increased. All populations moved along the same straight line described by log (total shoot weight per plant) against log (density of survivors), the slope of which is approximately $-3/2$. When populations reach this slope along which density declines and plant weight increases together, they are said to experience *self-thinning*. This relationship plant ecologists know as the *−3/2 power law* (Yoda et al. 1963). Many plants from herbaceous to woody appear to obey it (White 1980).

The interactions observed are described by the competition density equa-

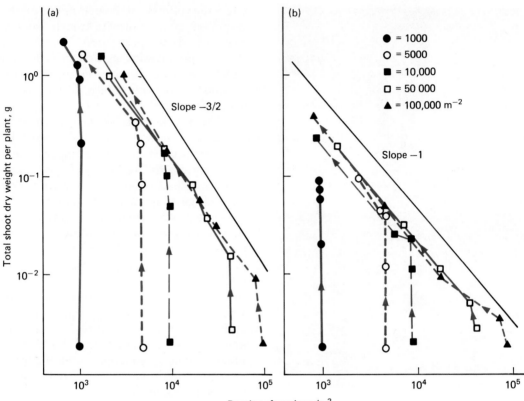

Figure 15.15 Self-thinning slopes for perennial ryegrass *Lolium perenne* sown at five different densities under two light conditions: (a) 0 percent shade, (b) 83 percent shade. Results of five successive harvests form a trajectory over time. Denser populations reached and followed the −3/2 slope of self-thinning much sooner. The last to reach it was the population with the thinnest density; but regardless of starting density, all achieved a similar final density. (From Lonsdale and Watkinson 1982: 433.)

tion $w = cd^{-a}$, where w = mean plant weight, d = plant density, and a and c are constants. The value c has the dimensions of g/m²/m. Known as the C-D index, a changes with time. With no competition the value of a is 0; under intense competition $a = 1$. At $a = 1$ growth of survivors balances the deaths of other individuals,

total weight per unit area remains constant, and biomass accumulation reaches the maximum for a particular species under given conditions. For the ryegrass grown under low light conditions, the slope was − 1. In this case $w = cd^{-1}$ is in contrast to $w = cd^{-3/2}$. The position of the − 3/2 line can be altered by light conditions, but usually not by nutrients. If the nutrient supply is increased, the line is reached more quickly and populations move along it faster because of an increased growth rate.

The − 3/2 power law is derived from the geometry of space occupied by the plant. The law assumes that all plants of any given species are similar geometrically, regardless of size and growing conditions. The average ground area or space (basal area) occupied by a plant will be proportional to the square of the linear dimension of the plant, and its weight will be proportional to the volume of space that the plant occupies. The canopy volume relates to a cube of some linear dimension, particularly height; thus the $-3/2$ relationship. Mortality in the population occurs when the percentage of canopy cover relative to ground area approaches 100 percent, blocking light below. What the $-3/2$ power law says is that the rate of growth (reaching the $-3/2$ line) and the rate of thinning depend on habitat conditions, especially nutrient availability and light; thinning proceeds in a density-dependent fashion; and the habitat rather than the density of the initial planting controls the amount of biomass that can be supported.

Plant shapes are variable, even within a species. The $-3/2$ law is based on isometric growth, but much plant growth is allometric, which should yield thinning exponents different from $-3/2$ (Weller 1987a, 1987b, 1989). Because thinning exponents vary according to plant geometry, they are not always near the idealized value of $-3/2$. Deviations in woody plants include height, average mass, average bole diameter at breast height, and average bole basal area. Weller points out that although the $-3/2$ power law may be weakened as a quantitative law, it does provide one means to study intraspecific competition in even-aged stands.

■ Density-Independent Influences

During the late 1950s and early 1960s ecologists diverged widely in their views on the relative importance of density-dependent and density-independent influences on populations. (For density-dependent viewpoints, see Nicholson 1954, 1956; Solomon 1957; Lack 1966; and Hairston, Smith, and Slobodkin 1960. For density-independent viewpoints see Andrewartha and Birch 1954; Milne 1957; and Thompson 1956. For a historical review see Krebs 1985.) The arguments were largely semantic, stemming from different approaches and different philosophies infused with a dose of advocacy. The density-independent school, for example, was dominated by insect population ecologists interested in proximate causes of population fluctuations and densities; whereas the density-dependent side was dominated by vertebrate ecologists more interested in natural selection and in evolutionary problems. Most ecologists now agree that the numbers of organisms are determined by an interaction between density-dependent and density-independent influences, which vary among and within populations.

By themselves, density-independent influences on the rate of increase do not regulate population growth. Regulation implies a homeostatic feedback that functions with density. However, density-independent influences can have considerable impact on changes in population size, and they can affect birthrates and death rates. Density-independent influences may so affect a population that they completely mask any effects of density-dependent regulation. A cold spring may kill the flowers of oaks, causing a failure of the acorn crop. Because of the failure, squirrels may experience widespread starvation the following winter. Although the proximate cause of starvation relates to the density of squirrels and available food supply, weather was the ultimate cause of the decline. Conditions beyond the organisms' limits of tolerance can have a disastrous impact, affecting growth, maturation, reproduction, survival, emigration, and dispersal within a population, and even eliminating local populations.

In general the influence of weather is stochastic: It is irregular and unpredictable, and it functions largely by influencing the availability of food. Pronounced changes in population growth often can be correlated directly with variations in moisture and temperature. For example, outbreaks of spruce budworm (*Choristoneura fumiferana*) are usually preceded by five or six years of anticyclonic weather, characterized by low rainfall and high evaporation, and end when wet weather returns. Such density-independent effects can take place on a local scale where topography and microclimatic conditions influence the fortunes of local populations.

Consider the San Francisco Bay checkerspot butterfly (*Euphydryas editha bayensis*). Adults emerge in a four to six week flight from mid-March to early May and lay their eggs at the base of the plantain *Plantago erectus*. Within two weeks the eggs hatch, and for three weeks the larvae feed on the plantain and owl's clover (*Orthocarpus densiflorus*) when they reach the third instar. Now the dry season sets in and the larvae go into diapause until December, when it is broken by the winter rains, which also stimulate the germination and growth of the food plants. The larvae go through three more instars before they pupate and emerge as adults. During a two-year period, 1983 and 1984, record rainfalls and cool weather retarded the development of the larvae, delaying the flights of the adults and the deposition of eggs.

Meanwhile the food plants *Plantago* and *Orthocarpus* were developing normally. By the time the eggs hatched, the host plants were already senescent or dying. Lacking food, the prediapause larvae faced heavy mortality from starvation.

Normally the warmer south-facing slope experiences earlier plant growth and faster development of larvae because of the warmer microclimate, while the north-facing slope experiences about a two-week delay in growth of food plants. Because of the delay in their development, the larvae faced senescent plants on the south-facing slope, while the host plants on the north-facing slope were not available for another week; and if they had been available, the plants would have died before the larvae reached the third instar. The following year only 18 adults were captured (Dobkin et al. 1987). In such a manner microclimatic changes can lead to sharp declines or extinctions in local populations.

Populations of deer in the northern part of their range, as well as moose, are sensitive to severe winters. In Minnesota Mech and others (1987) found a significant relationship between snow accumulation over the previous three years and viability of offspring, as indicated by fawn-doe ratios in deer, calf-cow ratios in moose, the twinning rate in moose, and the percent change in the deer population (Figure 15.16). The inverse effect relates to a winter-to-winter carryover of the nutritional effects on facundity, prenatal nutrition over winter, and in utero development in two-year-old deer. When the sum of a three-year average of snow accumulation exceeds 10.2 m and the average single-year accumulation exceeds 340 cm, we can expect an adverse effect on the ungulate population.

In desert regions a direct relationship exists between precipitation and rate of increase in certain rodents and birds (Figure 15.17). Merriam's kangaroo rat (*Dipodomys merriami*) occupies lower elevations of the Mojave Desert. Although the kangaroo rat has the physiological capacity to conserve water and survive long periods of aridity, it does require in its environment a level

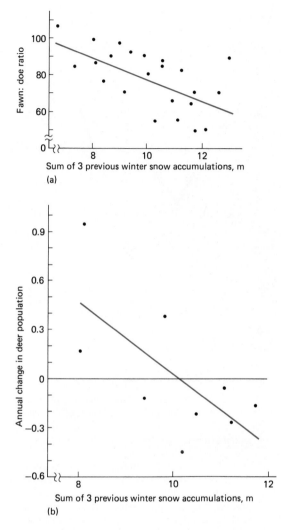

Figure 15.16 The relationship between the sum of the previous three winter monthly snow accumulations in northeastern Minnesota on (a) fecundity (fawn:doe ratio) and (b) percent annual change in next winter's population of white-tailed deer. (From Mech et al. 1987: 619, 622.)

of moisture sufficient to stimulate the growth of herbaceous desert plants in fall and winter. The kangaroo rat becomes reproductively active in January and February when plant growth, stimulated by fall rains, is green and succulent. Herbaceous plants provide a source of water, vitamins,

and food for pregnant and lactating females. If rainfall is scanty, annual forbs fail to develop and the production of kangaroo rats is low (Beatley 1969; Bradley and Mauer 1971). This close relationship to seasonal rainfall and relative success of winter annuals is also apparent in other rodents occupying similar desert habitats and in Gambel's quail and scaled quail (Francis 1970).

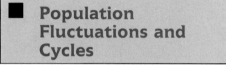

Population Fluctuations and Cycles

Populations tend to fluctuate about a long-term mean, which might be considered the asymptote of the logistic curve or *K* (Figure 15.18). These fluctuations come about because density-dependent mechanisms, mostly birthrates and mortality rates, tend to either undercompensate or overcompensate for population size. They also have built-in time lags and in addition are affected by extrinsic influences, particularly weather and predation. The nature of the fluctuations, based on census data, depends upon whether the populations are censused in the fall, winter, or spring. Population trends in winter may reflect neither the true carrying capacity of the breeding habitat nor the breeding population. Fall populations of some species, reflecting the success of recruitment, may be larger and more variable from year to year than spring breeding populations, which may show much less variability (Figure 15.19) and may more nearly reflect the true carrying capacity of the habitat.

Population fluctuations that are more regular than we would expect by chance are called oscillations or cycles. In a strict sense any fluctuations with peaks regularly separated by the same time intervals

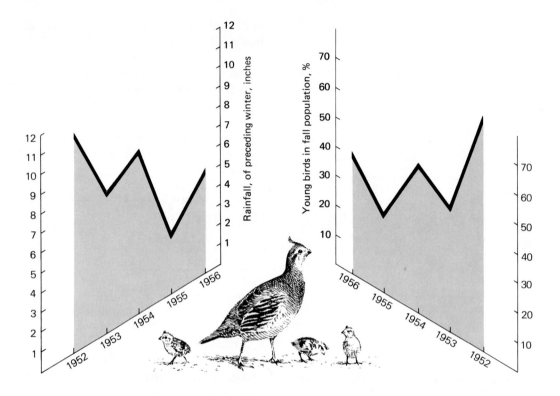

Figure 15.17 Relationship of winter rainfall to the percentage of young the following year of a Gambel's quail population in southern Arizona. Note how the production of young follows rainfall. Quail need an abundance of green vegetation in late winter and early spring to supply the nutrients and vitamins essential to reproduction. In years of low rainfall spring flush of vegetation in the semiarid southwest fails to appear, and many if not most of the birds fail to reproduce or even to break up winter coveys. Thus reproductive success in desert quail reflects a response to a density-independent influence, rainfall. (After Sowls 1960: 187.)

Figure 15.18 Fluctuation in a wintering population of black-capped chickadees *(Parus atricapillus)* in northwestern Connecticut about a mean long-term density of 160 birds. The short-term decline in the population in 1968–1969 was attributed to an influx of tufted titmice, decreasing survival rates and recruitment to the population; although the decline might have been due to chance, and the lower population density of chickadees exploited by the tufted titmouse. In spite of fluctuations the population exhibited no sustained increases or decrease. (Data from Loery and Nichols 1985.)

is a cycle. The fluctuations of Nicholson's blowfly populations are oscillations (see Figures 15.6 and 15.7) brought about by delayed density responses.

The two most common intervals between peaks are three to four years, typified by lemmings (Figure 15.20), and nine to ten years, typified by snowshoe hare

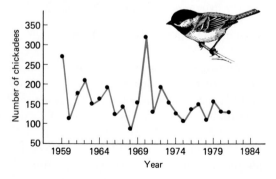

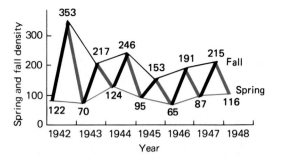

Figure 15.19 Trends of spring and fall populations in the bobwhite quail in Wisconsin. Note the seasonal fluctuations of population density from spring to fall and back to spring again. Fall densities fluctuate more than spring densities which changes in carrying capacity in the habitat which are fairly stable in this territorial species. (After Kabat and Thompson 1963: 78.)

and lynx (Figure 15.21). These cyclic fluctuations are largely confined to simpler ecosystems, such as the northern coniferous forest and the tundra. Usually only local or regional populations are affected, although there is some evidence to suggest broader synchronies.

A number of theories have been advanced to explain cycles. They can be divided into two main groups. One theory, represented by Cole (1951, 1954) and Palmgren (1949), holds that cycles cannot be distinguished statistically from random

Figure 15.20 The lemming cycle over a 20-year period in the coastal tundra at Barrow, Alaska, with three to six years between peaks. The periodicity is regular, but the amplitudes vary greatly. (From Batzli et al. 1980: 338.)

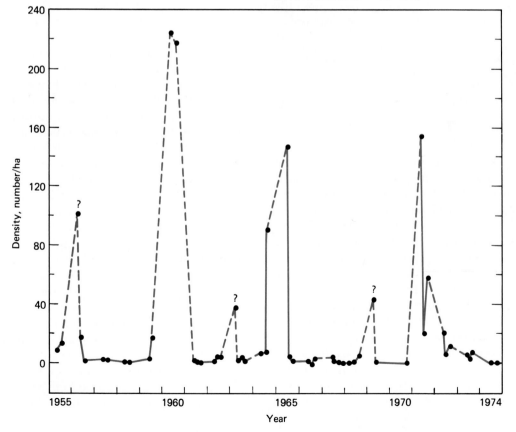

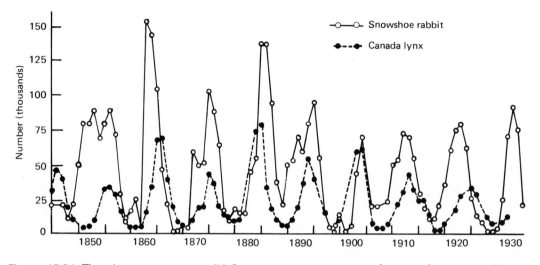

Figure 15.21 The nine- to ten-year cyclid fluctuation of snowshoe hare and lynx populations in northern North America. This cycle is based on fur returns from both the snowshoe hare and lynx. Although this graph supposedly shows lynx populations tracking the snowshoe hare populations, the superimposition of graphs is flawed because the data for the snowshoe hare are from eastern Canada, those for the lynx from western Canada. Standing alone, the graphs probably do depict the cyclic fluctuations of each species. (From MacCulich 1937.)

fluctuations. Populations reflect a variety of random oscillations or fluctuations in environmental conditions. However, the statistical reality of the cycles has been demonstrated by Bulmer (1974, 1975) for a number of northern animals (red fox, lynx, muskrat, snowshoe hare, horned owl, ruffed grouse) during the period 1951–1969 and by May (1976) for lemmings. The main features of these cycles is regularity of the period and the irregularity of the amplitude.

The other group maintains that something in the physical environment, in the ecosystem, or in the population itself causes cycles. Predation has been singled

out as a cause, but predators usually are not abundant enough when rodents are at a peak to bring about a decline. Endocrine malfunction has been invoked, as well as genetic changes in the quality of animals, aggressive behavior (Krebs 1985), and dispersal (Stanseth 1983). Food shortages have been implicated, and they may be a major part of the story (see Chapter 20). They are implicated in snowshoe decline (Keith 1974) and in cycles of lemmings in the arctic tundra. Schultz (1964, 1969) and Laine and Henttonen (1983) have proposed somewhat differing ideas on food quality and nutritional state of the vegetation. Pitelka (1973) and Batzli and others (1980) found that peaks of lemming populations built up under conditions of good snow cover, high quantity and quality of food, and low predation. Populations crashed when vegetation was depleted to the point that little grassy cover existed after snowmelt and lemmings were vulnerable to intense predation. Predators of lemmings, including owls, jaegers, foxes, and weasels, may track the cycles to a certain extent, but they play no important role in driving the cycles. Lynx populations supposedly track snowshoe hare populations, the basis of the classic snowshoe hare-lynx cycles, but the example is flawed because the snowshoe hare data are from eastern Canada, whereas the superimposed lynx data are from western Canada (Finerly 1979).

L. Keith and his associates (Meslow and Keith 1968; Keith and Windberg 1978; Keith et al. 1984) followed snowshoe hare populations through two periods of increase and three of decline in the Rochester district of central Alberta. The decline, which set in prior to the peak winter populations, was characterized by high winter-to-spring weight loss, a decrease in juvenile growth rate, decreased juvenile overwinter survival, reduction of adult survival beginning one year after the population peak and continuing to the low, and decreased reproduction (characterized by changes in ovulation rate, third and fourth litter pregnancy rates, and length of breeding season). Although the decline has been attributed to food shortages, an intensive study involving radio-collared hares revealed new insights into the decline (Keith et al. 1984). Starvation in winter is a short-term phenomenon, confined largely to the first few months of the major decline. It is followed by one to two winters in which malnutrition (determined by sampling bone marrow) and low temperatures (below $-30°$ C) interact strongly with heavy predation on nutritionally stressed hares by coyote, lynx, horned owl, and goshawk. Predation was the paramount immediate cause of mortality, accounting for the death of 80 to 90 percent of radio-collared hares. The upswing of the cycle, which set in about three years after the peak winter, was characterized by a lower winter-to-spring weight loss, increased juvenile growth rate, increased juvenile overwinter survival, and increased reproduction.

■ Key Factor Analysis

How do we determine what density-dependent influences are at work in a given population? One method is *key factor analysis*. A key factor is a biological or environmental condition associated with mortality that causes major fluctuations in population size. Key factor analysis is based on a k value derived from the life table. Related to the mortality rate q, k (sometimes called killing power) is defined as $\log_{10}l_x - \log_{10}l_{x+1}$. The k value has the advantage over q because it is additive. The summation of k values over age classes provides K, the total killing power, which reflects the rate or intensity of mortality. Like l_x it is comparable between populations. The use of k values and key factor analysis has been useful in the study of insect populations, which have discrete generations and life stages to which mortality can be assigned. It is more difficult to use in populations with overlapping generations, but with modifications it can be used to detect regulation in the life cycle and to search for the causes of mortality.

To demonstrate how key factor analysis works we can use the data from the life table of the gypsy moth (Table 14.2) The first figure in the k factor table (Table 15.4) is the maximum potential natality for each generation, determined by multiplying the number of females of reproductive age by the maximum number of eggs per female. For our example we will simply consider the maximum fecundity for one female gypsy moth in a sparse population as 800 eggs. The l_x value of each successive stage of the life cycle as it appears in the life table is entered and the values converted to logarithms. Each logarithm is subtracted from the previous one to give a *k* or mortality value for each age class. These values are added to give a total generation mortality, *K*. This work is done for a number of successive generations. To identify the key factor that influences trends in adult populations, the *k* values for each successive generation are plotted along with *K*. The plot shows whether the mortality rate for one particular stage or age classes consistently displays over the generations a strong correlation with total mortality (Figure 15.22). If there is some correlation between the *k* value of a particular stage and total *K*, then the analysis can be carried further to determine the *k* factor within that stage.

Houston (1982) used that approach in his study of the population dynamics of the northern Yellowstone elk. It was obvious from the life table (Table 14.5) that juvenile mortality was the major cause of population loss. Houston used five sets of data in his key factor analysis: annual census of elk, reduction in fertility (k_0) relative to potential maximum, neonatal mortality (k_1), calf hunting mortality (k_2), and overwinter mortality of calves (k_3). Reduction in fertility was measured as the difference between the number of calves born if all mature females produced calves and that of observed pregnancy. Neonatal mortality involved calves alive at 6 months; hunting mortality, calves alive at 9 months; and winter mortality, calves alive at 12 months. Total reduction in number of young, *K*, is equal to $k_0 + k_1 + k_2 + k_3$. Values for each were obtained by subtracting log (calves produced) from log (maximum potential calves) and so on. Houston assumed that each reduction acted in sequence with negligible overlap in time. Because of censusing and

Figure 15.22 Key factor analysis applied to the great tit *(Parus major)* and the winter moth *(Operophtera brumata)*. The key factor in the life cycle of the great tit in Marley Woods, Oxford, is k_4, mortality outside of the breeding season. In the rest of the annual life cycle density-dependent variations in clutch size (k_1) and hatching success (k_2) regulate the population. Key factors in the life cycle of the winter moth are k_1, overwintering loss of winter moth eggs and larvae before the first larval census in spring, and k_5, a density-dependent loss of the pupae in the soil due to predation in the spring. (From Podoler and Rogers 1975: 97, 101.)

TABLE 15.4 k *Values for a Gypsy Moth Population Based on Life Table for Sparse Population in Connecticut (Table 13.5)*

x	l_x	Logarithm of l_x	*k Value*
Maximum natality	800.0	2.903	—
Eggs laid	550.0	2.740	0.163
Larvae I-III	385.0	2.585	0.155
Larvae IV-VI	242.0	2.384	0.201
Prepupae	14.6	1.164	01.22
Pupae	11.7	1.068	0.096
Adult	1.4	0.146	0.922
			$K = 2.757$

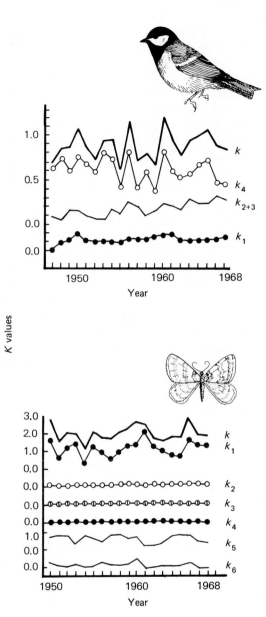

K values

Year

Year

source of loss in calves was neonatal natality (k_1).

■ Summary

When births and immigration exceed deaths and emigration, a population increases. The difference between the two (when measured as an instantaneous rate) is the population's rate of increase, r. In an unlimited environment a population expands geometrically, a phenomenon that may occur when a small population is released in a unfilled habitat. Geometric increase is characterized by a constant schedule of birth and death rates, an increase in numbers equal to the intrinsic rate of increase, and the assumption of a fixed or stable age distribution, which is maintained indefinitely. Because the environment is limited, such growth is not maintained indefinitely. Population growth eventually slows and arrives at a point of equilibrium with the environment's carrying capacity, K. However, natural populations rarely achieve such equilibrium levels; instead a population fluctuates in numbers. When populations become quite small, chance events, demographic or environmental, alone can lead to extinction.

Fluctuations about the equilibrium point suggest that populations possess some form of regulatory mechanism. Some ecologists have maintained that population fluctuations are most affected by density-independent influences. Others have argued that populations are regulated by density-dependent influences that relate to optimum population size. A more general idea is that local populations fluctuate between some upper and lower levels, and the fluctuations are brought about by an

other problems, the estimates of k were crude, but nevertheless reflected the true situation. Reduction in fertility k_0 was estimated essentially as a constant and hunting mortality was low. The major interaction of population density and influences extrinsic to the population, such as weather.

Density-independent influences affect but do not regulate populations. They can reduce local populations, even to the point of extinction, but their effects do not vary with population density. Regulation implies a homeostatic feedback that functions with density.

Intraspecific competition for resources in short supply is a density-dependent mechanism in the regulation of population numbers. There are two basic types of competition. In scramble competition all individuals have equal access to the resource and each attempts to get a part of it. In extreme cases scramble competition results in each individual obtaining insufficient amounts to survive or reproduce. In contest competition successful individuals divide the resource and the unsuccessful are denied access to it. Contest competition is characteristic of species whose individuals are able to defend a resource from others.

Intraspecific competition can result in density-dependent reduction in growth, delayed maturity and fecundity, and an increase in mortality, especially of the young. In plants response to density results in a decrease in the number of individuals and an increase in the mean weight of individuals. This relationship, known as self-thinning, is expressed in the $-3/2$ power law, $w = cd^{-3/2}$.

One approach to the study of density-dependent mortality in animal populations is k-factor analysis. A key factor is a biological or environmental condition associated with mortality that causes major fluctuations in population size.

Intraspecific Competition

■ **Density and Stress**

■ **Dispersal**

    Who Disperses?
    When and How Far?
    Why Disperse?
    Does Dispersal Regulate
      Population?

■ **Social Interactions**

    Social Dominance
    Territoriality
    Home Range

■ **Summary**

What form does intraspecific competition take, especially in situations in which resources such as food, space, and mates are not shared equally? How are resources allocated? What happens to individuals who do not receive their share? Intraspecific competition can operate in subtle and not-so-subtle ways. Its outcome affects the fitness of individuals and can influence population density.

■ Density and Stress

How do individuals respond as population density increases because of population growth or because of a decrease in living space through environmental degradation? Increased density can result in in-

creased social stress. Stress in animals, evidence suggests, can act on the individual through a physiological feedback involving the endocrine system (Figure 16.1). In vertebrates this feedback is most closely associated with the pituitary and adrenal glands (Christian 1963, 1978; Christian and Davis 1964; Davis 1978). Stress triggers hyperactivation of the hypothalmus-pituitary-adrenocorticular system, which in turn alters gonadotrophic secretions. Such profound hormonal changes result in a suppression of growth, a curtailment of reproductive functions, and delayed sexual activity. Further, these hormonal changes may suppress the immune system and cause breakdowns in white blood cells, increasing an individual's vulnerability to disease (for example, see Sinclair 1977). Social stress among pregnant fe-

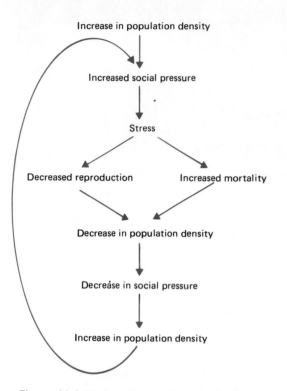

Increase in population density

Increased social pressure

Stress

Decreased reproduction   Increased mortality

Decrease in population density

Decrease in social pressure

Increase in population density

Figure 16.1 Christian's stress hypothesis of population regulation. Increased social pressure results in increased physiological stress. Stress in turn affects fecundity, mortality, and eventually population increase. As population density decreases, so does physiological stress, and the population increases. (From Krebs 1964.)

males may increase intrauterine mortality and cause inadequate lactation and subsequent stunting of nurslings. Thus stress can result in decreased births and increased mortality.

Such population-regulating effects have been confirmed in confined laboratory populations of several species of mice and to a lesser degree in enclosed wild populations of woodchucks *(Marmota monax)* (Lloyd et al. 1964) and Old World rabbits *(Oryctolagus cuniculus).* K. Myers and his associates (1967, 1971) held such rabbits at several densities in different living spaces within confined areas of natural habitat. Those living in the smallest space, in spite of the decline in numbers, suffered the most debilitating effects. Rates of sexual and aggressive behavior increased, particularly among females. Reproduction declined, fat about the kidneys decreased, and the kidneys exhibited inflammation and pitting on the surface. The weight of liver and spleen decreased and adrenal size increased.

Young rabbits born to stressed mothers were stunted in all body proportions and in organs. As adults they showed such behavioral aberrations as a high rate of aggressive and sexual activity and such physiological aberrations as abnormal adrenal glands, low body weight, and poor survival, all indicating a lack of fitness. Rabbits from low and medium densities had excellent health and survival.

Pheromones (chemical releasers that serve as communication among individuals of the same species) present in the urine of adult rodents may inhibit reproduction among members of a population. Such a function is suggested in a study involving wild female house mice *(Mus musculus)* living in high-density and low-density populations confined to grassy areas within a highway cloverleaf. Urine from females of a high-density population was absorbed onto filter paper. The paper was placed with juvenile wild female mice held individually in laboratory cages. Similarly, urine from females in low-density and sparse populations was placed with other juvenile test females. Juvenile females exposed to urine from high-density populations experienced delayed puberty, whereas females exposed to urine from low-density populations did not. The results suggest that pheromones present in the urine of adult females in high-density populations may delay puberty in juveniles and help slow population growth

(Massey and Vandenberg 1980). Juvenile female house mice exposed to urine of dominant adult males experienced an accelerated onset of puberty (Lombardi and Vandenbergh 1977).

Plants, too, respond to stress from crowding. How they respond is influenced by their adaptiveness to stress situations at the genet and modular levels. A major stress for plants is light (see Chapter 7). Under the light-intercepting canopy of other plants or similar low-light conditions, shade-adapted plants will respond to that stress by growing slowly, conserving energy by reducing the rate of photosynthesis, and forgoing the production of flowers and seed. In effect, they wait for a time when sufficient light will stimulate rapid growth. Plants adapted to other forms of stress follow a similar pattern (Grime 1977). Plants adapted to open light respond to low light intensities by growing rapidly in height (which under some situations will get them up into the light). However, they develop thin cell walls, which reduce the supporting ability of the stems, resulting in weak, spindly individuals highly vulnerable to fungal infections.

Plants growing under crowded conditions in which competition is high respond to conditions of low nutrients, low moisture, or other environmental stress in various ways. Individual plants react to increased density with decreased growth, as expressed by a reduction in mean plant weight. In spite of this reduction, yield per unit area is constant over a wide range of densities (the law of constant yield, Kira et al. 1953). Reduction in size often comes about in losses in modular units and changes in morphology. Genets may respond to high density by reducing the number of ramets, as perennial ryegrass (*Lolium perenne*) does, resulting in a lower density of tillers than genets in the population (Kays and Harper 1974). Plants also modify their morphology by reducing the number of nodes per stem, internode length, number of flowers and seeds, leaves per stem, and branches (for examples, see White 1984; Sarukhan et al. 1984). Such reductions, especially in leaf area, can increase mortality among individuals, especially those with reduced leaf area (Fowler and Antonovics 1981; Antonovics and Primack 1982).

Ruderal plants, those adapted to persistent and severe disturbance, respond to stress by producing seeds at the expense of vegetative development. Individual plants are small and poorly developed; yet the number of seeds relative to individual plant biomass is high. Seeds of such plants can survive buried in soil for long periods of time. They are capable of germinating rapidly when a disturbance exposes the seeds to light and fluctuating daily temperatures. Examples of such plants are annual weeds.

Such responses to stress do influence individual fitness among plants and the maintenance and expansion of populations, but there is little evidence that density-dependent stress in plants acts in any regulatory way. What evidence we do have comes from experimental populations, not ones growing under natural conditions. (For detailed review of density-dependent regulation in plants see Antonovics and Levin 1980.)

■ Dispersal

By late summer the reproductive season is over and juveniles have swelled the population. General belief is that the young will leave their natal area, driven out by competitive interactions with the adults and older subadults. Actually dispersal is more complex. Why disperse, when perhaps the animal might be better off staying put? Who disperses? When do dispersers leave, and where do they go? What motivates dispersers to leave? What effect does dispersal have on the fitness of the dispersers and on population regulation? Answers to such questions provide insights into one of the more conspicuous activities in woods and fields.

Who Disperses?

There is no hard and fast rule about who disperses. Howard (1960) defined dispersal as "the permanent movement an individual makes from its birth site to the place where it reproduces or would have reproduced if it had survived and found a mate." This definition suggests that juveniles and subadults are the dispersers. Among birds young are the major dispersers, and Greenwood (1980) suggests that leaving be called *natal dispersal*. The reason for the distinction is that adult birds may leave poorer reproductive sites for better ones. Greenwood (1980) terms this movement *breeding dispersal* (see also Greenwood and Harvey 1982).

Among rodents subadult males and females make up most of the dispersing individuals (Tamarin 1978; Krebs et al. 1976; Beacham 1981). In some situations dispersing females outnumber males; in other situations the opposite is true, or there is no difference at all (Johnson and Gaines 1987). Many dispersers, up to 40 percent in some cases (Beacham 1981), are in breeding condition. They tend to be lighter in weight than those that stay at home, and reach sexual maturity at a lower weight and younger age (Krebs et al. 1976).

Among many groups of insects, dispersal is undertaken by a polymorphic segment of normally flightless insects that acquire wings, notably aphids, leafhoppers, and water striders. These winged forms have the option of flying or not flying. Within insect populations in which all individuals are capable of flight, a proportion disperses (for review see Harrison 1980).

When and How Far?

Dispersal among all groups generally takes places in a prereproductive period. Among many birds dispersal takes place in spring, following the return of migrants or the breakup of winter flocks among resident species. In others, such as the ruffed grouse, young disperse in late fall and settle in breeding areas by early winter. Dispersal in rodents takes place during periods of increasing population and at the peak phase of population growth, and decreases during population decline (Krebs et al. 1976; Beacham 1981).

Lidicker (1985) has hypothesized two types of dispersal—presaturation and saturation. *Presaturation dispersal* takes place during the increase phase of population growth, before population reaches a peak or carrying capacity and before resources are depleted. The dispersers are in good condition, consist of any sex or age group, have a good chance of survival, and show a high probability of settling in a new area. These dispersals are density-independent. *Saturation dispersal* occurs when carrying capacity has been exceeded. The individuals, mostly juveniles and subdominants, have two options: to stay and perish or at the very best not breed, or to leave the area. If they move out, the odds are they will perish, although a few may arrive at some suitable area and settle down. The mass movements of squirrels in the 1800s (Allen 1962) and the muskrat dispersal described by Errington (1955) are examples of saturation dispersal. Such dispersal is density-dependent.

Because dispersers are seeking vacant habitat, the distance they travel will depend in part on the density of surrounding populations and the availability of suitable unoccupied areas. Murray (1967) has stated a rule of dispersal: Move to the first uncontested site you find and no further. Waser (1985) has developed models, based on Murray, for dispersal distance. The basic model, which assumes a straight-line dispersal, is

$$p_n = t(1 - t)^n$$

where t is the likelihood that the home range the disperser travels is unoccupied, $(1 - t)$ the probability that the animal has to disperse at all, times t the probability that the next home range it traverses is unoccupied, and $(1 - t)^n$ the probability that it has found no nearer home range. The models predict the distribution of dispersal as a function of the turnover of home ranges (number of home ranges vacated by death of previous owner), and the distance the disperser has to travel to find that area. Animals generally will either disperse in a straight line from their natal area or make exploratory forays into surrounding areas before leaving the natal site, and should settle in the first empty site. Waser measures dispersal distance in terms of the number of home ranges (based on average size of home range for the species) the disperser has to travel from its natal site (Figure 16.2).

Why Disperse?

Why should a young animal disperse at all, leaving its natal area for some strange place? Some do not. Whatever the choice, it carries certain costs and benefits (Table 16.1). A few dominant juveniles may establish themselves reproductively when the natal area is vacated through mortality of the adults; but because there is little probability they will reproduce in their natal area, most juveniles or subadults can maximize their fitness only if they leave the natal area, in spite of the risks. Where intraspecific competition is intense, dispersers can locate habitats where resources are more accessible, breeding sites are more available, and competition is less. Dispersers also increase the probability of encountering new individuals with which to mate, reducing probability of inbreeding and increasing fitness of offspring because of increased heterozygosity (see Chapter 13).

An animal has to have some motivation to disperse. Intraspecific competition of some sort may be a driving force behind some dispersals. Christian (1970) hypothesized that increased population density in rodents increased levels of aggression. Aggressive individuals forced subdominant ones to disperse. Under these conditions maximum dispersal would be at peak densities (saturation) and involve social subordinates, mostly males.

Social interaction prior to dispersal rather than aggressiveness is invoked by Bekoff (1977) as the mechanism behind dispersal. Asocial individuals, either dominant individuals avoided by their sibs or subdominants avoiding their sibs, are the mostly likely to disperse, because they fail to develop social ties. Because aggressive behavior is not an adequate stimulus for dispersal, no relationship exists between population density and dispersal.

Evidence on enclosed populations seems to point out that highest rates of dispersal occur before peak densities are

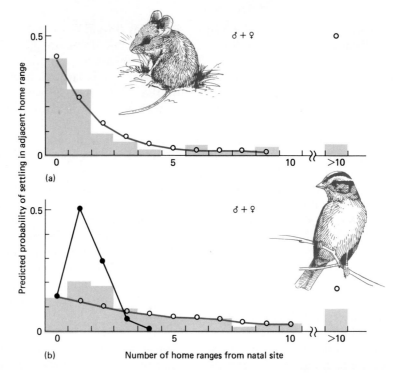

(a)

(b)

Number of home ranges from natal site

Predicted probability of settling in adjacent home range

Figure 16.2 Observed and expected dispersal distances (a) in the deermouse *(Peromyscus maniculatus)* and (b) in white-crowned sparrows *(Zonotrichia leucophrys)*. The bar graphs indicate the observed dispersal distances. The dashed line indicates dispersal distances expected if the search were along a radius away from the natal site. The solid line is the expected pattern if the search were spiral to the nearest empty site. Distributions are calculated using t, the observed probability of nondispersal. (From Waser 1985: 1173.)

reached, that not all dispersers are subdominants or the less aggressive, and that a large fraction of dispersing animals are females. Some animals may be born to disperse. Howard (1960) suggested a genetic basis for dispersal, and Myers and Krebs (1971) and Krebs et al. (1976) found that dispersers were not a random subsample from their control populations of the voles *Microtus pennsylvanicus* and *M. ochrogaster*.

Certain genotypes (leucine aminopeptidase) were more prone to dispersal than others when populations were increasing. However, there is no evidence on the heritability of tendencies to disperse.

A genetic-behavioral component is involved in Chitty's hypothesis (Chitty 1960; Chitty and Phipps 1967), later modified by Krebs (1973, 1976) to explain the two- to four-year cycle in microtine rodents (Figure 16.3). Chitty suggested that the decline in vole populations resulted from the selection pressures of mutual interference. When populations are low, mutually tolerant individuals would be most adaptable and have a selective advantage over aggressive individuals, because of high reproductive output. As the population reaches peak densities and begins to crowd in on available resources, particularly food, selection would shift in favor of aggressive individuals who control access

---

TABLE 16.1   Potential Costs and Benefits of Dispersal Choices

STAY AT HOME: PHILOPATRY

Costs	Benefits

Inbreeding depression (G)
Reduced fitness because of resource
    shortage (S)
Reduced indirect fitness: competition with
    kin (S)
	Optimal inbreeding: maintain locally adapted
	    genes (G)
	Reduced physical risk: increased survivor-
	    ship (S)
	Familiarity with local terrain: security (S)
	Familiar social environment (S)
	Adaptive local traditions (S)
	Maintain kin association (S)

DISPERSE

Costs	Benefits

Outbreeding depression: disrupt coadapted
    genes (G)
Hybrid young not well adapted (G)
Alleles less suited to the environment (G)
Greater risk in movement: predators, local
    diseases, unfamiliarity with terrain (S)
G = genetic
S = somatic
	Outbreeding enhancement (G)
	Avoid overcrowding (S)
	Avoid competing with kin (S)
	Improve fecundity (S)

Source: Adapted from Shields 1987.

---

to food, space, and mates. When the population reaches the bottom of a sharp decline, aggressive behavior in a low population is no longer advantageous. As the number of tolerant individuals in the population increases, the population builds up again. Krebs modified this hypothesis to emphasize competition among males for mates and aggressiveness on the part of females to defend nest and space, all forcing expulsion of subdominant individuals. These animals pay for aggressive behavior in time and energy, reducing their fitness, bringing on conditions conducive to population decline.

It is obvious that no one mechanism can be invoked to explain motivations for dispersal either within a population at a given time or in a given species. We know only that dispersal follows periods of reproduction and that mutual interference or aggressiveness may be involved.

Does Dispersal Regulate Population?

Much of what we know about dispersal and population interactions is based on comparisons of rodent populations in which dispersal could occur with those in which dispersal was prevented. Krebs and associates (1969) enclosed three populations in such a way that they could not immigrate or emigrate, but predators had access. They compared these populations with a control population whose members were able to disperse. The enclosed populations increased in size; in one plot the population was three times as high as the control. Overpopulation in the enclosures resulted in overgrazing, habitat deterioration, and starvation. Stress was relatively unimportant, because *Microtus* can exist at densities several times higher than those normally experienced by other voles.

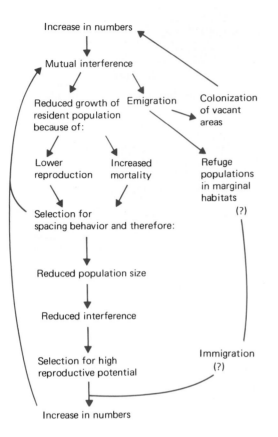

Figure 16.3 Modified version of Chitty's hypothesis to explain fluctuations in small rodents. Density-related changes come about through natural selection, especially through dispersal. Animals with the highest reproductive potential disperse; those that remain behind are less influenced by population density. Emigration or dispersal is the key to population regulation. (From Krebs et al. 1973: 40.)

In a later experiment Krebs and associates (1976) set up control and experimental areas populated with *Microtus townsendii*. They cleared experimental areas of voles, continually kept the areas vacant, and monitored recolonization. Colonization of experimental areas was most rapid when populations in the control areas were increasing. In fact, the experiments removed more animals from the experimental areas on which no litters could be produced than they could catch in the control area. In declining populations on the control area in which losses were due to emigration and not death, very little dispersal occurred.

These experiments represented a somewhat artificial situation. Lidicker (1973) and Tamarin (1978) compared the population dynamics of naturally enclosed populations of *Microtus* on islands where a water barrier prevented emigration with populations living on the mainland. Lidicker followed the growth of the California vole *Microtus californicus* on Brooks Island in San Francisco Bay (Figure 16.4). He found that population growth rates were lower in the mainland populations from which dispersal could take place than in the island population in which emigration was greatly limited. Tamarin found that populations of beach vole *(Microtus breweri)* on Muskeget Island near

Figure 16.4 Percentage increase (monthly intervals) from starting densities during periods of rapid increase to peaks in *Microtus californicus*. Note the rapid increase in the island population, which had no dispersal sinks, compared to the mainland population, which could disperse. (Adapted from Lidicker 1975.)

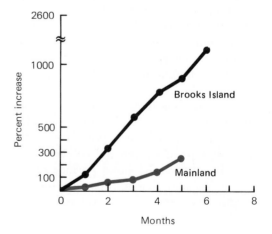

Nantucket Island, Massachusetts, did not undergo the degree of regular fluctuations characteristic of mainland populations of meadow vole (*M. pennsylvanicus*).

The results of the two studies seem contradictory, but the population of voles on Brooks Island represented only a period of rapid increase in the spring; the population was not followed over a period of several years. The Muskeget Island population was long established.

Lidicker hypothesized that presaturation dispersal could play a key role in population growth of voles. To disperse individuals must be motivated, be physically able to leave the area, and have available a "dispersal sink," an empty or unfilled habitat or even marginal or unsuitable habitat in which the animals could survive for a short time. The dispersal sink must permanently remove animals from the population. If a barrier exists to prevent dispersal or if dispersal sinks are unavailable, the dispersers must return to their home area. Because voles are not strongly territorial, these "frustrated dispersers" are allowed back into the population, adding to population growth. Tamarin (1978)

further hypothesizes that in situations in which emigration is inhibited the population would increase rapidly, as it did in the experiments of Krebs and Lidicker, overshoot the carrying capacity, and decline. Without normal regulatory process, that is, dispersal, the population will oscillate until it arrives at some equilibrium level with the environment, characterized by decreased birthrates and increased death rates as on Muskeget Island. In unenclosed populations, a large segment of the population leaves before the population reaches peak density, resulting in some cycling (Figure 16.5). Both the mainland population of voles in Tamarin's study

Figure 16.5 Population of female beach voles (*Microtus breweri*) on Muskeget Island compared to a population of female meadow voles (*M. pennsylvanicus*) on the mainland at Barnstable, Massachusetts. Note the relative stability of the "enclosed" island population, which achieved high density and then stabilized with increased mortality and decreased fecundity, brought about because of a lack of dispersal sinks. The "unenclosed" mainland population experienced dispersal which brought about cyclic fluctuations in population density. (Adapted from Tamarin 1978: 547.)

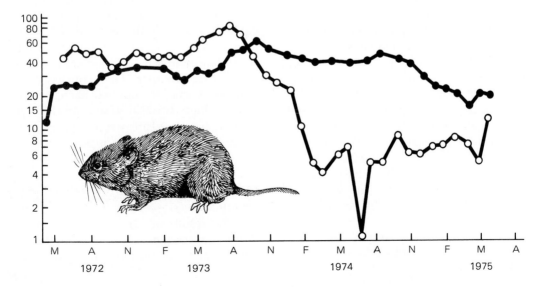

and Krebs' enclosed control population had a dispersal sink, and the populations responded with a high rate of increase and a high rate of dispersal.

These studies suggest that population dispersal involves two components, a source and a sink habitat. A *population sink habitat* is an area where conditions (such as predation, poor nesting sites, and lack of cover) are not sufficient for the species to carry out its life history. When local populations in the source habitat exceed the breeding sites available, surplus individuals leave, because potentially they can achieve a higher fitness by doing so. Most of these individuals settle into sink habitats in which local reproduction is not sufficient to balance local mortality. Populations in such habitats can be maintained only by immigration from more productive source areas. Sink habitats, however, can support very large populations, and the species may be more common in the sink habitat than in the source habitat. The sink may look like the prime habitat because of the large population.

As Gaines and McClenaghan (1980) point out in their survey of dispersal in small mammals, dispersal appears to be independent of population size in the source habitat. Although dispersal is positively correlated with population density and with the rate of population increase, there is no association between the proportions of the population leaving the area and the rate of population increase and decrease. Although dispersal may not function as a regulatory mechanism in a traditional sense, it can serve to expand populations, to aid in the persistence of local populations, and to function as a form of natural selection by sorting out phenotypes and genotypes.

Unanswered is the question of fitness of dispersers. A study of enclosed populations of the prairie vole *M. ochrogaster* by Johnson and Gaines (1987) provides some insights. They held populations of the vole in three 0.8 ha fenced exclosures. Dispersers from one exclosure (X) were marked and either placed in a vacant exclosure (Z) or returned to the original exclosure as "frustrated" dispersers. Residents of a third exclosure (Y) were free to disperse. Johnson and Gaines monitored the survival and reproduction of voles in all the exclosures for ten weeks and then replicated the experiment over a two-year period. They found no difference in sex ratio and reproductive condition between dispersers and residents, but subadults made up a greater proportion of dispersers than residents. They found that dispersers had the highest survival rate, followed by residents in exclosure X, and finally frustrated dispersers. Dispersers and frustrated dispersers exhibited the same reproductive activity, which was greater in both than in resident voles in exclosure X. Johnson and Gaines then combined survival and reproductive activity in a general fitness index. Dispersers had the highest relative fitness. Residents in exclosure X and the returned frustrated dispersers had lower and approximately equal fitness. Only the survival of females in exclosure Y, where all the dispersers were allowed to leave, increased as a result of decreased density. They concluded (1) that dispersers into optimal habitat experienced high survival and reproduction; (2) that frustrated dispersers experienced low survival, but those that became established in their natal exclosure showed high reproductive activity; and (3) that females gained the most from reduced population density.

The results of this experiment relate to the observations Tamarin made on his island and mainland populations of voles.

The island population, with its contingent of frustrated dispersers, undoubtedly experienced decreased survival of subadults and females and lowered reproductive activity, resulting in a stabilized population. On the mainland the vole population was open to dispersal. The subadults could move out, decreasing the density and stimulating bouts of high reproduction and population increase, followed by decline in populations, probably caused by high dispersal.

■ Social Interactions

A flock of mourning doves, two pairs of adults and their young, are feeding on grain you have scattered for them. Each maintains some distance from others and protects that distance by slight movement toward any individual that comes too close. Among them a dominant male commands the largest space, driving all others away. Several of the young are harassed, chased away from the grain. Feeling their frustration, you scatter more seed over a wider area, a space too large to be controlled by the dominants. The subdominant birds feed on the periphery with their own, and peace has settled on the flock for a time. A resource in a limited space causes intraspecific conflict among the birds. An increase in both the available food and feeding space reduces competitive interactions.

Aggressive or agonistic interactions are the basis of social organization, which takes two forms: dominance and territoriality. The difference between the two involves not only type of interaction but also utilization of space. Dominance, based on individual distance (the distance from another individual that provokes aggression or avoidance behavior) and a set of relationships, results in space being shared among individuals. Territoriality involves the division and exclusive occupation of space by a social unit or an individual with a defended boundary.

Social Dominance

Social dominance is based on intraspecific aggressiveness and intolerance and on the dominance of one individual over another. Two opposing forces are at work simultaneously: mutual attraction versus social intolerance, a negative reaction against crowding. Each individual occupies a position in the group or local population based on dominance and submissiveness.

In its simplest form an alpha individual is dominant over all others, a beta individual is dominant over all but the alpha, and so on to the omega, which is usually totally subordinate. This relationship was first described by Schjelderup-Ebbe (1922) for the domestic chicken. It is a straight-line or linear peck order, so named because pecking follows dominance—that is, birds peck others of lower dominance or rank (Figure 16.6). Even within a peck order complexities may exist, such as triangular or nonlinear hierarchies. In these triplets, the first individual is dominant over the second, the second is dominant over the third, and the third is dominant over the first. In such a situation, an individual of a lower rank can peck an individual of higher rank.

In some groups peck order is replaced by peck dominance, in which social rank is not absolutely fixed. Threats and pecks are dealt by both members during encounters, and the individual that pecks the most is regarded as dominant. The position of the individual in the social hierar-

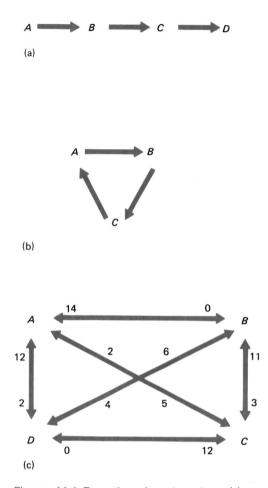

(a)

(b)

(c)

Figure 16.6 Examples of peck orders. (a) A straight-line peck order in which one animal is dominant over the animal below it. A is the alpha individual, D is the omega. (b) Triangular peck order in which A is dominant over B, B is dominant over C, yet C is dominant over A. (c) A more complex triangular peck order. The double arrows indicate encounters between individuals. The numbers represent the number of wins of one individual over another. For example, A is clearly dominant over B with 14 wins. B never dominated A. A is also dominant over D with 12 wins out of 14 encounters. C is also strongly dominant over D with 12 wins to 2 for D. B is also dominant over D with 6 wins to D's 4 wins. D clearly is the omega individual.

chy may be influenced by levels of male hormone, strength, size, weight, maturity, previous fighting experience, previous social rank, injury, fatigue, close associates, and environmental conditions.

In groups made up of both sexes separate hierarchies may exist for males and females with males dominant over females. In some groups females are equal to the males or dominant over them, and in others dominance is unrelated to sex. Once social hierarchies are well established within a group, newcomers and subdominant individuals rise in rank with great difficulty; and strangers attempting to join the group either are rejected or relegated to the bottom of the social order.

Rise in hierarchy often is related to breeding and sexual activity and hormones, particularly among those species that remain in flocks throughout the years. Individuals, male or female, that come into breeding condition early, even though subdominant in the winter group, rise in hierarchy through increased aggressiveness.

To the dominant individuals in such a social structure go most of the resources. Among many species the dominant male secures the most mates, thereby ensuring greater fitness at the expense of subdominant males. Dominant individuals have first choice of food, shelter, and space, and subdominant individuals obtain less than the despots. When shortages are severe, low-ranking individuals may be forced to wait until all others have fed, to take the leavings, to face starvation, or to leave the area.

Such social interactions cannot be called regulatory. Rather, the outcome of this type of contest competition simply ensures that the dominant animals in the population will continue to reproduce successfully. The fitness of such individuals is secured at the expense of subdominant individuals.

Social dominance, however, can influence population regulation if it affects reproduction and survival in a density-dependent manner. An example is the wolf. Wolves live in small groups of 6 to 12 or more individuals called packs. The pack is an extended kin group consisting of a mated pair, one or more juveniles from the previous year who do not become sexually mature until the second year, and several nonbreeding, related adults.

The pack has two social hierarchies, one headed by an alpha female and one headed by an alpha male, the leader of the pack to whom all other members defer. Below the alpha male is the beta male, closely related, often a full brother, who has to defend his position against pressures from other males below.

Mating within the pack is rigidly controlled. The alpha male (occasionally the beta male) mates with the alpha female. She prevents lower-ranking females from mating with the alpha and other males, while the alpha male inhibits mating attempts by other males. Therefore each pack has one reproducing pair and one litter of pups each year. These pups are reared cooperatively by all members of the pack. At low wolf densities, some packs may rear two litters of pups a year (Ballard et al. 1987).

The level of a wolf population in a region is governed by the size of the packs that hold exclusive areas. Regulation of pack size is achieved by events within the pack that influence the amount of food available to each wolf. The food supply itself does not affect births and deaths, but the social structure that leads to an unequal distribution of food does. The reproducing pair, the alpha female and the alpha male, has priority for food; they, in

effect, are independent of the food supply. The subdominant animals, male and female, with little reproductive potential, are affected most seriously. At high densities the alpha female will expel other adult females from the pack. Other individuals may leave voluntarily. Unless these animals have an opportunity to settle successfully in a new territory and form a new pack, they fail to survive.

The social pack, then, becomes important in population regulation. As the number of wolves increases, the size of the pack increases. Individuals are expelled or leave, and the birthrate relative to the population declines because most sexually mature females do not reproduce. Overall the percentage of reproducing females declines. When the population of wolves is low, sexually mature females and males leave the pack, settle in unoccupied habitat, and establish their own packs with one reproducing female. More rarely, the pack may produce two litters instead of one in a year (Ballard et al. 1987; Van Ballenberghe 1983). At very low densities, though, females may have difficulty locating males to establish a pack and so fail to produce or even survive. (For details on social regulation of population size in wolves, see Mech 1970; Zimen 1978; Fritts and Mech 1981; Ballard et al. 1987.)

Territoriality

The flock of mourning doves feeding on the ground will retain that social organization through the winter. In early spring the flock will break up and scatter across the countryside, form pairs, and establish another type of social organization, territories. A *territory* is a defended area, more or less fixed and exclusive, maintained by an individual or by a social group such as a wolf pack. Some would give the term an even broader definition and say that a territory exists when individuals or groups are spaced out more than you would expect from a random occupation of suitable habitat (Davies 1978).

Often it is difficult to draw a sharp distinction between social dominance and territoriality. Depending upon the season, the degree of crowding, and how resources are distributed, territory can grade into social hierarchy and vice versa. This mix is best expressed in the behavior of the feral domestic fowl (McBridge et al. 1969).

During the breeding season fowls' social behavior may range from extreme territorial to weak social hierarchy, as illustrated by six classes of males. The first class, dominant territorial males, restrict the movements of all other males to the territory and have females as constant companions. The second class of males possess subordinate territories, small ones defended against even dominant neighbors. The dominant neighbor may be able to penetrate the periphery of the territory, but not the whole extent. These males may have harem flocks. The third class, semiterritorial males, roost apart from the dominant males and make slight defense reactions against invasion by dominant males. However, the territories of these males are within the territory of a dominant male. They may be attended by females, but not constantly. The fourth class, subordinate males, roost apart from dominant males. The areas they occupy are also usually occupied by females, but the birds do not possess any. The fifth class is a group of subordinate males who roost with a dominant male and do not possess females. The last class consists of the runts, who generally leave the territory. Where more than one male occupies an area, territorial succession is a matter of peck order position. When male do-

mestic fowl are confined to smaller and smaller areas, dominant males adjust from territorial behavior to social dominance. When space is increased, the dominant males become territorial again (M. Schein).

Types of Territories. Types of territories vary according to the needs of the animals that defend them (Nice 1941). One type, common among songbirds and mammals such as muskrats, is a general purpose territory within which all activities take place. Late in the breeding season or soon after it, territorial defense breaks down. A second type, common among hawks, is a mating and nesting territory with feeding done elsewhere. Mating territories are exemplified by the leks of prairie grouse and the singing grounds of woodcock. Swallows and many colonial birds defend only a nesting territory, the size of which is often determined by the distance the bird can strike from its nest. Other animals, such as the hummingbirds and some squirrels, defend only a food resource, a feeding territory. Some birds defend a roosting territory where adequate roosting sites are scarce.

Means of Defense. Once an animal has established a territory, the owner must defend it against intruders. In the early period of territorial establishment, conflicts may be numerous. Territory-holding birds, frogs, and insects usually defend their claims vocally by singing from some conspicuous spot. Songs and calls advertise the fact that the area is already occupied. They are a long-distance warning that tells potential trespassers they should not waste their energies trying to settle there. Birds may shift their song perches throughout the territory (Figure 16.7) or vary their song patterns, perhaps in an effort to suggest that more than one male is

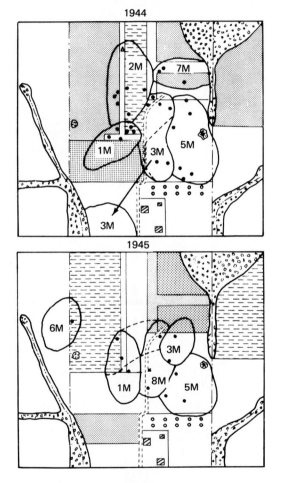

1944

1945

Figure 16.7 Territories of the grasshopper sparrow *(Ammodramus savannarum),* determined by observations of banded birds. Dots indicate song perches. Note how they are distributed near the territorial boundaries. The shaded areas represent crop field, the open areas hayfield. Dashed lines indicate boundary shifts in territory prior to second nesting. Note the return of the same males to nearly the same territorial area the second year. Such behavior is philopatry. (From Smith 1963: 160.)

torial male is removed but his song has been recorded and is played in the territory, other males will stay away (Carrick 1963).

If songs and calls fail and another individual does move into the area, the owner may confront the trespasser with visual display. Display may involve raising the crest, fluffing the body feathers, spreading the wings and tail, and waving wings among birds, or erecting the ears and baring the fangs among mammals (Figure 16.8). Such displays usually intimidate the invader, encouraging it to leave.

If intimidation displays fail, then the territorial owner is forced to attack and chase the intruder, an activity easily observed among many birds in spring and among dragonflies about a pond's edge. The gall-forming aphid *Pemphigus betae* (about which more will be said later) defends her territory by engaging in end-to-end kicking-shoving contests with the intruder, which may last more than two days and may result in the death of one or both aphids (Whitham 1987).

Some animals defend a territory in a more subtle manner by the use of scent markers. Wolf packs and coyotes mark territories with well-placed scent posts, frequently renewed by urine (Peters and Mech 1975; Barrette and Messier 1980). These scent marks warn neighboring members of their species about boundary rights. Just as important, these scent posts tell members of a wolf pack that they are

in the area. Song can be very effective in maintaining space between individuals. If a bird is removed from its territory, the space is quickly claimed by another deprived of a territory outright or forced to settle in some suboptimal area. If a terri-

Pectoral
sandpiper

Herring
gull

Hermit
thrush

Redpoll

(a)

Ring-necked
pheasant

(b)

(c)

Figure 16.8 Aggressive signals employed by some birds and mammals to defend territory or assert social dominance. (a) Agonistic displays among birds. Note the general similarity. (Pectoral sandpiper after Hamilton 1959; herring gull based on photographs in N. Tinbergen 1953; hermit thrush after Dilger 1956; redpoll after Dilger 1960; ring-necked pheasant after Collias and Taber 1951.) (b) Aggressive (upper) and submissive (lower) expressions in the American wolf, typical of canids. (c) Aggressive display in the blacktailed deer *(Odocoileus hemionus sitkensis)*. The illustration shows details of the head during the snort that occurs when the buck is circling in a crouch position. Note the widely opened preorbital gland, curled upper lip, and bulging neck muscles. The snort is a sibilant expulsion of air through the closed nostrils, causing them to vibrate. (After Cowan and Geist 1961.)

within their own territory and prevent accidental straying into hostile territory.

The use of scent and other chemical releasers, or pheromones, is widespread among animals, especially mammals and insects. They are secreted from endocrine glands, transmitted as a liquid or a gas, and smelled or tasted by others. They are important not only to mark territory and trails but also to convey such information as the identity of an individual, its sex and social rank, the location of food, and the presence of danger, and to attract mates (see Wilson 1971; Whittaker and Feeney 1971).

Why Defend a Territory? Why should an insect, bird, or mammal defend a territory? The reason varies. For some animals it is the acquisition and protection of a needed resource, such as food, or a reduction in the risk of predation. For others it is the attraction of a mate. Whatever the apparent reasons for defending a territory might be, the basic reason is the benefits derived, ultimately an increased probability of survival and improved reproductive success—in short, increased fitness. By defending a territory, the individual forces others into suboptimal habitat, reducing their fitness and at the same time increasing the proportion of its own offspring in the population.

Consider the territorial behavior of the gall-forming aphid *(Pemphigus betae)* on the leaves of its host plant, narrowleaf cottonwood *(Populus augustifolia)* (Whitham 1987). The female sexual aphid oviposits a single overwintering egg from which a wingless stem mother emerges in spring at the time of leaf break. The stem mothers migrate to and establish themselves on developing leaves. They probe the expanding tissues to form a small depression. In a short time leaf tissue will envelope the aphid, forming a hollow gall

in which she will parthenogenetically produce up to 300 progeny. The number of aphids invading the tree may be high, up to 850 per 1000 leaves.

The aphids have a preferred site, the base of the largest leaves. Such leaves have the lowest concentration of phenolics, which inhibit herbivorous insects (see plant defenses, Chapter 21) and they are the least susceptible to leaf fall. They are rare, only about 1.6 percent of available leaves, so there is a great deal of competition for them. Kicking-and-shoving contests between prospective colonists settle disputes for microterritories 3–5 mm in length. The winners usually are the largest stem mothers, who may or may not share a leaf with another. Whitham found that stem mothers who had a leaf to themselves had the highest reproductive success. When two aphids shared a leaf, the stem mothers that secured the basal portion of the leaf produced on the average 56 percent more progeny than those displaced to the inferior distal portion of the leaf (Figure 16.9). These differences in reproductive success result from microhabitat variations in the quality of the leaf. Only a few millimeters' difference in leaf position can affect aphid reproduction.

To determine what would happen to reproductive success, Whitham removed one member of a competing pair. Immediately the remaining aphid crossed the former territorial boundary and enlarged her own. If the remaining member was the distal stem mother, she increased the number of her progeny by an average of 48 percent. If she was the basal stem mother, she improved her reproductive success by only 18 percent.

This study points out several aspects of territoriality. The territorial animal that claims the better territory has the higher reproductive success. The reproductive cost of losing is lower fecundity, if the los-

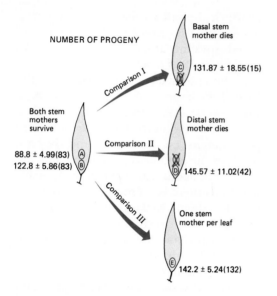

Figure 16.9 Comparisons of the reproductive success and fitness of territorial stem mothers of the gall-forming aphid *(Pemphigus betae)*. When either stem mother of a competing pair suffered an early death, the remaining stem mother produced significantly more progeny (A < C and B < D). Stem mothers occupying leaves singly from the beginning of gall formation produced more progeny than either member of a competing pair (A< B < E). When released from competition early in development, the surviving stem mothers on the average produced the same number of progeny as stem mothers solitary from the beginning of gall formation (C = D = E). Mean number of progeny from leaves of some quality and size ± 1 SE *(n)* are indicated for 355 surviving stem mothers. (From Whitham 1986: 140.)

ers reproduce at all. The winners are dominant individuals. Among aphids this is measured by size. Solitary stem mothers achieved the highest reproductive success, followed by the basal stem mother of a competing pair and then by the distal stem mother. Smaller stem mothers are forced to share leaves or have to settle on inferior smaller leaves. Thus the cost of losing a competitive interaction is reduced fitness.

Defending a territory can be a costly business, especially if optimal resources are limited. When the stakes of winning or losing in the aphids are high, contests over a position on a leaf may last two days and result in the death of one or both contestants. On a less intense scale, territorial defense uses energy, consumes time, and interferes with feeding, courtship, mating, and rearing of young.

Like all economic endeavors, territorial ownership has its costs and benefits, and the territorial owner has to balance the two (Figure 16.10) (Davies and Houston 1984). Some territories are economically defendable and some are not. A general prerequisite is a predictable resource somewhat dispersed. Acquisition of an area assures its owner of its resource needs, reducing foraging costs, and allows time for other activities. If resources are unpredictable and patchy, it may be advantageous for individuals to belong to a group and cooperatively seek needed resources without being restricted to any

Figure 16.10 A graphic model of cost-benefit ratio curves as they might apply to territorial defense. Costs of territorial defense increase as the size of territory increases. Defense is profitable only between X and Z. A maximum cost-benefit is at Y in each case. (See Schoener 1983.)

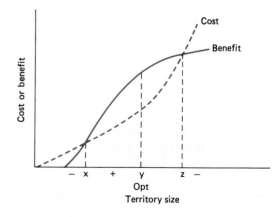

one area. Spotted hyenas *(Crocuta crocuta),* for example, live in clan territories in the Ngorongora Crater of Kenya, where resources are predictable, whereas on the Serengeti Plains, where food is seasonal, they range over wide areas and do not defend a territory (Kruuk 1972).

Territory Size. Closely associated with the cost-benefit ratio of territorial ownership is territorial size. As the size of a territory increases, the cost of territorial defense increases. Many male birds in spring attempt to claim more ground than they can economically defend. They are forced to draw in the boundaries to make the area more manageable to defend; but there is a minimum size area below which the territorial owner cannot go, because it will be too small to meet needs. The number of territorial owners an area can hold is determined by the total area available divided by the minimum size of the territory. This minimum size, however, can change from year to year, depending upon resource availability, habitat changes, adult mortality, settlement patterns of juveniles, and the like (Knapton and Krebs 1974). Somewhere along the gradient of too small and too large is an optimal size for a territory, one from which the owner gains maximum benefits for the costs incurred.

Some animals, notably certain spiders, have fixed territories (Riechert 1981). The size of the territories remains the same, regardless of spider numbers or resource availability. Such fixed sizes may be transmitted as part of an inherited behavior, or they may be transmitted culturally from one owner to the next, and between generations.

Other animals are more flexible. Optimal size will vary from year to year and from locality to locality. If a resource such as food is abundant, the territory may be small; and if resources are less abundant, it may be larger. In general, territory size tends to be no larger than required. For example, the territories of the golden-winged sunbird *(Nectarina reichenowi)* vary greatly in size and in floral composition, but each territory contains just enough of a nectar supply to meet an individual's daily energy requirement (Gill and Wolf 1975). Such flexibility in the size of territories has been likened by Julian Huxley (1945) to an elastic disk compressible to a certain size. Territory size decreases as density increases; but when the territory compresses to a certain size, the resident resists further compression and denies access to additional settlers. Because aggressive behavior varies among individuals, the most aggressive have the advantage and the less aggressive are forced to settle elsewhere.

For some animals—birds in particular—it is not the size of the territory that counts, but its quality (Figure 16.11). Some males are successful in claiming the best territories, usually measured by some feature of the vegetation that makes them superior nesting sites or food sources. For example, hummingbirds and Hawaiian honeycreepers seem to assess differences in standing crops of floral nectar and defend areas with richest nectar supplies, but cease to defend territories when food is superabundant (Carpenter 1987). The ovenbird *(Seiurus aurocapillus)* selects and defends areas with greatest prey abundance per unit area (Smith and Shugart 1987), and optimal territory for the dicksissel is measured by litter depth and vegetation density (Zimmerman 1971). Less successful males occupy suboptimal territories, and many males fail to secure a territory.

To successful males come the females, and the "wealthiest" males always obtain a mate. In fact among some species of birds,

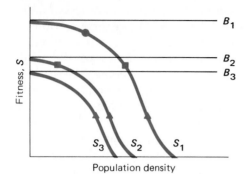

Figure 16.11 Relationship among habitat suitability, population density, and settling. The suitability of habitat is ranked from B_1 to B_3. The curves S_1, S_2, and S_3 show the fitness of individuals in habitats at different population densities. At low population density all individuals will settle in habitat B_1. At intermediate density some individuals will settle in habitat B_2. At high density all three habitats will be settled. The model assumes that each individual will select the optimum habitat, where chances of survival and success at reproduction will be the highest. If all individuals choose a habitat of highest suitability, then from the viewpoint of unsettled individuals, all occupied habitats will be equally suitable. When all optimum habitat is filled, then the next most suitable or suboptimal habitat becomes for the remaining individuals the optimum habitat, and they settle in it. Once that is filled, marginal habitat becomes suitable. The model implies that as density increases, habitat suitability decreases. (After Fretwell and Lucas 1969: 24.)

such as the dicksissel (Zimmerman 1971) of the midwestern grasslands and some of the marshland blackbirds (Orions 1969), females will accept a male already mated if he holds an outstanding territory. Apparently the quality of the territory is such that the odds of a successful nesting by the female are greater there, even though the female may have to share the area with another. Lenington (1980) found that female redwinged blackbirds (*Agelaius phoeniceus*) chose males based on territory quality. Early-arriving females were able to choose males on the basis of territory quality, whereas late-arriving females faced aggression from resident females and had to settle for males holding territories in poorer habitat. At the other extreme were male birds who owned such poor territories they were unable to attract a mate.

Floaters. As a result of contest competition among males for space, some individuals secure optimal territories, while at the other extreme, some individuals are denied territory. Thus a portion of the population does not reproduce because they are excluded from the suitable breeding sites by territorial individuals. They make up a surplus population, a floating reserve that would be able to reproduce if a territory became available to them.

Such a floating reserve of potentially breeding adults has been described for a number of species. The number of floaters may be great. Studies of a banded white-crowned sparrow population (*Zonotrichia leucophrys nuttalli*) in California indicated a nonbreeding surplus of potentially breeding birds (Petrinovich and Patterson 1982). In fact, 24 percent of the territorial holders entered the breeding population two to five years after banding, and 25 percent of the nestlings that acquired territories did so two to five years after their birth. Territory holders that disappeared during the breeding season were quickly replaced.

Although the existence of floating reserve populations is acknowledged, few data exist on the social organization and behavior of surplus birds. Floaters may live singly off the territories as the white-crowned sparrows do; they may spend time on the breeding territories of others; or they may form flocks with a dominance hierarchy on areas not occupied by territory holders. An example of the latter is the red grouse of the heather-dominated moors of Scotland.

The red grouse has three social classes: (1) territorial cocks and their hens; (2) nonterritorial surplus birds that live as a floating reserve on the periphery of the breeding ground; and (3) nonterritorial transient birds. In fall and spring red grouse experience a sharp decline in numbers associated with territorial behavior and loss of nonterritorial birds from predation. If one of the territorial birds dies or disappears, it place is taken by a bird from the floating reserve. By late winter all surplus birds are removed from the moors and the breeding population in spring is fixed by territorial behavior of the previous fall (Figure 16.12) (Watson and Jenkins 1968; Watson and Moss 1972). Competition for breeding territories sets an upper limit to the breeding numbers, but density of breeding popula-

tion varies from year to year because of secondary variables that influence territorial behavior. Especially important is the abundance of heather *(Calluna vulgaris)* and its nutritional quality as measured by its nitrogen content (Moss et al. 1972; Lance 1978). Size of grouse territories varies inversely with the proportion of ground occupied by patches of heather sward (Miller and Watson 1978).

Figure 16.12 Territoriality and social behavior as regulatory mechanisms in red grouse populations. In early winter the birds become divided into three classes, only one of which will breed the following spring. The other classes are eliminated. Breeding success in any year appears to be inversely proportional to breeding success the previous year: The greater the success in the previous year, the poorer it is in the current one. Breeding success and therefore population size is regulated through territorial behavior. (From Watson and Moss 1971: 97.)

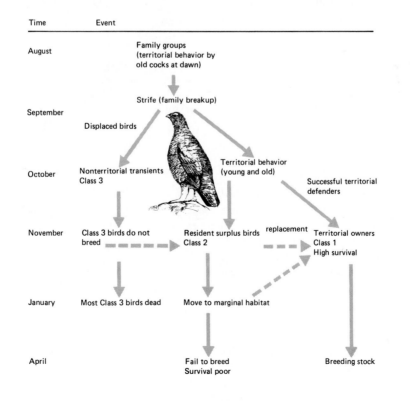

Another example of this strategy is provided by the detailed studies by S. Smith (1978) of the rufous-collared sparrow *(Zonotrichia capensis)* in Costa Rica. By observing banded birds, both territorial and nonterritorial, and by selectively removing certain individuals, Smith was able to determine the role of the floater or "underworld" bird. Territorial sparrows on her study area occupied small territories ranging from 0.05 to 0.40 ha and made up 50 percent of the total population. The other 50 percent, underworld birds consisting of both males and females, lived in well-defined restricted home ranges within other birds' territories (Figure 16.13). Male home ranges, often disjoined, embraced three or four territories. Female home ranges were usually restricted to a single territory. Because home range boundaries of both sexes coincided with territorial boundaries, each territory held two single-sex dominance hierarchies of floaters, one male and one female. When a territorial owner, male or female, disappeared, it was quickly replaced by a local underworld bird of appropriate sex on the territory. These floaters usually entered the territories as young birds hatched some distance away and were tolerated by the owners.

A similar situation exists among some territorial spiders (Reichert 1981) and the aphid *Pemphigus* (Whitham 1987). They, too, have a floating reserve of individuals who will quickly claim vacated sites. Among the spiders, the floaters live in cracks and crevices within occupied territories. Periodically, floaters will attempt unsuccessfully to take over an occupied habitat.

Population Regulation. A consequence of territoriality can be population regulation. If no limit to territorial size exists and all pairs that settle on an area get a territory, then territoriality only spaces out the population. No regulation of the population results. If territories have a lower limit in size, then the number of

Figure 16.13 Territorial boundaries of rufous-collared sparrows. Home ranges of two "underworld" females and males and superimposed on occupied territories. Eventually some of the females occupied a territory by replacing a missing bird. The male home range *a* and *b* are disjoined and each includes several territories. (Adapted from S. Smith 1978: 570.)

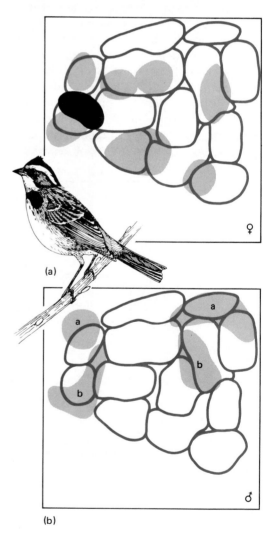

pairs that can settle on an area is limited. Those that fail to get a territory have to leave. Thus territoriality might regulate population density, but only under certain conditions.

For example, in the arctic ground squirrel *(Spermophilus undulatus)* all females are allowed to nest, but territorial polygamous males drive excess males from the colony into submarginal habitat, where they exist as a nonbreeding floating population. The number of breeding males remains constant because losses are continually replaced from the floating population, which in turn decreases drastically over the year largely from predation and weather. In this species territoriality stabilizes the number of breeding males but does not regulate the population, because only males appear to be surplus.

Krebs (1971) removed breeding pairs of great tits from their territories in an English oak woodland. The pairs were replaced by new birds, largely first-year individuals, that moved in from territories in hedgerows, considered suboptimal habitat (Figure 16.14). The vacated hedgerow territories, however, were not filled, suggesting that a floating reserve of nonterritorial birds did not exist. In this case territorial behavior limits density of breeding birds in optimal habitat, but does not regulate the population because all birds are breeding in some habitat.

Evolution of Territoriality. How territoriality might have evolved has been the subject of speculation (see Brown 1969). Several hypotheses have been advanced, including prevention of overpopulation (Wynne-Edwards 1962) and need for individual food exploitation. Defense of a food resource, or rather space containing a food resource, may be important in some species, such as the red grouse, the

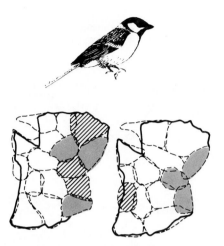

Figure 16.14 Replacement of removed individuals and settlement of vacated territory. Six pairs of great tits *(Parus major)* were removed between March 19 and March 24, 1969 (left stippled area). Within three days, four new pairs had taken up territories (right stippled area) and some residents had expanded their territories; so after replacement territories again formed a complete mosaic over the woods. (From Krebs 1971: 7.)

ovenbird (Stenger 1958), and the golden-winged sunbird (Gill and Wolf 1975), because in those species territorial size is inversely related to food density. Brown has proposed a general theory that for territoriality to evolve some resource must be in short supply, it must be defensible, and it must be "worth fighting for" (Figure 16.15). Because space is more defensible than food, aggressive behavior is directed toward excluding other animals from space containing food resources.

Closely linked to the evolution of territoriality is aggressive behavior. Aggressive behavior is not likely to benefit an individual unless it provides something worth the risk. If an individual can obtain a mate, food, or nesting space without defending a territory, then territorial behavior would be disadvantageous, but if a resource is in short supply, an animal has

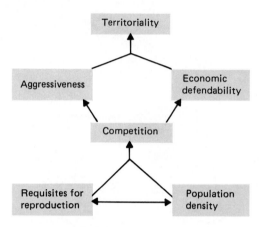

Figure 16.15 Outline of a general theory of evolution of diversity in intraspecific territorial systems. In this resource-centered theory, the value of territorial behavior for a species is determined by the supply of a critical resource and by the energy costs and gains associated with fighting for a share of the resource. (From Brown 1964: 161.)

much to gain in defense of a territory in spite of the risks of aggressive behavior.

Home Range

Territory should not be confused with home range. *Home range* is an area in which an animal normally lives. It is not necessarily associated with any type of aggressive behavior (Figure 16.16). A home range may be defended in part or in whole (in which case the home range is identical with a territory) and may overlap with those of other individuals of the same

Figure 16.16 Home ranges of female and male mountain lions of Idaho, determined by observations of marked animals. Females share the same home range. Males own exclusive home ranges. (Male 26 replaced male 7 and occupied the same home range.) (After Hornocker 1969: 407.)

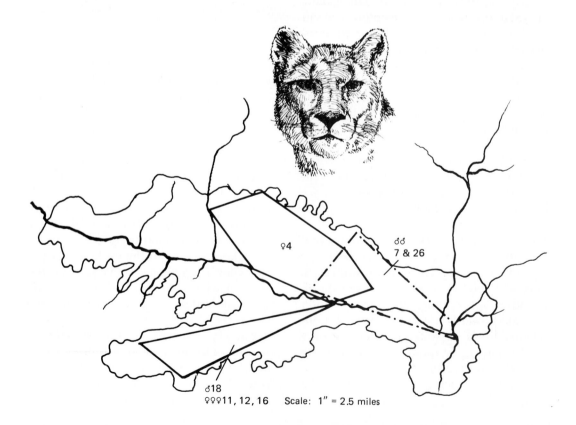

species. However, dominance may exist among individuals with overlapping home ranges, and subordinate individuals tend to avoid contact with dominant individuals. Both use portions of the same area by avoiding contact both spatially and temporally.

Generally home ranges do not have fixed boundaries. Seldom is a home range rigid in its use, size, and establishment. It may be compact, continuous, or broken into two or more discontinuous parts reached by trails and runways. Irregularities in distribution of food and cover produce corresponding irregularities in home range and in frequency of animal visitation. The animal does not necessarily visit every part daily. Its movements may be restricted to trails, and most of its activities may be concentrated in a smaller *core area* used more intensively than other parts.

The size of a home range relates to body size (McNab 1963; Harestad and Bunnell 1979). Large mammals have larger home ranges than smaller ones, and carnivores generally have larger home ranges than herbivores and omnivores of similar size. Males and adults have larger home ranges than females and subadults. Weight alone is sufficient to account for this difference within a species without invoking any competitive interactions. The home range of herbivores and omnivores increases at a nearly constant rate as body weight increases; among carnivores the home range increases at a greater rate as body weight increases (Figure 16.17).

Like territoriality, possession of a home range confers certain advantages. The animal becomes familiar with the local area, where to find food, shelter, and escape cover from enemies with a minimum expenditure of energy. It can define a series of escape routes to cover and travel routes to food sources throughout the year.

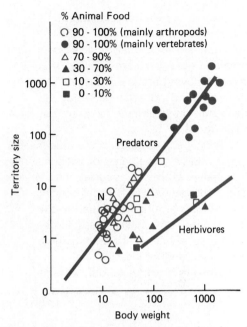

Figure 16.17 Relationship between the size of home range (ha) and body weight of North American mammals. (Adapted from Harestad and Bunnell 1979: 390.)

■ Summary

Some mechanisms of intraspecific competition involve physiological and behavioral interactions among members of the population. Increased density may affect population growth through physiological responses of individuals. Crowding may produce stresses that result in endocrine imbalances, especially in the adreno-pituitary complex, which result in abnormal behavior, abnormal growth, degeneration, and infertility.

Social pressure and crowding may also induce emigration or dispersal. Dispersal appears to be density-independent except at the highest population levels; most dispersal appears to take place when populations are increasing rather than at the peak population levels. Dispersal may function in population regulation by encouraging mostly subadults and subdominant individuals to leave their natal area and occupy vacant habitats. Although risks of moving are great, successful dispersing individuals, which may be genetically programmed to do so, improve their fitness.

Social behavior reflects the influence of density. Basic social behavior is agonistic or aggressive, expressed as social dominance and territoriality, which act as spacing mechanisms among individuals in the population. Territoriality, the defense of a fixed area, divides space among some individuals and excludes others. Territoriality is expressed only when a resource such as food is contained within an economically defensible space. Once established, territorial owners defend their space by aggressive signals such as song, display, or chemical marking. In contrast, social dominance results in the sharing of space and maintenance of individual distances, but with some individuals dominant. Subordinate individuals avoid contact with dominant ones by using different parts of a shared area or home range and by using the area at different times. Territoriality can function in population regulation only if it creates a surplus population consisting of sexually mature individuals prevented from breeding by territory holders.

Life History Patterns

- ## Reproductive Effort

    Energy Costs
    Parental Investment
    Age, Size, and Fecundity
    Reproductive Value
    Reproductive Costs
    Annual Versus Perennial
    Life Histories

- ## *r*-Selection and
  K-Selection

- ## Forms of Sexual
  Reproduction

- ## Mating Systems

    Monogamy
    Polygamy
    Plastic Mating Systems

- ## Sexual Selection

- ## Habitat Selection

- ## Summary

Life history patterns encompass many processes that affect fitness. They include, among other things, fecundity and survivorship, physiological adaptations, modes of reproduction, age at reproduction, number of eggs, young, or seeds produced, parental care, means of avoiding environmental extremes, size, and time to maturity. Success among individuals is measured only in terms of successful offspring. How this end is achieved is the organism's life history strategy.

■ Reproductive Effort

An ideal situation for any organism is to reproduce shortly after birth, produce a large number of highly adapted offspring, live for a long period of time, reproduce frequently, lavish parental care on the offspring, and expend the maximum amount of energy on reproduction. Realistically, any organism has to compromise considerably with this ideal. Let us examine the consequences of investing energy in the production of offspring.

Energy Costs

Investment in reproduction carries its costs. It collects its charges in terms of energy and survival. Parents have a finite amount of energy for reproduction at any given time. A certain minimum amount of energy is needed for viable offspring. As parents apportion available energy among an increasing number of offspring, the fitness of individual offspring declines. Eventually there comes a point at which increases in reproductive effort per offspring or an increasing number of offspring results in declining payoff in fitness (Smith and Fretwell 1974).

Parents have to apportion some of their assimilated energy to growth and maintenance and some to reproduction. How they allocate this energy to growth and reproduction is central to the reproductive strategy an organism employs. If an organism directs more energy to reproduction, it has less energy to allocate to growth and maintenance. As a result the individual grows more slowly to the next stage or it may fail to reproduce or to survive. For example Lawler (1976) found that reproductive females of the terrestrial isopod *Armadillidium vulgare* had a lower rate of growth than nonreproductive females and that nonreproductive females devoted as much energy to growth as the reproductive females devoted to both growth and reproduction. Tinkle (1969) has demonstrated that among lizards, species with larger fecundity have poorer individual survival because they channel more energy into egg production over a short period, resulting in heavy physiological stress. Many North American freshwater fish experience a considerable loss of size or growth from reproduction, up

to 25 percent or more in some trout *(Salmo)* (see Bell 1980). The female Allegheny salamander *(Desmognathus ochrophaeus)* invests 48 percent of its annual energy assimilation in reproduction, including energy stored in eggs and energy costs of brooding (Fitzpatrick 1973). In homoiotherms energy costs are more difficult to determine. The costs show up in a different manner, to be considered later.

Among plants energy costs vary. Plants, like animals, must allocate a portion of their net annual assimilation to reproductive effort. Primack (1979) found among 15 species of plantain *(Plantago)*, 6 perennials and 9 annuals, that annuals had higher reproductive efforts based on milligrams of seeds produced per square centimeter of leaf area than perennial species. Weedy perennial species had higher reproductive costs than native species; but perennials can allocate high amounts of yearly biomass to reproduction. For example, *Ranunculus bulbosum* and *R. acris* allocate up to 60 percent to reproduction, and the showy goldenrod *Solidago speciosa* of old fields and woods up to 35 percent (Abrahamson and Gadgil 1973). The goldenrod shows phenotypic plasticity in adjusting energy allocation to environmental situations. Plants that grow in adverse environments allocate more energy to reproduction than those in more moderate habitats (Figure 17.1).

The shortcomings of such estimates are many. In plants the estimates fail to take into account belowground energy allocations, including the production of bulbs, corms, and roots. Simply comparing aboveground biomass of vegetative parts to biomass of reproductive output may not be an accurate assessment of the energy costs of reproduction, as Reekie and Bazzaz (1987) found in intensive studies of reproductive costs in the grass *Agropyron repens*. They point out that re-

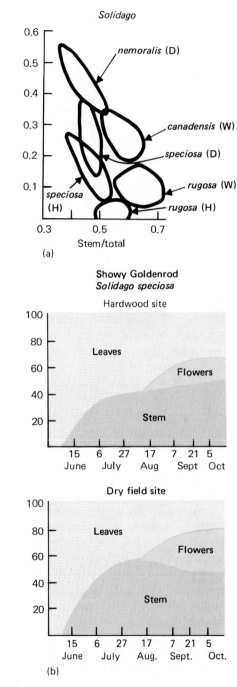

(a)

Showy Goldenrod
Solidago speciosa

Hardwood site

Dry field site

(b)

Figure 17.1 Allotment of energy to reproductive effort in goldenrods. The reproductive effort of six populations of goldenrods is measured by plotting the ratio of dry weight of reproductive tissue to total dry weight of aboveground tissue on the ordinate as a function of ratio of weight of stem tissue to total dry weight of aboveground tissue. Each enclosed curve embraces all points representing individuals included in a single population. D = dry field site; W = wet meadow; H = hardwood site. (b) the reproductive effort of showy goldenrod (*Solidago speciosa*) growing on a hardwood site and on a dry field site is measured by plotting the percent of total biomass of stems, leaves, and flowers as a function of time during the growing season. Note that goldenrods growing on the dry site allocate a greater portion of energy to reproductive effort, probably because density-independent mortality is higher there. (Adapted from Abrahamson and Gadgil 1973: 654.)

source allocation is not necessarily a reliable indicator of costs of reproduction, because plants experience different costs. A plant with low cost of reproduction could afford to allocate a larger proportion of its resources to reproduction. Much the same may be true in animals, particularly homoiotherms (Clutton-Brock 1984). Bell (1980) points out that defining costs of reproduction in terms of energy is irrelevant to the evolution of life histories.

Parental Investment

Of greater significance to evolution of life history is parental investment in offspring from hatching or birth of young to independence. Two basic choices are to apportion parental investment among many small young or to concentrate investment in a few larger ones. Within that range of choices the parents have to adjust the number of young they can rear without exceeding the resources available or significantly reducing the probability of their own survival.

Clutch Size. One method of apportioning energy to reproductive effort or parental investment is the adjustment of clutch size. Lack (1954) proposed that among most birds clutch size evolved through natural selection to correspond to the average largest number of young the parents can feed. There is indeed evidence that the optimal clutch size should be lower than that which would produce the maximum number of young each season. Individuals of some species may adjust parental effort to changing energy resources. For example, the common grackle *(Quiscalus guiscula)* begins incubation before its clutch of five eggs is complete, a pattern of hatching ensuring survival of some young under adverse conditions (Howe 1976). The eggs laid last are heavier and the young from them grow fast; but if food is scarce, the parents do not feed these late offspring, allowing them to die of starvation. Asynchronous hatching favors the early hatched young at the expense of those hatched later. Although parents attempt to assure the survival of all young, their investment in young is protected by starvation of late-hatched birds if a food shortage arises.

What happens if birds attempt to rear more than an average-sized brood has been demonstrated by the experimental enlargement of broods by adding young to the nest. The average weight of the young in these superbroods is reduced (Askenmo 1977; Westmoreland and Best 1987) and the survival of the overtaxed parents reduced. Askenmo (1979) compared the return rate of male pied flycatchers *(Ficedula hypoleuca)* who cared for enlarged broods with those that reared normal broods. The proportion of control males who returned was twice that of males who aided in the rearing of enlarged broods. Probably the parents with enlarged broods had to work harder, sacrificing maintenance energy for foraging energy.

Westmoreland and Best (1987) exchanged eggs of equal age among nests of mourning doves *(Zenaida macroura)* to create clutches of one and three eggs and compared the breeding success of those parents with ones incubating a normal clutch of two. Adults successfully incubated clutches of three eggs, but the nestlings grew more slowly than the two-brood young, had less full crops during the nestling period, took 1.3 days longer to fledge, and upon leaving the nest weighed 26 percent less than the control brood. Although the parents of enlarged broods fledged 30 percent more offspring, post-nesting survival may be lower. Both physiological and ecological factors probably restrict the mourning dove to a two-egg clutch, holding the birds to an optimum rather than a maximum clutch size, as Lack postulated.

Lack (1954) also proposed that clutch size is an adaptation to food supply. Temperate species have larger clutches because increasing daylength allows parents a longer time to forage for food to support larger broods. In the tropics, where daylength does not change, food becomes a limiting resource. Cody (1966) modified this concept by employing the principle of allocation of energy and proposed the hypothesis that clutch size results from different allocations of energy to egg production, avoidance of predators, and competitive ability. In temperate regions periodic local climatic catastrophes can hold the population below carrying capacity; natural selection favors an increase in r, and clutch size on the average is larger than in tropical regions. The stability and predictability of the climate in the tropics makes maintaining carrying capacity more important than increasing the production of offspring to replace losses due to environmental instability. More energy is expended in meeting intraspecific and interspecific competition, which is keener than in temperate regions.

The predictions of this theory are met by a number of field examples. Birds in temperate regions have larger clutch sizes than birds in the tropics (Figure 17.2), and mammals at higher latitudes have larger litters than those at lower latitudes (Lord 1960). Lizards exhibit a similar pattern. Those living at lower latitudes have smaller clutches, have higher reproductive success (or lower egg mortality), reproduce at an earlier age, and experience higher adult mortality than those living at higher latitudes (Tinkle and Ballinger 1972; Andres and Rand 1974).

Insects, too, support this theory. An example is the milkweed beetle (Landahl and Root 1969). When a population of the temperate species *Oncopeltus fasciatus* and one of the tropical species *O. unifasciatellus* were reared in the laboratory, both exhibited similar duration of egg stage, egg survivorship, developmental rate, and age of sexual maturity. Although clutch size for both species was the same, the temperate species produced eggs sooner and laid a larger number of eggs over time. The tropical species had fewer clutches, rather than smaller ones, and laid over a longer period. Its total egg production was only 60 percent of that of the temperate species.

Plants follow the general principle of allocation on a latitudinal basis, as do many animals. S. McNaughton (1975) investigated the allocation of resources to reproduction in a series of greenhouse studies of the cattail *(Typha)*. He measured reproductive effort by the growth of rhizomes, the cattail's principal means of population growth within a given habitat. Cattails shed their very small seeds to the wind as a means of colonizing a new habitat. McNaughton considered three species of cattails on a climatic gradient: the common cattail *(T. latifolia)*, a climatic generalist that grows from the Arctic circle to the equator; the narrowleaf cattail *(T. angustifolia)*, restricted to northern latitudes of North America; and *T. domingensis*, restricted to southern latitudes of North America. The cattails exhibited a climatic gradient in the allocation of energy to reproduction. *T. angustifolia* and northern populations of *T. latifolia* grew earlier and faster and produced a greater number of rhizomes than *T. domingensis* and southern populations of *T. latifolia*, but the southern plants produced larger

Figure 17.2 Relationship between clutch size and latitude in the subfamily Icteridae (blackbird, orioles, and meadowlarks) in North and South America and the worldwide genus *Oxyura* (ruddy and masked ducks) of the subfamily Anatinae. (Adapted from Cody 1966).

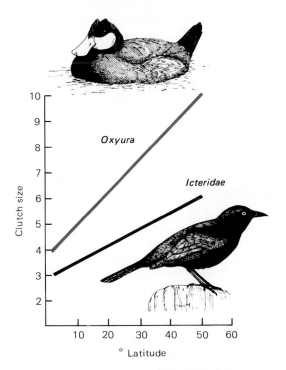

rhizomes. Faced with stronger intraspecific and interspecific competition, southern cattails invested more energy in vegetative growth. However, cattail species exhibited a gradient in trade-offs between colonization potential and competitive potential (Figure 17.3). An inverse relation between rhizome production and rate of foliage production allows adaptation to site conditions. Where competition is intense, plants can produce more and taller foliage, but reduce rhizome production. Where competition is low, plants can invest more energy in rhizome production.

Tropical members of the sunflower tribe Heliantheae in the aster family (Asteraceae) have significantly smaller clutch sizes (ovules per flowering head) than those of temperate and alpine regions, with an average of 51.2, 82.7, and 89.6 ovules respectively. Levin and Turner (1977) suggest that this may result from

greater risks of attack by herbaceous insect larvae. By spreading offspring into smaller cohorts, fewer would be destroyed in each attack.

Care of Young. Parental investment in animals includes care given the young from birth or hatching until independence (see Lloyd 1987). The extent and kind of care is influenced by the maturity of young at birth. Some birds incubate eggs for a much longer period of time, investing considerable energy prior to hatching; and some mammals have much longer gestation periods and give birth to young in a more advanced stage of development. This strategy, of course, reduces the proportion of parental investment in young after hatching or birth.

Basically birds and mammals are either precocial or altricial at hatching or birth. *Precocial* animals are able to move about at or shortly after birth, although some time may elapse before they can fly or move about as adults. *Altricial* animals are born helpless, naked or nearly so, often blind and deaf. Between the two extremes there is a wide variation in the stage and nature of maturity.

Nice (1962) has classified maturity at hatching in birds in a way that can be extended to mammals (Table 17.1). Precocial and semiprecocial birds, hatched with eyes open and completely covered with down, are mobile to some degree and leave the nest in a day or so. Most precocial birds are capable of feeding themselves on small invertebrates and seeds; others follow the parents and respond to their food calls; still others are fed by the parents. Semiprecocial birds are able to walk, but because of feeding habits of the parents, they are forced to remain in the nest. Semialtricial birds are hatched with a substantial covering of down and with eyes open or closed, but are unable to leave the nest. Altricial birds are com-

Figure 17.3 Relationship between rate of rhizome production and rate of foliage production in cattails *(Typha)*. Line is best least squares fit. Dark circles = *T. latifolia;* open circles = *T. angustifolia;* and dark squares = *T. domingensis.* (From S. MacNaughton 1975: 258.)

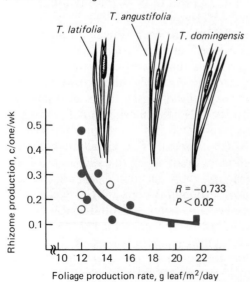

TABLE 17.1 Classification of Maturity at Hatching in Birds

Type	Characteristics	Example
FEED SELVES		
Precocials	Eyes open, down-covered, leave nest first day or two	
Precocials 1	Independent of parents	Megapods
Precocials 2	Follow parents but find own food	Ducks, shorebirds
Precocials 3	Follow parents and are shown food	Quail, chickens
FED BY PARENTS		
Precocials 4	Follow parents and are fed by them	Grebes, rails
Semiprecocials	Eyes open, down-covered, stay at nest though able to walk	Gulls, terns
Semialtricials	Down-covered, unable to leave nest	
Semialtricials 1	Eyes open	Herons, hawks
Semialtricials 2	Eyes closed	Owls
Altricials	Eyes closed, little or no down, unable to leave nest	Passerines

Source: Adapted from Nice 1962.

pletely helpless; their eyes are closed and they have little or no down.

A somewhat similar classification could be devised for mammals. Young mice, bats, and rabbits are born blind and naked and thus are altricial. Young of wolves, foxes, dogs, and cats are born with hair and are soon able to crawl about the nest or den, but are blind for several days. They might be called semialtricial. Deer, moose, and other ungulates, as well as horses and pigs, would fall into the semi-precocial category. They are very ungainly on their legs for several days after birth, and during this time they establish a nursing routine. They may be hidden alone by the mother or held in a "pool," characteristic of the elk (Altmann 1960). The young wait for the dam to return to the hiding place to nurse and to be licked. Most precocial of all mammals are seals, which might well be a "precocial 4" according to the classification in Table 17.1. Not only is delivery extremely rapid, approximately 45 seconds in the gray seal (Bartholomew 1959), but movements and vocalization appear very shortly after birth. Newly born fur seals are able to rise

up and call from 15 to 45 seconds after birth and are capable of shaky but effective locomotion a few minutes after birth (Bartholomew 1959). Even while the umbilical cord is still attached, the pups are able to shake off water, bite and nip at each other, and scratch dog-fashion with the hind flippers. They attempt to nurse within five minutes after birth. Although pups continue to nurse the cows for some time, the cows are protective and attentive toward their young only between parturition and estrus. Thus this behavior is conspicuous only for a few hours to a day after the cow has given birth to the pup.

Parental care is not well developed among invertebrates. Some retain eggs within the body until they hatch; others, such as crayfish, carry eggs externally. Invertebrate parental care is most highly developed in social ants, bees, and wasps. Social insects provide all five functions of parental care: food, defense, heat, sanitation, and guidance. (See Tallamy 1984 on insect parental care.)

In fish, the quality of parental care appears to be related to egg size (Sargent et al. 1987). The larger the egg size relative to the size of the fish, the greater the amount of parental care. This extra care apparently relates to the longer time offspring take to develop and absorb the yolk sac and become juveniles; but in turn juveniles that hatch from larger eggs experience lower mortality, faster growth, and earlier sexual maturity. Large-mouth bass, sticklebacks, and catfish lay relatively few large eggs, which the male actively defends; later he protects the young. Other fish, such as trout, lay more and smaller eggs and give them no care at all.

Parental care is usually poorly developed among amphibians, although some species of salamanders remain with the eggs. Some anurans, notably the male midwife toad, carry eggs and subsequent young on their bodies and eventually place them in a suitable environment for further growth.

Internal fertilization and terrestrial reproduction are fully developed among reptiles. Relatively few eggs well supplied with yolk may be carried inside the mother's body until they hatch; or they may be placed in nests buried in the ground and given little subsequent care. Crocodiles, however, actively defend the nest and young for a considerable time.

Age, Size, and Fecundity

A strong relationship exists between fecundity and size and age, especially in plants and poikilothermic animals, among which fecundity increases with size and age. Among the plantain (*Plantago*) species, for example, annuals have a higher seed output per unit of leaf area than perennial species; but perennials with a larger average leaf area produce a greater weight of seeds per plants (Primack 1979). Perennials delay flowering until they have attained a sufficiently large leaf area to support seed production. Annuals show

no such relationship between leaf area and reproductive output, which seems to be independent of plant size once reproduction starts; but size differences among annuals do result in differences in seed production. Small plants produce few seeds, even though the plants themselves may be contributing the same proportionate share to reproductive efforts. Annual jewelweeds (*Impatiens spp*) growing in optimal environments produce both outcrossed and self-fertilized flowers, the latter late in the season. Jewelweeds growing in stressed environments produce only a few self-fertilized flowers and seed early in the season (Waller 1982).

Age at maturity and seed production varies among trees. Some, like Virginia pine (*Pinus virginianus*), will start to produce seed at five to eight years when growing in open stands. In dense stands Virginia pine may delay seed production for 50 years. Quaking aspen (*Populus tremuloides*) produces seeds at the age of 20 years and white oak not until the age of 50 years. Acorn production varies with size. Trees 16 inches in diameter at breast height produce around 700 acorns; those 24 to 26 dbh produce 2000 or more acorns (Downs and McQuilkin 1944).

Similar patterns exist among poikilothermic animals. Fecundity in fish increases with size, which in turn increases with age. Because early fecundity reduces both growth and later reproductive success, there is a selective advantage to delaying sexual maturation until more body growth is achieved. Gizzard shad (*Dorosoma cepedianum*) reproducing at two years of age produce 59,000 eggs. Those delaying reproduction until the third year produce about 379,000 eggs. Egg production then declines to about 215,000 eggs in later years. Among these fish only about 15 percent spawn at two years of age and about 80 percent at three years of age.

Delayed maturation is characteristic of fish and other poikilotherms. Among 104 species of freshwater fish in the United States, only 15 percent mature the first year, and 21 percent the second year. Sixty-four percent of all species mature at three years of age or older. By contrast, 55 percent of 171 species of placental mammals reach reproductive maturity at one year of age, 20 percent the second year. Only 25 percent mature at the age of three or older (Bell 1980).

Among homoiotherms there is an apparent relationship between body size and fecundity, particularly as it relates to the age of first reproduction. Harvey and Zummato (1985), however, point out that when the effect of body size is removed, the relative age at which mammalian females first reproduce is strongly correlated with the expectation of life at birth (e_x). A possible reason may be that each bout of reproduction reduces subsequent reproductive success (by further increasing mortality or reducing fecundity) and that older mothers produce larger litters and more viable offspring than do younger mothers. As life expectancy decreases, selection will favor maturation at an earlier age. Among the red deer, young hinds and old hinds had the highest survival of their calves and middle-aged hinds the poorest (Clutton-Brock et al. 1982). This finding probably reflects the better body condition of the young hinds and a greater energy investment in offspring by old hinds. Such an investment by old hinds is not reflected in another measure of fecundity, reproductive value, which usually declines with age.

Reproductive Value

What is the average contribution to the next generation that members of a given age group in a population give between their current age and death? Such a contribution is termed *reproductive value,* a concept that was advanced by R. A. Fisher (1930). In a population of constant size, the reproductive value is defined as age-specific expectation of future offspring. It is given as

$$v_x = \sum_{y=x}^{\infty} \frac{l_y}{l_x} m_y$$

where l_y/l_x = probability of living from age x to age y, m_y = average reproductive success of individual at age y, and v = reproductive value.

For newborn individuals in a population that is neither increasing nor decreasing the reproductive value is the same as the net reproduction rate, $R = 1$. Populations change in size, however, so the definition of reproductive value becomes the present value of future offspring. This value represents the number of offspring an individual dying at age $x + 1$ would have to produce at age x to leave behind as many offspring had it not died at age $x + 1$ and survived to produce young thereafter. The general equation for reproductive value is

$$\frac{v_x}{v_o} = \frac{e^{rx}}{l_x} \sum_{y=x}^{\infty} e^{-ry} l_y m_y$$

For calculation of reproductive value see Appendix B.

The reproductive value of a newborn individual is influenced by the state of the population. In an expanding population the reproductive value of young is low for two reasons. First, in an increasing population the probability of death before reproduction also increases. Second, because the future breeding population of which it will be a part will be larger, the young will contribute less to the overall gene pool than offspring currently being born. Conversely, young born into a declining population will contribute more to the future than present progeny, so they are worth more.

Reproductive values are usually calculated for females because reproductive values for males are difficult to obtain, but it can be done if we know the number and fate of the offspring sired by the male. Clutton-Brock and his associates (1982) have done so for the red deer stag (Figure 17.4). In the red deer population on the Isle of Rhum, Scotland, the reproductive value for hinds declines between ages three and four. The reproductive value of stags initially follows that of hinds, levels, falls off later than the hinds, and drops sharply after the age of seven. Unlike old hinds, stags apparently do not invest heavily in breeding activities during the declining years.

Reproductive Costs

If an organism is to contribute its optimal fitness to future generations, it has to balance the profits of immediate reproductive investments against the costs to future prospects, including fecundity and its own survival, a trade-off between present progeny and future offspring (Willams 1966) (Figure 17.5a) These trade-offs involve *reproductive costs,* the effects of present breeding efforts on an individual's future survival and fecundity. The situation is well defined in a partitioning of the

Figure 17.4 An estimate of reproductive values for red deer hinds (dashed line) compared to reproductive values of red deer stags (solid line). For both sexes the reproductive value is calculated in terms of the number of female offspring surviving to one year old that parents of different ages can be expected to produce in the future. Measures of reproductive value of stags are approximate. Reproductive values so calculated are more realistic than ones for females based on age-specific fecundity, m_x. Note the gradual decline in the reproductive value of the hinds from four years, whereas the.reproductive value of the stag remains constant from three to seven years, then declines sharply. (From Clutton-Brock 1984: 154.)

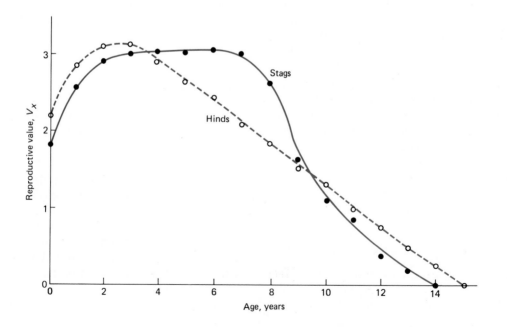

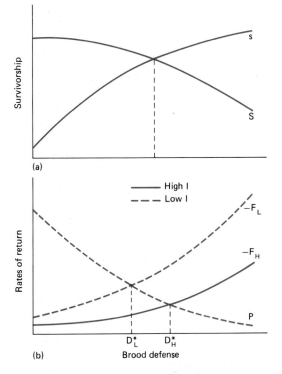

(a)

(b)
Brood defense

High I
Low I

Figure 17.5 A graphic analysis of William's (1966) trade-off model between fecundity and adult survival. (a) Survivorship of offspring and adults relative to parental investment in offspring as measured by brood defense. As parental investment in offspring increases, offspring survival increases. The point of intercept represents optimal parent investment relative to survivorship and thus future fecundity. Beyond this point parents are sacrificing future investment for present investment. (b) Parents achieve optimal brood defense and thus optimal return on parental investment when rates of return on brood defense for present reproduction (P) and future reproduction (F) are equal but opposite in sign (P = −F). Note the relationship between present reproduction and future reproduction relative to high and low investments in defense (D*). (After Sargent and Gross 1985:44.)

equation for reproductive value (Pianka and Parker 1975) into two parts:

$$V_x = m_x + \sum_{t=x+1}^{\infty} \frac{l_t}{l_x} t$$

In words, reproductive value at age x = present progeny (investment) + expected future progeny (future investment).

The equation clearly states the problem. Organisms operate under a fixed budget. Out of it so much energy has to be allocated to maintenance and survival and so much to reproduction. That is going to involve some trade-offs. For example, in anthropomorphic terms, the organism might opt for more growth, later maturity, later reproduction, and longer survival, or for less growth, earlier maturity, earlier reproduction, and reduced survival. Obviously, neither extreme is going to result in optimal fitness. Further, the organism can expend all of its energies in one major reproductive effort in its lifetime, or it can opt for repeated reproductive efforts over time, producing fewer young at each reproductive bout and saving some of its energies for future reproduction.

Those organisms that go for one major reproductive effort in a lifetime are termed *semelparous*. They invest all their energy in growth, development, and storage and then expend that energy in one massive, suicidal reproductive effort. Such a reproductive strategy is employed by most insects and other invertebrates, some species of fish, notably salmon, and by many plants—annuals, biennials, and some bamboos. Many semelparous plants and animals are small, short-lived, and occupy ephemeral or disturbed habitats, or are highly sensitive to seasonal environmental changes. For them it would not pay to be iteroparous, for the chances of future investment are slim. They can gain their maximum fitness by expending all their energies in one bout of reproduction. Other semelparous organisms are long-lived and delay reproduction. Mayflies may spend several years as larvae before emerging to the surface of the water for an adult life of several days devoted to reproduction. Periodical cicadas spend 13 to 17 years below ground before they emerge as adults to stage an outstanding exhibition of single-term reproduction. Pacific salmon may spend three to four years at sea building up biomass before returning to freshwater streams to spawn and die. The Hawaiian silverswords (*Argyroxiphium*) live 7 to 30 years before flowering and dying, and some species of bamboo delay flowering for 100 to 120 years, produce one massive crop of seeds, and die (Janzen 1976). The evolutionary reasons for delayed semelparity in long-lived organisms vary (for bamboos see Janzen 1976). The Haleakala silverswords are found on the summit of that Maui volcano where the environmental conditions are extreme: intense light, periods of strong heat and cold, and little moisture. The plant accumulates energy and moisture within its leaves, which it expends in a very short period of flowering once in its lifetime. Environmental conditions preclude iteroparity. In general, for a species to evolve semelparity, that mode of reproduction has to increase the reproductive effort enough to compensate for the loss of repeated reproductive efforts.

Organisms that choose to produce fewer young at one time and repeat reproductive efforts throughout a lifetime are termed *iteroparous*. For an iteroparous organism the problem is timing reproduction—early in life or at a later age. Early reproduction reduces survivorship and the potential for later reproduction. Later reproduction increases growth and improves survivorship but reduces fecundity. Energy expended in repeated reproduction weighs against future prospects, as measured by declining fecundity and reduced potential for survival (Figure 17.6). Thus a parent that has invested heavily in past will have less to invest in future fecundity and survival. Increased past investments will require that it increase its optimal level of present investment, because of decreased future survival. Somewhere in between lies the best strategy for optimal *lifetime reproductive success*, the sum of present and future reproductive success.

Achieving highest fitness involves aiming neither for maximum present reproduction nor for maximum future production. An optimal level of parental investment balances the rate of return from an investment in the present with the rate of return in investment in the future (Figure 17.7). Present investment should continue to grow through reproductive success of the offspring, while future investments continue to decline (see Sargent and Gross 1985; Coleman et al. 1986).

Annual Versus Perennial Life Histories

What are the evolutionary advantages of semelparity and iteroparity, of surviving to reproduce in a single season and die or of surviving to live and reproduce over a number of seasons? This question was addressed mathematically some years ago by Cole (1954). Cole calculated that an annual genotype could achieve an absolute gain in its intrinsic rate of population growth by becoming perennial, but it could also achieve the same result by remaining an annual and adding one individual to its litter size. Cole, however, considered no mortality during the life cycle. Charnov and Schaffer (1973) modified Cole's proposition somewhat by formulating a simple algebraic model that considers survivorship of both annual and perennial juveniles and survivorship of the adult perennial. They suggested that Cole's proposition be modified to read

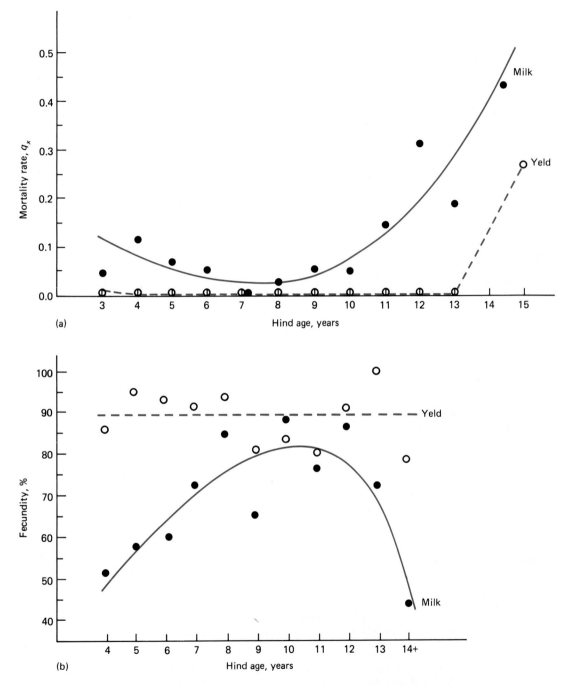

(a)

(b)

Figure 17.6 (a) Age-specific mortality of red deer milk hinds (ones who have reared a calf to weaning age) compared to age-specific mortality of yeld hinds (those that fail to do so). Milk hinds have a higher reproductive cost, which increases mortality. (b) The effect of that mortality and past investment in young is decreased future fecundity. (From Clutton-Brock 1984: 224, 216.)

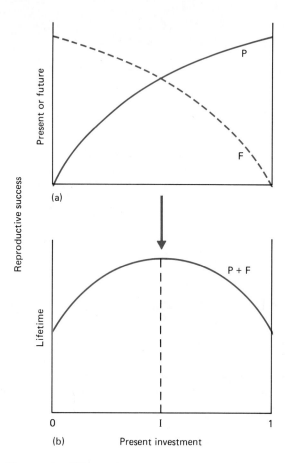

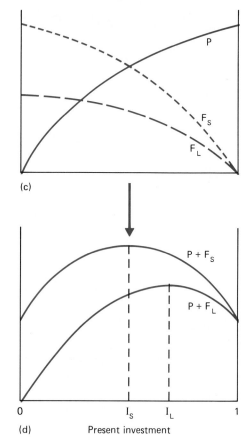

Figure 17.7 Effects of present investment on reproductive success. (a) Present reproductive success (P) is assumed to increase with diminishing returns (reproductive success of adult offspring) on present investment, while future reproductive success declines with present investment. (b) Lifetime reproductive success is the sum of present and future reproductive success (P + F). The optimal level of present investment (I) is that which brings about maximum lifetime reproductive success. (c) The effects of present investment (P) on future reproductive success after small (F_s) or large (F_l) past investments. An increase in past investments decreases future reproductive success from large (1) to small (s). (d) Thus an optimal present investment for a large past investment (I_l) is greater than that for a small past investment (I_s). Although theoretical, these models and that of Figure 17.5 provide some insights into reproductive behavior in the field—why birds may abandon a nest after expending some futile effort in defense or why grizzly bears aban-

don young as a reproductive tactic (see Tait 1980). (After Coleman et al. 1985: 60.)

For an annual species, the absolute gain in intrinsic population growth rate that can be achieved by changing to the perennial reproductive habit would be exactly equivalent to adding *P/C* individuals to the average litter size.

Let

Ba = number of annual offspring

Bp = number of perennial offspring

C = proportion of offspring surviving the first year

P = survival rate of perennial adults

N_t = number of organisms present year t just before reproduction but after mortality

Population growth for the annual genotype is

$$N_{t+1} = BaCN_t \text{ or } a = BaC$$

Population growth for the perennial genotype is

$$N_{t+1} = BpCN_t + PN_t \text{ or } p = BpC + P$$

Consider two genotypes, one annual and the other perennial, each of which produces 20 offspring with a survival rate of 50 percent. The perennial adults have a survival rate of 100 percent:

$$20 (0.5) < 20 (0.5) + 1$$
$$10 < 10 + 1$$

The perennial genotype has a reproductive advantage over the annual genotype.

Now suppose the annual genotype adds two additional offspring to the litter, with survival rates for all remaining the same:

$$22 (0.5) = 20 (0.5) + 1$$
$$11 = 10 + 1$$

By adding one additional *surviving* offspring, the annual has achieved the same rate of growth as the perennial.

If the population of both annual and perennial genotypes increase at the same rate, then $a = p$ and

Ba = Bp + P/C

Let the survival rate of both annual and perennial juveniles be the same, 25 percent, and the adult perennial survival rate be 100 percent. Because of its higher production, the annual genotype will have the larger number of offspring. Let the number of annual offspring prior to reproduction *Ba* be 25, and the number of perennial offspring 5. The adult peren-

nial survival rate/juvenile survival, P/C, will be 1.0/0.25. Then

$$25 > 5 + 1.0/0.25$$
$$25 > 5 + 4$$

Under these conditions, the annual genotype has the advantage.

If the survival rate of the juvenile perennials drops to 5 percent and the number of offspring remains the same as above, then

$$25 = 5 + 1.0/0.05$$
$$25 = 5 + 20$$

High adult survival and low juvenile survival of the perennial genotype balances the high juvenile survival of the annual. The annual, which in effect has increased its litter size by *P/C* or 20, would gain no advantage by becoming perennial. If the juvenile survival rate dropped lower to 4 percent, then $P/C = 1.0/0.04 = 25$ and $25 < 5 + 25$ and it would be advantageous for the annual genotype to assume a perennial habit.

In any organism there has to be a trade-off between survival and fecundity. A perennial life history strategy is favored in those environments in which juvenile survival is low, regardless of whether the survival is density-dependent or density-independent. The annual life history strategy is favored whenever adult survival is low or zero and juvenile survival is high.

■ *r*-Selection and *K*-Selection

One of the pervasive concepts in population ecology is *r*-selection and *K*-selection. The idea originated with MacArthur and Wilson (1967) in the development of their theory of island biogeography as a qualitative distinction in the types of selection pressures on island colonists. It was further elaborated by Pianka (1970). Uncrowded or empty environments favor colonists with the ability for rapid population growth in relatively unfavorable habitats. On the other hand, crowded environments favor colonists more able to compete for food and other resources. The former consist of species living in harsh or unpredictable (to the point of being ephemeral) environments in which mortality is largely independent of population density. They allocate more energy to reproduction and less to growth, maintenance, and adjustment to the environment. These species are known as r-strategists because environmental conditions keep growth of such populations on the rising part of the logistic curve. In such situations, genotypes with a higher r are favored. The latter involves species living in stable or predictable environments in which mortality results from density-related factors and competition is keen. They allocate more energy to nonreproductive activities. They are called K-strategists because they are able to maintain their densest populations at equilibrium (asymptote) or carrying capacity (K).

Species known as r-strategists are typically short-lived. They have high reproductive rates and produce a large number of offspring with low survival, but rapid developmental rates. They are opportunistic, that is, they have the ability to make use of temporary habitats, and often inhabit unstable or unpredictable environments where catastrophic mortality is environmentally caused and is relatively independent of the density of the population. For them environmental resources are not limiting and they are able to exploit relatively uncompetitive situations. Tough and adaptable, r-strategists have means of wide dispersal and are good col-

onizers; they respond rapidly to disturbances to the population. Such species are characteristic of early stages of succession.

K-strategists are competitive species with stable populations of long-lived individuals; they are relatively large and produce relatively few seeds, eggs, or young. Among animals parents care for the young; among plants seeds possess stored food that gives the seedling a strong start. K-strategists exist in environments in which mortality relates more to density-related causes than to unpredictability of conditions. They are specialists, efficient users of their particular environment, but their populations are at or near carrying capacity and are resource-limited. K-strategists are typically long-lived, requiring a relatively long time to mature. These qualities combined with their lack of means of wide dispersal make K-strategists poor colonizers. They are characteristic of later stages of succession.

K-strategists and r-strategists are under different selection pressures. Among r-species selection favors those genotypes that confer the highest possible intrinsic rate of increase, rapid development, small body size, early and single-stage reproduction, large number of offspring, and minimal parental care. Among K-species selection favors genotypes that confer the ability to cope with physical and biotic pressures, the ability to tolerate relatively high population densities, delayed reproduction, large body size, slower development, and repeated reproduction. r-selection favors productivity; K-selection favors efficient use of the environment (see Pianka 1970.)

However useful the terms r-selection and K-selection may be conceptually, they create problems. One rests with K. K is a parameter and not a function of life history traits. Individuals cannot be under direct selection pressure to increase K. Selection can only act on *r*, which is a function of life history traits: age, survivorship, and fecundity. Therefore *r* and *K* are not equivalent terms; they "cannot be reduced to units of common currency" (Stearns 1976).

The concepts of both *r*-selection and *K*-selection assume deterministic environments: one that is unpredictable, and another that is stable or predictable. However, environments are stochastic, fluctuating. Schaffer (1974), Stearns (1976) and others have advanced the "bet-hedging" hypothesis that if in a variable environment adult mortality is high and juvenile mortality is low, then selection should favor early maturity, larger reproductive effort, and more young. Under the same conditions if juvenile survival is high and adult mortality is low, then natural selection should favor late maturity. Thus a variable environment that affects juvenile survival more than adult survival should lead to lower annual reproduction and more reproductive seasons; conversely, if the variable environment affects adult survival more than juvenile survival, then it should lead to higher annual reproduction and reduced iteroparity. Thus we can arrive at the same strategies by considering the response of individuals to selection pressures only in terms of *r*.

Zammuto and Millar (1985) quantitatively tested various aspects of *r* and *K* theory versus bet-hedging for six populations of Columbian ground squirrels *(Spermophilus columbianus)* from Alberta, Canada. Three populations occupied less predictable environments at low elevations with more variable food resource levels; and three populations lived in a more predictable environment at higher elevations. The squirrels at the higher elevation had higher adult survival rates and later ages at maturity than squirrels at lower elevatins—in other words, they exhibited a *K*-strategy. Contrary to the predictions of

bet-hedging, juvenile survival, prereproductive survival, and percentage of squirrels surviving to maturity were positively associated with the degree of iteroparity and age at maturity, whereas adult/juvenile survival ratio and variance in reproductive output were negatively associated. In spite of the traditional theory, predictability of environment and the occurrence of *K*-strategists increased rather than decreased with movement upslope.

In another study of the Columbian ground squirrels in Alberta, Dobson and Murie (1987) found consistently different life history patterns at different elevations, which could not be explained either by *r* and *K* theory or by bet-hedging. They suggest a resource-limitation hypothesis that the ground squirrel responds plastically to food availability and other environmental influences on obtaining and utilizing resources.

r-selection and *K*-selection often are considered polar entities of a continuum from *r* to *K* (Pianka 1970). Such a view tempts a classification of species either as *r*-selected or *K*-selected, but it is difficult to force species into such a classification. The concept of *r*-selection and *K*-selection is most useful to compare organisms of the same type or to compare individuals within a population or populations within a species, as Zumato and Millar did with the ground squirrels. Under certain conditions individuals or populations will exhibit *r*-selected traits or *K*-selected traits. Meadow mice living in environments where dispersal can take place easily exhibit characteristics of *r*-selection, whereas those living under conditions in which there is no dispersal sink assume *K*-selected characteristics (Tamarin 1978) (see Chapter 16).

Because of the difficulty of forcing plants into *r* and *K* categories, J. Grime (1977, 1979) has proposed the idea of ruderal or *R*-strategists; competitive or *C*-strategists; and stress-tolerant or *S*-strategists. *R*-strategists reproduce early in life, possess high fecundity, experience lethal reproduction (semelparity), occupy uncertain or disturbed habitats, and have well-dispersed seeds. *C*-strategists and *S*-strategists occupy more stable environments and are relatively long-lived, often drastically reducing the opportunity of seedling establishment, resulting in high juvenile mortality. Beyond these two characteristics, the two have evolved quite different life-history strategies. *C*-strategists reproduce early and repeatedly utilize an annual expenditure of energy stored prior to seed production. *S*-strategists have delayed maturity, intermittent reproductive activity, and long-term energy storage.

Forms of Sexual Reproduction

Sexual reproduction is common to most multicellular organisms. Even those that rely primarily on asexual or vegetative reproduction will revert on occasions to sexual reproduction. Sexual reproduction allows the gene pool to become mixed, increasing the genetic variability necessary to meet changing selective pressures and to prevent an accumulation of harmful mutations. For the individual, though, sexual reproduction is expensive. Each individual can contribute only one-half of its genes to the next generation. The success of that contribution depends upon a member of the opposite sex. For that reason each individual must acquire the best possible mate.

Sexual reproduction can take a variety of forms. The most familiar is separate male and female individuals. Plants with that characteristic are called *dioecious,* and

examples are holly trees and stinging nettle *(Urtica).* The equivalent term for animals is *gonochoristic.* Some organisms possess both male and female organs. They may be *monoecious* or *hermaphroditic.* There is a difference. Hermaphrodites have both male and female organs in the same individual. Among plants that means flowers contain both stamens and ovules. Among animals individuals possess both testes and ovaries, a condition common to some invertebrates such as earthworms and some fish. Monoecy is restricted to plants, in which the individual possesses separate male and female flowers, such as birch *(Betula)* and aspen *(Populus)* trees. Some hermaphrodites are simultaneous, others sequential. The latter type are one sex when young and develop into the opposite sex when mature.

Sex reversal seems to be stimulated by a social change involving sex ratios in the population. Sex reversal among some species of marine fish can be initiated by the removal of one or more individuals of the other sex. Among some coral fish, removal of females from a social group stimulates an equal number of males to change sex and become females (Friche and Friche 1977). Among other species, removal of males stimulates a one-to-one replacement of males by sex-reversing females (see Shapiro 1980, 1987).

Plants also exhibit a gender change (Freeman et al. 1980). One such plant is jack-in-the-pulpit *(Arisaema triphyllum)* (Levett, Doust, and Cavers 1982; Bierzychudek, 1982, 1984a, 1984b; Policansky 1987), a clonal woodland herb whose genet is a perennial corm and whose ramet is a single annual shoot. Jack-in-the-pulpit produces staminate (male) flowers one year, an asexual vegetative shoot the next, and a carpellate (female) or monoecious shoot the next. Over its life span a jack-in-the-pulpit may produce both genders as well as an asexual vegetative shoot, but in no particular sequence. Usually an asexual stage follows a gender change. Gender changes in some plants are normally stimulated by environmental changes in moisture and light, but gender change in jack-in-the-pulpit appears to be triggered by an excessive drain on the photosynthate by female flowers. If the plant is to survive, one carpellate flowering could not follow another. To avoid death, the plant reduces its reproductive effort the next year by changing its gender or becoming vegetative. Among jack-in-the-pulpits, the size of the plant is more important than its sexual state in determining its future sex. Large males are more likely to be female the next spring than are small females (Policansky 1987). Dwarf ginseng *(Panax trifolium)* exhibits *diphasy*. Each individual plant switches sex in either direction between a male phase bearing only staminate flowers and a female phase bearing hermaphroditic flowers. Females experience a decrease in size following seed production. They have used up so much stored energy in reproduction that they convert to a male, which assimilates more resources than needed. After several years of restorative growth the plant may convert to a female again (Schlessman 1987).

Mating in hermaphroditic and monoecious species can involve either outcrossing, in which genetic variability is assured, or self-fertilization, which involves no genetic change. Apparently, most hermaphroditic animals are not self-fertilized. Earthworms mate with other individuals, and so do hermaphroditic coral reef fish. Some animal hermaphrodites are completely self-sterile; and only a few, like certain land snails, are self-fertilized.

Life can appear complex among animal hermaphrodites. How is reproductive effort to be allocated between egg and sperm and between male and female behaviors? The male side of behavior in a hermaphrodite involves attempts to fertilize the eggs of other hermaphrodites, increasing the reproductive success of the male; and the female side attempts to secure fertilization of her eggs to enhance her reproductive success. Should the hermaphroditic male defend the eggs of his partners or those of his own female side? Fischer (1981) investigated this problem in the hermaphroditic coral reef fish *Hypoplectrus nigracans*, the black hamlet. He found that the reproductive success of the hamlet as a male depended upon its ability to reproduce as a female, because spawning partners "traded eggs," giving up eggs to be fertilized in exchange for the opportunity to fertilize those of another individual. Courtship was largely a female function rather than a male function; it advertised that an individual has eggs. Each fish parcels out its daily clutch in four to five spawns, and it alternates sex roles with each spawn, taking turns giving up parcels to be fertilized. As long as there is a greater return to the individual per unit effort from the male than from the female function, the reproductive success of the hermaphrodite is greater than being purely female and is similar to that of a parthogen. Parthogens have twice the fecundity of sexual organisms because each of their offspring produce eggs; whereas sexual individuals have on the average half of their offspring males, which contribute little to fecundity (unless the male is involved in parental care; see Maynard Smith 1971). By trading eggs, hermaphrodites invest most of their reproductive efforts in female functions, and secure a fecundity advantage similar to that of a parthogen.

Many hermaphroditic plants are self-compatible but have evolved means to prevent self-fertilization. Anthers and pistils may mature at different times, or the pistil may extend well above the stamens. Other plants have evolved more effective means. One is a genetic mechanism that prevents the growth of a pollen tube down the style of the same individual. Other hermaphroditic species are divided into two or three morphologically different types between which pollination takes place.

Although many hermaphroditic organisms possess mechanisms to reduce self-fertilization, the capacity of self-fertilization in hermaphrodites does carry certain advantages, especially among plants. A single self-fertilized individual is able to colonize a new habitat and then reproduce itself, establishing a new population. Other hermaphrodites produce self-fertilized flowers under stressed conditions, ensuring a new generation. Jewelweed, for example, under normal and optimal envi-

ronmental conditions produces cross-pollinated *(chistogamous)* flowers. Under adverse environmental conditions or after the chistogamous flowers have set seed, jewelweed (as well as violets and other species) produces tiny, self-fertilized *(cleistogamous)* flowers that never open (Figure 17.8). They have vestigial petals, no nectar, and few pollen grains and remain in a green, budlike stage. Thus, if outcrossing fails or never develops, the plants have ensured a next generation by self-fertilization.

Figure 17.8 The cross-pollinated flowers of jewelweed (right) and violets (left) are termed chistogamous. They are conspicuous and produce obvious seeds. Hidden beneath the leaves of jewelweed and at ground level in violets are tiny self-fertilized cleistogamous flowers that never open but also produce seed. The seeds of cleistogamous flowers are important to the maintenance of locally adapted populations of these plants.

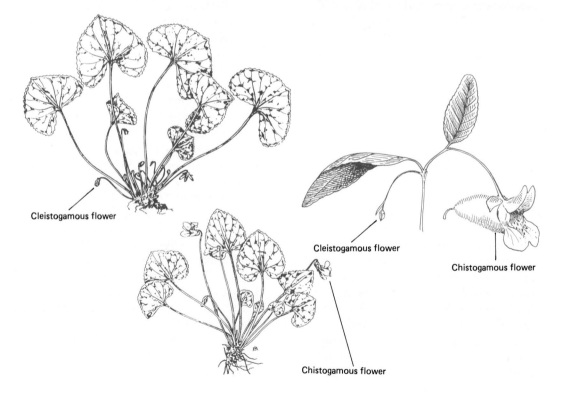

Cleistogamous flower

Chistogamous flower

Cleistogamous flower

Chistogamous flower

■ Mating Systems

The behavioral mechanisms and the social organization involved in an organism's obtaining a mate is called a *mating system*. A mating system includes such aspects as the number of mates acquired, the manner in which they are acquired, the nature of the pair bond, and the pattern of parental care provided by each sex. The structure of mating systems ranges from monogamy through many variations on polygamy. Mating systems employed may vary even within a species, involving different degrees of pair bonds and relationships between the pair (Table 17.2).

The nature and evolution of male-female relationships are influenced by

TABLE 17.2 An Ecological Classification of Mating Systems

Monogamy	Neither sex has opportunity of monopolizing additional members of the opposite sex. Fitness often maximized through shared parental care.
Polygyny	Individual males frequently control or gain access to multiple females.
Resource defense polygyny	Males control access to females *indirectly*, by monopolizing critical resources.
Female (or harem) defense polygyny	Males control access to females *directly*, usually by virtue of female gregariousness.
Male dominance polygyny	Mates or critical resources are *not economically monopolizable*. Males aggregate during the breeding season and *females select mates* from these aggregations.
Explosive breeding assemblages	Both sexes converge for a short-lived, highly synchronized mating period. The operational sex ratio is close to unity and sexual selection is minimal.
Leks	Females are less synchronized and males remain sexually active for the duration of the females' breeding period. Males compete directly for dominant status or position within stable assemblages. Variance in reproductive success and skew in operational sex ratio reach extremes.
Rapid multiple clutch polygamy	Both sexes have substantial but relatively *equal* opportunity for increasing fitness through multiple breedings in rapid succession. Males and females each incubate separate clutches of eggs.
Polyandry	Individual females frequently control or gain access to multiple males.
Resource defense polyandry	Females control access to males *indirectly*, by monopolizing critical resources.
Female access polyandry	Females do not defend resources essential to males but, through interactions among themselves, may limit access to males. Among phalaropes, both sexes converge repeatedly at ephemeral feeding areas where courtship and mating occur. The mating system most closely resembles an explosive breeding assemblage in which the operational sex ratio may become skewed with an excess of females.

Source: Emlen and Oring 1977.

ecological conditions, especially the availability and distribution of resources and the ability of individuals to control access to mates or resources. If the male has no role in the feeding and protection of young, no advantage accrues to the female to remain with the male. If the habitat is sufficiently uniform so that little difference in territorial quality exists, the number of young raised in the poorest habitat is only slightly less than the number reared in the best. Selection would favor monogamy because female fitness in both would be nearly the same. If the habitat is diverse with some parts more productive than others, competition among males may be intense and some males will settle on poorer territories. Under such conditions it may be more advantageous for a female to join another female in the territory of a male defending a rich resource than to expel the other female. Selection under those conditions will favor bigamous mating even though aid from the male in feeding the young is reduced or absent.

Monogamy

Monogamy is the formation of a pair bond between one male and one female. It is most prevalent among birds. It is relatively rare among mammals except for humans, several carnivores, such as the fox and mustelids, and a few herbivores, such as the beaver.

Monogamy results when neither of the two sexes has the opportunity to monopolize the other either directly or through control of resources. Monogamy is most prevalent among those species in which cooperation by both parents is required to rear young successfully. For one member of the pair to spend time and energy courting and mating with other individuals would result in the loss of individ-

ual fitness. Monogamy also prevails in situations where resources are rather uniformly distributed and little opportunity exists for an individual to monopolize them. Maximum fitness for both individuals of the pair is achieved when both parents share in the care of the young (see Coulson 1966).

Polygamy

Polygamy is the acquisition by an individual of two or more mates, none of which is mated to other individuals. A pair bond exists between the individual and each mate. When one member of the pair is freed from parental duty, partly or wholly, the more time and energy the emancipated member of the pair can devote to intrasexual competition for mates and resources. The more such critical resources as food or quality habitat are unevenly distributed, the greater is the opportunity for such an individual to control the resource and thus available mates.

How many members of the other sex an individual can monopolize depends upon the degree of synchrony in sexual receptivity. For example, if females in the population are sexually active for only a brief period, the number a male can monopolize is limited. If females are receptive over a long period of time, the number a male can control depends upon the availability of females and the number of mates the male can energetically defend. Such variability in environmental and behavioral conditions results in various types of polygamy. There are two basic forms: polygyny, in which an individual male gains control of or access to two or more females; and polyandry, in which an individual female gains control of or access to two or more males. A special form of polygamy is promiscuity, in which males and females copulate with one or many of the opposite sex but form no pair bonds. Emlen and Oring (1977) classify types of polygamous relationships according to the means by which individuals gain access to the limiting sex. This classification appears in Table 17.2.

Resource Defense Polygyny. Resource defense polygyny results when the male can defend resources essential to the female. How many females he can monopolize depends upon how the resources are distributed and how defendable they are. If resources are highly clumped, a male can control more of the resources and thus more of the females. In the choice of a male by the female the concomitant sexual selection pressure is influenced by the quality of the male-defended resource as well as by the quality of the male.

If both parents take care of the young, the female stands to lose if the male has more than one mate, unless the quality of the male or the quality of the resources he controls more than offsets the loss of his help at the nest. Under such a situation the female may have a greater chance of reproductive success and thus increase her fitness if mated polygamously to a male in high quality habitat than if mated monogamously to a male in poor quality habitat. Examples are the dickcissel (Zimmerman 1971) and the red-winged blackbird (Howard 1977), two species in which males defending optimal habitats, determined by volume of vegetation, attract multiple mates, while males in less suitable habitats are more apt to be monogamous or go unmated.

If the male takes no active interest in the nest and provides no care of the young and if the resources needed by the female are sufficiently clumped, he can spend his time defending the resources. Such a situation can lead to a strong development of polygyny, as it does in a number of species of hummingbirds (Stiles and Wolf 1970l; Gill and Wolf 1975). The males defend high nectar-producing flowers, rather than territory (Wolf et al. 1976), at which they allow the females but not other males to feed. By controlling a needed resource, food, the hummingbird also controls access to females, leading to polygyny.

Female Defense Polygyny. A second type of polygyny is defense of a group of females or a harem. This usually results when females are naturally gregarious and live in small herds for most of the year for reasons unrelated to reproduction. Such herds enable the dominant male to have access to and to control a number of females during the reproductive season. The male asserts his claim to the females by aggressive behavior toward other males and expends great amounts of energy defending the females from raiding activities of rivals. There is, of course, an upper limit to the number of females a male is able to defend; a male attempting to control too many will lose part of his harem to a rival.

Two types of harem defense exist. One, common to ungulates such as the wapiti and red deer and to wild horses, is defense of the females themselves. A second type, common to seals, is defense of access to areas where females congregate on the breeding ground.

Female defense polygyny leads to intense sexual selection and strong sexual dimorphism. The males of seals are many times larger than females; bull wapiti are considerably larger than cows and possess highly developed antlers. (For discussion of sexual dimorphism in this context, see Geist 1977.)

Male Dominance Polygyny. On the grasslands of middle North America where remnant populations of prairie grouse (prairie chickens and sharp-tailed grouse) still exist and on the sagebrush lands of western North America where sage grouse live, males of these species congregate in communal display areas called *leks* in order to attract and mate with females. These leks, called booming grounds for prairie chickens, dancing or parade grounds for prairie chickens and sharp-tailed grouse, and strutting grounds for sage grouse, are located on areas of open ground somewhat elevated and visible to the surrounding area. The cocks gather from daybreak on through the afternoon on the lek to display. Early in the season behavior is largely aggressive and fighting is frequent. Territories are established as a result of this fighting, with dominant and subdominant birds obtaining central positions on the lek. Once fighting has subsided, females begin visiting the lek, where they show a marked preference for centrally located males. When females arrive on the lek, courtship dominates with its bizarre displays and resonation of sounds from brightly colored air sacs (see Wiley 1974).

Lek behavior is an example of male dominance polygyny. The male does not defend females, nor does he defend a resource needed by the female. Rather, the males sort themselves out into dominance positions, excluding some or many from mating, and the females choose the male. Lek mating systems exist among a number of species of birds and some insects, frogs, and mammals.

Another type of male dominance polygyny, somewhat suggestive of lek behavior, is the explosive breeding assemblage. Males of such animals as singing insects and chorusing amphibians congregate on breeding sites to which females are attracted. During a short-lived, highly synchronous breeding period, mating is promiscuous. Breeding activity is frenzied, and because of female synchrony in breeding period, individual males have little opportunity to monopolize matings.

Male dominance polygyny results when individual males are completely freed from parental care, when the environment provides little potential for resource or mate control, and when it is economically (energetically) impossible for the male to control or monopolize resources necessary for acquiring females.

Rapid Multiple-Clutch Polygamy. Among some birds males assume the full burden of incubation and rearing of the brood. This frees the female from nesting duties and allows for more frequent matings. This reversal of parental duties could come about only if it increased male fitness. A lack of dependable breeding conditions could place a premium on the ability of the female to replace clutches lost to predation and other causes. From the male point of view, costs of assuming nesting duties are balanced by the ability of the female to produce new clutches rapidly. Among some shorebirds and gallinaceous birds the female lays the first clutch, which the male incubates, and then lays a second clutch, which she incubates. This doubles the reproductive effort in a short period of breeding time. Such an arrangement obviously increases the fitness of the female and it also increases male fitness if the female returns to her mate prior to laying of the second clutch and if the male remains sexually active as far into the breeding season as possible.

The assumption of incubation and brood rearing roles by the male is a step toward polyandry. The female increases her fitness by producing multiple clutches, but only to the point that males are sexually receptive and available to assume responsibility for additional clutches.

Resource Defense Polyandry. In resource defense polyandry the female competes for and defends resources essential for the male. As in polygyny, the degree of polyandry depends upon the distribution and defensibility of resources, especially quality habitat. The production by females of multiple clutches that all require the brooding services of a male leads to competition among females for access to available males. After a clutch is laid and a male begins incubation, he becomes sexually inactive and is effectively removed from the male pool, resulting in a scarcity of available males.

Polyandry usually occurs when nesting failure is frequent, requiring the female to be free to produce multiple clutches, when the breeding season is sufficiently long to allow renesting, and when the males arrive asynchronously on the nesting grounds (in contrast to synchrony in female breeding condition in polygyny).

Polyandry is best developed in two bird orders, Gruiformes (cranes, rails, etc.) and Charadriiformes (the shorebirds). An example in eastern North America is the spotted sandpiper (*Actitis macularia*), a species in which losses to egg predation are high and in which the female has an exceptional ability to lay multiple clutches (Hayes 1972; Oring and Knudsen 1972). An example of extreme polyandry is the American jacana of Central America. Because the breeding habitat of this bird is very limited, only a small fraction of the breeding population is active in any one year. Males defend small territories about ponds and lagoons, while females defend large territories that embrace the territories of several males. Females mate with more than one male, produce multiple clutches, and frequently have multiple mates incubating clutches simultaneously. Because nest predation is high, females provide replacement clutches for their males, who assume parental duties. In effect, female jacanas specialize in egg production, while the males specialize in brood rearing (Jenni and Collier 1972; Jenni 1974).

Among species exhibiting polyandry behavioral dimorphism and size dimorphism are evident. The females are larger, often considerably so, than the males (for example, the female jacana weights 50 to 75 percent more than the male), are dominant over males in aggressive interactions, and provide minimal care of eggs or young.

Female Access Polyandry. An exceptional type of polyandry in which no resource defense is involved at all is exhibited by phalaropes (Emlen and Oring 1977). Phalaropes, small shorebirds that nest on the arctic tundra and winter at sea in the Southern Hemisphere, are unique in that the female is more brightly colored than the male and the male performs all parental care. Because the food resource, aquatic larvae of mosquitos, midges, and beetles, is highly unpredictable and ephemeral, the birds shift courtship and nesting sites from year to year and even week to week. Males and females congregate at bodies of water, where they feed, display, and mate. Pair bonds are brief. After the female has completed the first clutch of eggs, she may attempt to increase fitness by mating again. This is especially true for females that complete the first clutch early. If nesting failures are frequent and if sufficient time remains in the breeding season to allow renesting, females compete for males. If the density of birds is high, female interactions may become severe and some females may be prevented from breeding. Thus females may limit access of other females to males, with which the female must remain until the clutch is complete and incubation begins.

Plastic Mating Systems

Mating systems often are not genetically fixed or species-specific. Considerable variability may exist in mating systems employed by individuals within a species, depending upon the availability and distribution of needed resources, the energy cost of defending resources and mates, and population density. If environmental conditions or population densities change from year to year, individuals within local populations may be able to respond to those changes by shifting from monogamy to polygamy or from one form of polygamy to another.

How mating systems may change under differing environmental conditions has been outlined in a general schema by Emlen and Oring (1977) (Figure 17.9). Ecological factors influence the environmental potential (resource availability and distribution) for polygamy. Phylogenetic factors influence the ability of the animal

to capitalize on environmental potential. The ability to capitalize involves the amount of parental care required for the successful rearing of young. If both parents are required, they must remain monogamous. If only one parent is required, the potential for polygamy exists.

■ Sexual Selection

The various mating systems result in a differing fraction of the male population acquiring mates. Mating becomes competitive. Members of one sex compete among themselves to mate with members of the other sex (intrasex competition), and members of the opposite sex (usually female) show a preference for those that win. Supposedly, males (and in some situations, females) that win the intrasex competition are the fittest, and by selecting a mate from among those males, females ensure their own fitness. Selection works against those males deprived of mates. Competitive mating results in differential production of progeny by different genotypes, or *sexual selection*. Ultimately, sexual selection results in the evolution of morphological and behavioral traits that influence both competitive mating and mating systems.

Figure 17.9 General determinants of a plastic mating system. (From Emlen and Oring 1977.)

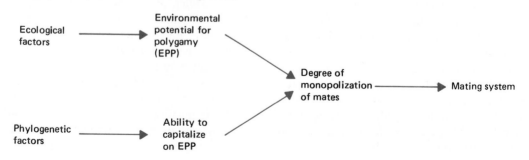

How does a female select a mate with greatest fitness? That is still a largely unanswered question. Actually, the female has little to go on. She might select a winner from among males that bested others in combat—as in bighorn sheep, elk, and seals—or ritualized display. She might select mates based on intensity of courtship display. Whatever the situation, the selection process comes down to salesmanship on part of the male and sales resistance on part of the female (G. C. Williams 1975). The female may attempt to elicit as much courtship behavior as possible from a potential mate. Females, for example, may force males to display in groups, as in prairie chickens and in some frogs and toads (see Wells 1977; Howard 1978a). They usually choose large and probably older males. In those species in which males monopolize a group of females, such as elk and seals, the females accept the dominant male. But even in that situation, the females are not placid and do have some choice. Protestations by female elephant seals over the attention of a dominant male may attract other large males nearby who may attempt to dislodge the male from the group. Such behavior ensures that females will mate only with the highest-ranking male (Cox and LeBoeuf 1977).

Often females chose males who offer the highest quality territory (for example, see Howard 1978b; Searcy 1979; Zimmerman 1971). In that situation the question is whether the female selects the male and accepts the territory that goes with him or whether she selects the territory and accepts the male that goes with it. In birds the intensity and repertoire of song has little to do with mate selection, other than calling the female's attention to his presence (see Searcy and Anderson 1986). By choosing a male with a high-quality territory, the female can best assure her own fitness.

Whatever the case, the ultimate strategy for both male and female is to assure maximum personal fitness; but what increases male fitness is not what necessarily improves female fitness. Sperm are cheap. With little investment involved, males should mate with as many females as possible to achieve maximum fitness. Because females invest considerably more in reproduction it is to their advantage to be more selective in choosing a mate. In general, they should refuse to mate with a promiscuous male or one with several mates. Instead they should favor one-to-one relationships in which the male helps care for the young. For males such a relationship carries certain risks. The male must guard the female from other males to ensure that the offspring he helps to raise are really his own. Females are always sure that the offspring they raise are their own.

Sexual selection is considered mostly an attribute of animals. Although Darwin recognized such selection in plants, we might have difficulty imagining how plants would engage in sexual selection. Such selection is associated with the evolution of dioecy from largely hermaphroditic plants (see Willson 1979, 1983; Bawa 1980; Ross 1982; and especially Willson and Burley 1983). It may involve largely intrasexual competition among hermaphroditic flowers for the dispersal of pollen. Selective advantage would go to those with larger flowers that provided the most pollen and nectar. The more energy these flowers put into pollen, the less energy is available for seed production by the female component. If some hermaphrodites carry a mutant gene for partial female sterility, all their energy could be directed toward pollen production. Such flowers could have a selective advantage over other hermaphrodites. At the same time, male-sterile flowers would not waste energy on pollen production. They would have a selective advantage over normal and female-sterile hermaphrodites. They could channel most of their resources to seed production and would not have to contend with their own pollen. Ultimately that type of sexual selection could lead to the evolution of dioecy.

Sexual selection in both plants and animals involves major selective processes acting on both sexes during the whole reproductive process. The role of intrasexual competition among males, however, is more easily observed and better understood than female sexual selection. We still do not know what criteria females use to make the important choice of the fittest male. We still have difficulty separating intramale competition for mates from female selection of males.

■ Habitat Selection

The importance of quality habitat to the fitness of individuals already has been emphasized in Chapter 16 (see Figure 16.11). How are individuals able to assess the quality of habitat? The answer to that long-standing question is important in these times of diminishing habitats for wildlife. To maintain and improve them, we need criteria important to the species occupying them (see Verner et al. 1986).

Habitat selection may involve *imprinting,* a form of associative learning characterized by a rapid establishment of a perceptual preference for an object. It seems to occur only at a specific and highly critical period in the life cycle of an animal and under a particular set of environmental conditions. Other evidence suggests that habitat selection among birds and

mice is little influenced by early experience and may have a genetic relationship (Wecker 1963; Klopfer 1963).

Habitat selection is partly a physiological process. Lack and Venables (1939) and Miller (1942) suggest that birds recognize their ancestral habitat by conspicuous though not necessarily essential features. The Nashville warbler *(Vermiforma ruficapilla),* a typical inhabitant of open heath edges of northern bogs, selects open stands of aspen and balsam fir and forest openings of blackberry and sweetfern in the southern part of its range. These habitats are visually suggestive of bog openings (Smith 1956). MacArthur and MacArthur (1961) and later many others (for example, MacArthur 1972; James 1971; Balda 1975) demonstrated a strong correlation between structural features of vegetation and the species of birds present.

Habitat selection probably involves a hierarchical approach (Hilden 1965; Wiens and Rotenberry 1981; Hutto 1985). Birds appear to assess initially the general features of the landscape—the type of terrain; presence of lakes, ponds, streams, and wetlands; gross vegetational features such as open grassland, shrubby areas, types and extent of forests; homogeneous or patchy vegetational distribution. Once in a broad general area, the birds respond to more specific features of habitats, such as the structural configuration of vegetation, particularly the density of leaves at various elevations above the ground and horizontal stratification (see Chapter 24). Although once regarded as unimportant in habitat selection and utilization, the floristics of the area also appear to be important. The structural characteristics of the trees and shrubs may affect the foraging activities of birds; and various species of plants influence the type of herbivorous arthropods, and thus the levels of prey abundance on different plants available for foraging birds (Holmes and Robinson 1981; Robinson and Holmes 1984; Wiens 1985).

Still other structural features determine a habitat's suitability (see Fish and Wildlife Service 1980; Verner et al. 1986). The lack of song perches may prevent some birds from colonizing an otherwise suitable habitat. The introduction of perches can result in the colonization of that area. For example, when telephone lines were strung across a treeless heath, tree pipits, birds that require an elevated singing perch, moved into the area (Lack and Venables 1939). Woodcock will not utilize a singing ground unless it allows sufficient room for flight (Sheldon 1967). A small opening surrounded by tall trees is not suitable, but an opening of the same size surrounded by low shrubs is.

An adequate nesting site is another requirement. Animals require sufficient shelter to protect parents and young against enemies and adverse weather. Selection of small island sites, such as muskrat houses, by geese provides protection against predators. Cavity-nesting animals require suitable cavities, dead trees, or another substrate in which they can construct such cavities. In areas where such sites are absent, populations of birds and squirrels can be increased dramatically by providing nest boxes and den boxes.

There appears to be a relationship between food and habitat selection (Smith and Shugart 1987, Hutto 1985), particularly in nonbreeding migratory birds. For them the availability of food is more important than more sharply defined structural features (Hutto 1985). They may use direct or indirect cues to assess quality of habitat and the use of space within a habitat.

The role of food availability and the use of direct structural cues indicating food in habitat selection has been demonstrated in the ovenbird *(Seiurus aurocapillus)* by Smith and Shugart (1987). They hypothesized that structural habitat cues are the proximate factor determining territory size by examining the relationships among habitat structure, prey abundance, and intrapopulation variation in the ovenbird in an eastern deciduous woods. They assessed vegetation structures, plotted territories, and determined prey abundance by sampling forest floor litter invertebrates from each of 115 sample vegetation plots contained within 23 territories of ovenbirds and from 100 plots outside of ovenbird territories during late April and early May (the time when arriving ovenbirds select their habitat within the woods). Smith and Shugart found a significant difference in prey abundance per unit area between territorial and nonterritorial sites. Seventy-five percent of the variation in prey abundance was correlated with habitat structure, particularly the nature of the canopy. Forest litter invertebrates were most abundant in those areas of the forest with large trees, a closed canopy, and a sparse understory, in contrast to fewer invertebrates in areas of open canopy and dense understory. The microclimate of the forest floor under a closed canopy, with its lower temperatures, less drastic diurnal fluctuations in temperature and moisture in forest litter, and increased relative humidity provided superior habitats for forest litter invertebrates, major food of the ovenbird. The ovenbird apparently cued in on areas of closed canopy and low understory as microhabitats within the woods offering the greater abundance of food. This hypothesis is supported by a significant partial correlation between predicted prey abundance and territory size, but not between actual prey abundance and territory size when the influence of habitat structure was statistically removed. Variation in territory size was related to structural features of the habitat rather than prey abundance or intraspecific competition.

Habitat selection is not restricted to bird and mammals, although they furnish most of the examples. Whitham (1980) found that the gall-forming aphid *Pemphigus,* which parasitizes the narrowleaf cottonwood, selects the largest leaves to colonize and discriminates against small leaves. Beyond that they select the best positions on the leaf (see Chapter 16). Occupancy of this particular habitat, which provides the best food source, produces individuals with the highest fitness.

Even though a given habitat may provide suitable cues, it still may not be selected. The presence of others of the same species may be necessary to attract more individuals. Among colonial and semicolonial birds, for example, the attractiveness of a colony is highly important. Only after several animals have settled on an area are others stimulated to do so. This factor is important in habitat selection by herring gulls nesting for the first time (Dorst 1958). On the other hand, the presence of some animals may inhibit others from occupying otherwise suitable habitat. Hooded warblers and American redstarts abandoned suitable habitat on a research area when their territories were colonized by a pair of canopy-nesting Copper's hawks (Kinsley and Smith pers obs.). Human activities on northern lakes inhibit the nesting of loons on otherwise suitable habitat (Ream 1976).

Habitat selection is not rigid. Most species exhibit some plasticity; otherwise these animals would not colonize new habitats. The ability of some members of a particular species to select habitats that deviate from that of others must exist on both a phenotypic and genetic level. Ovenbirds in the northern part of their range select woodland habitats that are exactly the opposite of those in Tennessee. The Ontario birds prefer a more open canopy and denser understory (Stenger and Falls 1959). The differences in habitat may result from differences in plant species associated with structural gradient and from their influence on composition, microclimate, and chemistry of the forest floor that affects prey abundance. Late successional stages in Ontario forests are dominated by mixed conifers and hardwoods, whereas those in Tennessee are deciduous. The contrast in ovenbird habitats suggests that species may have evolved a pattern of habitat selection to match patterns of productivity over their geographical range (Smith and Shugart 1987). Another example of strong contrast in habitat types is found in the black-throated green warbler *(Dendroica virens)*. It is associated with coniferous forests in the northern part of its range and with drier oak forests in the middle and southern Appalachians (Collins 1983).

■ Summary

Individuals achieve optimal fitness only by balancing energy allocated to present reproduction with that allocated to survival and future fecundity. One alternative is to invest a maximum amount of energy into a single reproductive effort in a lifetime, as exemplified by annual plants and many insects. The other alternative is to allocate less energy to a single reproduction and repeat reproductive efforts through a lifetime. Organisms may invest reproductive effort into many small offspring and provide minimal parental care; or they may invest a similar amount of energy into fewer, larger individuals and extend parental care. A single reproductive effort or production of many young with minimal energy invested in each is characteristic in environments and life histories in which mortality of young is very high. Because little difference in potential for population increase exists between single and repeated reproduction, repeated reproduction may be a response to an unpredictable survival of individuals from zygote to adult. By reproducing several times, an organism is more likely to assure reproductive success.

Reproduction has its costs. Individuals with repeated reproduction face the prospect of decreased survival and decreased future fecundity with each present investment in reproduction. Heavy early investment in reproduction can reduce individual survival and thus future fecundity. Light investment can improve survival and improve future fecundity, but the individual still faces the prospects of potential low life expectancy. Somewhere in between lies the best strategy for optimal lifetime reproductive success, the sum of present and future reproductive success.

Allocation of energy to reproduction is related in part to the nature of mortality, its relation to the population, and predictability of the environment. Organisms living in an unpredictable environment or subject to heavy environmentally induced mortality tend to allocate a greater proportion of energy to reproduction, and to expand rapidly when conditions are favorable. Such organisms are said to be r-selected because selection favors high productivity.

Organisms that occupy a more predictable environment and are more subject to density-related mortality tend to allocate less energy to reproduction. They are said to be K-selected because selection favors efficient use of the environment.

Associated with reproductive effort is a mating system, the behavioral mechanisms and social organization involved in the acquisition of mates. Mating systems, influenced by sexual selection and degree and nature of parental care given the young, include two basic types, monogamy and polygamy. Two general kinds of polygamy are polygyny, in which the male acquires more than one mate, and polyandry, in which the female acquires more than one mate. The potential for competitive mating is higher in polygamy than in monogamy. The nature of mating systems is influenced by both ecological and behavioral factors.

Because the quality of habitat is so important to reproductive success, habitat selection is an important part of an organism's life history pattern. How animals cue in on habitat quality is not clearly understood, but structural cues such as structure and diversity of vegetational cover, particularly as it relates to food, appear to be important. Quality may be modified by the availability of nesting sites, song perches, escape cover, and the like. Individuals across a species' range exhibit regional plasticity to selection of habitats.

Interspecific Competition

Individuals in single species population interact not only with individuals of their one species but with individuals of other species populations. Aphids suck juices from the leaves of plants and caterpillars feed on the leaves, mites feed on aphids, and caterpillars are parasitized by flies and consumed by birds. Bluebirds and tree swallows dispute ownership of nesting cavities. Within a community, populations of plants and animals interrelate in some manner. Some have minimal influence on one another. Other populations, such as parasites and their hosts and predators and their prey, have very direct and immediate relationships. From an individual's standpoint, these relationships can be detrimental or beneficial; from a population standpoint, they depress, stabilize, or increase a population's rate of growth.

■ Types of Population Interaction

The effects of interactions among populations can be positive, negative, or neutral in their effect on population growth (Table 18.1). Neutral interactions (designated 0 0) have no effect on the growth of interacting populations. Positive interactions (+ +) benefit both populations. These interactions may be nonobligatory, that is, not essential to either population, or obligatory and essential. Such relationships are termed *mutualism*. Negative interactions (− −) produce an adverse effect on both populations. Such a relationship is termed *competition*.

In some situations one population is favorably affected while the other is unaf-

fected (+ 0); or one population is negatively affected while the other remains unaffected (− 0). The former is *commensalism*, a one-sided relation between two species in which one benefits and the other is neither benefited nor harmed. Among such commensals are epiphytes, plants that grow in the branches of trees. They depend on trees for support only; their roots draw nourishment from humid air. The relationship in which one population is inhibited while the other remains unaffected is called *amensalism*. Amensalism probably involves some type of chemical interaction, such as the production of an antibiotic or allelochemical agent by one of the organisms involved. Amensalism is a nebulous relationship, most examples of which can probably be best described as a form of interspecific competition.

Other relationships positively affect one population and are detrimental to the other (+ −). Such relationships involve predation and parasitism. *Predation is* the killing and consumption of prey. *Parasitism*, which may be viewed as a weak form of predation, also involves the consumption of one organism by another, but either the host survives or it is killed slowly (Chapter 22). Parasites that cause the eventual death of the host are known as *parasitoids*. These relationships are summarized in Table 18.1. Three with important ecological consequences are mutualism (Chapter 23), predation (Chapter 19), and competition.

When two or more organisms in the same community seek the same resource, which is in short supply to the number seeking it, they compete with one another. If the competition is among members of the same species, it is called intraspecific (Chapter 16); if it takes place among individuals of different species, it is called *interspecific competition*. Individuals in pop-

TABLE 18.1 Population Interactions, Two Species System

Type of Interaction	Response A	B
Neutral	0	0
Mutualism	+	+
Commensalism	+	0
Amensalism	−	0
Parasitism	+	−
Predation	+	−
Competition	−	−

Note: 0 = no direct effect; + = positive effect on growth of population; − = negative effect on growth of population.

ulations experience both types of competition to a greater or lesser degree. To demonstrate that interspecific competition is important, we have to show that one species utilizes a resource to the extent that it limits the population size of another.

Interspecific competition may be exploitative or interference, two terms somewhat akin to scramble and contest types of intraspecific competition. In *exploitative competition* the individuals utilize the same resource. Prior utilization by one reduces the availability for another. The outcome is determined by how effectively each of the competitors utilizes the resource. It often results in reduced growth of all competitors. *Interference competition* involves a direct interaction between the competitors in which one interferes with access to the resource by another. In animals interference usually involves some form of aggressive behavior.

Among individual plants and sessile animals, both fixed in space, interspecific competition is influenced by the degree of proximity of individuals to one another. Each individual affects the environment of its neighbor by consuming resources in limited supply, by modifying environmental conditions (for example, shading and protecting plants from wind and predators), and by producing toxins. These changes can alter the rate of growth, biomass accumulation, and growth form of individual plants.

■ Classic Competition Theory

The Lotka-Volterra Model

Lotka and Volterra separately developed mathematical equations based on the logistic curve to describe the relationship between two species utilizing the same resource. They added to the logistic equation for the population of each species a constant to account for the interference of one species on the population growth of another. This constant, in effect, converts the number of members of one species population into the units of the other:

Species 1: $\dfrac{dN_1}{dt} = r_1 N_1 \left(\dfrac{K_1 - N_1 - \alpha N_2}{K_1} \right)$

Species 2: $\dfrac{dN_2}{dt} = r_2 N_2 \left(\dfrac{K_2 - N_2 - \beta N_1}{K_2} \right)$

where r_1 and r_2 = the rates of increase for species 1 and 2, respectively; K_1 and K_2 = equilibrium population size for each species in the absence of the other; α = a constant, characteristic of species 2, a measure of the inhibitory effect of one N_2 individual on the population growth of species 1; and β = a constant, characteristic of species 1, a measure of the inhibitory effect of one N_1 individual on the population growth of species 2.

Some major assumptions lie behind the Lotka-Volterra model: (1) the environment is homogeneous and stable, without any fluctuations; (2) migration is unimportant; (3) coexistence requires a stable equilibrium point; and (4) competition is the only important biological interaction (see Schoener 1982; Chesson and Case 1986; Roughgarden 1986).

In the absence of any interspecific competition—either α or N_2 = 0 in equation 1 and β or N_1 = 0 in equation 2—the population of each species grows logistically to equilibrium at carrying capacity. Inherent in the logistic equation is the inhibitory effect of intraspecific competition on each species' own population growth. This effect is represented by $1/K_1$ for species 1 and $1/K_2$ for species 2. In competing populations, the inhibitory effect of each N_2 individual on N_1 is α/K_1. Similarly, the inhibiting effect of each N_1 individual on the population growth of species 2 is β/K_2. The outcome of competition depends upon the relative values of K_1, K_2, α, and β. If $N_2 = K_1/\alpha$, N_1 can never increase; and if $N_1 = K_2/\beta$, N_2 can never increase.

What the equations describe can be better understood with some graphic models (Figure 18.1). In each case the ordinate will represent the population size of species 1 and the abscissa the population size of species 2. The two extreme cases, no competition and complete competition, are the ends of the diagonal line, K_1 and K_2, for carrying capacity of species 1 and 2 respectively, and K_1/α and K_2/β for the competitive effects on species 1 and species 2, respectively. The diagonal line represents equilibrium conditions; the space below it represents the area within which the population can increase. For species 1 in Figure 18.1a, equilibrium conditions are represented by the line K_1/α, K_1 obtained by plotting N_1 against N_2. The line represents a set of joint values of N_1 and N_2 along which the number of individuals in species 1 is neither increasing nor decreasing. This set of joint values is represented by $dN_1/dt = 0$. Populations of species 1 inside the line, indicated by the shaded area, will increase in size until they reach the diagonal line, which represents all points of equilibrium. Actual values can fall anywhere along this line. All populations outside the line (to the right) will decrease to the equilibrium points. A similar set of joint values exists in which N_2, the number of individuals in species 2, reaches equilibrium level (Figure 18.1b). Outside the diagonal equilibrium line, species 2 decreases; inside the line it increases. Values of N_1 and N_2 are considered jointly because of the effect of competition. An increase in N_1 diminishes the growth rate of species 2, and an increase in N_2 decreases the growth rate of species 1. In the presence of species 2, the higher the value of N_2, the lower is the value of N_1 at which species 1 stops growing.

What happens when species 1 and species 2 occupy the same space simultaneously as competitors? According to the Lotka-Volterra model, there are four possible outcomes: (1) species 1 wins and species 2 becomes extinct; (2) species 2 wins and species 1 becomes extinct; (3) either species can win, depending upon the ecological variables operative at any one time; or (4) neither species wins, and they eventually coexist, dividing the resources between them in some manner.

To illustrate the first two outcomes, if species 1 and species 2 occupy the same space as competitors and species 2 is the stronger competitor, species 2 slows down the population growth of species 1 and eventually wins, leading to the extinction of species 1 ($\alpha > K_1/K_2$ and $\beta < K_2/K_1$). In Figure 18.1c the plot of species 1 moves upward because the area on or above the line $K_1,K_1/\alpha$ (the carrying capacity of species 1) and the area below $K_2,K_2/\beta$ is below the carrying capacity of species 2. Species 2 will increase until it arrives at K_2; at that point only species 2 remains. Under a different set of conditions species 1 will win, as illustrated in Figure 18.1f. In each case no equilibrium exists because one species is able to increase in an area where the other species must decrease. Equilibrium can come about only when the diagonal lines cross one another.

In the third outcome (Figure 18.1d) the diagonal equilibrium lines cross each other. The equilibrium point is represented at their crossing, but it is unstable. The vectors are directed away from the equilibrium point, indicating that the true equilibrium points are K_1 and K_2. In this situation equilibrium between competing

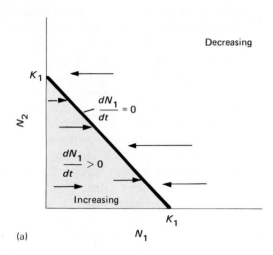

(a)

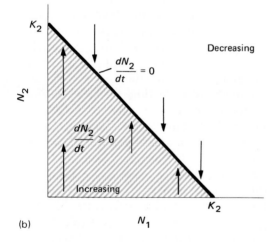

(b)

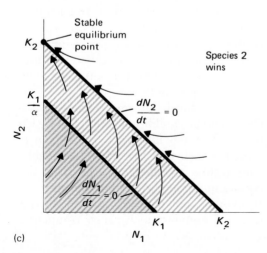

(c)

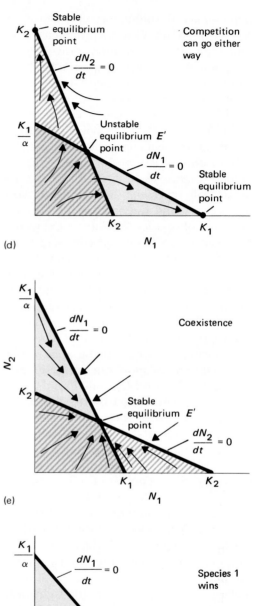

(d)

(e)

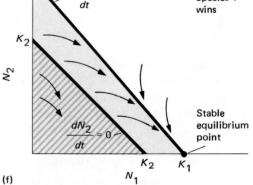

(f)

Figure 18.1 The Lotka-Volterra model of competition between two species. In (a) and (b) populations of species 1 and 2 in absence of competition will increase in size and come to equilibrium at some point along the diagonal line or isocline. The line represents all equilibrium conditions at which the population just maintains itself ($r = 0$). In the shaded area below the line r is positive and the population increases (as indicated by arrows). In (c) species 1 and 2 are competitive. Because the isocline, the zero growth curve of species 2, falls outside the isocline of species 1, species 2 wins, ultimately leading to the extinction of species 1. In (f) the situation is reversed and species 1 wins, leading to the exclusion of species 2. In (d) and (e) the isoclines cross. Each species, depending upon the circumstances, is able to inhibit the growth of the other. In (e) neither species can exclude the other. Each by intraspecific competition inhibits the growth of its own population more than it inhibits the growth of the other population. In (d) each species inhibits the growth of the other species more than it inhibits its own growth. What species wins often depends upon the initial proportion of the two species or changing environmental conditions.

species is unstable and either of the two species can win. Above the line $K_2,K_2/\beta$, species 2 is unable to increase; and above $K_1,K_1/\alpha$, species 1 is unable to increase. If the mix of the species is such that the point N_1,N_2 falls within the triangle $K_2,E'/K_1/\alpha$, species 1 is above its carrying capacity and species 2 is not. Species 2 will continue to increase and species 1 will decrease until it is gone. The reverse situation occurs in triangle $K_1,E',K_2/\beta$. What happens in parts of the diagram outside the triangles depends upon whether the starting value of N_1 is larger or smaller than that of N_2.

Finally, the two species might coexist with their populations in some sort of equilibrium (Figure 18.1e). As species 1 increases, species 2 may decrease and vice versa. Each species inhibits its own growth through density-dependent mechanisms more than it inhibits the growth of the

other species. Neither species reaches a high enough density to bring about any serious competition between them and the population growth of each is not strongly controlled by the same limiting conditions. As long as each species is limited by a different resource and both are only weakly competitive, then the two species will continue to coexist. Thus, in Figure 18.1e, species 2 has the advantage in the area $K_1,E',K_2/\beta$ and the plot moves up and to the left to the equilibrium point E'. The two competing species will reach a stable equilibrium point and persist indefinitely when

$$\alpha < \frac{K_1}{K_2} \text{ and } \beta < \frac{K_2}{K_1}$$

The Lotka-Volterra equation and graphical models based on them apply well to animal populations, but not to plants. The models are based on an animal's potential for increase in numbers. In many plant populations the potential is for increase in biomass. Accordingly, competitive relations among plants may be examined in terms of the influence of one species on the growth of another. Models developed from experimental studies of mixtures of two species sown at a variety of proportions detect changes in yield of dry matter, number of tillers, production of seed, and so on after a lapse of time. Proportions of the two species at the end of the period are plotted against proportions at the beginning. Such studies arrive at five basic types of interactions (Figure 18.2), suggestive of those predicted by the Lotka-Volterra equations (Harper 1977).

1. Neutral interaction. Proportion of two species remains unaltered after a period of growth together. The balance of the two species is subject only to random variation. Such interactions are rarely, if ever, seen in nature.

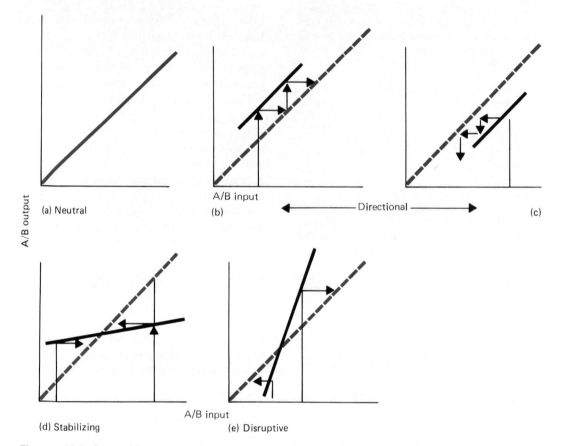

A/B output

(a) Neutral

A/B input

(b)

← Directional →

(c)

(d) Stabilizing

A/B input

(e) Disruptive

Figure 18.2 Competitive interactions among plants. Compare Figure 18.1, which is for animal populations. These models are based on replacement series experiments in which the ratio of species present after a period of time is plotted against the ratio of the species sown or planted. (a) The proportion of two species remains unaltered after a period of growth together. This graph represents an ideal situation. (b) Species 1 gains an advantage over species 2 in all situations and species 2 goes extinct. The degree of advantage is measured by the distance of the actual ratio line from the theoretical diagonal line of no advantage. If the lines are parallel, the advantage is independent of frequency. In (c) the situation is reversed and species 1 moves to extinction. (d) In this frequency-dependent situation the minority species is always at an advantage. Successional or selective advantage depends upon the frequency at which each species is sown or planted. For example, if a high population of species 1 is sown relative to species 2, species 2 gains in the mix. The mix tends toward a stable equilibrium frequency. (e) In this frequency-dependent situation the major species in the mix is at an advantage. If a high proportion of species 1 is sown, species 2 will go extinct. Thus the outcome depends upon the starting frequency of the species. There is no equilibrium mix. (After Harper 1977.)

2. Directional interaction in favor of species 1. Species 1 has the competitive advantage in the mixture at all proportions. If the advantage is carried from planting to harvest and back to plant-

ing again, generation after generation of species 1 ultimately would bring about the extinction of species 2.

3. Directional interaction in favor of species 2. Species 1 is eliminated by spe-

cies 2 at a speed dependent upon the distance between two parallel lines.

4. Stabilizing interaction. In this frequency-dependent situation the minority component is always at an advantage.

5. Disruptive interaction. In this frequency-dependent situation the majority component is at an advantage. This nonequilibrium mixture is unstabilized and disruptive.

Competitive Exclusion

The Lotka-Volterra equations predict, as graphs (c) and (f) in Figure 18.1 illustrate, that if one species in a competitive situation grows rapidly enough to prevent the population increase of another, it can reduce that population to extinction or exclude it from the area. This model led to the concept of competitive exclusion. This concept is called Gause's principle, because he demonstrated the concept experimentally with laboratory populations of *Paramecium*. It states that two species with identical ecological requirements cannot occupy the same environment.

The idea was far from original with Gause (and he laid no claim to it). For example, the ornithologist Joseph Grinnell in 1904 wrote: "Two species of approximately the same food habits are not likely to remain long evenly balanced in numbers in the same region. One will crowd the other out. The one longest exposed to local conditions, and hence best fitted, though ever so slightly, will survive to the exclusion of any less-favored would-be invader."

More recently the concept gained the name *competitive exclusion principle*. Hardin (1960) wrote: "Complete competitors cannot coexist. Two competing species with identical ecological requirements cannot occupy the same area." However, this so-called competitive exclusion principle

hardly rates as a principle (Cole 1960). It is little more than an ecological definition of a species. A corollary of the statement is that if two species coexist, they must possess ecological differences. Obviously, two separate species cannot have identical requirements; being different species, they must have somewhat different ecologies. However, two or more species can compete for some essential resource without being complete competitors.

Competition, for simplicity, is usually considered on a one-dimensional gradient; but in natural situations competition is spread over a number of resources. A high competitive interaction for one resource may be counterbalanced by low competitive interactions for other resources. In such situations, minimal competitive inhibitions on several gradients among a number of species can for some individual species be equivalent to strong competitive interaction for one resource from a single competing species. This relationship has been termed *diffuse competition* by MacArthur (1972). Theoretically, diffuse competition can exclude a species or greatly reduce its numbers through competitive interactions with a specific combination of other species, rather than just one strong competitor.

Pielou (1974) provides a set of conditions in addition to the utilization of resources in short supply that should be met for competitive exclusion to take place. These conditions include (1) competitors must remain genetically unchanged for a sufficiently long period of time for one species to exclude the other; (2) immigrants from areas with different conditions cannot move into the population of the losing species; (3) environmental conditions must remain constant; and (4) competition must continue long enough for equilibrium to be reached. In the absence of any of these requirements, species usually coexist.

■ Studies of Competition

Laboratory Studies

The best place to observe interspecific competition is in laboratory cultures of small invertebrates and microorganisms and in the greenhouse, isolated from environmental fluctuations and outside interferences. Gause (1934) set out to test the Lotka-Volterra equations experimentally. He used two species of *Paramecium*, *P. aurelia* and *P. caudatum*. *P. aurelia* has a higher rate of increase than *P. caudatum*. When both were introduced into one tube containing a fixed amount of bacterial food, *P. caudatum* died out. The population of *P. aurelia* interfered with the population growth of *P. caudatum* because of its higher rate of increase (Figure 18.3). In another experiment Gause used *P. cauda-*

Figure 18.3 Competition experiments with two species of *Paramecium*. The graphs show the growth of two related ciliated protozoans, *Paramecium aurelia* and *P. caudatum,* when grown separately and when grown in a mixed culture. In a mixed culture *P. aurelia* outcompetes *P. caudatum* and the result is competitive exclusion. (From Gause 1934.)

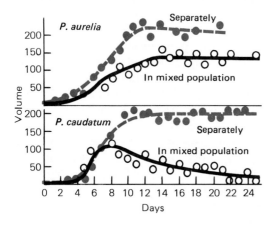

tum and *P. bursaria*. Both species reached stability, because *P. bursaria* confined its feeding to bacteria on the bottom of the tube, whereas *P. caudatum* fed on bacteria suspended in solution. Although the two used the same food supply, they occupied different parts of the culture. In effect, each utilized food unavailable to the other. Park (1948) and Crombie (1947) carried out competition experiments involving several species of flour beetle and obtained results similar to those of Gause.

Tilman and associates (Tilman et al. 1983) grew laboratory populations of two species of diatoms, *Asterionella formosa* and *Synedra ulna,* which require silica for the formation of cell walls. They monitored not only population growth and decline but also the level of silica. When grown alone in a liquid medium to which resource (silica) was continually added, both species kept silica at a low level (Figure 18.4 a, b). When grown together, *S. ulna* took silica to a level below which *A. formosa* could not survive and reproduce (Figure 18.4 c, d). In this experiment *S. ulna* competitively excluded *A. formosa* by reducing resource availability.

Field Studies

Demonstrating interspecific competition in the laboratory is relatively simple, but for experimental organisms life in the laboratory or even in outdoor enclosures is something like life in a jail or in a compound. The situation under which interactions take place is not normal, and dispersal is inhibited, placing unnatural restrictions on space and density. Attempting to demonstrate interspecific competition under truly natural conditions is difficult because of limited ability to establish adequate controls and to manipulate the populations. We can observe

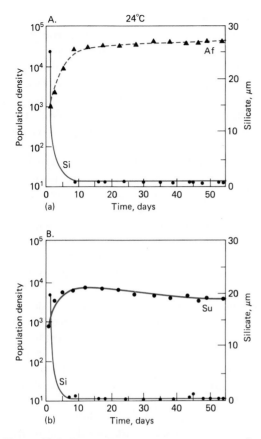

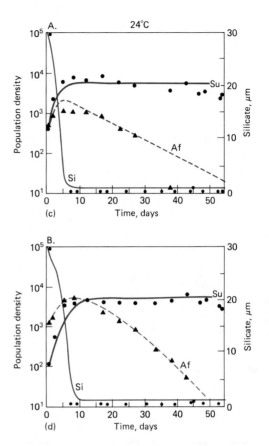

Figure 18.4 Competition between two species of diatoms, *Asterionella formosa* and *Synedra ulna,* both of which require silica for the formation of cell walls. Both are capable of reducing silica to low levels in a culture medium. (a) When grown alone in a culture flask, *A. formosa* reaches a stable population level at which it keeps the resource, silica, at a constant low level. (b) *S. ulna* does the same except that it draws the silica down to an even lower level (c, d, replicates). When the two species are grown together in a culture flask, *S. ulna* drives *A. formosa* to extinction, because *S. ulna* reduces the silica level to a point below which the other species cannot exist. All cultures illustrated were grown at the constant temperature of 24° C. (From Tilman et al. 1981: 1025, 1027.)

interspecific competition among birds at winter feeding stations and among bluebirds, tree swallows, and starlings over a limited number of nest boxes and cavities; but we have observations, not experimental evidence that such competition has a pronounced effect on population growth and on survival of individuals and populations. Although competition is important in the organization of community structure and in the evolution of organisms, the prevalence of interspecific competition in natural communities is not well understood.

There have been well over 100 field experiments attempting to demonstrate that interspecific competition has an effect on competing populations (Schoener 1983; Connell 1983); most results are inconclusive (Underwood 1986). The reasons include poor experimental design, lack of replication (see Chapter 2), problems with experimental controls, and confounding interspecific competition with possible environmental effects. The presence or absence of competitive interactions among species can be interpreted by the use of manipulative field experiments, if the hypotheses to be tested are well defined, if the experimental procedures represent the reality of the situation being investigated, and if the experiments are replicated.

A field experiment by Kenneth M. Brown (1982) with pond snails shows how laboratory studies aid in the design of a manipulative field experiment with spatial replications. First, Brown quantitatively sampled eight ponds with an Ekman bottom dredge to determine the relative abundance of pulmonate snails. He divided the ponds into four categories based on the time they remained full of water and the major food resource: (1) open temporary ponds that dried in early summer with a major food resource of periphyton and aquatic vegetation; (2) partially wooded ponds and (3) heavily wooded temporary ponds, both of which dried up somewhat later in summer and had a food base of decaying leaves; and (4) a permanent pond with a food base of periphyton. Brown found six species of snails, two of which were rare and occurred at low density. The four most abundant large pulmonate snails were (1) *Aplexa hypnorum*, a habitat specialist found in heavily wooded temporary ponds as well as a food specialist, consuming detritus; (2) *Lymnaea elodes*, a generalist feeding on periphyton and carrion; (3) *Physa gyrina*, a generalist with a broad range of habitats and niches; and (4) *Helisoma trivolvis*, a generalist in food preference but found only in the permanent pond.

Next Brown carried out experimental laboratory studies with all four species to determine their dietary food overlaps and

their reproductive overlap. If snails with similar food habits inhabited the same ponds, they might possibly compete if they produced new cohorts at the same time of year. The snails with the greatest dietary and reproductive overlap, thus potential competitors, were the two most abundant ones, the generalists *Physa gyrina* and *Lymnaea elodes*.

With this background information, Brown was ready to begin manipulative field experiments to determine (1) which species pair had the highest resource overlap in habitat and food and (2) whether the species suffered decreased growth or fecundity (fitness) due to competition.

He selected three different ponds: a partially wooded temporary pond A, and open temporary pond B, and a permanent pond F, for replicate experiments to compare the effects of pond permanency and productivity on competitive effects. Ranked for productivity based on periphyton biomass accumulation on slides submerged in the water, the permanent pond F was the most productive and the open temporary pond was the least productive: $F > A > B$.

One phase of the experiment involved rearing snails in specially constructed plastic containers inserted in the pond and replenished weekly with pond vegetation. In each pond the conspecific snails were reared at two densities in the container, two and four snails, to determine any density-dependent effects on growth and fecundity. Other containers held a mixed population of two individuals of each species. There were 15 replicates for each treatment. If the growth rates and fecundities were depressed in the mixed containers below averages for conspecific containers, interspecific competition between the two species could be considered stronger than density-dependent effects. Brown measured shell growth weekly in the container population, noted egg cases produced, and replaced any dead snails with live ones of the same size.

From his container experiments Brown found that both intraspecific and interspecific competition in *P. gyrina* lowered fecundity and growth rates in both temporary ponds but not in the permanent pond. Replacing two *P. gyrina* with two *L. elodes* significantly lowered fecundities for individuals of *P. gyrina* grown in all three ponds and almost eliminated reproductive activity in one temporary pond and the permanent pond (Figure 18.5a). Fecundities for this species were lowest in the permanent pond and highest in the temporary pond. *L. elodes* also experienced density-dependent effects on growth, and interspecific competition had less effect on growth increments (Figure 18.5b). However, *L. elodes* grew most rapidly in the productive, permanent pond and experienced its highest fecundity there. Replacing two individuals of this snail with two *P. gyrina* increased individual fecundities of *L. elodes* in the permanent pond, but did not affect fecundities in the two temporary ponds. Interspecific competition increased growth rates of both species, but decreased fecundities only in *P. gyrina*. Both species grew best in the more productive permanent pond.

Next Brown wanted to learn whether competitive effects would also occur in a larger pond environment. He conducted pen experiments in temporary pond A because it had large areas free of submerged logs and was intermediate in productivity. He constructed twelve 3 × 1 m pens, of which he placed six in the pond in the fall of 1979 and six more in the spring of 1980. Three of the pens served as controls. Fifty *P. gyrina* were added to each of the other nine pens on June 3, 1980. Of these nine pens, three had 50 *L. elodes*

added (high *Lymnaea* treatment), three had all *L. elodes* removed (*Lymnaea* removal treatment), and the last three received 50 additional *P. gyrina*. The number of *L. elodes* doubled the control densities based on sampling of *L. elodes*. After June 3 he made no effort to maintain densities to simulate natural conditions. Brown measured growth rates and fecundities of the snails in the pens.

Trends in the pen experiments paralleled those of the container experiments, although natural habitat heterogeneity among pens kept the treatment effects from being significant (Table 18.2). Summarizing, the pen experiments supported the hypothesis that interspecific competition is more severe for *P. gyrina* than for *L. elodes*. For *P. gyrina* both growth rates and fecundities tend to be lower when *L. elodes* is added in excess to pens. In contrast, only increases in conspecific densities decreased fecundity and growth rates in *L. elodes*. The pen experiments probably were more representative of competition in the pond itself.

Brown recognizes three deficiencies in his study: (1) high densities of *P. gyrina* were not maintained in the pens; (2) individual egg cases were not counted because they could not be removed from the screens of the pens, so fecundity was measured in terms of egg cases laid per snail (pen totals divided by pen densities); and (3) pens should have had more replications.

Nevertheless, these field manipulations support the hypothesis that a high overlap in resources resulted in competition in this guild of pond snails; that *P. gyrina* with a niche very similar to that of *L. elodes* suffers decreased fecundities under competition with that species, although the mechanism is not clear. The manipulative experiments do not clearly show that field distribution of pond snails

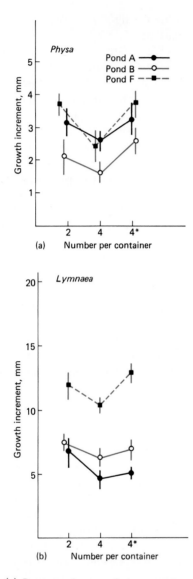

Figure 18.5 (a) Response in growth increments and fecundity of *P. gyrina* held in containers in three ponds and at three densities (two and four *Physa* per container and two *Physa* plus two *Lymnaea*). (b) Response in growth increments and fecundity of *L. elodes* in similar container experiments. Data in both are treatment means SE. *N* = 15 in both groups. (From Brown 1982: 418, 419.)

is the result of interspecific competition, but they do demonstrate that interspecific competition can occur in simple habitats such as temporary ponds.

Because plants are sedentary, and therefore more readily subject to manipulative experiments, studies of interspecific competition among plants are more adaptable to field experimentation. Interspecific competition is considered one of the driving forces behind plant succession

TABLE 18.2 Results ($\bar{x} \pm$ SE) For the Pen Experiment Averaged over the Whole Experiment for *Physa gyrina* and *Lymnaea elodes*

	PHYSA GYRINA			LYMNAEA ELODES		
Treatment	*No. Snails/Pen*	*Number of Egg Cases/Snail*	*Shell Length (mm)*	*No. Snails/Pen*	*Number of Egg Cases/Snail*	*Shell Length (mm)*
Control	1.56 ± .50	1.65 ± .72	14.83 ± .73	14.56 ± 1.86	1.89 ± .51	26.85 ± .48
High *Physa*	32.42 ± 11.99	2.71 ± 1.53	14.35 ± .91	9.33 ± 1.69	2.35 ± .62	26.38 ± .64
Lymnaea removal	19.33 ± 5.33	1.09 ± .56	14.20 ± .82	—	—	—
High *Lymnaea*	20.25 ± 5.56	.67 ± .21	13.61 ± 1.01	38.72 ± 3.48	1.29 ± .37	22.78 ± .52
F	3.85	<1	<1	35.85	<1	21.98
P	.016			<.001		<.001

F values and their significant levels are from a one-way analysis of variance. Density and fecundities were square root transformed.

Source: Brown 1982: 420.

(see Chapter 25). One very early successional species is mullein *Verbascum thapsus*. Mullein, a biennial, colonizes highly disturbed sites from seeds already in the soil and brought to the surface or from seeds carried to the area. The seeds germinate under a wide range of environmental conditions, requiring only light and moisture. Mullein's hold on the site, however, is tenuous, as it quickly gives way to grasses and perennials and becomes locally extinct.

Katherine L. Gross investigated the role of grasses and perennials on the local extinctions of mullein populations. Her study area was a 7-ha abandoned agricultural field planted to barley three years before. In it she located a 20 × 30-m area enclosing a reproducing mullein population of 600 adults, and established 56 1.2 × 1.2-m plots. Gross grouped the plants in the field into three categories: biennials and winter annuals (B), perennial dicotyledons (D), and grasses (G). Then she removed these three categories in eight types of removal-treatments, including controls: single removal (G, D, and B),

two-way combinations (GD, GB, and BD), and three-way combinations with all categories removed (GBD). These removals, carried out by cutting stems just below he ground, increased the amount of open space from 50 percent in the control and G-treatment plots to 100 percent open ground in the GBD plots. Gross replicated the eight treatments, including controls, seven times, assigned randomly to the 56 plots. She then followed the fates of 13,000 seedlings over a three-year period.

The seedlings emerged in three cohorts: mid-May, mid-June, and mid-August, usually three to four days after extended rains. The number of seeds germinating increased by one order of magnitude with each cohort. Both the number of seedlings of mullein that emerged and seedling survival was positively correlated with the number of plant categories removed from a plot (Figure 18.6). Seedling establishment was limited to the most open plots, and only in the completely open plots did individuals survive to flower and produce seed. Of these

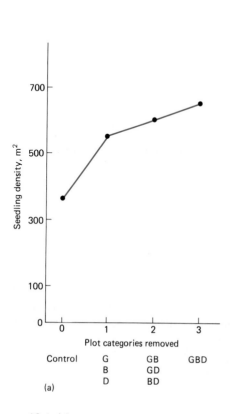

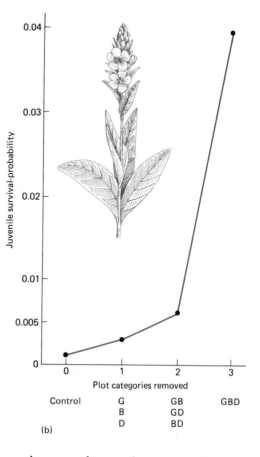

Figure 18.6 (a) Total number of seedlings in each of the vegetation-removal treatments for May, June and August cohorts emerging as a function of the vegetation categories removed from a plot. (b) Overall juvenile survival probability (from germination to mid-November of year 1) as a function of the number of vegetation categories removed from a plot. (From Gross 1980: 921, 923.)

duct good experiments on interspecific competition in the field, despite the difficulties.

■ **Coexistence**

the May cohort experienced the highest survival. All of the rosettes from the May cohort, 50 percent of the June cohort, and none of the August cohort survived to the June of the following year, and only the May cohort, the largest and most vigorous, reproduced.

These well-planned studies, among others, point out that it is possible to con-

Most interspecific competitive relations are probably expressed as some form of stable or unstable equilibrium, if indeed competition occurs at all. Classic competition theory assumes that the environment is stable and competition is continuous; but in reality interspecific competition is probably discontinuous (see Wiens 1977; Wiens et al. 1986; Chesson 1986), because of variable environments. Organisms using identical but limited resources coexist because of different responses to fluctuating environments and different life history traits.

In variable environments resource levels vary between superabundance and scarcity. Irregular periods of scarcity create ecological crises that can result in intense interspecific competition and act as a major selective force intermittently on competing species. On the small island of Daphne in the Galapagos Islands, Grant and his associates (Grant 1986) followed the populations of two species of Darwin's finches, *Geospiza fortis,* the medium ground finch, and *G. scandens,* the cactus ground finch, over a period of ten years, during which there was one long dry period from May 1976 to January 1978. During that period there was a precipitous decline in seed production, resulting in a period of food scarcity accompanied by a population crash of *G. fortis* and a less drastic decline of *G. scandens* (Figure 18.7a). Many seed-bearing plants died and were not replaced until the next wet year, 1983, keeping seed availability low for several years. Foraging diets and behavioral changes reduced overlaps in food. The diet of *G. scandens,* specializing on the seeds of cactus, became narrower, and that of the generalist *G. fortis,* feeding on a variety of seeds including those of cactus, became broader (Figure 18.7b). During the period of drought the diets of the two birds diverged and the overlap diminished, possibly because of the decline in jointly exploited foods.

■ Allelopathy

A particular form of interference competition among plants is *allelopathy,* the production and release of chemical substances by one species that inhibit the growth of other species. These substances range from acids and bases to simple organic compounds that reduce competition for nutrients, light, and space. Produced in profusion in natural communities as secondary substances, most compounds remain innocuous, but a few may influence community structure. For example, broomsedge *(Andropogon virginicus)* produces chemicals that inhibit the invasion of old fields by shrubs and thus maintains its dominance (Rice 1972). The allelopathic effects of goldenrods *(Solidago),* asters *(Aster),* and certain grasses prevent tree regeneration in glades (Horsley 1977). Likewise, the black walnut *(Juglans nigra)* of the eastern North American deciduous forest is antagonistic to many plants (Brooks 1951).

In desert shrub communities a number of shrubs *(Larrea, Franseria,* and others) release a variety of more or less toxic phenolic compounds to the soil through rainwater. Under laboratory conditions, at least, these substances inhibit germination and growth of seeds of annual herbs (McPherson and Muller 1969). Other desert shrubs *(Artemisia* and *Salvia)* that commonly invade desert grasslands release aromatic terpenes such as camphor to the air. These terpenes are adsorbed from the atmosphere onto soil particles. In certain clay soils these terpenes accumulate during the dry season in quantities sufficient to inhibit the germination and growth of herb seedlings. As a result invading patches of shrubs are surrounded by belts devoid of herbs and by wider belts in which the growth of grassland plants is reduced (Muller et al. 1968). Allelopathy may not be the only reason for these belts. Studies of plant-animal interactions suggest that the absence of plants may result from predation by hares and consumption of seeds by rodents, birds, and ants (Bartholomew 1970).

J. L. Harper (1977) suggests that toxic interactions in higher plants may not be as

Figure 18.7 Change in the population sizes (a) and breadth of diets (b) of two Darwin's finches, *Geospiza fortis* and *G. scandens,* in response to a drought on the island of Daphne. The biomass of small seeds at the beginning of the dry season shows a precipitous decline from 1976 to 1982. (c) In response to this change in food availability, *G. scandens* decreased and *G. fortis* increased its diet breadth. After rainfall broke the drought in 1978, the changes were reversed. (From Grant 1986: 180, 186.)

common as supposed. First, higher plants rapidly evolve tolerances to such environmental toxins as zinc, nickel, and copper, and have developed tolerances to herbicides. Second, complex organic molecules are broken down by soil microbes, and plant toxins probably experience the same fate.

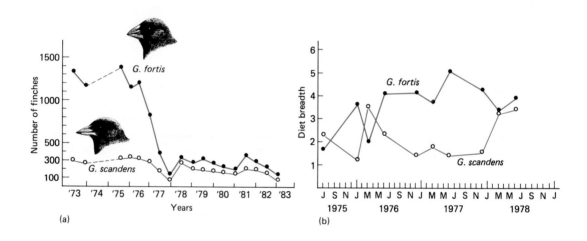

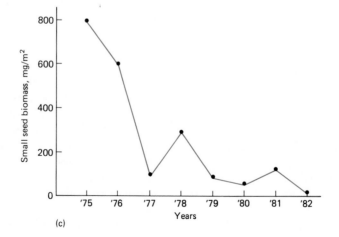

■ Diffuse Competition

If interspecific competition between two species is difficult to establish, how much more difficult it is to demonstrate diffuse competition. Davidson (1985) carried out a five-year experiment with colonies of three species of ants: *Pogonomyrmex desertorum, Po. regosus,* and *Pheidole xerophila.* Relative to populations of conspecifics on control plots, *Po. desertorum* increased on plots where its interference competitor, the large species *Po. rugosa,* was removed and a small species, *P. xerophila,* the exploitative competitor of *Po. desertorum,* declined. Davidson attributed the decline in the latter to the large species' indirect aid to the small species by suppressing populations of the intermediate-sized *Po. desertorum.*

■ Resource Partitioning

Observations of a number of species sharing the same habitat suggest that they coexist by utilizing different resources. Animals eat different sizes and kinds of food, or feed at different times or in different areas. Plants occupy a different position on a soil moisture gradient, require different proportions of nutrients, or have different tolerances for light and shade. Each species exploits a portion of the resources, which becomes unavailable or is unusable to others. Such resource partitioning or differential resource utilization is often regarded as an outcome of interspecific competition.

Theoretical Considerations

Consider species A, which in the absence of any competitor utilizes a range of dif-

ferent-sized food items (Figure 18.8). We can picture that utilization as a bell-shaped curve on a graph, with food as the ordinate and fitness as the abscissa. Most individuals feed about the optimum. Individuals at either tail feed on larger or smaller food items, respectively. As population size increases, the range of food taken may increase, as intraspecific competition forces some individuals to seek food at two extremes. Such intraspecific competition fosters increased genetic variability in the population.

Figure 18.8 Theoretical resource gradient utilized by three competing species, A, B, and C. A and B share the resource gradient with minimal overlap. A third species, C, whose optimal resource utilization lies between A and B, competes with them. In response to selection pressures, A and B narrow their range of resource utilization to the optimum, and C utilizes that portion at less than an optimal level for A and B.

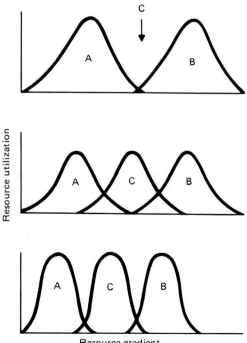

Now allow a second species, B, to enter the area. When its resource use curve is superimposed on the curve of species A, B shows considerable overlap. Selective pressure from interspecific competition forces both species A and species B to narrow their range of resource use. Natural selection will favor those individuals living in areas of minimal or no overlap. Ultimately the two species will narrow their ranges of resource use. They will diverge, moving to the left and the right on the graph. Direct interspecific competition will be reduced, and the two species will coexist. Thus, while intraspecific competition favors expansion of the resource base, interspecific competition narrows the range. The populations involved have to arrive at some balance between the two.

Now allow a third species, C, to invade this resource gradient at a point between the utilization curves of A and B. Species C can successfully invade if A and B are relatively rare, if they are below carrying capacity, and if resources are abundant. Under these conditions competition will force each of the three to become more specialized in their resource utilization, to utilize optimal resources, and to space themselves more narrowly on the resource gradient (for theory, see MacArthur and Levins 1967).

Field Examples

Intensive field studies have turned up numerous examples of presumed resource partitioning. Lack (1971) noted that nine species of the genus *Parus* living in the broadleaf woods and coniferous forests of Europe and Great Britain feed in different parts of the tree canopy and consume different food throughout the year. Three of them, the blue tit, the great tit, and the marsh tit, inhabit the broadleaf woods. The blue tit, the most agile of the three, works high up in the trees gleaning insects, mostly 2 mm in size or smaller, from leaves, buds, and galls. The great tit, which is large and heavy, feeds mostly on the ground and seeks prey in the canopy only when taking caterpillars to feed its young. Its food consists of large insects 6 mm and over, supplemented with seeds and acorns. The marsh tit feeds largely on insects around 3 to 4 mm, which it gleans in the shrub layer and in twigs and limbs below 6 m above the ground. It, too, feeds extensively on seeds and fruits. In the northern coniferous forests live the coal tit, the crested tit, and the willow tit. The more agile coal tit forages high up in the trees among the needles. There it seeks and feeds on aphids and spruce seeds. The willow tit consumes a high proportion of vegetable matter and feeds in the few available broadleaf trees. When in the conifers, it spends most of its time in the lower parts and on the branches rather than on the twigs. The crested tit is confined mostly to the upper and lower parts of the trees and the ground, but the bird does not feed in the herb layer. Thus by feeding in different areas and on different size insects, as well as different types of vegetable matter, these species divide the resources among them.

MacArthur (1958) observed a similar partitioning among five species of warblers inhabiting the spruce forests of the northeastern United States. Each fed in a different part of the canopy, and each was specialized behaviorally to forage in a somewhat different manner.

Intraspecific rather than interspecific competition appears to be most important among phytophagus (leaf-eating) insects. Although the assemblages or guilds of such insects are similar to those of other groups of species, there is no evidence that they result from interspecific competition (Lawton and Strong 1981). Interspecific competition for food is weak (except, perhaps, in outbreaks of gypsy moths, which consume most or all of the foliage in oak forests). Instead, environmental pressures such as harsh climatic conditions, phenology of host plants, seasonal changes in chemical composition of plant tissues (see Chapter 20), and patchy distribution of food plants overshadow the relationships among leaf-eating insects.

A similar partitioning of resources exists among plants. Plants experience strong abiotic and biotic selective pressures on gross morphology, both above ground and below ground, which result in differing methods of exploiting light, water, and nutrients. Once plants are committed to a particular life form, they are committed to a particular mode of resource utilization. For this reason intraspecific competition can be more influential than interspecific competition. This point is well illustrated in Cody's studies of life forms of desert plants. He found that belowground root morphologies were much more important than aboveground structures to coexistence. Conspecifics with similar root structures tended to be widely spaced, and their nearest neighbors were species with different and complementary root systems (Figure 18.9). Species such as *Echinocereus,* the hedgehog cactus, and *Yucca,* the yuccas, have spreading roots within 15 cm of the surface. Others have deep taproots that extend 2 m or more below the surface to reach water and nutrients. Still others, such as the *Opuntia* cactus, have deep spreading roots rather than taproots. The deep-rooted species would conflict with conspecific near neighbors but would allow shallow-rooted species to live as near neighbors. These shrub species coexist because they have different root systems that allow them to exploit water and nutrients in spatially separated areas.

■ Differential Resource Utilization

The usual approach to competition along a single resource gradient emphasizes population responses; but as populations use resources they are also depleting the resource base, which may or may not be renewed. What becomes important in interspecific competition, then, is rate of consumption versus the rate of renewal. Among plants, in particular, this competition is not just for one resource, such as food among animals, but for several resources simultaneously, such as light, soil nutrients, and water.

Tilman (1980, 1982, 1986) has presented a model of plant competition that considers both species growth and resource levels. It is based on the theory that each plant species is a superior competitor along a resource gradient of light and nutrients or light and moisture, and that changes in these resources should result in competitive interactions between plant species.

The theory is best illustrated graphically, first on a single-species basis, then on a two-species basis. Consider a single species using two essential resources, soil nitrogen and light. The resource availability can be plotted on two axes, with light on the X axis and soil nitrogen on the Y axis (Figure 18.10). The population response of the species at zero growth on two gradients is plotted as a solid line with a right-angle corner. It represents the resource-dependent growth isocline. For all points along the isocline population growth is zero; in other words, the plant population growth rate equals its loss rates.

For any one point along the zero net growth isocline there is also only one

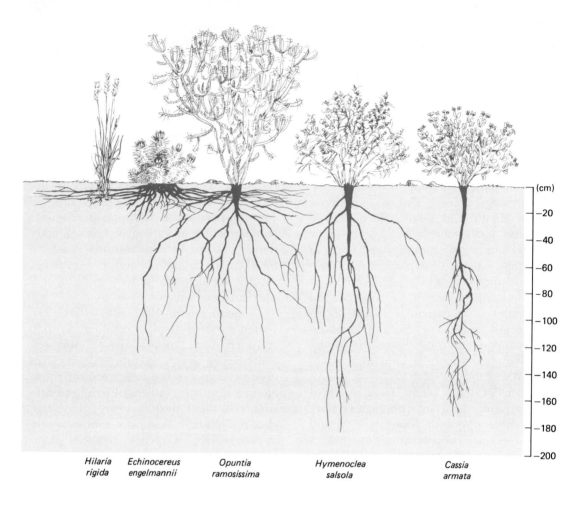

| | | | | (cm) |
| Hilaria rigida | Echinocereus engelmannii | Opuntia ramosissima | Hymenoclea salsola | Cassia armata |

Figure 18.9 Partitioning of the soil resource by a group of Mojave Desert plants. Root system morphology is species-specific. Species such as *Hilaria rigida* and *Echinocereus engelmannii* are shallow surface rooters, able to take up moisture quickly during occasional rains. *Opuntia* and *Hymenoclea* employ more spreading roots at various intermediate depths. Plants such as *Cassia* have deep taproots. Plants with the same root morphology are not near neighbors. (After Cody 1986: 386–387.)

point (supply point) where resource supplies are constant. If the species utilizes a resource at a rate faster than it can be renewed, then the population will decline along the zero isocline (toward the bottom left of the diagram) until it comes into a new equilibrium. Conversely, if renewal of the resource exceeds consumption, then the population will grow along the isocline

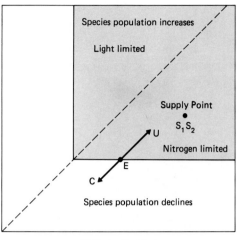

Light at soil surface

Figure 18.10 Zero net growth isocline for a plant species potentially limited by two resources, light at soil surface and available soil nitrogen. The solid line with the right-angle corner is the resource-dependent growth isocline. The broken line shows the proportions of the two resources, light and soil nitrogen, for which this population is equally limited by each. For zero net population growth, the plant population should utilize the two resources in the proportions shown. S_1, S_2 is the resource supply point. $\bar{C}$ is the consumption vector. It points away from the supply point because consumption draws down the resource supply. $\bar{U}$ is the resource supply vector. It points toward the supply vector because it renews the resource supply. The point on the zero net growth supply isocline is the equilibrium point associated with the supply point shown. At this point births equal deaths and resource supply balances resource consumption. The resulting vector parallels the broken line. (After Tilman 1986: 362.)

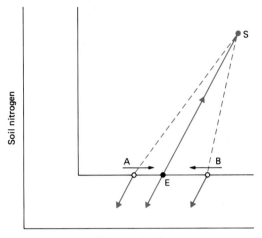

Light at soil surface

Figure 18.11 What happens if consumption vectors veer away from the equilibrium point? At A consumption rate is less than the renewal rate, so the population will grow toward the equilibrium point E. If consumption rate is greater than the renewal rate, as at B, then the population will decline to the equilibrium point.

until it establishes a new equilibrium point (Figure 18.11).

Now superimpose on the diagram the isocline of another species, which has a higher or lower requirement for one of the two resources than the other (Figure 18.12). If the isocline of species B falls inside the isocline of species A, the outcome

of interspecific competition depends upon the location of the supply point. If the supply point falls outside the isoclines of both species, then neither species can exist. If the supply point falls between the isoclines of species A and species B or within the isoclines of both species, then A reduces the resource concentrations along its own isocline to a point where species B cannot exist, and B is competitively excluded.

In the usual situation, one species has a greater requirement for a resource than another. In this case the isoclines will intersect (Figure 18.13). In our example A is the superior competitor for nitrogen and B is the superior competitor for light. The point at which their isoclines cross is the equilibrium point at which both species can coexist. Again, if the supply point falls outside either of the two isoclines, as

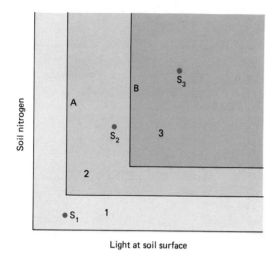

Figure 18.12 Competitive exclusion. The resource-dependent growth isocline of species B lies further away from the resource axes than that of species A. In area 1 neither species can exist because the supply point lies outside the isoclines of both species. In area 2 only A can exist because the supply point falls within the isocline of species A but outside that of species B. In area 3 A also wins because the supply point falls within its isocline, which is closer to the resource axes than that of species B. (After Tilman 1986.)

at point 1, neither species can survive. If the supply point falls at 2 between A and B, then A wins and B loses; and if it falls at 3 between B and A, then B wins. If the supply point falls somewhere within the common region enclosed by both, then the outcome depends upon the position of the supply point. If it falls at 4 then B, with low light requirements, should displace A from habitats with low light availability, because it reduces the light at the soil surface to a level below that required for species A to survive. In region 6, which has a low supply of nitrogen, both species will be nitrogen-limited and A, the superior competitor for nitrogen, should displace species B. In both situations the reason for displacement is the greater utilization by each species of the resource

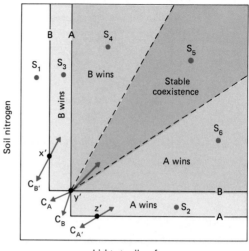

Figure 18.13 Competition for soil nitrogen versus light in two hypothetical species with different need of each resource. When the supply point is located at z, species A should displace species B because it has the lower requirement for nitrogen. A's equilibrium point is at z^1. At supply point x species B should displace species A because it has a lower requirement for light. Its equilibrium point is at x^1. Note that at each of the two equilibrium points the vectors $C_{a'}$ and $C_{b'}$ parallel the broken lines. The point at which the isoclines of each species cross is the two-species equilibrium point. When the supply point is at y, both species coexist because the intermediate habitat has supply rates of nitrogen and light for which each species is limited by a different resource. Note how the vector lines for consumption diverge at the equilibrium point. (After Tilman 1986: 363.)

that limits its own growth. In an intermediate habitat, 5, each species is limited by a different resource, B by nitrogen and A by light. Each species will consume relatively more of the resource that limits its growth and will be limited by the resource for which it is the poorer competitor. The equilibrium point of stable coexistence in habitat 5 will be that at which combined resource consumption by the two species equals resource supply. The concept can be extended to several species (Figure

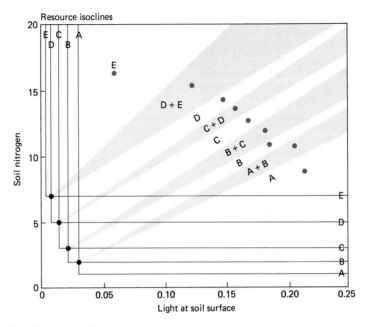

Figure 18.14 Model of resource competition involving several species. Species A to E are inversely ranked according to their competitive abilities for nitrogen and light. The amount of light and nitrogen used by each makes the four two-species equilibrium points stable. Depending upon where the supply point falls, it will support no species, one species, or the coexistence of two species. (Adapted from Tilman 1986: 366.)

18.14). (For details see Tilman 1980, 1982, 1986.)

This model of differential resource utilization appears to be more realistic than the classic competition model, because it takes into account not only population growth but also depletion and renewal of resources that affect that growth. Like other ecological models, though, it will require extensive field testing under natural conditions to support its validity.

■ The Niche

Closely associated with the concept of interspecific competition is the concept of the niche. *Niche* is one of those nebulous terms in ecology, its meaning colored by various interpretations that equate it with habitat, functional roles, food habits, and morphological traits. In everyday terms, a niche is a recess in a wall where you place something, usually an ornamental object;

or it is a place or position in life suitable or appropriate for a person. In ecology it means an organism's place and function in the environment—or does it?

One of the first to propose the idea of the niche was the ornithologist Joseph Grinnell (1917, 1924, 1928). In his study of the California thrasher *(Toxostoma redivivum)* and other birds, he suggested that the niche be regarded as a subdivision of the environment occupied by a species, "the ultimate distributional unit within which each species is held by its structural and functional limitations." Essentially Grinnell was describing the habitat of the species.

Figure 18.15 Models of niche dimension. Three elements comprise a hypothetical organism's niche: food size, foraging height, and humidity. Graph (a) represents a one-dimensional niche involving food size. In graph (b) a second dimension has been added, foraging height. Enclosing that space, we obtain a two-dimensional niche. Now suppose the organism can survive and reproduce only within a certain range of humidity, graphed as a third axis. By enclosing all those points, we arrive at (c), a three-dimensional niche space or volume for the organism.

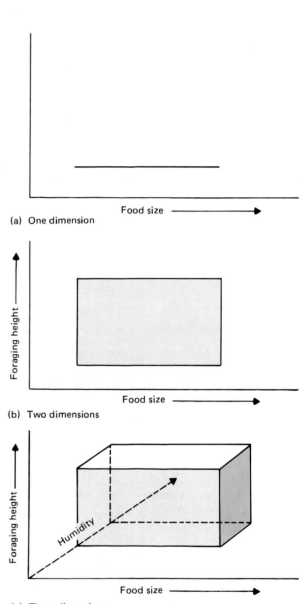

(a) One dimension

(b) Two dimensions

(c) Three dimensions

Charles Elton (1927), in his classic book *Animal Ecology*, considered the niche as the fundamental role of the organism in the community—what it does, its relation to its food and enemies. Basically this idea stresses the occupational status of the species in the community.

Other definitions are variations on the same theme. Odum (1959) considers the habitat as the animal's address and the niche as its occupation. Whittaker, Levin, and Root (1973) consider the niche only as a functional position, whereas the niche and habitat combined comprise the ecotope, "the ultimate evolutionary context of the species." Pianka (1978) regards the niche as embracing all the ways in which a given individual, population, or species conforms to its environment.

The definition that links the niche to competition was proposed by G. E. Hutchinson (1957). It is based on the competitive exclusion principle. According to the Hutchinsonian concept, an organism's niche consists of many physical and environmental variables, each of which can be considered a point in a multidimensional space. Hutchinson called that space the *hypervolume*.

We can visualize a multidimensional niche to a certain extent by creating a three-dimensional one. Consider three niche-related variables for a hypothetical organism: food size, foraging height, and humidity (Figure 18.15). Suppose the animal can handle only a certain range in food size. Food size, then, is one dimension of the niche. Add the foraging height, the area to which it is limited seeking food. If we graph that on the second axis and enclose the space, we have a rectangle, representing a two-dimensional niche. Suppose, too, that the animal can survive and reproduce only within a certain range of humidity. Humidity can be plotted on a third axis. Enclosing that space, we come up with a volume, a three-dimensional niche. Of course, a number of variables, both biotic and abiotic, influence a species' or an individual's fitness. A number of these dimensions, *n*—difficult to visualize and impossible to graph—make up the *n*-dimensional hypervolume that would be the species' niche. An individual or a species free from the interference of another could occupy the full hypervolume or range of variables to which it is adapted. That is the idealized *fundamental niche* of the species.

The fundamental niche of the species assumes the absence of competitors, but rarely is this the case. Competitive relationships may force the species to constrict a portion of the fundamental niche it could potentially occupy. In those parts its fitness might be reduced to zero. The conditions under which an organism actually exists are its *realized niche* (Figure 18.16), which may be further restricted by the absence of certain features of the niche at any given point in time and space. Like the fundamental niche, the realized niche is an abstraction. In their studies, ecologists usually confine themselves to one or two niche dimensions, such as a feeding niche, a space niche, or a tolerance niche.

Consider two examples. Root (1967) studied the exploitation of the feeding niche by the blue-gray gnatcatcher in California oak woodlands. He characterized the niche of the bird in part by the size of its food and by the height above the ground at which it captured food (Figure 18.17). Simplified for the sake of example, the bird's fundamental niche could be considered as described by a maximum

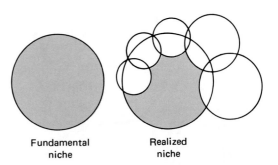

Fundamental niche Realized niche

Figure 18.16 Fundamental and realized niches. The fundamental niche of a species represents the full range of environmental conditions, biological and physical, under which it can successfully exist. However, under pressure of superior competitors whose niches overlap, the species may be completely displaced from part of its fundamental niche and forced to retreat to that portion of the fundamental niche hypervolume to which it is most highly adapted. The portion it occupies is its realized niche.

range of size of prey between 1 and 14 mm in length and by a foraging area of ground level to 10 m. The gnatcatcher's niche center, indicated by frequency of capture and stomach content analysis, consists of insects 3 to 5 mm taken 2.4 to 8.5m above the ground. The further the height and food dimensions diverge from this center, the more the gnatcatcher's niche may overlap those of other species. The boundaries of the realized niche could be defined by the boundaries of any one of the contour lines according to the degree of competition.

Putwain and Harper (1970) studied the population dynamics of two species of dock, *Rumex acetosa* and *R. acetosella*, each growing in hill grasslands in North Wales. *R. acetosa* grew in a grassland community dominated by velvet grass *(Holcus lanatus)* and red and sheep fescues *(Festuca rubra* and *F. ovina); R. acetosella* grew in a community dominated by sheep fescue and bedstraw *(Galium saxatile).* To determine interference and niches of the two dock

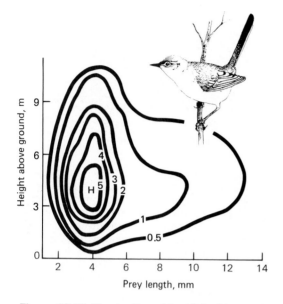

Figure 18.17 The feeding niche of the blue-gray gnatcatcher *(Polioptila caerulea)*, based on a combination of two variables, size of prey and feeding height above ground. The contour lines map the feeding frequencies (in terms of total diet for the two niche axes) for adult gnatcatchers during the incubation period during July and August in California oak woodlands. The maximum response level is at H. Contour lines spreading out from this optimum represent decreasing response levels. For illustrative purposes the outer contour line may be considered the outer boundary of the fundamental niche, although under field conditions in which the study was made, it probably represents the outer boundary of the realized niche for these two variables. (For a detailed discussion of such an analysis of the niche, see Maguire 1973; Hutchinson 1978.) (Diagram from Whittaker et al. 1974: 321, based on data from Root 1967.)

species, Putwain and Harper treated the flora with specific herbicides to remove selectively in different plots (1) grasses, (2) forbs except *Rumex* species, and (3) the *Rumex* species. All species except *R. acetosella* spread rapidly after the grasses were removed, but *R. acetosella* increased only after both grasses and nongrasses were removed.

The niches of these two plants are diagramed in Figure 18.18. The fundamental niche of *R. acetosella* (R) overlaps the fundamental niches of both grasses (G) and other forbs (D). Only when these competitors are eliminated does this dock realize its fundamental niche. *R. acetosa*, however, overlaps only with the grasses and only their removal is necessary to permit expansion of this dock throughout its fundamental niche.

Figure 18.18 Diagram of niche relationships of (a) *Rumex acetosa* and (b) *R. acetosella* in mixed grassland swards. In each diagram the fundamental niches of grass species (G) and nongrass species (D) overlap. The fundamental niche of *Rumex* (R) is shown as a continuous line and the realized niche is shaded. E is that part of the fundamental niche of *R. acetosa* which is expressed in the presence of nongrass species and does not overlap the fundamental niches of G and D. The fundamental niches of seedlings (S), shown by the small dark-colored circles, are contained within the fundamental niches of grasses, nongrasses, and mature *Rumex*. (From Putwain and Harper 1970.)

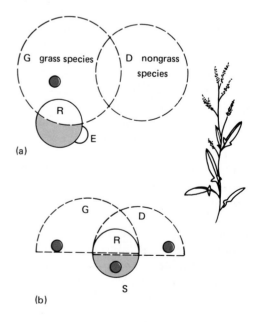

Note that the niches of seedlings differ from those of the mature plants. The fundamental and realized niches of an organism can change with its growth and development. Insects with a complex life cycle may occupy one niche as a larva and an entirely different niche as an adult. In other organisms niche space can change as the organism matures because food and cover requirements change as the organism grows larger.

Niche Overlap

The example of *Rumex* brings up the question of niche overlap. What happens when two or more species use a portion of the same resource, such as food, simultaneously? The theoretical model of the niche assumes that competition is intense, that only one species can occupy a niche space, and that competitive exclusion takes place in areas of overlap. The amount of niche overlap is assumed to be proportional to the degree of competition for that resource. In a condition of minimal or no competition, niches may be adjacent to one another with no overlap or they may be disjunct (Figure 18.19). At the other extreme, in a condition of intense competition, the fundamental niche of one species may be completely within or correspond exactly to another, as in the case of the seedling *Rumex*. In such instances there can be two outcomes. If the

niche of species 1 contains the niche of species 2 and species 1 is competitively superior, species 2 will be eliminated entirely. If species 2 is competitively superior, it will eliminate species 1 from the part of the niche space species 2 occupies. The two species then coexist within the same fundamental niche.

When fundamental niches overlap, some niche space is shared and some is exclusive, enabling the two species to coexist (Figure 18.19). Considerable niche overlap does not necessarily mean high competitive interaction. In fact, the reverse may be true. Competition involves a resource in short supply. Extensive niche overlap may indicate that little competi-

Figure 18.19 Niche relationships visualized as graphs on a resource gradient and as Venn diagrams (diagrams using overlapping circles to show relationships between sets, with each circle representing one set). Species A and B have overlapping niches of equal breadth but are competitive at opposite ends for the resource gradient. B and C have overlapping niches of unequal breadth. Species C shares a greater proportion of its niche with B than B does with C. (However, B also shares its niche with A at the other end.) C and D occupy adjacent niches with little possibility of competition. D and E occupy disjunct niches and no competition exists. Species F has a niche contained within the niche of E. If F is superior to E competitively, it persists and E shares that part of its niche with F. Compare this situation with Figure 18.17. (Adapted from Pianka 1978.)

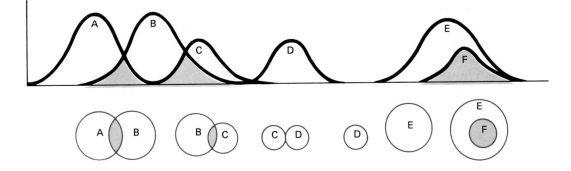

tion exists and that resources are abundant. Pianka (1972, 1975) has suggested that the maximum tolerable overlap in niches should be lower in intensely competitive situations than in environments with low demand/supply ratios.

In fact, both high niche overlap and the absence of overlap may reflect other environmental and behavioral influences and not interspecific competition at all. To attribute niche overlap or the lack of it to interspecific competition may be to ignore real reasons. Consider a study of habitat partitioning by three species of grassland sparrows: the grasshopper sparrow *(Ammodramus savannarum),* Henslow's sparrow *(A. henslowii),* and the savannah sparrow *(Passerculus sandwichensis),* on broad expanses of reclaimed surface mines in north-central Pennsylvania by Pieler and Whitmore (Pieler 1987). They mapped 65 sparrow territories (20 grasshopper, 3 savannah, and 22 Henslow's sparrows) and measured 14 structural variables of the vegetation, including basal and overhead cover, height, density, and litter depth. Territorial boundaries of all three species overlapped to a degree. Except for the semicolonial Henslow's sparrow, territorial boundaries within species were contiguous. Territorial boundaries among the three species showed varying degrees of overlap, the greatest being a 31 percent overlap between the grasshopper and savannah sparrows. The partitioning of the grassland habitat and thus habitat niche configurations could be explained by a gradient of vegetational structure (Figure 18.20) alone, based on increasing habitat richness (vegetation density, vegetation height, bare ground cover, and litter depth). Henslow's sparrows inhabited areas of tall, dense, thick grass. On the other end of the gradient, with some bare ground, thinner and shorter forb and grass cover, and less litter depth, was the

grasshopper sparrow. Intermediate between the two was the savannah sparrow. In addition to the wide differences in the vegetational cover they inhabited, the two congeneric species, grasshopper sparrow and Henslow's sparrow, were very similar in morphological characters of bill width and bill length, whereas the savanna sparrow had significantly smaller ($p < .05$) bill width and length.

For simplicity, niche overlap is usually considered one-dimensional or two-dimensional. In reality, a niche involves many types of resources: food, a place to feed, cover, space, and so on. Rarely do two or more species possess exactly the same requirements. Species overlap on one gradient, but not on another. Total competitive interactions may be less than the competition or niche overlap suggested by one gradient alone (Figure 18.21).

Niche Width

If we plotted the range of resources—for example, food size—used by an animal or the range of soil moisture conditions occupied by plants, the length of the axis intercepted by the curve would represent *niche width* (Figure 18.22). Theoretically, niche width (also called niche breadth and niche size) is the extent of the hypervolume occupied by the realized niche. A more practical definition is the sum total of the different resources exploited by an organism (Pianka 1975). Measurements of a niche usually involve the measure of some ecological variable such as food size or habitat space.

Niche widths are usually described as narrow or broad. The wider the niche, the more generalized the species is considered to be. The narrower the niche, the more specialized is the species. Most species have broad niches and sacrifice efficiency

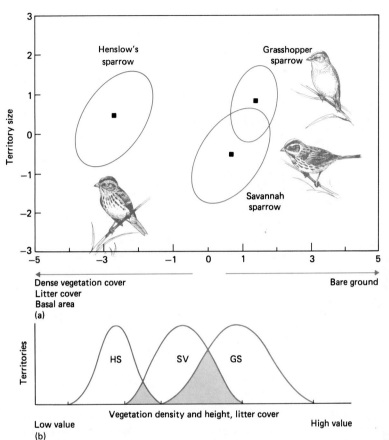

Dense vegetation cover
Litter cover
Basal area
(a)

Territories

HS SV GS

Vegetation density and height, litter cover

Low value   High value
(b)

Figure 18.20 Niche overlap and habitat partitioning among three species of grassland sparrows on a Pennsylvania study site. (a) A two-dimensional plot of the first (habitat) and second (territory size) discriminant axes. Note the wide separation between the two congeneric species, the grasshopper sparrow and Henslow's sparrow. (From Pieler 1987.) (b) The three sparrows arranged along a habitat resource gradient. There is a 1 percent overlap in habitat between Henslow's and grasshopper sparrows, a 14 percent overlap between Henslow's and savannah sparrows, and a 35 percent overlap between savannah and grasshopper sparrows.

in the use of a narrow range of resources for the ability to use a wide range of resources. As competitors they are superior to specialists if resources are somewhat undependable. Specialists, equipped to exploit a specific set of resources, occupy narrow niches. As competitors they are superior to generalists if resources are dependable and renewable. A dependable resource is closely partitioned among specialists with low interspecific overlap (Roughgarden 1974). If resource availability is variable, generalist species are subject to invasion and close packing with other species during periods of resource abundance.

Niche Change

Niche width provides some indication of resource utilization by a species. If a community made up of a number of species with broad niches is invaded by competitors, intense competition may force the

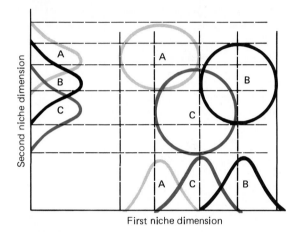

Figure 18.21 Niche relationships based on two gradients. Models of niche on a single-dimensional gradient do not indicate the true niche overlap where other gradients are involved. Two species may exhibit considerable overlap on one gradient and little or none on another. When niche dimensions are considered, niche overlap may be reduced considerably, as illustrated here. On resource gradient 1, A and B exhibit no overlap; and on resource gradient 2, they overlap equally and opposite. When both niches are considered (circles), A and B do not overlap. C on resource gradient 2 overlaps equally with B and very little but equally with A. On resource gradient 2, C overlaps with both A and B. When both gradients are considered, C overlaps mostly with B and very little with A. (Adopted from Pianka 1978.)

Figure 18.22 Hypothetical distribution of a species with a broad niche (A) and a species with a narrow niche (B) on a resource gradient. The niches overlap (shaded area). Species A overlaps species B more than species B overlaps species A.

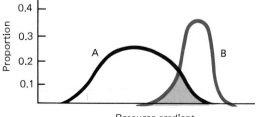

original occupants to restrict or compress their utilization of space and to confine their feeding and other activities to those patches of habitat providing optimal resources. Competition that results in the contraction of habitat rather than a change in the type of food or resources utilized is called *niche compressions* (MacArthur and Wilson 1967).

Conversely, if interspecific competition is reduced, a species may expand its niche by utilizing space previously unavailable to it. Niche expansion in response to reduced interspecific competition is called *ecological release*. Ecological release may occur when a species invades an island that is free of competitors, moves into habitats it never occupied on the mainland, and increases in abundance (Cox and Ricklefs 1977). Such expansion may also follow when a competing species is removed from a community, allowing a remaining species to move into a part of the habitat it previously could not occupy.

Associated with compression and release is another response, niche shift. *Niche shift* is the adoption of changed behavioral and feeding patterns by two or more competing populations to reduce interspecific competition. The shift may be a short-term ecological response or a long-term evolutionary response involving some change in a basic behavioral or morphological trait. With exception of evolutionary change, niche shift does not involve the establishment of a new niche, as the term might imply. Rather niche shift refers (or should refer) to a shifting of the realized niche within the range of the fundamental niche.

Werner and Hall (1976, 1977, 1979) demonstrated niche shift in three cogeneric species of sunfish (Centrarchidae): the bluegill *(Lepomis maerochirus)*, the pumpkinseed *(L. gibbosus)*, and the green sunfish *(L. cynellus)*, occupying experimental ponds 30 m in diameter supporting natural stands of emergent and submerged vegetation and their associated prey populations. Although this experiment lacked replication and the fish were stocked at higher than natural densities, it does provide an example of processes involved in niche shift. When stocked in the ponds alone, each species inhabited the vegetation zones supporting the largest prey. When the three species were stocked together in equal densities, all initially occupied the vegetated zones. As food resources declined, the bluegill and the pumpkinseed left for more open water, leaving the green sunfish, a more efficient forager, in the vegetation. Bluegills and pumpkinseeds concentrated their foraging efforts on bottom invertebrates, mainly Chironominae. The pumpkinseed, with its short, widely spaced gill rakers that do not become fouled when the fish sorts through the bottom sediments, exploited that food supply. The bluegill, with long, fine gill rakers that retain small prey, foraged in the open water.

The fish exhibited shifts in habitat utilization as well. When bluegill and green sunfish were confined together in equal densities in a pond, the bluegill shifted to open water; but in the absence of green sunfish, bluegills invaded the dense vegetation. Bluegills and pumpkinseeds have a dietary overlap of 50 to 55 percent. When the two are stocked together in a pond, the bluegill again moves to open water. In the end, however, the bluegill may be the superior competitor. Young sunfish of all three species feed on zooplankton in open water. This fact places the young of green sunfish and pumpkinseeds in direct competition for small-sized food with both young and adult bluegill. A more efficient open-water forager, the bluegill probably affects the recruitment of the other two species. Perhaps its generalized diet and

its ability to move into different habitats as the situation requires accounts for the bluegill's position as the most common sunfish in ponds.

■ Summary

Relations among species may be positive (+) or beneficial; negative (−) or detrimental; or neutral (0). There are six possible interactions: (0 0), neutral; (+ +), in which both populations mutually benefit each other (mutualism); (− −), in which both populations are affected adversely (competition); (0 +), in which one population benefits and the other is unaffected (commensalism); (0 −), in which one population is harmed and the other is unaffected (amensalism); and (+ −), in which one population benefits and the other is harmed (predation, parasitism).

Interspecific competition—the seeking of a resource in short supply by individuals of two or more species, reducing the fitness of both—may be one of two kinds, interference and exploitative. Exploitative competition depletes resources to a level of little value to either population. Interference involves aggressive interactions (passive in plants, active in animals). A particular form of interference competition is allelopathy, the secretion of chemical substances that inhibit the growth of other organisms.

As described by the Lotka-Volterra equations, four outcomes of interspecific competition are possible. Species 1 may win over species 2, species 2 may win over species 1. Both of these outcomes represent competitive exclusion. A third possibility is unstable equilibrium, in which the potential winner is determined by the one most abundant at the outside or with the most ability to respond to a changing environmental condition. A final possible outcome is stable equilibrium, in which two species coexist, but at lower population levels than if each existed in the absence of the other.

The competitive exclusion principle—two species with exactly the same ecological requirements cannot coexist—has conceptual difficulties. It has, however, stimulated critical examinations of other competitive relationships, especially how species coexist and how resources are partitioned. One way of looking at plant competition is differential resource utilization. It is based on the premise that species compete simultaneously for several resources, such as light and nutrients, but differ in their requirements for two limiting essential resources. Coexistence occurs when the combined resource consumption of the two species equals the resource supply. Groups of functionally similar species that share a spectrum of resources and interact strongly with each other are termed guilds.

Closely associated with the concept of interspecific competition is the concept of the niche. Basically a niche is the functional role of an organism in the community. It might be constrained by interspecific competition. In the absence of any competition, an organism occupies its fundamental niche. In the presence of interspecific competition, the fundamental niche is reduced to a realized niche, the conditions under which an organism actually exists. When two different organisms use a portion of the same resource, such as food, these niches are said to overlap. Overlap may or may not indicate competitive interaction.

The range of resources used by an organism suggests its niche width. Species with broad niches are considered generalist species, whereas those with narrow niches are considered specialists. Niche compression results when competition forces an organism to restrict its type of food or constrict its habitat. In the absence of competition, the organism may expand its niche and experience its ecological release. Organisms may also undergo niche shift by changing their behavioral or feeding patterns to reduce interspecific competition.

Predation

No type of population interaction is more misunderstood or stirs more emotional reaction than predation. Sympathy goes to the prey. Divorced from the reality of death, predation becomes a step in the transfer of energy in the ecosystem. It is commonly associated with the strong attacking the weak, the lion pouncing on the deer, the hawk upon the sparrow, but in the broadest sense, predation also includes parasitoidism, a case of the weak attacking the strong. In this situation, one organism, the parasitoid, attacks the host (the prey) by laying its eggs in or on the body of the host. After the eggs hatch, the larvae feed on the tissues of the host until it dies. Ultimately the effect is the same as that of predation. A special form of predation is *cannibalism*, in which the predator and the prey are the same species. The concept of predation has been extended further to include *herbivory*, in which grazing animals of all types feed on plants. Herbivores kill their prey when consuming seeds or the whole plant, or they function as parasites when they consume only part of the plant but do not destroy it. Thus predation in its broadest sense can be defined as one organism feeding on another living organism, or *biophagy*.

Ecologically, predation is more than just a transfer of energy. It represents a direct and often complex interaction of two or more species, of the eaters and the eaten. The numbers of some predators may depend upon the abundance of prey, and predation may be involved in the regulation of prey populations. These ideas are debatable; certainly the same generalizations cannot apply to all groups of predators and prey.

Models of Predation

The influence of predation on the population growth of a species received the attention of the mathematicians Lotka (1925) and Volterra (1926) (for an excellent discussion of their work see Hutchinson 1978; Kingsland 1984). Independently, they proposed mathematical equations to express the relationship between predator and prey populations. They attempted to show that as the predator population increases, the prey decreases to a point where the trend is reversed and oscillations are produced.

The Lotka-Volterra model involves one equation for the prey population and one for the predator population. The prey growth equation involves two components, the maximum rate of increase per individual and the removal of prey from the population by the predator:

$$\frac{dN_1}{dt} = r_1 N_1 - N_2$$

where N_1 = density of the prey population, r_1 = intrinsic rate of increase in absence of predation, N_2 = density of the predator population, and P = a coefficient of predation. The expression $N_1 N_2$ assumes the removal of prey from the population is proportional to the chance encounter between predator and prey. In turn growth of the predator population is influenced by the density of the prey population:

$$\frac{dN_2}{dt} = P_2 N_1 N_2 - d_2 N_2$$

where P_2 = a coefficient expressing effectiveness of the predator and d_2 = density-independent mortality rate of the predator.

The Lotka-Volterra model is graphically depicted in Figure 19.1. The ordinate H is the number of predators; the abscissa P is the number of prey. In the area to the right of the vertical line, predators increase; to the left they decrease. In the area below the horizontal line, prey increase, above it prey decrease. The circle of arrows represents the joint population of predators and prey, and the size of the population of each changes with it. If a point or arrow falls in the region left of the vertical line, the prey population is not large enough to support the predators and the predator population declines. If arrows fall left of the vertical and above the horizontal, both populations are declining; the predator population decreases enough to permit the prey population to increase, moving the arrows left of the vertical below the horizontal. The increase in prey population now permits predators to increase and the arrows move right of the vertical below the horizontal. As the predator population increases, depressing the prey population, the arrows right of the vertical move above the horizontal. This interaction between predator and prey populations will oscillate through time as indicated in the graph.

The Lotka-Volterra model is based on a number of simplifying assumptions that are difficult to justify in nature. It assumes

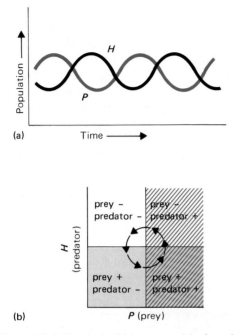

(a)

Population →

Time →

(b)

prey –
predator –

prey –
predator +

H (predator)

prey +
predator –

prey +
predator +

P (prey)

Figure 19.1 The Lotka-Volterra model of predator-prey interactions. The Lotka-Volterra equation can be depicted by the two graphs shown. (a) The abundance of each population is plotted as a function of time. The model shows the joint abundances of species. The zero growth curves of both predator and prey are straight and intersect at right angles. (b) The responses of the populations are indicated by a negative sign for population decline and a positive sign for population increase. Predators increase to the right of the vertical line; prey increases below the horizontal line.

that (1) predators move at random among a prey population that is distributed randomly; (2) every encounter of a predator with prey results in capture and consumption; (3) in absence of predation the prey exhibits exponential growth; and (4) all responses are instantaneous with no time lag for handling or ingesting prey. The model makes no allowance for age structure, for interaction of the prey with their own food supply, and for density-dependent mortality of the predator.

A decade later an ecologist, A. J. Nicholson, and a mathematician and engineer, V. A. Bailey, developed a mathematical model for a host-parasitoid relationship (Nicholson and Bailey 1935). Nicholson and Bailey also based their model on the assumption that predators search randomly, but they assumed prey populations to be uniformly distributed in a uniform environment. They assumed that predators are insatiable, regardless of prey density, and that predators "sample" a certain proportion of the total prey population.

This feature of the Nicholson-Bailey model allows an estimate of prey in the next generation. If the number of hosts that the parasitoid removes within its sampling area is equal to a fraction of prey that represents recruitment, then the base parental stock remains. If the parental stock remaining is sufficient to replace the individual prey taken and if it is sufficient to maintain the density of parasitoids, then the two populations remain stable indefinitely. But if the parasitoid removes part of the parental stock along with recruitment, the prey and ultimately the predator populations decline. If much of the recruitment is left untouched, prey increase and predators may not be up to the task of removing increasing recruitment. In either of these two cases the two interacting populations will undergo oscillation with increasing amplitude.

The Nicholson-Bailey model has its weaknesses. It assumes random searches by the predator for static prey homogeneously dispersed over a homogeneous landscape. The predators have a constant area of discovery and an insatiable appetite. Predators and prey have synchronized generations (same time span and same length) and predator mortality is independent of density. Of course, no such predator exists.

To make the Lotka-Volterra model more realistic, Rosenweig and MacArthur (1963) developed a series of graphic models that consider a wider range of predator-prey interactions. By modifying the zero growth isocline of the prey to account for a low rate of growth at low and high densities, they plot the growth curve as convex rather than horizontal. The basic components of increase and decrease described by the Lotka-Volterra equations remain the same. Thus in the Lotka-Volterra model the growth curve of the prey is a horizontal straight line and the growth curve of the predator is a vertical line (see Figure 19.1). In the Rosenweig-MacArthur model the growth curve of the prey is convex and the isocline of the predator is a vertical line (Figure 19.2). By moving the predator curve to the right or the left so that the curves no longer intersect at right angles, we can produce damped and unstable cycles.

Such models assume (1) an instantaneous change in reproduction of predators without a change in prey density; (2) a prey density unaffected by predator death rate; (3) predators spending the same amount of time in search of prey regardless of prey density; (3) predators

capturing and eating the same proportion of prey they encounter; (4) random distribution of prey and random hunting by predators; and (5) predators who do not compete for prey. Although models based on the Lotka-Volterra equations provide some insights into predator-prey relations, none of them mimic real world situations. Predators and prey simply do not conform to those mathematical models.

Figure 19.2 The Rosenweig-MacArthur model of the outcome of a predator-prey interaction, similar to that predicted by the Lotka-Volterra model but more realistic. The prey curve is convex rather than straight, involving density-dependent growth, but the growth of the predator population lacks density dependence. (a) In this stable cycle model, the prey population can increase if the joint abundances of predator and prey fall inside the area below the isocline of the prey. Prey population will decrease if it falls outside this area. The growth curves intersect at right angles, as they do in the Lotka-Volterra model in Figure 19.1. If the predator isocline is moved to the right or to the left, it will no longer intersect the prey isocline at right angles. The former situation will produced a damped cycle (b) and the latter an unstable cycle (c). (d) A prey refuge limits the amplitude of the oscillations and and tends to stabilize the system. (For further discussions of this model see MacArthur and Connell 1966: 154–156.)

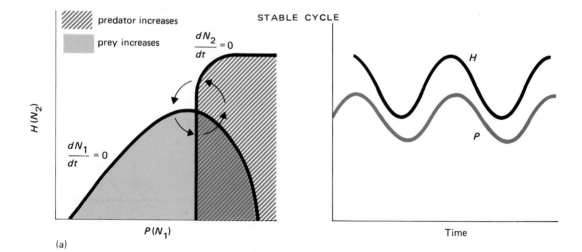

(a)

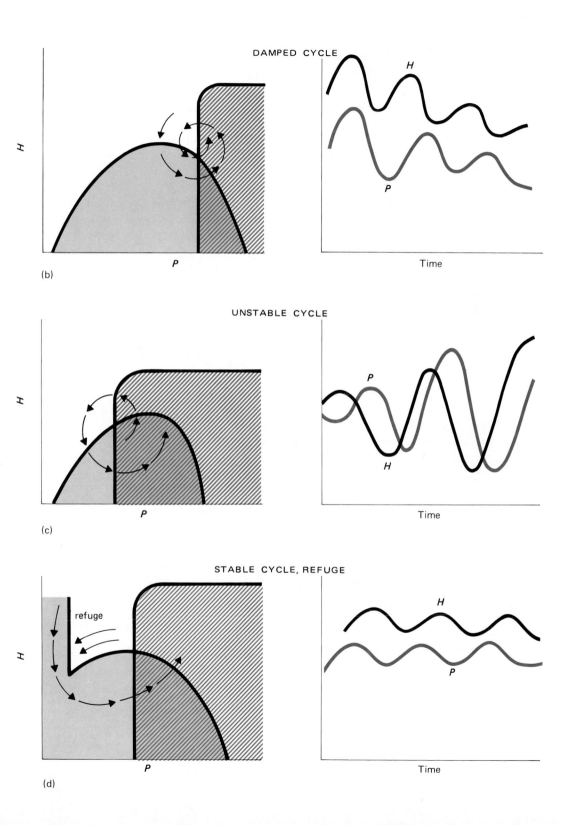

DAMPED CYCLE

H

P

Time

(b)

UNSTABLE CYCLE

H

P

Time

(c)

STABLE CYCLE, REFUGE

refuge

H

P

Time

(d)

■ Functional Response

The Lotka-Volterra type equations of predation hint at two distinct responses of predators to changes in prey density. As prey density increases, each predator may take more prey or take them sooner, a *functional response,* or predators may become more numerous through increased reproduction or immigration, a *numerical response.*

The idea of a functional response was introduced by Solomon (1949) and explored in detail by Holling (1959, 1961, 1966). He recognized three types of functional response (Figure 19.3): Type I, in which the number of prey eaten per predator increases linearly to a maximum as prey density increases; Type II, in which the number of prey eaten increases at a decreasing rate toward a maximum value; and Type III, in which the number of prey taken is low at first and then increases in a sigmoid fashion, approaching an asymptote.

Figure 19.3 Three types of functional response curves. (a) Type I, in which the number of prey taken per predator increases linearly to a maximum as prey density increases. Graphed as a percentage, predation declines relative to the growth of the prey population. (b) Type II, in which the number of prey taken rises at a decreasing rate to a maximum level. When considered as a percentage of prey taken, the rate of predation declines as the prey population grows. Type II predation cannot stabilize a prey population. (c) Type III, in which the number of prey taken is low at first, then increases in a sigmoid fashion approaching an asymptote. When plotted as a percentage, the functional response still retains some of the sigmoid features, but declines slowly as the prey population increases. Type III functional response has the potential of stabilizing prey populations. (After Holling 1959: 293.)

Type I Response

Type I response is a specialized type of functional response, the sort assumed in the simpler models just described (Lotka-Volterra, Rosenzweig-MacArthur, and others). In Type I predators of any given abundance take a fixed number of prey during the time they are in contact, usually enough to satiate themselves. Trout feeding on an evening hatch of mayflies would be an example of this type of functional response. Type I produces

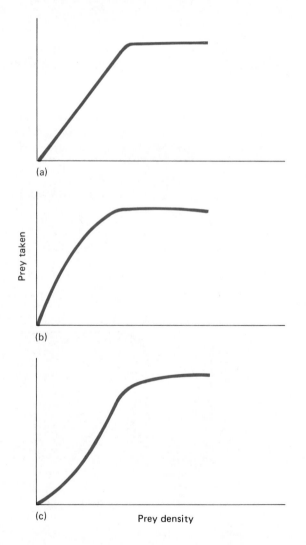

density-independent mortality up to satiation. Of more interest are Type II functional responses, which produce inverse density-dependent mortality in prey, and Type III, which produce changing density-dependent mortality.

Type II Response

The Type II response, generally but not exclusively associated with invertebrate predators, has attracted the most attention. It is described by the disk equation, named for an element in the experiment from which it was derived. In Holling's experiment the predator was represented by a blindfolded person and the prey by sandpaper disks 4 cm in diameter thumbtacked in different densities to a 3 ft square table. The predator tapped the table with a finger until a prey was found and then removed the disk. The predator continued the search and encounter (tapping, discovery, and removal) for one minute. Holling found that the number of disks the predator could pick up increased at a progressively decreasing rate as the density of disks increased. Predator efficiency rose rapidly as the density of disks increased until so much time was spent picking up and laying aside disks that the predator could handle only a maximum number at a time. The experiment demonstrated several important components of predation: density of prey, attack rate of predator, and handling time, including time spent pursuing, subduing, eating, and digesting prey.

Type II functional response is described by the disk equation

$$\frac{N_a}{P} = \frac{aNT}{1 + aT_hN}$$

where T is determined by the equation

$$T = T_s + T_hN_a$$

and

N_s/P = number of prey eaten per predator
N_a = number of prey or hosts killed or attacked
P = number of predators or parasitoids
a = a constant representing the attack rate of the predator or the rate of successful search
N = number of prey
T = total time predator and prey are exposed
T_n = handling time
T_s = time spent by predator in search of prey

Because handling time is the dominant component, rise in the number of prey taken per unit time decelerates to a plateau (Figure 19.4) while the number of prey is still increasing. For this reason Type II functional response cannot act as a stabilizing force on prey population unless the prey occurs in patches. Thus Type II response is destabilizing (for details see Murdock and Oaten 1975).

Type III Response

Type III functional response is more complex than Type II. It has been associated with vertebrate predators that can learn to concentrate on a prey when it becomes more abundant, but recent studies by Hassell et al. (1977) show that it can be found among invertebrate predators as well. Because some vertebrate predators, especially feeding specialists, may also show Type II response, it is much wiser not to attempt to assign types to either invertebrate or vertebrate predators.

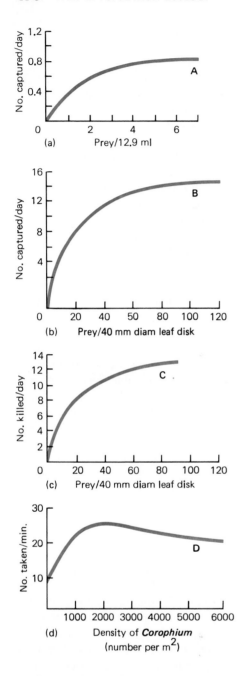

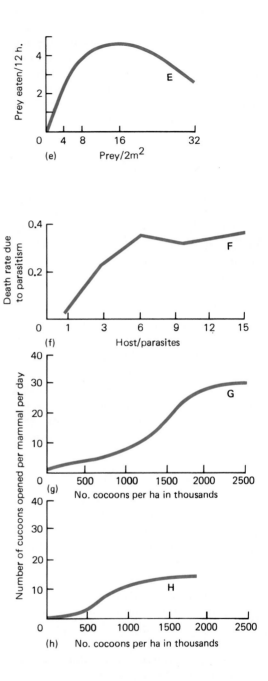

In Type III response the number of prey taken per predator increases with increasing density and then levels off to a plateau where the ratio of prey taken to prey available declines. Because the range of prey densities over which the death rate is imposed is an increasing function of density, that is, is density-dependent, Type III functional response is potentially stabilizing.

Figure 19.4 Examples of Type II and Type III functional response curves. Graphs (a) through (e) are Type II curves; graphs (f) through (h) are Type III. (a) First instar of *Linyphia triangulatus* (spider) feeding on *Drosophila*. (b) Second instar of coccinellid *Harmonia axyridis* feeding on aphids. (c) Adult female *Phytoseiulus persimilis* (mites) feeding on nymphs of *Tetranychus urticae*. (d) Numbers of amphipod crustacean *Corophium* taken per minute by redshank *Tringa totanus* in relation to density. (e) Dome-shaped Type II functional response curve of adult *Phytoseiulus persimilis* feeding on *Tetranychus urticae*. The dome-shaped curve probably results from interference among predators. Female parasitoids discover many hosts are already parasitized, or predators are discouraged from feeding in areas where a large number of individuals are already congregated and leave the area. (f) *Encarsia formosa* parasitizing *Trialeurodes vaporariorum*. (g) Shrew *Sorex* feeding on sawfly larvae. (h) *Peromyscus* (deer mice) preying on sawfly larvae. (a, b, c, e, f from Hassell et al. 1976: 138, 141, and Beddington et al. 1976: 196; d from Goss-Custard 1977: 22; g, h from Holling 1964.)

Threshold of Security

Type II responses occur in situations of varying densities of one prey species. Type III responses invariably involve two or more prey species; the predator has a choice of prey. In the presence of several prey species, the predator may distribute its attacks among the prey in response to the relative density of the prey species. Predators may take most or all of the individuals of a prey species that are in excess of a certain minimum number, determined, perhaps, by the availability of prey cover and the prey's social behavior. Errington (1946), drawing on his studies of predation in muskrat and bobwhite quail populations, called this prey population level, at which the predator no longer finds it profitable to hunt the prey species, its *threshold of security*. Type III responses have been called compensatory because as prey numbers increase above the threshold, the "doomed surplus" becomes vulnerable to predation through intraspecific competition (Chapter 16). Below the threshold of security the prey species compensates for its losses through increased litter size and greater survival of young.

Although intuitively the concept of the threshold of security appears to be sound, it has never been rigorously tested in the field. Based on intrinsic density-dependent population regulation through intraspecific competition, it minimizes the role of predators as an important extrinsic force in population regulation. This concept has been uncritically accepted as dogma by wildlife biologists. Upon it they have based their beliefs (1) that in good habitats wildlife populations cannot be overhunted in normal seasons and (2) that hunting take is compensatory and not additive to other population losses. What hunters remove is the "doomed surplus." Recently there is increasing evidence that this concept does not hold true and that even in good habitats some wildlife species can be reduced below the presumed threshold of security.

Aggregative Response

Although the functional response equations assume predators searching at random in a uniform prey population, the usual situation in nature is an uneven distribution of prey. This setup, in turn, may attract a number of predators to an area. One or two members of the predator species discover and begin to feed on the prey item; other members of the species observe the feeding response and follow suit (see Curio 1976). In this *aggregative response*, predators tend to congregate in patches of high prey density.

Intermediate levels of aggregation may increase the efficiency of predation, because a number of predators foraging together can locate areas of prey abundance more quickly than a few. At high levels of aggregation in areas of high prey density, interference among predators may be so great that the efficiency of predation declines. Encountering an individual of its own species, a predator may temporarily cease hunting or leave the area. Interference among predators reduces the proportion of total prey or hosts the predator or parasite encounters, because the predator's search time is reduced (Hassell et al. 1976).

Aggregative responses of predators to areas of high prey density may have a pronounced influence on stability of predator-prey interactions. Hassell and May (1974) presented an idealized general aggregative response curve for predators-to-prey distribution (Figure 19.5). The response curve exhibits a lower plateau of low prey density and an upper plateau of high prey density where predators do not distinguish between prey areas. The model predicts that predators discriminate markedly in areas of intermediate prey

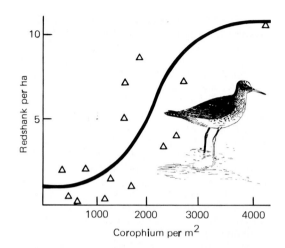

Figure 19.6 Aggregative response in the redshank. The curve plots the density of the predator (the redshank) in relation to the average density of arthropod prey *(Corophium)*. Compare this curve with the profitability curve in Figure 19.5. (After Hassel and May 1974: 569.)

density and tend to congregate in areas of higher density. An example of this type of distribution is the response curve for the redshank *(Tringa totanus),* a shorebird that tends to concentrate in areas of its preferred food, the amphipod crustacean *Corophium volutator* (Figure 19.6). On the other hand, predators tend to avoid areas of low prey density, making those individuals much less vulnerable. Low density areas provide the prey species with partial refuges against predation.

Switching

Involved in the Type III response curve is the role of the facultative predator and alternate prey. Although the predator may have a strong preference for a certain prey, it can turn to an alternate, more abundant prey species that provides more profitable hunting. If rodents, for example, are more abundant than rabbits and quail, foxes and hawks will concentrate on

Figure 19.5 Model of general aggregative response. At the lower plateau of prey density predators do not distinguish among relatively low (unprofitable) prey areas; at intermediate densities (shaded area) predators discriminate markedly; at the upper plateau predators do not discriminate among high density (profitable) areas. (Hassel and May 1974: 576.)

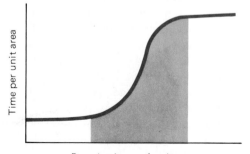

Prey density per 1 unit area

the rodents. This idea was advanced early by Aldo Leopold in his book *Game Management* (1933), in which he described alternate prey species as buffer species because they stood between the predator on one hand and game species on the other. If the population of buffer prey is low, the predators turn to the game species; the foxes and hawks will concentrate on the rabbits and quail. This turning by a predator to an alternate, relatively more abundant prey has more recently been termed *switching* by Murdoch (1969). In switching the individual predator concentrates a disproportionate amount of attacks on the more abundant species and pays little attention to the rarer species (Figure 19.7). As the relative abundance of the two prey species changes, the pre-

dator changes its diet and turns to the alternate prey when it becomes more abundant.

Switching, according to Murdoch and Oaten (1975), is caused by the predator's (1) changing its preference toward the more abundant prey as it eats it more frequently by choice; (2) ignoring rare prey; or (3) concentrating search in more rewarding areas. Any one of these three behaviors results in Type III response curves.

Search Image

The reason for the sigmoidal shape of Type III response is the subject of much study and debate (see Royama 1970; Croze 1970; Murdock and Oaten 1975; Curio 1976). One explanation has been advanced by L. Tinbergen (1960), based on his studies of the relation between woodland birds and insect abundance. According to Tinbergen's hypothesis, when a new prey species appears in a given area, its risk of becoming prey is low. The birds have not yet acquired a search image for the species. A *search image* is a perceptual change in the ability of a predator to detect a familiar cryptic prey. Once the predator has secured a palatable item of prey, the predator finds it progressively easier to find others of the same kind. The more adept and successful the predator becomes at securing a particular prey item, the longer and more intensely it concentrates on the item. In time the numbers of the prey species become so reduced or its population so dispersed that encounters between it and the predator lessen. The search image for that species begins to wane, and the predator begins to react to another species. The combination of increasing density of prey and establishment of a search image results in a sudden increase of the perceived prey species in the

Figure 19.7 A model of switching. The straight line represents a situation in which the proportion of a given prey is the same both in the environment and the diet of the predator. It represents a constant preference with no switch. The curved line represents the proportion of prey species actually taken. Up to a point the proportion taken is less than the proportion in the environment. Switching occurs at the point where the curved line crosses the straight. At that point the number of prey taken is disproportionate to the number in the environment.

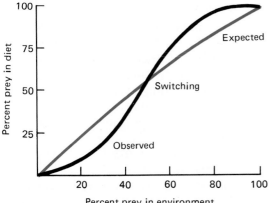

predator's diet, giving a sigmoid functional response curve.

The predator can acquire a search image from remarkedly few experiences (Croze 1970; Dawkins 1971; Curio 1976). In losing an image the predator may simply not respond to the perceived stimulus or may in fact no longer perceive it, that is, no longer distinguish the properties of the prey from the background. The search image is maintained by rewards in the form of the acquisition of food. When rewards are no longer there, the bird turns to another image. In effect, the predator responds to changes in rewards. The extinction of an image tends to occur more slowly than its acquisition, and among some predators the search image may be retained for some time, even in the absence of rewards. Croze (1970) found that carrion crows *(Corvis corone)* retained their search image for eight days without reward. After that time the search image declined rapidly, but was still retained.

Notwithstanding the apparent close relationship between switching and search image, Murdoch and Oaten (1975) emphasize they are not the same. Search image does not refer to a change in predator behavior in response to a change in the relative density of one prey in respect to another, a response that characterizes switching. However, Davies (1977), in a study of prey selection in wagtails *(Motacilla spp.)*, found that a change in diet of these birds over a ten-day period was related to changes in the absolute rather than the relative abundance of the preferred food, midges (Chironomidae). As numbers of this prey decreased, the wagtails incorporated more alternate prey, *Drosophila,* into their diet to maintain the feeding rate.

The search image may not be sufficient or necessary to produce the sigmoidal response curve, but it is still a valid observation of a behavioral phenomenon. It has been studied in some detail by a number of investigators (see Dawkins 1971; Murton 1971; Krebs 1973; Croze 1970). In his carrion crow studies Croze found that the bird did exhibit a search image and that it needed only a few experiences with camouflaged prey to find it. Croze placed meat bait under painted mussel and clam shells arranged on a sandy beach. The crows needed only 2.3 ± 0.5 experiences to become proficient in acquiring a search image for shells of a particular color. After discovery of one prey, the crow tended to concentrate its efforts in that area.

■ Numerical Response

In addition to functional responses predators may exhibit numerical responses. One type of numerical response is a change of predator density through the movement of predators in and out of areas in response to prey density. Such movements of predators represent an aggregative response to prey patchiness (Hassell 1966; Beddington et al. 1976), but can be considered a true numerical response only when predators move into an area from some distance. For example, Figure 19.8 shows a strong numerical response to increased prey density by sparrows feeding on larch sawflies in certain Canadian bogs. Sparrow populations increased largely by immigration involving family flocks, adult and subadult birds, and premigratory flocks (Buckner and Turnovk 1965). Aggregative response of local predators to local situations must be considered a part of functional response (Beddington et al. 1976).

Another and more important type of numerical response is the rate of change of a predator population (through birth

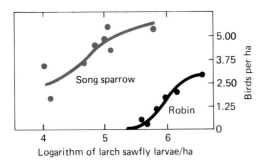

Figure 19.8 Numerical response of song sparrows and robins to larval larch sawflies. (From Buckner and Turnock 1965: 232.)

and death rates) in response to prey death rate. The nature of this type of response is determined by the kind of predator involved, whether it is a parasitoid or a true predator.

For true predators, those that require several prey to complete their development, and for arthropod predators in particular, numerical response or overall rate of increase involves three components (Beddington et al. 1976): (1) duration of each predator instar; (2) survival rate within instars or survival rate of young in nonarthropod predators; and (3) fecundity of adults. All of these components, of course, depend upon the rate at which predators are able to locate and consume suitable prey (Lawton et al. 1975).

Duration of the instar is not influenced by prey density if the parasitoid feeds on only one host. If the parasitoid feeds on more than one host during a developmental period, the amount of food intake and thus prey density influence development time. Growth takes place only after metabolic energy needs are met; if only minimal energy is available after metabolic needs are met, growth is minimal. The less food available, the longer the development time and ultimately the slower the numerical increase of the predator.

Survival rates of arthropod instars and the young of nonarthropod predators are directly dependent on prey density, the size and availability of food. Too few prey means a lack of food, a lack of food results in poor survival of young or within instars, and a poor survival rate has a direct bearing on the numerical increase of a predator population.

For parasitoids in which the complete development of each larva requires only one host, adult fecundity is not limited by nutritional needs, but by the number of hosts females can find. The relationship between prey density and fecundity is linear (Hassell and May 1973).

In all other situations nutrition controlled by the amount of prey eaten during the adult stage influences fecundity. Energy remaining after maintenance demands are met can be used in reproduction. If food and therefore energy is limited because of low prey density, fecundity is necessarily low (Figure 19.9) and positive numerical response is low. With increasing prey density fecundity increases, and the numerical response is proportionately higher.

For example, the population of the great horned owl in a 160 km^2 area in Alberta, Canada, increased over a three-year period, 1966 to 1969, from 10 birds to 18 as the population of its prey, the snowshoe hare, increased sevenfold (Rusch et al. 1972). The proportion of owls nesting increased from 20 percent to 100 percent as biomass of snowshoe hare in the owl's diet increased from 23 to 50 percent (Figure 19.10). Coyotes inhabiting a 756 km^2 area in Utah increased as the density of black-tailed jackrabbits increased, a species that made up three-fourths of the coyotes' diet (F. W. Clark 1972).

As we have seen, numerical response, positive or negative, is not immediate, especially in situations where fecundity depends upon the energy intake of adults. There is necessarily a time lag between

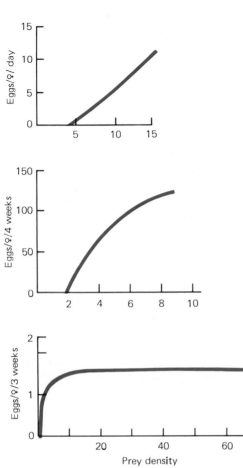

Figure 19.9 Fecundity in various predatory arthropods as a function of prey density. Note the shapes the curves can take. (From Beddington et al. 1976.)

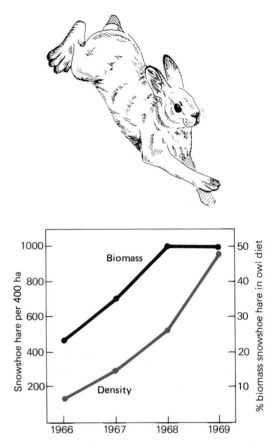

Figure 19.10 Numerical response of horned owls to increased abundance of snowshoe hare. Density of snowshoe hare in its habitat is plotted with the percentage of biomass of the hare in the diet of the great horned owl near Rochester, Alberta. (From Rusch et al. 1972: 291.)

adequate nutritional intake, development and birth of young, and their maturation to reproducing individuals. Examples of delayed numerical response can be found in a number of field studies.

Numerical response may involve both an aggregative response and an increase in fecundity. An example can be found among the "fugitive" warblers of northern forests, especially the Tennessee, Cape May, and bay-breasted warblers, whose abundance is dictated by outbreaks of

spruce budworm. During such periods populations of bay-breasted warblers have increased from 10 to 120 pairs per 40 ha (Mook 1963; Morris et al. 1958), and Cape May and bay-breasted warblers have larger clutches than associated warbler species (MacArthur 1958). In fact, Cape May and possibly bay-breasted warblers apparently depend upon occasional outbreaks of spruce budworm for their continued existence. At those times the two species are able to increase more rapidly than other warblers because of extra large clutches. Between outbreaks they are reduced in numbers and even become extinct locally.

In general numerical response takes three basic forms (Figure 19.11): (1) direct or positive response, in which the number of predators per unit area increases as the prey density increases; (2) no response, in which the predator population remains proportionately the same; and (3) inverse or negative response, in which the predator population declines in relation to prey population (Hassell 1966).

■ Total Response

In analyzing the relationship between predator density and prey density, functional and numerical responses may be combined to give a total response, and predation may be plotted as a percentage. Predation then falls into two types: (1) percentage of predation declines continuously as prey density increases (Figure 19.12), and (2) percentage of predation rises initially, then declines. The second type results in a dome-shaped curve (Figure 19.13), produced by the Type III functional response to prey density and by direct numerical response.

■ Foraging Theory

The Type III response curve may reflect in some predators the acquisition of a search image, but the same response curve could result from the manner in which predators must allocate energy to secure prey (Royama 1970). Because it is energetically unprofitable for predators to

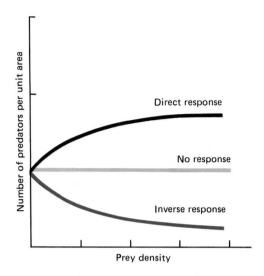

Figure 19.11 Basic forms of numerical response. No response means that the number of predators remains the same in the face of increasing prey density. A direct response implies that predators increase in response to an increasing prey density. Inverse response means that the number of predators per unit area declines as prey density increases. (From Hassell 1966: 67.)

spend time where prey density is low, predators must discover the most productive way to allocate their hunting time among different prey species of different abundances in different patches. Profitability is measured not by prey density, but by the amount of prey, preferably measured in terms of biomass, that a predator can harvest in a given time. This profitability of hunting is sufficient to produce Type III curves.

The profitability of hunting by a predator relates to the manner in which its prey is distributed. If prey were distributed in a fine-grained manner (see Chapter 14), the predator could select any food item in a coarse-grained manner. It could pick and choose among the prey, using a search image. In reality prey is distributed in a coarse-grained manner in patches

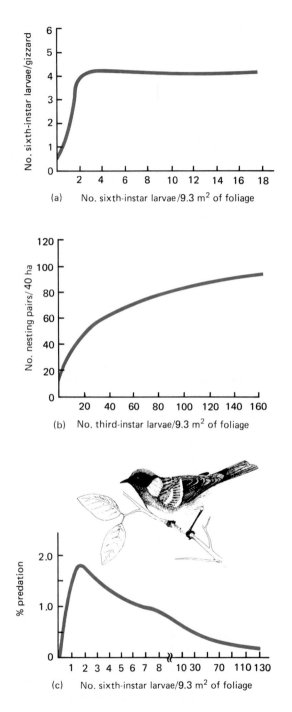

(a)

(b)

(c)

Figure 19.12 (a) Functional response, (b) numerical response, and (c) total response of the bay-breasted warbler to changes in abundance of spruce budworm in New Brunswick. Total response is based on the assumption that larvae are available for 30 days, the average feeding day is 16 hours, and the digestive period is 2 hours. (After Mook 1963.)

across the landscape. These patches vary in size and in the quality and quantity of resource, so the predator must be able to locate profitable patches. This problem gave rise to the concept of the optimal use of patchy environments advanced by MacArthur and Pianka (1966), which later evolved into optimal foraging theory. This theory forms a standard against which actual foraging strategies can be compared.

An *optimal foraging strategy* provides maximum net rate of energy gain, endowing the animal with the greatest fitness. It involves two separate but related components. One is optimal diet; the other is foraging efficiency. Theoretical ecologists have come up with certain rules (hypotheses) for optimal foraging.

Figure 19.13 Total response of predators to prey density expressed as a percentage of predation to prey density. Total response includes both functional response and numerical response. S = *Sorex* shrew; B = *Blarina* shrew; P = *Peromyscus* mouse. (After Holling 1964: 19.)

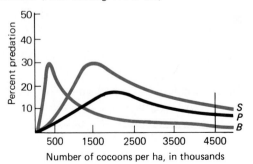

Number of cocoons per ha, in thousands

Optimal Diet

Suppose you had scattered black oilseed (a small type of sunflower seed) and millet, the former a large seed, the latter very small, for mid-fall birds. The oilseed was for cardinals, house finches, and nuthatches, who extract the meat; and the millet for the mourning doves, whose preferred fall foods, according to food studies, are the seeds of panic grass and foxtail (a wild millet). You would have soon discovered through frequent observations that the doves chose oilseed over millet. They did not crack the seeds as finches do—their thin pointed bills are not adequate to the task—but ate them whole. Only after the day's allotment of oilseed was consumed did the mourning doves turn to their supposedly preferred millet.

Obviously, the mourning doves found it energetically more profitable to take large oilseed rich in carbohydrates over the small millet seeds. Because of oilseed's larger size the doves could acquire more energy with less handling time, which meant remaining for a much shorter period in the food patch. When feeding on millet, the doves had to handle many more seeds, providing much smaller packets of energy per unit effort. In effect, the doves made an optimal economic decision. They chose larger, more profitable seeds over smaller ones and in their own way demonstrated some predictions of optimal foraging theory relative to optimal diet.

According to the "decision rules" the consumer should (1) prefer the more profitable prey; (2) feed more selectively when profitable prey or food items are abundant; (3) include less profitable items in the diet when the most profitable foods are relatively scarce; and (5) ignore unprofitable items, however common, when profitable prey are abundant. The mourning doves made all the "right" decisions;

but they were operating under ideal conditions unwittingly provided: a relative abundance of food with a choice of only two items. Natural conditions in which they had to locate food patches that provided much smaller and more diverse prey items might have produced a different outcome.

Although the theory of optimal diet makes practical sense, it is difficult to test under proper field conditions. Not only would we have to know exactly what items the animals were consuming; we would also have to know the relative availability of the food items in the habitat. Even then, the animals may make their decisions on criteria other than relative availability (for a short discussion on this point see Taylor 1984: 93–94). Some studies have been done under controlled conditions involving birds (great tit, Krebs et al. 1978), fish (brown trout *Salmo trutta*, Ringler 1979) and invertebrates (shore crabs *Carcinus maenas*, Elner and Hughes 1978). Werner and Hall (1974) presented groups of ten bluegill sunfish with three sizes of *Daphnia* in a large aquarium. They allowed the fish to forage for a period of time, then killed them and examined their stomachs to determine the number and size of *Daphnia* taken. When the density of the prey presented was low, the fish consumed the three sizes according to the frequency encountered. They showed no preference for size (Figure 19.14). When the prey population was dense, the fish consumed the largest prey items. When presented with an intermediate number, the fish took the two largest size classes. The results of these feeding trials support the optimal foraging theory. Feeding trials involving the shore crab and its prey, the blue mussel (*Mytilus edulis*), gave much the same results (Elner and Hughes 1978). The crab's diet extended to smaller mussels as the preferred size became scarce;

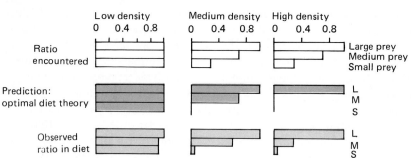

Figure 19.14 Optimal choice of diet in the blue-gill sunfish preying on different sizes of *Daphnia*. The histograms show the ratio of encounter rates with each size class at three different densities, the prediction of optimal ratios in the diet, and observed ratios in the diet. Note the bluegill's preference for large prey. (After Werner and Hall 1974: 1048.)

but at no time did the crabs exclude small mussels completely from their diet, even when larger mussels were more than abundant enough to fill them. The crabs did not pass up good food when encountered, even though it did come in smaller packages. That is probably the way animals in the wild respond.

Two field studies provide some insight into the way animals forage under natural conditions. Goss-Custard (1977a, 1977b) studied food selection by redshanks (*Tringa totanus*) on mudflats containing different sizes of polychaete worms. He found that as the number of large worms increased, redshanks became more selective and tended to ignore small worms, regardless of how common they were, as long as the density of large worms remained high. These field observations were supported by laboratory studies.

Davies (1977) studied the feeding behavior of the pied wagtail (*Montacilla alba*) and yellow wagtail (*M. flava*) in a pasture field near Oxford, England. The birds fed on various dung flies and beetles attracted to droppings. They had access to prey of several sizes: large, medium, and small flies and beetles. The wagtails showed a decided preference for medium-sized prey (Figure 19.15). The size of the prey corresponded to the optimum-sized prey the birds could handle profitably (Figure 19.16). The birds ignored small sizes. Although easy to handle, small prey did not return sufficient energy; large sizes required too much time and effort to handle.

Foraging Efficiency

Most animals live in a heterogeneous or patchy environment. In feeding, they have to seek out and concentrate their attention on the most productive food patches. This fact has given rise to an-

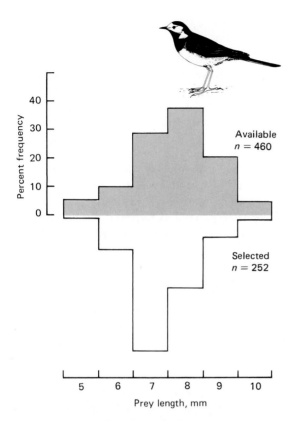

Figure 19.15 Pied wagtails show a definite preference for medium-sized prey, which are taken in disproportionate amounts compared to sizes of prey available in the environment. (Davies 1977: 48.)

Figure 19.16 Prey size chosen by pied wagtails is the optimal size for providing a maximum amount of energy per handling time. Small sizes provide too few calories. Large sizes require too much handling time. (Davies 1977: 48.)

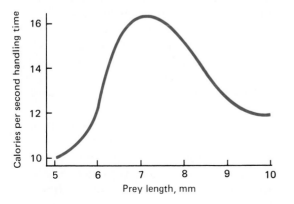

other set of decision rules in optimal foraging theory. The consumer should do the following: (1) concentrate foraging activity in the most productive patches; (2) stay with those patches until their profitability falls to a level equal to the average for the foraging area as a whole (Figure 19.17); (3) leave the patch once it has been reduced to a level of average productivity; and (4) ignore patches of low productivity.

These rules are covered by the *marginal value theorem* (Charnov 1976; Parker and Stewart 1976), which gives the length of time a forager should profitably stay in a resource patch before it seeks another. The length of stay relates to the richness of the food patch, the time required to get there, and the time required to extract the resource. When a forager arrives on a patch, it initially has a high rate of extraction and energy gain (Figure 19.17), but as time progresses the abundance of the resource and rate of extraction decline until on the average it is no longer profitable for the forager to remain. Too long a stay depletes the resource. Conversely, if the forager leaves a patch too soon, it does not utilize the resource efficiently. Ideally the forager should leave for another patch at the point (indicated by the intersect of the straight line tangent to the curve in Figure 19.17b) where energy gains start to diminish. The model predicts that the forager should remain in a rich food patch longer than in a poor one; and that as travel time between patches increases, it should remain in the patch longer to balance energy loss in travel. Overall the forager should leave all patches, regardless of their profitability, when they have been reduced to the same marginal value, which is average for the environment as a whole.

Whether animals go by these rules has been the object of experimentation both in the laboratory and in the field. Hub-

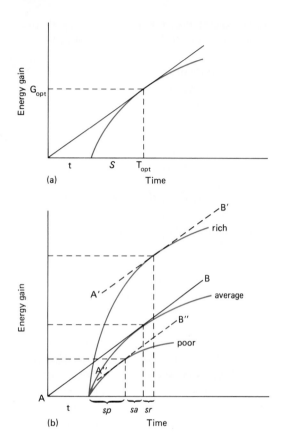

Figure 19.17 How long should a predator remain in a habitat patch seeking food? This graph provides a theoretical answer. Time spent in travel and time spent in a habitat patch are plotted against food depletion (net accumulation of gain in food, which declines as available food is depleted). (a) The curve represents the cumulative amount of food harvested relative to time in the patch (which is high early in time). The straight line represents average food intake per unit time for the habitat as a whole. Where the line touches the curve represents the point at which the predator has reached average cumulative net food gain for the habitat as a whole. Beyond this point net food gain declines below average. It is no longer profitable for the predator to remain. Thus the point represents the optimal time for the predator to leave and seek a more profitable food patch. (After Krebs 1978: 42.) (b) When a forager visits several different patch types, it should exploit each type until the patch drops to the average of the environment, line AB. The forager should spend less time in a poor patch than in a rich patch.

bard and Cook (1978) studied the foraging behavior of a parasitoid, the ichneumon wasp *(Memeritis canescens)*, in a laboratory arena containing patches of the host, larvae of *Ephestia cantella*. Hosts in densities of 64, 32, 16, 8, and 4 were placed in petri dishes filled to the brim with plaster of Paris, the hosts' substrate. Space between the larvae was filled with wheat bran. One result of the experiment was that all patches were depleted of larvae to a common level of host abundance. The richest patches suffered the greatest depletion. As exploitation proceeded, the amount of time spent by the wasp in patches of highest density declined, and the proportion of time spent in the next richest patch increased.

Zack and Falls (1976a, 1976b, 1976c) studied foraging strategy under more natural conditions. They exposed captive ovenbirds individually to a patchy food supply (mealworms) presented in natural outdoor pens in typical habitat.

In one experiment Zack and Falls presented four fixed patch locations in which they interchanged prey densities. They found that the ovenbirds increased their search path exponentially with prey density. The birds rapidly shifted their search efforts as prey densities were interchanged. Because less search path was required per prey found in patches of dense prey, the birds concentrated their efforts in areas of high profitability and took a higher percentage of prey available there. Ovenbirds did not always visit every patch location during the observation periods, but they always visited the high-density prey patches. This finding suggests that the birds' discovery of one or more profitable feeding areas discourages the sampling of other patches to assess their profitability. The ovenbird's tendency to quit searching after encountering one or more profitable patches and its ability to learn the location of and return to patches of high prey density may limit the number of patches the bird will exploit. The ovenbird may use other patches only if it discovers them by chance.

In another set of experiments Zack and Falls (1976b) exposed ovenbirds to various sets of patches. When they exposed birds to a single patch, the ovenbirds quickly concentrated foraging there. The birds took equal amounts of food at different prey densities and did not vary the amount of search path per visit. However, as the prey density in patches increased, ovenbirds decreased the number of visits to the patch, and reduced both the total search path per prey located and exploratory searches outside the patch at high prey densities. That finding suggests that with low depletion and high renewal rates, a single profitable feeding site might be enough.

In a third experiment Zack and Falls presented two prey patches successively. Prey in patch 1, presented on day 1, was not renewed on day 2. Instead a new patch with prey was presented. After an initial visit to patch 1, the birds quickly abandoned the first patch location, and after some exploratory behavior, they rapidly concentrated their search efforts in the new location.

Zack and Fall (1976c) also investigated when ovenbirds would give up the search in one area and move on to the next. They discovered that the birds did not take some optimal number based on previous experience. Instead the birds learned rapidly to find patches of prey; they chose feeding sites nonrandomly and avoided areas of no food and patches visited previously. Ovenbirds improved their foraging efficiency by searching nonrandomly within the patches, avoiding areas already exploited. By doing so, ovenbirds were less likely to deplete their prey. If the birds followed a systematic search pattern, they gave up and left when a patch was completely covered. The time at which they gave up the search was unrelated to prey density.

Observation and experimental studies support the hypothesis of optimal foraging up to a point. It is not surprising that they should. Much optimal foraging theory concerns actions you would expect any mobile animal to take: Forage in areas where food is abundant, leave when searching is no longer rewarding, select the larger and most palatable items of food, leave the poor items until last, and travel no further than necessary to feed. Where the theory breaks down is in the expectation that animals will choose patches in order of their profits or take only optimal food items first and ignore the rest. Such choices may be characteristic of animals foraging in a stable laboratory environment. It is not necessarily the way animals behave in the wild.

Being opportunists, animals will take some less than optimal food items upon discovery, and they may quit before food items are reduced to some minimal level. Nor will they pass up certain profitable patches because they do not meet some theoretical expectation. Animals quickly learn where food is and where food is not, and they do not waste much time on a patch after it is depleted. Foragers, however, will stay with a patch as long as the rate of replenishment exceeds the rate of depletion. Some animals are highly restricted in their choices. Sedentary animals, such as corals, barnacles, and blackfly larva, filter feeders all, have to take what food flows past them. Others have severely restricted foraging patterns that limit their choices of food.

Risk-Sensitive Foraging

The marginal value theorem model assumes that the animal knows the quality of the food patch and expects a constant reward on each visit. In reality the quality of the patch varies randomly over space and time. In this situation the animal has to decide whether to return to a patch that gives it a constant rate of return or visit a new patch where the resource may be in greater or lesser abundance. How animals make such a decision has been the subject of a number of behavioral experiments (Real and Caraco 1986). For example, Caraco and associates (1980) determined the daily energy requirements of yellow-eyed juncos (*Junco phaeonotuis*) and provided food (millet) at two feeding stations separated by a partition in their aviary cage. The experimenters could manipulate the energy budgets of the birds by depriving the birds of food prior to any trials. In a given experiment one feeding station always offered a constant reward; the other feeding station offered an unpredictable reward (no seed half the time; some seeds the other half of the time). Thus the birds faced choices between a constant number of seeds and a random number of seeds; but always the mean of the variable reward equaled the mean of the constant reward.

When the birds were deprived of food for one hour in experimental tests and were still in a positive energy balance, the juncos avoided risk by preferring the predictable site. When deprived of food for four hours the birds switched their preference to the variable reward. They changed from being risk-averse to risk-prone. Under energy stress the variable site offered the possibility of providing 50 percent more food, whereas the constant site would not provide sufficient food to meet energy needs. Of course, there was the 50 percent risk of finding no food. Nevertheless, in face of high energy demand, risk-prone behavior maximized daily survival. Animals living in natural conditions face such choices each day, particularly in winter. They may initially start out risk-prone and as time goes on become risk-averse. This behavior has given rise to the *expected energy budget rule:* Be risk-prone if the daily energy budget is negative; be risk-averse if it is positive (Stephens 1981).

There is yet another risk-sensitive approach available to animals: Sample a patch to assess its quality. The rewards obtained at the beginning of a patch visit can be used to estimate patch quality and how long the animal should stay. The longer the animal stays, up to a point, the more information it acquires about the patch; thus its stay in the patch should be longer than the marginal value model predicts (see Oaten 1977; Green 1980; McNamara 1982).

Foraging theory is an area of active interest among theoretical behavioral ecologists. They are attempting to develop models incorporating animal foraging decisions, resource quantity and quality, and environmental constraints to explain and to predict foraging behavior. Because of an animal's individuality in its decision making, it is very doubtful if foraging behavior can be reduced to sets of predictive mathematical equations, but foraging models can provide useful and valuable insights into ways in which animals utilize their environment. (For discussions see Krebs 1987; Kamil, Krebs, and Pullium, eds. 1986; Real and Caraco 1986.)

■ Summary

Predation is the consumption of one living organism by another, a relationship in which one organism benefits at the other's expense. In its broadest sense predation includes herbivory, parasitism, and cannibalism.

Interactions between predator and prey

have been described by the mathematical models of Lotka and Volterra and by subsequent modification of their model by others. Essentially all of these models predict oscillations of predator and prey populations. The oscillations may be stable, damped, or unstable. Relationships between predator and prey populations result in two distinct responses. As density of prey increases, predators may take more of the prey, a functional response, or predators may become more numerous, a numerical response.

There are three types of functional responses. In Type I the number of prey eaten per predator increases linearly to a maximum as prey density increases. In Type II the number of prey taken rises at a decreasing rate toward a maximum. In Type III the number of prey taken increases in a sigmoidal fashion.

Both invertebrate and vertebrate predation may exhibit Type II and Type III response curves. Type II occurs in situations of varying densities of one species of prey. Type III involves two or more species of prey. Inherent in Type III responses are a search image, in which the predator develops a facility for finding a particular prey item, and switching, in which the predator turns to an alternate, more abundant prey species for more profitable hunting. It takes that prey in a disproportionate amount relative to other prey species.

Functional response views predation in terms of the relation of predator attack rates to prey density. Numerical response refers to the increase of predators resulting from an increased food supply. Numerical response may involve an aggregative response, the influx of predators to a food-rich area, or more importantly, a change in the rate of growth of the predator population through changes in developmental time, survival rates, and fecundity. Such changes produce a delayed numerical response, for a time lag necessarily exists between birth of young and maturation of reproducing individuals.

Because prey occurs in patches, the predator finds it more efficient to spend time in areas not necessarily where prey is most abundant, but where hunting is most profitable in time allocated relative to net energy gained. Study of such behavior has given rise to the concept of optimal foraging, a strategy that obtains for the predator a maximum rate of net energy gain. There is a break-even point above which foraging in a particular patch is profitable and below which it is not. Optimal foraging involves an optimal diet, one that includes the most efficient size of prey for both handling and net energy return. Optimal foraging efficiency involves the concentration of activity in the most profitable patches of prey and the abandonment of those patches when they are reduced to the average of profitability of the area as a whole.

Plant-Herbivore Systems

■ Predator-Prey Systems

The interaction of predator and prey, particularly where an individual predator and an individual prey species are involved, is considered a predator-prey system. The predator directly influences growth and survival of the prey population and the density of the prey population influences growth and survival of the predator population. It is such a system that Lotka and Volterra attempted to describe mathematically and the biologist G. F. Gause (1934)

investigated experimentally. He reared together under constant environmental conditions a predator population *Didinium*, a ciliate, and its prey, *Paramecium caudatum* (Figure 20.1). The predator always exterminated the prey, regardless of the density of the two populations. After the prey was destroyed, the predator died of starvation. Only by periodic introductions of prey to the medium was Gause able to maintain the predator population and prevent it from dying out. In this manner he was able to maintain populations together and produce regular fluctuations in both as predicted by the Lotka-Volterra

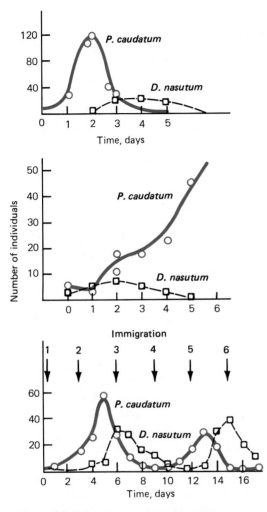

Figure 20.1 Outcome of Gause's experiments on predator-prey interactions between the protozoans *Paramecium caudatum* and *Didinium nasutum* in three microcosms: (a) oat medium without sediment; (b) oat medium with sediment; (c) with immigration in oat medium without sediment. (After Gause 1934.)

model. The predator-prey relation was one of overexploitation and annihilation, unless there was immigration from other prey populations.

In another experiment Gause introduced sediment in the floor of the tube.

Here prey could escape from the predator. When the prey was eliminated from the clear medium, the predators died from the lack of food. The paramecia that took refuge in the sediment continued to multiply and eventually took over the medium.

The Gause experiments took place in a relatively simple environment. In a different type of experiment C. Huffaker (1958) attempted to learn if an adequately large and complex laboratory environment could be established in which a predator-prey system would not be self-exterminating. Involved were the six-spotted mite (*Eotetranychus sexmaculatus*) and a predatory mite (*Typhlodromus occidentalis*). Whole oranges, placed on a tray among a number of rubber balls the same size, provided food and cover for the spotted mite. Such an arrangement permitted the experimenter to control both the total food resource available and the pattern of dispersion by covering the oranges with paper and sealing wax to whatever degree desired and by changing the general distribution of oranges among rubber balls. The experimenter could manipulate conditions to simulate a simple environment where the food of the herbivore was concentrated or a complex universe where food was widely dispersed, partially blocked by barriers, and where refuge areas were lacking.

In both situations the two species found plenty of food available at first for population growth. Density of predators increased as the prey population increased. In the environment where food was concentrated and dispersion of the prey population was minimal, predators readily found the prey, quickly responded to changes in prey density, and were able to destroy the prey rapidly. In fact, the situation was self-annihilative. In the environment where the primary food supply and the prey were dispersed, predator and prey went through two oscillations before the predators died out. The prey recovered slowly.

Several important conclusions resulted from the study. First, predators cannot survive when the prey population is low. Second, a self-sustaining predator-prey relationship cannot be maintained without immigration of prey. Third, the complexity of prey dispersal and predator-searching relationships, combined with a period of time for the prey population to recover from effects of predation and to repopulate the areas, had more influence on the period of oscillation than the intensity of predation.

The degree of dispersion and the area employed were too restricted in Huffaker's experiment to perpetuate the system. Pimental, Nagel, and Madden (1963) attempted to provide an environment with a space-time structure that would allow the existence of a parasite-host system. They chose as subjects a parasitic wasp (*Niasonia vitripennis*) and a host fly (*Musca domestica*) and provided for the environment a special population cage, a group of interconnected cells. A predator-prey system living in 16 cells died out, but a 30-cell system persisted for over a year. Increasing the system from 16 to 30 cells decreased the average density of parasties and hosts per cell and increased the chances for survival of the system. The lower density was due to the breakup and sparseness of both parasite and host populations. The greater number of individual colonies that remained following a severe decline of the host assured survival of the system, because these colonies provided a source of immigrants to repopulate the environment. Moreover, amplitude of the fluctuations of the host did not increase with time, as proposed by the model of Nicholson. Apparently the fluctuations were limited by intraspecific competition.

These laboratory experiments support studies made in the field. Sometime before 1839 prickly pear cactus (*Opuntia*) was introduced from America into Australia as an ornamental. As is often the case with introduced plants and animals, the cactus escaped from cultivation and rapidly spread to cover 60 million acres in Queensland and New South Wales. To combat the cacti, a South American cactus-feeding moth (*Cactoblastis cactorum*) was liberated. The moth multiplied, spread, and destroyed the cacti until plants existed only in small, sparse, widely distributed colonies.

The decline of the prickly pear also meant decline of the moth. Most of the caterpillars coming from moths that had bred on prickly pear the previous generation died of starvation. In areas where only a few moths survived, not many plants were parasitized. As prickly pear increased, so did the moth, until the cactus colony was again destroyed. In areas where no moths survived the colony spread once more, but sooner or later it was found by moths from other areas and was eventually destroyed. However, seed scattered into new areas established new colonies that maintained the existence of the species and thereby maintained the predator-prey system.

The rate of establishment of prickly pear colonies is determined by the time available for the colonies to grow before they are found by moths. As a result an unsteady equilibrium exists between cactus and moth. Any increase in the distribution and abundance of the cactus leads to an increase in the number of moths and subsequent decline in the cactus. The maintenance of this predator-prey, or more accurately herbivore-plant, system depends upon environmental discontinuity. The relative inaccessibility of host or prey in time and space limits the number of parasites and predators.

In further investigations of the moth and cactus relationship, J. Monro (1967) found that the moth may conserve food for succeeding generations of moths by limiting its own numbers. At high densities the moth clumps its egg sticks rather than laying them randomly on prickly pear. The clustering overloads certain plants of prickly pear with eggs, resulting in the destruction of the plants. In dense stands of prickly pear clustering initially has little influence on larval survival, for as an overloaded plant collapses, it falls on its neighbor and larvae can move to a new source of food. Later, as dense stands become broken up into isolated plants, the relatively sedentary larvae are unable to cross the wide gaps and die of starvation. As mean density increases, the proportion of eggs wasted by clumping increases.

However, because the eggs are clustered rather than widely distributed, more plants escape infestation altogether or are subject to a lighter infestation than would be expected if eggs were laid at random. Monro found that this mechanism, which is employed most in the center of the range of the moths, acts to conserve food supply for succeeding generations and to maintain a constant level of both the food resource and the moth population.

These examples to some extent illustrate predator-prey interactions both at the plant-herbivore and at the herbivore-carnivore level. Although for simplicity they are considered separately, predator-prey interactions at one trophic level influence predator-prey interactions at the next trophic level. Interactions at two or more trophic levels are often involved in predator-prey stability.

■ Effects of Plant Predation

Predation on plants by herbivores involves defoliation and consumption of fruits and seed. The results of the two forms of predation are different.

Defoliation is the destruction of plant tissue (leaf, bark, stem, and roots). Some plant predators, such as aphids, do not consume tissue directly, but, acting as parasites, tap plant juices without killing the plant. Other herbivores consume tissue directly, destroying parts or all of the plant. If grazers consume seedlings, they kill the plant outright. If they remove only part of a plant, its survival depends on the amount and continuation of grazing. Continued grazing may eventually kill the plant, but if grazing ceases, it may regenerate. Although grazed plants may persist and regenerate, defoliation still has an adverse effect. Plant biomass is decreased. Removal of leaves may damage the hierarchical position of the plant in the stand. Loss of foliage and subsequent death of some roots (root pruning) reduce the vigor of the plant, its competitive ability, and its fitness (Figure 20.2) (Harper 1977).

The impact of seed predation that results in the elimination of individuals is difficult to assess. If density-dependent processes let few seedlings survive, seeds removed by predators represent that portion of the population that has no future. In such instances seed predation has no real impact. If predators remove seeds from an expanding population or from areas being colonized, predation reduces the rate of increase. On the other hand, if consumption of seeds is a mechanism for seed dispersal, as when seeds are con-

tained within a palatable fruit and then carried in the gut of a fruit-consuming herbivore, predation can be to the plant's advantage (see Chapter 23).

Plant Fitness

Removal of plant tissue—leaf, bark, stems, roots, sap—affects a plant's fitness and its ability to survive, even though it may not be killed outright (see Dirzo 1984). Loss of foliage and loss of roots decrease plant bi-

Figure 20.2 (a) Intense predation on oaks by gypsy moth. Such defoliation can kill weaker trees in the stand and reduce the growth of others. Increased light and nutrient input from frass of the caterpillars can increase understory growth. (Photo by R. L. Smith.) (b) Heavy browsing on woody sprouts and herbaceous plants by white-tailed deer prevents any of the plants from escaping predation and achieving any significant growth. (Photo by R. L. Smith, Jr.)

omass, reduce the vigor of the plant, place it at a competitive disadvantage to surrounding vegetation, and lower its reproductive effort or fitness. This effect is especially strong in the juvenile stage, when the plant is most vulnerable to increased mortality and reduced competitive ability.

Although the plant may be able to compensate for the loss of leaves by increasing photosynthetic assimilation in the remaining leaves, it may be adversely affected by the loss of nutrients, depending on the age of the tissues removed. Young leaves are dependent structures, importers and consumers of nutrients drawn from reserves in roots and other plant tissues. As the leaf matures, it becomes a net exporter, reaching its peak before senescence sets in. Grazing herbivores such as sawfly and gypsy moth larvae, deer, and rabbits concentrate on more palatable, more nutritious leaves. They tend to reject older leaves because they are less palatable, being high in lignin and other secondary compounds (tannin, for example). If grazers concentrate on young leaves, they remove considerable quantities of nutrients.

Plants respond to defoliation with a flush of new growth that drains nutrients from reserves that otherwise would have gone into growth and reproduction. Defoliation also draws on the plants' chemical defenses, a costly response. Often the withdrawal of nutrients and phenols from roots exposes them to attack by root fungi while the plant marshals its defenses in the canopy (Parker 1981). If defoliation of trees is complete, as often happens during an outbreak of gypsy moths or fall cankerworms *(Alsophila pometaria)*, replacement growth differs from the primary canopy removed. The leaves are smaller and the total canopy area may be reduced by as

much as 30 to 60 percent (Heichel and Turner 1976). Defoliation in a subsequent year may cause an even further reduction in leaf size and number. Some trees may end up with only 29 to 40 percent of the original leaf area to produce food in a shortened growing season.

Severe defoliation and subsequent regrowth alter the tree physiologically. Growth regulators controlling bud dormancy are changed with the removal of leaves. The plant uses up reserve food to maintain living tissues until new leaves are formed. Buds for the next year's growth are late in forming. The refoliated tree is out of phase with the season; and the drain on nutrient reserves adversely affects the tree over winter, because twigs and tissues are immature at the onset of cold weather. Such weakened trees are more vulnerable to attacks by insects and disease the next year. Also, because the plant used nutrient reserves for regrowth and maintenance, it has no resources available for reproduction. Defoliation of coniferous trees results in death.

Some plant predators, such as aphids, tap plant juices on new growth and young leaves rather than consume tissue directly. Sap suckers can decrease growth rates and biomass of woody plants by 25 percent.

Damage to the cambium and growing tips (apical meristem) may be more important in some plants. Deer, mice, rabbits, and bark-burrowing insects feed on those parts, often killing the plant or changing its growth form.

Moderate grazing, even in a forest canopy, can have a stimulating effect, increasing biomass production, but at some cost to vigor and at the expense of nutrients stored in the roots. The degree of stimulation depends upon the nature of the plant, nutrient supply, and moisture. In general, the biomass of grass is increased by grazing up to a point; then biomass production declines. Adverse effects are greatest when new growth is developing. Defoliation then, as in deciduous trees, results in a loss of biomass, decreased growth, and delayed maturity (Andrzejewska and Gyllenberg 1980).

Grasses, however, are well adapted to grazing, and may benefit from it. Because the meristem is close to the ground, older rather than more expensive young tissue is consumed first. Grazing stimulates production by removing older tissue functioning at a lower rate of photosynthesis, reducing the rate of leaf aging, thus prolonging active photosynthetic production and increasing light intensity on underlying young leaves, among other things. Some grasses can maintain their fitness only under the pressure of grazing, even though defoliation reduces sexual reproduction (McNaughton 1979; Owen 1980; Owen and Wiegert 1981). The idea, however, that herbivory benefits grazed plants has been challenged (Belsky 1986). Certain grasses have traits that reduce the deleterious effects of grazing, but this fact may not mean that herbivory increases fitness.

Herbivore Fitness

Herbivory is a two-way street. Plants, although seemingly passive in the process, have a pronounced effect on the fitness of herbivores. For herbivores it is not the quantity of food that is critical—usually there is enough biomass—but the quality. Because of the complex digestive process needed to break down plant cellulose and convert plant tissue into animal flesh, high quality forage rich in nitrogen is necessary. Without that, herbivores can starve to death on a full stomach. Low quality foods are tough, woody, fibrous, and un-

digestable. High quality foods are young, soft, and green, or they are storage organs such as roots, tubers, and seeds. Most food is low quality, and herbivores forced to live on such resources experience high mortality or reproductive failure (Sinclair 1977). Added to the problem of quality is the task of overcoming the various defenses of plants that make food unavailable, hardly digestible, unpalatable, or even toxic.

Secondary plant substances can affect the reproductive performance of some mammals. Isoflavonoids in plants—usually concentrated in legumes, particularly alfalfa and ladino—mimic estrogenic hormones, especially progesterone. When consumed, these isoflavonoids exert an estrogenic effect and induce a hormonal imbalance in grazing herbivores that results in infertility, difficult labor, and reduced lactation. Secondary compounds also serve as reproductive cues for some voles (Berger et al. 1981). One compound, 6-methoxybenzoxolinone (6-MBOA), rapidly stimulates reproductive effort in montane voles (*Microtus montanus*). When voles feed on grass, they stimulate the injured plant tissue to release an enzyme that converts a precursor compound abundant in young growing tissue to 6-MBOA. The ingested chemical serves as a cue that the vegetative growing season has begun. Such a chemical cue allows the voles to produce offspring when food resources will be available to them. These chemical signals are important to the voles, because they live in an environment where food resources are unpredictable and depend upon the timing of snowmelt and other environmental conditions. Yearly differences in the appearance of new vegetative growth and of 6-MBOA may influence population fluctuations (Negus, Berger, and Forslund 1977).

■ Plant Defenses

Throughout their evolutionary history, plants have arrived at modes of defense against their herbivorous predators. These defenses range from chemical methods, widespread among plants, to mimicry and structural features.

Mimicry

Mimicry is usually considered an evolutionary response in animals (see Chapter 21), but animals in search for food may have stimulated mimicry in the plant kingdom. Gilbert (1975) found evidence of plant mimicry in his study of the passionflower butterfly *(Heliconius)* and its food plant, the passionflower *(Passiflora)*. *Passiflora*, a tropical vine of the New World comprising around 350 species, has a wide range of intraspecific and interspecific leaf and stipule shapes. The number of *Passiflora* species found in any one area is about 2 to 5 percent of the 350 species. Some 45 species of highly host-specific species of *Heliconius* use *Passiflora* species as an egg-laying site and as a source of larval food. Each species of *Heliconius* uses a limited group of plants. Visually sophisticated butterflies learn the position of the vines and return to them on repeated visits. Within a habitat the leaf shapes of passionflowers vary among species. Under visual selection by butterflies passionflowers apparently have evolved leaf forms that make them more difficult to locate. Because the larval food niche is broader than that of the ovipositing females, there has been selective pressure for divergence among *Passiflora* species (Figure 20.3). Probably because of these selection pressures, *Passiflora* leaf shapes converge on those of associated tropical plants that *Heliconius* finds inedible. So close are the convergences that plant taxonomists have named some *Passiflora* species after the genus they resemble.

In addition two *Passiflora* species, *P. cyanea* and *P. auriculata,* have evolved glandular outgrowths on the stipules that mimic the size, shape, and golden color of *Heliconius* eggs at the point of hatching. Because *Heliconius* females detect and then reject shoots that carry eggs and young of other females, *Passiflora* achieves a measure of protection by egg mimicry (Williams and Gilbert 1981).

Structural Defenses

Some of the least costly defenses available to plants are structures that make penetration by predators difficult, if not impossible. They include tough leaves, spines, and epidermal hairs on leaves, often hooked, which may trap, impale, or fence out insects, and discourage browsing by vertebrate herbivores. These structures may have evolved early in the history of the plants, when they might have been subject to even greater predatory pressure. Because they represent little investment, plants still retain them. Thick, hard seed coats provide protection from seed-eating animals. The problem with such seed defense is that the seeds need to be scarified—the hard seed coat softened—so the seedling itself can escape. If scarification is not achieved, the seedling is sealed in, never to germinate. Many plants, however, have turned seed predation into a mechanism for seed dispersal (see Chapter 23).

The role of structural defenses in plants is mostly presumed. Little experimental evidence exists to demonstrate the effectiveness of such apparent defensive

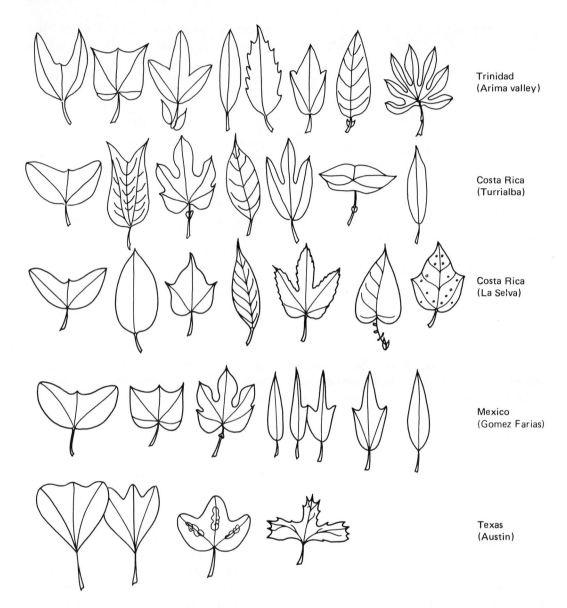

Trinidad
(Arima valley)

Costa Rica
(Turrialba)

Costa Rica
(La Selva)

Mexico
(Gomez Farias)

Texas
(Austin)

Figure 20.3 Variation in leaf shape among groups of sympatric species of *Passiflora*. The leaf shapes tend to converge with other common species of a number of genera, inedible to *Heliconius* butterflies. (After Gilbert 1975.)

structures against grazing herbivores. Cooper and Owen-Smith (1986) investigated experimentally the effects of plant spinescence on the feeding habits of three large browsing mammalian herbivores: the kudu *(Tragelapus strepsiceros)*, a large African antelope attaining female weights of 180 kg; the impala *(Aepyceros melampus)*, a medium-sized African antelope attaining a female body weight of 50 kg; and the Boer goat, a domestic ungulate weighing about 35 kg. The experimenters hand-reared the antelope from calves to allow close observation of feeding habits from very close range. In the rearing pens the antelopes and goats were introduced to plant species from the study area, the Nylsvley Nature Reserve in the northern Transvaal bushveld of South Africa. At six months of age, the antelope were released into a 213 ha enclosure, where they ranged freely and secured food from natural vegetation. Feeding animals were observed from distances of 1 to 5 m, and biting rates were determined by counting the number of sequential plucking actions made while feeding on a particular plant. Bites were converted to dry biomass by collecting samples of leaves and shoots of a size similar to those eaten and drying to a constant weight. Eating rate was calculated as the product of bite size (dry mass) and biting rate.

Cooper and Owen-Smith carried out two additional experiments to examine further the influence of spinescence. They

selected ten plants each of five species of trees at a height accessible to impalas outside of the enclosure. On each tree two branches were matched for size, shape, density of leaf cover, and ease of access to impala. They labeled the paired branches and removed the thorns from one of the pairs. They estimated the relative loss of foliage from browsing visually two months later. The other experiment involved feeding goats clipped and unclipped branches.

The woody plants (Table 20.1) exhibited three basic types of spinescence: (1) paired prickles or thorns situated in or close to the leaf axils; (2) short, sharp-tipped branchlets or spines, sometimes carrying small leaves; and (3) prickles of various kinds on leaves. Thorns were either straight and long, up to 70 mm, or short and sharply curved (hooked).

Results clearly showed that thorns and spines affected the feeding behavior of the three ungulates. These structures re-

TABLE 20.1 Characteristics of Spinescense of Selected Woody Plant Species Found on Nylsvley Nature Reserve, Transvaal, South Africa

Species	Abbrev.	Leaf Mass (g)	Features of Prickles Shape	Length (mm) (mm)	Prickliness Rating
Acacia burkei	Acbu	0.047	hooked thorns (paired)	3	*****
Acacia caffra	Acca	0.240	hooked thorns (paired)	2	**
Acacia nilotica	Acni	0.036	straight thorns (paired)	26	***
Acacia tortilis	Acto	0.018	mixed hooked and straight (paired)	41	*****
Dichorostachys cinerea	Dici	0.115	straight spines	28	**
Maytenus heterophylla	Mahe	0.022	straight spines	24	***
Strychnos cocculoides	Stco	0.122	curved thorns (opposite)	8	*
Struchnos pungens	Stpu	0.044	leaf tip spine		*
Ziziphus mucronata	Zimu	0.040	straight and curved thorns (paired)	8	*****

Source: Adapted from Cooper and Owen-Smith 1986:488.

stricted bite sizes mostly to single leaves or leaf clusters, and hooked thorns retarded biting rates. Acceptability of those plant species offering small leaf size along with prickles was lower, at least for kudu, than those of other palatable plant species. The inhibitory effect of prickles was less for impala and goats than for kudu (Figure 20.4). Nevertheless kudu bit off the shoot ends despite the prickles; and for certain straight-thorned species kudu compensated partially for their slow eating rates by extending feeding duration per encounter. Most spinescent species were similar to unarmed palatable species in their acceptability to the ungulates, even though the armed species had a higher

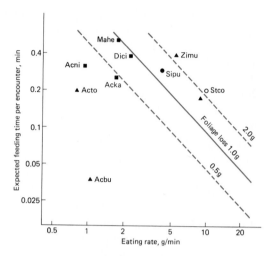

Figure 20.5 Expected feeding time per encounter (acceptability times mean feeding duration) by kudus, plotted on a log-log scale against eating rates for various spinescent and unarmed woody plants. Isoclines indicate the expected foliage losses per encounter of 2.0 g, 1.0 g, and 0.5 g. See Table 20.1 for plant codes. The most spinescent species are to the left of the 0.5 g isocline. The least spinescent fall between and on the 2.0 g isocline. (From Cooper and Owen-Smith 1986: 450.)

Figure 20.4 Kudu browsing on *Acacia*. (Photo by T. M. Smith.)

crude protein to their foliage. Probably these spinescent species, especially species of *Acacia*, would be preferred over unarmed species but for the thorns.

The main effect of these structural defense features is to restrict bite size, increasing handling time (Figure 20.5). Thorns, spines, and prickles reduce foliage losses to large herbivores. In addition, the animals may incur scar tissue in the esophagus and scratches in the buccal and esophageal mucosa.

Predator Satiation

A more subtle defense is the physiological mechanism of timing reproduction so that a maximum number of offspring is produced within one short period, thus satiating the predator and allowing a percentage of the offspring to escape.

Predator satiation is a major strategy against predation in plants. It is most prevalent in those species lacking strong chemical defenses. It involves four approaches (Janzen 1971). The first approach is to distribute seeds so that all of a seed crop is not equally available to all members of a seed predator population. Seeds of most trees are concentrated near the parent, and the number of seeds declines rapidly as the distance from the tree increases. The predators are attracted to the parent and many of the scattered seeds are missed by searching predators, in part because of search image and unprofitability (see Chapter 19). These survivors produce most of the recruitment. A second approach is to shorten the time of seed availability. If all seed matures and is available at one time, seed predators will be unable to utilize the entire crop before germination. A third approach is to produce a seed crop periodically rather than annually. The longer the time between seed crops, the less opportunity dependent seed predators have to maintain a large population between crops. Seed predators often experience local increases in density after good seed years, but decline rapidly when the food supply is depleted. This strategy reduces the number of predators available to exploit the next seed crop.

The production of a periodic seed crop depends upon synchronization of seed production among individuals. This synchrony is usually achieved by weather events, such as late frosts or protracted dry spells. As individuals of a tree species in a community become synchronized, strong selection pressures build up against nonsynchronizing individuals, because they experience heavy seed predation between peak seed years. Such individuals either drop out of the community over evolutionary time or become synchronized.

Predator satiation may be further assured if during any fruiting season the timing of the seed crop of one species is influenced by the presence of seed crops of others. If two or more species synchronize seed production and share seed predators, those predators may be attracted away from one species to another, reducing predatory pressure on both species.

Chemical Defenses

Chemical defense is another first line of defense of plants against herbivores (see Levin 1976). The basis of chemical defense is an accumulation of secondary products ranging from alkaloids to terpenes, phenolics, steroids, cyanogens, and mustard oil glycosides and tannins. Phenolics, a by-product of amino acid metabolism, are ubiquitous to seed plants. Alkaloids, also amino acid derivatives, occur in several thousand species, and cyanogenic glycosides in a few hundred species. The secondary products may be stored within the cells and released only when cells are broken, or they may be stored and secreted by epidermal glands to function as a contact poison or a volatile inhibitor.

Production and storage of such metabolites are expensive to the plant and seem to require a trade-off between defense and reproductive effort, but there is little evidence that plants evolved these metabolites for defense. During the evolutionary history of plants, these secondary compounds may have resulted as metabolic by-products or have served some past (or perhaps even present) physiological function (see Futuyma 1983). In time these compounds became useful deterrents to herbivore predation, although they are not able to defend against a full suite of enemies.

Chemical resistance to attack falls into two general types (Levin 1976). One involves accumulations and changes in metabolites of the host plant that act as toxins at the wound site. This response is commonly used to resist attacks from bacteria, fungi, and nematodes. The other type of resistance is based on the presence of inhibitors prior to attack. This approach is commonly employed against animals feeding on plants as well as against fungi.

Inhibitors may function as warning odors, repellents, attractants, or in some cases direct poisons. Volatile components advertise substances that insects and other herbivores would find repellent if they touched the plant. Bitter tastes imparted by tannins and cardiac glycosides can deter further consumption of both seeds (Janzen 1971) and foliage. Metabolites such as phenolic terpenes and saponins may be toxic and cause illness and occasionally death. Some, such as tannins, reduce digestibility of the plant materials consumed. Such repellents not only inhibit feeding on the plant possessing them, but also add a measure of protection to associated plants. For example, grazing by cattle on bentgrasses *(Agrostris)* and fescue *(Festuca)* is reduced considerably in the presence of buttercup *(Ranunculus bulbosus)*, which contains a powerful irritant of skin and mucous membranes (Phillips and Pfeiffer 1958). The presence of such plants can cause the herbivore to fail to locate the palatable plants or to reject them along with the repellent plant (Atsatt and O'Dowd 1976).

Plants containing secondary metabolites may also function as attractant decoys that cause the herbivore to feed on alternate prey. Many attractant plants are not what they advertise and cause mortality or reduced fecundity because of the presence of toxins or deficiency in certain nutritive materials (Atsatt and O'Dowd 1976). Coexisting toxic and nontoxic plants with similar attractant chemistry present a problem for host-specific herbivores. Some insects, for example, may be stimulated to lay eggs on a "wrong" but closely related plant, which eventually results in larval deaths.

The mode of defense varies with the nature of the plants. Feeney (1975) divided plants into two groups, apparent and unapparent; and Rhodes and Cates (1976) divided them into available and predictable and unavailable and unpredictable (Rhodes and Cates 1976). The two sets of terms are synonymous—the former from the viewpoint of the herbivore, the latter from the viewpoint of the plant.

Apparent plants are large, easy to locate, available to herbivores, usually long-lived, and woody. They possess the most expensive type of defense—quantitative or dosage-dependent. Such a defense, not easily mobilized, is most effective against herbivore specialists. The secondary compounds involved are mostly tannins and resins concentrated near the surface tissues of leaves, in bark, and in seeds. They form indigestible complexes with leaf proteins, reduce the rate of assimilation of dietary nitrogen, reduce the inability of microorganisms to break down proteins in herbivore digestive systems, and lower palatability. The problem with such a defense is slow response. A year after defoliation by gypsy moths, oaks increased tannin and phenolic content of their leaves and increased their toughness (Schultz and Baldwin 1982).

Unapparent plants are short-lived, mostly annuals and perennials, and scattered in space and time. The plants employ a qualitative, highly toxic defense involving secondary substances such as cardiac glucosides and alkaloids that interfere with metabolism or disrupt development of nonadapted insects. These substances can be synthesized quickly at little cost, are effective at low concentrations, are readily transported to the site of attack, and work quickly. They can be shuttled about the plant from growing tips to leaves, stems, roots, and seeds, and they can be transferred from seed to seedling. These substances protect mostly against generalist herbivores.

The distinction between quantitative and qualitative defense and between apparency and unapparency is not absolute. Although testing causal relationships of apparency is difficult (Fox 1981; Courtney 1985), the concept is useful in understanding some aspects of herbivore-plant relationships.

■ Herbivore Countermeasures

Although plants may possess powerful chemical defenses that work well against generalist herbivores, they can be breached, especially by specialists. The main mechanism involved is detoxification of secondary compounds. The major detoxifying system is mixed function oxidase, MFO. Possessed by all animals, MFO metabolizes foreign, potentially toxic substances. In vertebrates the MFO activity is located in the liver; in insects it is in the gut, fat bodies, and Malpighian tubules. By oxidation, reduction, and hydrolysis MFO converts fat-concentrating (lipophilic) foreign chemicals into water-soluble molecules that can be eliminated by the excretory system.

The MFO system is a general detoxifying agent, nonspecific in character and induced into activity by a wide array of toxic compounds. It probably evolved in animals to degrade toxic by-products of animal metabolism and harmful compounds ingested. Thus animals, especially the insects, are preadapted to handle many toxic, chemically unrelated compounds. Because of the ubiquitous occurrence of MFO, adaptations to new specific toxic compounds require little genetic change, as witnessed by the rapidity with which insect pests become adapted to new insecticides. Thus insects discovering an abundant new source of food can adapt quickly to novel toxic compounds and become feeding specialists on certain families of plants. For example, some species of butterflies of the family Pieridae, notably the cabbage butterfly, the large white, and small white, and cabbage aphids (*Brevicoryne brassicae*) feed on members of the Crucifer family. Its allyl glucosinolate is toxic to all noncruciferous feeders. Larvae of the monarch butterfly feed on the highly toxic milkweed and sequester its cardiogenic glucosides in their bodies as a chemical deterrent to predation. For such specialists, the volatile chemicals of the host plants act as an attractant rather than a deterrent. The females of specialists are programmed to seek out and lay eggs on plants on which the larvae are able to overcome chemical defense.

Other insects get around chemical defenses by cutting circular trenches in leaves before feeding, stopping the flow of chemical defenses to the leaf area on which they will feed (Carroll and Hoffman 1980). Beetles and certain caterpillars cut leaf veins on milkweeds and other latex-producing plants, blocking the flow of latex to the intended feeding site (Dussourd and Eisner 1987).

■ Models of Interaction

The interrelations of plants and herbivores have been examined theoretically by May (1973), Caughley (1976a, 1976b), and Noy-Meir (1975), all of whom present mathematical approaches and analyses.

The growth of vegetation as a function of plant biomass can be described by an expression comparable to the logistic growth equation

$$a V \left(1 - \frac{V}{K} \right)$$

where V = biomass of vegetation, K = maximum sustained biomass (carrying capacity), and a = rate of increase. The rate of increase slows as competition for sunlight, moisture, and nutrients and self-interference increases (Figure 20.6).

When ungrazed vegetation is subjected to grazing by a herbivore population, the vegetation's rate of growth is slowed by an amount proportional to the intensity of grazing or predation (number of herbivores consuming plants multiplied by the rate at which vegetation is consumed). When vegetation is at maximum sustained biomass *(K),* herbivores can eat all they want, although the quantity is limited by the herbivores' intake capacity. If vegetation increases while the herbivore population remains the same, grazers increase consumption up to a point of saturation. If the vegetation declines, the amount herbivores consume also declines, because intake is limited by the forage available. These conditions represent a Type II functional response curve (Figure 20.6a).

If herbivores increase, a numerical response, they may reach a level where they overgraze the vegetation, as frequently happens with deer, snowshoe hare, and lemmings. If the vegetation has no ungrazable reserve biomass or if it is grazed to a point where the green biomass is too sparse to maintain production, the plant population may go extinct (Figure 20.6c,d). If the vegetation has an ungrazed reserve, the reserve may be utilized to attain a low biomass steady state (Figure 20.6b). Depending upon the population density of the herbivore, the vegetation may stabilize at a high biomass *(V)* or reach an unstable equilibrium point at which the vegetation may be able to restore itself if grazing (predatory) pressure is reduced or at which the plant population may slip to extinction. In some situations, especially where a Type III functional response is involved, the vegetation may exhibit two stable steady points, one at a high plant biomass and another at a low plant biomass (Figure 20.6e). Between the two is an unstable equilibrium point.

Interactions between various vegetation growth curves and various herbivore densities can result in a number of plant-herbivore relations (for some detailed examples and discussion, see Noy-Meir 1975). As herbivores increase, vegetation declines. In turn the herbivore population declines (Figure 20.7). The vegetation recovers, the herbivore population increases, and the two populations approach equilibrium, the vegetation with grazing pressure and the herbivore with its food supply (Caughley 1976b).

A notable vegetation-herbivore interaction involves the ten-year cycle of the snowshoe hare (Figure 20.8). A decline in the peak densities of snowshoe hares is initiated by an overwinter shortage of its food, woody browse. In forest regions dominated by aspen *(Populus)* the hare's essential browse consists largely of aspen stems less than 3 mm in diameter, young growth with a concentration of nutrients

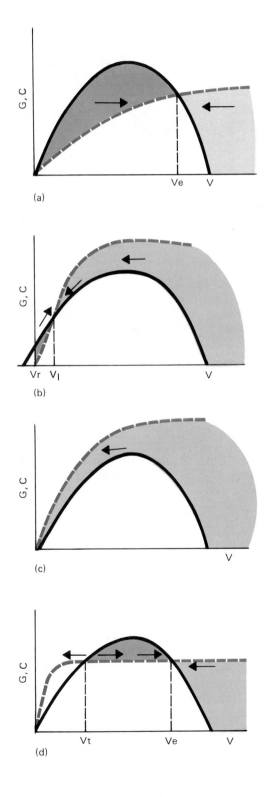

(a)

(b)

(c)

(d)

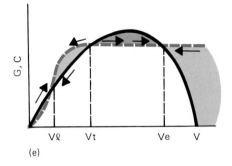

(e)

Figure 20.6 Logistic plant growth (G) as a function of plant biomass (V), over which is imposed consumption per animal (C) as a function of plant biomass. In dark-colored areas plant growth exceeds consumption; in light-colored areas consumption exceeds growth. (a) Consumption curve is below the growth curve at all biomass levels. Intersection of the two curves indicates point of stable equilibrium between plant growth and herbivore consumption. Deviation from it in either direction will cause net changes in V tending to restore stable equilibrium. In this undergrazed state plant growth, animal consumption, and secondary production are below maximum. (b) Overgrazing to low biomass steady state. Vegetation has some ungrazable reserve biomass that prevents complete extinction. (c) Overgrazing to extinction. The consumption curve exceeds the plant growth curve at all levels of V. If no inaccessible plant reserves capable of producing new plant growth exist, vegetation becomes extinct. (d) Steady state and unstable turning point to extinction. This situation occurs if the consumption curve is steeper than in (a). The two curves intersect at two points, one at a steady state at high biomass and the other at low biomass. Any deviation can lead to extinction if V becomes lower then V_t. (e) Two steady states, one at high biomass and the other at low biomass, occur when an ungrazable plant exists. The two curves intersect three times, producing a stable steady state at high plant biomass (V_e) and at low plant biomass (V_2) and an unstable equilibrium or turning point between them (V_t). (After Noy-Meir 1976.)

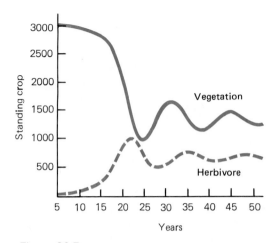

Figure 20.7 Trend of vegetation density and animal numbers after a herbivore eruption. Note that as the herbivore population increases, vegetation biomass decreases; and as herbivore decline, vegetation increases. Eventually vegetation growth and herbivore consumption reach sort of a steady state. (After Caughley 1976b: 103.)

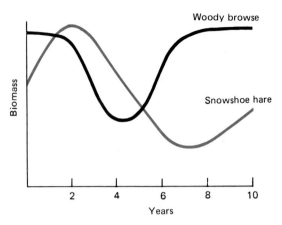

Figure 20.8 A vegetation-herbivore cycle involving woody vegetation, particularly aspen, and the snowshoe hare. Note the time lag between the cycle of vegetation recovery and the growth and decline of the snowshoe hare population. (Adapted from Keith 1974.)

(Keith 1974; Pease, Vowles, and Keith 1979; Wolf 1980). The hare-vegetation interaction becomes critical when essential browse falls below that needed to support the population overwinter, approximately 300 g per individual per day of stems 3 to 4 mm in diameter. Excessive browsing and girdling during population increase reduce subsequent increases of woody growth, bringing about a food shortage, which causes a high winter mortality of juvenile hares and lowered reproduction the following summer.

The decline in hares is also related to chemical defenses of alder, birches, and some willows, which strongly influence the selection of winter forage among many subarctic browsing vertebrates (Bryant and Kuropat 1980). Hares avoid these more nutritious plants, especially juvenile growth and buds, because they contain more resin and phenolic glucosides (Reichardt et al. 1984, Palo 1984, Sinclair and Smith 1984, Bryant et al. 1985). For example, hares avoid the juvenile internodes

of Alaskan green alder *(Alnus crispus)*. They contain three times the concentration of two deterrent secondary metabolites, pinosylvin and pinosylvin methyl ether, found in mature twigs, which the hares did consume (Bryant et al. 1983; Clausen et al. 1986). Mountain (Arctic) hares *(Lepus timidus)* selectively feed on mature over juvenile twigs of a number of species of willow and avoid low-growing species, which have high levels of secondary metabolites (Takvanainen 1985). The decline in secondary metabolites in mature woody growth suggests that juvenile resistance can be an adaptation against mammal browsing at the ground level. It could further reduce available winter food and help trigger cycling in hares.

Chemical defenses of the subarctic woody plants also interfere with or restrict the foraging patterns and winter food supply of other animals, particularly the ruffed grouse (Bryant and Kuropat 1980) and the moose (Bryant and Kuropat 1980; Reichardt et al. 1984).

■ Summary

Interaction of predators and their prey makes up a predator-prey system. Predators influence the growth and survival of prey populations, and the density of prey populations influences the growth and survival of predator populations. Predator-prey interactions at one trophic level influence predator-prey interactions at the next trophic level. Both plant-herbivore systems and herbivore-carnivore systems behave this way.

Plant predation involves defoliation by grazers and consumption of seeds and fruits. Interactions between changes in plant biomass and herbivores result in oscillations of both plant and herbivore populations or in an equilibrium situation.

Herbivory affects plant fitness by reducing the amount of photosynthate and the ability to produce more. Plants try to prevent losses by denying herbivores palatable or digestible food or by producing secondary compounds that interfere with growth and reproduction.

In response to the selection pressures of herbivory, plants have evolved measures of defense. They include mimicry among plants to hide from specialized herbivores or to attract seed dispersers, structural defenses such as hairs, thorns, and spines, and predator satiation. Reproduction is timed so that fruits and seeds are so abundant that seed predators can take only a fraction of them, leaving a number to escape and germinate. Widespread is chemical defense. It involves distasteful or toxic substances that repel, warn, or inhibit would-be attackers. These substances are secondary metabolic products such as alkaloids, phenolics, and cytogenic glucosides. Chemical defense is most successful against generalist herbivores. Certain specialists breach the chemical defense and detoxify secretions or sequester the toxins in their own tissues as defense against predators.

Herbivore-Carnivore Systems

Herbivory supports carnivory. Unlike herbivores, carnivores are not faced with a lack of quality in their food. The quality is there—all highly proteinaceous and easily digestible. It is quantity that is frequently lacking. That dictates a somewhat different relationship between eater and eaten. Numbers of prey become important. Fitness of the predator depends upon its ability to capture prey; and fitness of the prey depends upon its ability to elude predation and, if a herbivore, at the same time to overcome plant defenses. That combination puts a squeeze on herbivores.

■ Prey Defense

In an evolutionary context predator and prey play a game. Prey evolve often elaborate means of defense. To survive under selective pressure, the predator must come up with a way to breach the defense.

The relationship between the two involves a flux of adaptive genetic change in each.

Predators, however, do not seem to track closely changes in their prey. Over evolutionary time predators seem to have experienced an adaptive gap between themselves and their prey (Bakker 1983). Predators, as suggested by fossil records, did not evolve rapidly enough to track the escape adaptations of their prey. Thus predators possess a suboptimal and barely adequate efficiency in predation.

Chemical Defense

Chemical defenses are widespread among animals; and as with plants, they may have been borrowed from some other use. Venom, for example, protects snakes from enemies, but it is also the means by which the snakes capture prey.

There is an array of chemical defenses. Some species of fish release pheromones from the skin into the water that act as alarm substances, inducing fright in other members of the same or related species (Pfeiffer 1962). The fish produce the pheromone in specialized cells in the skin that do not open to the surface, so the pheromone is released only when the skin is broken. Fish in the vicinity receive the stimulus through the olfactory organs. Such alarm substances are most common among fish that are social, nonpredaceous, and lacking in defensive structures.

Arthropods, amphibians, and snakes employ secretions to repel predators. Strongly odorous, easily detected substances are produced in often copious amounts by arthropods (Eisner and Meinwald 1966; Eisner 1970). They produce the secretions in glands with large saclike reservoirs that are essentially infoldings of the body wall and discharge it through small openings. The secretions may ooze onto the animal's body surface, as in mil-

lipedes; be aired by the evagination of the gland, as in beetles; or be sprayed for distances of up to a meter, as in grasshoppers, earwigs, and stinkbugs. These secretions repel birds, mammals, and insects alike by their effect on the predator's face and mouth. Some mammals, such as shrews and skunks and other mustelids, also possess secretions that discourage attacks by would-be predators.

Active components in the defensive secretions of many arthropods occur as toxic secondary substances, such as saponins, glossypol, and cyanogenic glycosides, used as a chemical defense by plants. Although these toxins inhibit herbivores from feeding on the plants (see Chapter 20), some arthropods can incorporate toxic substances ingested from the plants into their own tissues. In turn the toxin protects the herbivore from its enemies. The monarch butterfly, for example, feeds on milkweeds that contain a cardiac glycoside, a substance that causes illness in birds that eat the monarch.

Warning Coloration and Mimicry

Animals with pronounced toxicity and other chemical defenses often possess warning coloration, bold colors with patterns that serve as warning to would-be predators. The black and white stripes of the skunk, the bright orange of the monarch butterfly, and the yellow and black coloration of many bees and wasps serve notice of danger to their predators. All their predators, however, must have had some unpleasant experience with the prey before they learn to associate the color pattern with unpalatability or pain.

Similarly, animals associated with inedible species sometimes evolve a similar mimetic or false warning coloration. That phenomenon was described some 100 years ago by the English naturalist H. W. Bates in his observations of tropical butterflies. The type of mimicry he described, now called Batesian, is the resemblance of an edible species, the mimic, to an inedible one, the model (Figure 21.1). Once the predator has learned to avoid the model, it avoids the mimic also. Batesian mimicry is disadvantageous to the model because the predator will encounter a number of tasty mimics and therefore take longer to avoid the model, which will suffer greater losses in the learning process. The greater the proportion of mimics to the models, the longer the learning time of the pred-

Figure 21.1 Mimicry in insects. The model, the distasteful pipevine swallowtail, has as its mimics the black swallowtail and the spicebush swallowtail. The black female tiger swallowtail is a third mimic. All these butterflies are found in the same habitat. The robber fly, a mimic of the bumblebee, illustrates aggressive mimicry. The drone fly is a mimic of the honeybee.

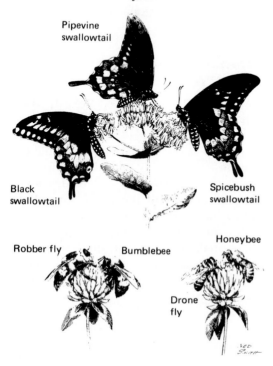

Pipevine swallowtail

Black swallowtail

Spicebush swallowtail

Robber fly   Bumblebee

Honeybee

Drone fly

ator. Usually the number of mimics is fewer than the model.

Among the North American butterflies, the palatable viceroy butterfly *(Basilarchia archippus)* mimics the monarch *(Danaus plexippus),* most of which are distasteful to birds (Brower 1958). Both model and mimic have an orange ground color with white and black markings; they are remarkably alike. The viceroy's nonmimetic relatives are largely blue-black in color.

A less common type of mimicry, called Mullerian, involves both unpalatable models and unpalatable mimics. Such mimicry is advantageous to both. The pooling of numbers between the model and the mimic reduces the losses of each, because the predator associates distastefulness with the pattern without having to handle both species. Mullerian mimicry differs from Batesian in that feedback from handling either species is negative, reinforcing the learning process in the predator. Batesian mimics and models belong to different phylogenetic lines, whereas Mullerian mimics include members of the same genus and family, probably because their ability to utilize and store poisons from plants has become fixed in an evolutionary line.

Cryptic Coloration

Certain color patterns and behaviors have evolved to hide prey from predators. Such cryptic colorations involve patterns, shapes, postures, movements, and behaviors that tend to make the prey less visible or break up their outlines. Some animals are protectively colored, blending into the background of their normal environment. Such protective coloration is common among fish, reptiles, and many groundnesting birds. Countershading or obliterative coloration, in which the lower part of the body is light and the upper part is dark, reduces the contrast between the unshaded and shaded areas of the animal in bright sunlight. Object resemblance is common among insects. For example, walkingsticks (Phasmatidae) resemble a twig, and some insects resemble leaves (for example, katydids). Some animals possess eyespot markings, which intimidate potential predators, attract their attention away from the animal, or delude them into attacking a less vulnerable part of the body. Eyespot patterns in Lepidoptera seem to intimidate predators by imitating the eyes of a large avian predator that attacks small, insectivorous passerine birds.

Associated with cryptic coloration is flashing coloration. Certain butterflies, grasshoppers, birds, and ungulates, such as the white-tailed deer, display extremely visible color patches when disturbed and put to flight. The flashing coloration may distract and disorient predators. When the animal comes to rest, the bright or white colors vanish, and the animal disappears into its surroundings (see Harvey and Greenwood 1978 for review).

Armor and Weapons

Some of the most effective means of defense involve protective armor. Clams, armadillos, turtles, and numerous beetles all withdraw into armor coats or shells when danger approaches. The associated problem is the animal's inability to assess the external environment. Is the predator large or small, still present or departed? How much foraging time should an animal sacrifice before daring to open up its defenses? Porcupines, hedgehogs, and echidnas have quills (modified hairs), which discourage predators.

Behavioral Defense

Some animals' defenses are behavioral. One is the alarm call, given at the moment of potential danger when a predator is sighted. High-pitched alarm calls are not species-specific. They are recognized by many different animals close by. An unanswered question is to whom the calls are directed—the predator or the conspecific prey. If directed toward potential prey, the alarm call could be either altruistic or selfish (see Chapter 13). If the alarm exposed the caller's position to the predator, the caller could draw the predator's attention away from conspecifics, including kin. On the other hand, it could attract more conspecifics for cooperative defense and lower its risk of being taken. Alarm calls do function to warn close relatives, as in the case of Belding ground squirrels (Sherman 1977). Highly sedentary, closely related females live in close proximity to each other. Adult and one-year-old females living with relatives respond quickly to danger and give most of the alarm calls, which warn offspring and other relatives. Beyond that there are few conclusive studies on the evolution and function of alarm calls.

Alarm calls often bring in numbers of potential prey that respond to the situation by mobbing or harassing the predator. An example is the harassment of an owl perched in a tree by many small birds attracted to the scene by general alarm calls. Mobbing may involve harassment at a safe distance or direct attack. The outcome for the prey is a reduction in the risk of predation to themselves and their offspring. As with alarm calls, the adaptive and evolutionary significance of mobbing are still obscure.

Distraction display diverts the attention of predators away from eggs or young. Distraction displays are most common among birds. Birds with precocious young, such as the killdeer, usually exhibit the most vehement distraction displays at the time the eggs hatch, and altricial birds, such as the vesper sparrow, at the time the young fledge. Because the beneficiaries of distraction display are the immediate offspring, the behavior probably evolved through kin selection.

Living in groups may be the simplest defense for some prey species. Groups, especially in mobbing situations, probably deter a predator that would not be so inhibited when facing only one or two prey individuals. Sudden explosive group flight can confuse a predator, unable to decide which individual to follow. The more prey are congregated, the less is any one individual's chances of being taken. By keeping close together, individuals reduce their chances of being the one closest to the predator. By maintaining a tight or cohesive group, prey make it difficult for a predator to snatch a victim.

Predator Satiation

A more subtle defense is the timing of reproduction so that most of the offspring are produced in a short period of time. Then food is so abundant that the predator becomes satiated, allowing a percentage of the offspring to escape. This strategy is employed by such ungulates as the wildebeest and caribou. It involves the synchronized birth of young; the young are so abundant in one short period of time that predators can take only a fraction of them, and the remaining young quickly grow beyond a size easily handled by predators (Schaller 1972; Bergerud 1971).

The 13-year and 17-year appearances of the periodical cicadas *(Magicicada spp.)* function in much the same manner. By appearing suddenly in enormous numbers, they quickly satiate predators and do not need to evolve costly defensive mechanisms. Although huge numbers of adults succumb to predators, the losses hardly dent the total population (Figure 21.2). In other years, predators must seek alternative prey. Thus the cicadas' major defense is to prevent predators from ever evolving any dependence upon them.

■ Predator Offense

As prey evolved ways of avoiding predators, predators by necessity had to evolve better ways of hunting and capturing prey (Bakker 1983).

Hunting Tactics

Predators have three general methods of hunting: ambush, stalking, and pursuit. Ambush hunting involves lying in wait for prey to appear. This method is typical among certain insects and some frogs, lizards, and alligators. Ambush hunting has a low frequency of success, but it requires a minimal expenditure of energy. Stalking, typical of herons and some cats, is a deliberate form of hunting with quick attack. The predator's search time may be great, but pursuit is minimal. Therefore it can afford to take smaller prey. Pursuit hunting, typical of many hawks, involves minimal search time. Because pursuit time is relatively great, these predators must secure relatively large prey to compensate for the energy expended. Searchers spend more time and energy to encounter prey. Pursuers, theoretically, spend more time to capture and handle prey once they notice it.

Predators have certain energy requirements that can be met only by profitable

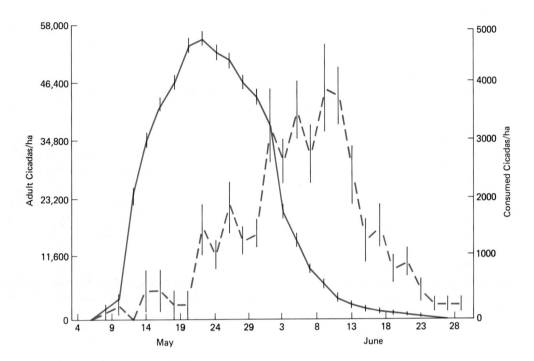

Figure 21.2 Estimated daily population density of periodical cicadas on a study site in Arkansas (left) and estimated daily bird predation rates based on cicada wing counts (right). Maximum cicada density occurred around May 24, and maximum predation occurred around June 10. Predatory pressures built up as birds apparently acquired a search image for the cicadas; but at the height of predation, the bulk of the cicadas had already emerged and escaped bird predation. (From Smith 1988.)

foraging. They cannot afford prey too small to meet their energy requirements unless that prey is abundant and can be captured quickly. Predators also have an upper size limitation. The prey must not be too large to consume or too difficult or dangerous to handle. In fact, some prey species become invulnerable to predation through body growth.

Some predators have evolved methods of killing prey much larger than themselves. Just as prey may live in groups to lessen the risk of predation, so predators may hunt in groups to lessen the risks they face in attacking prey. Wolves, African hunting dogs, jackals, and African lions are examples of predators that take large prey by hunting cooperatively. Certain snakes and arthropods may use venom to kill large prey.

Conversely, predators much larger than their prey have evolved ways of filtering organisms from their environments, particularly in aquatic communities. Examples are net-spinning caddisflies that feed on drift and baleen whales that feed on krill. Intermediate-sized predators are usually hunters, whereas very small predators are usually parasitoids. The boundaries between the various predator-prey sizes are usually set by some cost-benefit ratio.

Cryptic Coloration and Mimicry

If prey can use cryptic coloration, concealing coloration, and mimicry to their advantage, predators can do the same. Cryptic and concealing coloration enable them to blend into the background or break up their outlines. Predators also can deceive prey by resembling the host or prey. An example is the model bumblebee and the mimic robber fly *(Mallophora bomboides)* (Figure 21.1). Not only does the robber fly benefit from reduced predation, but it also exploits the model for food. The robber fly preys on Hymenoptera by preference, and its resemblance to its prey allows it to escape notice until the bee finds it too late to flee or defend itself (Brower and Brower 1962). The females of certain species of fireflies imitate the flashing of other species, attracting to them males of those species, which they promptly kill and eat (Lloyd 1980). That type of mimicry is called aggressive.

Adaptations for Hunting

Predators have acquired various adaptations that improve their hunting ability in addition to such weapons as fangs and claws. Bats, for example, produce ultrasonic sounds through the nose and mouth that enable them to detect prey by echolocation (for a good summary, see Vaughan 1986). Night-hunting owls can locate prey by hearing rather than by sight. Their feathered facial disks reflect the sounds of prey and direct them to the ears (Konishi 1973). The owl's large ear openings are positioned asymmetrically, enabling the bird to detect differences in the elevation of the prey. The owl's ability to fly noiselessly allows it to come upon the prey without alerting the victim. Day-hunting harriers flying over densely grown grass fields have similar facial disks and placement of ears that enable them to locate by sound voles hidden in the grass (Rice 1982).

■ Regulation

That predators have an adverse impact on the abundance of prey species is an ingrained idea hard to dislodge. Predators are quickly blamed for the decline in any species in which humans have a vested interest, such as game species. However, long-term studies on the effect of predation on vertebrate populations by Paul Errington (1943, 1945, 1946, 1963) have suggested that predators feed on the "doomed surplus" and have little impact on the productivity of prey populations. The unstated premise in Errington's studies was that the predator would have to reduce r, the rate of increase, of the prey population to regulate it. To accomplish that predators would have to remove a portion of the reproductive age classes (see Taylor 1984: 125–139), which are not part of the doomed surplus.

There is little evidence that natural predation accomplishes such regulation. Larger predators, such as African lions and wolves, take the most vulnerable individuals—the young, the old, and those in poor condition. Such predation is compensatory, that is, the losses are made up by future reproduction; the predators are not removing reproductive individuals (Figure 21.3). When the prey population declines sharply for reasons other than predation, large predators are forced to take reproductive individuals, in spite of the greater energy costs incurred. Such predation can drive prey to even lower levels. The lack of prey then reduces the predator population through malnutrition, reproductive failure, and social interactions. Freed from intensive predation and experiencing an abundance of food, prey becomes abundant again. Such

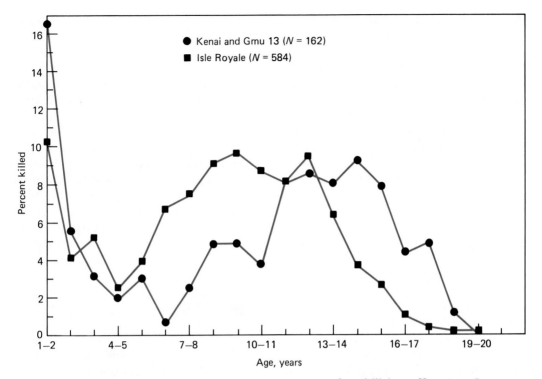

Figure 21.3 Compensatory predation. The ages of adult moose killed by wolves on the Kenai, Alaska, and on Isle Royale, Michigan. Yearlings and older individuals are most vulnerable, especially on the Kenai. A smaller moose population on Isle Royale apparently increased predatory pressure on younger moose, but heaviest predation still fell on the old. (Data from Peterson et al. 1984 and Ballard et al. 1987.)

loosely regulatory feedback in predator populations and the changing impact of predation on the prey result in wide fluctuations in both predator and prey populations.

These fluctuations can be influenced strongly by human exploitation of both predator and prey populations. Such exploitation is usually directed, intentionally or unintentionally, toward reproductive age classes and can have a strong regula-

tory or destabilizing effect (see Gasaway et al. 1983; Ballard et al. 1987).

Theory holds that predators interact with prey to produce either stable equilibria or cycles, but field evidence is wanting. Erlinge and associates (1984) studied the interactions of nine vertebrate predator species and their major prey, largely rodents and rabbits, in central Sweden. They found that the feeding habits of generalist predators in combination with their territorial behavior did prevent significant annual fluctuation in rodent numbers. Their field studies, supplemented by simulation models, suggest that generalist predators can maintain stable populations by shifting their diets (switching—see Chapter 19) in response to changing prey densities. Such regulation can occur only if alternative prey exist in abundance and predator populations are intrinsically regulated. By contrast, in large predator-ungulate systems the predators are loosely regulated by social interactions but lack alternative prey, so one prey species has to bear the brunt of predation. This situation can result in wide fluctuations in prey populations.

When one predator feeds on other predators, all of whom share prey species, the effect of the dominant prey species on population of the predaceous prey species can be pronounced. In the Carribean Islands lizards are the dominant vertebrate insectivore. They feed extensively on web spiders as well as the same prey taken by spiders. Lizards are both predators on and competitors with the spiders. Populations of web spiders are about ten times as dense on islands without lizards than on islands with lizards (Schoener and Toft 1983). To discover the process underlying this effect, Schoener and Spiller (1987) experimentally removed lizards from randomly selected plots on a very large island in the Bahamas. They found that spider densities in the removal plots were 2.5 times higher than on the control plots. The effects of removing the lizards showed down through the food web. Abundance of arthropod prey increased as did the consumption of prey. This and a similar but less extensive study (Palaca and Roughgarden 1984) demonstrate that lizards can significantly reduce spider populations on tropical areas. Contrary to most studies in which predation increases species diversity, predator removal resulted in an increase in the number of spider species.

Assessing the role of predation in the regulation of prey populations is difficult because human intrusion into predator-prey systems has so altered the relationships between predator and prey that few resemble the system under natural conditions. For example, human settlement in the Prairie Pothole Region of North Dakota severely reduced waterfowl breeding habitat and shifted a multispecies canid population to a single species, the red fox. Predation by red fox is more intense on nesting females than on males, distorting the sex ratio strongly toward males, 128 to 100, compared to the presumed pristine ratio of 110 males to 100 females. A reduction in the fox population resulted in a more even sex ratio on an experimental area, and a higher fox population shifted the ratio more strongly toward males (Johnson and Sargeant 1978). Fox predation accompanied by noncompensatory hunting mortality further distorts the sex ratio.

■ Predator-Prey Cycles

Cycles are one of the more pervasive ideas in population ecology, one that has risen to the level of ecological dogma. The concept of predator-prey cycles is an outgrowth of the neutral Lotka-Volterra predation equations, later modified as stable limit cycles, in which the prey cycle is driven by predation (Chapter 19). The classic example of a predator-prey cycle is that of the snowshoe hare and lynx. Using data obtained from the fur returns of the Hudson Bay Company, MacLuich (1937) first described the snowshoe hare-lynx cycle. The cycle was further analyzed by Elton and Nicholson (1942) and has become part of the ecological literature ever since. As described, the snowshoe hare and lynx cycled in tandem with a time lag between the two. The lynx population peaked as the snowshoe population declined (see Figure 15.20) and the snowshoe hare population recovered before that of the lynx—or so it appeared. The classic cycle, however, is based on erroneous data. The snowshoe hare cycle was plotted using data from the Hudson Bay region of eastern Canada, whereas the lynx data were from western Canada (Finerty 1980). The two cycles were not coupled.

As we saw in Chapter 20, the snowshoe hare cycle appears to be generated by a vegetation-hare interaction, which drives the predator cycle (to the extent that such a predator-prey cycle exists). The decline in hares brought about by a food shortage increases the ratio of predators to hares, intensifying predation on the hare. (Note that predators other than lynx also are exploiting the hare population.) This predation extends the period of decline and holds the population at a level much lower than the habitat could actually support. Facing a shortage of food, the predators (mostly lynx) fail to reproduce or to rear their young. With a decline in predatory pressure and a growing abundance of winter food, the hare population begins to rise sharply, starting another cycle (Figure 21.4).

Thus the cycle of the snowshoe hare and lynx is not a classic predator-prey cycle, in which the predator drives the cycle of the prey. Rather it is a vegetation-hare cycle tracked by the predator, whose major role in the cycle appears to be the suppression of the prey population at low densities. How effectively predators suppress is an unanswered question. Taylor (1984) suggests that if predators were not present, the hares might prevent the veg-

Figure 21.4 Model of the vegetation-herbivore-predator cycle involving the snowshoe hare and lynx. Note the time lag between the cycle of vegetation recovery, growth and decline of the snowshoe hare population, and the rise and fall of the predator population. (Adapted from Keith 1983.)

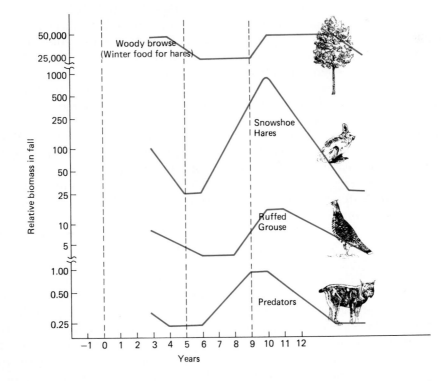

etation from recovering to a level that would support high hare populations.

The ten-year cycle of the snowshoe hare is characteristic of the boreal region. South of the boreal region, in mountain ranges and southern limits of its range, some populations of snowshoe hares exhibit cyclic fluctuations and some do not. Dolbeer and Clark (1975) and Tanner (1975), and Wolff (1980) have studied why. In coniferous forests and associated regions south of the boreal forests, snowshoe hares exhibit cyclic fluctuations only in a uniform environment of spruce and fir (which, of course, is also characteristic of the boreal forest). In regions where the environment is very patchy, where many kinds of vegetation patterns exist, cycles do not occur. There are several possible reasons. Fragmented habitats support a greater diversity of prey species, which adds stability to populations of facultative predators. These predators are able to maintain sustained predation on hares (Keith 1983). Hares occupying high quality habitats are protected from predation, whereas hares living in areas of poor cover are subject to predation. In effect, patches of high quality habitat act as refuges from which surplus animals are able to repopulate poorer habitat patches. These hares, in turn, are eliminated by predators. Such dispersal and predation tend to hold the population of hares in better habitats at a level at which they do not overutilize their food supply, thus damping cyclic behavior.

■ Cannibalism

A special form of predation is cannibalism, euphemistically called intraspecific predation. Cannibalism, more widespread and important in the animal kingdom than many ecologists admit, involves the killing and eating of an individual of the same species. Cannibalism is common to a wide range of animals, aquatic and terrestrial, from protozoans and rotifers through centipedes, mites, and insects to frogs, birds, and mammals, including humans. Interestingly, about 50 percent of terrestrial cannibals, mostly insects, are normally herbivorous species, the ones most apt to encounter a shortage of protein. In freshwater habitats, most cannibalistic species are predaceous, as they are in all marine ecosystems (Fox 1975a, 1975b, 1975c).

Cannibalism has been associated with stressed populations, particularly those facing starvation. Although some animals do not become cannibalistic until other food runs out, others do so when the relative availability of alternative foods declines and individuals in the population are disadvantaged nutritionally (Alm 1952). It is probably initiated when hunger triggers search behavior, lowers the threshold of attack, increases the time spent foraging, and expands the foraging area. It is consummated when the individual encounters vulnerable prey of the same species (see Polis 1981). Other conditions that may promote cannibalism are (1) crowded conditions or dense populations, even when food is adequate; (2) stress, especially when induced by low social rank; and (3) the presence of vulnerable individuals, such as nestlings, eggs, or runty individuals, even though food resources are adequate.

Whatever the cause, the intensity of cannibalism is influenced by local conditions and by the nature of local populations. In general cannibalism fluctuates greatly over both long and short periods of time. Among some predaceous fish, such as walleye (*Stizostedium vitreum*) (Fortney 1974), and insects, such as freshwater backswimmers (*Notonecta hoffmanni*) (Fox 1975b, 1975c), cannibalism is most prevalent in summer, which coincides with a decrease in normal prey and a reduction of spatial refuges for the young.

Not all individuals in a population become cannibals (see Polis 1981). Intraspecific predation is usually confined to older and larger individuals. Those receiving the brunt of cannibalism are the small and the young, but not always. In some situations groups of smaller individuals will attack and devour larger individuals.

Demographic consequences of cannibalism depend upon the age structure of the population and the feeding rates of various age classes. Even at very low rates, cannibalism can produce demographic effects. Three percent cannibalism in the diet of walleyes could account for 88 percent of mortality among young (Chevalier 1973). Cannibalism can account for 23 to 46 percent of the mortality among eggs and chicks of herring gulls (Parsons 1971), 8 percent of young Belding ground squirrels (Sherman 1981), and 25 percent of lion cubs (Bertram 1975). If a large proportion of either an entire population or a vulnerable age class is eaten, it can cause violent fluctuations in recruitment.

Cannibalism can become a mechanism of population control that rapidly decreases the number of intraspecific competitors as food becomes scarce. It is unlikely to bring about extinctions of local populations because of its short-term nature. It decreases as resources become more available to survivors and as vulnerable individuals become scarcer. By reducing intraspecific competition at times of resource shortages, cannibalism may actually reduce the probability of local extinction of a population. In the long term, however, cannibalism can be self-defeating because it runs counter to second law of thermodynamics and trophic level dynamics. The exceptions would be among those animals whose young feed at different trophic levels than the adults. Then cannibalism would involve the harvesting of young grazers.

Cannibalism can provide a selective advantage to survivors. Survivors gain a meal and eliminate both a potential competitor for food and a potential conspecific predator. With the population reduced, the survivor has more food, enhancing its chances of longer survival, rapid growth, and increased fecundity. Cannibalism may also be rewarding from a nutritional standpoint, leading to increased growth rates and reproduction of cannibalistic individuals over noncannibalistic ones. The individuals consumed contain the proper proportion of nutrients necessary for growth, maintenance, and reproduction.

Meefe and Crump (1987) tested this hypothesis by feeding females of the mosquitofish *Gambusia affinis*, which has cannibalistic tendencies, four dried diets, one of which contained conspecifics. Those females fed the cannibalistic diet exhibited increases in both body growth and reproduction relative to those on the other diets, and faster development of embryos. Both effects could increase fecundity. Hoogland (1985) noted that infanticide and cannibalism of closely related individuals in the prairie dog *(Cynomys ludovicianus)* provided nutritional benefits to lactating females.

Cannibals can also increase their own fitness by reducing the fitness of competitors. By killing and eating other individuals of the same sex, they reduce competition for mates. They also can eat the offspring of a competitor, as adults of the Belding ground squirrels do. Among some animals—insects, in particular—the females will kill and eat the male after mating, reducing the probability that other females will encounter a mate.

Cannibalism can be a selective disadvantage if individual survivors become too aggressive and destroy their own progeny or genotype completely, reduce their genotype faster than the genotypes of conspecific competitors, or reduce the chances of successful reproduction by eliminating suitable mates.

Selection can balance advantages against disadvantages. In some situations the disadvantages of cannibalism are less severe than starvation and reproductive failure caused by inadequate nutrition. For example, parents cannibalizing some of their own offspring can increase the probability of survival and fitness of either parents or surviving offspring or both (Polis 1981; Rohwer 1978) and utilize rather than waste energy already invested in them. If a population is reduced by starvation, the survivors may be nutritionally stressed. If individuals are removed by cannibalism, density is reduced early, and per capita food supply remains high. Survivors have improved their fitness because, being well fed as juveniles, they grow faster, survive better, and produce more young.

Cannibalism may be less costly to individuals, but it is disadvantageous from an evolutionary viewpoint. For this reason it is highly improbable that any strong selection exists for the trait. With a few exceptions, cannibalistic individuals do not distinguish between conspecific and other prey, but rather are opportunistic predators. Rarely is cannibalism a distinct behavioral trait.

■ Exploitation by Humans

A form of highly selective and intensive predation, often not related to the density of either predators or prey, is exploitation by humans. Overexploitation of wild populations, especially if coupled with the loss of habitat, has resulted in a serious decline and local or global extermination of some species. The overharvesting of buffalo, great auk, African ungulates, whales, and many pelagic fish are examples of shortsightedness on the part of humans. On the other hand, such wildlife populations as the white-tailed deer are underharvested in many places, particularly since their natural predators have been eliminated. In contrast to these examples of destructive exploitation, the objective of the wise exploitation of any natural population is the maintenance of some sort of equilibrium between recruitment and harvest.

Basic Concepts

Although some of the terms used in defining exploitation of populations are similar to those used in productivity, the meanings are somewhat different. When fishery and wildlife biologists speak of *yield,* they refer to the individuals or biomass removed when the population is harvested. *Biomass yield* is the product of the number harvested times the average weight. (Yield may indicate weight without numbers or vice versa.) The *standing crop* is the biomass present in a population at the time it is measured. *Productivity* is the difference between the biomass left in the population after harvesting at time t and the biomass present in the population just before harvesting at some subsequent time $t + 1$.

The objective of regulated exploitation of a population is *sustained yield:* The yield per unit time is equal to productivity per unit time. In its simplest form sustained yield is described by an equation first proposed by E. S. Russell for fishery exploitation. Although the equation was specifically developed for fisheries, it is applicable to any exploitable population.

With some minor modifications this equation is

$$B_{t+1} = B_t + (A_{br} + G_{bi}) - (C_{bf} + M_b)$$

where B_{t+1} = total biomass of exploitable stock just before harvesting, at time $t + 1$; B_t = total biomass of exploitable stock just after the last harvest, at time t; A_{br} = biomass gained by the younger recruits just grown to exploitable stock; G_{bi} = biomass added by the growth of individuals in both B_t and A_{br}; C_{bf} = any biomass exploitatively removed during the harvest period; and M_b = biomass lost from exploitable stock by natural causes during the time t to $t + 1$. The equation stresses the fact that productivity also includes individuals that were born and individuals that died during the time interval from the end of one harvest period to the beginning of the next.

The equation is highly simplified. In an unexploited fish population A_{br}, C_{bf}, and M_b are interdependent. For example, in a stable environment largely undisturbed by humans fish populations appear to be dominated by large species. In turn each species population appears to be dominated by large old fish (Johnson 1972). When humans start to exploit such a population significant changes take place. To compensate for exploitation directed first toward the largest members of the population (organisms that under natural conditions are normally secure from predation), the population exhibits an increased growth rate, a reduced age of sexual maturity, increased number of eggs per unit of body weight, and reduced mortality of small members of the population (Regier and Loftus 1972). Populations of other vertebrates react in a similar way.

Exploitation may also influence behavior of the species. Fishing or hunting techniques that are employed constantly because of their initial success gradually become less effective with time because of some conditioning or learning by members of the population. As harvest of the species begins to decline, the exploiters are forced to improve or change their methods of fishing and hunting. Also involved may be an interspecific competition. As both the numbers and larger members of a population decline, the niche may be occupied by unexploited, highly competitive, and closely related sympatric or introduced species. Thus, as the Pacific sardine populations declined, their place was taken by anchovies.

If exploitation is carried far enough, then the age classes in the population are too young to carry on reproduction and the population collapses. The principle behind sustained yield is to avoid that collapse of the population.

Exploitation and sustained yield of a population is clearly dependent on the rate of increase. Sustained yield does not imply holding a population at ecological carrying capacity (K), for obviously at that level the rate of increase equals zero. A population stable in the absence of harvesting can be harvested under sustained yield only after the plant-herbivore or the herbivore-carnivore system has been manipulated to raise r. This change can be made in two ways: (1) improve food and cover to increase the carrying capacity by increasing available resources, fecundity, and survival; (2) lower the density by removing a certain number and then stabilize the population at some lower density (Figure 21.5). Within limitations the lower the density of a population is below the carrying capacity, the higher is the rate of increase. Thus a higher rate of harvest is needed to hold the reproductive population stable at some desired lower density.

The idea is to have the rate of harvest

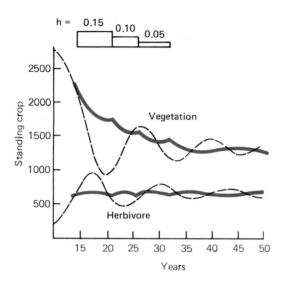

Figure 21.5 Effect of harvesting on a vegetation-herbivore system. Superimposed on the herbivore-vegetation interaction diagram in Figure 20.8 is the path of vegetation and herbivore standing crops after harvesting of herbivores over the intervals and at the rates per year diagramed in the rectangles above. As a result of management, fluctuations are reduced and the system moves to equilibrium. Such equilibrium results if harvest is initiated at a rate of about one-half of the population's intrinsic rate of increase when the animal population is well below the peak. This rate of harvest must be maintained until plant density levels off and begins to increase. (After Caughley 1976.)

H equal to the rate of increase r. In effect the rate of harvest should hold the rate of growth at zero. H would have to equal the rate at which the population would increase if the harvest were stopped (Caughley 1976). Consider a deer population increasing at a rate of 20 percent a year. It has a finite rate of increase, $e^r = 1.20$ and $r = 0.182$. To hold the population stable, the herd must be harvested at the instantaneous rate of $H = 0.182$. If the population is harvested only during a certain season of the year, as is usual with deer, then the sustained yield is calculated from

an isolated rate of harvest h, defined as $h = 1 - e^{-H}$, in this case 0.167. This isolated rate of harvest would have taken into account natural mortality to the population occurring between the period of birth and the time of harvest. Assuming a deer population after the fawning season of 1000 animals and allowing a natural finite rate of mortality of 0.25 per year, we could remove 151 animals in the fifth month. This action would allow the number at the next fawning season to climb back to 1000. The addition of 375 young would compensate for hunting and natural mortality (for details on carrying out calculations, see Caughley 1976).

Although often considered as such, sustained yield is not a particular value for a given population. There may be a number of sustained yield values corresponding to different population levels and different management techniques. The level of sustained yield at which the population declines if exceeded is known as *maximum sustained yield*. Maximum sustained yield is not always the most efficient harvest level because of other considerations such as species interactions, esthetics, land use problems, and the like. Harvesting may be aimed at the *optimum sustained yield,* the level of sustained yield determined by consideration of these other factors as well as maximum sustained yield.

The higher the r of a species, the higher will be the rate of harvest that produces the maximum amount of biomass production. Species characterized by scramble competition (r-strategists) have a high wastage of production. To manage a population influenced by density-independent variables, such as climate or temperature, the objective is to reduce wastage by increasing the rate of exploitation. The role of harvesting is to take all individuals that otherwise would be lost to natural mortality. This type of exploitation is described by the expression (K. E. F. Watt 1968)

$$\text{maximum yield} = B_t - \min(R_t)$$

where B_t = biomass at time t and $\min(R_t)$ = minimum number of reproducing individuals left at time t in order to ensure replacement of maximum yield at time $t + 1$.

Such a population is often difficult to manage because the stock can be severely depleted unless there is repeated reproduction. An example is the Pacific sardine (Murphy 1966, 1967), a species in which there is little relationship between breeding stock and the subsequent number of progeny produced. Exploitation of the Pacific sardine population in the 1940s and 1950s shifted the age structure of the population to younger age classes. Prior to exploitation 77 percent of the reproduction was distributed among the first five years. In the fished population, 77 percent of the reproduction was associated with the first two years of life. The population approached that of single-stage reproduction subject to pronounced oscillations (Figure 21.6). Two consecutive years of reproductive failures resulted in a collapse of the population from which it never recovered.

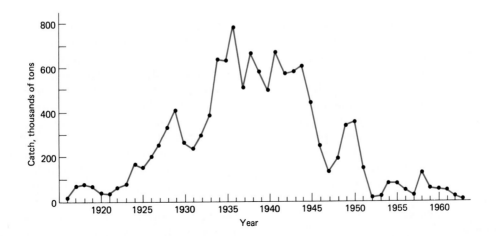

Figure 21.6 (a) Simulation of an exploited and an unexploited population of sardines, both subject to random environmental variation in reproductive success. The dotted line indicates the asymptotic population size. Note how exploitation adds to instability and how dangerously low the population can get. (From Murphy 1967: 734.) (b) Compare this simulation with the annual catch of the Pacific sardine along the Pacific coast of North America. The population collapsed from overfishing, environmental changes in the Pacific Ocean, and the increase in a competing fish, the anchovy, following a population decline of the sardine. (After Murphy 1966.)

In populations characterized by density-dependent regulation (K-strategists) the maximum rate of harvest depends on age structure, frequency of harvest, number left behind after harvest, fluctuations in environment, and fluctuations in fecundity. It also depends on density of the population to be harvested and the rate of harvest needed to stabilize the density at that level.

This type of harvest is described by the expression

$$P_b = \max B_{t+1} - B_t(X)$$

where P_b = biomass productivity from t to $t + 1$, B_t = biomass at time t, B_{t+1} = biomass at time $t + 1$, and X = the several variables that influence biomass production over time t to $t + 1$.

This relation is illustrated by Figure 21.7. For any position of the stock to the left of the 45° line there is a rate of exploitation that will maintain the stock at that position. Maximum sustained yield does not necessarily require a large standing crop. Let a be any position on the curve and c a perpendicular line that cuts the 45° line at b. At equilibrium the portion bc of the recruitment must be used for the maintenance of the stock, for $bc = ac$; ab can be harvested. There is, however, a limit to exploitation, a limit that is influenced by the inflection point of the curve. For curve C, the maximum rate of harvest is about 82 percent, for curve A 42 percent. In these curves the size of the reproductive stock that will give maximum sustained yield will not be greater than half of the replacement of the reproductive population. The greater the area of the reproduction curve above the 45° line, the greater the optimum rate of reproduction.

In summary, Caughley (1976) gives six points applicable to harvesting of populations:

1. A population stable in numbers must be reduced below a steady density to obtain a croppable surplus.
2. There is an appropriate sustained yield for each density to which a population is reduced.
3. For each level of sustained yield there are two levels of density from which this sustained yield can be harvested (Figure 21.8).
4. Maximum sustained yield can be harvested at only one density (Figure 21.8).
5. If a constant number is harvested from a population each year, the population will decline from steady density and stabilize at the upper population size for which that number is sustained yield. If this number exceeds the maximum sustained yield, the population declines to extinction.
6. If a constant percentage of the population is harvested each year (the percentage applying to the standing crop of that year), the population will decline and stabilize at a level at equilibrium with the rate of harvesting. This level may be above or below that generating maximum sustained yield.

One group of animals, notably fish and whales, is exploited commercially; another group is hunted for sport. Most game animals are characterized by contest competition and are largely, but not always, density regulated. It has been assumed by wildlife managers that regulated hunting and compensatory predation have similar effects on game animals. Just as a predator turns to another source of prey when the first source demands too great an expenditure of energy to hunt, so sport hunters abandon the field when hunting success (animals taken per unit time) drops below a certain level. Hunting mortality supposedly replaces natural mortality. If the surplus were not har-

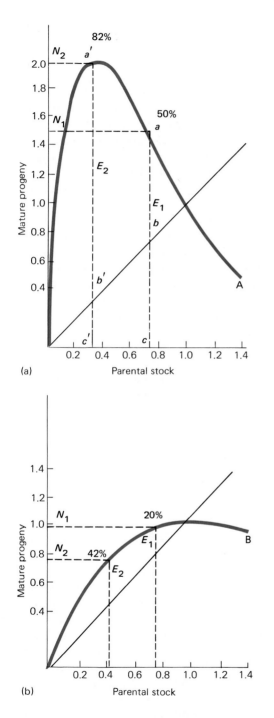

(a)

(b)

Figure 21.7 Reproduction curves illustrating rates of exploitation. Reproduction curves diagram the relationship between recruitment or net reproduction considered as mature progeny and the density of parental stock. The 45° line represents the replacement level of the stock, reproduction in which density dependence is absent. The dome-shaped curve is the plot of actual recruitment in relation to size of parental stock. The apex of the curve, lying above and to the left of the diagonal line, represents maximum replacement reproduction. The curve must cut the diagonal line at least once and usually only once. Where the curve and diagonal line intersect is the point at which parents are producing just enough progeny to replace current losses from reproductive units.

There are innumerable types of reproduction curves. The reproduction curve typical of r-strategists suggests that low parental stock can be very productive. On curve A a perpendicular line ac cuts the 45° at b. Segment ab (E) is harvest; bc is stock left for recruitment. Line ac represents the point on the curve at which 50 percent of the mature population is harvested each period. The line $a'c'$ is the 82 percent point, which represents the maximum surplus reproduction and the maximum rate of exploitation possible for this population. Curve B is typical of K-strategists, among which a low density of parental stock is not very productive. The two points represent levels at which 20 and 42 percent, the maximum, can be harvested.

Note the difference between curves A and B. In Curve A, $E_2 < E_1$ and $N_2 < N_1$. Under these conditions the greater the standing crop, the greater is sustained yield. In curve A, where $E_2 < E_1$ yet $N_1 < N_2$, a high standing crop does not result in greater sustained yield. A knowledge of parent-progeny relations is essential for exploitation of natural animal populations.

vested, the animals would succumb to disease, exposure, and the like. Thus for each individual removed by hunting natural mortality is reduced by one. It is assumed that population stability and sustained yield can be maintained if the population is harvested at the rate represented by the percentage of the young of the year or in some cases percentage of year-old animals in the population.

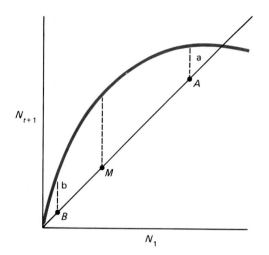

Figure 21.8 A reproduction diagram of an idealized population of *K*-strategists harvested for sustained yield under three regimes. In the first the population is harvested down from a steady state to a size $N_t = A$. The dashed line *a* represents the number that must be harvested or replaced each year to hold the population size stable at $N_t = A$. In the second region the population is reduced to $N_t = B$ and a sustained yield represented by line *b* could be harvested each year to hold the population stable at $N_t = B$. The yield in this case is a large proportion of a small population. Maximum sustained yield, *M*, is obtained at the population size where the diagonal line and the curve have maximum separation, $N_t = M$. (From Caughley 1976.)

The weakness of this assumption is that if hunting mortality and natural mortality replace each other and if the rate of harvest applied to the population equals the rate of mortality in the absence of hunting, then the only cause of death in the post-young population is hunting. Similarly, if the nonhunted population is below carrying capacity, no adult dies until *K* is reached (Caughley 1976). Obviously other mortality does take place among some hunted populations, and hunting and natural mortality are not compensatory. Part of the hunting mortality may be an addition to rather than a replacement of nonhunting mortality, as seems to be the case in some waterfowl populations (Geis et al. 1971).

Problems

Although sustained yield management (SYM), particularly multiple sustained yield (MSY), looks good on paper, it depends too heavily on the simplistic logistic equation. The problem is that SYM views management as a numbers game. The number or biomass removed theoretically is replaced by an equal number of new recruits to the harvestable component. Under SYM the "surplus," based on density compensation, supposedly is reduced; future production is optimized by optimizing harvest. Thus SYM assumes, as the models presented indicate, that a relationship exists between stock and recruitment. No proof of this theory exists. The usual approach to SYM fails to consider adequately size and age classes, differential rates of growth among them, sex ratio, survival, reproduction, and environmental uncertainties, all data difficult to obtain. Added to the problem is the common property nature of the resource. Attempting MSY without such information is balancing the population on the edge of catastrophe (see Figure 21.8). Overestimate the potential harvest in any one year and the population is headed for a sharp decline.

Several approaches to management of exploitable populations are in current use. One is the *fixed quota*, in which a certain percentage is removed each harvest period based on MSY estimates. Harvesting matches recruitment. Often used in fisheries, such an approach is risky because a fixed quota can drive a population to commercial if not actual extinction. Because populations fluctuate from harvest period to harvest period, the MSY will vary from year to year. If such fluctuations are not taken into account, there will be times of overharvest. Overharvest combined with environmental changes have been responsible for the demise of some fisheries, such as the Pacific sardine.

A second approach is *harvest effort*, often used in establishing seasons for sport hunting and fishing. The number of animals killed is manipulated by controlling hunting effort: the number of hunters in the field, the number of days of hunting (season length), and the size of the bag limit. To reduce the kill, hunting effort is decreased by reducing the bag limit, shortening the season, or closing the season entirely. The reverse approach is used to increase the kill. In general such a rule-of-thumb approach has been more successful in managing exploitable populations than the fixed quota approach.

The permit system is a special approach used to control more tightly the number of animals killed. It is based on the maximum number of animals to be removed from any one defined area and the amount of hunting effort needed to achieve that goal. For example, it is well established that approximately 15 percent of deer hunters are successful in any one hunting season. To achieve a desired level of kill the number of deer to be removed from any one area is used as the expansion ratio to set the number of permits issued. For example, if 100 animals is the desired harvest level, then approximately 700 permits should be issued.

A third approach is the *dynamic pool model*. It is essentially the one described in the opening part of this section. It assumes a constant natural mortality rate that is independent of density and is the same for all age classes. Growth rates are age-specific but unrelated to density. Fishing take replaces and is not additive to density-related natural mortality, which most fishery biologists believe is concentrated in the early life stages of the fish. The flaws in such an approach should be obvious, based on previous discussions of population dynamics. Fishing mortality can be additive to natural mortality; growth rates are affected by population density, as is recruitment to the population. Fishery biologists have no proof that natural mortality is concentrated only in the early life stages. In practice, the dynamic pool model translates fishing mortality into fishing effort, based on type of equipment such as size-selective gill nets that sort out age classes, efficiency of equipment, and the seasonal nature of exploitation. Few dynamic pool models have been developed, let alone put into practice. A general weakness of the model is the inability to estimate natural mortality accurately.

All of the three models are biological, are based on logistic growth, and have a major flaw. They fail to incorporate the most important component of population exploitation, economics (Figure 21.9) (see Hall et al. 1986; Walters 1986). Once exploitation of a natural resource becomes a commercial enterprise, the pressure is on to increase that exploitation to maintain the underlying economic infrastructure built upon the exploitable population. Once in place, any attempts to reduce the rate of exploitation meet strong opposition, supported by arguments that reduction will result in unemployment and industrial bankruptcy. This, of course, is a short-term argument, because in the long run the resource will be depleted and the economic collapse will occur. That fact is written across the country in abandoned fishery processing plants, rusting fishing fleets, and deserted logging towns. With conservative exploitation on lower economic and biological scales, the resource could still be exploited. Because of the failure of harvest regulations, nationally and internationally, and the common nature of the resource, exploitation efforts increase, even in the face of declining

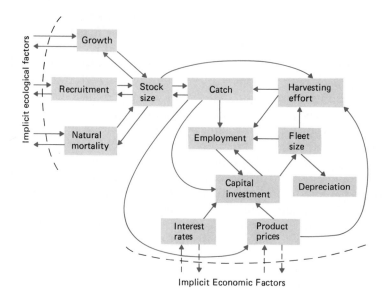

Figure 21.9 A model showing the interrelationship between sustained yield models and economic models related to harvesting. Sustained yield models ignore economic factors, which have a great impact. (From Walters 1984: 37.)

stocks. Instead of being reduced, harvest effort is increased by technological improvements in finding and harvesting the remaining resource to the point where it collapses.

An example is the whaling industry (Figure 21.10). Early whalers with hand lances and harpoons sought only those whales that could be overtaken and killed from small boats. As whales became scarce along the shore, whalers took to the high seas. In the sixteenth century whalers hunted off the Newfoundland coast and Iceland until stocks failed. The populations of Spitzenbergen and Davis Strait were next to go. The colorful New England whaling industry, which first exploited the stock of right whales, peaked in the first half of the nineteenth century, but as stocks of slow-moving, easily exploited species were depleted and as pe-

troleum replaced whale oil as fuel, the New England whaling industry died. Then two developments put international whaling into business. One was the invention of the explosive harpoon in 1865, and the other was the development of more powerful, faster boats that could overtake the swifter whales. The revitalized industry began to concentrate on the plankton-feeding blue whale and its relatives the fin and the sei whales. Again stocks were overexploited. Blue whale fishery in Norway ended in 1904, followed by a decline of the species off Iceland, the Faeroes, the Shetlands, the Hebrides, and Ireland. When these areas failed, whalers sailing free-ranging factory ships turned to the Antarctic and the Falklands to concentrate on smaller fin whales. The catches rose and the stocks again collapsed. Antarctic whaling was over and most of the whaling nations were out of business. Japan and the USSR developed a whaling industry and hunted the sei, sperm, and other whales wherever they could be found. Investments were made in large factory ships equipped with heli-

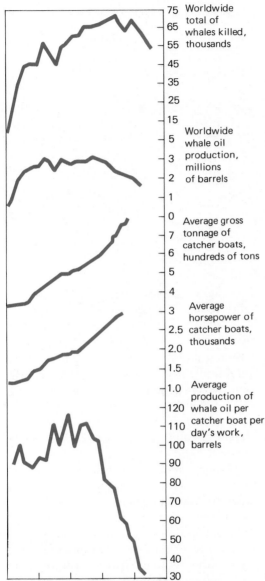

75 Worldwide
65 total of
55 whales killed,
   thousands
45
35
25
15
5 Worldwide
3 whale oil
  production,
2 millions
  of barrels
1
0 Average gross
7 tonnage of
  catcher boats,
6 hundreds of tons
5
4
3 Average
2.5 horsepower of
    catcher boats,
2.0 thousands
1.5
1.0 Average
    production of
120 whale oil per
110 catcher boat per
    day's work,
100 barrels
90
80
70
60
50
40
30

1945 1950 1955 1960 1965 1970

Figure 21.10 Relationship between economics, intensity of harvesting, and the declining stock of whales. These graphs pick up the data on the whaling industry following its resurgence after World War II in 1945. From 1945 on more and more whales were killed to produce less and less oil. At the same time boats became larger and more powerful, but their efficiency dropped greatly. Today by international agreement whaling has virtually ceased. (From D. C. Payne 1968.)

copters and accompanied by catcher boats that captured more whales but with less oil per effort and per ship.

Populations being overexploited exhibit certain easily discernible symptoms (K. E. F. Watt 1968). Up to a certain rate of exploitation, the stock is able to replace itself. Beyond this critical point certain changes point to impending disaster. Exploiters experience decreased catch per unit effort as well as a decreasing catch of one species relative to the catch of related species. There is a decreasing proportion of females pregnant, due both to sparse populations and to a high proportion of young nonreproducing animals. The species fails to increase its numbers rapidly after harvest. A change in productivity relative to age and age-specific survival shows that the ability of the population to replace harvested individuals has been impaired. An outstanding example is the blue whale (Figure 21.11). After 1860 the blue whale became the most important commercial species. Catches peaked in 1931 at 150,000 animals and declined to 1000 to 2000 in 1963. For the past 40 years the average age of the blue whale caught in the Antarctic has been six years, mostly immature females or females carrying their first calf. The species is near extinction.

An economic attitude also prevails in the management of some game species. In too many instances biologists have emphasized the increase of recreational opportunities for hunters rather than the welfare of the species involved. This bias is evident in the failure to reduce waterfowl bag limits when the population levels seem to require it, in late-season hunting of grouse well after breeding males have established their spring territories, and in the reluctance to restrict hunting of moose in Alaska. Hunting mortality is additive to predatory losses. Rather than re-

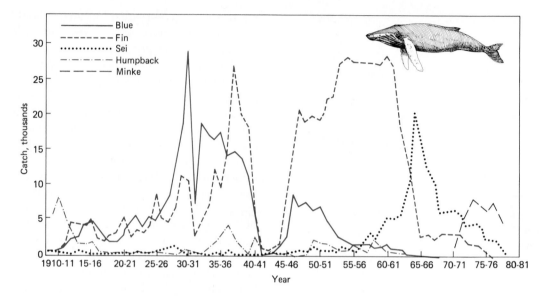

Figure 21.11 Catches of whales in the Southern Hemisphere, 1910–1981, the focal point of whaling after stocks in the Northern Hemisphere had been depleted. Note the virtual cessation of whaling during World War II, 1941–1945. The precipitous decline in the blue whale began before 1940. After World War II the fin whale bore the heaviest exploitation, but for a while blue whale take increased. The increase in harvesting intensity on the blue whale points out a truism in resource exploitation. A high harvesting effort on a declining stock can continue if some alternative resource is abundant enough to support that effort. The stock of fin whales supported the incidental harvesting of blue whales. Then the story was repeated with the fin whale. As that stock declined, whalers turned to the smaller sei and minke whales in an effort to maintain their investment in boats and equipment. Finally the whaling industry collapsed, but not until whales were close to the point of extinction. (Data from International Whaling Commission.)

strict human exploitation of moose populations, the approach is to kill wolves to reduce natural predatory loss (see Gasaway et al. 1983).

Reluctance to reduce seasons or to tighten bag limits relates in part to the economics of hunting. Most wildlife programs, directly or indirectly, depend upon hunting license revenue. Loss of such revenue reduces income to wildlife agencies. This reduction has a two-edged effect. It may reduce pressure on some hunted species, but it also reduces money to support wildlife habitat restoration and acquisition, which benefits all species, including endangered ones, and other programs essential to wildlife welfare. Obviously, a source of funding unrelated to hunting is needed.

Management of exploitable populations is too much crisis management. No steps are taken to rescue a species until its population has fallen so low that the species becomes endangered. Then it is protected and expensive programs are initiated with the hope that it will recover. The simpler solution is to manage populations more judiciously in the first place.

■ Summary

Herbivore-carnivore systems involve interactions between the second and third trophic levels that influence the fitness of both predator and prey. In response to selective pressures of predation, prey species have evolved measures of defense. Chemical defense involves a distasteful or toxic secretion that repels, warns, or inhibits a would-be attacker. Cryptic coloration and behavioral patterns enable the prey to escape detection or inhibit predators. Warning coloration signals to a predator that the intended prey is distasteful or disagreeable in some manner. Usually the predator has to experience at least one encounter with such a prey to learn the significance of a color pattern. Some palatable species will mimic unpalatable species, thus acquiring some protection from predators. Although warning and cryptic coloration and mimicry are usually associated with prey species, such mechanisms are also employed by predators to increase hunting efficiency.

Another form of defense is predator satiation. Reproduction is so timed that offspring are so abundant that predators can take only a fraction of the very young, leaving a number to escape by growing to a size too large for the predator to handle.

Predation may stabilize prey populations, but more often the relationship results in unstable fluctuations. Although ten-year predator-prey cycles typified by the snowshoe hare-lynx cycle were once regarded as driven by predation, the actual situation appears to be a vegetation-herbivore cycle that is tracked by the predator.

A unique form of predation is cannibalism, in which predator and prey are the same species. Often associated with stressed populations, cannibalism can result in pronounced demographic effects within a population, including the loss of younger age classes and lowered reproduction. It can become an important form of population control.

Commercial and sport hunting represents predation by humans on natural populations. Some attempt is made at sustained yield, in which the yield per unit time equals production per unit time. Such an approach to management of exploited populations necessarily differs between K-selected and r-selected species. Based on the logistic equation, sustained yield models fail to take into consideration all aspects of population dynamics, including natural mortality and environmental uncertainty. Also ignored is the economics of population exploitation. Economic pressures on a common resource encourage overexploitation, resulting in the commercial and even biological extinction of species.

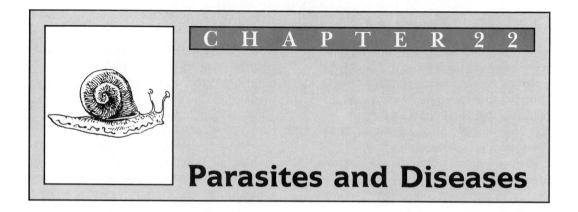

CHAPTER 22

Parasites and Diseases

Although parasites and pathogens are extremely important in interspecific relationships, ecologists have not given them enough attention. The effects of parasites and disease are most pronounced in host populations that have not evolved defenses against them. Then diseases sweep through and decimate the population. Examples are the outbreaks of rinderpest in African ungulates, myxomatosis in European rabbits, and blight in the American chestnut. In most natural populations, though, living in association with normal parasites and diseases, the effects of parasitism are elusive. Bodies of victims that die are quickly processed by detrivores and scavengers, eliminating traces of disease.

Aldo Leopold, in his classic book *Game Management* (1933), remarked that the role of disease in wildlife conservation has been radically underestimated, that it controls both predators and prey, that it may delimit the range of some species, limit population density, and become involved in population fluctuations. "Disease," he wrote, "does not yield to observational methods of study." Understanding begins only when field observations are combined with experimental and laboratory techniques. Recent years have seen advances in theoretical and experimental studies of disease in natural populations.

■ Definitions

Parasitism is a condition in which two organisms live together but one derives its nourishment at the expense of another. Parasites, strictly speaking, draw nourishment from the tissues of their larger hosts, a case of the weak attacking the strong. Typically parasites do not kill their hosts as predators do, although the host may die from secondary infection or suffer from stunted growth, emaciation, or sterility. Some parasitic larvae of insects draw nourishment from the tissues of their hosts and by the time of metamorphosis have completely consumed the soft tissues of the host. These parasites, known as *parasitoids*, essentially act as predators (Chapter 21).

Parasites include viruses, many bacteria, fungi, and an array of invertebrate taxonomic groups, including the arthropods. The presence of a heavy load of parasites is considered an infection, and the outcome of an infection is a disease. A *disease* is any change in the physical condition of an individual plant or animal that deviates from normal. Not all parasites are agents of disease. Many are mutualistic (see Chapter 23) and essential to the well-being of other organisms. Parasitic bacteria associated with digestion in the rumen of ungulates and the gut of termites are essential for the well-being of those herbivores.

Parasites exhibit a tremendous diversity in ways and adaptations to exploit their hosts (for an overview, see Croll 1966). Parasites may be plants or animals,

and they may parasitize plants or animals or both. They may live on the outside of the host (ectoparasites) or within its body (endoparasites). Some are full-time parasites, others only part-time. Part-time parasites may be parasitic as adults and free-living as larvae or the reverse. Parasites have developed numerous ways to gain entrance to their hosts, even to the point of using several hosts as dispersal agents. They have evolved various means and degrees of mobility, ranging from free-swimming ciliated forms to ones totally dependent upon other organisms for transport. They have developed diverse ways of securing themselves to the host to maintain position and means of counteracting the biochemical hazards of living inside a host. Parasites may be restricted to one or a limited number of species or genera of host. A number of parasites of birds, especially certain tapeworms, live only in one particular order or genera (see Baer 1951). Some parasites live their entire life cycle on one host, whereas others require more than one host.

The usual approach to typing parasites has been taxonomic: tapeworms, roundworms, and the like. More recently May and Anderson (1979) have suggested that parasites be distinguished on the basis of size, as microparasites and macroparasites. *Microparasites* include the viruses, bacteria, and protozoans. They are characterized by small size and a short generation time. They develop and multiply rapidly within the host and tend to induce immunity to reinfection in hosts that survive initial infections. The duration of the infection is short relative to the expected life span of the host. Transmission from host to host is direct, although they may involve some other species as a vector.

Macroparasites are relatively large in size and include parasitic worms, the platyhelminths, acanthocephalans, roundworms, flukes, lice, fleas, ticks, mites, fungi, rusts, smuts, and parasitic plants such as dodders, broomrape, and mistletoe. Macroparasites have a comparatively long generation time, and direct multiplication in the host is rare. The immune response macroparasites stimulate is of short duration and depends upon the number of parasites in the host. Macroparasites are persistent with continual reinfection. They may spread by direct transmission from host to host or by indirect transmission, involving intermediate hosts and vectors.

■ Hosts as Habitat

Hosts are homes for parasites, and parasites have exploited every conceivable habitat on and within them. Parasites live on the skin within the protective cover of feathers and hair. Some burrow beneath the skin. They live in the bloodstream, in the heart, brain, digestive tract, liver, spleen, mucosal lining of the stomach, spinal cord, and brain, in nasal tracts and lungs, in the gonads and in the bladder, pancreas, eyes, gills of fish, muscle tissue, to mention some sites among many. Parasites of insects live on the legs, on upper and lower body surfaces, and even on the mouthparts.

Like species of plants and animals in terrestrial and aquatic habitats, parasites within the host will colonize different sites in organ systems. Coccidian protozoans of the genus *Elmeria* occupy different regions: one in the duodenum, another in the lower duodenum and upper small intestine, a third in the lower small intestine, and a fourth in the caecum and rectum. There are similar divisions in the sharing of a particular organ system by closely related parasites in many animals.

Plant parasites, too, divide up the habitat. Some live on the roots and stems; others penetrate the roots and bark to live in the woody tissue beneath. Some live at the root collar where the plant emerges from the soil. Others live within the leaves, on young leaves, on mature leaves, on flowers, pollen, and fruits.

A major problem for parasites, especially parasites of animals, is gaining access to and escaping from the host. Parasites of the alimentary tract enter the host orally and escape through the rectal route, a path used by other parasites as well. Parasites of the lungs enter orally or penetrate the skin and travel to the lungs by the pulmonary system. They escape mainly by being coughed up and swallowed in the alimentary tract. Liver parasites, exploiting one of the richest habitats in the animal body, arrive there by way of the circulatory system, bile duct, and hepatic portal systems and escape through the same routes. Parasites of the urogenital system enter orally and travel through the gut to the site of residency and exit by the urinary system. Blood parasites enter and escape by way of the skin, but always with the aid of some obliging vector such as mosquitos and ticks. Parasites that end up in the muscle tissues, where they usually exist in capsules, reach a blind end. For them the only way out is for their host to be killed and consumed by a predator.

■ Life Cycles

Parasites face unique problems moving from host to host. Many parasites have evolved complex life cycles geared to escape and relocation. For most parasites the hosts function as islands from which they must escape to locate other host islands, some of which are occupied and some of which are not. For animal parasites the host "islands" are movable.

Parasites cannot move from host to host at will. They can escape only during a transmission stage, whether the means of dispersal is direct or indirect; and successful transmission depends upon contact between the host and the infective stage. The infective stage is essential.

All parasites reach a stage in their life cycle within the host when they can develop no further. The *definitive host* is the one in which the parasite becomes an adult and reaches maturity. All others are *intermediate hosts,* which harbor some developmental phase. Parasites may require one, two, or even three intermediate hosts. Each infective stage can develop only when it is independent of the definitive host, and it can continue its development only if it can find another intermediate or its definitive host. For this reason, many parasites employ animals as intermediate hosts to aid in the location of a definitive host or adapt to the host's habits. Thus the population dynamics of a parasite population is closely tied to the population dynamics of the host.

Direct Transmission

Direct transmission is the transfer of the parasite from one host to another by direct contact or through a carrier or vector. The life cycle of the parasite does not involve intermediate stages of development in a secondary host. Most microparasites are transmitted directly. One important viral disease of wild mammals and humans is rabies, transmitted through the saliva by the bite of a rabid animal. The rabies virus follows the nerves from the infection point to the spinal column and brain before symptoms appear. The symptoms are both behavioral and physical. Infected animals are excitable and restless, exhibit convulsions, wander aimlessly, and show no fear of humans. Physically they are emaciated, exhausted, and partially paralyzed. Among wild animals, rabies occurs most commonly in coyotes, foxes, skunks, raccoons, and bats (Figure 22.1). Foxes and dogs are the primary transmitters of the disease.

Microparasites that infect plants also involve transmission through direct contact with the virus or resting spores in the soil. In the presence of suitable hosts these spores germinate and penetrate the roots. Wind-carried spores come to rest on and infect leaves of the plant. Others are carried by insect vectors. One example is elm phloem necrosis caused by a mycoplasmal organism. It is transmitted by root grafts or by the widely distributed white-banded elm leafhopper *(Scaphoideus luteolus).* This insect ingests the mycoplasma while feeding on the sap of a diseased tree and transmits it to a new host tree.

Many important macroparasites of animals and plants move directly from infected to uninfected host by direct contact. Parasitic nematodes *(Ascaris)* live in the digestive tracts of mammals. Female roundworms lay thousands of eggs in the gut of the host, which are then expelled with the feces. If they are swallowed by the host of the correct species, the eggs hatch in the intestines of the host, bore their way into the blood vessels, and come to rest in the lungs. From here they ascend to the mouth, usually by causing the host to cough, and are swallowed again to reach the stomach, where they mature and enter the intestines.

Nematodes are important macroparasites of plants. The female of one species, the eelworm, *Heterodera schachti,* lives a parasitic existence in the roots of such

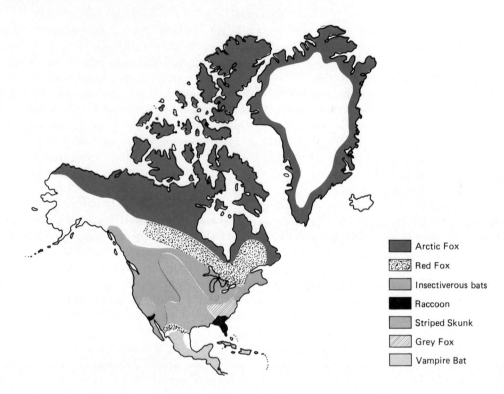

Arctic Fox
Red Fox
Insectiverous bats
Raccoon
Striped Skunk
Grey Fox
Vampire Bat

Figure 22.1 The distribution of wildlife rabies in North America, emphasizing the number of vector species. In Europe the only wildlife vector is the fox. In South and Central America the major vector is the vampire bat. (From Steck 1982: 58.)

plants as tomatoes and beets. Fertilized by the free-living male, the female sends thousands of larvae into the soil, ready to take up a parasitic existence in other plants.

Most important external and often debilitating parasites of birds and mammals are spread by direct contact. They include lice, mites that cause mange in mammals, particularly the canids, ticks, fleas, and botfly larvae. Many of these parasites lay their eggs directly on the host, but fleas lay their eggs and their larvae hatch in the nests and bedding of the host (even in the shag rugs of homes with dogs), and eventually leap onto nearby hosts.

Some fungal parasites of plants spread through root grafts, and others are carried by insect vectors. For example, an important fungal infection of white pine (*Pinus strobus*), *Fomes annosus*, spreads rapidly through pure stands of the tree by root graphs (a situation in which roots of one tree become grafted onto the roots of a neighbor). Oak wilt, a disease caused by the fungus *Ceratocystis fagacearium*, is spread both by root grafts and by nitidulid beetles, especially *Glischorochilus* and *Coleopterus* (True et al. 1960). The devastating Dutch elm disease caused by the introduced fungus *Ceratocystis ulmi* is spread from tree to tree by spore-carrying elm beetles *Scolytus multistriatus* and *Hylurgopinus rufipes*.

Plant macroparasites include a number of flowering plants. One group are *holoparasites*, plants that lack chlorophyll and draw their water, nutrients, and carbon from the roots of host plants. Notable among them are members of the broom-rape family, Orobanchaceae. Two are the familiar squawroot (*Conopholis americana*), which parasitizes the roots of oaks, and beechdrops (*Epifagus virginiana*), which parasitizes mostly beech trees. Another common plant macroparasite is dodder (*Cuscuta gronovii*), a member of the morning glory family. The seeds of dodder germinate in the soil, but when the seedlings attach themselves to plants and entwine about their hosts, dodders' roots degenerate and their contact with the ground ceases.

Another group are the *hemiparasites*. They are photosynthetic, but they draw water and nutrients from their host plant. The most familiar hemiparasites are mistletoes, whose sticky seeds attached to limbs (see Chapter 23) send out roots that embrace the limb and send roots into the sapwood. Mistletoe can reduce growth of its host.

Indirect Transmission

Many parasites, both plant and animal, utilize indirect transmission involving different hosts during different stages of the life cycle. The brainworm (*Parelaphostrongylus tenuis*) (in spite of its name the parasite is a lungworm) of the white-tailed deer has as its intermediate host during its larval stage a snail or slug that lives in the grass (Figure 22.2) (Anderson 1963, 1965). The deer picks up the infected snail while grazing. In the deer's stomach the larvae leave the snail, puncture the deer's stomach wall, enter the abdominal membranes, and by the way of the spinal cord reach spaces surrounding the brain. Here the worms mate and produce eggs. Eggs and larvae pass through the bloodstream to the lungs, where the larvae break into the air sacs, are coughed up,

swallowed, and passed out through the feces. The larvae are ingested by the snail, where they continue to develop to the infective stage.

Other parasites involve two intermediate hosts. An example is the black grub or black spot infection of minnows and sunfish (Figure 22.3), caused by a small white larva of a fluke, *Uvulifer ambloplitis*. Once infected, the fish host lays down a black pigment around the thick-walled cysts, causing black spots on the body wall. The adult stage is attached by suckers to the mucosa of the intestine of the primary host, the kingfisher. The eggs pass through the intestine of the kingfisher into the water, where the miracidia hatch. The miracidia, moving through the water by cilia, seek an intermediate host, a snail, to which they gain entrance by secreting a tissue-dissolving substance. Within the snail, the myracidia transform into saclike sporocysts, which eventually give rise to tailed larvae called cercaria. These free-swimming larvae emerge from the snail and seek the next intermediate host, a fish, which they must find within two days or die. The cercaria penetrate the fish and encyst. The fish, in turn, is eaten by the kingfisher; the cercaria become free and mature in the bird's intestinal tract. The new adults start laying eggs to repeat the cycle.

Indirect transmission among plant macroparasites is uncommon except among the rusts, which employ the wind to carry the infective stages from primary and intermediate hosts. An example is white pine blister rust (*Cronartium ribicola*) (Figure 22.4). The rust produces spores in the diseased bark of pine in the spring. These spores, which cannot reinfect pines, are carried by the wind. Some may land on species of *Ribes* or gooseberries. On the infected leaves of *Ribes* the rust produces two kinds of spores, early and late sum-

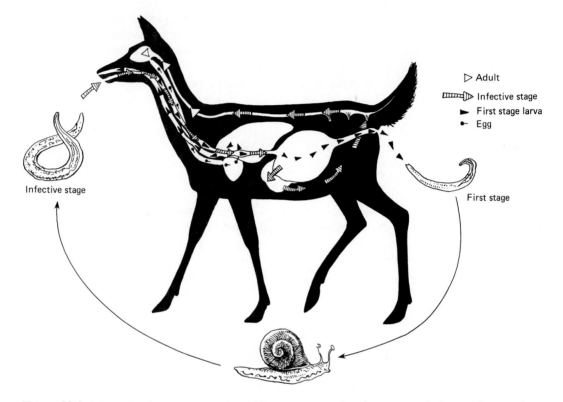

Figure 22.2 Life cycle of a macroparasite with indirect transmission, the brainworm (actually a lungworm) *Parelaphostrongylus tenuis* in white-tailed deer (but transmissible to moose and elk). Adult worms in the definitive host, the white-tailed deer, release eggs into the venous circulation. The eggs reach the lungs as emboli (undissolved material). In the lungs the eggs develop into first-stage larvae and hatch. The larvae move into air spaces in the lungs and pass up the bronchial tubes, are swallowed, and pass in the feces. In the external environment the first-stage larvae invade the foot of terrestrial snails as they move across the feces. The larvae grow in the snail, molt twice, and give rise to the third and infective stage. Deer acquire the third stage by ingesting the infected snails on vegetation. In the alimentary canal the larvae leave the tissues of the snail and penetrate the gastrointestinal wall, cross the peritoneal cavity, and follow the lumbar and other nerves to the vertebral canal, a journey of about ten days. The migrating third-stage larvae enter the dorsal horns of the gray matter of the brain, where they develop for about a month. Forty days after infection, the subadult worms enter the spinal subdural space. Here they mature and migrate anteriorly to the cranium, where they deposit eggs in the venous circulatory system. Some females may lay eggs in the brain, and the eggs develop to the first stage in the meninges. (After Strelive in Davidson et al. 1981: 141.)

mer forms. The early summer form infects other *Ribes* only, intensifying infection on these plants during the summer. Late summer spores infect only white pine needles. Carried by the wind, the spores germinate on the needles and grow until they reach the bark. There the rust produces spindle-shaped diseased areas, or cankers. Two to five years after infection and annually thereafter, spores produced in the bark are carried by the wind to nearby *Ribes*. Like other parasites that involve intermediate hosts, conditions for transmission are critical. They include a

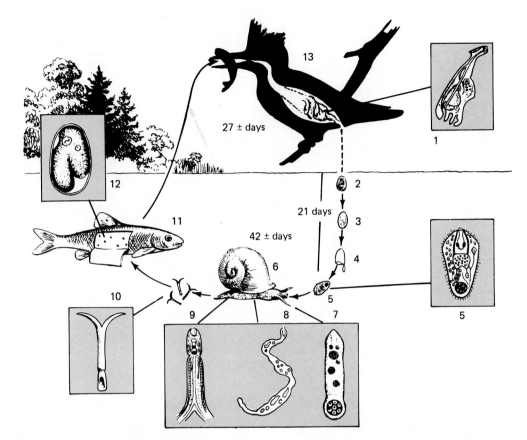

Figure 22.3 Life cycle of the black grub *(Uvulifer ambloplitis)*. (1) Adult trematode attached to the mucosa of the kingfisher intestine; (2) immature egg; (3) mature egg; (4) empty shell; (5) miracirdium; (6) the first intermediate host, the snail; (7) and (8) sporocysts; (9) cercaria; (10) free-swimming cercaria, which penetrate beneath the scales or into the fins and encyst, producing black spots in about 22 days; (11) minnow with skin cut away to show black grubs; (12) sketch of parasite within inner cyst; (13) kingfisher. (Adapted from Hunter and Hunter 1934, New York State Conservation Department.)

widespread infection of *Ribes,* an abundant production of pine-infecting spores, favorable temperature, moisture, and wind conditions, and hosts within 280 meters of the infected *Ribes,* the maximum wind-carried distance of spores on the wind. The disease has been controlled by eliminating *Ribes* within and near white pine stands.

Dynamics of Transmission

Transmission from host to host is the key to parasitic existence. It can take place only with the dispersal of an infective stage independent of the definitive host. Parasites requiring more than one host reach only a certain stage in their life cycle in each and they can complete the cycle only if they can infect another host. The sexual form of the malarial parasite cannot continue its existence in the mosquito. It must transfer to a vertebrate host to develop into the adult stage. The brainworm of deer has to locate a snail as an inter-

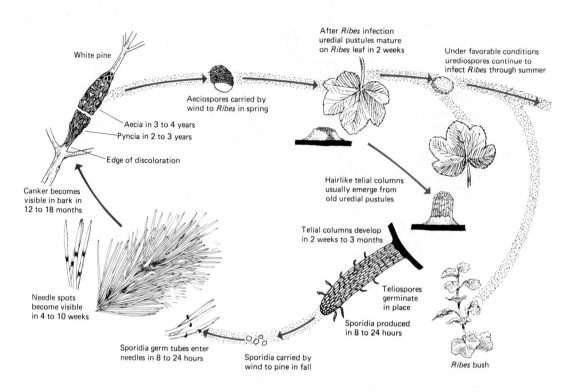

White pine

Aecia in 3 to 4 years
Pyncia in 2 to 3 years

Edge of discoloration

Canker becomes
visible in bark in
12 to 18 months

Needle spots
become visible
in 4 to 10 weeks

Sporidia germ tubes enter
needles in 8 to 24 hours

Aeciospores carried by
wind to *Ribes* in spring

Sporidia carried by
wind to pine in fall

After *Ribes* infection
uredial pustules mature
on *Ribes* leaf in 2 weeks

Under favorable conditions
urediospores continue to
infect *Ribes* through summer

Hairlike telial columns
usually emerge from
old uredial pustules

Telial columns develop
in 2 weeks to 3 months

Teliospores
germinate
in place

Sporidia produced
in 8 to 24 hours

Ribes bush

Figure 22.4 Life cycle of the white pine blister rust *(Cronartium ribicola)*. The rust infection on *Ribes* appears as yellowish to orange pustules of spores. An annual infection, it has little effect on *Ribes*. On the white pine the infection is perennial and appears as large orangish blisterlike cankers on the bark.

mediate host to continue its development to an infective stage that can be transmitted back to the deer. Even more comple‧ is the fluke infecting the kingfisher and other fish-eating birds. It has to develop three infective stages. Each stage has to be uniquely adapted to its host. It has to be able to access the host, overcome the immune responses of the host, and escape from the host. For many animal parasites this process means exploiting the feeding habits of the definitive host and adapting to the habits of the intermediate hosts.

The brainworm of the deer infects a snail as an intermediate host and exploits the snail's habit of crawling up grass stems where it risks being eaten along with the grass by a grazing deer. Unless the snail is swallowed by the deer, the parasite will perish.

Transmission of parasites is further complicated by the patchy or clumped distribution of the hosts. Only a few members of the host population harbor major parasite loads and act as reservoirs of the infection. Uninfected hosts are widely scattered or intermixed among populations of other species, so the probability of the parasite or its carrier coming in contact with susceptible individuals is low. In other words, transmission is highly dependent on both the density of and distance between parasite and potential hosts. Transmission of parasites is most successful when the population of potential hosts is dense, particularly if the parasites depend upon direct contacts among hosts. The spread of many viral and bacterial diseases is most rapid in dense populations, resulting in *epizootics* in animal populations (termed *epidemics* in humans). Such epizootics involve virulent forms that sweep through susceptible populations as an advancing front, exemplified by the spread of rabies through Europe (Figure 22.5).

Some parasites involve predator-prey relationships. The tapeworm *Echinococcus*

(Cestoda) lives in the intestines of carnivores. It spends its larval stage as hydatid cysts in the muscles and tissues of any organ in its intermediate hosts, herbivores, and under certain conditions humans closely associated with them. When the

Figure 22.5 The spread of rabies in Europe. Since originally diagnosed in Poland in 1939, it has spread westward over 1400 km of Europe at a rather constant rate of about 20–60 km per year. This epizootic developed independently of dogs and has spread across the continent with foxes as the principal vector. (From Steck 1982: 62.)

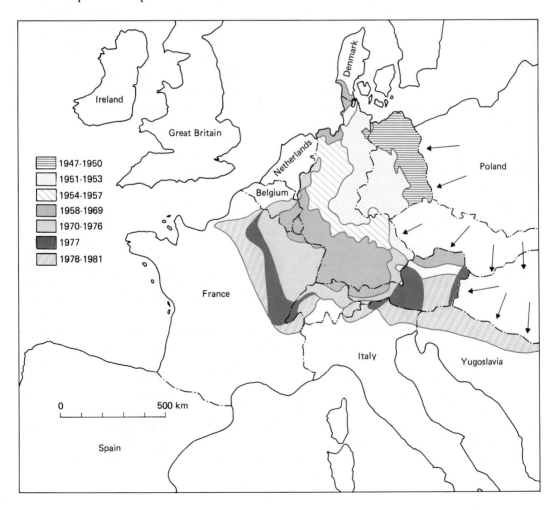

Ireland

Great Britain

| 1947-1950 |
| 1951-1953 |
| 1954-1957 |
| 1958-1969 |
| 1970-1976 |
| 1977 |
| 1978-1981 |

Denmark

Netherlands

Belgium

Poland

France

Italy

Yugoslavia

0   500 km

Spain

carnivore feeds on the flesh of herbivore prey, the hydatid cyst gains entrance to the digestive tracts of the carnivore and develops into adult tapeworms. One species, *E. granulosus*, the dog tapeworm, is found in dogs and wild canids, especially the wolf. A second species, *E. multilocularis*, has a restricted range in the northern regions of Asia, Europe, and North America. Its definitive host is the Arctic fox and its intermediate host several species of voles. Both species of tapeworm appear to change the behavior of its intermediate host to ensure transmission back to its determinate host. Wolves preferably select moose infected with hydatid larvae, which concentrate in their lungs; their presence may be indicated by some odor on the breath of the infected moose, by excretions, or by some aberrant behavioral activity. Voles heavily infected with hydatid larvae of *E. multilocularis*, which rapidly grow in and destroy the liver, appear large and fat, but they have difficulty moving and fall easy prey to the fox.

■ Host Response

Just as prey respond defensively to predators, so hosts show defensive reactions to the invasions of parasites. A first line of defense among animals is biochemical response. One response is inflammation, a generalized response to alien chemicals provoked by the death or destruction of host cells. The infection stimulates the secretion of histamines and increased blood flow to the site, bringing in phagocytes, lymphocytes, and leucocytes to attack the infection. Such reactions can result in scab formation on the skin by fibrosis, as in the case of heavy mange mite infestations on

red fox and other canids. Internal reactions can produce calcareous cysts in muscle or skin, which imprison the parasite. Examples are the cysts of the infective stage of the blackgrub fluke in its intermediate fish host (see Figure 21.3) and the cysts of the roundworm *Trichinella spiralis* (Nematoda), which cause trichinosis in humans, in the muscles of pigs and bears.

Some plants respond to bacterial and nematode infections in roots by forming cysts. Responses include scab formation in fruits and roots, cutting off contact of fungus with healthy tissue; root knot, a reaction to nematodes in tomatoes and other plants; black knot, a limb canker in black cherry *(Prunus serotina);* and root nodules in legumes.

The ultimate in biochemical response is the immune response of animals. Depending upon the parasite, it may produce short-term or lifelong immunity to the infection. When a foreign protein or antigen enters the bloodstream, it is taken up by lymphocytes, which produce a molecular template corresponding to the antigen. This template is known as an antibody. The antibodies in the bloodstream neutralize invading antigens.

The immune response, however, can be breached. Antibodies specific to an infection normally are composed of proteins. If the animal suffers from poor nutrition and its protein deficiency is severe, its normal production of antibodies is inhibited. This depletion of energy reserves breaks down the immune system and allows the virus and other parasites to become pathogenic. The ultimate breakdown in the immune system occurs in humans infected with the human immunodeficiency virus (HIV), the causal agent of the sexually transmitted AIDS. The virus attacks the immune system itself, exposing the host to a range of infections that prove fatal.

The host may also react with abnormal growth. Plants respond to the invasion of plant tissues—leaf, stem, and fruit or seed—by gall wasps, bees, and flies by forming abnormal growth structures unique to the particular gall insect involved (Figure 22.6). Production of galls does not benefit the plant. In fact, a heavy infestation of gall insects, particularly leaf galls, can weaken and ultimately kill the plant. However, gall formation can expose the larvae of some galls to predation. The conspicuous, swollen knobs of the goldenrod stem gall attract the downy woodpecker, who excavates and eats the larva within.

Malarial parasites in vertebrates can stimulate the spleen into a high production of red blood cells and antibodies, resulting in its enlargement. Certain helminth parasites can produce cancerous tumors in their mammalian hosts. Larvae of botfly developing beneath the skin of mammals produce abnormal swellings on the skin known as warbles. These growths subside after the transformed larvae escape through a hole cut in the skin. Invertebrates, especially mollusks, respond to irritation by parasitic larvae by pearl formation. Pearl formation is usually associated with some inanimate foreign body, but often it is the response of an invertebrate host to a parasitic infection.

Parasitic infection can also result in castration, especially of some invertebrate hosts. Shore crabs (*Carcinus maenas*) infected with a cirripede parasite (*Sacculina*) may experience hormonal changes that prevent molting of the shell. These infected crabs may be distinguished by weeds and barnacles growing on the shell, retained long enough to support such

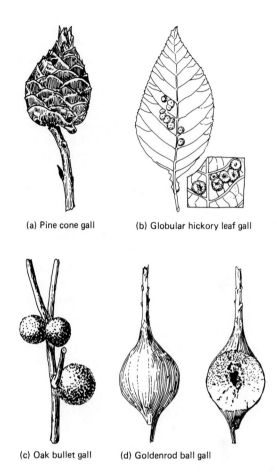

(a) Pine cone gall   (b) Globular hickory leaf gall

(c) Oak bullet gall   (d) Goldenrod ball gall

Figure 22.6 Galls represent a morphogenic response by plants to an alien substance in their tissues, in this case a parasitic egg. The response involves a genetic transformation of the host's cells. (a) The pine cone gall, a bud gall caused by the gall midge (*Rhabdophaga strobiloides*). (b) The hickory leaf gall, induced by a gall aphid (*Phylloxera canyaeglobuli*). (c) An oak fig root gall, a fleshy, fig-shaped white gall caused by a gall wasp (*Belanocnema treatae*). (d) The familiar goldenrod ball gall, a stem gall induced by a gall fly (*Eurosta solidagini*).

growth. More importantly, the parasite may grow into and destroy the gonads of male crabs, causing a sex reversal to females.

Parasitic infections can result in behavioral changes in the hosts. Rabbits infected with the bacterial disease tularemia *(Pasteurella tularensis),* transmitted by the rabbit tick *(Haemaphysalis leporis-paulstris),* are sluggish and difficult to move from cover. Rabid foxes may be overly aggressive or overly tame and unafraid of human contacts. Horsehair or gordiid worms (Nematomorpha) live in and lay their eggs in ponds and livestock troughs. The tiny motile armed larvae that hatch from the egg invade host insects, including wasps, bees, and beetles that visit the water. Approaching maturity, the larvae must return to water, and they can get there only if the insects carry them back. Parasites seemingly force their hosts to return to water, into which they often dive suicidally. Then the gordiid larvae burst out of the insect and swim free in the water to live brief lives as reproducing adults.

■ Population Dynamics

By now it should be apparent that interactions of parasites and hosts involve three or more populations. One is the host, the growth rate of which is influenced by intensity of parasitic infection. The second is the adult parasite in the definitive host, the growth rate of which is influenced by intraspecific competition within the host and the dispersion of parasites among the host individuals. The third is the free-living infective transmission stage, which experiences many of the same population limitations as the adult parasites. The growth of the adult population of the parasite in the definitive host is distinct from the production of the infective stage (or stages), which has its own birthrate and growth rate.

Models of Parasitism

Anderson and May (1978, 1979) and May and Anderson (1978, 1979) provide theoretical models for the growth of the three different populations. The assumptions are that parasites increase the rate of mortality of the host and that the rate of mortality is a function of the number of parasites per host. The basic models can be modified to consider direct and indirect transmission of parasites, parasites that reproduce within the definite host, time lags, and the like. The basic models are

Host: $dN/dt = (a - b)N - \alpha P$

Parasite: $dP/dt = \beta wN - (\mu + b + \alpha)P - \alpha(k - 1)P^2/kN$

Free-living infective stage:
$$dw/dt = \lambda P - cw - \beta wN$$

where

N = number of hosts
P = number of adult parasites
w = number of free-living infective stages
a = instantaneous host birthrate
b = instantaneous host death rate
α = instantaneous parasite-induced host death rate
β = rate of parasite transmission (hosts acquire individual parasites at a rate proportional to the number of contacts between host and parasitic infective stages βwN)
k = parasites distributed as a negative binomial (degree of overdispersion)
μ = natural mortality rate of adult parasites
λ = rate of production of infective stages by adult parasites
c = death rate of infective stages

The host model states that the growth of the host population is influenced by its own rate of increase reduced by parasite-caused mortality. The parasite model shows the relationship among the population size and transmissivity of the infective stage, reduction in adult parasite numbers caused by parasite-induced host mortality, natural host death rate, natural mortality of adult parasites, and parasite-induced host death rate as influenced by dispersion of intensity of infection of hosts by adult parasites. The infective stage model relates to production of infective or intermediate stages, which is decreased in part by loss of population due to invasion of the definitive host and transformation into adults.

Parasitism as a Regulatory Mechanism

The impact of parasites on host populations relates in part to the nature of the transmission and the density and dispersion of the host population. Microparasites, dependent for the most part on direct transmission, require a high host density to persist. For them ideal hosts are ones that live in groups or herds. To persist the parasites need a long-lived infective stage that does not ensure long-term immunity in the host population. Immunity reduces populations, if indeed it does not eliminate that particular parasite. An example of a parasite in wild populations that does not confer long-term immunity is the rabies virus; one that does confer immunity to animals that survive the disease is the distemper virus.

Indirect transmission, typical of macroparasites, is more complex. To persist the parasite requires a highly effective transmission stage, which often involves a close association with food webs. Parasites with indirect transmission exist at low population levels, but because of efficient transmissions they do well and persist for a long time in low density populations of hosts. The longevity of each parasitic stage varies in different hosts. Longevity is high in the definitive host, and much less in the intermediate hosts.

To regulate a host population, the parasite has to increase the host's death rate or decrease its reproductive capability. Ecologists know little about the regulatory effects of parasitism. They find it difficult to assess the regulatory effects of parasitic infections, let alone to determine the extent of infections in host populations. Infections, particularly of microparasites, are probably more widespread than is evident with a small reservoir of infected hosts in the population. When populations become dense and direct contact is frequent among individuals carrying the disease and susceptible hosts, disease breaks out and spreads rapidly through the population with high mortality.

Such effects are most evident when parasites are introduced into a population with no evolved defenses. In such cases the disease may be density-independent. It can reduce populations, exterminate them locally, or restrict the distribution of the host. For example, the fungus *Endothia parasitica* spread rapidly through the American chestnut (*Castanea dentata*), eliminating it as a commercial timber species and changing the composition of the eastern deciduous forest. Rinderpest, a viral disease whose natural host is cattle, spread swiftly through populations of native cattle, wildebeest, and African buffalo, decimating those species. It was introduced in Africa between 1884 and 1889 by cattle importations. After several periodic epizootics the disease was removed by vaccination of cattle. Rinderpest has had a pronounced effect on the ecology of the East African savanna ecosystem far beyond the ungulate species involved (see Sinclair 1977, 1979).

Another example is the effect of the accidental introduction of the night mosquito *Culex pipiens fatigans* to the Hawaiian Islands in 1826. The mosquitos spread rapidly throughout the lowland areas of the high islands and spread bird pox, malaria, and other diseases throughout the lowland bird populations. As a result several endemic species of birds disappeared from the lowlands to about a 600 m elevation (Warner 1968). A number of birds became extinct, and those of the species that still exist are restricted to regions above 600 m in spite of the availability of habitat at lower elevations. When they are carried to a lower elevation, they succumb to malaria or bird pox. Thus the high forests of the Hawaiian Islands are an ecological sanctuary from disease for the birds.

These are extreme cases and hardly regulatory. What about directly transmitted endemic diseases maintained in the population by a small reservoir of infected carrier individuals? Outbreaks of these diseases appear to occur when the density of the host population is high, and they tend to sharply reduce host populations. Examples are distemper in raccoons, which can be significant in controlling their populations (Gorham 1966; Haberman et al. 1958), and rabies in foxes. A more important regulatory parasite in foxes is a macroparasite, the mange mite *(Sarcoptes scabiei)*. Scarcoptic mange is highly density-dependent and is transmitted from fox to fox. Pups are usually infected in the den. Mange begins as flaking and cracking of the skin and loss of hair. Eventually the flaking skin areas develop into lesions, the skin becomes crusted and wrinkled, and mites invade the eyelids. Heavily infested foxes die. In New York

45 percent of red foxes found dead had mange (Tullar 1979), and up to 67 percent of red foxes in New Brunswick and Nova Scotia had mange mites (Smith 1978).

Do these parasitic infections in themselves result in density-dependent reductions of populations? They may be a proximate rather than the ultimate cause. An increasing population of a host may result in the depression of antibody formation and other body defenses, an increase in inflammatory processes, and an increased susceptibility to disease. Thus the disease may be a consequence of a high population rather than a cause of population decline.

The distribution of macroparasites, especially those with indirect transmission, is highly clumped or overdispersed (Figure 22.7) Some individuals in the host population carry a higher burden of parasites than others. These individuals are the ones that are most likely to succumb to parasite-induced mortality or suffer reduced reproductive rates, or both (Figure 22.8). Such parasitically induced deaths often are not caused directly by parasites but by some secondary infections. Herds of bighorn sheep *(Ovis canadensis)* may be infected with up to seven different species of lungworms (Nematoda) involving 100 percent of the animals. The level of infection varies with density of sheep and associated livestock, climate, range conditions, weather, and season. Highest infections occur in the spring, when the lambs are born. Young are often infected with lungworm as a result of transplacental migration of the third stage larvae in the fetal liver. After birth the nematodes migrate to the lungs. Heavy lungworm infestations in the young bring about a secondary infection, pneumonia, which kills the lambs. Such infections tend to stabilize or sharply reduce mountain sheep popu-

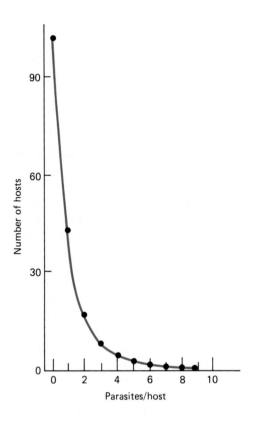

Figure 22.7 Overdispersion of the tick *Ixodes trianguliceps* (Birula) on a population of the European field mouse *Apodemus sylvaticus*. Most of the individuals in the host population carry no ticks and relatively few individuals carry most of the parasite load. (From Randolph 1975: 454.)

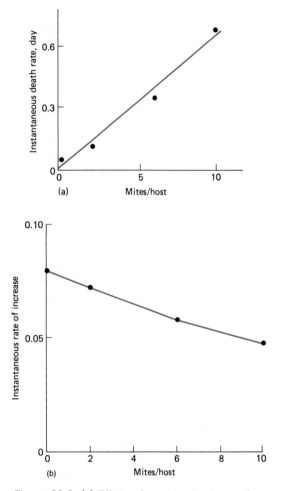

Figure 22.8 (a) Effects of parasite density on the rate of parasite-induced mortality of the host. This example shows the influence of parasitism by the mite *Hydryphantes tenuabilis* on the water measurer *Hydrometra myrae,* a common hemipteran on ponds. (b) The effect of parasitic infection by the same mite on the fecundity of the water measurer. Denser parasitic infections increased mortality and decreased fecundity. (After Lanciana 1975: 691.)

lations by reducing reproductive success. Wildlife biologists reduce the parasite load and increase lamb survival with chemotherapy by treating pregnant ewes with an antihelmenthic drug disguised in highly palatable fermenting apple mash.

High populations of hosts may result in a more uniform distribution of parasites (underdispersions), which may have some significance in population regulation. Carroll and Nichols (1986) monitored a high population of meadow voles *(Microtus pennsylvanicus)* and its important

ectoparasite, the larvae and nymphs of the American dog tick *(Dermacentor variabilis)*. During high vole density nearly all voles examined were parasitized. There was no significant difference ($P > 0.05$) in tick burdens between sexes on an annual basis, but juvenile and subadult voles harbored significantly fewer larval and nymphal ticks than adults. When the vole population declined over a two-year period—1983 and 1984—there was a sharp decline in adult ticks. Adult ticks in 1983 would have fed as nymphs on a declining population in the spring of the year when the vole population was at its lowest. The greater scarcity of adult ticks in 1984 was the result of low populations of both adult ticks and voles in 1983. This study suggests that the dynamics of host and parasite populations are closely related.

Infestations of a number of different species of both ectoparasites and endoparasites may have a combined effect, a case of "diffuse" parasitism. The eastern cottontail rabbit *(Sylvilagus floridanus)* has been declining over parts of its range, in spite of no apparent changes in its habitat. Jacobson, Kirkpatrick, and McGinnes (1978) investigated two local rabbit populations in Virginia occupying habitats similar in structure and vegetational composition, but differing in temperature, humidity, and soil conditions. One population was relatively abundant and the other, inhabiting a locality with higher summer temperatures and humidity, conditions favorable for many parasites, was chronically low for a number of years. The low population carried a high parasitic load. These parasites included a high infestation of rabbit ticks, carriers of the fatal bacterial disease tularemia, botflies, nematodes, a very high load of intestinal flukes, stomach worms inducing hemorrhagic gastritis, anemia, and weight loss, and staphylococcus and subcutaneous infections. Highest parasitic infections occurred in spring, resulting in low rabbit reproduction. The other population carried a much lower infestation of a some-

what different array of parasites. Field evidence seems to point to the role of parasites in holding a local population of rabbits at a level well below the carrying capacity of the habitat.

Parasite Population Dynamics

There is a similarity between predator-prey relationships and parasite-host relations. The parasite acts out in its own way the role of a predator, but with a major difference. The parasite lives inside the host. The host is at once food and habitat; so the parasite is much more intimately related to its "prey" than the predator. Its fate is closely related to the density of the host population and the distribution of the individual parasites among the host. A host can support only so many adult parasites before they experience intraspecific competition or parasite-induced mortality that takes the parasites within the host along with it.

The situation becomes magnified when the parasite possesses two or more infective stages. Each behaves as an individual population subject to its own population stresses. Growth and survival of the parasite population is affected by the immune responses of the host. A strong immune response can severely reduce or destroy a parasitic population. Growth is further influenced by the longevity of the host. The host has to live long enough for the parasite to complete its cycle within it and reach an infective stage. Premature death of the host results in the death of the parasites, unable to live an independent existence. Thus the parasite is faced with both the natural and parasite-induced mortality of the host.

Within the host the parasite faces its own intraspecific interactions, especially density-dependent interference competition. Growth of a parasitic population may be constrained by space within the host (Figure 22.9) as well as its fecundity (Figure 22.10). Ackert (1931) found a relationship between the amount of eggs and larvae taken in by chicks and hatching and rates of growth of the parasite. An intake of a smaller number of eggs gave a higher percentage of hatching and a greater growth rate of the worms, whereas a high infestation resulted in low percentage of hatching and slow growth rates.

Immune responses to high infestation can increase parasite mortality and reduce hyperinfestation. The nematode *Haemonchus contortus* occurs as an adult in the stomach of sheep and is transmitted in the larval stage found on grass. Entering the stomach, the infecting larvae burrow into the mucosal lining of the host's gut. They remain here for a short time before returning to the lumen of the stomach as adults. Before their stay ends, the larvae induce a rise in the histamine concen-

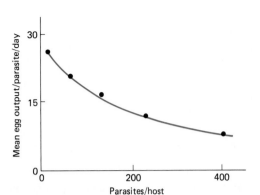

Figure 22.10 High density of parasites in a host, probably resulting in strong intraspecific competition, sharply reduces the production of transmission stages. The hosts are sheep infected with the liver fluke *Fasciola hepatica*. (From Boray 1969.)

tration in the sheep's blood and evoke an antibody response. In the stomach the adult nematodes take up feeding on blood. Continuous reinfection of the sheep brings in more larvae, which attach themselves to the mucosa. Their presence induces further production of histamine and antibodies, which circulate in the blood and are taken up by the blood-feeding adults, killing them. With the expulsion and death of the adult nematodes, the newly invading larvae find a place to live. Such a reaction prevents an overpopulation of parasites within the host.

The reproductive rate of a parasite population is given as R_o. Unlike the reproductive rate of nonparasites, it depends upon the host density. The population cannot grow until it reaches some transmission threshold N_T, at which R_o = 1. If R_o is less than 1, the parasite population will die out, and if the value is greater than 1, the parasite population increases (Figure 22.11) (Anderson and May 1979).

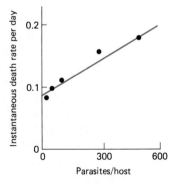

Figure 22.9 An example of the relationship between parasite density and instantaneous parasite death rate within a host. The hosts are chickens and the parasite is the gut nematode *Ostertagia ostertagi*. The rate of mortality is linear with density. (From Ackert et al. 1931.)

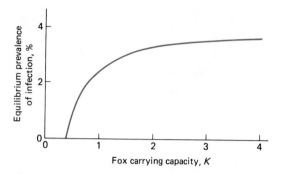

Figure 22.11 Theoretical relationship between the density of a host population, the fox, and the prevalence of infection of a microparasite, the rabies virus. The equilibrium prevalence of the infection does not rise until the host density rises above the threshold level N_T. After this point the level of infection increases rapidly and then tends to slow down and level off at high values of N. At high levels of infection and high levels of host mortality, the relationship can result in population regulation of the host by the parasite. Rabies outbreaks in high populations of foxes do reduce the fox population to much lower levels and then periodically reduce them again as the population again rises above the threshold level. The same type of curve is applicable to infections of macroparasites. (From Smith 1982: 137.)

Among microparasites R_o is the number of secondary infections produced by one infected host during its infective life when introduced into a large susceptible population. A large value of R_o means a high infection rate.

Among macroparasites R_o is the average number of offspring produced by a mature parasite throughout its reproductive lifespan that successfully complete their life cycle and attain reproductive maturity.

R_o is given as $\beta N/(a + b + v)$, where

N = host population size
β = transmission rate of the parasite
a = instantaneous host death rate when mortality is due to influence of the parasite

b = instantaneous host natural death rate
v = instantaneous birthrate of parasitic transmission stage

R_o equals 1 only when the population reaches a transmission threshold N_T. This threshold must be crossed if the disease or parasite is to spread. This threshold is given as

$$N_T = (a + b + v)/\beta$$

Thus $R_o = N/N_T$.

■ Evolutionary Responses

The relationship of parasites and their hosts involves an evolutionary response. As the host builds up defenses against parasitic infections, the parasite evolves new strains with reduced virulence or pathogenicity that enable it to survive within its host. The parasite gains no advantage if it kills its host. A dead host means dead parasites. A high virulence usually results in a high mortality of hosts. Low virulence ensures a long duration of infection. As an outcome both host and parasite evolve a mutual tolerance.

A dramatic example is the relation of hybrid corn, artificially developed or "evolved" by humans, and corn blight, a viral disease. In breeding varieties of hybrid corn, hybridizers succeeded in developing a uniform cytoplasm. They utilized factors within the cytoplasm that lead to male sterility, which eliminates the need for detasseling corn in seed fields. One particular corn cytoplasm, Texas male sterile, TMS, was widely incorporated into many lines; but TMS conferred susceptibility to a virulent race T of southern corn blight (Tatum 1971). When this blight appeared in the central corn belt and in Florida in 1969, it became epidemic, because 80 percent of field corn hybrids carried the highly susceptible TMS. The remaining hybrids, containing normal male-fertile cytoplasm, were resistant to race T. As a result corn breeders had to breed resistance into the susceptible hybrids. Such is the risk humans run with new varieties and races of cereal grains. The problem is to maintain a crop with a high degree of genetic diversity, so that when a disease develops a virulent strain, a genetic bank exists from which resistant forms can be bred.

Such genetic diversity was involved in the parasite-host interaction of the European rabbit and the viral infection myxomatosis. It is an example of evolutionary forces at work, the adaptation and counteradaptation of parasites and hosts (see Holmes 1983; May and Anderson 1983).

To control the introduced rabbit, the Australian government introduced the rabbit's viral parasite into the population. The first epidemic of myxomatosis was fatal to 97 to 99 percent of the rabbits. The second resulted in a mortality of 85 to 95 percent; the third, 40 to 60 percent (Fenner and Ratcliffe 1965). The effect on the rabbit population was less severe with each succeeding epizootic, suggesting that the two populations were adjusting to each other.

In this adjustment, attenuated genetic strains of virus, intermediate in virulence, tended to replace highly virulent strains. Too high a virulence killed off the host; too low a virulence allowed the rabbits to recover before the virus could be transmitted to another host. Also involved was a passive immunity to myxomatosis conferred upon the young born to immune

does. Finally a genetic strain arose in the rabbit population providing an intrinsic resistance to the disease.

The transmission of myxomatosis depends upon *Aedes* and *Anopheles* mosquitos, which feed only on living animals. Rabbits infected with the more virulent strain lived for a shorter period than those infected with a less virulent strain. Because the latter live for a longer period, the mosquitos have access to that virus for a longer time. That gives the less virulent strain a competitive advantage over the more virulent.

In those regions where the less virulent strains have the competitive advantage, the rabbits are more abundant because fewer die. That means more total virus is present in those regions than in comparable areas where the more virulent strains exist. Thus, the virus with the greatest rate of increase and density within the rabbit population is not the one with the selective advantage. Instead the virus whose demands are balanced against supply has the greatest survival value.

■ Social Parasitism

Another form of parasitic relationship is *social parasitism*, in which one organism is parasitically dependent on the social structure of another. Social parasitism may be temporary or permanent, facultative or obligatory, within a species or between species. Four forms of social parasitism can be defined (Wilson 1975).

The first is temporary, facultative parasitism within a species. This type is well developed among the ants and wasps. For example, a newly mated queen of the wasp genus *Polistes* or *Vespa* will attack established colonies of her own species and displace the resident egg-carrying queen (Wilson 1975). Females of a few species of birds, notably among the Galliformes, ostriches, and waterfowl, will parasitize the nests of others of the same species. For example, some female goldeneye ducks (*Bucephala clangula*) will lay their eggs in the nests of nearby females, who incubate them as part of their own clutch.

A second type is temporary, facultative parasitism between species. An example occurs among the formicine ant genus *Lasius*. A newly mated queen of the species *L. reginae* will enter the nest of a host species *L. alienus* and kill its queen. The *alienus* workers will care for the *reginae* queen and her brood. In time the *alienus* workers, deprived of their own queen and thus replacements, die out, and the colony then consists of *reginae* workers. A somewhat parallel situation exists among birds. Twenty-one species of ducks are known to lay eggs in nests other than their own (Weller 1959). An estimated 5 to 10 percent of female redhead ducks are nonparasitic and nest early. All others lay eggs parasitically at one time or another. More than half of them are semiparasites and nest themselves. The remainder are completely parasitic.

A third type of social parasitism is temporary, obligatory parasitism between species. Although common in ants, the most outstanding examples are obligatory egg or brood parasitism in birds. Brood parasitism has been carried to the ultimate by the cowbirds and Old World cuckoos, both of which have lost the act of nest building, incubating the eggs, and caring for young. They are obligatory parasites who pass off these duties to the host species by laying eggs in their nests. The brown-headed cowbird of North America removes one egg from the nest of the intended host, usually the day she is to lay,

and the next day lays one of her own as replacement. Some host birds counter by ejecting the egg from the nest. Others hatch the egg and rear the young cowbird, usually to the detriment of their own offspring. The host's young may be pushed from the nest or die from lack of food because of the more aggressive nature and larger size of the young cowbird.

A fourth type is permanent, obligatory parasitism between species. The parasitic form spends its entire life cycle in the nest of the host (Wilson 1975). This type of social parasitism is common among ants and wasps. In most cases the species are workerless and queens have lost the ability to build nests and care for young. The queen gains entrance to the nest of the host and either dominates the host queen or kills her outright and takes over the colony.

Like parasitism, social parasitism can adversely affect a host experiencing its first contact with social parasite. Such a situation has developed with the Kirtland warbler (*Dendroica kirtlandii*), a relict species that inhabits extensive jack pine stands in a compact central homeland of about 250 km^2 in northern lower Michigan (Mayfield 1960). Before white settlers arrived, Kirtland's warbler apparently was isolated by 320 k of unbroken forest from the parasitic brown-headed cowbird of the central plains, a bird closely associated with grazing animals. When settlers cleared the forest and brought grazing animals with them, cowbirds spread eastward and northward into jack pine country. Never associated with the cowbird, Kirtland's warbler has not evolved de fenses against the social parasite, such as egg ejection, building a new nest over the cold, or rearing young successfully along with the cowbird. The warbler has a short nesting season and an incubation period a day or two longer than other songbirds, so the young cowbird is already out of the egg when the warblers hatch. This parasitism resulted in an alarming decline in the number of warblers fledged, a trend that has been reversed by a strong cowbird control program in the Kirtland warbler's breeding range.

The outcome of social parasitism may be neutral to beneficial for the host and deleterious for the parasite, a switch in the relationship. Being indeterminate layers, female goldeneye ducks who have their nests parasitized by their own species accept the eggs as their own and adjust their own final clutch accordingly (Andersson and Eriksson 1982). The reproductive success of the host is reduced by the proportion of parasitized eggs. The selection pressure for such parasitism is probably the lack of suitable nest sites. Goldeneyes nest in hollows of trees. The parasitized females do not defend their easily located nests, and they respond to the parasitic eggs as their own. Because female waterfowl have a very strong tendency to return to their own natal place (philopatry), the host and the parasitic bird are probably genetically related, so the effects on overall fitness may not be great.

The situation is somewhat different when goldeneye nests are parasitized by another waterfowl species. Eadie and Lumsden (1985) studied the effects of nest parasitism of goldeneyes by a smaller cavity-nesting species, the hooded merganser (*Lophodytes cucullatus*). In this case the young ducklings are decidedly different in appearance from young goldeneyes and experience higher mortality from brood predation. In effect, the young mergansers act as buffers for the young goldeneyes. The reason for high predation on the mergansers may be a possibly slower reaction to parental alarms; or the difference in size and color may encourage predators to select the odd prey. In either case, the host seems to gain, because loss rate of goldeneye young is less in parasitized broods. Thus the improved survival of her own young may compensate for her reduced clutch and the cost of hatching anc caring for the parasitic young (which is not great because the young are precocial). What advantages accrue for the parasitic merganser are questionable. For her parasitism may be a salvage strategy, opting for some reproductive success in the face of a shortage of nesting sites.

Social brood parasitism can in some situations be mutually beneficial, as in the often cited case of the giant cowbird (*Scaphidura oryzivora*) and the hosts oropendola (*Zarhynchus*) and cacique (*Cacicus*), grackle-like birds of the family Icteridae (N. Smith 1968). Under certain conditions a greater number of host offspring survive from cowbird-parasitized nests than from unparasitized ones. A botfly that burrows into the chick's body to feed and crawls out again to pupate on the bottom of the nest is the major cause of mortality among the nestlings. Host chicks in nests parasitized by the cowbird are almost never bothered by botflies, whereas chicks in unparasitized nests are and sustain heavy mortality. The reason is that the aggressive parasitic cowbird nestling actually removes bots and eggs from its nestmates and at the same time protects itself by snapping at any moving object, including adult botflies.

There is more to the story that can only be briefly told. Cowbird eggs vary in coloration and markings from colony to colony. At some colonies giant cowbirds produce eggs whose color and markings mimic the eggs of the host. In other colonies the cowbird lays nonmimetic eggs. Those cowbirds laying mimetic eggs secretly deposit only one egg in a host nest that already contains eggs. Cowbirds laying nonmimetic eggs openly leave two to three in an empty nest. Hosts for mimetic eggs discriminate against mismatched ones and throw them out. Hosts for nonmimetic eggs do not discriminate. A study showed that three-fourths of the nests of nondiscriminating hosts contained cowbird chicks; only one-fourth of the nests of discriminators had cowbirds.

Oropendula and cacique colonies are often clustered near nests of wasps and stingless but biting bees, which in some way or another repel botflies. Nests near bees and wasps have a lower incidence of botfly parasitism. Interestingly, colonies protected by wasps and bees are composed of discriminator hosts and mimetic cowbirds, whereas unprotected colonies consist of nondiscriminating hosts and nonmimetic cowbirds.

The outcome of these complex interrelations, which involve social parasitism, commensalism (birds and wasps), and protocooperation (nonmimetic cowbirds and nondiscriminatory hosts), is mixed. Reproductive success of discriminator oropendulas and caciques, who build nests near wasps and bees, is reduced by giant cowbird parasitism and gains no advantage from cowbird nestlings. However, nondiscriminatory hosts gain an advantage from an association with cowbirds. The average number of oropendulas and cacique chicks fledged per nest in colonies near wasps and bees was 0.39 and for nests distant from the insects was 0.43. The average number of cowbird chicks fledged in the two types of colonies was 0.76 and 0.73, respectively. The reproductive success of both oropendulas and caciques and their cowbird parasites is about the same under both circumstances.

It is evident that oropendulas and caciques maintain polymorphisms in sets of populations, which allow part of the population to accept cowbird eggs and others to reject them. At the same time cowbirds maintain polymorphisms of egg mimicry and of aggressive and secretive behavior to take full advantage of the host (for details, see N. Smith 1968).

■ Summary

Parasitism is a situation in which two organisms live together, but one derives its nourishment at the expense of the other. A parasitic infection can result in disease, a condition of a plant or animal worse than normal well-being. Parasites may be divided into microparasites and macroparasites. Microparasites include the viruses, bacteria, and protozoa. They are small in size, have a short generation time, multiply rapidly in the host, tend to produce immunity, and spread by direct transmission. They are usually associated with dense populations of the host. Macroparasites, relatively large in size, include parasitic worms, lice, ticks, fleas, rusts, smuts, fungi, and other forms. They have a comparatively long generation time and rarely multiply directly in the host, are persistent with continual reinfection, and spread by both direct and indirect transmission.

Hosts are the habitat of parasites. The problem of parasites is to gain entrance into and escape from the host. The life cycles of parasites revolve about these two problems. The adult stages live in the definitive host, from which they escape by means of direct contact with other hosts or by means of vectors. Vectors are organisms that carry or transmit the parasite from one organism to another. Many vectors are hosts for some developmental or infective stage of the parasite. These vectors become intermediate hosts of the parasite, and transmission from definitive to intermediate and back to definitive hosts is considered indirect. Indirect transmission often involves the food chain.

Transmission of parasites, direct or indirect, is complicated by patchy or clumped distribution of the host. As a result parasites become overdispersed. The greatest load of parasites is carried by relatively few individuals in the population, and most remain free of infection. Hosts respond to parasitic infection by biochemical responses, including the inflammatory process and immune reactions. Parasitic infections may result in abnormal growth and in behavioral changes.

Interactions of parasites and host involve up to three populations, each subject to its own rate of growth and fecundity. Because of the close relationship between parasite and host, the population dynamics of one is influenced by the other. A heavy parasitic load can increase

mortality and decrease fecundity of the host population. Under certain conditions parasitism can function as population regulatory mechanism. Conversely, a high population of parasites within an individual can experience both intraspecific competition and immune responses from the host. These factors interact to increase mortality of parasites and reduce their fecundity. Because the death of a host does not benefit a parasite that depends on its host for both food and shelter, natural selection favors less virulent forms of the parasite that can live in the host without killing it. Both hosts and parasites evolve a mutual tolerance with a low-grade widespread infection.

Another type of parasitism is social parasitism. It may be temporary, permanent, facultative, or obligatory. A common example is brood parasitism among some species of birds and among some ants and wasps.

CHAPTER 23

Mutualism

Coevolution

The relationships of predators with their prey and parasites with their hosts share one characteristic. As selection pressures of predation increased upon the prey, the prey responded by evolving some means of defense, such as increased speed, warning coloration, and chemical defenses among animals and chemical and structural defenses among plants. In turn predators, always a few steps behind, responded by evolving their own means of improving their hunting ability or breaching chemical defenses. Parasites evolved ways of gaining access to and exits from

their hosts and means of counteracting immune responses, the major defensive mechanisms of their hosts. Such reciprocal selection pressures on two interacting populations result in *coevolution*. Certain traits of each species evolve in response to the traits of the other (Ehrlich and Raven 1965; Janzen 1980). Any evolutionary change in one member changes the selection forces acting on the other. Coevolution is basically a game of adaptation and counteradaptation to changing selection pressures imposed on one taxon by another.

Such a restricted definition seems to imply that interacting species grew up together over evolutionary time. No proof exists that they did so. The chances are that many supposedly coevolved pairs did not grow up together and that today they are not inhabiting the environments in which they evolved (Howe 1985; Herrara 1985). Instead, the organisms probably evolved in a mix of habitats through time, each experiencing somewhat different selection pressures. When these plants or animals invaded different habitats, they adjusted to the organisms at hand. If the traits they had already acquired fit the situation, then the two interacted in a manner that suggested long-term coevolution. If the relationship meshed, then further evolutionary changes were minor. What appears to be a highly coevolved system may not have involved an evolutionary change in either partner. The relationship would then continue to be selected for by current interaction.

Another approach to coevolution is to consider it as a less restrictive, more general response of one group of species to another. A particular trait evolves in several species in one taxon in response to a trait or a suite of traits in several species in another taxon. Plants might have evolved chemical and physical defenses against a diverse array of herbivorous insects. In turn, many insects evolved the ability to detoxify a wide range of plant chemicals. Similarly, animals might have evolved a generalized immune system in response to a wide range of parasites. Plants adapt to nonspecific pollinators that visit their flowers. Such interactions are termed *diffuse coevolution* because the adaptive responses spread over many interacting species. They contrast with a pairwise response between one species and another, such as a specialized parasite and its host. Among interspecific relationships, coevolution is most closely related to mutualism in its various forms.

■ Types of Mutualism

Mutualism is a positive, reciprocal relationship at the individual or population level between two different species (Boucher et al. 1982). Out of this relationship, most obvious at the individual level, both species enhance their survival, growth, or fitness (Holmes 1983). Evidence suggests that mutualism is more a reciprocal exploitation than the cooperative effort that the definition seems to imply (Barrett 1983).

Like competition, mutualism involves interactions between two populations, but the interactions are positive rather than negative. Mutualism has not received as intensive mathematical treatment as interspecific competition. The general approach has been a modification of the terms of the Lotka-Volterra equations for competition in which the negative alphas of competition become positive; and because the maximum values of K do not enter into the relationship, K is changed to X (Pianka 1988):

$$\frac{dN_1}{dt} = r_1 N_1 \left(\frac{X_1 - N_1 + \alpha_{12} N_2}{X_1} \right)$$

$$\frac{dN_1}{dt} = r_2 N_2 \left(\frac{X_2 - N_2 + \alpha_{21} N_1}{X_2} \right)$$

where

X_1 and X_2 = equilibrium density of species 1 and species 2 in the absence of the other species;

α_{12} = the coefficient of mutualism, the beneficial effect of one individual of N_1 on N_2;

α_{21} = beneficial effect of one individual of N_2 on N_1.

The pair of linear equations predicts that the population equilibrium density of each species is increased by the density of the other species. Such equations suggest that mutualism could produce a runaway positive feedback that would be unstabilizing for both populations unless a strong negative feedback of density-dependent population regulation checked it.

Mutualistic relationships obviously are much more complex than the two basic equations describe. More complex mathematical formulations have been developed, especially ones relative to stabilizing the population dynamics of the mutualistic species (see Vandermeer and Boucher 1979; Vandermeer 1980; Travis and Post 1979; Wolin and Lawlor 1984; Post, Travis, and DeAngelis 1985; Wolin 1985). Stability may be realized by the introduction of a third species that is a competitor or a predator of one of the mutualists (see Heithaus, Culver, and Beattie 1980).

Obligate Symbiotic Mutualism

Mutualism may be symbiotic or nonsymbiotic, facultative or obligate. *Symbiotic* refers to two organisms living together in close association. Parasites and their hosts are a form of symbiosis. In symbiotic mutualism individuals interact physically and their relationship is obligate. At least one member of the pair cannot live without the other.

Some forms of the relationship are so permanent and obligatory that the distinction between the two interacting populations becomes blurred. A good example is the fungal-algal symbiosis in the lichens. The basic structure of the lichen is a mass of fungal hyphae. Within this formation is a thin zone of algae, which usually forms colonies of 2 to 32 cells (Figure 23.1). Although some 26 different genera of algae

have been associated with lichens, the most common genus is *Trebouxia,* the only one not found in the free-living state (Holmes 1983). Most of the fungae involved are close relatives of free-living ascomycetes. Many lichens produce and disperse spores, which form mats of mycelia that may live an independent, saprophytic existence for a short time. Then they become associated with compatible algal cells.

Although algae and fungi supposedly exist together for each other's benefit, they may not really. Algal cells may leak metabolites to the surrounding soil rather than passing them on the fungus; and there is no evidence that the fungus provides anything for the alga other than protection from damaging solar radiation and desiccation (Admadjian 1970).

A better example of coevolved mutualism is the association of plant roots with the mycelia of fungi. Common to many trees of temperate and tropical forests, one form, *ectomycorrhizae,* produces shortened and thickened roots that suggest coral. The hyphae work between the root cells, and on the outside of the root they develop into a network that functions as extended root hairs. Another form is *endomycorrhizae,* notably the vesicular arbuscular mycorrhizae. The plant's roots are infected by mycelia from the soil. They penetrate the cells of the host and form a finely bunched network called an arbuscle. The mycelia act as extended roots for the plant, drawing in phosphorus at distances beyond those reached by the roots and root hairs, but they do not change the shape or structure of the root. Mycorrhizae, especially important in nutrient-poor soils, aid in the decomposition of litter and translocation of nutrients, especially nitrogen and phosphorus, from the soil into root tissue (Zak 1964; Marx

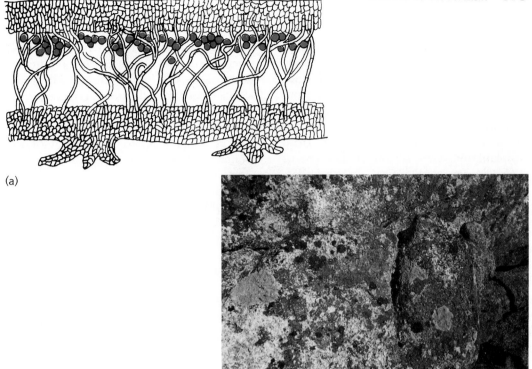

(a)

(b)

Figure 23.1 (a) Section through a lichen body, showing the algae within the fungal mass. (b) Crustose lichens growing on a boulder. (Photo by R. L. Smith.)

1971). Mycorrhizae increase the capability of roots to absorb nutrients, provide selective ion accumulation and absorption, mobilize nutrients in infertile soil, and make available certain nutrients bound up in silicate minerals (Voigt 1971). In addition, mycorrhizae reduce susceptibility of their hosts to invasion of pathogens by utilizing root carbohydrates and other chemicals attractive to pathogens. They provide a physical barrier to pathogens and stimulate the roots to elaborate chemical inhibitory substances (Marx 1971). In return, the roots of the host provide support and a constant supply of carbohydrates.

The association between the two can be tenuous. Any alteration in the availability of light or nutrients for the host creates a deficiency of carbohydrates and thiamine for the fungi. Interruption of photosynthesis causes a cessation of fruiting by mycorrhizae.

Similar mutualistic relationships permitting organisms to exploit nutrient-poor environments may be found in the sea. The oceans offer innumerable examples of all sorts of mutualistic interactions, especially among the coral reefs. Coral reefs, calcareous substrates formed and occupied by anthozoans, are found largely in warm nutrient-poor tropical waters. Living within the cells of the endoderm layer of the oral cavity of the coralline anthozoans are photosynthetic dinoflagellate algae (zooxanthellae). The heterotrophic anthozoans utilize the photosynthetic products of the algae. In turn the coral anthozoans remove, retain, and recycle essential nutrients from the water (Muscatine and Porter 1977). Although they are carnivorous suspension feeders capturing zooplankton from the surrounding water, anthozoans derive only about 10 percent of their daily energy requirement from zooplankton. They obtain 86 percent of their energy and caloric requirements from algal fixation of C and N and are able to survive and flourish in their nutrient-poor environment by recycling nutrients with their symbiont algae (Figure 23.2). In addition algal photosynthesis accelerates the ability of the anthozoans to lay down calcified coral structures, enabling them to build reefs fast enough both to counteract destruction and to increase their benthic cover.

Obligate Nonsymbiotic Mutualism

More common is obligate nonsymbiotic mutualism, in which the mutualists live independent lives yet cannot survive without each other. One example is the mutualistic relationship of the yucca and the yucca moth. The yucca depends upon yucca moths for pollination, and the lar-

vae of yucca moth are obligate predators of yucca seeds. In return for pollination yucca must pay out a certain amount of seeds as larval food, without which the yucca moth could not survive. Overall yucca moth larvae decrease viable seed production by as much as 19 percent, and yuccas expend up to 30 percent of the benefits gained from pollination to support yucca moth larvae.

The approximately 900 species of the widespread tropical genus of figs *(Ficus)* have a complex obligatory relationship with pollinating agaonid fig wasps (Wiebes 1979). They, too, lay their eggs in the developing seeds upon which the larvae feed. Figs experience 44 to 77 percent seed mortality, a high cost for pollination (Janzen 1979).

Figure 23.2 A model of the potential pathway for recycling of carbon and nitrogen in a mutualistic symbiosis between heterotrophic coral anthozoans and phototrophic zooxanthellae. Evidence to support all parts of the cycle is not available. Apparently nitrogen deficiency is avoided by fixation of nitrogen by algae, its translocation from algae to the coral, and possible uptake and retention of ammonium excreted by the heterotroph. (Adapted from Muscatine and Porter 1977: 458.)

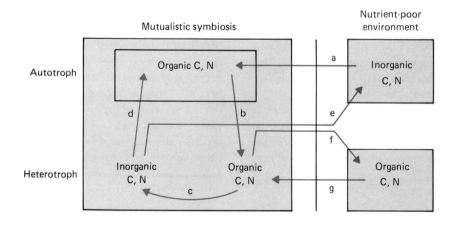

Other such obligatory relationships involve shelter, protection against predators, and reproduction. Some of the most interesting exist between ants and plants, from the relationship between fungus-growing attine ants and a slow-growing fungus that cannot survive without them (Martin 1970) to the ant-acacia relationship (Janzen 1966; Hocking 1975). The Central American ants live in the swollen thorns of acacia *(Acacia spp.)*, from which they derive shelter and a balanced and almost complete diet for all stages of development. In turn the ants protect the plants from herbivores. At the least disturbance the ants swarm out of their shelters, emitting repulsive odors and attacking the intruder until it is driven away. Neither the ants nor the acacias can survive in the absence of the other.

Some mutualistic relationships involve a third member. Some ectomycorrhizae are epigenous; that is, they produce their sporocarps above ground and forcibly discharge their spores to the air. Hypogenous mycorrhizae produce their sporocarps below ground in structures popularly known as truffles. In the coniferous forests of western North America, these mycorrhizae depend upon chipmunks and voles to disperse their spores (Maser, Trappe, and Nussbaum 1978). Attracted to the fruiting bodies by species-specific odors, rodents eat the sporocarps, which make up a significant part of their diet. When they defecate, the rodents spread the viable spores necessary for the survival and health of the conifers through the forests.

Facultative Mutualism

Most mutualisms are nonobligatory and facultative, at least on one side. Such mutualisms are diffuse, involving interactions among guilds of species. They include seed dispersal and pollination. The benefits are spread over many plants, pollinators, and seed dispersers.

Seed Dispersal. Plants with seeds too heavy to be dispersed by wind depend upon animals to carry the seeds some distance from the parent plant and deposit them in sites favorable for seedling establishment. Some seed-dispersing animals upon which the plant depends may be seed predators, consuming the seeds for their own nutrition. Plants depending on such animals must produce a tremendous number of seeds over their reproductive lifetimes and sacrifice most of them to ensure that a few will survive, come to rest on a suitable site, and germinate.

An example is the almost obligatory relationship between the Clark nutcracker *(Nucifraga columbiana)* and the whitebarked pine *(Pinus albicaulis)*. Whitebarked pine and a few others such as pinyon pine *(Pinus edulis)* produce large wingless seeds that can be dispersed from the parent trees only by animals. The seeds of whitebarked pine are eaten and hoarded by several species of rodents, including chipmunks, and by Steller's jays and Clark nutcrackers. Only the nutcracker possesses the behavior appropriate to disperse the seed systematically and successfully away from the tree (Hutchins and Lanner 1982; Tomback 1982). The bird carries seeds in cheek pouches and caches them deep enough in the soil of forests and open fields to reduce predation by rodents. The number of seeds cached per individual per year is enormous, an estimated 98,800. The nutcrackers fail to retrieve enough of these seeds to allow a large number of them to survive and establish seedlings. Although the cost is high, whitebarked pine is virtually dependent on the bird for seed dispersal.

Some plants use a seed predator not

only to disperse the seeds but also to protect them from other predators. In the deserts of southwestern United States (O'Dowd and Hay 1980), in the sclerophyllous shrublands of Australia (Berg 1975), and in the deciduous forests of eastern North America (Beattie and Culver 1981; Handel 1978) a number of herbaceous plants, including many violets *(Viola spp.)*, depend upon ants to disperse their seeds. Such plants, called *myrmecochores,* have an ant-attracting food body on the seed coat called an *elaiosome.* Appearing as shiny tissue on the seed coat, the elaiosome contains lipids and sterols essential to certain physiological functions in insects. The ants carry seeds to their nests, where they sever the elaiosome and eat it or feed it to their larvae. The ants discard the intact seed within abandoned galleries of the nest. Ant nests, whose substrate is richer in nitrogen and phosphorous than surrounding soil (Culver and Beattie 1978), provide a suitable substrate for seedling emergence and establishment. Further, by removing seeds from the parent plant, the ants significantly reduce rodent predation on them (O'Dowd and Hay 1980; Heithaus 1981). Heithaus (1981) found that rodents, particularly the white-footed mouse *(Peromyscus leucopus),* removed 84 percent of the seeds of bloodroot *(Sanguinaria canadensis)* when ants were excluded from the parent plants, but only 13 to 43 percent when ants were allowed access.

Plants have an alternative approach for seed dispersal: to enclose the seed in a nutritious fruit attractive to fruit-eating animals, the frugivores. Frugivores are not seed predators but consume only the endocarp surrounding the seed; with some exceptions, they do not impair the vitality of the seed. Most frugivores do not depend exclusively on fruits, because fruits tend to be deficient in certain nutrients such as protein, and because they are only seasonally available.

To use frugivorous animals as agents of dispersal, plants must attract them and at the same time discourage the consumption of unripe fruit. Plants accomplish that by cryptic coloration, such as green, unripened fruit among green leaves, and by unpalatable texture, repellent substances, and hard outer coats. When seeds mature, plants can attract fruit-eating animals by presenting attractive odors (Howe 1980), altering the texture of fruits and seeds, improving succulence, acquiring a high content of sugar and oils, and "flagging" their fruits with colors—red, black, blue, yellow, white—to catch frugivores' attention (Stiles 1982). Wheelwright and Janson (1985) report that among 383 bird-dispersed plant species in the tropical forests of Costa Rica and Peru, over 60 percent bore either black or red fruits.

Plants have two alternative approaches to seed dispersal by frugivores. One is to become opportunistic and evolve fruits that can be exploited by a large number of dispersal agents. Such plants opt for quantity dispersal, the scattering of a large number of seeds with the chance that a diversity of consumers will drop some seeds in a favorable site.

Such a strategy is typical of but not exclusive to plants of the temperate regions, where most fruit-eating birds and mammals are opportunistic consumers, rarely specializing in any one kind of fruit and not depending exclusively on fruit for their basic sustenance. The fruits are usually succulent and rich in sugars and organic acids and contain small seeds that pass through the digestive tract unharmed (Stiles 1980). To accomplish this passage such plants have evolved seeds with hard coats resistant to digestive enzymes. Seeds of some such plants may not germinate unless they have been conditioned or scarified by passage though the digestive tract. Large numbers of small seeds may be so dispersed, but relatively few are deposited on suitable sites. The lenght of time of such seeds within the digestive tracts of some small birds may be no more than 30 minutes, so the distance dispersed depends on how far and where the birds go after eating (Stiles 1980). Such dispersal is a lottery in the truest sense (Figure 23.3).

In temperate regions fruits ripen in early and late summer, when the young of the year are no longer dependent on highly proteinaceous food and both adults and young can turn their attention to a growing abundance of fruits. The fruits may be small-seeded or large-seeded, high or low in nutrients, sweet or otherwise. The fruits of early summer ripen before migratory birds pass through. Small-seeded fruits—blackberries, blueberries, and mulberries—are sweet, fragrant, easily invaded by fungi and yeasts, and highly appealing to small mammals and birds. Seeds easily pass through the gut, but dispersal distance is restricted. Large-seeded fruits, mostly cherries (*Prunus*), hang high, where they are taken by raccoons, opossums, turkeys, and robins, and the fruits that drop are eaten by foxes and skunks. These animals distribute seeds

Figure 23.3 Red cedar (*Juniperus virginiana*) possesses a fleshy, dark blue, highly aromatic, berrylike cone highly attractive to birds and small mammals, the major dispersal agents for the tree. The pattern of invasion of this field by red cedar reflects dispersion of its seeds by animals. (Photo by R. L. Smith.)

more widely, occasionally into suitable habitats such as old field, but the seeds are left in piles of dung, so that any emerging seedlings face heavy competition from conspecifics.

Fruits of late summer and fall ripen when migrant birds come through. They congregate in flowering dogwoods, spicebush, and wild grape to feed on high quality fruits with relatively large seeds and nutrient-rich flesh. Such fruits do not last long, and their seeds are scattered widely. Fruits of lower quality have less fats and sugars, are unlikely to be invaded by microorganisms, and hang on well into winter (Stiles 1980). They are available to birds over a longer period of time and are sure to be consumed when more palatable, short-lived succulent fruits are gone. In such a manner those fruits may avoid competition for dispersers in early fall.

The second approach is to depend upon a small number of birds and mammals that are exclusively consumers of fruit. Such plants are mostly tropical forest species, 50 to 75 percent of which produce fleshy fruits whose seeds are dispersed by animals (Howe and Smallwood 1982). Rarely are frugivores obligates of particular fruits on which they feed. Exceptions include the oilbirds and a large number of tropical fruit-eating bats. Among them, the flying foxes of the Old World eat the fruits in place and drop the seeds beneath the tree. The smaller spearnosed fruit bats of the New World pluck the fruits and fly to a safe perch some distance away to avoid predators waiting for them in the fruiting trees. Most trees, though, attract frugivores that consume many different fruits. Dispersers of the seeds of one plant are also dispersers for others, for several reasons. Fruits vary widely in their nutritional value; by eating a variety of them, frugivores tend to balance their diets. Plants have few means to restrict consumption of their fruits to a certain few frugivores. However, because plants do have fruits of various sizes, shapes, colors, aromas, nutrient contents, and palatability, some are consumed chiefly by mammals and others by birds (for review see Howe 1986).

Like other areas of mutualism, the study of the relationships among plants and their frugivores and seed dispersal is still embryonic, lacking in empirical studies in spite of the relatively large literature available (see Howe 1986).

Pollination. The "goal" of seed dispersal by plants is to get the seeds away from the parent plant and settled in some site favorable for seedling establishment. The goal in pollination is much more specific and direct. The plant must transfer its pollen from the anthers of one plant to the stigma of a conspecific. Some plants simply disperse their pollen to the wind. This method works well and costs little when the plants grow in large homogeneous stands, as grasses and pine trees do. Wind dispersal is unreliable when conspecifics are scattered individually or in patches across a field or forest. These plants depend upon animals for pollen transfer, mostly insects with some assistance from nectivorous birds and bats.

Like frugivores, nectivorous animals visit the plants to exploit them as a source of food, not to function as pollinators. With a few exceptions, the nectivores are generalists; like the frugivores they find little advantage specializing, except as temporary facultatives. Further, because of the short seasonal flowering of each species, often shorter than the availability of fruits, nectivores depend on a progression of flowering plants through the season. Nectivores cannot afford to commit themselves to one flower, but they do concentrate on one species while its flowers are available.

Plants are the ones that have to specialize, to entice animals to them by color, fragrances, and odors, dust them with pollen, and then reward them with a rich source of food: sugar-rich nectar, protein-rich pollen, and fat-rich oils. Providing such rewards is expensive for plants. Nectar and oils are of no value to the plant except as attractants for potential pollinators. They represent energy that the plant otherwise might expend in growth.

Many species of plants, such as blackberries, elderberries, cherries, and goldenrods, are generalists themselves. They flower profusely and provide a glut of nectar that attracts a diversity of pollen-carrying insects, from bees and flies to beetles. Other plants are more selective and screen their visitors to ensure some efficiency in pollen transfer. These plants may have long corollas, allowing access only to insects and hummingbirds with long tongues and bills and keeping out small insects that eat nectar but do not outcross the plants. Some, such as the closed gentian, have petals that only large bees can pry open.

Orchids, whose individuals are scattered widely through their habitats, have evolved extreme precision in pollen transfer and reception so that pollen is not lost when the insect visits flowers of other species. New World tropical orchids attract the solitary euglossine or golden bees, specialists in the pollination of orchids. Extremely fast fliers, these bees cover long distances between orchids in the tropical forests. However, they do not depend upon orchids as an energy source; the orchids could not afford to meet the bees' energy demands. Instead the bees depend on other, more rewarding flowers for their energy, although some orchids do

provide a fragrance that the bees collect from secretory cells on the lips of the flowers. Males use this fragrance as pheromones to attract females, who ignore orchids altogether. Most orchids offer no reward at all. To attract the euglossines, orchids mimic other flowers that do; or they have flowers with the scent, shape, or color patterns that mimic females, with which the males attempt to copulate, picking up pollen.

The orchids contain their pollen in a single mass, the pollinium, with an attachment device. When the bee brushes against it, the pollinium becomes attached to the bee at a specific location on its body. It remains there, for days or weeks if necessary, until the bee visits another orchid of the same species, whose stigma retrieves it. Bees may visit more than one species of orchid and pick up pollinia from each of them. To prevent wrong deliveries, each species of orchid has its pollinium so located within the flower that it adheres at its own specific location on the bee's body. It becomes detached only when the bee enters the correct orchid. This fact suggests that coevolution has occurred between bees and orchids and that the bees have been a strong selection force in the evolution of orchids (see Heinrich 1979; van der Pijl and Dodson 1966; Feinsinger 1983; Faegri and van der Pijl 1979).

In addition to nectar, some plants provide oil as a reward (Vogel 1969; Buchmann 1987). Many genera in a number of families, including Iridaceae, Orchidaceae, Scrophulariaceae, Concurbitaceae, Solanaceae, and Primulaceae, mostly in neotropical savannas and forest, have specialized oil-secreting organs, called *elaiophores*. One type is epithelial elaiophores, which consist of small areas of secretory epidermal cells beneath a protective cuticle on the petals, in which secreted lipids accumulate. A second type are trichome elaiophores, made up of hundreds to thousands of glandular trichomes that secrete lipids in a thin film of oil exposed to the air or protected within deep floral spurs. The oil flowers are visited by highly specialized bees in four families that use the energy-rich floral oils in place of or along with pollen as provisions for developing larvae. These bees possess modified cuticular and setal structures designed for mopping up, storing, and transporting oil to the nest (see Buchmann 1987).

The relationship between the common Central American plants *Heliconia* or wild plaintain and the hummingbird illustrates the many factors that may be involved in such mutualisms. Growing in openings of tropical forests or along the forest edge, *Heliconia* propagates vegetationally by rhizomes and usually forms large clumps. When two years of age or older, each individual *Heliconia* plant in the clump blooms. The bloom or inflorescence consists of several showy bracts, each of which encloses several flowers. The bracts open one after the other over a period of days or weeks with each flower within the bract lasting only a day. The flowers are tubular and vary in length and curvature depending upon the species of *Heliconia*. Some have long, curvaceous corollas, 33 mm or more in length; others have short, straight corollas, 32 mm or less (Stiles 1975). Some species bloom either in the wet season or the dry season, whereas others bloom throughout the year but have a wet or dry seasonal peak in flowering. All are pollinated by insects or birds and offer a supply of sugar-rich nectar as an inducement to their pollinators.

Stiles (1975) found that in his Costa Rican study area nine species of hummingbirds visit the nine species of *Heliconia*. Just as the flowers of *Heliconia* have straight or curved corollas, Stiles found that the hummingbirds, too, could be divided into two groups: hermit hummingbirds with long, curved bills and nonhermits with shorter, straight bills. Stiles observed that the five *Heliconia* species with long, curved corollas are visited to a significantly greater extent by hermits than nonhermits, whereas three of the species with short corollas are visited mostly by straight-billed nonhermits.

The hermit and nonhermit groups differ in another way. The nonhermits frequently hold territories about clumps of *Heliconia* with short corollas. The hermits seldom hold flower-centered territories, and those that do defend them inconsistently. The difference in territoriality is directly related to a difference in feeding strategy that is in turn directly related to the energy-supplying attributes of different *Heliconia* species. The nonhermits defend large clumps of *Heliconia* that at the peak of flowering are able to supply the birds' total energy needs. The hermits feed on *Heliconia* that grow in scattered clumps, have a lower rate of flowering per inflorescence, and have a lower rate of nectar production. Even at the height of seasonal bloom, clumps of these species cannot provide sufficient nectar to sustain a single bird. The hermits' feeding strategy involves "traplining." The birds travel between clumps of flowers, often following a definite route and visiting clumps in a particular sequence. Each clump produces sufficient nectar to warrant repeated visits, but not enough to be worth defending.

In return for nectar hummingbirds pollinate the respective flowers. *Heliconia* depend upon hummingbirds for pollen transfer. Hermit hummingbirds, probing into long, curved corollas, carry pollen at the base of the bill or on the head. The nonhermits or straight-billed hummingbirds carry pollen on the chin or mandibles. If the short corolla flowers are somewhat curved but the path to the nectar is short and straight, allowing easy access, the hummingbird has pollen deposited on the bill.

Because of the number of types of plants and birds involved, some isolating mechanisms are essential. *Heliconia* select against hybridization by sequential and nonoverlapping peaks in flowering (wet or dry season), by spatial isolation (shady or sunny habitats), and by behavioral differences in hummingbirds. The behavioral differences include responses of hummingbirds to visual cues of flowers and to caloric content of nectar. The mutualism thus depends not only on the morphological fit between bird bill and flower corolla, but also on several other complementary relationships involving flowering phenology, energy content of nectar, and energy demands and behavioral responses of hummingbirds.

Defensive Mutualism

A major problem for many livestock producers is the toxic effects on cattle of certain grasses, particularly perennial ryegrass and tall fescue. These grasses are infected by certain fungal endophytes that live inside plant tissue. The fungi (Clavicipitaceae Ascomycetes) produce physiologically active alkaloids in the tissue of the host grasses. These alkaloids, which impart a bitter taste to the grass, are toxic to grazing mammals, particularly domestic animals, and a number of insect herbivores. In mammals the alkaloids constrict small blood vessels to the brain, causing convulsions, tremors, stupor, gangrene of the extremities, and death. At the same time these fungi seem to stimulate plant growth and seed production. This symbiotic relationship suggests a defensive mutualism between plant and fungi in which the fungi defend the host plant against grazing (Clay 1988).

There are costs to the plant. The fungal infection causes sterility in the host plant by inhibiting flowering or aborting seeds. Some plants have a few counter-adaptations that restore fertility; but in most plants the loss of sexual reproduction is balanced by the greater vegetative growth of the infected plants and enhanced growth in the absence of herbivory.

Indirect Mutualism

The mutualistic relationships described so far are direct; one species positively benefits the other. Other mutualistic relationships may be indirect. For example, suppose that prey species A, consumed by predator X, strongly inhibits prey species B, consumed by predator Y. If the effect of predator X on prey A permits the expansion of prey B, X in effect provides more food for predator Y. The activities of X then indirectly benefit species Y.

A similar mutualistic effect may occur in competitive situations. Suppose species A competes strongly with species B, but mildly with C, and that species C competes strongly with B but only mildly with A. The combined effects of A and C on B reduce competition for species C (but are detrimental to species B). A number of such indirect mutualisms could influence community organization (Chapter 24). (For more discussion of indirect mutualism see Vandermeer 1980; Waser and Real 1979; Lane 1985.)

■ Origins of Mutualism

The examples suggest that mutualism probably evolved from predator-prey, parasite-host, or commensal relationships. Initially one member of the relationship increased the stability of a resource level for the second. In time energy benefits accrued to the second member, and perhaps its activities began to improve the fitness of the first. For example, a host tolerant of a parasitic infection may exploit the relationship. In time the two exploit each other, as in the mycorrhizal-plant mutualisms. Selection then favors mutual interaction to the point that two become totally dependent on each other, as in obligate symbiotic mutualism. At the extreme the two function as one individual, as the lichens do.

Among nonsymbiotic mutualists, obligate or facultative, the relationship may have begun with exploitation. Birds or insects came to plants to feed on pollen and nectar and frugivores to feed on fruit. In the process they accidentally carried pollen to similar plants or dispersed seeds away from the parent plant. Such plants experienced improved fitness and ultimately adapted to exploit the visitors as a means of dispersing pollen and seeds.

Such relationships did not evolve on purpose. Pollinators came to the plant to collect pollen or tap nectar supplies for food, not to aid the plant in completing its life cycle. Frugivores visited plants for fruits, not to disperse seed. Neither did the mutualisms arise as one-to-one relationships. Plants were visited by an array of hungry insects, birds, and mammals. Except in rare situations, these animals did not and still do not specialize on one plant; so groups of plants found themselves visited by groups of various taxons of animals, which differed locally. The result was diffuse coevolution. Guilds of related or unrelated taxa evolved the capacity to use a range of plant resources, resulting in novel ecological associations and diffuse rather than paired coevolution.

A number of constraints favor diffuse coevolution. One is multiple relationships.

A flower may be served best by one or two species of pollinators, but it is visited by many. An insect such as a bee may visit only one or two species of flowers, but that same flower will be visited by other insects as well. A plant may have its seeds dispersed most efficiently by one type of frugivore, but the frugivore will eat a variety of different fruits seasonally. Therefore it is unlikely that close evolution occurs between just one plant and one pollinator or seed disperser. A second constraint is asymmetrical evolution. For instance, woody plants are much older geologically and have evolved more slowly than any of their pollinators or seed dispersers (Herrera 1985). Over millions of years these plants have seen species and even taxons of animals come and go. As one animal group disappeared from the scene, its place was taken up by new groups that exploited the plants in a similar manner. Plants, however, retained any traits that encouraged pollination or fruit consumption. Further restricting tight coevolution are variations in the distribution and abundance of plants and animals, seasonal differences in fruiting and flowering, and seasonal changes in animal abundance and distribution.

■ Population Effects

Mutualism is most easily appreciated at the individual level. It is fairly easy to comprehend the interaction between an ectomycorrhizal fungus and its oak or pine host, to observe and quantify the dispersal of oak seeds by squirrels and jays, and to measure the cost of dispersal to oaks in terms of seeds consumed. Mutualism improves the fitness of an oak and the consumption of acorns improves the fitness of its seed predators; but what are the consequences at the level of the population?

The population consequences of mutualism are considerably more difficult to define than those of predation and parasitism, because the relationship is more nebulous and harder to model. Mutualism exists at the population level only if the growth rate of species A increases with the increasing density of species B, and vice versa.

The question is most relevant to obligate and facultative nonsymbiotic mutualists. For obligate symbiotic mutualists the relationship is straightforward: Remove species A and the population of species B no longer exists. If ectomycorrhizal spores fail to infect the rootlets of young pines, they will not develop, and if the young pines invading a nutrient-poor old field fail to acquire mycorrhizal symbionts, they will not grow well. (Foresters and nurserymen inoculate nursery-grown pine seedlings with the appropriate mycorrhizae.)

For nonsymbionts, obligate or facultative, the effect on populations may be limited to the extent that one species benefits another and to the proportion of each other's life history cycle in which the mutualistic relationship is involved. Consider the yucca and the yucca moth on which it depends for pollination. Aker (1982) studied the relationship of *Yucca whipplei* and its moth *Tegeticula maculata*. Throughout the flowering season the number of adult yucca moths were distributed evenly among the available flowers, so the number of pollinators on the inflorescences or flower heads was directly proportional to the number of flowers available. In turn the number of fruits set on the plant was directly proportional to the number of flowers produced. If too few moths were available, many flowers would go unpollinated. Most flowers were not fertilized, and the fruit production of some plants

was limited by pollination. In general, the relative abundance of moths to flowers was low enough that most plants were pollinator-limited.

Yuccas regulate mature fruit production by aborting excess fruit, which affects the survival of the seed-consuming moth larvae and their emergence from the capsules at the end of the summer. Thus the size of the adult moth population the following year is determined by the number of plants that flowered and matured fruit. The abcission of immature fruit (along with the larvae) matches the moth larval abundance to the number of plants reproducing in the current year.

These observations suggest that the population of the yucca and the yucca moth have reciprocal influences; but whether the yucca moth limits or influences population recruitment and growth of yucca is questionable. The yucca moth may limit seed production (Addicott 1985), but recruitment of yuccas may be influenced more by such extrinsic conditions as insufficient rainfall for seed germination, seed predation, and animal browsing on seedlings. The number of moths visiting the yuccas can be affected by weather conditions and by the spatial distribution of the plants. The vigor of the yucca as evidenced by the size of the basal rosettes influences the total number of flowers produced (Aker 1982). In other words, the population growth and density of yuccas and moths may be influenced by situations unrelated to their strong mutualistic relationship.

A more definitive example of the population consequences of mutualism is provided by a demographic analysis of an ant-seed mutualism by Hanzawa, Beattie, and Culver (1988). It involves a guild of ants and golden corydalis *(Corydalis aurea)*, an annual or biennial widely distributed in open or disturbed sites in the northeastern and western continental United States, Canada, and Alaska. The three compared the survivorship of both seeds and plants, fecundity, reproduction, and growth rates of two seed cohorts of the plant, one relocated to an ant nest by undisturbed ant foragers and a control cohort of equal numbers hand-planted near each nest. The ant-handled cohort had significantly higher survivorship than the control cohort (Figure 23.4). The ant-handled cohort produced 90 percent more seeds than the control cohort; and its net reproductive rate R_o was 8.0 and that of the control 4.2. The finite rate of increase of the ant-handled cohort was 2.83 per year, compared to 2.05 per year for the control. The ant-handled cohort experienced greater reproductive success not because of any great difference in the fecundity of the reproductive plants but because of a significantly higher survival to reproductive age. Survival was due to the superior microsites of the ant nests, not the removal of seeds from the vicinity of parent plants or the distance moved.

In recent years theoretical ecologists have been attempting to model mutualism in much the same manner as predation, parasitism, and competition, some involving Lotka-Volterra equations (Whittaker 1975; Christiansen and Fenchel 1977; Vandermeer and Boucher 1978; Heithaus, Culver, and Beattie 1980; May 1981; Keeler 1981; Dean 1983). Because each species benefits the other, an increase in one population directly influences and is directly influenced by an increase in the other. The population growth equation of each must include a term for the rate of increase of the other. However, because there is an upper limit to the population growth based on carrying capacity, the influence of each has to have a saturation point. Addicott (1979), for example, observed that in ant-aphid

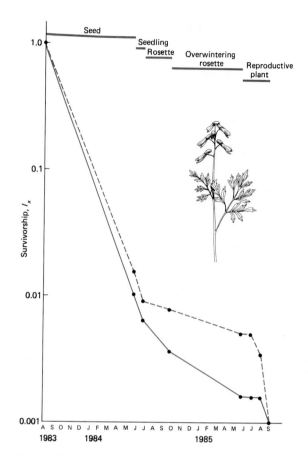

Figure 23.4 Survivorship curves for ant-handled (dashed line) and control (solid line) cohorts of golden corydalis (*Corydalis aurea*). The mutualism between ants and the plant results in a higher survivorship of overwintering rosettes and reproductive plants. (After Hanzawa, Beattie, and Culver 1988: 7.)

mutualisms, the response of the aphids to tending was greatest when the local populations were small, and declined as density of local populations increased. Most models, however, fail to include any variable to stop unbounded growth. To cap such growth and bring stability to the system, some outside regulatory force such as predation must be brought in (Heithaus, Culver, and Beattie 1980), as often happens in real-life mutualisms involving seed dispersal. Further, the environment may impose its own limit on population growth. In addition, in a plant-pollinator system a certain population density of both interacting species is necessary before any equilibrium is possible and both populations reach maximum stable densities. If plant density is too low and pollinators have difficulty finding plants, the pollinators may decline below replacement level. Finally, it is difficult to develop a realistic two-species model analogous to the predator-prey and interspecific competition models, because so many mutualistic systems are diffuse, involving arrays of species in different taxons rather than one-to-one relationships.

■ Summary

When predators and parasites seek prey and hosts, prey and hosts seek escape; each is imposing selection pressure on the other. These reciprocal selection pressures result in coevolution. Certain traits of each interacting species evolves a response to the trait of another. A positive reciprocal relationship between two species is mutualism, which may have evolved from predator-prey, host-parasite, and commensal relationships.

Mutualism may be symbiotic (living together in close relationship) or nonsymbiotic and obligatory or nonobligatory. Obligate symbiotic mutualists are physically dependent on each other, one usually living within the tissues of the other. Obligate nonsymbiotic mutualists depend on each other but they live independent lives. Nonobligate, faculative mutualists include interactions among guilds of species involved in seed dispersal and pollination. In exchange for dispersal of pollen and seeds, plants reward animals with food—fruit, nectar, and oil. To reduce wastage of pollen, some plants possess morphological structures that permit only certain animals to reach the nectar.

Population effects of mutualism are difficult to model because so many mutualistic systems are diffuse, involving arrays of species in different taxons rather than one-to-one relationships. Because an increase in the population of the one species results in an increase in the population of the other, models must incorporate some variable to halt unbounded growth.

PART IV

THE COMMUNITY

Community Basics

- **Physical Structure**

    Vertical Stratification
    Horizontal Heterogeneity
    Edges and Ecotones

- **Biological Structure**

    Species Dominance
    Species Diversity
    Species Abundance

- **Island Biogeography**

    Theory
    Application

- **Interactions and Community Structure**

    Competition
    Predation
    Parasites and Diseases
    Mutualism

- **Community Pattern**

- **Classification Systems**

- **Summary**

Forest, grassland, hedgerow, lawn, stream, marsh—these and other habitats in the landscape are easily recognized as unique groupings of organisms or biotic communities. A *community* is a naturally occurring and interacting assemblage of plants and animals living in the same environment and fixing, utilizing, and transferring energy in some manner. Like many ecological terms, the word *community* has several meanings. Some ecologists use it to describe groups of similar organisms occu-

pying the same habitat, such as a community of desert lizards or grassland birds. Others use the term in an even more restricted sense to refer to specialized groups within a specific habitat, such as insect-feeding birds in the forest canopy. In this case, ecologists are referring to *guilds,* well-defined groups interrelated in the manner in which they exploit the environment. In all cases, however, the term *community* does refer to an assemblage of interacting species.

Our broad definition of a community encompasses all populations of plants and animals with varying degrees of interaction. Within any community the interaction among many groups is minimal or at best indirect. For example, the assemblage of organisms occupying a fallen log on the forest floor may interact with each other, but they have minimal relationships with organisms living in the upper canopy of the forest. Within the community relationships involve a matter of scale in size, time, and space.

A community may be autotrophic in the sense that it includes photosynthetic plants and gains its energy from the sun. Other communities, such as springs and caves, are heterotrophic; they depend upon the input of fixed energy, such as detritus, from the outside. Autotrophic communities usually contain a number of heterotrophic microcommunities such as fallen logs.

The nature of the community is determined by the adaptations of its organisms to the physical environment—soil, temperature, moisture, light, nutrients, and the like—and by interactions among its organisms—competition, predation, parasitism, and mutualism. The adaptations and interactions influence such attributes of the community as structure, dominance, species diversity, and niches. The community, in effect, is a product of the evolutionary processes that have adapted individual species to the environment and to each other.

■ Physical Structure

The most easily observed feature of a given community is its physical structure. There are pronounced differences between a grassland and a forest and between a stream and a lake. Differences in terrestrial communities are determined largely by vegetation. Differences in aquatic communities are determined largely by the depth and flow of water.

The form and structure of terrestrial communities are determined by the nature of the vegetation. Vegetation may be classified according to growth form. The plants may be tall or short, evergreen or deciduous, herbaceous or woody. We might speak of trees, shrubs, and herbs, and then further subdivide the categories into needle-leafed evergreens, broad-leafed evergreens, evergreen sclerophylls, broad-leafed deciduous, thorn trees and shrubs, dwarf shrubs, ferns, grasses, forbs, and lichens.

A more useful system is the one designed in 1903 by the Danish botanist Christen Raunkiaer. Instead of considering plants' growth form, he classified plants by life form, the relation of their height above ground to their perennating tissue (tissue that survives from one growth season to the next, remaining inactive over winter or dry periods). Perennating tissue is the embryonic or meristemic tissue of buds, bulbs, tubers, roots, and seeds. Raunkiaer recognized five principal life forms (Table 24.1 and Figure 24.1). All the species in a region can be grouped into six life form classes: therophytes, cryptophytes, hemicryptophytes, chamaephytes, phanerophytes, and epiphytes. The ratio among these life form classes expressed as a percentage provides a life form spectrum for the area that reflects the plants' adaptations to the environment, particularly climate (see Table 24.2 and Figure 24.2). A community with a high percentage of perennating tissue well above the ground (phanerophytes) would be characteristic of warm climates; a community consisting mostly of chamaephytes and hemicryptophytes would be

TABLE 24.1 Raunkiaer's Life Forms

Name	Description
Therophytes	Annuals survive unfavorable periods as seeds. Complete life cycle from seed to seed in one season.
Geophytes (Cryptophytes)	Buds buried in the ground on a bulb or rhizome.
Hemicryptophytes	Perennial shoots or buds close to the surface of the ground; often covered with litter.
Chamaephytes	Perennial shoots or buds on the surface of the ground to about 25 cm above the surface.
Phanerophytes	Perennial buds carried well up in the air, over 25 cm. Trees, shrubs, and vines.
Epiphytes	Plants growing on other plants; roots up in the air.

Figure 24.1 Raunkiaer's life forms. (1) phanerophytes; (2) chamaephytes; (3) hemicryptophytes; (4) geophytes (cryptophytes); (5) therophytes. The parts of the plant that die back are unshaded; the persistent parts with buds (or seeds in the case of therophytes) are dark.

TABLE 24.2 Analysis of Life-Form Spectra of Two Plant Communities: a New Jersey Pine Barren and a Minnesota Jack Pine Forest

Basis of Spectrum	Community	Number of Species	Percentage				
			Ph	Ch	H	G	Th
Species list	New Jersey	19	84.2	0	10.5	5.2	0
	Minnesota	63	23.8	4.7	60.3	7.9	3.1
Cover	New Jersey	19	98.1	0	1.9	0.	0.
	Minnesota	63	11.8	2.5	55.6	28.7	1.4

Source: Stern and Buell 1951.

characteristic of cold climates; and a community dominated by therophytes would be characteristic of deserts.

Vertical Stratification

A distinctive feature of the community is vertical stratification (Figure 24.3; see Chapters 30–33). Stratification of a community is determined largely by the life form of plants—their size, branching, and leaves—which in turn influences and is influenced by the vertical gradient of light. The vertical structure of the plant community provides the physical structure in which many forms of animal life are adapted to live.

A well-developed forest ecosystem has a highly stratified structure with a large variety of components. It consists of several layers of vegetation, each of which provides a habitat for animal life in the forest. From top to bottom these layers are the *canopy,* the *understory tree layer,* the *shrub layer,* the *herb* or *ground layer,* and the *forest floor.* We can continue down into the root layer and the soil strata. The canopy is the major site of primary production and has a pronounced influence on the structure of the rest of the forest. If it is fairly open, considerable sunlight reaches the lower layers, and the shrub and understory tree strata are well developed. If the canopy is closed, the understory trees, the shrubs, and even the herbaceous layer are poorly developed. The nature of both the shrub and herb layers depends upon soil moisture conditions, slope position, density of the overstory, and aspect of slope, all of which can vary from place to place throughout the forest. The final layer, the forest floor, depends again on all these factors and in

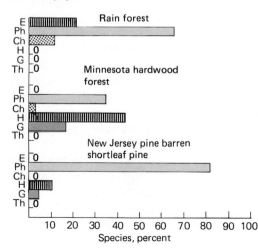

Figure 24.2 Life form spectra of a tropical rain forest (adapted from Richards 1952), a Minnesota hardwood forest (data from Buell and Wilbur 1948), and a New Jersey pine barren (data from Stern and Buell 1951). Note the absence of hemicryptophytes, geophytes, and therophytes from the tropical rain forest and the prominence of epiphytes. The pine barrens are dominated by phanerophytes.

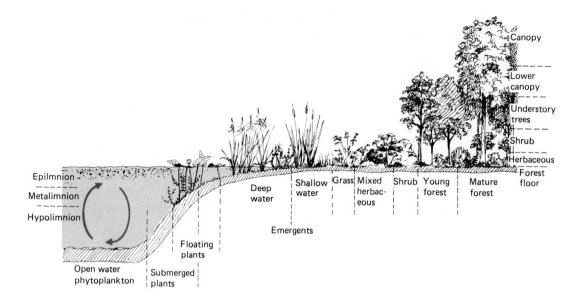

Figure 24.3 Vertical structure of communities from aquatic to terrestrial. In all, the zone of decomposition and regeneration is the bottom stratum and the zone of energy fixation is the upper stratum. In the sequence of stages from aquatic to terrestrial, stratification and complexity of the community become greater. Stratification as it affects the distribution of species and their niches is largely physical, influenced by gradients of oxygen, temperature, and light. Development of stratification in terrestrial communities is largely biological. Dominant vegetation affects the physical structure of the community and microclimate conditions of temperature, moisture, and light. Because the forest has four to five strata, it can support a greater diversity of life than a grassland with basically two strata. Floating and emergent aquatic plant communities can support a greater diversity of life than open water.

turn determines how and what nutrients are released for recycling for the growth of the other layers.

Other communities have a similar, if not as highly stratified, structure. Grasslands have a herbaceous layer that changes through the seasons, a ground or mulch layer, and a root layer. The root layer is more pronounced in grasslands than in any other ecosystem, and the mulch layer has a pronounced influence on plant development and animal life, especially insects and small mammals.

The strata of aquatic communities are determined by light penetration, temperature profile, and oxygen profile (for details, see Chapter 10). Well-stratified lakes in summer contain the epilimnion, a layer of freely circulating surface water; the metalimnion, characterized by a thermocline; the hypolimnion, a deep layer of dense water about 4° C, often low in oxygen; and a layer of bottom mud. In addition, two structural layers based on light penetration are recognized—an upper zone roughly corresponding to the epilimnion, dominated by plant plankton and the site of photosynthesis, and a lower layer in which decomposition is most active. The lower layer roughly corresponds to the hypolimnion and bottom mud.

Communities, whether terrestrial or aquatic, have similar biological structures. They possess an autotrophic layer, which fixes the energy of the sun and manufactures food from inorganic substances. It consists of the area where light is most available: the canopy of the forest, the herbaceous layer of the grassland, and the upper layer of water of the lake and sea. Communities also possess a heterotrophic layer, which utilizes the food stored by the autotrophs, transfers energy, and circulates nutrients by means of predation in the broadest sense and decomposition.

Each vertical layer in the community is inhabited by its own more or less characteristic organisms. Although considerable interchange takes place among several strata, many highly mobile animals confine themselves to only a few layers, particularly during the breeding season. Occupants of the vertical strata may change during the day or season. Such changes reflect daily and seasonal variations in humidity, temperature, light, oxygen content of water, and other conditions or the different requirements of organisms for the completion of their life cycles. For example, D. L. Pearson (1971) found that birds occupying the upper strata of a tropical dry forest in Peru moved to the lower strata during the middle of the day for several reasons: to secure food (insects move to lower levels), to escape heat stress, to escape a high degree of solar radiation, and to conserve moisture.

In general, the finer the vertical stratification of a community, the more diverse is its animal life. The variety of life in a terrestrial community is in part a function of the number and development of layers of vegetation. If a certain layer is absent, then the animal life it normally shelters

and supports is also missing. Thus grassland with few strata is poorer in species than a highly stratified forest ecosystem (see Karr and Roth 1971). The distribution of life and biological activity in aquatic systems is to a large extent governed by vertical gradients of light, temperature, and oxygen; the greater the variation along these gradients, the greater the diversity of life the system supports.

Horizontal Heterogeneity

Walking across a typical old field, you move through patches of open grass, clumps of goldenrods, tangles of blackberry, and small thickets of sumac and other tall shrubs. Continue into an adjacent woodland and you may cross through open understory, patches of shade-tolerant undergrowth of laurel and viburnum, and gaps in the canopy where dense thickets of new growth have claimed the sunlit openings. The mosaic of vegetation patches form a quiltwork across the landscape. These patches, each spatially separated from one another, produce a horizontal heterogeneity, which adds to the physical complexity of the environment.

Horizontal heterogeneity results from an array of environmental and biological influences (Figure 24.4). Soil structure, soil fertility, moisture conditions, and aspect influence the microdistribution of plants. Patterns of light and shade shape the development of understory vegetation. Runoff and small variations in topography and microclimate produce well-defined patterns of plant growth. Grazing animals have subtle but important effects on the spatial patterning of vegetation, as do abiotic disturbances such as windthrow and fire (see Chapter 26). Among plants

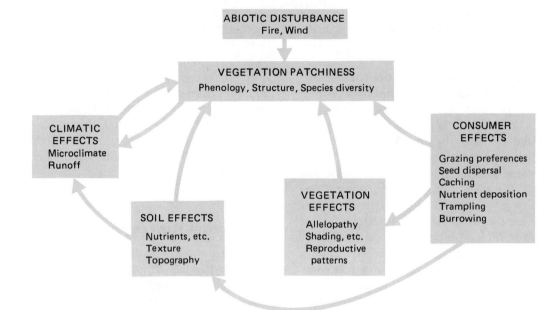

Figure 24.4 General relationships of some of the major influences that govern vegetational patchiness in terrestrial environments. (From Wiens 1976.)

opportunities for recruitment, growth, reproduction, and survival vary spatially, as do variations in biological interactions such as competition for space. The mode of plant reproduction and availability of propagules over time affect vegetation patterns. Plants with wind-dispersed and animal-dispersed seeds have a wider distribution across the landscape than plants with heavy seeds. Vegetative or clonal reproduction produces distinctive clumps or patches of certain plants in an otherwise homogeneous environment. Allelopathic effects and shading lead to the suppression of some plant species and to the development and growth of others. A patchy environment (see Wiens 1976) in turn influences the distribution of animal life across the landscape.

Edges and Ecotones

Closely associated with horizontal stratification are edges and ecotones. Although the two terms are often used synonymously, they are different. An *edge* is where two or more different vegetational communities meet. An *ecotone* is where two or more communities not only meet but also intergrade (Figure 24.5).

Some edges result from abrupt changes in soil type, topographic differences, geomorphic differences (such as rock outcrops), and microclimatic changes. Because adjoining vegetation types are determined by long-term natural features, such edges are usually stable and permanent and are considered *inherent* (Thomas et al. 1979). Other edges result from such natural disturbances as fires, storms, and floods or from such human-induced disturbance as livestock grazing, timber harvesting, land clearing, and agriculture. The adjoining vegetational types are suc-

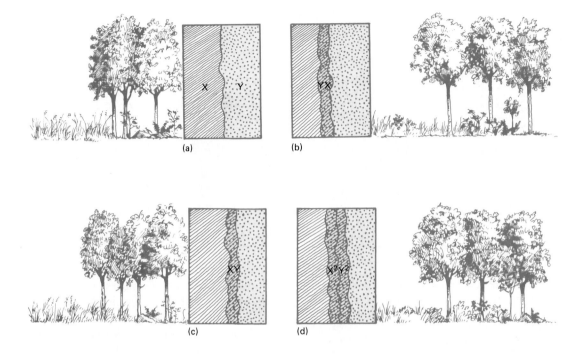

(a)

(b)

(c)

(d)

Figure 24.5 Edge and types of ecotones. (a) Abrupt, narrow edge with no development of an ecotone. (b) Narrow ecotone developed by advancement of community Y into community X to produce ecotone YX. (c) Community X advances into community Y to produce ecotone XY. (d) Ideal ecotone development in which plants from both communities invade each other to create a wide ecotone X^2Y^2. This type of ecotone will serve the most edge species.

cessional or developmental (see Chapter 25) and will change or disappear with time. Such edges are termed *induced*. They can be maintained only by periodic disturbance (see Chapter 26). Induced edges, too, may be abrupt or transitional, resulting in an ecotone.

Ecotones arise from the blending of two or more vegetational types. Plants competitively superior and adapted to en-vironmental conditions existing in the edge advance as far into either community as their ability to maintain themselves will allow (Figure 24.5). Beyond this point interior plants of adjacent communities maintain themselves. As a result, the ecotone exhibits a shift in dominants of certain species of each community.

The ecotone also involves a number of highly adaptable species that tend to colonize such areas. Plants of the edge tend to be opportunistic (see Chapter 25) and shade-intolerant. They grow well in a relatively xeric environment, including a high rate of evapotranspiration, reduced soil moisture, and fluctuating temperatures. Animal species of the edge are usually those that require two or more vegetational communities. For example, one edge species, the ruffed grouse (*Bonasa umbellatus*), requires forest openings with

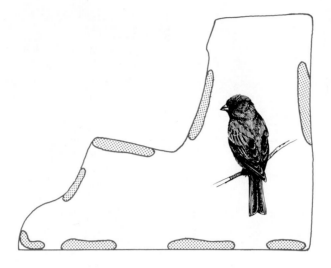

Figure 24.6 Map of territories of a true edge species, the indigo bunting *(Passerina cyanea)*, which inhabits woodland edges, larger gaps in forests creating edge conditions, hedgerows, and roadside thickets. The male requires tall, open song perches and the female a dense thicket in which to build a nest. (After Whitcomb et al. 1981: 143.)

an abundance of herbaceous plants and low shrubs, dense sapling stands, pole timber for nesting cover, and mature forests for winter food and cover. Because the ruffed grouse spends its life in an area of 5 to 10 hectares, this amount of land must provide all of its seasonal requirements. Some species, such as the indigo bunting, are restricted exclusively to edge situations (Figure 24.6). Because of species responses, the variety and density of life are often greatest in and about edges and ecotones. This phenomenon has been called the *edge effect* (Leopold 1933).

Edge effect is influenced by the amount of edge available—its length, width, and degree of contrast between adjoining vegetational communities (Patton 1975). The greater the contrast between

adjoining plant communities, the greater the species richness should be (Figure 24.7). An edge between a forest and grassland should support more species than an edge between a young and a mature forest. The larger the adjoining communities, the more opportunity exists for flora and fauna of adjoining communities as well as species that favor edge situations to occupy the area. If patches of vegetation are too small to support species characteristic of larger patches of that habitat, then the area becomes a homogeneous community dominated by edge species.

The edge effect comes about because environmental conditions differ from those of adjacent vegetational communities, especially adjoining forests (Ranney 1977). Environmentally, such edges reflect steep gradients of wind flow, moisture, temperature, and solar radiation between the extremes of open land and forest in-

Figure 24.7 Inherent and induced edges. Edges of high contrast involve widely different adjacent communities, such as shrub and mature forest. Edges of low contrast involve two closely related successional stages, such as shrubs and sapling growth. (After Thomas et al. 1977.)

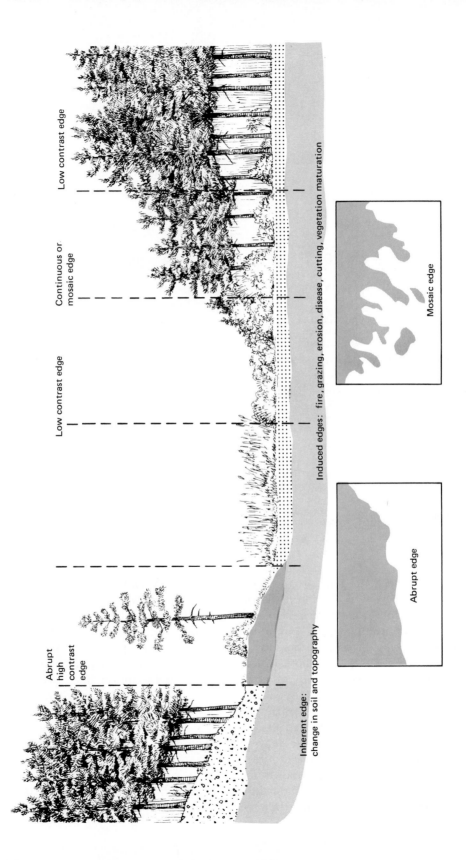

Abrupt high contrast edge

Low contrast edge

Continuous or mosaic edge

Low contrast edge

Inherent edge: change in soil and topography

Induced edges: fire, grazing, erosion, disease, cutting, vegetation maturation

Mosaic edge

Abrupt edge

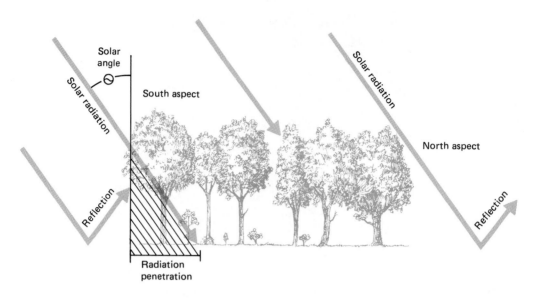

Figure 24.8 Influence of solar radiation on the nature of the edge. Solar radiation does not affect all edge aspects equally. North-facing edges receive almost no direct sunlight and limited reflection from nearby fields, whereas south-facing edges receive considerable solar radiation. The depth to which solar radiation penetrates forest edge on level ground depends upon the height of the canopy and solar angle. Within the zone of radiation penetration light intensities and summer daytime temperatures are higher than in forest interiors sheltered by tree canopies. (After Ranney 1977.)

terior. Wind velocity is greater at the forest's edge, helping to create xeric conditions in and around the edge. Evaporation increases, placing increased demands on soil moisture by plants. Variations in solar radiation, both directed and reflected, probably have the greatest influence on forest edge, especially as it relates to north-facing and south-facing edges (Figure 24.8). A south-facing edge may receive three to ten times more hours of sunshine a month during midsummer than a north-facing edge, making it much warmer and drier. Also affecting the phys-

ical environment of the edge is the depth to which solar radiation penetrates the vertical edge of the forest. This penetration depends upon solar angle, edge aspect, density and height of vegetation, latitude, time of year, and time of day.

Because of increased exposure to wind and direct radiation, vegetation of the edge is subject to different environmental stresses than forest interior species. Trees on the forest edge are under greater heat stress and dissipate proportionately more heat by evaporation than by radiation; interior forest trees dissipate more heat by radiation. Higher light intensities result in sun scald of some tree species suddenly exposed to direct solar radiation and increased crown expansion and epicormic branching (development of new limbs on the bole of a tree) in others. Some mesic, shade-tolerant trees succumb. High light intensity and xeric conditions favor those plant species that are tolerant of light and xeric soil moisture conditions, are capable of high root competition for moisture and nutrients, reproduce vegetatively, and depend on birds and mammals for distribution of seeds. Density and frequency of saplings and stump sprouts are greater in border areas.

As the edge canopy thickens, shade-tolerant plants appear. Eventually the edge becomes a mixture of both interior and edge species. Generally more interior-related species are found in the edge than edge species are found in the interior, although the interior holds a greater proportion of rare species (see Gysel 1951; Wales 1972; Ambuel and Temple 1983; Blake and Karr 1984; Freemark and Merriam 1986).

Certain attributes influence the amount of edge and its richness. The amount of edge available, its length, width, and configuration (abrupt or mosaic), is often influenced by adjoining land use practices (Figure 24.9). Richness is a product of the combination of flora and fauna from the adjoining communities and the addition of species that favor edge situations. The degree of richness depends upon the contrast between adjoining plant communities and stand size. The greater the contrast between plant communities, the greater may be the richness. Edge between forest and grassland, for example, should be richer than edge between shrub stand and sapling forest. As will be emphasized later in this chapter in the section on island biogeography, the size of the stand between certain upper and lower limits also influences richness. There is a limit to how small a vegetation patch can be and still support a rich edge. A large homogeneous stand of vegetation can be so broken up by edge into smaller and smaller stands or patches that heterogeneity once provided by edge becomes itself a homogeneous community dominated by edge species. Small woodlands, for example, set in grasslands or other types of vegetation are essentially edge habitats rather than forest and edge. A minimum size of habitat block is needed to achieve a maximum number of species. The point at which maximum diversity tends to decrease is when the average size of the habitat block becomes smaller than the size required to support the maximum number of species present.

The amount of edge relative to the size of forest interior or vegetation patch may have some impact on future trends in vegetation development. Exposed as they are to animals and wind, plant propagules of edge species have a higher probability of dispersal than forest interior species. As the ratio of cleared to forest land increases with a proportionate increase in edge, forest edge species may contribute more heavily to propagules available for transport between stands. Thus edge species have a selective advantage in forest regeneration, which could shift the composition of future forest stands toward edge and xeric species.

■ Biological Structure

The community is influenced not only by physical or abiotic conditions, but also by biological conditions: the abundance and the diversity of species, and the interactions among them. Biologically controlled communities are often influenced by a single species or by a group of species that modify the environment. These organisms are called *dominants*.

Species Dominance

It is not easy to determine what constitutes a dominant species or, in fact, to determine the dominant species. The dominants in a community may be the most numerous, possess the highest biomass,

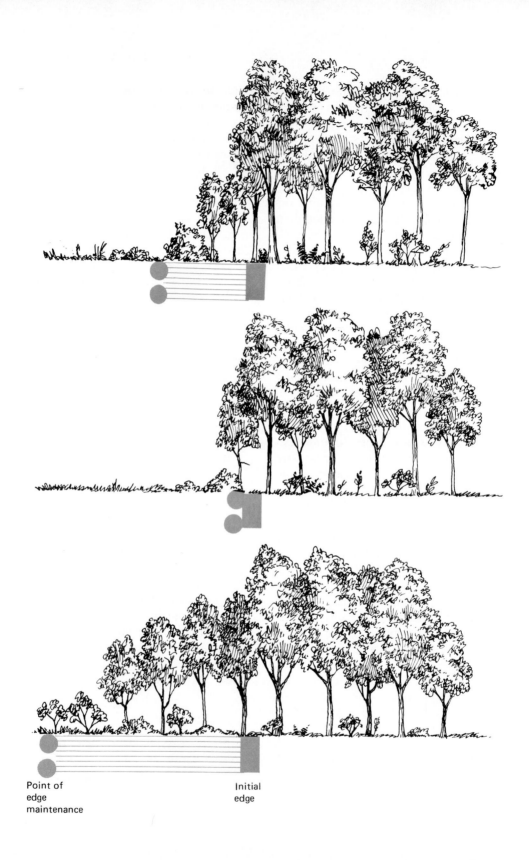

Point of
edge
maintenance

Initial
edge

Figure 24.9 Edge structure as influenced by human disturbance. When the disturbance line is adjacent to the tree line, the edge is abrupt, with an overlapping or cantilevered canopy. Such an edge is of minimal value to wildlife. If the disturbance line lies outside the overhanging canopy and away from the tree trunks, woody vegetation and perennial herbaceous plants develop beneath the drip line of the canopy. If the disturbance is minimal, woody vegetation extends well away from the initial edge. (After Ranney 1977.)

preempt the most space, make the largest contribution to energy flow or mineral cycling, or by some other means control or influence the rest of the community.

Some ecologists assign the role of dominance to numerically superior organisms, but numerical abundance alone is not sufficient. A species of plant, for example, may be widely represented yet exert little influence on the community as a whole. In the forest small or understory trees may be numerically superior; yet the nature of the community is controlled by a few large trees that overshadow them. In such a situation dominance is measured not by number but by biomass or basal area.

The dominant organism may be scarce, yet by its activity control the nature of the community. The predatory starfish *Piaster,* for example, preys on a number of species similar in habit and thereby reduces competitive interactions among them, so these different prey species coexist (Paine 1966). If the starfish is removed, a number of prey species disappear and one of them becomes dominant. In effect, the predator controls the structure of the community and so must be regarded as the dominant or keystone species.

The degree of dominance expressed by any one species appears to depend in part on the position it occupies on a phys-

ical or chemical gradient. A species becomes dominant because it can exploit a range of environmental conditions more efficiently and adapt to a wider range of ecological tolerances than associated species. For example, at one point on a moisture gradient species A and species B may be the dominants. As the gradient becomes drier, species B may assume a subdominant position in the community, and its place might be taken by species C. Nutrient enrichment, too, can change the structure of a community. Lakes receiving excess sewage discharges shift from a diverse assemblage of nutrient-thrifty diatoms to a few blue-green algae that can exploit a nutrient-rich system (see Edmundson 1970).

To determine dominance ecologists use several approaches. They measure relative abundance of species, comparing the numerical abundance of one species to the total abundance of all species (see Appendix A). They measure relative dominance, a ratio of basal area occupied by one species to total basal area, or they use relative frequency as a measure. Often all three measurements are combined to arrive at an importance value for each species. Most species do not arrive at a high level of importance in the community, but those that do serve as index species. Once species in a community have been assigned importance values, the stands can be grouped by their leading dominants. Such techniques are useful in the study and ordination of communities on some environmental gradient, such as moisture (see Appendix A).

Species Diversity

Species Richness and Evenness.
Among the array of species that make up a community, few are abundant. Individ-

uals of most species make up only a small proportion of the total population in the community. Consider the structure of the tree component of a mature woodland consisting of 24 species over 10 cm dbh (diameter breast height), as presented in Table 24.3. Individuals of two tree species, yellow-poplar and white oak, make up nearly 44 percent of the stand. The next most abundant trees—black oak, sugar maple, red maple, and American beech—make up from under 7 percent to just over 5 percent of the stand each. Nine species range from 1.2 to 4.7 percent of the stand, and the eight remaining species

as a group represent about 5 percent of the stand. Data in Table 24.4 show a woodland sample of somewhat different composition. That community consists of ten species of which two, yellow-poplar and sassafras, make up almost 84 percent of the stand.

These two forest communities illustrate a typical pattern in temperate regions—a few common species associated with less abundant ones. They also illustrate two other characteristics of distribution of species within a community—*species richness*, the number of species, and *evenness*, the relative abundance of individ-

TABLE 24.3 Structure of Vegetation of a Mature Deciduous Forest in West Virginia

Species	Number	Percentage of Stand	Species	Number	Percentage of Stand
Yellow-poplar (*Liriodendron tulipifera*)	76	29.7	Bitternut hickory (*Carya cordiformis*)	5	2.0
White oak (*Quercus alba*)	36	14.1	Pignut hickory (*Carya glabra*)	3	1.2
Black oak (*Quercus velutina*)	17	6.6	Flowering dogwood (*Cornus florida*)	3	1.2
Sugar maple (*Acer saccharum*)	14	5.4	White ash (*Fraxinus americana*)	2	.8
Red maple (*Acer rubrum*)	14	5.4	Hornbeam (*Carpinus caroliniana*)	2	.8
American beech (*Fagus grandifolia*)	13	5.1	Cucumber magnolia (*Magnolia grandiflora*)	2	.8
Sassafras (*Sassafras albidum*)	12	4.7	American elm (*Ulmus americana*)	1	.39
Red oak (*Quercus rubra*)	12	4.7	Black walnut (*Juglans nigra*)	1	.39
Mockernut hickory (*Carya tomentosa*)	11	4.3	Black maple (*Acer nigrum*)		
Black cherry (*Prunus serotina*)	11	4.3	Black locust (*Robinia pseudoacacia*)	1	.39
Slippery elm (*Ulmus rubra*)	10	3.9	Sourwood (*Oxydendrum arboreum*)	1	.39
Shagbark hickory (*Carya ovata*)	7	2.7	Tree of heaven (*Ailanthus altissima*)	1	.39
				256	100.00

TABLE 24.4 Structure of Vegetation of a Second Deciduous Forest in West Virginia

Species	Number	Percentage of stand
Yellow-poplar (*Liriodendron tulipifera*)	122	44.5
Sassafras (*Sassafras albidum*)	107	39.0
Black cherry (*Prunus serotina*)	12	4.4
Cucumber magnolia (*Magnolia grandiflora*)	11	4.0
Red maple (*Acer rubrum*)	10	3.6
Red oak (*Quercus rubra*)	8	2.9
Butternut (*Juglans cinerea*)	1	.4
Shagbark hickory (*Carya ovata*)	1	.4
American beech (*Fagus grandifolia*)	1	.4
Sugar maple (*Acer saccharum*)	1	.4
	174	100

uals among the species. The more equitable the distribution of individuals among the species, the greater is evenness. Species diversity increases as the number of species increases and as the number of individuals in the total population are more evenly distributed among them. The stand described in Table 24.3 is richer and has greater evenness than the stand in Table 24.4.

In order to quantify species diversity for purposes of comparison, a number of indexes have been proposed (see Appendix B; Lloyd and Ghelardi 1964; Pielou 1971, 1975). One of the most widely used is the Shannon Index, which includes both richness and equitability.

The Shannon Index measures diversity by the formula

$$H = -\sum_{i=1}^{s} (p_i)(\log p_i)$$

where
H = diversity index
s = number of species
i = species number
p_i = proportion of individuals of the total sample belonging to the ith species

This index, based on information theory, is a measure of uncertainty. The higher the value of H, the greater is the probability, or uncertainty, that the next individual chosen at random from a collection of species containing N individuals (in our example the species of tree) will not belong to the same species as the previous one. The lower the value, the greater the probability that the next individual encountered will be the same species as the previous one. Thus in the woodland with low diversity (84 percent yellow-poplar and sassafras) the probability is high that in sampling the trees the next tree picked at random will be a yellow-poplar or a sassafras. In the woodland with higher diversity, the chances that the next tree encountered at random will be a yellow-poplar or a white oak are considerably less.

The more abundant species are not necessarily the most influential. In communities embracing organisms with a wide range of sizes, an index may lead us to underestimate the importance of fewer but larger individuals and overestimate that of the more common species. One of the distinctive failures of the indexes is the inability to distinguish between the abundant and the important species. Nevertheless, diversity indexes do provide one measure of community differences. (For discussion of indexes, see Hurlbert 1971; Peet 1974.)

Diversity indexes may be used to compare species diversity within a community (alpha diversity, α), between communities or habitats (beta diversity, β), and among communities over a geographical area (gamma diversity, γ) (Whittaker 1972).

Local versus Regional Diversity. The richness of species in a local area reflects the uniqueness of the local habitat, which is exposed to specific sets of conditions. The presence or absence of species in a local area is influenced by availability of nutrients, moisture, and food, disturbance, and the physical condition of the environment, all of which change over space and time. The death of one or several large trees in the forest can open up a microsite and bring in shrubby growth and associated wildlife. The influence of that disturbance is temporary and will decline as the vegetation grows. Certain species may persist in a given locality only because immigrants move in from more productive areas. Some plant species may occur locally only because persistent disturbances maintain potential habitats.

Local diversity reflects and is influenced by regional diversity, which is a product of climatic history, historical accidents, and the geographical position of dispersal barriers (Ricklefs 1987). The plowing of the Great Plains, a historical event, reduced diversity of grassland wildlife. The massive clearing of the eastern deciduous forest in North America, which reduced a dispersal barrier, and the subsequent conversion of much of the area to farmland allowed midwestern grassland species to move eastward, increasing diversity in the deciduous forest region.

Global Diversity. Global diversity is influenced by both latitudinal and altitudinal gradients. Species of nesting birds (Fischer 1960), mammals (Simpson 1964), fish (Lowe-McConnell 1969), lizards (Pianka 1967), and trees (Monk 1967) decrease latitudinally from the tropics to the arctic. Species diversity of marine life increases from the continental shelf, where food is abundant but the environment is changeable, to deep water, where food is less abundant but the environment is more stable. Mountain areas generally support more species than flatlands because of topographic diversity, and peninsulas have fewer species than adjoining continental areas. From east to west in North America the number of species of breeding land birds (MacArthur and Wilson 1967) and mammals (Simpson 1964) increases. This increased diversity on an east-west gradient relates to an increased diversity of the environment both horizontally and altitudinally (Figure 24.10). Eastern North America has more uniform topography and climate and therefore fewer species than western North America. However, because of more favorable moisture conditions, amphibians are more abundant and diverse in eastern North America than in the western part of the continent, whereas reptiles are more diverse in the hot, arid regions of western North America (Kiester 1971) (Figure 24.11).

Species Diversity Hypotheses. Many hypotheses have been proposed to explain why the tropics should hold a greater abundance of species than the temperate regions or why one island and locality should hold more species than another.

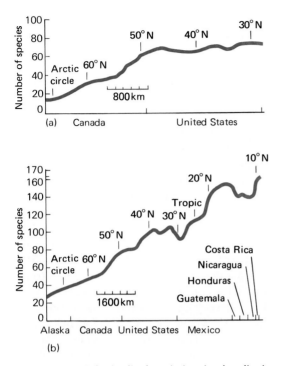

(a) Canada   United States

(b) Alaska Canada United States Mexico

Figure 24.10 Latitudinal variation in the distribution of mammals. Diversity of mammals reflects both latitudinal and altitudinal variations. (a) Species density of North American mammals from the Arctic to the Mexican border along the 100th meridian. (b) Species density from the Arctic through Central America. (After Simpson 1964.)

required for a species to disperse into unoccupied areas of suitable habitat. Because there has not been enough time for many species to move into temperate zones, these areas are unsaturated by the total number of species they now support. Many cannot move until barriers to dispersal are broken; others are moving out

Figure 24.11 Latitudinal variation in the distribution of reptiles (a) and amphibians (b). The most pronounced latitudinal variations occur among amphibians and reptiles. Being poikilothermic and ectothermic, reptiles have their greatest diversity in hot desert regions and lower latitudes of North America. Being not only poikilothermic but also highly sensitive to moisture conditions, amphibians reach their greatest diversity in the central Appalachians and decrease northward, southward, and westward. Species numbers are lowest in dry and cold regions. (From Keister 1971.)

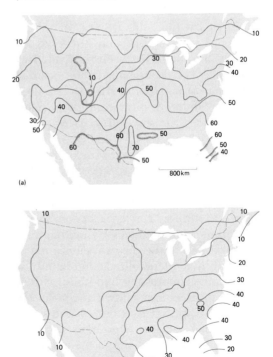

Many of these hypotheses are similar, but not identical.

The *evolutionary time hypothesis* (Fisher 1960; Simpson 1964) proposes that diversity relates to the age of the community. Old communities (in an evolutionary sense) hold a greater diversity than young communities. Tropical communities are older and evolve and diversify faster than temperate or arctic communities, in part because the environment is more constant and climatic catastrophes less likely.

Considering a shorter time scale, the *ecological time theory* is based on the time

of the tropics into the temperate regions. Examples are the natural spread of the cattle egret into North America from Africa by way of South America and the northward spread of the armadillo.

The *spatial heterogeneity theory* (Simpson 1964) holds that the more complex and heterogeneous the physical environment, the more complex will its flora and fauna be. The greater the variation in topographic relief, the more complex the vertical structure of the vegetation, and the more types of habitats the community contains, the more kinds of species it will hold. This theory is supported by the fact that forests with marked vertical structure hold more species of birds (MacArthur 1972; Pearson 1971).

The *climatic stability theory* (Fischer 1960; Connell and Orias 1964) holds that the more stable the environment, the more species will be present. Through evolutionary time the tropics, of all regions of Earth, has probably remained the most constant and has been relatively free from severe environmental conditions that could have catastrophic effects on a population. Under tropical conditions selection favors organisms with narrow niches and specialized feeding habits. Because each species uses a smaller fraction of the total resources, more species are able to exist in regions of constant climate. At higher latitudes where the climate is severe and unpredictable, selection favors organisms with broad limits of tolerance for variations in the physical environment and with more generalized food habits.

Another interesting hypothesis closely related to climatic stability is the *productivity theory* advanced by Connell and Orias (1964). In brief, this hypothesis proposes that the level of diversity of a community is determined by the amount of energy flowing through the food web. The rate of energy flow is influenced by the limitation

of the ecosystem and by the degree of stability of the environment. The productivity theory in effect says that the more food produced, the greater the diversity. Although it is true in a general sense, there are too many exceptions. In some aquatic systems, for example, increased enrichment from sewage and other nutrient sources results in a decrease in diversity. Marine benthic regions of low productivity have a higher abundance of species than areas of high productivity (Sanders 1968). In tropical bird communities it appears that the number of ways energy is packaged rather than the total energy is best correlated with diversity (Karr 1975).

Sanders (1968) combined the environmental stability hypothesis and the time hypothesis into still another one, the *stability-time hypothesis*. This hypothesis assumes that two contrasting types of communities exist: the physically controlled and the biologically controlled. In physically controlled communities organisms are subjected to physiological stress by fluctuating physical conditions. Organisms in time evolve adaptive mechanisms to meet these conditions; but at least some of the time the organisms are subjected to severe physiological stress and the probabilities of reproductive success and survival are low. As a result diversity is low. In the biologically controlled community, physical conditions are relatively uniform over long periods of time and are not critical in controlling the species. Evolution proceeds along the lines of interspecific competition, one species adapting to the presence of another and sharing resources with it. The environment is more predictable, the physiological tolerances are low, and diversity is high. However, there is no wholly physically controlled or wholly biologically controlled community. Rather, any community is influenced by the interaction of the two.

The *competition theory* (Dobzhansky 1951; Williams 1964) states that in environments of high physical stress, such as the Arctic with its frigid cold and the temperate regions with their wide fluctuations in annual temperatures, selection is controlled largely by the physical variables. In more benign climatic regions biological competition becomes more important in the evolution of species and the specialization of niches.

A *predation theory* has been proposed as a source of species diversity, particularly on a more local and regional basis (Paine 1966). This theory proposes that a higher species diversity exists in those communities in which predators reduce the numbers of prey to a level where interspecific competition among them is greatly reduced, allowing the coexistence of a number of prey species.

All these hypotheses are based on the assumption that communities exist at competitive equilibrium: The rate of change of competitors is zero. Rarely are natural communities at equilibrium. Fluctuations in the physical environment, predation, herbivory, and all sorts of density-independent mortality keep natural communities in a state of nonequilibrium. Huston (1979) proposed a *dynamic equilibrium hypothesis* based on the differences in rates at which populations of competing species reach competitive equilibrium. The major determinant of diversity in nonequilibrium situations is the population growth rates of competitors. Most communities fail to achieve equilibrium because of a fluctuating environment and periodic reductions in populations. In the absence of disturbance, an increase in population growth of major competitors results in a low diversity. Low diversity also results if disturbances are so frequent that many populations fail to increase. Greatest diversity occurs at intermediate

levels of disturbance, in which a dynamic balance may become established between competitive displacement and frequency of population reductions. Low availability of basic nutrients would reduce growth rates of organisms requiring them, resulting in higher diversity. Communities with high productivities would have lower diversity because of high growth rates of superior competitors.

However intriguing these hypotheses may be, they are difficult to test experimentally. Nevertheless, it is apparent that species diversity can be related to a number of variables, such as the structure of the habitat, the diversity of microhabitats, the nature of the physical environment, climate and protection from its adverse effects, competition, predation, availability of food, nutrient availability, time, disturbances, historical accidents, and geographical barriers.

Species Abundance

Associated with species diversity, in fact central to it, is species abundance, the relationship between abundance of individuals within a species and the number of species having similar abundances. Theoretically, it represents the manner in which species divide up the niche space (Chapter 18). There are three hypotheses of species abundance. All involve plotting the relative abundance of each species in the community in order of its rank from the most to the least abundant.

The *random niche hypothesis* or broken stick model (MacArthur 1960) suggests that the boundaries of the niches of species can be treated as points cast at random on a line. The points then represent the niche boundaries. The model assumes that species in the community utilize the critical resource with no overlap between species. If the segments, representing the importance value of the species (the percentage of total density, biomass, and so on, of all species in a community that a single species represents), are plotted in a sequence from the longest to the shortest, then a curve like curve A in Figure 24.12 will result. This hypothesis is realistic only in rare instances (Hairston 1969). Although considered obsolete, it is still employed. The curve is approached only by some small samples of taxonomically related animals with stable populations and relatively long life cycles occupying a small homogeneous community, such as nesting birds in a forest.

The *niche preemption hypothesis* or geometric distribution supposes that the most successful or dominant species preempts the most space, the next most successful claims the next largest share of space, and the least successful occupies what little space is left. If the relative importance of each species is plotted in species sequences on a log scale, the result is a straight line (curve B in Figure 24.12) and the distribution of the species forms a geometric series. Such a distribution is achieved only by a few plant communities containing relatively few species and occupying severe environments such as a desert. In most plant and animal communities, species overlap in the use of space and resources.

The *log-normal hypothesis* (Preston 1962) supposes that niche space occupied by a species is determined by a number of conditions, such as resources, space, microclimate, and other variables, that affect the success of one species in competition with another. The relative importance of each species is determined by the way these variables affect it (see May 1979, 1981). This results in a bell-shaped curve or normal distribution of importance values (Figure 24.13). As in the broken stick model, the line can be divided into segments, but in this case each segment represents ranges in importance values arranged into frequency distribution classes called octaves. (For methodology see Ludwig and Reynolds 1988; Krebs 1989; Poole 1974.) The octaves represent a doubling of individuals—1–2, 2–4, 4–8, and so on. This doubling translates into taking logarithms of abundance to the base 2. The octave with the greatest number of individuals is equal to 0. The rest of the octaves are numbered in plus and minus directions away from the mode. If this dis-

Figure 24.12 Graphic representation of species abundance hypotheses. In these graphs, the importance values (the percentage of all species that a particular species represents), expressed as total density, total productivity, or some other measurement, are plotted against a sequence of species. In the random niche hypothesis (curve A), the boundaries are located at random positions along the line. In the niche preemption hypothesis or geometric distribution (curve B), the relative importance of each species is plotted in sequence on a log scale, resulting in a straight line. The log-normal hypothesis (curve C) supposes that species distribution is determined by a large number of variables that affect the competitive abilities of the species involved. (After Whittaker 1965.)

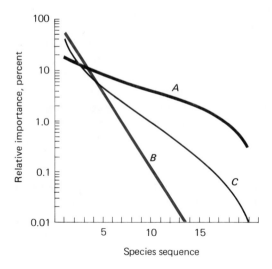

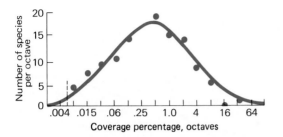

Figure 24.13 A bell-shaped curve of importance values resulting from the log-normal distribution of plant species. The importance value in this example is determined by the percentage ground surface covered by the species. It is represented on the horizontal scale (logarithmic) by octaves in which the species are grouped by doubling the units of percentage of cover. The largest number of species occurs in the middle octave. (After Whittaker 1965.)

tribution is plotted logarithmically, the curve produced will fall somewhere between the random niche and geometric distribution curves (Figure 24.12). The log-normal distribution most closely approximates the distribution of importance values obtained from communities rich in species. It is most useful in summarizing observed abundance relationships within and among communities. All of these distributions describe species abundances, but they are of little value in determining the underlying causes for the observed abundance relationships.

■ Island Biogeography

Theory

Large land masses hold more species than small land masses. This fact was not overlooked by early naturalist-explorers and biogeographers. Much later the zoogeographer P. Darlington (1957) suggested a rule of thumb: A tenfold increase in area leads to a doubling of the number of species. In 1962 F. W. Preston formalized the relationship between area and species diversity. The number of species on an island area increased linearly with island size (Figure 24.14):

$$S = cA^z$$

where
S = the number of species
A = area of island
c = a constant measuring the number of species per unit area
z = a constant measuring the slope of line relating S and A

MacArthur and Wilson (1963) proposed that the number of species on an island represents a dynamic equilibrium between the immigration rate of new species to the island and the extinction among those species already present on the island. They formally presented the concept as a theory of island biogeography. Immigration rate of new species to an island from a pool of colonists will decrease as the number of species on the island increases and the pool of available new species declines. Because at first few or no species inhabit the island, the immigration curve initially will be high. As the number of colonizing species increases, the number of immigrants arriving on the island decreases for several reasons. Fewer new immigrants are available from the source pool, and later immigrants may be unable to establish populations because available habitats are filled or resources are already utilized. With more species on the island the extinction rate increases. The rate at which one species is lost and a replacement gained is the *turnover rate*. The point where the two curves intersect represents the equilibrium

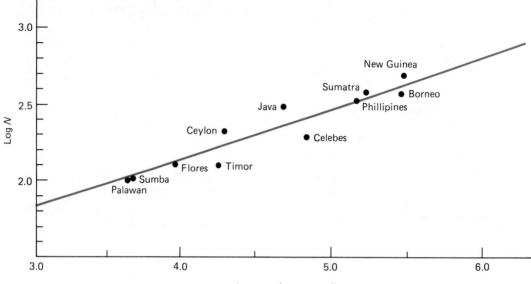

Figure 24.14 Number of bird species on the islands of the East Indies in relation to area. The abscissa gives areas of various islands. The ordinate is the number of bird species breeding on each island. (From Preston 1962: 195.)

number of species for the island (Figure 24.15). At equilibrium between immigration and extinction the number of species remains stable, although the composition of species may change. The equilibrium level is influenced by the size of the island and the distance of the island from a pool of potential immigrants. Small islands hold fewer species than large ones, and islands most distant from a pool of colonists (for example, a continental land mass) hold fewer species than islands close to a pool of colonists (Figure 24.16).

Testing hypotheses on island biogeography theory and verifying the model is difficult. An intensive study of the colonization of very small islands was undertaken by Simberloff and Wilson (1969; Wilson and Simberloff 1969; Simberloff 1969). This experiment involved the de-

faunation of four minute islands of mangroves at varying distances from a large mangrove island in the Florida keys. The arthropod fauna was removed by covering the islands with sheeting and applying insecticides. The sheeting was then removed

Figure 24.15 Equilibrium model of species on a single island. The point at which the curve for the rate of immigration intersects the curve for rate of extinction determines the equilibrium number of species in a given taxon on an island.

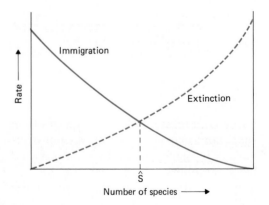

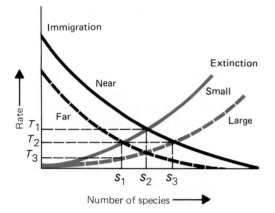

Figure 24.16 Graphical representation of the island biogeography theory involving both distance and area. Immigration rates decrease with increasing distance from a source area. Thus distant islands attain species equilibrium with fewer species than near islands, all else being equal ($S_3 < S_2$ for large islands; $S_2 < S_1$ for small islands). Extinction rates increase as the size of the island becomes smaller. As a result large islands have lower extinction rates than small islands. Considering both immigration and extinction rates, we can hypothesize that near islands reach equilibrium with more species than distant islands ($S_3 < S_1$ for small islands; $S_3 < S_2$ for large islands). Small islands reach equilibrium with fewer species than large islands, all else being equal (S_2 for small islands, S_3 for large islands); and large far islands have an equilibrium density equal to that of near small islands. Turnover rates are greater for near small islands than for distant small islands ($T_1 < T_2$) because near islands are closer to a source of immigrants. Similarly, turnover rates are greater for near large islands than for distant large islands ($T_2 < T_3$). Turnover rates are greater for near small islands than for near large islands because extinction rates on small islands are greater than on large islands ($T_1 < T_2$); and all else being equal, turnover rates are greater on small islands than on large islands ($T_1 < T_3$).

and certain groups of arthropods were censused at frequent intervals during the following three years. The islands richest and poorest in species prior to defaunation were also the richest and poorest af-

ter recolonization. Islands with the greatest number of species were those nearest to the main stands of mangroves. The number of anthropod species on all islands, however, continued to increase beyond the original number, and species found were frequently not the same as those present before the experiment. Species turnover was high, in part because of the difficulty of separating transient from breeding colonists. These experiments tended to verify the dynamic equilibrium model.

Rey (1981; Strong and Rey 1982) removed the *Spartina alterniflora* arthropod fauna from six small islets at Oyster Bay, Florida. Rey (1981) defaunated six islets in the spring, then followed recolonization weekly for one year, along with monitoring the arthropod populations on two control islets and one mainland plot. Recolonization and extinction rates corresponded to the island biogeography curves in general, with immigration rates negatively correlated and extinction rates positively correlated with species number. However, great variability in both rates, especially extinctions, made correspondence to the MacArthur-Wilson curves statistically insignificant, suggesting that both immigration and extinction are influenced by life history and trophic characteristics of the species, predation, host plant chemistry, and other factors.

Equilibrium size and turnover rates on already occupied islands can be assessed only by long-term annual censuses on each island. An example is a seasonal census of confirmed nesting birds over a 26-year period in the 16 ha Eastern Wood of Bookham Common, an isolated woodland near Surrey, England. Over the 26-year period, 44 species of birds appeared in the woods. Of these species, 6 apparently did not nest, leaving 38 breeding species. Of these breeders 4 species had territories extending beyond the woods, and 9 species never had more than two pairs nesting in the woods itself. Eleven more species nested in fewer than 5 years during the 26 years. They had to be considered casual species, counted as immigrants and extinctions in calculating annual turnover. Only 14 species were regular breeders. During the 26 years Eastern Wood experienced a considerable turnover of species with an average of three immigrations and three extinctions annually. Species equilibrium as determined by immigration and extinction graphs (Figure 24.17) is 32 species, somewhat higher than the 27 species that on the average Eastern Woods supported over the years (Williamson 1981).

The failure of most studies to support or validate the model points out a weakness of island biogeography theory (Gilbert 1980). The MacArthur-Wilson model is deterministic. It makes no allowances for chance arrivals and extinctions of spe-

cies on islands. It ignores the effects of turnover on the composition of species. Species richness may remain constant, but species composition changes. On any given island the balance between immigration and extinction is explained as a function of size and degree of isolation. The degree of isolation is relative, depending upon the taxon involved. What is a short distance for a bird may be a insurmountable distance for a mammal or a lizard. The assumption that extinctions relate to an island's area overlooks the fact that immigrations and extinctions may not be independent. Extinction of a dwindling population may be slowed or even halted by an influx of immigrants, the *rescue effect* (Brown and Kodrich-Brown 1977). Extinctions are influenced by life history traits and are not due only to isolation. The model assumes that an island's area determines the number of species, and overlooks the role of habitat heterogeneity. There is considerable evidence that habitat heterogeneity overrides island size in some situations (Rigby and Lawton 1981; Blake and Karr 1984; Freemark and Merriam 1986). The equilibrium number of species, S, is considered a constant when it can be variable. Further, the model treats all species in a taxon as equals, with the probability of extinctions and immigrations the same for all species; it ignores population dynamics and life history requirements of the species involved.

Because of its apparent simplicity and applicability, island biogeography theory was accepted prematurely and uncritically. In spite of its shortcomings, island biogeography theory has stimulated new insights and research into the distribution, diversity, and conservation of species. (See, for example, Williamson 1981; Harris 1984; Diamond 1975, 1986.)

Figure 24.17 Immigration and extinction curves for Eastern Wood, Bookham Common, Surrey, England, a 16 ha oak woodland. The immigration curve is a regression line (the points have been omitted) that cuts the abscissa at 39 species. The maximum number of species that bred at one time or another in Eastern Wood was actually 44. The extinction line is at 45°, indicating one extinction for every species present over 29. The equilibrium point, where immigration intersects extinction, is 32 species. The lines are straight rather than curved because the data are discrete, not continuous. (Adapted from Williamson 1981: 101.)

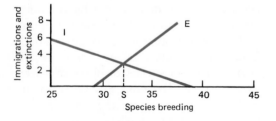

Application

Although the equilibrium theory originally applied to oceanic islands, not all islands are oceanic. Mountaintops (Brown 1971, 1978), bogs, ponds, dunes, areas fragmented by human land use, host plants and their insects (Janzen 1968; Strong et al. 1984) and animals and their parasites—all are essentially island habitats. As Simberloff (1974: 162) put it: "Any patch of habitat isolated from similar habitat by different, relatively inhospitable terrain traversed only with difficulty by organisms of the habitat patch may be considered an island." Investigations of these island situations, all stimulated by island biogeography theory, have led to new insights and modifications of the theory.

It is not difficult to understand why, among all ecological theories, island biogeography theory seems appealing in the management of wild plants and animals, particularly the problems engendered by the increasing fragmentation of habitats. It is relevant to such problems as the size, shape, number, and distribution of reserves and the degree to which fragmentation of habitat leads to species extinction. Some of the theory does seem to apply to habitat fragmentation. Large areas in general do support more species than small areas. Habitat parcels more distant from a large pool of immigrants hold fewer species than near ones. Species richness tends to reach some level of equilibrium. However, these apparent truisms are not wholly supported by data available.

In spite of the lack of sufficient supporting data, island biogeography theory too often has been uncritically accepted in attempts to develop some rules and recommendations for the establishment of nature reserves and the management of wildlife refuges. There is vigorous debate over an important question: Which is preferable, one large reserve or several smaller reserves that add up to the same area? One group Abele and Connor 1979; Higgs 1981; Jarvinen 1982, 1984; (Simberloff and Abele 1976, 1982, 1984) argues that two or more smaller reserves will hold more species than one large reserve of the same area. Data on species richness seem to support that argument. Others (Diamond and May 1976; Wilson and Willis 1974; Willis 1984; Whitcomb et al. 1976) argue that a large island is preferable because that area will not only support more rare species but also large populations of other species, making them less vulnerable to random extinction.

The problem with these arguments is their emphasis on area and their failure to consider all aspects of the situation. To begin with, there are considerable differences between oceanic islands, on which the theory was developed, and habitat islands (Wilcox 1980).

Oceanic islands are isolates. They are ecological units surrounded by an aquatic barrier to dispersal. They are inhabited by organisms or various taxa that arrived there by chance dispersal over a long period of time or represent remnant populations that existed on the area long before isolation. By contrast, habitat islands are samples of populations of a much larger area. These samples contain fewer species, fewer individuals within a species, and more species represented by only a few individuals. As a large area of habitat is fragmented, the total habitat area is reduced and what is left is distributed in disjointed fragments of varying size (Figure 24.18). These fragments are separated by other types of habitats, particularly urban and suburban developments and agricultural lands. As more land area is carved out of the original habitat, the distinctiveness of the habitat patch becomes accentuated. Surrounding areas represent more than barriers to dispersal for habitat island inhabitants; they are terrestrial habitats with their own sets of species, including domestic animals such as cats, dogs,

Figure 24.18 At the time of European settlement, this section in central Maryland (USA) was completely forested. Over the course of several hundred years much of the land was cleared for agriculture, fragmenting the remaining forestland into isolated patches. In the past quarter century urban and suburban developments have encroached upon the area, further fragmenting and isolating the forest, and doing the same to agricultural lands. (Adapted from Whitcomb et al. 1981: 127.)

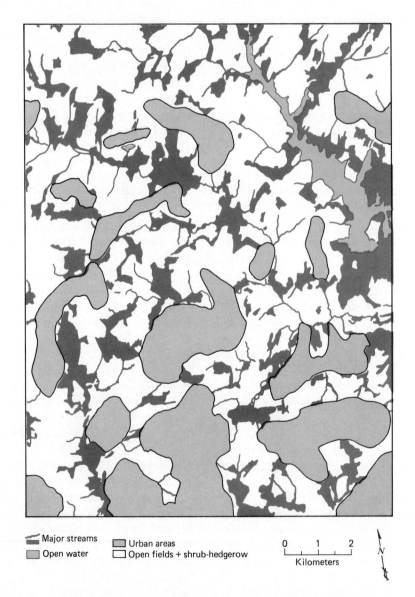

Major streams
Open water
Urban areas
Open fields + shrub-hedgerow

0 1 2
Kilometers

N

sheep, cattle, and species of wildlife highly adaptable to human habitations. These species invade the edges and move into the interior; cats, weasels, raccoons, and crows, for example, increase predatory pressure on interior species (Gates and Gysel 1978; Chusko and Gates 1982; Wilcove 1985; Ratti and Reese 1988) and the brown-headed cowbird encounters new naive hosts for nest parasitism (Brittingham and Temple 1983). Exposure to wind and solar radiation causes mortality in plant species on the edge, further reducing the integrity of the fragment (see Wales 1972; Ranney 1977; Lovejoy et al. 1986).

Large areas are whittled away piecemeal by logging, surburban developments, and road construction, a process conspicuous in any part of the world. Up to a point no species are lost, but as fragmentation continues the remaining area is reduced to a critical size below which the habitat will not provide the requirements of many of the original species, and a number of them disappear (Figure 24.19). The first to go are the susceptible species—those on the higher trophic levels, habitat and food specialists, and ones that require large areas of habitat to maintain viable populations. McLellen and associates (1986) designed a computer simulation to mimic the effects of fragmentation on two species pools, one with large area requirements and low vagility and the other with smaller area requirements and greater vagility. When the habitat was reduced to 50 percent of its original area and fragmented into a large number of small tracts, the area-sensitive pool began losing species. When the habitat was reduced 95 percent, only 30 percent of the original pool remained. Species of the second pool did not decline until the area was reduced to 20 percent of its original, but they were vulnerable to collapse when

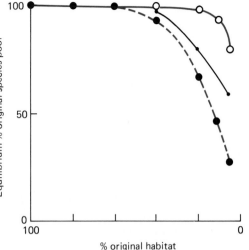

Figure 24.19 A model of the number of species remaining in each species pool as fragmentation proceeds. Large, solid circles show the pool of species with larger area requirements and low vagility; open circles show the species pool with less stringent area requirements. The small solid dots depict the proportion of the first species pool that would be present when the habitat is minimally fragmented. (From McLellen et al. 1986: 311.)

the habitat became even more highly fragmented. Species typical of the second pool may hang on because they are able to use resources just outside the fragment (for example, the Yellowstone elk). Habitats or reserves not too distant from a replacement pool may hold on to species whose resident populations can be supplemented by immigrants.

Another problem with small habitat islands is the potential for experiencing all the problems of small populations—random sampling of gene frequency, genetic drift, erosion of genetic variation, increased inbreeding and accompanying loss of fitness—all increasing the probability of species extinction (see Chapter 13).

On the other hand, dependence on one or two large habitats alone would increase the probability of extinction of some species. With populations subdivided into smaller reserves, the chance of the entire species going extinct would be reduced. Consider the fate of the heath hen, the whole population of which was confined to the island of Martha's Vineyard off the coast of Cape Cod, Massachusetts. Would the species have survived if the population had been scattered in several reserves?

What size patches of remnant vegetation are needed to maintain regional populations and satisfy habitat requirements of the species concerned? What size of island maintains greatest species diversity? At what size of habitat patch do interior or area-sensitive species disappear? Such questions have stimulated a number of studies of the response of both plants and animals to habitat fragmentation (see Wilcove 1986; Lovejoy et al. 1986; McLellan 1986; Higgs 1981).

The minimum size of forest habitat needed to maintain interior species differs with plants and animals. For forest interior plants the minimum area depends upon the size at which moisture and light conditions become sufficiently mesic and shady to support shade-tolerant species. Size alone is not as important in species persistence and extinction as environmental conditions (Weaver and Kellman 1981). Area depends in part on the edge about the stand (Levenson 1981; Raney, Bruner, and Levenson 1981)—whether it is open or closed, reducing the penetration of light and wind; it depends also on canopy closure and on the ratio of edge to interior (Figure 24.20). If the stand is too small or too open, the interior environment becomes so xeric that it prevents reproduction by mesic species, both herbaceous and woody. As a result, when mature residual mesic species such as

sugar maple and beech die, they are replaced by less mesic species such as oak.

Levinson (1981) found that the species richness of plants was greatest in edge situations where xeric species coexist with some interior species (Figure 24.21). The total number of woody species in Wisconsin woodlots increased with woodlot size up to approximately 2.3 ha. At that point vegetation achieved a maximum balance between edge and residual interior species. Beyond that size species richness declined and leveled off at 9.3 ha (23 acres), as mesic conditions and shade-tolerant species persisted. Thus there appears to be a negative correlation between edge species and the size of forest islands and a positive correlation between interior species and increased area.

The nature of species replacement, turnover, and immigration in isolated Wisconsin woodlands was influenced by the distance of forest islands from seed sources. Seed dispersal to woodlands was aided by hedgerows and other narrow belts of vegetation linking one forest island with another. Species that are bird-dispersed are more successful immigrants than species dispersed by mammals and the wind. The latter two types tend to become local in distribution, leading to their extinction in many forest islands.

Among birds a similar pattern exists. In general, larger forests have more species than smaller forests, the latter occupied by edge and ubiquituous species at home in any size forest tract (Moore and Hooper 1975; Galli et al. 1976; Ambuel and Temple 1983; Blake and Karr 1984; Freemark and Merriam 1986). A New Jersey study suggested that maximum bird diversity was achieved with woodlands 24 ha in size. However, those woodlands held no true forest interior species such as the worm-eating warbler (*Helmitheros vermivo-*

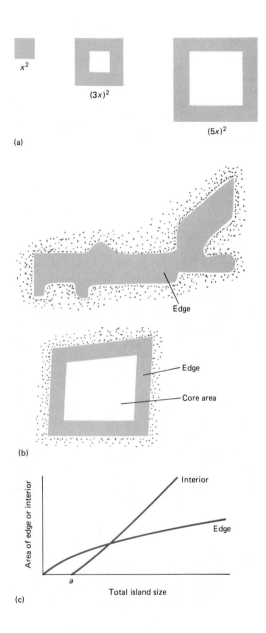

(a)

(b)

(c)

Figure 24.20 Relationship of island or fragment size to edge and interior conditions. (a) Functionally all islands are edge. Allowing the depth of the edge to remain constant, the ratio of edge to interior decreases as island size increases. When the island size is large enough to maintain mesic and interior conditions, an interior begins to develop. (b) In an application of this concept, it is evident that size alone is not a primary determinant of interior-edge conditions. Configuration or shape of the island is also critical. In this example the rectangular wooded island contains 39 ha; yet it is entirely edge, because its width does not exceed the depth of its edge. Out of its 16 species, none are interior. The square woodland of 47 ha has a core interior area of 20 ha, and 6 of its 16 species are interior species sensitive to fragmentation. (c) The graph shows this relationship. Below point *a,* the size at which interior species can exist, the woodland is edge. As its size increases, interior area increases and the ratio of edge to interior decreases. This relationship of size to edge holds for circular or square islands. Long narrow woodland islands as in (b), whose width does not exceed the depth of the edge, would be edge communities, even though their area might be the same as that of square or circular ones. (a and c after Levenson 1981: 32; b after Temple 1986: 304.)

(1986), investigating bird species composition of large and small forest habitat islands (from 3 to 7620 ha) in agricultural regions in Illinois USA and Ontario, Canada, respectively, found that although two or more smaller forest habitats supported more species, long-distance migrants and interior species, typical of larger tracts, were missing or poorly represented. Large forest tracts were important for increasing the number of forest interior neotropical migrants and certain resident species, such as the hairy woodpecker. Large forest tracts with a high degree of heterogeneity held the most species, supporting birds of both edge and interior. The presence of forest interior species in smaller woodlands depended upon the nearness

rus) and the ovenbird *(Seiurus aurocapillus),* which are highly sensitive to forest fragmentation and require extensive areas of woods (Whitcomb et al. 1981; Lynch and Whitcomb 1977). Both Blake and Karr (1984) and Freemark and Merriam

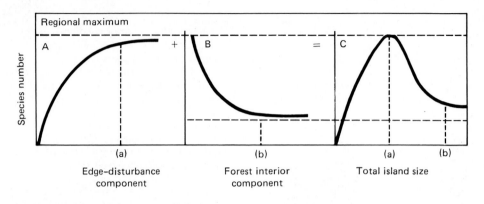

Figure 24.21 Species richness of a woodland relates to the ratio of edge to interior, area, and a variety of conditions. Under ideal conditions of both sufficient area and variation in environmental conditions, all woody species of a region could coexist (curve A). The number would be asymptotic to the number of woody species found in the region. At some point (a) along the curve interior conditions would prevail over edge. (b) As mesic, low light conditions develop, light-demanding species will decline (curve B) and the interior population will stabilize at the number of mesic, shade-tolerant species found in the region (point c). Considering edge and interior species together, species richness would follow the curve in C. Smaller islands would hold edge species exclusively up to some point a. Then, as interior conditions develop, edge species would drop out. Mostly interior species would remain at point b. (After Levenson 1981: 34.)

of those habitat islands to a pool of replacement individuals.

To what degree heterogeneity associated with the creation of gaps and edge conditions can be introduced into a forest and not affect interior species is not well defined. Yahner (1982, 1986) investigated the effects of patch clearcutting in 50- to 60-year-old stands of aspen and mixed oak in central Pennsylvania for ruffed grouse management compared with untreated stands. Management involved a mosaic of 1 ha clearcuts, which amounted to 36 percent of the stand. The treated section had a greater species richness and more total individuals combined than the untreated section. The treated section held seven species of birds associated with early succession; yet the degree of fragmentation was not sufficient to discourage the presence of interior species.

The question, then, is not whether to create large versus small reserves but how to mix both small and large habitat parcels, both heterogeneous and homogeneous habitats. An overemphasis on diversity, so prevalent in species management, can lead to a decline and ultimate extinction of interior or area-sensitive species that need more homogeneous habitats.

■ Interactions and Community Structure

Competition

Since Darwin (1878) ecologists have considered interspecific competition, especially competitive exclusion, as the cornerstone of community structure. Lack (1954, 1971), Hutchinson (1959), and MacArthur (1960, 1972), among others, have stressed the role of competition in shaping communities, as reflected in species distributions, resource allocations, niche segregation, and the like (see Chapter 18). Salisbury (1929) early pointed out the role of competition in shaping plant communities.

Numerous studies have been undertaken to demonstrate the role of interspecific competition in community structure, with emphasis on the animal component (see Connell 1983; Schoener 1983). Many of these studies suggest, if they do not prove with certainty, that interspecific competition exists between certain species pairs; but none demonstrate that interspecific competition has any significant influence in shaping the whole assemblage of organisms making up the community.

This statement is not to imply that interspecific competition does not exert an influence on community structure. Anyone who has a garden experiences the effects of competition between garden plants and weeds. A great deal of competition, both exploitative and interference, for light, moisture, nutrients, and space exists among plants from forests to deserts. The outcome of competitive interaction among plants has a pronounced influence on the physical structure of the terrestrial community in which they belong and in turn on distribution and abundance of animal life in the community.

Yodzis (1986) breaks down competition within the community into two major types: consumptive competition and competition for space. In consumptive competition one competitor consumes resources that others might have consumed and in some way affects their welfare. Such exploitative competition is influenced by the nature of the niche each species occupies. For example, in autumn many animals, especially jays, wild turkey, deer mice, squirrels, black bear, and white-tailed deer, feed on acorns. If the acorn crop is poor, these animals compete strongly for acorns. Because squirrels depend heavily on acorns, they may feel the effects of exploitative competition more keenly than the more generalist consumers (Allen 1943). For another example, mortality of oak trees defoliated by gypsy moths is greater on nutrient-rich mesic sites than on xeric sites (Statler and Serro 1983; Hicks and Fosbroke 1987). One hypothesis for this response is that yellow-poplar and sugar maple, usually ignored by oak-feeding gypsy moth larvae, maintain their canopy and utilize the moisture and nutrient resources that normally would be used by the now defoliated oaks. Oaks on mesic sites, forced to draw on the nutrient reserves in their roots and lacking foliage to carry on transpiration, are unable to compete strongly for moisture and nutrients. Unable to refoliate successfully, many oaks succumb. Oaks dominate the more xeric sites. Because all the oaks are defoliated, exploitative competition with foliated trees does not exist, and nutrients and moisture are available to the refoliating oaks.

Yodzis recognizes two types of competition for space: dominance and founder control. In the former, some organisms become dominant and control the resource. Among plants interspecific competition interacts with resource abundance, local environmental heterogeneity, and disturbances to control community structure. Fast-growing trees, for example, intercept light and shade plants beneath their canopy. Mussels completely cover rocky substrates along the shore. Founder control involves no clear dominants. Space belongs to those species that colonize the area first and hold on to it. In this case, disturbance may be important in determining the colonization by different species (see Chapter 26).

The clearest examples of competition influencing community structure are the effects of introductions or invasions of exotic species. Exotic species of birds introduced into the Hawaiian islands have not only excluded native species with similar ecological requirements but have also excluded or prevented the successful establishment of other exotics with similar morphological characteristics (Moulton and Pimm 1986). When two introduced species of plankton-feeding fish, the alewife and rainbow smelt, proliferated in Lake Michigan, seven native species of fish with similar food habits declined drastically (Crowder et al. 1981). Japanese honeysuckle, a garden escapee, and multiflora rose, widely planted in the past for soil conservation purposes, have invaded old fields and the forest edge, crowding out native plants and affecting the structure and composition of animal life (in some cases improving wildlife habitat).

The role and importance of competition will vary from community to community and among groups within the community. Competition may be more important or more pronounced among some groups, especially sessile ones such as plants and mussels, or members of the same guilds. Other groups, such as phytophagous insects, experience no interspecific competition (Strong et al. 1984). The major impact of competition in a community may come from competition among plants, which can affect the structural aspects of the community. At best it is difficult to establish the role of competition in the community, because its apparent outcomes may have alternative explanations. The major forces in structuring communities may be other influences that affect populations, especially weather, climate, and predation (Andrewarthe and Birch 1984), all of which can influence or modify competitive relationships.

Predation

Although the influence of competition on community structure is somewhat obscure, the influence of predation is more obvious and demonstrable. Part of the reason is that competition occurs among species on the same trophic level. Because predation affects two or more trophic levels, its influence can be more readily noticed throughout a community, particularly as it influences competitive relationships among species.

The role of herbivores in influencing plant community structure had been well demonstrated (see Chapter 20). The influence of rabbits on the diversity of species in English pastures is a classic example (Tansley and Adamson 1925; Zeevalking and Fresco 1977). Often this effect works its way through the community. For example, a sharp reduction in the rabbit population in southern England from myxomatosis resulted in an aggressive growth of meadow grass in fields inhabited by the spectacular large blue butterfly *Maculinea arion*. Heavy grass resulted in the extinction of open-ground ant colonies, the nests of which were utilized by large blue caterpillars. As a result, the large blue is nearly extinct. The loss of one keystone grazing herbivore, the rabbit, resulted in the local extinctions of two other species as well.

Grazing herbivores, from prairie dogs (Whicker and Detling 1988) to pocket gophers (Huntley and Inouye 1988) to large ungulates (McNaughton, Ruess, and Seagle 1988), influence the structure of grassland communities from the North American plains to East Africa. Grazing ungulates, in particular, reduce canopy height, stimulate tiller growth, and favor low-growing grass and herbaceous plants at the expense of taller grasses. Intensive grazing favors plants that are prostrate, small-leafed, or in rosettes close to the ground and selects against taller forms that can easily become dominant. When grazing pressure is relaxed, tall-growing grasses assume dominance at the expense of low-growing forms, reducing plant diversity. The effect often is evident with experimental exclosures or removal of the grazing herbivores. McNaughton (1984a, b) found that on the Serengeti mean plant height within fenced plots averaged 6 times and maximum height 16 times that of plants within unfenced plots, although biomass was twice as high on unfenced plots.

Browsing ungulates, such as deer and moose, can change the structure and composition of forest vegetation. Browsing by

white-tailed deer in forests of eastern North America has eliminated understory woody plants, including tree reproduction, converted the understory to fern, and created browse lines on the trees (Marquis 1981; Marquis and Grisez 1978). Heavy browsing by moose on Isle Royale favored spruce over deciduous trees. The change in quantity and nature of litterfall resulted in a decrease in microbial biomass and nitrogen availability (McInnes et al. 1988).

The rocky intertidal region of the New England coast is inhabited by a number of marine algae, including the ephemeral species *Enteromorpha, Ulva,* and *Porphyra* and the perennial fucoid algae *Ascophyllum Nodosum* and *F. distichus,* and by a grazing herbivore, the snail *Littorina littorea* (Figure 24.22). Lubenchenco (1983, 1984) excluded snails from rocks relatively sheltered against wave action and discovered that the ephemeral species of algae colonized and become established on the rocks. They monopolized the space and inhibited colonization by the perennial fucoid algae. Where snails were present, grazing pressure reduced competition for space among the three ephemerals and

Figure 24.22 (a) The common periwinkle *(Littorina littorea)* is the most common grazing herbivore on the rocky intertidal zone of northeastern North America. (b) Rockweed *(Ascophyllum Nodosum)* is the dominant alga of the rocky intertidal zone in the presence of the common periwinkle. (Photos by R. L. Smith.)

(a)

(b)

prevented any one from becoming dominant, although each has the potential of excluding the other. Instead the perennial *Fucus* was able to colonize the rocks and assume dominance. Although small *Fucus* are eaten by the snails, *Littorina* avoids larger individuals if ephemerals are present. *Fucus* has chemical and structural defenses against snails, but the ephemerals have no known defenses. They persist where grazers are absent or exist at very low levels, or are inactive (during winter months). Because snails are rare at sites exposed to heavy wave action, exposed rocky intertidal sites become havens for ephemerals. These experiments suggest that by controlling competitive interactions, herbivory increases species richness and influences community structure of rocky intertidal shores.

Such studies, and there are many of them, suggest some basic principles on the influence of grazing on plants. If the herbivore selectively grazes the major dominant, then plant diversity increases. Overgrazing in winter and spring followed by undergrazing in summer produces maximum floral richness. (This case is usual in natural situations because deer, elk, and others can overutilize their food in winter, but undergraze it in summer.) The most species-rich communities are developed by continuous grazing with a maintained population of herbivores. If the dominant plant is highly palatable, overgrazing will reduce it and allow other, less palatable species to occupy the area. If the dominant species are unpalatable, then grazing only serves to consolidate their dominance. If the herbivore is regulated by food supply and not by predators, then it can reduce a plant species to a minor component and allow invasion by other species. As the intensity of grazing increases, more and more unpalatable species are grazed until eventually, only unpalatable

species remain or move into the area. Thus the complex balance of species in a plant community is sensitive to some control by feeding activities of grazing animals.

By selectively eating seeds of certain plants over others, seed-eating predators can influence plant composition. Again using exclosures and plots sown with known quantities of seeds, Borchert and Jain (1978) found that meadow mice and house mice consumed 75 percent of wild oat (*Avena fatua*) seed, 44 percent of wild barley (*Hordeum leporinum*) seed, and 37 percent of ripgut brome (*Bromus diandrus*) seed. Showing a strong preference for *Avena* and *Hordeum*, the mice reduced the numbers of these plants by 62 percent and 30 percent respectively, stimulating competitive release of *Bromus*. Mice can also influence the population of plants by cutting mature plants and eating seedlings (Batzli and Pitelka 1970).

Similar influence of predation on community structure may occur at the next trophic level. An example involves intertidal invertebrate predators and prey. The rocky intertidal community on the Pacific Coast of Washington State (USA) consists of four species of algae, a sponge, filter-feeding barnacles and mussels, browsing limpets and chitons, a predatory whelk, and the predatory starfish *Pisaster ochraceus*. Both barnacles and mussels, when given the opportunity, aggressively and tenaciously take over space and exclude other sessile organisms. Paine (1966) removed the top carnivore, the starfish that feeds on sessile barnacles and mussels as well as on limpets and chitons, and excluded it for two years from an 8 m long, 2 m deep swath on a rocky intertidal area. On the control areas nothing changed. On the swath where the starfish was removed, barnacles settled successfully but were soon crowded out by the mussels. All but one species of alga disappeared for the lack of space, browsers moved away for the lack of space and food, and the number of species dropped from 15 to 8. Apparently the predaceous starfish by feeding on barnacles and mussels made space available for competitively subdominant species, helping to maintain species diversity and a more complex intertidal community structure.

These experiments on rocky intertidal habitats appear to support the hypothesis that herbivores and predators influence community structure. However, other influences may be involved. Availability of planktonic colonists, vagaries of larval settlement, settling ability, and environmental fluctuations may lessen the importance of predation and competition in structuring intertidal communities (Underwood and Denley 1986). From these experiments ecologists should not develop sweeping generalizations on the effects of predation on other communities.

Predation can have its greatest impact when an exotic predator is turned loose in a natural community. An excellent example is the effect of releasing the predaceous peacock bass *(Cichla ocellaris)* into a tributary of Panama's Lake Gatun around 1967. This voracious predator eliminated 8 of the 11 principal native fish species and reduced 3 others by 75 to 90 percent. With these fish gone, another species, *Cichlasoma maculicauda*, whose juveniles formed the prey of some of the species eliminated, increased. With plankton-feeding fish eliminated, zooplankton that had been their preferred prey increased; and with their prey species largely eliminated, herons, terns, kingfishers, and tarpon decreased (Zaret and Paine 1973).

Sih and associates (1985) reviewed an array of studies on the effects of the interaction of predation and competition on prey communities and arrived at certain trends reflected in the studies. Herbivores have a stronger effect on plant communities than carnivores have on animal communities. Overall effects of predators on a prey community are heavily influenced by the predator's impact on competitive dominants or middle-level predators (second-level consumers). The effects may be direct or indirect. One exploitative competitor affects another by reducing the level of a shared resource, a third species. Reduction of one or both competitors increases the third, an indirect effect. Predation may also influence lifestyles of the prey, including morphology, physiology, and behavior such as habitat use, foraging mode, and the like. These factors in turn determine prey encounter rates with both competitors and predators. In all, the interactions between predation and competition and community structure involve complexities at which experimental studies to date can only hint.

Parasites and Diseases

Parasites, which include disease-producing organisms, are an integral part of natural communities. Their overall effect on community structure becomes most apparent when an outbreak of disease decimates or reduces an affected population. With the clearing of forests, the brown-headed cowbird, a nest parasite, has expanded its range through North America. It has invaded forest edges and even the forest interior, where it parasitizes birds that were never exposed to nest parasitism in their evolutionary history. Lacking antiparasitic behavior, these birds are experiencing parasite-induced declines in their populations (Goldwasser et al. 1980; Walkinshaw 1983). The introduction of rinderpest to East African ungulates (see Chapter 22) decimated herds of African buffalo and wildebeest. The introduction of avian malaria carried by introduced mosquitos eliminated most native Hawaiian birds below 1000 m in altitude, above which is the mosquito-free zone, whereas introduced bird species are resistant to malaria (Warner 1968). Outbreaks of mange periodically reduce red fox populations in New York and other areas.

How parasitism can influence interactions within a community beyond a simple host-parasite relationship is illustrated in the moorland ecosystem of Scotland. Heather *(Calluna vulgaris)* and bracken *(Pteridum aquilinium)* (see Chapter 30) dominate the moorlands of Scotland, the major habitat of the red grouse *(Lagopus lagopus scoticus)* and grazing land for domestic sheep. The red grouse, which feeds on young shoots of heather, is the host for several parasites, internal and external. The parasitic nematode *Trichostongyle tenuis* reduces the breeding success of grouse and may account for the short-term population cycles of the bird (Potts et al. 1984; Hudson 1986). Red grouse, the chicks in particular, are also parasitized by the tick *Ixodes ricinus,* which carries a tick-borne virus, louping ill. The virus normally is associated with sheep, but ticks transmit the disease to highly susceptible grouse chicks (Duncan et al. 1978). The problem is intensified by management schemes. Sheep grazing on the moorlands provide the reservoir of the virus. Bracken, once managed and harvested for livestock bedding, is spreading into and replacing heather. Mature heather resists invasion; but burning of moorlands to provide young nutritious shoots for the grouse reduces mature heather and its ability to resist invasion of bracken with its fast-growing rhizome system. Bracken produces a mat of humid litter that provides an ideal habitat for ticks. Spread of bracken has increased exposure of the grouse to the tick and subsequently to the louping ill virus,

which is associated with the long-term decline in grouse densities (Dobson and Hudson 1986). Because the virus is so pathogenic in grouse, the birds cannot maintain the virus without the presence of sheep, the vaccination of which should theoretically eliminate the disease.

Virulent tree diseases have markedly changed the composition of North American forests. The chestnut blight, introduced into North America from Europe, nearly exterminated the American chestnut and removed it as a major component of the forests of eastern North America. With its demise oaks and birch increased. Dutch elm disease has nearly removed the American elm from the forest and landscape of North America and the English elm from Great Britain.

Mutualism

Mutualism is too often overlooked as a mechanism in community structure. Although difficult to demonstrate, it may be more significant than either competition or predation. The importance of mutualistic relationships among conifers, mycorrhizae, and voles in the forests of the Pacific northwest is one example (see Chapter 23). Temple (1977) suggests that the tree *Calvaria major* on the island of Mauritius has failed to produce seedlings in spite of adequate seed production because its seeds have to pass through the gut of the now-extinct dodo *(Raphus cucullatus)*. Relationships between pollinators and certain flowers are so close that loss of one could result in the extinction of the other (Pimm and Pimm 1982).

There is some evidence that a relationship exists between mutualism and interspecific competition, just as there seems to be a relationship between interspecific competition and predation. This relationship could come about through indirect mutualism, in which the affected species never come into contact, but affect each other's fitness or population growth rate (Boucher, James, and Keeler 1982). Suppose that species A and species C both compete with species B for resources partially shared. The interaction between A and C on species B in effect reduces competition for each and prevents competitive exclusion from the resources they consume (Figure 24.23). The interaction between competitors reduces niche overlap, an effect similar to that induced by a keystone predator. In another situation species A and C may be mutualistic with B, in which case A and C benefit each other indirectly.

Boucher, James, and Keeler (1982) provide some possible examples of indirect mutualism. One is the study of the in-terrelationships between two species of herbivorous *Daphnia* and their predators, a midge larvae *(Chaoborus)* and a larval salamander *(Ambystoma)* in subalpine ponds in Colorado (Dodson 1970). *Ambystoma* eats large species of *Daphnia,* the midge larvae small species of *Daphnia.* Where salamander larvae were present, the number of large *Daphnia* was low and the number of small *Daphnia* high; but in ponds in which the salamander larvae were absent, small *Daphnia* were absent and midges could not survive. The two species of *Daphnia* apparently compete for the same resources, whereas the midge and the salamander use different resources. The midge apparently depends upon the presence of salamander larvae for its survival in the pond. This possible indirect mutualism can be demonstrated only under controlled experiments involving population manipulations of the four species.

The idea of indirect mutualism is highly speculative, still to be demonstrated experimentally in natural communities. It does suggest that mutualism, as well as competition and predation, can be an integrating force in the structuring of natural communities. Mutualism, long overshadowed by interspecific competition, is now receiving the attention it deserves (see Howe and Westley 1988).

Figure 24.23 How indirect mutualism might work. In this hypothetical situation species A and species C compete moderately, and both compete strongly with B. When all three species occur together, A and C exert strong competition on B, reducing the intensity of competition by B on each other. The presence of B indirectly benefits A and C by reducing the fitness of a strong competitor of the other species. (This situation could also be an example of diffuse competition on B.) (From Pianka 1981.)

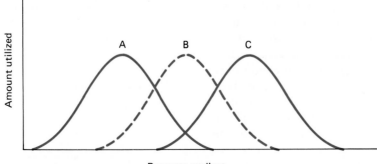

■ Community Pattern

The composition of any one community is determined in part by the species that happen to be distributed in the area and can grow and survive under prevailing conditions. Seeds of many plants may be carried by the wind and animals, but only those adapted to grow in the habitat where they are deposited will take root and thrive. The element of chance also is involved. One adapted species may colonize an area and prevent others equally well adapted from entering. Wind direction and velocity, size of the seed crop, disease, insect and rodent damage all influence the establishment of vegetation. Thus the exact species that settle an area and the number of individual species that succeed are situations that seldom if ever are repeated in any two places at any two times. Nevertheless, there is a certain pattern, with more or less similar groups recurring from place to place. Only a relatively small group of species are potential dominants because a limited number are well adapted to the overall climate and soils of the region they occupy.

Communities are often regarded as distinct natural units or associations, especially for practical reasons of description and study, but more often than not community boundaries are hard to define. Some, such as ponds, tidal beaches, grassy balds, islands of spruce and fir within a hardwood forest, old fields and burns, have sharply defined boundaries. Here the vegetational pattern is discontinuous. Most often, however, one community type blends into another. The species comprising the community do not necessarily associate only with one another, but are found with other species where their distribution overlaps (Figures 24.24, 24.25).

Some organisms will succeed only under certain environmental conditions and tend to be confined to certain habitats. Other tolerate a wider range of environmental conditions and are found over a wider area. Species shift in abundance and dominance, because of change in altitude, moisture, temperature, and other physical conditions. One species may be dominant in one group, an associated species in an-

Figure 24.24 Four models of species distribution along environmental gradients. (a) The abundance of one species on an environmental gradient is independent of the others. The association of several species along the gradient changes with the response of individual species to that gradient. (b) The abundance of one species is associated with the abundance of another. Two or more species are always found in association with each other. (c) The distribution of one species is independent of another on the environmental gradient, but the distribution of each species is sharply restricted at some point on the gradient by interspecific competition. (d) The distribution of a species is sharply restricted by a change in some environmental variable. This case is characteristic of an ecotone or edge.

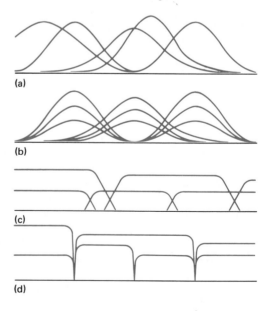

(a)

(b)

(c)

(d)

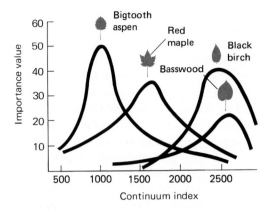

Figure 24.25 Distribution of some forest trees on a continuum index. Note that the smaller the index value, the earlier the species appears in forest succession. (Adapted from Curtis and McIntosh 1951.)

other. This sequence of communities showing a gradual change in composition is called a *continuum* (Curtis 1959). Each community is similar to, but slightly different from its neighbor, the difference increasing roughly with the distance between them. Even when the dominant plants change completely, the community may integrate in the understory vegetation. The continuum is much like a light spectrum. The end colors, red and blue, and other primary colors in the middle are distinguishable, but the boundaries grade continuously in either direction. Eventually the continuum must end, when environmental conditions favor a completely different group of organisms. A community in such a gradient can be described as a discrete area or point in the continuum, the point being defined by some given criteria.

The distribution of species along an environmental gradient is not confined to plants alone. The same phenomena also have been found in insects (Whittaker 1952) and birds (Bond 1957).

Classification Systems

To give some order to the study of communities, we need a system of classification, even though we may not be able to place the communities of a region into discrete categories. There are a number of approaches to community classification, each arbitrary and each suited to a particular need or viewpoint (Mueller-Dombois and Ellenberg 1974). The most common classification systems are based on physiognomy, species composition, dominance, and habitat.

Physiognomy, or general appearance, is a highly useful method of naming and delineating communities, particularly in surveying large areas and in subdividing major types into their component communities. Because animal distribution appears to be more closely correlated to the structure of vegetation than to species composition, classification by physiognomy will relate both to plant and animal life. Communities so classified are often named after the dominant form of life, usually plants, such as a coniferous or deciduous forest, sagebrush, short-grass prairie, and tundra. A few are named after animals, such as the barnacle-blue mussel (*Balanus-Mytilus*) community of the rocky intertidal zone. One, of course, may grade into another, so the classification may be based on arbitrary, although specific, criteria. In places where habitat boundaries are well defined, communities may be classified by physical features such as tidal flats, sand dunes, cliffs, ponds, and streams.

Finer subdivisions may be based on species composition, a system that works much better with plants alone than with animals. Such a classification requires a detailed study of the individual community (see Appendix B; Mueller-Dombois and Ellenberg 1974), which involves the concepts of frequency, dominance, constancy, presence, and fidelity—the occurrence of a species in only a few community types.

In this system a group of stands in which similar combinations of species occur can be classified as the same community type, named after the dominant organisms or the ones with the highest frequency. Examples are the *Quercus-Carya* association or oak-hickory forest, the *Stipa-Bouteloua* association or mixed prairie.

European ecologists have developed a floristic classification with emphasis on dominance, constancy, and diagnostic species. They group communities into classes, orders, alliances, and associations (see Poore 1962; Whittaker 1962; Mueller-Dombois and Ellenberg 1974). Such a classification involves *fidelity,* the faithfulness of a species to a community type. Species with low fidelity occur in a number of different communities and those with a high fidelity in only a few. Seldom, if ever, are the latter found away from certain closely associated plants and animals. The greater the ratio of constant species to the total number of species, the more homogeneous is the community and the more sharply it can be delineated. Often, however, this close association merely reflects a group of species unable to grow successfully under a wide range of ecological conditions or with other species. Species can be grouped as *exclusive,* those completely or nearly confined to one type of community; *characteristic,* those most closely identified with a certain community; and *ubiquitous,* those with no particular affinity to any community. The species grouped as characteristic, high in constancy and dominance, are the ones that define the community type.

A different approach to classifying vegetation involves ordinational techniques. An early version was the *continuum index* (Curtis and McIntosh 1951; Curtis 1959), a synthetic index ordering sites by changing vegetational composition (Figure 24.25). As originally conceived, the continuum index was based on a linear, unidimensional ordination of stands from early to late stages of succession. Tree species were rated with a climax adaptation number, from 10 for a climax species such as sugar maple or beech to 1 for an early successional species such as aspen. A continuum index was obtained by multiplying the climax adaptation number of each species in a stand by its importance value *(see Appendix B of the Student Resource Manual)* to obtain a continuum index for the stands. The weighted total was then used as a basis for placing each stand in relation to other stands on a linear ordination.

The continuum index only relates the vegetative component to a successional sequence; but an objective in vegetation analysis and classification is the correlation of vegetation with the environment. One approach is gradient analysis (Whittaker 1956, 1960, 1967), which plots the pattern of plant responses to changes along a particular environmental gradient such as moisture, temperature, and altitude. *Direct gradient analysis* uses a graphical technique for displaying patterns of species distribution from data obtained by sampling the vegetation along some environmental gradient. The resulting bell-shaped curves of plant species on the gradient (Figure 24.26) are suggestive of the utilization curves or niche dimensions on the resource gradients of animal ecologists (Austin 1984).

An advance over the continuum index and direct gradient analysis was the development of community ordination, a form of *indirect gradient analysis. Ordination,* in

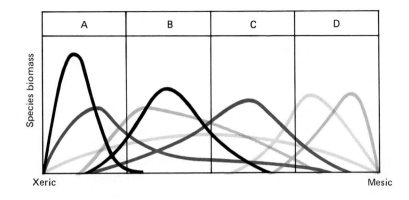

Figure 24.26 Vegetation distributed along a moisture gradient from xeric to mesic. Each species responds in its own individual way to moisture; yet sufficient overlap in response exists among species to allow a number of them to associate with each other on the gradient. The nature of the community, its dominants and associated species, depends upon the point at which community boundaries are placed. In the hypothetical gradient, each demarked community is characterized by its own dominants, although some species are shared with other communities. The communities at either end of the gradient are distinct, although they share one ubiquitous species. Artificially shifting the community boundaries along the gradient (as often happens in sampling communities on a gradient) would result in changes in community composition and dominants.

plant ecology, is the technique of arranging vegetational samples or species in relation to one or more ecological gradients or to one or more abstract coordinate axes that represent such gradients. The relative positions of the sample units on the axes and to each other provide maximum information about their ecological similarities (Austin 1984; Ludwig and Reynolds 1988). Ordination is an exploratory data-analysis technique designed to seek patterns or trends in the data.

The earliest of these ordination methods in plant ecology was the polar ordination of Bray and Curtis (1957), a technique specifically developed to analyze plant community data. It involves the use of community similarities or dissimilarities (see Appendix B) to determine endpoints on the axis, followed by a geometric positioning of the remaining stands in between (see Reynolds and Ludwig 1988; Beals 1984). Although subjective, polar ordination still remains a useful method.

Since then more mathematically sophisticated and complex multivariate methods have been developed to analyze the community. One is principal components analysis; developed for vegetational studies, this method has become widely used in other areas of ecology. Others include cluster analysis, discriminate analysis, correspondence analysis, and nonlinear ordinations. The value of these methods is their ability to summarize a large, complex set of variables that enable ecologists to develop hypotheses about environmental effects on species distribution and community patterns.

■ Summary

A biotic community is a naturally occurring assemblage of plants and animals living in the same environment, mutually sustaining, and interacting directly or indirectly with one another.

Communities exhibit some form of layering or vertical stratification, which in terrestrial communities largely reflects the life forms of plants and influences the nature and distribution of animal life in the community. In addition to vertical stratification, communities may also exhibit horizontal heterogenity produced by clumped distribution of vegetation, resulting in a patchy environment. The communities that are most highly stratified, both vertically and horizontally, hold the richest variety of animal life, for they contain a greater assortment of microhabitats and available niches.

The place where two different communities meet is an edge. The area where two communities blend is an ecotone. An edge may be inherent, produced by a sharp environmental change such as a topographical feature; or an edge may be induced, created by some form of disturbance and change through time. Because it supports not only selected species of adjoining communities but also a group of opportunistic edge species, an ecotone has a high species richness.

Communities may be organized about dominant species, especially in the temperate zone. The dominant species may be the most numerous, possess the highest biomass, preempt the most space, or make the largest contribution to energy flow. It is not necessarily the most important species in the community.

Communities differ in their richness and diversity of species. Species diversity implies both a richness in the number of species and evenness of distribution of individuals among the species. The more evenly the individuals are distributed, the higher is species diversity. A number of hypotheses have been proposed to explain species diversity among different communities, especially on a latitudial basis, but they are difficult to test. A species diversity index is useful only on a comparative basis, either within a single community over time or among communities. Species diversity within a community is called alpha diversity; among communities, beta diversity; and in a geographical area, gamma diversity.

A relationship exists between species diversity and area. In a general way, large areas support more species than small areas. This species-area relationship is involved in the theory of island biogeography, which holds that the number of species an island holds represents a balance between immigration and extinction. Immigration rates in an island are influenced by the distance of an island from a mainland or source of potential immigrants. Thus, islands most distant from a mainland would receive the fewest immigrants and the ones closest to the mainland, the most. Extinction rates are influenced by the area of the island. Because small islands hold smaller populations and have less variation in habitat, they experience higher extinction rates than large islands.

The theory of island biogeography is relevant to the fragmentation of natural habitats such as forests that become isolated by surrounding agricultural and urban lands. Although smaller fragments hold fewer species than larger fragments, size is not the only criterion for the value of such fragments for wildlife. Important is the ratio of core or interior area to edge. Many species are area-sensitive, and unless the habitat parcel is large enough to hold interior species, the area is inhabited only by edge species and area-insensitive species. Therefore species diversity is not a reliable indicator of ecological conditions.

The structure of a community may also be influenced by interaction among its member species. Interspecific competition may exclude or reduce the abundance of species or the number of individuals within a species. Predators may reduce the population sizes of competing prey species, permitting their coexistence. When key predators are eliminated, a certain few prey species may assume dominance. Parasites and diseases may reduce or eliminate certain species from the community, opening it for invasion by others. Finally, mutualistic relationships, both direct and indirect, may integrate community structure in ways that are now only beginning to be recognized.

The composition of a community is determined in part by the species that happen to be distributed in the area and that can grow and survive under prevailing conditions. The exact species that happen to settle in an area and the number that survive are rarely repeated in any two places at the same time, but there is a certain recurring pattern of more or less similar groups. Rarely can different groups of communities be sharply delimited, because they tend to form a continuum along a gradient of environmental resources.

Nevertheless, there is a need to classify communities by some criteria for convenience in field work and other purposes. The simplest classification scheme is based on physiognomy and dominant species. More detailed schemes involve detailed studies of floristics. One scheme, direct gradient analysis, has been more or less replaced by indirect gradient analysis, which involves such multivariate statistical approaches as principal components analysis and discriminant analysis.

C H A P T E R 2 5

Succession

Abandoned cropland is a common sight in agricultural regions, particularly in areas originally covered with forests. No longer tended, the lands grow up in grasses, gold-enrod, and other herbaceous plants. Only the most unobservant would fail to notice that in a few years these same weedy fields are invaded by shrubby growth—blackberries, sumac, and hawthorn. These shrubs are followed by fire cherry, pine, and aspen. Many years later this abandoned cropland supports a forest of maple, oak, cherry, or pine. Thus over a period of years one community replaces another until what appears to be a stable forest occupies the area (Figure 25.1).

■ Succession Defined

The changes involved in the return of the forest are not haphazard, but orderly, and barring disturbance by humans or natural events, the reappearance of a forest is predictable. This change in species composition and community structure and function over time is *ecological succession*. More precisely, succession may be defined as a unidirectional, sequential change in the relative dominance of species (as measured by biomass) in a community.

The series of communities from grass to shrub to forest that terminates in a relatively stable community is called a *sere,* and each of the changes that takes place is a *seral stage*. Although each seral stage is a point in a continuum of vegetation through time, each is recognizable as a distinct community with its own characteristic structure and species composition, especially at the point of optimal development. Seral stages may last a short time or for many years. Some stages may be missed completely, or they may appear only in abbreviated or altered form. For

example, when an old field grows up immediately in forest trees (Figure 25.1e), the shrub stage appears to be bypassed, but structurally its place is taken by the incoming young trees.

Eventually succession slows down, and the plant community achieves some degree of equilibrium or steady state with the environment. This mature, relatively self-maintaining seral stage traditionally has been called the *climax* community, and the vegetation supporting it the climax vegetation.

Succession that takes place on areas devoid of or unchanged by organisms is called *primary* (Figure 25.2). Succession that proceeds from a state in which other organisms were already present is called *secondary* (Figure 25.1). Secondary succession arises on sites in which the vegetational cover has been disturbed by humans, animals, or natural forces such as fires, wind storms, and floods. Its development may be controlled or influenced by the activities of humans or domestic and wild animals. Barren areas, whether they are natural primary sites, such as rock outcrops, sand dunes, and alluvial deposits, or disturbed areas, such as abandoned cultivated fields or roadbanks, are a natural biological vacuum eventually to be filled by living organisms. Plants that colonize such sites comprise the *pioneer species.*

■ A Descriptive Approach

Most successional studies are largely descriptive, describing the process rather than elucidating the mechanisms of succession. Descriptive approaches have their place because they set the pattern of

(a)

(d)

(b)

(e)

(c)

(f)

Figure 25.1 Successional changes in an old field over 45 years. (a) The field as it appeared in 1942, when it was moderately grazed. (b) The same area in 1963. (c) A close view of the rail fence in the left background of (a). (d) The same area 20 years later. The rail fence has rotted and white pine and aspen grow in the area. (e) The same field in 1972, when aspen has claimed much of the ground. (f) The same field in 1987. The open field is now covered with a young forest dominated by quaking aspen and red maple. (Photos by R. L. Smith.)

succession, illustrate the process, and provide useful points from which to discuss what happens during succession.

Terrestrial Primary Succession

On terrestrial primary sites no soil exists initially, and successive communities become more complex as soil develops. There are descriptive studies of primary succession on rocks, outcrops, and cliffs (Burbanck and Platt 1964; Shure and Ragsdale 1977), on sand dunes (Cowles 1899; Olson 1958), and on glacial till (Lawrence 1958).

Vasek and Lund (1980) investigated primary plant succession on a Mojave Desert dry lake bed, a harsh site, with special emphasis on changing soil characteristics. The substrate is characterized by high levels of Na, Cl, and NO_3 and low levels of

Ca, Mg, and K. Summer cypress *(Kochia)* and seepweed *(Suaeda)* colonize the mud cracks in the playa and accumulate about them small mounds of silty-sandy soil. High levels of nutrients of NO_3, Cl, K, Ca, and Mg accumulate in these small mounds and their levels decrease beneath the mounds. Saltbush *(Atriplex)* and goldenweed *(Haplopappus)* invade the mounds, increase their size by holding wind-deposited sandy soil, and add NO_3, Na, Cl, Ca, and Mg to the mounds through accumulating leaves. Eventually the large mounds coalesce to form giant mounds, in which the level of the nutrients is much less than in the playa-level soil beneath the mounds. Following the death of the shrubs, the giant mounds erode and the soil is deposited between them. At this point the overall deposited soil is similar to that originally found in the giant mounds, and Na and Cl are leached down to playa level. These early pioneer plants accumulate soil and redistribute the nutrients, allowing primary succession to proceed. This example emphasizes one aspect of primary succession: that the colonizing species ameliorate the environment, paving the way for invasion of other species.

Figure 25.2 Primary succession in the subalpine zone of the Wasatch Mountains, Utah. Here the early stages include trees; the climax is a mixed herb community. Note the changes in soil depth from a rocky surface with fine soil only in crevices to a well-defined solum essentially free from rocks. (Based on data from Ellison 1954.)

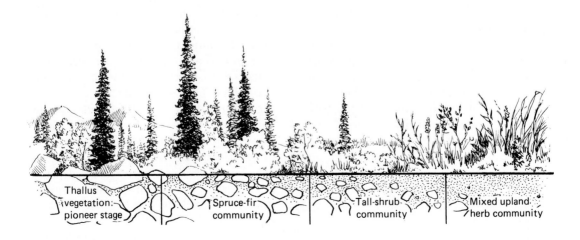

Thallus vegetation: pioneer stage | Spruce-fir community | Tall-shrub community | Mixed upland herb community

Newly deposited alluvial soil on flood plains represents another barren primary site. Walker and associates (1986) studied primary succession on an alluvial flood plain in Alaska. Seeds of all colonizers—willow, alder, balsam poplar *(Populus balsamifera)*, and white spruce—arrived on the site more or less simultaneously and gave rise to seedlings. Their appearance, however, was determined by life history traits. Willow with its light, wind-dispersed seeds was most abundant and the three other species were less widely distributed. Willow and alder grew rapidly, but the willow, naturally short-lived, was heavily browsed by snowshoe hares. Nitrogen-fixing alder then became dominant, eventually to be replaced by balsam poplar and long-lived white spruce. Thus primary succession may be influenced by stochastic events, such as seed production, and life-history traits of the colonists.

Primary succession does not always involve colonization of the site by species with the highest environmental tolerances and their amelioration of the site. In other words, primary succession may not be a lichens to moss to annuals sort of progression nor one in which the most vagile species are the early colonizers. Wood and del Moral (1987) followed early primary succession in subalpine habitats on Mount St. Helens for the first six years following its latest eruption. They found that species of high environmental tolerance, such as *Aster ledophyllus,* dispersed only short distances; most seedlings were within 3 m of the conspecific adults and therefore were unable to colonize much of the barren substrate. Species that dispersed the farthest were incapable of growing on barren ash; they required some amelioration of the site. They could survive only in patches of original substrate that remained uncovered by volcanic ash. Thus primary succession may be impeded by the lack of vagility or inability of adapted pioneering species, preventing rapid colonization of the site.

Terrestrial Secondary Succession

Secondary succession is most commonly encountered on abandoned farmland and noncultivated ruderal sites (waste places) such as fills, spoil banks, railroad grades, and roadsides, all artificially disturbed and frequently subject to erosion and settling movements.

Species most likely to colonize such places are the so-called weeds, species out of place from a human perspective. Although hard to define, weeds have two characteristics in common. They invade areas modified by human action—in fact, a few are confined to such artificially modified habitats—and they are exotics, not native to the region. Once native species move in, these "weed" plants disappear.

Annual, biennial, or perennial, all plants that successfully colonize disturbed areas possess tolerance for soil disturbance and partial defoliation. Their seeds remain viable for a long time and may remain in the soil for a number of years until conditions are right for germination. Some weeds require an open seedbed and exposed mineral soil for germination. Their rapid and successful colonization is aided by an efficient means of dispersal. Some have light seeds that are carried by the wind; others spread by underground rhizomes. These vigorous pioneer plants grow rapidly under favorable conditions; in less favorable habitats they set seed, even when small (Sorensen 1954). In spite of their vigor these plants cannot maintain dominance for long—two to three years at the most—if all disturbance ceases. Short life cycles, advantageous at first, are not adaptable to conditions imposed by incoming plants with longer life cycles, ones that begin growth early in the spring and persist throughout the summer (see Peterson and Bazzaz 1978).

One of the classic examples of secondary succession is Keever's (1950) study of old field succession in the Piedmont of North Carolina. The year a crop field is abandoned, the ground is claimed by annual crabgrass *(Digitaria sanguinalis)* whose seeds, lying dormant in the soil, respond to light and moisture and germinate. The crabgrass's claim to the ground is short-lived. In late summer the seeds of horseweed, a winter annual, ripen. Carried by the wind, they settle on the old field, germinate, and by early winter have produced rosettes. The following spring horseweed, off to a head start over crabgrass, quickly claims the field. During the summer the field is invaded by other plants—white aster *(Aster ericoides)* and ragweed *(Ambrosia artemissifolia)*. Competition from aster and inhibiting effects of decaying horseweed roots on horseweed itself allow aster to achieve dominance.

By the third summer broomsedge *(Andropogon virginicus),* a perennial bunch-grass, invades the field. Abundant organic matter and the ability to exploit soil moisture more efficiently permit broomsedge to dominate the field. About this time pine seedlings, finding room to grow in open places among the clumps of broomsedge, invade the field. Within five to ten years the pines are tall enough to shade the broomsedge. A layer of poorly decomposed pine needles (duff) that prevents most pine seeds from reaching mineral soil, dense shade, and competition for moisture among successfully germinating seedlings and shallow-rooted parent trees

inhibit pines from regenerating themselves on the site. Hardwoods grow up through the pines and, as the pines die (if they are not cut), take over the field. Development of the hardwood forest continues as shade-tolerant trees and shrubs—dogwood, redbud, sourwood, hydrangea, and others—fill the understory. The sere (sequence of communities) has arrived at the mature or tolerant stage in which only the dominant species of the crown can replace themselves in their own shade.

As plants change, so do the animals. Early stages are characterized by such arthropods as crickets, grasshoppers, and spiders and by such seed-eating birds as mourning doves. Broomsedge brings in meajdowlarks and meadow mice, and the low pine growth brings in rabbits. Mature pines shelter pine warblers and sparrows. As pines decline and hardwoods claim the area, downy woodpeckers, flycatchers, and hooded and Kentucky warblers appear.

Aquatic Succession

The transition from a pond to terrestrial community involves primary succession. Succession starts with open water and a bottom barren of life. The first forms of life to colonize the pond or lake are plankton, which may become so dense they cloud the water. If the plankton growth is rich enough, the pond may begin to support other forms of life—caddisflies, bluegills, green sunfish, and large-mouthed bass.

At the same time the pond acts as a settling basin for inputs of sediment from the surrounding watershed. These sediments form an oozy layer that provides a substrate for rooted aquatics such as the branching green algae, *Chara,* and pondweeds. These plants bind the loose matrix and add materially to the accumulation of organic matter. Rapid addition of organic matter and sediments reduces water depth and increases the colonization of the basin by emergent and submerged vegetation. That, in turn, enriches the water with nutrients and organic matter, further stimulating pelagic production and sedimentation, and expands the surface area available for colonization by macrophytes (see Carpenter 1981). Thus aquatic succession goes from an oligotrophic to a eutrophic state. Eventually the substrate, supporting emergent vegetation such as sedges and cattails, develops into a *marsh*. As drainage improves and the land builds higher, emergents disappear, the soil rises above the water table, and organic matter, exposed to air, decomposes more rapidly. Meadow grasses invade to form a marsh meadow in forested regions and prairie in grass country. Depending upon the nature of the environment, the area passes into grassland, forest, or peat bog.

Is the chronosequence of aquatic succession self-driven (autogenic), or does it result from outside (allogenic) influences and disturbance? Paleoecological investigations of aquatic succession in the Indiana sand dunes by Jackson and associates (1988) suggest that the classical sequence of supposedly autogenic succession observed today is the result of the differential effects of disturbances, not of gradual, self-induced successional changes. The dune ponds remained more or less in floating and submerged stages until times of human disturbance, when changes were rapid. Increased sedimentation and changes in water chemistry probably hastened the extinction of certain macrophytes and permitted invasion by others. Thus observed differences in open water or hydrarch succession over a relatively short time reflect differences in disturbances that distort vegetational patterns rather than autogenic successional changes.

Intertidal Succession

There is evidence that succession is involved in the patterns of intertidal vegetation. The subtidal kelp forests off the California coast, for example, exhibit three major vertical layers: a *Macrocystis pyrifera* surface canopy, a dense subsurface canopy of *Pterygophora californica*, and an understory of coralline algae. When Reed and Foster (1984) experimentally removed the canopy and subcanopy layers, annual brown algae *(Desmarestia ligulata)* and a moderate recruitment of the canopy species responded to increased light. This finding suggests that, unless disturbed, these perennial kelp species inhibit their own recruitment as well as the invasion of other species.

Disturbance alters the situation. Harris and associates (1984) observed that a severe storm denuded a southern California marine reef, exposing large areas of virgin rock and decimating the herbivorous sea urchins that graze on the kelp. They followed the sequence of recolonization. Immediately after the storm diatoms colonized the surface, and they were soon followed by filamentous and leafy seaweeds. Dense stands of filamentous brown algae preceded the appearance of small sporophytes of kelp. Where growing in the open, small kelp experienced heavy grazing by two herbivorous fishes, whereas those growing within stands of brown algae escaped serious predation. The brown algal refuge permitted the kelp to become established and grow large enough to withstand predation.

Predation, however, can have a major influence on successional development in subtidal regions. In the Pacific Northwest sea urchins prefer early successional kelp *(Nereocystis leutkeana)* over late successional kelp *(Agarum cribrosum)*. When urchins are present, late successional *Agarum* occurs. If the urchins are naturally rare or removed, the early successional *Nereocystis* inhibits the establishment of *Agarum* (Vadas 1977). In the rocky intertidal zone of New England, littorine snails prefer the early successional green alga *Ulva lactuca,* which can delay the recruitment of less preferred *Fucus vesiculosus.* When the snails are abundant, *Fucus* become established much more quickly (Lubchenco 1978). In southern California sea urchins may overgraze the middle and late successional perennial red and brown algae and permit the early successional ephemeral species to persist. The marine successional example emphasizes another important point, that succession can be strongly influenced by the grazing action of herbivores.

■ Models of Succession

H. C. Cowles (1899) demonstrated succession convincingly in his study of vegetational development on the sand dunes of Lake Michigan, and F. E. Clements (1916) advanced the concept as an ecological hypothesis. His concept of succession had a powerful influence (or inhibition, some critics say) on ecological thought. In fact, Clements' theory of succession dominated ecology for years, almost becoming dogma, because it provided an orderly, logical explanation for the development of plant communities.

Pioneering Concepts

Clements viewed succession as a process involving several phases, to each of which he gave his own terminology. Succession began with a bare site, called *nudation.* Nudation was followed by *migration,* the arrival of propagules. Migration was followed by the establishment and growth of vegetation. Clements called this phase *ecesis.* As vegetation became well established, grew, and spread, various species began to compete for space, light, and nutrients. This phase Clements called *competition,* which was followed by *reaction,* self-induced *(autogenic)* effects of plants on the habitat. The outcome of this reaction was the replacement of one plant community by another, resulting in the persistence of one species complex or community, or *stabilization.* In spite of vigorous criticism of Clements' concept of succession, the basic processes he outlined are still valid and accepted subconsciously, even by his critics.

The problem with Clements' view of succession lies outside the outline of the process. To Clements each stage of succession represented a step in the development of a superorganism, the climax. The climax was an assemblage of vegetation that belonged to the highest type of life form possible under the prevailing climate. "Each climax is the direct expression of its climate; the climate is the cause, the climax is the effect, which in turn reacts upon the climate" (Weaver and Clements 1938: 479). The climax, according to Clements, is able to reproduce itself, "repeating with essential fidelity the stages of its development." Each seral stage so modifies the environment that plants of that stage can no longer exist there. They are replaced by plants of the next stage until the vegetation arrives at the self-reproducing climax. That marked the end of succession.

Henry Gleason (1917) challenged Clements' view of succession. Gleason regarded a plant association or climax not as a superorganism, but as a community that developed from a random process in which short-lived colonizing species were replaced eventually by longer-lived species. Plants involved in succession were those that arrived first on the site and were able to establish themselves under prevailing environmental conditions. Competitive and other interactions among the species involved the plants' modification of the microenvironment, and the ability of the plants to exploit nutrients, moisture, and other environmental inputs on the site determine the final outcome.

Later Frank Egler (1954) recast these views of succession in terms of *relay floristics* and *initial floristic composition*. Relay floristics most nearly corresponds to Clements' idea of succession—groups of associated species marching together and disappearing through time as one group replaces the other. Initial floristics, involved only in secondary succession, proposes that the propagules of most species, both pioneer and late stage, are initially on the site. Which species becomes dominant depends upon life history characteristics and competitive interactions. Short-lived species are eventually replaced by long-lived, but not necessarily climax, species.

These views find further expression in two divergent approaches to succession: a holistic or ecosystem approach and a reductionist or population approach.

The Ecosystem Concept

The ecosystem approach is a direct descendent of Clements' organismal theory of succession. It views succession as a property of the community driven by changes in attributes between youthful and mature systems. It eventually leads to the formation of an emergent entity with unique characteristics involving nutrient flow, biomass accumulation, and species diversity (Odum 1969, 1983). Succession begins with the developmental stages of short-lived, intolerant plant species and terminates with a mature stage dominated by long-lived species. Young developmental stages so modify the environment that the existing community is replaced by a more mature one, better able to exploit the changed environment.

As succession proceeds, ecosystem attributes change. Succession begins with an unbalanced community metabolism. Net community production is greater than respiration and biomass accumulates over time. As the ecosystem matures, respiration begins to equal production. Ultimately, in the terminal stage, production balances respiration. The ratio of biomass accumulation increases to a point at which maximum biomass is maintained per unit of energy flow. Nutrient cycling, more open in earlier stages, becomes closed and internal, with minimal dependence on external inputs. Storage and turnover times of nutrients within the system increase during the development of the sere, as do controls of nutrient retention and conservation. As species composition changes through time, species diversity increases. This approach views succession as an orderly transition from one seral stage to another, driven by self-induced (autogenic) modification of the environment.

The Population Concept

Reductionists reject the ecosystem approach, arguing that successional changes can be explained by population dynamics, especially competition, regeneration, and mortality (Peet and Christensen 1980; Noble and Slatyer 1980) and by physiology and life history strategies (Bazzaz 1979). Noble and Slatyer (1980) proposed a set of vital attributes applicable to secondary plant succession. Such attributes include the method of arrival on a disturbed site, persistence of a species on the site before and after disturbance, the ability of a species to establish itself and grow to maturity in a developing community, and the time a species requires to reach its critical life stage and replacement. What species become established is largely subject to chance. Species composition over time is determined by development, longevity, and response to competition. There is no need to invoke emergent properties or to seek such indices of succession as changes in the ratio of gross production to biomass.

This approach to succession emphasizes the role of population dynamics and life history patterns. Succession in its very earliest stages begins with an open site (nudation) that is eventually colonized by early successional species, variously called pioneer, opportunistic, or fugitive. On a primary site these species have to be carried in by wind, water, or animals (migration). On secondary sites, residual propagules—seeds lying dormant in the soil, roots, and rhizomes—are important in colonization. Having arrived on the site, the colonizers have to establish themselves and grow (ecesis).

Successful early colonists have characteristics that enable them to establish themselves quickly on open sites. They are generally small and low-growing, have short life cycles, and reproduce annually by seeds or send out new shoots from buds near the ground (see Chapter 17). They produce large numbers of easily dispersed small seeds, which can remain dormant for a long time in the soil waiting for favorable conditions to germinate. These plants respond quickly to disturbance, es-

pecially exposure of mineral soil. They grow rapidly, put most of their production into photosynthetic tissue, and attain dominance quickly by suppressing for a while the growth of any later-stage plants that might exist as seedlings beneath them. They are tolerant of fluctuating environmental conditions, particularly a wide range of daily temperatures on the soil's surface, alternate wetting and drying, and intense light.

Early colonists soon lose their temporary dominance—in part because they have to renew their photosynthetic biomass each year—to the taller, more vigorous growth of later-stage plants, which carry over biomass from year to year. These plants gradually assume dominance and place a much greater demand on resources such as nutrients, light, and moisture. As availability of resources decreases and demand increases, competition within and among populations of plants becomes more intense. Mortality, reflected in self-thinning, increases. As the population declines, resources are released for use by plants of still later stages (Peet 1981).

Species characteristic of later stages of succession grow more slowly and are relatively long-lived. They are able to dominate the site over a much longer period of time. Much of their production goes into storage and maintenance. They produce relatively few heavy seeds, dispersed mostly by animals and gravity. Their seeds are large, providing an abundance of nutrients to get new seedlings started, but the seeds' vitality and longevity are low. These species mostly are specialists, adapted to a narrow range of environmental conditions in which the plants either hoard resources or use them more effectively.

Thus reductionists consider the successional process as involving differences in colonizing ability, growth, and longevity of species adapted to grow along a gradient of changing environmental conditions. As environmental conditions change, species composition also changes gradually. The replacement of one or several species or groups of species by others results in part from interspecific competition, which permits one group of plants to suppress slower growing species. As plants of earlier stages are supplanted by species of later stages, the structure of the community as dictated by growth forms and longevity of plants also changes. Eventually succession arrives at a point where long-lived species create a relatively stable community called the climax.

Connell-Slatyer Models

Connell and Slatyer (1977) proposed three different models of succession: facilitation, tolerance, and inhibition. All three models emphasize some points of Clements' original theory: nudation, migration, ecesis, competition, reaction, and stabilization.

The *facilitation model,* basically a Clementsian approach, is autogenic; changes are brought about from within by the organisms themselves. Early stage species modify the environment and prepare the way for later stage species, facilitating their success.

The *tolerance model* involves the interaction of life history traits, especially competition. It suggests that later successional species are neither inhibited nor aided by species of earlier stages. Later stage species can invade a site, become established, and grow to maturity in the presence of those preceding them. They can do so because these later species have a greater tolerance to a lower level of resources than the earlier ones. Such interactions lead to communities composed of those species most efficient in exploiting re-

sources either by interference competition or by using sources unavailable to other species.

The *inhibition model* is purely competitive. No species is completely superior to another. The site belongs to those species that become established first and are able to hold their position against all invaders. They make the site less suitable for both early and late successional species. As long as they live, they maintain their position; but the ultimate winners are the long-lived plants, even though early successional species may suppress them for a long time.

The Resource-Ratio Model

Recent approaches to succession combine aspects of both the holistic and reductionist points of view. One is the resource-ratio hypothesis advanced by Tilman (1985). This hypothesis involves two components: (1) interspecific competition for resources and (2) a long-term pattern of a supply of limiting resources, especially soil nutrients and light. Succession comes about as the relative availability of those resources changes through time. The gradient ranges from habitats with soils poor in nutrients but with a high availability of light at the soil surface to habitats with nutrient-rich soils and low availability of light. Community composition changes along that gradient as the availability of two or more limiting resources, particularly nitrogen and light, changes. Species reach an equilibrium with the supply rates of the limiting resources. In doing so they lower the available resources to a point at which other species cannot invade.

In early primary succession, the colonizing species are those adapted to a low soil nutrient and high light regime. As biogeochemical processes make more soil nutrients available, plant growth increases, reducing the availability of light at the soil surface. The changing ratio of soil nutrients to light leads to the replacement of one plant species by another, favoring through time plants with high nutrient and low light availability at the soil surface. Because of the relative slowness of primary succession, succession along the gradient arrives at various plateaus of equilibrium, dominated by species competitively superior at each ratio.

Secondary succession flows in much the same pattern, but at a more rapid rate. Species composition and rapidity of change depend upon the point on the gradient at which the species colonize the area. For example, old field succession normally involves a relatively low level of soil nutrients and high light on the soil surface; but where a seed source is available and a site exhibits a point on the gradient with high soil nutrients and high light, the area may be colonized by such tree species as pines or yellow-poplar and exclude earlier successional species.

The Nonequilibrium Model

The population approach proposes that succession is a population process involving competition that results from inversely correlated traits: a high utilization of light associated with poor utilization of nutrients. Those species that do best in high light conditions outcompete shade-tolerant species but lose their competitive advantage under lower light and higher nutrient conditions. Competition is viewed as taking place among species.

However, competition is an individual process. The ability of an individual to compete is restrained by individual traits based on a suite of life history attributes common to the species: height, shade tolerance, rapidity of growth, size, and longevity. There are two components to plant interactions. One is the response of the individual to the prevailing environment; the other is the influence of the plant on the environment. Thus the outcome of plant interactions is a result of plants modifying the environment.

Huston and Smith (1987) have proposed a nonequilibrium model of succession that involves both competition and autogenesis. It uses competition among individual plants to explain species replacement on a spatial and evolutionary gradient. It assumes that both environmental resources and intensity of competition change through time and among communities. The hypothesis is based on three premises. (1) As plants grow, they alter the environment in such a way that the relative availability of resources changes, altering the rules for competitive success. (2) The physiological traits of plants prevent any one species from achieving maximum competitive ability under all circumstances. (3) The interaction between premises 1 and 2 produces an inverse correlation between certain groups of traits, so species that are good competitors under one suite of environmental conditions are poor competitors under another.

■ The Climax

According to classical ecological theory, succession stops when the community or sere has arrived at an equilibrium or steady state with the environment. At this point the community is stable and self-replicating and, barring major disturbances, will persist indefinitely. This end point of succession is termed the *climax*.

The climax, theoretically, takes on certain characteristics. The vegetation is tolerant of the environmental conditions it has imposed upon itself. It is characterized by an equilibrium between gross primary production and total respiration, between energy utilized from sunlight and energy released by decomposition, between the uptake of nutrients and the return of nutrients by litterfall. It has a wide diversity of species, a well-developed spatial structure, and complex food chains.

Ideally, every individual in the climax stage is replaced by another of the same kind; average species composition reaches an equilibrium. If offspring of the same species are favored over others, then a dead mature individual may be replaced by a plant of its own kind. If the offspring are concentrated about a mature parent, it may be replaced by its own progeny. This case is most likely if the replacement is essentially the same unit—a root or stump sprout from the dead individual. If conditions beneath the mature tree are less favorable for its own species than for other species, it will be replaced by associated species. If conditions are neither more or less favorable for the offspring than for other species, replacement individuals will be influenced by the relative abundance of propagules arriving on the site, suppressed individuals already present, and competitive interactions among them.

In reality, self-destructive changes are continually taking place in the climax community. Trees grow old and die and more often than not are replaced by individuals of a different species. Changes are constantly occurring in patches across the community (see Chapter 26), slowly altering the average species composition. Although succession may slow down, it never ceases; and the stability and persistence of the climax reflects mostly the fact that the dominant species are long-lived in terms of a human time frame.

There are three theoretical approaches to the climax. One is the *monocli-max theory* advanced by Clements (1916). This theory recognizes only one climax, whose characteristics are determined solely by climate. Successional processes and modifications of the environment overcome the effects of differences in topography, parent material of the soil, and the like. Given sufficient time, all seral communities in a given region converge to a single climax and the whole landscape will be clothed with a uniform plant community. All communities other than the climax are related to the climax by successional development and are recognized as subclimax, postclimax, disclimax, and so on.

Tansley (1935) advanced the *polyclimax theory*, which proposes that the climax vegetation of a region consists of a mosaic of vegetational climaxes controlled by soil moisture, soil nutrients, topography, slope exposure, fire, and animal activity (see Daubenmire 1968a, b; Whittaker 1974). The spatial pattern of habitats influences the spatial pattern of climax communities.

A third is the *climax pattern theory* (Whittaker 1953, 1954; McIntosh 1958; Selleck 1960). The total environment of the ecosystem determines the composition, species structure, and balance of a climax community. Involved in the environment are the characteristics of species populations, their biotic interrelationships, availability of flora and fauna to colonize the area, chance dispersal of seeds and animals, and soils and climate. The mosaic of climax vegetation will change as the environment changes. The climax community then represents a pattern of populations that corresponds to and changes with the pattern of environmental gradients to form ecoclines. The central and most widespread community is the prevailing or climatic climax.

Vegetation is dynamic, and mortality of individual trees is a feature of the climax. As old trees die, new individuals and species take their place. Thus species composition changes, even though physiognomy remains the same. Gross production is high, but so is respiration because of the decomposition of woody material. Net production is relatively low, but production never achieves equilibrium with respiration. Biomass continues to accumulate on individual trees, however slowly. Bormann and Likens (1971) have suggested that rather than achieving equilibrium, a mature stage of forest vegetation achieves a *shifting-mosaic steady state* in which the standing crop of living and total biomass (living + dead) fluctuates about a mean. Such an ecosystem consists of patches in various stages of seral development, from ones exhibiting high net production and biomass accumulation to ones of senescence and downed timber in which respiration exceeds production. The total production of these patches remains more or less the same throughout time.

Studies of old or mature ecosystems to test the climax theory are rare, especially in the eastern deciduous forest where extremely few old untouched stands exist. One is the Dick Cove Natural Area, a mesic forest on the western slope of the Cumberland Plateau in Franklin County, Tennessee. It is a 100-acre tract surrounded by second growth forest. The forest dominants are oaks. Northern red oak, white oak, chestnut oak, hickories, sugar maple, and yellow-poplar are associated with some 17 other species. Age is beginning to tell on the old trees. McGee (1984) studied the mortality and succession in this old-age forest and found that during a nine-year period from 1972 through 1981, 26 percent of the hickory and 18 percent of white oak and red oak over 17 inches dbh died from a combination of senescence, drought, insect damage, and blowdown. These dead trees ranged in age from 90 to 375 years. Dead red oak averaged 135 years and dead

white oak, chestnut oak, and hickory averaged 210 years old. The demise of the oaks after 350 years appears to be accelerating, bringing about a change in structure and species composition of the stand from the original apparent climax of oak and hickory to one of the slow-growing, shade-tolerant sugar maple and the fast-growing, shade-intolerant yellow-poplar with a strong component of hickory. Red oak lacks replacement trees and yellow-poplar is filling in the gaps (see Chapter 26). This study emphasizes that even old-growth apparent climax forests are in a state of flux that is observable only over periods of several hundred years. What appears permanent to the observer is slowly transitory.

This point is even more applicable to the old-growth 450 to 1000-year-old stands of Douglas-fir of the Pacific Northwest. Technically, these old Douglas-fir forests are not climax, because Douglas-fir is a pioneer species that happens to be very long-lived (Franklin and Hemstrom 1981). The climax species that might replace Douglas-fir in another 500 years are western hemlock (*Tsuga heterophylla*) and Pacific silver fir (*Abies amabilis*). Only by human standards of permanency are these old stands "climax." Moreover, these old Douglas-fir stands are uneven-aged when, theoretically, they should be even-aged. Age classes range from 145 years on up within the stand, suggesting occasional disturbances and replacement, or perhaps failure of canopy closures over a period of time because of variable seed crops.

The variation in age classes within the stand suggests some agreement with the shifting-mosaic steady state. Gross production is high, but so is respiration, and net production is low. Ratio of live biomass to dead biomass peaks at 300 to 400 years. Total biomass is around its highest at about 750 years and total dead biomass is greatest at 800 to 1000 years. The forest is a complex system of many species of saprophytes and epiphytic lichens. Nutrient retention within the system involves complex detrital pathways. Nitrogen is fixed by lichens in tree crowns and by microflora in decomposing logs. Structural diversity is high, with a large range in the size of individual living trees and numerous large standing dead trees, large downed trees, and decomposing logs. Because of their great age, these old growth forests have seen the evolution of species that depend upon them for their continued existence. Among these species are the spotted owl *(Strix occidentalis)* and the tree vole *(Phenacomys longicaudus)*.

Cycles and Fluctuations

Cyclic Replacement

Successional stages that appear to be directional are often, in fact, phases in a cycle of vegetational replacement. Such cyclic replacement results from destruction of existing vegetation by some characteristic of the dominant organisms or by periodic disturbances that start regeneration again at some particular stage. Such changes are part of community dynamics, usually occur on a small scale within the community, and are repeated over the whole of the community. Each successive community or phase is related to the others by orderly changes in the upgrade and downgrade series. Such cyclic replacements contribute to community persistence.

Watt (1947) described phasic cycles in Scottish heaths. Scottish heather *(Calluna)* represents the peak of the upgrade series. After the death of heather, a lichen *(Cladonia silvatica)* becomes dominant and covers the dead heather stems. Eventually the lichen disintegrates to expose bare soil, the last of the downgrade series. The bare soil is colonized by bearberry to initiate the upgrade series. Heather then reclaims the area and dominates again. There are other, shorter phasic cycles, one involving heather, lichen, bare soil, and back to heather again.

Cyclic replacement is a relatively common and important phenomenon in different ecosystems. Cyclic succession is frequently initiated by ants or ground squirrels in old field communities in Michigan (Evans and Cain 1952). Here bare areas at the bottom of the downgrade series are invaded by mosses to start the upgrade series (Figure 25.3). The mosses are invaded by Canada bluegrass and dock. The accumulation of dead culms of these plants is covered by lichens of several species and together they crowd out the grass. Rain, frost, and wind destroy the lichens, and the bare soil is left open to invasion by mosses again. In the desert scrub of Texas creosote bush *(Larrea tridentata)* exchanges place with Christmas tree cholla *(Opuntia leptocaulis)* with an interval of bare ground between (Yeaton 1978). The overriding pattern of successional development on the coastal tundra of Alaska is cyclic, controlled primarily by changes in microrelief and drainage regimes (Webber et al. 1980). Cyclic replacement retains the long-term stability of pothole marshes in north-central North America (Figure 25.4). During periods of drought—about every 5 to 20 years—shallow marshes dry. Organic debris accumulated on the bottom decays rapidly, releasing nutrients for recycling and stimulating the germination of seeds (Weller and Fredrickson 1973). The upgrade of the cycle begins with seed germination on exposed mud. It is followed by a newly flooded stage with sparse, often well-dis-

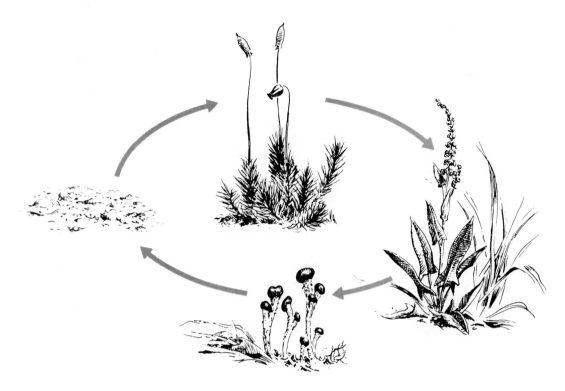

Figure 25.3 Cyclic replacement in an old field community. The cycle moves from dock to lichen to bare ground to moss and back to dock again.

persed vegetation, dominated by annuals and immature perennials; a flooded, dense marsh dominated by perennials; and a deep open marsh rimmed with emergents, in part caused by the feeding activities of muskrats. The cycle begins anew when the marsh dries. Although these short-term cycles give the shallow marsh the appearance of an unstable ecosystem, such cyclic replacements ensure the long-term stability of the marsh ecosystem.

Fluctuations

Fluctuations are nonsuccessional or short-term reversible changes (Rabotnou 1974).

Fluctuations differ from succession in that the floristic composition over time is stable: No new species invade the site and changes in dominants may be reversible. These changes in floristic composition result from such environmental stresses as soil moisture fluctuations, wind, grazing, and the like.

Fluctuations in forest communities may involve an alternation of species in canopy gap replacements (see Chapter 26). In such forests there appears to be a tendency for each species to be replaced by its competitor (Fox 1977). If one species becomes moderately abundant in the canopy, alternate species may be abundant beneath it. Thus over time dominance in the canopy may eventually shift in favor of the temporarily disadvantaged species. For example, Fox (1977) found that in old northern hardwood stands, sugar maple tends to replace beech in

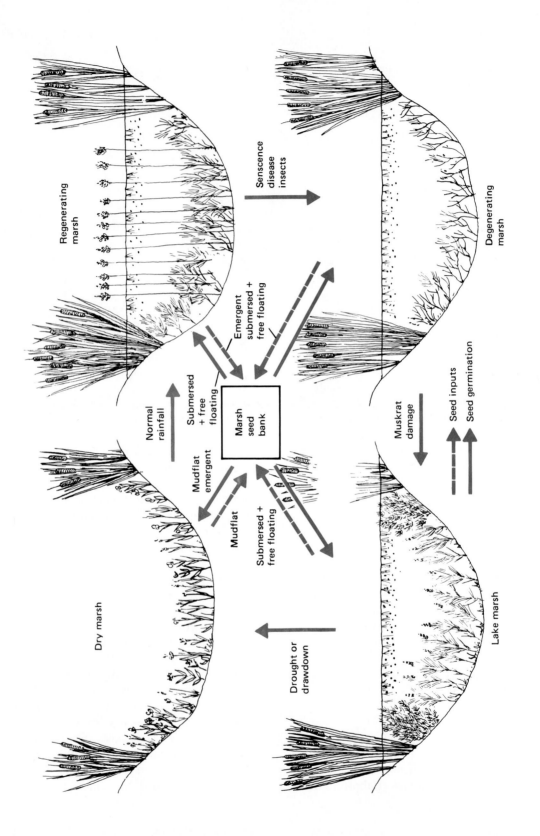

Regenerating marsh

Senscence
disease
insects

Degenerating marsh

Emergent
submersed +
free floating

Normal
rainfall

Submersed
+ free
floating

Mudflat
emergent

Marsh
seed
bank

Muskrat
damage

Seed inputs

Seed germination

Mudflat

Submersed +
free floating

Dry marsh

Drought or
drawdown

Lake marsh

Figure 25.4 Cyclic replacement of vegetation in a prairie glacial marsh, influenced by the input and output of seeds. The cycle is initiated by periods of drought followed by periods of normal rainfall, but the key to replacement is the seed bank in the marsh mud. (After van der Valk and Davis 1978:333.)

small openings and beech replaces sugar maple. In general, the tendency is for the dominant tree to be replaced more than half the time by its competitor, for two reasons. First, the dominant tree usurps the site, concentrating the bulk of biomass at one particular place in a single tree. Its conspecifics are thinned out more severely than its competitors. Second, because of its influence on nutrient regeneration and light and moisture regimes, the canopy tree creates a somewhat species-specific microhabitat for seeds and seedlings beneath it. With an alternate species favored in the understory, the forest maintains a species equilibrium.

Fluctuations may also involve replacement of one age class by another within a

Figure 25.5 A regeneration wave in a balsam fir forest. The wave is initiated at the location of standing dead trees with mature trees beyond it and an area of vigorous reproduction below it. In the area where dead trees have fallen a crop of young fir seedlings is developing. Beyond them is a dense stand of fir saplings, followed by a mature fir forest and then by a second wave of dying trees. (After Sprugle 1976.)

species. Such fluctuations are important in maintaining certain forest ecosystems, particularly coniferous forests (Korchagin and Karpov 1974). Sprugle (1976) describes a wave regeneration pattern in balsam fir *(Abies balsamae)* forests in the northeastern United States. Trees die off continually at the edge of a "wave" and are replaced by vigorous stands of young balsam fir (Figure 25.5). The cycle is initiated when an opening occurs in the forest, exposing trees to the wind on the leeward side of the opening. Desiccation of the canopy foliage by winter winds, the loss of branches and needles in winter from rime frost forming on them, and decreased primary production due to cooling of needles in summer cause the death of trees. Their death exposes the trees behind them to the same lethal conditions and they die too. This process continues, so a wave of dying trees through the forest is followed by a wave of vigorous reproduction.

These regeneration waves follow at intervals of about 60 years. The process is so regular that all stages of degeneration and regeneration are found in the forest at all times, provided the stand is not cut. The phasic cycle results in a steady state, because the degenerative changes in one part of the forest are balanced by regenerative stages in another. The wave regeneration process ensures the stability of the forest and prevents its advancement into a hardwoods stage.

■ Changes in Ecosystem Attributes

Although a reductionist approach minimizes the influence of changing ecosystem structure and function as a driving force behind succession, undeniably certain changes take place.

In 1942 R. Lindeman postulated that productivity should increase as succession proceeds, that succession improves the efficiency of transfer and use of energy and nutrients. This postulate was modified and expanded by Margelef (1963, 1968) and E. P. Odum (1969) into the concept of maturity. They held that structural complexity and organization of an undisturbed ecosystem increase and mature with time (Table 25.1). Succession may begin with a bare area colonized by small plants and terminate with large plants whose growth form results in increased vertical stratification and increased influence on environmental variables within the community. These changes and others, according to Margelef and Odum, are typical.

Early successional stages, according to the ecosystem models of succession, are characterized by few species, low biomass, and dependence on an abiotic source of nutrients. Net community production is greater than respiration (see Table 25.1), resulting in increased biomass over time. Energy is channeled through relatively few pathways to many individuals of a few species, and production per unit biomass is high. Food chains are short, linear, and largely grazing.

The mature stages in succession are characterized by a greater diversity of species, high biomass, a nutrient source largely organic in nature, and gross production that about equals respiration.

Table 25.1 Expected Trends in Ecological Succession from Development to Mature Systems

Attribute	Trend	Accept or Reject
Biomass	Increases	A
GPP/ER	Approaches 1	R (plants)
GPP/B	Decreases	A
B/ER	Increases	A
Net community production	Decreases	R (plants)
		A (animals)
Total organic matter	Increases	A
Inorganic nutrient input	External to internal	?
Species richness	Increases	A
Species equability	Increases	R
Stratification	Increases	R
Size of organisms	Increases	R
Niche specialization	Broad to narrow	?
Role of detritus	Increases in importance	?
Growth form	r to K	R
Nutrient conservation	Increases	R

Note: GPP is gross primary production; ER is ecosystem respiration; B is standing crop biomass.

Accept or Reject: If trends of attributes are considered hypotheses, they can be tentatively accepted or rejected based on current data. A is Accept; R is Reject.

Source of attributes: Odum 1981:446.

Food chains are complex and largely detrital. Inorganic nutrients accumulate in the soil and vegetation, and considerable quantities are locked or hoarded in plant tissue. Fundamental niches shift from broad and general to narrow and specialized. Accompanying these changes is an increase in species diversity.

How well does succession conform to these proposed attributes? Studies of ecosystem functions provide some insights into the validity of ecosystem theory. Consider species diversity. Diversity in succession can be considered from two points of view: across an environmental gradient and across a temporal gradient (Auclair and Goff 1971). Because the two are inversely related, no sweeping generalizations can be made concerning diversity and maturity. Diversity may or may not increase with advancing successional stages (Figure 25.6). Some early stages of succession may have a greater diversity of vegetation than later stages (Bazzaz 1975). Some of the later stages may be dominated by plants with strong allelopathic interference, which reduces species diversity. Sassafras (*Sassafras albidum*), for example, maintains itself in relatively pure stands by releasing into the soil at different times of the year phytotoxins that inhibit germination of seeds and growth of other plants (Gant and Clebsch 1975). A great deal depends upon the initial state of the site. Auclair and Goff (1971) concluded that pioneering forest communities on xeric sites increase in diversity through time (Figure 25.7), whereas successional forest communities on mesic sites decline in diversity from late successional to equilibrium forest. On intermediate sites diversity approaches an asymptote late in succession. Even within this framework diversity of a given site will be influenced by soil, microtopography, disturbance, and grazing by herbivorous animals. Diversity

of bird species and trees increases with succession but reaches a maximum before the mature stage.

Other attributes also fail to match the proposed ones, according to results obtained from 15 years of data collected from developing a hardwood forest at Hubbard Brook, New Hampshire. In this research Bormann and Likens (1979) divided the seral stages of forest recovery following clear-cutting into four phases: a *reorganization,* or young seral stage; an *aggradation,* or developmental seral stage in which the growing forest is passing through the shrubs and pole timber; a *transition* phase, when growth is beginning to slow and a number of trees die from competition and self-thinning; and finally a mature, *steady state* phase. Data for ecosystem function in the reorganization and early aggradation phase were actual; those for later stages were simulations based on actual data using the JABOWA model.

The ratio of production to respiration (P/R) did not follow a straight route from less than 1 to 1. Instead, the P-R ratio for the reorganization stage and the transition stage was less than 1 and for the steady state, equilibrium. Only in the aggradation phase was the P-R ratio greater than 1. These changes in the P-R ratio reflected the loss of biomass through decomposition of dead organic matter following clear-cutting in the reorganization phase and death of trees in the transition phase.

Gross production increased through all stages, and net production was highest in the aggradation phase, which led to a rapid accumulation of biomass. Thus biomass accumulation was highest in the aggradation phase, declined during the transition stage, and leveled off during the steady state phase (Figure 25.8). Nutrient cycling was loosest during the reorganization phase, when quantities of nutrients were being exported from the system (see

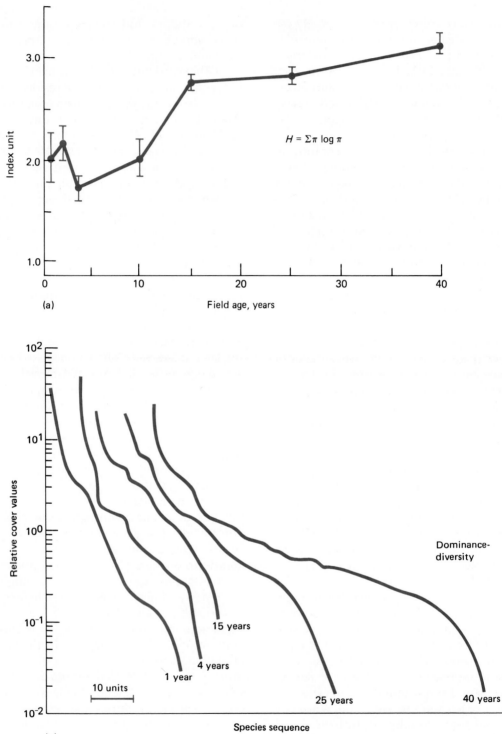

$H = \Sigma \pi \log \pi$

(a)

Index unit

Field age, years

(b)

Relative cover values

10^2

10^1

10^0

10^{-1}

10^{-2}

Dominance-diversity

15 years

4 years

1 year

25 years

40 years

10 units

Species sequence

Figure 25.6 Relationship between plant species diversity and succession. (a) Plant species diversity generally increases with succession and may reach a maximum in the forest stage (see Figure 25.8), when shade-tolerant and shade-intolerant guilds are present together in the community. Relatively low species diversity in a successional community (as in the example graphed here) may result from development of strong dominance by a species with allopatric interference. High species diversity may result from a high degree of vertical and horizontal microenvironmental heterogeneity. (b) Dominance-diversity curves of a successional community are geometric at first, suggesting the niche-preemption theory (see Figure 24.12). Dominance curves become less steep with time as more species are added and gradually a log-normal distribution with an increase in species with intermediate relative importance values develops. (From Bazzaz 1975:486, 487.)

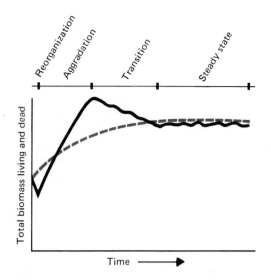

Figure 25.8 Model of biomass accumulation during successional development of a forest after clear-cutting. The dashed line is an asymptotic model of net biomass accumulation (living and dead) until a steady state is achieved, as predicted by the ecosystem theory of succession. The solid line is biomass accumulation as predicted and observed by a shifting-mosaic steady state model. (From Bormann and Likens 1979:166.)

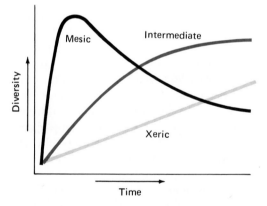

Figure 25.7 Species diversity in relation to time for xeric, intermediate, and mesic moisture conditions. Note that diversity increases with time on xeric and intermediate sites but decreases on mesic sites, which reach their maximum diversity in earlier stages of succession. Diversity in later stages is maximal on intermediate sites. (From Auclair and Goff 1971:525.)

most highly regulated of all stages. Habitat diversity was highest during the aggradation phase, with its mix of shade-intolerant and shade-tolerant and intermediate species.

Chapter 26). The reason was rapid decomposition of detrital material and the inability of incoming vegetation to utilize all the resources. Nutrient cycling was tightest during the aggradation phase, the

■ Time and Direction

Time is an integral component of succession; but time is relative, measured in terms of human experience. Theoretically, climax vegetation is stable with a degree of permanency. What is permanent? Vegetation that remains the same over several human lifetimes, like the old-growth forests? By that standard, old field vegetation could be a climax community to ants, birds, and meadow mice, because the vegetation would extend over several lifetimes of those organisms.

Nevertheless, successional communities have their life spans, governed in part by the longevity of the plants that comprise the seral stages. The annual weed stage in an old field may last no longer than one or two years. Pioneer lichens and moss on a granite outcrop may remain for hundreds of years. Seral grass stages may last 10 to 15 or even fewer years before being overtaken by woody growth. Woody growth in a shrub stage—whether true shrubs or incoming tree growth below a height of 6 m—may last an additional 10 to 15 years until pioneering trees take over or the canopy closes. If the trees are pioneer or shade-intolerant species such as aspen, the stage may last 25 to 40 years before shade-tolerant species become dominant and hold the site 250 to 1000+ years.

The time line implies that shade-tolerant species replace shade-intolerant ones and that trees replace shrubs. It is not always so. A dense growth of shrubs such as meadowsweet, mountain laurel, or Saint-John's-wort may claim a site for 60, 70, 80 years without any indication of change. Shade-intolerant species, such as yellow-poplar and red pine, can hang on to a site and maintain their positions for many years. Douglas-fir, a pioneer species in western North America, may claim a site for more than 1000 years before giving way to western hemlock, which may need another 500 years to achieve dominance.

How long a seral stage lasts could be simply an academic question, except that time has implications in forestry and wildlife management. Certain types of wildlife habitat are seral and thus ephemeral. Their maintenance may be critical to the welfare of certain species, which requires human interference with the successional process. Certain commercially valuable timber trees are pioneering species that require periodic disturbance to ensure regeneration.

Classic succession theory holds that succession is directional and therefore predictable. Succession has to head somewhere, even if in a circle (cyclic succession); but if it implies unidirection—a one-way trip—the answer to whether succession is directional would have to be a qualified no.

We could predict with a high degree of probability that an old field in eastern North America will, barring further disturbance, return to forest. We would have much more difficulty predicting the kind of forest, even if we had knowledge of previous vegetation. Each successional community, climax and otherwise, is individualistic, a one-time product of abiotic and biotic forces operating during the time of its development. The exact interaction of these forces will not be repeated again. Any new successional community will be molded by current abiotic and biotic inputs. The exact original composition of the species will not be duplicated. Thus, we might predict successfully the physiognomy over a broad region but miss the successional communities on a local level (see Horn 1975, 1981). There are too many side roads succession can take, influenced by environmental conditions and biotic interactions that do not lead to the supposed climax (Figures 25.9, 25.10).

Figure 25.9 Soil conditions, particularly soil moisture, influence species composition and pattern of succession. This diagram illustrates plant succession on moderately drained to well-drained soils in a high elevation hanging valley, Canaan Valley in West Virginia. Succession proceeds along several directions, resulting in different "climax" communities. The original climax in the valley was spruce, to which it probably will never return. (After Fortney 1975.)

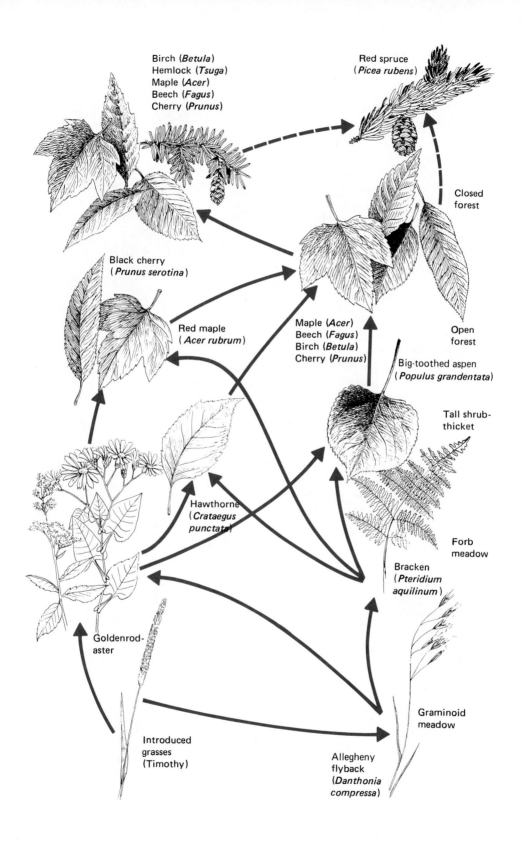

Birch (*Betula*)
Hemlock (*Tsuga*)
Maple (*Acer*)
Beech (*Fagus*)
Cherry (*Prunus*)

Red spruce
(*Picea rubens*)

Closed forest

Black cherry
(*Prunus serotina*)

Red maple
(*Acer rubrum*)

Maple (*Acer*)
Beech (*Fagus*)
Birch (*Betula*)
Cherry (*Prunus*)

Open forest

Big-toothed aspen
(*Populus grandentata*)

Tall shrub-thicket

Hawthorne
(*Crataegus punctata*)

Bracken
(*Pteridium aquilinum*)

Forb meadow

Goldenrod-aster

Introduced grasses
(Timothy)

Allegheny flyback
(*Danthonia compressa*)

Graminoid meadow

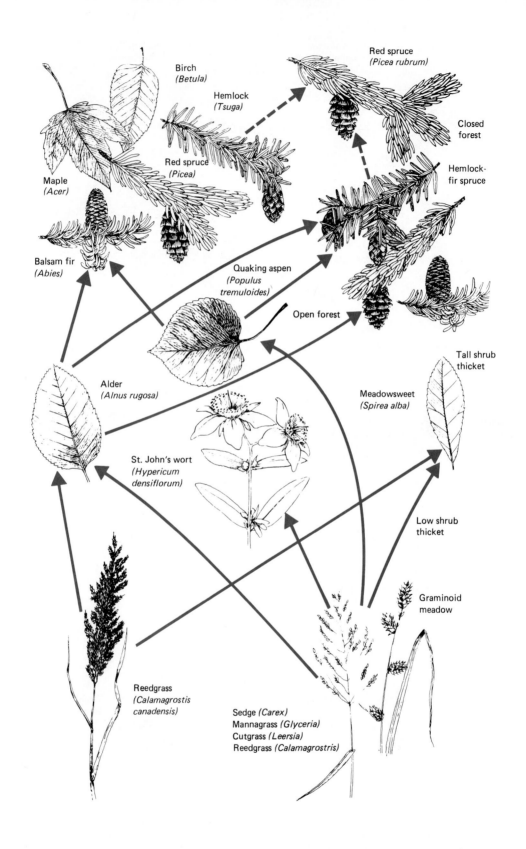

Birch
(Betula)

Hemlock
(Tsuga)

Red spruce
(Picea)

Maple
(Acer)

Balsam fir
(Abies)

Red spruce
(Picea rubrum)

Closed
forest

Hemlock-
fir spruce

Quaking aspen
(Populus
tremuloides)

Open forest

Tall shrub
thicket

Alder
(Alnus rugosa)

Meadowsweet
(Spirea alba)

St. John's wort
(Hypericum
densiflorum)

Low shrub
thicket

Graminoid
meadow

Reedgrass
(Calamagrostis
canadensis)

Sedge (Carex)
Mannagrass (Glyceria)
Cutgrass (Leersia)
Reedgrass (Calamagrostris)

Figure 25.10 Plant succession on wet mineral soil in Canaan Valley, West Virginia, advances differently from succession on well-drained soils. The terminal community includes a mixed forest of hemlock, balsam fir, red spruce, and maple. Succession to forest can be blocked by terminal shrub communities of either meadowsweet or Saint-John's-wort. (After Fortney 1975.)

Successional direction is implicit in two popular models of forest succession and their variations: JABOWA (Botkin, Janak, and Wallis 1972; Aber, Botkin, and Melillo 1978) and FORET (Shugart, Crow, and Hett 1973). These models involve explicit variables, including temperature, moisture, light, nutrients, life history patterns of species involved, mortality, replacement growth rates, and competition. These models are predictive over large areas as long as conditions con-

form to the boundaries set by the variables. Such models are useful in planning timber harvests and predicting changes in wildlife habitat and the effects of pollution and fire on forest ecosystems.

■ Animal Succession

As seral stages change, animal life changes with them (Figure 25.11). Because animal life is influenced more by structural char-

Figure 25.11 Wildlife succession in large conifer plantations in central New York State. Note how some species appear and disappear as vegetation density and height change. Some species are common to all stages. (After Smith 1960.)

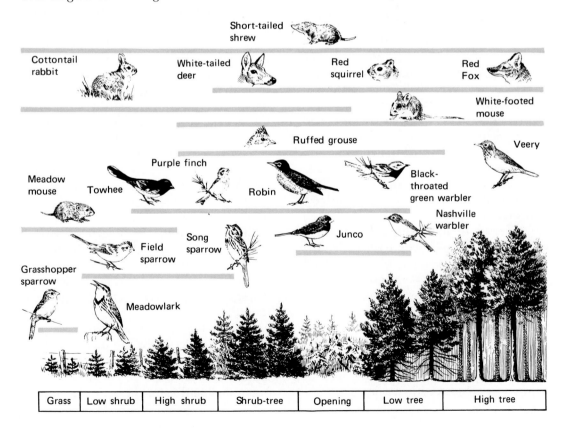

| Grass | Low shrub | High shrub | Shrub-tree | Opening | Low tree | High tree |

acteristics than by species composition, each stage has its own distinctive group of animals. For this reason successional stages of animal life might not correspond to the successional stages identified by plant ecologists. Animal ecologists would classify a young stand of yellow-poplar or balsam fir under 6 m tall as a shrub stage; a plant ecologist would consider yellow-poplar an intolerant tree stage and fir a tolerant tree stage.

Early terrestrial successional stages support animals of grasslands and old fields, such as meadowlarks, meadow voles, and grasshoppers. When woody plants invade, a new structural element is added. Meadowlarks decline and field sparrows and song sparrows appear. When woody vegetation, whether tall shrubs or young trees, eventually claims the area, shrubland animals move in. Field sparrows decline and the thickets are claimed by towhees, catbirds, brown thrashers, and goldfinches. Meadow mice give way to white-footed mice. When woody growth exceeds 6 m in height and the canopy closes, shrub-inhabiting species decline, to be replaced by birds and insects of the canopy. As the community matures and more structural elements are added, new species appear, such as tree squirrels, woodpeckers, and birds of the forest understory like hooded warblers and wood thrushes.

Diversity of animal life across a range of seral stages varies with each individual community. In a general way, shrubland and edge communities and mature stands have a greater diversity of animal life than young forest stands. The key to diversity of wildlife is a heterogeneous landscape with patterns of habitat patches of successional stages and a maintenance of this pattern in an area sufficiently large to support its characteristic species.

■ Microcommunity Succession

Within each major community and dependent upon it for an energy source are a number of microcommunities. Dead trees, animal carcasses and droppings, plant galls, tree holes, all furnish a substrate on which groups of plants and animals live, succeed each other, and eventually disappear, becoming in the final stages a part of the nutrient base of the major community itself. In these instances succession is characterized by early dominance of heterotrophic organisms, maximum energy available at the start, and a steady decline in energy as the succession progresses.

An acorn supports a tiny parade of life from the time it drops from the tree until it becomes a part of the humus (Winston 1956). Succession often begins while the acorn still hangs on the tree. The acorn may be invaded by insects, which carry to the interior pathogenic fungi fatal to the embryo. Most often the insect that invades the acorn is the acorn weevil (*Curculio rectus*). The adult female burrows through the pericarp into the embryo and deposits its eggs. Upon hatching, the larvae tunnel through to the embryo and consume about half of it. If fungi (*Penicillium* and *Fusarium*) invade the acorn simultaneously with the weevil or alone, they utilize the material. The embryo then turns brown and leathery and the weevil larvae become stunted and fail to develop. These organisms represent the pioneer stage.

When the embryo is destroyed, partially or completely, by the pioneering organisms, other animals and fungi enter the acorn. Weevil larvae cut through the outer shell and leave the acorn. Through this exit hole fungi-feeders and scavengers enter. Most important is the moth *Valentinia glandenella*, which lays its eggs on or in the exit hole, mostly during the fall. Upon hatching, the larvae enter the acorn, spin a tough web over the opening, and proceed to feed on the remainder of the embryo and the feces of the previous occupant. At the same time several species of fungi enter and grow inside the acorn, only to be utilized by another occupant, the cheese mites *(Tryophagus* and *Rhyzoglyphus)*. By the time the remaining embryo tissues are reduced to feces, the acorn is invaded by cellulose-consuming fungi. The fruiting bodies of these fungi, as well as the surface of the acorn, are eaten by other mites and collembolans and, if moist, by cheese mites too. At this time predaceous mites enter the acorn, particularly *Gamasellus,* which is extremely flattened and capable of following smaller mites and collembola into crevices within the acorn. Outside on the acorn, cellulose and lignin-consuming fungi soften the outer shell and bind the acorn to twigs and leaves on the forest floor.

As the acorn shell becomes more fragile, holes other than the weevils' exits appear. One of the earliest appears at the base of the acorn where the hilum (the scar marking the attachment point of the seed) falls out. Through this hole, larger animals such as centipedes, millipedes, ants, and collembolans enter, although they contribute nothing to the decay of the acorn. The amount of soil in the cavity increases and the greatly softened shell eventually collapses into a mound and gradually becomes incorporated into the humus of the soil.

Thus microcommunities illustrate one concept of succession: that the change in the substrate is brought about by the organisms themselves. When organisms exploit an environment, their life activities make the habitat unfavorable for their own survival and create a favorable environment for different groups of organisms. Those responsible for the beginning of succession are all quite specialized for feeding in acorns, the later forms are less so, and the final group are generalized soil animals, such as earthworms and millipedes.

■ Human Communities

No land-use change is more drastic than industrialization and urbanization, a "climax" type in human succession. Natural vegetation is destroyed by humans and replaced by ecologically permanent bare areas of concrete, asphalt, and steel. Even here a diversified group of animals and some plants are able to exist. Norway rats, common rock pigeons, starlings, English sparrows, cockroaches, and flies are common to this environment, as well as some grasses, algae, and other plants able to gain a foothold in cracks in concrete and vacant lots. Peregrine falcons have substituted the artificial canyons created by tall buildings for natural cliffs. After their disappearance from eastern North America because of pesticide poisoning (see Chapter 12), the peregrine falcon has been successfully reintroduced to both urban and wilderness habitats.

Fumes from factories, coke ovens, and smelters destroy the vegetation of surrounding areas. Even after the cause has been eliminated, many years are required before the vegetation begins to return. Pollution of streams by sewage, industrial wastes, and siltation eliminates oxygen-demanding fish like trout; they are replaced by carp and bullheads, able to adapt to polluted conditions. Dam construction for power drowns terrestrial communities and converts part of the river community to a deep lake. Migrant fish, particularly the salmon, may be blocked from reaching their spawning grounds in the headwater streams.

Human settlement of an area from past to present has undergone a sort of succession. The first to live in or penetrate a region, the pioneers, are hunters and trappers, who, aside from harvesting animals, leave little mark on the land. They are followed by a subsistence farming or grazing culture, which can completely change a natural community. Some plants and animals may be destroyed, succession set back to an earlier and more economically productive stage, and new animals introduced. Land too poor or too abused to support human society economically may be abandoned and revert to natural vegetation. Traces of old settlements and abandoned land can be found throughout the country. Ghost towns, old stone and rail fences, house and barn foundations, lilac bushes, wells and springs hidden back in the woods, all attest to former human occupancy of the land.

Industrial and urban settlements might be considered climax stages of human succession. The tremendous growth of suburban settlements onto fertile farmland marks well this type of succession. Just as there are successional trends in the various later stages of forest development, so there are successional trends in the stages of urban development. The urban community begins as a small central core and grows outward into the surrounding countryside. Just as the forest invades an old field, so this outward expansion or invasion takes over the surrounding country.

Initially the center of the city is the most desirable place to locate. Because all establishments cannot locate there, a sorting process takes place, resulting in the segregation of both functional units and social units based on socioeconomic status, culture, and race (Mayer 1969). As the city grows, the pressure for other central locations forces outward expansion. This expansion or invasion does not take place at equal rates because of resistance (competition) from land-use functions, people, and transportation facilities. As one zone exerts pressure on the adjacent outer zone, it eventually replaces the outer zone. At the same time the outer zone tends to invade the next or colonize a new area. Thus one successional stage replaces another (Figure 25.12).

With the passage of time, the mature community deteriorates, and the zones about the central core become less desirable. Buildings deteriorate, industrial and commercial complexes become less efficient and far removed from the source of productivity—the young vigorous communities on the outer zones in earlier stages of succession. There new central core areas develop, determined by transportation links, shopping centers, and industrial parks. As these new core areas become firmly established, they in turn initiate their own successional forces and patterns. In time they will be absorbed by the expanding urban area.

Meanwhile, in the original central core, homes once occupied by higher socioeconomic classes have been abandoned to new immigrants, the original residents having moved to outer zones. As the new inhabitants of the core better their socioeconomic status, they too move out to the adjacent zones, while their occupants move still further out. Eventually the structure of the inner city becomes unattractive, the density of the population declines, and the core becomes a blighted area unable to pay any sort of economic

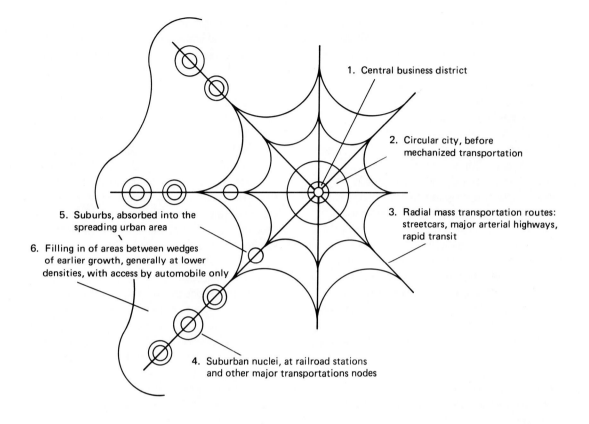

1. Central business district

2. Circular city, before mechanized transportation

3. Radial mass transportation routes: streetcars, major arterial highways, rapid transit

5. Suburbs, absorbed into the spreading urban area

6. Filling in of areas between wedges of earlier growth, generally at lower densities, with access by automobile only

4. Suburban nuclei, at railroad stations and other major transportations nodes

Figure 25.12 Expansion of urban areas suggests succession in old fields. Just as forests move out into old fields (see Figure 25.1), so urban and suburban areas move into the countryside. Their spread follows dispersal routes created by avenues of transportation, as suggested by this model. (From Mayer 1969.)

rent. Subsidized public intervention becomes necessary. Just as fire or some other natural event tends to renew the aged, declining natural community, so urban renewal represents an attempt to put the core city back to an earlier, more productive stage of succession.

Unlike other organisms, humans have the power to direct the pattern of succession through zoning, city planning, and other practices aimed at protecting the integrity of the environment. Such activities require a strong political will and motivation to do so.

■ Summary

With the passing of time natural communities change. Old fields of today return to forests tomorrow; weedy fields in prairie country revert to grasslands. This gradual sequential change in the relative abundances of dominant species in a community is succession. It is characterized by the replacement of opportunistic, early successional stage species by late stage species, a progressive change in community structure, an increase in biomass and organic matter accumulation, and an approach toward a balance between community production and community respiration. Succession that begins on sites devoid of or unchanged by organisms is termed primary; and succession that proceeds from a state in which other organisms were already present is called secondary.

The process and mechanisms involved in succession have been the subject of considerable study, controversy, and review, with succession being relegated either to the community or species level. Succession may begin with one organism or a group of organisms capable of growing on an open site successfully and arriving there early either as seeds, spores, or residual propagules. They preempt space and continue to exclude or inhibit the growth of others until they die or are damaged, releasing resources and allowing new, longer-lived species to enter. In another view, succession comes about because of changes induced by the organisms themselves. As they exploit the environment, their life activities make the habitat unfavorable for their own continued survival and create a favorable environment for different species.

Although a conflict exists between these two views of succession, both processes are involved to a greater or lesser degree. Although any point on the time gradient may give the impression that one discrete community replaces another, seral stages actually represent a pattern of species replacement driven by chance, differential longevity, and competition. Species, not discrete communities, are replacing one another, but the end result is seral or community change.

Competition and plant replacement out in the field take place not between species but between individuals. Another approach to succession makes competition among individuals the mechanism behind species replacement during plant succession.

Various combinations of life history and physiological traits, characteristics of species but expressed by individuals, interact with environmental conditions to produce the variety of successional patterns at both primary and secondary levels.

Eventually communities arrive at some form of steady state with the environment. This stage, usually called the climax, is more or less self-sustaining, largely through small-scale disturbances as individuals die and are replaced. The degree of change in a climax depends upon whether the replacements are of the same or different species. If replacements are the same species, stability is assured; but because of different microenvironments, progeny of the same species are usually less favored than those of other species. As a result the climax is usually a mosaic of regenerating patches in which new growth may not be the same species as the individuals being replaced.

Some changes are not truly successional. They include cyclic replacement and fluctuations. Cyclic replacement results from the destruction of vegetation induced by some periodic biotic or environmental disturbance that starts regeneration again at some particular stage. Each phase is related to the one before it in the rise and decline of successive communities. Such cyclic succession aids in community persistence. Nonsuccessional, reversible changes or fluctuations in communities result from environmental stresses such as changes in soil moisture, wind, and grazing.

Perhaps the most outstanding characteristic of natural communities is their dynamic nature. They are constantly changing through time—rapidly in early stages of development, more slowly in later stages. Even those communities that are seemingly the most stable slowly change through time.

Disturbance

- **Agents of Disturbance**

  Natural Disturbance
  Human Disturbance

- **Scale of Disturbance**

  Small-Scale Disturbances
  Large-Scale Disturbances

- **Frequency of Disturbance**

- **Disturbance and Nutrient Cycling**

- **Animal Response**

- **Ecosystem Stability**

- **Summary**

We tend to think of familiar natural landscapes as frozen in time, like the image in a photograph, to remain unchanged. To preserve the area as we see it, we protect it against fire, insect attack, and other events we consider harmful. In doing so we change it, because we fail to comprehend that nature is not constant, that disturbance is the means by which landscape diversity is maintained.

Consider a photograph of western U.S. land in the late 1800s. This scene is a heterogeneous landscape with a diversity of patches: grassland, aspen, Douglas-fir, pine (Figure 26.1). The patchy landscape resulted from periodic disturbances by fires. Protected from fire for 90 years, the vista has changed. The open country is grown to Douglas-fir and the diverse patches of vegetation are gone. A patchy landscape protected from disturbance has changed to a homogeneous one.

In Switzerland the scenic Alpine landscape associated with Heidi changed when cattle and sheep no longer grazed the high mountain meadows. Spruce forests began to claim the meadows, changing the landscape familiar to tourists. Only by reintroducing grazing, even on a subsidy basis, could the traditional Alpine landscape be maintained (Wiegandt 1976). The point of both examples is that diver-

(a)

(b)

Figure 26.1 These two photos of the same area show the effects of fire suppression and grazing disturbance on landscape diversity. (a) A photo taken in 1888 shows a north-northwest view in the vicinity of Fort Maginnis (U. S. Army troops in the foreground), located in the Judith Mountains about 20 miles northeast of Lewistown, Montana. Ground cover in the ridge and lower slopes in the distance apparently are dominated by perennial grasses. Conifers on the far slopes are confined largely to localized areas, suggesting that wildfires burned the slopes several decades earlier. (b) The same site on July 18, 1980, 92 years later. The foreground shows the effects of current livestock grazing. Shrubs, including currant, chokeberry, rose, shrubby cinquefoil, and common juniper are more conspicuous in the rocks, as are shrubs and conifers in the left and central midground. The stream course in the distance at center left of photo, which was formerly treeless, now supports large cottonwoods, Douglas-fir, and ponderosa pine. Absence of fire has allowed a profuse growth of Douglas-fir and ponderosa on the far slopes. (Photo (a) by W. H. Culver and photo (b) by W. J. Reich, courtesy of Montana Historical Society. Interpretative text based on description written by George E. Gruell in Fire and vegetative trends in the northern Rockies: Interpretations from 1871-1982. U.S.D.A. Forest Service General Technical Report INT-158, page 54.)

sity in the landscape comes about only through disturbance, which is a normal part of ecosystem functioning.

Disturbance is a relatively discrete event in time coming from the outside that disrupts ecosystems, communities, or populations, changes substrates and resource availability, and creates opportunities for new individuals or colonies to become established. Disturbances have both spatial and temporal characteristics: size of the area disturbed; frequency, the mean number of events per unit time; turnover, the mean time between disturbances; intensity, the physical force of the event per area per time; and severity, the impact on the organism or the community.

Disturbances are not specific to any one level of organization; and a disturbance at one level is not necessarily a disturbance at another (Allen and Hoekstra 1984). A disturbance to a herbaceous layer on the forest floor may have little effect on the whole system, although it may be a major disturbance to the populations of woodland herbs. Conversely, a fire in a forest may have a major effect on the whole system, but act only as a temporary stress on individual animals that escape the fire or on herbaceous plants that are able to respond by sending out new growth quickly.

■ Agents of Disturbance

Natural Disturbance

Wind. Wind is a subtle agent of disturbance to vegetation. It shapes the canopies of trees exposed to prevailing winds, affects their growth of wood, and uproots them from the ground. Especially vulnerable to windthrow are mature trees, whose trunks lack the suppleness of youth, and trees growing on shallow and poorly drained soils in which the roots, spreading along the ground, cannot get a firm grip in the soil. Trees weakened by fungal disease, insect damage, and lightning strikes and tropical forest trees carrying a heavy load of epiphytes in their crowns are also candidates for windthrow. The impact of wind is accentuated when strong winds accompany heavy snowfall that weighs down trees or heavy rains that soften the soil about the roots. Both hurricanes and tornados can cause massive blowdowns.

The position trees occupy in the forest also affects their vulnerability to wind damage. Trees bordering ragged forest gaps and trees growing along forest edges, roads, and power lines are more likely to blow down than trees in the forest interior.

Moving Water. Moving water is a powerful agent of disturbance. Storm floods scour stream bottoms, cut away banks, change the courses of streams and rivers, move and deposit sediments, and bury or carry away aquatic organisms. Strong waves on rocky intertidal and subtidal shores overturn boulders and dislodge sessile organisms. This action clears patches of hard substrate available for recolonization and maintains local diversity (see Chapter 24) (Sousa 1979, 1985; Connell and Keough 1985). High storm tides break down barrier dunes, allowing seawater to invade behind the dunes, and changing the geomorphology of barrier islands.

Drought. Prolonged drought can have a pronounced effect on vegetation composition and structure. On the grasslands of western Kansas during the drought years of the 1930s, blue grama with its physiological ability to resist dry conditions became twice as dense as the drought-sensitive buffalo grass. When the rains came, buffalo grass quickly responded. In two years it became five times as dense as blue grama. After ten years of favorable conditions the two species shared codominance (Coupland 1958). In temperate forests prolonged drought can result in heavy mortality of shallow-rooted tree species such as hemlock and yellow birch and understory trees and shrubs (Hough and Forbes 1943; Bjorkbom and Larson 1977). Drought dries up wetlands, causing crisis conditions among waterfowl and other wetland birds and muskrats (Errington 1963), seriously reducing their populations.

Fire. Fire is a major natural disturbance, influencing species composition and shaping the character of a community. It has long played an important role in vegetational development worldwide. Globally large regions are characterized by vegetation that evolved under fire: the grasslands of North America, the chaparral of the southwestern United States, the maquis of the Mediterranean, the South African fynbos (see Chapter 30), African grasslands and savannas (Batchelder 1965), the southern pinelands of the United States, and even-aged stands of coniferous forests of western North America. Up to 95 percent of the virgin forests of Wisconsin were burned during the five centuries before the land was settled by white Europeans (Curtis 1959). These fires not only enabled such species as yellow birch, hemlock, pines, and oaks to persist, but also were normal and necessary to perpetuate those forests (Maisurow 1941; Curtis 1959). In Alaska fires have converted white spruce stands into treeless herbaceous and shrub communities of fireweed and grass or dwarf birch and willow (Lutz 1956), whose growth is so thick that forest trees cannot become established. Thus fire is a powerful selective and regulatory force on the evolution and maintenance of many types of ecosystems.

Four conditions are necessary for fire to assume ecological importance: (1) an accumulation of organic matter sufficient to burn; (2) dry weather conditions to render the material combustible; (3) a landscape conducive to spread of fire; and (4) a source of ignition. The only two important sources of ignition are lightning and humans.

Globally certain regions possess conditions conducive to burning and the spread of fires. One condition is a fire climate: extended dry periods during which fuel accumulated during wetter periods can burn and the prevalence of lighting storms not universally accompanied by precipitation. Such conditions prevail in certain parts of North America, Africa, the Mediterranean, and Australia. In the western United States 70 percent of forest fires are caused by dry lightning during the summer. Because of the seasonal nature of lightning, fires so caused are more numerous during the growing season, from April through August. At that time fires are the least severe but have the greatest impact as a selective force.

Fires may be surface, ground, or crown. Their type and behavior depend upon the kind and amount of fuel, moisture, wind, other meteorological conditions, season, and the nature of the vegetation. *Surface fire*, the most common type, feeds on the litter layer. In grassland it consumes dead grass and mulch, converting organic matter to ash. Usually fire does not harm the basal portions of root stalks, tubers, and underground buds, but it does kill most of the invading woody vegetation. In the forest surface fires consume leaves, needles, woody debris, and humus. They kill herbaceous plants and seedlings and scorch the bases and occasionally the crowns of trees. Damage to trees depends upon the intensity of the fire and the susceptibility of trees to heat. Surface fires may kill thin-barked trees by scorching the cambium layer. Thick-barked trees are better protected, but they can be scarred, allowing fungal infection (Figure 26.2). Shallow-rooted trees are more vulnerable to surface fires than deep-rooted ones such as oaks and hickories.

If the fuel load is high and the wind strong, surface fires may leap into the forest canopy to cause a *crown fire*. Crown fires are most prevalent in coniferous forests because of the flammability of the fo-

Figure 26.2 This old yellow-poplar has experienced several surface fires in its lifetime, as evidenced by the fire scar at its base, mostly healed over, and by the blackened ridges of bark. The tree was relatively fire-resistant, but the scars, although walled off by subsequent growth, did allow fungal infection to gain entrance into the heartwood. (Photo by R. L. Smith.)

liage. If the canopy is unbroken, the fire may sweep across it and tops and branches fall to the ground to further feed the fire. Crown fires kill most aboveground vegetation; yet certain forest types, such as jack pine and lodgepole pine, require all-consuming fires to regenerate the stand.

Ground fire that consumes organic matter down to the mineral substrate or bare rock is the most destructive (Figure 26.3). It is most prevalent in areas of

Figure 26.3 Fires of great intensity can have a profound influence on the ecosystem. After the spruce forest was cut, intense ground fires, fed by piles of logging debris, burned over the Allegheny Plateau area in West Virginia known as Dolly Sods. Fire consumed the peatlike ground layer to bedrock and mineral soil. The forest never recovered from the Civil War years' burn. The plateau is now a boulder-strewn landscape with intermittent patches of blueberry, dwarfed birches, mountain ash, and bracken fern. (Photo by R. L. Smith.)

deep, dried-out peat and of extremely dry, light organic matter. Such a fire is flameless, keeps extremely hot, and persists until all available fuel is consumed. In spruce and pine forests, with their heavy accumulation of fine litter, a ground fire can burn down to expose rocks and mineral soil, eliminating any opportunity for that vegetation type to return.

Animals. A walk into a forest inhabited by a high population of deer or across an overgrazed grassland provides visual evidence of the impact that herbivorous animals can have on communities. Overgrazing native rangelands of southwestern United States, for example, has reduced the organic mat and thus the incidence of fire. Because of infrequent fires, reduced competition from grasses, and dispersal of

seed through cattle droppings, mesquite and other shrubs have invaded the grasslands (Phillips 1963; Box et al. 1967). In many parts of eastern North America, large populations of white-tailed deer have eliminated certain trees such as white cedar and American yew from the forest, destroyed forest reproduction, and developed a browse line—the upper limits on a tree at which deer can reach foliage. In cutover areas of hardwood forests on the Allegheny Plateau in Pennsylvania, deer greatly reduced pin cherry and blackberry, selectively reduced sugar maple, and favored the expansion of ferns and grass (Marquis 1974, 1981).

The African elephant influences the nature of the savanna ecosystem. When their numbers are in balance with the vegetation and their movements are not restricted, elephants have an important role in creating and maintaining the forest. When elephants exceed the capacity of their habitat to support them, their feeding habits combined with fires devastate flora, fauna, and soils. Elephant depredation on trees (Figure 26.4) acts as a catalyst to fires, which are the primary cause of converting forest to grassland.

Beaver create major disturbances in many forested regions of North America and Europe. By damming streams they alter the structure and dynamics of flowing water ecosystems (see Chapter 32) (Naiman et al. 1986). Pools behind dams become catchments for sediments and sites for organic decomposition. By flooding lowland areas, beaver convert forested stands into wetlands. By feeding on aspen, willow, and birch, beaver maintain stands of these trees, which otherwise would be replaced by later successional species.

Outbreaks of insects such as introduced gypsy moth and balsam aphid and spruce budworm defoliate large areas of

Figure 26.4 By consuming great quantities of woody vegetation and uprooting trees, elephants of the African savanna have an important influence on ecosystem succession and stability. The life cycles of certain trees and the maintenance of the savanna ecosystem depend in part upon disturbance by elephants. Too many elephants result in destruction of the ecosystem; too few elephants results in bush encroachment. (Photo by T. M. Smith.)

forest, which results in the death or reduced growth of affected trees. The degree of mortality may range from 10 to 30 percent in hardwood forests infested by gypsy moth to 100 percent in spruce and fir stands. Outbreaks of bark beetles have much the same effect in pine forests. The impact of spruce budworm, bark beetles, and other major forest insects is most intense in large expanses of homogeneous, forested landscape where natural fires have been suppressed for long periods of time, allowing late-stage stagnated stands, highly susceptible to spruce budworm outbreaks, to develop (Wulf and Cates 1987).

In their own way these insects act to regenerate senescent or stagnated forests.

Human Disturbance

Humans have a more profound impact on ecosystems than natural causes, because we have the ability to change the natural environment radically. We have removed forests, overgrazed and plowed grasslands, converted natural communities to cultivated cropland, peeled away layers of Earth to reach mineral deposits, dammed, channeled, and changed the courses of rivers, filled marshes, paved enormous acreages of land under concrete and asphalt, through land mismanagment hastened the encroachment of deserts, and polluted aquatic ecosystems (see Chapter 12).

Cultivation. Cultivated plant communities are simple, highly artificial, and consist mainly of introduced species well adapted to grow on disturbed sites. Because of the very simple and homogeneous ecosystem involved, tillage brings with it new pests destructive to both cultivated and natural vegetation. Tillage disturbs the structure of the soil and exposes it to water and wind erosion. In tropical regions cultivation associated with fire can permanently change vegetation upon land abandonment.

Ionizing Radiation. Intense radiation such as produced by unshielded nuclear reactors, by atomic blasts, and by fallout can have a pronounced effect on plant communities (Woodwell 1963). Early successional stages, dominated by grasses, hemicryptophytes, and cryptophytes, are the least sensitive to ionizing radiation. Radiosensitive species such as pines are destroyed by chronic irradiation of 20 to 60 roentgens, and hardwood species by over 360 roentgens. If a complex ecosystem is subject to low exposure, the growth of sensitive species is inhibited temporarily. The effect is comparable to that of wind or frost damage. Recovery would be rapid, and the direction of succession would not be changed. At higher levels of dosage, radiation would cause selective mortality of sensitive populations. Its impact on succession is similar to that of logging and grazing. Under intense radiation the basic structure of the ecosystem and its capacity for future support might be reduced.

Surface Mining. Surface mining accounts for a high percentage of coal production and for most other extracted mineral such as iron and gold ores. The impact and magnitude of damage vary with the region and the degree and success of reclamation efforts. The impact is most pronounced in mountainous regions, as in the Appalachians, where surface mining is on the contour or involves mountaintop leveling and valley fill. Whatever the method, surface mining does violence to the land. Deep, unweathered rock strata are broken and brought to the surface, where the material is subject to rapid weathering, releasing toxic nutrients required in small amounts. Carried away in high concentration by water coming off mined sites, these elements reduce water quality downstream for both aquatic life and humans. Unless expensive precautions are taken, sediment deposition in streams can be extremely high.

Surface mining alters the ground-water regime. Water tables, once deep in the underlying rock strata, are exposed and flow freely to the newly created surface. Large quantities of water that would have been taken up by trees and lost to the atmosphere by trees are added to the amount of runoff, which during heavy storms intensifies the height and damage of flash floods.

Pollution. The settlement of a region by humans subjects aquatic ecosystems to increased inputs of natural and unnatural substances. Agriculture and construction of roads and buildings have clogged rivers and streams with incalculable tons of silt, filled lakes, dams, and estuaries. Humans have added a heavy load of nutrients, especially nitrogen, phosphorus, and organic matter from sewage and industrial effluents (see Chapter 12). We have poured in thousands of different chemicals and wastes, including pesticides, which natural ecosystems are ill-adapted to handle. The outcome of this has been excessive nutrient enrichment of aquatic ecosystems, producing significant chemical, biological, and ecological changes in streams, lakes, and estuaries (see Chapter 32).

Timber Harvesting. Disturbance by logging depends upon the methods employed: selection cutting or some form of even-aged management or clear-cutting. In selection cutting single trees or groups of trees are removed, based on their position in the stand and possibilities of future growth. Selection cutting produces only gaps in the forest canopy and favors reproduction of tolerant over intolerant trees. Forest composition essentially remains unchanged (Trimble 1973; Johnson 1984; Lorimer 1989).

Even-aged management involves removal of the forest and reversion to an earlier stage of succession (Figure 26.5a, b). Unless followed by fire or badly disturbed by erosion and logging activities (Figure 26.5c), the cutover area fills in rapidly with herbs, shrubs, sprout growth, and seedlings of trees present as advanced regeneration in the understory. The area passes quickly through the shrub stage to an even-aged pole forest. Because many of the most valuable timber trees are intolerant to mid-tolerant species, they can be regenerated only by removal of mature trees, exposing the ground to sunlight.

There are three approaches to even-aged management (D. Smith 1986). One is clear-cutting 11 to 44 ha blocks of timber within large forest tracts. A second method is strip-cutting, the removal of all merchantable timber and remaining trees in strips 15 to 30 m wide. Every third strip is removed, followed by the remaining strips in two cuttings two to four years apart. A third method is shelter wood cutting, which leaves 10 to 70 percent of the stand remaining after cutting. When new growth is well under way, the remaining trees are removed. The first two methods of even-aged management favor regeneration of intolerant tree species. Shelter wood cutting retains some of the characteristics of the original forest yet permits intolerant species to regenerate.

Foresters often modify the regenerating forest to meet their requirements, introducing a form of disturbance during the development of the stand. Early in the life of a new forest, foresters may remove tree species not desired for timber or individuals of poor form. This alteration improves, by economic but not necessarily ecological standards, the composition of the stand and the quality of the trees. The maximum growth of crop trees can be encouraged by thinning. Increased space between the trees stimulates crown expansion and increases growth.

■ Scale of Disturbance

Disturbance involves a matter of scale in size of area, intensity, severity, and frequency. It ranges from very small, frequent disturbances such as the death of a single tree in a forest to large-scale, rare disturbances which embrace extensive areas swept by fire, buried under volcanic ash, torn by landslides, or denuded by human land-clearing schemes.

Small-Scale Disturbances

What constitutes a small-scale disturbance depends upon the scale of the landscape. The loss of a group of trees in a small woodland would have more impact than the same loss in a large forest. Small-scale disturbances are caused by the deaths of individuals or groups, which open the canopy or the substrate. Abrasive action of waves tears away mussels and algae from hard rocky tidal substrates. In grasslands digging by badgers and groundhogs exposes patches of mineral soil for colonization by herbaceous plants. A tree dies or falls in a forest, opening up the canopy. In these instances the effect of the disturbance is to create a *gap*, a term originally

(a)

(b)

(c)

Figure 26.5 (a) A new clear-cut in the dormant season. The large and small materials have been completely removed and 1- to 5-inch stems have been felled. Such a cutting represents a major disturbance to a forest ecosystem. (b) Response to clear-cutting sometimes is rapid, as this seven-year-old stand of hardwood reproduction illustrates. (Photos courtesy of U.S. Forest Service.) (c) Unless carefully managed, clearcutting can result in severe disturbance to a forest ecosystem. (Photo courtesy of Soil Conservation Service.)

Figure 26.6 Treefall gap in the tropical rain forest in Brazil. Treefalls, which peak during periods of rain and gusty wind, are essential for the regeneration of many tropical forest plants, both primary and pioneer species, and provide a diversity of habitats for wildlife within the forest. (Photo by Alexine Kergoulian.)

applied by Watt (1947) to a patch in a forest created by the death of a canopy tree. The meaning has been extended to other localized sites of regeneration and growth caused by death or loss of original occupants.

Consider a forest. A single tree struck by lightning or killed by a fungal infection that remains standing as a snag creates a minimal gap. A canopy tree uprooted by a windstorm creates a larger, more complex gap (Figure 26.6). As it falls it opens a space in the canopy it occupied. Its upturned roots tear up the soil and form a pit and a mound of exposed mineral soil open for colonization. Its falling trunk slices through its neighbors, snapping the limbs of nearby canopy and understory trees. The fallen crown crushes the understory beneath it. Once created, some gaps in the forest continue to grow as peripheral trees succumb to winds and insect damage.

Gaps, large or small, in the forest are sites of increased light, soil temperature, and nutrients, and decreased soil moisture and relative humidity. How woody plants respond to gap formation is suggested by studies on the effects of different timber harvesting practices on forest regeneration. Natural gap formation can be simulated by single tree and group selection cuttings. If the gap is small, such as one produced by the death or removal of an

individual tree or a small group of trees, the response is typically vegetational reorganization. Trees about the edges of the gap expand their crowns to fill in the opening. Their closure of the canopy inhibits any initial response understory plants may have made (Trimble and Tyron 1966; Lorimer 1989). Larger gaps provide the environmental conditions and opportunities for tolerant species in temperate hardwood forests (Trimble et al. 1973) and primary species in tropical forests (Brokaw 1984), so replacement vegetation is likely to be similar to that of the canopy. Few trees in small gaps in the forest reach the canopy without the aid of another disturbance (Runkle and Yetter 1987; Tryon and Trimble 1969). These disturbances need not occur within the gap but only among neighboring trees. Incoming trees need at least two disturbances to reach a position in the canopy. In such gaps tolerant understory trees such as flowering dogwood may respond more quickly and successfully than replacement canopy saplings (Erenfeld 1980). They fill in the lower canopy and for a time, at least, inhibit the growth of young canopy-species trees.

Response to small gaps in northern hardwoods forests appear to involve an alternation of species (Fox 1977; Forcier 1975; Runkle 1984; Hibbs 1982). Forcier (1977) reports that in northern hardwood forests of New Hampshire beech seedlings and saplings are positively associated with yellow birch and sugar maple and negatively with their own canopy. When a gap develops, the first to fill the space is usually the widespread, opportunistic yellow birch. It grows rapidly, shading an understory of sugar maple seedlings and saplings. When the short-lived yellow birch dies, its place is taken by sugar maple, beneath which beech seedlings now grow. When sugar maple dies, young beech fill the gap. In forests with lower diversity sugar maple tends to replace beech and beech replaces sugar maple (Fox 1977). In this manner the northern hardwoods forest is able to maintain a quasi-equilibrium after small-scale disturbances.

Much larger gaps encourage the growth of intolerant species. Invasion or response by intolerant species takes place only in gaps with a minimum size of 0.1 ha or twice the height of the surrounding trees. Even then the most responsive growth is in the center of the gap (Tyron and Trimble 1969). Intolerant species

such as yellow-poplar and black cherry respond to larger gaps. Their ability to do so accounts for their conspicuous and continued presence in mature forests dominated by tolerant species. Formation of larger gaps up to 0.4 ha provides the greatest diversity in the forest. Overall, gap-phase replacement results in patches of different stages of successional or compositional maturity. Chronic small-scale disturbances are important in the maintenance of species richness and structural diversity within a mature forest ecosystem (see Shugart 1984; Platt and Strong eds. 1989).

Patches of disturbed soil created by fallen trees provide a variety of microreliefs, ranging from pits to mounds, about upturned roots (Figure 26.7). Beatty (1984) studied the relationship between such microreliefs and herbaceous vegetation pattern in a hemlock-hardwood forest. She found that eight herbaceous species, including asters *(Aster spp.)* and tall white lettuce *(Prenanthes altissima)*, were exclusive colonizers of mounds of disturbed soils. Six species, among them ramps *(Allium tricoccum)*, toothwort *(Dentaria diphylla)*, and jack-in-the-pulpit *(Arisaema atrorubens)*, were associated exclusively with pits, and five species grew only on undisturbed portions of the forest floor. Thompson (1980) noted that in Illinois woodlands upturned roots were colonized mostly by species growing within 1 meter of the disturbed area of woodland soil and spread there largely by vegetative means. Among the species growing some distance from the microsite, only those with an effective means of dispersal, such as violet and aniseroot *(Osmorhiza longistylis)*, colonized the disturbed areas.

Herbaceous plants responding to gap formation can influence the success or failure of woody seedlings. Maguire and Forman (1983) found that the density of tree seedlings was much greater in places

Figure 26.7 Trees uprooted by ice and wind storms create new microreliefs on the forest floor to be colonized by herbaceous woodland plants. (Photo by R. L. Smith.)

on the forest floor with little herbaceous cover than in patches of herbaceous growth. The inhibitory effect of dense patches of herbs and ferns on forest tree regeneration is well known among foresters (Horsley 1977; Ferguson and Boyd 1988).

Large-Scale Disturbances

Large-scale disturbances induced by fire, logging, land-clearing, and other such events results in responses that go beyond vegetational reorganization and involve colonization by opportunistic species. What species colonize the disturbed area is influenced by available seeds, seedlings, and saplings, soil conditions, amount of competition, and other ecological conditions.

Successful colonization of a disturbed site is influenced by the availability of seeds on the site as well as seeds from outside. The amount of seed already available on the site often is enormous. Wendel (1987) collected and moved to a greenhouse square-foot samples of the forest floor from four hardwood forest stands in West Virginia. Seeds that germinated and root stocks that sprouted gave rise to 44 different plant species, woody and herbaceous. The most abundant of the nine tree species was sweet birch (Betula lenta) and yellow-poplar was second. Blackberry (Rubus) was the most abundant shrub, followed by wild grape; and violets were the most abundant herbaceous species. Under ideal conditions some 800,000 stems per 0.40 ha could occupy the disturbed site the second year following clear-cutting.

Removal of Canopy. Response to the sudden removal of a forest canopy is often rapid, as seeds and seedlings of intolerant woody plants take advantage of changed environmental conditions (Bormann and Likens 1979). One example is pin cherry *(Prunus pensylvanica)*, whose seeds are carried to a forest by birds and small mammals or are deposited on the forest floor by an earlier stand of cherry. Pin cherry seeds can remain dormant for up to 50 years. When the forest canopy is removed and moisture, temperature, and light conditions become favorable, pin cherry seeds germinate and young trees quickly dominate the site, crowding out the associated blackberry *(Rubus spp.)* that also colonizes the area (Marks 1974). If the seedling growth is dense, a pin cherry canopy can close in four years, eliminating other species except highly shade-tolerant seedlings of sugar maple or beech (Figure 26.8). If seedling growth is moderately dense, species with wind-disseminated seeds, such as yellow birch and paper birch, will also occupy the site. Within 30 to 40 years pin

cherry dies out, allowing birch, sugar maple, and beech to dominate the gap; but during its tenure pin cherry contributed numerous seeds to the forest floor, ready to reclaim the site when another disturbance provides the opportunity.

Sites may be colonized by species seeding in from outside the area. One such species is yellow-poplar. The nature and success of its colonization depends

Figure 26.8 The importance of different species along a gradient of time following disturbance of a typical northern hardwoods forest. Immediately after disturbance blackberries (a) dominate the site, but they quickly give way to yellow birch (b), quaking aspen (c), and pin cherry (d). Intolerant pin cherry assumes dominance early but within 30 years fades from the forest. Yellow birch, an intermediate species, assumes early dominance, which it retains into mature or climax stand. Trembling aspen, an intolerant species, begins to drop out after 50 years. Meanwhile, sugar maple (e) and beech (f), highly tolerant species, slowly gain dominance through time. In about 100 years the mature forest is dominated by beech, sugar maple, and birch. (After Marks 1974: 75.)

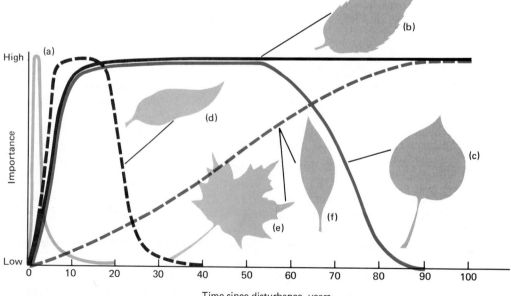

Time since disturbance, years

upon a number of conditions, including distance from the site, size of the seed crop, timing of seed arrival, a sufficient number of seeds to ensure germination and seedling survival, and exposed mineral soil as a seedbed.

Although the response of woody vegetation to large-scale removal of timber is rather well understood, little attention has been given to the response of the shade-tolerant understory herbaceous plants. Ash and Barkham (1974) studied the response of the herbaceous understory layer of an English coppice forest after cutting. (A coppice forest is one in which the stump sprouts or root suckers are maintained as a main source of regeneration, with cutting rotations between 20 and 40 years.) Cutting coppice involves complete canopy removal, resulting in increased surface temperature on the forest floor and full exposure to light. Typically, a number of open habitat or opportunistic species germinate and become established (Figure 26.9) but are soon excluded by the developing canopy cover. In spite of the disturbance, characteristic woodland species persist throughout the cycle. Adapted to a high light regime in spring before the leaves are out, these plants have the ability to tolerate high light intensity. At the same time they are able to coexist with annuals and open-habitat perennials, which cast a shade on the ground like tree cover. As opportunistic species disappear, woodland herbs again assume dominance, often developing into monospecific stands. Probably much the same response occurs in North American forests subject to similar large-scale disturbance.

Fire. Fire, a semirandom, recurring event, is the major natural large-scale disturbance to ecosystems. It induces certain synergistic effects not common to other types of large-scale disturbances. It raises the temperature of the soil, but the degree depends upon soil moisture present. Soil temperature does not rise above 100° C until all the moisture is evaporated, and even under hot fires in semiarid country temperatures rarely exceed 200° C at depths of 2.5 cm. However, the heat consumes organic matter and breaks down soil aggregates, increasing bulk density of soil and decreasing its permeability. This change reduces infiltration of water into the soil, increases surface runoff, and pro-

Figure 26.9 Response of understory herbaceous vegetation to disturbance in a coppice stand in England. The graphs show changes in the percentage of cover produced by growth of the canopy and the number of herb and shrub species in the field or ground layer at different times after coppicing. Not all species were present during the 30-year period of growth. Total numbers for each type are indicated. Note the rapid decline of annuals, biennials, and open site perennials as the canopy closes. Open-site perennials dominate the field layer shortly after coppicing. Perennials of shaded sites show a sigmoidal growth response as the canopy closes. (From Ash and Barkham 1976: 706.)

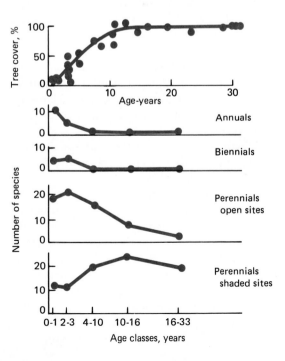

motes erosion and soil slippage on steep slopes.

As an agent of disturbance fire sets the process of stand regeneration into motion by stimulating sprouting from roots and germination of seeds. Fire prepares the seedbed for some species of trees by exposing mineral soil, eliminating competition from fire-sensitive and shade-tolerant species for soil moisture and nutrients. Periodic surface fires thin some coniferous stands, such as ponderosa and longleaf pine. Importantly, fire acts as a sanitizer, terminating outbreaks of insects and such parasites as mistletoe by destroying senescent stands and deadwood and providing conditions for regeneration of vigorous young trees.

Some vegetation types require the rejuvenating effects of periodic fires. Such vegetation evolved in a fire environment, according to a hypothesis advanced by Mutch (1970), and acquired certain characteristics that enhance fire spread and increase flammability. These characteristics may be chemical, physical, or physiological (Philpot 1977). They include a high content of flammable resins, waxes, terpenes, and volatile products, finely branched needle-like leaves that carry fire from plant to plant, and accumulation of dead material as the plants age. Plants of these fire-prone ecosystems, such as chaparral, lodgepole and other pines, and eucalyptus, possess certain adaptive traits that permit them to respond to fire in one of three general ways.

Plants can survive as mature adults with little or no damage. Their mode of defense is bark thick enough to insulate the cambium from the heat of surface fires. Because such protection is not 100 percent effective and the heat of fire is not uniform about the tree, one side of the tree may burn, creating a fire scar on the annual ring, while the other side does not. A second defense is rapid growth, self-pruning the lower branches and raising the crown high enough above the ground to escape surface fire and to reduce the danger of the fire leaping into the crown. A third defense is to allow the accumulation of a mat of needles that supports frequent low-intensity surface fires. These frequent surface fires prevent the buildup of a heavy fuel load that once ignited could destroy the forest.

A second adaptation involves the death of mature plants as a means of regenerating and perpetuating the stand. Such destruction results in even-aged stands in which all the trees arise and become senescent simultaneously, promoting severe but infrequent fires. Such a response works only if frequency between fires is long enough to allow the plants to mature and produce seed. The seed may be stored in the soil or on the plant, awaiting the fire to release seeds or stimulate germination.

Jack pine and lodgepole pine are two coniferous species that retain unripened cones for many years on trees. Seeds remain viable within the cones until a crown fire destroys the stand. Then the heat opens the cones and releases the seeds *(serotiny)* to a newly prepared seedbed well fertilized with ash.

Other species rely on stand destruction and fire-stimulated germination of seeds stored in the soil. Involved is *hard-seededness.* The seed coat is impervious to water and other softening agents. The increased temperature of fire-heated soil cracks the hard seed coat or releases the seed from soil-stored chemical inhibitors imposed by living overhead vegetation. Some of the most abundant shrubs in the chaparral of California, ceanothus *(Ceanothus)* and manzanita *(Arctostaphylos)* are *obligate seeders,* plants that regenerate by seeds only rather than by root or stem sprouting. When mature and senescent these shrubs possess an abundance of dead stems that feed intense fires and few potential resprouters.

A third response to fire is resprouting. Although fire kills the tops and foliage, new growth appears as bud sprouts and root sprouts. Certain trees, particularly a number of *Eucalyptus* species in Australia, possess buds protected beneath the thick bark of larger branches. The buds survive crown fires and break out to develop new foliage (Figure 26.10). Other plants sprout from buds on roots, rhizomes, root collars, and specialized structures called *lignotubers*. These are basal burls from which latent axillary buds are released from inhibition when fire removes the region of growing cells at their tips (apical meristems). Ferns have subterranean buds on rhizomes that respond to loss of aboveground foliage. Shrubs such as blackberries and blueberries and trees such as aspen *(Populus)* sprout vigorously from roots. Trees such as oaks and hickories sprout from buds that develop at the root collar just below the ground. Certain Mediterranean-type shrubs, including the North American chamise *(Adenostoma fasciculatum)* and species in at least seven genera in Australia, have lignotubers. Such sprouting is an effective means of survival and increase.

Figure 26.10 Eucalyptus forests are subject to periodic fires. The trees are rarely burned, but the bark may be scorched and charred and the trees defoliated by radiant heat from the ground. The trees responded by resprouting not from roots but from trunk and branches, as in this recently burned area. Foliage regenerates from dormant shoots, either along the trunk, on main branches, only in the upper branches, or in some combination, depending upon the species. (Photo by R. L. Smith.)

■ Frequency of Disturbance

Disturbances on a given area occur as separate events in a given period of time. The mean number of events per period of time is the *frequency* of disturbance. The *return interval* or turnover time is the inverse of frequency or the mean time between disturbances on the same piece of ground.

In systems such as temperate and tropical forests, in which natural disturbances involve gap formation, the frequency of disturbance within the stand is high, but the rate of disturbance is low, from 0.5 to 2.0 percent per year. The return interval is between 50 and 200 years (Brokaw 1985; Runkle 1985). Runkle (1985) estimates that 4 to 14 percent of the total land area of the Great Smoky Mountains National Park is subject to disturbance over a decade. It is this slow rate of disturbance and replacement that maintains diversity in a mature forest over many years and allows the coexistence of species with different life history characteristics. This type of disturbance gives the time-locked observer the impression that the forest is unchanging.

In natural large-scale disturbances, the frequency is low but the magnitude is high. If the frequency of disturbance in such systems is related to the life span of the longest living species and the plants are adapted to disturbance, then the event probably is internally triggered and maintained (Allen and Hofstra 1984). Consider a boreal forest of black spruce (*Picea mariana*). Longevity of black spruce is about 200 years, but few stands ever survive beyond 70 years. In fact, it is difficult to find an old stand unburned (Yarie 1981). In the boreal forest black spruce, which grows on cool, wet ground, experiences fire frequencies of about 43 years. Dense stands of spruce are prone to lightning fires that burn off all vegetation, but revegetation is rapid. Moss and light-seeded herbs provide the seedbed and cover for black spruce seedlings, which grow above the shrubs and establish closed dense stands in 40 to 60 years. At about this age productivity of the stand declines and the microenvironment becomes cooler and wetter. Lacking a method of self-perpetuation without fire, the spruce forest changes into a treeless moss and lichen bog.

How frequently a fire burns over an area—its return rate—is influenced by the occurrence of droughts, accumulation and inflammability of the fuel, the resulting intensity of the burn, and the human interference. In the grasslands of presettlement North America fires occurred about every two to three years, the time needed for sufficient mulch, dead stems, and leaves to accumulate. In forest ecosystems the frequency of fires varies greatly, depending upon the type of forest (Heinselman 1981a). Frequent light surface fires may have a return interval of 1 to 25 years, whereas crown fires may have a return interval of 25, 100, or even 300 years. Usually light and severe fires occur in combination. A red pine forest may experience a light to moderate surface fire every 5 to 30 years and a crown fire every 100 to 300 years.

Various forest ecosystems appear to burn and develop under certain fire frequencies. Frequent low-intensity surface fires every 5 to 20 years were typical in presettlement forests of ponderosa pine (*Pinus ponderosa*) in western North America. Such fires reduced the needle layer periodically, prevented the buildup of a heavy fuel load, thinned the stand, eliminated incoming shade-tolerant conifers that could act as a fire ladder to the crown, and encouraged an open, grassy understory. Red pine and white pine forests in the Great Lakes region experienced infrequent surface fires of low intensity with a return interval of about 20 to 38 years that scarred individual trees but killed few. They were punctuated at longer intervals of 150 to 300 years by severe surface or crown fires that destroyed the stand (Heinselman 1981b).

Systems dependent on infrequent, large-scale disturbances for regeneration respond negatively to deviations from natural disturbance intervals. If disturbance comes at too frequent intervals, the system becomes degraded or converted into some other system. For example, Zedler and associates (1983) report on the effects of short intervals between fires on chaparral vegetation in California. In 1979 a fire burned over a large area of chaparral. Although such burned-over areas are quickly recolonized by seedlings and sprouts, they typically are reseeded artificially to ryegrass to control erosion. Fine-leaved ryegrass itself is fire-prone. In the following year a portion of the same area burned again, which resulted in a drastic change in vegetation. The two most abundant shrubs, *Ceanothus olignathus* and *Adenostoma fasciculatum,* were reduced in density by 97 percent. Because the seed reserves were depleted by response to the previous fire, the area was changed suddenly to a relatively permanent, degraded shrubland and annual grassland. The increased frequency of fires brought about by the planting of grass and by the prevalence of human-set fires is having a similar effect on large areas of California chaparral.

Suppression of disturbance can lead a disturbance-controlled system into a more fragile, less resilient one, susceptible to destruction. In a lodgepole pine forest or mixed coniferous forest of western North America, a natural fire regime involves recurring ground fires of intervals of 7 to 25 or more years. These fires have a significant role in maintaining conditions for tree regeneration and nutrient cycling and creating and maintaining openings in the forest canopy, in effect creating natural fire breaks. Fire suppression (fire management) has allowed the accumulation of fuel, crown closure, development of an understory that acts as a fire ladder to the crown, and senescence and death of individual trees, all adding to the fuel load and greatly lengthening the fire cycle. It makes for an intense fire that can cause extensive tree damage and mortality over a very large area (Holling 1980; Kilgore 1973). The long years of fire suppression in the Yellowstone National Park until 1975 (Romme and Knight 1982) accompanied by an unexpected period of drought set the stage for extensive fires in Yellowstone National Park in 1988. The long-term impact will be a rejuvenated ecosystem.

A similar effect occurs when natural insect predators of trees are suppressed to protect stands of susceptible timber. Spruce budworm infestations, like short-interval ground fires, create a patchy environment; but extensive protection of spruce and fir forest from short-term timber losses sets the stage for a widespread outbreak of the insects, covering hundreds of thousands of acres.

Kessell and Fischer (1981) developed a model that predicts the effects of high frequency and low frequency fire disturbance in Douglas-fir stands (Figure 26.11). High frequency fires reduce the system to one dominated by opportunistic species. Low frequency fires result in the establishment of climax Douglas-fir forest. Intermediate frequency fires, characterized by ground fires and low intensity surface

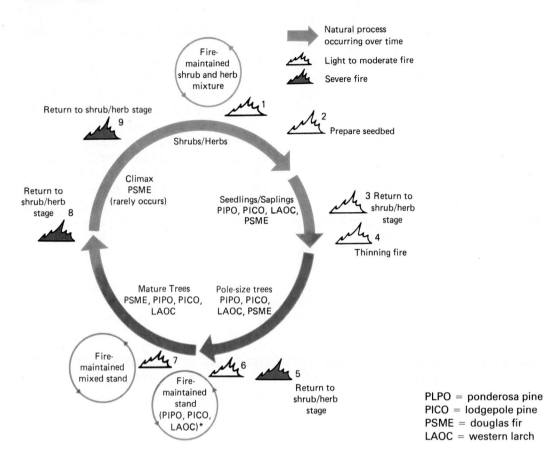

Figure 26.11 A model of forest succession in a moist Douglas-fir habitat influenced by fire. Following a fire disturbance, secondary succession begins with a mixture of shrubs and herbs (1). Fire creates a seed bed conducive to seedling germination. However, frequent fires will kill off conifer regeneration and maintain the shrub and herb community. If a fire should recur during the seedling-sapling stage the site again will return to the shrub-herb community (3). In somewhat older stands fire will eliminate accumulated fuel and thin the conifer stand (4). If a severe fire occurs in a pole-stage stand, the site will go back to the shrub-herb stage. However, surface fires of lower intensity maintain open seral stands, often dominated by ponderosa and lodgepole pines and larch (6). Where Douglas-fir is more aggressive, it will dominate the mixture of seral trees. Periodic light surface fires maintain this mixed stand (7). Long periods of fire suppression permit the establishment of a dense understory of Douglas-fir, which acts as a fuel ladder to the overstory. A severe stand-destroying crown fire in a closed mature stand will recycle the site to a shrub-herb stage (8). Where ponderosa pine, lodgepole pine, and larch are prominent components, the interval between fires is shorter than the life span of seral trees, and Douglas-fir cannot develop. On some sites Douglas-fir may achieve near-climax status, especially if it is the dominant tree in early succession and the intervening fires are ground-clearing light to moderate ones. A severe fire in a climax Douglas-fir stand would return the vegetation to a shrub-herb community. This model graphically shows how fire intensity and fire frequency can influence vegetation. (After Kessell and Fischer 1981:4.)

fires, result in more open stands with greater species diversity and a mix of tolerant and intolerant species.

■ Disturbance and Nutrient Cycling

As succession proceeds, more and more nutrients become locked up in woody biomass and are not immediately available for recycling through the system. Litterfall, detritus, and the death of individual trees provide an opportunity for the mobilization of some of these reserves through decomposition. Major disturbances such as fire, timber harvesting, and insect outbreaks cause a sudden release of the nutrient capital and provide the opportunity for reestablishing a cycle of change, initiating succession, and increasing productivity. The key is the retention of nutrients. Without uptake by rejuvenating vegetation, activity of soil microorganisms, and the colloidal properties of soil, nutrients could be lost to the system.

Normal populations of foliage-consuming insects account for up to 40 percent of the input of nitrogen and phosphorus to the litter, strongly influencing nutrient cycling in the forest (Kitchell et al. 1979). During an outbreak of foliage-consuming insects such as gypsy moths, fall cankerworm, and loopers, great amounts of nutrients contained in the leaves are returned to the litter in the form of frass, the fecal material of caterpillars.

During an outbreak, the California oak moth *Phryganidia californica* removed all foliage from both the evergreen coast live oak *(Quercus agrifolia)* and the deciduous valley oak *(Quercus lobata)*, both of which contribute quantities of litter to the forest floor during normal years. The caterpillars doubled the amount of nitrogen and phosphorus input to the ground. Seventy percent of total nitrogen and phosphorus from the live oak and 60 percent of the nitrogen and 40 percent of the phosphorus from the valley oak flowed to the ground through frass. Because of the high solubility of frass, loss of nutrients was rapid.

Swank and associates (1981) investigated the effects of an epidemic outbreak of fall cankerworm, *Alsophila pometaria,* a spring defoliator, on nitrate export from three mixed hardwood forests in North Carolina. The larvae consumed approximately 33 percent of the foliage biomass and produced considerable quantities of frass. This deposition resulted in an accelerated leaching and subsequent stream export of nitrate nitrogen, NO_3-N, although there were no changes in the concentrations of other nutrients. The overall effect of such defoliation was a large increase in leaf litterfall, in total litter, in soil metabolism, in standing crop of total microbes and nitrifying bacteria, and in pools of available nutrients. The trees, in response to defoliation, increased their uptake of nutrients made accessible by frass and more rapid litter turnover and shifted production from wood to leaf to compensate for the loss of leaves to insects.

Fire is the great regenerator of nutrients (Ahlgren and Ahlgren 1960), provided they stay in place and the ash is not eroded away. Nutrient recycling by fire is influenced by the nutrient content of the fuel, the temperature and duration of the fire, nutrient release from the ash, nutrient storage capacity of the soil, surface runoff, soil erosion, leaching, and the response of rejuvenating vegetation (Stark 1976). Burning of heather in Scottish highlands reduced annual primary production and tended to make calcium and magnesium less soluble. Young growth of heather, experiencing a postburn flush, contained increased concentrations of K, Mg, P, and N for the first several years; then the concentration declined rapidly back to the prefire level (Miller and Watson 1974).

Fire reduces soil organic matter to ash. Temperatures of 200° to 300° C destroy 85 percent of organic matter; release CO_2, nitrogen, and ash to the atmosphere; and deposit minerals in the form of ash on the soil. Volatilized nitrogen and potassium are released and lost by distillation at temperatures over 200° C, but little is lost under 200° C (Knight 1966). Much of the nitrogen lost, however, is in a form unavailable to plants and is replaced by nitrogen-fixing legumes. Their growth is stimulated by fire, by increased activity of soil microorganisms, including free-living nitrogen-fixing bacteria favored by increased pH following fire, and by weathering of soil materials.

Timber harvesting differs from other types of disturbance because it involves removal of biomass and the nutrients it contains from the ecosystem. The quantity of nutrients removed depends upon the type of harvest involved. Stem-only harvesting leaves tops and foliage behind. Whole tree harvesting leaves only some foliage, with all woody material being reduced to chips. Most exploitative of all is complete tree harvesting, which involves removal of the total tree, including roots. Block clearcutting involving stem-only removal on Hubbard Brook Forest removed about 28 percent of the calcium contained in aboveground biomass, 28 percent of potassium, 24 percent of nitrogen, 34 percent of sulfur, and 19 percent of phosphorus (Hornbeck et al. 1987).

Timber harvesting immediately affects nutrient cycling within the disturbed forest. Loss of the trees blocks uptake of nutrients. The sudden introduction of large quantities of detritus and loss of canopy cover, allowing increased precipitation and sunlight to reach the forest floor, increases decomposition and accelerates nitrification. Increased concentrations of nutrients in the dissolved organic and inorganic fractions are easily leached. With minimal uptake by vegetation, nutrients are exported by surface water and groundwater to streams. This flux of nutrients represents a loss to the disturbed site and a gain to the stream ecosystem. Such an input to a stream may have positive or negative effects, depending upon its nutrient status (see Chapter 32) (Bormann and Likens 1973; Hornbeck et al. 1987; Waide et al. 1988; Vitousek and Melillo 1979; Martin et al. 1986). Input of nutrients into receiving streams is greater in the second year following cutting than during the first, and by the fourth year stream ion concentration usually returns to precutting levels.

The reason may relate to the means of nutrient conservation available to disturbed ecosystems. One is immobilization of nutrients by decomposer organisms. Bormann and Likens (1979) suggest that in the first year after harvest, when the C:N and element:P ratios of decaying organic matter are high, dissolved nutrients in the soil are taken up and immobilized by the increased numbers of microorganisms (see Chapter 3). As the ratios decrease, the demands by microorganisms decline, and export of nutrients to streams increases. During this time, however, rapidly growing, early successional vegetation begins to recycle nutrients efficiently, retaining most of the nutrients on site. Boring and associates (1981) found that incoming woody and herbaceous plants on a clear-cut Appalachian hardwood forest during the first year held 29 to 44 percent of the amount of N, P, K, Mg, and Ca found in the net primary production of the control. Uhl and Jordan (1984) found similar responses in a cutover tropical forest. By the end of the fifth year, live plant regeneration held 15 percent of precut levels of N, 23 percent of P, 39 percent of K, 48 percent of Ca, and 45 percent of Mg. Thus microbial activity, soil processes, and rapid recovery of vegetation contribute to the resilience of forest ecosystems following logging.

■ Animal Response

The impact of a disturbance on animals is both short-term and long-term, negative or positive, depending upon the species involved. With disturbance it is clearly a case of "one man's meat is another man's poison." Small-scale disturbances, particularly treefall gaps, have a positive response. The new vegetative growth filling in the gap provides low ground cover attractive to such gap species as hooded warblers and Kentucky warblers. The canopy gaps provide open areas needed for sallying by flycatchers; yet the openings have no effect on canopy-dwelling species. Clear-cut areas create the early successional habitats needed by opportunistic or ephemeral species such as prairie warblers, chestnut-sided warblers, and woodcock. Large clear-cut areas, on the other hand, eliminate habitat of canopy-dwelling species, vertebrate and invertebrate, that will not return for several decades (Smith 1988). For the most part small ground-dwelling mammals are little affected by timber harvesting. Elimination of old growth forest can permanently eliminate old-growth dependent species such as spotted owls and red-backed voles of the Pacific Northwest.

A view of its immediate aftermath may give the impression that fire is a major agent of destruction to wildlife. On a short-term basis fire destroys or partially destroys habitat and causes some injury and death, either directly by fire or indirectly by predators who take advantage of prey suddenly driven from or deprived of cover. On the African savanna kites and other birds are grass fire followers, hunting insects driven to flight by the advancing fires. Many flying insects, such as grasshoppers and moths, fly in front of the flames and are often engulfed by wind-driven gas clouds, but a surprising number go through the fire unscathed. Unless nesting, birds are rarely directly affected by fire other than short-term loss of habitat. Some species of birds may decrease following fires, but ground-foraging birds seem to increase (Wirtz 1977; Wright and Bailey 1982).

Many mammals, especially those that live in burrows, survive fires. Large mammals are adept at keeping ahead of flames and working their way back through gaps and unburned patches to burned-over areas behind the flames (Main 1981). The major problem faced by these mammals is the short-term lack of food and cover. High populations of postfire grazing herbivores feeding on newly regenerating plants may overgraze and eliminate palatable species.

Severe fires alter the habitat and eliminate for a time those species dependent on it. At the same time fire creates new habitat for a different set of species, especially those that favor open and shrubby land. Many animals favor both prefire and postfire conditions and in fact are dependent on such fluctuations in habitat. Fire produces a mosaic of shrubs, timber, and open land (see Figure 26.1). Such patchy environment is essential for snowshoe hare, black bear, white-tailed deer, ruffed grouse, and many other species.

A few species are wholly dependent on the disturbance of fire. One is the endangered Kirtland's warbler (*Dendroica kirtlandii*). Restricted to the jack pine forests of the lower peninsula of Michigan, the warbler requires large blocks (40+ ha) of even-aged stands of pine 1.5 to 4.5 m tall with branches close to the ground. Smaller or larger trees are unacceptable. Intervals of fire sufficient to maintain blocks of these young jack pines are necessary to maintain the habitat of this warbler.

Another group of fire-dependent birds is the *Sylvia* warblers that occupy the Mediterranean shrublands of Sardinia (Walter 1977). Of the five species, two are fire-dependent. One species, the Dartford warbler *(S. sarda),* can occupy a habitat patch that has been burned within six years. Another species, Marmora's warbler *(S. undata),* occupies only 18- to 20-year-old tall, shrubby growth. Two of the remaining species are fire-adapted, and one is fire-tolerant.

■ Ecosystem Stability

We view disturbance as an intrusion into the stability of a system. If an ecosystem maintains its integrity and returns to its original condition in the face of a short-term disturbance, it is considered stable. The faster it returns to its original condition and the less it varies from its original state, the more stable the system.

Disturbance may be called *perturbation;* the two terms are often used synonymously, but there is a subtle difference. Disturbance is a cause of or action involved in a change, such as grazing or fire. It is the external input or driving variable. Perturbation is the effect of or reaction to disturbance. The changes produced are measured by the deviation from a steady state reference point.

Stability is the tendency of an ecosystem to reach and maintain an equilibrium condition of either a steady state or a stable oscillation (Holling 1973, 1984). If the system is highly stable, it resists departure

from that condition; and if it happens to be directed away from that condition, it returns rapidly with the least fluctuation.

Stability may be local or global. *Local stability* is the tendency of a system to return to its original state from a small perturbation. Forest gaps filling in with similar tree species are examples of local stability. *Global stability* is the tendency of a community to return to its original condition from all possible perturbations. Chaparral or eucalyptus forest returning quickly to its original condition and species composition after a fire represents global stability. Such systems exhibit low variability and resistance to change.

Ecosystems exhibit resistance and resilience. *Resistance* is a measure of the degree to which a system is changed from an equilibrium state following a disturbance. Ecosystems most resistant to change characteristically have a large biotic structure, as trees do, and have nutrients and energy stored in standing biomass. A forest ecosystem is relatively resistant to disturbance. It can withstand such environmental disturbances as sharp temperature changes, drought, and insect outbreaks because the system is able to draw on stored reserves of nutrients and energy. For example, a late spring frost may kill the new leaves of trees, but the forest is able to draw on energy reserves to replace leafy growth. If the forest is highly disturbed by fire or logging, though, its return to its original condition is slow. The system exhibits low resilience.

Resilience is the speed with which a perturbed system returns to an equilibrium value. A rapid return is evidence of high resilience, and a slow return indicates low resilience. For example, in the spruce-fir forests of northern North America the spruce budworm population under certain environmental conditions increases rapidly, escapes control of predators and parasites, and feeds heavily on balsam fir, killing many trees and leaving only the less susceptible spruce and birch. After the spruce budworm population collapses because of the exhaustion of the food supply, young balsam fir grows back in thick stands with spruce and birch. Between budworm outbreaks balsam fir outcompetes spruce and birch, but during outbreaks spruce and birch are favored over balsam fir. Thus some time after the outbreak of budworm, the system returns to balsam fir. The system is resilient even though some of the interacting populations have low resistance.

Aquatic ecosystems, which lack any long-term storage of energy and nutrients in biomass, exhibit little resistance but are resilient. An influx of pollutants such as sewage effluents disturbs the system, adding more nutrients than it can handle; but because the system is limited in its capacity to retain and recycle nutrients, it returns to its original condition relatively soon after the perturbation is reduced or removed.

For example, Lake Washington, near Seattle, was used as a basin for sewage disposal. It received a large input of excessive nutrients, especially phosphorus. The input eliminated certain diatom and algal populations and encouraged others, especially filamentous algae, to increase, changing the clear-water lake ecosystem. Once the sewage input was diverted from the lake, phosphorus levels in the lake declined, algal populations shrank, and the lake cleared. The lake had high resilience but low resistance.

A strong disturbance may carry an ecosystem into a different level of stability. The system may be so greatly disturbed that it is unable to return to its original state, and a different ecosystem with a different domain of stability takes its place (Figure 26.3). Perry and associates (1989) provide an excellent example of such a response to disturbance in 10 to 15 ha high

elevation clear-cuts in white fir (*Abies concolor*) forests of the Siskiyou Mountains of southwestern Oregon and northern California. Following clear-cutting the areas were sprayed with herbicides to eliminate incoming hardwood growth preparatory to replanting the sites with conifers. Hardwood growth, however, is a natural successional recovery process after a disturbance such as fire in these forests. The rapid response of hardwood growth shades the soil, ameliorates surface temperatures and moisture regimes, and maintains the integrity and stability of the belowground soil and rhizosphere organisms.

The loss of woody growth greatly weakened the close link between plants and soils and altered the belowground system. Bacteria, mostly actinomycetes, inhibited the growth of other microbes and plants. The failure of the original vegetation to recover from the disturbance further reduced the belowground mutualists. The loss of the belowground microbial community in turn caused the loss of soil aggregation and subsequently a reduction in soil pore space and water holding capacity. The original vegetation was replaced by brome grass, bracken fern, and manzanita, species that have a competitive advantage when mycorrhizal fungi and rhizosphere organisms decline. A white fir forest can return to the site only after its mutualistic soil microbial community is restored.

■ Summary

Disturbance is a relatively discrete event in time coming from outside the system that disrupts ecosystems, communities, and populations, changes substrates and resource availability, influences species composition and system structure, initiates succession, and adds diversity to the landscape. Major agents of disturbance are wind, moving water, weather, fire, timber harvesting, and animal activity, including insect outbreaks.

Important to the nature of disturbances is scale in size and time. Small-scale disturbances are typical of rocky intertidal shores and temperate and tropical forests. Wave action and moving water create gaps among sessile organisms on rocky substrates. Treefall and removal of individual trees by logging create small gaps in forests. Response to gap formation involves canopy closure, invasion by opportunistic species, and growth of tolerant species, depending upon biotic conditions. Large-scale disturbances induced by major events such as fire, logging, or insect outbreaks, result in responses that go beyond vegetational organization and involve colonization by opportunistic species. Such disturbances can modify the system, favoring certain species and eliminating others, or they can ensure regeneration of the system.

Fire is the major natural large-scale disturbance to terrestrial ecosystems. It has both beneficial and adverse affects. It results in loss of soil nutrients but also makes nutrients available. It sets into motion regeneration of fire-adapted systems by stimulating root sprouting and germination of seeds. It can favor fire-resistant species and eliminate fire-sensitive ones, thereby influencing the composition and structure of forest systems.

Of great ecological importance is the frequency and return interval of disturbances. Small-scale disturbances have a high frequency within the system, but the rate of disturbance is low and the return interval is between 50 and 200 years. Natural large-scale disturbances have low frequency and a return interval of 25 to several hundred years. Frequency of large-scale disturbances generally is related to the life span of the longest living species. Too frequent disturbances can eliminate certain species by destroying the plants before they have had time to mature and seed. Too long a time between disturbances can result in the elimination of midtolerant species, reduce system diversity, and set the stage for highly destructive disturbances.

Disturbances suddenly release nutrients locked up in biomass. Mechanisms exist with the system to counter the blocked uptake of nutrients. Although considerable amounts of nutrients may be lost to the system following major disturbances, much of the nutrient pool is immobilized by increased populations of soil microorganisms and incorporated into plant biomass by the resurging growth of vegetation.

Response of the animal component of the systems depends upon the species involved. Short-term impacts involve the loss of food and cover. Long-term effects may be the loss of habitat for some species and the gain of habitat for others. Some species depend upon disturbance for the maintenance of their habitat, especially those associated with the more ephemeral stages of early succession. Other species depend upon periodic fires to maintain their habitat and to provide a mosaic of vegetation types required in their life cycle. A few fire-dependent species would go extinct without periodic fires to maintain their habitat.

Disturbance is involved in the stability of ecosystems. Stability is the tendency of a system to return to and maintain equilibrium after perturbation. If the system is highly stable it resists departure from that condition or returns rapidly to it. Stable systems are highly resistant to disturbance but are not necessarily resilient. Resilience is the speed with which a system returns to an equilibrium state. Resistance is the ability of the system to absorb changes and persist.

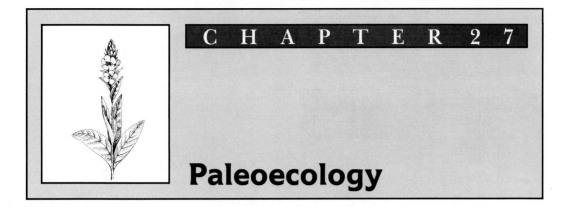

C H A P T E R 2 7

Paleoecology

- **Continental Drift**

- **Climatic Changes and Dispersal of Species**

- **The Pleistocene Epoch**

Pre-Pleistocene Development
Glacial Periods and Vegetation
Speciation
Extinction

- **Summary**

Succession as experienced by humans takes place over a short period in terms of Earth's history. Changes also take place on a grander time scale. Earth is dynamic, and since its inception it has changed profoundly. Land masses broke into continents. Mountains emerged, seas rose and fell, ice sheets covered large expanses of the Northern and Southern hemispheres. All these changes affected the climate and other environmental conditions from one region of Earth to another. Many species of plants and animals disappeared and were replaced by others, and there were major shifts in vegetational patterns. What life was like in the past and under what conditions it existed we infer from present-day conditions. The past, the geologist Lyell remarked, is the key to the present. Conversely, the key to distributions of animals and plants today may be found in the past.

Records of these past communities, their animals and plants, lie buried as fossils: bones, insect exoskeletons, plant parts, and pollen grains. These fossils enable us to determine plant and animal associations of the past and in a broad way the climatic changes that brought about the gradual destruction of one type of community and the emergence of another. Such interpretation is based on the assumption that organisms of the past possessed ecological requirements similar to those of related species living today. For example, if modern palms and broad-leaf evergreens are tropical plants, we assume that their ancient prototypes also lived in a tropical climate. The study of relationships of ancient flora and fauna to their environment is *paleoecology*.

Earth, geologists estimate, is some 4600 million years old. Life formed in the late Precambrian era about 4000 million

TABLE 27.1 Geological Time Scale

Era	Period	Epoch	Age (Millions of Years)	Dominant Life — Plants	Dominant Life — Animals
CENOZOIC: the age of mammals	Quaternary	Recent	0.01	Agricultural plants	Domesticated animals
		Pleistocene	2		Ice Age—First true humans; mixture and then thinning out of mammalian faunas
	Tertiary	Pliocene	10	Herbaceous plants rise; forests spread	Culmination of mammals; radiation of apes
		Miocene	25	First extensive grasslands	Modernization of mammals; mammals become dominant
		Oligocene	35		Mammals become conspicuous
		Eocene	55		Expansion of mammals; extinction of dinosaurs
		Paleocene	70		
MESOZOIC: the age of reptiles	Cretaceous		135	Angiosperms rise; gymnosperms decline	Dinosaurs reach peak; first snakes appear
	Jurassic		180	Cycads prevalent	First birds and mammals appear
	Triassic		230	Gymnosperms rise; seed ferns die out	First dinosaurs; reptiles prominent
PALEOZOIC	Permian		280	Conifers become forest trees; cycads important	Great expansion of primitive reptiles
	Carboniferous			Lepidodendron, sigillaria, and calamites dominant; the swamp forest	Age of cockroaches; first reptiles
	Pennsylvanian		310		
	Mississippian		345	Lycopods and seed ferns abundant	Peak of crinoids and bryozoans
	Devonian		405	First spread of forests	First amphibians; insects and spiders
	Silurian		425	First known land plants	First land animals (scorpions)
	Ordovician		500	Algae, fungi, bacteria	Earliest known fishes; peak of trilobites
	Cambrian		600	Algae, fungi, bacteria; lichens on land	Trilobites and brachiopods; marine invertebrates
PRECAMBRIAN	Late			Algae, fungi, bacteria	First known fossils
	Early		4500	Bacteria	No fossils found

years ago and developed through geological ages with the evolution, emergence, and extinction of numerous groups of plants and animals (see Table 27.1).

■ Continental Drift

The distribution of life in the fossil record suggests that the distribution of land masses during early geological ages was much different from that today. The flora that covered the southern land masses during the age of plants, the Permian period of the Paleozoic era, differed from that of the northern land mass. This fact and the discovery that glaciers covered the southern continents could be explained only if the land masses of the earth were joined into a supercontinent.

In 1924 the German meteorologist and astronomer Alfred Wegener observed that if the continental shelves of South America and Africa on either side of the Atlantic Ocean were fitted together like pieces in a jigsaw puzzle, many geological features that ended abruptly on continental edges became continuous. Fossils of similar plants and animals once separated by oceans and realms of fossils of different ages were brought together again. Wegener proposed that in the Permian and earlier periods the land masses of the earth were compressed into one large continent, Pangaea, surrounded by one ocean, Panthalassa. Over eons of time this land mass broke apart and drifted into the positions the continents occupy today. Because Wegener was not a geologist and little was known of the earth's surface and ocean floors, geologists rejected Wegener's theory. As additional evidence on the nature of the earth's interior and crust, fossils, and paleomagnetism (magnetic directions of ancient rocks) accumulated, the theory

of continental drift resurfaced; it is widely accepted today.

Some evidence seems to indicate that in the lower Paleozoic three separate land masses existed: Asia, North America and Europe, and Gondwanaland, which included modern-day Africa, South America, Australia, New Zealand, and Antarctica. During the Permian these three blocks joined, raising the Caledonian, Ural, and Appalachian mountain ranges and forming the single land mass Pangaea. During the Paleozoic, 420 million years ago, North America and Africa lay close together around the South Pole and the rest of Gondwanaland lay on the far side of the South Pole, pointing toward the equator (Figure 27.1). Slowly the land mass moved northward. By the Carboniferous age, 340 million years ago, the whole of Africa had moved across the South Pole, and Antarctica lay in the region of the South Pole. Glaciers covered southern South America, South Africa, India, and Australia; and Europe and North America lay along the equator.

As Pangaea moved northward it began to break apart slowly, 5 to 10 cm a year (Figure 27.2). The first break in the single land mass apparently took place in the mid-Mesozoic age, when North America and Africa separated to form the first narrow strip of the Atlantic. Africa and South America were still connected by the end of the Jurassic, but drift was taking place in the South Atlantic region. Africa was separated from the Antarctic between the middle Jurassic and middle Cretaceous. By the middle Cretaceous Africa and South America had split apart. By the late Cretaceous southern Greenland began to separate from the British Isles and to move northward. South America and Antarctica still clung together until sometime in the early Cenozoic, finally becoming separated in the Eocene.

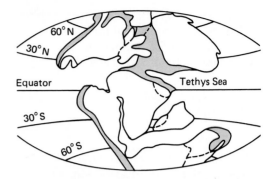

Figure 27.2 Earth in the mid-Cretaceous period, 105 million years ago. Pangaea not only moved northward, breaking away from Gondwanaland, but also began to break up into separate continents. Dotted lines indicate present-day continental coastlines; shaded areas indicate epicontinental seas. (From Cox et al. 1973.)

Figure 27.1 Map of Pangaea in the Permian and Triassic (without southeast Asia, because its position at this time is uncertain). Dashed lines indicate the outlines of present-day continental coastlines. Laurasia, consisting of North America and Eurasia, is attached to Gondwanaland, the continents of Africa, South America, Antarctica, and Australia. The series of dark dots indicates the positions of the South Pole from the Cambrian to the Jurassic. (Redrawn from Cox et al. 1973.)

■ Climatic Changes and Dispersal of Species

The breakup and northward drift of continents resulted in broad climatic changes that affected the evolving plant and animal life. During the Carboniferous period Laurasia (North America and Europe), having moved north to the equator, had a warm, humid, and seasonless climate. A large part of the area was covered with swamps and tropical rain forests dominated by *Lepidodendron, Sigillaria,* and the forerunner of coniferous trees, *Cordaites.* Throughout the Mesozoic no distinct floral or faunal regions existed. For most of the Mesozoic the land, even though partially separated, was one. There were no effective barriers to the dispersal of plants and animals. Mountain ranges lay about the edges of Pangaea. Although shallow seas invaded the land, especially in the late Mesozoic, the invasions were short-lived, geologically speaking, and did not effectively disturb the distribution of animals. The climate, too, was warm and equitable, mostly tropical to subtropical, even to the coast of Alaska. These conditions allowed the great reptiles and early mammals to roam freely over the continental land masses.

In the late Cretaceous conditions began to change as the continental land masses drifted apart. The warm, subtropical middle Cretaceous climate was replaced in the late Cretaceous by a warm, temperate climate. The cooling climate marked the end of the great reptiles. As the Cretaceous period moved into the Paleocene period of the Cenozoic, greater changes took place. Between the lower Cretaceous and the Eocene, the single, connected land mass inhabited by gymnosperms and reptiles was replaced by a divided land mass inhabited by flowering plants (angiosperms) and mammals. Continental drift effectively separated plant and animal populations and hastened the development of different floras and faunas.

The climate changed even more. Land areas along the seas experienced mild, less variable climates. New mountain ranges interrupted the pattern of rainfall, encouraging the development of deserts or grasslands on the lee side. The uplift of the Andes in South America resulted in a dry, cooler climate that favored the development of grassland on the eastern side of the continent, a situation that exists to the present day. As land masses moved northward, the leading edges reached high latitudes and the land was covered by huge ice sheets. As Gondwanaland moved south, the southern limits of that land mass became buried under a layer of permanent ice. In the late Cretaceous Gondwanaland was broken up; the only major intact land mass was Laurasia.

Before Gondwanaland had broken up, the angiosperms experienced a sudden explosive evolutionary radiation. They supplanted the gymnosperms and, apparently before the breakup, spread throughout the supercontinent. As a result four families of angiosperms—Compositae, Graminae, Leguminosae, and Cyperaceae—are abundant throughout the world. When Gondwanaland broke apart, the resulting continental land masses of the southern hemisphere carried with them similar flora.

Floral evolution in Laurasia was somewhat different. Laurasia retained the conifers that had arisen from the Cordaites. The conifers could not spread across the hotter equatorial regions, a barrier reinforced by the periodic invasion of shallow seas that covered southern Europe and northern Africa during the Jurassic and Cretaceous, separating Laurasia and Gondwanaland. Because of the colder climate in the northern part of Laurasia, a still different flora evolved in northern and central Eurasia. Later a shallow epicontinental sea, the Turgai Straits east of the Ural Mountains, separated eastern North America and central Europe from Asia and western North America, causing the development of two separate floral realms.

Although plants achieved worldwide distribution before the continents broke up, the mammals did not. They were not significant in the fauna until the dinosaurs disappeared, and by then the continents were well on their way to separation. One of the first groups of mammals, the marsupials, were confronted with competition from placental animals. Had the placental mammals achieved dominance before the breakup of the continents, the marsupials might never have survived; but marsupials apparently had spread to Antarctica and Australia before these land masses separated. Antarctica moved south, where the cold eliminated terrestrial mammalian life, while Australia, separate from all other land masses, became a final refuge for marsupial life. Free from competing placental forms, marsupials were able to radiate into a variety of forms and to fill niches similar to those occupied by placental animals elsewhere. Except for the opossum, marsupial mammals in North

America and South America were rapidly replaced by placental mammals.

South America, isolated from North America during the lower Cenozoic (Tertiary period), supported diverse and unique forms of placental mammalian life, including the primitive ungulates *Thoatherium* and *Toxodon*. After a land bridge, the Isthmus of Panama, became exposed between the two continents, these animals were unable to compete with the more advanced placentals that moved down from the north.

Competition resulted in the elimination of certain kinds of animals, but the major influence on developing fauna was the rapidly changing climate of the northern land mass. The fauna of the Old World tropics remained free from the influences of great temperature changes, although the increasing aridity in Africa that began in the Oligocene and Miocene epochs of the Cenozoic era brought about the replacement of tropical forests by grasslands, on which evolved the huge herds of diverse ungulate fauna and eventually humans.

Europe, North America, and Asia were still connected in a manner that permitted animals to move from one to the other. Until the lower Eocene North America was still connected to Europe by Greenland and Scandinavia. At that time North America appeared to be the center of the early evolution of placental mammals. From there they spread to Europe. During that time Asia was separated from Europe by the Turgai Straits but was connected to North America by the Bering land bridge between Siberia and Alaska. However, the cold climate inhibited the movement of mammalian fauna from North America to Asia.

The mid-Eocene was a time of further changes. The North Atlantic joined the Arctic Ocean and separated Europe from North America (Figure 27.3). New mammalian groups that evolved in Europe could not cross to North America, but at this period in the earth's history the Turgai Straits dried up, allowing European mammals access to Asia. The same period saw the return of a warm, moist climate; a semitropical rain forest extended to Alaska, and tropical conditions extended as far north as England. The climate of the Bering land bridge was benign enough to encourage the passage of mammals from Asia to North America.

In the Oligocene the climate cooled again, restricting the migration of mammals to those tolerant of cold temperatures and eliminating tropical forms of plants from northern lands. From the Oligocene through the Pliocene the cooling trend continued. The movement of animals from Eurasia to North America was restricted to such cold-tolerant species as mammoths and humans, and the flora of the continent was basically a modern one. The cooling of the Pliocene continued until the Pleistocene ushered in the Ice Age.

Figure 27.3 Earth during the lower Tertiary period (upper Eocene), 50 million years ago. Dashed lines indicate present-day continental coastlines; shaded areas indicate shallow epicontinental seas. North America has separated from Eurasia; South America has moved well away from Africa; and Australia has separated from Antarctica. (Redrawn from Cox et al. 1973.)

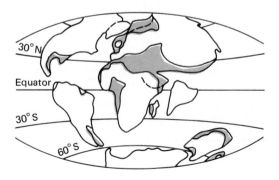

■ The Pleistocene Epoch

Of particular interest to the paleoecologist are the climatic and vegetational changes that followed the advance and retreat of glaciers in the Pleistocene. As the glaciers advanced and retreated at least four times, vegetation was destroyed and the relief or physiognomy changed radically. The climate about the edges of the glacier supposedly was cold and optimal only for tundra and taiga vegetation.

Changes in postglacial vegetation and climate are recorded in the bottoms of lakes and bogs. As the glaciers retreated, they left scooped out holes and dammed up rivers and streams, which filled with water to form lakes. Organic debris accumulated on the bottom to form peat, marl, and muds. Pollen, spores, and small invertebrates that blew in from adjacent vegetation settled on the water and sank. Microscopic examination of samples of organic bottom deposits obtained at regular intervals reveals the fossil remains of these organisms. Various genera of fossil pollen can be identified by comparisons with pollen growing today. The relative abundance of pollen of several genera indicates the predominant vegetation at the specified depth of deposition (Figure 27.4).

Pollen investigation can indicate only trends in vegetation and climate through the past. At present it is impossible to determine the exact structure and composition of prevailing vegetation at any one time period. Tree species that produced more pollen than others may appear more abundant than they really were. Some pollen might have been carried some distance by the wind or perhaps buried deeper by soil invertebrates. Many pollen grains can be identified only to genera and not to species. For example, the pollen of different types of oak cannot be distinguished to give a clue to the particular type of oak forest existing at a particular time. However, modern paleoecological techniques that take these problems into account and improved identification procedures are enabling paleoecologists to provide a rather accurate picture of postglacial vegetation.

The Pleistocene, which began some two million years ago, marked the end of the Tertiary period and the Pliocene epoch by ushering in the Ice Age. However, recent studies of deep sea sediments, the geophysics of the ocean bottoms, and the Antarctic ice suggest that ice caps have been part of the earth's geological and ecological history at least since the Miocene and even earlier. Thus the Pleistocene is simply a stage on a continuum of ice and vegetation through the Cenozoic. The present distribution of plants and animals can be appreciated only in the context of longer successional development in the past.

Pre-Pleistocene Development

At the beginning of the Cenozoic era, some 70 million years ago, most of present-day continental North America and Europe was land. By the beginning of the Miocene epoch forests closely related to the present-day deciduous forest existed with little variation across the northern continents. Known as the Arcto-Tertiary forest, it was a mixture of broad-leafed and coniferous species roughly divided into boreal and temperate elements. The temperate or deciduous element was much like the mixed mesophytic forest of the central Appalachians. The boreal element, consisting of pines, spruces, cypress, birches, and willows, bore little resemblance to today's uniform northern coniferous forests. Because tropical and subtropical climates existed far north of the present positions, neotropical and Pa-

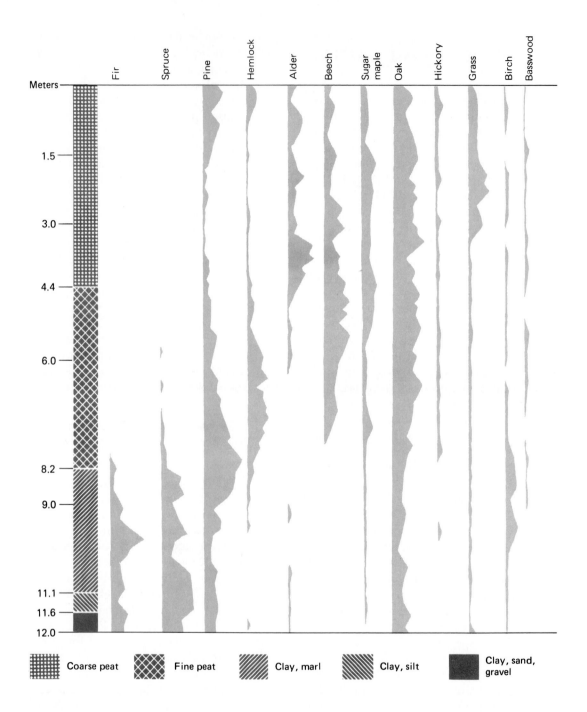

leotropical-Tertiary forest, ancestors of today's tropical forests, covered most of central and southern Europe and central North America. Probably in the Eocene, a mixed woodland, the Madro-Tertiary flora, developed on the Mexican plateau.

From the Miocene on, the climate began to deteriorate. The western mountain system in North America rose; climatic zones and their biota were pushed southward, and tropical forests were driven into Central America. The American portion of the Arcto-Tertiary forest was separated from that of Europe and Asia by continental drift, and certain species, such as *Metasequoia, Aliarthus,* and *Ginkgo,* became extinct in North America. As the mountains rose the broad rain shadow on their lee side wiped out the Arcto-Tertiary forest and stimulated the development of grasslands in the central part of North America. A relict Arcto-Tertiary forest, poor in species, but including the sequoias and redwoods, was left in the Pacific northwest. Elements of the Madro-Ter-

tiary forest moved northward to occupy dry lands vacated by the Arcto-Tertiary forest and eastward to form a sclerophyllous-pine woodlands ancestral to the Southern oak-pine woodlands of today. In the late Pliocene a continuing climatic deterioration accompanied by mountain building resulted in the development of continental glaciers.

Glacial Periods and Vegetation

The Pleistocene was an epoch of great climatic fluctuations throughout the world. At least four times during the Pleistocene ice sheets advanced and retreated in North America (Figure 27.5) and at least three times in Europe (Figure 27.6). Four times in North America and three times in Eurasia the biota retreated and advanced,

Figure 27.5 Glaciation in North America. In the Pleistocene northern North America was covered by at least four ice sheets, the limits of each usually marked by a terminal moraine. The last and most significant was the Wisconsin ice sheet.

Figure 27.4 Pollen diagram for Crystal Lake, Hartstown Bog, Crawford County, Pennsylvania, which was at the edge of the maximum advance of the Wisconsin glaciation. The graphs indicate the percentage of various genera based on counts of 20 pollen grains for each spectrum level. Grass pollen counts are expressed as percentages of total free pollen. Note five major forest successions from bottom to top. (1) Initial spruce-fir forest together with some pine and oak invaded as the glacier retreated. Alder, which today precedes spruce, followed spruce. (2) As the climate warmed, spruce and fir gave way to jack pine forest with some oak and birch. (3) Later forest surrounding the pond was dominated by oak with hickory, beech, and hemlock. (4) Oaks next dominated the forest, with some hickory, sugar maple, and beech. Grass became important component in forest openings. (5) Forests were dominated by oak and pine (white?) with sugar maple and beech. Hemlock again became important. Compare this diagram with Figures 27.10 and 27.11. (Adapted from Walker and Hartman 1960: 463.)

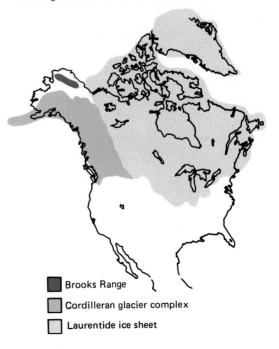

Brooks Range

Cordilleran glacier complex

Laurentide ice sheet

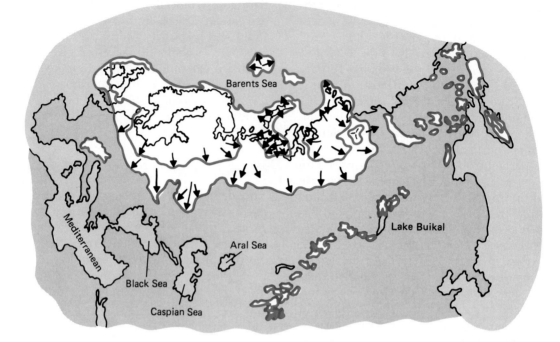

each advance having a somewhat different mix of species.

Each glacial period was followed by an interglacial period (Table 27.2 and Figure 27.7). The climate oscillations in each interglacial period had two major stages, cold and temperate. During the cold stage tundra-like vegetation dominated the landscape. As the glaciers retreated and

Figure 27.6 Glaciation in Europe. In the Pleistocene, northern Eurasia was covered by ice sheets similar to those covering North America. The most important was the last, known as the Weischselian. Note the disjunct glacier in the region of the Alps. (After Flint 1971: 545.)

Figure 27.7 Schematic representation of the glacial-interglacial cycle in northwestern Europe. (After van der Hammen, Wijmstra, and Zagwinj 1974.)

TABLE 27.2 Glacial and Interglacial Stages

Britain	Northern Europe	North America	Climate
Flandrian (postglacial)	Weichselian		temperate
Devensian (last glaciation)	Weichselian	Wisconsin	cold glacial
Ipswichian (last interglacial)		Sangamon	temperate
Wolstonian	Saalian	Illinoian	cold glacial
Hoxnaian	Holstein	Yarmouth	temperate
Anglian	Elsterian	Kansas	cold glacial
Comerian		Aftonian	temperate
Beestonian		Nebraskan	cold
Pastonian			temperate
Baventian			cold

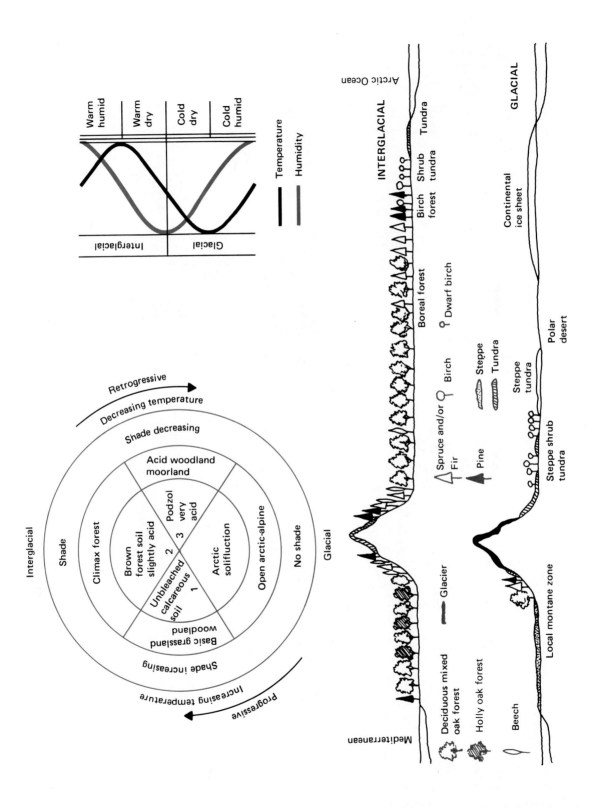

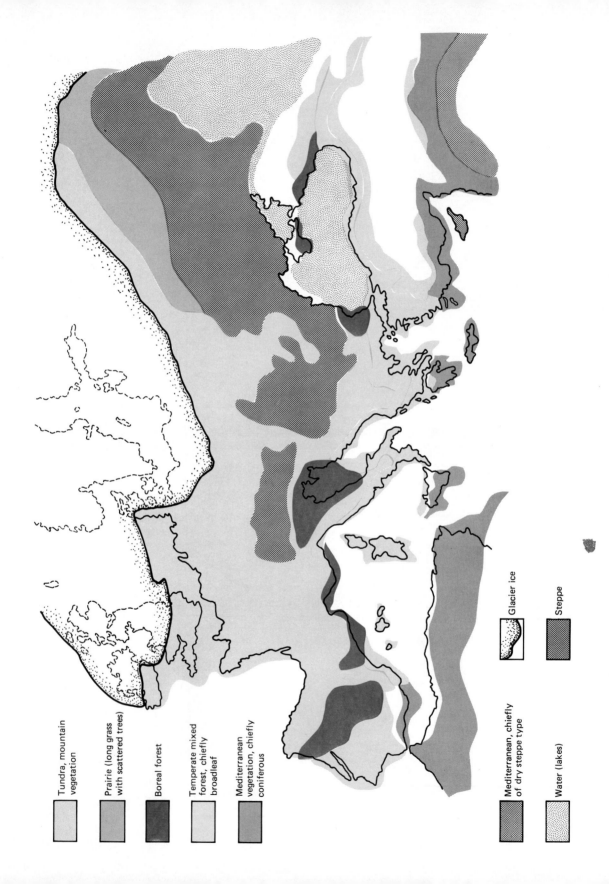

Tundra, mountain
vegetation

Prairie (long grass
with scattered trees)

Boreal forest

Temperate mixed
forest, chiefly
broadleaf

Mediterranean
vegetation, chiefly
coniferous

Mediterranean, chiefly
of dry steppe type

Water (lakes)

Glacier ice

Steppe

the climate ameliorated, light-demanding forest trees such as birch and pine advanced. As the soil improved and the climate continued to warm, these trees were replaced by more shade-tolerant species, such as oak and ash. As the next glacial period began to develop, species such as firs and spruces dominated the forest, changing the soil from mull to acid mors. As both climate and soil began to deteriorate, heaths began to dominate the vegetation and forest species disappeared.

Major differences in vegetation bordering the glacier existed between Europe and North America. In Britain and Europe a wide belt of tundra edged the glacier. The Alps, with their large ice cap, created an air flow pattern in which the westerly flow of warm air was diverted southward. As a result the Arcto-Tertiary flora was decimated by the early cold stages of the Pleistocene. Temperate genera were forced southward, but the southward retreat was blocked by mountains, deserts, and seas (Figure 27.8). Only the hardy boreal genera could survive. In North America no such barriers existed. The glacier extended further south into a warmer zone across which the flow of warm air was unimpeded (Figure 27.9). As a result spruce forests grew virtually to the edge of the ice, and Tertiary precursors of the present-day deciduous forest survived to spread north.

The last great ice sheet, the Laurentian, reached its maximum advance about 20,000 BP to 18,000 BP during the Wisconsin glaciation stage in North America

Figure 27.8 Assumed distribution of vegetation in Europe at the Weichsel/Wurm maximum. The Black Sea and Caspian Sea are interconnected lakes. Note the predominance of tundra vegetation, the patches of boreal forest, and the highly restricted distribution of temperate deciduous forest. (After Flint 1971.)

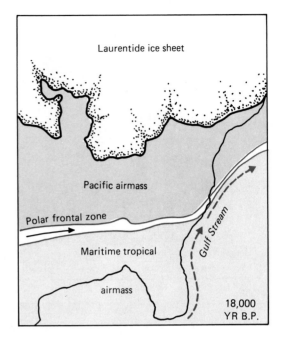

Figure 27.9 At the full glacial interval at 18,000 BP in North America, the boreal forest was separated from the temperate deciduous forest further south by a narrow belt of cool temperate coniferous and deciduous tree species. This narrow belt of vegetation probably represents the mean annual position of a strong zonal circulation of air, the Polar Frontal Zone, which extended eastward across the western North Atlantic Ocean. This narrow, stable climatic boundary separated the Pacific Air Mass immediately to the north of the Polar Frontal Zone from the Maritime Air Mass to the South. Compare this map with Figure 27.10b. (Adapted from Delcourt and Delcourt 1984: 277.)

(Figure 27.10). Canada was under ice. A narrow belt of tundra about 60 to 100 km wide bordered the edge of the ice sheet and probably extended southward into the high Appalachians; a few relict examples exist today (Wright 1970, 1971; Delcourt and Delcourt 1981, 1985). Boreal forest, dominated by spruce and jack pine, covered most of eastern and central United States as far as western Kansas. Its

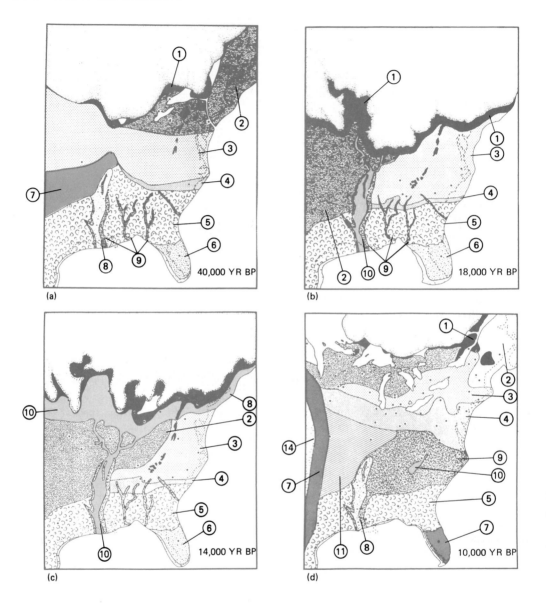

(a) 40,000 YR BP

(b) 18,000 YR BP

(c) 14,000 YR BP

(d) 10,000 YR BP

southern limit was about 1200 km south of the modern southern border of boreal forest in Canada. West of Kansas lay uninterrupted sand dunes, a treeless landscape that resulted from intense winds generated by the nearby ice sheet (see COHMAP 1988). South of the boreal for-

est was a transition belt of conifers and mixed hardwoods that separated the boreal forest from the oak-pine forests to the south. Mesic, temperate hardwood species—beech, sugar maple, basswood, walnut, buckeye, yellow-poplar, chestnut, hickory, and oak—found a refuge in the

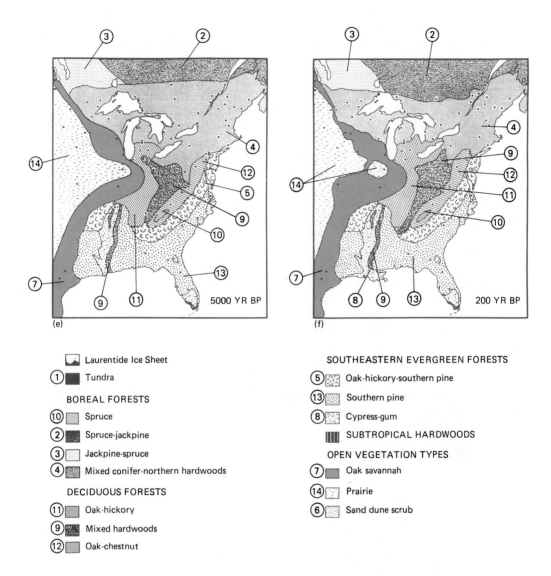

Figure 27.10 Changes in vegetation during and following the retreat of the Wisconsin ice sheet, reconstructed from pollen analysis at sites throughout eastern North America. (Adapted from Delcourt and Delcourt 1981.)

loess-capped uplands of the Mississippi Valley, in dissected valley slopes along major southern river systems, ravines, and irregular topography of karst terrain, and perhaps along the southern Atlantic and Gulf coastal regions exposed by seas that were 300 m lower than today (Figure 27.10) (Delcourt and Delcourt 1985).

The climatic changes that accelerated the retreat of the Laurentian ice sheet in the late Wisconsin also brought about a sudden end to the boreal forest in unglaciated North America. In the western part of the glacial region spruce was replaced by prairie grass. Further east, closer to the edge of the present prairie region, spruce gave way to pine, birch, and alder (Amundson and Wright 1979). In the southern Appalachians oak and pine replaced spruce, except for relict stands at high elevations. Pines moved northward rapidly from their Appalachian refugia in the Carolinas and dominated much of the region about the newly formed Great Lakes (Figure 27.11) (Davis 1981, 1983). Hemlock and other species appeared in the southern Appalachians, from which they invaded the deglaciated areas. Other species, such as chestnut, moved much more slowly, taking 3000 years to reach New England from the central Appalachians (Davis 1981).

In the Western Cordillera (parallel chains of mountains) during the Wisconsin glaciation the tree line and tundra vegetation moved downslope 800 to 1000 m lower than today. Many of the modern desert basins were shrub steppes dominated in part by sagebrush. At the end of the Wisconsin period the valley glaciers melted, the climate warmed, and the tundra and coniferous forest ascended the mountains.

Speciation

The expansion of vegetation during the interglacial periods and the destruction and isolation of large areas of vegetation during the full glacial periods undoubtedly set the stage for speciation of many forms of life, such as insects (Ross 1970), birds (Mengel 1964, 1970; Selander 1965), mammals (Flerow 1971), and amphibians (Blair 1965). In North America much of this speciation apparently took place in the forest regions, for they were subject to long periods of geographical isolation during the glacial periods. Little speciation took place on grasslands (Ross 1971; Mengel 1964) because they were

not divided into isolated units as were the forests. Rather the central grasslands acted as a persistent although changeable "sea" that separated the deciduous forest "continent" from "islands" of western montane forests and woodlands (Mengel 1970). Acting as a land bridge that periodically united the two was the spruce forest, expansive during interglacial stages and narrow and broken during glacial periods (Figure 27.12). During full glacial times species occupying the spruce forest would be isolated in eastern and western refugia and could begin to differentiate. During the interglacial periods the western forms would be further isolated when the montane forest moved upslope as the climate warmed.

Figure 27.11 Postglacial migration of three tree species. The isopleths represent the leading edge of the expanding populations. (a) An aggressive pioneer, spruce moved quickly into the tundra and took over the landscape newly exposed by the melting ice sheet. Its speed of northward movement varied. It moved into the Great Lakes region shortly after the ice receded, but 2000 years intervened until it arrived in New England. Spruce preceded alder, the opposite of successional sequence on glacial material today (see Figure 27.4). (b) White pine found a refuge during the height of the ice sheet along the East Coast and on the foothills of the Appalachians, where it was mixed with stands of hardwoods. The species was absent from the full glacial sites in the Mississippi Valley and Florida. Paleoecological evidence of white pine appeared first in Virginia, from where it expanded rapidly northward and westward to Minnesota 7000 years ago. Its westward expansion was blocked by dry climate. White pine expanded northward, reaching sites north of its present range. There it occurred briefly in large populations before competitive reduction by hardwoods. (c) Oaks, widespread in the southern United States during full glaciation, spread rapidly northward between 10,000 and 9000 BP and reached the full limits of their range by 7000 BP. The northern limit of oaks coincides with the northern limit of white pine. (Adapted from Davis 1981: 138, 144, 145.)

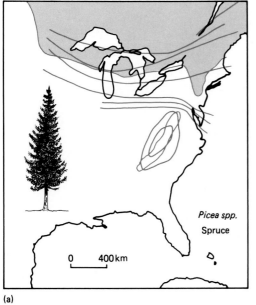

(a)

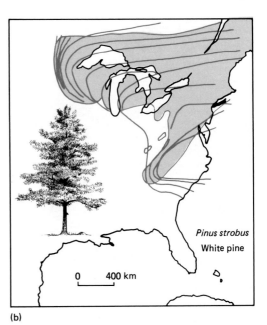

(b)

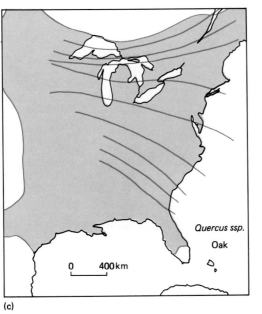

(c)

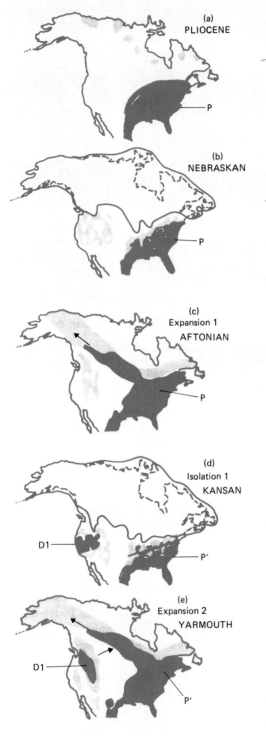

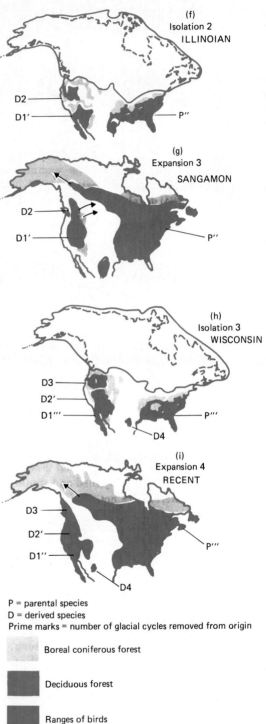

(a) PLIOCENE — P

(b) NEBRASKAN — P

(c) Expansion 1 AFTONIAN — P

(d) Isolation 1 KANSAN — D1, P′

(e) Expansion 2 YARMOUTH — D1, P′

(f) Isolation 2 ILLINOIAN — D2, D1′, P″

(g) Expansion 3 SANGAMON — D2, D1′, P″

(h) Isolation 3 WISCONSIN — D3, D2′, D1‴, P‴, D4

(i) Expansion 4 RECENT — D3, D2′, D1″, P‴, D4

P = parental species
D = derived species
Prime marks = number of glacial cycles removed from origin

Boreal coniferous forest

Deciduous forest

Ranges of birds

Figure 27.12 Model of the influence of the Pleistocene glaciation on speciation as exemplified by wood warblers in North America. With the first (Nebraskan) glaciation components of the boreal coniferous Arcto-Tertiary forest invaded deep into the Arcto-Tertiary deciduous forest of southeastern North America. Some warblers adapted to coniferous forests or their seral stages. In the next interglacial period a broad transition zone of coniferous forest developed, permitting warblers to establish a continent-wide range and providing conditions for separation of stocks into eastern and western segments during subsequent glacial advances. In the west montane forest replaced glaciers during the interglacial periods, creating island habitats in which differentiation could proceed. Repetition of the process during four glacial cycles resulted in the differentiation of the 12 endemic western species of warblers. (After Mengel 1964.)

R. Mengel (1964) developed a model to show how an ancestral species invaded the spruce after the first glaciation when the boreal Arcto-Tertiary forest was forced south by glaciation. Upon retreat of the glacial ice and the northward expansion of the boreal forest, the species would spread throughout the transcontinental range. At the second glaciation the species would be disjoined and would subsequently differentiate not only into eastern and western species, but also into varying forms in the western montane islands. In the next interglacial stage the eastern species would expand again and the entire process of expansion and subsequent disjunction during the advance of the ice sheet would be repeated through the third and fourth glaciations.

An example that fits the model well is the black-throated green warbler complex (Figure 27.13). The parent species would be an ancestral black-throated green warbler *(Dendroica virens)*. Nearly continental in distribution, the black-throated green warbler is found to a limited extent in the central Appalachian deciduous forest and more extensively in deciduous forest containing some hemlock *(Tsuga)*.

In the west are three and possibly four related species. Townsend's warbler *(D. Townsendi)*, an inhabitant of the coniferous forest of the northwest Pacific coast, is similar to the black-throated green in plumage pattern and in quality of song. It was probably derived from a westward invasion of the ancestral black-throated green warbler. Another species, the hermit warbler *(D. occidentalis)*, found in the tall conifers of the Cascades and Sierra Nevadas, also is only slightly different from the black-throated green, and may stem from the same invasion. Both Townsend's and the hermit warbler hybridize in a narrow zone of sympatry. Overlapping the ranges of both is the black-throated gray warbler *(D. nigrescens)*, which occupies shrubby openings in the northwest coniferous forests, shrubby openings in western mixed woods, and dry slopes covered with chaparral. Small in size and completely lacking yellow pigments, the bird nevertheless is remarkably similar in color pattern to Townsend's warbler. It may have descended from a western colonization more remote in time than that postulated for the other two warblers. In the Edwards Plateau of Texas is the golden-cheeked warbler *(D. chrysoparia)*, which inhabits the relict deciduous forest. Appearing as a dark form of the black-throated green warbler and possessing a similar song, this species is regarded as an offshoot of the black-throated green warbler or its ancestor, which may have reached Texas during the Wisconsin glaciation.

Extinction

The Pleistocene was not only a time of widespread speciation; it was also a time of extinctions. At the end of the Pliocene a number of warm climate forms—southern species of deer, tapirs, elephants, and apes—disappeared from northern Asia

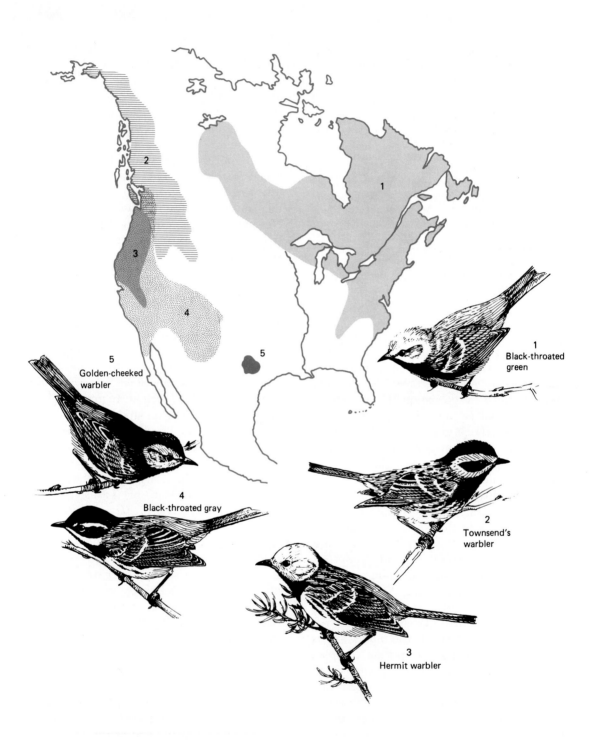

1
Black-throated green

5
Golden-cheeked warbler

4
Black-throated gray

2
Townsend's warbler

3
Hermit warbler

Figure 27.13 Approximate breeding distribution of the black-throated green warbler group. Compare the present-day ranges of these species with the maps in Figure 27.12. (Adapted from Mengel 1964.)

and Europe. As these forms disappeared, new groups evolved. By the middle of the Pleistocene a number of arctic and subartic genera of mammals evolved at the edges of the ice sheet—reindeer, musk-ox, woolly rhinoceros, long-horned bison, saber-toothed tiger, wolverine, and various lemmings. By the end of the Pleistocene most of the cold climate forms had disappeared and those that survived were limited to restricted ranges.

The Pleistocene also saw the expansion of *Homo sapiens* across the North American continent. Because of the lowered sea levels during glacial times humans were able to move across the land bridge from Asia to Alaska. Much of Alaska was unglaciated, and during the glacial retreat 11,000 to 12,000 years ago, new immigrants were able to move into North America through an unglaciated corridor that led down through Canada. There human colonists found an abundance of large mammals. The spread of humans over the continent was spurred by a population increase and the need to move southward as resources were depleted (Martin 1973).

Accompanying this explosive increase in humans was the extinction of many species, such as the woolly mammoth, mastodon, woolly rhinoceros, camel, horse, and royal bison. Some students of the Quaternary Period believe that the reason for the extinction of certain large mammals was overkill by humans. P. S. Martin (1973) points out that overkill was most pronounced in North and South America. According to his hypothesis humans swept through North and South America in a series of advancing fronts of

dense population. A front remained stationary for a decade or less until humans reduced or eliminated populations of big game. Then the front advanced to a new area and remained there until the animal resource was exhausted. In time as food resources were depleted, human populations, too, declined.

The hypothesis of mammalian extinctions by early human populations has its critics. Some argue that it is difficult to imagine that primitive humans could have destroyed a whole species by killing off individuals, even if the methods of taking animals were wasteful. (A common method of capturing or killing large game was to drive herds over a cliff or to surround them with fire.) Others believe that although hunting by humans was extensive, climatic changes and changes in food supply helped to cause extinctions (Bryson 1970; Flerow 1971). Climatic changes that replaced browse with short grass over large areas of the continent could have reduced the food supply for large browsers such as the woolly mammoth (note the location of the grasslands in the last glaciation in Figure 27.9). This argument is further supported by the fact that the mammoth became extinct in the Soviet Union about the same time it became extinct in North America (Flerow 1971). If the highest density of human population were associated with extinctions, we might have to conclude that Europe was less densely populated than Siberia, because such Pleistocene species as big-horned deer and aurocks became extinct in Siberia, but the wisent or European bison still survives. This conclusion is highly improbable. Another descendent of the Pleistocene fauna, the American bison, did not face demise until white Europeans deliberately went about its extermination to subjugate the descendants of Pleistocene human settlers.

Extinction probably resulted from a number of causes, differing for each species. Extinction for one set of large grazing herbivores might have resulted from a combination of a declining food base and intense predation from both humans and large carnivores. There is considerable anthropological evidence that the native Americans of the central United States depended heavily on white-tailed deer for meat (Smith 1975). So, too, did the wolf and the mountain lion. Wolves and to an extent mountain lions take mostly young of the year and old and sick individuals. Humans, as excavations of old settlements reveal, took mostly individuals one to four years old. Thus humans and wolves were noncompetitive complementary predators that together held deer herds to some level below the carrying capacity of the habitat. If humans, saber-toothed tigers, dire wolves, and others were complementary predators, all of them acting together could have exerted more predatory pressure than the large herbivores, faced with a changing vegetation because of a warming climate, could withstand. As a result the larger mammals disappeared, human populations sharply declined, and large predators became extinct.

■ Summary

Earth is a dynamic planet. Over the course of its existence, it has experienced the breakup of large land masses into continents that drifted apart, the building up and wearing away of mountains, the rise and fall of sea levels, and the advance and retreat of ice sheets. With these physical changes came climatic changes—warming and cooling, aridity and heavy precipitation. Plant and animal species appeared and disappeared as new forms evolved. Patterns of vegetation changed as climates changed.

Some of the most pronounced changes, as evidenced by pollen profiles from bogs and pond and lake bottoms, occurred during the Pleistocene. At that time several advances and retreats of ice sheets eliminated vegetation over much of the northern part of the Northern Hemisphere and pushed plants and animals southward. With each northward retreat of the ice sheet trees moved north. Trees experienced different survival patterns from one glaciation to the next, producing different forest communities during each glacial interval. In Europe tree species found their southward retreats blocked by the glacial ice of the Alps. Many species became extinct, reducing present-day species diversity of the European deciduous forest. In North America no such barriers existed. During the last great ice sheet of the Wisconsin glacial period, the Laurentide, pockets of tundra existed adjacent to the ice sheet, boreal forests extended south to mid-continent, oak-pine forests covered the south, and the species-rich mesic forest was confined to the Mississippi Valley, the Gulf Plain, and other scattered refugia. As the ice sheet retreated, the climate warmed rapidly and trees moved northward. Boreal forest shifted into Canada, replaced to the south by oak-pine forest and mixed deciduous forest. Mesic species moved north and northeastward out of their confined refuges at different rates and arrived at the northern limits of their ranges at different times.

Isolation and environmental changes during the glacial and interglacial stages encouraged the speciation of some forms of plant and animal life and the extinction of others. The nature of vegetational and faunal communities today reflects the evolutionary impact of changing conditions during the Pleistocene.

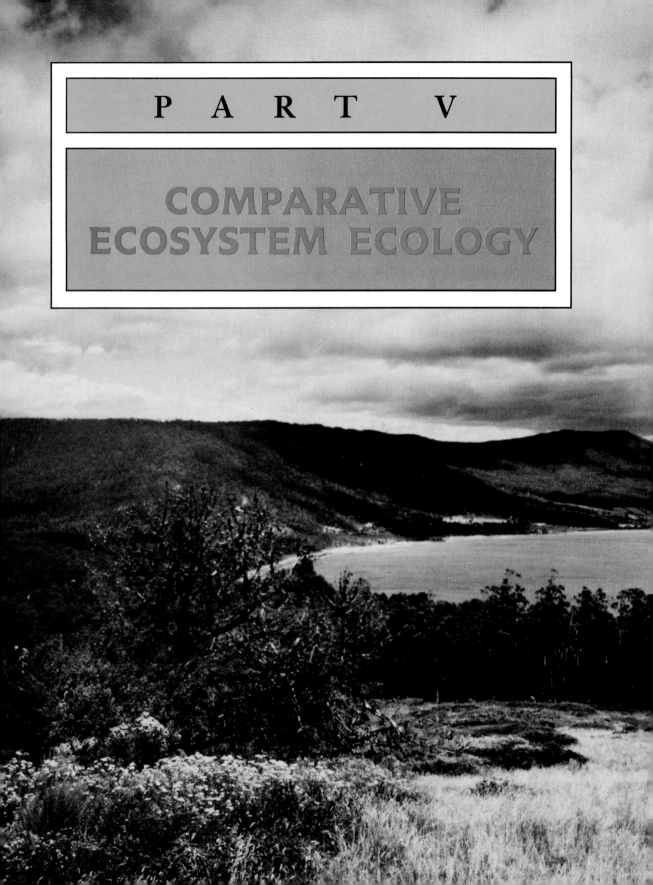

PART V

COMPARATIVE ECOSYSTEM ECOLOGY

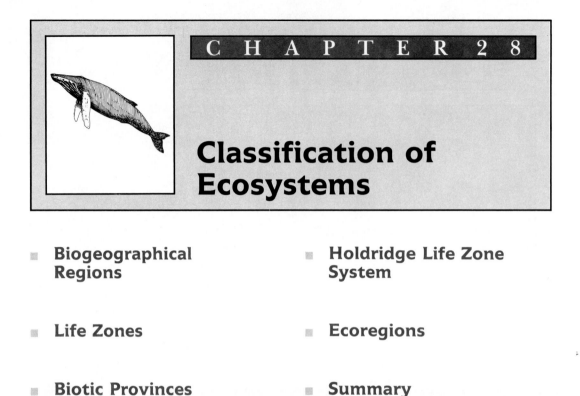

CHAPTER 28

Classification of Ecosystems

- **Biogeographical Regions**

- **Life Zones**

- **Biotic Provinces**

- **Biomes**

- **Holdridge Life Zone System**

- **Ecoregions**

- **Summary**

Human beings have always been interested in the plants and animals around them, but their knowledge for centuries was limited to life in their own immediate area. As adventurous explorers expanded horizons to new lands around the world, they brought back stories and specimens of new and often strange forms of life. As naturalists joined the explorers, they became more and more familiar with plant and animal life around the world, and began to note similarities and differences.

Botanical explorers and zoological explorers looked at the world differently. Botanists in time noted that the world could be divided into great blocks of vegetation—deserts and grasslands, coniferous, temperate, and tropical forests.

These divisions they called *formations,* even though they had difficulty drawing sharp lines between them. In time plant geographers attempted to correlate the formations with climatic differences, and found that blocks of climate reflected the blocks of vegetation and their life form spectra.

At the same time zoogeographers studying the distributions of animals found them much more difficult to map. By the beginning of the twentieth century, naturalists had accumulated the basic facts of worldwide animal distribution. All they needed to do—a great enough task—was to arrange the facts and draw some general conclusions. In 1878 Philip Sclater mapped birds into six regions roughly corresponding to continents. Several different schemes of classification have been devised since then.

■ Biogeographical Regions

The master work in zoogeography was done by Alfred Wallace, who is also known for reaching the same general theory of evolution as Darwin. The *realms* of Wallace, with some modification, still stand today. There are six biogeographical regions, or realms, each more or less embracing a major continental land mass and each separated by oceans, mountain ranges, or desert (inside cover). They are the Palearctic, Nearctic, Neotropical, Ethiopian, Oriental, and Australian. Because some zoogeographers consider the Neotropical and the Australian regions to be so different from the rest of the world, they group the four other regions together to make a more general classification: Neogea (Neotropical), Notogea (Australia), and Metagea (the rest of the world). Each region possesses certain distinctions and uniformity in the taxonomic units it contains, and each to a greater or lesser degree shares some of the families of animals with other regions. Each at one time or another in the history of Earth has had some land connection with another across which animals and plants could pass, Australia more briefly than the others.

Two regions, the Palearctic and the Nearctic, are closely related; in fact, the two are often considered as one, the Holarctic. The Nearctic contains the North American continent south to the Tropic of Cancer. The Palearctic region contains the whole of Europe, all of Asia north of the Himalayas, northern Arabia, and a narrow strip of coastal North Africa. The regions are similar in climate and vegetation. They are alike in their faunal composition, and they share, particularly in the north, similar animals, such as the wolf, the hare, the moose (called elk in Europe), the stag (called elk in North America), the caribou, the wolverine, and the bison.

Below the coniferous forest belt the two regions become more distinct. The Palearctic is not rich in vertebrate fauna, of which few are endemic. Palearctic reptiles are few and usually are related to those of the African and Oriental tropics. The Nearctic, in contrast, is the home of many reptiles and has more endemic families of vertebrates. The Nearctic fauna is a complex of New World tropical and Old World temperate families; the Palearctic is a complex of Old World tropical and New World temperate families.

South of the Nearctic lies the Neotropical, which includes all of South America, part of Mexico, and the West Indies. It is joined to the Nearctic by the Central American isthmus and is surrounded by the sea. Isolated until 15 million years ago, the fauna of the Neotropical is most distinctive and varied. In fact, about half of the South American mammals, such as the tapir and llama, are descendants of North American invaders, whereas the only South American mammals to survive in North America are the armadillo, opossum, and porcupine. Lacking in the Neotropical is a well-developed ungulate fauna of the plains, so characteristic of North America and Africa. The Neotropical, however, is rich in endemic families of vertebrates. Of 32 families of mammals, excluding bats, 16 are restricted to the Neotropical. In addition, 5 families of bats, including the famous vampire, are endemic.

The Old World counterpart of the Neotropical is the Ethiopian, which includes the continent of Africa south of the Atlas Mountains and Sahara Desert and the southern corner of Arabia. It embraces tropical forests in central Africa and savanna, grasslands, and desert in the mountains of East Africa. During the Miocene and the Pliocene, Africa, Arabia, and India shared a moist climate and a continuous land bridge, which allowed the animals to move freely between them. This connection accounts for some similarity in the fauna between the Ethiopian and the Oriental regions. Of all the regions the Ethiopian contains the most varied vertebrate fauna; it is second only to the Neotropical in endemic families.

Lush forests cover much of the Oriental region, which includes India, Indochina, south China, Malaysian, and the western islands of the Malaysian Archipelago. It is bounded by the Himalayas, the Indian Ocean, and the Pacific Ocean. On the southeast corner, where the islands of the Malaysian Archipelago stretch out toward Australia, there is no definite boundary, although a line drawn by Wallace is often used to separate the Oriental from the Australian region. This line runs between the Philippines and the Moluccas in the north, bends southwest between Borneo and the Celebes, then turns south between the islands of Bali and Lombok. A second line, Weber's, has been drawn to the east of Wallace's line; it separates the islands with a majority of Oriental animals from those with a majority of Australian ones. Because the islands between these two lines are a transition between the Oriental and the Australian regions, some zoogeographers call the area Wallacea (see inside cover).

Of the tropical regions the Oriental possesses the fewest endemic species and lacks a variety of widespread families. It is rich in primate species, including two families confined to the region, the tree shrews and the tarsiers.

The most interesting and the strangest region, and certainly the most impoverished in vertebrate species, is the Australian. It includes Australia, Tasmania, New Guinea, and a few smaller islands of the Malaysian Archipelago. New Zealand and the Pacific Islands are excluded, for they are regarded as oceanic islands, separate from the major faunal regions. Partly tropical and partly south temperate, the Australian is noted for its lack of a land connection with other regions, the poverty of freshwater fish, amphibians, and reptiles, the absence of placental mammals, and the dominance of marsupials. Included are the monotremes with two egg-laying species, the duck-billed platypus and the spiny anteater. The marsupials have become diverse and have evolved ways of life similar to those of the placental mammals of other regions.

Life Zones

By the turn of the century some biologists were attempting to combine the regional distributions of both plants and animals into one scheme. C. Hart Merriam, then chief and founder of the United States Bureau of Biological Survey (later to become the Fish and Wildlife Service), proposed the idea of *life zones* (1894a, 1894b). Merriam divided the North American continent into three primary transcontinental regions, the Boreal, the Austral, and the Tropical. Differences among transcontinental belts running east to west, expressed by the animals and plants living there, supposedly are controlled by temperature. The Boreal region extends from the northern polar seas south to southern Canada, with extensions running down the three great mountain chains, the Appalachians, the Rockies, and the Cascade-Sierra Nevada Range. The Austral region embraces most of the United States and a large part of Mexico. The Tropical region clings to the extreme southern border of the United States and includes some of the lowlands of Mexico and most of Central America. Each of these regions Merriam further subdivided into life zones.

The Boreal region he subdivided into three zones. The Arctic-Alpine zone, characterized by arctic plants and animals, lies north of the tree line and includes the arctic tundra as well as those parts of mountains further south that extend above the timber line. The Hudsonian zone, the land of spruce, fir, and caribou, embraces the northern coniferous forest and the boreal forests covering the high mountain ranges to the south. The Canadian zone includes the southern part of the boreal forest and the coniferous forests that cloak the mountain ranges extending south.

The Austral region is split into five zones. First is the Transition zone (called the Alleghenian in the east) which extends across the northern United States and runs south on the major mountain ranges. It is a zone in which the coniferous forest and the deciduous forest intermingle. Extending in a highly interrupted fashion across the country from the Atlantic to the Pacific is the upper Austral zone. It is further subdivided into the Carolinian area in the humid east and the upper Sonoran of the semiarid west. The lower Austral embraces the southern United States from the Carolinas and the Gulf states to California. In the humid southeast it is known as the Austroriparian area and in the arid west as the Lower Sonoran.

Once widely accepted, life zones are rarely used today, although they creep now and then into the literature on the vertebrates. In the first place a life zone is not a unit that can be recognized continent-wide by a characteristic and uniform faunal or vegetational component. For example, the Transition zone includes the hardwoods of the east, the yellow pine of the Rocky Mountains, and the redwoods of California. Thus it covers too many types of vegetation and too many different animals to be useful. The life zones south of the Arctic and Canadian are not transcontinental. The Transition and Upper and Lower Austral of the east are totally different from those of the west. Then the temperatures at times of the year other than the season of growth and reproduction influence the distribution of plants and animals. Nevertheless there is something evocative about the life-zone terminology. The Arctic-Alpine zone recalls the cold, windswept mountains above the timber line; and the name Sonoran, slowly spoken, sings of the sun-baked desert, cactus, mesquite, horned lizards, and roadrunners.

■ Biotic Provinces

A third approach to the subdivision of the North American continent into geographical units of biological significance was the biotic provinces concept, defined and mapped by Dice in 1943. It differs from the others in that a province embraces a continuous geographic area that contains ecological associations distinguishable from those of adjacent provinces, especially at the species and subspecies level. Each biotic province is further subdivided into ecologically unique subunits, districts, or life belts, based largely on altitude, such as grassland belts and forest belts.

Basically the biotic province concept is an attempt to classify the distribution of plants and animals, especially the latter, on the basis of ranges and centers of distribution of the various species and subspecies. The regions themselves and their subdivisions are largely subjective. The boundaries more often than not coincide with physiographic barriers rather than with vegetation types, and the regions never occur as discontinuous geographic fragments. Although a number of species may be confined in some biotic province, others occur over several provinces, because their distribution is determined more by the presence of suitable habitat, which is rarely restricted to a single region. Because the boundaries of biotic provinces and the ranges of subspecies of animals with a wide geographic distribution do coincide, this system is used at times by mammalogists, ornithologists, and herpetologists in the study of a particular group.

Biomes

As attempts at combining plant and animal distribution into one system, all of the above units of classification are unworkable, for plant and animal distributions do not coincide. Another approach, pioneered by Victor Shelford, was simply to accept plant formations as the biotic units and to associate animals with plants. This approach works fairly well, because animal life does depend upon a plant base. These broad natural biotic units are called *biomes* (Figure 28.1). Each biome consists of a distinctive combination of plants and animals in a climax community; and each is characterized by a uniform life form of vegetation, such as grass or coniferous trees. It also includes stages in the development of the community toward its final form, which may be dominated by other life forms. Because the species that dominate the developmental or seral stages are more widely distributed than those of the climax, they are of little value in defining the limits of the biome.

On a local and regional scale, communities are considered as gradients in which the combination of species varies as the individual species respond to environmental gradients. On a larger scale we can consider the terrestrial and even some aquatic ecosystems as gradients of worldwide communities and environments. Such gradients of ecosystems are *ecoclines*.

If we were to sample vegetation and associated animal life along a transect cutting across a number of ecosystems, such as up a mountain slope or north to south across a continent, we would discover that communities change gradually just as species change along community gradients. If we were to run this transect across midcontinent North America beginning in the moist, species-rich forests of the Appalachians and ending in the desert, the transect would follow a gradient of ecosystems on a climatic moisture gradient—from the mesophytic forest of the Appalachians through oak-hickory forests, oak woodlands with grassy understory, tall-grass prairies (now cornland), mixed prairies (wheatland), short-grass plains, desert grasslands, and desert shrublands (Figure 28.2a). Likewise, a transect from southern Florida to the Arctic would move along a climatic temperature gradient from a subtropical forest through temperate deciduous forest, temperate mixed forest, to boreal coniferous forest and tundra (Figure 28.2b).

In addition to gradual changes in vegetation, there are gradual changes in other ecosystem characteristics. As we go from highly mesic and warm temperatures to xeric situations or cold temperatures, productivity, species diversity, and the amount of organic matter decreases. There is a corresponding decline in the complexity and organization of ecosystems, in the size of the plants, and in the number of strata to the vegetation.

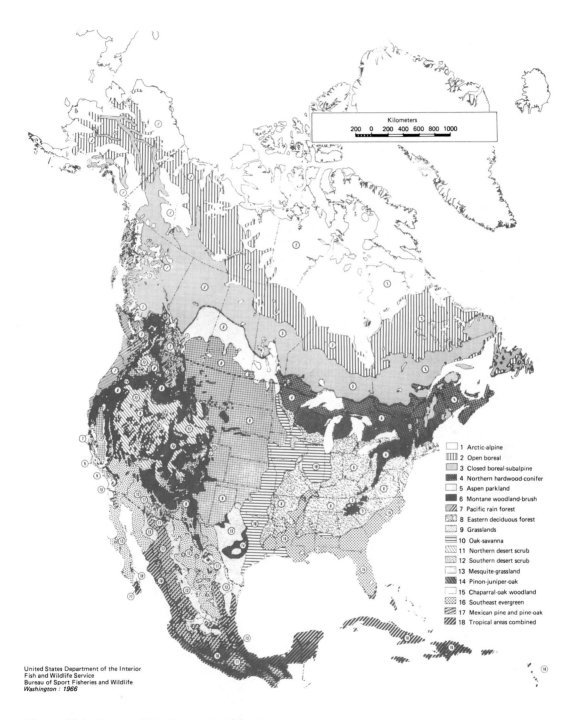

United States Department of the Interior
Fish and Wildlife Service
Bureau of Sport Fisheries and Wildlife
Washington : 1966

1 Arctic-alpine
2 Open boreal
3 Closed boreal-subalpine
4 Northern hardwood-conifer
5 Aspen parkland
6 Montane woodland-brush
7 Pacific rain forest
8 Eastern deciduous forest
9 Grasslands
10 Oak-savanna
11 Northern desert scrub
12 Southern desert scrub
13 Mesquite-grassland
14 Pinon-juniper-oak
15 Chaparral-oak woodland
16 Southeast evergreen
17 Mexican pine and pine-oak
18 Tropical areas combined

Figure 28.1 Biomes of North America. (Map by
John Aldrich, courtesy U.S. Department of Interior, Fish and Wildlife Service.)

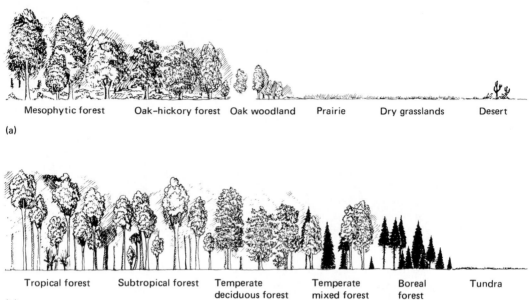

(a)

Mesophytic forest   Oak–hickory forest Oak woodland   Prairie   Dry grasslands   Desert

(b)

Tropical forest   Subtropical forest   Temperate deciduous forest   Temperate mixed forest   Boreal forest   Tundra

Figure 28.2 Gradients of vegetation in North America from east to west and north to south. (a) The east-west gradient runs from the mixed mesophytic forest of the Appalachians through oak-hickory forests of the central states, the ecotone of bur oak and grasslands, to the prairie, short-grass plains, and desert. The transect does not cut across the Rocky Mountains. This gradient reflects precipitation. (b) The north-south gradient reflects temperatures. The transect cuts across tundra, boreal coniferous forest, the mixed mesophytic forests of the Appalachians, the subtropical forests of Florida and Mexico, and the tropical forests of southern Mexico. (After Whittaker 1970: 164.)

Growth forms change. The tropical rain forest is dominated by phanerophytes and epiphytes; the arctic tundra by hemicryptophytes, geophytes, and therophytes. Wherever similar environments exist on Earth, the same growth forms exist, even though the species differences may be great. Thus different continents tend to have communities of similar physiognomy.

The various biomes of the world fall into a distinctive pattern when plotted on a gradient of mean annual temperature and mean annual precipitation (Figure 28.3). The plots are obviously rough. Many types intergrade with one another, and adaptations of various growth forms differ on several continents. Soil and exposure to fire can influence which one of several biomes will occupy an area. Structure of biomes is further influenced by the nature of the climate, whether marine or continental. The same amount of rain, for example, can support either shrubland or grassland.

Holdridge Life Zone System

Another useful approach to classification and study of ecosystems is the Holdridge life zone system (Holdridge 1964, 1967; Holdridge et al. 1971). It is based on the assumptions that (1) mature, stable plant formations represent physiognomically

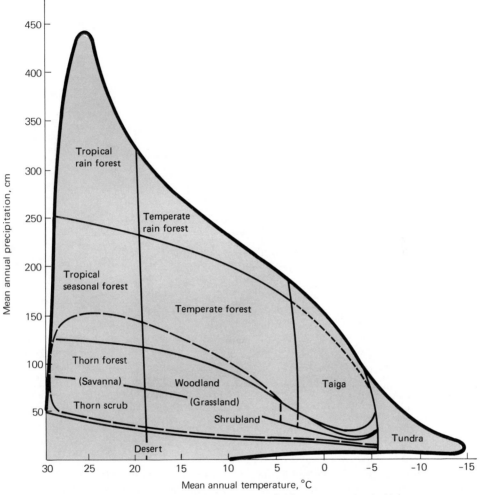

Figure 28.3 Pattern of world plant formation in relation to temperature and moisture. Where climate (maritime versus continental) varies, soil can shift the balance between such types as woodland, shrubland, and grass. The dashed line encloses a wide range of environments in which either grassland or one of the types dominated by woody plants may form the prevailing vegetation in different areas. (After Whittaker 1970: 167.)

discrete vegetation types recognizable throughout the world and (2) geographical boundaries of vegetation correspond closely to boundaries between climatic zones (Holdridge et al. 1971). Vegetation is determined largely by an interaction of temperature and rainfall.

The Holdridge system divides the world into life zones arranged according to latitudinal regions, altitudinal belts, and humidity provinces (Figure 28.4). The boundaries of each zone are defined by

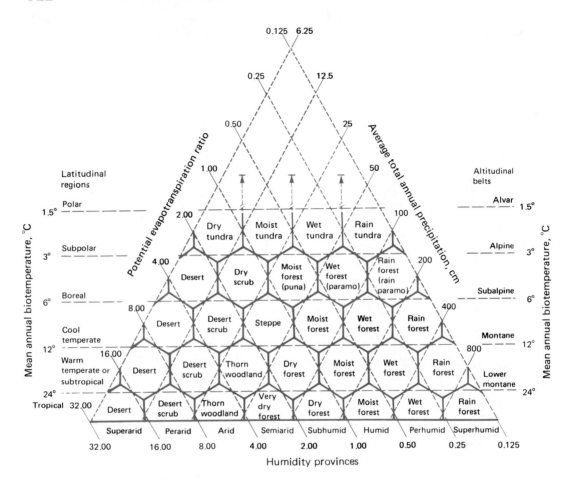

Figure 28.4 The Holdridge life zone system for classifying plant formations. The life zones are determined by a gradient of mean annual biotemperatures by latitude and altitude, potential evapotranspiration ratio, and an average total annual precipitation. (After Holdridge 1967.)

mean annual precipitation and mean annual biotemperature. Altitudinal belts and latitudinal regions are defined in terms of mean biotemperatures and not in the usual terms of latitudinal degrees or meters of elevation.

The Holdridge system is based on three levels of classification: (1) climatically defined life zones; (2) associations, which are subdivisions of life zones based on local environmental conditions; and (3) further local subdivisions based on actual cover or land use. The term *association,* as used by Holdridge, means a unique ecosystem, a distinctive habitat or physical environment and its naturally evolved community of plants and animals.

The subdivision into zones or associations is based on an interaction between environment and vegetation as defined by biotemperature, precipitation, moisture availability, potential evapotranspiration, and potential evapotranspiration ratio.

Biotemperature, as defined by Holdridge, is the mean of unit period temperatures with the substitution of zero for all unit period values below 0° C and above

30° C; temperatures below 0° and above 30° limit growth and physiological activities of plants. Mean biotemperature on a daily basis for an area is calculated by summing the hourly temperatures above 0° C and below 30° C and dividing by 24.

Precipitation is the total amount of precipitation for a region rather than seasonal distribution. Long-term average annual precipitation tends to be partially correlated with its seasonal distribution (in terms of soil moisture availability) and life zone boundaries as reflected in vegetation. This fact leads to a classification at the association level.

Moisture availability is the supply of moisture available for plant growth. A function of the relationship between precipitation and potential evapotranspiration, it determines the biologically significant condition of humidity. For all major components of vegetation, soil rather than direct precipitation supplies water necessary for physiological activity. In any given climatic situation moisture available to plants varies with those soil characteristics that influence filtration, storage, and so on. These variations determine association classifications.

Potential evapotranspiration from land vegetation is essentially a function of biologically positive heat balance or biotemperature. Potential evapotranspiration from land vegetation is a hypothetical figure, not a directly measurable figure against which other moisture values may be compared. It is the amount of water that would be transpired under constantly optimal conditions of soil moisture and plant cover. It is calculated by multiplying mean annual temperature by an experimentally derived constant, 59.93.

Potential evapotranspiration ratio is the ratio of the mean annual potential evapotranspiration to average total annual precipitation. It provides an index of biological humidity conditions. A ratio of 1.00 indicates precipitated moisture is exactly equal to potential evapotranspiration for a long-term average year. As the ratio increases, precipitation is progressively less than water lost and climates are increasingly arid. As the ratio decreases, the climates become progressively more humid.

■ Ecoregions

A recent attempt at the classification of the biotic world is the ecoregion approach, developed largely by R. G. Bailey (1976) of the U.S. Forest Service. It is based on the ecoregion concept of J. M. Crowley (1967). The classification scheme involves a synthesis of climate types, vegetation associations, and soil types into a single geographic, hierarchical classification, which reflects both ecological properties and spatial patterns (Tables 28.1).

An *ecoregion* is a continuous geographical area across which the interaction of climate, soil, and topography are sufficiently uniform to permit the development of similar types of vegetation (Figure 28.5 on page 726). Boundaries between ecoregions, however, are arbitrary, drawn where the associations of the two regions cover approximately the same area. With in an ecoregion certain ecological communities may be similar to those of other regions.

The ecoregion classification relates more to management of forest, range, and wildlife resources than to ecology. Its purpose is to provide a foundation for ecological management of resources. Because each region has its own distinctive flora, fauna, climate, soil, and landform, each requires its own approach to management. The ecoregion approach provides a means of studying the problems of resource management on a regional basis. It

TABLE 28.1 Characteristics of Some First-Order and Second-Order Ecoregions

Domain	Division	Temperature	Rainfall	Vegetation	Soil
Polar	Tundra	Mean temperature of warmest month <10° C	Water deficient during the cold season	Moss, grasses, and small shrubs	Tundra soils (entisols, inceptisols, and associated histosols)
	Subarctic	Mean temperature of summer is 10° C; of winter, −3° C	Rain even throughout the year	Forest, parklands	Podzols (spodosols and associated histosols)
Humid temperate	Warm continental	Coldest month below 0° C, warmest month <22° C	Adequate throughout the year	Seasonal forests, mixed coniferous-deciduous forests	Gray-brown podzolic (spodosols, alfisols)
	Hot continental	Coldest month below 0° C, warmest month >22° C	Summer maximum	Deciduous forests	Gray-brown podzolic (alfisols)
	Subtropical	Coldest month between 18° C and −3° C, warmest month >22° C	Adequate throughout the year	Coniferous and mixed coniferous-deciduous forests	Red and yellow podzolic (ultisols)
	Marine	Coldest month between 18° C and −3° C, warmest month <22° C	Maximum in winter	Coniferous forests	Brown forest and gray-brown podzolic (alfisols)

	Temperature	Precipitation	Vegetation	Soil
Prairie	Variable	Adequate all year, excepting dry years; maximum in summer	Tall grass, parklands	Prairie soils, chernozems (mollisols)
Mediterranean	Coldest month between 18° C and −3° C, warmest month >22° C	Dry summers, rainy winters	Evergreen woodlands and shrubs	Mostly immature soils
Dry — Steppe	Variable, winters cold	Rain <50 cm/yr	Short grass, shrubs	Chestnut, brown soils, and sierozems (mollisols, aridisols)
Desert	High summer temperature, mild winters	Very dry in all seasons	Shrubs or sparse grasses	Desert (aridisols)
Humid tropical — Savanna	Coldest month >18° C, annual variation <12° C	Dry season with <6 cm/yr	Open grassland, scattered trees	Latosols (oxisols)
Rain forest	Coldest month >18° C, annual variation <3° C	Heavy rain, minimum 6 cm/month	Dense forest, heavy undergrowth	Latosols (oxisols)

Note: Names in parentheses in the soil column are soil taxonomy orders (USDA Soil Survey Staff 1975).

Source: Bailey 1978, map.

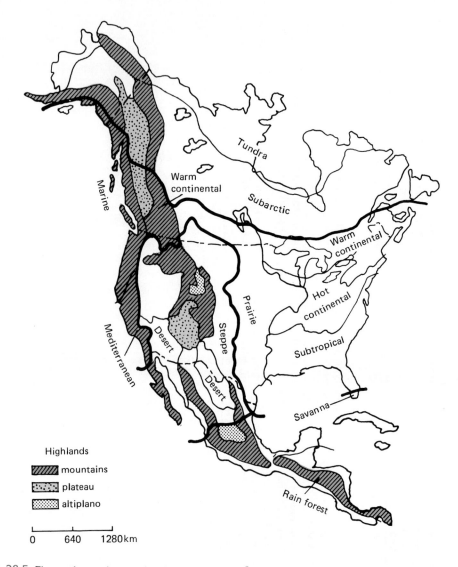

Figure 28.5 First-order and second-order ecoregions of North America. The heavy lines outline the domains: polar, humid, temperate, and humid tropical. The thin lines demark the divisions (see Table 29.1). (From Bailey 1978.)

provides a framework for the organization and retrieval of data gathered in resource inventory data. The concept has been widely accepted in the area of applied ecology.

The classification involves a hierarchy for ecosystems (Table 28.2 on page 728). The largest category is the *domain*, a subcontinental area of broad climatic similarity. Domains are divided into *divisions*, determined by isolating areas of differing vegetation and regional climates. Divisions are divided into *provinces*, which correspond to broad vegetational regions having a uniform regional climate and the same type or types of zonal soils (Figure 28.6). Provinces are subdivided into sections and smaller categories useful in local areas.

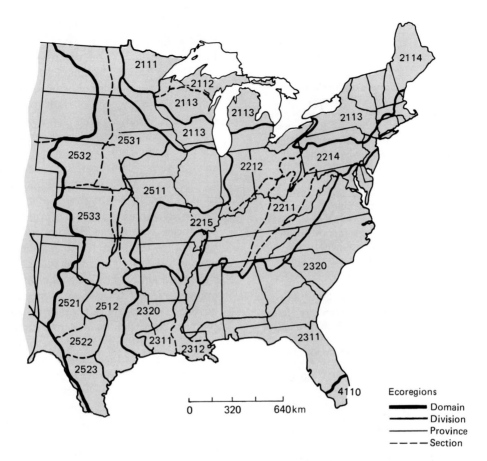

Figure 28.6 The subdivision of ecoregions in eastern and central United States to the edge of the humid temperate domain. (From Bailey 1978.)

2000 Humid temperate domain
  2100 Warm continental division
    2110 Laurentian mixed forest province
      2111 Spruce forest
      2112 Northern hardwood-fir forest
      2113 Northern hardwood forest
      2114 Northern hardwood-spruce forest
  2200 Hot continental division
    2210 Eastern deciduous forest
      2211 Mixed mesophytic forest
      2212 Beech-maple forest
      2213 Maple-basswood forest
      2214 Appalachian oak forest
      2215 Oak-hickory forest

2300 Subtropical division
  2310 Outer coastal plain forest
    2311 Beech-sweet gum-magnolia-pine-oak forest
    2312 Southern floodplain forest
  2320 Southeastern mixed forest
2500 Prairie division
  2510 Prairie parkland province
    2511 Oak-hickory-bluestem parkland
    2512 Oak-bluestem parkland
  2520 Prairie brushland
    2521 Mesquite-buffalo grass
    2522 Juniper-oak-mesquite
    2523 Mesquite-acacia
  2530 Tall-grass prairie
    2531 Bluestem prairie
    2532 Wheatstem-bluestem-needlegrass
    2533 Bluestem-grama
4000 Humid tropical domain
  4100 Savanna division
    4110 Everglade province

TABLE 28.2 A Hierarchy for Ecosystems

Name	Character
1. Domain	Subcontinental area of related climates
2. Division	Single regional climate at the level of Köppen's types (Trewartha 1943)
3. Province	Broad vegetation region with the same type or types of zonal soils
4. Section	Climatic climax at the level of Küchler's potential vegetation types (1964)
5. District	Part of a section having uniform geomorphology at the level of Hammond's land-surface form regions (1964)
6. Landtype association	Group of neighboring landtypes with recurring pattern of landforms, lithology, soils, and vegetation associations
7. Landtype	Group of neighboring phases with similar soil series or families with similar plant communities at the level of Daubenmire's habitat types (1968)
8. Landtype phase	Group of neighboring sites belonging to the same soil series with closely related habitat types
9. Site	Single soil type or phase and single habitat type or phase

Source: From R. G. Bailey 1978, accompanying map; adapted from Crowley 1967 and Wertz and Arnold 1972.

■ Summary

Through the years a number of attempts have been made to classify life into meaningful distributional units. Most of these attempts have been faunistic in approach.

The first division of the world into distributional units was the biogeographical or faunal realms. Three realms are subdivided into six regions. Each region is separated by a barrier of oceans, mountain ranges, or deserts that prevent the free dispersal of animals, and each possesses its own distinctive forms of life. Each region is further subdivided by secondary barriers such as vegetation types and topography.

Various classifications of vegetation and topography within the more general regions have been labeled life zones, biotic provinces, and biomes. The life zone concept, restricted to North America, divides the continent into broad transcontinental belts. The plant and animal differences between them are governed chiefly by temperature. The biotic province approach divides the North American continent into continuous geographic units that contain ecological associations different from those of adjacent units, especially at the species and subspecies level. The biome system groups plants and animals of the world into integral units characterized by distinctive life forms in a climax community, the stage of development at which the community is in approximate equilibrium with its environment. Boundaries of biomes, or major life zones as they are known in Europe, coincide with the boundaries of major plant formations of the world. By including both plants and animals as a total unit that evolved together, the biome permits recognition of the close relationship among all living things. More refined is the Holdridge life zone system, which divides the world into zones defined by mean annual precipitation and mean annual biotemperature.

The more recent ecoregion scheme classifies ecosystems on the basis of large geographical areas in which the interaction of climate, soil, and topography are sufficiently uniform to permit the development of similar types of vegetation. The underlying purpose of this classification is to provide an ecological foundation for the management of regional resources.

Each system has its advantages and disadvantages. All reflect the fact that natural ecosystems are complex and difficult to describe and classify.

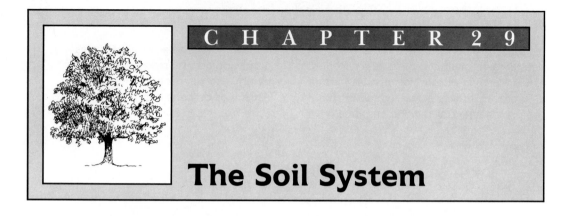

CHAPTER 29

The Soil System

Basic to all terrestrial ecosystems is soil (see Chapter 9). Populated by high numbers of species as well as individuals, the soil embraces another world with its whole chain of life, its predators and prey, its herbivores and carnivores, and its fluctuating populations. Because of their abundance, feeding habits, and ways of life, these small organisms have an important influence on the world above them. Because for all practical purposes it is a community separate from the one above, soil has been considered an ecosystem or biocenose, but it is not. Its energy source comes from dead bodies and feces from the community above (Figure 29.1). It is but a stratum of the whole ecosystem (see Castri 1970; Kuhnelt 1970; Ghilarov 1970).

■ Soil as an Environment

Soil is a radically different environment for life than the one above the surface, yet the essential requirements do not differ. Like animals that live outside the soil, soil fauna require living space, oxygen, food, and water.

To soil fauna, the soil in general possesses several outstanding characteristics as a medium for life. It is relatively stable, both chemically and structurally. Variability in the soil climate is slight compared to above-surface conditions. The atmosphere remains saturated or nearly so, until soil moisture drops below a critical point. Soil

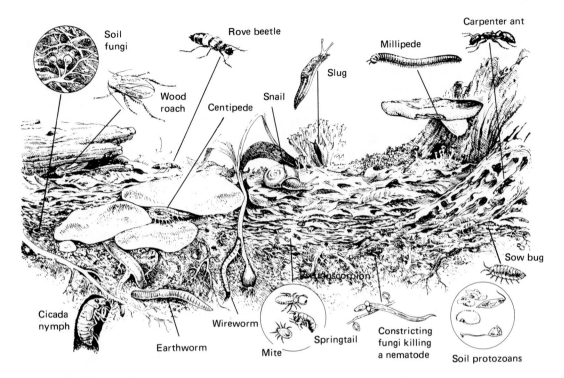

Soil fungi
Rove beetle
Carpenter ant
Millipede
Slug
Wood roach
Snail
Centipede
Cicada nymph
Wireworm
Earthworm
Mite
Springtail
Constricting fungi killing a nematode
Pseudoscorpion
Sow bug
Soil protozoans

Figure 29.1 Life in the soil. This drawing shows only a small fraction of the kinds of organisms that inhabit the soil and litter. Note the fruiting bodies of fungi, which are consumed by animals, invertebrate and vertebrate.

affords a refuge from high and low extremes in temperature, wind, evaporation, light, and dryness. This fact permits soil fauna to make relatively easy adjustments to unfavorable conditions.

On the other hand, soil has low penetrability. Movement is greatly hampered. Except to such channeling species as earthworms, soil pore space is important, for it determines the nature of the living space, humidity, and gaseous condition of the environment. Variability of these conditions creates a diversity of habitats, which is reflected by the diversity of species found in the soil (Birch and Clark

1953). The number of different species, representing practically every invertebrate phylum, found in the soil is enormous.

Only a part of the soil litter is available to most soil animals as living space. Spaces between surface litter, cavities walled off by soil aggregates, pore spaces between individual soil particles, root channels and fissures—all are potential habitats. Most of the soil fauna are limited to pores and cavities larger than themselves. Distribution of these forms in different soils is often determined in part by the structure of the soil (Weis-Fogh 1948), for there is a relationship between the average size of soil spaces and the fauna inhabiting them (Kuhnelt 1950). Large species of mites inhabit loose soils with crumb structure, in contrast to smaller forms inhabiting compact soils. Larger soil species are confined to upper layers, where soil interstices are the largest (Haarløv 1960).

Water in the spaces is essential, because the majority of soil fauna are active only in water. Soil water is usually present as a thin film lining the surface of soil particles. This film contains, among other things, bacteria, unicellular algae, protozoa, rotifers, and nematodes. Most of them are restricted in their movements by the thickness and shape of the water film in which they live. Nematodes are less restricted, for they can distort the water film by muscular movements and thus bridge the intervening air spaces. If the water film dries up, these species encyst or enter a dormant state. Millipedes and centipedes, on the other hand, are highly susceptible to desiccation and avoid it by burrowing deeper into the soil (Kevan 1962).

Excess water and lack of aeration are detrimental to many soil animals. Excessive moisture, which typically occurs after heavy rains, often is disastrous to some soil inhabitants. Air spaces become flooded with deoxygenated water, producing a zone of oxygen shortage for soil inhabitants. If earthworms cannot evade this zone by digging deeper, they are forced to the surface, where they die from excessive ultraviolet radiation. The snowflea (a collembolan) comes to the surface in the spring to avoid excess soil water from melting snow (Kuhnelt 1950). Many small species and immature stages of larger species of centipedes and millipedes may be completely immobilized by a film of water and unable to overcome the surface tension imprisoning them. Adults of many species of these organisms possess a waterproof cuticle that enables them to survive some temporary flooding.

Soil acidity long has been regarded as having an important effect on soil fauna. Because pH is readily measured, it has been overplayed in an attempt to correlate soil characteristics with the fauna. Bornebusch (1930) regarded a pH of 4.5 as in-imical to earthworms; yet some earthworms, such as *Lumbricus rubellus*, are quite tolerant of relatively acid conditions. Although every species of earthworm has its optimum pH, and although some species, such as *Dendrobaenus*, are characteristic of acid conditions, most of them seem to be able to settle in most soils, provided they contain sufficient moisture (Petrov 1946). In northern hardwood forests earthworms are most abundant both in species and in numbers when the pH is between 4.1 and 5.5 (Stegeman 1960). Mites and springtails (Collembola) can exist in very acid conditions (Murphy 1953).

■ Structure

The interrelations of organisms living in the soil are complex, but within the upper layers of the soil energy flows through a series of trophic levels similar to those of surface communities.

The primary source of energy in the soil community is the dead plant and animal matter and feces from the ground layer above. These materials are broken down by the microbial life—bacteria, fungi, protozoans. Upon this base rest phytophagous consumers, which obtain nourishment from assimilable substances of living plants, as do the parasitic nematodes and root-feeding insects; from fresh litter, as do the earthworms; and from exploitation of the soil microflora. Some members of this consumer level, such as some protozoa and freeliving nematodes, feed selectively on the microflora. Others, including most earthworms, pot worms, millipedes, and small soil arthropods, ingest large quantities of organic matter and utilize only a small fraction of it, chiefly the bacteria and fungi, as well as any protozoans and small invertebrates contained within the material. (For a review see Petersen and Luxton 1982.)

On the next trophic level are the predators—the turbellaria, which feed on nematodes and pot worms, the predatory nematodes and mites, insects, and spiders. In such a manner does the community in the soil operate on an energy source supplied by the unharvested organic material of the world above.

Prominent among the larger soil fauna are the Oligochaetes, which include two common families, the Lumbricidae (earthworms) and the Enchytraeidae (white or pot worms). The small, whitish pot worms abound in the upper 8 cm of the soil if humidity is fairly constant. They are able to live under a greater variety of conditions than earthworms, but their numbers undergo violent fluctuations. Populations are at a maximum in winter and at a minimum in summer. They are not extensive burrowers and appear to divide the earth and humus more finely than earthworms. Little is known about their feeding biology beyond that they ingest organic debris, from which they may digest bacteria, protozoans, and other microorganisms (C. O. Nielson 1961).

Earthworm activity in the soil consists mainly of burrowing, of ingestion and partial breakdown of organic matter, and of subsequent egestion in the form of surface or subsurface casts. Ingested soil is taken during burrow construction, mixed with intestinal secretions, and passed out either as aggregated castings on or near the surface or as a semiliquid in intersoil spaces along the burrow. Earthworms pull organic matter into their burrows and ingest some of it; it is then partially or completely digested in the gut. Casts of soil passed through the alimentary canal contain a larger proportion of soil particles less than 0.005 cm in diameter than uningested soil and a higher total nitrogen, organic carbon, exchangeable calcium and magnesium, available phosphorus, and pH.

Surface casting and burrowing slowly overturn the soil. Subsurface soil is brought to the top and organic matter is pulled down and incorporated with the subsoil to form soil aggregates. These aggregates result in a more open structure in heavy soil and bind particles of light soil together.

Millipedes probably are the next most important group of litter-feeders. They and their somewhat similar associates, the centipedes, are essentially animals of the woodland floor. Millipedes occupy three woodland habitats: the floor and aerial parts of vegetation, the litter and upper soil layer, and the areas beneath bark and stones and in rotten logs and stumps. The three most common forms are glomerids, or oval pill millipedes; flat-backed polydesmids with flattened lateral expansions; and large iuloids. The glomerids and polydesmids are not adapted to burrowing and must find refuge against both floods and drought in surface retreats. Iuloids, however, burrow extensively in the soil. Millipedes ingest leaves, particularly those in which some fungal decomposition has taken place; for lacking the enzymes necessary for the breakdown of cellulose, they live on the fungi contained within the litter. Different species of millipedes ingest varying quantities of litter, depending upon the tree species (van der Drift 1951). *Iulus* consumes more red oak litter, *Cylindroiulus* more pine.

The chief contribution of millipedes to soil development and to the soil system is the mechanical breakdown of litter, making it more vulnerable to microbial attack, especially by the saprophytic fungi.

Litter-feeders of importance are snails and slugs, which among the soil invertebrates possess the widest range of enzymes to hydrolyze cellulose and other plant polysaccharides, possibly even lignin (C. O. Nielsen 1962). In Australian rain forests amphipods are a conspicuous part of the fauna and play a major part in the disintegration of the leaf litter (Birch and Clark 1953).

Not to be ignored are termites (Isoptera), white, wingless, social insects. The termite, together with some dipteran and beetle larvae, is the only larger soil inhabitant that is able to break down the cellulose of wood. It accomplishes this task with the aid of a symbiotic protozoan living in its gut. The termite has a mouth structure adapted to ingest wood; the protozoan produces the enzymes that digest cellulose into the simple sugars that the termite can use. Together, the two organisms function perfectly. Without the protozoan, the termite could not exist; without the termite the protozoan could not gain access to wood.

Termites do not play a major role in the temperate soils, but in the tropics they dominate the soil fauna. In these regions they are responsible for the rapid removal of wood and other cellulose-containing materials, twigs, leaves, dry grass, structural timbers, and so on from the surface. In addition to removal of organic matter, termites are important soil churners. They move considerable quantities of soil, perhaps as much as 12.500 t/ha, in constructing their huge and complex mounds. In semidesert country the openings and galleries of subterranean termites allow the infrequent rains to penetrate deep into the subsoil rather than to run off the surface (Kevan 1962).

Of all soil animals, the most abundant and widely distributed are the mites (Acarina) and the springtails (Collembola). Both occur in nearly every situation where vegetation grows, from tropical rain forest to tundra. Flattened dorsoventrally, they are able to wiggle, squeeze, and even digest their way through tiny caverns in the soil. Here they browse on fungi or search for prey in the dark interstices and pores of the organic mass.

The more numerous of the two, both in species and in numbers, are the mites, tiny, eight-legged arthropods from 0.1 to 2.0 mm in size. The most common mites in the soil and litter are the Orbatei. In the pine-woods litter of Tennessee, for instance, they make up 73 percent of all the litter mites (Crossley and Bohnsack 1960). These mites live largely on fungal hyphae that attack dead vegetation as well as on the sugars digested by this microflora from evergreen needles.

The Collembola are the most generally distributed of all insects. Typically, they are brightly colored or completely white. Their common name, springtail, is descriptive of the remarkable springing organ at the posterior end, which enables them to leap comparatively great distances. The springtails are small, from 0.3 to 1 mm. They consist of two groups: the round springtails, or Symphypleona, and the long springtails, or Arthropleona. Neither has specialized feeding habits. They consume decomposing plant materials, largely for the fungal hyphae they contain.

Small arthropods are the principal prey of spiders, beetles (especially the Staphylinidae), the pseudoscorpions, the mites, and the centipedes. The centipedes are one of the major invertebrate predators. The two most common groups are the nonburrowing type and those that burrow, earthwormlike, into the soil. Predacious Mesostigmata mites prey on herbivorous mites, nematodes, enchytraeid worms, small insect larvae, and other small soil animals.

Most of the microorganisms of the soil, the protozoans and rotifers, myxobacteria and nematodes, feed on bacteria and algae. Nematodes are ubiquitous, found wherever their need for a film of water in which to move is met. Soil and freshwater nematodes form one ecological group, with many species in common. In the soil they exist at much higher densities than in fresh water, up to 20 million/m^2. They are most abundant in the upper 5 cm in the vicinity of roots, where they feed on plant juices, soil algae, and bacteria. A few are predaceous.

These bacteria-feeders and algae-feeders, in turn, are consumed by various predacious fungi. Among them there are three groups: (1) the Zoopagales, an order of Phycomycetes that preys chiefly on protozoans, although a few species prey on nematodes; (2) the endozoic Hyphomycetes; and (3) the ensnaring Hyphomycetes. The latter two capture and digest nematodes, crustaceans, rotifers, and to an extent, protozoans (Maio 1958; Doddington, in Kevan 1955). Zoopagales possess sticky mycelia, which capture the prey like flypaper. Endozoic Hyphomycetes release spores, which stick to the integument of nematodes. Germ tubes penetrate the tube of the animal and develop into internal mycelium.

The nematode-trapping Hyphomycetes possess unique morphological adaptations that enable them to capture their prey. One of the most common forms of traps is a network of highly adhesive loops, which catch and hold nematodes on contact. Others possess sticky, knoblike processes to which the nematodes adhere. The most unusual of all is the rabbit-snare trap, of which there are two types, nonconstricting and constricting. Both possess rings of filaments attached by short branches to the main filament. Each ring trap consists of three curved cells; and its inside diameter is just large enough to permit a nematode attempting to pass

through to become wedged and unable to withdraw. In the constricting ring the friction of the nematode's body stimulates the ring cells to inflate to about three times their former volume and to grip the nematode in a stranglehold. The response is rapid: complete distention of the cells is accomplished within one-tenth of a second.

Other groups of animals, although feeding largely on the surface and contributing little to litter breakdown, are important as soil mixers. Ants are especially important as soil animals, for they are widely distributed, pioneer new sites, and bring up large quantities of soil from below ground. Prairie dogs raise earth from lower levels and deposit it at the surface, where it is broken down by weathering and incorporated with organic material. They carry surface soil down to plug passageways, and on clay soils they increase the proportion of fine soil particles on the surface. Moles, too, move considerable quantities of earth, although the amount has not been calculated. Their varied influences include improving the natural drainage and aeration of soil and increasing organic matter by burying surface vegetation and litter under their hills.

■ Function

The soil community is largely heterotrophic, dependent upon energy fixed by the green plants it supports (Figure 29.2). Eighty to 90 percent of the energy bound in the litter and available to the community is captured by the microfloral decomposers, mostly the fungi, the most important functional group in the soil community. The remaining 10 to 20 percent is divided among the numerous and highly diverse groups of soil fauna.

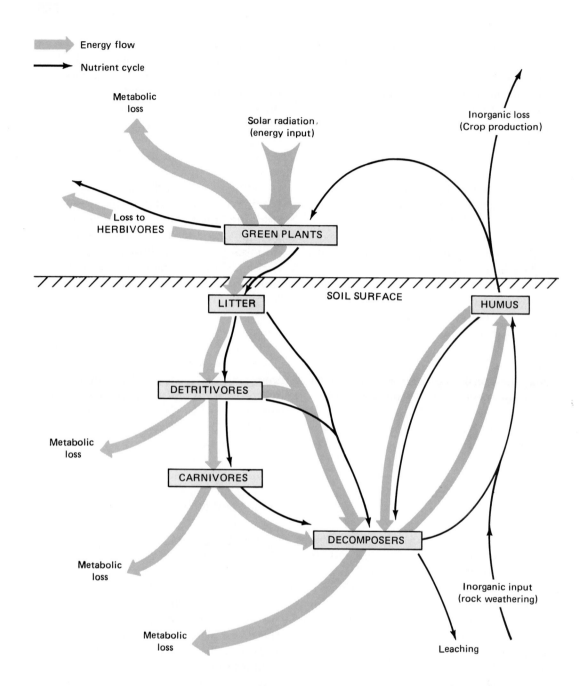

Figure 29.2 Energy flow and nutrient cycling in the soil ecosystem. Note that the soil ecosystem, if it can be properly called one, is heterotrophic, with its food web dependent upon the auto- trophic community above it. This diagram emphasizes the role of soil as the site of decomposition and nutrient exchange. (Adapted from Wallwork 1973.)

How that energy is distributed is largely unknown, and because of the complexity of studying the soil fauna, it will probably remain unknown for a long time. To discover how energy flow is partitioned we need to know at least population density, biomass, fluctuations in both, and feeding, assimilation, and respiration rates for each population. Engelmann (1961) investigated the functional role of soil arthropods, especially the herbivorous orbatid mites in an old field community. By gathering data on biomass, respiration, and caloric flow, he estimated that in 1 m^2 of soil 12.5 cm deep the mites consumed 10,248 calories of food a year and assimilated 2085 calories or 20 percent of the food ingested (Figure 29.3). Respiration accounted for 96 percent of the energy assimilated, leaving little for production. Assuming the population was in a steady state, Engelmann found that the mite biomass was replaced each year. The main role of the orbatid mites was to control the fungal and bacterial populations breaking down the dead litter.

Figure 29.3 Annual energy budget for a mite population in a square meter of an old field grassland in Michigan. (From Engelmann 1961.)

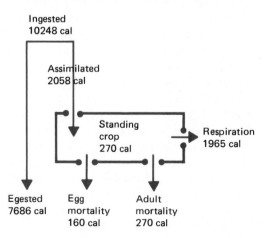

Feeding upon the soil herbivores are soil carnivores. Less is known of the flow of energy from herbivores to carnivores than from detritus to herbivores. Engelmann attempted to estimate this flow. He restricted his analysis to several groups of herbivores, two groups of carnivorous mites, and Japygidae (primitive, wingless insects with forcepslike structures at the caudal end of the body), which may be omnivorous or carnivorous. Using data on respiration rates for the various groups of herbivores and carnivores, Engelmann estimated ecological efficiency, the flow of energy from herbivores to carnivores, to range between 8 and 30 percent.

The role of soil arthropods in nutrient cycling is difficult to estimate and more difficult to follow, although the basic pattern is similar to that of aboveground cycling (Crossley 1976; Petersen and Luxton 1982). Raw litter is fragmented by wind and trampling and by the activity of macroarthropods, which process 20 to 30 percent of annual dead organic matter input to the soil. The increased surface area speeds the leaching of soluble nutrients by rainwater and makes the litter more accessible for colonization by bacteria, fungi, and microarthropods (Figure 29.4). Soil fauna consume microbes and dead organic matter, incorporating considerable amounts of nutrients into biomass and accelerating mineralization by converting part of the litter to feces and body secretions. Saprovores concentrate all elements except potassium at levels above that found in the substrate; and fungi, rich in nitrogen, calcium, and sodium, become a major source of those nutrients for saprovores and fungivores. The latter consume a high proportion of fungal net production. Total invertebrate production accounts for 8.1 percent of calcium in litterfall, 10 percent of potassium, and 15.6 percent of manganese, not a significant

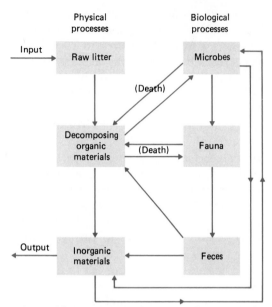

Figure 29.4 A model of nutrient flow through the soil decomposition system. Input is raw litter; output is inorganic materials. Interactions among various compartments keep nutrients cycling within the system. (From Crossley 1977: 53.)

contribution to the cycling of those nutrients. Saprovores appear to be less efficient than fungivores in cycling nutrients, but fungivores consume nutrient-rich fungal hyphae (Table 29.1).

With nitrogen and phosphorus the situation is different (Table 29.1). Soil invertebrates immobilize about 70 percent of N in the litter, much of which they release at the beginning of the growing season. McBrayer (1976) has estimated that biomass of soil invertebrates holds approximately 4.68 g/m^2 of nitrogen and that the soil fauna releases 0.91 g/m^2 of N as ammonia in early spring when the populations experience a sharp decline. This amount is equivalent to 14 percent of N in litterfall made available quickly at the onset of early spring herbaceous growth.

For phosphorus the situation is more dramatic. Total annual production of soil invertebrates utilizes an equivalent of 118 percent of P contained in litterfall, which implies tight internal cycling; but the spring decline in populations of soil invertebrates releases 1.38 mg/P/m^2 or 22 percent of P returned as litter. It seems that for N and P, two critical elements in ecosystem functioning, soil invertebrates have an important input in nutrient cycling. Overall, soil fauna appear to provide a chemical homeostasis mechanism that prevents excessive fluctuations in nutrient availability through the seasons (Petersen and Luxton 1982; Ingham et al. 1985).

■ Soil Mistreated

Although topsoil is the foundation upon which all terrestrial life and human welfare and survival depends, we treat the soil like dirt. Civilizations have risen and fallen on the exploitation and poor management of topsoil. Clearing of forests, slash-and-burn and tillage agriculture, overgrazing, and road building, dam building and construction of all kinds have exposed our essential soil resource to the ravages of wind and water.

Disturbance of the topsoil by tillage and other activities expose it to erosion. Stripped of its protective vegetation and litter by plow, ax, grazing, and major construction, soil is removed by wind and water faster than it can be formed. Loss of the upper layers of humus-charged, granular, highly absorptive topsoil exposes the humus-deficient, less stable, less absorptive, and highly erodable layers beneath. If the subsoil is clay, it absorbs water so slowly that heavy rains produce a highly abrasive and rapid runoff. Soil of the uplands ends up in muddied rivers where it settles out in dams, shortening their useful life, and builds up in river deltas.

TABLE 29.1 Elemental Fluxes (g/m²/yr) Through a Cryptozoan Food Web

Element		Saprovores	Fungivores	Predators	Sum
Ca	C	.976	6.292	3.762	11.03
	E	.101	1.178	.740	2.02
	E/C	.10	.19	.20	—
Mg	C	.085	.362	.447	.89
	E	.003	.137	.095	.24
	E/C	.03	.38	.21	—
K	C	.087	.229	.359	.68
	E	.001	.119	.053	.17
	E/C	.02	.52	.15	—
Na	C	.004	.076	.479	.56
	E	.001	.179	.110	.29
	E/C	.24	1.0+	.23	—
N	C	.525	5.338	8.211	14.07
	E	.076	2.309	2.293	4.68
	E/C	.14	.43	.28	—
P	C	.039	.458	1.377	1.87
	E	.008	.415	.314	.74
	E/C	.20	.91	.23	—
S	C	.100	.763	.424	1.29
	E	.004	.119	.118	.24
	E/C	.04	.16	.28	—

Note: C is ingestion; E is losses ($\cong$ productivity).

Bare soil, finely divided, loose, and dry, as it often is after tillage, is ripe for wind erosion. Very fine particles of dust are picked up by the wind and carried on as dust clouds, as occurred during the Dust Bowl days in the Great Plains during the drought years of the 1930s and is happening on an increasing scale today. Often dust particles are lifted high in the atmosphere and carried for hundreds and even thousands of miles.

Erosion by wind and water impoverishes the land. It carries away organic layers, exposes the subsoil, depletes nutrients, changes soil structure, deposits soil elsewhere, increases runoff, and causes land ruin and abandonment. Land abandoned because of mismanagement is usually so degraded that natural vegetation has difficulty colonizing the area. Erosion worsens, gullies deepen, and conditions become progressively worse, unless drastic steps are taken to stop erosion and restore vegetation and rebuild the living soil.

■ Summary

Terrestrial ecosystems are supported by the living soil, an underground system influenced by and influencing the terrestrial community above. Organisms in the soil, like all others, reflect their environment.

Their abundance and composition in an area depend upon the physical nature of the soil, its nutrient status, the vegetation present, the kind of litter the vegetation produces, and the ability of plants to return calcium and other nutrients to the soil. Soil fauna influence the future development of upper soil layers.

Life in the soil requires the same essentials as aboveground life: living space, oxygen, food, and water. Soil furnishes a relatively stable environment, chemically, physically, and structurally. Important in the soil environment, influencing both diversity of belowground habitats and species, are pore spaces, which provide living space for soil organisms and affect their movement. Water film about pore spaces provides habitats for microfauna; yet at the same time excess water can be detrimental to soil fauna, flooding pore spaces and reducing oxygen.

The soil system is heterotrophic, depending on organic detritus from above. This energy and nutrient source supports complex food webs belowground with their own components of predators, parasites, and mutualists. These food webs tie in with aboveground food webs, linking belowground to aboveground systems.

The direct decomposition of plant litter is accomplished by microflora, bacteria, and fungi. Soil invertebrate fauna make organic matter more readily available to microflora by breaking down litter mechanically, by exposing new areas for fungal invasion, by spreading fungal spores through their feces, and by increasing surface area exposed to attack by bacteria and fungi. Soil fauna also consume great quantities of fungi and depress bacterial and fungal populations. Predaceous species in turn influence the population levels of litter-feeding and decomposer organisms. Such is the chain of life in the world beneath the ground.

Grasslands, Savannas, Shrublands, and Deserts

Grasslands

When the explorers looked out across the prairies for the first time, they witnessed a scene they had never before experienced. Nowhere in all western Europe had they seen anything similar. Lacking any other name to call them, the explorers named these grasslands "prairie," from the French, meaning "grassland." This land was the North American prairie and plains, the climax grassland that occupied the midcontinent (Figure 30.1). It was one of several great grassland regions in the world, including the steppes of Russia, the pusztas of Hungary, the South African veld, and the South American pampas (see inside cover). In fact, at one time grasslands covered about 42 percent of the land surface of the world, but today much of it is under cultivation. All have in

Figure 30.1 The extensive, unbroken grasslands of North America supported vast herds of bison and other ungulates such as the pronghorn antelope, and their associated predators. The grasslands of East Africa in places still support great herds of a diverse group of ungulates. (Photo courtesy South Dakota Fish and Game Commission.)

common a climate characterized by high rates of evaporation, periodic severe droughts, a rolling-to-flat terrain, and animal life that is dominated by grazing and burrowing species. They occur largely where rainfall is between 25 and 75 cm/yr, too light to support a heavy forest growth and too heavy to encourage a desert. Grasslands, however, are not exclusively a climatic formation. Most of them require periodic fires for maintenance and renewal and for the elimination of incoming woody growth.

Types of Grasses

Grasses that make up the haylands, pastures, and prairies are either sod formers or bunch grasses. As the names imply, the former develop a solid mat of grass over the ground, and the latter grow in bunches (Figure 30.2), the space between which is occupied by other plants, usually

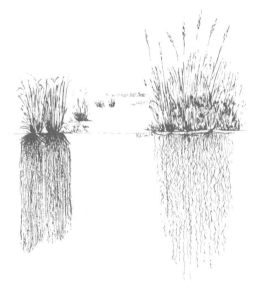

Figure 30.2 Growth forms and root penetration (maximum depth of about 2.5 m) of a sod grass (right) and a bunch grass (left).

herbs. Orchardgrass, broomsedge, crested wheatgrass, and little bluestem are typical bunch grasses, which form clumps by the erect growth of all shoots and spread at the base by tillers. Sod-forming grasses, which include such species as Kentucky bluegrass and western wheatgrass, reproduce and spread by underground stems. Some grasses may be either sod or bunch, depending upon the local environment. Big bluestem will develop a sod on a rich, moist soil and form bunches on a dry soil.

Associated with grasses are a variety of legumes and forbs. Cultivated haylands and pastures usually are planted to a mixture of grasses and such legumes as alfalfa and red clover. With them may grow unwanted plants, such as mustard, dandelion, and daisy. Seral grasslands often consist of a mixture of native grasses, such as timothy and bluegrass, and an assortment of herbaceous plants, including cinquefoil, wild strawberry, daisy, dewberry, and goldenrod. On the prairie legumes and forbs, particularly the composites, are important components of the climax grassland (J. E. Weaver 1954). From spring to fall the color and aspect of the grassland change from pasqueflower and buttercups to goldenrod.

Types of Grasslands

Tame and Successional Grasslands. Grasslands in normally forested regions are either tame or successional. In highly developed agricultural areas, such as eastern and central North America, Great Britain, and Europe, tame or cultivated grasslands are the major representatives of their class, although a few natural types do exist. By clearing the forests, humans developed tame grasslands principally as a source of food for livestock. In some agricultural regions, especially New England and the Lake states, grasslands were

planted and then abandoned to revert to forest. In other regions, especially Britain, some grasslands have existed for centuries, becoming a sort of climax community supporting its own distinctive vegetation (see Duffey et al. 1974).

Tame grasslands can be classified as *permanent*, in grass over seven years and managed for hay or pasture; *temporary* or *rotational*, plowed every three to five years for crop production; and *rough*, marginal, unimproved, semiwild lands used principally for grazing. Many seral or successional grasslands fit the last category.

Ecologically permanent hay fields and grazing lands differ from rotational hay fields. Permanent haylands, more common in Britain than in North America, and permanent grazing lands consist of species adapted to periodic defoliation by cutting and grazing. They consist chiefly of hemicryptophytes that produce their maximum leaf area in the early part of the season.

Rotational or temporary hay fields are dominated by two or three cultivated species, usually two grasses and a legume. Such hay fields are denser and ranker in growth than are permanent and seral grasslands. Management consists of fertilization, mowing, and plowing at regular intervals for growing other crops in the rotation. Such hay fields can provide an excellent habitat for grassland wildlife, but early season mowing destroys cover at the beginning or the height of the nesting season and exposes the ground surface to high solar radiation in late spring and early summer.

Tall-Grass Prairie. Tall-grass prairie occupies, or rather occupied, a narrow belt running north and south next to the deciduous forest (see Figure 28.1). In fact, it was well developed within a region that could support forests. Oak-hickory forests did extend into the grasslands along streams and rivers, on well-drained soils, sandy areas, and hills. Prairie fires, often set by Indians in the fall, stimulated a vigorous growth of grass and eliminated the encroaching forest. When settlers eliminated burning, oaks invaded and overtook the grasslands (Curtis 1959).

Big bluestem was the dominant grass of moist soils and occupied the valleys of rivers and streams and the lower slopes of the hills. Associated with bluestem were a number of forbs, goldenrods, compass plants, snakeroot, and bedstraw. Although grasses dominated the biomass, they were not numerically superior. Studies on remnant prairies in Wisconsin (Curtis 1959) show the legumes comprised 7.4 percent of all species, grasses 10.2, composites 26.1. The high percentage of nitrogen-fixing legumes accounts in part for the annual production of 8500 k/dry matter/ha.

Drier uplands in the tall-grass country were dominated by the bunch-forming needlegrass, side-oats grama, and prairie dropseed. Like the lowland, the drier prairie contained many species other than grass. In Wisconsin composites accounted for 27.5 percent of all species, butterfly-weed and legumes 4.6 percent each, and grasses 13.7 (Curtis 1959). The suggestion has been made that perhaps the xeric prairie might be more appropriately called "daisy-land."

Mixed-Grass Prairie. West of the tall-grass prairie region is the mixed-grass prairie, in which midgrasses occupy the lowland and short grasses the higher elevations. The mixed prairie, typical of the northern Great Plains, embraces largely the needlegrass-grama grass community, with needlegrass-wheatgrass dominating gently rolling soils of medium texture (Coupland 1950). Because the mixed prairie is characterized by great annual ex-

tremes in precipitation, its aspect varies widely from year to year. In moist years midgrasses are prevalent, whereas in dry years short grasses and forbs are dominant. The grasses here are largely bunch and cool-season species, which begin their growth in early April, flower in June, and mature in late July and August.

Short-Grass Plains. South and west of the mixed prairie and grading into the desert are the short-grass plains or steppe, a country too dry for most midgrasses. The short-grass plains reflect a climate where the rainfall is light and infrequent (25 to 43 cm in the west, 51 cm in the east), the humidity low, the winds high, and the evaporation rapid. Shallow-rooted, the short grasses utilize moisture in the upper soil layers, beneath which is a permanent dry zone into which the roots do not penetrate. Sod-forming blue grama and buffalo grass dominate the short-grass plains. On wet bottomlands, switchgrass, Canada wild rye, and western wheatgrass replace grama and buffalo grass. Because of the dense sod, fewer forbs grow on the plains, but prominent among them is purple lupine.

Just as the tall-grass prairie was destroyed by the plow, so has much of the short-grass plains area been ruined by overgrazing and by plowing for wheat, which, because of low available moisture, the land could not support. Drought, lack of a tight sod cover, and winds turned much of the southern short-grass plains into the Dust Bowl, the recovery from which has taken years.

Desert Grassland. From southeastern Texas to southern Arizona and south into Mexico lies the desert grassland, similar in many respects to the short-grass plains except that triple-awn grass replaces buffalo grass (Humphrey 1958). Composed largely of bunch grasses, the desert grasslands are widely interspersed with other vegetation types, such as oak savanna and mesquite. The climate is hot and dry. Rain falls only during two seasons, summer (July and August) and winter (December to February), in amounts that vary from 30 to 41 cm in the western parts to 51 cm in the east; but evaporation is rapid, up to 203 cm/year. Vegetation puts on most of its annual growth in August. Annual grasses germinate and grow only during the summer rainy season, whereas annual forbs grow mostly in the cool winter and spring months.

Annual Grassland. Confined largely to the Central Valley of California is annual grassland. It is associated with mediterranean-type climate, characterized by winter precipitation and hot, dry summers. Growth occurs during early spring and most plants are dormant in summer, turning the hills a dry tan color accented by the deep green foliage of scattered California oaks. The original vegetation was perennial grasses dominated by purple needlegrass (*Stipa pulchra*), but since settlement these grasses have been replaced by vigorous annual species well adapted to mediterranean-type climate. Dominant species are wild oats (*Avena fatua*) and slender oats (*Avena barbara*) grass.

Tropical Grasslands. Around the world is a belt of tropical monsoon grasslands that extend from western Africa to eastern China and Australia. Within this belt monsoon grasslands fall into ecoclimatic gradients of arid to semiarid grasslands; medium rainfall grasslands found mainly in India, Burma, and northern Australia; the high-rainfall monsoonal or equatorial grasslands of southeast Asia; and grasslands whose species are adapted to a hot monsoonal summer and cool-to-cold winters (Whyte 1968). The tropical grasslands of South America do not fall into this group because geographical conditions do not promote true or false monsoonal conditions. Instead the grasslands of Latin America are largely steppe, consisting almost entirely of bunch grass with no legumes and very few herbs, bushes, or trees (McIlroy 1972).

Much tropical grassland exists because of fires that prevent the intrusion of woody vegetation. Grass in turn reacts to burning by putting out new shoots and drawing on reserves of moisture and food in the rhizomes and roots, which then are depleted before the arrival of rain. If the grass is overgrazed, the plants are weakened and deteriorate and may be replaced by annual or unpalatable species. On the other hand, the elimination of fire is just as disastrous.

There are a number of differences between tropical and temperate grasses (Steward 1970). Tropical grasses are lower in crude protein and higher in crude fibers. Tropical grasses have maximum photosynthesis at 30° to 35° C, temperate grasses at 15° to 20° C. Tropical grasses have maximum photosynthetic rates of 50 to 70 mg/hr, compared with 20 to 30 mg/hr for temperate grasses. Individual leaves of temperate grasses reach light saturation at relatively low levels, whereas leaves of tropical grasses become saturated only at much higher levels. Temperate grasses have high rates of photorespiration; tropical plants do not.

Structure

Vegetation. Grasslands possess essentially three strata—roots, ground layer, and herbaceous layer (Figure 30.3). The root layer is more pronounced in grasslands than in any other major community. Half or more of the plant is hidden be-

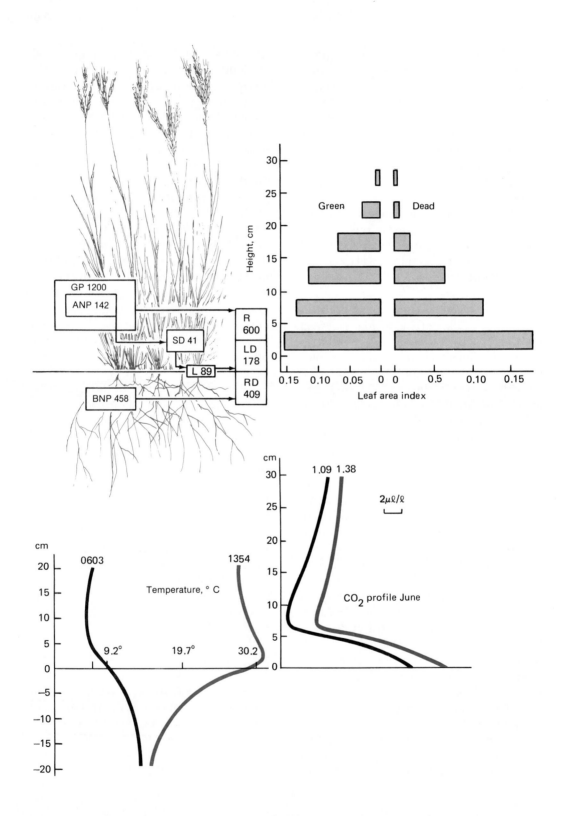

GP 1200
ANP 142

SD 41

L 89

BNP 458

R 600

LD 178

RD 409

Height, cm

30
25
20
15
10
5

Green   Dead

0.15 0.10 0.05 0   0   0.5   0.10 0.15
Leaf area index

cm
30
25
20
15
10
5
0

1.09 1.38

2μℓ/ℓ

CO_2 profile June

cm
20   0603   1354
15
10
5   9.2°   19.7°   30.2
0
−5
−10
−15
−20

Temperature, ° C

Figure 30.3 Profile of a grassland, showing energy flow, structure, and stratification of the physical environment. The energy flow diagram (upper left) indicates primary production in an ungrazed short-grass prairie. Figures in the compartments represent net production during the growing season in g/m^2. ANP = aboveground net production; BNP = belowground net production; SD = standing dead; R = roots; L = litter; RD = root disappearance; LD = litter disappearance. (Data from Sims and Singh 1971.) Bars in the graph (upper right) indicate leaf area indexes at different levels for green and dead plant structures. Note the differences. In the temperature profile (bottom left) the actual temperature at soil surface is shown on each curve. In the CO_2 profile (bottom right) note the concentration of CO_2 near the soil surface, indicating flux from soil, litter, and lower canopy. (Adapted from Ripley and Redmann 1977.)

neath the soil; in winter almost the total grass plant is hidden, a sharp contrast to the leafless trees of the forest. The bulk of the roots occupy rather uniformly the upper 1.6 dm of the soil profile and decrease in abundance with depth. The depth to which the roots of grasses extend is considerable. Little bluestem reaches 1.3 to 1.7 m and forms a dense mat in the soil to 0.8 m (J. E. Weaver 1954). Roots of blue grama and buffalo grass penetrate vertically to 1 m. In addition, many grasses possess underground stems, or rhizomes, that serve both to propagate the plants and to store food. On the end of the rhizome, which has both nodes and scalelike leaves, is a terminal bud, which develops into aerial stems or new rhizomes. Rhizomes of most species grow at shallow depths, not over 10 to 14 cm deep. The exotic quack grass is notorious for its tough rhizomes. Forbs such as goldenrod, asters, and snakeroot possess large, woody rhizomes and fibrous roots that add to the root mat in the soil. Some, such as snakeroot, have extensive taproots 5 m long, and rushlike lygodesmia, common in

many prairies, extends down over 6 m in mellow soil. Among hayland plants, alfalfa possesses a taproot that grows to a considerable depth.

All the roots of grassland plants are not confined to the same general area of the soil but develop in three or more zones. Some plants are shallow-rooted and seldom extend much below 6 dm. Others go well below the shallow-rooted species, but seldom more than 1.5 m. Deep-rooted plants extend even further into the soil and absorb relatively little moisture from the surface soils. Thus plants roots absorb nutrients from different depths in the soil at different times, depending upon moisture (J. E. Weaver 1954).

The ground layer is characterized by low light intensity during the growing season and by reduced windflow. Light intensity decreases as the grass grows taller and furnishes shade. Temperatures decrease as solar insolation is intercepted by a blanket of vegetation, and windflow is at a minimum. Even though the grass tops may move like waves of water, the air on the ground is calm. Conditions on grazed lands are different. Because the grass cover is closely cropped, the ground layer receives much higher solar radiation and is subject to higher temperatures and to greater wind velocity near the surface.

The herbaceous layer may vary from season to season and from year to year, depending upon the moisture supply. Essentially the layer consists of three or more strata, more or less variable in height, according to the grassland type (Coupland 1950). Low-growing and ground-hugging plants, such as wild strawberry, cinquefoil, violets, dandelions, and mosses, make up the first stratum. All of them become hidden, as the season progresses, beneath the middle and upper layers. The middle layer consists of shorter grasses and such herbs as wild mustard, coneflower, and daisy fleabane. The upper layer consists of tall grasses and forbs, conspicuous mostly in the fall.

Mulch. Grasslands, unmowed, unburned, and ungrazed, accumulate a layer of mulch on the ground surface. The oldest layer consists of decayed and fragmented remains of fresh mulch. Fresh mulch consists of residual herbage, leafy and largely undecayed. Three or four years must pass before natural grassland herbage will decompose completely (H. H. Hopkins 1954). Not until mulch comes in contact with mineral soil does the decomposition process, influenced by compaction and depth, proceed with any rapidity. As the mat increases in depth, more water is retained, creating favorable conditions for microbial activity (McCalla 1943).

The amount of accumulated mulch often is enormous. On a tall-grass prairie it may be two to three times the amount of annual production (Knapp and Seastedt 1986). On a relict of a climax prairie organic matter and other humic materials amounted to 885.4 g/m^2, 581.1 g of it fresh mulch and 50.1 g fresh herbage (Dhysterhaus and Schmutz 1947). Another prairie supported 461.2 g/m^2 of fresh mulch and 830.1 g/m^2 of humus (Dix 1960).

Grazing reduces mulch, as do fire and mowing. Light grazing tends to increase the weight of humic mulch at the expense of fresh (Dix 1960); moderate grazing results in increased compaction, which favors an increase in microbial activity and a subsequent reduction in both fresh and humic mulch. Heavy grazing of a stand of bluestem reduced mulch from 1014.7 g/m^2 to 100.8 g/m^2 (Zeller 1961). An ungrazed North Dakota prairie averaged 441.2 g/m^2 of mulch, compared to 241.5 g/m^2 for a grazed prairie. Burning reduces both, but the mulch structure re-

turns two to three years after a fire on lightly grazed and ungrazed lands (Tester and Marshall 1961; Hadley and Kieckhefer 1963). Mowing greatly reduces fresh mulch and in a matter of time humic mulch also. An unmowed prairie accumulated 10 metric tn of humic mulch/ha; a similar prairie, mowed, had less than 4 metric tn (Dhysterhaus and Schmutz 1947).

Animal Life. All types of grasslands support similar forms of life. Much of the animal life exists within the several strata of vegetation: the roots, ground layer, and herb cover. Invertebrates, particularly insects, occupy all strata at some time during the year. During winter insect life is confined largely to the soil, litter, and grass crowns as pupae or eggs. In spring soil occupants are chiefly earthworms and ants. Ants are the most prevalent in some eastern North American meadows, constituting 26 percent of the insect population (Walcott 1937). Ground and litter layers harbor scavenger carabid beetles and predaceous spiders. Because this layer contains only limited supports for webs, there are more hunters than web builders among spiders.

Life in the herbaceous layer varies as the strata become more pronounced from summer to fall. Here invertebrate life is most diverse and abundant. Homoptera, Coleoptera, Orthoptera, Diptera, Hymenoptera, and Hemiptera are all represented. Insect life reaches two highs during the year, a major peak in summer and a less well-defined one in the fall.

Mammals are the most conspicuous vertebrates of the grasslands and the majority of them are herbivores. A large and rich ungulate fauna evolved on the grasslands. The bison and antelope (*Antilocarpa americana*) of North America were equaled only by the richer and more diverse ungulate fauna of the East African plains. Today herds of cattle have replaced the buffalo and rodents and rabbits have the distinction of being the most abundant native vertebrate herbivores.

Grassland animals share some outstanding traits. Hopping or leaping is a common method of locomotion. Strong hind legs enable a variety of animals, such as grasshoppers, jumping mice, jackrabbits, and gazelles, to rise above the grass where visibility is unimpeded. Speed, too, is well developed. Some of the world's fastest mammals, such as antelopes and cheetahs, live in grasslands. Many of the rodents and rabbits are adapted to digging or burrowing. Because of dense grass and lack of trees for singing perches, some grassland birds have conspicuous flight songs that advertise territory and attract mates.

Animal life in tame and successional grasslands depends upon human management for the maintenance of habitat. Mowing for hay, a major management tool, results in the destruction of habitat at a critical time of year. Nests of rabbits, mice, and birds are exposed at the height of the nesting season. Losses from both mechanical injury and predation are often heavy, although most species remain on the area to complete or reattempt nesting activity.

Pasturelands more often than not are so badly overgrazed that they support little vertebrate life. Rotation pastures may support more grassland life.

Successional grassland, because of infertility and plant cover of poverty grass *(Danthonia spp.)* and broomsedge, does not usually support as wide a variety of life as hayland. Poverty grass-dewberry fields are inhabited by grasshopper sparrows *(Ammodramus savannarum)*, vesper sparrows *(Pooecetes gramineus)*, and meadow mice; but deep grass species, such as meadow-

larks *(Sturnella spp.)* and bobolinks *(Dolichonyx oryzivorus)*, are few if not entirely absent. Broomsedge fields contain grasshopper sparrow, meadowlarks, and cotton rats. Both successional types offer poor quality food for such herbivores as cottontail rabbits and deer, although deer do feed on young broomsedge sprouts in early spring. Otherwise these dominant grasses are unpalatable to cattle and native herbivores alike.

Function

Grasslands vary in nature, type, species composition, and productivity as the pattern of rainfall changes (Table 30.1). Primary production is influenced by a number of environmental variables, particularly temperature and moisture. Primary production is lowest where precipitation is lowest and temperatures are high, and highest where mean annual precipitation is greater than 800 mm and mean annual temperature is above 15° C (see data in Coupland 1979; Breymeyer and Van Dyne 1980). Production, however, is most directly related to precipitation (Figure 30.4 and Table 30.1). The greater the mean annual precipitation, the greater is aboveground production, in part because increased moisture reduces water stress and enhances the uptake of nutrients and in part because of water use efficiency.

Grasses of arid regions use water inefficiently, although most of the water is evaporated from the bare surface of the soil. Adaptations among desert grasses are not for the efficient use of water but for survival and persistence by other means. In semiarid country where water is limiting, plants have evolved adaptations to make the maximum use of water, so water efficiency is high (see Chapter 5). In humid, tall-grass country the vegetative can-

TABLE 30.1   Mean Standing Crop of Selected Grasslands, Growing Season (g/m^2, oven dry)

Standing Crop	DESERT		SHORT-GRASS PLAINS	
	Ungrazed	Grazed	Ungrazed	Grazed
Above ground	81.2	49.5	69.0	42.9
Cool season grass	0	0	6.6	2.6
Warm season grass	51.3	9.4	42.6	27.1
Old dead	0.9	1.1	25.9	18.3
Mulch	68.7	52.3	97.5	89.3
Below ground, alive and dead	197.0	185.0	1600.0	1800.0
Rainfall (cm)	23		30	

Standing Crop	MIXED PRAIRIE		TRUE PRAIRIE	
	Ungrazed	Grazed	Ungrazed	Grazed
Above ground	154.2	94.6	256.1	220.8
Cool season grass	101.6	9.5	4.6	24.2
Warm season grass	45.8	82.1	227.1	152.6
old dead	80.2	58.5	204.0	44.0
mulch	448.3	239.2	111.2	206.5
Below ground, alive and dead	1213.0	2086.0	893.0	781.0
Rainfall (cm)	38		94	

Source: Adapted from Lewis 1971.

opy is dense and intercepts most of the solar radiation, so adaptations of plants are for efficient capture of light at the expense of efficient use of water.

Production is mirrored in biomass accumulation (Table 30.1). Mean green shoot aboveground biomass among the array of grasslands studied in the International Biological Program (IBP) ranges from 50 g/m^2 in arid grasslands to 827 g/m^2 in subhumid grasslands. Mean underground biomass ranges from 45 to 4707 g/m^2. In most tropical grasslands, that biomass is usually less than 1000 g/m^2 (median 200 g/m^2). Among grasslands the ratio of root-to-shoot (or belowground to aboveground biomass) ranges from 0.2 to 10.3. The ratio is smallest in tropical grasslands, a mean of 0.8. The ratio for seminatural grasslands is a mean of 3.3 and for temperate grasslands, 4.4. Except

for a short period of maximum aboveground biomass, belowground biomass is two to three times that above ground. Primary productivity ranges from 82 g/m^2/yr in semiarid regions to 3396 g/m^2 in subhumid tropical grasslands. In natural and seminatural temperate grasslands, production ranges from 98 to 2430 g/m^2/yr. Much of the net production of grassland does not appear as aboveground biomass. Seventy-five to 85 percent of the photosynthate is translocated to the roots for storage.

Grasslands have the highest index of productivity per unit of standing crop, 20 to 55 percent, among all terrestrial ecosystems. The tundra during its short growing season has an index of 10 to 20 percent. The index for the boreal forest is only 2 to 5 percent and for the temperate deciduous forest, 8 to 10 percent. Of course, a

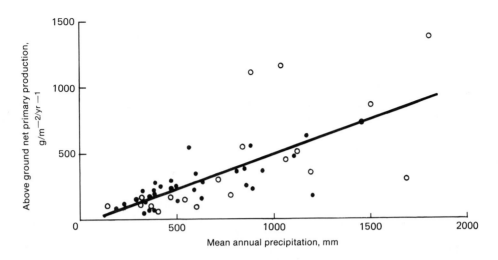

Figure 30.4 Relationship between aboveground primary production and mean annual precipitation for 52 grassland sites around the world. North American grasslands are indicated by dots. Annual net production −0.5 (annual precipitation) −29; $r^2 = 0.51$. (From Lauenroth 1979:10.)

great deal of energy in woody ecosystems goes into the maintenance of high standing crops, which grasslands do not support. Efficiency of energy capture by grasslands is 0.75 percent for net primary production. Light to moderate grazing increases that efficiency to 0.87 percent.

Grasslands evolved under grazing pressures of ungulates since the Cenozoic. Grasses have adapted to grazing by compensatory growth of tissue damaged or removed by grazing (Sims and Singh 1971; McNaughton 1979; Risser 1985; for extensive discussions see Coupland ed. 1979; Breymeyer and Van Dyne eds. 1980). Such compensation involves increased photosynthetic rates in residual tissue; reallocation of nutrients and photosynthates from one part of the plant to another, especially from roots to stems; removal of photosynthetically inefficient mature tissue to make room for young tissue and subsequent increase in light inten-

sities to stimulate new growth; reduction of leaf senescence, prolonging the active photosynthetic period of remaining tissues; hormonal control of meristems, stimulating tillering and leaf growth; conservation of soil moisture by reducing transpiration surface; and recycling of nutrients through dung and urine (McNaughton 1979).

Numerous experimental studies of herbivore-grass interactions by plant ecologists and range managers have demonstrated that moderate grazing stimulates primary production of grasslands. McNaughton (1984) points out that moderate grazing on the Serengeti grasslands stimulated production up to twice the levels in ungrazed control plots. IBP grassland studies examined grazed and ungrazed systems from a functional point of view. In general, total net primary production was greater on grazed than on ungrazed sites. However, grazed grasslands channeled more organic matter below ground than did ungrazed grasslands (see Table 30.2). Grazed grasslands allocated about 63 percent of total net primary production to below ground net production and about 37 percent to aboveground net production. Ungrazed systems sent about

TABLE 30.2 Net Primary Productivity and Efficiency of Energy Capture by Ungrazed and Grazed Grasslands within the Growing Season

Type	ABOVE GROUND		BELOW GROUND		TOTAL	
	Net Production (kcal/m^2)	Efficiency (%)	Net Production (kcal/m^2)	Efficiency (%)	Net Production (kcal/m^{20})	Efficiency (%)
GRAZED						
Desert	456	.06	625	.09	1081	.13
Short-grass plains	476	.10	3285	.70	3761	.80
Mixed prairie	444	.10	1810	.41	2254	.51
True prairie	2048	.42	1701	.35	3749	.77
UNGRAZED						
Desert	688	.09	489	.07	1177	.16
Short-grass plain	568	.12	2153	.45	2721	.57
Mixed prairie	788	.18	1264	.29	2052	.47
True prairie	1348	.27	872	.17	2220	.44

Source: Adapted from Sims and Singh 1971.

52 percent of total net primary production below ground and about 42 percent above. Ungrazed treatments had an above ground production efficiency of 0.26 percent and the grazed 0.33 percent (see Sims and Singh 1971; Singh et al. 1980).

Another response of grasslands to grazing is a change in species composition. Some grasses and forbs are sensitive to intensive grazing pressure and tend to disappear while other species increase (see Table 30.3). On desert grasslands of North America black grama is replaced by weedy species; on short-grass plains, which are the most stable under grazing pressure, blue grama and prickly pear increase; on mixed-grass prairies midgrasses decrease and short grasses and sedges increase. On tall-grass sites tall grasses disappear and little bluestem and tall dropseed increase; if grazing pressure is heavy the site may be invaded by the weedy Japanese chess (Lewis 1971).

Natural grasslands experienced grazing pressure from free-ranging and often migratory populations of mammals. When wild ungulates were replaced by domestic ones, humans confined them with fences and overstocked the ranges, which resulted in serious deterioration of grassland systems (Figure 30.5). Overgrazing desert grasslands increases the spread of mesquite because of lessened competition from grass and the spread of seed by livestock. On other grasslands mulch deteriorates and disappears, because only a small amount of litter is added to the ground. Water flows over the surface, taking topsoil with it. Lacking moisture and nutrients, the original species cannot maintain themselves and the vegetation cover continues to decrease until only an erosion pavement remains.

Although large herbivores are the most conspicuous consumers, invertebrates are the most important grazers. The aboveground biomass of invertebrates—including plant consumers, saprovores, and predators—ranges from 1 to 50 g/m^2, and grazing mammals amount to about 2 to 5 g/m^2. Belowground invertebrates exceed 135 g/m^2. The major above-

TABLE 30.3 Response of Grassland Species to Grazing Disturbance, Grassland Biome, 1970

| Grassland | Species | % COMMUNITY STANDING CROP | |
		Ungrazed	Grazed
Desert	Black grama *(Bouteloua eriopoda)*	53	8
	Broom snakeweed *(Xanthocephalum sarothrae)*	15	26
	Russian thistle *(Salsola kali)*	3	25
Short-grass plains	Blue grama *(Bouteloua gracilis)*	63	72
	Prickly pear *(Opuntia polyacantha)*	5	12
Mixed prairie	Wheat grass *(Agropyron smithii)*	43	9
	Blue grama *(Bouteloua gracilis)*	9	37
	Buffalo grass *(Buchloe dactyloides)*	28	44
True prairie	Little bluestem *(Andropogon scoparius)*	69	23
	Indian grass *(Sorghastrum nutans)*	7	t
	Japanese chess *(Bromus japonicus)*	0	22
	Tall dropseed *(Sporobolus asper)*	2	37

Source: Lewis 1971.

Figure 30.5 The effects of overgrazing by cattle are apparent on this short-grass range. Prickly pear and mesquite have replaced grass. The ground is bare, most of the moisture from rains is lost by runoff and evaporation, and erosion is serious. (Photo courtesy Soil Conservation Service.)

ground consumers are grasshoppers and the major belowground consumers are nematodes. Nematodes account for 90, 95, and 93 percent of all belowground herbivory, carnivory, and saprophagous activity, respectively. They account for 46 to 67 percent of root and crown consumption, 23 to 85 percent of fungal consumption, and 43 to 88 percent of belowground predation. Aboveground herbivores consume 2 to 7 percent of primary production, or 3 to 10 percent if both the amounts consumed and wasted are considered. Belowground consumption, including wastage, amounts to 13 to 46 percent.

Not only is a large proportion of primary production consumed below ground, but a greater proportion is utilized at each trophic level (Figure 30.6). Efficiency of transfer, however, is about the same. Invertebrates convert about 9 to 25 percent of ingested energy to animal tissue, whereas homoiotherm grazers, such as sheep and cattle, convert 3 to 15

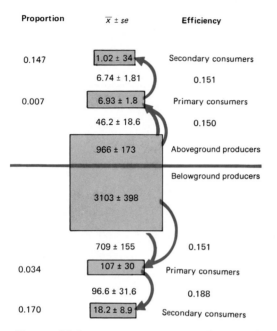

Figure 30.6 Grassland production (in boxes) and consumption (between boxes), representing means and standard errors for desert, tall-grass, mixed, and short-grass sites in North American grasslands. Production in one trophic level is represented as a proportion of production in the next trophic level and efficiency as the ratio of production to consumption. (From French 1979: 185.)

percent to animal tissue. Some invertebrate consumers are wasteful. The amount of aboveground vegetation detached or otherwise killed about equals that consumed by vertebrate grazing herbivores.

Energy flow through small mammal populations in grasslands may be high relative to their biomass but low relative to the standing crop biomass (Risser et al. 1981). Small mammals consume only about 4 percent of the herbage available. Energy flow through breeding bird populations may be even smaller, about 1.01 to 2.33 kcal/m^2. Birds probably exert little in-

fluence on ecosystem structure, function, or dynamics through their direct effects on the flow and storage of energy and nutrients (Wiens 1973).

Most of the primary production, above ground especially, goes to the decomposers, dominated by fungi whose biomass is 2 to 7 times that of bacteria. Overall the decomposer biomass exceeds that of invertebrates.

Green plant consumers, however, are important in the cycling of nutrients in grassland ecosystems. Invertebrate consumers are highly inefficient in assimilating ingested material. Much of their intake is deposited as feces or frass. Because the nutrients they contain are in a highly soluble form, they are returned rapidly to the system. Large grazing herbivores return a portion of their intake as dung, which becomes a major pathway of nutrient cycling (Figure 30.7). They harvest nutrients over a large area and concentrate them in a small area (McNaughton 1985). A well-developed corprophagous fauna speeds the decay of manure and accelerates the activity of bacteria in feces.

Nitrogen provides an example of nutrient cycling in grasslands (Figure 30.8). About 90 percent of nitrogen in grasslands is tied up in soil organic matter, 2 percent in litter, 5 percent in live and dead plant cover, 1 percent in dead shoots, and about 0.8 percent in soil microflora, and a very small amount in aboveground and belowground invertebrates. When nitrogen is limiting, most of it is shunted to green herbage, where it remains during the growing season. Some of it is consumed by herbivores and another fraction is translocated to roots. Much of this nitrogen is moved above ground to new growth the following season. A fraction is retained in standing dead leaves and returned to the litter and

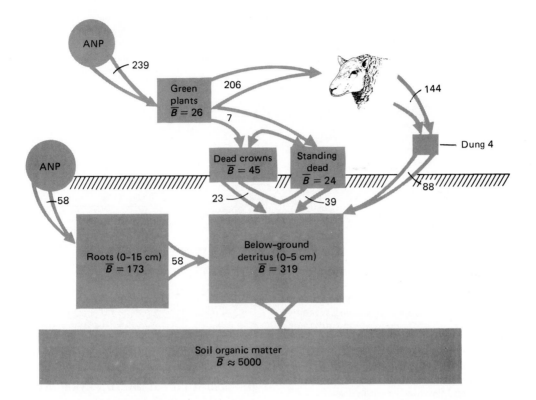

Figure 30.7 Carbon (energy) flow in a grazed pasture, indicating the role of grazing herbivores in the functioning of grassland ecosystems. Boxes represent the retention of carbon (g/m^2). Arrows represent the flow of carbon (g/m^2/yr). ANP = annual net productivity; B = biomass. Note the high proportion of energy that moves through the grazing herbivore. (From Breymeyer 1980: 803.)

soil surface, where it is acted upon by fungi and bacteria. Part becomes immobilized in microbial biomass but is quickly mineralized upon the death of microbes and reenters the nitrogen cycle. Still another fraction is tied up in humus and resists decomposition. Turnover, however, is fairly rapid, and most of the nitrogen that enters a green plant one growing season will reenter another the following year.

Central to nutrient cycling in grasslands is the accumulation of mulch or detritus. An accumulation of mulch can have both positive and negative effects. Mulch increases soil moisture through its influence on infiltration and evaporation; it decreases runoff, stabilizes soil temperatures, may improve conditions for seed germination, provides nesting habitat for grassland birds such as the meadowlark and dickcissel, and offers cover for small mammals. On the other hand, a large standing crop of detritus can have detrimental effect on nitrogen cycling, particularly in tall-grass ecosystems (Knapp and Seastedt 1986). Detritus intercepts rainfall from which microbes can assimilate inorganic nitrogen directly before it reaches plant roots, while the mulch itself inhibits nitrogen fixation by blue-green and free-

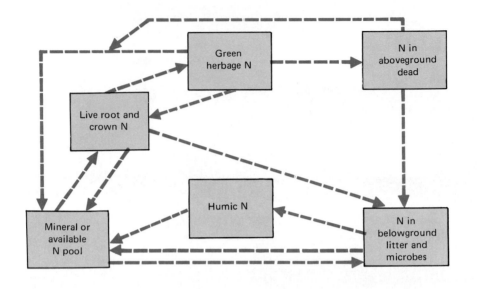

Figure 30.8 Internal flows of nitrogen in a *Andropogon gerardi-Andropogon scoparius* prairie in Missouri. The values for pools are in tm/ha and the values for flows are in tm/ha/yr. Little nitrogen is transferred from the aboveground biomass to below ground. Much of the nitrogen is returned by decomposition of detrital material. The relationships and flows between the various pools suggest a conservative cycling of nitrogen in the grassland ecosystem.

living nitrogen-fixing microbes. An accumulation of mulch reduces available light for and alters both the microclimate and the physiology of emerging shoots, so CO_2 uptake is reduced. By insulating the soil surface from solar radiation, mulch reduces root productivity and inhibits the activity of soil microbes and invertebrates. Periodic grassland fires clear away the mulch layer and release nutrients in detritus to the soil, but nitrogen, equal to about two years of nitrogen inputs to the system through rainfall, is lost to the atmosphere. Fires, however, stimulate the growth of nitrogen-fixing leguminous forbs and improve conditions for earthworms. Annual burning could eventually deplete soil organic nitrogen.

■ Tropical Savannas

The one ecosystem that defies general description is the tropical savanna. The problem is an old one, involving even its name (Hill 1965; Sarmiento and Monasterio 1975; Bourliere and Hadley 1983). The word in its several origins, largely Spanish, referred to grasslands or plains. Over time the word was applied to an array of vegetation types representing a continuum of increasing cover of woody vegetation, from open grassland to widely spaced shrubs or trees to closed woodland with a grass understory (Figure 30.9). Most are dominated by Andropogonoid grasses.

Moisture appears to be the major determinant of savannas (Figure 30.10), a function of both rainfall (amount and distribution) and soil texture (Sarmiento and Monasterio 1975; Tinley 1982). Fire, soil nutrients, and herbivores are important modifiers of the basic vegetation that the water regime promotes (Huntley and Walker 1982). Savannas cover much of central and southern Africa, western In-

(a)

(b)

(c)

(d)

Figure 30.9 Savanna ecosystems in southern Africa: (a) grass savanna (note the termite mound); (b) shrub savanna; (c) tree savanna; (d) savanna woodland dominated by *Combreteum*. (Photos by T. M. Smith.)

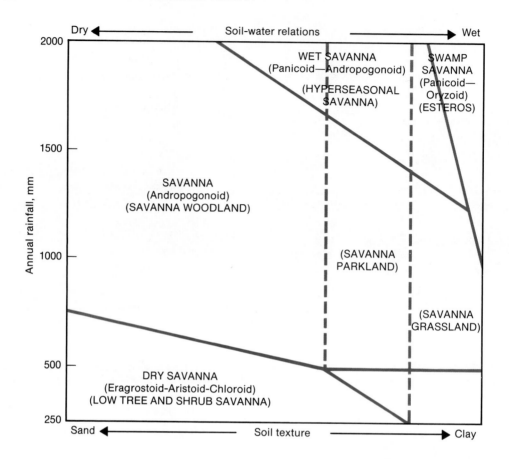

Figure 30.10 Classification of savannas of the world, based on annual rainfall and soil moisture as influenced by soil texture. Dominant grasses include the genera *Andropogon, Panicum, Oryza,* and *Eragrostis* and genera related to them. (After Johnson and Tothill 1985.)

dia, northern Australia, large areas of northern and east-central South America, and some of Malaysia. Some savannas are natural. Others are seminatural or anthropogenic, brought about and still maintained by centuries of human interference.

Characteristics

Savannas, in spite of their vegetational differences, exhibit a certain set of char-acteristics. Savannas occur on land surfaces of little relief, often old alluvial plains. The soils are low in nutrients, and this fact and other conditions, including moisture availability, mean that they cannot support a forest. Precipitation exhibits extreme seasonal fluctuations; in South American savannas, in particular, the soil water regime may fluctuate from excessively wet to extremely dry, often below the permanent wilting point. Savannas are subject to recurrent fires, and the dominant vegetation is fire-adapted. Grass cover with or without woody vegetation is always present. When present, the woody component is short-lived, with individuals seldom surviving for more than several decades (except for the African baobab trees). Finally, detrital-processing termites are a conspicuous component of savanna animal life.

Structure

The major and most essential stratum of the savanna ecosystem is grass, mostly bunch or tussock, with no vertical structure. The woody component adds one or two more vertical layers, ranging from about 50 to 80 cm when small woody shrubs are present to about 8 m in the tree savannas. Highly developed root systems make up the larger part of the living herbaceous biomass. The root system is concentrated in the upper 10 cm but extends down to about 30 cm. Savanna trees have extensive horizontal roots that go below the layer of grass roots. They have a high root-to-shoot ratio (R/S) in sharp contrast to typical forest trees, which have a low R/S ratio. Competition may exist between grass and woody vegetation for soil moisture, but more intense competition takes place between trees, accounting for the spacing patterns of woody vegetation (Gutierrez and Fuentes 1979; Smith and Walker 1983; Smith and Grant 1986).

In contrast to the rather poorly developed vertical structure is a well-developed, although often unapparent, horizontal structure. The tussock grasses form an array of clumps set in a matrix of open ground, creating patches of low vegetation with frequent changes in microclimatic conditions. The addition of woody growth, the widely spaced shrubs and trees, lends horizontal structure to the soil. Trees add some organic matter and nutrients to the soil beneath them, reduce evapotranspiration, resulting in increased herbaceous and woody shrub growth, and provide patches of shade. On the African savanna in particular, large grazing herbivores rest in the shade during the heat of the day and concentrate nutrients from dung and urine beneath the trees.

Savannas support or are capable of supporting a large and often varied assemblage of herbivores, both grazing and browsing, particularly in Africa. This element of large grazing herbivores is missing from the South American savannas, where it is replaced in part by a number of deer, tapirs, and capybara (*Hydrochoerus*), the largest living rodent (Ojasti 1983). In spite of their visual dominance, large ungulates consume only about 10 percent of primary production. Dominant herbivores are the invertebrates, including acrid mites, acridid grasshoppers, seed-eating ants, and detrital-feeding dung beetles and termites (Gillon 1983).

Function

Primary production in savannas is not well documented. Because of the wide diversity of savanna types, it is difficult to make any strong generalizations. Probably a wide range of production exists between grass savanna on one end of the gradient and tree savanna and woodlands on the other (see Lamotte 1975; Rutherford 1978; Lamotte and Bourliere 1983). Primary production is initiated at the beginning of the wet season when moisture releases nutrients from materials accumulated in the dry season and stimulates nutrient translocation from the roots. A quick flush of growth into grass and woody plants follows. Large root systems efficiently transfer nutrients from the soil to aboveground biomass.

The nutrient pool, especially nitrogen, is low and its circulation in a tree savanna (Bernhard-Reversat 1981; Sarmiento 1984) provides some insight into the function of savanna ecosystems. Nitrogen content and organic matter of the soil are low. Both are concentrated largely in the upper 10 cm of soil, with the highest concentration beneath the scattered trees. The first rains of the wet season trigger mineralization.

Its rate, 5 to 8 percent of the total N per year, is fairly high. It is the function of the number of days the soil was wet for a given temperature and N content. Bernhard-Reversat (1981) found that the nitrogen flux between soil and vegetation was higher under the trees than in the open, because of reduced evaporation, which kept the soil moist, more organic matter accumulation, and more herbaceous growth. Savanna trees, especially the African acacias, exhibited tight internal cycling. Nitrogen concentration in the leaves decreased as the dry season approached, with maximum withdrawal before leaf fall. The trees transferred some of the nitrogen into new woody growth, but much of it went to the root reserve, where N accumulated for the flush of new season growth.

A similar tight circulation exists in neotropical savannas (Sarmiento 1984). An important portion of nitrogen, 66 percent of it, is translocated to the perennial roots, while another portion remains in the dry standing biomass (Figure 30.11). Most of the nitrogen in the dry aboveground biomass is lost to the atmosphere by volatilization if fire sweeps the savanna; otherwise a fraction will be transferred to the soil through leaching effects of rainwater.

■ Shrublands

Covering large portions of the arid and semiarid world is climax shrubby vegetation (see inside cover). In addition climax shrubland exists in parts of temperate regions, because historical disturbances of landscapes have seriously affected their potential to support forest vegetation (Eyre 1963). Among such shrub-dominated plagioclimaxes are the moors of

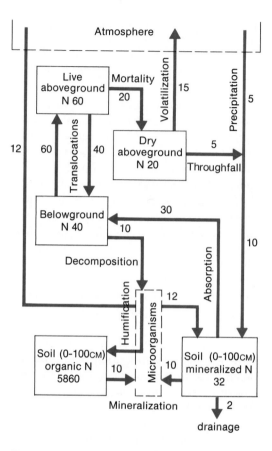

Figure 30.11 The annual nitrogen cycle in the seasonal savanna of *Axonopus purpusii-Leptocoryphiium lanatum* at Barinas, Venezuela. The aboveground biomass at maximum density accumulates 60 kg/ha/N. As the aboveground growth dries, about 66 percent of the nitrogen translocates to the roots; the remainder remains in dry standing crop. Of this N an estimated 25 percent is incorporated into the mineral nitrogen pool of the soil by throughfall. Some of the nitrogen in the dry aboveground material is lost through volatilization when burned. In the roots 25 percent of the biomass is recycled through decomposition. In this situation about 10 kg/ha/yr passes through soil microorganisms to the organic nitrogen pool. Compare this figure with Figure 30.8 for a grassland. The value of nitrogen accumulated in the pools are in tm/ha, and the values of the flows are in tm/ha/yr. (After Sarmiento 1984: 191.)

Scotland, the macchia of South America. Outside these regions, shrublands are seral, a stage in land's progress back to forest. There they exist as second-class citizens of the plant world (McGinnes 1972), given little attention by botanists, who tend to emphasize dominant plants. Too little work has been done on the seral shrub communities.

Characteristics of Shrubs

Shrubs are difficult to characterize. They have, as McGinnes (1972) points out, a "problem in establishing their identity." They constitute neither a taxonomic nor an evolutionary category (Stebbins 1972). One definition is that a shrub is a plant with woody persistent stems, no central trunk, and a height of up to 4.5 to 6 m. Size does not set shrubs apart, though, because under severe environmental conditions many trees will not exceed that size. Some trees, particularly coppice stands, are multiple-stemmed, and some shrubs have large single stems. Shrubs may have evolved either from trees or herbs (for detailed discussion on evolution of shrubs, see Stebbins 1972).

The success of shrubs depends largely on their abilities to compete for nutrients, energy, and space (West and Tueller 1972). In certain environments shrubs have many advantages. They have less energetic and nutrient investment in aboveground parts than trees. Their structural modifications affect light interception, heat dissipation, and evaporative losses, depending upon species and environments involved. The multistemmed forms influence interception and stemflow of moisture, increasing or decreasing infiltration into the soil (Mooney and Dunn 1970b). Because most shrubs can get their roots down quickly and form extensive root systems, they can utilize soil moisture

deep in the profile. This feature gives them a competitive advantage over trees and grasses in regions where the soil moisture recharge comes during the nongrowing season. Because they have a low shoot-to-root ratio, shrubs draw less nutrient input into aboveground biomass and more into roots. Their perennial nature allows immobilization of limiting nutrients and slows the nutrient recycling, favoring further shrub invasion of grasslands.

Subject to strong competition from herbs, some climax shrubs, such as chamiso *(Adenostoma fasciculatum)*, inhibit the growth of herbs by allelopathy (see Chapter 18), the secretion of substances toxic to other plants (McPherson and Muller 1969). Only when fire destroys mature shrubs and degrades the toxins do herbs appear in great numbers. As the shrubs recover, herbs decline. Seeds of herb species affected apparently have evolved the ability to lie dormant in the soil until released from suppression by fire.

Types of Shrublands

Mediterranean-Type Shrublands. In five regions of the world lying for the most part between 32° and 40° North and South of the equator are areas characterized by a mediterranean climate: the semiarid regions of western North America, the regions bordering the Mediterranean, central Chile, the Cape region of South Africa, and southwestern and southern Australia. The mediterranean climate is characterized by hot, dry summers with at least one month of protracted drought and cool, moist winters. At least 65 percent of the annual precipitation falls during the winter months and for at least one month the temperature remains below 15° C (Aschmann 1973).

All five areas (see inside cover) support physiognomically similar communities of xeric broadleaf evergreen shrubs and dwarf trees known as broad sclerophyll vegetation with herbaceous understory (for a detailed description, see McKell et al. 1972; de Castri and Mooney 1973; de Castri, Goodall, and Specht 1981). Although vegetation in all the mediterranean-type ecosystems shares certain characteristics, exhibiting strong convergence in vegetation forms (see Mooney 1977), each has evolved its own distinctive flora and fauna. In the Northern Hemisphere the vegetation evolved from tropical floras and developed in dry summer climates that did not exist until the Pleistocene (Raven 1973).

In addition to similar forms, vegetation in each of the mediterranean systems also shows similar adaptations to fire (see Mooney and Conrad 1978) and to low nutrient levels in the soil. In the mediterranean systems of the Northern Hemisphere annuals make up 50 percent of the species and 10 percent of the plant genera; 40 percent of the species are endemics.

There are variations of the basic mediterranean-type ecosystems (see de Castri 1981). In the Mediterranean region shrub vegetation often results from forest degradation and falls into three major types. The *garrigue*, resulting from degradation of pine forests, is low, open shrubland on well-drained to dry calcareous soil. The *maqui*, replacing cork forests, is higher thick shrubland in areas of more rainfall. The *mattoral*, further subdivided into high, middle, low, and scattered (Tomaselli 1981), appears to be equivalent to the North American chaparral (Soriano 1972; Tomaselle 1981). The Chilean mediterranean system, also called *mattoral*, varies across topographic positions from the coast to slopes of coastal ranges and the foothills of the Andean cordillera (Rundel 1981).

Much of the South African mediterranean shrubland is heathlands known as *fynbos*, discussed later. The more typical mediterranean-type shrubland, dominated by a broad sclerophyll woody shrub, goes by the names of *strandveld, coastal renosterveld*, and *inland renosterveld*.

In southwest Australia the mediterranean shrub country, known as *mallee*, is dominated by low-growing *Eucalyptus*, 5 to 8 m high with broad sclerophyllous leaves (Figure 30.12). There are six types of mallee ecosystems, which intergrade. Three fall into a mediterranean-type ecosystem with a typically grassy and herbaceous understory. The other three types occur on oligotrophic soils and fall under the category of heathland shrubs. Razed by fire at irregular intervals, the mallee ecosystem differs markedly from typical mediterranean-type ecosystems in its summer growth rhythm. Mediterranean-type shrublands typically initiate new growth during the spring; but the mallee retains its summer growth rhythm from the subtropical Tertiary, out of phase with the mediterranean climate of the area. Growth takes place during the summer, the driest part of the year, drawing on water conserved in the soil during wet winter and spring showers (Specht 1981).

In North America the sclerophyllic shrub community is known as *chaparral*, a word of Spanish origin meaning a thicket of shrubby evergreen oaks. California chaparral is dominated by shrub oak (*Quercus dumosa*) and chamise (*Adenostoma fasciculatum*). Another shrub type, also designated as chaparral, is associated with the Rocky Mountain foothills in Arizona, New Mexico, Nevada, and elsewhere. It differs from California chaparral in two ways. It is dominated by Gambel oak (*Quercus gambelii*) and other species and lacks chamise; and it is summer-active and winter-deciduous, whereas California chapparal is ev-

Figure 30.12 Tall shrub mallee in Victoria, Australia, is an example of a mediterranean-type shrubland dominated by *Eucalyptus*. Note the canopy structure and the open understory of grass at the beginning of the spring rains. This type of vegetation supports a rich diversity of bird life. (Photo by R. L. Smith.)

ergreen, winter-active, and summer-dormant (see Hanes 1981).

For the most part mediterranean-type shrublands lack an understory and ground litter, are highly inflammable, and are heavy seeders. Many species require the heat and scarring action of fire to induce germination. Others sprout vigorously after a fire.

For centuries periodic fires have roared through mediterranean-type vegetation, clearing away the old growth, making way for the new, and recycling nutrients through the ecosystem. When humans intruded into this type of vegetation, they changed the fire regime, either by attempting to exclude fire completely or by overburning (see Trabaud 1981). In the absence of fire, chaparral grows tall and dense and yearly adds more leaves and twigs to those already on the ground. During the dry season the shrubs, even though alive, nearly explode when ignited. Once set on fire by lightning or humans, an inferno follows.

After fire the land returns either to lush green sprouts coming up from buried root crowns or to grass, if a seed source is nearby. Grass and vigorous young sprouts are excellent food for deer, sheep, and cattle. Then as the sprout growth matures, chaparral becomes dense, the canopy closes, the litter accumulates, and the stage is set for another fire.

Because of the rough terrain characteristic of the mediterranean-type ecosystem some areas have remained relatively undisturbed, especially in California and South Africa. In Australia and the Mediterranean basin, human activity, especially livestock grazing and fruit and vegetable farming, have degraded the broadleaf sclerophyllous vegetation (see Trabaud 1981).

Northern Desert Scrub. In the Great Basin of North America, the northern, cool, arid region lying east of the Rocky Mountains, is the northern desert scrub. The climate is continental, with warm summers and prolonged cold winters. Although this region is perhaps more appropriately considered a desert, it is one of the most important shrublands in North America (Figure 30.13). Its physiognomy differs greatly from the southern hot desert and the dominant vegetation is shrub. The vegetation falls into two main associations: one is sagebrush, dominated by *Artemisia tridentata* which often forms pure stands, and the other is shad-

Figure 30.13 This northern desert shrubland in Wyoming is dominated by sagebrush. Although classified as cold desert, sagebrush forms one of the most important shrub types in North America. (Photo by R. L. Smith.)

scale, *Atriplex confertifolia*, a C$_4$ species, and other chenopods (see Tiedemann et al. 1983; Caldwell 1985), halophytes tolerant of saline soils. Inhabiting this shrubland are pocket and kangaroo mice, lizards, and sage grouse, sage thrasher, sage sparrow, and Brewer's sparrow, four birds that depend upon sagebrush for their continued existence.

A similar type of shrubland exists in the semiarid inland of southwestern Australia. Numerous chenopod species, particularly the saltbushes of the genera *Atriplex* and *Maireana*, form extensive low shrublands on low riverine plains there (Figure 30.14).

Heathlands. Typically, heathlands have been associated with cool to cold temperate climatic regions of northwestern Europe. It was probably coincidental that the original name came to refer to land dominated by Ericaceae. The word "heath" comes from the German word *Heide*, meaning "an uncultivated stretch of land," regardless of the vegetation. It just so happened that in parts of Germany uncultivated or waste land was dominated by Ericaceae. In reality heathlands are found in all parts of the world, from the tropics to polar regions and from lowland to subalpine and alpine altitudes.

Heathland flora probably evolved in the Mesozoic in the eastern to central portion of Gondwanaland. It survived the climatic changes of the Tertiary and Quaternary and retained most of its morphological and physiological characteristics. It expanded out of the fragments of Gondwanaland from Africa into western

Figure 30.14 Saltbush shrubland in Victoria, Australia, is dominated by *Atriplex*. It is an ecological equivalent to the shrublands of the Great Basin in North America. (Photo by R. L. Smith.)

the deciduous forest, as shrubby understory. Although associated with shrubs of the family Ericaceae, notably heather *(Calluna)*, several families of heaths are prominent, including Empetraceae, Vacciniaceae, Epacridaceae, and Prionotaceae. In the Southern Hemisphere heathlands have many primitive and ancient dicotyledonous and monocotyledonous families.

Heathlands invariably occur on oligotrophic soils especially deficient in phosphorus and nitrogen. Heathlands are most extensive in the arctic regions. Other extensive areas occur in the mediterranean-type regions of South Africa, where they are known as *fynbos* (Kruger 1979), and in southeastern and western Australia (see George et al. 1979). In subtropical to

Figure 30.15 Heathlands, dominated by ericaceous shrubs, have a similar physiognomy around the world. Typical is this heathland in the Grampian Mountains of Australia. (Photo by R. L. Smith.)

Europe and the northern part of Eurasia and North America, from India into southeastern Asia, and from Australasia into the Malaysian Archipelago.

Heathland vegetation is an assemblage of dense to mid-dense growth of ancient or primitive genera of angiosperms: evergreen sclerophyllous shrubs and subshrubs together with hemicryophytes and thereophytes, all adapted to fire (Figure 30.15). Heathland shrubs have leaves with thick cuticles, sunken stomata, thick-walled cells, and hard and waxy upper surfaces. Many species have leaves with small surface area—less than 25mm and termed leptophyll—and others roll their edges in toward the midrib. Although mostly associated with heathlands, many heathland shrubs are usually present, at least locally, in other ecosystems, such as

tropical climates true heathlands are confined to alpine areas and to lowland, oligotrophic soils subject to seasonal waterlogging. Some heathlands, such as the heather-dominanted moors of Scotland, are human-induced and can be maintained only by periodic fires (Gill and Groves 1981; Gimingham 1981).

There are two distinct heathland ecosystems: dry heathlands and wet heathlands. Dry heathlands are on well-drained soils and subject to seasonal drought. Wet heathlands are subject to seasonal waterlogging. In wet heathlands graminoids (grasses and sedges) may become codominant with heathland shrubs; and in extreme wet heathlands the grass component is suppressed by *Sphagnum* moss (see Chapter 32).

Both height and foliage cover of heathlands vary considerably with the habitat. They are divided according to height of the uppermost stratum: shrubs > 2m, scrub; shrubs 1–2 m, tall heathland; shrubs 25–100 cm, heathland; and shrubs < 25 cm, dwarf heathland.

Successional Shrublands. On drier uplands shrubs rarely exert complete dominance over herbs and grass. Instead the plants are scattered or clumped in grassy fields, the open areas between filled with the seedlings of forest trees, which in the sapling stage of growth occupy the same ecological position as tall shrubs (Figure 30.16). Typical are the hazelnut, forming thickets in places, sumacs, chokecherry, and shrub dogwoods.

On wet ground the plant community often is dominated by tall shrubs and contains an understory intermediate between that of a meadow and a forest (Curtis 1959). In northern regions the common tall shrub community found along streams and around lakes is the alder thicket, composed of alder or alder and a mixture of

Figure 30.16 In eastern North America and in northern and western Europe shrublands are usually successional communities; but if the vegetation is dense shrub, these communities may persist for many years. This slope is claimed by Saint-John's-wort *(Hypericum virginicum)* and wild indigo *(Baptisa tinctoria)*. (Photo by R. L. Smith.)

other species such as willow and red osier dogwood. Alder thickets are relatively stable and remain for some time before being replaced by the forest. Out of the alder country, the shrub or carr community (*carr* is an English name for wet-ground shrub communities) occupies the low places. Dogwoods are some of the most important species in the carr. Growing with them are a number of willows, which as a group usually dominate the community.

Shrub thickets are valued as food and cover for game; and many shrubs, such as hawthorn, blackberry, sweetbrier, and dogwoods, rank high as game food. However, the overall value of different types of shrub cover, its composition, quality, and the minimum amounts needed, have never been assessed. There is some evidence (Egler 1953; Niering and Egler 1955) that even where the forest is the normal end of succession, shrubs can form a stable community that will persist for many years. If incoming tree growth is removed either by selective spraying or by cutting, shrubs eventually form a closed community resistant to further invasion by trees. This fact could have wide application to the management of power line rights-of-way.

Structure

Shrub ecosystems, seral or climax, are characterized by woody structure, increased stratification over grasslands, dense branching on a fine scale, and low height, up to 8 m. Typically there are three layers: a broken upper canopy, an irregular low shrub canopy, and a grass/herbaceous layer. This structure, of course, may vary. Dense shrubland may have only a canopy layer; and stratification often decreases as the shrubs reach maximum height, particularly in seral shrublands. Horizontal stratification varies with the vegetation type. Heathlands may exhibit minimal horizontal stratification. Mediterranean-type shrublands, notably the mattoral and the mallee, may have pronounced horizontal stratification. Greatest patchiness probably occurs in seral shrublands with scattered clumps of invading shrubs and trees.

Shrub communities have their own distinctive animal life. Seral shrub communities support not only species common to shrubby edges of forest and shrubby borders of fields but a number of species dependent upon them, such as bobwhite quail, cottontail rabbit, prairie warbler *(Dendroica discolor)*, and yellow-breasted chat *(Ictera virens)*. In Great Britain some shrub communities, especially hedgerows, have been stable for centuries and many forms of animal life, invertebrate and vertebrate, have become adapted to or dependent upon them. Among these species are the whitethroat *(Sylvia communis)*, linnet *(Acanthis cannabina)*, and blackbird *(Turdus merula)*.

Climax shrub communities have a complex of animal life that varies with the region (see, for example, Bigalke 1979; Dwyer et al. 1979; Newsome and Catling 1979; Blondel 1981; Cody 1973; Schodde 1981). Within the mediterranean-type shrublands and heathlands similarity in habitat structure and in the nature and number of niches has resulted in pronounced parallel and convergent evolution among bird species (see Blondel 1981; Cody 1973).

Function

The functional aspects of shrubland ecosystems are poorly studied, but data from a few mediterranean-type systems provide some interesting insights into nutrient cycling.

In the California chaparral precipitation falls mostly in the cool winter months. About 8 percent of this precipitation is intercepted and evaporated, 15 percent flows into streams, 40 percent penetrates the substrate to flow as groundwater, and about 33 percent is lost as evapotranspiration (Mooney and Parsons 1973). Most of the plant growth and flowering is concentrated in spring; 75 percent of the flowering plants, half of them annuals, bloom in May at the end of the rainy season. However, the evergreen dominant can fix CO_2 throughout the year. The greatest daily carbon gain is in the spring; the lowest during summer drought; and in winter with its short photoperiod, carbon fixation is minimal (Mooney 1981; Mooney and Miller 1985). The amount fixed in summer depends upon the amount of rainfall received during the previous winter. Yearly accumulation of aboveground biomass for a sclerophyllous plant of intermediate age is about 1000 kg/ha (Mooney and Parsons 1973).

The soils of mediterranean-type ecosystems are low in nutrients and are especially deficient in nitrogen and phosphorus. The cycling of these nutrients appears to be tight, conservative, and seasonal (Mooney and Miller 1985; Groves 1981), reflecting rainfall (or the lack of it) and associated microbial activity, high during the wet season. A flush of microbial activity, which involves decomposition of humus and mineralization of nitrogen and carbon, follows wetting of dry soil. Nitrate accumulation depends upon a progressive drying period after rains, when the topsoil gradually dries out to an increasing depth. Improved soil aeration and increasing soil temperature favor rapid bacterial nitrification and the retention of nitrate ions in the soil. As the dry cycle continues, nitrates accumulate and remain fixed in the topsoil along with other nutrients (Schaefer 1973). When the rains arrive and wet the soil, the concentration of nutrients stimulates a flush of growth. If heavy rains suddenly enter dry topsoil, quantities of nutrients may be lost by leaching. In California chaparral much of the nitrogen returned to the soil is lost through erosion (Mooney and Parsons 1973).

Some plants of the mediterranean systems exhibit some nutrient conservation mechanisms. *Ceanothus*, an early successional species in the California chaparral, is a nonleguminous nitrogen fixer (Mooney and Parsons 1973). In the Australian mallee *Atriplex vesticana*, a dominant plant, lowers the nitrogen content of the surrounding soil and concentrates nitrogen through litterfall in the soil directly beneath the plant. More nitrogen, however, is withdrawn from the soil than is returned by litterfall, which represents about 10 percent of the total plant nitrogen. Litter has lower nitrogen and phosphorus content than fresh leaves, which suggests that the plant withdraws nitrogen and phosphorus from the leaf into the stem before the leaf falls.

Australian sclerophyllous shrubs also exhibit phosphorus conservation, involving three mechanisms. First, phosphorus, like nitrogen, in aging leaves is recirculated through the plant with minimal losses to litter. Second, a fine mat of proteoid or mycorrhizal roots penetrates the decomposing litter and takes up phosphorus. Third, polyphosphate forming and hydrolyzing enzymes are present in the roots of sclerophyllous plants. As orthophosphate is released from decomposing litter in spring, it is stored in roots as long-chain polyphosphate. When growth begins, the polyphosphate seems to be hydrolyzed back to orthophosphate and transported to the growing shoots (Specht 1973).

The best available data for mineral cycling in a mediterranean-type ecosystem are for a 17-year-old *Quercus coccifera* garrigue in southern France (Figure 30.17). The total aboveground mineral mass is 773 kg/ha; 629 kg is in wood, and 144 kg in leaves. Calcium is the most important element, amounting to 485 kg/ha, followed by nitrogen, 159 kg/ha, and potassium, 85 kg/ha. The mean annual incorporation of biogenic elements into perennial organs amounts to 37 kg/ha, of which two-thirds is calcium. Annual return of elements through the litter amounts to 75.2 kg/ha. Net primary productivity is on the order of 3.4 tn/ha/yr

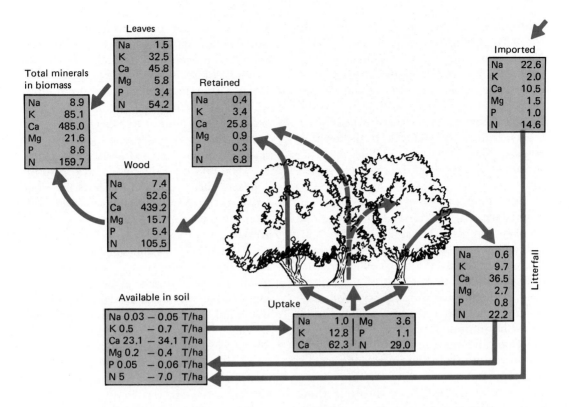

Figure 30.17 Annual turnover of macronutrients in the *Quercus coccifera* garrigue at Le Puech de Juge near Montpellier in southern France. (After Lossaint 1973.)

with a mean annual production of litter of 2.3 tn/ha (Lossaint 1973; Rapp and Lossaint 1981).

Deserts

Geographers define deserts as land where evaporation exceeds rainfall. No specific rainfall serves as a criterion, but deserts may range from extremely arid ones to those with sufficient moisture to support a variety of life. Deserts have been classified according to rainfall into *semideserts*, with precipitation between 150 and 300 to 400 mm per year; *true deserts*, with precipitation below 120 mm per year; and *extreme deserts*, areas with rainfall below 70 mm per year (Shmida 1985). Deserts, which occupy about 26 percent of the continental area, occur in two distinct belts between 15° and 35° latitude in both the Northern and Southern hemispheres—around the Tropic of Cancer and the Tropic of Capricorn.

Deserts are the result of several forces. One force that leads to the formation of deserts and the broad climatic regions of Earth is the movement of air masses over Earth (see Chapter 4). High-pressure areas alter the course of rain. The high-pressure cell off the coast of California and Mexico deflects rainstorms moving south from Alaska to the east and prevents moisture from reaching the southwest. In winter high-pressure areas move southward, allowing winter rains to reach southern California and parts of the North American desert. Winds blowing over cold waters become cold also; they carry little moisture and produce little rain. Thus the west coast of California and Baja California, the Namib desert on coastal southwest Africa, and the coastal edge of the Atacama in Chile may be shrouded in mist, yet remain extremely dry.

Mountain ranges also play a role in desert formation by causing a rain shadow on their lee side. The High Sierras and the Cascade Mountains intercept rain from the Pacific and help maintain the arid conditions of the North American desert. The low eastern highlands of Australia block the southeast trade winds from the interior. Other deserts, such as the Gobi and the interior of the Sahara, are so remote from the source of oceanic moisture that all the water has been wrung from the winds by the time they reach those regions.

Characteristics of Deserts

All deserts have in common low rainfall, high evaporation (7 to 50 times as much as precipitation), and a wide daily range in temperature, from hot by day to cool by night. Low humidity allows up to 90 percent of solar insolation to penetrate the atmosphere and heat the ground. At night the desert yields the accumulated heat of the day back to the atmosphere. Rain, when it falls, is often heavy, and unable to soak into the dry earth, it rushes off in torrents to basins below.

The topography of the desert, unobscured by vegetation, is stark and, paradoxically, partially shaped by water. Unprotected, the soil erodes easily during violent storms and is further cut away by the wind. Alluvial fans stretch away from eroded, angular peaks of more resistant rocks. They join to form deep expanses of debris, the *bajadas*. Eventually the slopes level off to low basins, or *playas*, which receive waters that rush down from the hills and water-cut canyons, or *arroyos*. These basins hold temporary lakes after the rains, but water soon evaporates and leaves behind a dry bed of glistening salt.

The aridity of the desert may seem inimical to life; yet in the deserts life does exist, surprisingly abundant, varied, and well adapted to withstand or circumvent the scarcity of water.

Deserts are not the same everywhere. Differences in moisture, temperature, soil drainage, topography, alkalinity, and salinity create variations in vegetation cover, dominant plants, and groups of associated species. There are hot deserts and cool deserts, extreme deserts and semideserts, ones with sufficient moisture to verge on being grasslands or shrublands, and gradations between those extremes within continental deserts. However, there is a certain degree of similarity among hot deserts and cold deserts of the world. The cold deserts, including the Great Basin of North America, the Gobi, Takla Makan, and Turkestan deserts of Asia, and high elevations of hot deserts are dominated by *Artemisia* and chenopod shrubs, and may be considered shrub steppes. The northern part of the North American Great Basin is dominated by nearly pure stands of big sagebrush (see Figure 30.13) and the southern part by shadscale and bud sage *(Artemisia)*. The hot deserts range from no or scattered vegetation to some combination of chenopods, dwarf shrubs, and succulents. (For a detailed review of world deserts see Evanari, Noy-Meir, and Goodall eds. 1985, 1986). The deserts of southwestern North America—the Mojave, the Sonoran, and the Chihuahuan—are dominated by creosote bush *(Larrea di-*

varicata) and creosote bush *(Franseria spp.)*. Areas of favorable moisture support tall growths of *Acacia*, saguaro *(Cereus giganteus)*, palo verde *(Cercidium spp.)*, and ocotillo *(Fouquieria spp.)*.

Structure

Woody-stemmed and soft brittle-stemmed shrubs are characteristic desert plants (Figure 30.18). In a matrix of shrubs grows a wide assortment of other plants, the yucca, cacti, small trees, and ephemerals. In the Sonoran, Peru-Chilean, South African Karoo, and southern Na-

Figure 30.18 Two hot deserts dominated by woody, brittle-stemmed shrubs. (a) Edge of the Great Victorian desert in Australia. (b) Chihauhuan Desert in Nuevo Leon, Mexico. The substrate of this particular desert is sand-sized particles of gypsum. The sparseness and spacing of the shrubs reflects extreme aridity. (Photos by R. L. Smith.)

mib deserts, large succulents rise above the shrub level and change the aspect of the desert far out of proportion to their numbers (Figure 30.19). The giant saguaro, the most massive of all cacti, grows on the bajadas of the Sonoran desert. Ironwood, smoketree, and palo verde grow best along the banks of intermittent streams, not so much because their moisture requirements demand it, but because their hard-coated seeds must be scraped and bruised by the grinding action of sand and gravel during flash floods before they can germinate.

Both plants and animals are adapted to the scarcity of water either by drought evasion or by drought resistance (see Chapter 5). Plant drought-evaders flower only in the presence of moisture. They persist as seeds during drought periods, ready to sprout, flower, and produce seeds when moisture and temperature are favorable. There are two periods of flower-

(a)

(b)

Figure 30.19 Saguaro dominates the aspect of this portion of the Sonoran Desert in the southwestern United States. Note the water-shaped topography. (Photo courtesy Arizona Fish and Game Department.)

ing in the North American deserts, after winter rains come in from the Pacific Northwest and after summer rains move up from the southwest out of the Gulf of Mexico. Some species flower only after winter rains, others only after summer rains, and a few bloom during both seasons. If rains fail, these ephemeral species do not bloom. Drought-evading animals, like their plant counterparts, adopt an annual lifestyle or go into estivation or some other stage of dormancy during the dry season. If extreme drought develops during the breeding season, birds fail to nest and lizards do not reproduce.

Belowground biomass in the desert can be as patchy as the aboveground biomass. Desert plants may be deep-rooted woody shrubs, such as mesquite (*Prospis spp.*) and *Tamarix*, whose taproots reach the water table, rendering them independent of water supplied by rainfall. Some, such as *Larrea* and *Atriplex*, are deeprooted perennials with superficial laterals that extend as far as 15 to 30 m from the stems. Other perennials, such as the various species of cactus, have shallow roots often located no more than a few centimeters below the surface. Ephemerals have shallow and poorly branched roots reaching a depth of about 30 cm, where they pick up moisture quickly from light rains (see Drew 1979).

The desert floor is stark, a raw mineral substrate devoid of a continuous litter layer. Dead leaves, bud scales, and dead twigs, mostly associated with drought-resistant species that shed them to reduce transpiring surfaces, accumulate in wind-protected areas beneath the plants and in depressions in the soil.

Function

The desert ecosystem differs from other ecosystems in at least one important way: The input of precipitation is highly discontinuous. It comes in pulses as clusters of rainy days 3 to 15 times a year, of which only 1 to 6 may be large enough to stimulate biologic activity. Thus the desert ecosystem experiences periods of inactive steady states broken by periods of production and reproduction. Both processes are stimulated by rain and continue through short periods of adequate moisture; when water is scarce again, both production and biomass return to some low steady state.

Primary production in the desert depends upon the proportion of available water used and the efficiency of its use. Data from various deserts in the world (for a summary, see Noy-Mier 1973, 1974) suggest that annual net primary production of aboveground vegetation varies from 30 to 200 g/m^2.

The amount of biomass that accumulates and rate of turnover (ratio of production to biomass) depend upon the dominant type of vegetation. In deserts such as the Sonoran, where trees, shrubs, and cacti dominate, annual productivity is about 10 to 20 percent of the aboveground standing crop biomass of 300 to 1000 g/m^2. In deserts with perennial type vegetation, annual production is 20 to 40 percent of biomass of 150 to 600 g/m^2. Annual or ephemeral communities have a 100 percent turnover of both roots and aboveground foliage; annual production is the same as peak biomass. These turnover rates are higher than those of forests and tundra. The ratio of belowground biomass to aboveground (stems and foliage) for perennial grasses and forbs is between 1 and 20 and for shrubs between 1 and 3. In general, hot desert plants do not have a high root-to-shoot ratio. The root biomass is relatively small. Cold desert plants have a higher root-to-shoot ratio.

Adding to primary production in the desert are lichens and green and blue-green algae, abundant as soil crusts. Blue-green algal crusts, whose biomass ranges up to 240 kg/ha, have an unusually high rate of nitrogen fixation, 10 to 20 g/m^2/yr. In spite of high fixation rates, only 5 to 10 g/m^2 of total nitrogen input becomes part of higher plants (Figure 30.20). Approximately 70 percent of the nitrogen is short-circuited back to the atmosphere as volatilized ammonia and as N$_2$ from denitrification, speeded by dry alkaline soils (Reichle 1975; West 1979).

Nutrient cycling in arid ecosystems is tight, and two major nutrients, phosphorus and nitrogen, are in short supply (Figure 30.21). Much of the nutrient supply is tied up in plant biomass, living and dead; the tissues of desert plants have higher concentrations of nutrients than plants of mesic environments, and they tend to retain certain elements before shedding any parts. For example, the nitrogen and phosphorus content of fallen phyllodes (the enlarged and commonly flattened leaf stalks that function as leaves) of *Acacia aneura* in the northern Australian desert decreases markedly at the time of major phyllode fall, but the content of potassium, calcium, and magnesium show little change. The desert plant *Artemensia* translocates phosphorus and potassium back to the twig before shedding its leaves. The nutrients retained in the shed parts collect and decompose beneath the plants, where microclimate conditions created by the shrubs favor biological activity. The soil is further enriched by animals attracted to the shade (Binet 1981; West 1979, 1981; West and Skugins 1978). The plants, in effect, create islands of fertility beneath themselves (Figure 30.21).

In spite of its aridity, desert ecosystems support a surprising diversity of animal life, notably herbivorous species (Figure 30.22). Grazing herbivores of the desert tend to be generalists and opportunists in their mode of feeding. They consume a wide range of species, plant types, and parts. Desert sheep feed on succulents and ephemerals when available and then switch to woody browse during the dry period. As a last resort herbivores consume dead litter and lichens. Small herbivores—the desert rodents, particularly the family Heteromyidae, and ants—tend to be granivores, feeding largely on seeds. They are important in the dynamics of desert ecosystems. One of the notable small herbivores of the Sonoran Desert is the harvester ant *(Pogonomyrmex occiden-*

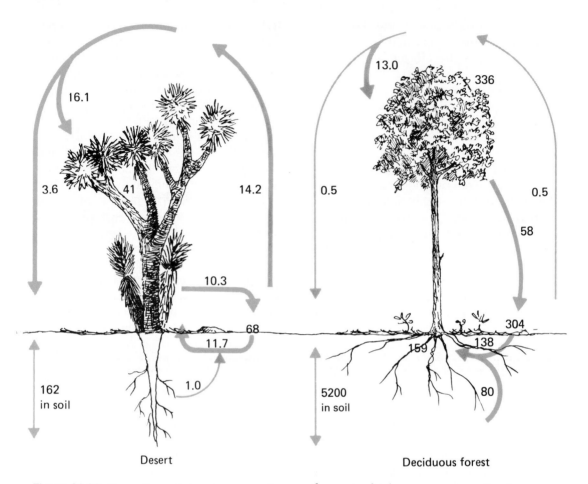

16.1

3.6   41   14.2

10.3

68

11.7

1.0

162
in soil

Desert

13.0   336

0.5   0.5

58

304

159   138

5200   80
in soil

Deciduous forest

Figure 30.20 Comparison of the nitrogen cycle of the desert with that of the temperate deciduous forest. In the forest a considerable portion of the nitrogen is cycled through the plants. Although desert algae fix considerable quantities of nitrogen, most of it is lost through denitrification. (After Reichle 1975.)

talis), which lives on seeds gathered from the desert floor and stored in underground granaries. During periods of drought these ants gather mainly the seeds of several species of perennials. During winter rains, when annual plants flower and seed, ants gather the seeds of these plants. By utilizing different plants,

the ant obtains a constant food source throughout the desert year.

Herbivores can have a pronounced impact on primary producers, especially if they are more abundant then the range's capacity to support them. Once grazers have utilized annual production, they consume plant reserves, especially in long dry periods. Overbrowsing can so weaken the plant that vegetation is destroyed or irreparably damaged. Areas protected from grazing, especially by goats and cattle, have higher biomass and a greater percentage of palatable species than grazed areas.

Herbivores in a shrubby desert, under most conditions, consume only a small

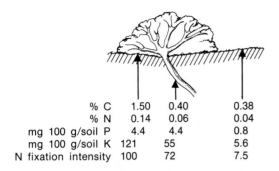

% C	1.50	0.40	0.38
% N	0.14	0.06	0.04
mg 100 g/soil P	4.4	4.4	0.8
mg 100 g/soil K	121	55	5.6
N fixation intensity	100	72	7.5

Figure 30.21 Organic content and carbon content of desert soil beneath and away from the base of an Egyptian desert shrub, *Anabasis aretioides*. The values from top to bottom indicate the percent of carbon, percent of nitrogen, amount of phosphorus (mg 100 g/soil), amount of potassium (mg 100 g/soil), and intensity of nitrogen fixation, the value of 100 being attributed to the point situated at the soil surface, under the plant canopy. (From Binet 1981: 327.)

Figure 30.22 Although the typical herbivores of the desert are rodents, some large herbivores such as these gemsbok in the Kalahari Desert are well adapted to arid desert conditions. (Photo by T. M. Smith.)

part of the aboveground primary production, but seed-eating herbivores can consume most of the seed production. In one of the few studies of herbivory in the desert, Chew and Chew (1970) found that small grazing herbivores (the jackrabbit and kangaroo rat) used only about 2 percent of the aboveground net primary production, but consumed 87 percent of the seed production, a rate of consumption that could have a pronounced effect on plant composition and plant populations. Fifty-five percent of energy flow through small mammals in the shrub desert passed through the kangaroo rat, 22 percent through the browsing jackrabbit, and 6.5 percent through the insectivorous grasshopper mouse *(Onychomys torridus)*.

Carnivores, like the herbivores, are opportunistic feeders, with few specialists. Most desert carnivores, such as foxes and coyotes, have mixed diets that include leaves and fruits, and even insectivorous birds and rodents consume herbivorous foods. Omnivory rather than carnivory

and complex food webs seems to be the rule in the desert ecosystem.

The detrital food chain seems to be less important in the desert than in other ecosystems. Although most functional and taxonomic groups of soil microorganisms exist in the desert, fungi and actinomycetes are prominent. Microbial decomposition, like the blooming of ephemerals, is limited to short periods when moisture is available. For this reason dry litter tends to accumulate until the detrital biomass is greater than aboveground living biomass. Most of the ephemeral biomass disappears through grazing, weathering, and erosion. Decomposition proceeds mostly through detritus-feeding anthropods such as termites that ingest and break down woody tissue in their guts. In some deserts considerable amounts of nutrients may be locked up in termite structures, to be released when the structure is destroyed. Other important detritivores are acarids and various isopods.

■ Summary

Natural grasslands occupy regions where rainfall is between 25 cm and 70 cm a year, but they are not exclusively climatic. Many persist through the intervention of fire and human activity. Once covering extensive areas of the globe, grasslands have shrunk to a fraction of their original size because of conversion to cropland and grazing lands. Disappearing along with the native grasslands were the native grazing herbivores, replaced in part by domestic livestock. Conversion of forests into agricultural lands, the planting of hay and pasture fields, and development of successional grasslands on disturbed sites have extended the range of some grassland animals into once forested regions. Successional and climax grasslands consist of sod formers, bunch grasses, or both. Depending upon their fire history and degree of grazing, grasslands accumulate a layer of mulch that retains moisture, influences the character and composition of plant life, and provides shelter and nesting sites for some animals.

Productivity varies considerably, influenced greatly by precipitation. It ranges from $82g/m^2/yr$ in semiarid grasslands to 30 times that much in subhumid, tame, and cultivated grasslands. The bulk of primary production goes underground to the roots. To a point grazing stimulates primary production. Although the most conspicuous grazers are the large herbivores, the major consumers are invertebrates. The heaviest consumption takes place below ground, where the dominant herbivores are nematodes. Most of the primary production goes to decomposers. Nutrients are recycled rapidly. A significant quantity goes to the roots, to be moved above ground to next year's growth.

Savannas are grasslands with woody vegetation. They are characteristic of regions with alternating wet and dry seasons. Difficult to characterize precisely, savannas range from grass with an occasional tree to shrub and tree savannas. The latter grade into woodland and thornbush with an understory of grass. Much of the nutrient pool is tied up in plant and animal biomass, but nutrient turnover is high with little accumulation of organic matter.

Shrublands, which go by different names in various parts of the world, dominate regions with a mediterranean-type climate in which winters are mild and wet and summers are long, hot, and dry. Successional shrublands occupy land in transition from grassland to forest. Such shrubland may remain stable for years.

Shrublands characteristically have a densely branched woody structure and low height. The success of shrubs depends upon their ability to compete for nutrients, energy, and space. In semiarid situations shrubs have numerous competitive advantages, including structural modifications that affect light interception, heat losses, and evaporative losses. Growth in mediterranean-type shrublands is concentrated at the end of the wet season, when nutrients in solution and a relative abundance of moisture produce a flush of vegetation. Nutrient cycling, especially of nitrogen and phosphorus, is tight. Many plants translocate nutrients from leaves to stem and roots before leaf fall; others concentrate nitrogen in litterfall, which the plants take up again quickly in the wet season.

Deserts occupy about one-seventh of Earth's land surface and are largely confined to two worldwide belts, around the Tropic of Cancer and the Tropic of Capricorn. Deserts result largely from the climatic patterns of Earth, rain-blocking mountain ranges, and remoteness from sources of oceanic moisture. Two types of deserts exist, cool deserts, exemplified by the Great Basin of North America, and hot deserts.

The desert is a harsh environment in which plants and animals have evolved ways of circumventing aridity and high temperature by becoming either drought-evaders or drought-resistors. Functionally deserts are characterized by low net production, by opportunistic feeding patterns for herbivores and carnivores, and by a detrital food chain that is less important than in other ecosystems. Crustlike growths of blue-green algae on the desert floor fix quantities of nitrogen, but most of it is lost to the atmosphere.

Forests and Tundras

Of all the vegetation types of the world probably none is more widespread or more diverse than the forest (see inside cover). A map of world vegetation shows forest growth in distinct bands. Starting below the tundra in the Northern Hemisphere are consecutive belts of coniferous, temperate deciduous, and tropical forests, most extensive in the Southern Hemisphere. Within these bands of global forests are a diversity of forest types.

Regardless of their types, all forests possess large aboveground biomass. This biomass creates several layers or strata of vegetation, all of which influence both the vertical structure and the environmental conditions within the stand: light, moisture, temperature, wind, and carbon dioxide. Many differences in physical and environmental structures will emerge as we discuss representative forest types.

■ Coniferous Forests

Types

Boreal Forests or Taiga. The taiga forms a circumpolar belt of coniferous forest about the Northern Hemisphere. Its northern limit is roughly along the July 13 isotherm, the southern extent of the Arctic front in summer. Its southern limit, much less abrupt, is more or less marked by the winter position of the Arctic front, roughly just north of the 58° N latitude. In Eurasia it begins in Scandinavia and extends across the continent to northern Japan. In Europe the forest is dominated by Norway spruce *(Picea abies);* in Siberia by Siberian spruce *(P. obovata),* Siberian stone pine *(Pinus sibirica),* and larch *(Larix sibirica);* and in the Far East by Yeddo spruce *(P. jezoensis).* In North America the taiga extends from Labrador across Canada and through Alaska to the Brooks Range. It is dominated by four genera of conifers, *Picea, Abies, Pinus,* and *Larix,* and two genera of deciduous trees, *Populus* and *Betula* (Figure 31.1). Dominant trees include black spruce *(P. mariana)* and jack pine *(Pinus banksiana).* Although the taiga throughout may have the same general appearance, its vegetation exhibits important regional differences.

The boreal forest comprises four major vegetation zones: the forest-tundra ecotone, characterized by open stands of stunted spruce, lichens, and moss; the open boreal woodland, characterized by lichen-black spruce woodland; the main boreal forest with continuous stands of coniferous trees and a moss and low shrub understory broken up by poplar and birch on disturbed areas; and the boreal-mixed forest ecotone, where the boreal forest

Figure 31.1 The boreal forest extends from the taiga, dominated by black spruce, at its northern limits to the spruce-birch forest at its southern limits. This forest of red spruce and gray birch in eastern United States is an ecotone between the spruce and northern hardwoods forest. (Photo by R. L. Smith.)

grades into the mixed forest of southern Canada and the northern United States (Oechel and Lawrence 1985).

Throughout much of its extent the boreal forest experiences great seasonal fluctuations of temperature, freeze-thaw cycles that affect shallow-rooted trees, nutrient-poor podzolic soils, and depending upon the region, the presence or absence of permafrost. Occupying, for the most part, glaciated land, the taiga is also a region of cold lakes, bogs, rivers, and alder thickets.

Temperate Rain Forests. South of Alaska the coniferous forest differs from the northern boreal forest, both floristically and ecologically. The reasons for the change are both climatic and topographic.

Moisture-laden winds move in from the Pacific, meet the barrier of the Coast Range, and rise abruptly. Suddenly cooled by this upward thrust into the atmosphere, the moisture in the air is released as rain and snow in amounts up to 635 cm/yr. During the summer, when winds shift to the northwest, the air is cooled over chilly northern seas. Although rainfall is low, cool air brings in heavy fog, which collects on the forest foliage and drips to the ground to add 127 cm or more of moisture. This land of superabundant moisture, high humidity, and warm temperatures supports the temperate rain forest, a community of luxuriant vegetation dominated by a variety of conifers well adapted to wet, mild winters, dry warm summers, and nutrient-poor soils (see Franklin and Dyrness 1973; Waring and Franklin 1979; Franklin and Waring 1979; Lassoi, Hinckley, and Grier 1985). The forest (Figure 31.2) is dominated by western hemlock *(Tsuga heterophylla)*, mountain hemlock *(T. mertensiana)*, Pacific silver fir *(Abies amabilis),* and Douglas-fir, all trees high in foliage and stem biomass. Further south, where precipitation still is high, grows the redwood *(Sequoia sempervirens)* forest, occupying a strip of land about 724 km long.

Western Montane Forests. The air masses that drop their moisture on the western slopes of the Coast Range descend the eastern slopes, heat, and absorb moisture, creating the conditions that produce the Great Basin desert (see Chapter 30). The same air rises up the western slopes of the Rockies, cools, and drops moisture again, although far less than on the Coast Range. Here in the Rocky, Wasatch, Sierra Nevada, and Cascade Mountains develop several coniferous forest associations (see Franklin and Dyrness 1973; Pfister et al. 1977). In the southwestern

Figure 31.2 The temperate rain forest embraces the redwood and sequoia forests of California and the Douglas-fir-western hemlock-Sitka spruce forests of the northwest Pacific Coast. Typical is this Douglas-fir stand with an abundant western hemlock understory. (Photo courtesy U. S. Forest Service.)

United States these coniferous forests occur between 2500 and 4200 m elevation and in the northern United States and Canada between 1700 and 3500 m elevation. At high elevations in the Rocky Mountains, where winters are long and snowfall is heavy, grows the subalpine forest, dominated by Engelmann spruce *(Picea engelmannii)* and subalpine fir *(Abies lasiocarpa)* (Figure 31.3). Mid-elevations have stands of Douglas-fir and lower elevations are dominated by open stands of ponderosa pine *(Pinus ponderosa)* and open to thick stands of the early successional pioneering conifer, lodgepole pine *(P. contorta).*

Similar forests grow in the Sierras and Cascades. There high elevation forests consist largely of mountain hemlock, red

(a)

(b)

Figure 31.3 (a) A montane coniferous forest in the Rocky Mountains. The drier lower slopes support ponderosa pine; higher elevations are cloaked with Douglas-fir. (b) Subalpine forest dominated by subalpine fir *(Abies lasiocarpa)*. This tree grows with Engelmann spruce and mountain hemlock. (Photos by R. L. Smith.)

fir *(Abies magnifica)*, and lodgepole pine. Among them grow sugar pine *(Pinus lambertiana)*, incense-cedar *(Libocedrus decurrens)*, and the largest tree of all, the giant sequoia *(Sequoiadendron giganteum)*, which grows only in scattered groves on the west slopes of the California Sierras.

A deciduous seral stage species, but occasionally permanent, common both to the montane and the boreal forest is trembling aspen *(Populus tremuloides)*, the most widespread tree of North America (Figure 31.4).

Montane coniferous forests dominated by pine also occur in Mexico and

Figure 31.4 Quaking aspen *(Populus tremuloides)* is the dominant deciduous tree in the western montane forest. A successional species, it often forms long-lived patchy stands that turn golden yellow in autumn. (Photo R. L. Smith.)

Guatemala. In Europe forests of Scots pine *(Pinus sylvestris)* extend across the mountain ranges from Norway and Siberia to the Alps and Spanish Sierra Nevada.

Southern Pine Forests. The pine forests of the coastal plains of the South Atlantic and Gulf states (Figure 31.5) are usually considered part of the temperate deciduous forest because they represent a seral rather than final stage. These pines maintain their presence by possessing a competitive advantage over hardwoods on nutrient-poor, dry sandy soil and by their

Figure 31.5 Homogeneous stands of pine cover much of the lowland coastal plain and piedmont of the southern United States. Many of these stands are dominated by loblolly pine. (Photo courtesy U.S. Forest Service.)

adaptation to a fire regime. At the northern end of the coastal pine forest in New Jersey pitch pine *(Pinus rigida)* is the dominant species. Further south, loblolly pine *(P. taeda)*, longleaf pine *(P. palustris)*, and slash pine *(P. caribaea)* are most abundant. Where fires are allowed to burn or controlled burning is practiced, the understory is open and dominated by wiregrass *(Aristida)*.

Structure

Coniferous forests can be divided into three broad classes by growth form and general growth behavior: (1) pines with straight, cylindrical trunks, whorled spreading branches, and a crown density that varies with the species from the dense crowns of red and white pine to the relatively open thin crowns of Virginia pine, jack pine, Scots pine, and lodgepole pine; (2) spire-shaped evergreens, including spruce, fir, Douglas-fir, and (with some exceptions) the cedars, with more or less tall pyramidal crowns, gradually tapering trunks, and whorled, horizontal branches; (3) deciduous conifers such as large and bald cypress, with pyramidal, open crowns that shed their needles annually. Growth form and behavior influence animal life and other aspects of coniferous ecosystems.

Vertical stratification in coniferous forests is not well developed. Because of a high crown density, the lower strata are poorly developed in spruce and fir forests and the ground layer consists largely of ferns and mosses with few herbs. The maximum canopy development in spire-shaped conifers is about one-third down from the open crown, which gives such forests a profile different from that of pines. Pine forests with a well-developed high canopy lack lower strata. However, old stand, open-crowned pines may have three strata: an upper canopy, a shrub

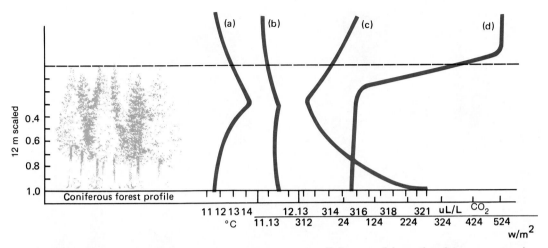

Figure 31.6 Mean hourly profiles of temperature (a), humidity (b), carbon dioxide (c), and net radiation (d) in a coniferous forest, a Sitka spruce plantation in Scotland. CO_2 increases below the canopy and above the forest floor because the soil is a source of CO_2. The hollow in the humidity profile in the lower canopy and beneath it probably represents a horizontal humidity flux caused by increased air flow in the open spire-shaped canopy. Height of 12 m is scaled as reduced height for comparison with the deciduous profile in Figure 31.12.

layer, and a thin herbaceous layer. The litter layer in coniferous forests is usually deep, poorly decomposed, and on top of instead of mixing in with the mineral soil.

Vertical stratification influences the environmental stratification within the stand (Figure 31.6). Light intensity is progressively reduced though the canopy to only a fraction of full sunlight (see Chapter 7). The upper crown of spruce and firs, a zone of widely spaced narrow spires, is open and well lighted, whereas the lower crown is dense and intercepts most of the solar radiation. Most pines form a dense upper canopy that excludes so much sunlight that lower strata cannot develop. Open-crowned pines allow more light to reach the forest floor, stimulating a grassy or shrubby understory. Because conifers retain their foliage through the

year, light reaching the lower strata in coniferous forests is about the same throughout the year; but illumination is usually the greatest during midsummer, when the sun's rays are most direct, and the lowest in winter, when the intensity of incident sunlight is the lowest.

The temperature profile of a coniferous forest varies with the growth form. In forests of sprucelike trees temperatures tend to be coolest in the upper canopy, perhaps because of greater air circulation, and reach their highest in the lower canopy (see Figure 31.6).

Animal life in the coniferous forest varies widely, depending upon the nature of the stand. Soil invertebrate litter fauna is dominated by mites. Earthworm species are few and their numbers low. Insect populations, although not diverse, are high in numbers, and encouraged by the homogeneity of the stands, are often destructive. Spruce budworm (*Choristoneura fumiferana*), found throughout the boreal forest, attacks balsam fir and spruces. Related species attack jack pine and red pine. Sawflies (*Neodriprion*) attack a wide variety of pines, including pitch, Virginia, shortleaf, and loblolly.

A number of bird species are closely associated with coniferous forests. In North America they include chickadees,

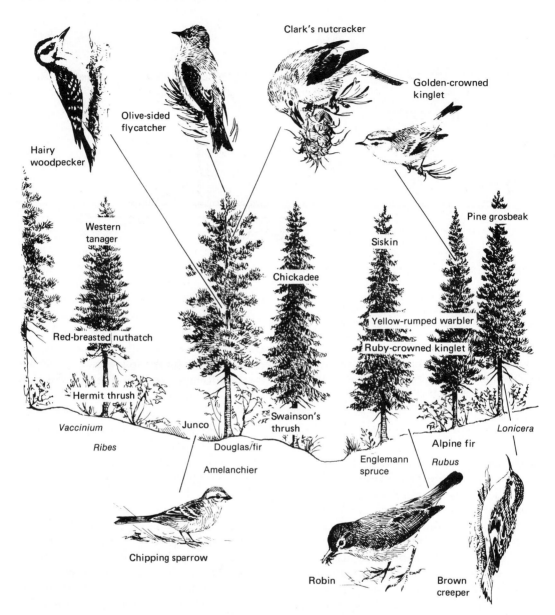

Figure 31.7 Vertical distribution of some birds in a spruce-fir forest in Wyoming. (After Salt 1967.)

kinglets, pine siskins, crossbills, purple finches, and hermit thrush (Figure 31.7). Related species, the tits and grosbeaks, are common to European coniferous forests.

Species diversity varies. In general, coniferous forests of northeastern and southeastern North America and the Sierra Nevadas support the richest avifaunas (Wiens 1975). Seventeen to 24 percent of all individuals present belong to a single dominant species. Bird densities are highest in the Pacific Northwest and lowest in the immature northeastern coniferous forest. There more than 50 percent of the individuals are warblers, whereas in western coniferous forest less than 10 percent are warblers. Foliage-feeding insectivorous birds are dominant in all types of coniferous forests; in North America they are more dominant in the north, northeast, and southeast than in the west and northwest.

Except for strictly boreal species, such as the pine martin and lynx, mammals have much less affinity for coniferous forests. Most are associated with both coniferous and deciduous forest; the white-tailed deer, moose, black bear, and mountain lion are examples. Their north-south distribution seems to be limited more by climate, especially temperature, than by vegetation. The red squirrel, commonly associated with coniferous forests, is common in deciduous woodlands in the southern part of its range.

Function

Ecosystem function has been studied most extensively in forest ecosystems (see Reichle 1981). One type of coniferous forest ecosystem studied in detail is Douglas-fir. Nutrient budgets and cycling have been prepared for both young and old growth stands (Cole et al. 1969; Cole and Rapp 1981; Johnson et al. 1982). Differences between a 36-year-old Doulgas-fir stand and a 450-year-old stand are summarized in Table 31.1 and Table 31.2). The distribution of organic matter and nitrogen differs between the two stands. Although the old stand had considerably larger total biomass than the young stand, and both had similar foliage biomass, the young stand had the higher percentage of its living biomass in foliage and roots. Most of the living biomass in both age classes was in branches and bole. Litter accumulation accounted for the greater part

TABLE 31.1 Organic Matter Distribution, kg/ha (% in parentheses) for Young and Old-Growth Douglas-Fir, and Young Loblolly Pine Ecosystems

Component	Douglas-Fir 36-Yr-Old	Douglas-Fir 450-Yr-Old	Loblolly Pine 16-Yr-Old
Overstory			
Foliage	9,097 (4)	8,906 (2)	9,700 (4)
Branches	22,031 (11)	48,543 (8)	27,900 (12)
Stemwood	121,687 (60)	472,593 (78)	132,180 (57)
Bark	18,728 (9)	*	18,460 (8)
Roots	32,986 (16)	74,328 (12)	44,550 (19)
Total	204,529	604,370	232,490
Subordinate vegetation	1,010	9,864	
Forest Floor			
Wood	6,345 (1)	55,200 (1)	
Litter &	16,427 (18)	43,350 (55)	
humus Total	22,772	98,550	
Soil	111,552 (82)	79,250 (45)	
Total Ecosystem	339,863	792,034	

Source: Data for Douglas-fir from Johnson et al. 1982; for loblolly pine from Jorgensen and Wells 1986.

*Value for bark included in stemwood.

TABLE 31.2 Nitrogen Distribution (% in parentheses), kg/ha, for Young and Old-Growth Douglas-Fir and Young Loblolly Pine Ecosystems

Component	Douglas-Fir 42-Yr-Old	Douglas-Fir 450-Yr-Old	Loblolly Pine 16-Yr-Old
Overstory			
Foliage	102 (32)	75 (16)	81 (26)
Branches	61 (19)	49 (10)	60 (19)
Stemwood	77 (24)	189 (40)	78 (24)
Bark	48 (15)	162 (34)	36 (11)
Roots	32 (10)	475	64 (20)
Total	320		319
Subordinate	6		
Vegetation	58		
Forest Floor			
Wood	14	132	
Litter &	161	434	
humus	175	566	307
Total			
Soil	2809	4300	1750
Total Ecosystem	3310	5399	2376

Source: Data for Douglas-fir from Johnson et al. 1982; for loblolly pine from Jorgensen and Wells 1986.

of organic detrital matter in the old stand; in the young stand, soil held most of the detrital organic matter. Litter organic matter exceeded soil organic matter in the old stand because of the accumulation and slow decomposition of needles and long-term decomposition of large fallen trunks and limbs.

The nitrogen budget emphasizes the importance of internal cycling and storage within the system. The young stand had most of its N in foliage and bole whereas the old stand had most of its N in bole and root. The 450-year-old stand accumulated considerably more nitrogen, but had a lower percentage of it in the soil. The young stand had 98 percent of its detrital nitrogen accumulated in the soil. Input of nitrogen into both systems was low and approximately the same, about 2 kg/ha/yr. The old stand returned somewhat more nitrogen to the forest floor than the young stand, mostly because of greater stemflow and throughfall. The young

stand had a considerably greater N requirement and a larger uptake than the old stand, 25.1 versus 14.5 kg/ha/yr. This uptake amounted to about 55 percent of the young stand's requirement and 44 percent of the old stand's needs. In both stands the deficiency of uptake was made up by recycling elements within the biomass. This internal cycling is also typical of other nutrients except for calcium and magnesium (Table 31.3). Uptake of those nutrients far exceeds requirement, so that cycling within the tree is not necessary.

Contrasting with the Douglas-fir stands of the Pacific Northwest is loblolly pine of the coastal plains of the southeastern United States. Loblolly pine grows on a wide range of soils, from poorly drained lowland sites where it does best to dry, poor upland sites. This pine grows rapidly, achieving a diameter of 20 cm and a height of 18 m by age 20. At this age loblolly pine ties up a large portion of the site's nutrients (Table 31.3) (a trait that

TABLE 31.3 Nitrogen Transfers (kg/ha/yr) in Young and Old-Growth Douglas-Fir and Young Loblolly Pine Ecosystems

Component	Douglas-Fir 42-Yr-Old	Douglas-Fir 450-Yr-Old	Loblolly Pine 16-Yr-Old
Input	1.67	2.0	5.4
Return to Forest Floor			
Throughfall	0.53	3.4	4.1
Litterfall	25.4	25.6	33.4
Total	25.93	29.0	37.5
Within Vegetation			
Requirement	45.8	33.3	66.0
Redistribution	20.7	18.5	11.6
Uptake	25.1	14.5	54.4

Source: Data for Douglas-fir from Johnson et al. 1982; for loblolly pine from Jorgensen and Wells 1986.

gives conifers the reputation of being accumulator plants), with maximum accumulation by age 30 (Figure 31.8).

Some coniferous ecosystems scavenge nutrients directly from rainfall through microcommunities of algae and lichens colonizing canopy leaves in balsam fir forests (Lang, Reiners, and Heier 1976) and western coniferous forests (Johnson et al. 1982; Carroll 1979). Old growth Douglas-fir supports in its canopy a complex biological community, including primary producers, consumers, and decomposers (Figure 31.9). Cyanophycophilous lichens fix atmospheric nitrogen. Organic nitrogen lost through leaching from lichens combines with canopy moisture to form a dilute organic solution that in turn is taken up by microorganisms and other canopy epiphytes. Part of this microbial production is consumed by canopy arthropods. These nutrient minicycles in the canopy tend to influence and even restrict the amount of nutrients that reach the forest floor by throughfall and stemflow.

Ground moss is an important but overlooked component of some coniferous forests. Binkley and Graham (1981) found that in an old-growth Douglas-fir stand, ground-layer mosses, which ac-count for only 0.13 percent of the aboveground biomass, add 5 percent to the estimated aboveground primary production,

Figure 31.8 Accumulation of macronutrients in whole trees in loblolly pine stands on good sites. Note the rapid accumulation of nitrogen and calcium. (Adapted from Jorgensen and Wells 1986: 9.)

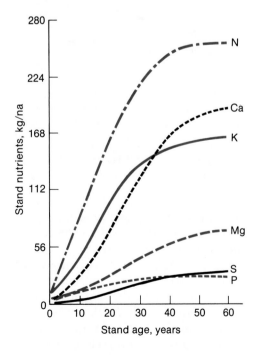

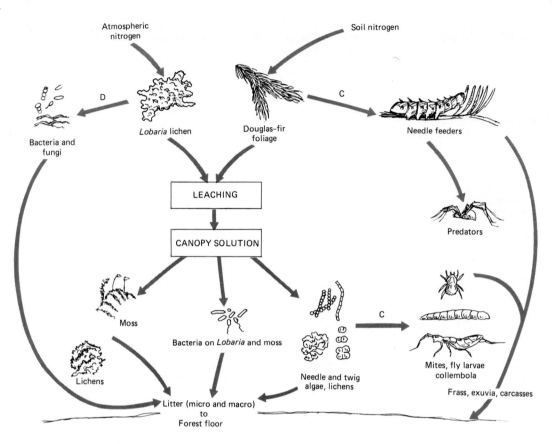

Figure 31.9 Nitrogen cycle in the canopy of old-growth Douglas-fir. Microecosystems exist in high canopy of old Douglas-fir and probably in many other forests, including tropical rain forest. The communities consist of primary producers (lichens) and biophage and saprophage consumers. These microecosystems conserve and recycle nutrients such as nitrogen and influence nutrient return to the forest floor by leaching and throughfall. (After Johnson et al. 1982: 193.)

5 percent to the uptake of calcium and potassium, and 10 percent to the uptake of nitrogen and phosphorus.

Roots, fungi, and mycorrhizae (see Chapter 23) have a role in nutrient cycling in coniferous stands, the extent of which is not well understood. Stark (1972, 1973) provides some interesting insights into the role of fungi in mineral cycling in a xeric stand of Jeffrey pine (*Pinus Jeffreyi*) in Nevada. Litterfall is largely pine needles, which are infected by fungi a few months before they fall. During the first winter under snow on the ground, basidiomycete fungi remove cell contents and most of the nitrogen and phosphorus, leaving mostly cellulose behind. In the second year the remains of needles are incorporated in the fermentation zone of litter, to be acted upon by other fungi.

Stimulating fungal activity in the fermentation zone is the rain of pollen in spring. This pollen rain, from 0.9 to 3.0 kg/ha/yr, supplies among other nutrients 38.0 g N, 15 g K, 8.5 g P, and 0.56 g Ca. Although this level of nutrient input is inconsequential to tree growth, it is essential to fungi in the fermentation zone in summer. Nitrogen and phosphorus apparently stimulate the fungi to complete litter decay and release other elements to tree growth. Because of dry conditions during the summer, minimal amounts of nitrogen and phosphorus are leached from the fermentation layer; pollen is the main nutrient source for fungi.

Fungal hyphae or rhizomorphs in turn concentrate nutrients in their tissues, especially the fruiting body. In comparison to the amount of nutrients found in pine needles, the basidiomycete fruiting bodies are low in calcium, magnesium, and manganese, but high in nitrogen, phosphorus, sodium, and several trace elements. The rhizomorph tissue of fungi is high in calcium, nitrogen, phosphorus, sodium, and zinc. Thus fungi act as a living sink of nutrients. Resistant to leaching, the fungal rhizomorphs hold biologically important elements in the litter. In fact, in an experimental study, rhizomorph tissues held 99.9 percent of 10 elements measured against a leaching force equivalent to one year's precipitation. Unprotected against leaching, most elements would be lost to the ecosystem. Stark suggests that fungi may release the elements slowly to the soil through exudates for recycling.

One component of forest ecosystems too often overlooked is dead wood. The mass of dead wood waxes and wanes with tree mortality and disturbances (Lang 1985) and never achieves equilibrium within a stand (Lang 1985; Long 1982). In general, wood litterfall increases through succession (Long 1982) and becomes most conspicuous in old-age temperate forests, both deciduous and coniferous, where logs and large limbs decay slowly. Fallen trees and large limbs made up over 71 percent of the forest floor mass in a 250-year-old oak forest in New Jersey (Lang and Forman 1975) and over 60 percent in old growth Douglas-fir stands. Volume of woody debris is much greater in old-growth coniferous forests of the Pacific Northwest

than in Eastern deciduous forests (Grier and Logan 1977; Sollins et al. 1980).

Dead wood in the form of large standing dead trees or snags (Figure 31.10) or downed trunks and limbs makes up a unique and critical component of the forest ecosystem (Maser and Trappe 1984; Maser et al. 1979). Standing dead trees provide essential nesting and den sites for cavity-nesting birds and tree-dwelling mammals, food, and foraging areas. Downed dead trees, which may make up 10 to 20 percent of the ground surface in forests, provide food, protection pathways for small mammals, and reproductive sites for certain woody plants. The elimination of standing and fallen dead trees can greatly impoverish animal life in the forest.

■ Temperate Broadleaf Forests

Temperate forests, in spite of their name, do not exist in a temperate environment. They face extreme fluctuations in daily and seasonal temperatures which stresses the physiological activity of plants and animals. Deciduous forests are leafless during the winter and in northern regions remain leafless for the greater part of the year. They are exposed to droughts and in places flooding. In spite of their intemperate environment, temperate forest ecosystems are able to maintain high productivity.

Types

Temperate Deciduous Forests. The temperate deciduous forest once covered large areas of Europe and China, parts of South America and the middle American highlands, and eastern North America (see inside cover). The deciduous forests of Europe and Asia have largely disappeared, cleared for agriculture. The dom-

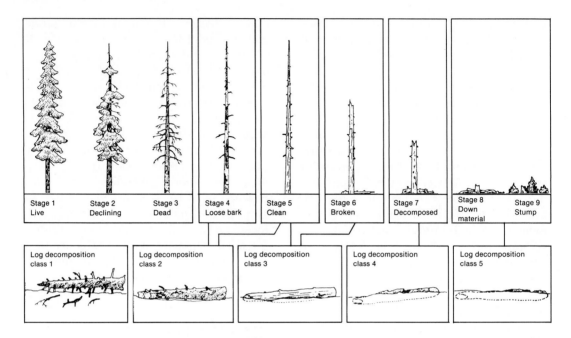

Figure 31.10 An important component of all forest ecosystems is dead wood, both standing and down. Snags or dead trees go through a successional process of decay, each stage of which supports its own group of animal life. When dead trees fall, they enter one of the first four decomposition classes. Fallen logs provide habitats for an array of animal life and germination sites for forest tree seedlings. (After Thomas et al. 1979: 64.)

inant trees include European beech (*Fagus sylvatica*), pendunculate oak (*Quercus robor*), ashes (*Fraxinus spp.*), birches (*Betula spp.*), and elms (*Ulmus spp.*). Because of glacial history (see Chapter 27) the diversity of European deciduous forests does not compare with that of North America or China.

In eastern North America the temperate deciduous forest consists of a number of forest types, which intergrade into one another. The northern segment of the deciduous forest complex is the hemlock-white pine-northern hardwoods forest, which occupies southern Canada and extends southward through northern United States and along the high Appalachians into North Carolina and Tennessee. Beech, sugar maple, basswood, yellow birch, black cherry, red oak, and white pine are the chief components. White pine was once the outstanding tree of the forest, but because most of it was cut before the turn of the century, it now grows only as a successional tree on abandoned land and as scattered trees through the forest.

On relatively flat, glaciated country with its deep, rich soil grow two somewhat similar forests, the beech-sugar maple forest, restricted largely from southern Indiana north to central Minnesota and east to western New York; and the sugar maple-basswood forest, found from Wisconsin to Minnesota south to northern Missouri.

Further south is the extensive central hardwood forest. The central hardwood can be divided into three major types. (1) The cove, or mixed mesophytic, forest consists of an extremely large number of species, dominated by yellow-poplar. This forest, which reaches its best development on the northern slopes and deep coves of the southern Appalachians, is one of the most magnificent in the world. Much of its original grandeur has been destroyed by high-grading and fire, but even in second- and third-growth stands, its richness is apparent. (2) On more xeric sites, the southern slopes and drier mountains, grows the oak-chestnut forest. The chestnut, killed by blight, has been replaced by additional oaks. (3) The western edge of the central hardwoods in the Ozarks and the forests along the prairie river systems are dominated by oak and hickory.

Temperate Woodlands. In western parts of North America where the climate is too dry for montane coniferous forests, we find the temperate woodlands. These ecosystems are characterized by open growth small trees with a well-developed understory of grass or shrubs. There are a number of types of temperate woodlands, which may consist of needle-leaved trees, deciduous broad-leaved trees, or sclerophylls, or any combination of these. An outstanding example is the pinyon-juniper woodland in which two dominant genera, *Pinus* and *Juniperus,* are always associated. This ecosystem is found from the front range of the Rocky Mountains to the eastern slopes of the Sierra Nevada foothills. One of the best examples stands on the Kaibab Plateau of northern Arizona. In southern Arizona, New Mexico, and northern Mexico occur oak-juniper and oak woodlands, and in the Rocky Mountains, particularly in Utah, there are oak-sagebrush woodlands. In the Great Valley of California grows still another type—evergreen-oak woodlands with a grassy undergrowth.

Temperate Evergreen Forests. In several subtropical areas of the world are extensive mixed forests of both broadleaf evergreen and coniferous trees. Such forests include the eucalyptus forests in Australia, paramo forests and anacardia gallery forests of South America and New Cale-

donia, and false beech *(Nothofagus spp.)* forests in Patagonia. Representatives of temperate evergreen forests also occur in the Caribbean and on the North American continent along the Gulf Coast, in the hummocks of Florida Everglades, and in the Florida Keys. Depending upon location, these forests are characterized by oaks, magnolias, gumbo-limbo *(Bursera simaruba),* and royal and cabbage palms.

Structure

Highly developed, uneven-aged, temperate deciduous forests usually consist of four strata (Figures 31.11, 31.12). The upper canopy consists of dominant and co-dominant trees, below which is the lower tree canopy and then the shrub layer. The ground layer consists of herbs, ferns, and mosses.

Even-aged stands, the results of fire, clear-cut logging, and other large-scale disturbances (see Chapter 26), often have poorly developed strata beneath the canopy because of dense shade. The low tree and shrub strata are thin and the ground layer also is poorly developed, except in small, open areas.

In general, the diversity of animal life is associated with stratification and the growth forms of plants (see Chapter 24). Some animals, particularly forest arthropods, are associated with or spend the major part of their life in a single stratum; others range over two or more strata. The greatest concentration and diversity of life in the forest occurs on and just below the ground layer. Many animals, the soil and litter invertebrates in particular, remain in the subterranean stratum. Others, such as mice, shrews, ground squirrels, and forest salamanders, burrow into the soil or litter for shelter and food. Larger mammals live on the ground layer and feed on herbs, shrubs, and low trees. Birds move freely among several strata, but favor one layer

Figure 31.11 A virgin stand of mixed mesophytic forest in the central Appalachians, composed of white oak, beech, yellow-poplar, and other species. Note the well-developed understory in this uneven-aged climax stand. (Photo courtesy U.S. Forest Service.)

over another. Ruffed grouse, hooded warbler, and ovenbird occupy essentially the ground layer but move up into the upper strata to feed, roost, or advertise territory.

Other species occupy the upper strata—the shrub, low tree, and canopy. The red-eyed vireo, the most abundant bird of the eastern deciduous forest, inhabits the lower tree stratum and the wood pewee the lower canopy. The black-throated green warbler and scarlet tanager live in the upper canopy. Squirrels are mammalian inhabitants of the canopy, and woodpeckers, nuthatches, and creepers live in the open space of tree trunks between shrubs and the canopy.

The physical stratification of the forest influences the microclimate within the forest. The highest temperatures are in the upper canopy, because this stratum intercepts solar radiation. Temperatures tend to decrease through the lower strata (Figure 31.12). The most rapid decline takes place from the leaf litter down through the soil.

The temperature profile changes through the 24-hour period. At night temperatures are more or less uniform from the canopy to the floor. Radiation takes place most rapidly in the canopy; as the air cools it sinks and becomes slightly heated by the warmer air beneath the canopy. During the day the air heats up, and by midafternoon temperature stratification be- comes most pronounced. On rainy days the temperatures are more or less equalized because water absorbs heat from warmer surfaces and transfers it to the cooler surfaces.

Temperature stratification varies seasonally (Christy 1952). In fall, when the leaves drop and the canopy thins, temperatures fluctuate more widely at the various levels. Maximum temperatures decrease from the canopy downward, but rise again at the litter surface. The soil, no longer shaded by an overhead canopy, absorbs and radiates more heat than in summer.

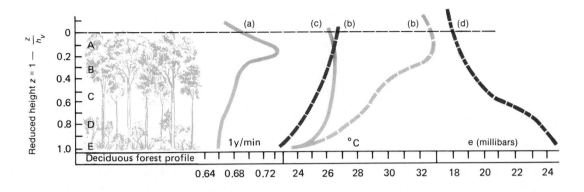

Figure 31.12 Microclimate profiles of a mixed oak, linden, and maple forest in central Russia. (a) Long-wave radiation within the forest during summer. (b) Air temperature at midmorning during a sunny summer day. (b') Air temperature at midafternoon. (c) Idealized temperature profile in summer. (d) Water vapor pressure within the forest during warm weather. Height is scaled as reduced height determined by $Z = 1 - Z/h_v$ where Z is some determined level and h_v is the maximum height of the stand. In this example the height of the canopy is 12 meters. (After Raunder 1972.)

Below the insulating pavement of litter, temperatures decrease again through the soil. Thus there may be two temperature maximums in the profile, one in the canopy, the other on the surface of the litter. Winter temperatures decrease from the canopy down to the small tree layer, where in some forests they rise and then drop at the litter surface. From here the temperature increases rapidly down through the soil. During spring conditions are highly variable. Maximum temperatures are found on the leaf-litter surface, which at this season of the year intercepts solar radiation, and temperatures decrease upward toward the canopy.

Humidity in the forest interior is high in summer because of plant transpiration and poor air circulation. During the day, when the air warms and its water-holding capacity increases, relative humidity is lowest. At night, when temperature and moisture-holding capacities are low, relative humidity rises. The lowest humidity in the forest is a few feet above the canopy, where air circulation is best. The highest humidity is near the forest floor, the result of evaporation of moisture from the ground and settling of cold air from the strata above.

Variation of humidity within the forest is influenced in part by the degree to which the lower strata are developed. Leaves add moisture to the immediate surrounding air; well-developed strata with more leaves have higher humidity. Thus layers of increasing and decreasing humidity may exist from the floor to the canopy.

Bathed in full sunlight, the uppermost layer of the canopy is the brightest part of the forest. Down through the forest strata light intensity dims to only a fraction of full sunlight. In an oak forest only about 6 percent of the total midday sunlight reaches the forest floor; brightness of light at the forest floor is about 0.4 percent of that of the upper canopy.

Light intensity within the forest varies seasonally (see Chapter 7). The forest floor receives its maximum illumination during early spring before the leaves appear; a second lower peak of maximum illumination during the growing season occurs in the fall. The darkest period is midsummer. Light intensity during summer is highly variable from point to point and time to time as sun shines through gaps in the canopy. Sun flecks can influence the distribution of herbaceous vegetation on the forest floor.

Function

Energy flow and nutrient cycling have been assessed for several deciduous forest stands, including the Hubbard Brook northern hardwoods forest in New Hampshire, Walker Branch mesic hardwoods forest at Oak Ridge, Tennessee, and an oak-pine forest at Brookhaven, Long Island.

Energy flow is most conveniently measured in terms of carbon pools and fluxes. The carbon budget for the mesic yellow-poplar *(Liriodendron tulipifera)* and oak *(Quercus spp.)* at Oak Ridge is summarized in Table 31.4 and Figure 31.13. Estimates of aboveground and central root carbon pool are 8.03 kg with an annual aboveground accumulation rate of 0.166 kg $C/m^2/yr$ (Reichle et al. 1973). Mean standing belowground lateral root carbon

amounts to 0.76 kg C/m^2. Increment to lateral root biomass amounts to 8 percent per year. Root turnover from death occurs largely in roots less than 0.5 cm in diameter. Much of this root death takes place in late spring and late autumn. Similarly, most root production, which occurs mostly in late winter and midsummer, also takes place in roots less than 0.5 cm. Mean annual standing crop in the O_1 and O_2 litter layer is 237 g C/m^2. Total amount of soil carbon is 12.3 kg C/m^2 to a depth of 75 cm. Soil organic matter decreased from 4.6 percent dry weight of soil in the upper 10 cm to 1.3 percent at 21 cm depth. Litter invertebrates amounted to 520 mg C/m^2 and soil invertebrate fauna, largely earthworms, 6.4 mg C/m^2. Soil microflora, 65 percent fungi and 35 percent bacteria, totaled 58 g C/m^2.

Fluxes of carbon in photosynthesis and autotrophic respiration determined by means of gas exchange include an estimated gross carbon uptake (gross primary production) of 2.15 kg $C/m^2/yr$. Net primary production is 0.73 kg $C/m^2/yr$. The autotrophic respiration, including contributions from the forest plants, is 1.44 kg $C/m^2/yr$. Respiration from the forest floor

TABLE 31.4 Metabolism of Two Deciduous Forests

	Parameter	*Liriodendron* Forest	*Quercus-Pinus* Forest
Total standing crop	TSC	8.76	5.96
Net primary production	NPP	0.73	0.60
Relative production	NPP/TSC	8.3%	10%
Autotroph respiration	R_A	1.44	0.68
	R_A/TSC	0.16	0.11
Heterotroph respiration	R_H	0.67	0.29
Ecosystem respiration	$R_E = R_A + R_H$	2.11	1.01
Net ecosystem production	NEP = NPP − R_H	0.06	0.28
Annual decay		0.70	0.36

Note: Units of measure are kg C/m^2 and kg $C/m^2/yr^1$ for compartments and fluxes.

Source: Liriodendron data after Reichle et al. 1973a; *Quercus-Pinus* data from Woodwell and Botkin 1970; in National Academy of Science 1974.

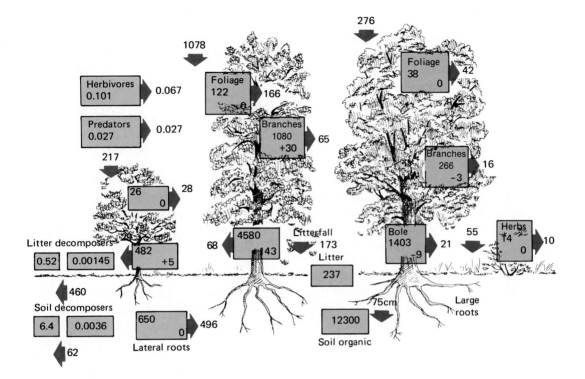

Figure 31.13 Carbon budget in a mesic hard-wood forest ecosystem at Oak Ridge, Tennessee. From left to right: the trees represent the understory, dominant yellow-poplar (*Liriodendron tulipifera*) and other overstory trees. Structural components of the ecosystem have been abstracted as compartments with major fluxes. Vertical arrows represent photosynthetic fixation. Lateral arrows represent respiratory losses. Units of measure are g C/m^2 and g C/m^2/yr for compartment increments and fluxes. (From Reichle et al. 1973.)

was 1.04 kg C/m^2/yr; decomposer respiration amounted to 0.21 kg C and summed respiration of canopy insects 0.094 kg C/m^2/yr.

Annual litterfall amounts to 229 g C/m^2, of which 78 percent is leaves. Tree mortality accounts for 50 g C/m^2/yr. Loss of photosynthetic surface through insect consumption varies from 1.9 percent to 3.4 percent, while actual reduction in photosynthetic surface amounts to 5.6 to 10.1

percent. Carbon flux due to actual consumption is 4.5 g C/m^2/yr.

Another comprehensively analyzed temperate forest ecosystem is a young oak-pine forest at Brookhaven, New York, a relatively xeric stand of pitch pine and scarlet, white, and bear oak with an understory of blueberry and huckleberry (Whittaker and Woodwell 1969; Woodwell and Botkin 1970). In some respects the oak-pine stand and the yellow-poplar stand are similar, even though they have different standing crops, 8.76 kg C/m^2 for yellow-poplar–oak compared to 5.96 kg C/m^2 for the oak-pine (Table 31.4). Both have comparable net primary production and autotrophic respiration. Both lose approximately the same amount of photosynthetic surface to insects, about 3 percent of net annual production.

However, the two stands differ considerably at the heterotrophic level. Heterotrophic respiration for yellow-poplar is more than twice that of the oak-pine forest. This large heterotrophic respiration plus the large autotrophic respiration result in total ecosystem respiration for yellow-poplar forest over two times as great as that of the oak-pine forest. The main difference in heterotrophic respiration is in annual decomposition of 712 g C/m^2 for yellow-poplar compared to 360 g C/m^2 for the oak-pine. Much of this difference comes from a high root turnover rate in the yellow-poplar stand, as well as the resistance of oak leaves and pine needles to rapid decomposition. Comparison of the ratios of net ecosystem production (NEP) to total standing crop (0.05 for oak-pine and 0.007 for yellow-poplar) indicates that the young oak-pine forest is accumulating carbon seven times as fast as the yellow-poplar stand.

Mineral cycling has been studied extensively in several temperate forest ecosystems, including an oak-pine forest (Whittaker and Woodwell 1969), a northern hardwoods forest (Likens et al. 1971; Likens 1976; Bormann and Likens 1979), a European oak forest (Duvigneaud and Denaeyer-De Smet 1970), and many others (see Cole and Rapp 1981). Nutrient cycling may be considered as a balance between inputs to the biological system and outputs or losses from the system in streamflow through the watershed. The difference represents the amount of recharge to the system from the soil pool. In the oak-pine forest at Brookhaven, Long Island, New York, total input from precipitation for the four cations potassium, calcium, magnesium, and sodium amounts to 2.48 g/m^2, while losses to the water table range from 3.68 to 5.18 g/m^2 (Woodwell and Whittaker 1968). Output from the Hubbard Brook forest in New Hampshire, a northern hardwoods system, summarized in Table 31.5, exceeds the input of calcium, magnesium, and sodium, while potassium shows a gain through input (Likens et al. 1967).

These and other budgets suggest several characteristics of mineral cycling in temperate deciduous forests. One is that the uptake of nutrients does not meet requirements (Table 31.6). For example, in the Walker Branch yellow-poplar–oak stand, the uptake of nutrients does not meet requirements. Uptake of nitrogen amounts of 58.1 kg/ha/yr, only 66 percent of requirement. Deficiencies have to be met by cycling nutrients within the tree biomass. Nutrient accumulations in tree biomass (Table 31.7) form a considerable pool of nutrients, most of which is unavailable for short-term recycling. Forty-nine percent of N, 35 percent of P, 42 percent of K, 62 percent of Ca, and 49 percent of Mg are incorporated into woody biomass; 31, 51, 43, 28, and 24 percent respectively in root biomass; and 20, 14, 15, 10, and 27 percent respectively in foliage.

Considering the whole forest ecosystem, 7 percent of N, 3 percent of P, 12 percent of Ca, and 1 percent each of K

TABLE 31.5 Input and Losses of Nutrients of the Hubbard Brook Forest (kg/ha)

	Ca	Mg	Na	K
Input	3.0 ± 0	0.7 ± 0	1.0 ± 0	2.5 ± 0
Output	8.0 ± 0.5	2.6 ± 0.06	5.9 ± 0.3	1.8 ± 0.1
Loss	5.0 ± 0.5	1.9 ± 0.06	4.9 ± 0.3	0.7 ± 0.1

Source: Data from Likens et al. 1967.

TABLE 31.6 Annual Element Balance of a 30- to 80-Year-Old Yellow-Poplar–Oak Forest, Oak Ridge, Tennessee (kg/ha/yr)

	N	P	K	Ca	Mg
Requirement	87.9	6.3	47.5	82.6	21.7
Uptake	58.1	3.4	40.0	87.8	12.4
Internal recycling	29.8	2.9	7.5	(−5.2)	9.3

Note: Requirement = annual increment of elements associated with bole and branch wood plus current foliage production.
Uptake = annual increment of elements associated with bole and branch wood plus annual loss through litterfall, leaf wash, and stem flow.
Internal recycling = Requirement − Uptake: deficiency in uptake made up by recycling elements within plant biomass.
Source: Data from Cole and Rapp 1981: 359.

TABLE 31.7 Nutrient Balance Sheet for 30- to 80-Year-Old Yellow-Poplar–Oak Forest, Oak Ridge, Tennessee (kg/ha)

	N	P	K	Ca	Mg
Input/output					
Atmosphere	8.7	0.54	1.0	9.1	1.1
Leaching to watershed	1.8	0.02	6.8	147.5	77.1
Loss/gain	+6.9	+0.52	−5.8	−138.4	−76
Internal cycling					
Litterfall	36.2	2.7	19.1	58.3	8.3
Throughfall	12.0	0.4	18.4	21.9	3.4
Total	48.2	3.1	37.5	80.2	11.7
Live accumulation					
Woody biomass	189	15	127	462	38
Roots	122	22	132	211	19
Foliage	78	6	45	75	21
Total	389	43	304	748	78
Detrital accumulation					
Forest litter	187	11	14	294	22
Soil rooting zone	7,300	1,400	36,000	6,300	8,700
Total	7,487	1,411	36,014	6,594	8,722

Source: Data from Cole and Rapp 1981: 394.

and Mg are stored in vegetation. The litter layer is the most important nutrient pool, because it is quickly decomposed (average turnover time of four years) and recycled, although the bulk of the nutrient pool is in mineral soil. Nutrients stored in living biomass, especially in roots, are translocated and recycled through the living biomass, particularly the foliage. The foliage, in turn, translocates a considerable portion of its nutrients back to roots before leaf fall. However, mineral cycling can be maintained only if nutrients are pumped from soil reserves or are released through the weathering of parent materials.

The role of the various components in nutrient cycling is illustrated by long-term studies of the nitrogen cycle in deciduous forests at Hubbard Brook, New Hampshire (Bormann et al. 1977) and Walker Branch, Oak Ridge, Tennessee (Henderson et al. 1973). Both studies point out that natural forest ecosystems tend to accumulate and cycle large amounts of nitrogen. At both sites most of the nitrogen (87 to 90 percent) is incorporated in the mineral soil horizons. The remaining nitrogen is in vegetation and forest floor (Figure 31.14). The most important mechanism for cycling nitrogen from vegetation to soil is lateral root turnover. In the Walker Branch forest it amounts to 56 kg/ha/yr. This amount includes only root mortality and not exudates. As measured at Hubbard Brook, root exudates release about 1 percent of the inorganic nitrogen made available by net mineralization. The second most important mechanism is litterfall, which at Walker Branch accounts for 37 kg/ha/yr and at Hubbard Brook 54.2 kg/ha/yr. Of this 91 percent consists of leaves and reproductive parts; the remaining 9 percent is bole and branch fall

Figure 31.14 Representation of the nitrogen cycle on a mixed mesophytic forested watershed at Oak Ridge, Tennessee. Nitrogen pools in ecosystem components are shown on the right and annual transfers on the left. (After Auerbach et al. 1974; courtesy Oak Ridge National Laboratory.)

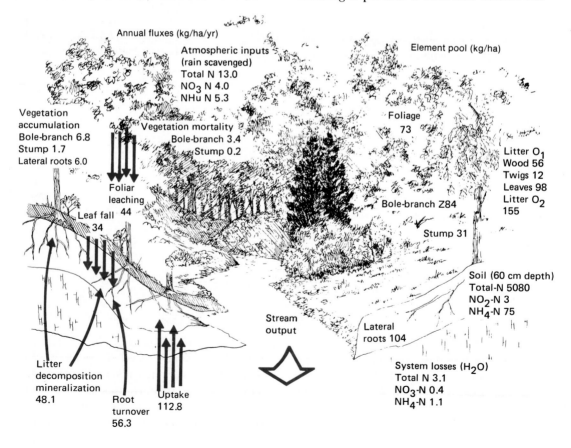

Annual fluxes (kg/ha/yr)

Atmospheric inputs (rain scavenged)
Total N 13.0
NO_3 N 4.0
NHu N 5.3

Element pool (kg/ha)

Vegetation accumulation
Bole-branch 6.8
Stump 1.7
Lateral roots 6.0

Vegetation mortality
Bole-branch 3.4
Stump 0.2

Foliage 73

Litter O_1
Wood 56
Twigs 12
Leaves 98
Litter O_2 155

Foliar leaching 44

Bole-branch Z84

Leaf fall 34

Stump 31

Litter decomposition mineralization 48.1

Root turnover 56.3

Uptake 112.8

Stream output

Lateral roots 104

Soil (60 cm depth)
Total-N 5080
NO_2-N 3
NH_4-N 75

System losses (H_2O)
Total N 3.1
NO_3-N 0.4
NH_4-N 1.1

due to tree mortality. Foliar leaching, a third mechanism, contributes 4.4 kg/ha at Walker Branch, 9.3 kg/ha at Hubbard Brook. Vegetation retains 14.8 kg/ha of which bole and branch, stump, and lateral root components account for 47, 12, and 41 percent, respectively. Nitrogen released from forest litter decomposition amounts to an estimated 48 kg/ha. Atmospheric input accounts for 6.5 kg/ha/yr at Hubbard Brook, 23.5 kg/ha at Walker Branch. An additional estimated input of 14.2 kg/ha is added by nitrogen fixation, an input not estimated at Walker Branch. Uptake from mineral soil amounts to 112.8 kg/ha at Walker Branch, 79.6 kg/ha at Hubbard Brook.

The Hubbard Brook study emphasizes other important aspects of the nitrogen cycle in the forest. Of the nitrogen added to long-term storage about 54 percent is accumulated in living matter, and 46 percent is stored in organic matter in the forest soil. Of an estimated 20.7 kg/ha entering the system each year about 81 percent is held within the system. Of the estimated 119 kg/ha of nitrogen used in plant growth, 33 percent is withdrawn from storage in the living plant and utilized in growth. A like amount is withdrawn from leaves before leaf fall and stored in the stems. In both forest ecosystems only a small fraction of the nitrogen added to the inorganic pool within the ecosystem is lost from the system in streamflow. At Hubbard Brook this leakage amounts to 5 percent; at Walker Branch, 1.9 percent. Thus an internal source of nitrogen not subject to loss through streamflow (providing the vegetation with a buffer against short-term fluctuations in available soil nitrogen), annual uptake by living vegetation, and annual additions of nitrogen to wood biomass are important in promoting a tight cycling of nitrogen.

How the substrate can influence mineral cycling is illustrated by the difference between the cation budgets of two forests of similar vegetation, one growing on a site underlain by granitic bedrock and one growing on a site underlain by dolomitic rocks rich in calcium and magnesium (Table 31.8). The forest growing at Walker Branch on dolomitic rocks shows an output of calcium and magnesium more than 20 times greater than the output from the Coweeta, North Carolina, forest on granitic substrate. Net losses of the other two cations, sodium and potassium, are small and about the same magnitude.

Coniferous versus Deciduous Forests

Deciduous forests appear to differ functionally from coniferous forests in the magnitude and nature of nutrient cycling. The differences are illustrated graphically in Figure 31.15, comparing nutrient cycling in a mixed oak forest and spruce forest in Belgium, both of which have an annual production of 14.6 tn/ha/yr. From these data it is evident that considerably more nutrients are cycled through the deciduous forest than through the coniferous forest and that the spruce retains relatively more nutrients in its biomass than the oak. Spruce retains more nitrogen, sulfur, and phosphorus than it returns through litterfall, throughfall, stemflow, and dead parts of the herbaceous layer. The oak forest, on the other hand, returns more of all elements than it retains. The deciduous forest is much more efficient at recycling calcium. The oak forest recycles calcium largely through litterfall, whereas the spruce forest returns considerable quantities through stemflow and throughfall. Although the deciduous forest takes up more potassium than the spruce forest, it retains less. The spruce forest retains

TABLE 31.8   Average Annual Cation Budgets for Two Undisturbed Watersheds in the Appalachian Highlands, 1969–1972 (kg/ha)

	COWEETA (NORTH CAROLINA) Mature Hardwoods Granitic Rock	WALKER BRANCH (TENNESSEE) Mature Hardwoods Dolomitic Rock
Calcium + +		
Input	6.16	28.6
Output	6.92	138.4
Net loss or gain	− 0.76	− 103.8
Magnesium + +		
Input	1.26	3.2
Output	3.09	69.6
Net loss or gain	− 1.82	− 66.4
Potassium +		
Input	3.16	4.8
Output	5.17	5.6
Net loss or gain	− 2.02	− 0.8
Sodium +		
Input	5.40	9.2
Output	9.74	5.3
Net loss or gain	− 4.34	+ 3.9

Source: National Academy of Science 1974.

nearly half its uptake of potassium. This difference confirms the idea that conifers tend to be accumulators.

■ Tropical Forests

Types

Not all tropical forests are rain forests. There are two major tropical forest groups, the tropical rain forest and tropical seasonal forest, each with many subtypes.

Tropical Rain Forests. The tropical rain forest, named in 1898 by the German botanist A. F. W. Schimper (see Chapter 1) as *tropische Regenwald*, comes in at least 30 to 40 types, which include the monsoon forest, the evergreen savanna forest, the evergreen mountain forest, and the tropical evergreen alluvial forest, as well as the true equatorial lowland tropical rain forest (see Odum 1970; Richards 1972; Whitmore 1984; Mabberley 1983). They form or once formed a worldwide belt about the equator. The largest continuous rain forest is found in the Amazon basin of South America (Figure 31.16). West and central Africa and the Indo-Malaysian regions (Figure 31.17) contain other major tropical rain forests. Smaller types of rain forest, now virtually gone, occur on the eastern coast of Australia (Figure 31.18), on the windward side of the Hawaiian Islands (Figure 5.4b), and on the east coast of Madagascar. All these rain forests grade into temperate and subtemperate rain forest.

Tropical rain forests grow where seasonal changes are minimal. The mean annual temperature is about 26° C, the mean minimum rarely goes below 25° C, and the difference in temperature through the

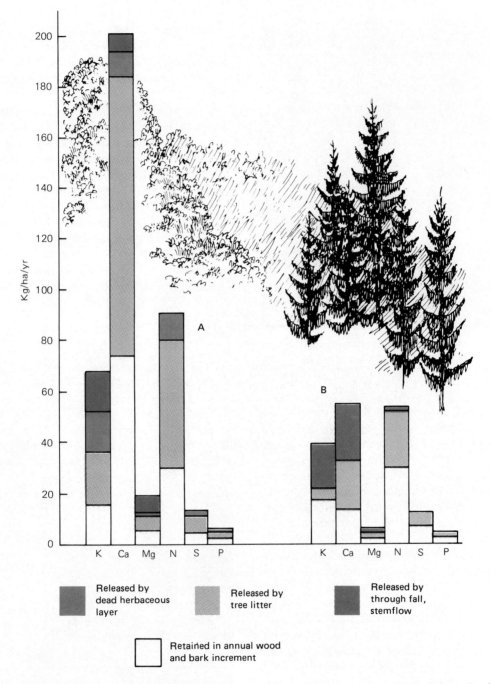

Figure 31.15 Simplified representation of the biological cycle of nutrients in two forests: (a) a mixed oak forest *(Quercus)* at Virelles, Belgium, and (b) a spruce forest *(Picea)* at Mitwart, Belgium. Productivity for both forests is 14.6 tn/ha/yr. (After Duvungneaud and Denaeyer-De Smet 1975.)

Figure 31.16 Amazonian tropical rain forest. Note the canopy emergents. (Photo by Alexine Keuroghlian.)

of which 27 are endemic. (The Asian diptocarps have 12 genera and 470 species.) The few rain forest communities with single dominants are limited to areas of particular combinations of soils and topography.

The tree trunks are straight, smooth, and slender, often buttressed, and reach 25 to 30 m before expanding into crowns with large, leathery, simple leaves (for the complex architecture of tropical trees see Hallee, Oldeman, and Tomlinson 1978). Climbing plants, the *lianas*, long, thick, and woody, hang from trees like cables, and epiphytes grow on trunks and limbs. Undergrowth of the dark interior is sparse and consists of shrubs, herbs, and ferns. Litter decays so rapidly that the clay soil,

Figure 31.17 Lowland tropical rain forest in peninsular Malaysia. The prominent emergent canopy trees are *Shorea* representatives of the dipterocarp family. (Photo by R. L. Smith.)

year is less than 4° C. Heavy rainfall occurs throughout the year, not less than 100 mm in any month for two out of every three years. What constitutes heavy rainfall varies with the region. The tropical rain forest regions in Latin America receive about 4000 mm rain annually and those in Africa about 1500 mm. Under such perpetual midsummer conditions plant activity continues uninterrupted, resulting in luxurious growth.

Tree species number in the thousands. A 10 square kilometer area of tropical rain forest may contain 1500 species of flowering plants and up to 750 species of trees. The richest is the lowland tropical forest of the Malaysian Peninsula, which contains some 7900 species. There, one of the major groups, the Dipterocarpa, contains 9 genera and 155 species,

Figure 31.18 Rain forest in Australia is confined to the eastern coast. The most extensive areas are in Queensland with scattered pockets to the south. Most of the rain forest has been destroyed. This stand of southern closed forest is dominated by mountain ash *(Eucalyptus regnans)* and myrtle beech *(Nothofagus cunninghamii)*; it has a luxuriant understory of tree fern *(Dicksonia antarctica)*. (Photo by T. M. Smith.)

more often than not, is bare. The tangled vegetation, popularly known as jungle, is secondary (second growth) forest that develops where primary forest has been disturbed. Tropical rain forests account for several million species of flora and fauna, one-half of all known plant and animal species, and 20 to 25 percent of all known arthropods.

Tropical Seasonal Forests. Tropical rain forests grade into seasonal forests, characterized by less rainfall, more variable temperatures, and a dry season during which many of the trees lose their leaves.

Such forests are most common in southeastern Asia, India, South America, and Africa, and along the Pacific side of Mexico and Central America.

Structure

The tropical rain forest can be divided into five general layers, but stratification is often poorly defined because many tree species have the same growth plan but differ in size (Hallee, Oldeman, and Tomlinson 1978; Tomlinson 1983; Brunig 1983). Stratification is most apparent in the undisturbed forest. The uppermost or emergent layer consists of trees over 60 to 80 m high whose deep crowns rise above the rest of the forest to form a discontinuous canopy. The second layer, consisting of trees about 50 m high, forms another lower discontinuous canopy. Not clearly separated from one another, these two layers form an almost complete canopy. The third layer is the lowest tree stratum; it is continuous, often the deepest layer, and is well defined. The fourth layer, usually poorly developed in deep shade, consists of shrubs, young trees, tall herbs, and ferns. The fifth stratum is the ground layer of tree seedlings and low herbaceous plants.

A conspicuous part of the rain forest is plant life dependent on trees for support. Such plants include epiphytes, climbers, and stranglers. Epiphytes such as orchids and members of the Ericaceae are hemiparasites. They attach themselves to a tree and take up water, nutrients, and some photosynthate. Some of the epiphytes are important in recycling minerals leached from the canopy. Climbers are vinelike plants that reach to the tops of trees and expand into the form and size of a tree crown. Climbers grow prolifically in openings, giving rise to the image of the impenetrable jungle. Stranglers start life as epiphytes. They send roots to the ground and increase in number and girth until they eventually encompass the host tree and claim the crown limbs as support for their own leafy growth.

The mature tropical forest, like the mature temperate forest, is a mosaic of continually changing vegetation. Death of tall trees, brought about by senescence, lightning, wind storms, hurricanes, defoliation by caterpillars, and other causes, creates gaps (see Chapter 27) that shade-intolerant pioneer species quickly fill (Poore 1968; Hartshorn 1978; Doyle 1981). These trees are replaced eventually by shade-tolerant late successional species; but continuous random disturbances across the forest ensure persistence of the species in the mature forest. A high frequency of tree fall—Poore (1968) estimated that in his 12 ha Malaysian study area, one-half of it would have experienced gap formation—may account for the low density of large trees (1 m+ dbh) in mature rain forests.

Most tropical rain forest trees reach full height when they have achieved only about one-third to one-half of their final bole diameter. Thus stratification or layering results when a group of species of similar mature height dominate a stand (Whitmore 1984). Layering is also influenced by crown shape, which in turn is correlated with tree growth. Young trees still growing in height have a single stem and a tall narrow crown (monopodial). Mature trees have a number of large limbs diverging from the upper stem or trunk (sympodial). The limbs continue to grow, adding to crown width after the tree has reached mature height. Thus a pattern of gap formation, succession, and mature phases in a stand results in a poorly defined stratification in many tropical rain forests.

Layering of vegetation influences the internal microclimate of the forest (Figure 31.19). The crowns of emergent trees experience conditions similar to open land. Through the canopy the level of CO_2 and amount of humidity increases and temperature and evaporation decrease. From the ground to 1 m the levels of CO_2 are high and humidity stands at 90 percent; temperature on the average is 6° C cooler than outside forest cover, and it experiences a strong nocturnal inversion (Bourgeron 1983). Light decreases rapidly down through the canopy. The amount of light that reaches the floor of a Malaysian rain forest is about 2 to 3 percent of incident radiation, and half of that comes from sun flecks, shafts of light that pass through the leaves and change through the day; about 6 percent comes from breaks in the canopy, and 44 percent from reflected and transmitted light (Mabberley 1983).

Stratification of animal life in the tropical rain forest, however, is pronounced (Figure 31.20). Harrison (1962) recognized six distinct feeding strata. (1) A group feeding above the canopy consists largely of insectivorous and some carnivorous birds and bats. (2) A top of the canopy group comprising a large variety of birds, fruit bats, and other species of mammals feeds on leaves, fruit, and nectar. A few are insectivorous and mixed feeders. (3) Below the canopy, a zone of tree trunks, is a world of flying animals— birds and insectivorous bats. (4) Also in the middle canopy are scansorial mammals, which range up and down the trunks, entering the canopy and the ground zone to feed on the fruits of epiphytes growing on tree trunks, on insects, and on other animals. (5) Large ground animals make up the fifth feeding group. It includes large mammals and a few

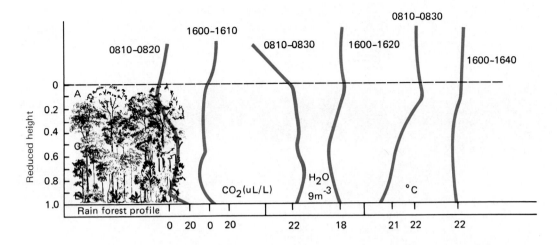

Figure 31.19 Microclimatic profiles in a tropical rain forest in November at Bosque de Florencia, Costa Rica. The height at the top of the canopy, 40 meters, is scaled as reduced height. CO_2 is expressed in relative values of 0 to 20; 0 represents a 0 gradient and values are deviations from an arbitrary standard. Note the increase in CO_2 near the ground and in the morning and the decrease in the afternoon, except near the ground. Temperature increases up through the canopy in the morning; in late afternoon the temperature gradient is nearly 0. (After Allen et al. 1976.)

birds, living on the ground and lacking climbing ability, that are able to reach up into the canopy or cover a large area of forest. They include the large herbivores and their attendant carnivores. (6) The final feeding stratum includes the small ground and undergrowth animals, birds and small mammals capable of some climbing, that search the ground litter and lower parts of tree trunks for food. This stratum includes insectivorous, herbivorous, carivorous, and mixed feeders.

Animal life in the tropical rain forest is largely hidden, either by the dense foliage of the upper strata or by the cover of night. Birds are largely arboreal, and although brightly colored, remain hidden in the dense foliage. Ground birds are small and dark-colored, difficult to see. Mammals appear scarcer than they really are, for they are largely nocturnal or arboreal. Ground-dwelling mammals are small and secretive. Tree frogs and insects are most conspicuous at evening, when their tremendous choruses are at full volume. Insects are most diverse at forest openings, along streams, and at forest margins, where light is more intense, temperatures fluctuate, and air circulates freely. Highly colored butterflies, beetles, and bees are common. Among the unseen invertebrates, hidden in loose bark and in axils of leaves, are snails, worms, millipedes, centipedes, scorpions, spiders, and land planarians. Termites are abundant in the rain forest and play a vital role in the decomposition of woody plant material. Together with ants, they are the dominant insect life. Ants are found everywhere in the rain forest, from the upper canopy to the forest floor, although in common with other rain forest life, the majority tend to be arboreal.

Many specialized interactions among plants and animals exist in tropical rain forests. Plants, often widely dispersed in

Figure 31.20 Stratification of vegetation and animal life in a Malaysian rain forest.

the forest, depend for pollination upon birds, bats, and insects, especially beetles, bees, moths, and butterflies (see Procter and Yeo 1973; Baker et al. 1983; Howe and Westley 1988) and for seed dispersal on fruit-eating birds, bats, rodents, and primates. Heavy predation on seeds by insects may result in wide dispersal of tree species (Janzen 1971). Other interactions involve repellent toxins to discourage predation by herbivores and even insect-plant mutualisms in which insects such as ants live in hollow stems and prevent other insects from gaining entrance or feeding on the plant (Janzen 1967) (see Chapter 23).

Function

The most intensely studied tropical forest is the one at the Puerto Rico Nuclear Center. H. T. Odum (1970) and his associates worked up an energy budget for that forest. Incoming solar radiation amounted to 3830 kcal/m^2/day. Gross production amounted to 131 kcal/m^2/day, of which 116 kcal was used in respiration, leaving a net production of 15.2 kcal/m^2/day as determined by gas analysis. Roots were responsible for 60 percent of the respiration, leaves for 33 percent, and trunks, branches, and fruit for the remainder. Net productivity as measured by biomass accumulation was 16.31 kcal/m^2/day. Cumulative biomass addition through wood growth was 0.72 kcal/m^2/day or 3.8 percent of net production. The remainder of net production passed through the grazing and detrital food chains.

Odum's energy budget is for a specific rain forest. Because site, soil, and other conditions of different tropical forests vary widely, productivity also varies widely (Figure 31.21). However, a few generalizations can be made about their energy budgets. Tropical forests use 70 to 80 percent of their energy intake in maintenance and 20 to 30 percent for net production. Average gross primary production is about 67 mtn/ha/yr or 28×10^3 kcal/m^2/yr (Golley 1972). Mean annual net production is about 21.6 mtn/ha. This amount exceeds temperate forests, averaging 13 mtn/ha/yr, by a factor of 1.7 and boreal forests, averaging 8 mtn/ha/yr, by a factor of 2.7 (Golley and Farnsworth 1973). However, tropical and temperate forests differ somewhat in their efficiency of production. Efficiency in this case is defined as the sum of energy stored in wood, leaves, fruit, and litter divided by total solar energy available to the community. Jordan (1971, 1983) found that high overall productivity of tropical forests relates more to foliage than to wood production. The rate of wood production was not any greater than in temperate hardwood forests, but the rate of leaf and litter production was higher in the tropics. However, the efficiency of wood production was higher in temperate forests, probably because more selective pressure exists in temperate forests, where solar energy is not as abundant, to produce the maximum amount of wood.

High year-round temperatures and abundant rainfall in tropical rain forest areas produce rapid geological cycling. Because geological cycling is accelerated, biological cycles apparently are modified to keep nutrients in the living portion of the system. Nutrients may be stored in living biomass where they are protected from leaching, or the time the nutrient elements remain in the soil may be reduced to a minimum. These strategies result in some basic differences between tropical forest and temperate forest ecosystems (Figure 31.22).

A large standing biomass is typical of tropical ecosystems. The tropical rain forest averages about 300 tn/ha, compared to 150 tn/ha for a temperate forest. Tropical rain forests tend to concentrate proportionately more calcium, silica, sulfur, iron,

Figure 31.21 Variation in production in tropical rain forests, shown by autotrophic respiration (R) as a percent of gross primary production (GPP). All types of vegetation are represented in this figure. Variation among stands makes it hard to generalize about primary production in tropical forests. (From Golley 1972.)

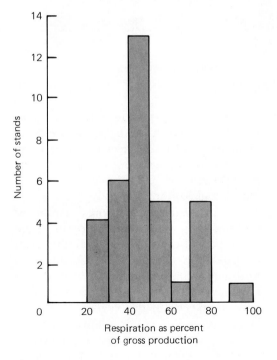

Number of stands

Respiration as percent of gross production

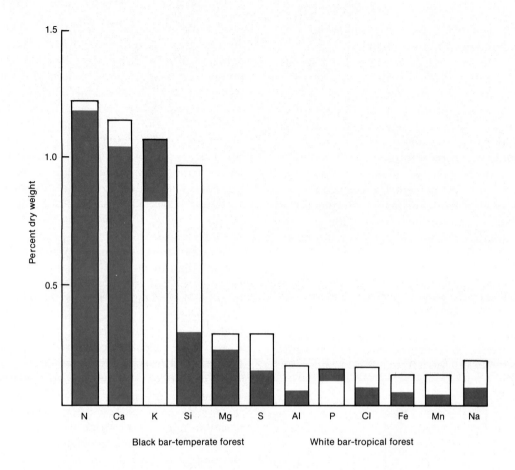

Black bar-temperate forest   White bar-tropical forest

Figure 31.22 Comparison of concentration of elements in temperate and tropical forests. Tropical forests have a higher concentration of nitrogen in their biomass. Temperate hardwood forests have higher concentrations of phosphorus and potassium. (From Golley 1975.)

magnesium, and sodium and less potassium and phosphorus than temperate forests. In spite of these differences, however, rain forests hold larger quantities of nutrients simply because of their much larger biomass. Mineral concentrations do vary widely among tropical forests, influenced by site, soil, and climate (Golley 1975). For example, Amazon rain forests

are low in nutrient concentrations compared to Panamanian tropical forests. A great range of differences in nutrient cycling exists among tropical moist forests on different soil types (for review see Vitousek and Sanford 1986).

As in the temperate forest a ratio of standing crop of available nutrients in the active part of the soil to the standing crop in vegetation provides some insight into nutrient cycling (Figure 31.23). In five Panamanian forests described by Golley and others (1975) the vegetation held a large percentage of phosphorus and potassium. Much of the mineral recycling takes place through litterfall. The ratio of

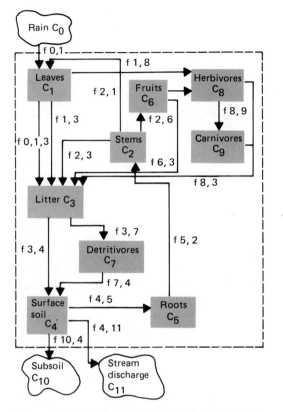

Figure 31.23 Model of mineral cycling in a tropical rain forest. Dotted lines indicate system boundary. Boxes identify system components. Arrows indicate transfer functions between components and sources. (From Golley 1975.)

TABLE 31.9 Turnover of the Mineral Inventory in Vegetation by Litterfall in Tropical Forests (yr)

Forest	P	K	Ca	Mg
Tropical moist, Panama	25	37	22	25
Premontane wet, Panama	9	28	17	10
High forest, Ghana	12	10	9	6
Deciduous evergreen, Thailand	5	6	9	8
Montane, Puerto Rico	—	84	9	10

Source: F. B. Golley 1975.

tropical forest of Panama indicate that inputs of phosphorus and potassium nearly balance outputs (Table 31.10), but more calcium and magnesium are lost from the system than are gained by rainfall input. This finding suggests considerable input from the soil reservoir. Annual uptake of phosphorus, potassium, and calcium far exceeds the stream discharge, while the reverse is true for magnesium. The budget also suggests phosphorus and potassium might be limiting and that the elements are conserved by a rapid internal cycling.

Internal cycling may be aided by (1) the rapid return of nutrients leached by throughfall; (2) retention of nutrients by fungal rhizomorphs; and (3) uptake by mycorrhizal fungi. Data on throughfall in the rain forests of the Ivory Coast of Africa indicate that more than 60 percent of the potassium recycled and 15 to 56 percent of other nutrients such as calcium, magnesium, and nitrogen are supplied by throughfall. Nutrients leached from the leaves, especially at the end of the dry season when leaching is greatest, are apparently taken up efficiently by the soil-root system (Bernhard-Reversat 1975).

In the temperate forest bacteria and fungi, the main agents of decay, release

mineral elements held in biomass to the amount returned to the soil by litter provides some estimate of turnover time (Table 31.9). Phosphorus, magnesium, calcium, and potassium all appear to have turnover times of less than 100 years, and most are recycled in 20 years.

Nutrient budgets, like those of the temperate forest, can be assessed by comparing inputs from rainfall with outputs by stream discharge. The difference between the two represents inputs from weathering of the substrate. Data for a

TABLE 31.10 Comparison of Biological and Geological Cycles in a Tropical Moist Forest

	Geological Cycle		Biological Cycle
Element	Input Rain (kg/ha/yr)	Stream Output (kg/ha/yr)	Annual Uptake (kg/ha/yr)
P	1.0	0.7	11
K	9.3	9.5	187
Ca	29.0	163.0	270
Mg	5.0	44.0	30

Source: F. B. Golley 1975.

nutrients directly into the mineral soil, where they are subject to leaching. Feeder roots of these trees are woven into the matrix of mineral soil. Feeder roots of tropical trees, however, are concentrated in the well-aerated upper 2 to 15 cm of humus and only a few penetrate the upper layer of mineral soil (Cornforth 1970; Jordan 1979, 1982). Symbiotically associated with the roots are mycorrhizal fungi. Some mycorrhizal fungi live around the roots. Others live partly within the cells of the root (see Chapter 23). Although their role as symbionts is not clearly defined, it appears the mycorrhizae are capable of digesting organic matter and passing phosphates and other minerals from the soil to the roots. During the process the mycorrhizae extract sugar and growth substances from the roots.

The roots of tropical forests support an abundance of mycorrhizal fungi that attach the feeder roots to dead organic matter by hyphae and rhizomorph tissue (Figure 31.24). As a result, considerable nutrient cycling in tropical rain forests appears to be through mycorrhizae (Went and Stark 1968). They appear to cycle nutrients directly from dead organic matter to living roots with only a minimum of leakage into mineral soil. In such a direct mineral cycling, minerals remain tied up in living and dead organic matter and are

transferred through hyphae from dead branches or leaves to living roots. Very little mineral matter becomes soluble and moves into the soil. This fact, according to Went and Stark (1968), may explain why feeder roots are concentrated mainly in the humus layer and why more mycorrhizal roots occur in poorer tropical soils.

Fate of Tropical Forests

In the 1800s North America, especially the United States, experienced the most rapid and massive deforestation of a continent by humans in Earth's history. Well over 80 percent of the rich hardwoods were destroyed, and the great pine forests of the Great Lakes region experienced a devastation from which the land has never recovered. At the same time humans eliminated all but a minuscule fraction of the prairie ecosystem. This destruction resulted in the demise of some wildlife species and the extinction of the passenger pigeon and the ivory-billed woodpecker. The extent of this deforestation is masked today by the return of the forest to previously cutover and abandoned agricultural land, although the replacement forests lack the diversity and grandeur of the original.

Today, the latter part of the 1900s, the tropical forest regions of the planet

(a) Direct litter-humus-root

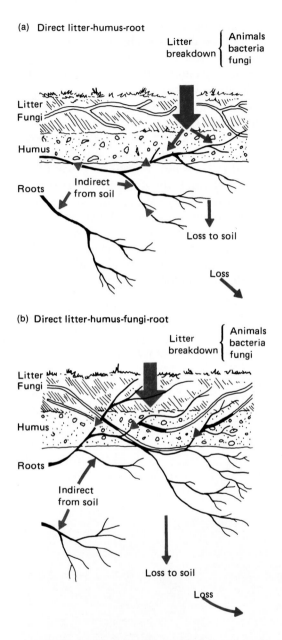

Litter
breakdown { Animals
bacteria
fungi

Litter
Fungi

Humus

Roots

Indirect
from soil

Loss to soil

Loss

(b) Direct litter-humus-fungi-root

Litter
breakdown { Animals
bacteria
fungi

Litter
Fungi

Humus

Roots

Indirect
from soil

Loss to soil

Loss

Figure 31.24 Mechanisms of mineral cycling from soil in tropical forests. (a) One type of direct nutrient cycling is characterized by a transfer of elements from litter to humus to roots without the aid of mycorrhizal fungi (except at root ends). This type of transfer occurs in tropical latisols where water penetration is slow. (b) Another type of direct cycling involves a litter-humus-fungi-root pathway. It is characterized by the breakdown of litter by fungi, bacteria, and animals, uptake from humus by mycorrhizal fungi, and transfer of materials to living roots. (After Stark 1973.)

25 years little will remain of the Amazonian rain forest. What makes the deforestation of the tropical areas so critical, compared to the deforestation of North America, is that much of the rain forest is poorly adapted to agriculture; and once cleared, burned, planted, and abandoned, it is slow to return to forest, if ever.

Lost in the destruction of the tropical rain forest is its high diversity of plant and animal life. Although occupying only 7 percent of the Earth's surface, the tropical rain forest regions hold 50 to 80 percent of the world's plant species. Most of the

Figure 31.25 The tropical rain forest faces destruction by large land clearing schemes, as is the fate of this Amazonian tropical rain forest. (Photo by Alexine Keuroghlian.)

are experiencing an even faster rate of massive deforestation (Figure 31.25). By the end of this century at the present rate of destruction most of the rain forests of southeast Asia will be gone, together with the still remaining patches in Australia, West Africa, and Madagascar; and within

animal species are endemic, resident, and nonmigratory. Once the habitat is destroyed, these species go extinct. So do hundreds of plant species that depend upon them for pollination and seed dispersal. Tropical forests are a source of food, medicinal plants little studied and utilized, genetic diversity, and economic opportunity that are being poorly managed for the future (see Myers 1983; Ehrlich and Ehrlich 1981; Wilson 1988).

■ Tundras

North of the coniferous forest belt lies a frozen plain, clothed in sedges, heaths, and willows, which encircles the top of the world (see inside cover). It is the arctic tundra—the word comes from the Finnish *tunturi*, meaning a treeless plain (Figure 31.26). At lower latitudes similar landscapes, the alpine tundra, occur in the mountains of the world. In the Antarctic a

well-developed tundra is lacking. Arctic or alpine, the tundra is characterized by low temperatures, a short growing season, and low precipitation (cold air can carry very little water vapor).

The tundra is a land dotted with lakes and transected by streams. Where the ground is low and moist, extensive bogs exist. On high, drier areas and places exposed to the wind, vegetation is scant and scattered, and the ground is bare and rock-covered. These regions are the fell-fields, an anglicization of the Danish *fjoeld-mark*, or rock deserts. Lichen-covered, the fell-fields are most characteristic of the highly exposed alpine tundra. Bliss (1981) divides the arctic into two major types:

Figure 31.26 Wide expanse of the arctic tundra. This photo shows an area near Sadlerochit River on the Arctic National Wildlife Refuge, 8 km from the Arctic Ocean. Note the frost polygons in the foreground. The caribou, a major arctic herbivore, are part of the porcupine herd. (Photo courtesy U. S. Fish and Wildlife Service.)

tundra with 100 percent cover and wet to moist soil, and *polar desert* with less than 5 percent cover and dry soil.

Characteristics

Frost molds the tundra landscape. Alternate freezing and thawing and the presence of a permanent frozen layer in the ground, the *permafrost,* create conditions unique to the arctic tundra. The sublayer of soil is subject to annual thawing in spring and summer and freezing in fall and winter. The depth of thaw may vary from a few centimeters in some places to half a meter in others. Below the thaw depth the ground is always frozen solid and is impenetrable to both water and roots. Because the water cannot drain away, flatlands of the Arctic are wet and covered with shallow lakes and bogs. This reservoir of water lying on top of the permafrost enables plants to exist in the driest parts of the Arctic.

The symmetrically patterned landforms so typical of the tundra result from frost. The fine soil materials and clays, which hold more moisture than the coarser materials, expand while freezing and then contract upon thawing. This action tends to push the larger material upward and outward from the mass to form the patterned surface.

Typical nonsorted patterns associated with seasonally high water tables are frost hummocks, frost boils, and earth stripes (Figure 31.27). Frost hummocks are small earthen mounds up to 1.5 m in diameter and 1.3 m high, which may or may not contain peat. Frost boils are formed when the surface freezes across the top, trapping the still unfrozen muck beneath. As this bulge chills and expands, the mud is forced up through the crust. Raised earth stripes, found on moderate slopes, appear as lines or small ridges flowing downhill.

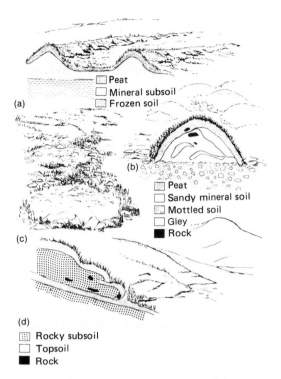

Figure 31.27 Patterned ground forms of the tundra region: (a) unsorted earth stripes; (b) frost hummocks; (c) sorted stone nets and polygons; (d) solifluction terrace. (Diagrams adapted from Johnson and Billings 1963.)

They apparently are produced by a downward creep or flow of wet soil across the surface of the permafrost.

Sorted patterns are characteristic of better-drained sites. The best known are the stone polygons, the size of which is related to frost intensity and the size of the material (Johnson and Billing 1962). The larger stones are forced out to a peripheral position, and the smaller and finer material, either small stones or soil, occupies the center. The polygon shape may result from an accumulation of rocks in desiccation cracks formed during drier periods. These cracks appear as the surface of the soil dries out, in much the same way as cracks appear in bare, dry, compacted clay surfaces in temperate regions. On the slopes, creep, frost-thrusting, and downward flow of soil change polygons into sorted stripes running downhill. Mass movement of supersaturated soil over the permafrost forms solifluction terraces, or "flowing soil." This gradual downward creep of soils and rocks eventually rounds off ridges and other irregularities in topography. This molding of the landscape by frost action is called *cryoplanation* and is far more important than erosion in wearing down the arctic landscape.

In the alpine tundra permafrost exists only at very high elevations and in the far north, but the frost-induced processes—small, solifluction terraces and stone polygons—are still present. The lack of a permafrost results in drier soils; only in alpine wet meadows and bogs do soil moisture conditions compare with the Arctic. Precipitation, especially snowfall and humidity, is higher in the alpine than in the arctic tundra, but the steep topography results in a rapid runoff of water.

The arctic and the alpine regions share many features that characterize the tundra biome. The vegetation of the tundra is structurally simple (Figure 31.28). The number of species tends to be small, growth is low, and most of the biomass and functional activity are confined to a relatively few groups. Growing season and reproductive season are short. Most of the vegetation is perennial and reproduces vegetatively rather than by seed. Although it appears homogeneous, the pattern of vegetation is diverse. Small variations in microtopography result in a steep gradient of moisture. The combination of microrelief, snowmelt, frost heaving, and aspect, among other conditions, produces an almost endless change in plant associations from spot to spot (Polunin 1955).

Although the environment of the arctic and alpine regions is somewhat similar,

the vegetation of the two differs in species composition and in adaption to light. Alpine sorrel *(Oxyria digyna),* found both in the arctic and alpine tundras, exhibits increased production of flowers and decreased production of rhizomes in the southern portion of its range (Mooney and Billings 1960). Northern populations of the plant have a higher photosynthetic rate at lower temperatures and attain a maximum rate at a lower temperature. Alpine plants reach the saturation point for light at higher intensities than arctic plants, which are adapted to lower light intensities (Mooney and Billings 1961). Arctic plants require longer periods of daylight than alpine plants. The further north the geographic origin of the plant, the more slowly the plant grows under short photoperiod.

At the tree line, where the forest gives way to the tundra, lies an area of stunted, wind-shaped trees, the Krummholz or "crooked wood" (Figure 31.29). The Krummholz in the North American alpine region is best developed in the Appalachians. In the west it is much less marked, for there the timber line ends almost abruptly with little lessening of height; the trees for the most part are flagged, that is, branches remain only on the lee side. On the high ridges of the Appalachians, particularly in the White mountains, and on the Adirondack mountains, the trees begin to show signs of stunting far below the timber line. As the trees climb upward, stunting increases until spruces and birches, deformed and semiprostrate, form carpets 0.6–1 m high, impossible to walk through but often dense enough to walk upon. Where strong winds come in from a constant direction, the trees are sheared until the tops resemble close-cropped heads, although the trees on the lee side of the clumps grow taller than those on the windward side.

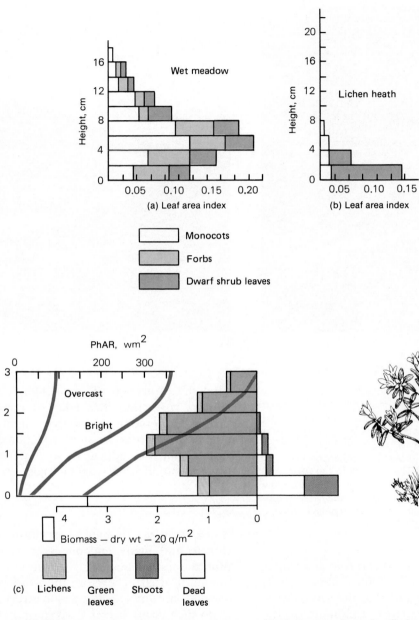

(a) Leaf area index

(b) Leaf area index

Monocots

Forbs

Dwarf shrub leaves

(c) Lichens Green Shoots Dead
leaves leaves

Figure 31.28 Structure of vegetation in arctic tundra. (a) Maximum leaf area indices (m²/m²) at different plant heights (single surface) measured by inclined point quadrants for a wet meadow. (b) Leaf area indices for a lichen heath. Note the structural simplicity. (After Berg et al. 1975.) (c) Vertical structure and radiation profile of the *Loiseleuria* heath. The instantaneous values of photosynthetically active radiation (W/m²) are plotted against stand height for a bright day and an overcast day. The air and soil temperatures are for a hot summer day. (After Larcher et al. 1975.)

(a)

(b)

Figure 31.29 The Krummholz. (a) In the Rocky Mountains the tree line is sharply defined. Dwarf spruce and fir grow in narrow pockets that hold snow in winter. Although the low growth form is partly genetic, leaves, twigs, and branches exposed above the snow are broken or killed by wind and cold. (b) The tree line in the Australian Alps in the Brindabella Range is marked by low twisted snow gum *(Eucalyptus pauciflora)* in protected pockets that gives way to low-growing heaths. (Photos by R. L. Smith.)

Though the wind and cold and winter desiccation (Wardle 1968; Lindsay 1971) generally are regarded as the cause of the dwarf and misshapen condition of the trees, Clausen (1965) has demonstrated that the ability of some species of trees to show a Krummholz effect is genetically determined. Eventually conditions become too severe even for the prostrate forms, and the trees drop out completely except for those that have taken root be-

hind the protection of some high rocks. Tundra vegetation then takes over completely. On some slopes the trees might be able to grow on better sites at higher elevations, but they appear to be eliminated by competition from sedges (Griggs 1946).

Structure

Arctic Tundra. In spite of its distinctive climate and many endemic species, the tundra does not possess a vegetation type unique to itself. Therefore the word *tundra* does not imply a vegetation structure as do such terms as *prairie, deciduous forest,* or *tropical rain forest.* In effect, the tundra is structurally a grassland.

In the Arctic only those species able to withstand constant disturbance of the soil, buffeting by the wind, and abrasion from wind-carried particles of soil and ice can survive. On well-drained sites heath shrubs, dwarf willows and birches, dry-land sedges and rushes, herbs, mosses, and lichens cover the land. On the driest and most exposed sites—the flat-topped domes, rolling hills, and low-lying terraces, all usually covered by coarse, rocky material and subject to extreme action by the frost—vegetation is sparse and often confined to small depressions. Plant cover consists of scattered heaths and mats of mountain avens, as well as crustose and foliose lichens growing on the rocks. Willows, birch, and heath occupy well-drained soils of finer material, and between them grow grasses, sedges, and herbs.

Over much of the Arctic the typical vegetation is a cotton grass-sedge-dwarf heath complex (H. C. Hanson 1953). Hummocks may support growths of lichens, willow, blueberry, and heaths. Depressions are covered with sedge-marsh vegetation, and over the rest grow tussocks of cotton grass. The spaces between the tussocks may be filled with sphagnum, on top of which dwarf shrubs grow; in other places sphagnum may overgrow the sedges and cotton grass. On mounds and hummocks in freshwater marshes, on well-drained knolls and slopes, in areas of late-melting snowbanks, along streams, and on sandy and gravelly beaches, grassland types develop.

Topographic location and snow cover delimit a number of arctic plant communities. Steep, south-facing slopes and river bottoms support the most luxurious and tallest shrubs, grasses, and legumes, whereas cotton grass dominates the gentle north-facing and south-facing slopes, reflecting higher air and soil temperatures and greater snow depth. Pockets of heavy snow cover create two types of plant habitats, the snow patch and the snowbed. Snow-patch communities occur where wind-driven snow collects in shallow depressions and protects the plants beneath. Snowbeds, typical of both arctic and alpine situations, are found where large masses of snow accumulate because of certain topographic peculiarities. Not only does the deep snow protect the plants beneath, but the meltwater from the slowly retreating snowbank provides a continuous supply of water throughout the growing season. Snowbed plants, usually found only here, have an extremely short growing season but are able to break into leaf and flower quickly because of the advanced stage of growth beneath the snow.

The conditions unique to the Arctic result in part from three interacting forces: permafrost, vegetation, and the transfer of heat. Permafrost is sensitive to temperature changes. Any natural or human disturbances, however slight, can cause the permafrost to melt. Because the permafrost itself is impervious to water, it forces all the water to move above it. Thus the surface water becomes quite conspicuous even though precipitation is low (see Brown and Johnson 1964; Brown 1970). Vegetation protects the permafrost by shading, which reduces the heating of the soil. It retards the warming and thawing of the soil in summer and increases the average temperature in winter. If the vegetation is removed, the depth of thaw is 1.5 to 3 times that of the area still retaining the vegetation. Accumulated organic matter and dead vegetation further retard the warming of the soil in summer, even more than a vegetative cover. Thus vegetation and its organic debris impede the thawing of the permafrost and act to conserve it (see also Pruitt 1970; Bliss et al. 1973).

In turn permafrost chills the soil, retarding the general growth of both aboveground and belowground parts of plants and the activity of soil microorganisms. It also impoverishes the aeration and nutrient content of the soil (Tyrtikov 1959). The effect is more pronounced the closer the permafrost comes to the surface, where it contributes to the formation of shallow root systems. The effect of permafrost on vegetation is so pronounced that vegetation can be used to map areas of permafrost.

The tundra world holds some fascinating animals, even though the diversity of species is low. The animals of the Arctic are mostly circumpolar. They include the caribou or reindeer *(Rangifer tarandus)*, muskox *(Ovibos moschatus)*, arctic hare *(Lepus arcticus)*, and arctic ground squirrel *(Alopex lagopus)*. Some 75 percent of the birds of the North American tundra are common to the European tundra (Udvardy 1958).

Muskox and caribou are the dominant large herbivores. Muskox are intensive grazers with low herd movements. In summer they feed on grasses, sedges, and dwarf willow in the valleys and plains; in winter they move up to the windswept ridges where snow cover is scant. Caribou are extensive grazers, spreading over the tundra in summer to feed on sedges and grasses, and in winter migrating southward to the taiga to feed on lichens.

Intermediate-sized herbivores include the arctic hares, which feed on willows. In winter the hares disperse over the range; in summer they tend to congregate in more restricted areas. The smallest and dominant herbivore over much of the arctic tundra is the lemming *(Lemmus spp.)* which feeds on fresh green sedges and grasses. Breeding throughout the year beneath the snow, the lemming has a three- to five-year population cycle. During the highs this rodent can reach densities as great as 125 to 250 per hectare and during the lows as few as 3 to 5 per hectare. Herbivorous birds are relatively few, dominated by ptarmigan and migratory geese.

The major arctic carnivore is the wolf *(Canis lupus)*. (The polar bear is a marine predator.) The wolf preys on muskox, caribou, and when they are abundant, lemmings. Medium-sized predators include the arctic fox, which feeds on the arctic hare. The smallest mammalian predators, the least weasel *(Mustela rixosa)* and the short-tailed weasel *(Mustela erminea)*, feed principally on lemmings and the eggs and nestlings of birds. Major avian predators, the snowy owl *(Nyctea scandiaca)* and the hawklike jaeger *(Stercorarius)*, feed heavily on lemmings. Except for the wolf, whose populations remain relatively stable when free from human persecution, the fortunes of most arctic predators rise and fall with the flood and ebb of lemming life.

The arctic tundra, with its wide expanse of ponds and boggy ground, is the haunt of myriads of waterfowl, sandpipers, and plovers, which arrive when the ice is out, nest, and return south before winter sets it.

Invertebrate life is scarce, as are amphibians and reptiles. Some snails are found in the arctic tundra about the Hudson Bay. Insects, reduced to a few genera, are nevertheless abundant, especially in mid-July. The insect horde is composed of black flies, deerflies, and mosquitos (Shelford and Twomey 1941).

Animal activity in the arctic tundra is geared to short summers and long winters. The only hibernator is the ground squirrel, although the female polar bear does den in the snow, where she gives birth to her cubs. The ground squirrel is active only from May to September. It mates almost as soon as it emerges from the burrow in spring, and the young are born in mid-June, after a 25-day gestation period. The young are self-sufficient by mid-July, attain adult weight, and are ready to hibernate by late September and early October (W. Mayer 1960). A similar speedup in the life cycle exists among some of the arctic birds. Because of the long days, the northern robin feeds the young for 21 hours a day, and the young may leave the nest when they are slightly over 8 days old, in contrast to the 13 or more days when born in the temperate region (Karplus 1949). The species that are unable to withstand severe cold migrate to warmer or more protected areas.

Alpine Tundra. In general the alpine tundra is a more severe environment for plants than the arctic tundra, and the adaptation of plants to the physical environment is probably more important than the interrelations of one species with another. The alpine tundra is a land of strong winds, snow, and cold and widely fluctuating temperatures. During the summer the temperature on the surface of the soil ranges from 40° to 0° C (Bliss 1956). The atmosphere is thin, so light intensity, especially ultraviolet, is high on clear days.

The alpine tundra is a land of rock-strewn slopes, bogs, alpine meadows, and shrubby thickets (Figure 31.30). In spite of the similarity of conditions, only about 20 percent of the plant species of the arctic and of the Rocky Mountain alpine tundra are the same, and they are different ecotypes. Heaths are lacking in the tundras of the Rockies, as well as the heavy growth of lichens and mosses between other plants. Lichens are more or less confined to the rocks, and the ground is bare between plants.

Cushion- and mat-forming plants, rare in the arctic, are important in the alpine tundra. Low and hugging the ground, they are able to withstand the buffeting of the wind. The cushionlike blanket traps heat and the interior of the cushion may be 20° warmer than the surrounding air, a microclimate that is uti-

Figure 31.30 Alpine tundra in (a) the Rocky Mountains and (b) the Australian Alps. Although the species of vegetation are different, the growth forms are convergent. The physiognomy of both alpine areas is similar. (Photos by R. L. Smith.)

(a)

(b)

lized by insects. Thick cuticles, which increase plants' resistance to desiccation, and an abundance of epidermal hairs and scales are characteristic of alpine plants. The significance of this feature is still debated. The hairs appear to absorb and reflect the bright light of the alpine environment. At the same time they may act as a heat trap, perhaps preventing cold injury when air temperatures drop to freezing (Krogb1955) and enabling the plantsto develop and bloom while the air is still cold.

Alpine vegetation and its associated soils vary on a complex gradient of topographic site and snow cover, both of which interact with wind (Figure 31.31). The vegetational pattern worked out for the Beartooth Plateau of Wyoming (Johnson and Billings 1962) serves as an example of the high alpine areas of the Rocky Mountains. The high, windswept areas, rocky and free of snow, support only lichens, which may completely cover the sheltered side; but on the windward side they are short, no higher than the depth of snow, and they may be completely lacking on the most exposed sites. Below the lichen growth are the xeric cushion-plant communities, which extend further downslope on the windward side than they do on the lee. This land of rock, lichens, and cushion plants is the alpine rock desert. In somewhat more protected sites grows the geum turf, a sodlike covering of geum and associated plants, such as sedges, lupines, polygonums, and mountain avens. Alpine meadows develop on well-drained soils of sheltered uplands and lower mesic slopes and basins. Hairgrass (*Deschampsia*), often growing in pure stands, is the dominant species. These meadows are subject to considerable disturbance both from frost activity and from pocket gophers. Alpine bogs, communities quite similar to those of the arctic tundra, support a growth of sedge and cottongrass. Willow

Figure 31.31 Relationship of major vegetation types to slope cover and slope position in the western alpine tundra on Bearfoot Plateau, Wyoming. (After Johnson and Billings 1962.)

thickets, dense and uniform in height, grow along drainage channels and in alpine valley bottoms.

The alpine tundra of the high Appalachians is not nearly so cold and windswept as that of the Rockies. Tundra areas are small and lack the diversity of species found in the western mountains. Indeed, little floristic similarity exists between the regions. There is a much closer affinity between the flora of the eastern alpine tundra and that of the arctic and the alpine communities of Scandinavia and central Europe.

Nine plant communities are recognized in the tundras of the Presidential Range in New Hampshire (Bliss 1963). They occur on two gradients, one of increasing snow depth, the other of increasing moisture (Figure 31.32). On exposed windswept sites where winter snow cover is thin or nonexistent, *Diapensia*, a dwarf, tussock-forming shrub, grows. Over those widespread areas where snow cover is variable a dwarf heath-rush community oc-

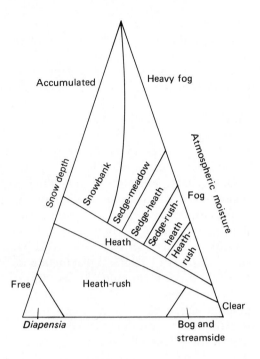

Figure 31.32 Relationship of alpine communities to snow and atmospheric moisture in the Presidential Range in New Hampshire, USA. (After Bliss 1963.)

cupies the sites. Dwarf heaths—bearberry, bilberry, Lapland rosebay—dominate where the deep snow cover melts early. Snowbank communities are most prevalent on the east- and southeast-facing slopes, in the lee of the prevailing winds.

The second gradient, one of increasing summer atmospheric and soil moisture and fog, is largely restricted to north- and west-facing slopes on the higher peaks. Sedge meadows at the highest elevations give way downslope to a sedge-dwarf heath community. At lower elevations this is replaced by a sedge-rush-dwarf heath type. Two other communities, the streamside and the bog, are common at low elevations.

The alpine tundra, which extends upward like islands in the mountain ranges of the world, is small in area and contains few characteristic species. The alpine regions of western North America are inhabited by pikas, marmots, mountain goat (not a goat at all, but related to the South American chamois), mountain sheep, and elk. Sheep and elk spend their summers in the high alpine meadows and winter on the lower slopes. The marmot, a mountain woodchuck, hibernates over winter, whereas the pika cuts grass and piles it in tiny haycocks to dry for winter. Some rodents, such as the vole and the pocket gopher, remain under the ground and snow during winter. The pocket gopher is important, for its activities influence the pattern of alpine vegetation. The tunneling gophers kill the sedge and cushion plants by eating the roots and by throwing the soil to the surface, smothering plant life. Other plants then take over on the windblown, gravelly soil. These pioneering plants are rejected by the gopher. The rodent moves on, the cushion plants move back in, organic matter accumulates again. Slowly the sedges recover, and when they do, the gophers return once more.

The alpine regions contain a fair representation of insect life. Flies and mosquitos are scarce, but springtails, beetles, grasshoppers, and butterflies are common. Because of the ever-present winds, butterflies fly close to the ground; other insects have short wings or no wings at all. Insect development is slow; some butterflies may take two years to mature, and grasshoppers three.

Function

Low temperature, a short growing season ranging from 50 to 60 days in the high arctic tundra to 180 days in low altitude alpine tundra, and low availability of nutrients interact to keep primary production on the tundra low.

Photosynthesis. Plants are photosynthetically active on the arctic tundra about three months out of the year. The onset of photosynthesis occurs in spring as quickly as snow cover disappears, but it is limited initially because plant leaves are poorly developed. Alpine species, however, exhibit a rapid burst of growth following snowmelt, but at the expense of belowground root and rhizome carbohydrate reserves (Hadley and Bliss 1964).

Throughout the growing season photosynthesis is influenced by light and temperature (Johansson and Linder 1975). Across the tundra plants exhibit ecotypic variation in their adjustments to these two influences. Maximum photosynthesis takes place when ambient temperatures are 10 to 15° C (Mayo et al. 1977; Tieszen et al. 1980). Photosynthesis is inhibited when leaf temperatures rise above 40°, which they can reach on bright, clear arctic summer days (Bliss 1975). Tundra plants, which possess a C_3 photosynthetic pathway (see Chapter 3), have adapted to short growing seasons and low intensities of the arctic summer in three ways: (1) they carry on photosynthesis throughout the 24-hour daylight period, even at midnight when the light is one-tenth that of noon; (2) the plants, especially the monocots, possess a high leaf area index (0.5 to 1.0); (3) the leaves are rarely light-saturated. The nearly erect leaves of some arctic plants permit the almost complete interception of the slanting rays of arctic solar radiation (Bunnell et al. 1975; Berg et al. 1975). Some arctic plants, however, particularly mosses, are poorly adapted to 24-hour light. Under this condition they turn yellowish or brownish (Kallio and Valanne 1975).

Much of the photosynthate goes into the production of new growth, but about one month before the growing season ends, plants cease to allocate photosynthate to aboveground biomass. They withdraw photosynthate from the leaves and move it to the roots and belowground biomass, sequestering ten times the amount stored by temperate grasslands (see Chapin et al. 1980).

Primary Production. Net annual primary production varies markedly across the tundra, depending upon the plant community considered (Bliss 1975; Wielgolaski 1975). Because of microenvironmental conditions influenced by soil, slope, aspect, exposure to wind, snow depth, and drainage conditions plant communities change rapidly over relatively short distances. Primary production for selected tundra communities is summarized in Tables 31.11, 31.12, and 31.13. In general alpine tundras are more productive than arctic tundras (Table 31.11). At the Devon Island IBP site in Canada total primary production above and below ground for a hummocky sedge-moss meadow was 148.3 g/m²/yr; for a wet sedge meadow, 175.4 g/m²/yr; and for a raised beach cushion plant-lichen community, 30.4 g/m²/yr (Bliss 1975). In contrast, net primary production above ground only for two communities on an Austrian alpine heath, a *Vaccinium* heath and a *Loiseleuria* heath, was 422 g/m²/yr and 277 g/m²/yr, respectively (Larcher et al. 1975). Total aboveground and belowground productivity for three Norwegian alpine tundra communities, a lichen heath, a wet sedge-moss meadow, and a dry meadow dominated by mountain avens, *Dryas*, forbs, and grass, was 188 g/m²/yr, 1527 g/m²/yr, and 786 g/m²/yr respectively (Ostbye et al. 1975).

Different components of the plant community make different contributions to net productivity, depending upon the site (compare Tables 31.11 and 31.12). In the hummocky sedge-moss meadow of the Devon Island IBP site, Canada, aboveground net primary production of vascular plants amounted to 44.7 g/m²/yr, while mosses contributed 33 g/m²/yr and lichens contributed nothing. In the raised beach community dominated by lichens and cushion plants vascular plants contributed 17.8 g/m²/yr, mosses 2 g/m²/yr, and lichens 25 g/m²/yr. In an alpine wet meadow vascular plants contributed 254 g/m²/yr and mosses 173 g/m²/yr.

Much of the primary production of vascular plants is below ground rather than above (Table 31.11). The data emphasize this important functional aspect of the tundra. The net annual aboveground

TABLE 31.11 Primary Production and Biomass of Vascular Plants, Selected Tundra Ecosystems (g/m²/yr)

Location and Type	Vegetation	Primary Production	Biomass	A/B Ratio Biomass	Reference
Devon Island, Canada					
arctic	hummocky sedge-moss meadow	44.7 (103.6)	86 (1085)	1:12.6	Bliss 1975
arctic	wet sedge-moss meadow	45.7 (129.7)	78 (691)	1:8.9	Bliss 1975
arctic	raised beach, cushion plant-lichen	17.8 (2.6)	89 (57)	1:.06	Bliss 1975
Taimyr, USSR					
arctic	herb and grass meadow	68.4	71.5		Matveyera et al. 1975
arctic	mossy *Salix polaris* mesic	31.8	20.2		Matveyera et al. 1975
arctic	frost boil dry frost boil	23.8	21.0		Matveyera et al. 1975
Hardangervidda, Norway					
alpine	lichen heath	88 (100)	62 (191)		
alpine	wet meadow	254 (1316)	147 (410)	1:3.08	Ostbye et al. 1975
alpine	dry meadow	241 (545)	161 (245)	1:8.95	Ostbye et al. 1975
	1:3.38	Ostbye et al. 1975	1:8.95	Ostbye et al. 1975	
Mt. Patscherkofel, Austria					
alpine	*Loiseleuria* heath	277	1084 (2213)	1:2.04	Larcher et al. 1975
alpine	*Vaccinium* heath	422	1013 (2206)	1:2.17	Larcher et al. 1975
Macquarie Island, Tasmania					
subantarctic	grassland	1890 (3670)	912 (1690)	1:1.85	Jenkin 1975
	herbfield	314 (550)	139 (670)	1:4.82	Jenkin 1975

Note: Numbers in parentheses are belowground production and biomass; without parentheses, aboveground production and biomass.

TABLE 31.12 Primary Production and Biomass of Mosses and Lichens, Selected Tundra Ecosystems (g/m 2/yr)

Location and Type	Vegetation	Primary Production		Biomass		Reference
		Mosses	Lichens	Mosses	Lichens	
Devon Island, Canada						
arctic	hummocky sedge-moss	33.0	0	908	0	Bliss 1975
arctic	meadow	102.6	0	1097	0	Bliss 1975
arctic	wet sedge-moss meadow raised beach, cushion plant-lichen	2.0	25	15	49	Bliss 1975
Hardangervidda, Norway						
alpine	lichen heath	10.0	78	7	370	Ostbye et al. 1975
alpine	wet meadow	173.0	0	175	0	Ostbye et al. 1975
alpine	dry meadow	48.0	4	31	19	Ostbye et al. 1975
Mt. Patscherkofel, Austria						
alpine	*Loiseleuria* heath	nd		0	136	Larcher et al. 1975
	Vaccinium heath	nd		44	22	Larcher et al. 1975

primary production of vascular plants in a wet sedge meadow at Devon Island was 45.7 g/m^2 and belowground 129.7 g; a wet meadow at an alpine tundra at the Hardangervidda IBP site in Norway had a biomass of 254 g/m^2 above ground and 1316 g/m^2 below ground. On more xeric sites dominated by lichens there is less belowground biomass than above ground (Table 31.13). (For summaries see Bliss et al. 1973; Wielgolaski et al. 1981; Miller et al. 1980.)

Although total production of the tundra is low because of a short growing season, daily primary production rates of 0.9 to 1.9 g/m^2 in arctic tundra and 2.2 g/m^2 in some alpine tundras are comparable to some temperate grasslands. The efficiency of primary production of the Devon Island tundra (Table 31.14) ranges from 0.03 percent for the polar desert to 1.03 percent for the hummocky sedge-moss meadow (Bliss 1975). Efficiency of a Norway tundra ranges from 0.6 percent for a dwarf heath community to 2.4 percent for a willow thicket (Wielgolaski and Kjelvik 1975). By comparison, efficiency of temperate grassland ecosystems ranges from 0.33 to 3.8 percent.

Net radiation in the tundra is somewhat higher than radiation in temperate regions, but photosynthetically active radiation appears to be lower, increasing the percentage efficiency of primary production of tundra plants (Bliss 1975).

Biomass. Biomass reflects the pattern of net production. At Devon Island, for example, biomass for the polar desert amounts to 270 g/m^2 and for the hummocky sedge-moss meadow 3208 g/m^2. More live biomass exists below ground than above in most tundra communities (Table 31.11). Live roots make up about 60 percent of the total belowground biomass. The ratio of aboveground to be-

TABLE 31.13 Live Standing Crop and Net Annual Production of Communities on Truelove Lowland and Plateau (energy constant, kJ/m^2)

	Vascular Plants					
	Above Ground	Below Ground	Mosses	Lichens	Algae	Total
STANDING CROP						
Cushion plant-lichen	1,824	1,167	272	841	0	4,104
Cushion plant-moss	2,581	1,025	10,920	393	nd	14,919
Frost-boil sedge-moss	1,088	6,720	10,012	0	17	17,837
Dwarf shrub-heath	3,397	18,912	7,389	410	nd	30,108
Hummocky sedge-moss	1,674	20,656	16,527	0	67	38,924
Wet sedge-moss	1,519	13,154	19,966	0	33	34,672
Polar desert (plateau)	301	75	4,222	0	0	4,598
Semidesert (raised beach)	2,130	1,113	4,531	661	0	8,435
Wet sedge tundra (all meadows)	1,402	14,008	13,351	0	42	28,803
Total Lowland	1,602	8,422	9,506	205	21	19,756
NET PRODUCTION						
Cushion plant-lichen	326	63	38	50	0	477
Cushion plant-moss	552	105	364	33	nd	1,054
Frost-boil sedge-moss	565	1,121	146	0	17	1,849
Dwarf shrub heath	423	1,904	364	59	nd	2,750
Hummocky sedge-moss	858	1,979	602	0	67	3,506
Wet sedge-moss	895	2,477	1,874	0	33	5,279
Polar desert (plateau)	42	8	92	0	0	142
Semidesert (raised beach)	418	79	167	46	0	710
Wet sedge tundra (all meadows)	724	1,619	460	0	42	2,845
Total Lowland	544	1,004	331	17	21	1,917

Source: Bliss 1975.

lowground biomass depends upon the nature of the tundra community. On Devon Island the hummocky sedge-moss community has an A/B ratio of 1:12.6, the cushion plant-lichen community 1:.06; the wet sedge communities of both the arctic and the alpine tundra 1:8.9.

Biomass on the tundra also accumulates below rather than above ground. The entire monocot portion of the vegetation grows and dies each season. Most of the herbaceous and woody plants accumulate little above ground and rapidly turn over the energy in the living portion of the system. The belowground portion is more persistent, with roots lasting for two to ten years, reflecting low mortality and slow decomposition. Because decomposition is slowest in wet sites, the difference between aboveground and belowground biomass is greatest there. Decomposition is faster on well-drained and on nutrient-rich sites. In fact, greatest accumulation of belowground biomass is associated with those sites where net productivity is the lowest.

Decomposers. Most of the production of the tundra goes by the way of the decomposers (Figure 31.33). At Devon Island, for example, grazing by lemmings at the most accounts for only 3 to 4 percent of aboveground standing crop. On the hummocky sedge-moss meadow 2 percent of

TABLE 31.14 Net Annual Production and Efficiency for Various Tundra Plant Communities and Components (Growing Season of 50 to 60 Days)

Component	Net Production (kJ/m^2)	Total Radiation (kJ/m^2)	Efficiency (%)	
			Total Radiation	PAR*
Polar desert (plateau)	138	11.72 × 10^5	0.01	0.03
Polar semidesert (raised beach)	711	11.13 × 10^5	0.06	0.16
Sedge-moss meadows (all meadows)	2845	9.04 × 10^5	0.31	0.79
Hummocky sedge-moss meadow	3506	9.04 × 10^5	0.39	1.03
Total lowland	1916	9.62 × 10^5	0.20	0.50

*PAR (photosynthetically active radiation) = 40 percent of total radiation; efficiency was calculated on the basis of radiation received during the growing season for each component part and the contribution (%) which that component provides to the total lowland (raised beach types, meadow types, dwarf shrub heath, and so on).
Source: Bliss 1975.

the plant production is utilized by herbivores; 98 percent of the primary production is channeled to microbivores and saprovores. Comprising the latter is a wide variety of soil invertebrates, dominated by protozoans. Among important soil invertebrates are the Enchytraeidae or pot worms, nematodes, fly larvae, and various crustacea (McLean 1980). Annual production and consumption among these animals is greater than the herbivore system even in a lemming high (Bunnel et al. 1975).

However, the major detrital consumers are bacteria and fungi (Bunnell et al.

1980; Flanagan and Bunnell 1980). Tundra soils contain a diversity of soil bacteria—iron oxidizers, nitrogen fixers, ammonia oxidizers, sulfate reducers, and fermenters. Bacteria occur in the same abundance as in temperate soils. The same is true for fungi. The amount of my-

Figure 31.33 Energy flow diagram for a sedge-moss meadow system. Standing crop (boxes) and energy flow (arrows) are expressed in kJ/m^2. Respiration (R) is given for all components along with energy estimates for rejecta. Ivores are fungal, bacterial, and protozoan feeders. Insectivores are seed and insect-feeding birds. (From Bliss 1975.)

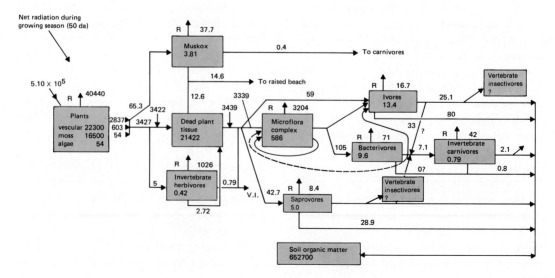

celia in tundra soils is as great as or often greater than in temperate mull or mor soils. As in other terrestrial systems, fungi are more important than bacteria. Like the primary producers, the decomposers are restricted in their activities by the cold. Studies of decomposition rates for sedge indicate a dry weight loss of 19 percent for the first year, 12 percent for the second year. For moss the rate was 1.3 percent per year (see Bliss et al. 1973; Bliss 1975; Bunnell et al. 1975; Rosswall 1975).

Consumers. Major herbivores of the tundra include waterfowl, ptarmigan, lemming, hares, muskox, caribou, and reindeer. In some parts of the arctic tundra lemmings are the dominant herbivores. Adult lemmings consume about 0.32 g dry material/g body weight and juveniles 0.53 g. Adults consume 170 percent and juveniles 200 percent of their body weight in summer with an assimilation efficiency of 47 percent. Lemmings during a high may consume over 25 percent of the above ground primary production or 10 percent of total plant production. Consumption, however, is seasonal; heaviest consumption takes place in winter when during a high, lemmings may consume the entire aboveground vascular plant biomass. Of the food consumed lemmings return about 70 percent as feces, which accumulate in winter and release quantities of nutrients during the spring melt (Bunnell et al. 1975). During periods of highs, lemmings can hold primary production to 3 to 48 g/m^2. By reducing the litter layer, these animals reduce insulation of the soil and increase the depth of thawing. In doing so, lemmings can influence the composition and nature of the tundra plant community.

In other parts of the arctic tundra lemmings are not significant grazing herbivores. Their place may be taken by the muskox and caribou. These two ungulates have an average standing crop of 0.17 kg/ km^2 in the Canadian and Alaskan tundra, low compared to the ungulate biomass of 140 kg/km^2 of the African savanna. In summer muskox consume approximately 30 to 34 grams of vegetation per kilogram of body weight and have an assimilation efficiency of 56 percent. Muskox remove less than 1 percent of potential primary production. On the range at Devon Island, muskox grazed 14 percent of the total area available and removed 15 percent of the total herbage available. However, muskox are selective grazers, and on the restricted areas they grazed the animals removed 80 to 85 percent of the herbage available. Dung decomposition is slow, requiring 5 to 12 years to be recycled (Bliss 1975).

Preying on the herbivores are a number of carnivores. Arctic fox consume 23.3 to 44.7 kcal/kg body weight a day in summer and 54.5 to 72.2 kcal/kg body weight in winter. An efficient assimilator, the fox rarely passes as feces more than 5 percent of the total energy ingested. One investigator (Speller, cited by Bliss et al. 1973) estimated that between the first of May and the last of September, one fox would require the equivalent of 400 28-g lemmings, 400 30-g snow buntings, and 20 3.46-kg arctic hares. Weasels consume more than 1 28-g lemming a day. Extremely effective as predators, weasels may consume up to 20 percent of the lemming population. Snowy owl, another lemming predator, has a winter food requirement of four to seven lemmings a day. Thus not only can the predators act as a force to drive the lemming population to a low, but predators themselves are strongly affected by a scarcity of lemmings. When lemming populations are low, arctic fox experience reproductive failures and snowy owls cannot exist on the tundra. When forced southward, the owls face an uncertain fate.

Nutrients. Arctic and alpine tundras are low in nutrients because of a short growing season, cold temperatures, low precipitation, and restricted decomposition. Of all terrestrial ecosystems, the arctic tundra has the smallest proportion of its nutrient capital in live biomass. Dead organic material functions as a nutrient pool. Most of this nutrient capital is not directly available to plants. Pools of soluble soil nutrients, especially nitrogen and phosphorus, are small relative to exchangeable pools, which in turn are small compared to the nonexchangeable pools (Figure 31.34) (Bunnell et al. 1975). As a result nutrient cycling in the tundra is conservative. Vascular plants have a strong internal cycling. They retain and reincorporate nutrients, especially nitrogen, phosphorus, potassium, and calcium, in their tissues rather than release them to decomposers. In forest ecosystems more nutrients are retained in new growth than released through litterfall, but in the tundra ecosystem death of tissues compensates for new growth. The system arrives at a steady state in which return equals uptake.

Leaching or removal of nutrients is minimal, occurring mostly in spring at the beginning of the growing season. As the snow melts, runoff increases, carrying with it animal debris that accumulates over winter. Spring rains leach senescent plant material produced the previous growing season. Rising temperatures stimulate decomposition. Because 60 percent of the active roots are in the upper 5 cm of soil, the root mass once thawed takes up most of the nutrients. However, early in the season, plants have to compete with microbes, whose uptake of nutrients can exceed that of plant roots (Chapin et al. 1980). In summer vascular plants leak nutrients, especially phosphorus and potassium, from the cuticle of the leaves to the surface, where they are washed off by summer rains. These nutrients are often picked up by bryophytes, which absorb or adsorb nutrients before they reach the soil. The nutrients incorporated into the mosses are slowly released by them. Thus bryophytes function as a temporary nutrient sink.

In some tundra ecosystems lemmings become involved in nutrient cycling. Foraging in meadows and ridges, but building their nests and defecating in troughs, lemmings transport nutrients from one site to another. Higher levels of soil nutrients, especially phosphorus, in polygon troughs, and higher rates of decomposition in these microsites may reflect the activity of lemmings (Bunnell et al. 1975).

Two nutrients, nitrogen and phosphorus, are most limiting. Compared to temperate forest and grassland, the input of nitrogen is small. Two major sources are precipitation and biological fixation. Nitrogen fixation is accomplished by both anaerobic and aerobic free-living bacteria; by blue-green algae in soil, in water, and especially in the moss where they live epiphytically; and by lichens, particularly in the Fennoscandian tundra (Granhall and Lid-Torsvik 1975; Kallio and Kallio 1975). A rough estimate places nitrogen fixation up to 1 or 2 $g/m^2/yr$, although values for blue-green algae range up to 5 to 6 $g/m^2/yr$. Nitrogen fixation is controlled by light, temperature, and moisture.

Precipitation is the second important input of nitrogen. The input from precipitation is usually lower than the input from biological fixation, but at some sites the two are equal. At Hardangervidda, Norway, for example, input from precipitation equals that of fixation, about 200 $mg/m^2/yr$ (Granhall and Lid-Torsvik 1975).

Nitrogen accumulates in tissues in greater quantities than other nutrients, making much of it unavailable for recycling. At the Barrow IBP site in Alaska about 65 percent of the gross input, 59.2 mg/m^2, is stored in the system as living

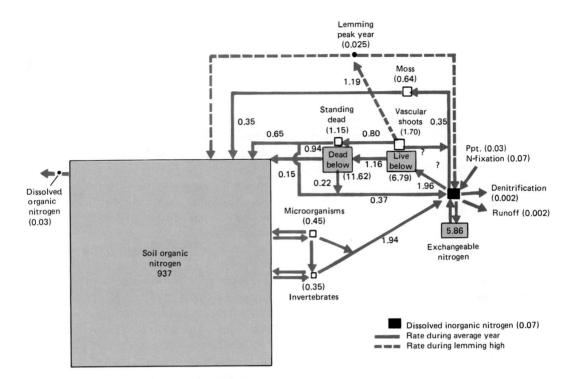

Figure 31.34 Annual nitrogen budget in a moist meadow on arctic coastal tundra to the depth of 20 cm. The area of each box is proportional to its compartment size in g/m². Values next to the arrows indicate annual fluxes in g/m²/yr. The budget represents that for years of low lemming populations. Years of high lemming populations are indicated by dashed lines to account for increased flux through the rodents. Amount of leaching from vascular plants is unknown; and mosses are assumed to get all of their nutrients from the soil (which they probably do not). Note the relatively large amount of N in the soil organic matter and the small amounts in other compartments. (After Chapin et al. 1980: 471.)

and dead organic matter (Bunnell 1975). Circulation in the system is restricted by decomposers, whereas flux rates are more or less controlled by the nitrogen-fixing microorganisms. Relatively small quantities of nitrogen are lost through leaching, stream flow, and denitrification. The most significant leaching takes place during snowmelt in the spring.

Phosphorus can be very limiting in the tundra ecosystem. Phosphorus apparently controls the rate of production of new leaves, primarily by controlling the rate at which nutrients are removed from older leaves and translocated into new leaves (Bunnell et al. 1975). The effect of phosphorus limitation is not nutrient deficiency but a limitation of leaf area, which lowers production. Although the accumulation of nitrogen, potassium, and calcium is similar from site to site across the tundra, the accumulation of phosphorus is highest at the most productive sites.

Because pools of both nitrogen and phosphorus are small, a constant turnover between the exchangeable and soluble pools is necessary to replenish the quantity absorbed by plants. Bunnell et al. (1975) estimate that at the Barrow site soluble and exchangeable nitrogen must turn over 11 times during a growing season to meet plant needs, and phosphorus must be replenished 200 times a season or 3 times a day during the growing season. Because the tundra soil does not store available nutrients in any great quantity, primary productivity depends upon release of nutrients from decomposition, the uptake of which is often aided by mycorrhizae. In this respect the tundra ecosystem is more similar in function to the tropical forest ecosystem than to the temperate forest ecosystem.

■ Synthesis

A simple comparison of ecosystem functions provides a first step toward seeking some common properties or metabolic patterns in ecosystems and developing some general principles of ecosystem functioning. Table 31.15 presents comparative metabolic parameters for four types of terrestrial ecosystems: a deciduous forest (mesic hardwood), a coniferous forest, a short-grass prairie, and a tun-

dra. The values are general and, of course, selected ecosystems within the four types might present values different from those used. Data for the deciduous forest and the short-grass prairie are probably more accurate than data for the other two.

Although gross production of the deciduous forest ecosystem is 1.6 times that of the coniferous forest and 1.7 times that of the short-grass prairie, net primary production among the three is comparable, although the coniferous forest is somewhat less. Much of the GPP of the forest ecosystems goes to the maintenance of structure. This fact is reflected in a comparison of the deciduous forest with the short-grass prairie. Forest structure requires 4.5 times greater allocation of photosynthate to maintain structure than the grassland. Heterotrophic respiration in the deciduous forest and short-grass prairie is similar, but values for the coniferous forest are considerably lower, reflecting either an inaccurate data base or more likely the lower activity of soil microflora and microfauna, allowing accumulation of litter on the forest floor. All four systems are

TABLE 31.15 Comparative Metabolic Parameters and Metabolic Ratios of Four Contrasting Ecosystems ($gC/m^2/yr$)

		Deciduous Forest	Coniferous Forest	Grassland	Tundra
Metabolic parameters					
Gross primary production	GPP	2150	1320	983	208
Autotrophic respiration	R_A	1440	680	430	120
Net primary production	NPP	720	600	840	88
Heterotrophic respiration	R_H	660	370	670	85
Net ecosystem production	NEP	150	270	180	3
Ecosystem respiration	R_E	2105	1050	1090	205
Metabolic ratios					
Production efficiency	R_A/GPP	0.67	0.52	0.30	0.57
Effective production	NPP/GPP	0.33	0.45	0.70	0.42
Maintenance efficiency	R_A/NPP	2.0	1.13	0.51	1.36
Respiration allocation	R_H/R_A	0.46	0.54	1.55	0.71
Ecosystem productivity	NEP/GPP	0.02	0.20	0.14	0.01

Source: Deciduous forest and grassland data from Burgess and O'Neill 1976. Coniferous forest and tundra data from National Academy of Science 1974.

characterized by a large pool of organic detritus, suggesting that a large, stable pool of organic matter is typical of ecological systems. Three of the systems are characterized by annual accumulation of large amounts of organic matter below ground.

This accumulation of organic matter is one aspect of capture, distribution, and conservation of elements essential to system persistence. Currency for translocation and retention of nutrient elements is carbon allocated for maintenance of primary producers and decomposers (Burgess and O'Neill 1976). Retention and distribution of elements within an ecosystem may be dependent on the maintenance of the detrital pool in litter and soil and in the decomposer community. The cost to the ecosystem is the loss of carbon to heterotrophic respiration. Microbial immobilization of nutrients appears to be synchronized with likely periods of maximum elemental uptake by roots, which results in a reduced amount of nutrients being lost from the ecosystem. Maximum growth of roots appears to take place at times of minimum nutrient immobilization by decomposers. This reciprocal interaction between plants and decomposers may be a major mechanism underlying nutrient conservation in ecosystems.

Among the four types of ecosystems the tundra stands out as possessing extremely low metabolic parameters. Gross production of the tundra, for example, is 10 times less than that of the deciduous forest. However, when all four systems are compared in terms of ratios of metabolic parameters, especially production efficiencies and effective production, ecosystems appear more uniform. Respiratory expenditures per unit production are about the same regardless of the environment. Thus certain parameters of ecosystem functioning analyzed in terms of ratios (such as the relative apportioning of resources among components of various ecosystems) yield some basic patterns common to most ecosystems.

■ Summary

Coniferous, deciduous, and tropical rain forests are three of the dominant types of forest. The coniferous forest, which forms a vast belt encircling the northern portion of the Northern Hemisphere, is typical of regions where summers are short and winters are long and cold. The deciduous forest, richly developed in North America, western Europe, and eastern Asia, grows in a region of moderate precipitation and mild temperatures during the growing season. The tropical rain forest grows in equatorial regions where humidity is high, the rainfall is heavy, seasonal changes are minimal, and annual mean temperature is about 28° C. All three are more or less stratified into layers of vegetation. Accompanying this vegetative stratification is a stratification of light, temperature, and moisture. The canopy receives full impact of climate and intercepts light and rainfall; the forest floor is shaded through the year in most coniferous and tropical rain forests and in late spring and summer in the deciduous forest.

The coniferous and deciduous forests hold different species of animal life, but animal adaptations are similar. The greatest concentration and diversity of life are on and just below the ground layer. Other animals live in various strata from low shrubs to the canopy. The tropical rain forest has pronounced feeding strata from above the canopy to the forest floor, and many of its animals are strictly arboreal. Whatever the forest, the different trees that compose it create different environments that ultimately dictate the kinds of plants and animals that can live within it.

Although gross primary productivity of forest ecosystems is high, so much of the GPP is allocated to the maintenance of forest structure that effective production is low: deciduous forest, 33 percent; coniferous forest, 45 percent; and tropical rain forest, 12 percent. Mineral cycling is tight. Nutrients accumulate in woody biomass to form a pool unavailable for short-term cycling. Although the bulk of nutrients is in mineral soil, the most important pools in mineral cycling are root mortality, litterfall, and foliar leaching. Internal cycling of some nutrients, particularly nitrogen, is important in nutrient conservation. In most forest systems only a small fraction of nutrients is lost from the system through streamflow.

Coniferous forests exhibit short-term cycling between litterfall and uptake by trees. At the same time conifers appear to be accumulators, removing elements from the soil in quantities large enough to upset nutritional balances in ecosystems. Nutrient cycling in conifers appears to depend on mycorrhizae, fungi, and root activity. Fungi act as nutrient sinks, but a symbiotic relationship exists between mycorrhizae and roots in the uptake of nutrients.

Tropical rain forests possess a large standing crop biomass that ties up great quantities of nutrients. Much of the mineral cycling takes place between a rapid decomposition of litterfall and rapid uptake of nutrients it contains. Roots are concentrated in the top of the ground in close contact with the litter, where a symbiotic relation between roots and mycorrhizae facilitates nutrient uptake.

The tropical rain forest and coniferous forest contrast with the deciduous forest, in which microbial decomposition releases nutrients to mineral soil, in turn providing the pathway through the roots for nutrient uptake.

The alpine tundra of the high mountain ranges in lower latitudes and the arctic tundra that extends beyond the tree line of the far north are at once similar and dissimilar. Both have low temperatures, low precipitation, and a short growing season. Both possess a frost-molded landscape and plant species whose growth rates are slow. The arctic tundra has a permafrost layer; rarely does the alpine tundra. Arctic plants require longer periods of daylight than alpine plants and reproduce vegetatively, whereas alpine plants propagate themselves by seed. Over much of the Arctic, the dominant vegetation is cotton grass, sedge, and dwarf heaths. In the alpine tundra cushion and mat-forming plants, able to withstand buffeting by the wind, dominate exposed sites, while cotton grass and other tundra plants are confined to protected sites. Net primary production is low because of a short growing season, although daily primary production rates are comparable to those of temperate grasslands. Biomass accumulates below rather than above ground. In spite of an assemblage of grazing ungulates and rodents, most production goes to decomposers. Major detrital consumers are bacteria and fungi. Nutrient levels are low, and nutrients tend to accumulate and become stored in living and dead plant material unavailable for recycling. Circulation is restricted by limited activity of decomposers, while the flux rates are controlled by nitrogen-fixing organisms. Because pools of both nitrogen and phosphorus are small, constant turnover between exchangeable and soluble pools is necessary to replenish the quantity absorbed and retained by plants.

Freshwater Ecosystems

Global aquatic ecosystems fall into two broad classes defined by salinity—the freshwater ecosystem and the saltwater ecosystem. The latter may include inland brackish water, as well as marine and estuarine habitats. Freshwater ecosystems, the study of which is known as limnology, are conveniently divided into two groups, *lentic* or standing water habitats and *lotic* or running water habitats. Both can be considered on an environmental gradient. The lotic follows a gradient from springs to mountain brooks to streams and rivers. The lentic follows a gradient from lakes to ponds to bogs, swamps, and marshes.

Lentic Ecosystems

Lakes are inland depressions containing standing water. They vary in size from small ponds of less than a hectare to large seas covering thousands of square kilometers. They range in depth from one meter to over 2000 meters. *Ponds* are defined as small bodies of standing water so shallow that rooted plants can grow over most of the bottom. Most ponds and lakes have outlet streams; and both may be more or less temporary features on the landscape.

Lakes and ponds arise in many ways. Some North American lakes were formed by glacial erosion and deposition. Glacial abrasion of slopes in high mountain valleys carved basins, which filled with water from rain and melting snow to produce tarns. Retreating valley glaciers left behind crescent-shaped ridges of rock debris, which dammed up water behind them. Numerous shallow kettle lakes and potholes were formed on the glacial drift sheets that cover much of northeastern North America and northwestern Europe.

Lakes are also formed by the deposition of silt, driftwood, and other debris in the beds of slow-flowing streams. Loops of streams that meander over flat valleys and floodplains often become cut off, forming crescent-shaped oxbow lakes.

Shifts in Earth's crust, either by the uplifting of mountains or by the breaking and displacement of rock strata, causing part of the valley to sink, develop depressions that fill with water. Craters of extinct volcanos may fill with water; and landslides can block off streams and valleys to form new lakes and ponds. In a given area all natural lakes and ponds have the same geological origin and the same general characteristics; but because of varying depths at the time of origin, they may represent several stages of development.

Many lakes and ponds are formed by nongeological activity. Beavers dam up streams to make shallow but often extensive ponds. Humans intentionally create artificial lakes by damming rivers and streams for power, irrigation, and water storage or by constructing small ponds and marshes for water, fishing, and wildlife. Quarries and strip mines fill with water to form other ponds.

Structure

Unlike most terrestrial ecosystems, lentic ecosystems have well-defined boundaries—the shoreline, the sides of the basin, the surface of the water, and the bottom sediment. Within these boundaries environmental conditions vary. Light penetrates to a depth determined by turbidity produced by sediments and phytoplankton. Temperatures vary seasonally and with depth. Because only a small proportion of the water is in direct contact with the air and because decomposition takes place on the bottom, the oxygen content of lake water is low compared to that of running water (see Chapter 10). In some lakes oxygen decreases with depth, but there are many exceptions. These gradations of oxygen, light, and temperature profoundly influence life in the lake, its distribution and adaptations (Figure 32.1).

Lentic ecosystems can be subdivided into vertical and horizontal strata based on photosynthetic activity. The *littoral* or shallow-water zone is the one in which light penetrates to the bottom. This area is occupied by rooted plants such as water lilies, rushes, and sedges. Beyond it is the *limnetic* or open-water zone, which extends to the depth of light penetration. It is inhabited by plant and animal plankton and the *nekton*, free-swimming organisms such as fish, which are capable of moving about voluntarily. Beyond the depth of effective light is the *profundal* zone, which depends upon the rain of organic material from the limnetic as the energy source. Common to both the profundal and littoral zones is the *benthic* zone or bottom region, which is the zone of decomposition. Although these zones are named and often described separately, all are closely dependent upon one another in nutrient and energy flow.

Limnetic Zone. The open water is a world of minute, suspended organisms, the plankton. Dominant is the phytoplankton, a group of plant organisms containing the diatoms, desmids, and filamentous green algae. Because photosynthesis in open water is carried on only by these

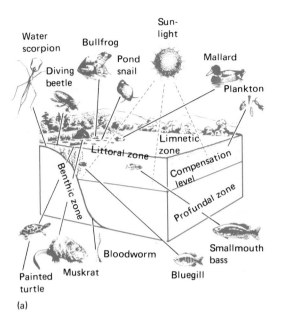

(a)

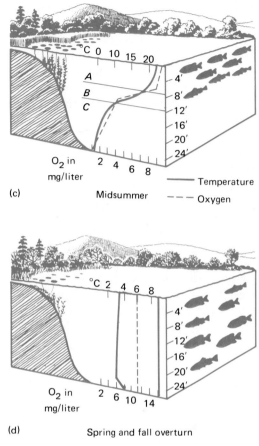

(c)   Midsummer
——— Temperature
- - - Oxygen

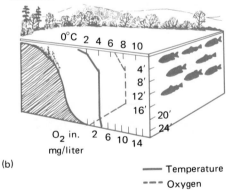

(b)
——— Temperature
- - - Oxygen

(d)   Spring and fall overturn

Figure 32.1 Seasonal variations in the stratification of oxygen and temperature and the distribution of aquatic life in lake ecosystems. This generalized picture of a lake in midsummer shows the major zones—littoral, limnetic, profundal, and benthic. The compensation level is the depth at which light is too low for photosynthesis. Surrounding the lake is a variety of organisms typical of a lake community. The distribution of oxygen and temperature in a lake during the different seasons affects the distribution of fish life. The narrow fish silhouettes represent trout, or cold water species. The wider silhouettes are bass, or warm water species. Note the pronounced horizontal stratification in midsummer and the nearly vertical oxygen and temperature curves during the spring and fall overturns. (See also Chapter 10.)

tiny plants, they are the base upon which the rest of limnetic life depends. Suspended with phytoplankton is the zooplankton, which grazes upon the minute plants. These animals form an important link in the energy flow in the limnetic zone.

Vertical distribution or stratification of plankton organisms is influenced by the physiochemical properties of water, especially temperature, oxygen, light, and current. Light, of course, sets the lower limit at which phytoplankton can exist. Because zooplankton feeds on these minute plants, it, too, is concentrated in the trophogenic zone. Phytoplankton, by its own growth, limits light penetration and thus reduces the depth at which it can live. As the zone becomes more shallow, the phytoplankton can absorb more light, and organic production is increased. Within these limits the depths at which the various species live is influenced by the optimum conditions for their development. Some phytoplankton live just beneath the water's surface; others are more abundant a meter beneath, while those requiring colder temperatures live deeper still. Cold-water species, in fact, are restricted to those lakes in which phytoplankton growth is scarce in the upper region and in which the oxygen

content of the deep water is not depleted by the decomposition of organic matter. Many of these cold-water species never move up through the metalimnion.

Because many species of zooplankton are capable of independent movement, animal plankton exhibits stratification that often changes seasonally. In winter some plankton forms are spread evenly to considerable depths; in summer they concentrate in the layers most favorable to them and to their stage of development. At this season animal plankton undertakes a vertical migration during some part of the 24-hour period. Depending upon the species and stage of development, zooplankton spends the night or day in the deep water or on the bottom and moves up to the surface during the alternate period to feed on phytoplankton (Figure 32.2).

In the limnetic zone fish make up the

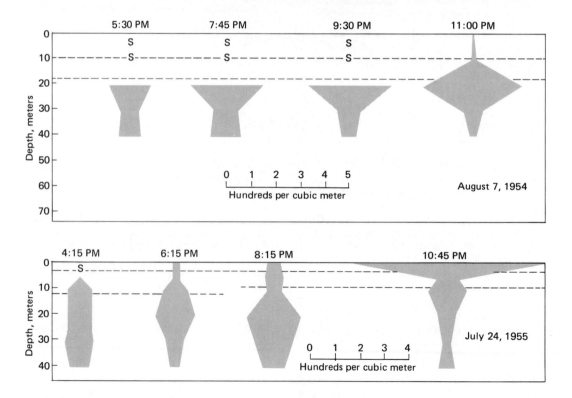

bulk of the nekton. Their distribution is influenced mostly by food supply, oxygen, and temperature. During the summer, largemouth bass, pike, and muskellunge inhabit the warmer epilimnion waters, where food is abundant. In winter they retreat to deeper water. Lake trout, on the other hand, move to greater depths as summer advances. During the spring and fall overturn, when oxygen and temperature are fairly uniform throughout, both warm-water and cold-water forms occupy all levels.

Profundal Zone. Diversity and abundance of life in the profundal zone are influenced by oxygen, temperature, and the amount of organic matter and nutrients supplied from the limnetic zone above. In highly productive waters decomposer organisms so deplete the profundal waters of oxygen that little aerobic life can exist there. The profundal zone of a deep lake is relatively much larger, so the productivity of the epilimnion is low in comparison to the volume of water and decomposition does not deplete the oxygen. Here the profundal zone supports some life, particularly fish, some plankton, and such organisms as certain cladocerans, which live in the bottom ooze. Other zooplankton may occupy this zone during some part of the day, but migrate upward to the surface to feed. Only during the spring and autumn overturns, when organisms from the upper layers enter this zone, is life abundant in the profundal waters.

Easily decomposed substances floating

Figure 32.3 Littoral vegetation. The heavy emergent growth of the littoral zone adds large amounts of detrital material important to the functioning of lake ecosystems. (Courtesy Soil Conservation Service.)

down through the profundal zone are partly mineralized while sinking (Kleerekoper 1953). The remaining organic debris, dead bodies of plants and animals of the open water and decomposing plant matter from shallow-water areas, settles on the bottom. Together with quantities of material washed in by inflowing water they make up the bottom sediments, the habitat of benthic organisms.

Littoral Zone. Aquatic life is richest and most abundant in the shallow water about the edges of lakes (Figure 32.3). Plants and animals found here vary with water depth, and a distinct zonation of life exists from the deeper water to shore (Figure 32.4). A blanket of duckweed (*Lemna spp.*), supporting a world of its own, may cover

Figure 32.2 Vertical distribution of the planktonic copepod *Limnocalanus marcurus* on two midsummer days. The maximum number reach the surface 1.5 to 4 hours after sunset. Note that this organism inhabits deeper water. The dashed lines represent the metalimnion. (After Wells 1961.)

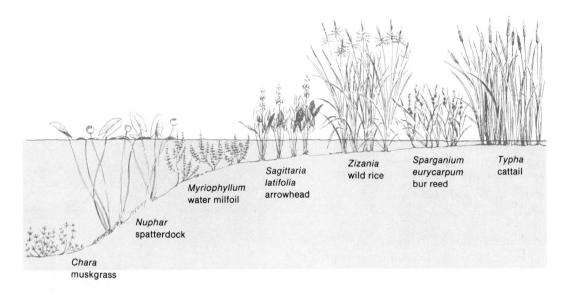

Zizania
wild rice

*Sparganium
eurycarpum*
bur reed

Typha
cattail

*Sagittaria
latifolia*
arrowhead

Myriophyllum
water milfoil

Nuphar
spatterdock

Chara
muskgrass

Figure 32.4 Zonation of emergent, floating, and submerged vegetation at the edge of a lake or pond. Such zonation does not necessarily reflect successional stages, as is often inferred, but response of vegetation types to water depth.

the surface of the littoral waters. Submerged plants, such as pondweeds (*Potamogeton spp.*), grow at water depths beyond that tolerated by emergent vegetation. As sedimentation and accumulation of organic matter increase toward the shore and the depth of water decreases, floating rooted aquatics, such as pond lilies and smartweed (*Polygonum spp.*), appear. Floating plants offer food and support for numerous herbivorous animals that feed both on phytoplankton and the floating plants. In the shallow water beyond the zone of floating plants grow the emergents, plants whose roots and stems are immersed in water and whose upper stems and leaves stand above the water. Among these emergents are plants with narrow, tubular or linear leaves, such as the bulrushes, reeds, and cattails. The distribution and variety of plants vary with the water depth and the fluctuation of the water level. Within the sheltering beds of emergent plants animal life is abundant. Although largely ignored in limnological studies, the littoral zone contributes heavily to the productivity of lentic ecosystems and provides a large input of organic matter to the system (see pages 839–849).

Benthic Zone. The bottom ooze is a region of great biological activity, so great that oxygen curves for lakes and ponds show a sharp drop in the profundal water just above the bottom (see Figure 10.9). Because organic muck lacks oxygen completely, the dominant organisms there are anaerobic bacteria. Under anaerobic conditions, decomposition cannot proceed to inorganic end-products. When the amounts of organic matter reaching the bottom are greater than can be utilized by the bottom fauna, odoriferous muck rich in hydrogen sulfide and methane results. Thus lakes and ponds with highly productive limnetic and littoral zones have an impoverished fauna on the profundal bottom. Life in the bottom ooze is most abundant in lakes with a deep hypolimnion in which oxygen is still available.

As the water becomes more shallow, the benthos changes. The bottom materials—stones, rubble, gravel, marl, clay—are modified by the action of water, by plant growth, by drift materials, and by recent organic deposits. Increased oxygen, light, and food result in a richness of species and an abundance not found on the profundal bottom.

Closely associated with the benthic community are the *periphyton* or *aufwuchs*, those organisms that are attached to or move upon a submerged substrate but do not penetrate it. Small aufwuchs communities are found on the leaves of submerged aquatics and on sticks, rocks, and other debris. The organisms found there depend upon the movement of the water, temperature, kind of substrate, and depth.

Periphyton found on living plants are fast-growing and lightly attached. They consist primarily of algae and diatoms. Because the substrate is so short-lived, they rarely exist for more than one summer. Aufwuchs on stones, wood, and debris form a more crustlike growth of blue-green algae, diatoms, water mosses, and sponges.

Function

In many ways a lake might be considered a self-contained ecosystem, but in fact, lentic ecosystems are strongly influenced by inputs of nutrients from sources outside of the basin (Figure 32.5). Nutrients and other substances move across the boundaries of the lentic system along biological, geological, and meteorological pathways (Likens and Bormann 1975).

Wind-borne particulate matter, dissolved substances in rain and snow, and atmospheric gases represent meteorological inputs to the system. Meteorological outputs are small, mainly spray aerosols and gases such as carbon dioxide and methane. Geological inputs include nutrients dissolved in groundwater and inflowing streams and particulate matter washed into the basin from the surrounding terrestrial watershed. Geological outputs include dissolved and particulate matter carried out of the basin by outflowing waters and nutrients incorporated in deep sediments which may be removed from circulation for a long period of time. Biological inputs and outputs, relatively small, include animals such as fish that move into and out of the basin. Energy input is largely sunlight, and energy output is heat. The lentic ecosystem receives its hydrological input from precipitation and the drainage of surface waters. Outputs involve seepage through walls of the lake basin, subsurface flows, evaporation, and evapotranspiration. Within the lentic ecosystem nutrients move among three compartments, dissolved organic matter, particulate organic matter, and primary and secondary minerals. Nutrients and energy move through the system by way of the grazing and detrital food chains.

Although studies of lake metabolism have emphasized the phytoplankton-zooplankton grazing food chain, in reality lakes, like terrestrial communities, are dominated by the detrital food chain (Figure 32.6). The reason for an emphasis on the grazing food chain in lakes has been a preoccupation with the open-water zone and a disregard of the littoral zone, which adds significantly to lake productivity and supplies substantial quantities of detritus to the system. Detritus is all dead organic carbon. It includes particulate and dissolved organic carbon (POC and DOC) from external sources that enter and cycle within the system, and organic matter lost to a particular trophic level by such nonpredatory losses as egestion, excretion, and secretion (Rich and Wetzel 1978; Wetzel 1975).

Lake ecosystems function mostly within a framework of organic carbon

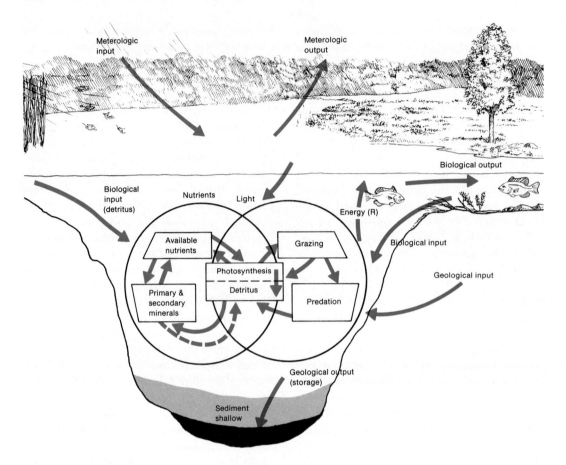

Figure 32.5 Model for nutrient cycling and energy flow in a lake ecosystem. Meteorological, geological, and biological inputs enter the lentic system from the watershed that contains it. The nutrient and energy inputs as well as the nutrients and energy generated within the system move through a number of pathways. Part of the nutrients and energy fixed accumulate in bottom sediments. (After Likens and Bormann 1974.)

transfer. The central pool, which comes from both internal *(autochthonous)* and external *(allochthonous)* sources, represents the major flow through the system. Particulate organic carbon comes from three sources: (1) imports into the system from the outside; (2) the littoral zone; (3) the lentic zone. Most of the detrital metabo-

lism takes place in the benthic zone, where particulate matter is decomposed, and in the open-water zone during sedimentation.

Primary production is carried out in the limnetic zone by phytoplankton and in the littoral zone by macrophytes. The ratio of these contributions varies among lentic systems. Phytoplankton production is influenced by nutrient availability in the water column. If nutrients are not limiting and the only losses are respiratory, the rate of net photosynthesis and biomass accumulation is high. In fact, a linear relationship exists between phytoplankton production and phytoplankton biomass (Brylinsky 1980). However, as phyto-

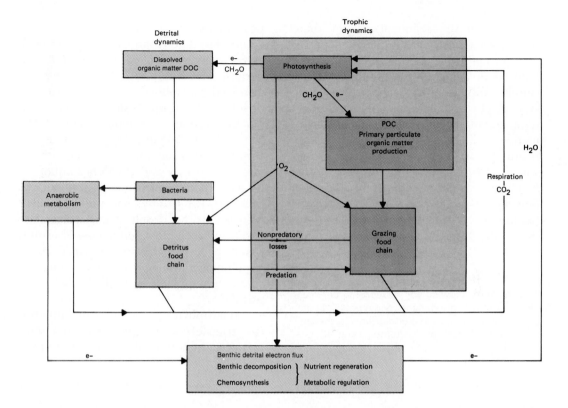

Figure 32.6 Energy flow and nutrient cycling in a lake ecosystem. Heterotrophic, autotrophic, and detrital pathways are necessary. This model shows the integration of detrital and trophic structure involving herbivores, bacteria, and anaerobic metabolism. Herbivores and bacteria represent the grazing and detrital food webs, which depend upon each other for prey and nonpredatory losses—the movement of organic matter to decomposers. Anaerobic metabolism is a pathway for energy leaving organic substrates and the biota to interact more directly with abiotic factors. Export and uncoupled oxidation represent energetic losses, whereas import and net photosynthesis represent energy gains. CO_2 = oxidized carbon; H_2O = reduced oxygen: e^- = electron. (From Rich and Wetzel 1978: 68.)

plankton biomass increases, shading increases, respiration per unit surface increases, and net photosynthesis and thus production declines. When nutrients are low, respiration and mortality increase, re- ducing net photosynthesis and thus biomass. However, if zooplankton grazing and bacterial decomposition are high, nutrients are recycled rapidly, resulting in a high rate of net photosynthesis even though the concentration of nutrients and biomass accumulation are low.

Macrophytes also contribute heavily to lake production. Their maximum biomass is close to annual cumulative net production. The ratio of macrophytic production to microphytic production is influenced by the fertility of the lake. Highly fertile lakes support a heavy growth of phytoplankton that shades out macrophytes and reduces their contribution. In less fertile lakes where phytoplankton production is low, light penetrates the water and rooted aquatics grow. Macrophytes are little affected by nutrient exchange in the water column. Rooted aquatics draw on nutrients from the sediments rather than from open water.

Nutrient transfers within lentic ecosystems take place largely between the water column and sediments, and involve uptake by phytoplankton, zooplankton, bacteria, and other consumers as well as sedimentation in both the water column and benthic muds (Figure 32.7). In spring when phytoplankton bloom is at its height, nitrogen and phosphorus become depleted in the trophogenic zone, in part because of the high rate of photosynthesis, the high rate of sinking of dead

phytoplankton, and the high rate of sedimentation. The trophogenic zone experiences a decrease in particulate N and P because of decomposition, an increase in dissolved phosphorus, and a decrease in dissolved N because of denitrification.

In summer conditions change. Because of a decline in phytoplankton in the trophogenic zone and a slower sinking rate, as much N and P enters solution as is taken up by phytoplankton in photosynthesis. In the trophogenic zone, N and P increase in the dissolved and particulate pools and in bottom sediments. Phosphorus, in particular, becomes trapped in the hypolimnion and remains there, unavailable to phytoplankton until the fall overturn.

Macrophytes, however, can influence this transfer of phosphorous from sediments to the water column and on to phy-

Figure 32.7 A model of nitrogen cycling in a lentic ecosystem, showing the relationship between water column and sediments. Sediments are both a storehouse and a source of nutrients in the lentic system. (Based on Golterman and Kouwe 1980: 138.)

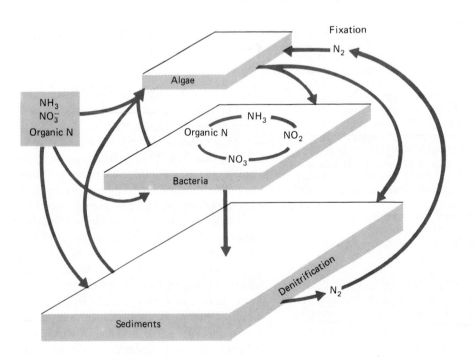

toplankton, and at the same time increase the accretion of bottom sediments. Rooted macrophytes draw phosphorus from the sediment. Carpenter (1981) in a study in Lake Wingra, Wisconsin, found that macrophytes doubled the amount of sedimentary phosphorus made available to phytoplankton that it otherwise would obtain from direct release from the sediment. Macrophytes obtained 73 percent of their shoot phosphorus from sediments, eventually making available to phytoplankton some 550 kg of P compared to 470 kg available from sediments through overturns and disturbance and resuspension of sediments. This uptake stimulates both the production of macrophytic biomass and phytoplankton biomass, adding to sediment accumulation on the bottom. Sediment accumulation provides new areas for colonization and increases the availability of additional phosphorus for phytoplankton. Macrophytes enhance the recycling of phosphorus by mobilizing it from the sediments. Such mobilization and recycling accelerate the enrichment or eutrophication of lakes.

Mirror Lake, studied by G. Likens and his associates (1985), provides an example of the functioning of a lentic ecosystem, limited here to a review of the carbon cycle. Mirror Lake is a kettle hole lake that appeared in the postglacial tundra landscape 14,000 BP. It has an average depth of 5.75 m, a maximum depth of 11 m, and an area of 15 ha, and it is fed by numerous streams. Both the lake and its surrounding watershed have undergone considerable change. The lake shore is rocky, sparsely vegetated with submerged macrophytes and benthic algae. The waters are clear, well illuminated, warm, and well oxygenated. The profundal zone is anoxic in summer.

Carbon fluxes occur both across the system, with inputs from the surrounding watershed, and within the system. Seven percent of the organic carbon in the water of the lake and 0.6 percent in the top 10 cm of the lake occur in living organisms. The rest of the carbon is detrital, both particulate and dissolved. Dissolved organic carbon (DOC) is 11 times greater than POC. Inputs of carbon amount to 60.5 g/C/m^2/yr (Table 32.1). Of this input 79 percent is internal or autochthonous. Phytoplankton is the major producer, accounting for 88 percent of the autochthonous input. Inflowing water (fluvial) accounts for 14 percent of the allochthonous input, litterfall 5 percent, and precipitation 2 percent (Figure 32.8). Outputs are largely respiration (Table 32.2), which accounts for 57.0 g/C/m^2/yr or 70 percent including plant respiration and 35.6 g/C excluding plants. Twenty-one percent (12.8 g/C/m^2/yr) is permanently buried in the sediments; 14 percent of in-

TABLE 32.1 Inputs of Organic Carbon (mg C/m^2/yr) for Mirror Lake

Autochthnous (gross)	
Phytoplankton·(POC and DOC)	56,500
Epilithic algae	2,500
Epipelic and epiphytic algae	>1,000
Macrophytes	2,500
Dark CO$_2$ fixation	2,100
	Sum 64,600
Allochthonous	
Precipitation	1,400
Shoreline litter	4,300
Fluvial: DOC	10,500
FPOC (0.45 m-1mm)	300
FPOC (>1mm)	50
CPOC	800
	Sum 17,350
Total Inputs	81,950

Source: Adapted from Jordan, Likens, and Peterson 1985:294.

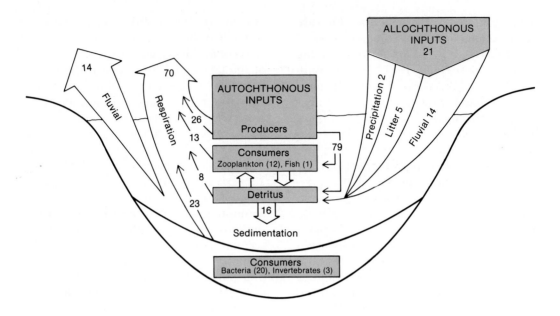

TABLE 32.2 Outputs of Organic Carbon (mg C/m²/yr) for Mirror Lake

Respiration		
Phytoplankton	19,100	
Zooplankton	10,000	
Epilithic algae	1,000	
Epipelic and epiphytic algae	400	
Macrophytes	1,000	
Benthic invertebrates	2,800	
Fish	200	
Salamanders	0	
Bacteria on rocks and sand	3,250	
Bacteria in gyttja	12,700	
Bacteria in plankton	6,550	
	Sum 57,000	
Exports		
Permanent sedimentation	12,800	
Ground water (DOC)	4,800	
Fluvial: (DOC)	5,990	
(FPOC)	780	
Insect emergence	500	
	Sum 24,950	
Total Outputs	81,950	

Source: Adapted from Jordan, Likens, and Peterson 1985:299.

Figure 32.8 A model of a food web for Mirror Lake, Hubbard Brook Forest, New Hampshire, indicating the carbon flows. Numbers represent percent of annual organic carbon flux. Note the importance of internal or autochthonous inputs, the photosynthetic activity of phytoplankton and macrophytes. Most of the outside or allochthonous inputs are fluvial, largely subsurface water flowing into the lake system. Compare Figure 32.14. (From Likens 1985: 339.)

puts leaves as fluvial outflow and 1 percent leaves as insect emergence.

Algal growth and decomposition are the major sources of DOC for the sediment and water column. This internal input of DOC (20 g/C/m²/yr) is greater than allochthonous input of DOC (12 g). Most of the DOC of algal origin is metabolized rapidly and never accumulates in the water. The amount of DOC in the water column, in fact, is relatively stable. Superficially, the grazing food chain and the microbial food web (DOC and dissolution of detrital POC) are separate and independent. Zooplankton speeds the release of DOC from algal cells by mechanical disruption and makes the nutrients available to microbes. Bacteria grow at the expense of algal DOC.

Although the productivity of a lake is related to the nutrient richness of its waters, other internal forces influence actual productivity. Any two lakes with a similar nutrient load may differ in productivity, a difference that relates to the allocation of chlorophyll *a* among the major primary producers, the phytoplankters, which vary in their rates of metabolic activity and of nutrient recycling, both size-dependent (Carpenter and Kitchell 1984). Small phytoplankters, for example, have higher maximum growth rates overall, higher maximum growth at low levels of nutrients, and a lower sinking rate than large phytoplankton species.

Feeding on phytoplankters are the zooplankton, essential to the recycling of nutrients, particularly N and P (see Chapter 12). Zooplankton of various sizes graze on different sizes of phytoplankton. Depending on the size relationship of the dominant zooplankters, they influence the species composition and size structure of the phytoplankton community. Zooplankters, in turn, are consumed by invertebrate planktivores (insect larvae and crustaceans) and vertebrate planktivores (minnows and small spiny fish). These predators, too, are size-dependent in their food selection (see Chapter 19). Invertebrate planktivores are also subject to predation by vertebrate planktivores, which in turn become prey for fish-eating predators (piscivores).

Interactions among these feeding groups flows down through the food web, influencing productivity at each trophic level. A rise in the biomass of predatory fish (bass, pike, trout) can result in changes in the density, species composition, and behavior of zooplanktivorous fish and that in turn on invertebrate planktivores. Vertebrate planktivores, taking the largest available prey, reduce the density of large zooplankters and invertebrate planktivores select smaller species. Any changes in the relative densities of these two groups of planktivores influence the structure and density of zooplankton, affecting grazing intensity and nutrient recycling rates. A decline in herbivore biomass results in an increase in phytoplankton biomass. Each change in biomass at one trophic level has the opposite response at the next trophic level. Throughout the food web, however, maximum production is achieved at intermediate levels of density. Alteration of biomass at various feeding levels and trophic interactions can influence and regulate productivity of lake ecosystems (Carpenter, Kitchell, and Hodgson 1985).

Nutrient Status

A close relationship exists between land and water ecosystems. Primarily through the hydrological cycle one feeds upon the other. The water that falls on land runs from the surface or moves through the soil to enter streams, springs, and eventually lakes. The water carries with it silt and nutrients in solution, all of which enrich aquatic ecosystems. Human activities, including road construction, logging, mining, construction, and agriculture, add an additional heavy load of silt and nutrients, especially nitrogen, phosphorus, and organic matter. The outcome of these inputs is nutrient enrichment of aquatic systems. This enrichment is termed *eutrophication*.

The term *eutrophy* (from the Greek *eutrophus*, "well nourished") means a condition of being nutrient-rich. The opposite of eutrophy is *oligotrophy*, the condition of being nutrient-poor. The terms were introduced by the German limnologist C. A. Weber in 1907, when he applied them to the development of peat bogs. E. Naumann later associated the terms with phytoplankton production in lakes: Eutrophic lakes found in fertile lowland regions hold high populations of phytoplankton; oligotrophic lakes, common to regions of primary rocks, contain little plankton. This concept of oligotrophy and eutrophy ignores the input of highly productive littoral zones.

Eutrophic Systems. A typical eutrophic lake (Figure 32.9) has a high surface-to-volume ratio; that is, the surface area is large relative to depth. It has an abundance of nutrients, especially nitrogen and phosphorus, that stimulate a heavy growth of algae and other aquatic plants.

Increased photosynthetic production leads to increased regeneration of nutrients and organic compounds, stimulating further growth. Phytoplankton becomes concentrated in the upper layer of the water, giving it a murky green cast. The turbidity reduces light penetration and restricts bi-

Figure 32.9 Comparison of oligotrophic and eutrophic lakes. (a) The oligotrophic lake is deep and has relatively cool water in the epilimnion. The hypolimnion is well supplied with oxygen. Organic matter that drifts to the bottom falls through a relatively large volume of water. The watershed surrounding the lake is largely oligotrophic, dominated by coniferous forests on thin and acid soil. (b) The eutrophic lake is shallow and warm, and oxygen is nearly depleted in the deeper water. The amount of organic detritus is large in relation to the volume of water. The watershed surrounding the lake is eutrophic, consisting of a nutrient-rich deciduous forest and farmland.

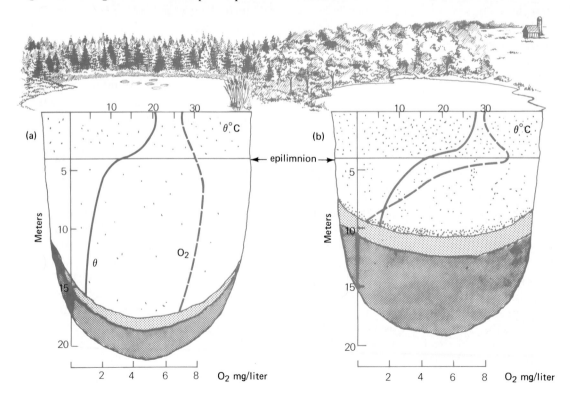

ological productivity to a narrow zone of surface water. Algae, inflowing organic debris and sediment, and the remains of rooted plants drift to the bottom, adding to the highly organic sediments. On the bottom bacteria partially convert dead matter into inorganic substances. The activities of these decomposers deplete the oxygen supply of the bottom sediments and deep water to a point where the deeper parts of the lake are unable to support aerobic forms of life. The number of species declines, although the numbers and biomass of organisms may remain high. As the basin continues to fill, the volume decreases and the resulting shallowness speeds the cycling of available nutrients and further increases plant production. Positive feedback carries the system to eventual extinction—the filling in of the basin and the development of a marsh, swamp, and ultimately a terrestrial community.

Oligotrophic Systems. Oligotrophic lakes are characterized by a low surface-to-volume ratio, water that is clear and appears blue to blue-green in the sunlight, bottom sediments that are largely inorganic, and a high oxygen concentration through the hypolimnion. The nutrient content of the water is low; although nitrogen may be abundant, phosphorus is highly limiting. Low nutrient availability results from a low input of nutrients from external sources. This factor in turn causes a low production of organic matter, particularly phytoplankton. Low organic production results in low rate of decomposition and high oxygen concentration in the hypolimnion. These oxidizing conditions are responsible for low nutrient release from the sediments. The lack of decomposable organic substances results in low bacterial populations and slow rates of microbial metabolism. Although the num-

ber of organisms in oligotrophic lakes may be low, the diversity of species is often high. Fish life is dominated by members of the salmon family.

When nutrients in moderate amounts are added to oligotrophic lakes they are rapidly taken up and circulated. According to Vallentyne (1974), tissues of aquatic algae contain phosphorus, nitrogen, and carbon in the ratio of 1 P : 7 N : 40 C per 500 g wet weight. If nitrogen and carbon are in excess and phosphorus is limiting, the addition of phosphorus would stimulate growth; if nitrogen is limiting, the addition of that element would do the same. In most oligotrophic lakes phosphorus rather than nitrogen is limiting. Because of its low ratio, 1 P per 500 g, an addition of even moderate amounts of phosphorus generates considerable growth of algae (for detailed discussions see Vallentyne 1974; Wetzel 1976). As increasing quantities of nutrients are added to the lake, it begins to change from oligotrophic to *mesotrophic* (having a moderate amount of nutrients) to eutrophic. This change has been happening at an increasing rate to clear oligotrophic lakes around the world.

In fact, this "galloping eutrophication" has been changing naturally eutrophic lakes into *hypertrophic* ones. An excessive nutrient content results from a heavy influx of wastes, raw sewage, drainage from agricultural lands, river basin development, runoff from urban areas, and burning of fossil fuels. This accelerated enrichment, which results in chemical and environmental changes in the system and causes major shifts in plant and animal life, has been called *cultural eutrophication* (Hasler 1969).

Dystrophic Systems. Lakes that receive large amounts of organic matter from surrounding watersheds, particularly in the form of humic materials that stain the water brown, are called *dystrophic*. Although the productivity of dystrophic lakes is considered low, this refers only to planktonic production. Dystrophic lakes generally have highly productive littoral zones, particularly those that develop bog flora. This littoral vegetation dominates the metabolism of the lake ecosystem, providing a source of both dissolved and particulate organic matter (see Wetzel and Allen 1970; Wetzel 1975).

Marl Systems. A fourth type of system, the *marl* lake, contains extremely hard water due to inputs of calcium over a long period of time. A hard-water lake is relatively unproductive and remains so because of the reduced availability of nutrients, even though carbonates remain high. Under certain conditions in these lakes, phosphorus, iron, magnesium, and other nutrients form insoluble compounds and are lost to the system. Sodium and potassium are low, but nitrogenous compounds are high. Because of low photosynthetic productivity, nitrogen remains unutilized. Calcium is often supersaturated, and carbonates, especially calcium carbonate, and humic dissolved organic matter precipitate to form marl deposits on the bottom. If carbonate inputs from drainage are reduced or depleted or if sediments build up high enough to support littoral vegetation and the growth of *Sphagnum*, the marl lake gradually develops into a bog (for details, see Wetzel 1972, 1975.)

■ Lotic Ecosystems

Continuously moving water is the outstanding feature of streams and rivers. Current cuts the channel, molds the character of the stream, and influences the life

and ways of organisms inhabiting flowing waters.

Streams may begin as outlets of ponds or lakes or they may arise from springs and seepage areas. Added in varying quantities is surface runoff, especially after heavy or prolonged rains and rapid snowmelt. Because precipitation, the source of all runoff and subsurface water, varies seasonally, the rate and volume of streamflow may fluctuate widely from flood conditions to dry channels.

As water drains away from its source, it flows in a direction dictated by the lay of the land and the underlying rock formations. Its course may be determined by the original slope and its regularities; or the water, seeking the least resistant route to lower land, may follow the joints and fissures in bedrock near the surface and shallow depressions in the ground. Whatever its direction, water is concentrated into rills that erode small furrows, which soon grow into gullies. Water, moving downstream, especially where the gradient is steep, carries with it a load of debris that cuts the channel wider and deeper and that sooner or later is deposited within or along the stream. At the same time erosion continues at the head of the gully, cutting backward into the slope and increasing its drainage area. Just below its source, the stream may be small, straight, and often swift, with waterfalls and rapids. Further downstream, where the gradient is less and the velocity decreases, meanders become common. They are formed when the current, deflected by some obstacle on the floor of the channel, by projecting rocks and debris, or by the entrance of swifter currents, strikes the opposite bank. As the water moves downstream, it is thrown back to the other side again. These abrasive forces create a curve in the stream, on the inside of which the velocity is slowed and the water drops its load. Such cutting and deposition often cause valley streams to change course and to cut off the meanders to form oxbow lakes. When the water reaches level land, its velocity is greatly reduced, and the load it carries is deposited as silt, sand, or mud.

At flood time the material carried by the stream is dropped on the level lands, over which the water spreads to form floodplain deposits. These floodplains, which humans have settled so extensively, are a part of the stream or river channel used at the time of high water, a fact that few people recognize. The current at flood time is swiftest in the normal channel of the stream and slowest on the floodplain. Along the margin of the channel and the floodplain, where the rapid water meets the slow, the current is checked and all but the fine sediments are dropped on the edges of the channel. Thus the deposits on the floodplain are higher on the immediate border and slope off gradually toward the valley side.

When a stream or river flows into a lake or sea, the velocity of the water is suddenly checked and the load of sediment is deposited in a fan-shaped area at the inlet point to form a delta. Here the course of the water is broken into a number of channels, which are blocked or opened with subsequent deposits. As a result the delta is characterized by small lakes and swampy or marshy islands. Material not deposited at the mouth is carried further out to open water, where it settles on the bottom. Eventually the sediments build up above the water to form a new land surface.

The character of a stream is molded by the velocity of the current. This velocity varies from stream to stream and within the stream itself, and it depends upon the size, shape, and steepness of the stream channel, the roughness of the bottom, the depth, and the rainfall.

The velocity of flow influences the degree of silt deposition and the nature of the bottom. The current in the riffles is too fast to allow siltation, but coarser silt particles drop out in the smooth or quiet sections of the stream. High water increases the velocity; it moves bottom stones, scours the streambed, and cuts new banks and channels. In very steep streambeds, the current may remove all but very large rocks and leave a boulder-strewn stream.

Flowing water also transports nutrients to and carries waste products away from many aquatic organisms and may even sweep them away. Balancing this depletion of bottom fauna, the current continuously reintroduces bottom fauna from areas upstream. Similarly, as nutrients are washed downstream, more are carried in from above. For this reason, the productivity of primary producers in streams is 6 to 30 times that of those in standing water (Nelson and Scott 1962). The transport and removal action of flowing water benefits such continuous processes as photosynthesis.

Structure

Fast Water. Fast or swiftly flowing streams are, roughly, all those whose velocity of flow is 50 cm/sec or higher (A. Nielsen 1950). At this velocity the current will remove all particles less than 5 mm in diameter and will leave behind a stony bottom. The fast stream is often a series of two essentially different but interrelated habitats, the turbulent riffle and the quiet pool (Figure 32.10). The waters of the pool are influenced by processes occurring in the rapids above, and the waters of the rapids are influenced by events in the pool.

Riffles are the sites of primary pro-

Figure 32.10 Two different but related habitats in a stream, the riffle (foreground) and the pool (background). (Photo by R. L. Smith.)

duction in the stream (see Nelson and Scott 1962). Here the aufwuchs assume dominance and occupy a position of the same importance as the phytoplankton of lakes and ponds. The aufwuchs consist chiefly of diatoms, blue-green and green algae, and water moss. Extensive stands of algae grow over rocks and rubble on the streambed and form a slippery covering. Growth during favorable periods may be so rapid that the stream bottom is covered in ten days or less (Blum 1960). Many small algal species are epiphytes and grow on the tops of or in among other algae.

The outstanding feature of much of this algal growth is its ephemeral nature. Scouring action of water and the debris it carries tears away larger growth, epiphytes and all, and sends the algae downstream. As a result there is a constant contribution from upstream to the downstream sequence.

Above and below the riffles are the

pools. Here the environment differs in chemistry, intensity of current, and depth. Just as the riffles are the sites of organic production, so the pools are sites of decomposition. They are the catch basins of organic materials, for here the velocity of the current is reduced enough to allow a part of the load to settle out. Pools are the major sites for free carbon dioxide production during the summer and fall.

Overall production in a stream is influenced in part by the nature of the bottom. Pools with sandy bottoms are the least productive because they offer little substrate for either aufwuchs or animals. Bedrock, although a solid substrate, is so exposed to currents that only the most tenacious organisms can maintain themselves (Figure 32.11). Gravel and rubble bottoms support the most abundant life because they have the greatest surface area for the aufwuchs, provide many crannies and protected places for insect larvae, and are the most stable (Figure 32.12). Food production decreases as the

Figure 32.11 A fast mountain stream in a deep woods. The bottom is largely bedrock. (Photo by R. L. Smith.)

particles become larger or smaller than rubble. Insect larvae, on the other hand, differ in abundance on the several substrates. Mayfly nymphs are most abundant on rubble, caddisfly larvae on bedrock, and Diptera larvae on bedrock and gravel (Pennak and Van Gerpen 1947).

The width of the stream also influences overall production. Bottom production in streams 6 m wide decreases by one-half from the sides to the center; in streams 30 m wide it decreases by one-third (Pate 1933). Streams 2 m or less in width are four times as rich in bottom organisms as those 6 to 7 m wide. This is one reason why headwater streams make such excellent trout nurseries.

Slow Water. As the current slows, a noticeable change takes place in streams (Figure 32.13). Silt and decaying organic matter accumulate on the bottom, and fine detritus from upstream is the main

Figure 32.12 Comparison of life in a fast stream (a) and a slow stream (b). (1) Blackfly larvae (Simuliidae); (2) net-spinning caddisfly (*Hydropsyche* spp.); (3) stone case of caddisfly; (4) water moss *(Fontinalis)*; (5) algae *(Ulothrix)*; (6) mayfly nymph *(Isonychia)*; (7) stonefly nymph *(Perla* spp.); (8) water penny *(Psephenus)*; (9) hellgrammite (dobsonfly larva, *Corydalis cornuta)*; (10) diatoms (Diatoma); (11) diatoms *(Gomphonema)*; (12) cranefly larva (Tipulidae); (13) dragonfly nymph (Odonata, Anisoptera); (14) water strider *(Gerris)*; (15) damselfly larva (Odonata, Zygoptera); (16) water boatman (Corixidae); (17) fingernail clam *(Sphaerium)*; (18) burrowing mayfly nymph *(Hexegenia)*; (19) bloodworm (Oligochaeta, *Tubifex* sp.); (20) crayfish *(Cambarus spp.)*. The fish in the slow stream are left to right: northern pike, bullhead, and smallmouth bass.

(a)

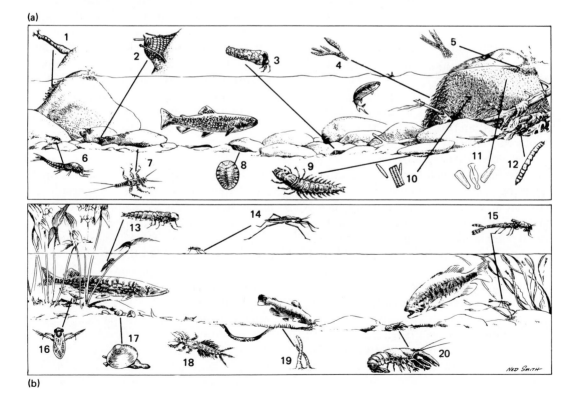

(b)

Figure 32.13 A slow stream reflects the sky of a summer afternoon. (Photo by R. L. Smith.)

source of energy. Faunal organisms are able to move about to obtain their food, and a plankton population develops. The composition and configuration of the stream community approaches that of standing water.

With increasing temperatures, decreasing current, and accumulating bottom silt, organisms of the fast water are replaced gradually by organisms adapted to these conditions. Brook trout and sculpin give way to smallmouth bass and rock bass, the dace to shiners and darters. With current at a minimum many resident fish lack the strong lateral muscles typical of the trout and have compressed bodies that permit them to move with ease through masses of aquatic plants. Mollusks, particularly *Sphaerium* and *Pisidium*, and pulmonate snails, crustaceans, and burrowing mayflies replace the rubble-dwelling insect larvae. Only in occasional stretches of fast water in the center of the stream are remnants of headwater-stream organisms found.

As the volume of water increases, as

the current becomes even slower, and as the silt deposits become heavier, detritus-feeders increase. Rooted aquatics appear. Emergent vegetation grows along the riverbanks, and duckweeds float on the surface. Indeed, the whole aspect approaches that of lakes and ponds, even to zonation along the river margin.

The higher water temperature, weak current, and abundant decaying matter promote the growth of protozoan and other plankton populations. Scarce in fast water, plankton increases in numbers and species in slow water. Rivers have no typical plankton of their own. Those found there originate mainly from backwaters and lakes. In general, plankton populations in rivers are not nearly as dense as those in lakes. Time is too short for much multiplication of plankton because relatively little time is needed for a given quantity of water to flow from its source to the sea. Also occasional river rapids, often some distance in length, kill many plankton organisms by violent impact against suspended particles and the bottom. Aquatic vegetation filters out this minute life as the current sweeps it along.

Function

The lotic or flowing water system is open and largely heterotrophic, especially in headwater streams. A major source of energy and nutrients is detrital material carried to the streams from the outside (Figure 32.14). Much of this organic matter comes in the form of leaves and woody debris dropped from streamside vegetation, collectively called coarse particulate organic matter (CPOM) (particles larger than 1 mm). Another type of organic input is fine particulate organic matter (FPOM), material less than 1 mm, including leaf fragments, invertebrate feces, and precipitated dissolved organic matter. A

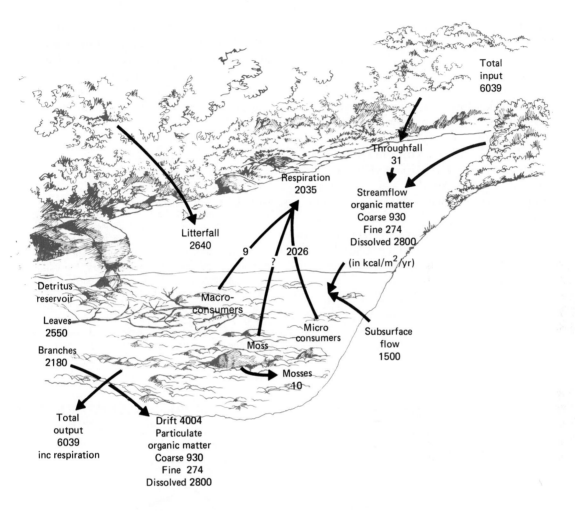

Figure 32.14 Energy flow in a stream ecosystem. Streams are open ecosystems with much of the energy input coming from outside sources. Note the great dependence on materials from terrestrial sources and inflow from upstream and the role of coarse particulate organic matter (CPOM), fine particulate organic matter (FPOM), and dissolved organic matter (DOM). Primary production contributes little to energy flow. Energy values are based on Bear Brook, Hubbard Forest, New Hampshire. (Data from Fisher and Likens 1973.)

third input is dissolved organic matter (DOM), material less than 0.5 micron in solution. One source of DOM is rainwater dropping through overhanging leaves, dissolving the nutrient-rich exudates on them. Other DOM input comes from subsurface seepage, which brings nutrients leached from adjoining forest, agricultural, and residential lands. Many streams receive inputs from dumping of industrial and residential effluents. Supplementing this detrital input is autotrophic production in streams (autochthonous) by diatomaceous algae growing on rocks and by rooted aquatics such as water moss (*Fontinalis*).

The processing of this organic matter involves both physical and biological mechanisms. In fall, leaves drift down from overhanging trees, settle on the water, float downstream, and lodge against banks, debris, and stones. Soaked with water, the leaves sink to the bottom, where they quickly lose 5 to 50 percent of their dry matter as water leaches soluble organic matter from their tissues. Much of this DOM is either incorporated onto detrital particles or precipitated to become part of FPOM. Another part is incorporated into microbial biomass. Once softened, leaves and other debris are processed by a number of species of invertebrates, which can be placed into several functional groups (Figure 32.15).

Functional Groups. Within a week or two, depending upon the temperature, the surface of the leaves is colonized by bacteria and fungi, largely aquatic hypho-

mycetes. Fungi are more important on CPOM because large particles offer more surface for mycelial development (Cummins and Klug 1970). Bacteria are associated more with FPOM. Microorganisms degrade cellulose and metabolize lignin.

Figure 32.15 Model of a stream ecosystem emphasizing structure and function. It illustrates the processing of leaves and other particulate matter and dissolved organic matter. Leaves, branches, seeds, bark, and flowers represent the coarse particulate organic matter, which is colonized by bacteria and aquatic fungi. They constitute most of the nutrient and energy input into the stream. Algae and moss, the microproducers and macroproducers, are responsible for primary production in the stream. At any one point there is additional input of dissolved organic matter from upstream. Processing these inputs are the functional groups of organisms, mostly invertebrates: the shredders, collectors, grazers, and predators. (Adapted from Cummins 1974: 663.)

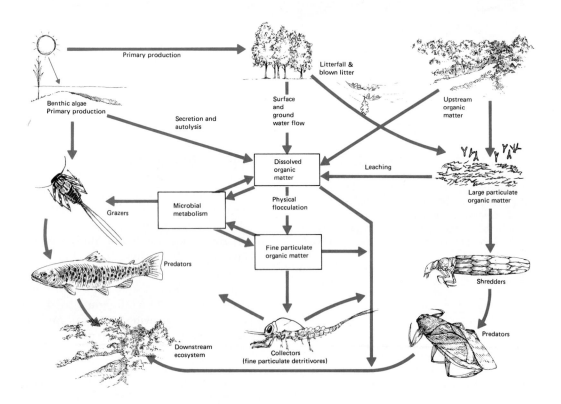

Their populations form a layer on the surface of leaves and detrital particles that is much richer nutritionally than the detrital particles themselves (Anderson and Cummins 1979). Leaves and other detrital particles are attacked by a major feeding group, the *shredders*, invertebrates that feed on leaves and other large organic particles. Among the shredders are the larvae of craneflies (Tipulidae) and caddisflies (Trichoptera), the nymphs of stoneflies (Plecoptera), and crayfish. They break down the CPOM, feeding on the material not so much for the energy it contains but for the bacteria and fungi growing on it (Cummins 1974; Cummins and Klug 1979). In doing so they skeletonize the leaf by feeding on the softer portions. Shredders assimilate about 40 percent of the material they ingest and pass off 60 percent as feces.

Broken up by shredders and partially decomposed by microbes, the leaf material along with fecal material becomes part of the FPOM, which also includes some precipitated DOM. Drifting downstream and settling on the stream bottom, FPOM is picked up by another feeding group of stream invertebrates, the *filtering* and *gathering collectors*. The filtering collectors include, among others, the larvae of blackflies (Simulidae) with filtering fans, and net-spinning caddisflies, including *Hydropsyche*. Gathering collectors, such as the larvae of midges, pick up particles from stream bottom sediments. Collectors obtain much of their nutrition from bacteria associated with fine detrital particles.

Whereas shredders and collectors feed on detrital material, another group feeds on algal coating of stones and rubble. They are the *scrapers*, which include the beetle larvae, popularly known as the water penny (*Psephenus spp.*), and mobile caddisfly larvae. Much of the material they scrape loose enters the drift as FPOM. Be-

haviorally and morphologically, scrapers are adapted to maintain their position in the current either by flattening to avoid the main force of flow or by weighting down with heavy mineral cases. Feeding on mosses and filamentous algae are the *piercers*, a feeding group made up largely of microcaddisflies. Another group, associated with woody debris, are the *gougers*, invertebrates that burrow into waterlogged limbs and trunks of fallen trees.

Feeding on the detrital feeder and scrapers are predaceous insect larvae such as large stoneflies, the powerful dobsonfly larvae (*Corydalus cornutus*), stream salamanders, and fish such as sculpin (*Cottus* spp.) and trout. Invertebrate predators employ either ambush or searching strategies. Some engulf the prey whole or in pieces; others pierce the body of prey and suck out all or part of the contents. Predaceous aquatic insects detect prey largely by mechanical cues (Peckarsky 1982), and are size-selective opportunists. Prey species in turn have evolved morphological and behavioral defenses, including flattened body shapes, protective cases (caddisflies), and cryptic coloration. Predators, especially those in headwater streams, do not depend solely on aquatic insects; they also feed heavily on terrestrial invertebrates that fall or are washed into the stream.

Because of current, quantities of CPOM and FPOM and invertebrates tend to drift downstream to form a traveling benthos. This process is normal in streams, even in the absence of high water and abnormal currents (for reviews see Hynes 1970; Waters 1972). Drift is so characteristic of streams that a mean rate of drift can serve as an index of the production rate (Pearson and Kramer 1972). This drift is essential to the production processes of downstream systems (Wallace et al. 1982).

Energy Flow and Nutrient Cycling.
Energy flow in lotic ecosystems has been documented for only a few streams. One energy budget is for the well-studied, small, forested Bear Brook in Hubbard Forest of northern New Hampshire (Fisher and Likens 1973). That budget is summarized in Table 32.3 and Figure 32.14. Over 90 percent of the energy input came from the surrounding forested watershed or from upstream areas. Primary production by mosses accounted for less than 1 percent of the total energy supply. Algae were absent from the brook. Inputs from litter and throughfall accounted for 44 percent of the energy sup-

ply, and geological inputs accounted for 56 percent. Energy was introduced in three forms: CPOM represented by leaves and other debris; FPOM represented by drift and small particles; and DOM. In Bear Brook 83 percent of the geologic input and 47 percent of the total energy input was in the form of DOM. Sixty-six percent of the organic input was exported downstream, leaving 34 percent to be utilized locally.

Nutrient cycling is more difficult to assess, because of the open nature of the lotic system. Triska and associates (1984) estimated the nitrogen budget of a small stream draining a 10.1 ha watershed

TABLE 32.3 Annual Energy Budget for Bear Brook

Item	kg (whole stream)*	kcal/m^2	Percentage
Inputs			
Litter fall			
Leaf	1990	1370	22.7
Branch	740	520	8.6
Miscellaneous	530	370	6.1
Wind transport			
Autumn	422	290	4.8
Spring	125	90	1.5
Throughfall	43	31	0.5
Fluvial transport			
CPOM	640	430	7.1
FPOM	155	128	2.1
DOM, surface	1580	1300	21.5
DOM, subsurface	1800	1500	24.8
Moss production	13	10	0.2
Input total	8051	6039	99.9
Outputs			
Fluvial transport			
CPOM	1370	930	15.0
FPOM	330	274	5.0
DOM	3380	2800	46.0
Respiration			
Macroconsumers	13	9	0.2
Microconsumers	2930	2026	34.0
Output total	8020	6039	100.2

Note: CPOM is coarse particulate organic matter; FPOM is fine particulate organic matter; DOM is dissolved organic matter.

*Budget in kg does not balance because of different caloric equivalents of budgetary components.

in the old-growth Douglas-fir H. J. Andrews Experimental Forest in Oregon (Table 32.4). The major annual input of dissolved nitrogen, amounting to 15.25 g/N/m^2, came from two major sources: hydrological and biological. Hydrological inputs, largely subsurface flow in the form of seeps, accounted for most of the input, 11.06 g/m^2. Biological inputs accounted for 4.19 g/m^2. Particulate organic nitrogen accounted for 11.93 g, of which CPOM made up 7.16 g and FPOM 4.77 g. Total output of nitrogen was 11.36 g/m^2. Of this output 2.53 g was particulate organic matter dominated by FPOM (1.66 g) and 8.81 g was dissolved N, less than the input because of biological uptake. The stream intercepted and transported nearly all of the nitrogen loss from the 10 ha watershed, yet retained and processed a considerable

portion of it. Nitrogen transported downstream would be consumed there by microbes.

A major problem for flowing water ecosystems is retention of nutrients upstream to reduce losses to downstream. Although nutrient cycling is downhill in all ecosystems, the problem in lotic systems is the constant movement of the substrate—water—away from the system. Nutrients in terrestrial and lentic systems are recycled more or less in place. An atom of a nutrient passes from soil or water column to plants and consumers back to soil or water in the form of detrital material or exudates and then is recycled within the same segment of the system, although losses do occur. Cycling is essentially temporal. Lotic systems have an added spatial cycle. Nutrients in the form of DOM and

TABLE 32.4 Nitrogen Budget (/m^2) for a Small Coniferous Forest stream (Watershed 10, H. J. Andrews Experimental Forest, Oregon)

Inputs			
Dissolved organic nitrogen pool			15.25
Hydrological		11.06	
DON	10.56		
NO$_3$–N	0.50		
Biological		4.19	
N-fixation	0.76		
Throughfall	0.30		
Litterfall	1.35		
Lateral movement of leaves	1.78		
Particulate organic nitrogen pool			11.93
FPOM	4.77		
CPOM	7.16		
Outputs			
Total nitrogen output			11.36
Dissolved nitrogen		8.81	
DON	8.38		
NO$_3$–N	0.43		
Particulate organic nitrogen		2.53	
FPOM	1.66		
CPOM	0.87		
Insect emergence		0.02	

Source: Triska et al. 1984.

POM undergo constant downstream displacement as a function of physical processes. The degree of displacement is determined by water flow and physical and biological retention in place. The greater the flow, the more rapid the displacement. Physical retention involves storage in wood detritus such as logs and snags in the stream, accumulation of debris in pools formed behind boulders, leaf sediments, and beds of macrophytes. Recycling is biological, controlled by biological uptake and storage in animal and plant tissue.

The process of recycling, retention, and downstream displacement may be pictured as a spiral lying longitudinally in a stream (Figure 32.16). *Spiraling* involves

Figure 32.16 Nutrient spiraling between particulate organic matter, including microbes, and water column in a lotic ecosystem. Uptake and turnover take place as nutrients flow downstream. (a) represents tight spiraling, and (b) more open spiraling. The tighter the spiraling, the longer the nutrient remains in a particular segment of a stream. (Adapted from Newbold et al. 1982: 630.)

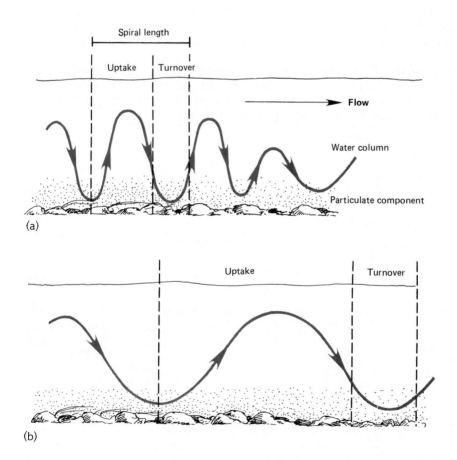

(a)

(b)

the combined processes of nutrient cycling and downstream transport. One cycle in the spiral involves the uptake of an atom or nutrient from DOM, its passage through the food chain, and its return to water where it is available for reutilization. Thus one cycle or loop begins and ends with the water compartment. Spiraling is measured as the distance needed for completion of one cycle. The longer the distance required, the more open the spiral; the shorter the distance, the tighter the spiral.

Tight spiraling of nutrients in flowing water ecosystems depends upon the retention of leafy detritus in place long enough to allow the biological component of the stream, especially the shredders and microbes, to process the organic matter. This delay is especially important in high gradient headwater streams, which can rapidly transport unprocessed particulate organic matter downstream. Logs, snags, other woody debris in the channel and along the banks of streams, and large rocks act as dams, intercepting leafy detritus and forming pools that collect sediments. These debris dams create a diversity of physical habitats within the stream and influence the nature of the invertebrate community (Figure 32.17). On some small streams (orders 0–2) wood and debris and sediments stored behind them

Figure 32.17 Large woody debris modifies the stream habitat, slowing the flow, retaining organic debris for processing by invertebrates, and adding to the diversity of the stream biota. Note the addition of another functional group, the gougers, where woody debris is prominent. Woody debris habitat favors collectors and shredders, especially on small-order streams.

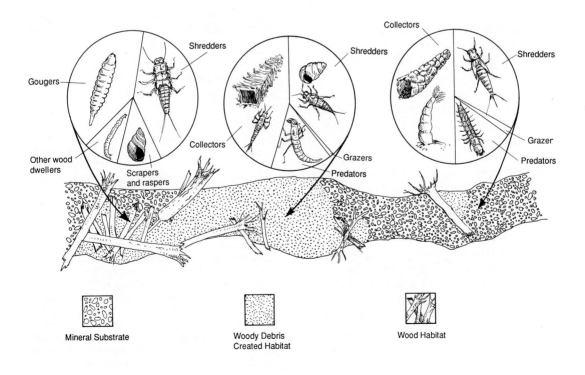

Mineral Substrate

Woody Debris Created Habitat

Wood Habitat

may make up 50 percent of the stream area, and in larger streams (orders 3–4) 25 percent of the stream area (Triska et al. 1983).

Bilby (1989) demonstrated the effectiveness of organic debris dams in regulating the export of dissolved and particulate organic matter from headwater streams. He removed organic debris dams from a stretch of a small stream in the Hubbard Brook forest and followed the effects over a period of a year. Removal of the dams resulted in only a slight increase (6 percent) in the export of DOM, but it caused a 500 percent increase in the export of CPOM and FPOM. Removal was most dramatic at periods of high discharge.

Newbold and his associates (1983) investigated the spiraling of exchangeable phosphorus in the form of $^{32}PO_4$ in Walker Branch at Oak Ridge, Tennessee, a first-order stream. The tagged P moved downstream at the rate of 10.4 m a day and cycled once every 18.4 days. Thus the average downstream distance of one spiral was 190 m. In other words, one atom of P on the average completed one cycle from the water compartment back to the water compartment again for every 190 m of downstream travel. The spiraling length was partitioned into an uptake length of 165 m, associated with transport in the water column, mostly as DOM; a particulate turnover length of 25 m, associated with FPOM; and a consumer turnover length of 0.05, associated with consumer drift. CPOM accounted for 60 percent of the uptake; FPOM 35 percent; and aufwuchs, 5 percent. Turnover time of P in CPOM ranged from 5.6 to 6.7 days; and FPOM 99 days. Only 2.8 percent of P uptake from particulate matter was transferred to consumers; most of the P was transferred back to water. About 30 percent of the consumer uptake was transferred to predators.

The River Continuum

Lotic ecosystems involve a continuum of physical and biological conditions from the headwaters to the mouth (Vannote et al. 1980; Minshall et al. 1983) (Figure 32.18). The upper reaches of the headwaters (stream orders 1–3) are usually swift, cold, and in forested regions shaded. Riparian vegetation reduces light, inhibiting autotrophic production (Hawkins et al. 1982), and contributes more than 90 percent of organic input into streams as terrestrial detritus. Even when headwater streams are exposed to sunlight and autotrophic production exceeds heterotrophic inputs, organic matter produced invariably enters detrital food chains (Minshall 1978). Dominant organisms are shredders, processing large litter and feeding on CPOM, and collectors, processors of FPOM. Populations of grazers are minimal, reflecting the small amount of autotrophic production, and predators are mostly small fish—sculpins, darters, and trout. Headwater streams are accumulators, processors, and transporters of particulate organic matter from terrestrial systems. As a result the ratio of gross pri-

Figure 32.18 The lotic system is essentially a continuum from headwaters to the river's mouth. The headwater stream is strongly heterotrophic, dependent on terrestrial input of detritus, and the dominant consumers are shredders and collectors. As stream size increases, the input of organic matter shifts from particulate organic matter to primary production from algae and rooted vascular plants. The major consumer groups are now collectors and grazers. The zone at which the shift occurs is influenced by the degree of shading. As the stream increases to river, the lotic system shifts back to heterotrophy. It is dependent upon inputs of fine particulate organic matter and dissolved organic matter. A phytoplankton population may develop, its extent influenced by depth and turbidity. The bottom consumers are collectors, mostly bottom organisms.

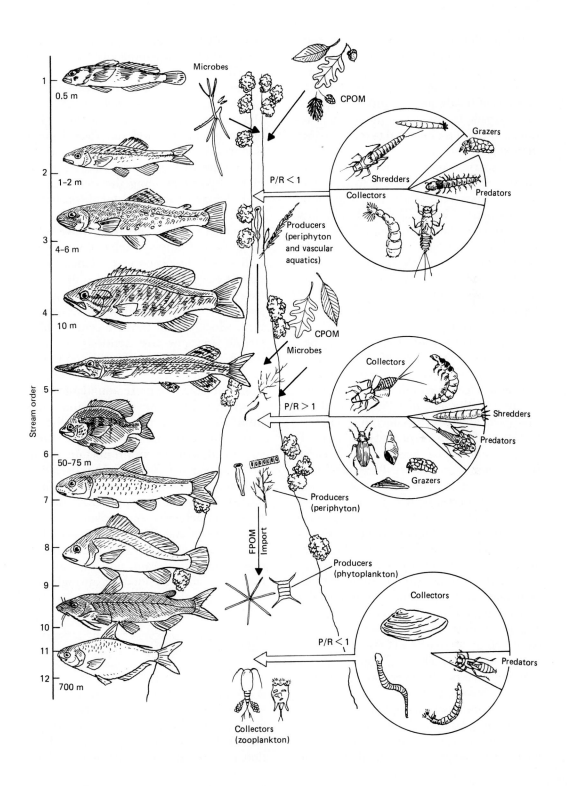

Microbes

CPOM

Grazers

Shredders

Collectors

Predators

P/R < 1

Producers
(periphyton
and vascular
aquatics)

CPOM

Microbes

P/R > 1

Collectors

Shredders

Predators

Grazers

Producers
(periphyton)

FPOM
Import

Producers
(phytoplankton)

Collectors

P/R < 1

Predators

Collectors
(zooplankton)

Stream order

1   0.5 m

2   1–2 m

3   4–6 m

4   10 m

5

6   50–75 m

7

8

9

10

11

12   700 m

mary production to community respiration is less than 1. Consumer organisms utilize long-chain and short-chain organic compounds for transport downstream. Organisms of headwater streams are adapted to a narrow temperature range, to a reduced nutrient regime, and to maintenance of their position in the current.

As streams increase in width to medium-sized creeks and rivers (orders 4 to 6), the importance of riparian vegetation and its detrital input decreases. The lack of shading results in increased temperature of stream water. As the gradient declines, the current slows and becomes more variable. The diversity of microenvironments supports a greater diversity of organisms. An increase in light and temperature and a decrease in terrestrial input encourage a shift from heterotrophy to autotrophy, relying on algal and rooted plant production. Gross primary production now exceeds community respiration. Because of the lack of CPOM, shredders disappear, and collectors, feeding on FPOM transported downstream, and grazers, feeding on autotrophic production, become the dominant consumers. Predators show little increase in biomass but shift from cold-water species to warm-water species.

As the stream order increases from 6 through 10 and higher, riverine conditions develop. The channel is wider and deeper. The volume of flow increases, and the current becomes slower. Sediments accumulate on the bottom. Both riparian and autotrophic production decrease, with a gradual shift back to heterotrophy. A basic energy source is FPOM, utilized by bottom-dwelling collectors, now the dominant consumers. However, slow, deep water and dissolved organic matter (DOM) support a minimal phytoplankton and associated zooplankton population.

Large rivers (and many small ones) have been highly modified, altering the nature of the river continuum (Cummins 1988). Biological complexity has been reduced by massive snag removal projects. Snags retarded flow and created stable habitats in active channels. Constructed channels and levees have straightened the channels and isolated the river from its floodplain. Impoundments have altered the nature of flows, produced large sediment collections along the river's course, and changed the nature of the fish community (Bain, Finn, and Booke 1988). Regulated streams and rivers have highly variable and unpredictable flow regimes influenced by economic demands for power, navigation, and flood control. Species and size classes of fish that are restricted to shallow depths and slow current velocities and thus concentrate along stream margins are reduced in abundance. Fish species and size classes that occupy a broad range of habitats, deep fast water, or both are more abundant.

Throughout the downstream continuum, the lotic community capitalizes on upstream feeding inefficiency. Downstream adjustments in production and the physical environment are reflected in changes in consumer groups (Figure 32.18). Through the continuum the lotic ecosystem achieves some balance between the forces of stability, such as natural obstructions in flow that aid in the retention of nutrients upstream, and the forces of instability, such as flooding, drought, and temperature fluctuations.

■ Wetlands

Associated with lentic and lotic ecosystems are wetlands, areas where water is near, at, or above the level of the land (Figure 32.19). However utilitarian such a definition may be, it is not sufficient to define

(a)

(b)

Figure 32.19 Two examples of wetlands. (a) A glacial prairie marsh, dominated by emergent vegetation. (b) A cypress deepwater swamp in the Southern United States. (Photos courtesy U.S. Fish and Wildlife Service.)

such heterogeneous communities. The U.S. Fish and Wildlife Service (1979) has defined wetlands as

> lands transitional between terrestrial and aquatic systems where the water table is usually at or near the surface or the land is covered by shallow water. . . . Wetlands must have one or more of the following three attributes: (1) at least periodically, the land supports predominately hydrophytes; (2) the substrate is predominately undrained hydric soil; and (3) the substrate is nonsoil and is saturated with water or covered by shallow water at some time during the growing season of the year.

The U.S. Army Corps of Engineers (1984) defines wetlands as

> those areas that are inundated or saturated by surface or ground water at a frequency and duration sufficient to support, and that under normal circumstances do support, a prevalence of vegetation typically adapted for life in saturated soil conditions. Wetlands generally include swamps, marshes, bogs, and similar areas.

Biologically, wetlands are among the richest and most interesting ecosystems; yet they are among the least appreciated and the first to be destroyed by filling and drainage. Wetlands are a halfway world between terrestrial and aquatic ecosystems and exhibit some of the characteristics of each. A wide variety of wetlands exists, and classifying them for management and conservation has presented some problems. An older but still useful classification appears in Table 32.5 on pages 864 and 865. A more recent, much more comprehensive hierarchical classification is *Classification of Wetlands and Deepwater Habitats of the United States* (Cowardin et al. 1979).

(text continues on page 866)

TABLE 32.5 Types of Wetlands

Type	Site Characteristics	Plant and Animal Populations
Inland Fresh Areas		
Seasonally flooded basins or flats	Soil covered with water or waterlogged during variable periods, but well drained during much of the growing season; in upland depressions and bottomlands	Bottomland hardwoods to herbaceous growth
Fresh meadows	Without standing water during growing season; waterlogged to within a few inches of surface	Grasses, sedges, rushes, broadleaf plants
Shallow fresh marshes	Soil waterlogged during growing season; often covered with 15 cm or more of water	Grasses, bulrushes, spike rushes, cattails, arrowhead, smartweed, pickerelweed; a major waterfowl–production area
Deep fresh marshes	Soil covered with 15 cm to 1 m of water.	Cattails, reeds, bulrushes, spike rushes, wild rice; principal duck-breeding area
Open fresh water	Water less than 3 m deep	Bordered by emergent vegetation such as pondweed, naiads, wild celery, water lily; brooding, feeding, nesting area for ducks
Shrub swamps	Soil waterlogged; often covered with 15 cm or more of water	Alder, willow, buttonbush, dogwoods; nesting and feeding area for ducks to limited extent
Wooded swamps	Soil waterlogged; often covered with 0.3 m of water; along sluggish streams, flat uplands, shallow lake basins	North: tamarack, arborvitae, spruce, red maple, silver maple, south: water oak, overcup oak, tupelo, swamp black gum, cypress
Bogs	Soil waterlogged; spongy covering of mosses	Heath shrubs, *Sphagnum,* sedges
Coastal fresh areas		
Shallow fresh marsh	Soil waterlogged during growing season; at high tide as much as 15 cm of water; on landward side, deep marshes along tidal rivers, sounds, deltas	Grasses and sedges; important waterfowl areas
Deep fresh marshes	At high tide covered with 15 cm to 1 m of water; along tidal rivers and bays	Cattails, wild rice, giant cutgrass

Type	Site Characteristics	Plant and Animal Populations
Open fresh water	Shallow portions of open water along fresh tidal rivers and sounds	Vegetation scarce or absent; important waterfowl areas
Inland Saline Areas		
Saline flats	Flooded after periods of heavy precipitation; waterlogged within few inches of surface during the growing season	Seablite, salt grass, saltbush; fall waterfowl-feeding areas
Saline marshes	Soil waterlogged during growing season; often covered with 0.61 to 1 m of water; shallow lake basins	Alkali hard-stemmed bulrush, wigeon grass, sago pondweed; valuable waterfowl areas
Open saline water	Permanent areas of shallow saline water; depth variable	Sago pondweed, muskgrasses; important waterfowl-feeding areas
Coastal Saline Areas		
Salt flats	Soil waterlogged during growing season; sites occasionally to fairly regularly covered by high tide; landward sides or islands within salt meadows and marshes	Salt grass, seablite, saltwort
Salt meadows	Soil waterlogged during growing season; rarely covered with tide water; landward side of salt marshes	Cord grass, salt grass, black rush, waterfowl-feeding areas
Irregularly flooded salt marshes	Covered by wind tides at irregular intervals during the growing season; along shores of nearly enclosed bays, sounds, etc.	Needlerush, waterfowl cover area
Regularly flooded salt marshes	Covered at average high tide with 15 cm or more of water; along open ocean and along sounds	Atlantic: salt-marsh cord grass; Pacific: alkali bulrush, glassworts; feeding area for ducks and geese
Sounds and bays	Portions of saltwater sounds and bays shallow enough to be diked and filled; all water landward from average low-tide line	Wintering areas for waterfowl
Mangrove swamps	Soil covered at average high tide with 15 cm to 1 m of water; along coast of southern Florida	Red and black mangroves

Source: Adapted from Shaw and Fredine 1956.

Wetlands dominated by emergent vegetation, plants with roots in soil covered part or all of the time by water and leaves held above the water, are *marshes*. Growing to reeds, sedges, grasses, and cattails, marshes are essentially wet prairies. They develop along margins of lakes (lentic) or in shallow basins with an inflow and outflow of water (lotic), along slow-moving rivers (riverine), and on tidal flats (tidal marshes). Wetlands in which considerable amounts of water are retained by an accumulation of partially decayed organic matter are *peatlands* or *mires*. Mires fed by water moving through mineral soil and dominated by sedges are known as *fens*. Because most of their nutrients come from mineral soil, they are called minerotrophic or *rheotrophic* mires. Mires dominated by *Sphagnum* mosses and depending largely on precipitation for their water supply are *bogs*. Mires that develop on upland situations where decomposed, compressed peat forms a barrier to the downward movement of water, resulting in a perched water table above mineral soil, are *blanket mires* and *raised bogs*. Blanket bogs are more popularly known as *moors*. Because bogs depend upon precipitation for nutrient inputs, they are highly deficient in mineral salts and low in pH. Such bogs are called *ombrotrophic*. Bogs may also develop by the filling in of a lake basin from above rather than from below. Because mire vegetation often grows on a floating mat of peat over open water, such bogs are often termed *quaking*— more descriptively, *schwingmoor* in German. Wooded wetlands are *swamps*. They may be deepwater swamps dominated by cypress, tupelo, and swamp oaks; or they may be shrub swamps dominated by alder and willows. Shrubby swamps are often called *carrs*. Along large river systems may occur extensive tracts of *riparian woodlands*, which are occasionally flooded by river waters but are dry for most of the growing season. In the United States such wetlands are called *bottomland hardwood forests*.

Only in recent years have ecologists given serious attention to the structure and function of freshwater wetlands (Good, Whigham, and Simpson 1978; Greeson, Clark, and Clark 1979; Weller 1981; Mitsch and Gosselink 1986; Gore 1983; Odum et al. 1984). An impetus for increased studies is the potential of wetlands as filtering systems to reduce heavy metals such as iron and aluminum from mine drainage, sewage, and industrial wastes. Two wetland types will serve as examples, a freshwater marsh and peatlands.

Freshwater Marshes

Freshwater marshes occur both inland and along the coast. Inland freshwater marshes are a diverse lot, ranging from glacial prairie potholes to extensive tracts covering many hectares, such as the playas of the Great Basin of the western United States and marshes in the region of the Great Lakes. Inland marshes are fed by inflowing water, seepage, and precipitation. Because of the variability of water supply, inland marshes usually experience flooding during periods of high rainfall and drawdown during dry periods, phenomena which appear to be essential to their long-term existence (see Chapter 26). Dominant species are mostly graminoid; depending on the marsh system they may include cattails, wild rice, sedges (*Carex spp.*), bulrushes (*Scirpus spp.*), spike rushes (*Eleocharis spp.*), smartweeds (*Polygonum spp.*), and others.

Tidal freshwater marshes are close enough to the coast to be affected by daily tides but distant enough to be unaffected by any intrusion of salt water. Because of their nutrient subsidy from tides and relatively stable water levels, coastal marshes do not undergo such variable fluctuations in water levels as inland marshes. The deposition of sediments during periods of high tides creates a gradient of microelevations across the marsh, which influences the pattern of vegetation. Deeper water is occupied by such submerged plants as spatterdock (*Nuphar advena* and pondweeds (*Potamogeton spp.*). Creek banks scoured clean each fall by strong tidal currents are colonized by annuals such as smartweeds and bur marigold (*Bidens spp.*). The low marsh with its deeper water holds broadleaf monocotyledons such as arrow arum (*Peltandra virginica*) and pickerelweed (*Pontederia cordata*). The high marsh is a mixture of annual and perennial species, ranging from smartweeds and rose mallow (*Hibiscus coccineus*) to cattails and big cordgrass (*Spartina cynosuroides*).

Wetlands achieve an importance far beyond their areal extent as breeding grounds for a diverse array of ducks, geese, herons, shorebirds, and furbearers such as muskrats, as stopover points for migrating waterfowl and shore birds, and as wintering grounds for many of the same. Coastal freshwater marshes function as nurseries for many nektonic fish, such as herring, shad, and striped bass. In spite of their great ecological and economic importance, wetlands continue to be drained for agricultural, suburban, recreational, and industrial development.

Productivity of freshwater marshes is influenced by hydrological regimes: groundwater, surface runoff, precipitation, drought cycles, and the like. It is also affected by the nature of the watershed in which the wetland lies, soils, nutrient availability, types of vegetation, and the life history of the plant species.

Aboveground biomass may vary with the proportionate abundance of annual

and perennial species, which change in dominance through the growing season. Annual emergents exhibit a linear increase in biomass through the growing season, reaching a maximum in late summer (Figure 32.20). Perennial biomass increases during the first part of the growing season, then declines or levels off as leaves become senescent (Whigham et al. 1978). In general, however, the average maximum standing crop of biomass matches annual aboveground productivity. For example, a sedge wetland had a maximum standing crop biomass greater than 1000 g/m^2 and a net primary productivity greater than 1000 g/m^2/yr. If the life history of the plants—including mortality, regeneration, and winter activity—was considered, net productivity was 1600 g/m^2/yr (Bernard and Gorham 1978).

Belowground production, much more

Figure 32.20 Pattern of aboveground biomass accumulation through the growing season for a freshwater annual, *Lythrum* (loosestrife), and a perennial, *Typha* (cattail). Note the linear increase in biomass of the annual and the sigmoid growth of the perennial. (From Whigham et al. 1978: 12.)

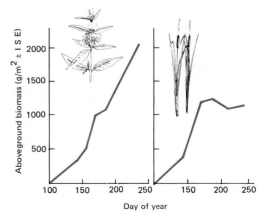

difficult to estimate, appears for some species to be highest in summer, at the same time the peak aboveground biomass is achieved. Others, such as *Typha* and *Scirpus*, reach peak production in the fall, when nutrients are stored in the roots. Such species may have minimal root biomass in the summer because of nutrient transfer to aboveground biomass.

The cycling of phosphorus in a cattail marsh on Lake Mendota, Wisconsin (Prentki et al. 1978) provides an example (Figure 32.21). In spring growing new shoots of cattail draw on the phosphorus reserves in the rootstalks for their initial growth before mobilizing P from the soil. By June cattails are accumulating P at a rate higher than they are accumulating bio-

mass. By midsummer *Typha* has accumulated about 4 g/P/m² in the shoots, 78 percent of total P in biomass. At the same time the belowground pool is minimal. As the season progresses P begins to accumulate in the rhizomes until December, but at a rate slower than accumulation of belowground biomass. In fall large amounts of P begin to disappear. Most of it is lost through leaching and death of the shoots. Only about 28 percent of the summer accumulation is returned to the rhizomes, to be depleted rapidly again by spring growth.

To balance losses, cattails and other emergent plants must draw on phosphorus from the soil (Figure 32.22). By doing so, the plants act as a nutrient pump, drawing nutrients from the soil, translocating them into the shoots, and then during the growing and postgrowing season releasing them to the surface soil by leaching and death of the shoots. In this way marsh plants make nutrients sequestered in the soil available for growth.

Nutrients accumulated in plant biomass become available for use through decomposition. Initial decomposition in wetlands is high as leaching removes soluble compounds that enter DOM (Figure 32.23). After initial leaching, decomposition proceeds more slowly. Permanently submerged leaves decompose more rapidly than those on the marsh surface because they are more accessible to detritivores (Smock and Harlowe 1983) and because the stable physical environment is more favorable for microbes.

Peatlands

Peatlands are ecosystems in which the rate of production of organic matter by living organisms exceeds the rate at which the compounds are respired and degraded

Figure 32.21 Seasonal stocks of phosphorus in cattail *(Typha latifolia)* plant parts and belowground deficit in a marsh at Lake Mendota, Wisconsin. Note the sharp rise in P in the shoots during the summer months and its just as rapid decline in the fall when the shoots die. Much P is lost through leaching, but note the accumulation of some P in the current year rhizomes. Belowground deficit is highest during the summer, when most of the P is in the shoots, but the deficit is erased by winter. (After Prentki, Gufason, and Adams 1978: 176.)

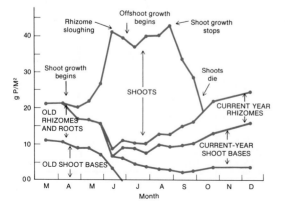

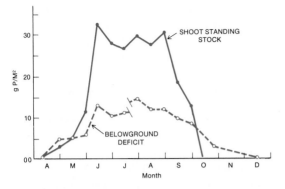

Figure 32.22 Flow of phosphorus through a stand of river bulrush (Scirpus fluviatilis). Flows are g/m²/yr and compartments are g/m² standing crop. Note that two-thirds of the phosphorus in the aboveground standing crop is lost by leaching. The loss is made up by takeup from the soil. In this manner the plants function as a nutrient pump, mobilizing P sequestered in the soil and releasing it to the marsh surface when they die. (After Klopatek 1978: 207.)

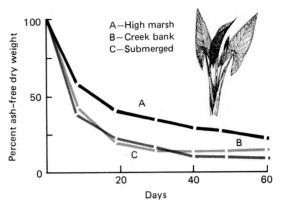

Figure 32.23 Decomposition of leaves of *Peltandra virginica*, arrow arum, in a tidal freshwater marsh, measured by the percentage of original ash-free dry weight remaining in litter bags under three conditions: irregularly flooded high marsh exposed to alternate wetting, creek bed flooded two times daily, and permanently submerged. Note that the detrital material consistently wet showed the highest rate of decomposition, although the overall pattern of decomposition is similar. (From Odum and Haywood 1978: 92.)

(Moore and Bellamy 1974). As a result part of the production accumulates as organic deposits or peat. As the peat blanket thickens, the surface vegetation becomes insulated from the mineral soil. Environmental conditions change, bringing about a change in plant and animal life.

Although peatlands are most commonly associated with northern regions of the world, they occur worldwide wherever humidity and precipitation are high or hydrological situations encourage an accumulation of partly decayed organic matter. (For world examples see Gore 1983.)

Formation. Peat forms when the movement of water through a low wetland area becomes partially blocked. The inflowing water no longer carries sediments out of the basin and much of the water is retained in the system. When dead organic matter, its decomposition slowed by immersion in water, begins to act as an inert body displacing its own volume of water, peat formation begins (Moore and Bellamy 1974). Peat formation continues until it reaches a level at which water drains from its surface. At this point peat no longer acts as an inert body but as its own reservoir, holding quantities of water from drainage.

In tropical and subtropical regions peatlands are found mostly in mountains or in lowland estuarine and delta regions where enough inflowing water keeps the basin filled, retarding decomposition. Examples are the Everglades, the tropical coal swamps of the Carboniferous period, and the pocasins of the southern United States coastal plains.

Peatlands are more characteristic of northern regions where inflow and precipitation balance outflow and retention, or where permafrost impedes drainage of water and the bulk of summer precipitation is retained. In northern regions peatlands are dominated by sphagnum moss and develop not only in basins, but also on mineral soil, producing blanket mires, raised bogs, and patterned fens (see Moore and Bellamy 1974; Heinselman 1970).

The formation and types of mires are complex. Peatland may develop from an aquatic ecosystem through the filling of the basin by sediments or from a terrestrial system through the swamping of wooded areas (paludification) (Malmer 1975). In the process the original systems, characterized by rapid and fairly complete turnover of dead organic matter and a high productivity, are slowly converted to a common ecosystem of low productivity, low plant biomass, low diversity, a slow and incomplete turnover of organic matter, and a storage of nutrients.

The classic example of bog development is the filling of a lake basin with sediment and organic matter carried into the area by inflowing water. As the open water, dominated by cations of calcium and carbonates, becomes more shallow, emergent vegetation begins to encroach upon and cover the basin. Excess organic matter produced by the marsh vegetation accumulates, building the bottom up to and eventually above the water table. The accumulated organic matter is continually saturated, inhibiting decomposition. If eutrophic conditions persist, as they often do, terrestrial vegetation may invade the area. Under oligotrophic and dystrophic conditions, calcium and sulphate ions dominate the water and humic compounds dissolved from organic matter increase and pH decreases. The marsh vegetation is replaced by acid-tolerant sedges and grasses.

The buildup of sediment and peat diverts inflowing water first to the periphery of the basin and eventually away from the basin altogether. Now the only source of water is rainfall. Under conditions of high humidity and precipitation sphagnum mosses colonize the area. Sphagnum has the ability to absorb ions out of nutrient-poor water by exchanging hydrogen ions in its tissues for cations in solution. This exchange, which results in an increase in hydrogen and sulphate ions in the water, increases the acidity of the system, further decreasing the rate of decomposition and increasing the net accumulation of organic matter. Sphagnum, which has a spongelike ability to hold water, increases water retention on the site, and its habit of adding new growth onto the accumulating remains of past moss generations aids in thickening the organic mat. In time the water-saturated mat of peat and sphagnum expands and rises above the general water table level to form a raised bog (Figure 32.24). Between the mat of peat and surrounding higher land is a moatlike area of shallow water, often the remnant of flowing groundwater around the peat mat. These water areas, dominated by sedges, are called *laggs*.

Bogs may also develop by the filling of a basin from above rather than from below. Many of the sediments coming into the lake, especially the finer materials, remain suspended as colloids during their movement from the water's surface to the bottom. Often they remain in suspension until precipitated by bacterial, chemical, or photosynthetic action to form a fine, soft deposit called a false bottom. Mean-

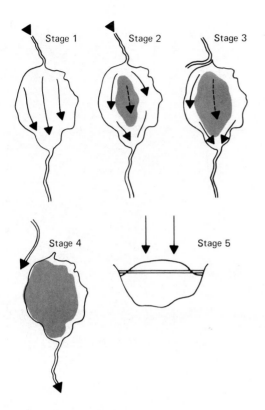

Figure 32.24 Stages in the development of a mire. Stage 1: A flow of water brings allochthonous material into the system. If the water flow is large, the amount of outside material is heavy and the rate of peat growth is slow, producing heavy peat above which the water flows. If the rate of water flow is slow, growth of peat is rapid, producing a light peat beneath which the water flows (see Figure 32.25). Stage 2: The accumulation of peat tends to channelize the main flow of water. Under certain conditions, the whole peat mass may be inundated. Stage 3: Continued peat growth diverts inflow from the basin. The major source of water is now precipitation, surface runoff, or seepage from the surrounding area. Stage 4: Additional accumulation of peat leaves large areas of mire unaffected by moving water, but subject to inundation during periods of heavy rainfall. Stage 5: Continued peat growth raises the surface of the mire above the influence of groundwater. The mire now possesses its own water table fed by rain. (After Moore and Bellamy 1975: 57.)

while peat develops around the edges of the basin. Sphagnum invades the area and fills the open spaces between sedges and other emergent plants. When a consolidated mass of peat develops or when jutting rocks, logs, or even the leading edges of low-growing plants allow a foothold, a mat extends outward over the water (Figure 32.25). As the mat thickens and advances toward the center of the lake, the older peat mat is colonized first by such shrubs as bog rosemary and leatherleaf, followed by forest trees like pine, spruce, or tamarack, producing the concentric concave rings of vegetation so characteristic of northern bog lakes.

Beneath the horizontal structure on the surface are horizontal zones of peat accumulation and basin filling. Kranz and DeWitt (1986) in a study of the internal mechanisms influencing the development of a northern bog recognized three horizontal zones within the basin, from the open water of the lake to the original shore: (1) a zone near the lake's edge where the vertical accumulation of peat thickens into a floating mat; (2) a midzone of compaction where the vertical accumulation of peat compacts underlying peat; and (3) an equilibrium zone bordering the shore where no peat accumulates and peat reaches its maximum density. The deposition and accumulation of peat in the bog Kranz and Dewitt studied was influenced by the spatial configuration of the basin, and in turn peat accumulation influenced the configuration.

Although basin filling is the textbook model of bog formation, most of the boreal peatland develops on higher ground on mineral soil. Covering large areas, such mire development creates high moors and blanket mires (Figure 32.26). Such upland bogs start in one of several ways. Sphagnum and other mosses may invade higher

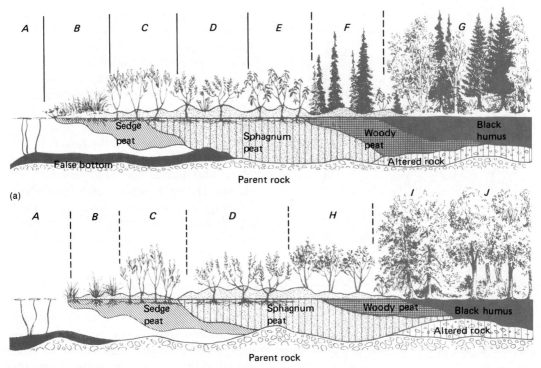

(a)

(b)

Figure 32.25 (a) Transect through a quaking bog, showing zones of vegetation, sphagnum mounds, peat deposits, and floating mats. A, pond lily in open water; B, buckbean and sedge zone; C, sweet gale zone; D, leatherleaf; E, Labrador tea; F, black spruce; G, birch-balsam fir-black spruce forest. (b) Alternative vegetational sequence: H, alder; I, aspen, red maple; J, mixed deciduous forest. (Adapted from Dansereau and Segadas-Vianna 1952.)

ground surrounding a lake basin by creeping into depressions. Beaver dams backing up water over streamside forests and along stream courses may initiate mire development. Logging of forests on wet sites where humus decomposition is incomplete may provide sites for colonization by mosses. Fresh mosses and sedges on the surface of developing peatland are porous and permeable to water. Partially decomposed remains of these plants be-

neath the new growth become compressed, increasing bulk density and reducing pore volume. Eventually these decomposed, compressed peats act much like impermeable clays. Water unable to move through the mass to mineral soil remains as a perched water table near peat surface. Far removed from flowing waters, these blanket mires and raised bogs depend wholly upon precipitation for water and minerals. As a result the water is highly deficient in mineral salts and low in pH. Such bogs are ombrotrophic.

Upland bogs or paludified landscapes are not uniform, but rather are a mosaic of vegetation types ranging from blanket mires and raised bogs to fens fed by water that has moved through mineral soil and thus supports a richer vegetation of sedges and associated plants (see Moore and Bellamy 1974; Heinselman 1970,

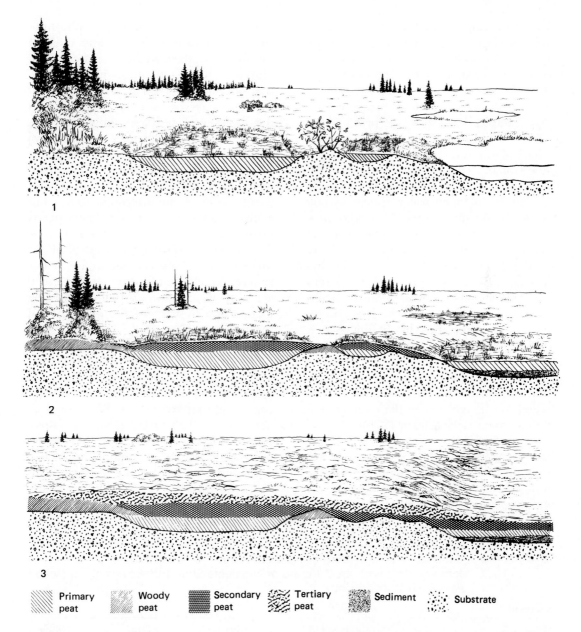

▨ Primary peat	▨ Woody peat	■ Secondary peat	▨ Tertiary peat	▨ Sediment	⋰ Substrate

Figure 32.26 Development of a raised bog and blanket mire. In the first stage primary peat forms in basins and depressions. It reduces the surface retention of the reservoir. After peat fills the basin, it may continue to develop beyond the confines of the basin to form a raised bog. The raised bog consists of secondary peat, which acts as a reservoir, increasing surface retention of water on the area. Peat that develops above the physical limits of the groundwater and blankets even terrestrial situations is tertiary peat. It acts as a reservoir, holding a volume of water by capillary action above the level of the main groundwater drainage. The landscape acts as a perched water table, fed by precipitation falling on it. As the area becomes more wet and acidic, tree growth dies out and the landscape is covered by sphagnum and sedges.

1975). Although peatlands have been considered climax vegetation, it seems inaccurate to do so because their development or succession appears to be subject to random changes induced by such local events as fire, drought, flooding, and the like. For example, occasionally the peat mat becomes infiltrated with alkaline water, which allows fermentation and decomposition of the peat. Pockets form beneath the surface; the peat mat sags and caves in. Patches of open water appear as bog pools. In the bog forest the evaporative power of the vegetation may become greater than the water-retaining capacity of the peaty material, upsetting the hydrological balance necessary for formation and maintenance of peatlands. Then succession tends toward a drier, more mesic condition; but lumbering, windthrow, or any action that destroys tree growth can reverse this process and convert the area to an open bog. Thus succession in peatlands may simply involve ceaseless change rather than directional development (Heinselman 1975).

Function. Because bog vegetation is not in contact with mineral soil and because inflow of groundwater is blocked, nutrient input is largely by precipitation. Nutrient availability, especially nitrogen, phosphorus, and potassium, is low, and most of the nutrients fixed in plant tissue are removed from circulation in the accumulation of peat (Table 32.6). The amount of nutrients received by a bog depends upon the location of peatland. Bogs near the sea, such as English blanket mires, receive considerably more nutrients, especially magnesium and potassium, than inland bogs (Table 32.7). Few data exist on nutrient cycles in ombrotrophic bogs, but evidence suggests that some bog plants possess mechanisms to conserve nutrients.

In general nitrogen is available from three sources: (1) precipitation, the major source; (2) nitrogen-fixation by blue-green algae living in close association with bog mosses and by bog myrtle, if the pH of the substrate is at least 3.5, below which nitrogen-fixing root nodule bacteria are inhibited; and (3) carnivorous habit, by which certain plants such as sundews and pitcher plants extract nitrogen from captured and digested insects. Blue-green algae may fix 0.23 to 0.9 $g/N/m^2/yr$. Rosswall has calculated a nitrogen budget and nitrogen cycle for an ombrotrophic bog dominated by the trailing ground plant *Rubus chamaemorus* (Figure 32.27). Yearly demand is about 1 to 2 g/m^2.

TABLE 32.6 Cumulative Loss or Gain of Nutrients over the Period February 1969 to January 1972 with an Estimate of Nutrients Stored Irrecoverably in Deep Peat Each Year (mg/m^2)

Element	1969	1970	1971	Total	Stored
Ca	−525	−329	−1207	−2061	6364
Mg	+429	+1950	+964	+2343	9091
K	−52	+361	−84	+225	4546
NH_4	−567	−588	−537	−1692	3200
NO_3	−369	−427	−356	−752	
P	−5	−2	−13	−20	2045
Fe	−201	−77	−94	−371	nd*
Cu	+1	+5	−2	+4	nd

Source: Moore et al. 1975.

TABLE 32.7 Annual Nutrient Input by Rainwater at Various Sites (kg/ha)

Site	Annual Rainfall	Inorganic N	P	Na	K	Ca	Mg
Lerwick, Shetland Isles	57.3	2.10	—	133.00	5.52	6.70	19.20
Lancashire, UK	161.7	6.28	0.43	35.34	2.96	7.30	4.63
Pennines, UK	186.1	6.89	0.27	32.14	2.27	9.53	4.48
Kent, UK	84.0	—	<0.4	19.30	2.80	10.70	<4.20
Hubbard Brook, USA	129.0	—	—	1.5	1.40	2.60	0.70

Source: Moore and Bellamy 1974.

The phosphorus cycle holds losses to a minimum and is more closed than the nitrogen cycle. *Rubus chamaemorus* increases its uptake of phosphorus prior to bud break. After bud break the plant increases phosphorus in stem, leaf, and root. This increase correlates with an increase in peat temperature. After the plant completes shoot growth in summer,

Figure 32.27 Nitrogen budget for an ombrotrophic mire (Stordalen, Sweden). Quantities of nitrogen are expressed as $g/N/m^2$, flows as $g/N/m^2/yr$. The vegetation has been divided into aboveground parts exemplified by *Rubus chamaemorus* and lichens (1.6 g N/m^2), belowground parts of vascular plants (3.0 g N/m^2), aboveground litter (3.2 g N/m^2), and mosses (4.4 g N/m^2). N_{org} = organic nitrogen; N_{acc} = nitrogen accumulated; N_{part} = particulate nitrogen (dry deposition). (After Rosswall 1975.)

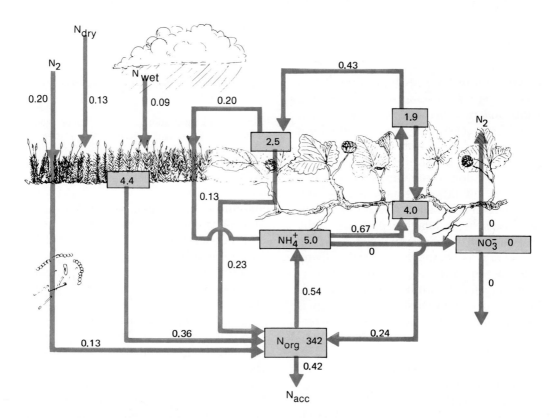

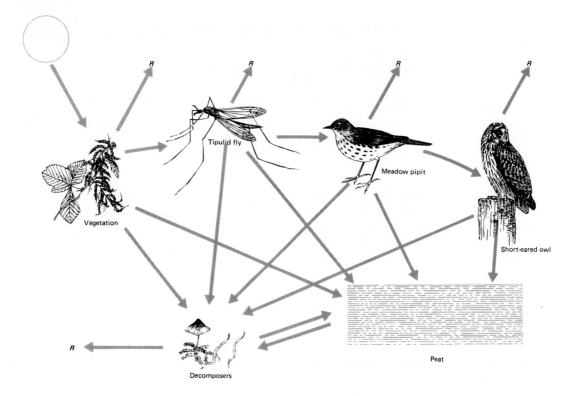

Figure 32.28 Energy flow in a mire ecosystem. Note the distinct difference between this energy flow pattern and other ecosystems. Instead of being lost as respiration, a portion of the energy is stored as peat, which accumulates in the system. Such an accumulation in the geological past resulted in the eventual formation of coal, oil, and gas deposits. (After Moore and Bellamy 1975.)

its phosphorus level declines in the shoots as it mobilizes phosphorus in developing fruits and in roots and rhizomes. In advanced senscence the plant rapidly loses phosphorus from the shoots and accumulates it in roots and winter buds.

Energy flow in the mire system differs from that of other systems, because decomposition is impaired (Figure 32.28). In most ecosystems material that enters the detrital food web eventually is recycled, and energy is liberated or stored in living material. In mire systems about 10 percent of primary production accumulates in an undecomposed state, and energy is locked up in peat until environmental conditions change, favoring decomposition, or the material is burned.

Because of low temperatures, acidity, and nutrient immobilization primary production in peatlands is low. In the Stordalen mire in Sweden primary production

amounted to 70 $g/m^2/yr$ by bryophytes and 83 $gm/m^2/yr$ by dwarf shrubs, forbs, and monocots collectively (Rosswall 1975). In an English blanket mire average production was 635 $g/m^2/yr$, with sphagnum on wet sites contributing 300 $g/m^2/yr$ (Heal et al. 1975).

Herbivorous utilization of bog vegetation is low in part because of unpalatability of the vegetation. Herbivores may include red grouse, willow grouse, hares, bog lemmings, and sheep. With these herbivores consumption is selective and usually confined to current shoots, opening buds, and fruits. In most bogs the dominant herbivores are insects. In an English blanket mire psyllids and tipulids were the main invertebrate herbivores, the former feeding on the phloem of heather and the latter feeding on liverworts (Heal et al. 1975; Moore et al. 1975). Predators include rodent-consuming weasels, harriers, short-eared owls, insect-consuming frogs, shrews, pipits (Europe), Nashville warblers (North America), and invertebrates such as spiders and ground beetles.

Decomposer organisms are low in number and the dominant species vary from year to year. In a Swedish mire total fungal biomass was 58 g/m^2 and the bacterial biomass, largely anaerobic, was 22 g/m^2 (Rosswall et al. 1975). Invertebrate detritivores include rotifers, tartigrades, mites, and nematodes. In an English mire a single species of enchytraeid, *Cognettia sphagnetorum*, accounted for 70 to 75 percent of the total energy assimilated by the detritivores (Heal et al. 1975).

Rates of decomposition vary among bog vegetation. Litter of *Rubus chamaemorus* lost 2 percent of its weight in one year and 50 percent in three years. Shrub litter decomposes more slowly, while sphagnum decomposes hardly at all.

In an English mire (Moor House National Nature Reserve) the vegetational standing crop has an annual turnover of about 0.3 percent, and only about 1 percent of production is consumed by herbivores. Only 14 percent of the herbivore fauna is assimilated by predators. The remainder of the vegetation enters the decomposer food web, where about 5 percent is assimilated by decomposers and 10 percent passes below the water table. About 85 percent of the production is decomposed by microflora; but the rate of decomposition is slow, with 95 percent turnover time of about 3000 years. In the top 20 cm, 95 percent turnover time is about 70 years (Heal et al. 1975).

■ Summary

Two distinctive types of freshwater ecosystems are lentic or still water, and lotic or flowing water. Lentic ecosystems are lakes and ponds, standing bodies of water that fill a depression in the landscape. Geologically speaking, lakes and ponds are ephemeral features of the landscape. In time they fill in, draw smaller, and may finally be replaced by a terrestrial community.

A lake exhibits gradients in light, temperature, and dissolved gases, resulting in seasonal stratification. These gradients influence biological stratification. The area where light penetrates to the bottom of the lake, a zone called the littoral, is occupied by rooted plants. Beyond is the open water or limnetic zone, inhabited by plant and animal plankton and fish. Below the depth of effective light penetration is the profundal region, where the diversity of life varies with temperature and oxygen supply. The bottom or benthic zone is a place of intense biological activity, for here decomposition of organic matter takes place. Anaerobic bacteria are dominant on the bottom beneath the profundal water, whereas the benthic zone of the littoral is rich in decomposer organisms and detritus feeders. Although lake ecosystems are often considered as autotrophic systems dominated by phytoplankton and the grazing food web, lakes are strongly dependent on the detrital food web. Much of that detrital input comes from the littoral zone.

Lakes may be classified as eutrophic or nutrient-rich, oligotrophic or dystrophic, acidic and rich in humic material. Most lakes are subject to cultural eutrophication, which is the rapid addition of nutrients, especially nitrogen and phosphorus, from sewage and industrial wastes. Cultural eutrophication has produced significant biological changes, mostly detrimental, in many lakes.

Lotic ecosystems are characterized by inputs of detrital material from terrestrial sources and currents of varying velocities that carry nutrients and other materials downstream. Lotic ecosystems exhibit a continuum of physical and ecological variables from source to mouth. There is a longitudinal gradient in temperature, depth, and width of the channel, velocity of the current, and nature of the bottom. Changes in physical conditions are reflected in biotic structure. Headwater streams in forested regions are shaded and strongly heterotrophic, dependent on inputs of detritus. This detrital material is processed by a number of invertebrates, classified functionally as shredders, collectors, grazers, and gougers. Larger streams, open to sunlight, shift from heterotrophic to autotrophic. Primary production from algae and rooted aquatics becomes an important energy source. Large rivers return to a heterotrophic condition. They are dependent on fine particulate organic matter and dissolved organic matter as sources of nutrients and energy. Downstream systems depend upon the inefficiencies of energy and nutrient processing upstream.

Closely associated with lakes and streams are wetlands, areas where water is at, near, or above the level of the ground and occupied by hydrophytic vegetation. Wetlands dominated by grasses are marshes; those dominated by woody vegetation are swamps. Wetlands characterized by an accumulation of peat are mires. Mires fed by water moving through the mineral soil (rheotrophic) and dominated by sedges are fens; those dominated by sphagnum moss and dependent on precipitation for moisture and nutrients (ombrotrophic) are bogs. Bogs are characterized by blocked drainage, accumulation of peat, and low productivity. A significant portion of nutrients fixed in plants is removed from circulation and stored in accumulated peat. The stored energy remains locked in peat until environmental conditions change to favor decomposition or until the material is burned. In contrast, marsh ecosystems are heavily influenced by nutrients stored in the soil organic material. Marsh vegetation acts as a nutrient pump, drawing nutrients from the substrate, translocating them to the shoots, and depositing them on the surface or losing them through leaching to the water.

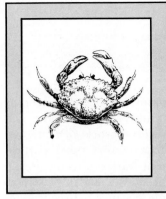

Marine Ecosystems

Freshwater rivers empty into the oceans, and terrestrial ecosystems end at the edge of the sea. For some distance there is a region of transition. Rivers enter the saline waters of the ocean, creating a gradient of salinity. That gradient provides a habitat for organisms uniquely adapted to the half-world between salt water and fresh. The coastal regions exposed to the open sea are inhabited by other organisms, able to live in the often severe environments dominated by tides. Beyond them lies the open ocean—shallow seas overlying continental shelves and then the deep oceans.

The marine environment occupies 70 percent of Earth's surface. The volume of surface area lighted by the sun is small compared to the volume of water. Dimness and the dilute solution of nutrients limit production. It is deep, in places nearly 7 kilometers. All of the seas are interconnected by currents, dominated by waves, influenced by tides, and saline (see Chapter 10).

Salinity imposes certain restrictions on life in the oceans (see Kinne 1971). Fish and marine invertebrates that inhabit marine, estuarine, and tidal environments have to maintain osmotic pressure under changing salinities. Most marine species are adapted to live in high salinities, and the number of species declines as salinity is reduced.

Another aspect of the marine environment is pressure. Pressure in the ocean varies from 1 atmosphere at the surface to 1000 atm at its greatest depth. Pressure changes are many times greater in the sea than in terrestrial environments and have a pronounced effect on the distribution of life. Certain organisms are restricted to surface waters, whereas others are adapted to pressure at great depths. Some marine organisms, such as sperm whales and certain seals, can dive to great depths and return to the surface without difficulty.

■ Tides

One of the fundamental laws of physics is Newton's law of universal gravitation. The law states that every particle of matter in the universe attracts every other particle with a force that varies directly as the product of their masses and inversely as the square of the distance between them. The gravitational pull of the sun and the moon each cause two bulges in the waters of the oceans. The moon, being much closer to Earth than the sun, exerts a tidal force twice as great as the sun's. The two bulges caused by the moon occur at the same time on opposite sides of Earth on an imaginary line extending from the moon through the center of the earth. The tidal bulge on the moon side is due to gravitational attraction; the bulge on the opposite side occurs because the gravitational force there is less than at the center of the earth. As Earth rotates eastward on its axis, the tides advance westward. Thus any given place on Earth will in the course of one daily rotation pass through two of the lunar tidal bulges, or high tides, and two of the lows or low tides, at right angles to the high. Because the moon revolves in a 29.5-day orbit around Earth, the average period between successive high tides is approximately 12 hours, 25 minutes.

The sun also causes two tides on opposite sides of Earth, but because they are less than half as high as the lunar tides, solar tides are usually partially masked by the lunar tides. Twice during the month, when the moon is full and when it is new, Earth, moon, and sun are nearly in line and the gravitational pulls of the sun and the moon are additive. The intensified force causes what are known as the spring tides, tides of the maximum rise and fall. When the moon is at either quarter, the gravitational pulls of the sun and moon interfere with each other, creating the neap tides of minimum difference between high and low tides.

Tides are not entirely regular, nor are they the same all over Earth. They vary from day to day in the same place, following the waxing and the waning of the moon. They may act differently in several localities within the same general area. In the Atlantic semidaily tides are the rule. In the Gulf of Mexico the alternate highs and lows more or less efface each other, and flood and ebb follow one another at about 24-hour intervals to produce one daily tide. Mixed tides, combinations of semidaily and daily tides, are common in the Pacific and Indian oceans, with different combinations at different places.

Local inconsistencies of tides are due to many variables. The elliptical orbits of Earth about the sun and of the moon about Earth influence the gravitational pull, as does the declination of the moon—the angle of the moon in relation to the axis of Earth. Latitude, barometric pressure, offshore and onshore winds, depth of water, contour of shore, and internal waves modify tidal movements.

■ Estuaries

Waters of all streams and rivers eventually drain into the sea; the place where this fresh water joins the salt is called an *estuary*. Estuaries are semienclosed parts of the coastal ocean where river water mixes with and measurably dilutes sea water (Ketchum 1983). Estuaries differ in size, shape, and volume of water flow, all influenced by the geology of the region in which they occur. As the river reaches the encroaching sea, the stream-carried sediments are dropped in the quiet water. They accumulate to form deltas in the up-

per reaches of the mouth and shorten the estuary. When silt and mud accumulations become high enough to be exposed at low tide, tidal flats develop, which divide and braid the original channel of the estuary. At the same time, ocean currents and tides erode the coastline and deposit material on the seaward side of the estuary, also shortening the mouth. If more material is deposited than is carried away, barrier beaches, islands, and brackish lagoons appear.

Structure

Current and salinity, both complex and variable, shape life in the estuary. Estuarine currents result from the interaction of a one-direction stream flow, which varies with the season and rainfall, with oscillating ocean tides and with the wind (Ketchum 1951, 1983; Burt and Queen 1951; Smayda 1983). Because of the complex nature of the currents, generalizations about estuaries are difficult to make (see Lauff 1967).

Salinity varies vertically and horizontally, often within one tidal cycle. Vertical salinity may be the same from top to bottom, or it may be completely stratified, with a layer of fresh water on top and a layer of dense saline water on the bottom. Salinity is homogeneous when currents, particularly eddy currents, are strong enough to mix the water from top to bottom. The salinity in some estuaries may be homogeneous at low tide, but unstable at high tide. As the tide floods, a surface wedge of seawater moves upstream more rapidly than the bottom water, creating a density inversion of salinity stratification. Seawater on the surface tends to sink as lighter fresh water rises, and mixing takes place from the surface to the bottom. This phenomenon is known as tidal overmixing. Strong winds, too, tend to mix salt water with the fresh (Barlow 1956) in some estuaries, but when the winds are still, the river water flows seaward on a shallow surface over an upstream movement of seawater that only gradually mixes with the salt.

Horizontally, the least saline waters are at the river entrance, and the most saline at the mouth of the estuary (Figure 33.1). The configuration of the horizontal zonation is determined mainly by the deflection caused by the incoming and outgoing currents (see Officer 1983). In all estuaries of the Northern Hemisphere, outward-flowing fresh water and inward-flowing seawater are deflected to the east because of Earth's rotation. As a result, salinity is higher on the western side.

Salinity also varies with changes in the quantity of fresh water pouring into the estuary through the year. Salinity is highest during the summer and during periods of drought, when less fresh water flows into the estuary. It is lowest during the winter and spring, when rivers and streams are discharging their peak loads. This change in salinity may happen rather

Figure 33.1 Vertical and horizontal stratification of salinity from the river mouth to the estuary at both low and high tide. At high tide the incoming seawater increases the salinity toward the river mouth; at low tide salinity is reduced. Note also how salinity increases with depth, because lighter fresh water flows over denser salt water.

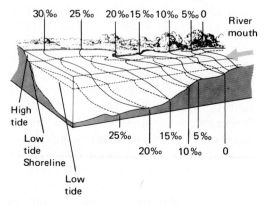

rapidly. For example, early in 1957 a heavy rainfall broke the most severe drought in the history of Texas. The resultant heavy river discharge reduced the salinities in Mesquite Bay on the central Texas coast by over 30 ppt in a two-month period. At the height of the drought salinities ranged from 35.5 to 50.0 ppt, but after the break in the drought they ranged from 2.3 to 2.9 ppt (Hoese 1960). Such rapid changes have a profound impact on the life of the estuary.

The salinity of seawater is about 35 ^{o}/oo; that of fresh water ranges from 0.065 to 0.30 ^{o}/oo. Because the concentration of metallic ions carried by rivers varies from drainage to drainage, the salinity and chemistry of estuaries differ. The proportion of dissolved salts in the estuarine waters remains about the same as that of seawater, but the concentration varies in a gradient from fresh water to sea.

Exceptions to these conditions exist in regions where evaporation from the estuary may exceed the inflow of fresh water from river discharge and rainfall (a negative estuary). This situation causes the salinity to increase in the upper end of the estuary, and horizontal stratification is reversed.

Temperatures in estuaries fluctuate considerably diurnally and seasonally. Waters are heated by solar radiation and inflowing and tidal currents. High tide on the mud flats may heat or cool the water, depending on the season. The upper layer of estuarine water may be cooler in winter and warmer in summer than the bottom, a condition that, as in a lake, will result in a spring and autumn overturn.

Mixing waters of different salinities and temperatures acts as a nutrient trap (Officer 1983). Inflowing river waters more often than not impoverish rather than fertilize the estuary, except for phosphorus. Instead, nutrients and oxygen are carried into the estuary by the tides. If vertical mixing takes place, these nutrients are not soon swept back out to sea, but circulate up and down among organisms, water, and bottom sediments (Figure 33.2).

Organisms inhabiting the estuary are faced with two problems—maintenance of position and adjustment to changing salinity (Vernberg 1983). The bulk of estuarine organisms are benthic and are securely attached to the bottom, are buried in the mud, or occupy crevices and crannies about sessile organisms. Motile inhabitants are chiefly crustaceans and fish, largely young of species that spawn offshore in high-salinity water. Planktonic organisms are wholly at the mercy of the currents. Because the seaward movement of stream flow and ebb tide transports plankton out to sea, the rate of circulation or flushing time determines the nature of the plankton population. If the circulation is too vigorous, the plankton population may be small. Phytoplankton in summer is most dense near the surface and in low-salinity areas. In winter phytoplankton is more uniformly distributed. For any planktonic growth to become endemic in an estuary, reproduction and recruitment must balance the losses from physical processes that disperse the population (Barlow 1955).

Changing salinity dictates the distribution of life in the estuary. Essentially, the organisms of the estuary are marine, able to withstand full seawater. Except for anadromous fishes no freshwater organisms live there. Some estuarine inhabitants cannot withstand lowered salinities, and these species decline along a salinity gradient. Sessile and slightly motile organisms have an optimum salinity range within which they grow best. When salinities vary on either side of this range, populations decline. Two animals, the clam worm and the scud, illustrate this situation. Two species of clam worm, *Nereis occidentalis* and *Neanthes succinea*, inhabit the estuaries of the southern coastal plains of

Figure 33.2 Circulation of fresh and salt water in an estuary creates a nutrient trap. Note a salt wedge of intruding seawater on the bottom, producing a surface flow of lighter fresh water and a counterflow of heavier brackish water. This countercurrent serves to trap nutrients, recirculating them toward the tidal marsh. The same countercurrent also sends phytoplankton up the estuary, repopulating the water. When nutrients are high in the upper estuary, they are taken up rapidly by tidal marshes and mud flats. These areas tend to trap particulate nitrogen and phosphorus, convert them to soluble forms, and export them back to open waters of the estuary. Plants on the tidal marshes and mud flats act as nutrient pumps between bottom sediments and surface water. (From Correll 1978.)

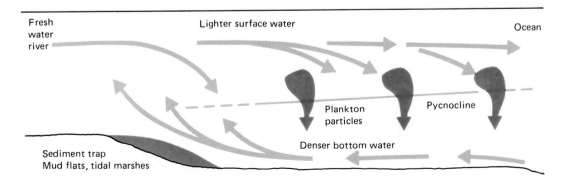

North America. *Nereis* is more numerous at high salinities and *Neanthes* at low salinities. In European estuaries the scud *Gammarus* is an important member of the bottom fauna. Two species, *G. locusta* and *G. marina,* are typical marine species and cannot penetrate far into the estuary. Instead they are replaced by a typical estuarine species, *G. zaddachi.* This species, however, is broken down into three subspecies, separated by salinity tolerances. *G. zaddachi* lives at the seaward end, *G. z. salinesi* occupies the middle, and *G. z. zaddachi,* which can penetrate up into fresh water for a short time, lives on the landward end (Spooner 1947; Segerstrale 1947).

Stage of development influences the range of motile species within the estuarine waters. This influence is particularly pronounced among estuarine fish. Species such as the striped bass spawn near the interface of fresh and low-salinity water (Figure 33.3). The larvae and young fish move downstream to more saline waters as they mature. Thus for the striped bass the estuary serves both as a nursery and feeding ground for the young. Anadromous species, such as the shad, spawn in fresh water; the young fish spend the first summer in the estuary, then move out to the open sea. Other species, such as the croaker, spawn at the mouth of the estuary, and the larvae are transported upstream to feed in the plankton-rich low-salinity areas. Others, such as the bluefish, move into the estuary to feed. In general, marine species drop out toward fresh water and are not replaced by freshwater forms. In fact, the mean number of species progressively decreases from the mouth of the estuary to upstream stations (H. W. Wells 1961).

Salinity changes often affect larval forms more severely than adults. Larval veligers of the oyster drill *Thais* succumb to low salinity more easily than the adults. A sudden influx of fresh water, especially after hurricanes or heavy rainfall, sharply lowers the salinity and causes a high mortality of oysters and their associates. When the drought-breaking heavy rainfall sharply reduced the salinities of Mesquite Bay in Texas, the highly salt-tolerant marine sessile and infaunal mollusks were completely wiped out. The high-salinity community of the oyster *Ostrea equestris* and the mussel *Brachidontes exustus* was replaced by the oyster *Crassotrea virginica* and the mussel *Brachidontes recurvus.* The rapid lowering of the salinity did not kill fish or other motile forms, which apparently moved out of the area (Hoese 1960).

The oyster bed and the oyster reef are the outstanding communities of the estuary. The oyster is the dominant organism about which life revolves. Oysters may be attached to every hard object in the intertidal zone or they may form reefs, areas where clusters of living oysters grow cemented to the almost buried shells of past generations. Oyster reefs usually lie at right angles to tidal currents, which bring planktonic food, carry away wastes, and sweep the oysters clean of sediment and debris.

Closely associated with oysters are encrusting organisms such as sponges, barnacles, and bryozoans, which attach themselves to oyster shells and are dependent

Figure 33.3 Relationship of a semianadromous fish, the striped bass, to the estuary. Adults live in the marine environment, but young fish grow up in the estuary. (From Cronin and Mansueti 1971.)

Fresh water   Estuarine   Marine

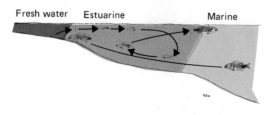

on the oyster or algae for food. The oyster crab strains food from the oyster's gills (Christensen and McDermott 1958), and a pramidellid snail lives an ectoparasitic life by feeding on body fluids and tissue debris from the oyster's mouth (Hopkins 1958). Beneath and between the oysters live polychaete worms, decapods, pelecypods, and a host of other organisms. In fact, 303 different species have been collected from an oyster bed (H. W. Wells 1961).

Function

Estuarine systems function on both plankton-based and detrital-based food webs. The producer component, particularly in the middle and lower estuary, consists of dinoflagellates and diatoms. The latter convert some of the carbon intake to high-caloric fats and lipids rather than low-energy carbohydrates typical of most green plants. This fat provides a high-energy food base for higher trophic levels.

Although inflowing waters from rivers and coastal marshes carry nutrients into the estuary and can stimulate an increase in phytoplankton production, phytoplankton production is regulated more by internal nutrient cycling than by external sources. This internal cycling involves excretion of mineralized nutrients by herbivorous zooplankton (see Chapter 12), and release of nutrients remineralized by invertebrates of the bottom sediments, by the roiling of sediments, and by steady-state exchanges between nutrients present in the particulate and dissolved phases (see Smayda 1983).

Nutrient buildup over winter in temperate estuaries stimulates a winter-spring bloom. As the nutrients become depleted and the phytoplankton experience intensive predation by zooplankton, the bloom collapses and falls to the bottom sediments, where it is fed upon by bivalves and other filter-feeding invertebrates (see Wolff 1983). In well-mixed estuaries nutrients remineralized in the benthos are released to the water column, stimulating a summer bloom.

Estuarine zooplankton, dominated by copepods (for ecology and life history, see Miller 1983), undergo their own seasonal fluctuations. Although expansive growth and subsequent declines in zooplankton populations can be associated with phytoplankton blooms, population dynamics are determined by many physical and biotic influences, including flushing rates of the estuary. The rate of increase for many zooplankton populations must balance the rate of loss from river flood and tidal flushing. Because of the unstable physical environment, zooplankton never have evolved species endemic to the estuary.

In shallow estuarine waters rooted aquatics such as widgeon grass and eelgrass (*Zostera marina*) assume major importance (for review see Phillips and McRoy 1980; Thayer, Adams and LaCroix 1984; Zieman 1984). They are complex systems supporting a large number of epiphytic and epizoic organisms. Such communities are important to vertebrate grazers such as brant, Canada geese, the black swan in Australia, and sea turtles, and provide a nursery ground for shrimp and bay scallops.

An example of energy flux in an estuary is provided by data for Pamlico River estuary in North Carolina (Copeland et al. 1974). The estuary is characterized by low salinity, high turbidity, and shallow water. The shallow water supports dense stands of widgeon grass, with associated attached algae and animals such as the scuds and grass shrimp. The benthos is dominated by the clams *Rangia* and *Macoma*. The euphotic zone, the upper 2 m, is dominated by dinoflagellates. Grazing on the phytoplankton are the zooplankters *(Acartia tonsa)* and the harpacticoid copepods that move up to the surface at night. Ctenophores crop each day 30 percent of the zooplankton population. At the same time considerable detrital material enters the estuarine system as dissolved organic matter and particulate organic matter. Because the phytoplankton depends heavily on organic detritus, the complex food web can be considered as primarily detrital. Both the detritus and the phytoplankton base support a number of trophic levels that lead eventually to fish and humans (Figure 33.4).

Figure 33.4 Simplified estuarine food web based on an estuary in the southeastern United States. The energy base is largely heterotrophic, supported by particulate matter and dissolved organic matter. (Adapted from Copeland et al. 1974.)

Rivers inflowing into estuaries carry along with the sediment a load of inland pollutants: domestic wastes, drainage from agricultural lands with nutrients and pesticides, and industrial effluents carrying toxic elements. Industrial, shipping, and housing developments along the estuarine coast add their own load of contaminants, alter flow of tidal water, and increase the anoxic condition of the estuarine bottom sediments. These conditions favor only a few bottom organisms, such as polychaete worms, and reduce the production of fish, clams, and oysters or make them inedible because of the accumulation of toxic elements and pesticides in their tissues.

■ Tidal Marshes

On the alluvial plains about the estuary and in the shelter of the spits and offshore

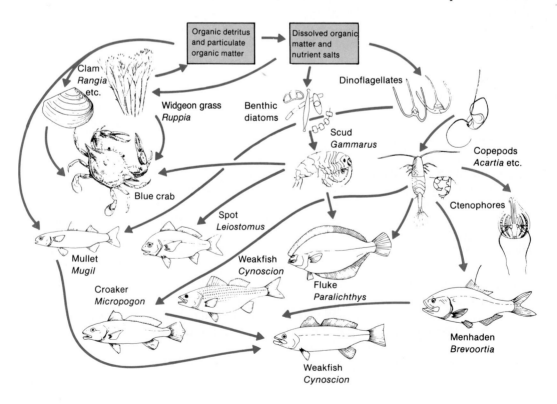

bars and islands exists a unique community, the tidal marsh (Figure 33.5). Although to the eye tidal marshes appear as waving acres of grass, they are a complex of distinctive and clearly demarked plant associations. The nature of the complex is determined by tides and salinity. The tides play the most significant role in plant segregation, for twice a day the salt marsh plants on the outermost tidal flats are submerged in salty water and then exposed to the full insolation of the sun. Their roots extend into poorly drained, poorly aerated soil, in which the soil solution contains varying concentrations of salt. Only plant species with a wide range of salt tolerance can survive such conditions. Thus from the edge of the sea to the highlands, zones of vegetation, each recognizable by its own distinctive color, develop.

Figure 33.5 A salt marsh on the Virginia Coast, showing the pattern of salt marsh vegetation. The mosaic is influenced by both water depth and salinity. The three dominant grasses are coarse-leaved salt marsh cordgrass *Spartina alternifolia*, the fine-leafed *Spartina patens*, salt marsh hay grass, and spike grass *Distichlis spicata*. (Photo by R. L. Smith.)

Structure

Plants. Tidal salt marshes begin in most cases as sands or mudflats, first colonized by algae and, if the water is deep enough, by eelgrass. As organic debris and sediments accumulate, eelgrass is replaced by the first salt marsh colonists—the sea poa *(Puccinellia)* on the European coast and saltwater cordgrass *(Spartina alterniflora)* on the eastern coast of North America. Stiff, leafy, up to 3 m tall, and submerged in salt water at every high tide, saltwater cordgrass forms a marginal strip between the open mudflat to the front and the higher grassland behind. As a wet grassland, the tall form of *Spartina alterniflora* is unique. No litter accumulates in the stand. Strong tidal currents sweep the floor of the *Spartina alterniflora* clean, leaving only black, thick mud.

As fine organic debris carried in and deposited by tides is buried by further deposition on top, an anaerobic environment develops. Here bacteria and nematodes live on organic matter, utilizing it by parallel oxidizing and reducing reactions (Teal and Kanwisher 1961). Such end products as methane, hydrogen sulfide,

and ferrous compounds accumulate. Increasing degrees of reduction suppress biological activity. In fact, if the bacteria of the mud are supplied with oxygen, their rate of energy degradation increases 25 times. Thus the tidal marsh is a vertically stratified system in which free oxygen is abundant in the surface and absent in the mud. Between these two extremes is a zone of diffusion and mixing of oxygen.

Spartina alterniflora is well adapted to grow on the intertidal flats of which it has sole possession. It has a high tolerance for salt water and is able to live in a semisubmerged state. It can live in a saline environment by selectively concentrating sodium chloride in its cells at a level higher than the surrounding salt water, thus maintaining its osmotic integrity. To rid itself of excessive salts, *Spartina alterniflora* has special salt-secreting cells in the leaves. Water excreted with the salt evaporates, leaving behind sparkling crystals on the surface of leaves to be washed off by tidal water. To get air to its roots buried in anaerobic mud, *Spartina alterniflora* has hollow tubes leading from the leaf to the root through which oxygen diffuses.

Across the marsh *Spartina alterniflora* has two or three growth forms, tall (100–300 cm), short (17–80 cm), and in places intermediate. Tall *Spartina* occupies that portion of the tidal marsh between mean low water and mean high water, mostly along tidal creeks (Figure 33.6). At the level of mean high water tall *Spartina* gives way sharply to short *Spartina*, yellowish, almost chloritic in appearance, contrasting sharply with the dark green tall form. The difference in growth has been the subject of considerable study and debate. Are the growth forms ecotypes, or are they ecophenes, plants exhibiting a plastic, phenotypic response to differences in tidal water inundation, nutrient availability, and salinity? The general consensus is that the

short forms are ecophenes, reflecting differences in plant-soil-microbial processes and interactions (Valiela, Teal, and Deuse 1978; Anderson and Treshow 1980; Haines and Dunn 1985). The short form reflects higher salinity of the high marsh, decreased input of nutrients, accumulation of toxic wastes, higher leaf temperatures because of decreased height growth, and other characteristics. Some electrophoresis studies show no genetic differences between the two growth forms (Shea, Warren, and Niering 1975), but tall and short *Spartina* subjected to a long-term garden transplant study under the same environmental conditions retained their morphological and physiological differences, suggesting a genetic difference (Gallagher et al. 1988).

The low marsh is monocultural, dominated by tall *Spartina alternifolia* that quickly fills in any disturbance. The high marsh with higher salinity, lower tidal exchange rates, a shorter, more open canopy, higher soil temperatures, and higher evaporation rates provides an opportunity for the growth of other plants. Here are the fleshy, translucent glassworts (*Salicornia spp.*) that turn bright red in the fall, sea lavender (*Limonium carolininum*), spearscale (*Atriplex patula*), and sea blite (*Suaeda maritima*).

Where the microelevation is about 5 cm above mean high water, short *Spartina alterniflora* and its associates are replaced by *Spartina patens,* salt marsh hay cordgrass, and an associate, spike grass, *Distichlis spicata. Spartina patens* is a fine, small grass that grows so densely and forms such a tight mat that few other plants can grow with it. Dead growth of the previous year lies beneath current growth, shield-

Figure 33.6 A stylized transect through part of a salt marsh, showing the relationship of plant distribution to microrelief and tidal submergence.

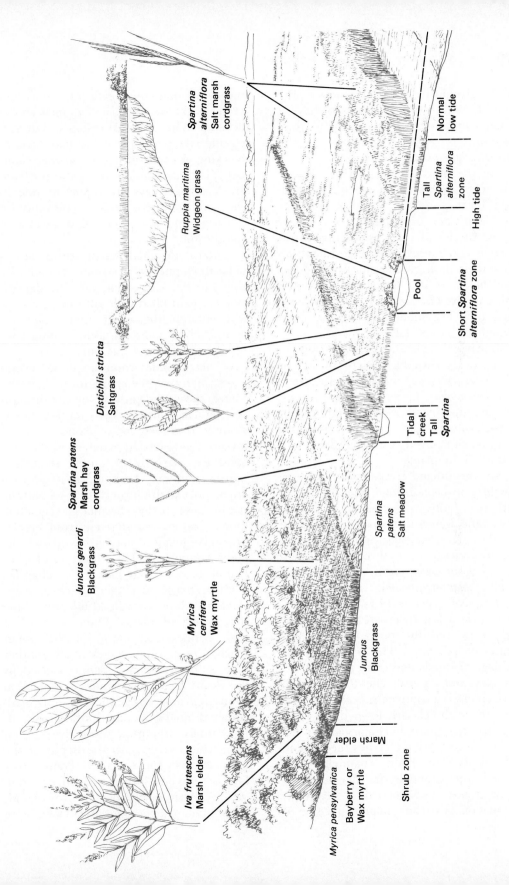

Iva frutescens
Marsh elder

Myrica cerifera
Wax myrtle

Juncus gerardi
Blackgrass

Spartina patens
Marsh hay
cordgrass

Distichlis stricta
Saltgrass

Ruppia maritima
Widgeon grass

*Spartina
alterniflora*
Salt marsh
cordgrass

Myrica pensylvanica
Bayberry or
Wax myrtle

Marsh elder

Juncus Blackgrass

*Spartina
patens*
Salt meadow

Tidal
creek
Tall
Spartina

Short *Spartina
alterniflora* zone

Pool

Tall
*Spartina
alterniflora*
zone

High tide

Normal
low tide

Shrub zone

ing the ground from the sun and keeping it moist. Where the soil tends to have a higher salinity or to be waterlogged *Spartina patens* is replaced or shares the site with *Distichlis spicata*. (Figure 33.6).

As the microelevation rises several more centimeters above mean high tide and if there is some intrusion of fresh water, *Spartina* and *Distichlis* may be replaced by two species of black grass *(Juncus roemerianus* and *Juncus gerardi)*, so called because their dark green color becomes almost black in the fall. Rarely are the rushes covered by ordinary high tides, but often they are submerged by the neap tides of spring and fall.

Beyond the black grass and often replacing it is shrubby growth of marsh elder *(Iva frutescens)* and groundsel *(Baccharis halimifolia)*. These shrubs tend to invade the high marsh where a slight rise in microelevation exists, but such invasions are often short-lived, as storm tides sweep in and kill the plants.

On the upland fringe grow bayberry *(Myrica pensylvanica)* and the pink-flowered sea hollyhock *(Hibiscus palustris)*. Where the water is fresh to brackish we find the reed *(Phragmites communis)*, spikerush *(Eleocharis spp.)*, three-square bullrush *(Scirpus americanus)*, and narrowleaf cattail *(Typha angustifolia)*.

Two conspicuous physiographic features of the salt marshes are the meandering creeks and the pond holes, called pannes or salt pans. The creeks are the drainage channels that carry the tidal waters back out to sea. The formation of these creeks is a complex process. In some cases the channels are formed by water deflected by minor irregularities on the surface. In estuarine marshes the river itself forms the main channel. Once formed, the channels are deepened by scouring and heightened by a steady accumulation of organic matter and silt. At

the same time the heads of the creeks erode backward and small branch creeks develop. Where lateral erosion and undercutting take place, the banks may cave in, blocking or overgrowing the smaller channels. The distribution and pattern of the creek system plays an important role in the drainage of the surface water and the drainage and movement of the water in the subsoil.

Across the tidal marsh are many circular to elliptical depressions. At high tide they are flooded; at low tide the depressions remain filled with salt water. If shallow enough the water may evaporate completely, leaving an accumulating concentration of salt on the mud.

These pannes come about in several ways. Many of them are formed as the marsh develops. Early plant colonization is irregular, and bare spots on the flat become surrounded by vegetation. As the level of the vegetated marsh rises, the bare spots lose their water outlet. If such a panne eventually becomes attached to a creek, normal drainage is restored and the panne eventually becomes vegetated. Other pannes are derived from creeks. Marsh vegetation may grow across the creek bottom and dam a portion of it, or lateral erosion may block the channel. With drainage no longer effective, water remains behind after flood tide, inhibiting the growth of plants. Often a series of such pannes may form on the upper reaches of a single creek. Still another type is the rotten-spot panne, caused by the death of small patches of vegetation from one cause or another, such as inadequate drainage or a concentration of salt.

Pannes support a distinctive vegetation, which varies with the depth of the water and salt concentration. Pools with a firm bottom and sufficient depth to retain tidal water support dense growths of widgeon grass *(Ruppia maritima)*, with long threadlike leaves and small, black, triangular seeds relished by waterfowl. The pools are usually surrounded by forbs such as sea lavender that add so much color to the marsh.

Shallow depressions in which water evaporates are covered with a heavy algal crust and crystallized salt. The edges of these "salt flats" may be invaded by *Salicornia, Distichlis,* or even *Spartina alterniflora.*

The exposed banks of tidal creeks that braid through the salt marshes support a dense population of mud algae, the diatoms and dinoflagellates, photosynthetically active all year. Photosynthesis is highest in summer during high tides and in winter during low tides when the sun warms the sediments (Pomeroy 1959). Some of the algae are washed out at ebb tide and become part of the estuarine plankton available to such filter feeders as oysters.

The salt marsh described is typical of the North American Atlantic Coast, but many variations exist locally and latitudinally around the world (Chapman 1977). North America has several distinctive types: Arctic salt marshes with few species, East and Gulf Coast marshes dominated by *Spartina* and *Juncus,* and Pacific Coast marshes, poorly developed and dominated by *Salicornia.* Some salt marshes on the western coast of Europe and in Britain are similar to those of the northeastern coast of North America, but great differences exist among them (see Chapman 1977). Tidal marshes of Europe and the east coast of North America form on a gently sloping continental coastal shelf. In western North America tidal marshes are confined to narrow river mouths, because the rivers flow directly onto a steep continental slope.

Although the species differ, *Spartina* is the dominant plant of salt marshes. From New England to New Jersey salt marshes exhibit a rather clear-cut zonation, with much of the area in high marsh, growing to *Spartina patens* and associated plants. From New Jersey to North Carolina tidal marshes exhibit less distinctive zonation and much of the high marsh is dominated by the short *Spartina alterniflora.* From North Carolina south the salt marsh reaches its best development on the heavy silt deposits with large expanses of tall *Spartina alterniflora* (for a detailed review, see Cooper 1974; Pomeroy and Wiegert 1981). On the west coast tidal land between mean sea level and mean high water is dominated by *Spartina foliosa,* and the land between mean high water and the highest tide is almost completely occupied by the perennial glasswort, *Salicornia virginica.*

Consumers. Animal life of the marsh, if not noted for its diversity, is certainly outstanding for its interest. Some of the inhabitants are permanent residents in sand or mud, others are seasonal visitors, and most are transients coming to feed at high or low tide.

Three dominant animals of the tall *Spartina alterniflora* are ribbed mussel (*Modiolus demissus*), fiddler crab (*Uca pugilator* and *Uca pugnax*), and marsh periwinkle (*Littorina spp.*) The marsh periwinkle, related to the periwinkles of the rocky shore, moves up and down the stems of *Spartina* and onto the mud as the tidal cycle changes. At low tide the periwinkle moves down the lower stems of *Spartina* and onto the mud to feed on algae and detritus.

Buried halfway in the mud is the ribbed mussel. At low tide the mussel is closed; at high tide the mussel opens to filter particles from the water, accepting some and rejecting others in a mucous ribbon known as pseudofeces.

Running across the marsh at low tide like a vast herd of tiny cattle are fiddler crabs. They earn their scientific name from their aggressive behavior and their common name from the highly developed single claw on the male. Among marsh animals fiddler crabs are the most adaptable. They have both gills and lungs. They can endure periods of high tides and cold winters without oxygen. They have a salt and water control system that enables them to move from diluted seawater to briny pools. The crabs are omnivorous feeders, consuming animal remains, plant remains, algae, and small animals. Fiddler crabs live in burrows, marked by mounds of freshly dug, marble-sized pellets. The burrowing activity of the crabs is similar to that of the earthworms because in overturning the mud they bring nutrients to the surface.

Prominent about the base of *Spartina* stalks and under debris are sandhoppers *(Orchestia)*. These detrital-feeding amphipods may be very abundant and are important in the diet of some of the marshland birds.

Two conspicuous vertebrate residents of the intertidal marsh of eastern North America are the diamond-backed terrapin *(Malaclemys terrapin)* and the clapper rail *(Rallus longirostris)*. The diamond-backed terrapin, which hibernates in the marsh mud, is carnivorous, feeding on fiddler crabs, small mollusks, marine worms, and dead fish. The rail finds its diet of fiddler crabs and sandhoppers along the marsh banks at ebb tide and in the tall grass. It builds its nest in *Spartina alterniflora,* keeping it just above the level of high tide. Less conspicuous is the seaside sparrow *(Ammospiza maritima)*, which, like the clapper rail, builds its nest just above the normal summer high-tide mark and feeds on the sandhoppers and other small invertebrates.

On the high marsh animal life changes almost as suddenly as the vegetation. The pulmonate marsh snail *Melampus* replaces the marsh periwinkle of the low marsh. At low tide the small, coffee-bean colored *Melampus* may be found by the thousands under the low grass where the humidity is high. Before high tide *Melampus* moves up the fine stalks of *Spartina patens*. Within the matted growth of *Spartina patens* is a maze of runways made by another high marsh inhabitant, the meadow mouse *(Microtus)*, which feeds heavily on the grass. Replacing the clapper rail and the seaside sparrow on the high marsh are the willet *(Catoptrophorus semipalmatus)* and the seaside sharp-tailed sparrow *(Ammospiza caudacuta)*.

Along the shrubby fringes of the marsh dense growths of marsh elder and groundsel give nesting cover for blackbirds and provide sites for heron rookeries. Remote stands of these shrubs support the nests of smaller herons and egrets, whereas the tall dead pines and man-made structures support the nests of the fish-eating osprey.

Low tide brings a host of predaceous animals onto the marsh to feed. Herons, egrets, gulls, terns, willets, ibis, raccoons, and others spread over the exposed marsh floor and the muddy banks of tidal creeks to feed. At high tide the food web changes as tide waters flood the marsh. Such fish as the killifish *(Fundulus heteroclitus)*, silversides *(Menidia menidia)*, and four-spined stickleback *(Apeltes quadracus)*, restricted to channel waters at low tide, spread over the marsh at high tide, as does the blue crab *(Callinectes sapidus)*.

Function

Comparative Studies. Two studies of salt marsh ecosystems, one done in Georgia by Teal (1962) and the other done in New England by Nixon and Oviatt (1973), provide contrasting insights on the functioning of salt marshes.

Teal confined his study to that area of salt marsh of which 42 percent was in short *Spartina alterniflora* and 58 percent was in the tall form. Average annual gross production of the marsh averaged about 34,600 kcal/m²/yr, resulting in an efficiency of about 6.1 percent (Figure 33.7). However, respiration was high, averaging 28,000 kcal/m²/yr, over 75 percent of gross production. It reduced net production to 1.4 percent of incoming solar radiation. Average net production for *Spartina alterniflora* was 6585 kcal/m²/yr (1600

Figure 33.7 Energy flow through a Georgia salt marsh. Note the importance of the detrital food web. (From Teal 1962.)

g/m²/yr), about 19 percent of gross production. In addition to *Spartina*, mud algae contributed 1620 kcal/m²/yr. Gross production amounted to 1800 kcal/m²/yr and respiration 180 kcal/m²/yr. Total net production for the marsh averaged over 8200 kcal/m²/yr (approximately 2000 g/m²/yr).

Feeding on *Spartina alterniflora* were two groups of grazing herbivores: the salt marsh grasshopper *(Orchelimum)*, which eats the plant tissues, and the plant hopper *(Prokelisis)*, which sucks plant juices. Neither harvested significant portions of the standing crop. Grasshoppers ingested about 3 percent of net production and assimilated 1 percent. Total annual net energy flow through the grasshoppers was

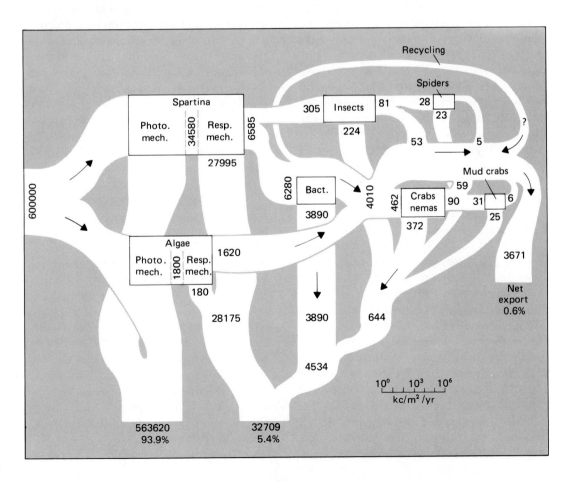

28 kcal, of which production was 10.8 kcal. Plant hoppers accounted for an energy flow of 275 kcal/m^2/yr, of which 70 kcal was production. Thus most of the *Spartina* production went to the detrital food chain.

Before *Spartina* is available to most detrital feeders it must be broken down by bacterial action. Part of the *Spartina*, particularly that in short growth, decomposes in place in the marsh. Most of the detritus from the tall *Spartina* is washed out to sea by tidal currents. Although bacterial decomposition consumes 82 percent of the organic matter, the remainder is enriched as animal food. Teal found that bacterial respiration accounted for 3890 kcal/m^2/yr or 59 percent of the available *Spartina*.

Feeding on decomposing detrital material and the bacteria it supports are fiddler crabs. Where population densities are high, fiddler crabs can sweep over the surface of the marsh between successive tides. They accounted for an assimilation of 206 kcal/m^2/yr. Mussels accounted for an assimilation of 56 kcal. Feeding on the herbivores were secondary consumers which largely went unstudied. The mud crab *(Eurytium)* assimilated 27 g/m^2/yr, of which 5.3 kcal was production.

Nixon and Oviatt (1973) studied a whole marsh system (16,800 m^2) and its embayment, Bissel Cove (6680 m^2). Tall *Spartina alterniflora* made up about 7 percent of the marsh, short *Spartina alterniflora* 78 percent, and high marsh 15 percent. The New England marsh was considerably less productive than the Georgia marsh. Production was 840 kcal/m^2/yr for the tall *Spartina alterniflora* and 432 g dry wt/m^2 for the short *Spartina*. Efficiency of production for the tall *Spartina* was 0.51 percent, one-half that of the tall *Spartina* of the Georgia marsh.

Growth in the New England marsh ceased in late fall (the Georgia marsh has some production in the winter) and dead grass remained until broken up by ice in late winter. Thus in New England marsh ice performed the task carried out in part by fiddler crabs in the Georgia marsh. Fiddler crabs are rare in New England salt marshes, as are the ribbed mussel and marsh snail, and there is no significant population of grasshoppers. Thus the entire production of the marsh goes into the detrital food chain.

The associated embayment supported dense, but patchy populations of widgeon grass *(Ruppia)* and sea lettuce *(Ulva)* that went through several periods of rapid growth, death, decay, and export. Maximum biomass of widgeon grass was twice that of tall *Spartina alterniflora*. Major consumers in the embayment were grass shrimp *(Palaemonetes pugio)* and fish, some 20 species including several mummichogs *(Fundulus)*, silversides, bluefish *(Pomatomus saltatrix)*, winter flounder, and stickleback. The fish were largely predatory; the grass shrimp were detritus-feeding herbivores. Feeding on fish were terns, gulls, black ducks, and mallards. The major fish predator, the herring gull, consumed less than 0.5 percent of the standing crop of fish. The ducks fed extensively on widgeon grass and sea lettuce.

The New England salt marsh is characterized by sharp seasonal changes in light, temperature, and salinity, all of which lower production. Most of the production of the salt marsh goes into embayment as detritus to enter a large sedimentary organic storage compartment as an energy source. Although the embayment is capable of considerable primary production over a short period of time, it is a semiheterotrophic system that depends upon the input of organic matter produced by the marsh grasses and algae. Over the year the annual energy budget for the embayment shows that consumption exceeds production.

The grass shrimp plays an important

role in the embayment system at Bissel Cove by feeding on detrital material washed in from the salt marsh. It plucks away at the surface of dead leaves and stems, breaking them into smaller pieces that are colonized by bacteria and diatoms. Inefficient assimilation repackages the detrital material into fecal pellets and adds large quantities of ammonia and phosphates to the water. The grass shrimp also prevents the buildup of detritus from widgeon grass and sea lettuce by feeding on it and reducing it to fine sediment. By its action on the detrital material the grass shrimp makes nutrients and biomass available for other trophic levels. Although the grass shrimp is eaten by the mummichogs, the shrimp's ability to live in a low oxygen environment limits predation and competition and allows its population to reach the high levels necessary to function as the major detritivore.

Production of west coast salt marshes and British salt marshes is comparable to that of east coast salt marshes. Production of *Spartina foliosa* on California marshes ranged from a high of 1200 to 1700 $g/m^2/yr$ (Cameron 1973) to 270 to 690 $g/m^2/yr$ (Mahall and Park 1976), with about 56 percent of primary production exported as detritus (Cameron 1973). Production of *Spartina townsendii* on British marshes ranged from 760 $g/m^2/yr$ (Ranwell 1961) to 840 $g/m^2/yr$ (Jefferies 1972). *Salicornia virginica* on California marshes produced 550 to 960 $g/m^2/yr$ (Mahall and Park 1976); production of the annual *Salicornia strictissima* on British marshes was comparable, amounting to 876 $g/m^2/yr$ (Jefferies 1972).

Synthesis. The salt marsh is one of the most productive ecosystems, exceeding most others, including intensive agriculture. The reason is a tidal subsidy. Tidal flushing brings in new nutrients, sweeps out accumulated salts, metabolites, sul-

fides, and toxic wastes, and replaces anoxic interstitial water with oxygenated water. Added to this advantage is tight internal nutrient cycling. Algal and bacterial populations turn over rapidly, and detrital material is fragmented and decomposed. Up to 47 percent of net primary production is respired by microbes; part is grazed by nematodes and microscopic benthic organisms, both of which are consumed by deposit feeders. Depending upon the nature of the salt marsh, between 11 and 66 percent of decomposed marsh grass is converted to microbial biomass. Belowground production enters an anaerobic food web of fermentation, reducing sulfates and producing methane.

Sulfur, present in tidal water, appears to be significantly involved in energy flow in the salt marsh (Howarth and Teal 1979; Howarth et al. 1983). In the anaerobic environment of the salt marsh soil bacteria convert the seawater sulfates to sulfites by oxidizing organic compounds. In doing so, bacteria trap a portion of the energy, with the remainder residing sulfide radicals. These stored sulfides are reoxidized over the year by oxygen diffusing from the roots of salt marsh plants and become available for further growth of sulfur-oxidizing bacteria. An estimated 70 percent of the aboveground net primary production in a New England salt marsh flows through reduced inorganic sulfur compounds, and an equivalent of 20 percent of aboveground net primary production is exported from the marsh as reduced sulfur compounds.

What happens to excess carbon production in tidal marsh is not well understood (Table 33.1). Each salt marsh apparently differs and what happens to its excess carbon production is influenced by its route of transformation though a food web, and its route and importance of export. Some salt marshes are dependent on tidal exchanges and import more than

TABLE 33.1 Simulated Annual Carbon Budget for the Duplin RIver Marshes, Sapelo Island, Georgia

Source or Process	Net Balance [a] (g C/m²/yr)	
Production		
S. alterniflora	1575	
Algae	131	
Total production		1706
Loss		
Respiration (CO₂ + CH₄)		
in soil	− 623	
in water	− 222	
in air	− 68	
	− 586	
Total loss		− 1499
Net change		207
Sedimentation	29	
Unexplained	178	

Note: All values prorated over the entire marsh system—soil, creeks, and Duplin River.
Source: Pomeroy and Wiegert 1981: 225.

they export (Woodwell et al. 1979), whereas others export more than they import (Valiela et al. 1978; Valiela and Teal 1979). Some of the excess production goes into sediments; some may be transformed microbially in the water in the marsh and tidal creeks. A portion may be exported to the estuary physically as detritus, as bacteria, or as fish, crabs, and intertidal organisms through the food web.

The importance of tidal marshes as a nutrient source and sink for the estuary has been a question of long standing. An insight into this relationship is partially provided by phosphorus and nitrogen cycling in the marsh.

The major flux of phosphorus in and out of salt marshes and estuarine waters is dissolved phosphate. Over the year input equals output, and usually phosphorus is surplus to the requirements in both soil and water. Excess phosphorus is mineralized and utilized by autotrophs and is

moved back through the system by heterotrophs.

Nitrogen is a different story (Table 33.2 and Figure 33.8). Denitrification often exceeds nitrogen fixation, and the marsh depends upon inputs into the system. The amount of nitrogen cycled is determined by tidal input, physical and chemical exchanges with air and water, and biological fluxes. In some salt marshes, exemplified by Great Sippewissett marsh in Massachusetts, groundwater inflow brings in nitrates, some of which percolates through the peat and is exported to the estuary as organic nitrogen, ammonium, and nitrates. Of the total influx, one-third is exported by denitrification and two-thirds by tides (Valiela et al. 1978). The much larger salt marshes of Sapelo Island, Georgia, depend upon inputs of nitrogen from associated river and tides. Only in summer does the system appear to export any nitrogen. Mostly the marsh is a net sink for nitrogen.

The nitrogen cycle in the salt marsh is not well understood, but a study of nitro-

TABLE 33.2 Provisional Nitrogen Budget for the Duplin River Watershed at Sapelo Island, Georgia

Flux Components	Nitrogen Flux g N/m²/yr
Inputs	
Rain	0.3
Sedimentation	3.3
Nitrogen fixation	14.8
Tidal exchange	46.6
Losses	
Denitrification	65.
Internal Cycles	
Primary production	70.
Soil remineralization	70.

Note: Tidal exchange is depicted as seeking a steady state, offsetting differences in nitrogen fixation and denitrification.
Source: Pomeroy and Wiegert 1981: 180.

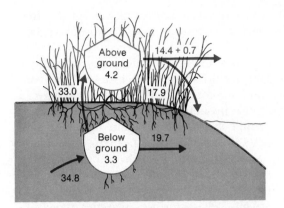

Figure 33.8 Nitrogen cycle (g/m^2/yr) in a Georgia salt marsh. Note both the degree of internal cycling and the amount exported to the mud and to the estuary. (From Hopkinson and Schubauer 1984: 966.)

gen cycling in *Spartina alternifolia* by Hopkinson and Schubauer (1984) in the Sapelo marshes, Georgia (Figure 33.8) provides some insight into the mechanisms. Aboveground concentration of total nitrogen was two times that below ground. The concentration was highest in young stems but it decreased with age. Maximum accumulation above ground was in midsummer and below ground in midwinter, when 83 percent of nitrogen was in roots and rhizomes. Total uptake of nitrogen was 34.8 g/m^2/yr. Total transfer of nitrogen from belowground to aboveground tissue was 33 g/m^2/yr. Forty-six percent of new nitrogen was taken up from the soil. Of N transferred to aboveground biomass, 14.4 g/m^2/yr was lost as detritus upon culm death, 0.7 g was leached from the living culm, and 17.9 g was translocated to rhizomes. The movement of N from aboveground to belowground biomass during the period of active growth and the dependence of new growth in spring on N stored in the rhizomes suggests strong nutrient conservation and recycling.

Maintenance of this cycling of N within *Spartina* may involve an interaction with the plant's animal associate the ribbed mussel, whose strong bissal threads bind sediments and prevent their erosion. Jordan and Valiela (1982) found that ribbed mussels pump tidal water in excess of the tidal volume of their New England salt marsh and deposited as much particulate matter in the form of feces and pseudofeces on the surface mud as was exported. This nitrogen is immediately available for *Spartina*. Investigating further, Bertness (1984) by manipulating density of ribbed mussels experimentally demonstrated that this invertebrate apparently stimulated production of *Spartina*. Increased height, biomass, and flowering of *Spartina* and soil nitrogen levels were positively associated with mussel density.

■ Mangrove Swamps

Replacing salt marshes on tidal flats in tropical regions are *mangrove swamps* or *mangal*. They develop in coastal areas where wave action is absent, sediments accumulate, and the muds are anoxic. The dominant plants are mangroves, which include 12 genera dominated by *Rhizophora*, *Avicennia*, *Brugivera*, and *Sonneratia*. Growing with them are other halophytic species, mostly shrubs. Mangrove trees are characterized by shallow, widely spreading roots, or by prop roots coming from the trunk (Figure 33.9). Many species have root extensions called *pneumatophores* that take in oxygen for the roots. The leaves, although tough, are often succulent and may have salt glands (see Chapter 10). Mangroves have a unique method of reproduction. Seeds germinate on the tree, grow into a seedling with no resting stage between, drop to the water, and float up-

Figure 33.9 Mangrove swamps replace tidal marshes in tropical regions. (Photo by R. L. Smith.)

right until they reach water shallow enough for their roots to penetrate the mud.

Structure

The formation and physiognomy of mangrove swamps is strongly influenced by the range and duration of tidal flooding and surface drainage (Lugo and Snedaker 1974). One of the features of this response is zonation, the changes in vegetation from seaward edge to true terrestrial environment. Although often used as an example, the mangrove swamps of the Americas, particularly Florida, have the least pronounced zonation, largely because of the few species involved. The pioneering red mangrove (*Rhizophora mangle*) occupies the seaward edge and experiences the deepest tidal flooding. Red mangroves are backed by pneumatophore-possessing black mangroves (*Avicennia germinanas*), shallowly flooded by high tides. The landward edge is dominated by white mangroves (*Laguncularia racemosa*) along with buttonwood (*Conocar-*

pus erectus), a nonmangrove species that acts as a transition to terrestrial vegetation.

In the Indo-Pacific region mangrove forests are much better developed, contain up to 30 to 40 species, and have a more pronounced zonation. The seaward fringe is dominated by one or several species of *Avicennia* and perhaps trees of the genus *Sonneratia,* which do not grow well in the shade of other mangrove species. Behind the *Avicennia* is a zone of *Rhizophora,* characteristic mangrove species with prop roots and pneumatophores, which grow in areas covered by daily high tide up to the point covered only by the highest spring tides. At and beyond the level of high spring tides is a broad zone of *Bruguiera.* The final and often indefinite mangrove zone is an association of small shrubs, mainly *Ceriops.*

Associated with mangroves are a mix of marine organisms that occupy prop roots and the mud, and those that live in the trees. As in the salt marsh, *Littorina* snails live on the roots and trunks of mangrove trees. Attached to stems and prop roots are barnacles and oysters. Fiddler crabs burrow into the mud during low tide and live on prop roots and high ground during high tide. In the Indo-Pacific mangrove swamps live mudskippers, fish of the genus *Periophthalmus* with modified eyes set high on the head. They live in burrows in the mud, spend time out of the water crawling about the mud, and in many ways act more like amphibians than fish. Herons and other wetland birds nest in the mangrove trees; and alligators, crocodiles, bears, pumas, and wildcats inhabit the forest interior.

Function

Net productivity of mangrove swamps is variable, ranging in Florida mangroves from about 450 g C/m^2/yr to 2700 g C/m^2/yr. Productivity is influenced by tidal inflow and flushing, water chemistry, salinity, and soil nutrients, much as in the salt marsh. Highest rates of productivity occur in the mangrove forests that are under the influence of daily tides (Lugo and Snedaker 1974; Careter et al. 1973). For example, gross primary productivity of red mangrove, flooded daily by high tides, decreases with increased salinity, whereas gross primary productivity of white and black mangroves increases with increasing salinity. In areas of intermediate salinity, white mangroves had twice the productivity of red mangroves. Zonation of mangrove swamps appears to reflect the optimal productivity niches of the species involved rather than physical or successional conditions (Lugo 1980). Like the salt marsh, mangrove swamps export a considerable portion, up to 50 percent, of their aboveground primary productivity to the adjacent estuary. This detrital input is important to the commercial and sport fisheries of the Gulf of Mexico (Odum 1970).

■ Rocky Shores

Where land and sea meet exists the fascinating and complex world of the seashore. Rocky, sandy, muddy, protected or exposed to the pounding of incoming swells, all shores have one feature in common: They are alternatingly exposed and submerged by the tides. Roughly, the region of the seashore is bounded by the extreme high-water mark and the extreme low-water mark. Within these confines, conditions change from hour to hour with the ebb and flow of the tides. At flood tide the seashore is a water world; at ebb tide it belongs to the terrestrial environment, with its extremes in temperature, moisture, and solar radiation. Nevertheless, the seashore inhabitants are essentially marine, adapted to withstand some degree of exposure to the air for varying periods of time.

Structure

As the sea recedes at ebb tide, life hidden by tidal water emerges, layer by layer. The uppermost layers, those near the high-water mark, are exposed to air, wide temperature fluctuations, intense solar radiation and desiccation for a considerable period of time, whereas the lowest fringes on the intertidal shore may be exposed only briefly before the flood tide submerges them again. These varying conditions result in one of the most striking features of the rocky shore, the zonation of life (Figure 33.10). Although this zonation may be

Figure 33.10 The broad zones of life exposed at low tide on the rocky shore of the Bay of Fundy. Note the heavy growth of *Fucus* on the lower portion and the white zone of barnacles above. (Photo by R. L. Smith.)

strikingly different from place to place as a result of local variations in aspect, substrate, wave action, light intensity, shore profile, exposure to prevailing winds, climatic differences, and the like, it possesses everywhere the same general features. All rocky shores have three basic zones, characterized by the dominant organisms occupying them (Figure 33.11).

Where the land ends and seashore begins is difficult to fix. The approach to a rocky shore from the landward side is marked by a gradual transition from lichens and other land plants to marine life, dependent in part at least on the tidal waters (Figure 33.12). The first major change from land shows up on the *supralittoral fringe*, where the salt water comes only every fortnight on the spring tides. It is marked by the black zone, a patchy or beltlike encrustation of Verrucaria-type li-

chens and Myxophyceae algae such as *Calothrix* and *Entrophysalis*. Capable of existing under conditions so difficult that few other plants could survive, these blue-green algae, enclosed in slimy, gelatinous sheaths, and their associated lichens represent an essentially nonmarine community, on which graze basically marine animals, the periwinkles (Doty, in Hedgpeth 1957). Common to this black zone is the rough periwinkle that grazes on the wet algae covering the rocks. On European shores lives a similarly adapted species, the rock periwinkle, the most highly resistant to desiccation of all the shore animals.

Below the black zone lies the *littoral zone,* a region covered and uncovered daily by tides. The littoral tends to be divided into subzones. In the upper reaches, barnacles are most abundant. The oyster, blue mussel, and limpet appear in the middle and lower portions of the littoral, as does the common periwinkle.

Occupying the lower half of the littoral zone (midlittoral) of colder climates and in places overlying the barnacles is an ancient group of plants, the brown algae, more commonly known as rockweeds, or wrack (*Fucus spp.*) (Figure 33.13). Rockweeds attain their finest growth on protected shores, where they may grow 2 m long; on wave-whipped shores they are considerably shorter.

The lower reaches of the littoral zone may be occupied by blue mussels (Figure 33.14) instead of rockweeds, particularly on shores where hard surfaces have been covered in part by sand and mud. No other shore animal grows in such abundance; the blue-black shells packed closely together may blanket the area.

Near the lower reaches of the littoral zone, mussels may grow in association with the red alga *Gigartina*, a low-growing, carpetlike plant. Algae and mussels together often form a tight mat over the

Figure 33.11 Basic or universal zonation on a rocky shore. Use this diagram as a guide when studying the subsequent drawings of zonation and discussion in the text. (Adapted from Stephenson 1949.)

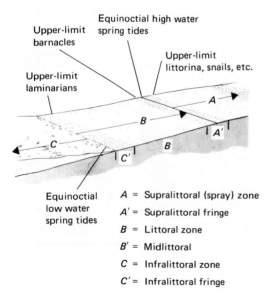

A = Supralittoral (spray) zone
A' = Supralittoral fringe
B = Littoral zone
B' = Midlittoral
C = Infralittoral zone
C' = Infralittoral fringe

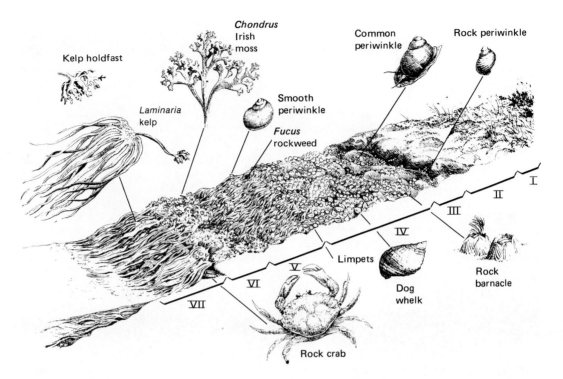

Figure 33.12 Zonation on a rocky shore along the North Atlantic. Compare Figure 33.11. I, land: lichens, herbs, grasses; II, bare rock; III, zone of black algae and rock periwinkles; IV, barnacle zone: barnacles, dog whelks, common periwinkles, mussels; V, fucoid zone: rockweed *(Fucus)* and smooth periwinkles; VI, *Chondrus* zone: Irish moss; VII, kelp *(Laminaria)* zone. (Based on data from Stephenson and Stephenson 1954, author's photographs and observations.)

rocks. Here, well protected in the dense growth from waves, live infant starfish, sea urchins, brittle stars, and bryozoan sea mats or sea lace *(Membranipora).*

The lowest part of the littoral zone, uncovered only at the spring tides, and not even then if wave action is strong, is the *infralittoral fringe.* This zone, exposed for short periods of time, consists of forests of the large brown alga *Laminaria,* one of the kelps, with a rich undergrowth of smaller plants and animals among the holdfasts.

Beyond the infralittoral fringe is the *sublittoral zone,* the open sea. This zone is principally neritic and benthic and contains a wide variety of fauna, depending upon the substrate, the presence of protruding rocks, gradients in turbulence, oxygen tensions, light, and temperature.

Although waves, tides, exposure, dessication, and temperature influence the intertidal zonation, the pattern of life on rocky shores is heavily influenced, too, by biotic interactions of grazing, predation, and competition, and by larval settlement (see Underwood, Denley, and Moran 1983). Where wave action is heavy on New England intertidal rocky shores, periwinkles are rare, which permits more vigorous growth of algae. Lack of grazing favors ephemeral algal species such as *Ulva* and *Enteromorpha,* whereas grazing allows

Figure 33.13 Knotted wrack *(Ascophyllum nodosum)* exposed at low tide on the Bay of Fundy. (Photo by R. L. Smith.)

Figure 33.14 A tidal flat covered with blue mussels *(Mytilus edulis)*. (Photo by R. L. Smith.)

the perennial *Fucus* to become established (see Lubchenco 1980, 1983; Chapter 26). On the New England coast the mussel *Mytilus edulis* outcompetes barnacles and algae; but predation by the starfish *Asterias spp.* and the snail *Nucella lapillus* prevents dominance by mussels except on the most wave-beaten areas. A similar situation exists on the Pacific coast. Barnacles of several species tend to outcompete and displace algal species, but in turn the competitively dominant mussel *M. californianus* destroys the barnacles by overgrowing them (Dayton 1971). Where present the predatory starfish *Pisaster ochraceus* prevents the mussel from completely overgrowing barnacles (Paine 1966). The result of such interactions of the physical and the biotic is a patchy distribution of life across the rocky intertidal shore (for a good overview see Nybakken 1988: 255–283).

Function

The rocky shore is both autotrophic and heterotrophic. Many organisms, such as barnacles, depend upon tides to bring them food; others such as periwinkles, graze on algal growth on the rocks. In fact, the functioning of the rocky seashore ecosystem involves complex interactions between physical and biotic aspects of the ecosystem. Like salt marshes, which receive a daily energy subsidy from tidal flooding, rocky shores receive an energy subsidy, although indirectly, from the waves. Leigh and associates (1987) estimate that wave-generated energy impinging on a rocky shore is roughly twice that of solar radiation, amounting to 0.045 watts/cm^2 compared to 0.017 to 0.025 watts/cm^2 of solar radiation. Energy input is reflected in productivity, which in the kelp beds of the Pacific Northwest exceeds that of the tropical rain forest. Part of this productivity relates to the much higher leaf surface per square meter—about 20 m^2/m^2 of growing surface compared to about 8 m^2 for a tropical forest. The highest standing crops of kelp, mussels, and other organisms were restricted to areas receiving the heaviest wave action.

Obviously the intertidal organisms do not use the wave energy directly. Rather the force of the waves favorably affects the environmental conditions that promote increased productivity. For one, heavy wave action reduces the activity of such predators of sessile intertidal invertebrates as starfish and sea urchins (see Chapter 20). Waves bring in a steady supply of nutrients and carry away products of metabolism. And they keep in constant

motion the fronds of various seaweeds, moving them in and out of shadow and sunlight, allowing for more even distribution of incident light and thus more efficient photosynthesis. By dislodging organisms (both plants and invertebrates, from the rocky substrate), waves open up space for colonization by algae and invertebrates and reduce strong interspecific competition. In effect, disturbance (see Chapter 26), which influences community structure, is the root of intertidal productivity.

Tide Pools

The ebbing tide leaves behind pools of water in rock crevices, in rocky basins, and in depressions (Figure 33.15). They are tide pools, "microcosms of the sea," as Yonge (1949) describes them. They represent distinct habitats, which differ considerably from the exposed rock and the open sea, and even differ among themselves. At low tide all the pools are subject to wide and sudden fluctuations in temperature and salinity, but these changes are most marked in shallow pools. Under the summer sun the temperature may rise above the maximum many organisms can tolerate. As the water evaporates, especially in the smaller and more shallow pools, salinity increases and salt crystals may appear around the edges. When rain or land drainage brings fresh water to the pool, salinity may decrease. In deep pools such fresh water tends to form a layer on the top, developing a strong salinity stratification in which the bottom layer and its inhabitants are little affected. If algal growth is considerable, oxygen content of the water varies through the day. Oxygen will be high during the daylight hours but will be low at night, a situation that rarely occurs at sea. The rise of carbon dioxide at night means a lowering of pH.

Obviously, pools near low tide are influenced least by the rise and fall of the

(a)

(b)

Figure 33.15 Tidal pools. (a) A rock pool nestles in a canyon on a rocky shore. (b) A small rock pool on the ledge of a large rock. Note the rockweed and barnacles and the transparency of the water. (Photos by R. L. Smith.)

tides; those that lie near and above the high-tide line are exposed the longest and undergo the widest fluctuations. Some may be recharged with seawater only by the splash from breaking waves or occasional high spring tides. Regardless of their position on the shore, most pools suddenly return to sea conditions on the rising tide and experience drastic and instantaneous changes in temperature, salinity, and pH. Life in the tidal pools must be able to withstand wide and rapid fluctuations in the environment.

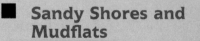

■ Sandy Shores and Mudflats

Both the sandy shore and the mudflat at low tide appear barren of life, a sharp

contrast to the life-studded rocky shores; but beneath the wet and glistening surface life exists, waiting for the next high tide.

Both in some ways are harsh environments, but the sandy shore is especially so. The very matrix of this seaside environment is a product of the harsh and relentless weathering of rock, both inland and along the shore. Through eons the ultimate products of rock weathering are carried away by rivers and waves to be deposited as sand along the edge of the sea. The size of the sand particles deposited influences the nature of the sandy beach, water retention during low tide, and the ability of animals to burrow through it. Beaches with relatively steep slopes usually are made up of larger sand grains and are subject to more wave action. Beaches exposed to high waves are generally flattened, for much of the material is transported away from the beach to deeper water, and fine sand is left behind (Figure 33.16). Sand grains of all sizes, especially the finer particles in which capillary action is greatest, are more or less cushioned by a firm of water about them, reducing further wearing away.

Figure 33.16 A long stretch of sandy beach washed by waves on Maui, Hawaiian Islands. Although the beach appears barren, life is abundant beneath the sand. (Photo by R. L. Smith.)

The retention of water by the sand at low tide is one of the outstanding environmental features of the sandy shore.

Structure

Life on the sand is almost impossible. Sand provides no surface for attachments of seaweeds and their associated fauna; and the crabs, worms, and snails so characteristic of rock crevices find no protection here. Life, then, is forced to exist beneath the sand.

Life on sandy and muddy beaches consists of the *epifauna*, organisms living on the surface, and the *infauna*, organisms living within the substrate. Most infauna either occupy permanent or semipermanent tubes within the sand or mud or are able to borrow rapidly into the substrate. Multicellular infauna obtain oxygen either by gaseous exchange with the water through their outer covering or by breathing through gills and elaborate respiratory siphons.

Within the sand and mud live vast numbers of *meiofauna* with a size range between 0.5 mm and 62 μm, including copepods, ostracods, nematodes, and gastrotrichs. Interstitial fauna are generally elongated forms with setae, spines, or tubercles greatly reduced. The great majority do not have pelagic larval stages. These animals feed mostly on algae, bacteria, and detritus. Interstitial life, best developed on the more sheltered beaches, shows seasonal variations, reaching maximum development in summer months.

Sandy beaches also exhibit zonation related to tidal influences (Figure 33.17), but it must be discovered by digging. Sandy and muddy shores can be divided roughly into supralittoral, littoral, and infralittoral zones, based on animal organisms, but a universal pattern similar to that of the rocky shore is lacking. Pale, sand-colored ghost crabs and beach hop-

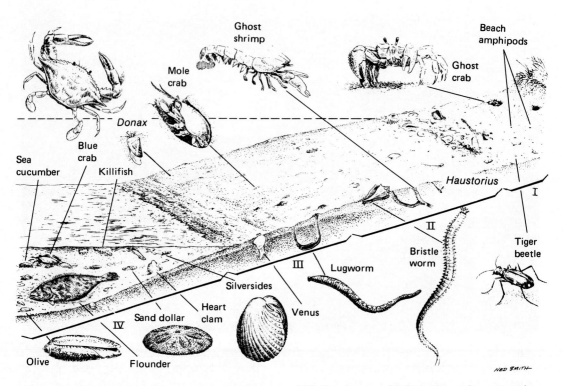

NED SMITH

Figure 33.17 Life on a sandy ocean beach along the Atlantic Coast. Although strong zonation is absent, organisms still change on a gradient from land to sea. I, supratidal zone: ghost crabs and sand fleas; II, flat beach zone: ghost shrimp, bristle worms, clams; III, intratidal zone: clams, lugworms, mole crabs; IV, subtidal zone. The dashed line indicates high tide.

pers occupy the upper beach, the supralittoral. The intertidal beach, the littoral, is a zone where true marine life appears. Although sandy shores lack the variety found on rocky shores, the populations of individual species of largely burrowing animals often are enormous. An array of animals, among them starfish and the related sand dollar, can be found above the low-tide line and in the infralittoral.

Organisms living within the sand and mud do not experience the same violent fluctuations in temperature as these on the rocky shores. Although the surface temperature of the sand at midday may be

10° C or more higher than the returning seawater, the temperature a few inches below remains almost constant throughout the year. Nor is there a great fluctuation in salinity, even when fresh water runs over the surface of the sand. Below 25 cm, salinity is little affected.

Function

For life to exist on the sandy shore, some organic matter has to accumulate. Most sandy beaches contain a certain amount of detritus from seaweeds, dead animals, feces, and material blown in from shore. This organic matter accumulates within the sand, especially in sheltered areas. In fact, an inverse relationship exists between the turbulence of the water and the amount of organic matter on the beach, with accumulation reaching its maximum on the mudflats. Organic matter clogs the space between the grains of sand and binds them together. As water moves down through the sand, it loses oxygen from both the respiration of bacteria and the oxidation of chemical substances, especially ferrous compounds. The point within the mud or sand at which water loses all its oxygen is a region of stagnation and oxygen deficiency, characterized by the formation of ferrous sulfides. The iron sulfides cause a zone of black whose depth varies with the exposure of the beach. On mudflats such conditions exist almost to the surface.

The energy base for sandy beach and mudflat fauna is organic matter. Much of it becomes available through bacterial decomposition, which goes on at the greatest rate at low tide. The bacteria are concentrated around the organic matter in the sand, where they escape the diluting effects of water. The products of decomposition are dissolved and washed into the sea by each high tide, which, in turn, brings in more organic matter for decomposition. Thus the sandy beach is an important site for biogeochemical cycling, supplying offshore waters with phosphates, nitrogen, and other nutrients.

At this point energy flow in sandy beaches and mudflats differs from that in terrestrial and aquatic systems because the basic consumers are bacteria. In other energy flow systems bacteria act largely as reducers responsible for conversion of dead organic matter into a form that can be utilized by producer organisms. In sandy beaches and mudflats bacteria not only feed on detrital material and break down organic matter, but they are also a major source of food for higher level consumers.

A number of deposit-feeding organisms ingest organic matter largely as a means of obtaining bacteria. Prominent among them on the mudflats are numerous nematodes and copepods (Harpacti-coida), the polychaete clam worm *(Nereis)*, and the gastropod mollusk *(Hydrobia)*. Deposit-feeders on sandy beaches obtain their food by actively burrowing through the sand and ingesting the substrate to obtain the organic matter it contains. The most common among them is the lugworm *(Arenicola)*, which is responsible for the conspicuous coiled and cone-shaped casts on the beach.

Other sandy beach animals are filter feeders, obtaining their food by sorting particles of organic matter from tidal water. Two of these "surf fishers" who advance and retreat up and down the beach with the flow and ebb of tide are the mole crab and coquina clam.

Associated with these essentially herbivorous animals are the predators, always present whether the tide is in or out. Near and below the low tide line live predatory gastropods, which prey on bivalves beneath the sand. In the same area lurk predatory portunid crabs such as the blue crab and green crab, which feed on mole crabs, clams, and other organisms. They move back and forth with the tides. The incoming tides also bring other predators, such as killifish and silversides. As the tide recedes, gulls and shorebirds scurry across the sand and mudflats to hunt for food.

Because of their dependence upon imported organic matter and their essentially heterotrophic nature, the sandy beaches and mudflats should be considered not as separate ecosystems, but as a part of the whole coastal ecosystem (Figure 33.18). Except in the cleanest of sands, some primary production does take place in the intertidal zone. The major primary producers are the diatoms, confined mainly to fine grain deposits of sand containing a high proportion of organic matter. Productivity is low; one estimate places productivity of moderately exposed sandy beaches at 5 g C/m^2/yr. Production

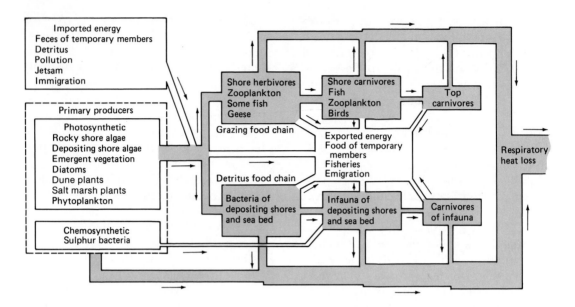

Figure 33.18 The coastal ecosystem, a supra-ecosystem consisting of the shore, the fringing terrestrial regions, and the sublittoral zones. It involves two food webs: (1) that of the rocky shore, with its algae, herbivores, and zooplankton; and (2) the detrital food webs, involving the bacteria of the depositing shore and sublittoral muds and the dependent detritivores and carnivores. Coastal ecosystems are extremely productive; because energy imports exceed exports, the system is continuously gaining energy. (From S. K. Eltringham 1971.)

may be temporarily increased by phytoplankton carried in at high tide and left stranded on the surface. Again these organisms are mostly diatoms. More important as producers are the sulfur bacteria in the black sulphide or reducing layer. These chemosynthetic bacteria use energy released as ferrous oxide is reduced. Some mudflats are covered with the algae *Enteromorpha* and *Ulva*, whose productivity can be substantial.

In effect, the shore and mudflats are part of a larger coastal ecosystem involving the salt marsh, the estuary, and coastal waters. They act as sinks for energy and nutrients because the energy they utilize comes not from primary production, but from organic matter that originates outside of the area. Many of the nutrient cycles are only partially contained within the borders of the shores.

■ The Open Sea

Beyond the rocky and sandy shores lies the open sea (Figure 33.19). Compared to terrestrial ecosystems, the pelagic ecosystem lacks well-defined communities, largely because it lacks the supporting structures and framework of large dominant plant life. Nevertheless, differences based on physical characteristics and life forms permit a division of the ocean into different regions. The Arctic Ocean comprises marine waters that lie north of the land masses in the Northern Hemisphere and are open only to the Atlantic Ocean. The Southern or Antarctic Ocean lies about the continent of Antarctica and is open to three oceans, the Atlantic, Pacific,

Figure 33.19 The open sea is the realm of fishes, dolphins, whales, and sea birds that rarely visit land. (Photo by R. L. Smith.)

and Indian. Warm oceanic waters making up the Atlantic and Pacific have some of their own distinctive communities. The Deep Sea benthic ecosystems are quite distinct from the lighted waters. Other important distinctive marine ecosystems include the coral reefs (see page 919), and upwelling systems off the coasts of California, Peru, Northwest Africa, Southwest Africa, India, and Pakistan. Also important and distinctive are the shelf-sea ecosystems; shallow, productive, and nutrient-rich, they support a diversity of fish and invertebrate marine life.

Just as lakes exhibit zonation, so do the seas. The ocean is divided into two main regions (Figure 33.20), the *pelagic* or whole body of water, and the *benthic* or bottom. The pelagic is further subdivided into the *neritic province*, or water that overlies the continental shelf, and the *oceanic province*. Because conditions change with depth, the pelagic is divided into a number of vertical layers or zones. From the surface to about 200 m is the *photic zone*, in which there are sharp gradients in light, temperature, and salinity. From 200 to 1000 m is the *mesopelagic* region, where little light penetrates and where the temperature gradient is more even, gradual, and without seasonal fluctuations. In the tropical region the lower boundary is considered as the 10° C isotherm. It contains an oxygen-minimum layer and often the maximum concentrations of nitrate and phosphate. Below the mesopelagic is the *bathypelagic*, falling between the 10° and 4° C isotherms, and depending upon its global locations, lying between 700–100 m and 2000–4000 m. Lying over the major plains of the ocean is the *abyssal pelagic* or *benthipelagic* zone, down to about 6000 m. Water of the deep oceanic trenches is called the *hadalpelagic*. The sea bottom is called the *benthic* region. It, too, can be divided into three zones, which differ in the nature of their bottom life. The *bathyal* zone covers the continental shelf and down to 4000 m. The *abyssal* zone embraces the broad ocean plains down to 6000 m, and the *hadal* zone is the bottom of the deep oceanic trenches. The benthic zone underlying the neritic zone of the continental shelf is the *sublittoral* or *shelf* zone. The intertidal zone, like the margins of lakes, is called the *littoral*.

Structure

Because the global regions vary from oligotrophic to eutrophic and from cold to tropical conditions, a brief discussion of structure must be general and emphasize the major groups. (For a good introduction to the structure and function of marine ecosystems see Nybakken 1988.)

Phytoplankton. As in lakes, the dominant form of plant life or primary producers in the open sea is phytoplankton. Seawater is so dense that oceanic phytoplankton do not need well-developed sup-

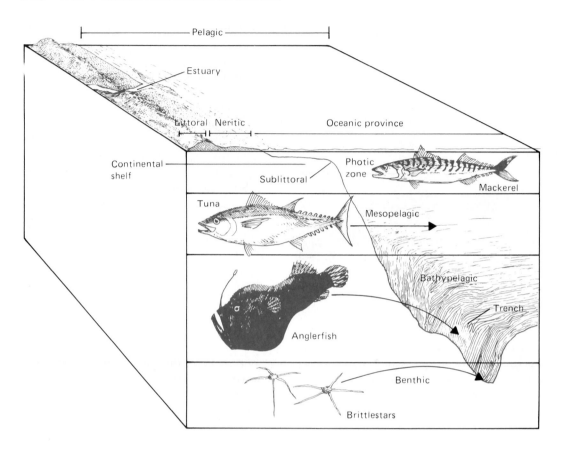

Figure 33.20 Regions of the ocean.

porting structures. In coastal waters and areas of upwelling phytoplankton 100 microns or more in diameter may be common, but in general plant life is much smaller and widely dispersed.

Because of its requirement for light, phytoplankton is restricted to the upper surface waters. The depth of its occurrence is determined by the depth of light penetration and so may range from tens to hundreds of meters. Because of seasonal, annual, and geographic variations in light, temperature, nutrients, and grazing by zooplankton, the distribution and composition of phytoplankton change with time and place.

Each ocean or region within an ocean appears to have its own dominant forms. Littoral and neritic waters and regions of upwelling are richer in plankton than midoceans. In regions of downwelling, dinoflagellates concentrate near the surface in areas of low turbulence. They attain their greatest abundance in warmer waters. In summer they may so concentrate in the surface waters that they color it red or brown. Often toxic to other marine life, such concentrations of dinoflagellates are responsible for red tides.

In regions of upwelling the dominant forms of phytoplankton are diatoms. Enclosed in a silica case, diatoms are particularly abundant in arctic waters.

Smaller than diatoms are the Coccolithophoridae, so small they pass through plankton nets (and so are classified as nan-noplankton). Their minute bodies are protected by calcareous plates or spicules embedded in a gelatinous sheath. Universally distributed in all waters except the polar seas, the Coccolithophoridae possess the ability to swim. They are characterized by droplets of oil, which aid in buoyancy and storage of food.

In the equatorial currents and in shallow seas the concentration of phytoplankton is variable. Where both lateral and vertical circulation of water is rapid, the composition reflects in part the ability of the species to grow, reproduce, and survive under local conditions.

Zooplankton. Grazing on the phytoplankton is the herbivorous zooplankton, consisting mainly of copepods, planktonic arthropods that are the most numerous animals of the sea, and the shrimplike euphausiids, commonly known as krill. Other planktonic forms are the larval stages of such organisms as gastropods, oysters, and cephalopods. Feeding on the herbivorous zooplankton is the carnivorous zooplankton, which includes such organisms as the larval forms of comb jellies (Ctenophora) and arrowworms (Chaetognatha).

Because its food in the ocean is so small and widely dispersed, zooplankton has evolved ways more efficient and less energy-demanding to harvest phytoplankton than filtering water through pores. It has webs, bristles, rakes, combs, cilia, sticky structures, and even bioluminescence.

The composition of zooplankton, like that of phytoplankton, varies from place to place, season to season, year to year. In general zooplankton falls into two main groups, the larger forms characteristic of shallow coastal waters and the smaller forms characteristic of the deeper open

ocean. Zooplankton forms of the continental shelf contain a large proportion of larvae of fish and benthic organisms. They have a greater diversity of species, reflecting a greater diversity of environmental and chemical conditions. The open ocean, being more homogeneous and nutrient-poor, supports less diverse zooplankton. In polar waters zooplankton species spend the winter in a dormant state in the deep water and rise to the surface during short periods of diatom blooms to reproduce. In temperate regions distribution and abundance depend upon temperature conditions. In tropical regions, where temperature is nearly uniform, zooplankton is not so restricted, and reproduction occurs throughout the year.

Also like phytoplankton, zooplankton lives mainly at the mercy of the currents; but many forms possess sufficient swimming power to exercise some control. Most species of zooplankton migrate vertically each day to arrive at a preferred level of light intensity. As darkness falls, zooplankton rapidly rises to the surface to feed on phytoplankton. At dawn the forms move back down to preferred depths.

By feeding in the darkness of night and hiding in the darkened waters by day, zooplankton avoids heavy predation; and by remaining in cooler water by day, it conserves energy during the resting period. Surface currents move zooplankton away from its daytime location during feeding, but it can return home by countercurrents present in the deeper layers. Response to changing light conditions is useful in another way. As clouds of phytoplankton pass over the water above, zooplankton responds to the shadow of food by moving upward. This motion takes it out of the deep current drift and nearer the surface. At night it can move directly up to the food-rich surface water.

Zooplankton that lacks a vertical migration, and even some of those that do, drifts out of breeding areas with surface currents. Survival of a breeding population is assured by a complex cycle of seasonal migrations.

Nekton. Feeding on zooplankton and passing energy along to higher trophic levels is the nekton, swimming organisms that can move at will in the water column. They range in size from small fish to large predatory sharks and whales, seals, and a number of marine birds, such as penguins. Some of the predatory fish, such as herring and tuna, are more or less restricted to the photic zone. Others are found in the deeper mesopelagic and bathypelagic zones or can move between them and the surface, as the sperm whale does. Although the ratio in size of predator to prey falls within certain limitations, some of the largest nekton organisms in the sea, the baleen whales, feed on disproportionately small prey, euphasiids (popularly called krill). By contrast, the sperm whale attacks very large prey, the giant squid.

Living in a world that lacks any sort of refuge against predation or site for ambush, inhabitants of the pelagic zone have evolved various means of defense and of securing prey. Among them are the stinging cells of jellyfish, the remarkably streamlined shapes that allow speed both for escape and for pursuit, unusual coloration, advanced sonar, a highly developed sense of smell, and a social organization involving schools or packs. Some animals, such as the baleen whale, have specialized structures that permit them to strain krill and other plankton from the water. Others, such as the sperm whale and certain seals, have the ability to dive to great depths to secure food. Phytoplankton lights up darkened seas, and fish take advantage of that bioluminescence to detect their prey.

The dimly lighted regions of the mesopelagic and the dark regions of the bathypelagic zone depend upon a rain of detritus as an energy source. Such food is limited. The rate of descent of organic matter, except for larger items, is so slow that it is consumed, decayed, or dissolved before it reaches the deepest water or the bottom. Other sources include saprophytic plankton, which exist in the darker regions, particulate organic matter, and import of such material as wastes from the coastal zone, garbage from ships, and large dead animals.

Residents of the deep have special adaptations for securing food. Some, like the zooplankton, swim to the upper surface to feed by night. Others remain in the dimly lighted or dark waters. Darkly pigmented and weak-bodied, many of the deep-sea fish depend upon luminescent lures, mimicry of prey, extensible jaws, and expandable abdomens (which enable them to consume large items of food) as means of obtaining sustenance. Although most fish are small (usually 15 cm or less in length), the region is inhabited by rarely seen large species such as the giant squid. In the bathypelagic region bioluminescence reaches its greatest development. Two-thirds of the species produce light. Bioluminescence is not restricted to fish. Squid and euphausiids possess searchlightlike structures complete with lens and iris, and squid and shrimp discharge luminous clouds to escape predators. Fish have rows of luminous organs along their sides and lighted lures that enable them to bait prey and recognize other individuals of the same species.

Benthos. There is a gradual transition of life from the benthos on the rocky and sandy shores and that in the ocean's depths. From the tide line to the abyss, organisms that colonize the bottom are influenced by the nature of the substrate. If the bottom is rocky or hard, the populations consist largely of organisms that live on the surface of the substrate, the epifauna and the epiflora. Where the bottom is largely covered with sediment, most of the inhabitants, chiefly animals, are infauna and live within the deposits. Particle size of the substrate determines the type of burrowing organisms in an area, because the mode of burrowing is often specialized and adapted to a certain type of substrate.

The substrate varies with the depth of the ocean and the relationship of the benthic region to land areas and continental shelves. Near the coast bottom sediments are derived from the weathering and erosion of land areas along with organic matter from marine life. Sediments of deep water are characterized by fine-textured material, which varies with depth and the type of organisms in the overlying waters. Although these sediments are termed organic, they contain little decomposable carbon, consisting largely of skeletal fragments of planktonic organisms. In general, with regional variations, organic deposits down to 4000 m are rich in calcareous matter. Below 4000 m hydrostatic pressure causes some form of calcium carbonate to dissolve. At 6000 m and lower sediments contain even less organic matter and consist largely of red clays, rich in aluminum oxides and silica.

Within the sediments are layers that relate to oxidation-reduction reactions. The surface or oxidized layer, yellowish in color, is relatively rich in oxygen, ferric oxides, nitrates, and nitrites. It supports the bulk of the benthic animals, such as polychaete worms and bivalves in shallow water, flatworms, copepods, and others in deeper water, and a rich growth of aerobic bacteria throughout. Below the oxidized layer is a grayish transition zone to the black reduced layer, characterized by a lack of oxygen, iron in the ferrous state, nitrogen in the form of ammonia, and hydrogen sulfide. It is inhabited by anaerobic bacteria, chiefly reducers of sulfates and methane.

In the deep benthic regions variations in temperature, salinity, and other conditions are negligible. In this world of darkness there is no photosynthesis, so the bottom community is strictly heterotrophic, depending entirely for its source of energy on what organic matter finally reaches the bottom. Estimates suggest that the quantity of such material amounts to only 0.5 g/m^2/yr (H. B. Moore 1958). Bodies of dead whales, seals, and fish may contribute another 2 or 3 g.

Bottom organisms have four feeding strategies: (1) they may filter suspended material from the water, as do stalked coelenterates; (2) they may collect food particles that settle on the surface of the sediment, as do the sea cucumbers; (3) they may be selective or unselective deposit-feeders, such as the polychaetes; or (4) they may be predators, as are the brittle-stars and the spiderlike pycnogonids.

Important in the benthic food chain are bacteria of the sediments (see Figure 33.21). Common where large quantities of organic matter are present, bacteria may reach several tens of grams per square meter in the topmost layer of silt. Bacteria synthesize protein from dissolved nutrients and in turn become a source of protein, fats, and oils for deposit-feeders.

In spite of living in a world of darkness and pressure, the benthic fauna exhibits a surprising diversity. Faunal composition and diversity increase with depth down to the mid or lower bathyal region and decreases toward the abyssal plain (Rex 1981). Diversity along a gradient of

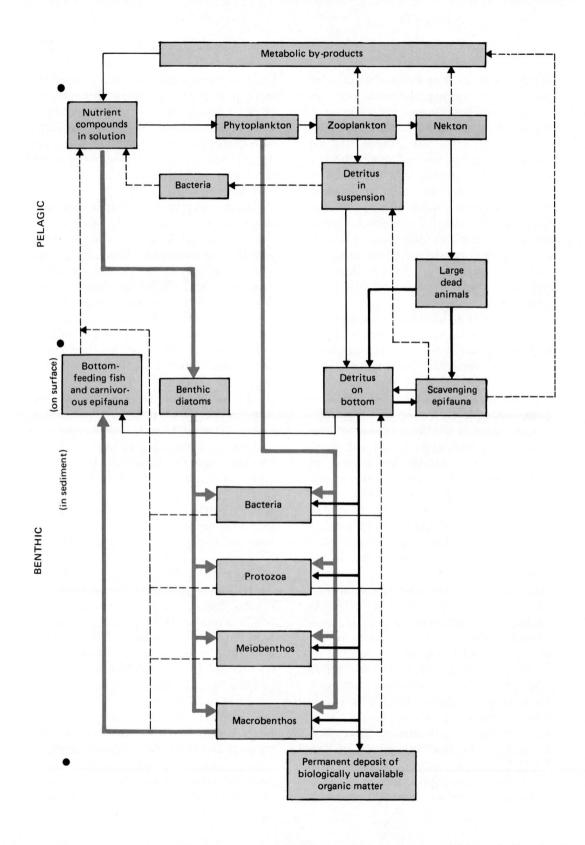

PELAGIC

BENTHIC

(on surface)

(in sediment)

Metabolic by-products

Nutrient compounds in solution

Phytoplankton

Zooplankton

Nekton

Bacteria

Detritus in suspension

Large dead animals

Bottom-feeding fish and carnivorous epifauna

Benthic diatoms

Detritus on bottom

Scavenging epifauna

Bacteria

Protozoa

Meiobenthos

Macrobenthos

Permanent deposit of biologically unavailable organic matter

depth appears to be related to the productivity of the waters above, competition among bottom fauna for food, and predation.

Hydrothermal Vents. In 1977 oceanographers discovered along volcanic ridges in the ocean floor of the Pacific near the Galagapos Islands high temperature deep sea springs never known before. These springs vent jets of hydrothermal fluids that heat the surrounding water to 8°–16° C, considerably higher than the 2° C ambient water. Since then oceanographers have discovered similar vents on volcanic ridges along fast-spreading centers of ocean floor, particularly in the eastern Pacific (for the geology of such vents see Hayman and McDonald 1983).

Vents form when cold sea water flows down through fissures and cracks in the basaltic lava floor deep into the underlying crust. The waters react chemically with the hot basalt, giving up some minerals but becoming enriched with others, such as copper, iron, sulfur, and zinc. Heated to a high temperature, the water reemerges through the flood through mineralized chimneys rising up to 13 meters above the forest floor. Among the various types of chimneys are white smokers and black smokers (Figure 33.22). White smoker chimneys rich in zinc sulfides issue a milky fluid under 300° C. Black smokers, narrower chimneys rich in copper sulfides, issue jets of clear water of 300° to over 450° C that is soon blackened by precipitation of fine-grained sulfur-mineral particles.

Associated with these vents is a rich diversity of newly discovered deep sea forms of life confined within a few meters

Figure 33.21 Simplified marine food web. (After Raymont 1963, from M. Gross 1972.)

of the vent system. They include giant clams, mussels, polychaete worms that encrust the white smokers, crabs, and vestimentifera worms lacking a digestive system. The primary producers are chemosynthetic bacteria that oxidize the reduced sulfur compounds, such as H_2S, to release energy, which they use to form organic matters from carbon dioxide. Primary consumers, the clams, mussels, and worms, filter bacteria from the water and graze on bacterial film on rocks. The giant clam *Calyptogena magnifica* and the large vestimentifera worm *Riftia pachyptila* contain symbiotic chemosynthetic bacteria in their coelomic tissues. These bacteria need a reduced sulfide source, which is carried to them by the blood of these animals. Aix and Childress (1983) report that *Riftia* has in its blood a sulfide-binding protein that concentrates sulfide from the environment and transports it to the bacteria. Such concentrations would poison normal animals, but the sulfide-bearing protein of the worm and apparently of the clam has a high affinity for free sulfides, preventing them from accumulating in the blood and entering the cells (Powell and Somero 1983).

Function

The oceans occupy 70 percent of Earth's surface and their average depth is around 4000 m; yet their primary production is considerably less than that of Earth's land surface. They are less productive because only a superficial illuminated area up to 100 m deep can support plant life; and that plant life, largely phytoplankton, is patchy because most of the open sea is nutrient-poor (Table 33.3). Much of this nutritional impoverishment results from a limited to almost nonexistent nutrient reserve that can be recirculated. Phytoplankton, zooplankton, and other organisms

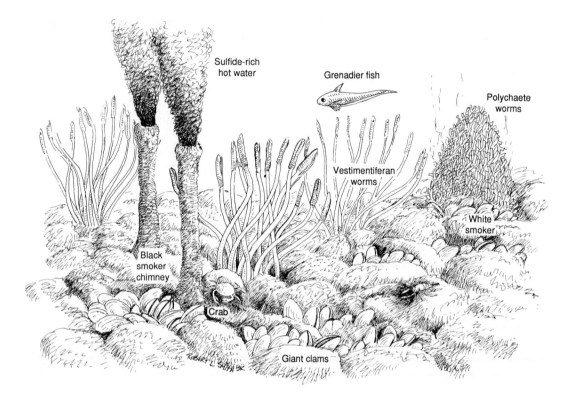

Figure 33.22 A typical hydrothermal vent mound resting on flows of black basaltic lava has a cluster of physical structures and an array of associated life. Black smoker chimneys form around jets of hot water well over 300° C. When the jets cool in the surrounding 2° C water, they precipitate calcium sulfate and sulfide minerals. White smokers encrusted with polychaete worms emit streams of milky-white fluids with temperatures less than 300° C. Elsewhere on the mounds fluids less than 100° C discharge from fissures and cracks. Newly discovered animal life occupies various temperature regions. These organisms include tube worms, giant clams, and crabs. (Adapted from Haymon and McDonald 1985 and other sources.)

and their remains sink below the lighted zone into the dark benthic water. While this sinking supplies nutrients to the deep, it robs the upper layers.

The depletion of nutrients is most pronounced in tropical waters. The permanent thermal stratification, with a layer of warmer, less dense water lying on top of a colder, denser deep water, prevents an exchange of nutrients between the surface and the deep. Thus in spite of high light intensity and warm temperatures, tropical seas are the lowest in production, about 18–50 g C/m²/yr.

The temperate oceans are more productive, largely because a permanent thermocline does not exist. During the spring and to a limited extent during the fall temperate seas, like temperate lakes, experience a nutrient overturn. The recirculation of phosphorus and nitrogen from the deep stimulates a surge of spring phytoplankton growth. As spring wears on the temperature of the water becomes stratified and a thermocline develops, preventing a nutrient exchange. The phytoplankton growth depletes the nutrients and the phytoplankton population suddenly declines. In the fall a similar over-

TABLE 33.3 Division of the Ocean into Provinces According to Level of Primary Organic Production

Province	Percentage of Ocean	Area (km^2)	Mean Productivity (g C/m^2/yr)	Total Productivity (10^9 tn C/yr)
Open ocean	90.0	326.0×10^6	50	16.3
Coastal zone*	9.9	36.0×10^6	100	3.6
Upwelling area	0.1	3.6×10^5	300	0.1
Total				20.0

*Includes offshore area of high productivity.
Source: J. Ryther 1969. © 1969 American Association for the Advancement of Science.

turn takes place, but the rise in phytoplankton production is slight because of decreasing light intensity and low winter temperatures. Reduced production in winter holds down annual productivity of temperate seas to a level a little above that of tropical seas, 70–110 g C/m^2/yr.

Most productive are coastal waters and regions of upwelling, whose annual production may amount to 1000 g C/m^2/yr. Major areas of upwelling are largely on the western sides of continents: off the coasts of southern California, Peru, northern and southwestern Africa, and the Antarctic. Upwellings result from the differential heating of polar and equatorial regions that produces the equatorial currents and the winds. As the water is pushed northward or southward toward the equator by winds, it is deflected away from the coasts by the Coriolis force. As the deflected surface water moves away, it is replaced by an upwelling of colder, deeper water that brings a supply of nutrients into the sunlit portions of the sea. As a result regions of upwellings are highly productive and contain an abundance of life. Because of their high productivity, upwellings support (or did support until overfished) important commercial fisheries such as the tuna fishery off the California coast, the anchoveta

fishery off Peru, and the sardine fishery off Portugal.

Other zones of high production are coastal waters and estuaries, where productivity may run to 380 g C/m^2/yr. Turbid, nutrient-rich waters are major areas of fish production. Thus between upwellings and coastal waters, the most productive areas of the seas are the fringes of water bordering the continental land masses. A great deal of the measured productivity of the coastal fringes comes from the benthic as well as the surface waters because these seas are shallow. Benthic production, largely unavailable, is not considered in the productivity of the open sea. Recent estimates (Koblentz-Mishke et al. 1970) of the total production for marine plankton is 50 Gt dry matter per year; if benthic production is considered, total production may be 55 Gt of dry matter per year.

Carbohydrate production by phytoplankton, largely diatoms, is the base upon which life of the seas exist. Conversion of primary production to animal tissue is accomplished by zooplankton, the most important of which are the copepods. To feed on the minute phytoplankton, most of the grazing herbivores must also be small, measuring between 0.5 and 5.0 mm. Most of the grazing herbivores in the oceans are the genera *Calanus, Acartia, Temora,* and *Metridia,* probably the most abundant animals in the world. The single most abundant copepod is *Calanus finmarchicus* and its close relative *Calanus helgolandicus.* In the Antarctic krill, fed upon by the blue whale, are the dominant herbivores.

The copepods then become the link in the food chain between the phytoplankton and the second-level consumers, as illustrated in the North Sea food web (Figure 33.23). The dominant primary producer is the diatom *Skeletoma,* which is grazed by *Calanus.* The major predator on *Calanus* is the semiplanktonic sand eel *Ammodytes,* which in turn is food for herring. The herring, however, can shorten the food chain by bypassing the sand eel and feeding directly on *Calanus.* In addition to this

major food chain, the herring is also involved in a number of side food chains that add stability to its food supply.

This classic food chain of the sea, however, overlooks the part of the food chain that begins not with the primary producers, the phytoplankton, but with organisms even smaller. Recent investigations show that bacteria and protists, both heterotrophic and photosynthetic, make up one-half of the biomass of the sea, and are responsible for the largest part of energy flow in pelagic systems (Fenchel 1987, 1988). Photosynthetic nanoflagellates (2–20µ) and cyanobacteria (1–2µ) are responsible for a large part of photosynthesis in the sea. These phytoplankton cells excrete a substantial fraction of their photosynthate in the form of dissolved organic material that is utilized by heterotrophic bacteria. Populations of such bacteria are dense, around 1 million cells per milliliter of sea water. Production by these heterotrophic bacteria accounts for about 20 percent of primary production. Bacterial growth efficiency does not exceed 50 percent, so one-half of phytoplankton primary production in the form of dissolved organic material is consumed by bacteria. Bacterial numbers in the sea remain relatively stable, suggesting predation; but filter-feeding zooplankton cannot retain particles of bacterial size. Therefore, the consumption of bacteria by heterotrophic nanoflagellates, experimentally demonstrated, accounts for the disappearance of bacterial production. This uptake by heterotrophic bacteria of dissolved organic matter produced by the plankton and the subsequent consumption of bacteria by nonoplankton introduces a feeding loop, termed the *microbial loop* (Azam et al. 1983) (Figure 33.24), and adds several more trophic levels to the plankton food chain.

Figure 33.23 Food web and energy flow from an area of the North Sea. Note the low energy yield to humans, 6.6 kcal/m^2/yr from a primary production of 900 kcal/m^2/yr. (From Steele 1971.)

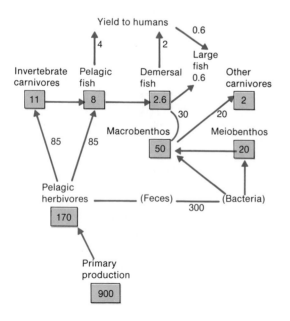

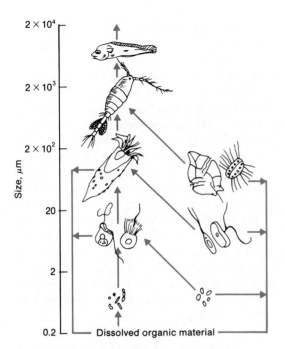

Figure 33.24 A microbial loop that feeds into the classic marine food web. On the right are representative photoautotrophs, nanoflagellates, and cyanobacteria and on the left are associated heterotrophs, bacteria, and heterotrophic nanoflagellates. (From Fenchel 1988: 24.)

■ Coral Reefs

In subtropical and tropical waters around the world are structures of biological rather than geological origin: coral reefs. Coral reefs are built by carbonate-secreting organisms, of which coral (Cnidaria, Anthozoa) may be the most conspicuous, but not always the most important. Also contributing heavily are the coralline red algae (Rhodaphyta, Corallinaceae), foraminifera, and mollusks. Built only under water at shallow depths, coral reefs need a stable foundation upon which to grow. Such foundations are provided by shallow continental shelves and submerged volcanoes.

Coral reefs are of three types, with many gradations among them. (1) *Fringing reefs* grow along the rocky shores of islands and continents. (2) *Barrier reefs* parallel shore lines along continents. (3) *Atolls* are coral islands that begin as horseshoe-shaped reefs surrounding a lagoon. Such lagoons are about 40 m deep and are usually connected to the open sea by breaks in the reef. Reefs build up to sea level. To become islands or atolls, the reefs have to be exposed by a lowering of the sea level or be built up by the action of wind and waves.

Structure

Coral reefs are complex ecosystems involving close relationships between coral and algae. In the tissues of the gastrodermal layer of coral live *zooxanthellae*, symbiotic endozoic dinoflagellate algae, both the encrusting red and green coralline species and filamentous species, including turf algae. Associated with coral growth are mollusks, such as giant clams *(Tridacna, Hippopus)*, echinoderms, crustaceans, polychaete worms, sponges, and a diverse array of fishes, both herbivorous and predatory.

Zonation and diversity of coral species are influenced by an interaction of depth, light, grazing, competition, and disturbance (Huston 1985). The basic diversity gradient is established by light. Lowest at the crest, where only species tolerant of intense or frequent disturbance can survive, diversity increases with depth to a maximum of about 20 m and then decreases as light becomes attenuated, eliminating shade-intolerant species (see Wellington 1982a). Imposed upon this gradient are a variety of biotic and abiotic disturbances that vary in intensity and decrease with depth. Growth rates of photosynthetic coral are highest in shallow depths and a few species, especially the branching corals, can easily dominate the reef by overgrowing and shading the crustose corals and algae. Disturbances by wave action, storms, and grazing reduce rate of competitive displacement among corals. Heavy grazing of overgrowing algae by sea urchins and fish, such as parrotfish, increases encrusting coralline algae. Light grazing allows rapidly growing filamentous and foliose algal species to eliminate crustose algae (see Wellington 1982b).

Function

Coral are partially photosynthetic organisms and partially heterotrophic. During the day zooxanthellae carry on photosynthesis and directly transfer organic material to coral tissue. At night coral polyps feed on zooplankton, securing phosphates and nitrates and other nutrients needed by the anthozoans and their symbiotic algae. Thus nutrients are recycled in place between the anthozoans and the algae (Johannes 1967; Pomeroy and Kuenzler 1969). In addition, carbon dioxide concentrations in animal tissue enable the coral to extract the calcium carbonate needed to build the coral skeletons.

Adding to the productivity of the coral are crustose coralline algae, turf algae, macroalgae, seagrass, sponges, and phytoplankton, and a large bacterial population. Coral reefs are among the most highly productive ecosystems on Earth. Net productivity ranges from 1500 to 5000 g C/m^2/yr compared to 15 to 50 g C/m^2/yr for the surrounding ocean. Because of the ability of the coralline community to retain nutrients within the system and to act as a nutrient trap, coral reefs are

oases of productivity within a nutrient-poor sea.

This high productivity and the diversity of habitats within the reef supports a high diversity of life. There are thousands of kinds of exotic invertebrates, some of which, such as sea urchins, feed on coral animals and algae, hundreds of kinds of herbivorous fish that graze on algae, and many predatory species (see Yonge 1963; Haitt and Strasburg 1960; Sale 1980). Some of these predators, such as the puffers and file fish, are corallivores, feeding on coral polyps. Others lie in ambush for prey in coralline caverns. In addition, there is a wide array of symbionts, such as cleaning fish and crustaceans that pick parasites and detritus from larger fish and invertebrates.

In spite of its vastness, the ocean is fragile and in jeopardy. It has become the dumping ground for humans inhabiting the continents it surrounds. Sewage, industrial wastes, agricultural runoff, and pesticides and other toxic materials flow into its shallow estuarine and offshore waters. There these materials contaminate shellfish, cause cancerous growths in fish, bacterial infections in marine mammals, and deformed growth of marine birds. Toxic substances move up through the food chain making economically important fish and shellfish inedible for humans. The wastes of cities—garbage and trash—are dumped in offshore waters to wash up on beaches. Even nuclear wastes have been jettisoned into the ocean. Oil from supertankers broken apart at sea and run aground on reefs and rocky shores becomes fouling slicks on the surface, killing marine birds and mammals, washes up on beaches, and settles as a life-smothering blanket on the bottom. Humans have plundered the ocean of its fishes and whales, reducing many to the point of extinction and radically altering species composition and food chains of the ocean. Unless all nations take the care of the sea and its life seriously and act cooperatively, the ocean, like much of the land, will be marked with ruin.

■ Summary

The marine environment, which occupies 70 percent of Earth's surface, is characterized by salinity, waves, currents, tides, and vastness. Border ecosystems between terrestrial and marine environments exist at estuaries, tidal marshes, and rocky, sandy, and muddy shores.

In estuaries where fresh water meets the sea and their associated tidal marshes and swamps, the nature and distribution of life are determined by salinity. As salinity declines from the mouth up through the river, so do estuarine fauna, chiefly marine species. The estuary serves as a nursery for marine organisms, for here the young can develop protected from predation and from competing species unable to withstand lower salinity. Tidal marshes are dominated by salt-tolerant plants and flooded by daily tides. They are highly productive because tidal flushing brings in new nutrients and sweeps away wastes. Most of the primary production goes unharvested by herbivores and is converted to microbial biomass. Some production goes into sediments; another portion may be exported to the estuary. If vertical mixing between fresh water and salt water occurs in the estuary, the nutrients are circulated up and down between the organisms and bottom sediments. In tropical regions mangrove swamps replace salt marshes.

Sandy shores and rocky coasts are places where the sea meets the land. The drift line marks the farthest advance of tides on the sandy shore. On the rocky shore the tide line is marked by a zone of black algal growth. The most striking feature of the rocky shore, its zonation of life, results from alternate exposure and submergence of the shore by the tides. The black zone marks the supralittoral, the upper part of which is flooded only every two weeks by spring tides. Submerged daily by tides is the littoral, characterized by barnacles, periwinkles, mussels, and fucoid seaweeds. Uncovered only at spring tides is the infralittoral, which is dominated by large brown laminarian seaweeds, Irish moss, and starfish. Although the basic pattern of zonation results from tidal flooding and exposure, distribution and diversity of life across the rocky shore is influenced by disturbance from wave action, competition, herbivory, and predation. Left behind by outgoing tides are tidal pools. These distinct habitats are subject over a 24-hour period to wide fluctuation in temperature and salinity and inhabited by varying numbers of organisms, depending upon the amount of emergence and exposure.

By contrast, sandy and muddy shores appear barren of life at low tide; but beneath the sand and mud conditions are more amenable to life than on the rocky shore. Zonation of life is hidden beneath the surface. The energy base for sandy and muddy shores is organic matter made available by bacterial decomposition. They are important sites for biogeochemical cycling, supplying nutrients for offshore waters. The basic consumers are bacteria, which in turn are a major source of food for both deposit-feeding and filter-feeding organisms. Sandy shores and mudflats are a part of the larger coastal ecosystem, including the salt marsh, estuary, and coastal waters.

Beyond the estuary and rocky and

sandy shores lies the open sea. There the dominant plant life is phytoplankton and the chief consumers are zooplankton. Also important in the marine food web are photosynthetic nanoflagellates and cyanobacteria, and heterotrophic bacteria. Depending upon this base are the nekton organisms dominated by fish.

The open sea can be divided into three main regions. The bathypelagic is the deepest, void of sunlight and inhabited by darkly pigmented, weak-bodied animals characterized by luminescence. Above it lies the dimly lit mesopelagic region, inhabited by its own characteristic species, such as certain sharks and squid. Both the mesopelagic and bathypelagic regions depend upon a rain of detritus from the upper region, the epipelagic, for their energy source. Along the volcanic oceanic ridges, especially in the mid-Pacific, are hydrothermal vents inhabited by unique and newly discovered forms of life, including clams, worms, and crabs. The source of primary production for these hydrothermal vent communities are chemosynthetic bacteria that use sulfates as an energy source.

The impoverished nutrient status of ocean water makes productivity low. Nutrient reserves in the upper layer of water are limited, phytoplankton and other life sink to the deep water, and a thermocline, permanent in deep water, prevents the recirculation of deep water to the upper layer. Most productive are shallow coastal waters and upwellings, where nutrient-rich deep water comes to the surface. The most productive fisheries are confined to these areas, but pollution and overexploitation are reducing the productivity of the seas.

Nutrient-rich oases in the nutrient-poor tropical seas are coral reefs. Coral reefs are complex ecosystems based on anthozoan corals and their symbiotic endozoan dinoflagellate algae and corraline algae. Recycling nutrients within the system and functioning as nutrient sinks, coral reefs are among the most productive ecosystems in the world. Their productive and varied habitats support a high diversity of colorful invertebrate and vertebrate life.

GLOSSARY

This glossary of 660 items contains most of the italicized terms in the text as well as a number of other terms commonly used throughout the book.

A horizon surface stratum of mineral soil characterized by maximum accumulation of organic matter, maximum biological activity, and loss of such materials as iron, aluminum oxides, and clays

abiotic non-living component of the environment including soil, water, air, light, nutrients, and the like

abyssal relating to the bottom waters of oceans, usually below 1000 m

acclimation alteration of physiological rate or other capacity to perform a function through long-term exposure to certain conditions

acclimatization changes or differences in a physiological state that appear after exposure to different natural environments

acid rain precipitation with an extremely low pH; brought about by a combination of water vapor in the atmosphere with hydrogen sulfide, and nitrous oxide vapors released to the atmosphere from the burning of fossil fuels; the result is sulfuric and nitric acid in rain, fog, and snow

active temperature range range of body temperatures over which ectotherms carry out their daily activities

active transport movement of ions and molecules across a cell membrane against a concentration gradient involving an expenditure of energy; movement of the ion or molecule is in a direction opposite to direction taken under simple diffusion

adaptation genetically determined characteristic (behavioral, morphological, physiological) which improves an organism's ability to survive and successfully reproduce under prevailing environmental conditions

adaptive radiation from a common ancestor, evolution of divergent forms adapted to distinct ways of life

adiabatic cooling resulting decrease in air temperature when a rising parcel of warm air cools by expansion (which uses energy) rather than by losing heat to the outside surrounding air; rate of cooling is approximately 1°C/100 m for dry air and 0.6°C/100 m for moist air

adiabatic lapse rate rate at which a parcel of air loses temperature with elevation if no heat is gained from or lost to an external source

adiabic process operation in which heat is neither lost to nor gained from the outside

aerenchyma plant tissue with large air-filled intercellular spaces; usually found in roots and stems of aquatic and marsh plants

aerobic living or occurring only in the presence of free uncombined molecular oxygen, either as a gas in the atmosphere or dissolved in water

aestivation dormancy in animals through a drought or dry season

aggrading gradually increasing in structure or biomass

aggregative response reaction in which consumers spend most of the time in food patches with the greatest density of prey

aggressive mimicry resemblance of a predator or parasite to a harmless species to deceive potential prey

agonistic behavior all types of hostile response to other organisms, ranging from overt attack to overt escape

albedo percentage of solar radiation reflected by the earth's surface

alfisol soil characterized by an accumulation of iron and aluminum in lower or B horizon *allele* one of two or more alternative forms of a gene that occupies the same relative position or locus on homologous chromosomes

allelopathy effect of metabolic products of plants (excluding microorganisms) on the growth and development of other nearby plants

Allen's rule trend among homoiotherms for limbs to become longer, and extremities (such as ears) to become less compact, in warmer climates than in colder ones

allochthonous food material reaching an aquatic community in the form of organic detritus

allogenic succession ecological change or development of species structure and community composition brought about by some externally generated force such as fire or storms

allopatric having different areas of geographical distribution; possessing nonoverlapping ranges

alluvial soil soil developing from recent alluvium (material deposited by running water); exhibits no horizon development; typical of floodplains

alpha diversity the variety of organisms occupying a given place or habitat

altricial condition among birds and mammals of being hatched or born usually blind and too weak to support their own weight

altruism form of behavior in which an individual increases the welfare of another at the expense of its own welfare

ambient refers to surrounding, external, or unconfined conditions

amensalism relationship between two species in which one is inhibited or harmed by the presence of another

ammonification breakdown of proteins and amino acids especially by fungi and bacteria with ammonia as the excretory by-product.

anabolism metabolic reactions usually requiring energy provided by ATP in which molecules are linked together to form more complex compounds

anadromous refers to fish that typically inhabit seas or lakes but ascend streams to spawn; for example, salmon

anaerobic adapted to environmental conditions devoid of oxygen

antibiotic substance produced by a living organism which is toxic to organisms of different species

aposematism possession of warning coloration; conspicuous markings on animals which are poisonous, distasteful, or possess some unpleasant defensive mechanism

apparent plants large, easy to locate plants possessing quantitative defenses not easily mobilized at the point of attack; for example, tannins

aridosol desert soils characterized by little organic matter and high base content

asexual reproduction any form of reproduction, such as budding, which does not involve the fusion of gametes

assimilation transformation or incorporation of a substance by organisms; absorption and conversions of energy and nutrient uptake into constitutents of an organism

association natural unit of vegetation characterized by a relatively uniform species composition and often dominated by a particular species

aufwuchs community of plants and animals attached to or moving about on submerged surfaces; also called *periphyton,* but that term more specifically applies to organisms attached to submerged plant stems and leaves

autecology ecology of individual species in response to environmental conditions

autochthonus organic matter produced and used within the system

autogenic self-generated

autotrophy ability of an organism to produce organic material from inorganic chemicals and some source of energy

available water capacity supply of water available to plants in a well-drained soil

B horizon soil stratum beneath the A horizon characterized by an accumulation of silica, clay, and iron and aluminum oxides and possessing blocky or prismatic structure

basal metabolic rate minimal amount of energy expenditure needed by an animal to maintain vital processes

Batesian mimicry resemblance of a palatable or harmless species, the mimic, to an unpalatable or dangerous species

bathyal pertaining to anything, but especially organisms, in the deep sea (below the photic or lighted zone, and above 4000 m)

benthos animals and plants living on the bottom of a lake or sea (from high water mark to the deepest depths)

Bergman's rule populations of homoiotherms living in cooler climates tend to have a larger body size and a smaller surface-to-volume ratio than related populations living in warmer climates

beta diversity variety of organisms occupying a number of different habitats over a region; regional diversity compared to very local or alpha diversity

biennial plant that requires two years to complete a life cycle, with vegetative growth the first year and reproductive growth (flowers and seeds) the second

biochemical oxygen demand (BOD) measure of the oxygen needed in a specified volume of water to decompose organic materials; the greater the amount of organic matter in water, the higher the BOD

biogeocenosis translated Germanic and Slavic language equivalent of ecosystem; more exact than ecosystem, the term emphasizes both the biological and geological aspects of the ecosystem

biogeochemical cycle movement of elements or compounds through living organisms and nonliving environment

biological clock internal mechanism of an organism that controls circadian rhythms without external time cues

biological magnification process by which pesticides and other substances become more concentrated in each successive link of the food chain

biological species group of potentially interbreeding populations reproductively isolated from all other populations

bioluminescence production of light by living organisms

biomass weight of living material, usually expressed as dry weight-per-unit area

biome major regional ecological community of plants and animals; usually corresponds to plant ecologists' and European ecologists' classification of plant formations and classification of life zones

biophage organisms that feed on living material

biosphere thin layer about the earth in which all living organisms exist

biotic community any assemblage of populations living in a prescribed area or physical habitat

blanket mire large areas of upland dominated by sphagnum moss and dependent upon precipitation for a water supply; a moor

bog wetland ecosystem characterized by an accumulation of peat, acid conditions, and dominance of sphagnum moss

boreal forest needle-leaved evergreen or coniferous forest bordering subpolar regions; also called *taiga*

bottleneck evolutionary term for any stressful situation that greatly reduces a population

brackish water that has salt concentration greater than fresh water (>0.5 o/oo) and less than seawater (35 o/oo)

browse part of current leaf and twig growth of shrubs, woody vines, and trees, available for animal consumption

bryophyte member of the division in the plant kingdom of nonflowering plants comprising mosses (Musci), liverworts (Hepaticae), and hornworts (Anthocerotae)

buffer chemical solution that resists or dampens change in pH upon addition of acids or bases; also used in population biology to refer to any prey species that tends to dampen the impact of predation on another

C horizon soil stratum beneath the solum (A and B horizons) relatively little affected by biological activity and soil forming process

C_3 *plant* any plant that produces as its first step, in photosynthesis, a three-carbon compound phosphoglyceric acid

C_4 *plant* any plant that produces as its first step, in photosynthesis, a four-carbon compound malic or aspartic acid

calcicole plants susceptible to aluminum toxicity, acidity, and other factors influenced by the absence of calcium

calcification process of soil formation characterized by accumulation of calcium in lower horizons

calcifuge plants with low calcium requirement that can live in soils with pH of 4.0 or less

caliche alkaline, often rock-like salt deposit on the surface of soil in arid regions; forms at the level where leached Ca salts from the upper soil horizons are precipitated

calorie amount of heat needed to raise 1 gram of water 1°C (usually from 15°C to 16°C)

Cambrian earliest period of geological time in the Paleozoic era from abut 570 to 500 million years ago; algae and many marine invertebrates appeared

cannibalism killing and consumption of one's own kind; intraspecific predation

canonical distribution particular configuration of the log normal distribution of species abundance

capillary water that portion of water in the soil held by capillary forces between soil particles

Carboniferous second period of the Upper Paleozoic from about 345 to 280 million years ago named after extensive coal deposits formed at this time; warm, humid period characterized by forests and swamps, appearance of early amphibians, reptiles, and giant ferns

carnivore organism that feeds on animal tissue; taxonomically, a member of the order Carnivora (Mammalia)

carr vegetation dominated by alder and willow and occupying eutrophic peat

carrying capacity (K) number of individual organisms the resources of a given area can support usually through the most unfavorable period of the year; this term has acquired so many meanings it is almost useless

catabolism metabolic breakdown of complex molecules to simpler ones

catadromous refers to fish that feed and grow in fresh water, but return to the sea to spawn (see *anadromous*)

catastrophic extinction major episode of extinction involving many taxa occurring fairly suddenly in the fossil record

catena group of related soils

cation part of a dissociated molecule carrying a $^+$ electrical charge

cation exchange capacity ability of a soil particle to absorb $^+$ charged ions

Cenozoic era major division of geological time extending from the Mesozoic era some 65 million years ago to the present

chamaephyte perennial shoots or buds on the surface of the ground to about 25 cm above the surface

chaos apparently random, recurrent, and unpredictable behavior in a simple deterministic system

chaparral vegetation, consisting of broadleaved evergreen shrubs, found in regions with a mediterranean climate of hot, dry summers and mild, wet winters

character convergence evolution of similar appearance or behavior in unrelated species

character displacement divergence of characteristics in two otherwise similar species occupying overlapping ranges; brought about by selective effects of competition

character divergence evolution of behavioral, physiological, or morphological differences among species occupying the same area; brought about by the selective pressure of competition

chasmogamy production of flowers that open to expose reproductive organs, allowing cross-pollination

chilling tolerance ability of a plant to carry on photosynthesis within a range of 5° to 10°C

circadian rhythm endogenous rhythm of physiological or behavioral activity of approximately 24 hours in duration

cleistogamy self-pollination within a flower that does not open

climax stable end community of succession that is capable of self-perpetuation under prevailing environmental conditions

climograph diagram describing a locality based on the annual cycle of temperature and precipitation

cline gradual change in population characteristics over a geographical area; usually associated with changes in environmental conditions

clone population of genetically identical individuals resulting from asexual reproduction

coarse-grained qualities, aspects, or characteristics of an environment (occurring over large patches) with respect to the activities of organisms involved

coevolution joint evolution of two or more noninterbreeding species having a close ecological relationship; through reciprocal selective pressures, the evolution of one species in the relationship is partially dependent on the evolution of the other

coexistence two or more species living together in the same habitat, usually with some form of competitive interaction

cohort group of individuals of the same age

cold resistance ability of a plant to resist low temperature stress without injury

colluvium mixed deposits of soil material and rock fragments accumulated near the base of steep slopes through soil creep, landslides, and local surface runoff

commensalism relationship between species which is beneficial to one, but neutral or of no benefit to the other

community group of interacting plants and animals inhabiting a given area

compensation intensity light intensity at which photosynthesis and respiration balance each other so that net production is 0; in aquatic systems, usually the depth of light penetration at which oxygen utilized in respiration equals oxygen produced by photosynthesis

competition any interaction that is mutually detrimental to both participants; occurs between species that share limited resources

competitive exclusion hypothesis which states that when two or more species coexist using the same resource, one must displace or exclude the other

conduction direct transfer of heat from one substance to another

connectance ratio of potential links or interactions in a food web to those that actually exist

consumer any organism that lives on other organisms, either dead or alive

continuum gradient of environmental characteristics or changes in community composition

continuum index measure of a position of a community on a gradient defined by species composition

convection transfer of heat by the circulation of fluids, liquid or gas

convergent evolution development of similar characteristics in different species living in different areas but under similar environmental conditions

coprophagy feeding on feces

Coriolis effect physical consequences of the law of conservation of angular momentum; as a result of the earth's rotation, a moving object veers to the right in the Northern Hemisphere and to the left in the Southern Hemisphere

countercurrent circulation anatomical and physiological arrangement by which heat exchange takes place between outgoing warm arterial blood and cool venous blood returning to the body core; important in maintaining temperature homeostasis in many vertebrates

Cretaceous final period of the Mesozoic from about 136 and 65 million years ago when much of the present land area was covered by shallow seas; rise of angiosperms and continued dominance of dinosaurs until mass extinction at the end of the period; evolution of modern birds and fish; appearance of primitive mammals

critical thermal maximum temperature at which an animal's capacity to move is so reduced that it cannot escape from thermal conditions that will lead to death

crude density number of individuals per unit area

cryptic coloration coloration of organisms that makes them resemble or blend into their habitat or background

cryptophyte buds buried in the ground on a bulb or rhizome

cyclic replacement type of succession in which the sequence of seral stages is repeated by imposition of some disturbance so that the sere never arrives at a climax or stable sere

dark reactions group of light-independent reactions following the light reactions in photosynthesis; they reduce carbon dioxide to produce glucose and other carbohydrates

day-neutral plant plant that does not require any particular photoperiod to flower

death rate number of individuals in a population dying during a given time interval divided by the number alive at the midpoint of the time interval

deciduous (of leaves), shed during a certain season (winter in temperate regions, dry seasons in the tropics); (of trees), having deciduous parts

decomposer organism that obtains energy from breakdown of dead organic matter to more simple substances; most precisely refers to bacteria and fungi

decomposition breakdown of complex organic substances into simpler ones

deductive method in testing hypotheses, going from the specific to the general

definitive host one in which a parasite becomes an adult and reaches maturity

deme local population or interbreeding group within a larger population

demographic stochasticity random variations in population birth and death rates due entirely to chance differences experienced by individuals

denitrification reduction of nitrates and nitrites to nitrogen by microorganisms

density dependent varying in relation to population density

density independent unaffected by population density

dependent variable variable *y*, which yields the second of two numbers in an ordered pair *(x, y);* the set of all values taken on by the dependent variable is called the *range* of the function; see *independent variable*

deterministic model mathematical model in which all relationships are fixed and a given input produces one exact prediction as an output

detritivore organism that feeds on dead organic matter; usually applies to detritus-feeding organisms other than bacteria and fungi

detritus fresh to partly decomposed plant and animal matter

Devonian period between the Ordovician and the Carboniferous some 405 to 355 million years ago; the age of fish; evolution of early amphibians; appearance of vascular plants, insects, and spiders

dewpoint temperature at which condensation of water in the atmosphere begins

diameter breast height (dbh) diameter of tree measured 4 feet, 6 inches (1.4 m) from ground level

diapause period of dormancy, usually seasonal, in the life cycle of an insect in which growth and development cease and metabolism is greatly decreased

diffuse coevolution coevolution involving the interactions of many organisms in contrast to pairwise interactions

diffuse competition type of competition in which a species experiences interference from numerous other species that deplete the same resources

dimorphism existing in two structural forms, two color forms, two sexes, and the like

dioecious plants in which male and female reproductive organs are borne on separate individuals

diploid having chromosomes in homologous pairs or twice the haploid numbers of chromosomes

directional selection selection favoring individuals at one extreme of the phenotype in the population

disease any deviation from normal state of health

dispersal leaving an area of birth or activity for another area

dispersion distribution of organisms within a population over an area

disruptive selection selection in which two extreme phenotypes in the population leave more offspring than the intermediate phenotype, which has lower fitness

diversity abundance in number of species in a given location

dominance (ecological) control within a community over environmental conditions influencing associated species by one or several species, plant or animal, enforced by number, density, or growth form; (social) behavioral and hierarchical order in a population that gives high-ranking individuals priority of access to essential requirements; (genetic) ability of an allele to mask the expression of an alternative form of the same gene in a heterozygous condition

dominant population possessing ecological dominance in a given community and thereby governing type and abundance of other species in the community

dormant state of cessation of growth and suspended biological activity during which life is maintained

drought avoidance ability of a plant to escape dry periods by becoming dormant or surviving the period as a seed

drought resistance sum of drought tolerance and drought avoidance

drought tolerant ability of plants to maintain physiological activity in spite of the lack of water or to survive the drying of tissues

dynamic pool model optimum yield model using growth, recruitment, mortality, and fishing intensity to predict yield

dystrophic term applied to a body of water with a high content of humic organic matter, often with high littoral productivity and low plankton productivity

ecesis establishment of the first stage in succession

ecocline geographical gradient of communities or ecosystems produced by responses of vegetation to environmental gradients of rainfall, temperature, nutrient concentrations, and other factors

ecological density density measured in terms of the number of individuals per area of available living space

ecological efficiency percentage of biomass produced by one trophic level that is incorporated into biomass of the next highest trophic level

ecological release expansion of habitat or increase in food availability resulting from release of a species from interspecific competition

ecologism linkage of ecological ideas with energy economics; the environmental movement

ecosystem the biotic community and its abiotic environment functioning on a system

ecotone transition zone between two structurally different communities; see also *edge ecotype* subspecies or race adapted to a particular set of environmental conditions

ectothermy determination of body temperature primarily by external thermal conditions

edaphic relating to soil

edge place where two or more vegetation types meet

edge effect response of organisms, animals in particular, to environmental conditions created by the edge

effective population size size of an ideal population that would undergo the same amount of random genetic drift as the actual population; sometimes used to measure the amount of inbreeding in a finite, randomly mating population

egestion elimination of undigested food material

elaiosome shiny, oil-containing, ant-attracting tissue on the seed coat of many plants

emigration permanent movement out of an area by part of a population

endemic restricted to a given region

endogenous any process that arises within an organism

endothermy regulation of body temperature by internal heat production; allows maintenance of appreciable difference between body temperature and external temperature

energy capacity to do work

entrainment synchronization of an organism's activity cycle with environmental cycles

entropy transformation of matter and energy to a more random, more disorganized state

environment total surroundings of an organism, including other plants and animals and those of its own kind

environmental stochasticity unpredictable environmental changes, such as bad weather or food failure, that affect some aspect of population growth

ephemeral organisms with a short life cycle; lasting only a season or a few days

epidemic rapid spread of a bacterial or viral disease in a human population; compare with *epizootic*

epifauna benthic organisms that live on or move across the surface of a substrate

epilimnion warm, oxygen-rich upper layer of water in a lake or other body of water, usually seasonal

epiphyte organism that lives wholly on the surface of plants, deriving support but not nutrients from the plants

epizootic rapid spread of a bacterial or viral disease in a dense population of animals

equilibrium species species whose population exists in equilibrium with resources and at a stable density

equilibrium turnover rate change in species composition per unit time when immigration equals extinction

equitability evenness of distribution of species abundance patterns; maximum equitability is the same number of individuals in all species

estivation dormancy in animals during a period of drought or a dry season

estuary partially enclosed embayment where fresh water and sea water meet and mix

ethology study of animal behavior

eukaryotic cell that has membraneous organelles, notably the nucleus

euphotic zone surface layer of water to the depth of light penetration where photosynthetic production equals respiration

eutrophic refers to a body of water with high nutrient content and high productivity

eutrophication nutrient enrichment of a body of water; called cultural eutrophication when accelerated by introduction of massive amounts of nutrients by human activity

evapotranspiration sum of the loss of moisture by evaporation from land and water surfaces and by transpiration from plants

evenness degree of equitability in the distribution of individuals among a group of species; see *equitability*

evolution change in gene frequency through time, resulting from natural selection and producing cumulative changes in characteristics of a population

evolutionary time period during which a population evolves and becomes adapted to an environment by means of genetic change

exothermic chemical reaction that releases heat to the environment

exploitative competition competition among members of different or the same species for the same limited resource

exponential growth instantaneous rate of population growth expressed as proportional increase per unit of time

extinction coefficient point at which the intensity of light reaching a certain depth is insufficient for photosynthesis; ratio of intensity of light at a given depth to the intensity at the surface

F_1 *generation* first generation of offspring from a cross between individuals homozygous for contrasting alleles; F_1 is necessarily heterozygous.

F₂ generation offspring produced by selfing or by allowing the F_1 generation to breed among themselves

facilitation model model of succession in which previous community prepares or "facilitates" the way for a succeeding community

fecundity potential ability of an organism to produce eggs or young; rate of production of young by a female

fen wetlands dominated by sedges in which peat accumulates

fermentation breakdown of carbohydrates and other organic matter under anaerobic conditions

field capacity amount of water held by soil against the force of gravity

fine-grained qualities, aspects, or characteristics of an environment, occurring in patches so small relative to the activity of organisms that they do not distinguish among them

fitness genetic contribution by an individual's descendants to future generations

fixation process in soil by which certain chemical elements essential for plant growth are converted from a soluble or exchangeable form to a less soluble or nonexchangeable form

floating reserve individuals in a population of a territorial species that do not hold territories and remain unmated, but are available to refill territories vacated by death of an owner

flux flow of energy from a source to a sink or receiver

foliage height diversity measure of the degree of layering or vertical stratification of foliage in a forest

food chain movement of energy and nutrients from one feeding group of organisms to another in a series that begins with plants and ends with carnivores, detrital feeders, and decomposers

food web interlocking pattern formed by a series of interconnecting food chains

foraging strategy manner in which animals seek food and allocate their time and effort in obtaining it

forb herbaceous plant other than grass, sedge, or rush

formation classification of vegetation based on dominant life forms

founder's effect population started by a small number of colonists which contain only a small and often biased sample of genetic variation of the parent population; may result in a markedly different new population

fragmentation reduction of a large habitat area into small, scattered remnants; reduction of leaves and other organic matter into smaller particles

free-running cycle length of a circadian rhythm in absence of an external time cue

frost pocket depression in the landscape into which cold air drains, lowering the temperature relative to the surrounding area; such pockets often support their own characteristic group of cold tolerant plants

frugivore organism that feeds on fruit

fugitive species species characteristic of temporary habitats

functional response change in rate of exploitation of a prey species by a predator in relation to changing prey density

fundamental niche total range of environmental conditions under which a species can survive

fynbos sclerophyllous shrub community occurring in regions of South Africa with a mediterranean climate

gamma diversity diversity differences between similar habitats in widely separated geographic regions

gap opening made in a forest canopy by some small disturbance such as wind fall, death of an individual or group of trees that influences the development of vegetation beneath

gap phase replacement successional development in small disturbed areas within a stable plant community; filling in of a space left by a disturbance, but not necessarily by the species eliminated by the disturbance

garrigue scrub woodland characteristic of limestone areas with low rainfall and thin, poor dry soils; widespread in the Mediterranean countries of southern Europe

gene unit material of inheritance; more specifically, a small unit of DNA molecule coded for a specific protein to produce one of the many attributes of a species

gene flow exchange of genetic material between populations

gene frequency actually allele frequency; relative abundance of different alleles carried by an individual or a population

gene pool sum of all the genes of all individuals in a population

genet genetic individual that arises from a single fertilized egg

genetic drift random fluctuation in allele frequency over time due to chance occurrence alone without any influence by natural selection; important in small populations

genetic feedback evolutionary response of a population to adaptations of predators, parasites, or competitors

genotype genetic constitution of an organism

geometric rate of increase factor by which size of a population increases over a period of time

gley soil soil developed under conditions of poor drainage resulting in reduction of iron and other elements, and in gray colors and mottles

global stability ability of a community to withstand large disturbances and return to its original state

gouger general group of stream invertebrates that live and feed on woody debris

granivore organism that feeds on seeds

greenhouse effect selective energy absorption by carbon dioxide in the atmosphere which allows short wavelength energy to pass through but absorbs longer wavelengths and reflects heat back to the earth

gross production energy fixed per unit area by photosynthetic activity of plants before respiration; total energy flow at the secondary level is not gross production, but rather assimilation, because consumers use material already produced with respiratory losses

group selection elimination of one group of individuals by another group of individuals possessing superior genetic traits; not a widely accepted hypothesis

growth form morphological category of plants, such as tree, shrub, and vine

guild group of populations which utilizes a gradient of resources in a similar way

gyre circular motion of water in major ocean basins

gyttja mud rich in organic matter found at the bottom of or near the shore of some lakes

habitat place where a plant or animal lives

hadal that part of the ocean below 6000 m

halocline changes in salinity with depth in the oceans

halophyte plant able to survive and complete its life cycle at high salinities

haploid having a single set of unpaired chromosomes in each cell nucleus

Hardy-Weinberg law proposition that genotypic ratios resulting from random mating remained unchanged from one generation to another, provided that natural selection, genetic drift, and mutation are absent

heliothermism acquisition of heat energy by ectotherms by basking in the sun

hemicryptophyte perennial shoots or buds close to the surface of the ground; often covered with litter

herbivore organism that feeds on plant tissue

hermaphrodite organism possessing the reproductive organs of both sexes

heterogenity state of being mixed in composition; can refer to genetic or environmental conditions

heterosis situation in which the heterozygote is more fit than either homozygote

heterotherm organism that during part of its life history becomes either endothermic or ectothermic; hibernating endotherms become ectothermic, and foraging insects such as bees become endothermic during periods of activity; characterized by rapid, drastic, repeated changes in body temperature

heterotrophic requiring a supply of organic matter or food from the environment

heterozygous containing two different alleles of a gene, one from each parent, at the corresponding loci of a pair of chromosomes

hibernation winter dormancy in animals characterized by a great decrease in metabolism

hierarchy sequence of sets made up of smaller subsets

histosol soil characterized by high organic matter content

homeostasis maintenance of nearly constant conditions in function of an organism or in interaction among individuals in a population

home range area over which an animal ranges throughout the year

homoiohydric ability to maintain a stable internal water balance independent of the environment

homoiotherm animal with a fairly constant body temperature; also spelled homeotherm and homotherm

homoiothermy regulation of body temperature by physiological means

homologous chromosomes corresponding chromosomes from male and female parents which pair during meiosis

homozygous containing two identical alleles of a gene at the corresponding loci of a pair of chromosomes

horizon major zones or layers of soil each with its own particular structure and characteristics

host organism that provides food or other benefit to another organism of a different species; usually refers to an organism exploited by a parasite

humus organic material derived from partial decay of plant and animal matter

hybrid plant or animal resulting from a cross between genetically different parents

hydrosphere body of water on or near the earth's surface

hyperthermia rise in body temperature to reduce thermal differences between an animal and a hot environment, thus reducing the rate of heat flow into the body

hypha filament of a fungus thalli or vegetative body

hypolimnion cold, oxygen-poor zone of a lake that lies below the thermocline

hypothesis proposed explanation for a phenomenon; it should be testable for acceptance or rejection by experimentation

immigration arrival of new individuals into a habitat or population

immobilization conversion of an element from inorganic to organic form in microbial or plant tissue, rendering the nutrient relatively unavailable to other organisms

importance value sum of relative density, relative dominance, and relative frequency of a species in a community

imprinting type of rapid learning at a particular and early stage of development in which the individual learns identifying characteristics of another individual or object

inbreeding mating among close relatives

inbreeding depression detrimental effects of inbreeding

incipient lethal temperature temperature at which a stated fraction of a population of poikilothermic animals (usually 50 percent) will die when brought rapidly to it from a different temperature

inclusive fitness sum of the total fitness of an individual and the fitness of its relatives, weighted according to the degree of relationship

independent variable variable *x*, which yields the first of two numbers in an ordered pair *(x, y)*; the set of all values taken on by the independent variable is called the *domain* of the function; see *dependent variable*

inductive method in testing hypotheses, going from the specific to the general

infaunal organisms living within a substrate

infralittoral region below the littoral region of the sea

inhibition model model of succession proposing that the dominant vegetation occupying a site prevents colonization of that site by other plants of the next successional community

instar form of insect or other arthropod between successive molts

interdemic selection group selection of populations within a species

interference competition competition in which access to a resource is limited by the presence of a competitor

intermediate host hosts which harbor developmental phases of parasites; the infective stage or stages can develop only when the parasite is independent of its definitive host; see *definitive host*

interspecific between individuals of different species

intraspecific between individuals of the same species

intrinsic rate of increase (r) intrinsic growth rate of a population under ideal conditions without inhibition from competition

introgression incorporation of genes of one species into the gene pool of another

inversion (genetic) reversal of part of a chromosome so that genes within that part lie in reverse order; (meteorological) increase rather than decrease in air temperature with height caused by radiational cooling of the earth (radiational inversion), or by compression and consequent heating of subsiding air masses from high pressure areas (subsidence inversion)

island biogeography study of distribution of organisms on islands

isolating mechanism any structural, behavioral, or physiological mechanism that blocks or inhibits gene exchange between two populations

isotherm lines drawn on a map connecting points with the same temperature at a certain period of time

Jurassic middle period of the Mesozoic some 190 to 135 million years ago; dinosaurs abundant; bony fishes evolving rapidly; gymnosperms dominant; warm, humid climate

K-selection selection under carrying capacity conditions and high level of competition *key factor analysis* statistical analysis of population processes accountable for changes in population size

kin selection differential reproduction among groups of closely related individuals

kinetic energy energy in motion

krummholtz stunted form of trees characteristic of transition zone betwen alpine tundra and subalpine coniferous forest

lagg moat-like area of shallow water, often dominated by sedges, surrounding a peat mat

landscape ecology study of the structure, function, and change in a heterogeneous landscape composed of interacting ecosystems

Langmuir cell localized area of water a few meters wide and hundreds of meters long created by the wind; involves a local circulation pattern of upwelling water that diverges from the center, converges, and sinks at the boundaries

latent heat of fusion amount of heat given up when a unit mass of a substance converts from a liquid to a solid state, or the amount of heat absorbed when a substance converts from the solid to liquid state

laterization soil-forming process in hot, humid climates characterized by intense oxidation resulting in loss of bases and in a deeply weathered soil composed of silica, sesquioxides of iron and aluminum, clays, and residual quartz

leach dissolving and removal of nutrients by water out of soil, litter, and organic matter

leaf area index ratio of area of canopy foliage to ground area

lek communal courtship area used by males to attract and mate with females

lentic pertaining to standing water, as lakes and ponds

life expectancy average number of years to be lived in the future by members of the population

life table tabulation of mortality and survivorship schedule of a population

life zone major area of plant and animal life equivalent to a biome; transcontinental region or belt characterized by particular plants and animals and distinguished by temperature differences; applies best to mountainous regions where temperature changes accompany changes in altitude

light reaction light-dependent sequences of photosynthetic reactions

lignotuber specialized structure on the roots of certain fire-adapted trees, particularly *eucalyptus*, from which new growth sprouts following a fire

limit cycle stable oscillation in the population levels of a species, usually involving predator and prey interactions

limiting resource resource or environmental condition that most limits the abundance and distribution of an organism

limnetic pertaining to or living in the open water of a pond or lake

limnetic zone shallow water zone of lake or sea in which light penetrates to the bottom

lithosol soil showing little or no evidence of soil development; consisting mainly of partly weathered rock fragments or nearly barren rock

lithosphere rocky material of the earth's outer crust

littoral shallow water of lake in which light penetrates to the bottom, permitting submerged, floating, and emergent vegetative growth; also shore zone of tidal water between high-water and low-water marks

local stability ability of a system to return to its initial conditions following a small disturbance

locus site on a chromosome occupied by a specific gene

loess soil developed from wind-deposited material

logistic curve S-shaped curve of population growth which slows at first, steepens, and then flattens out at asymptote, determined by carrying capacity

logistic equation mathematical expression for the population growth curve in which rate of increase decreases linearly as population size increases

log-normal distribution frequency distribution in which the horizontal or x axis is expressed in a logarithmic scale; in the frequency distribution of species, the x axis represents the number of individuals and the y axis represents the number of species

long-day organism plant or animal that requires long days (days with more than a certain minimum of daylight) to flower or come into reproductive condition

lotic pertaining to flowing water

macronutrients essential nutrients needed in relatively large amounts by plants and animals

macroparasite parasitic worms, lice, fungi, and the like; have comparatively long generation time; spread by direct or indirect transmission, and may involve intermediate hosts or vectors

maquis sclerophyllous shrub vegetation in the Mediterranean region

marl unconsolidated deposit formed in freshwater lakes that consists chiefly of calcium carbonate mixed with clay and other impurities

marsh wetland dominated by grassy vegetation, such as cattails and sedges

mattoral sclerophyllous shrub vegetation in regions of Chile with mediterranean climate

mediterranean-type climate semiarid climate characterized by a hot, dry summer and a wet, mild winter

meiofauna benthic organisms within the size range of 1 to 0.1 mm; interstitial fauna

meiosis two successive divisions by gametic cells, with only one duplication of chromosomes so that the number of chromosomes in daughter cells is one-half the diploid number

mesic moderately moist habitat

Mesozoic middle era in the geological time scale some 230 to 70 million years ago; Age of Reptiles; three main periods are Triassic, Jurassic, and Cretaceous

metalimnion transition zone in lake between hypolimnion and epilimnion; region of rapid temperature decline

micella soil particle of clay and humus, carrying $^+$ electrical charges at the surface

microbivore organism that feed on microbes, especially in the soil and litter

microclimate climate on a very local scale that differs from the general climate of the area; influences the presence and distribution of organisms

microflora bacteria and certain fungi inhabiting the soil

microhabitat that part of the general habitat utilized by an organism

micronutrients essential nutrients needed in very small quantities by plants and animals

microparasite viruses, bacteria, and protozoans, characterized by small size, short generation time, and rapid multiplication

migration intentional, directional, and usually seasonal movement of animals between two regions or habitats; involves departure and return of the same individual; a round trip movement

mimicry resemblance of one organism to another or to an object in the environment evolved to deceive predators

mineralization microbial breakdown of humus and other organic matter in soil to inorganic substances

minimum viable population size of a population which, with a given probability, will ensure the existence of the population for a stated period of time

mire wetland characterized by an accumulation of peat

mitosis cell division involving chromosome duplication resulting in two daughter cells with full complement of chromosomes, genetically the same as parent cells

model in theoretical and systems ecology, an abstraction or simplification of a natural phenomenon developed to predict a new phenomenon or to provide insights into existing ones; in mimetic association, the organism mimicked by different organism

moder type of forest humus layer in which plant fragments and mineral particles form a loose net-like structure held together by a chain of small arthropod droppings

mollisol soil formed by calcification characterized by accumulation of calcium carbonate in lower horizons and high organic content in upper horizons

monoecious in plants, occurrence of reproduction organs of both sexes on the same individual, either as different flowers (hermaphroditic) or in the same flower (dioceous)

monogamy mating of an animal and maintenance of a pair bond with only one member of the opposite sex at a time

moor a blanket bog or peatland

mor type of forest humus layer of unincorporated organic matter usually matted or compacted, or both, and distinct from mineral soil; low in bases and acid in reaction

morphology study of the form of organisms

mortality rate number of individuals in a population dying during a given time interval divided by the number alive at the beginning of the time interval

motane pertaining to mountains

mull humus which contains appreciable amounts of mineral bases and forms a humus-rich layer of forested soil consisting of mixed organic and mineral matter; blends into the upper mineral layer without abrupt changes in soil characteristics

Mullerian mimicry resemblance of two or more conspicuously marked distasteful species which increases predator avoidance

multivoline organisms that have several generations during a single season

mutation transmissable changes in structure of gene or chromosome

mutualism relationship between two species in which both benefit

mycelium mass of hyphae that make up the vegetative portion of a fungus

mycorrhizae association of fungus with roots of higher plants which improves the plants' uptake of nutrients from the soil

natality production of new individuals in a population

natural selection differential reproduction and survival of individuals that results in elimination of maladaptive traits from a population

negative feedback homeostatic control in which an increase in some substance or activity ultimately inhibits or reverses the direction of the processes leading to the increase

nekton aquatic animals that are able to move at will through the water

neritic marine environment embracing the regions where land masses extend outward as a continental shelf

net production accumulation of total biomass over a given period of time, after respiration is deducted from gross production in plants and from assimilated energy in consumer organisms deducted from gross production

net reproductive rate number of young a female can be expected to produce during a lifetime

net aboveground production accumulation of biomass in aboveground parts of plants over a given period of time

net belowground production accumulation of biomass in roots, rhizomes, and the like

neutrophilus pertains to plants not restricted to high calcium or to acidic soils

niche functional role of a species in the community, including activities and relationships

niche breadth range of a single niche dimension occupied by a population

nitrification breakdown of nitrogen-containing organic compounds into nitrates and nitrites

nitrogen fixation conversion of atmospheric nitrogen to forms usable by organisms

null hypothesis statement of no difference between sets of values formulated for statistical testing

numerical response change in size of a population of predators in response to change in density of its prey

nutrient substance required by organisms for normal growth and activity

nutrient cycle pathway of an element or nutrient through the ecosystems from assimilation by organisms to release by decomposition

old growth forest forest that has not been cut for decades nor disturbed by humans for hundreds of years

oligotrophic term applied to a body of water low in nutrients and in productivity

omnivore an animal that feeds on both plant and animal matter

ombrotrophic condition in bogs or mires in which water is highly deficient in mineral salts and has a low pH

opportunistic species organisms able to exploit temporary habitats or conditions

optimal foraging tendency of animals to harvest food efficiently, to select food sizes or food patches that will result in maximum food intake for energy expended

optimum yield amount of material that can be removed from a population which will result in production of maximum amount of biomass on a sustained yield basis

ordination process by which communities are positioned graphically on a gradient of one to several axes so that the distances between them reflect differences in composition

Ordovician second oldest period of the Paleozoic era some 510 to 440 million years ago; marine invertebrates abundant

oscillation regular fluctuation in a fixed cycle above or below some set point

osmosis movement of water molecules across a differentially permeable membrane in response to a concentration or pressure gradient

outbreeding production of offspring by the fusion of distantly related gametes

overturn vertical mixing of layers in a body of water brought about by seasonal changes in temperature *oxisol* soil developed under humid semitropical and tropical conditions characterized by silicates and hydrous oxides, clays, residual quartz, deficiency in bases and low plant nutrients; formed by the process of laterization

paleoecology study of ecology of past communities by means of fossil record

Paleozoic geological era of 620–230 million years ago, starting with the Cambrian period and ending with the Permian, during which land plants, insects, amphibians, and the first reptiles evolved

parapatric having ranges coming into contact but not overlapping by much more than the dispersal range of an individual in its lifetime

parasitism relationship between two species in which one benefits while the other is harmed (although not usually killed directly)

parasitoid insect larva that kills its host by consuming completely the host's soft tissues

peat unconsolidated material consisting of undecomposed and only slightly decomposed organic matter under conditions of excessive moisture

ped soil particles held together in clusters of various sizes

pedalfer podzolic soil possessing a layer of iron accumulation (hardpan) that impedes free circulation of air and water

pelagic referring to the open sea

periphyton in freshwater ecosystems, organisms that are attached to submerged plant stems and leaves; see *aufwuchs*

permafrost permanently frozen soil

permanent wilting point point at which water potential in the soil and conductivity assume such low values that the plant is unable to extract sufficient water to survive and wilts permanently

Permian most recent period of the Paleozoic about 280–230 million years ago; dominance of reptiles; appearance of modern insects and gymnosperm plants

perturbation another word for disturbance; borrowed from physics to suggest an event that alters the state of or direction of change in a system

phagioclimax vegetation type maintained as climax over a long period of time by continued human activity

phanerophyte perennial buds of trees, shrubs, and vines carried high up in the air

phenology study of seasonal changes in plant and animal life and the relationship of these changes to weather and climate

phenotype physical expression of a characteristic of an organism as determined by genetic constitution and environment

phenotypic plasticity ability to change form under different environmental conditions

pheromone chemical substance released by an animal that influences behavior of others of the same species

photic zone lighted water column of a lake or ocean inhabited by plankton

photoperiodism response of plants and animals to changes in relative duration of light and dark

photophosphorylation production of ATP from ADP and inorganic phosphorus using light energy in photosynthesis

photorespiration respiration that occurs in light in C_3 plants and is not coupled to oxidative phosphorylation and does not generate ATP; a wasteful process decreasing photosynthetic efficiency

photosynthesis synthesis of carbohydrates from carbon dioxide and water by chlorophyll using light as energy and releasing oxygen as a by-product

phreatophyte type of plant that habitually obtains its water supply from zone of groundwater

physiognomy outward appearance of the landscape

physiological longevity maximum lifespan of an individual in a population under given environmental conditions

phytoplankton small, floating plant life in aquatic ecosystems; planktonic plants

pioneer species plants that are initial invaders of disturbed sites or early seral stages of succession

plankton small, floating or weakly swimming plants and animals in freshwater and marine ecosystems

Pleistocene geological epoch extending from about 2 million to 10,000 years ago, characterized by recurring glaciers; the Ice Age

pneumatophore erect respiratory root that protrudes above waterlogged soils; typical of bald cypress and mangroves

podzolization soil-forming process resulting from acid leaching of the A horizon and accumulation of iron, aluminum, silica, and clays in lower horizon

poikilohydric internal water state matches that of the environment

poikilothermy variation of body temperature with external conditions

polyandry mating of one female with several males

polygamy acquisition by an individual of two or more mates, none of which is mated to other individuals

polygyny mating of one male with several females

polymorphism occurrence of more than one distinct form of individuals in a population

polyploid having three or more times the haploid number of chromosomes

positive feedback control in a system which reinforces the process in the same direction

potential energy energy available to do work

potential evapotranspiration amount of water that would be transpired under constantly optimal conditions of soil moisture and plant cover

Precambrian earliest and longest of the geological time periods some 4600 and 570 million years ago; precedes the Paleozoic era; appearance of blue-green algae and fungi

precocial young birds hatched with down, eyes open, and able to move about; also young mammals born with eyes open and able to follow their mother after birth (for example, fawn deer, calves).

predation one living organism serves as a food source for another

preferred temperature range of temperatures within which poikilotherms function most efficiently

primary production production by green plants

primary succession vegetational development starting from a new site never before colonized by life

producer green plants and certain chemosynthetic bacteria that convert light or chemical energy into organismal tissue

production amount of energy formed by an individual, population, or community per unit time

productivity rate of energy fixation or storage per unit time; not to be confused with production

profundal deep zone in aquatic ecosystems below the limnetic zone

proximate factor any characteristic of the environment that an organism used as a cue to behavioral or physiological responses; mechanisms behind the operation of an adaptation; opposite of *ultimate factors*

pycnocline layer of water that exhibits a rapid change in density

Quanternary geological period from 2 million years ago to the present; includes the Pleistocene and recent epochs

r-selection selection under low population densities; favors high reproductive rates under conditions of low competition

rain forest tropical forests with regular heavy rain, constantly high temperatures, and prolific plant growth

rain shadow dry area on lee side of mountains

raised bog bog in which the accumulation of peat has raised its surface above both the surrounding landscape and the water table; it develops its own perched water table

ramet any individual belonging to a clone

realized niche portion of fundamental niche space occupied by a population in face of competition from populations of other species; environmental conditions under which a population survives and reproduces in nature

recombination exchange of genetic material resulting from independent assortment of chromosomes and their genes during gamete production, followed by a random mix of different sets of genes at fertilization

recruitment addition of reproduction of new individuals to a population

regolith mantle of unconsolidated material below the soil from which soil develops

relict surviving species of a once widely dispersed group

reproductive isolation separation of one population from another by the inability to produce viable offspring when the two populations are mated

reproductive value potential reproductive output of an individual at a particular age (x) relative to that of a newborn individual at the same time

residual reproductive value reproductive value of an individual reduced by its expected present reproduction

resilience ability of a system to absorb changes and return to its original condition *resistance* ability of a system to resist changes from a disturbance

resource environmental component utilized by a living organism

resource allocation action of apportioning the supply of a resource to a specific use

respiration metabolic assimilation of oxygen accompanied by production of carbon dioxide and water, release of energy, and breaking down of organic compounds

restoration ecology study of the application of ecological theory to the ecological restoration of highly disturbed sites

rete a large network or discrete vascular bundle of intermingling small blood vessels carrying arterial and venous blood; acts as a heat exchanger in mammals and certain fish and sharks.

rheotrophic applies to wetlands, especially bogs, that obtain much of their nutrient input from groundwater

rhizobia bacteria capable of living mutualistically with higher plants

rhizome horizontally growing underground stem that through branching gives rise to vegetative structures

rhizoplane root surface

rhizosphere soil region immediately surrounding roots

richness component of species diversity; the number of species present in an area

riparian along banks of rivers and streams; riverbank forests are often called gallery forests

ruminant ungulates with a three- or four-chambered stomach; the large first chamber is known as the rumen, in which bacterial fermentation of plant matter is consumed

saprophage organism that feeds on dead plant and animal matter

savanna tropical grassland usually with scattered trees or shrubs

sclerophyll woody plant with hard, leathery, evergreen leaves that prevents moisture loss

scraper aquatic insects that feed by scraping algae from a substrate

search image mental image formed in predators, enabling them to find more quickly and to concentrate on a common type of prey

secondary production production by consumer organisms

secondary substances organic compounds produced by plants that are utilized in chemical defense

secondary succession plant succession taking place on sites that have already supported life

seiche oscillation of a structure of water about a point or node

semelparity having only one terminal reproductive effort in a lifetime

semiarid region of fairly dry climate with precipitation between 25 and 60 cm a year; with an evapotranspiration rate high enough so the potential loss of water to the environment exceeds inputs

senescence process of aging

seral series of stages that follow one another in succession

sere series of successional stages on a given site that lead to a terminal community

serotinous cones cones of pine that remain on the tree several years and require the heat of fire to open them and release the seeds

serpentine soil soils derived from ultrabasic rocks that are high in iron, magnesium, nickel, chromium, and cobalt and low in calcium, potassium, sodium, and aluminum; support distinctive communities

sessile not free to move about; permanently attached to a substrate

shade tolerant plants that are able to grow and reproduce under low light conditions

short-day organisms plants and animals that come into reproductive condition under conditions of short days (days with less than a certain maximum length)

sibling species species with similar appearance but unable to interbreed

sigmoid curve S-shaped curve of logistic growth

Silurian period between the Ordovician and Devonian periods some 400 to 405 million years ago; characterized by early land plants, invertebrates, primitive jawless fish

site combination of biotic, climatic, and soil conditions which determine an area's capacity to produce vegetation

snag dead or partially dead tree at least 10.2 cm dbh and 1.8 m tall; important for cavity-nesting birds and mammals

social parasite animal that uses other individuals or species to rear its young; for example, cowbirds

soil association group of defined and named soil taxonomic units occurring together in an individual and characteristic pattern over a geographic region

soil horizon developmental layer in the soil with its own characteristics of thickness, color, texture, structure, acidity, nutrient concentration, and the like

soil profile distinctive layering of horizons in the soil

soil series basic unit of soil classification consisting of soils which are essentially alike in all major profile characteristics except texture of the A horizon; soil series are usually names for the locality where the typical soil was first recorded

soil structure arrangement of soil particles and aggregates

soil texture relative proportions of the three particle sizes (sand, silt, and clay) in the soil.

soil type lowest unit in the natural system of soil classification, consisting of soils which are alike in all characteristics including texture of the A horizon

speciation separation of a population into two or more reproductively isolated populations

species diversity measurement which relates density of organisms of each type present in a habitat to the number of species in a habitat

species packing increase in species diversity within a comparatively narrow range of resource variation

species richness number of species in a given area

species selection form of group selection in which sets of species with different characteristics increase by speciation or decrease by extinction at different rates because of differences in adaptive characteristics

specific heat amount of energy that must be added or removed to raise or lower temperature of a substance by a specific amount

spiraling mechanism of retention of nutrients in flowing water ecosystems involving the interdependent processes, nutrient recycling, and downstream transport

spodosol soil characterized by the presence of a horizon in which organic matter and amorphous oxides of aluminum and iron have precipitated; includes podzols

stability ability of a system to resist change or to recover rapidly after a disturbance; absence of fluctuations in a population

stabilizing selection selection favoring the middle in the distribution of phenotypes

stable age distribution constant proportion of individuals of various age classes in a population through population changes

stand unit of vegetation that is essentially homogeneous in all layers and differs from adjacent types qualitatively and quantitatively

standard deviation statistical measure defining the dispersion of values about the mean in a normal distribution

standing crop amount of biomass per unit area at a given time

stationary age distribution special form of stable age distribution in which the population has reached a constant size, the birth rate equals the death rate, and age distribution remains fixed

stochastic patterns arising from random factors

stochastic model mathematical model based on probabilities; predictions of the model are not fixed but variable; opposite of *deterministic model*

stratification division of an aquatic or terrestrial community into distinguishable layers on the basis of temperature, moisture, light, vegetative structure, and other such factors creating zones for different plant and animal types

sublittoral lower division of sea from about 40 m to 60 m to below 200 m

subsidence inversion atmospheric inversion produced by sinking air movement from aloft

subspecies geographical unit of a species population distinguishable by certain morphological, behavioral, or physiological characteristics

succession replacement of one community by another; often progresses to a stable terminal community called the climax

sun plant plant able to grow and reproduce only under high light conditions

supercooling in ectotherms, lowering of body temperature below freezing without freezing body tissue; involves the presence of certain solutes, particularly glycerol

sustained yield yield per unit time from an exploited population equal to production per unit time

swamp wooded wetland in which water is near or above ground level

switching predator changing its diet from a less abundant to a more abundant prey species; see *threshold of security*

symbiosis living together of two or more species

sympatric living in the same area; usually refers to overlapping populations *synecology* study of groups of organisms in relation to their environment; community ecology

system set or collection of interdependent parts or subsystems enclosed within a defined boundary; the outside environment provides both inputs and receives attributes transmitted to it by the system

taiga the northern circumpolar boreal forest

temperate rain forest forests in regions characterized by a relatively mild clime and heavy rainfall that produces lush vegetative growth; one example is the coniferous forest of the coastal Pacific Northwest of North America

territory area defended by an animal; varies among animals according to social behavior, social organization, and resource requirements of different species

Tertiary first period of the Cenozoic era about 65 to 2 million years ago; composed of Paleocene, Eocene, Oligocene, Miocene, and Pliocene epochs; characterized by emergence of mammals

thermal conductance rate at which heat flows through a substance

thermal neutral zone among homoiotherms, the range of temperatures at which metabolic rate does not vary with temperature

thermal tolerance range of temperatures in which an aquatic poikilotherm is most at home

thermocline layer in a thermally stratified body of water in which temperature changes rapidly relative to the remainder of the body

thermogenesis increase in production of metabolic heat to counteract the loss of heat to a colder environment

therophyte life form of plants that survives unfavorable conditions in the form of a seed; annual and ephemeral species

thinning law,-3/2 self-thinning plant populations, sown at sufficiently high densities approach and follow a thinning line with a slope of roughly -3/2; thus, in a growing population, plant weight increases faster than density decreases to a point where the slope changes to a -1

threshold of security point in local population density at which the predator turns its attention to other prey (see *switching*) because of harvesting efficiency; the segment of prey population below the threshold is relatively secure from predation

time lag delay in a response to change

tolerance model model of succession that proposes that succession leads to a community composed of those species most efficient in exploiting resources; colonists neither increase or decrease the rate of recruitment or growth of later colonists

toposequence pattern of local soils whose development was controlled by topography of the landscape

torpidity temporary condition of an animal involving a great reduction in respiration; results in loss of power of motion and feeling; usually occurs in response to some unfavorable environmental condition such as heat or cold to reduce energy expenditure

trace element element occurring and needed in small quantities; see *micronutrient*

translocation transport of materials within a plant; absorption of minerals from soil into roots and their movement throughout the plant

transpiration loss of water vapor by land plants

Triassic oldest period of the Mesozoic some 230 to 195 million years ago; increase in primitive amphibians and reptiles

trophic related to feeding

trophic height considered 1 plus the length in the sense of ecological efficiencies of the shortest food chain linking a species to a primary producer

trophic level functional classification of organisms in an ecosystem according to feeding relationships, from first level autotrophs through succeeding levels of herbivores and carnivores

trophic structure organization of a community based on the number of feeding or energy transfer levels

trophogenic zone upper layer of the water column in lakes, ponds, and oceans in which light is sufficient for photosynthesis

tropholytic zone area in lakes and oceans below the compensation point

tundra areas in arctic and alpine (high mountains) regions characterized by bare ground, absence of trees, and growth of mosses, lichens, sedges, forbs, and low shrubs

turnover rate rate of replacement of a substance or a species when losses to a system are replaced by additions

ultimate factor survival value of an adaptation; evolutionary reason for an adaptation *ultimate incipient lethal temperature* upper or lower limit of temperature at which an acclimatized organism will succumb

univoltine breeding only once a season

upwelling areas in oceans where currents force water from deep within the ocean into the euphotic zone

vacuole fluid-filled cavity within the cytoplasm

vagile free to move about

validation an explicit and objective test of the basic hypothesis

vapor pressure the amount of pressure water vapor exerts independent of dry air

variance the square of the standard deviation

vector organism that transmits a pathogen from one organism to another

vegetative reproduction asexual reproduction in which plants propagate themselves by means of specialized multicellular organs such as bulbs, corms, rhizomes, stems, and the like

verification process of testing whether or not a model is a reasonable representation of a real-life system being investigated

viscosity property of a fluid that resists the force within the fluid that causes it to flow

Wallace's line biogeographic line between the islands of Borneo and the Celebes which marks the eastward boundary of many land-locked Eurasian organisms and the boundary of the Oriental region

water potential measure of energy in an aqueous solution needed to move water molecules across a semipermeable membrane; water tends to move from areas of high or less negative to areas of low or more negative potential

watershed entire region drained by a waterway which drains into a lake or reservoir; total area above a given point on a stream that contributes water to the flow at that point; the topographic dividing line from which surface streams flow in two different directions

wilting point moisture content of soil on an oven-dry basis at which plants wilt and fail to recover their turgidity when placed in a dark, humid atmosphere

xeric dry conditions, especially relating to soil

xerophyte plants adapted to life in a dry or physiologically dry (saline) habitat

Zeitgeber time-setter, usually light, that entrains a circadian rhythm to environmental rhythms

zoogeography study of the distribution of animals

zooplankton floating or weakly swimming animals in freshwater and marine ecosystems; planktonic animals

BIBLIOGRAPHY

ABELE, L. G., AND E. F. CONNER. 1979. Application of island biogeography theory to refuge design: Making the right decisions for the wrong reasons. In *Proceedings of First Conference on Scientific Research in National Parks*, ed. R. M. Linn, 89–94, Vol. 1. U.S. Department of Interior, Washington, D.C.

ABER, J. S., D. B. BOTKIN, AND J. M. MELILLO. 1978. Predicting the effects of different harvesting regimes on forest floor dynamics in northern hardwoods. *Can. J. For. Res.*, 8:305–315.

ABRAHAMSON, W. G., AND M. D. GADGIL. 1973. Growth form and reproductive effort in goldenrod (*Solidago, Compositae*). *Am. Nat.*, 107:651–661.

ACKERT, J. E., G. L. GRAHAM, L. O. NOLF, AND D. A. PORTYER. 1931. Quantitative studies on the administration of variable numbers of nematode eggs (*Ascaridia lineata*) to chickens. *Trans. Am. Microscopical Soc.*, 50:206–214.

ADDICOTT, J. F. 1979. A multispecies aphid–plant association: A comparison of local and metapopulations. *Can. J. Zool.*, 56:2554–2564.

———. 1985. On the population consequences of mutualism. In T. Case and J. Diamond (eds.), *Community Ecology*. Harper & Row, New York, pp. 425–436.

———. 1986. Variation in the costs and benefits of mutualism: The interaction between yuccas and yucca moths. *Oecologica*, 70:486–494.

ADKISSON, P. L. 1966. Internal clocks and insect diapause. *Science*, 154:234–241.

AHLGREN, I. F., AND C. H. AHLGREN. 1960. Ecological effects of forest fires. *Bot. Rev.*, 26:483–533.

AHMADJIAN, V., AND J. B. JACOBS. 1982. Algal–fungal relationships in lichens: Recognition, synthesis, and development. In L. J. Goff (ed.), *Algal Symbiosis: A Continuum of Interaction Strategies*. Cambridge University Press, Cambridge, England.

AHMED, A. K. 1976. PCB's in the environment. *Environment*, 18(2):6–11.

AKER, C. L. 1982. Spatial and temporal dispersion patterns of pollinators and their relationship to the flowering strategy of *Yucca whipplei* (Agavaceae). *Oecologica*, 54:243–252.

ALCOCK, J. 1973. Cues used in searching for food by red-winged blackbirds (*Agelaires Phoeniceus*). *Behaviour*, 17:130–233.

ALEXANDER, M. 1965. Nitrification. In W. V. Barthelomew and F. F. Clark (eds.), *Soil Nitrogen. Am. Soc. Agron. Monogr.* No. 10, pp. 307–343.

ALEXANDER, M. M. 1958. The place of aging in wildlife management. *Am. Sci.*, 46:123–137.

ALEXANDER, R. D., AND G. BORGIA. 1978. Group selection, altruism, and levels of organization of life. *Ann. Rev. Ecol. Syst.*, 9:449–474.

ALEXANDER, R. D., AND D. W. TINKLE (EDS.). 1981. *Natural Selection and Social Behavior: Recent Research and New Theories*. Chiron, New York.

ALLEE, W. C., A. E. EMERSON, O. PARK, T. PARK, AND K. P. SCHMIDT. 1949. *Principles of Animal Ecology*. Saunders, Philadelphia.

ALLEN, D. L. 1943. Michigan fox squirrel management. Mich. Dept. Cons. Game Div. Pub., 100.

———. 1962. *Our Wildlife Legacy*, 2nd ed. Funk and Wagnalls, New York.

ALLEN, L. N., AND E. R. LEMON. 1976. Carbon dioxide exchange and turbulence in a Costa Rican tropical rain forest In. J. C. Monteith (ed.), *Vegetation and the Atmosphere*. Vol. 2 *Case Studies*. Academic Press, London, pp. 265–308.

ALLEN, T. F. H., AND T. W. HOEKSTRA. 1984. The abuse of the concept "disturbance": A scaling problem. Unpublished manuscript.

ALLEN, T. F. H., AND T. B. STARR. 1982. *Hierarchy: Perspectives for Ecological Complexity*. University of Chicago Press, Chicago.

ALM, G. 1952. Year class fluctuations and span of life of perch. *Rept. Inst. Freshwater Res. Drottningholm*, 33:17–38.

ALTMANN, M. 1960. The role of juvenile elk and moose in the social dynamics of their species. *Zoologica*, 45:35–40.

AMBUEL, B., AND S. A. TEMPLE. 1983. Area-dependent changes in the bird communities and vegetation of southern Wisconsin forests. *Ecology*, 64:1057–1068.

AMERICAN CHEMICAL SOCIETY. 1969. *Cleaning Our Environment: The Chemical Basis for Action*. American Chemical Society, Washington, D.C.

AMUNDSON, D. C., AND H. E. WRIGHT, JR. 1979. Forest changes in Minnesota at the end of the Pleistocene. *Ecol. Monogr.*, 49:109–127.

ANDERSON, C. C., AND M. TRESHOW. 1980. A review of environmental genetic factors that affect height in *Spartina alternifolia* Loisel (salt marsh and grass). *Estuaries*, 3:168–176.

ANDERSON, M., AND M. O. ERICKSON. 1982. Nest parasitism in goldeneyes, *Bucephala clangula*: Some evolutionary aspects. *Am. Nat.*, 120:1–16.

ANDERSON, N. H., AND K. W. CUMMINS. 1979. Influences of diet on the life histories of aquatic insects. *J. Fish. Res. Bd. Can.*, 36:335–342.

ANDERSON, R. C. 1963. The incidence, development, and experimental transmission of *Pneumostrongylus tenuis* Dougherty (Metastrongyloidae: Protostrongyliidae) of the meninges of the white-tailed deer (*Odiocoilus virginianus borealis*) in Ontario. *Can. J. Zool.*, 41:775–792.

———. 1965. Cerebospinal nematodiasis (*Pneumostrongylus tenuis*) in North American cervids. *Trans. N. Am. Nat. Res. Conf.*, 13:156–167.

ANDERSON, R. C., AND A. K. PRESTWOOD. 1981. Lungworms. In W. R. Davidson (ed.), *Diseases and Parasites of White-tailed Deer*, Mscl. Publ. No. 7. Tall Timbers Research Station, Tallahassee, Fla., pp. 266–317.

ANDERSON, R. M. 1981. Population ecology of infectious diseases. In R. M. May (ed.), *Theoretical Ecology: Principles and Applications*, 2nd ed. Sinauer Associates, Sunderland, Mass., pp. 318–355.

ANDERSON, R. M., AND R. M. MAY. 1978. Regulation and stability of host–parasite population interactions. I. Regulatory processes. *J. Animal Ecol.*, 47:219–247.

———. 1979. Population biology of infectious diseases. Part I. *Nature,* 280:361–367.

ANDREWARTHA, H. G. 1961. *Introduction to the Study of Animal Populations.* Methuen, London.

ANDREWARTHA, H. G., AND L. C. BIRCH. 1954. *The Distribution and Abundance of Animals.* University of Chicago Press, Chicago.

———. 1984. *The Ecological Web.* University of Chicago Press, Chicago.

ANDREWS, R., AND A. S. RAND. 1974. Reproductive effort in anoline lizards. *Ecology,* 55:1317–1327.

ANDREWS, R. D., D. C. COLEMAN, J. E. ELLIS, AND J. S. SINGH. 1975. Energy flow relationships in a short grass prairie ecosystem. *Proc. 1st Inter. Cong. Ecol.,* 22–28. W. Junk Publishers, The Hague.

ANDRZEJEWSKA, L., AND G. GYLLENBERG. 1980. Small herbivore subsystem. In A. I. Bretmeyer and G. M. Van Dyne (eds.), *Grasslands, Systems Analysis and Man,* pp. 201–268. Internationall Biological Programme No. 19, Cambridge University Press, Cambridge, England.

ANTOINE, L. H., JR. 1964. Drainage and the best use of urban land. *Public Works,* 95:88–90.

ANTONOVICS, J., A. N. BRADSHAW, AND R. G. TURNER. 1971. Heavy metal tolerance in plants. *Advances in Ecological Research,* 71:1–85.

ANTONOVICS, J. A., AND D. A. LEVIN. 1980. The ecological and genetical consequences of density-dependent regulation in plants. *Ann. Rev. Ecol. Syst.,* 11:411–452.

ANTONOVICS, J. A., AND R. B. PRIMACH. 1982. Experimental ecology and genetics in *Plantago.* VI. The demography of seedling transplants of *P. lanceolata. J. Ecol.,* 70:55–75.

ARP, A. J., AND J. J. CHILDRESS. 1983. Sulfide binding by the blood of the hydrothermal vent tube worm *Riftia pachyptila. Science,* 219:295–297.

ASCHMANN, H. 1973. Distribution and pecularity of Mediterranean ecosystems. In F. di Castri and H. A. Mooney (eds.), *Mediterranean Type Ecosystems, Origin and Structure.* Springer-Verlag, New York, pp. 11–19.

ASCHOFF, J. 1958. Tierische Periodik unter dem Einfluss von zeitgebern. *Z. F. Tierpsychol.,* 15:1–30.

———. 1966. Circadian activity pattern with two peaks. *Ecology,* 47:657–662.

ASH, J. E., AND J. P. BARKHAM. 1976. Changes and variability in the field layer of a coppiced woodland in Norfolk, England. *J. Ecol.,* 64:697–712.

ASKENMO, C. 1977. Effects of addition and removal of nestlings and nestling weight, nestling survival, and female weight loss in the pied flycatcher *Ficedula hypoleuca* (Pallas). *Ornis. Scand,* 8:1–8.

———. 1979. Reproductive effort and return rate of male pied flycatchers. *Am. Nat.,* 114:748–752.

ATSATT, P. R., AND D. J. O'DOWD. 1976. Plant defense guilds. *Science,* 193:24–29.

AUCLAIR, A. N., AND F. G. GOFF. 1971. Diversity relations of upland forests in the western Great Lakes area. *Am. Nat.* 105:499–528.

AUERBACH, S. I., D. J. NELSON, AND E. G. STRUXNESS. 1974. Environmental Sciences Division Annual Progress Report Period ending Sept. 30, 1973. *Environ. Sci. Div. Pub.* No. 575. Oak Ridge National Laboratory, Oak Ridge, Tenn.

AUMANN G. D., AND J. T. EMLEN. 1965. Relationship of population density to sodium availability and sodium selection by microtine rodents. *Nature,* 208:198–199.

AUSMUS, B. S., N. T. EDWARDS, AND M. WITKAMP. 1975. Microbial immobilization of carbon, nitrogen, phosphorus, and potassium: Implications for forest ecosystem processes. *Proc. Brit. Ecol. Soc. Symp. on Decomposition.* Blackwell, Oxford, England.

AUSTIN, M. P. 1985. Continuum concept, ordination methods, and niche theory. *Ann. Rev. Ecol. Syst.,* 16:39–62.

AUSTIN, M. P., R. B. CUNNINGHAM, AND R. B. GOOD. 1983. Altitudinal distribution in relation to other environmental factors of several Eucalypt species in southern New South Wales. *Aust. J. Ecol.,* 8:169–180.

AUSTIN, M. P., R. B. CUNNINGHAM, AND P. M. FLEMING. 1984. New approaches to direct gradient analysis using environmental scalars and statistical curve-fitting procedures. *Vegetatio,* 55:11–27.

AZAM, F., T. FENCHEL, J. D. FIELD, L. A. MEYER-REIL, AND F. THINGSTAD. 1983. The ecological role of water-column microbes in the sea. *Mar. Ecol. Prog. Ser.,* 10:257–263.

BACON, P. J. (ED.). 1985. *Population Dynamics of Rabies in Wildlife.* Academic Press, London.

BAER, J. G. 1951. *Ecology of Animal Parasites.* University of Illinois Press, Urbana.

BAES, C. F., JR., H. E. GOELLER, J. S. OLSON, AND R. M. ROTTY. 1977. Carbon dioxide and the climate: The uncontrolled experiment. *Am. Sci.* 65:310–320.

BAILEY, P. C. E. 1986. The feeding behavior of a sit-and-wait predator *Ranatra dispar* (Heteroptera: Nepidae): Optimal foraging and feeding dynamics. *Oecologica,* 68:291–293.

BAILEY, R. W. 1978. *Description of the Ecoregions of the United States.* U.S.D.A. Forest Service Intermountain Region, Ogden, Utah.

BAIN, M. B., J. T. FINN, AND H. E. BOOKE. 1988. Streamflow regulation and fish community structure. *Ecology,* 69:382–392.

BAKER, H. G., K. S. BAWA, G. W. FRANKIE, AND P. A. OPLER. 1983. Reproductive biology of plants in tropical forests. In F. B. Golley (ed.), *Tropical Forest Ecosystems: Structure and Function.* Elsevier, Amsterdam, pp. 183–215.

BAKER, M. C., L. M. MEWALDT, AND R. M. STEWART. 1981. Demography of white-crowned sparrows *(Zonotricia leucophrys nuttalli). Ecology,* 62:636–644.

BAKKEN, G. S. 1976. A heat transfer analysis of animals: Unifying concepts and the application of metabolism chamber data to field ecology. *J. Theor. Biol.,* 60:337–384.

BAKKER, R. T. 1983. The deer flees, the wolf pursues: Incongruencies in predator–prey evolution. In D. Futuyma and M. Slakin (eds.), *Coevolution.* Sinauer Associates, Sunderland, Mass., pp. 350–382.

BALDA, R. P. 1975. Vegetation structure and breeding bird diversity. In D. R. Smith (ed.), *Symposium on Management of Forest and Range Habitats for Nongame Birds,* U.S.D.A. For. Serv. GTR WB-1.

BALLARD, W. B., J. S. WHITMAN, AND C. L. GARDNER. 1987. Ecology of an exploited wolf population in south-central Alaska. *Wildl. Monogr.,* 98, 54 pp.

BARKALOW, F. S., JR., R. B. HAMILTON, AND R. F. SOOTS, JR. 1980. The vital statistics of an unexploited gray squirrel population. *J. Wildl. Manage.,* 34:489–500.

BARLOW, J. P. 1955. Physical and biological processes determining the distribution of zooplankton in a tidal estuary. *Biol. Bull.,* 109:211–225.

———. 1956. Effect of wind on salinity distribution in an estuary. *J. Marine Res.,* 15:192–203.

BARRETT, J. A. 1983. Plant–fungus symbioses. In D. Futuyma and M. Slakin (eds.), *Coevolution.* Sinauer Associates, Sunderland, Mass., pp. 137–160.

BARRETTE, C., AND MESSIER. 1980. Scent marking in free-ranging coyotes *canis latrans. Anim. Behav.,* 28:814–819.

BARTHOLOMEW, G. A. 1959. Mother–young relations and the maturation of pup behaviour in the Alaskan fur seal. *Anim. Behav.,* 7:163–171.

———. 1970. Bare zone between California shrub and grassland communities: The role of animals. *Science,* 170:1210–1212.

———. 1981. A matter of size: An examination of endothermy in insects and terrestrial vertebrates. In B.

B-2

Heinrich (ed.), *Insect Thermoregulation.* Wiley-Interscience, New York, pp. 45–78.

BASSAM, J. A. 1965. Photosynthesis. In J. Bonner and J. E. Varna (eds.), *Plant Biochemistry.* Academic Press, New York, pp. 875–902.

———. 1977. Increasing crop production through controlled photosynthesis. *Science,* 197:630–638.

BASSAM, J. A., AND B. B. BUCKANAN. 1982. Carbon dioxide fixation pathways in plants and bacteria. In Govindjee (ed.), *Photosynthesis, Development, Carbon Metabolism and Plant Productivity,* Academic Press, New York.

BATCHELDER, R. R. 1967. Spatial and temporal patterns of fire in the tropical world. *Proc. 6th Tall Timber Fire Ecol. Conf.,* pp. 171–208.

BATZLI, G. O., AND F. A. PITELKA. 1970. Influence of meadow mouse populations on California grassland. *Ecology,* 51:1027–1039.

BAWA, K. S. 1980. Evolution of dioecy in flowering plants. *Ann. Rev. Ecol. Syst.,* 11:15–39.

BAZZAZ, F. A. 1975. Plant species diversity in old field successional ecosystems in southern Illinois. *Ecology,* 56:485–488.

———. 1979 The physiological ecology of plant succession. *Ann. Rev. Ecol. Syst.,* 10:351–371.

BEACHAM, T. D. 1980. Dispersal during population fluctuations of the vole *Microtus townsendii. J. Anim. Ecol.,* 49:867–877.

BEALS, E. W. 1968. Spatial pattern of shrubs on a desert plain in Ethopia. *Ecology,* 49:744–746.

———. 1985. Bray-Curtis ordination: An effective strategy for analysis of multivariate ecological data. *Adv. Ecol. Res.,* 14:1–55.

BEATLEY, J. C. 1969. Dependence of desert rodents on winter annuals and precipitation. *Ecology,* 50:721–724.

BEATTIE, A. J., AND D. C. CULVER. 1981. The guild of myrmechores in the herbaceous flora of West Virginia forests. *Ecology,* 62:107–115.

BEATTY, S. W. 1984. Influence of microtopography and canopy species on spatial patterns of forest understory plants. *Ecology,* 65:1406–1419.

BECK, S. D. 1980. *Insect Photoperiodism,* 2nd ed. Academic Press, New York.

BEDDINGTON, J. R., M. P. HASSELL, AND J. H. LAWTON. 1976. The components of arthropod predation. II. The predator rate of increase. *J. Anim. Ecol.,* 45:165–185.

BELLROSE, F. C., T. G. SCOTT, A. S. HAWKINS, AND J. B. LOW. 1961. Sex ratios and age ratios in North American ducks. *Illinois Nat. Hist. Surv. Bull.,* 27:391–474.

BEKOFF, M. 1977. Mammalian dispersal and the ontogeny of individual behavioral phenotypes. *Am. Nat.,* 111:715–732.

BELL, A. D. 1974. Rhizome organization in relation to vegetative spread in *Medeola virginiana. J. Arnold Arb.,* 55:458–468.

BELL, G. 1980. The costs of reproduction and their consequences. *Am. Nat.,* 116:45–76.

BELLOWS, T. S., JR. 1981. The descriptive properties for some models for density-dependence. *J. Anim. Ecol.* 50:139–156.

BELOVSKY, G. E. 1981a. Food plant selection by a generalist herbivore: The moose. *Ecology,* 62:1020–1030.

———. 1981b. A possible population response of moose to sodium availability. *J. Mamm.,* 63:631–633.

BELOVSKY, G. E., AND P. F. JORDAN. 1981. Sodium dynamics and adaptations of a moose population. *J. Mamm.,* 63:613–621.

BELSKY, A. J. 1986. Does herbivory benefit plants? A review of the evidence. *Am. Nat.,* 127:870–892.

BENTLEY, B. L. 1977. Extrafloral nectaries and protection by pugnacious bodyguards. *Ann. Rev. Ecol. Syst.,* 8:407–427.

BERGER, P. J., N. C. NEGUS, E. H. SANDERS, AND P. D. GARDNER. 1981. Chemical triggering of reproduction in *Microtus montanus. Science,* 214:69–70.

BERGERUD, A. T. 1971. The population dynamics of Newfoundland caribou. *Wildl. Monogr.* No. 25.

BERGERUD, A. T., AND F. MANUEL. 1969. Aerial census of moose in central Newfoundland. *J. Wildl. Manage.,* 33:910–916.

BERGERUD, A. T., W. WYETH, AND B. SNIDER. 1983. The role of wolf predation in limiting a moose population. *J. Wildl. Manage.,* 47:977–988.

BERNARD, J. M., AND E. GORHAM. 1978. Life history aspects of primary production in sedge wetlands. In R. E. Good, D. F. Whigham, and R. L. Simpson (eds.), *Freshwater Wetlands.* Academic Press, New York, pp. 39–51.

BERNHARD-REVERSAT, F. 1975. Nutrients in throughfall and their quantitative importance in rain forest mineral cycles. In F. Golley and E. Medina (eds.), *Tropical Ecological Systems Trends in Terrestrial and Aquatic Research.* Springer-Verlag, New York, pp. 153–159.

———. 1982. Biogeochemical cycle of nitrogen in a semi-arid savanna. *Oikos,* 38:321–332.

BERRY, J. A., AND O. BJORKMAN. 1980. Photosynthetic response and adaptation to temperature in higher plants. *Ann. Rev. Ecol. Syst.,* 31:491–553.

BERRY, J. A., AND W. J. S. DOWNTON. 1982. Environmental regulation of photosynthesis. In Govindjee (ed.), *Photosynthesis Vol II, Development, Carbon Metabolism, and Plant Productivity.* Academic Press, New York, pp. 263–343.

BERRY, J. A., AND J. K. RAISON. 1981. Responses of macrophytes to temperature. In O. Lange, P. S. Nobel, C. B. Osmund, and H. Zeigler (eds.), *Physiological Plant Ecology I, Vol. 12A Encyclopedia of Plant Physiology, New Series.* Springer-Verlag, New York, pp. 237–338.

BERRYMAN, A. A. 1981. *Population Systems: A General Introduction.* Plenum, New York.

BERTHOLD, P. 1974. Circannual rhythms in birds with different migratory habits. In E. T. Pengelley (ed.), *Circannual Clocks: Annual Biological Rhythms.* Academic Press, New York, pp. 55–94.

BERTNESS, M. D. 1984. Ribbed mussels and *Spartine alterniflora* production on a New England marsh. *Ecology,* 65:1794–1807.

BERTRAM, B. C. R. 1975. Social factors influencing reproduction in wild lions. *J. Zool. Lond.,* 177:463–482.

BETZER, P. R., R. H. BYRNE, J. G. ACKER, C. S. LEWIS, R. R. JOLLY, AND R. A. FEELY. 1984. The oceanic carbonate system: A reassessment of biogenic controls. *Science,* 226:1074–1077.

BIEL, E. R. 1961. Microclimate, bioclimatology, and notes on comparative dynamic climatology. *Am. Sci.,* 49:326–357.

BIERZYCHUDEK, P. 1982a. The demography of Jack-in-the-pulpit, a forest perennial that changes sex. *Ecol. Monogr.,* 52:335–351.

———. 1982b. Life history and demography of shade tolerant temperate herbs: A review. *New Phytol.,* 190:757–776.

BIGALKE, R. C. 1979. Aspects of vertebrate life in Fynbos, South Africa. In R. L. Specht (ed.), *Heathlands and Related Shrublands.* (*Ecosystems of the World* Vol. XX). Elsevier, Amsterdam, pp. 81–96.

BILBY, R. E. 1981. Role of organic debris dams in regulating the export of dissolved and particulate matter from a forested watershed. *Ecology,* 62:1234–1243.

BILBY, R. E., AND G. E. LIKENS. 1980. Importance of organic debris dams in the structure and function of stream ecosystems. *Ecology,* 61:1107–1113.

BINET, P. 1981. Short-term dynamics of minerals in arid ecosystems. In D. W. Goodall and R. A. Perry (eds.), *Arid-land Ecosystems: Structure, Functioning, and Management. II.* Cambridge University Press, Cambridge, England, pp. 325–356.

BINKLEY, D., AND R. L. GRAHAM. 1981. Biomass, production,

and nutrient cycling of mosses in an old-growth Douglas-fir forest. *Ecology,* 62:1387–1389.

BIRCH, L. C., AND D. P. CLARK. 1953. Forest soil as an ecological community with special reference to the fauna. *Quart. Rev. Biol.,* 28:13–36.

BIRDSALL, C. W., C. E. GRUE, AND A. ANDERSON. 1986. Lead concentrations in bullfrog *Rana catesbeiana* and green frog *R. clamatans* tadpoles inhabiting highway drainages. *Environmental Pollution (Series A),* 43:233–248.

BITMAN, J. 1970. Hormonal and enzymatic activity of DDT. *Agr. Sci. Rev.,* 7(4):6–12.

BJORKMAN, O., AND J. BERRY. 1973. High efficiency photosynthesis. *Sci. Am.,* 229(4):80–93.

BLACKMAN, F. F. 1905. Optima and limiting factors. *Ann. Bot.,* 19:281–298.

BLAIR-WEST, J. R., J. A. COGHLAN, D. A. DENTON, J. F. NELSON, ET AL. 1968. Physiological, morphological, and behavioral adaptations to a sodium-deficient environment by wild native Australian and introduced species of animals. *Nature,* 217:922–928.

BLAKE, D. R., AND F. S. ROWLAND. 1988. Continuing worldwide increase in trophospheric methane, 1978–1987. *Science,* 239:1129–1131.

BLAKE, J. G., AND J. R. KARR. 1984. Species composition of bird communities and the conservation benefit of large versus small forests. *Biol. Cons.,* 30:193–187.

BLISS, L. C. 1956. A comparison of plant development in microenvironments of arctic and alpine tundras. *Ecol. Monogr.,* 26:303–337.

———. 1963. Alpine plant communities of the Presidential Range, New Hampshire. *Ecology,* 44:678–697.

———. 1975. Devon Island, Canada. In T. Rosswall and O. W. Heal (eds.), *Structure and Function of Tundra Ecosystems.* Swedish National Science Research Council, Stockholm, pp. 17–60.

———. 1981. North American and Scandinavian tundras and polar deserts. In L. C. Bliss, O. W. Heal, and J. J. Moore (eds.), *Tundra Ecosystems: A Comparative Analysis.* Cambridge University Press, Cambridge, England, pp. 8–24.

BLISS, L. C., G. M. COURTIN, D. L. PATTIE, R. R. WIEWE, D. W. A. WHITFIELD, P. WIDDEN. 1973. Arctic tundra ecosystems. *Ann. Rev. Ecol. Syst.,* 4:359–399.

BLISS, L. C., O. W. HEAL, AND J. J. MOORE (EDS.). 1981. *Tundra Ecosystems: A Comparative Analysis.* International Biological Programme No. 25, Cambridge University Press, Cambridge, England.

BLONDEL, J. 1981. Structure and dynamics of bird communities in Mediterranean-type habitats. In F. di Castri, D. W. Goodall, and R. Specht (eds.), *Mediterranean-type Shrubland (Ecosystems of the World* 14), Elsevier, Amsterdam, pp. 361–386.

BLUM, J. L. 1960. Algal populations in flowing waters. In *Ecology of Algae. Spec. Pub. No. 2,* Pymatuning Lab. of Field Biology, pp. 11–21.

BOND, R. R. 1957. Ecological distribution of breeding birds in the upland forests of southern Wisconsin. *Ecol. Monogr.,* 27:351–384.

BONNELL, M. C., AND R. K. SELANDER. 1974. Elephant seals: Genetic variation and near extinction. *Science,* 184:908–909.

BONT, DE, R. G., J. J. VANGELDER, AND J. H. J. OLDERS. 1986. Thermal ecology of the smooth snake *Coronella austriaca* Laurenti during spring. *Oecologica,* 69:72–78.

BORCHERT, M. I., AND S. K. JAIN. 1978. The effect of rodent seed predation on four species of California annual grasses. *Oecologica,* 33:101–113.

BORING, L. R., C. D. MONK, AND W. T. SWANK. 1981. Early regeneration of a clearcut southern Appalachian forest. *Ecology* 62:1244–1253.

BORMANN, F. H., AND G. E. LIKENS. 1979. *Pattern and Process in a Forested Ecosystem.* Springer-Verlag, New York.

BORMANN, F. H., G. E. LIKENS, AND J. H. MELILLO. 1977. Nitrogen budget for an aggrading northern hardwood forest ecosystem. *Science,* 196:981–983.

BORNEBUSCH, C. H. 1930. *The Fauna of Forest Soils.* Nielsen and Lydiche, Copenhagen.

BOTKIN, D. B. 1977. Forest, lakes, and the anthropogenic production of carbon dioxide. *Bioscience,* 27:325–331.

BOTKIN, D. B., J. F. JANAK, AND J. R. WALLIS. 1972. Some ecological consequences of a computer model of forest growth. *J. Ecol.,* 60:948–972.

BOTT, T. L., AND T. D. BROCK. 1968. Bacterial growth rates above 90°C in Yellowstone hot springs. *Science,* 164:1411–1412.

BOUCHER, C., AND E. J. MOLL. 1981. South African Mediterranean shrublands. In F. di Castri, D. W. Goodall, and R. L. Spech (eds.), *Mediterranean-type Ecosystems (Ecosystems of the World* 11). Elsevier, Amsterdam, pp. 233–248.

BOUCHER, D. H., S. JAMES, AND H. D. KELLER. 1982. The ecology of mutualism. *Ann. Rev. Ecol. Syst.,* 13:315–347.

BOURGERON, P. S. 1983. Spatial aspects of vegetation structure. In F. B. Golley (ed.), *Tropical Rain Forest Ecosystems (Ecosystems of the World* 14a). Elsevier, Amsterdam, pp. 29–47.

BOURLIERE, F. (ED.). 1983. *Tropical Savannas (Ecosystems of the World* 13). Elsevier, Amsterdam.

BOURLIERE, F., AND M. HADLEY. 1983. Present-day savannas: An overview. In F. Bourliere (ed.), *Tropical Savannas.* Elsevier, Amsterdam, pp. 1–18.

BOWDEN, W. B. 1986. Nitrification, nitrate reduction, and nitrogen immobilization in a tidal freshwater marsh sediment. *Ecology,* 67:88–99.

BOWDEN, W. B., AND F. H. BORMANN. 1986. Transport and loss of nitrous oxide in soil water after clear-cutting. *Science,* 233:867–869.

BOX, T. W., J. POWELL, AND D. L. DRAWE. 1967. Influence of fire on south Texas chapparal. *Ecology,* 48:955–961.

BRADLEY, W. G., AND R. A. MAUER. 1971. Reproduction and food habits of Merriam's kangaroo rat, *Dipodomys merriami. J. Mamm.,* 52:497–507.

BRAMWELL, A. 1989. *Ecology in the 20th Century.* Yale University Press, New Haven, Conn.

BRAY, J. R., AND J. T. CURTIS. 1957. An ordination of the upland forest communities of southern Wisconsin. *Ecol. Monogr.,* 27:325–349.

BREY, J. R., AND E. GORHAM. 1964. Litter production in the forests of the world. *Adv. Ecol. Rec.,* 2:101–157.

BREYMEYER, A. I. 1980. Trophic structure and relationships. In A. I. Breymeyer and G. M. Van Dyne (eds.), *Grasslands, Systems Analysis, and Man.* Cambridge University Press, Cambridge, England.

BREYMEYER, A. I., AND G. M. VAN DYNE (EDS.). 1980. *Grasslands, Systems Analysis, and Man.* International Biological Programme No. 19, Cambridge University Press, Cambridge, England.

BRIAND, F. 1983a. Biogeographic patterns in food web organization. In D. L. DeAngelis, W. M. Post, and G. Sugihara (eds.), *Current Trends in Food Web Theory.* ORNL 5983, Oak Ridge National Laboratory, Oak Ridge, Tenn., pp. 41–44.

———. 1983b. Environmental control of food web structure. *Ecology,* 64:253–263.

BRIAND, F., AND J. E. COHEN. 1984. Community food webs have scale-invariant structure. *Nature,* 307:264–266.

BRITTINGHAM, M. C., AND S. A. TEMPLE. 1983. Have cowbirds caused forest songbirds to decline. *Bioscience,* 33:31–35.

BROCK, T. R. 1979. *Biology of Microorganisms.* Prentice-Hall, Englewood Cliffs, N.J.

BROECKER, W. S. 1970. Man's oxygen reserves. *Science,* 168:1537–1538.

B-4

BROECKER, W. S., T. TAHAKASHI, H. J. SIMPSON, AND T. H. PING. 1979. Fate of fossil fuel carbon dioxide and the global carbon budget. *Science,* 206:409–418.

BROKAW, N. V. L. 1985. Gap-phase regeneration in a tropical forest. *Ecology,* 66:682–687.

BROOKS, M. B. 1951. Effect of black walnut trees and their products on vegetation. *W. Va. Univ. Agr. Exp. Stat. Bull.,* 347:1–31.

BROWER, J. E., AND J. H. ZAR. 1984. *Field and Laboratory Methods for General Ecology,* 2nd ed. Brown, Dubuque, Iowa.

BROWER, J. V. Z. 1958. Experimental studies of mimicry in some North American butterflies: I. The monarch, *Danaus plexippus,* and Viceroy, *Limenitis archippus; 2, Battus philenor* and *Papilio troilus, P. polyxenes* and *P. glaucus; 3. Danaus glippus berenice* and *Limenitis archippus floridensis. Evolution,* 12:32–47, 123–136, 273–285.

BROWER, J. V. Z., AND L. P. BROWER. 1962. Experimental studies of mimicry: 6. The reaction of toads *(Bufo terrestris)* to honeybees *(Apis mellifera)* and their drone mimics *(Eristalis vinetorum). Am. Nat.,* 97:297–307.

BROWN, E. R. 1961. The black-tailed deer of western Washington. *Washington State Game Dept. Biol. Bull. No. 13,* Olympia.

BROWN, J. H. 1971. Mammals on mountaintops: Nonequilibrium insular biogeography. *Am. Nat.,* 105:467–478.

———. 1978. The theory of insular biogeography on the distribution of boreal birds and mammals. *Great Basin Nat. Mem.* 2:209–227.

———. 1981. Two decades of homage to Santa Rosalia: Toward a general theory of diversity. *Am. Zool.,* 21:877–888.

BROWN, J. H., AND A. C. GIBSON. 1983. *Biogeography.* Mosby, St. Louis, Mo.

BROWN, J. H., AND A. KODRICH-BROWN. 1977. Turnover rates in insular biogeography: Effect of immigration on extinction. *Ecology,* 58:445–449.

BROWN, J. L. 1964. The evolution of diversity in avian territorial systems. *Wilson Bull.,* 76:160–169.

———. 1969. Territorial behavior and population regulation in birds: A review and reevaluation. *Wilson Bull.,* 81:292–329.

BROWN, K. M. 1982. Resource overlap and competition in pond snails: An experimental analysis. *Ecology,* 63:412–422.

BROWN, L. 1976. *British Birds of Prey,* Collins, London.

BROWN, R. J. E. 1970. *Permafrost in Canada.* University of Toronto Press, Toronto.

BROWN, R. J. E., AND G. H. JOHNSON. 1964. Permafrost and related engineering problems. *Endeavour,* 23:66–73.

BROWN, S., AND A. E. LUGO. 1984. Biomass of tropical forests: A new estimate based on forest volumes. *Science,* 223:1290–1293.

BRUNIG, E. F. 1983. Vegetation structure and growth. In F. Golley (ed.), *Tropical Rain Forest Ecosystems: Structure and Function.* Elsevier, Amsterdam, pp. 49–75.

BRYANT, J. A. 1987. Feltleaf willow–snowshoe hare interactions: Plant carbon/nutrient balance and flood plain succession. *Ecology,* 68:1319–1327.

BRYANT, J. P., AND P. J. KUROPAT. 1980. Selection of winter forage by subarctic browsing vertebrates: The role of plant chemistry. *Ann. Rev. Ecol. Syst.,* 11:261–285.

BRYANT, J. P., G. D. WIELAND, T. CLAUSEN, AND P. J. KUROPAT. 1985. Interactions of snowshoe hares and feltleaf willow *(Salix alaxensis)* in Alaska. *Ecology,* 66:1564–1573.

BRYLINSKY, M. 1980. Estimating the productivity of lakes and reservoirs. In E. D. Le Cren and R. H. Lowe-McConnell (eds.), *The Functioning of Freshwater Ecosystems.* (International Biological Programme No. 22.) Cambridge University Press, Cambridge, England, pp. 411–418.

BRYSON, R. A., D. A. BAERRIS, AND W. M. WENDLAND. 1970. The character of late glacial and post glacial climatic changes. In W. Dort and J. K. Jones (eds.), *Pleistocene and Recent Environments of the Great Central Plains.* Spec. Publ. No. 3, Department of Biology, University of Kansas, Lawrence, pp. 53–74.

BRYSON, R. A., AND R. A. RAGOTZKIE. 1960. On internal waves in lakes. *Limnol. Oceanogr.,* 5:397–408.

BUCKMANN, S. L. 1987. The ecology of oil flowers and their bees. *Ann. Rev. Ecol. Syst.,* 18:343–369.

BUCKNER, C. H., AND W. J. TURNOCK. 1965. Avian predation on the larch sawfly, *Pristiphora erichsonii* (Hymenoptera: Tenthredinidae). *Ecology,* 46:223–236.

BUDYKO, M. I. 1963. *The Heat Budget of the Earth.* Hydrological Publishing House, Leningrad.

BUELL, M. F., AND R. E. WILBUR. 1948. Life form spectra of the hardwood forests of the Itaska Park region, Minnesota. *Ecology,* 29:352–359.

BULMER, M. G. 1974. A statistical analysis of the 10-year cycle in Canada. *J. Anim. Ecol.,* 43:701–718.

———. 1975. Phase relations of the 10-year cycle. *J. Anim. Ecol.,* 44:609–621.

BULL, J. J., AND R. C. VOGT. 1979. Temperature-dependent sex determination in turtles. *Science,* 206:1186–1188.

BULL, K. R., W. J. EVERY, P. FREESTONE, J. R. HALL, AND D. OSBORN. 1983. Alkyl lead pollution and bird mortalities in the Mersey Estuary, UK, 1979–1981. *Environmental Pollution A,* 31:239–259.

BUNNELL, F. L., S. F. MCCLAEAN, JR., AND J. BROWN. 1975. Barrow, Alaska, U.S.A. In T. Rosswall and O. W. Heal (eds.), *Structure and Function of Tundra Ecosystems.* Swedish Natural Science Institute, Stockholm.

BUNNELL, F. L., O. K. MILLER, P. W. FLANAGAN, AND R. E. BENOIT. 1980. The microflora: Composition, biomass, and environmental relations. In J. Brown, P. C. Miller, L. L. Tieszen, and F. L. Bunnell (eds.), *An Arctic Ecosystem: The Coastal Tundra at Barrow, Alaska.* Dowden, Hutchinson, and Ross, Stroudsburg, Pa., pp. 255–290.

BUNNING, E. 1964. *The Physiological Clock,* 2nd. ed. Academic Press, New York.

BURBRANCH, M. P., AND R. B. PLATT. 1964. Granite outcrop communities of the Piedmont Plateau in Georgia. *Ecology,* 45:292–306.

BURGESS, R. L., AND R. V. O'NEILL. 1976. Eastern deciduous forest biome progress report September 1, 1974–August 31, 1975, EDFB/IBP 76/5. *Env. Sci. Div. Pub. No. 871,* Oak Ridge National Laboratory, Oak Ridge, Tenn.

BURT, W. V., AND J. QUEEN. 1957. Tidal overmixing in estuaries. *Science,* 126:973–974.

BUSKIRK, R. E., AND W. H. BUSKIRK. 1976. Changes in arthropod abundance in a highland Costa Rica forest. *Am. Midl. Nat.,* 95:288–298.

BUTCHER, G. S., W. A. NIERING, W. J. BARRY, AND R. W. GOODWIN. 1981. Equilibrium biogeography and the size of nature preserves: An avian case study. *Oecologica,* 49:29–37.

BUTLER, P. A. 1964. Commercial fisheries investigations. In Pesticide-Wildlife Studies, *U.S. Serv. Circ. 226,* pp. 11–25.

BUTTEMER, W. A. 1985. Energy relations of winter roost-site utilization by American goldfinches *(Carduelis tristis). Oecologioca,* 18:126–132.

CALDWELL, M. M. 1971. Solar ultraviolet radiation as an ecological factor for alpine plants. *Ecol. Monogr.,* 38:243–268.

———. 1981. Plant responses to solar ultraviolet radiation. In O. L. Lange, P. S. Nobel, C. B. Osmond, and H. Zeigler (eds.), *Physiological Plant Ecology I, Vol. 12A, Encyclopedia of Plant Physiology.* Springer-Verlag, New York, pp. 120–197.

CALDWELL, M. M., R. ROBBERRECHT, AND W. D. BILLINGS. 1980. A steep latitudinal gradient of solar ultraviolet-B radiation in the arctic-alpine life zone. *Ecology*, 61:600–611.

CAMERON, G. N. 1972. Analysis of insect trophic diversity in two salt marsh communities. *Ecology*, 53:58–73.

CAMPBELL, C. A., E. A. PAUL, D. A. RENNIE, AND K. J. MCCALLUM. 1967. Applicability of the carbon-dating method of analysis to soil humus studies. *Soil Sci.*, 104:217–224.

CAMPBELL, R. W. 1969. Studies on gypsy moth population dynamics. In *Forest Insect Population Dynamics, USDA Res. Paper NE-125*, pp. 29–34.

CANHAM, C. D., AND O. L. LOUCKS. 1984. Catastrophic windthrow in presettlement forests of Wisconsin. *Ecology*, 65:803–809.

CARBYN, L. N. 1974. Wolf predation and behavioral interactions with elk and other ungulates in a high prey diversity *Can. Wildl. Serv. Rept.*, 233 pp.

———. 1982. Coyote population fluctuations and spatial distribution in relation to wolf territories in Riding Mountain National Park. *Can. Field Nat.*, 96:176–183.

———. 1983. Wolf predation on elk in Riding Mountain National Park, Manitoba. *J. Wildl. Manage.*, 47:963–976.

CARBYN, L. N., AND M. C. S. KINGSLEY. 1979. Summer food habits of wolves with emphasis on moose in Riding Mountain National Park. *Proc. North Am. Moose Conf. Workshop*, 15:349–361.

CAREY, F. C. 1982. A brain heater in the swordfish. *Science*, 216:1327–1329.

CARL, E. 1971. Population control in arctic ground squirrels. *Ecology*, 52:395–413.

CARLISLE, A., A. H. F. BROWN, AND E. J. WHITE. 1966. The organic matter and nutrient elements in the precipitation beneath a sessile oak canopy. *J. Ecol.*, 54:87–98.

CARPENTER, F. L. 1987. Food abundance and territoriality: to defend or not to defend? *Am. Zool.*, 27:401–409.

CARPENTER, S. R. 1980. Enrichment of Lake Wingra, Wisconsin, by submerged macrophyte decay. *Ecology*, 61:1145–1155.

CARPENTER, S. R., AND J. F. KITCHELL. 1984. Plankton community structure and limnetic primary production. *Am. Nat.*, 124:159–172.

CARPENTER, S. R., J. F. KITCHELL, AND J. HODGSON. 1985. Cascading trophic interactions and lake productivity. *Bioscience*, 35:634–639.

CARRICK, R. 1963. Ecological significance of territory size in the Australian magpie *Gymnorhina tibiten*. *Proc. Int. Orn. Cong.*, 13:740–753.

CARROLL, G. C. 1979. Forest canopies: Complex and independent subsystems. In R. H. Waring (ed.), *Forests: Fresh Perspectives from Ecosystem Analysis. Proc. Ann. Biol. Coll.*, Oregon State University Press, Corvallis, pp. 87–107.

CARROLL, C. R., AND C. A. HOFFMAN. 1980. Chemical feeding deterrent mobilized in response to insect herbivory and counter adaptation by *Epilachnia tridecimnoata*. *Science*, 209:414–416.

CARROLL, J. F., AND J. D. N. NICHOLS. 1986. Parasitization of meadow voles *Microtus pennsylvanicus* (Ord) by American dog ticks *Dermacenter variabilis* (Say) and adult tick movement during high host density. *J. Entomol. Sci.*, 21:102–113.

CARSON, R. 1962. *Silent Spring*. Houghton Mifflin, Boston.

CASE, T. J., AND M. L. CODY. 1984. Testing theories of island biogeography. *Am. Sci.*, 75:402–411.

CASTRI, F. D. I. 1970. Les grands problems qui se posent aux ecologistes pour l'etude des ecosystemes du sol. In J. Phillipson (ed.), *Methods of Study in Soil Biology*. UNESCO, Paris, pp. 15–31.

———. 1981. Mediterranean-type shrublands of the world. In Castri, F. di, D. W. Goodall, and R. L. Specht (eds.), *Mediterranean-type Shrublands* (Ecosystems of the World 11), Elsevier, Amsterdam.

CASTRI, F. DI, AND H. A. MOONEY (EDS.). 1973. *Mediterranean Type Ecosystems, Origin and Structure*. Springer-Verlag, New York.

CAUGHLEY, G. 1966. Mortality patterns in mammals. *Ecology*, 47:906–918.

———. 1970. Eruption of ungulate populations with emphasis on Himalayan thar in New Zealand. *Ecology*, 51:53–72.

———. 1976a. Wildlife management and the dynamics of ungulate populations. *Appl. Biol.*, 1:183–246.

———. 1976b. Plant and herbivore systems. In R. B. May (ed.), *Theoretical Ecology: Principles and Applications*. Saunders, Philadelphia, pp. 94–113.

———. 1977. *Analysis of Vertebrate Populations*. Wiley, New York.

CERNUSA, A. 1976. Energy exchange within individual layers of a meadow. *Oecologia* 23:141–149.

CHAFFEE, R. R. J., AND J. C. ROBERTS. 1971. Temperature acclimation in birds and mammals. *Ann. Rev. Physiol.*, 33:155–202.

CHANGNON, S. A. 1968. La Porte weather anomaly: Fact or fiction? *Bull. Am. Meteorol. Soc.*, 49:4–11.

CHAPIN, F. S., P. C. MILLER, W. D. BILLINGS, AND R. I. COYNE. 1980. Carbon and nutrient budgets and their control in coastal tundra. In J. Brown, P. C. Miller, L. L. Tieszen, and F. L. Bunnell (eds.), *An Arctic Ecosystem: The Coastal Tundra at Barrow, Alaska*. Dowden, Hutchinson, and Ross, Stroudsburg, Penn., pp. 458–482.

CHAPIN, F. S., L. L. TIESZEN, M. C. LEWIS, P. C. MILLER, AND B. H. MCCOWEN. 1980. Control of tundra plant allocation patterns and growth. In J. Brown, P. C. Miller, L. L. Tieszen, and F. L. Bunnell (eds.), *An Arctic Ecosystem: The Coastal Tundra at Barrow, Alaska*. Dowden, Hutchinson, and Ross, Stroudsburg, Penn., pp. 140–185.

CHAPMAN, J. A., AND R. P. MORGAN III. 1973. Systematic status of the cottontail complex in western Maryland nearby West Virginia. *Wildl. Monogr. 36*, Wildlife Society, Washington, D.C.

CHAPMAN, V. J. 1960. *Salt Marshes and Salt Deserts of the World*. Leonard Hill, London.

———. 1976. *Coastal Vegetation*, 2nd ed. Pergamon Press, Oxford.

CHARNOV, E. L. 1976. Optimal foraging: The marginal value theorem. *Theor. Pop. Biol.*, 9:129–136.

CHARNOV, E. L., AND W. M. SCHAFFER. 1973. Life history consequences of natural selection: Cole's results revisited. *Am. Nat.*, 107:791–793.

CHARLEY, J. C. 1972. The role of shrubs in nutrient cycling. In C. M. McKell (ed.), *Wildland Shrubs: Their Biology and Utilization. USDA Forest Service Gen. Tech. Rept.* INT-1, pp. 182–203.

CHASKO, G. G., AND J. E. GATES. 1982. Avian habitat suitability along a transmission line corridor in an oak-hickory forest region. *Wildl. Monogr.*, 82.

CHAZDON, R. L., AND R. W. PEARCY. 1986. Photosynthetic responses to light variation in rainforest species. II. Carbon gain and photosynthetic efficiencies during light flecks. *Oecologica*, 69:524–531.

CHEATUM, E. L., AND C. W. SEVERINGHAUS. 1950. Variations in fertility of white-tailed deer related to range conditions. *Trans. North Am. Wildl. Conf.*, 15:170–189.

CHERNOV, YU I., ET AL. 1975. Tareya, USSR. In T. Rosswall and O. W. Heal (eds.), *Structure and Function of Tundra Ecosystem*. Swedish Natural Science Research Council, Stockholm, pp. 159–181.

CHESSON, P. L. 1986. Environmental variation and coexistence of species. In J. Diamond and T. Case, (eds.), *Community Ecology*. Harper & Row, New York, pp. 240–256.

CHESSON, P. L., AND T. CASE. 1986. Overview: Nonequilibrium community theories: Chance, variability, history, and coexistence. In J. Diamond and T. Case (eds.), *Community Ecology.* Harper & Row, New York, pp. 229–239.

CHESSER, R. K. 1983. Isolation by distance: Relationship to the management of genetic resources. In C. Schonewald-Cox, S. Chambers, B. MacBryde, and W. Thomas (eds.), *Genetics and Conservation; A Reference for Managing Wild Animal Plant Populations.* Benjamin/Cummings, Menlo Park, Calif., pp. 51–65.

CHESTNUT, A. F. 1974. Oyster reefs. In H. T. Odum, B. J. Copeland, and E. A. McMahan (eds.), *Coastal Ecological Systems of the U.S. Vol. 2.* Conservation Foundation. Washington, D.C., pp. 171–203.

CHEVALIER, J. R. 1973. Cannibalism as a factor in the first year survival of walleye in Oneida Lake. *Trans Am. Fish. Soc.,* 102:739–744.

CHEW, R. M., AND A. E. CHEW. 1970. Energy relationships of the mammals of a desert shrub *(Larrae tridentatia)* community. *Ecol. Monogr.,* 40:1–21.

CHITTY, D. 1960. Population processes in the vole and their reference to general theory. *Can. J. Zool.,* 38:99–113.

CHITTY, D., AND E. PHIPPS. 1966. Seasonal changes in survival in mixed populations of two species of vole. *J. Anim. Ecol.,* 35:313–331.

CHOW, T. J. 1970. Lead accumulation in roadside soil and grass. *Nature,* 225:295–296.

CHOW, T. J., AND J. L. EARL. 1970. Lead aerosols in the atmosphere: Increasing concentrations. *Science,* 169:577–580.

CHRISTENSEN, A. M., AND J. J. MCDERMOTT. 1958. Life history of the oyster crab *Pinnotheres ostreum. Biol. Bull.,* 114:146–179.

CHRISTIAN, J. J. 1963. Endocrine adaptive mechanisms and the physiologic regulation of population growth. In W. V. Mayer and R. G. Van Gelder (eds.), *Physiological Mammalogy,* Vol. 1, Mammalian Populations. Academic Press, New York, pp. 189–353.

———. 1971. Fighting, maturity, and population density in *Microtus pennsylvanicus. J. Mammal.,* 52:556–567.

———. 1978. Neurobehavioral endrocrine regulation of small mammal populations. In D. P. Snyder (ed.), *Populations of Small Mammals under Natural Conditions,* Pymatuning Symposia in Ecology, Vol. 5. University of Pittsburgh Press, Pittsburgh, Penn., pp. 143–158.

CHRISTIAN, J. J., AND D. E. DAVIS. 1964. Endocrines, behaviour and populations. *Science,* 146:1550–1560.

CHRISTIANSEN, F. B., AND T. M. FENCHEL. 1977. *Theories of Population in Biological Communities.* Springer-Verlag, New York.

CHRISTY, H. R. 1952. Vertical temperature gradients in a beech forest in central Ohio. *Ohio J. Sci,* 52:199–209.

CICERONE, R. J. 1987. Changes in stratosphere ozone. *Science,* 237:35–42.

CLARK, F. E. 1969a. The microflora of grassland soils and some microbial influences on ecosystems functions. In R. L. Dix and R. G. Beidleman (eds.), The grassland ecosystem: A preliminary synthesis. *Range Sci. Dept. Science Ser.* No. 2. Colorado State University, Fort Collins, pp. 361–376.

———. 1969b. Ecological associations among soil microorganisms. In *Soil Biology.* UNESCO, Paris, pp. 125–161.

CLARK, F. W. 1972. Influence of jackrabbit density on coyote population change. *J. Wildl. Manage.,* 36:343–356.

CLARKE, J. F. 1969. Nocturnal urban boundary layer over Cincinnati, Ohio. *Monthly Weather Rev.,* 97:582–589.

CLARKSON, D. T., AND J. B. HANSON. 1980. The mineral nutrition of higher plants. *Ann. Rev. Plant Physiology,* 31:239–298.

CLAUSEN, J. 1965. Population studies of alpine and subalpine races of conifers and willows in the California High Sierra Nevada. *Evolution,* 19:56–68.

CLAUSEN, T. P., J. P. BRYANT, AND P. B. REICHARDT. 1986.

Defense of winter-dormant green alder against snowshoe hare. *J. Chem. Ecol.,* 12:2117–2131.

CLAY, K. 1988. Fungal endophytes of grasses: A defensive mutualism between plants and fungi. *Ecology,* 69:10–16.

CLEMENTS, F. E. 1916. *Plant Succession. An analysis of the Development of Vegetation.* Carnegie Inst. Wash. Publ. 242.

CLEMENTS, F. E., AND V. E. SHELFORD. 1939. *Bio-ecology.* McGraw-Hill, New York

CLOUDSLEY-THOMPSON, J. L. 1956. Studies in diurnal rhythms: VII. Humidity responses and nocturnal activity in woodlice (Isopoda). *J. Exp. Biol.,* 33:576–582.

———. 1960 Adaptive functions of circadian rhythms. *Cold Spring Harbor Symp. Quant. Biol.,* 255:345–355.

CLUTTON-BROCK, T. H. 1984. Reproductive effort and terminal investment in iteroparous animals. *Am. Nat.,* 123:212–229.

CLUTTON-BROCK, T. H., F. E. GUINNESS, AND S. D. ALBON. 1982. *Red Deer: Behavior and Ecology of the Two Sexes.* University of Chicago Press, Chicago.

———. 1983. The costs of reproduction to red deer hinds. *J. Anim. Ecol.,* 367:52.

CLUTTON-BROCK, T. H., M. MAJOR, AND F. E. GUINNESS. 1985. Population regulation in male and female red deer. *J. Anim. Ecol.,* 54:831.

CODY, M. L. 1966. A general theory of clutch size. *Evolution,* 20:174–184.

———. 1969 Convergent characteristics in sympatric species: A possible relation to interspecific competition and aggression. *Condor,* 71:222–239.

———. 1973. Parallel evolution and bird niches. In F. di Castri and H. A. Mooney (eds.), *Mediterranean-type Ecosystems* (Ecological Studies 7). Springer-Verlag, New York, pp. 307–338.

———. 1986. Structural niches in plant communities. In J. Diamond and T. Case (eds.), *Community Ecology.* Harper & Row, New York, pp. 381–405.

CODY, M. L., AND J. DIAMOND (EDS.) 1975. *Ecology and Evolution of Communities.* Harvard University Press, Cambridge, Mass.

CODY, M. L., AND H. A. MOONEY. 1978. Convergence versus nonconvergence on Mediterranean-climate ecosystems. *Ann. Rev. Ecol. Syst.,* 9:265–321.

COHEN, J. E. 1978. *Food Webs and Niche Space.* Princeton University Press, Princeton, N.J.

———. 1989. Food webs and community structure. In J. Roughgarden, R. May, and S. Levin (eds.), *Perspectives in Ecological Theory.* Princeton University Press, Princeton, N.J., pp. 181–202.

COHEN, J. E., AND F. BRIAND. 1984. Trophic links of community food webs. *Proc. Natl. Acad. Sci., USA,* 81:4105–4109.

COHMAP MEMBERS. 1988. Climate changes of the last 18,000 years: Observations and model simulations. *Science,* 241:1043–1052.

COKER, R. E. 1947. *This Great and Wide Sea.* University of North Carolina Press, Chapel Hill.

COLE, D. W., S. P. GESSEL, AND S. F. DICE. 1967. Distribution and cycling of nitrogen, phosphorus, potassium and calcium in a second-growth Douglas-fir ecosystem. In *Symposium on Primary Productivity and Mineral Cycling in Natural Ecosystems.* University of Maine Press, Orono, pp. 197–232.

COLE, D. W., AND M. RAPP. 1981. Elemental cycling in forest ecosystem. In D. E. Reichle (ed.), *Dynamic Properties of Forest Ecosystems.* Cambridge University Press, New York, pp. 341–409.

COLE, K. 1982. Late quaternary zonation of vegetation in eastern Grand Canyon. *Science,* 217:1142–1145.

COLE, L. C. 1946. A study of the crypotozoa of an Illinois woodland. *Ecol. Monogr.,* 16:49–86.

———. 1951. Population cycles and random oscillations. *J. Wildl. Manage.,* 15:233–252.

————. 1954a. The population consequences of life history phenomena. *Quart. Rev. Biol.*, 29:103–137.

————. 1954b. Some features of random cycles. *J. Wildl. Manage.*, 18:107–109.

————. 1960. Competitive exclusion. *Science*, 132:348–349.

COLEMAN, R. M., M. R. GROSS, AND R. C. SARGENT. 1985. Parental investment decision rules: A test in bluegill sunfish. *Behav. Ecol. Sociobiol.*, 18:59–66.

COLGAN, M. W. 1987. Coral reef recovery on Guam (Micronesia) after catastrophic predation by *Acabthaster planci. Ecology*, 68:1592–1605.

COLLIAS, N. E., AND R. D. TABER. 1951. A field study of some grouping and dominance relations in ring-necked pheasants. *Condor*, 53:265–275.

COLLINS, S. C. 1983. Geographic variation in habitat structure of black-throated green warbler *(Dendroica virens). Auk*, 100:382–398.

COMMONER, B. 1970. Threats to the integrity of the nitrogen cycle: Nitrogen compounds in soil, water, atmosphere, and precipitation. In S. F. Singer (ed.), *Global Effects of Environmental Pollution.* Springer-Verlag, New York, pp. 70–95.

CONNELL, J. H. 1961 The influence of interspecific competition and other factors on the distribution of the barnacle *Chthamalus stellatus. Ecology*, 42:710–723.

————. 1975 Some mechanisms producing structure in natural communities: A model and evidence from field experiments. In M. Cody and J. Diamond (eds.), *Ecology and Evolution of Communities.* Harvard University Press, Cambridge Mass., pp. 460–480.

————. 1983. On the prevalence and relative importance of interspecific competition: Evidence from field experiments. *Am. Nat.* 122:661–696

CONNELL, J. H., AND M. J. KEOUGH. 1985. Disturbance and patch dynamics of subtidal marine animals on hard substrate. In P. White and S. T. Pickett (eds.), *The Ecology of Natural Disturbance and Patch Dynamics.* Academic Press, Orlando, Fla., pp. 125–151.

CONNELL, J. H., AND E. ORIAS. 1964. The ecological regulation of species diversity. *Am. Nat.*, 98:399–414.

CONNELL, J. H., AND R. O. SLATYER. 1977. Mechanisms of succession in natural communities and their role in community stability and organization. *Am. Nat.*, 111:1119–1144.

COOKE, G. D. 1967. The pattern of autotrophic succession in laboratory microcosms. *Bioscience*, 17:717–721.

COOPER, A. W. 1974. Salt marshes. In H. T. Odum, B. J. Copeland, and E. A. McMahan (eds.), *Coastal Ecosystems of the U.S.*, Vol 2. Conservation Foundation, Washington, D.C., pp. 55–98.

COOPER, S. M., AND N. OWEN-SMITH. 1986. Effects of plant spinescence on large mammalian herbivores. *Oecologica*, 68:446–455.

COPE, O. B. 1971. Interaction between pesticides and wildlife. *Ann. Rev. Entomol.*, 16:325–364.

COPELAND, B. J., K. R. TENORE, AND D. B. HORTON. 1974. Oligohaline regime. In H. T. Odum, B. J. Copeland, and E. A. McMahan (eds.), *Coastal Ecosystems of the U.S.*, Vol. 2. Conservation Foundation, New York, pp. 315–358.

CORBETT, E. S., AND R. P. CROUSE. 1968. Rainfall interception by annual grass and chaparral. *USDA Forest Serv. Res. Paper PSW 48.*

CORNFORTH, I. S. 1970a. Leaf-fall in a tropical rain forest. *J. Appl. Ecol.*, 7:603–608.

————. 1970b. Reafforestation and nutrient reserves in the humid tropics. *J. Appl. Ecol.*, 7:609–615.

CORRELL, D. L. 1978. Estuarine productivity. *Bioscience*, 28:646–650.

COTTAM, C., AND E. HIGGINS. 1946. DDT: Its effects on fish and wildlife. *USDI Fish and Wildlife Circ. 11.*

COULSON, J. C. 1968. Differences in the quality of birds nesting in the center and on the edges of a colony. *Nature*, 217:478–479.

COUPLAND, R. T. 1950. Ecology of mixed prairie in Canada. *Ecol. Monogr.*, 20:217–315.

————. 1958. The effects of fluctuations in weather upon the grassland of the Great Plains. *Botan. Rev.*, 24:273–317.

COUPLAND, R. T. (ED.). 1979. *Grasslands of the World: Analysis of Grasslands and Their Uses.* Internation National Biological Programme No. 18. Cambridge University Press, New York.

COUTANT, C. 1970. Biological aspects of thermal pollution: 1. Entrainment and discharge canal effects. *CRC Critical Reviews in Environ. Control*, November 1970, pp. 341–381.

COWAN, I., AND V. GEIST. 1961. Aggressive behavior in deer of the genus *Odocoileus. J. Mammal*, 42:522–526.

COWAN, R. L. 1962. Physiology of nutrition as related to deer. *Proc. 1st Natl. White-tailed Deer Disease Symp.*, pp. 1–8.

COWARDIN, L. M., V. CARTER, F. C. GOLET, AND E. T. LAROE. 1979. *Classification of Wetlands and Deep-water Habitats of the United States.* U.S. Fish and Wildlife Service, Washington, D.C.

COWLES, H. C. 1899. The ecological relations of the vegetation on the sand dunes of Lake Michigan. *Botan. Gaz.*, 27:95–117, 167–202, 281–308, 361–391.

COWLES, R. B., AND C. M. BOGERT. 1944. A preliminary study of the thermal requirements of desert reptiles. *Bull. Am. Mus. Natur. Hist.*, 83:265–296.

COX, C. B., I. N. HEALEY, AND P. D. MOORE. 1973. *Biogeography, An Ecological and Evolutionary Approach.* Blackwell, Oxford.

COX, C. R., AND B. J. LE BOEUF. 1977. Female incitation of male competition: A mechanism of mate selection. *Am. Nat.*, 111:317–355.

COX, G. W. 1985. *Laboratory Manual of General Ecology*, 5th ed. Wm. Brown, Dubuque, Iowa.

————. 1985. The evolution of avian migration systems between temperate and tropical regions of the New World. *Am. Nat.*, 126:451–474.

COX, G. W., AND R. E. RICKLEFS. 1977. Species diversity and ecological release in Caribbean land bird faunas. *Oikos*, 28:113–122.

CRITCHFIELD, W. B. 1971. Profiles of California vegetation. *USDA Forest Serv. Res. Paper PSW-76.*

CROCH, J. H., J. E. ELLIS, AND J. D. GOSS-CUSTARD. 1976. Mammalian social systems: Structure and function. *Anim. Behav.*, 24(2):261–274.

CROLL, N. A. 1966. *Ecology of Parasites.* Harvard University Press, Cambridge, Mass.

CROMBIE, A. C. 1947. Interspecific competition. *J. Anim. Ecol.*, 16:44–73.

CRONIN, L. E., AND A. J. MANSUETI. 1971. The biology of the estuary. In P. A. Douglas and R. H. Stroud (eds.), *A Symposium on the Biological Significance of Estuaries.* Sport Fishing Institute, Washington, D.C., pp. 14–39.

CROSBY, G. T. 1972. Spread of the cattle egret in the Western hemisphere. *Bird-banding*, 43:205–212.

CROSSLEY, D. A., JR. 1977. The roles of terrestrial saprophagous arthropods in forest soils: Current status of concepts. In W. J. Mattson (ed.), *The Role of Arthropods in Forest Ecosystems.* Springer-Verlag, New York, pp. 49–59.

CROSSLEY, D. A., JR., AND K. K. BOHNSACK. 1960. Long-term ecological study in the Oak Ridge area: 3. Oribatid mite fauna in pine litter. *Ecology*, 41:628–639.

CROW, J. F., AND M. KIMURA. 1970. *An Introduction to Genetics Theory.* Harper & Row, New York.

CROWDER, L. B., J. J. MAGNUSON, AND S. B. BRANT. 1981. Complementarity in the use of food and thermal habitat by Lake Michigan fishes. *Can. J. Fish. Aquatic Sci.*, 38:662–668.

CROWLEY, J. 1967. Biogeography. *Can. Geog.*, 11:312–326.

CROZE, H. 1970. *Searching Image in Carrion Crows.* Paul Parey, Berlin.

CULVER, D. D., AND A. J. BEATTIE. 1978. Myrmecochory in *Viola:* Dynamics of seed–ant interactions in some West Virginia species. *J. Ecol.,* 66:53–72.

———. 1980. The fate of *Viola* seeds dispersed by ants. *Am. J. Bot.,* 67:710–714.

CUMMINS, K. W. 1974. Structure and function of stream ecosystems. *Bioscience,* 24:631–641.

———. 1988. The study of stream ecosystems: A functional view. In L. R. Pomeroy and J. J. Alberts (ed.), *Concepts of Ecosystem Ecology.* Springer-Verlag, New York, pp. 247–262.

CUMMINS, K. W., W. P. COFFMAN, AND P. A. ROFF. 1966. Trophic relationships in a small woodland stream. *Verheindlung der Internationalen Vereinigung für Theoretische und Angewandte Limnologie,* 16:627–637.

CUMMINS, K. W., AND M. J. KLUG. 1979. Feeding ecology of stream invertebrates. *Ann. Rev. Ecol. Syst.,* 10:147–172.

CURIO, E. 1976. *The Ethology of Predation.* Springer-Verlag, New York.

CURTIS, J. T. 1959. *The Vegetation of Wisconsin.* University of Wisconsin Press, Madison.

CURTIS, J. T., AND R. P. MCINTOSH. 1951. An upland forest continuum in the prairie-forest border region of Wisconsin. *Ecology,* 32:476–496.

CUSHING, E. J., AND H. E. WRIGHT, JR. (EDS.). 1967. *Quaternary Paleoecology.* Yale University Press, New Haven, Conn.

DAHL, E. 1953. Some aspects of the ecology and zonation of the fauna of sandy beaches. *Oikos,* 4:1–27.

DANSEREAU, P. 1945. Essae de correlation sociologique entre les plantes superieures et les poissons de la Beine du Lac Saint-Louis. *Rev. Can. Biol.,* 4:369–417.

———. 1959. Vascular aquatic plant communities of southern Quebec: A preliminary analysis. *Trans. 10th Northeast Wild. Conf.,* pp. 27–54.

DANSEREAU, P., AND F. SEGADAS-VIANNA. 1952. Ecological study of the peat bogs of eastern North America. *Can. J. Bot.,* 30:490–520.

DARLINGTON, P. J., JR. 1957. *Zoogeography: The Geographical Distribution of Animals.* Wiley, New York.

DARNELL, R. M. 1961. Trophic spectrum of an estuarine community based on studies of Lake Pontchartrain, Louisiana. *Ecology,* 42:553–568.

DARWIN, C. 1859. *The Origin of Species.* Murray, London.

———. 1897. *The Descent of Man, and Selection in Relation to Sex.* Murray, London.

DAPSON, R. W., P. R. RAMSEY, M. H. SMITH, AND D. F. URBSTAN. 1979. Demographic differences in contiguous populations of white-tailed deer. *J. Wildl. Manage.,* 43:889–898.

DASH, M. C., AND A. K. HOTA. 1980. Density effects on the survival, growth rate, and metamorphosis of *Rana tigrina* tadpoles. *Ecology,* 61:1025–1028.

DAUBENMIRE, R. F. 1959. *Plants and Environment: A Textbook of Plant Autecology.* Wiley, New York.

———. 1966. Vegetation: Identification of typal community. *Science,* 151:291–298.

———. 1968a. Soil moisture in relation to vegetation distribution in the mountains of northern Idaho. *Ecology,* 49:431–438.

———. 1968b. Ecology of the fire in grasslands. *Adv. Ecol. Res.,* 5:208–266.

———. 1968c. *Plant Communities: A Textbook of Plant Synecology.* Harper & Row, New York.

DAVIDSON, C. I., J. R. HARRINGTON, M. J. STEVENSON, M. C. MONAGHAN, W. R. PUDYKEIWICZ, AND W. R. SCHNELL. 1987. Radioactive cesium from the Chernobyl accident in the Greenland ice sheet. *Science,* 237:633–634.

DAVIDSON, D. W. 1985. An experimental study of diffuse competition in a desert ant community. *Am. Nat.,* 125:500–506.

DAVIDSON, W. R. (ED.). 1981. *Diseases and Parasites of White-tailed Deer.* Southeastern Cooperative Wildlife Disease Study, Mscl. Pub. No. 7. Tall Timbers Research Station, Tallahassee, Fla.

DAVIES, N. B. 1977. Prey selection and social behaviour in wagtails (Aves: Moticillidae). *J. Anim. Ecol.,* 46:37–57.

———. 1978. Ecological questions about territorial behaviour. In J. V. Krebs and N. D. Davies (eds.), *Behavioural Ecology: An Evolutionary Approach.* Blackwell, Oxford, pp. 317–350.

DAVIES, N. B., AND A. I. HOUSTON. 1984. Territory economics. In J. R. Krebs and N. B. Davies (eds.), *Behavioral Ecology: An Evolutionary Approach,* 2nd ed. Blackwell, Oxford, pp. 148–169.

DAVIS, C. D., AND A. G. VAN DER VALK. 1978. Litter decomposition in glacial prairie marshes. In R. E. Good, D. F. Whigham, and R. L. Simpson (eds.), *Freshwater Wetlands.* Academic Press, New York, pp. 99–144.

DAVIS, D. E. 1978. Physiological and behavioral responses to the social environment. In D. P. Snyder (ed.), *Populations of Small Mammals under Natural Conditions,* Vol. 5, Special Publication Ser. Pymatuning Laboratory of Ecology, University of Pittsburgh Press, Pittsburgh, pp. 84–91.

DAVIS, M. B. 1981. Quaternary history and the stability of forest communities. In D. C. West, H. H. Shugart, and D. B. Botkins (eds.), *Forest Succession: Concepts and Application.* Springer-Verlag, New York, pp. 132–153.

———. 1983. Holocene vegetational history of the eastern United States. In H. E. Wright, Jr. (ed.), *Late-Quaternary Environments of the United States. Vol. II. The Holocene.* University of Minnesota Press, Minneapolis, pp. 166–188.

DAWKINS, M. 1971. Perceptual change in chicks: Another look at the "search image" concept. *Anim. Behav.,* 19:566–574.

DAWSON, W. R., AND G. A. BARTHOLOMEW. 1968. Temperature regulation and water economy of desert birds. In G. W. Brown (ed.), *Desert Biology.* Academic Press, New York.

DAY, F. P., JR., AND D. T. MCGINTY. 1975. Mineral cycling strategies of two deciduous and two evergreen tree species on a southern Appalachian watershed. In F. C. Howell, J. B. Gentry, and M. H. Smith, (eds.), *Mineral Cycling in Southeastern Ecosystems.* National Technical Information Service, U.S. Department of Commerce, Washington, D.C., pp. 736–743.

DAYTON, P. 1971. Competition, disturbance, and community organization: The provision and subsequent utilization of space in a rocky intertidal community. *Ecol. Monogr.,* 41:351–389.

———. 1975. Experimental evaluation of ecological dominance in a rocky intertidal algal community. *Ecol. Monogr.,* 45:137–159.

DEAN, A. M. 1983. A simple model of mutualism. *Am. Nat.,* 121:409–417.

DEAN, R., L. E. ELLIS, R. W. WHITE, AND R. E. BEMERET. 1975. Nutrient removal by cattle from a short-grass prairie. *J. Appl. Ecol.,* 12:25–29.

DECOURSEY, P. J. 1960a. Phase control of activity in a rodent. *Cold Spring Harbor Symp. Quant. Biol.,* 25:49–54.

———. 1960b. Daily light sensitivity rhythm in a rodent. *Science,* 131:33–35.

———. 1961. Effect of light on the circadian activity rhythm of the flying squirrel, *Glaucomys volans. Z. Vergleich. Physiol.,* 44:331–354.

DELAUNE, R. D., AND W. H. PATRICK. 1980. Nitrogen and phosphorus cycling in a Gulf Coast salt marsh. In U. S. Kennedy (ed.), *Estuarine Perspectives.* Academic Press, New York, pp. 143–151.

DEEVEY, E. S. 1947. Life tables for natural populations of animals. *Quart. Rev. Biol.,* 22:283–314.

DELCOURT, H. R., AND P. A. DELCOURT. 1985. Quaternary palynology and vegetational history of the southeastern United States. In W. M. Bryant, Jr., and R. G. Holloway (eds.), *Pollen Records of Late-Quaternary North American Sediments.* American Association of Stratigraphic Palynologists Foundation, Washington, D.C., pp. 1–37.

DELCOURT, P. A., AND H. R. DELCOURT. 1981. Vegetation maps for eastern North America, 40,000 yr BP to present. In R. Romans (ed.), *Geobotany.* Plenum Press, New York, pp. 123–166.

———. 1984. Late quaternary paleoclimates and biotic responses in eastern North America and the western North Atlantic Ocean. *Paleogeography, Paleoclimatology, Paleopecology,* 48:263–284.

DHYSTERHOUS, E. J., AND E. M. SCHMUTZ. 1947. Natural mulches or "litter of grasslands"; with kinds and amounts on a southern prairie. *Ecology,* 28:163–179.

DIAMOND, J. 1975. The island dilemna: Lessons of modern biogeographic studies for the design of natural preserves. *Biol. Cons.,* 7:129–146.

———. 1986. Overview: Laboratory experiments, field experiments, and natural experiments. In J. Diamond and T. Case (eds.), *Community Ecology.* Harper & Row, New York, pp. 3–22.

DIAMOND, J., AND R. M. MAY. 1976. Island biogeography and conservation: Strategy and limitations. In R. M. May (ed.), *Theoretical Ecology: Principles and Application.* Saunders, Philadelphia.

DICE, L. R. 1943. *The Biotic Provinces of North America.* University of Michigan Press, Ann Arbor.

DICKERMAN, J. A., AND R. G. WETZEL. 1985. Clonal growth in *Typha latifolia:* Population dynamics and demography of the ramets. *J. Ecol.,* 73:535–552.

DILGER, W. C. 1960. Agonistic and social behavior of captive redpolls. *Wilson Bull.,* 72:115–132.

DIX, R. L. 1960. The effects of burning on the mulch structure and species composition of grassland in western North Dakota. *Ecology,* 41:49–56.

DIX, R. L., AND R. G. BIEDLEMAN (EDS.). 1969. The grassland ecosystem: A preliminary synthesis. *Range Sci. Dept. Sci. Ser. No. 2.* Colorado State University, Fort Collins.

DIX, R. L., AND F. E. SMEINS. 1967. The prairie, meadow, and marsh vegetation of Nelson County, North Dakota. *Can. J. Bot.,* 45:21–58.

DOBKIN, D. S., I. OLIVIERI, AND P. R. EHRLICH. 1987. Rainfall and the interaction of microclimate with larval resources in the population dynamics of checkerspot butterflies *(Euphydryas editha)* inhabiting serpentine grassland. *Oecologica,* 71:161–166.

DOBSON, A. P., AND P. J. HUDSON. 1986. Parasites, disease, and structure of ecological communities. *Trends Ecol. Evol.,* 1:11–15.

DOBSON, F. S., AND J. O. MURIE. 1987. Interpretation of intraspecific life history patterns: Evidence from Colombian ground squirrels. *Am. Nat.,* 129:382–397.

DOBSON, F. S., AND W. T. STONE. 1985. Multiple causes of dispersal. *Am. Nat.,* 126:855–858.

DOBZHANSKY, T. 1951. *Genetics and the Origin of Species,* 3rd ed. Columbia University Press, New York.

DOBZHANSKY, T., AND S. WRIGHT. 1941. Genetics of natural populations. V. Relations between mutation rates and accumulation of lethals in populations of *Drosophila pseudoobscura. Genetics,* 26:23–51.

DOLBEER, R. A., AND W. R. CLARK. 1975. Population ecology of snowshoe hares in the central Rocky Mountains. *J. Wildl. Manage.,* 39:535–549.

DORST, R. 1958. Über die Ansiedlung von jung ins Binnerien verfrachteten Silbernowen *(Larus argentatus). Vogelwarte,* 17:169–173.

DORT, W., JR., AND J. K. JONES, JR. (EDS.). 1970. *Pleistocene and Recent Environments of the Central Great Plains. Spec. Publ. No. 3.* Dept. Biology, University of Kansas, Lawrence.

DOTY, M. S. 1957. Rocky intertidal surfaces. In J. W. Hedgpeth (ed.), *Treatise in Marine Ecology and Paleoecology 1. Ecology.* Memoir 67, Geological Soc. Am., pp. 535–585.

DOWDY, W. W. 1944. The influence of temperature on vertical migration of invertebrates inhabiting different soil types. *Ecology,* 25:449–460.

———. 1951. Further ecological studies on stratification of the arthropods. *Ecology,* 32:37–52.

DOWNS, A. A., AND W. E. MCQUILKIN. 1944. Seed production of southern Appalachian oaks. *J. For.,* 42:913–920.

DOYLE, T. W. 1981. The role of disturbance on the gap dynamics of a montane rain forest: An application of a tropical forest succession model. In D. C. West, H. H. Shugart, and D. B. Botkin (eds.), *Forest Succession: Concepts and Application.* Springer-Verlag, New York, pp. 56–73.

DRAKE, J. A. 1983. Invasability in Lotka-Volterra interaction webs. In D. C. DeAngelis, W. M. Post, and G. Sugihara (eds.), *Current Trends in Food Web Theory,* ORNL 5983. Oak Ridge National Laboratory, Oak Ridge, Tenn., pp. 83–90.

DREW, M. C. 1979. Root development and activities. In D. W. Goodall and R. A. Perry (eds.), *Arid Land Ecosystems: Structure, Functioning, and Management,* Vol. 1. Cambridge University Press, Cambridge, pp. 573–608.

DREW, M. C., M. B. JACKSON, AND S. GIFFORD. 1979. Ethylene-promoted adventitious roots and development of cortical air spaces (aerenchyma) in roots may be adaptive responses to flooding in *Zea mays* C. *Planta,* 147:83–88.

DRIFT, J. VAN DER. 1951. Analysis of the animal community in a beech forest floor. *Tijdschrift voor Entomologie,* 94:1–168.

———. 1971. Production and decomposition of organic matter in an oakwood in the Netherlands. In P. Duvigneaud (ed.), *Productivity of Forest Ecosystems* (Proc. Brussels Symposium 1969). UNESCO, Paris, pp. 631–634.

DRIZO, R. 1984. Herbivory: A phytocentric overview. In R. Drizo and J. Sarukhan (eds.), *Perspectives in Plant Population Ecology.* Sinauer Associates, Sunderland, Mass., pp. 141–165.

DRURY, W. H., JR., AND I. C. T. NISBET. 1973. Succession. *J. Arnold Arboretum,* 54:331–368.

DUDDINGTON, C. L. 1955. Interrelations between soil microflora and soil nematodes. In D. Kevan, *Soil Zoology.* Butterworth, Washington, D.C., pp. 284–301.

DUFFEY, E. 1974. *Grassland Ecology and Wildlife Management.* Chapman and Hall, London.

DUGDALE, R. C., AND J. J. GOERING. 1967. Uptake of new and regenerated forms of nitrogen in primary productivity. *Limnol. Oceanogr.,* 12:196–206.

DUNCAN, J. S., H. W. REED, R. MOSS, J. P. P. PHILLIPS, AND A. WATSON. 1978. Ticks, louping ill, and red grouse on moors in Speyside, Scotland. *J. Wildl. Manage.,* 42:500–505.

DUSSOURD, D. E., AND T. EISNER. 1987. Vein-cutting behavior: Insect counterploy to the latex defense of plants. *Science,* 237:898–901.

DUVIGNEAUD, P., AND S. DENAEYER-DESMET. 1967. Biomass, productivity and mineral cycling in deciduous forests in Belgium. In *Symposium on Primary Productivity and Mineral Cycling in Natural Ecosystems.* University of Maine Press, Orono, pp. 167–186.

———. 1970. Biological cycling of minerals in temperate deciduous forests. In D. Reichle (ed.), *Analysis of Temperate Forest Ecosystems. Ecological Studies 1.* Springer-Verlag, New York, pp. 199–225.

————. 1975. Mineral cycling in terrestrial ecosystems. In National Academy of Science, 1975, pp. 133–154.

DWYER, P. D., J. KIKKAWA, AND G. J. INGRAM. 1979. Habitat relations of vertebrates in subtropical heathlands of coastal southeastern Queensland. In R. L. Specht (ed.), *Heathlands and Related Shrublands* (Ecosystems of the World, 9A). Elsevier, Amsterdam, pp. 281–300.

DYER, M. I. 1980. Mammalian epidermal growth factor promotes plant growth. *Proc. Natl. Acad. Sci.*, 77:4836–4837.

EARP, R. 1974. Tidepools. In H. T. Odum, B. J. Copeland, and E. A. McMahan (eds.), *Coastal Ecological Systems of the U.S.* Vol. 2. Conservation Foundation, Washington, D.C., pp. 1–29.

EADIE, J., AND H. G. LUMSDEN. 1985. Is nest parasitism always deleterious to goldeneyes? *Am. Nat.*, 126:859–866.

EASTWOOD, E. 1971. *Radar Ornithology.* Methuen, London.

EATON, J. S., G. E. LIKENS, AND F. H. BORMANN. 1973. Throughfall and stemflow chemistry in a northern hardwood forest. *J. Ecol.*, 61:495–508.

ECKHARDT, F. E. (ED.). 1968. *Functioning of Terrestrial Ecosystems at the Primary Production Level*, Proc. Copenhagen Symposium, Natural Resources Research. UNESCO, Paris.

EDMONDSON, W. T. 1956. The relation of photosynthesis by phytoplankton to light in lakes. *Ecology*, 37:161–174.

————. 1969. Eutrophication in North America. In *Eutrophication: Causes, Consequences, Correctives.* National Academy of Sciences, Washington, D.C., pp. 124–149.

————. 1970. Phosphorus, nitrogen, and algae in Lake Washington after diversion of sewage. *Science*, 169:690–691.

EDWARDS, C. A., AND G. W. HEATH. 1963. The role of soil animals in the breakdown of leaf material. In J. Doeksen and J. van der Drift (eds.), *Soil Organisms.* North Holland, Amsterdam, pp. 76–84.

EGLER, F. E. 1953. Vegetation management for rights-of-way and roadsides. *Smithsonian Inst. Ann. Rept. 1953*, pp. 299–322.

————. 1954. Vegetation science concepts. 1. Initial floristic composition—a factor in old field vegetation development. *Vegetatio*, 4:412–417.

EHLERINGER, J. R. 1980. Leaf morphology and reflectance in relation to water and temperature stress. In N. C. Turner and P. J. Kramer (eds.), *Adaptations of Plants to Water and Temperature Stress.* Wiley/Interscience, New York, pp. 295–308.

EHLERINGER, J. R., AND I. FORSETH. 1980. Solar tracking by plants. *Science*, 210:1094–1098.

EHRENFELD, J. G. 1980. Understory response to canopy gaps of varying size in a mature oak forest. *Bull. Torrey Bot. Club*, 107:29–41.

EHRLICH, P. R., AND A. H. EHRLICH. 1981. *Extinction: The Causes and Consequences of the Disappearance of Species.* Random House, New York.

EHRLICH, P. R., AND P. H. RAVEN. 1965. Butterflies and plants: A study in coevolution. *Evolution* 18:586–608.

EHRLICH, P. R., AND J. ROUGHGARDEN. 1987. *The Science of Ecology.* Macmillan, New York.

EISNER, T. 1970. Chemical defense against predation in arthropods. In E. Sondheimer and J. B. Simeone (eds.), *Chemical Ecology.* Academic Press, New York, pp. 157–217.

EISNER, T., AND J. MEINWALD. 1966. Defensive secretions of arthropods. *Science*, 153:1341–1350.

ELLISON, L. 1954. Subalpine vegetation of the Wasatch Plateau, Utah. *Ecol. Monogr.*, 24:89–184.

————. 1960. Influences of grazing on plant succession on rangelands. *Bot. Rev.*, 26:1–78.

ELSTER, H. J. 1965. Absolute and relative assimilation rates in relation to phytoplankton populations. In C. R. Goldman (ed.), *Primary Productivity in Aquatic Environments*, Mem. 1st Ital. Idrobiol. 18 Suppl. University of California Press, Berkeley, pp. 79–103.

ELTON, C. S. 1927. *Animal Ecology.* Sidgwick & Jackson, London.

ELTON, C., AND M. NICHOLSON. 1942. The ten-year cycle in numbers of lynx in Canada. *J. Anim. Ecol.*, 11:215–244.

ELNER, R. W., AND R. N. HUGHES. 1978. Energy maximization in the diet of the shore crab *Carcinus maenas. J. Anim. Ecol.*, 47:103–116.

ELTRINGHAM, S. K. 1971. *Life in Mud and Sand.* Crane Russak, New York.

ELWOOD, J. W., AND G. S. HENDERSON. 1975. Hydrologic and chemical budgets at Oak Ridge, Tenn. In A. D. Hasler (ed.), *Coupling of Land and Water Systems* (Ecological Studies 10). Springer-Verlag, New York, pp. 31–51.

EMANUEL, W. R., H. H. SHUGART, AND M. L. STEVENSON 1985. Climate change and the broad scale distribution of terrestrial ecosystem complexes. *Climate Change*, 7:29–43.

EMANUELSSON, A., E. ERIKSSON, AND H. EGNER. 1954. Composition of atmospheric precipitation in Sweden. *Tellus*, 3:261–267.

EMLEN, J. M. 1973. *Ecology: An Evolutionary Approach.* Addison-Wesley, Reading, Mass.

EMLEN, J. T., JR. 1940. Sex and age ratios in the survival of California quail. *J. Wildl. Manage.*, 4:2–99.

EMLEN, S. T., AND L. W. ORING. 1977. Ecology, sexual selection, and the evolution of mating systems. *Science*, 197:215–223.

ENGELMANN, M. D. 1961. The role of soil arthropods in the energetics of an old field community. *Ecol. Monogr.*, 31:221–238.

ENGLE, L. G. 1960. Yellow poplar seedfall pattern. *Central States Forest. Expt. Stat. Note 143.*

ENRIGHT, J. T. 1970. Ecological aspects of endogenous rhythmicity. *Ann. Rev. Ecol. Syst.*, 1:221–238.

————. 1975. Orientation in time: Endogenous clocks. In O. Kinne (ed.), *Marine Ecology: Vol. 2, Physiological Mechanisms, Part 2*, pp. 917–944.

ERIKISSON, E. 1962. Composition of atmospheric precipitation: 1. Nitrogen compounds. *Tellus*, 4:214–232.

————. 1963. The yearly circulation of sulfur in nature. *J. Geophys. Res.*, 68:4001–4008.

ERKERT, H. G. 1982. Ecological aspects of bat activity rhythms. In T. H. Kunz (ed.), *Ecology of Bats.* Plenum Press, New York, pp. 201–320.

ERKERT, H. G., AND S. KRACHT. 1978. Evidence for ecological adaptation of circadian systems: Circadian activity rhythms of neo-tropical and their re-entrainment after phase shifts of the Zeitgeber-LD. *Oecologia*, 32:71–78.

ERLINGE, S., G. GORANSSON, G. HOGSTEDT, G. JANSSON, O. LIBERG, J. LORMAN, I. N. NILSSON, T. VON SCHANTZ, AND M. SYLVEN. 1984. Can vertebrate predators regulate their prey? *Am. Nat.*, 12:125–133.

ERRINGTON, P. L. 1943. Analysis of mink predation upon muskrats in the north-central U.S. *Iowa Agr. Exp. Sta. Res. Bull.*, 320:797–924.

————. 1945. Some contributions of a fifteen-year local study of the northern bobwhite to a knowledge of population phenomena. *Ecol. Monogr.*, 15:1–34.

————. 1946. Predation and vertebrate populations. *Quart. Rev. Biol.*, 21:144–177, 221–245.

————. 1963. *Muskrat Populations.* Iowa State University Press, Ames.

ETHERINGTON, J. R. 1976. *Environment and Plant Ecology.* Wiley, New York.

EVANS, F. C., AND S. A. CAIN. 1952. Preliminary studies on the vegetation of an old-field community in southeastern Michigan. *Contrib. Lab. Vert. Biol. University of Michigan*, 51:1–17.

EVENARI, M. 1985. The desert environment. In M. Evenari, I. Noy-Mier, and D. W. Goodall (eds.), *Hot Deserts and*

Arid Shrublands A (Ecosystems of the World 12A). Elsevier, Amsterdam, pp. 1–22.

EVENARI, M., I. NOY-MEIR, AND D. GOODALL (EDS.). 1985. *Hot Deserts and Arid Shrublands A* (Ecosystems of the World 12A). Elsevier, Amsterdam.

———. 1986. *Hot Deserts and Arid Shrublands B* (Ecosystems of the World 12B). Elsevier, Amsterdam.

EYRE, S. R. 1963. *Vegetation and Soils: A World Picture.* Aldine, Chicago.

FAEGRI, K., AND L. VAN DER PIJL. 1979. *The Principles of Pollination Ecology,* 3rd ed. Pergamon, Oxford.

FARNER, D. S. 1955. The annual stimulus for migration: Experimental and physiologic aspects. In A. Wolfson (ed.), *Recent Studies in Avian Biology,* pp. 198–237.

———. 1959. Photoperiodic control of annual gonadal cycles in birds. In R. B. Witherow (ed.), *Photoperiodism and Related Phenomena.* American Association for the Advancement of Science, Washington, D.C., pp. 717–758.

———. 1964a. The photoperiodic control of reproductive cycles in birds. *Am. Sci.,* 52:137–156.

———. 1964b. Time measurement in vertebrate photoperiodism. *Am. Nat.,* 98:375–386.

FARNER, D. S., AND R. A. LEWIS. 1971. Photoperiodism and reproductive cycles in birds. In A. C. Giese (ed.), *Photophysiology,* Vol. 6. Academic Press, London, pp. 325–364.

FARNSWORTH, E. G., AND F. B. GOLLEY. 1974. *Fragile Ecosystems: Evaluation of Research and Applications in the Neotropics.* Springer-Verlag, New York.

FEENEY, P. 1975. Biochemical coevolution between plants and their insect herbivores. In L. E. Gilbert and P. H. Raven (eds.), *Coevolution of Animals and Plants.* University of Texas Press, Austin, pp. 3–19.

FEHRIG, L., AND G. MERRIAM. 1985. Habitat patch connectivity and population survival. *Ecology,* 66:1762–1768.

FEINSINGER, P. 1983. Coevolution and pollination. In D. Futuyma and M. Slatkin (eds.), *Coevolution.* Sinauer Associates, Sunderland, Mass., pp. 282–310.

FELTON, P. M., AND H. W. LULL. 1963. Suburban hydrology can improve watershed conditions. *Public Works,* 94:93–94.

FENCHEL, T. 1987. *Ecology—Potential and Limitations* (Excellence in Ecology 1.) Ecology Institute, Oldendorf/Luhe, Federal Republic of Germany.

———. 1988. Marine plankton food chains. *Ann. Rev. Ecol. Syst.,* 19:19–38.

FENNER, F. 1953. Host–parasite relationships in myxomotosis of the Australian wild rabbit. *Cold Spring Harbor Symp. Quant. Biol.,* 18:291–294.

FENNER, F., AND F. N. RATCLIFFE. 1965. *Myxamatosis.* Cambridge University Press, Cambridge, England.

FINERTY, J. P. 1980. *The Population Ecology of Cycles in Small Mammals: Mathematical Theory and Biological Fact.* Yale University Press, New Haven, Conn.

FISCHER, A. G. 1960. Latitudinal variation in organic diversity. *Evolution,* 14:64–81.

FISH AND WILDLIFE SERVICE. 1980. Habitat as a basis for environmental assessment. *Ecological Services Manual 101.* USDI Fish and Wildlife Service, Division of Ecological Services, Washington D.C.

FISCHER, R. A. 1930. *The Genetical Theory of Natural Selection,* 2nd rev. ed., Dover, New York.

FISHER, S. G., AND G. E. LIKENS. 1973. Energy flow in Bear Brook, New Hampshire: An integrative approach to stream ecosystem metabolism. *Ecol. Monogr.,* 43:421–439.

FITZPATRICK, L. C. 1973. Energy allocation in the Allegheny mountain salamander, *Desmognathus ochrophaeus. Ecol. Monogr.,* 43:43–58.

FLANAGAN, P. W., AND F. L. BUNNELL. 1980. Microflora activities and decomposition. In J. Brown, P. C. Miller, L. L. Tieszen, and F. L. Bunnell (eds.), *An Arctic Ecosystem:*

The Coastal Tundra at Barrow, Alaska. Dowden, Hutchinson, and Ross, Stroudsburg, Penn., pp. 291–334.

FLEROW, C. C. 1971. The evolution of certain mammals during the late Cenozoic. In K. K. Turekian (ed.), *The Late Cenozoic Glacial Ages.* Yale University Press, New Haven, Conn., pp. 479–485.

FLINT, H. L. 1974. Phenology and genecology of woody plants. In H. Leith (ed.), *Phenology and Seasonality Modeling.* Springer-Verlag, New York, pp. 83–97.

FLINT, R. F. 1970. *Glacial and Quaternary Geology.* Wiley, New York.

FLOOK, D. R. 1970. Causes and implications of an observed sex differential in the survival of wapiti. *Can. Wildl. Serv. Rept. Ser. No. 11.*

FOLTZ, D. W., AND J. L. HOOGLAND. 1983. Genetic evidence of outbreeding in the black-tailed prairie dog *(Cynomys ludovicianus). Evolution,* 37:273–281.

FONS, W. L. 1940. Influence of forest cover on wind velocity. *J. Forest.,* 38:481–486.

FOOSE, T. J. 1983. The relevance of captive populations to the conservation of biotic diversity. In C. M. Schonewald-Cox, S. M. Chambers, B. MacBryde, and W. L. Thomas (eds.), *Genetics and Conservation.* Benjamin/Cummings, Menlo Park, Calif., pp. 374–401.

FOOSE, T. J., AND E. FOOSE. 1984. Demographic and genetic status and management. In B. Beck and C. Wemmer (eds.), *Pere David's Dee: The Biology and Conservation of an Extinct Species.* Noyes, Park Ridge, N.J.

FORBES, S. A. 1887. The lake as a microcosm. *Bull. Peoria Sci. Assoc.;* reprinted 1925, *Ill. Natural History Surv. Bull.,* 15:537–550.

FORCIER, L. K. 1975. Reproductive strategies and cooccurrence of climax tree species. *Science,* 189:808–810.

FOREL, F. A. 1901. *Handbuch der Seenkunde, Allgemeine Limnologie.* J. Engelhorn, Stuttgart.

FORMAN, R. T. T., A. E. GALLI, AND C. F. LECK. 1976. Forest size and avian diversity in New Jersey woodlots with some land use implications. *Oecologia,* 26:1–8.

FORMAN, R. T. T., AND M. GODRON. 1986. *Landscape Ecology.* Wiley, New York.

FORSETH, I. N., AND J. R. EHLERINGER. 1983. Ecophysiology of two solar tracking desert annuals. IV. Effects of leaf orientation on calculated water gain and water use efficiency. *Oecologica,* 58:10–18.

FORTESQUE, J. A. C., AND G. C. MARTIN. 1970. Micronutrients: Forest ecology and systems analysis. In D. C. Reichle (ed.), *Analysis of Temperate Forest Ecosystems.* Springer-Verlag, New York, 1970, pp. 173–198.

FORTNEY, J. L. 1974. Interactions between yellow perch abundance, walleye predation, and survival of alternate prey in Oneida Lake, New York. *Trans. Am. Fish Soc.,* 103:15–24.

FORTNEY, R. H. 1975. The vegetation of Canaan Valley: A taxonomic and ecological study. PhD thesis, West Virginia University, Morgantown.

FOSTER, M. S. 1974. A model to explain molt, breeding overlap, and clutch size in some tropical birds. *Evolution,* 28:182–190.

FOWELLS, H. A. 1948. The temperature profile in a forest. *J. Forest.,* 46:897–899.

FOWELLS, H. A. (ED.). 1965. *Silvics of Forest Trees of the United States. USDA Agricultural Handbook 271.* Washington, D.C.

FOWLER, C. W. 1981. Density dependence as related to life history strategy. *Ecology,* 62:602–610.

FOWLER, N. L., AND J. ANTONOVICS. 1981. Small scale variability in the demography of transplants of two herbaceous species. *Ecology,* 62:1450–1457.

FOWLER, T. D., AND C. W. SMITH. 1981. *Dynamics of Large Animal Populations.* Wiley, New York.

FOX, B. J. 1981. Niche parameters and species richness. *Ecology,* 62:1415–1425.

FOX, F. M., AND M. M. CALDWELL. 1978. Competitive

interaction in plant populations exposed to supplementary ultraviolet-B radiation. *Oecologica,* 36:173–190.

FOX, J. F. 1977. Alternation and coexistence of tree species. *Am. Nat.,* 111:69–89.

FOX, L. R. 1975a. Some demographic consequences of food shortage for the predator *Notonecta hoffmanni. Ecology,* 56:868–880.

———. 1975b. Factors influencing cannibalism, a mechanism of population limitation in the predator *Notonecta hoffmanni. Ecology,* 56:933–941.

———. 1975c. Cannibalism in natural populations. *Ann. Rev. Ecol. Syst.,* 6:87–106.

———. 1981. Defense and dynamics in plant-herbivore systems. *Am. Zool.* 21:853–864.

FOX, M. W. 1970. A comparative study of the development of facial expressions in canids: Wolf, coyote, and foxes. *Behaviour,* 36:4–73.

FRANCIS, W. J. 1970. The influence of weather on population fluctuations in California quail. *J. Wildl. Manage.,* 34:249–266.

FRANKEL, O. H., AND M. E. SOULÉ. 1981. *Conservation and Evolution.* Cambridge University Press, Cambridge, England.

FRANKIE, G. W., H. G. BAKER, AND P. A. OPLER. 1974. Tropical plant phenology: Applications for studies in community ecology. In H. Leith (ed.), *Phenology and Seasonality Modeling.* Springer-Verlag, New York, pp. 287–296.

FRANKLIN, J. F., AND C. T. DYRNESS. 1973. *Natural Vegetation of Oregon and Washington.* U.S.D.A. Forest Service Gen. Tech. Rept. PNW 8. U.S.D.A. Forest Service, Corvallis, Oregon.

FRANKLIN, J. F., AND R. H. WARING. 1979. Distinctive features of the northwestern coniferous forest: Development, structure, and function. In R. H. Waring (ed.), *Forests: Fresh Perspectives from Ecosystem Analysis.* Proc. 40th Ann. Biol. Coll. Oregon State University Press, Corvallis.

FREDRIKSEN, R. L. 1972. Nutrient budget of a Douglas-fir forest on an experimental watershed in western Oregon. In J. F. Franklin, L. J. Dempser, and R. H. Waring (eds.), *Proceedings Research on Coniferous Forest Ecosystems, a Symposium.* Pacific Northwest Forest and Range Experiment Station, Portland, Oregon, pp. 115–131.

FREEMAN, C. L., K. T. HARPER, AND E. L. CHARNOV. 1980. Sex change in plants: Old and new observations and new hypotheses. *Oecologica,* 47:222–232.

FREEMARK, K. E., AND H. G. MERRIAM. 1986. Importance of area and habitat heterogeneity to bird assemblages in temperate forest fragments. *Biol. Cons.,* 36:115–141.

FRENCH, A. R. 1988. The patterns of mammalian hibernation. *Am. Sci.,* 76:569–575.

FRENCH, C. E., L. C. MCEWEN, N. C. MAGRUDER, R. H. INGRAM, AND R. W. SWIFT. 1955. Nutritional requirements of white-tailed deer for growth and antler development. *Penn. State Univ. Agr. Exp. Sta. Bull. No. 600.*

FRENCH, N. R. (ED.). 1971. *Preliminary analysis of structure and function in grasslands.* Range Sci. Dept., Sci. Ser. No. 10. Colorado State University, Fort Collins.

FRENCH, N. R., W. E. GRANT, W. E. GRODZINSKI, AND D. M. SWIFT. 1976. Small mammal energetics in grassland ecosystems. *Ecol. Monogr.* 46:201–220.

FRETWELL, S. D., AND H. L. LUCAS. 1969. On territorial behavior and other factors influencing habitat distribution in birds. *Acta Biotheoretica,* 19:16–36.

FRICHE, H., AND S. FRICHE. 1977. Monogamy and sex change by aggressive dominance in coral reef fish. *Nature,* 266:830–832.

FRINK, C. R. 1971. Plant nutrients and water quality. *Agr. Sci. Rev.,* 9(2):11–25.

FRITTS, S., AND D. MECH. 1981. Dynamic movements and feeding ecology of a newly protected wolf pack. *Wildl. Monogr. 80.*

FRY, F. E. J. 1947. Effects of the environment on animal activity. *Univ. Toronto Stud. Biol.,* 55:1–62.

FRY, F. E. J., J. S. HART, AND K. F. WALKER. 1946. Lethal temperature relations for a sample of young speckled trout *Salvelinus fontinalis. Univ. Toronto Studies, Fish. Res. Lab. No. 66.*

GAINES, M. S., AND L. R. MCCLENAGHAN, JR. 1980. Dispersal in small mammals. *Ann. Rev. Ecol. Syst.,* 11:163–198.

GALLAGHER, J. L., G. F. SOMERS, D. M. GRANT, AND D. M. SELISKAR. 1988. Persistent differences in two forms of *Spartina alterniflora:* A common garden experiment. *Ecology,* 69:1005–1008.

GALLI, A. E., C. F. LECK, AND R. T. T. FORMAN. 1976. Avian distribution patterns in forest islands of different sizes in central New Jersey. *Auk,* 93:356–364.

GALLOWAY, J. N., G. E. LIKENS, AND M. N. HAWLEY. 1984. Acid precipitation: Natural versus anthropogenic causes. *Science,* 226:829–831.

GANT, R. E., AND E. C. CLEBSCH. 1975. The allelopathic influences of *Sassafras albidum* in old field succession in Tennessee. *Ecology,* 56:604–615.

GASAWAY, W. C., R. STEPHENSON, J. DAVIS, P. SHEPHARD, AND O. BURRIS. 1983. Interrelationships of wolves, prey, and man in interior Alaska. *Wildl. Monogr.,* 84:50 pp.

GATES, D. 1962. *Energy Exchange in the Biosphere.* Harper & Row, New York.

———. 1965. Radiant energy: Its receipt and disposal. *Meteorol. Monogr.,* 6:1–26.

———. 1966. Spectral distribution of solar radiation at the earth's surface. *Science,* 151:523–528.

———. 1968. Energy exchange between organisms and environment. In W. P. Lowry (ed.), *Biometeorology,* Proc. 28th Ann. Biol. Colloq. Oregon State University Press, Corvallis.

———. 1985. *Energy and Ecology.* Sinauer Associates, Sunderland, Mass.

GATES, J. E., AND L. W. GYSEL. 1978. Avian nest dispersion and fledgling success in field-forest ecotone. *Ecology,* 59:871–883.

GAUSE, G. F. 1934. *The Struggle for Existence.* Williams & Wilkins, Baltimore.

GEIS, A. D., R. I. SMITH, AND J. P. ROGERS. 1971. Black duck distribution, harvest characteristics, and survival. *U. S. Fish and Wildl. Serv. Spec. Sci. Rept., Wildl. No. 139.*

GEIST, V. 1963. On the behaviour of the North American moose (*Alces alces andersoni,* Peterson, 1950) in British Columbia. *Behaviour,* 20:377–416.

———. 1977. A comparison of social adaptations in relation to ecology in gallinaceous birds and ungulate societies. *Ann. Rev. Ecol. Syst.,* 8:193–207.

GEMMEL, R. P., AND G. T. GOODMAN. 1980. The maintenance of grassland on smelter wastes in the lower Swansea Valley. III. Zinc smelter wastes. *J. Appl. Ecol.,* 17:461–468.

GEORGE, A. S., A. J. M. HOPKINS, AND N. G. MARCHANT. 1979. The heathlands of western Australia. In R. L. Specht (ed.), *Heathlands and Related Shrublands* (Ecosystems of the World 9A). Elsevier, Amsterdam, pp. 211–230.

GHILAROV, M. S. 1970. Soil biocoensis. In J. Phillipson (ed.), *Methods of Study in Soil Ecology.* UNESCO, Paris, pp. 67–77.

GHISELIN, M. T. 1974. *The Economy of Nature and the Evolution of Sex.* University of California Press, Berkeley.

GILBERT, L. E. 1975. Ecological consequences of a coevolved mutualism between butterflies and plants. In L. E. Gilbert and P. H. Raven (eds.), *Coevolution of Animals and Plants.* University of Texas Press, Austin, pp. 210–240.

GILBERT, F. S. 1980. The equilibrium theory of biogeography: Fact or fiction? *J. Biogeography,* 7:209–235.

GILBERTSON, C. B., ET AL. 1970. The effect of animal density and surface slope on the characteristics of runoff, solid waste, and nitrate movement on unpaved feedyards. *Nebraska Agr. Exp. Sta. Bull. 508.*

GILBERTSON, M., R. D. MORRIS, AND R. A. HUNTER. 1976. Abnormal chicks and PCB residue levels in eggs of colonial birds on the lower Great Lakes (1971–1973). *Auk*, 93:434–442.

GILL, R. A., AND R. H. GROVES. 1981. Fire regimes in heathlands and their plant ecological effects. In R. L. Specht (ed.), *Heathlands and Related Shrublands* (Ecosystems of the World 9A). Elsevier, Amsterdam, pp. 61–84.

GILL, D. E. 1975. Spatial patterning of pines and oaks in the New Jersey pine barrens. *J. Ecol.,* 63:291–298.

GILL, F. B., AND L. L. WOLF. 1975. Economics of feeding territorality in the golden-winged sunbird. *Ecology,* 56:333–345.

GILLION, D. 1983. The fire problem in tropical savannas. In F. Bourliere (ed.), *Tropical Savannas* (Ecosystems of the World 13). Elsevier, Amsterdam, pp. 617–641.

GILLION, Y. 1983. The invertebrates of the grass layer. In F. Bourliere (ed.), *Tropical Savannas* (Ecosystems of the World 13). Elsevier, Amsterdam, pp. 289–311.

GILPIN, M. E., AND M. E. SOULÉ. 1986. Minimum viable populations: The processes of species extinctions. In M. E. Soulé (ed.), *Conservation Biology: The Science of Scarcity and Diversity.* Sinauer Associates, Sunderland, Mass., pp. 13–34.

GIMINGHAM, C. H. 1979. Conservation: European heathlands. In R. L. Specht (ed.), *Heathlands and Related Shrublands* (Ecosystems of the World 9A). Elsevier, Amsterdam, pp. 249–260.

GIMINGHAM, C. H., S. B. CHAPMAN, AND N. R. WEBB. 1979. European heathlands. In R. L. Specht (ed.), *Heathlands and Related Shrublands* (Ecosystems of the World 9A). pp. 365–414.

GISBORNE, H. T. 1941. How the wind blows in the forest of northern Idaho. *Northern Rocky Mt. Forest Range Expt. Sta.*

GIST, C. S., AND D. A. CROSSLEY, JR. 1975. A model of mineral-element cycling for an invertebrate food web in a southeastern hardwood forest litter community. In F. Howell, J. B. Gentry, and M. H. Smith (eds.), *Mineral Cycling in Southeast Ecosystems.* National Technical Information Service, U.S. Dept. Commerce, pp. 84–106.

GLEASON, H. A. 1917. The structure and development of the plant association. *Bull. Torrey Bot. Club,* 44:463–481.

———. 1926. The individualistic concept of the plant association. *Bull. Torrey Bot. Club,* 53:7–26.

GOFF, L. J. 1982. Symbiosis and parasitism: Another viewpoint. *Bioscience,* 32:255–256.

GOLDSMITH, J. R. 1969. Epidemiological bases for possible air quality criteria for lead. *Air Poll. Contr. Assoc. J.,* 19:714–719.

GOLDSMITH, J. R., AND A. C. HEXTER. 1967. Respiratory exposure to lead: Epidemiological and experimental dose-response relationship. *Science,* 158:132–134.

GOLDWASSER, S., D. GAINES, AND S. R. WILBUR. 1980. The Least Bell's Vireo in California: A *de facto* endangered race. *Am. Birds,* 34:742–745.

GOLLEY, F. B. 1960. Energy dynamics of a food chain of an old field community. *Ecol. Monog.,* 30:187–206.

———. 1975. Productivity and mineral cycling in tropical forests. In National Academy of Science, *Productivity of World Ecosystems.* National Academy of Science, Washington, D.C., pp. 106–115.

GOLLEY, F. B. (ED.). 1983. *Tropical Rain Forest Ecosystems* (Ecosystems of the World 14A). Elsevier, Amsterdam.

GOLLEY, F. B., J. T. MCGINNIS, R. C. CLEMENTS, G. S. CHILD, AND M. J. DUEVER. 1975. *Mineral Cycling in a Tropical Moist Forest Ecosystem.* University of Georgia Press, Athens.

GOLLEY, F. B., K. PETRUSEWICZ, AND L. RYSZKOWSKI. 1975. *Small Mammals: Their Productivity and Population Dynamics.* Cambridge University Press, Cambridge, England.

GOLLEY, F. B., AND E. MEDINA (EDS.). 1975. *Tropical Ecological Systems Trends in Terrestrial and Aquatic Research.* Springer-Verlag, New York.

GOLLEY, P. M., AND F. B. GOLLEY (EDS.). 1972. *Tropical Ecology, with an Emphasis on Organic Production.* University of Georgia Press, Athens.

GOLTERMAN, H. L., AND F. A. KOUWE. 1980. Chemical budgets and nutrient pathways. In E. D. LeCren and R. H. Lowe-McConnell (eds.), *The Functioning of Freshwater Ecosystems* (International Biological Programmes No. 22). Cambridge University Press, Cambridge, England, pp. 88–140.

GOOD, R. E., D. F. WHIGHAM, AND R. L. SIMPSON (EDS.). 1978. *Freshwater Wetlands.* Academic Press, New York.

GORDON A. G., AND E. GORHAM. 1963. Ecological aspects of air pollution from an iron-sintering plant at Wana, Ontario. *Can. J. Bot.,* 41:1063–1078.

GORE, A. J. P. (ED.). 1983. *Mires: Swamp, Bog, Fen, Moor* (Ecosystems of the World 13). Elsevier, Amsterdam.

GORHAM, J. R. 1966. The epizootiology of distemper. *J. Am. Med. Assoc.,* 149:610–622.

GORDON, M. S. 1972. *Animal Physiology,* 2nd ed. Macmillan, New York.

GOSS, R. J., C. E. DINSMORE, L. N. GRIMES, AND J. K. ROSEN. 1974. Expression and suppression of circannual antler growth cycle in deer. In E. T. Pengelley (ed.), *Circannual Clocks: Annual Biological Rhythms.* Academic Press, New York, pp. 393–422.

GOSS-CUSTARD, J. D. 1970. The responses of redshank *Tringa totanus* (L) to spatial variation in the density of their prey. *J. Anim. Ecol.,* 39:91–113.

———. 1977a. The energetics of prey selection by redshank *Tringa totanus* (L) and a preferred prey *Corophium volutator* (Pallas). *J. Anim. Ecol.,* 46:21–35.

———. 1977b. The energetics of prey selection by redshank *Tringa totanus* (L) in relation to prey density. *J. Anim. Ecol.,* 46:1–19.

GOSZ, J. R., E. LIKENS, AND F. H. BORMANN. 1976. Organic matter and nutrient dynamics of the forest and forest floor in the Hubbard Brook Forest. *Oecologica,* 22:305–320.

GOSZ, J. R., G. E. LIKENS, J. S. EATON, AND F. H. BORMANN. 1975. Leaching of nutrients from leaves of selected tree species in New Hampshire. In F. C. Howell, J. B. Gentry, and M. H. Smith (eds.), *Mineral Cycling in Southeastern Ecosystems.* National Technical Information Service, U.S. Dept. Commerce, pp. 630–641.

GRACE, J. 1977. *Plant Response to Wind.* Academic Press, London.

GRAHAM, D., AND B. D. PATTERSON. 1982. Responses of plants to low, non-freezing temperatures: Protein metabolism and acclimation. *Ann. Rev. Plant Physiol.* 33:347–372.

GRANHALL, R., AND V. LID-TORSVIK. 1975. Nitrogen fixation by bacteria and free-living blue-green algae in tundra ecosystem. In F. E. Wielgolaski (ed.), 1975a, pp. 305–315.

GRANT, P. R. 1972. Interspecific competition among rodents. *Ann. Rev. Ecol. Syst.,* 3:79–106.

———. 1986. Interspecific competition in fluctuating environments. In J. Diamond and T. J. Case (eds.), *Community Ecology.* Harper & Row, New York, pp. 173–191.

GREEN, R. F. 1980. Baysean birds: A simple example of Oaten's stochastic model of optimal foraging. *Theor. Popul. Biol.,* 18:244–256.

GREENWAY, H., AND R. MURRS. 1980. Mechanisms of salt tolerance in nonhalophytes. *Ann. Rev. Plant Physiol.,* 31:149–190.

GREENWOOD, P. J. 1980. Mating systems, philopatry, and dispersal in birds and mammals. *Anim. Behav.,* 28:1140–1162.

B-14

GREENWOOD, P. I., AND P. H. HARVEY. 1982. The natal and breeding dispersal of birds. *Ann. Rev. Ecol. Syst.*, 13:1–21.

GREESON, P. E., J. R. CLARK, AND J. E. CLARK (EDS.). 1979. *Wetland Functions and Values: The State of Our Understanding.* American Water Resource Association, Minneapolis.

GRIER, C. C., AND D. W. COLE. 1972. Elemental transport changes occurring during development of a second growth Douglas-fir ecosystem. In J. L. Franklin, L. J. Dempster, and R. H. Waring (eds.), *Proceedings, Research on Coniferous Forest Ecosystems, a Symposium.* Pacific Northwest Forest and Range Experiment Station, Portland, Oregon, pp. 103–113.

GRIER, C. C., AND R. S. LOGAN. 1977. Organic matter distribution and net production in plant communities of a 400-year old Douglas-fir ecosystem. *Ecol. Monogr.,* 47:373–400.

———. 1978. *Pseudotsuga menziesii* communities of a western Oregon watershed: Biomass, distribution, and production budgets. *Ecol. Monogr.,* 47:373–400.

GRIER, C. C., C. E. MEIER, AND R. L. EDMONDS. 1982. Mycorrhizal role in net primary production and nutrient cycling in *Abies amabilis* ecosystems in western Washington. *Ecology,* 63:370–380.

GRIER, J. W. 1982. Ban of DDT and subsequent recovery of reproduction in bald eagles. *Science,* 218:1232–1234.

GRIGGS, R. F. 1946. The timberlines of northern America and their interpretation. *Ecology,* 27:275–289.

GRIME, J. P. 1966. Shade avoidance and shade tolerance in flowering plants. In F. Bainbridge. (ed.), *Light as an Ecological Factor.* Blackwell, Oxford, pp. 187–207.

———. 1977. Evidence for the existence of three primary strategies in plants and its relevance to ecological and evolutionary theory. *Am. Nat.,* 111:1169–1194.

———. 1979. *Plant Strategies and Vegetative Processes.* Wiley, New York.

GRINNELL, J. 1904. The origin and distribution of the Chestnut-backed Chickadee. *Auk,* 21:364–382.

———. 1917. The niche relationships of the California thrasher. *Auk,* 34:427–433.

———. 1924. Geography and evolution. *Ecology,* 5:225–229.

———. 1928. Presence and absence of animals. *University of California Chronicle,* 30:429–450.

GROSS, K. L. 1980. Colonization by *Verbascum thapsus* (Mullien) of an old-field in Michigan: Experiments in the effects of vegetation. *J. Ecol.,* 68:919–927.

GROSS, M. 1972. *Oceanography.* Prentice-Hall, Englewood Cliffs, N.J.

GRUBB, P. J. 1977. The maintenance of species richness in plant communities: The importance of the regeneration niche. *Biol. Rev.,* 52:107–145.

GRUE, C. E., D. J. HOFFMANM, W. D. BEYER, AND L. P. FRANSON. 1986. Lead concentrations and reproductive success in European starlings *Sturnus vulgaris* nesting within highway roadside verges. *Envir. Pollut.* (Series A), 42:157–182.

GRUELL, G. E. 1983. *Fire and Vegetative Trends in the Northern Rockies: Interpretations from 1871–1982 Photographs.* U.S.D.A. For. Serv. Gen. Tech. Rept. INT-158.

GULLON, D. 1983. The fire problem in tropical savannas. In F. Bourliere (ed.), *Tropical Savannas* (Ecosystems of the World 13). Elsevier, Amsterdam, pp. 617–642.

GUNTER, G. 1961. Some relations of estuarine organisms to salinity. *Limno. Oceanogr.,* 6:182–190.

GUTIERREZ, J. R., AND E. R. FUENTES. 1979. Evidence for intraspecific competition in the *Acacia cazen* (Leguminosae) savanna of Chile. *Oecol. Plant.,* 14:151–158.

GWINNER, E. 1978. Effects of pinealectomy on circadian locomotor activity rhythms in European starlings, *Sturnus vulgaris. J. Comp. Physiol.,* 126:123–129.

GYLLENBERG, G. 1980. Bioenergetic parameters of the main group of herbivores. In A. I. Breymeyer and G. M. Van Dyne (eds.), *Grassland, Systems Analysis and Man* (International Biological Programme No. 19). Cambridge University Press, Cambridge, England, pp. 238–256.

HAARLOV, N. 1960. Microarthropods from Danish soils. *Oikos, Suppl. No. 3,* pp. 1–176.

HABERMANN, R. T., C. M. HERMAN, AND F. P. WILLIAMS. 1958. Distemper in raccoons and foxes suspected of having rabies. *J. Am. Vet. Med. Assoc.,* 132:31–35.

HACKETT, C. 1965. Ecological aspects of the nutrition of *Deschampsia flexuosa* (L): The effects of Al, Ca, Fe, K, Mn, P, and pH on the growth of seedlings and established plants. *J. Ecol.,* 53:315–333.

HACSKAYLO, E. 1971. Metabolite exchanges in ectomycorrhizae. In E. Hacskaylo (ed.), *Mycorrhizae, U.S.D.A. Forest Serv. Misc. Pub. No. 1189.*

HADLEY, E. B., AND L. C. BLISS. 1964. Energy relationships of alpine plants on Mount Washington, New Hampshire. *Ecol. Monogr,* 34:331–357.

HADLEY, E. B., AND B. J. KIECKHEFER. 1963. Productivity of two prairie grasses in relation to fire frequency. *Ecology,* 44:389–395.

HADLEY, J. L., AND W. K. SMITH. 1986. Wind effects on needles of timberline conifers: Seasonal influence on mortality. *Ecology,* 67:12–19.

HAECKEL, E. 1869. Euber Entwichelunge Gang 4. *Aufgabe de Zoologie Jemaische z.,* 5:353–370.

HAINES, B. L., AND E. L. DUNN. 1985. Coastal marshes. In B. F. Chabot and H. A. Mooney (eds.), *Physiological Ecology of North American Plant Communities.* Chapman and Hall, New York, pp. 323–347.

HAIRSTON, N. G. 1969. On the relative abundance of species. *Ecology,* 50:1091–1094.

HAIRSTON, N. G., F. E. SMITH, AND L. B. SLOBODKIN. 1960. Community structure, population control, and competition. *Am. Nat.,* 94:421–425.

HALDANE, J. S. B. 1932. *The Causes of Evolution.* Longsman, Green, London.

———. 1954. The measurement of natural selection. *Proc. 9th Int. Congr. Genetics.,* pp. 480–487.

HALE, N. 1971. *Biology of Lichens.* Edward Arnold, London.

HALL, C. A. S., C. J. CLEVELAND, AND R. KAUFMAN. 1986. *Energy and Resource Quality: The Ecology of the Economic Process.* Wiley, New York.

HALL, R. J., G. E. LIKENS, S. B. FIANCE, AND G. R. HENDREY. 1980. Experimental acidification of a stream in the Hubbard Brook Experimental Forest, New Hampshire. *Ecology,* 61:976–989.

HALLE, F. R., A. A. OLDMEMAN, AND P. B. TOMLINSON. 1978. *Tropical Trees and Forests: An Architectural Analysis.* Springer-Verlag, New York.

HALLIDAY, T. R. 1978. Sexual selection and mate choice. In J. R. Krebs and N. D. Davies (eds.), *Behavioural Ecology: An Evolutionary Approach.* Blackwell, Oxford, England, pp. 180–213.

HAMILTON, W. D. 1964. The genetical evolution of social behavior I, II. *J. Theoret. Biol,* 82:1–16, 17–52.

———. 1972. Altruism and related phenomena, mainly in social insects. *Ann. Rev. Ecol. Syst.,* 3:193–232.

HAMILTON, W. J., III. 1959. Aggressive behavior in migrant pectoral sandpipers. *Condor,* 61:161–179.

HAMMOND, A. L. 1972. Chemical pollution: Polychlorinated biphenyls. *Science,* 175:155–156.

HAMNER, W. M. 1968. The photorefractory period of the house finch. *Ecology,* 49:211–227.

HANDEL, S. N. 1978. The competitive relationship of three woodland sedges and its bearing on the evolution of ant dispersal of *Carex penduculata. Evolution,* 32:151–163.

HANES, T. L. 1981. California chaparral. In F. di Castri and H. A. Mooney (eds.), *Mediterranean-type Shrublands: Origin and Structure.* Springer-Verlag, New York, pp. 139–174.

HANSON, H. C. 1953. Vegetation types in northwestern

Alaska and comparisons with communities in other arctic regions. *Ecology,* 34:111–140.

HANSON, W. C. 1971. Seasonal patterns in native residents of three contrasting Alaskan villages. *Health Physics,* 20:585–591.

HANSON, W. C., D. G. WATSON, AND R. W. PERKINS. 1967. Concentration and retention of fallout radionuclides in Alaska arctic ecosystems. In B. Aberg and F. P. Hungate (eds.), *Radioecological Concentration Processes.* Pergamon Press, London, pp. 233–245.

HANZAWA, F. M., A. J. BEATTIE, AND D. C. CULVAR. 1988. Directed dispersal: Demographic analysis of an ant-seed mutualism. *Am. Nat.* 131:1–13.

HARDIN, G. 1960. The competitive exclusion principle. *Science,* 131:1292–1297.

HARESTAD, A. S., AND E. L. BUNNELL. 1979. Home range and body weight—a reevaluation. *Ecology,* 60:389–402.

HARMEL, D. E. 1983. Effects of genetics on antler quality and body size in white-tailed deer. In R. D. Brown (ed.), *Antler Development in Cervidae.* Caeser Kleberg Wildlife Research Institute, Kingsville, Texas.

HARPER, J. L. 1977. *Population Biology of Plants.* Academic Press, New York.

HARPER, J., AND A. D. BELL. 1979. The population dynamics of growth form in organisms with modular construction. In R. M. Anderson (ed.), *Population Dynamics, 20th Symposium, British Ecological Society.* Blackwell, Oxford, England, pp. 29–52.

HARPER, J. L., AND J. WHITE. 1974. The demography of plants. *Ann. Rev. Ecol. Syst.,* 5:419–463.

HARRIS, G. G., A. W. EBLING, D. R. LAUR, AND R. J. ROWLEY. 1984. Community recovery after storm damage: A case of facilitation in primary succession. *Science,* 224:1136–1138.

HARRIS, L. D. 1984. *The Fragmented Forest.* University of Chicago Press, Chicago.

HARRIS, W. F., P. SOLLINS, N. T. EDWARDS, B. E. DINGER, AND H. H. SHUGART. 1975. Analysis of carbon flow and productivity in a temperate deciduous forest ecosystem. In *Productivity of World Ecosystems.* National Academy of Science, Washington D.C.

HARRISON, J. L. 1962. Distribution of feeding habits among animals in a tropical rain forest. *J. Anim. Ecol.,* 31:53–63.

HART, D. D. 1985. Causes and consequences of territoriality in a grazing stream insect. *Ecology,* 66:404–414.

HART, J. S. 1951. Photoperiodicity in the female ferret. *J. Expt. Biol.,* 28:1–12.

HARTSHORN, G. S. 1978. Tree falls and tropical forest dynamics. In P. B. Tomlinson and M.H. Zimmerman, *Tropical Trees as Living Systems.* Cambridge University Press, Cambridge, England, pp. 617–638.

HARVEY, P. H., AND P. J. GREENWOOD. 1978. Antipredator defense strategies: Some evolutionary problems. In J. R. Krebs and N. B. Davies (eds.), *Behavioural Ecology: An Evolutionary Approach.* Blackwell, Oxford, England, pp. 129–151.

HARVEY, P. H., AND R. M. ZAMMUTO. 1985. Patterns of mortality and age at first reproduction in natural populations of mammals. *Nature,* 315:319–320.

HASLER, A. D. 1969. Cultural eutrophication is reversible. *Bioscience,* 19:425–431.

HASLER, A. D. (ED.). 1975. *Coupling of Land and Water Systems,* Ecological Studies, Vol. 10. Springer-Verlag, New York.

HASSELL, M. P. 1966. Evaluation of parasite or predator response. *J. Anim. Ecol.,* 35:65–75.

HASSELL, M. P., J. H. LAWTON, AND J. R. BEDDINGTON. 1976. The components of arthropod predation. 1. The prey death rate. *J. Anim. Ecol.,* 45:135–164.

———. 1977. Sigmoid functional responses by invertebrate predators and parasitoids. *J. Anim. Ecol.,* 46:249–262.

HASSELL, M. P., AND R. M. MAY. 1973. Stability in insect host-parasite models. *J. Anim. Ecol.,* 42:693–726.

———. 1974. Aggregation in predators and insect parasites and its effect on stability. *J. Anim. Ecol.,* 43:567–597.

HAWKINS, C. P., M. L. MURPHY, AND N. H. ANDERSON. 1982. Effects of canopy, substrate composition, and gradient on the structure of macroinvertebrate communities in the Cascade Range streams of Oregon. *Ecology,* 63:1840–1856.

HAWTHORNE, W. R., AND P. B. CAVERS. 1978. Population dynamics of the perennial herbs *Plantago major* and *P. rugelli* Decne. *J. Ecol.,* 64:511–527.

HAYES, H. 1972. Polyandry in the spotted sandpiper. *Living Bird,* 11:43–57.

HAYMAN, R. M., AND K. C. MCDONALD. 1985. The geology of deep sea hot springs. *Am. Sci.,* 73:441–449.

HAYNE, D. W., AND R. C. BALL. 1956. Benthic productivity as influenced by fish predation. *Limnol. Oceanogr.,* 1:162–175.

HEAL, O. W., H. E. JONES, AND J. B. WHITTAKER. 1975. Moore House, U.K. In T. Rosswall and O. W. Heal (eds.), *Structure and Function of Tundra Ecosystems,* pp. 295–320.

HEATH, J. E. 1965. Temperature regulation and diurnal activity in horned lizards. *Univ. Cal. Pub. Zool.,* 64:97–136.

HEDGPETH, J. W. (ED.). 1957. Sandy beaches. In Hedgpeth (ed.), *Treatise in Marine Ecology and Paleoecology:* 1. *Ecology.* Memoir 67, Geological Soc. Amer., pp. 587–608.

HEGGLESTAD, H. E. 1969. Consideration of air quality standards for vegetation with respect to ozone. *Air. Pollut. Assoc. Cont. J.,* 19:424–426.

HEICHEL, G. H., AND N. C. TURNER. 1976. Phenology and leaf growth of defoliated hardwood trees. In J. F. Anderson and H. K. Karpus (eds.), *Perspectives in Forest Entomology.* Academic Press, New York, pp. 31–40.

HEINRICH, B. 1976. Heat exchange in relation to blood flow between thorax and abdomen in bumblebees. *J. Exp. Biol,* 64:567–585.

———. 1979. *Bumblebee Economics.* Harvard University Press, Cambridge, Mass.

HEINSELMAN, M. L. 1970. Landscape evolution, peatland types, and the environment in Lake Agassiz peatlands natural area, Minnesota. *Ecol. Monogr.,* 33:327–374.

———. 1975. Boreal peatlands in relation to environment. In A. D. Hasler (ed.), *Coupling of Land and Water Systems.* Springer-Verlag, New York, pp. 93–103.

HEINSELMAN, M. L., AND T. M. CASEY. 1981a. Fire and succession in the conifer forests of northern North America. In D. E. West, H. H. Shugart, and D. B. Botkin (eds.), *Forest Succession: Concepts and Applications.* Springer-Verlag, New York, pp. 374–405.

———. 1981b. Fire intensity and frequency as factors in the distribution of northern ecosystems. In *Fire Regimes and Ecosystem Properties,* U.S. For. Serv. Rept. Wo-26. U.S. Department of Agriculture, Washington, D.C., pp. 7–57.

HEITHAUS, E. R. 1981. Seed predation by rodents on three ant-dispersed plants. *Ecology,* 63:136–145.

HEITHAUS, E. R., D. C. CULVER, AND A. J. BEATTIE. 1980. Models of some ant–plant mutualisms. *Am. Nat.,* 116:347–361.

HENDERSON, G. S., AND W. H. HARRIS. 1975. An ecosystem approach to the characterization of the nitrogen cycle in a deciduous forest watershed. In B. Bernier and C. W. Wingel (eds.), *Forest Soils and Land Management.* Le Presse de l' Universido, Lavel Quebec.

HENDERSON, L. J. 1913. *The Fitness of the Environment.* Macmillan, New York.

HENDRIKSSON, E., AND B. SIMU. 1971. Nitrogen fixation by lichens. *Oikos,* 22:119–121.

HERRERA, C. M. 1982. Defense of ripe fruit from pests: Its significance in relation to plant disperser interactions. *Am. Nat.,* 120:218–241.

———. 1985. Determinants of plant-animal coevolution: The case of mutualistic dispersal of seeds by vertebrates. *Oikos,* 44:132–141.

HETT, J., AND O. L. LOUCKS. 1976. Age structure models of balsam fir and eastern hemlock. *J. Ecol.*, 64:1029–1044.

HIBBS, D. E. 1982. Gap dynamics in a hemlock-hardwood forest. *Can. J. For. Res.*, 12:522–527.

———. 1983. Forty years of forest succession in central New England. *Ecology*, 64:1394–1401.

HICKS, R. R., JR., AND D. E. FOSBROKE. 1987. Stand vulnerability: Can gypsy moth damage be predicted? In S. Fosbroke and R. Hicks (eds.), *Proc. Cooperative Workshop*, pp. 73–80.

HIGGS, A. T. 1981. Island biogeography theory and nature reserve design. *J. Biogeogr.*, 8:117–124.

HIGGS, A. T., AND M. B. USHER. 1980. Should nature reserves be large or small? *Nature* (London), 285:568–569.

HILDEN, O. 1964. Ecology of duck populations in the island group of Valassarret, Gulf of Bothnia. *Ann. Zool. Fennica*, 1:153–279.

———. 1965. Habitat selection in birds: A review. *Ann. Zool. Fennica*, 2:53–75.

HILL, E. P., III. 1972. Litter size in Alabama cottontails as influenced by soil fertility. *J. Wildl. Manage.*, 36:1199–1209.

HILL, R. W. 1976. *Comparative Physiology of Animals: An Environmental Approach.* Harper & Row, New York.

HILLBRICHT-ILKOWSKA, A. 1974. Secondary productivity in freshwaters, its value and efficiencies in plankton food chain. In *Proc. 1st Inter. Congr. Ecol.*, pp. 164–167.

HILLS, T. L. 1965. Savannas: A review of a major research problem in tropical geography. *Can. Geog.*, 9:216–228.

HINNERI, S., M. SONESSON, AND A. K. VEUM. 1975. Soils of Fennoscandian I.B.P. Tundra ecosystems. In F. E. Wielgolaski (ed.), *Fennoscandian Tundra Ecosystems. Part 1. Plants and Microorganisms.* Springer-Verlag, New York, pp. 31–40.

HIRSHFIELD, M. F., AND D. W. TINKLE. 1975. Natural selection and the evolution of reproductive effort. *Proc. Nat. Acad. Sci.*, 72(6):2227–2231.

HOBBIE, J., J. COLE, J. DUNGAN, R. A. HOUGHTON, AND B. PETERSON. 1984. Role of biota in global CO_2 balance: The controversy. *Bioscience*, 34:492–498.

HOCK, R. J. 1960. Seasonal variations in physiological functions of arctic ground squirrels and black bears. In C. P. Lyman and A. R. Dawe (eds.), *Mammalian Hibernation.* Museum Comparative Zoology, Harvard University, Cambridge, Mass., pp. 155–169.

HOCKING, B. 1975. Ant–plant mutualism: Evolution and energy. In L. E. Gilbert and P. H. Raven (eds.), *Coevolution of Animals and Plants.* University of Texas Press, Austin, pp. 78–90.

HOESE, H. D. 1960. Biotic changes in a bay associated with the end of a drought. *Limnol. Oceanogr.*, 5:326–336.

HOFFMAN, K. 1965. Clock mechanisms in celestial orientation of animals. In J. Aschoff (ed.), *Circadian Clocks.* North Holland, Amsterdam, pp. 87–94.

HOLDEN, C. 1974. Fish flour: Protein supplement has yet to fulfill expectations. *Science*, 173:410–412.

HOLDRIDGE, L. R. 1947. Determination of wild plant formations from simple climatic data. *Science*, 105:367–368.

———. 1967. Determination of world plant formations from simple climatic data. *Science*, 130:572.

HOLDRIDGE, L. R., W. C. GRENKE, W. H. HATHEWAY, T. LIANG, AND J. A. TOSI, JR. 1971. *Forest Environments in Tropical Life Zones, A Pilot Study.* Pergamon Press, New York.

HOLLING, C. S. 1959. The components of predation as revealed by a study of small mammal predation of the European pine sawfly. *Can. Entomol.*, 91:293–320.

———. 1961. Principles of insect predation. *Ann. Rev. Entomol.*, 6:163–182.

———. 1965. The functional response of predators to prey density and its role in mimicry and population regulation. *Mem. Entomol. Soc. Can. No. 45.*

———. 1966. The functional response of invertebrate predators to prey density. *Mem. Entomol. Soc. Can. No. 48.*

———. 1973. Resilience and stability of ecological systems. *Ann. Rev. Ecol. Syst.*, 4:1–23.

———. 1980. Forest insects, forest fires, and resilience. In H. Mooney, J. M. Boninicksen, N. L. Christensen, J. E. Latan, and W. A. Reiners (eds.), *Fire Regimes and Ecosystem Properties.* U.S.D.A. For. Serv. Gen. Tech Rept., pp. 20–26.

HOLLINGER, D. Y. 1986. Herbivory and the cycling of nitrogen and phosphorus in isolated California oak trees. *Oecologica*, 70:291–297.

HOLMES, J. 1983. Evolutionary relationships between parasitic helminths and their hosts. In D. J. Futuyma and M. Slatkin (eds.), *Coevolution.* Sinauer Associates, Sunderland, Mass.

HOLMES, R. T., AND S. K. ROBINSON. 1981. Tree species preferences of foraging insectivorous birds in a northern hardwoods forest. *Oecologica*, 48:31–35.

HOLMES, W. G. 1984. Sibling recognition in thirteen-lined ground squirrels: Effects of genetic relatedness, rearing association, and olfaction. *Behav. Ecol. Sociobiol.*, 14:225–233

HOLMES, W. T., AND P. W. SHERMAN. 1982. The ontogeny of kin recognition in two species of ground squirrel. *Am. Zool.*, 22:491–517.

———. 1983. Kin recognition in animals. *Am. Sci.*, 7:46–55.

HOOGLAND, J. C. 1982. Prairie dogs avoid extreme inbreeding. *Science*, 215:1639–1641.

———. 1985. Infanticide in prairie dogs: Lactataing females kill offspring of close kin. *Science*, 230:1037–1040.

HOOK, D. D. 1984. Adaptation to flooding with fresh water. In T. Kozlowski (ed.), *Flooding and Plant Growth.* Academic Press, Orlando, Fla., pp. 265–294.

HOPKINS, H. H. 1954. Effects of mulch upon certain factors of the grassland environment. *J. Range Manage.*, 7:255–258.

HOPKINS, S. H. 1958. The planktonic larvae of *Polydora websteri* Hartman (Annelida, Polychaeta) and their settling on oysters. *Bull. Marine Sci. of Gulf and Caribbean*, 8:268–277.

HOPKINSON, C. S., AND J. P. SCHUBAUER. 1984. Static and dynamic aspects of nitrogen cycling in the salt marsh graminoid *Spartine alternifolia*. *Ecology*, 65:961–969.

HORN, H. S. 1975. Markovian properties of forest succession. In M. L. Cody and J. Diamond (eds.), *Ecology and Evolution of Communities.* Harvard University Press, Cambridge, Mass., pp. 196–211.

———. 1981. Some causes of variety of patterns of secondary succession. In D. E. West, H. H. Shugart, and D. B. Botkin (eds.), *Forest Succession: Concepts and Applications.* Springer-Verlag, New York, pp. 24–35.

HORN, H. S., H. H. SHUGART, AND D. URBAN. 1989. Simulators as models of forced dynamics. In J. Roughgarden, R. May, and S. Levin (eds.), *Perspectives in Ecological Theory.* Princeton University Press, Princeton, N. J., pp. 256–267.

HORNBECK, J. W., G. E. LIKENS, AND J. S. EATON. 1976. Seasonal patterns in acidity of precipitation and implications for forest-stream ecosystems. In L. S. Dochinger and T. A. Seligar (eds.), *Proc. 1st. International Symp. on Acid Precipitation and the Forest Ecsystem.* U.S.D.A. For. Serv. Gen. Tech. Rept. NE-23.

HORNBECK, J. W., C. W. MARTIN, R. S. PIERCE, F. H. BORMANN, G. E. LIKENS AND J. S. EATON. 1987. The northern hardwood forest ecosystem: Ten years of recovery from clearcutting. U.S.D.A. For. Serv. NERP-576.

HORNOCKER, M. G. 1969. Winter territoriality in mountain lions. *J. Wildl. Manage.*, 33:457–464.

———. 1970. An analysis of mountain lion predation upon mule deer and elk in the Idaho primitive area. *Wildl. Monogr. No. 21.*

HORSLEY, S. B. 1977. Allelopathic inhibition of black cherry by ferns, grass, goldenrod, and aster. *Can. J. For. Res.*, 7:205–216.

HOUGH, A. F., AND R. D. FORBES. 1943. The ecology and silvics of forests in the high plateaus of Pennsylvania. *Ecol. Monogr.*, 13:299–320

HOUGHTON, R. A., J. E. HOBBIE, J. M. MELILLO, B. MORE, B. J. PETERSON, G. R. SHAVER, AND G. M. WOODWELL. 1983. Changes in the carbon cycle in terrestrial biota and soils between 1860 and 1980: A vert release of CO_2 to the atmosphere. *Ecol. Monogr.*, 53:235–262.

HOUSTON, D. B. 1982. *The Northern Yellowstone Elk: Ecology and Management.* Macmillan, New York.

HOWARD, D. V. 1974. Urban robins: A population study. In J. Noyes and D. R. Progulske (eds.), *Wildlife in an Urbanizing Environment.* Massachusetts Cooperative Extension Service, Amherst, pp. 67–75.

HOWARD, R. D. 1978a. The evolution of mating strategies in bullfrogs, *Rana catesbeiana. Evolution*, 32:850–871.

———. 1978b. The influence of male defended oviposition sites on early embryo mortality in bullfrogs. *Ecology*, 59:789–798.

———. 1979. Estimating reproductive success in natural populations. *Am. Nat.*, 114:221–231.

HOWARD, W. E. 1960. Innate and environmental dispersal of individual vertebrates. *Amer. Mid. Natur.*, 63:152–161.

HOWARD-WILLIAMS, C. 1970. The ecology of *Becium homblei* in Central Africa with special reference to metaliferous soils. *J. Ecology*, 58:745–763.

HOWARTH, R. W., AND J. M. TEAL. 1979. Sulfate reduction in a New England salt marsh. *Limno. Oceanogr.*, 24:999–1013.

HOWE, H. F. 1976. Egg size, hatching asynchrony, sex, and brood reduction in the common grackle. *Ecology*, 57:1195–1207.

———. 1980. Monkey dispersal and waste of a neotropical fruit. *Ecology*, 61:944–959.

———. 1985. Comphothere fruits: A critique. *Am. Nat.*, 125:853–865.

———. 1986. Seed dispersal by fruit-eating birds and mammals. In D. R. Murray (ed.), *Seed Dispersal.* Academic Press, Sydney, pp. 123–190.

HOWE, H. F., AND J. SMALLWOOD. 1982. Ecology of seed dispersal. *Ann Rev. Ecol. Syst.*, 13:201–238.

HOWE, H. F., AND L. C. WESTLEY. 1982. *Ecological Relationships of Plants and Animals.* Oxford University Press, New York.

HOWELL, F. G., J. B. GENTRY, AND M. H. SMITH (EDS.). 1975. *Mineral Cycling in Southeastern Ecosystems* (ERDA Symposium Series). National Technical Information Science, U.S. Dept. Commerce.

HOWES, B. L., J. W. H. DACEY, AND J. M. TEAL. 1985. Annual carbon mineralization and belowground production of *Spartine alterniflora* in New England. *Ecology*, 66:595–605.

HUBBARD, S. F., AND R. M. COOK. 1978. Optimal foraging by parasitoid wasps. *J. Anim. Ecol.*, 47:593–604.

HUDSON, H. J., AND J. WEBSTER. 1958. Succession of fungi on decaying stems of *Agropyron repens. Brit. Mycol. Soc. Trans.*, 41:165–177.

HUDSON, J. W. Evolution of thermal regulation. In G. W. Whittow (ed.), *Comparative Physiology of Thermoregulation*, Vol. 3. Academic Press, New York, pp. 97–165.

HUDSON, P. J. 1986. The effect of a parasitic nematode on the breeding production of red grouse. *J. Anim. Ecol.*, 55:85–92.

HUFFAKER, C. B. 1958. Experimental studies on predation: Dispersion factors and predator-prey oscillations. *Hilgardia*, 27:343–383.

HUMPHREY, R. R. 1958. The desert grassland, a history of vegetational changes and an analysis of causes. *Bot. Rev.*, 24:193–252.

HUNN, J. B., L. CLEVELAND, AND E. E. LITTLE. 1987. Influence of pH and aluminum on developing brook trout in low calcium water. *Environmental Pollution*, 43:63–73.

HUNT, E.G., AND A. I. BISCHOFF. 1960. Inimical effects on wildlife of periodic DDT applications to Clear Lake. *Calif. Fish Game*, 46:91–106.

HUNTER, G. W., III, AND W. S. HUNTER. 1934. Further studies on bird and fish parasites. *Supp. 24th Ann. Rept. N.Y. State Dept. Cons., No. 9, Rept. Biol. Surv. Mohawk-Hudson Watershed*, pp. 267–283.

HUNTLEY, B. J., AND B. H. WALKER (EDS.). 1982. *Ecology of Tropical Savannas.* Springer-Verlag, New York.

HUNTLEY, M., AND R. INOUYE. 1988. Pocket gophers in ecosystems: Patterns and mechanism. *Bioscience*, 38:786–793.

HURLBERT, S. H. 1971. The nonconcept of species diversity: A critique and alternative parameters. *Ecology*, 52:577–586.

———. 1984. Pseudoreplication and the design of ecological field experiments. *Ecol. Monogr.*, 54:187–211.

HUSTON, M. 1979. A general hypothesis of species diversity. *Am. Nat.*, 113:81–101.

———. 1985. Patterns of species diversity on coral reefs. *Ann. Rev. Ecol. Syst.*, 16:149–177.

HUSTON, M., AND T. SMITH. 1987. Plant succession: Life history and competition. *Am. Nat.*, 130:168–198.

HUTCHINS, H. E., AND R. M. LANNER. 1982. The central role of Clark's nutcracker in the dispersal and establishment of white-bark pine. *Oecologica*, 55:192–201.

HUTCHINSON, G. E. 1957. *A Treatise on Limnology, Vol. 1, Geography, Physics, Chemistry.* Wiley, New York.

———. 1958. Concluding remarks. *Cold Spring Harbor Symp. Quant. Biol.*, 22:415–427.

———. 1959. Homage to Santa Rosalia, or why are there so many kinds of animals? *Am. Nat.*, 93:145–159.

———. 1969. Eutrophication, past and present. In *Eutrophication: Causes, Consequences, Correctives.* National Academy Science, Washington, D.C., pp. 12–26.

———. 1978. *An Introduction to Population Ecology.* Yale University Press, New Haven. Conn.

HUTTO, R. L. 1985. Habitat selection by nonbreeding migratory land birds. In M. Cody (ed.), *Habitat Selection in Birds.* Academic Press, Orlando, Fla., pp. 455–476.

HYDER, D. N. 1969. The impact of domestic animals on the structure and function of grassland ecosystems. In R. L. Dix and R. G. Biedleman (eds.), *The Grassland Ecosystem: A Preliminary Synthesis.* Range Sci. Dept. Ser. No. 2. Colorado State University, Fort Collins, pp. 243–260.

HUXLEY, J. S. 1934. A natural experiment on the territorial instinct. *Brit. Birds*, 27:270–277.

HYNES, H. 1970. *Biology of Running Water.* University of Toronto Press, Ontario.

IKUSIMA, I. 1965. Ecological studies on the productivity of aquatic plant communities: 1, Measurement of photosynthetic activity. *Bot. Mag. Tokyo*, 78:202–211.

ILLIES, J., AND L. BOTOSANEANU. 1963. Problemes et methodes de la classification et de la zonation ecologique des eaux courantes, considerées surtout du point de vue faunistique. *Mitt. int. Verein theor. Angew. Limnol.*, 12:1–57.

INGHAM, R. E., J. A. TROFYMOW, E. H. INGHAM, AND D. C. COLEMAN. 1985. Interaction of bacteria, fungi, and their nematode grazers: Effects on nutrient cycling and plant growth. *Ecol. Monogr.*, 55:119–140.

INOUYE, R. S., G. S. BYERS, AND J. H. BROWN. 1980. Effects of predation and competition on survivorship, fecundity, and community structure of desert animals. *Ecology*, 61:1344–1351.

IRVING, I. 1960. Birds of Anaktuviik Pass, Kovuk and Old Crow: A study in arctic adaptation. *U.S. Nat. Mus. Bull.*, 217.

IRVING, I., AND J. KROGH. 1954. Body temperatures of arctic and subarctic birds and mammals. *J. Appl. Physiol.*, 6:667–680.

IWAKI, H. 1974. Comparative productivity of terrestrial ecosystems in Japan, with emphasis on the comparison between natural and agricultural ecosystems. *Proc. 1st Int. Cong. of Ecol.*, pp. 40–45.

JACKSON, J. K., AND S. G. FISCHER. 1986. Secondary production, emergence, and export of aquatic insects of a Sonoran Desert stream. *Ecology*, 67:629–638.

JACKSON, M. B. 1982. Ethylene as a growth promoting hormone under flooded conditions. In P. F. Wareing (ed.), *Plant Growth Substances*. Academic Press, London, pp. 291–301.

JACKSON, M. B., AND M. C. DREW. 1984. Effects of flooding on growth and metabolism of herbaceous plants. In T. T. Kozlowski (ed.), *Flooding and Plant Growth*. Academic Press, Orlando, Fla., pp. 47–128.

JACKSON, S. T., R. P. FUTUYMA, AND D. A. WILCOX. 1988. A paleoecological test of a classical hydrosere in the Lake Michigan dunes. *Ecology*, 69:928–936.

JACOBSON, H. A., R. L. KIRKPATRICK, AND B. S. MCGINNES. 1978. Disease and physiologic characteristics of two cottontail populations in Virginia. *Wildl. Monogr.*, 60:53 pp.

JACOBSON, J. S., AND A. C. HILL. 1970 *Recognition of Air Pollution Injury to Vegetation: A Pictorial Atlas*. Air Pollution Control Association, Pittsburgh, Penn.

JACOT, A. P. 1930. Reduction of spruce and fir litter by minute animals. *J. Forestry*, 37:858–860.

———. 1940. The fauna of the soil. *Quart. Rev. Biol.*, 15:38–58.

JAFFE, L. S. 1970. The global balance of carbon monoxide. In F. S. Singer (ed.), 1970, *Global Effects of Environmental Pollution*. Springer-Verlag, New York, pp. 34–49.

JAMES, F. 1971. Ordination of habitat relationships among birds. *Wilson Bull.*, 83:215–236.

JAMES, F. C., AND H. SHUGART. 1974. The phenology of the nesting season of the American robin *(Turdus migratorius)* in the United States. *Condor*, 76:159–168.

JANZEN, D. H. 1966. Coevolution of mutualism between ants and acacias in Central America. *Evolution*, 20:249–275.

———. 1967. Synchronization of sexual reproduction of trees within the dry season in Central America. *Evolution*, 21:620–637.

———. 1968. Host plants as islands in evolutionary and contemporary time. *Am. Nat.*, 102:592–595

———. 1971. Seed predation by animals. *Ann. Rev. Ecol. Syst.*, 2:465–492.

———. 1976. Why bamboos wait so long to flower. *Ann. Rev. Ecol. Syst.*, 2:465–492.

———. 1979. How to be a fig. *Ann. Rev Ecol Syst.*, 10:13–51.

———. 1980. When is it coevolution? *Evolution*, 34:611–612.

JÄRVINEN, O. 1982. Conservation of endangered plant populations: Single large or several small reserves. *Oikos*, 38:301–307.

———. 1984. Dismemberment of facts: A reply to Willis on subdivision of reserves. *Oikos*, 42:402–403.

JARVIS, P. G., G. B. JAMES, AND J. J. LANDSBERG. 1976. Coniferous forest. In J. L. Monteith (ed.), *Vegetation and the Atmosphere. Vol. 2 Case Studies*. Academic Press, London, pp. 171–240.

JAWORSKI, N. A., AND L. J. HELLING. 1970. Relative contribution of nutrients to the Potomac River Basin from various sources. In *Relationship of Agriculture to Soil and Water Pollution*. Cornell University Press, Ithaca, N.Y.

JEFFERIES, R. L. 1972. Aspects of salt marsh ecology with particular reference to inorganic plant nutrients. In R. S. K. Barnes and J. Green (eds.), *The Estuarine Environment*. Applied Science Publishers, London, pp. 61–85.

JEFFERS, J. N. R. 1988. *Practitioner's Handbook on the Modelling of Dynamic Change in Ecosystems* (Scope 34). Wiley, New York.

JENKIN, J. F. 1975. Macquarie Island, Subantarctic. In T. Rosswall and O. W. Heal (eds.), *Structure and Function of Tundra Ecosystems*, Swedish Natural Science Research Council, pp. 375–397.

JENKINS, D., A. WATSON, AND G. R. MILLER. 1963. Population studies on red grouse *Lagopus lagopus scoticus* (Lath) in northeast Scotland. *J. Anim. Ecol.*, 32:317–376.

JENKINS, D. W. 1944. Territory as a result of despotism and social organization in geese. *Auk*, 61:30–47.

JENNI, D. A. 1974. Evolution of polyandry in birds. *Am. Zool.*, 14:129–144.

JENNY, H. 1933. Soil fertility losses under Missouri conditions. *Missouri Agri. Exp. Sta. Bull.*, 324.

———. 1980. *The Soil Resource.* Springer-Verlag, New York.

JENSEN, S. 1966. A new chemical hazard. *New Scientist*, 32:612.

JOHANNES, R. E. 1964. Uptake and release of dissolved organic phosphorus by representatives of a coastal marine ecosystem. *Limnol. Oceanogr.*, 9:224–234.

———. 1965. Influence of marine protozoa on nutrient regeneration. *Limnol. Oceanogr.*, 10:434–442.

———. 1967. Ecology of organic aggregates in the vicinity of coral reef. *Limnol. Oceanogr.*, 12:189–195.

———. 1968. Nutrient regeneration in lakes and oceans. In M. R. Droop and E. J. Ferguson (eds.), *Advances in the Microbiology of the Sea*, Vol. 1. Academic Press, New York, pp. 203–213.

JOHANNES, R. E., AND K. L. WEBB. 1970. Release of dissolved organic compounds by marine and fresh water invertebrates. *Proc. Conf. on Organic Matter in Natural Waters, Occ. Pub. No. 1.* Inst. Marine Sci., University of Alaska.

JOHANSSON, L. G., AND S. LINDER. 1975. The seasonal pattern of photosynthesis of some vascular plants on a subarctic mire. In F. E. Wielgolaski (ed.), pp. 194–200.

JOHNSON, A. H. 1983. Red spruce decline in the northeastern US: Hypotheses rearding the role of acid rain. *J. Air Pollution Control Assoc.*, 33:1949–1054.

JOHNSON, A. H., AND T. G. SICCAMA. 1983. Acid deposition and forest decline. *Environ. Sci. Technol.*, 17:249A–304A.

JOHNSON, C. G. 1969. *Migration and Dispersal of Insects by Flight.* Methuen, London.

JOHNSON, D. H., AND A. B. SARGEANT. 1977. Impact of red fox predation on the sex ratio of prairie mallards. *Wildl. Res. Rept. 6.* U.S. Fish and Wildlife.

JOHNSON, D. W. 1968. Pesticides and fishes: A review of selected literature. *Trans. Am. Fish. Soc.*, 97:398–424.

JOHNSON, D. W., D. W. COLE, C. S. BLEDSOE, K. CROMACK, R. L. EDMONDS, S. P. BESSEL, C. C. GRIER, B. N. RICHARDS, AND K. A. VOGT. 1982. Nutrient cycling in forests of the Pacific Northwest. In R. L. Edmonds (ed.), *Analysis of Coniferous Forest Ecosystems in the Western United States*. Hutchinson, Ross, Stroudsburg, Penn., pp. 186–232.

JOHNSON, D. W., D. O. RICHTER, H. MIEGROET, AND D. W. COLE. 1983. Contribution of acid deposition and natural processes to cation leaching from forest soils: A review. *J. Air Pollution Control Assoc.*, 33:1036–1041.

JOHNSON, F. S. 1970. The oxygen and carbon dioxide balance in the earth's atmosphere. In F. S. Singer (ed.), *Global Effects of Environmental Pollution*. Springer-Verlag, New York, pp. 4–11.

JOHNSON, H. B. 1975. Plant pubescence: an ecological perspective. *Bot. Rev.*, 41:233–258.

JOHNSON, J. A. 1984. Small woodlot management by single-tree selection: 21-year results. *Northern J Appl. For.*, 1:69–71.

JOHNSON, M. C., AND M. S. GAINES. 1987. The selective basis for dispersal of the prairie vole *Microtus ochrogaster*. *Ecology*, 68:684–694.

JOHNSON, N. M., R. C. REYNOLDS, AND G. E. LIKENS. 1972. Atmospheric sulphur: Its effect on the chemical weathering of New England. *Science*, 177:514–515.

JOHNSON, P. L., AND W. O. BILLINGS. 1962. The alpine vegetation of the Beartooth Plateau in relation to cryopedogenic processes and patterns. *Ecol. Monogr.*, 32:105–135.

JOHNSON, P. L., AND W. T. SWANK. 1973. Studies of cation budgets in the southern Appalachians on four experimental watersheds with contrasting vegetation. *Ecology*, 54:70–80.

JOHNSON, R. W., AND J. C. TOTHILL. 1985. Definition and broad geographic outline of savanna lands. In J. C. Tothill and J. J. Mott (ed.), *Ecology and Management of the World's Savannas.* Commonwealth Agricultural Bureau, Australian Academy of Science, Canberra, pp. 1–13.

JOHNSON, W. K., AND G. J. SCHROEPFER. 1964. Nitrogen removal by nitrification and denitrification. *J. Water Pollut. Control Fed.*, 36:1011–1036.

JOHNSTON, J. W. 1936. The macrofauna of soils as affected by certain coniferous and hardwood types in the Harvard forest. PhD. Dissertation, Harvard University Library, Cambridge, Mass.

JOHNSTON, R. F. 1956a. Predation by short-eared owls in a *Salicornia* salt marsh. *Wilson Bull.*, 68:91–102.

———. 1956b. Population structure in salt marsh song sparrows: Part 2, Density, age structure and maintenance. *Condor*, 58:254–272.

JONES, H. G. 1983. *Plants and Microclimate.* Cambridge University Press, Cambridge, England.

JONES, H. LEE, AND J. M. DIAMOND. 1976. Short time base studies of turnover in breeding bird populations on the California channel islands. *Condor*, 78:526–549.

JORDAN, C. F. 1971a. Productivity of a tropical forest and its relation to a world pattern of energy storage. *J. Ecol.*, 59:127–142.

———. 1971b. A world pattern in plant energetics. *Am. Sci.*, 59:425–433.

JORDAN, C. F. 1981. Do ecosystems exist? *Am. Nat.*, 106:237–253.

———. 1982a. Amazon rain forest. *Am. Sci.*, 70:394–401.

———. 1982b. The nutrient balance of an Amazonian rain forest. *Ecology*, 63:647–654.

———. 1985. *Nutrient Cycling in Tropical Forest Ecosystems: Principles and Their Application in Management and Conservation.* Wiley, New York.

JORDAN, C. F., F. GOLLEY, J. D. HALL, AND J. HALL. 1979. Nutrient scavenging of rainfall by the canopy of an Amazonian rain forest. *Biotropica*, 12:61–66.

JORDAN, C. F., J. R. KLINE, AND D. S. SASSIER. 1972. Relative stability of mineral cycles in forest ecosystems. *Am. Nat.*, 106:237–253.

JORDAN, W. R., III, M. E. GILPIN, AND J. O. ABER. 1987. *Restoration Ecology: A Synthetic Approach to Ecological Research.* Cambridge University Press, Cambridge, England.

JORDEN, T. E., AND I. VALIELA. 1982. The nitrogen cycle of the ribbed mussel *Geukinsea demissa* and its significance in nitrogen flow in a New England salt marsh. *Limno. Oceanogr.*, 27:75–90.

JORGENSEN, J. R., AND C. G. WELLS. 1986. *Forester's Primer in Nutrient Cycling.* U.S.D.A. For. Serv. Gen. Tech. Rept. SE-37.E.

KABAT, C., AND D. R. THOMPSON. 1963. *Wisconsin Quail, 1834–1962: Population Dynamics and Habitat Management.* Tech. Bull. No. 30. Wisconsin Conserv. Dept., Madison.

KAJAH, Z., G. BRETSCHKO, F. SCHIEMER, AND C. LEVEGUE. 1980. Zoobenthos. In E. D. LeCren and R. H. Lowe-McConnell (eds.), *The Functioning of Freshwater Ecosystems* (International Biological Programme No. 22). Cambridge University Press, Cambridge, England, pp. 285–307.

KAKENOSKY, G. 1972. Wolf predation on wintering deer in east central Ontario. *J. Wildl. Manage.*, 36:357–367.

KALININ, G. R., AND V. D. BYKOV. 1969. The world's water resources, present and future. *Impact of Science on Society*, 19:135–150.

KALLE, K. 1971. Salinity: General introduction. In O. Kinne (ed.), *Marine Ecology: Vol. 1, Environmental Factors, Part 2.* Wiley, New York, pp. 683–688.

KALLIO, P. 1974. The ecology of the nitrogen fixation in subarctic lichens. *Oikos*, 25:194–198.

KALLIO, S., AND P. KALLIO. 1975. Nitrogen fixation in lichens at Kevo, North Finland. In F. E. Wielgolaski (ed.), 1975, pp. 292–304.

KALLIO, P., AND N. VALANNE. 1975. On the effect of continuous light on photosynthesis in mosses. In F. E. Wielgolaski (ed.), pp. 149–162.

KAMIL, A. C., J. R. KREBS, AND H. R. PULLIAN (EDS.). 1987. *Foraging Behavior.* Plenum, New York.

KAMMER, A. E. 1981. Physiological mechanisms of thermoregulation. In O. L. Lange, P. S. Nobel, C. B. Osmond, and H. Zeigler (eds.), *Physiological Plant Ecology. I. Responses to the Physical Environment.* Springer-Verlag, Berlin, pp. 115–148.

KANWISHER, J. W. 1955. Freezing in intertidal animals. *Biol. Bull. Woods Hole*, 109:56–63.

———. 1959. Histological metabolism of frozen intertidal animals. *Biol. Bull. Woods Hole*, 116:285–264.

KAPPER, L. 1981. Ecological significance of resistance to high temperature. In O. Lange, P. S. Nobel, C. B. Osmond, and H. Zeigler (eds.), *Physiological Plant Ecology, I. Responses to the Environment.* Springer-Verlag, New York.

KARPLUS, M. 1949. Bird activity in continuous day-light of arctic summer. *Bull. Ecol. Soc. Am.*, 30:60.

KARR, J. R. 1971. Structure of avian communities in selected Panama and Illinois habitats. *Ecol. Monogr.*, 41:207–233.

———. 1975. Production and energy pathways, and community diversity in forest birds. In F. Golley and E. Medina (eds.), *Tropical Ecological Systems: Trends in Terrestrial and Aquatic Research.* Springer-Verlag, New York, pp. 161–176.

———. 1976. Seasonality, resource availability and community diversity in tropical bird communities. *Am. Nat.*, 110:973–994.

KARR, J. R., AND R. R. ROTH. 1971. Vegetation structure and avian diversity in several new world areas. *Am. Nat.*, 105:423–435.

KASTINGS, J. F., AND T. P. ACKERMAN. 1986. Climatic consequences of very high carbon dioxide levels in the earth's early atmosphere. *Science*, 234:1383–1385.

KATZ, B. A., AND H. LEITH. 1974. Seasonality of decomposers. In H. Leith (ed.), *Phenology and Seasonality Modeling.* Springer-Verlag, New York, pp. 163–184.

KAYS, S., AND J. L. HARPER. 1974. The regulation of plant and tiller density in a grass sward. *J. Ecol.*, 62:97–105.

KEELER, K. 1981. A model of selection for facultative mutualism. *Am. Nat.*, 118:488–498.

KEELING, C. D. 1968. Carbon dioxide in surface ocean waters. *J. Geophys. Res.*, 73:4543.

KEEVER, C. 1950. Causes of succession in old fields of the Piedmont, North Carolina. *Ecol. Monogr.*, 20:229–250.

KEISTER, A. R. 1971. Species density of North American amphibians and reptiles. *Syst. Zool.*, 20:128ff.

KEITH, L. B. 1963. *Wildlife's Ten-year Cycle.* University of Wisconsin Press, Madison.

———. 1974. Some features of population dynamics in mammals. *Proc. Inter. Congr. Game Biol. Stockholm*, 11:17–58.

———. 1983. Role of food in hare population cycles. *Oikos*, 40:385–395.

KEITH, L. B., J. R. CARY, O. J. RONGSTAD, AND M. C. BRITTINGHAM. 1984. Demography and ecology of declining snowshoe hare population. *J. Wildl. Manage.*, 90:1–43.

KEITH, L. B., AND L. A. WINDBERG. 1978. A demographic analysis of the snowshoe hare cycle. *Wildl. Monogr.* 58.

KELLOGG, C. E. 1936. *Development and Significance of the Great Soil Groups of the United States. USDA Misc. Pub. 229.*

KELLOGG, W. W., ET AL. 1972. The sulfur cycle. *Science,* 175:587–599.

KERR, R. A. 1987. Halocarbons linked to ozone hole. *Science,* 236:1182–1183.

KESSEL, S. R., AND W. C. FISHER. 1981. Predicting post fire plant succession for fire management planning. *U.S.D.A. For. Serv., Gen. Tech. Rept. Int-94.*

KETCHUM, B. H. 1967. The phosphorus cycle and productivity of marine phytoplankton. In *Symposium on Primary Productivity and Mineral Cycling in Natural Ecosystems.* University of Maine Press, Orono, pp. 32–51.

KETCHUM, B. H. (ED.). 1983. *Estuaries and Enclosed Seas* (Ecosystems of the World 26). Elsevier, Amsterdam.

KETCHUM, B. H., AND N. CORWIN. 1965. The cycle of phosphorus in a plankton bloom in the Gulf of Maine. *Limnol. Oceanogr. Suppl. to Vol. 10.,* pp. R148–R161.

KETTLEWELL, H. B. D. 1961. The phenomenon of industrial melanism in Lepidoptera. *Ann. Rev. Entomol.,* 6:245–262.

———. 1965. Insect survival and selection for pattern. *Science,* 148:1290–1296.

KEVAN, D. K. MCE. 1955. *Soil Zoology.* Butterworth, Washington, D.C.

———. 1962. *Soil Animals.* Philosophical Library, New York.

KHALIL, M. A. K., AND R. A. RASMUSSEN. 1984. Carbon monoxide in the earth's atmosphere: Increasing trend. *Science,* 224:54–56.

KIESTER, A. R. 1971. Species density of North American amphibians and reptiles. *Syst. Zool.,* 20:127–137.

KILGORE, B. M. 1973. Impact of prescribed burning on a sequoia-mixed conifer forest. *Proc. Tall Timbers Fire Ecol. Conf.,* 12:345–375.

———. 1976. Fire management in the national parks: An overview. *Proc. Tall Timbers Fire Ecol. Conf.,* 14:45–57.

KILRICH, M. I., J. A. MACMAHON, R. R. PARAMETER, AND D. V. SISSON. 1986. Native seed preferences of shrub-steppe rodents, birds, and ants: The relationship of seed attributes and seed use. *Oecologica,* 68:327–337.

KINGSLAND, S. 1985. *Modelling Nature.* University of Chicago Press, Chicago.

KINGSTON, T. J., AND M. J. COE. 1977. The biology of the giant dung beetle *(Heliocopris dilloni)* (Coleptera: Scaraboeidae). *J. Zool.* (London), 181:243–263.

KINNE, O. (ED.). 1970. *Marine Ecology: A Comprehensive Integrated Treatise on Life in Oceans and Coastal Waters: Vol. 1, Environmental Factors.* Wiley, New York.

KIRA, T., AND T. SHIDEI. 1967. Primary production and turnover of organic matter in different forest ecosystems of the western Pacific. *Japanese J. Ecol.,* 17:70–87.

KIRA, K., H. OGAWA, AND K. SHINOZOKI. 1953. Intraspecific competition among higher plants. I. Competition and density yield interrelations in regularly dispersed populations. *J. Polytechnic Institute,* Osata City University, 4(4):1–16.

KIRK, J. T. O. 1983 *Light and Photosynthesis in Aquatic Systems.* Cambridge University Press, New York.

KIRTPATRICK, M. 1981. Spatial and age dependent patterns of growth in New England black birch. *Am. J. Bot.,* 68:535–543.

KITCHELL, J. F., R. V. O'NEILL, D. WEBB, G. A. GASLLEPP, S. M. BARTELL, J. F. KOONCE, AND B. S. AUSMUS. 1979. Consumer regulation of nutrient cycling. *Bioscience,* 29:28–34.

KLEEREKOPER, H. 1953. The mineralization of plankton. *J. Fish. Res. Board Can.,* 10:283–291.

KLOPATEK, J. M. 1978. Nutrient dynamics of freshwater riverine marshes and the role of emergent macrophytes. In R. E. Good, D. F. Whigham, and R. L. Simpson, *Freshwater Wet Ones.* Academic Press, New York, pp. 195–216.

KLOPFER, P. H., AND J. U. GANZHORN. 1985. Habitat selection: Behavioral aspects. In M. Cody (ed.), *Habitat Selection in Birds.* Academic Press, Orlando, Fla. pp. 436–454.

KLUGE, M. 1982. Crassulacean acid metabolism (CAM). In Govindjee (ed.), *Development, Carbon Metabolism, and Plant Productivity.* Academic Press, New York. pp. 232–264.

KNAPP, A. K., AND T. R. SEASTEDT. 1986. Detritus accumulation limits productivity of tall grass prairie. *Bioscience* 36:662–668.

KNAPTON, R. W., AND J. R. KREBS. 1976. Settlement patterns, territory size, and breeding density in the song sparrow *(Melospiza melodia). Can. J. Zool.,* 52:1413–1420.

KNIGHT, R. R., AND L. L. EBERHARDT. 1985. Population dynamics of Yellowstone grizzly bears. *Ecology,* 66:323–334.

KNOX, E. G. 1952. Jefferson County (N.Y.) *Soils and Soil Map.* N.Y. State College Agr., Cornell University, Ithaca, N.Y.

KNUTSON, R. M. 1974. Heat production and temperature regulation in eastern skunk cabbage. *Science,* 186:745–747.

KOBLENTZ-MISHKE, J. J., V. V. VOLKOVINSKY, AND J. G. KABANOVA. 1970. Plankton primary production of the world's oceans. In W. S. Wooster (ed.), *Scientific Exploration of the South Pacific.* National Academy of Science, Washington D.C.

KOK, B. 1965 Photosynthesis: The pathway of energy. In J. Bonner and J. E. Varner (eds.), *Plant Biogeochemistry.* Academic Press, New York, pp. 904–960.

KONISHI, M. 1973. How the owl tracks its prey. *Am. Sci.,* 61:414–424.

KOPEC, R. J. 1970. Further observations of the urban heat island of a small city. *Bull. Am. Meteorol. Soc.,* 51:602–606.

KORCHAGIN, A. A., AND V. G. KARPOV. 1974. Fluctuations in coniferous tiaga communities. In R. Knapp (ed.), *Vegetation Dynamics. Part 8. Handbook of Vegetation Science.* Junk, The Hague, Netherlands, pp. 225–231.

KOZLOVSKY, D. C. 1968. A critical evaluation of the trophic level concept: 1. Ecological efficiencies. *Ecology* 49:48–60.

KOZLOWSKI, T. T. 1984. Response of woody plants to flooding. In T. T. Kozlowski (ed.), *Flooding and Plant Growth.* Academic Press, Orlando, Fla., pp. 129–163.

———. 1984. Plant responses to flooding of soil. *Bioscience,* 34:162–168.

KOZLOWSKI, T. T., AND S. G. PALLARDY. 1984. Effects of flooding on water, carbohydrate, and mineral relations. In T. T. Kozlowski (ed.), *Flooding and Plant Growth.* Academic Press, Orlando Fla., pp. 165–194.

KRAMER, P. J. 1981. Carbon dioxide concentration, photosynthesis, and dry matter production. *Bioscience,* 31:29–33.

———. 1983. *Water Relations of Plants.* Academic Press, Orlando, Fla.

KRAMM, K. R. 1975a. Entrainment of circadian activity rhythms in squirrels. *Am. Nat.,* 109:379–389.

———. 1975b. Circadian activity of the red squirrel *Tamiascuirius hudsonicus* in continuous darkness and continuous illumination. *Int. J. Biometeor.,* 19:232–245.

———. 1976. Phase control of circadian activity in the antelope ground squirrel. *J. Interdicip. Cycle Res.,* 7:127–138.

KRATZ, T. K., AND C. D. DEWITT. 1986. Internal factors controlling peatland-lake ecosystem development. *Ecology,* 67:100–107.

KREBS, C. J. 1963. Lemming cycle at Baker Lake, Canada, during 1959–62. *Science,* 146:1559–1560.

———. 1964. The lemming cycle at Baker Lake, Northwest Territories, during 1959–1962. *Tech. Paper No. 15, Arctic Institute of North America.*

———. 1978. A review of the Chitty hypothesis of population regulation. *Can. J. Zool.,* 56:2463–2480.

———. 1985. *Ecology: The Experimental Analysis of Distribution and Abundance*. 3rd ed., Harper & Row, New York.

———. 1989. *Ecological Methodology*. Harper & Row, New York.

KREBS, C. J., M. S. GAINES, B. L. KELLER, J. H. MYERS, AND R. H. TAMARIN. 1973. Population cycles in small rodents. *Science*, 179:35–41.

KREBS, C. J., I. WINGATE, J. LEDUC, J. A. REDFIELD, M. TAITT, AND R. HILBORN. 1976. *Microtus* population biology: Dispersal in fluctuating populations of *M. townsendii*. *Can. J. Zool.*, 54:79–95.

KREBS, J. R. 1971. Territory and breeding density in the great tit *Parus major*. *Ecology*, 52:2–22.

———. 1973. Behavioral aspects of predation. In P. P. Bateson and P. H. Klopfer, Plenum, New York.

KREBS, J. R., AND N. B. DAVIES (EDS.). 1978. *Behavioral Ecology: An Evolutionary Approach*. Blackwell, Oxford, England.

———. 1984. *Behavioral Ecology: An Evolutionary Approach*, 2nd ed. Blackwell, Oxford, England.

KREBS, J. R., A. KACELNIK, AND P. TAYLOR. 1978. Optimal sampling by foraging birds: An experiment with great tits *(Parus major)*. *Nature* 275:27–31.

KREMER, J. N., AND S. W. NIXON. 1978. *A Coastal Marine Ecosystem*. Springer-Verlag, New York.

KROG, J. 1955. Notes on temperature measurements indicative of special organization in arctic and subarctic plants for utilization of radiated heat from the sun. *Physiol. Plantarum*, 8:836–839.

KRUBECKER, A. R. 1954. The ecology of serpentine soils. III. Plant species in relation to serpentine soils. *Ecology*, 35:267–287.

KRUG E. C., AND C. R. FRINK. 1983. Acid rain on acid soil: A new perspective. *Science*, pp. 520–525.

KRUGER, F. J. 1979. South African heathlands. In R. L. Specht (ed.), *Heathlands and Related Shrublands: Descriptive Studies* (Ecosystems of the World 9A). Elsevier, Amsterdam.

KRUUK, H. 1972. *The Spotted Hyena*. University of Chicago Press, Chicago.

KUCERA, C. L., R. C. DAHLMAN, AND M. R. KOELLING. 1967. Total net productivity and turnover on an energy basis for tallgrass prairie. *Ecology*, 48:536–541.

KÜCHLER, A. W. 1964. *Potential Natural Vegetation of the United States*. U.S. Geological Survey National Atlas, U.S. Geological Survey, Washington, D.C.

KUENZLER, E. J. 1958. Niche relations of three species of Lycosid spiders. *Ecology*, 39:494–500.

———. 1961. Phosphorus budget of a mussel population. *Limnol. Oceanogr.*, 6:400–415.

KUHNELT, W. 1950. *Bodenbiologie mit besonderer Berücksichtigung der Tierwelt*. Herold, Vienna.

———. 1970. Structural aspects of soil-surface-dwelling biocoenoses. In J. Phillipson (ed.), *Methods of Study in Soil Biology*. UNESCO, Paris, pp. 45–56.

KURCHEVA, G. F. 1964. Wirbellose Tiere abd Faktor der Zersetzung von waldstreu. *Pedobiologia*, 4:8–30.

LACK, D. L. 1947. The significance of clutch size. *Ibis*, 89:30–52; 90:25–45.

———. 1953. *The Life of the Robin*. Penguin Books, London.

———. 1954. *The Natural Regulation of Animal Numbers*. Clarendon Press, Oxford.

———. 1964. A long term study of the great tit *(Parus major)*. *J. Anim. Ecol. Suppl.*, 33:159–173.

———. 1966. *Population Studies of Birds*. Clarendon Press, Oxford.

———. 1971. *Ecological Isolation in Birds*. Harvard University Press, Cambridge, Mass.

LACK, D. L., AND L. S. V. VENERABLES. 1939. The habitat distribution of British woodland birds. *J. Anim. Ecol.*, 8:39–71.

LAMARCHE, V. C., D. H. GRAYBILL, H. C. FRITTS, AND M. R. ROSE. 1984. Increasing atmospheric carbon dioxide tree ring evidence for growth enhancement in natural vegetation. *Science*, 225:1019–1021.

LA MOTTE, M. 1975. The structure and function of a tropical savannah ecosystem. In F. B. Golley and E. Medina (eds.) *Tropical Ecological Systems Trends in Terrestrial and Aquatic Research*. Springer-Verlag, New York, pp. 179–222.

LA MOTTE, M., AND F. BOURLIERE. 1983. Energy flow and nutrient cycling in tropical savannas. In F. Bourliere (ed.), *Tropical Savannas* (Ecosystems of the World No. 13.). Elsevier, Amsterdam.

LANCE, A. N. 1978. Territories and the food plant of individual red grouse: 2. Territory size compared with an index of nutrient supply in heather. *J. Anim. Ecol.*, 47:307–313.

LANCINANI, C. A. 1975. Parasite-induced alterations in host reproduction and survival. *Ecology*, 56:689–695.

LANDAHL, J., AND R. B. ROOT. 1969. Differences in the life tables of tropical and temperate milkweed bugs Genus *Oncopeltus* (Hemoptera Lygaerdae). *Ecology*, 50:734–737.

LANDSBERG, H. E. 1970. Man-made climatic changes. *Science*, 170:1265–1274.

LANE, P. A. 1985. A food web approach to mutualism in lake communities. In D. H. Boucher (ed.), *The Biology of Mutualism*. Oxford University Press, New York, pp. 344–374.

LANG, G. E. 1985. Forest turnover and the dynamics of bole wood litter in subalpine balsam fir forest. *Can. J. For. Res.*, 15:262–288.

LANG, G. E., AND R. T. FORMAN. 1978. Detrital dynamics in a mature forest: Hutchinson Memorial Forest. *Ecology*, 59:580–595.

LANG, G. E., AND D. H. KNIGHT. 1983. Tree growth, mortality, recruitment, and canopy gap formation during a 10-year period in a tropical moist forest. *Ecology*, 64:1075–1080.

LANG, G. E., D. H. KNIGHT, AND D. A. ANDERSON. 1971. Sampling the density of tree species in a species-rich tropical forest. *For. Sci.*, 17:395–400.

LANG, G. E., W. A. REINERS, AND R. R. HEIER. 1976. Potential alterations of precipitation chemistry by epiphytic lichens. *Oecologica*, 25:229–241.

LANG, G. E. W. A. REINERS, AND L. H. PIKE. 1980. Structure and biomass dynamics of epiphytic lichen communities of balsam fir forests in New Hampshire. *Ecology* 63:541–550.

LARCHER, W., ET AL. 1975. Mt. Patscherfofel, Austria. In T. Rosswall and O. W. Heal (eds.), *Structure and Function of Tundra Ecosystems*. Swedish Natural Science Research Council, Stockholm, pp. 125–139.

LARCHER, W., AND H. BAUER. 1981. Ecological significance of resistance to low temperature. In O. L. Lange, P. S. Nobel, C. B. Osmond, and H. Zeigler (eds.), *Physiological Plant Ecology. I. Responses to the Physical Environment*. Springer-Verlag, Berlin, pp. 403–437.

LASSOIE, J. P., T. M. HINCHLEY, AND C. C. GRIER. 1985. Coniferous forests of the Pacific Northwest. In B. F. Chabot and H. A. Mooney (eds.), *Physcological Ecology of North American Plant Communities*. Chapman & Hall, New York, pp. 127–161.

LAUFF, G. (ED.). 1967. *Estuaries*. American Association Advancement Science, Washington, D.C.

LAVE, L. B., AND E. P. SESKIN. 1970. Air pollution and human health. *Science*, 169:723–733.

LAWLOR, L. R. 1976. Molting, growth and reproductive strategies in the terrestrial isopod *Armadillidium vulgare*. *Ecology*, 57:1179–1194.

LAWRENCE, D. B. 1958. Glaciers and vegetation in southeastern Alaska. *Am. Sci.*, 46:89–122.

LAWTON, J. H., M. P. HASSELL, AND J. R. BEDDINGTON. 1975.

Prey death rates and rate of increase of arthropod predator populations. *Nature,* 255:60–62.

LAWTON, J. H., AND D. R. STRONG. 1981. Community patterns and competition in folivorous insects. *Am. Nat.,* 118:371–388.

LEAN, D. R. S. 1973a. Movements of phosphorus between its biologically important forms in lake water. *J. Fish. Res. Board Can.,* 30:1525–1536.

———. 1973b. Phosphorus dynamics in lake water. *Science,* 179:678–680.

LEE, R. 1978. *Forest Microclimatology.* Columbia University Press, New York.

LEES, D. R., AND E. R. CREED. 1975. Industrial melanism in *Biston betularia:* The role of selective predation. *J. Anim. Ecol.,* 44:67–83.

LEIGH, E. G., JR. 1975. Structure and climate in tropical rain forest. *Ann. Rev. Ecol. Syst.,* 6:67–86.

———. 1987. Wave energy and intertidal productivity. *Proc. Natl. Acad. Sci. USA,* 84:1314.

LEITH, H. 1960. Patterns of change within grassland communities. In J. L. Harper (ed.), *The Biology of Weeds.* Oxford University Press, Oxford, pp. 27–39.

———. 1963. The role of vegetation in the carbon dioxide content of the atmosphere. *Geophys. Res.,* 68:3887–3898.

———. 1973. Primary production: Terrestrial ecosystems. *Human Ecol.,* 1:303–332.

———. 1975. Primary production of the major vegetation units of the world. In R. H. Whittaker and H. Leith, *Primary Productivity of the Biosphere.* Springer-Verlag, New York, pp. 203–205.

LEITH, H. (ED.). 1974. *Phenology and Seasonality Modeling.* Springer-Verlag, New York.

LEMON, E. R. (ED.). 1983. *CO₂ and Plants: The response of Plants to Rising Levels of Atmospheric Carbon Dioxide.* Westview Press, Boulder, Colo.

LENINGTON, S. 1980. Female choice and polygymy in redwinged blackbirds. *Behaviour,* 28:347–361.

LEOPOLD, A. 1933. *Game Management.* Scribner, New York.

LESSELLS, C. M., AND M. I. AVERY. 1987. Sex ratio adjustment in *Odocoileus:* Does local resource competition play a role? *Am. Nat.,* 129:452.

LETT, P. F., R. K. MOHN, D. F. GRAY. 1981. Density-dependent processes and management strategy for the northwest Atlantic Harp seal populations. In C. W. Fowler and T. D. Smith (eds.), *Dynamics of Large Mammal Populations.* Wiley, New York, pp. 135–158.

LEVENSON, J. B. 1981. Woodlots as biogeographic islands in southeastern Wisconsin. In R. L. Burgess and D. M. Sharpe (eds.), *Forest Island Dynamics in Man-Dominated Landscapes.* Springer-Verlag, New York, pp. 13–39.

LEVERICH, W. J., AND D. A. LEVIN. 1979. Age specific survivorship and reproduction in *Phlox drummondii. Am. Nat.,* 113:881–903.

LEVIN, D. A. 1971. Plant phenolics, an ecological perspective. *Am. Nat.,* 105:151–181.

———. 1973. The role of trichomes in plant defense. *Quart. Rev. Biol.,* 48:3–15.

———. 1976. The chemical defenses of plants to pathogens and herbivores. *Ann. Rev. Ecol. Syst.,* 7:121–159.

LEVIN, D. A., AND H. W. KERSTER. 1974. Gene flow in seed plants. *Evolutionary Biology,* 7:139–220.

LEVINS, R. 1968. *Evolution in Changing Environments.* Princeton University Press, Princeton, N.J.

LEWIS, J. K. 1971. The grassland biome: A synthesis of structure and function, 1970. In N. R. French (ed.), *Preliminary Analysis of Structure and Function in Grasslands.* Range Sci. Dept. Sci. Ser. No. 10. Colorado State University, Fort Collins, pp. 317–387.

LIDICKER, W. Z., JR. 1973. Regulation of numbers in an island population of the California voles, a problem in community dynamics. *Ecol. Monogr.,* 43:271–302.

———. 1975. The role of dispersal in the demography of small mammals. In F. B. Golley, K. Petrusewicz, and L. Ryszkowski (eds.), *Small Mammals, Their Productivity and Population Dynamics.* Cambridge University Press, Cambridge England, pp. 103–128.

LIGHT, L. E. 1967. Growth inhibition in crowded tadpoles: Intraspecific and interspecific effects. *Ecology,* 48:736–745.

LIGNON, J. D. 1968. Sexual differences in foraging behavior in two species of *Dendrocopos* woodpeckers. *Auk,* 85:203–215.

———. 1981. Demographic patterns and communal breeding in the green hoopoo *Phoeniculus purpureus.* In R. D. Alexander and D. W. Tinkle (eds.), *Natural Selection and Social Behavior: Recent Research and New Theories.* Chiron, New York, pp. 231–246.

LIKENS, G. E. 1975. Nutrient flux and cycling in freshwater ecosystems. In F. G. Howell, J. B. Gentry, and M. H. Smith (eds.), *Mineral Cycling in Southeastern Ecosystems.* National Technical Information Service, U.S. Dept. Commerce, pp. 314–348.

LIKENS, G. E., ET AL. 1977. *Biogeochemistry of a Forested Ecosystem.* Springer-Verlag, New York.

LIKENS, G. E., AND F. H. BORMANN. 1974. Acid rain: A serious regional environmental problem. *Science,* 184:1176–1179.

———. 1974. Effects of forest clearing on the northern hardwood forest ecosystem and its biochemistry. *Proc. 1st Int. Cong. Eco.,* pp. 330–335.

———. 1974. Linkages between terrestrial and aquatic ecosystems. *Bioscience,* 24(8):447–456.

———. 1975. Nutrient-hydrologic interactions (eastern United States). In A. D. Hasler (ed.), *Coupling of Land and Water Systems.* Springer-Verlag, New York, pp. 1–5.

———. 1976. Effects of forest clearing on the northern hardwood forest ecosystem and its biogeochemistry. In *Proc. 1st Int. Cong. Ecol.: Structure, Functioning, and Management of Ecosystems.* Center for Agricultural Publ. and Documentation, Wageninger, Netherlands.

LIKENS, G. E., F. H. BORMANN, AND N. M. JOHNSON. 1969. Nitrification: Importance to nutrient losses from a cutover forest ecosystem. *Science,* 163:1205–1206.

LIKENS, G. E., F. H. BORMANN, N. M. JOHNSON, AND R. S. PIERCE. 1967. The calcium, magnesium, potassium, and sodium budgets for a small forested ecosystem. *Ecology,* 38:46–49.

———. 1970. Effects of forest cutting and herbicide treatment on nutrient budgets in the Hubbard Brook watershed ecosystem. *Ecol. Monogr.,* 40:23–47.

LIKENS, G. E., F. H. BORMANN, R. S. PIERCE, AND J. S. EATON. 1985. The Hubbard Brook Valley. In G. E. Likens (ed.), *An Ecosystem Approach to Aquatic Ecology.* Springer-Verlag, New York, pp. 9–39.

LIKENS, G. E., F. H. BORMANN, R. S. PIERCE, AND D. W. FISHER. 1971. Nutrient hydrologic cycle interaction in small forested watershed ecosystems. In P. Duvigneaud (ed.), *Productivity of Forest Ecosystems.* UNESCO, Paris, pp. 553–563.

LILLYWHITE, H. B. 1970. Behavioral temperature regulation in the bullfrog, *Rana catesbeiana. Copeia,* 1970:158–168.

LINDQUIST, B. 1942. Experimentelle Untersuchingen über die Bedeutung einiger Landmollusken für die zersetzung der Waldstreu. *Kgl Fysiograf Sallskap Lund Forh.,* 11:144–156.

LINDSAY, J. H. 1971. Annual cycle of leaf water potential in *Picea engelmannii* and *Abies lasiocarpa* at timberline in Wyoming. *Arctic and Alpine Research,* 3:131–138.

LINHART, Y. B. 1974. Intra-population differentiation in annual plants. 1. *Veronica peregrina* L. raised under non-competitive conditions. *Evolution,* 28:232–243.

LITTLE, S., AND H. A. SOMES. 1965. Atlantic white cedar being eliminated by excessive animal damage in south Jersey. *U.S.D.A. For. Serv. Res. Note. NE-33.*

B-23

LITUIATIS, J. A., J. A. SHERBURNE, AND J. A. BISSONETTE. 1985. Influence of understory characteristics on a snowshoe hare habitat use and density. *J. Wildl. Manage.,* 49:866–873.

LLOYD, D. G. 1987. Selection of offspring size at independence and other size vs. number strategies. *Am. Nat.,* 129:800–817.

LLOYD, M., AND H. S. DYBAS. 1966. The periodical cicada problem. 1. Population ecology. *Evolution,* 20:133–149.

LLOYD, M., AND R. J. GHERARDI. 1964. A table for calculating the "equitability" component of species diversity. *J. Anim. Ecol.,* 33:217–225.

LOACH, K. 1967. Shade tolerance in tree seedlings. 1. Leaf photosynthesis and respiration in plants raised under artificial shade. *New Phytol.,* 66:607–621.

———. 1970. Shade tolerance in tree seedlings. 2. Growth analysis of plants raised under artificial shade. *New Phytol.,* 69:273–286.

LOEHLE, C., AND J. H. K. PECKMANN. 1988. Evolution: The missing ingredient in systems ecology. *Am. Nat.,* 132:884–899.

LOERY, G., AND J. D. NICHOLS 1985. Dynamics of a black-capped chickadee population, 1958–1983. *Ecology,* 66:1203.

LOMBARDI, J. R., AND J. B. VANDENBERGH. 1977. Pheromonially induced sexual maturation in females: Regulation by the social environment of the male. *Science,* 196:545–546.

LONG, G. 1982. Productivity of western coniferous forests. In R. L. Edmonds (ed.), *Analysis of Coniferous Forest Ecosystems in Western United States.* Hutchinson and Ross, Stroudsburg, Penn., pp. 89–125.

LONSDALE, W. M., AND A. R. WATKINSON. 1982. Light and self-thinning. *New Phytol.,* 90:431–435.

LOOMIS, R. S., ET AL. 1967. Community architecture and the productivity of terrestrial plant communities. In A. San Pietro et al. (eds.), *Harvesting the Sun.* Academic Press, New York, pp. 291–308.

LORD, R. D. 1960. Litter size and latitude in North American mammals. *Am. Midl. Nat.,* 64:488–499.

LORIMER, C. G. 1989. Relative effects of small and large disturbances on temperate hardwood forest structure. *Ecology,* 70:565–567.

LOSSAINT, P. 1973. Soil-vegetation relationships in Mediterranean ecosystems of southern France. In F. di Castri and H. A. Mooney (eds.), 1973, pp. 199–210.

LOTKA, A. J. 1925. *Elements of Physical Biology.* Williams & Wilkins, Baltimore.

LOVEJOY, T. E., R. O. BIERREGAARD, JR., H. B. RYLANDS, J. R. ALCORN, C. E. QUINTELA, L. H. HARPER, K. S. BROWN, JR., A. H. POWELL, G. V. N. POWELL, H. O. R. SCHUBERT, AND M. B. HAYS. 1986. Edge and other effects of isolation on Amazonian forest fragments. In M. Soulé (ed.), *Conservation Biology.* Sinauer Associates, Sunderland, Mass., pp. 257–285.

LOVETT, G. M., W. A. REINERS, R. K. OLSON. 1982. Cloud droplet deposition in subalpine balsam fir forests: Hydrological and chemical input. *Science,* 218:1303–1304.

LOVETT DOUST, J., AND P. B. CAVERS. 1982. Sex and gender dynamics in jack-in-the-pulpit *Arisaema triphyllum* (Araceae). *Ecology,* 63:797–808.

LOWE, V. P. Q. 1969. Population dynamics of red deer (*Cervus elaphus* L.) on Rhum. *J. Anim. Ecol.,* 38:425–457.

LOWE-MCCONNELL, R. H. 1969. Speciation in tropical freshwater fishes. *Biol. J. Linn. Soc.,* 1:51–75.

LUBCHENCO, J. 1978. Plant species diversity in a marine intertidal community: Importance of herbivore food preferences and algal competitive abilities.

———. 1980. Algal zonation in the New England rocky intertidal community: An experimental analysis. *Ecology,* 61:333–344.

———. 1983. *Littorina* and *Fucus:* Effects of herbivores, substratum heterogeneity, and plant escapes during succession. *Ecology,* 64:1116–1123.

———. 1986. Relative importance of competition and predation: Early colonization by seaweeds in New England. In J. Diamond and T. Case (eds.), *Community Ecology.* Harper & Row, New York, pp. 537–555.

LUCUS, W., AND J. A. BERRY. 1985. *Inorganic Carbon Uptake by Aquatic Photosynthetic Organisms.* American Society of Plant Physiology, Washington, D.C.

LUDWIG, J. A., AND J. F. REYNOLDS. 1988. *Statistical Ecology.* Wiley, New York.

LUGO, A. E. 1980. Mangrove ecosystems: Successional or steady state? *Biotropica,* 12:65–72.

LUGO, A. E., AND S. BROWN. 1986. Steady-state terrestrial ecosystem and the global carbon cycle. *Vegetatio,* 68:83–90.

LUGO, A. E., AND S. C. SNEDAKER. 1974. The ecology of mangroves. *Ann. Rev. Ecol. Syst.,* 5:39–64.

LULL, H. W. 1967. Factors influencing water production from forested watersheds. *Municipal Watershed Mgmt. Symp. Proc. 1965, Univ. Mass Coop. Ext. Serv. Pub.,* 446:2–7.

LULL, H. W., AND W. E. SOPPER. 1969. Hydrologic effects from urbanization of forested watersheds in the northeast. *U.S.D.A. Forest Serv. Res. Paper, NE-146.*

LUTZ, H. J. 1956. Ecological effects of forest fires in the interior of Alaska. *U.S.D.A. Tech. Bull. No. 1133.*

LUTZ, H., AND R. F. CHANDLER. 1954. *Forest Soils.* Wiley, New York.

LYMAN, C. P. 1982. *Hibernation and Torpor in Mammals and Birds.* Academic Press, Orlando, Fla.

LYMAN, C. P., AND A. R. DAWE (EDS.). 1960. *Mammalian Hibernation.* Museum Comparative Zoology, Harvard University, Cambridge, Mass.

LYNCH, J. F., AND N. K. JOHNSON. 1974. Turnover and equilibria in insular avifaunas with special reference to California Channel Islands. *Condor,* 76:370–384.

LYNCH, J. F., AND R. F. WHITCOMB. 1977. Effects of insularization of the eastern deciduous forest and avifaunal diversity and turnover. In *Classification, Inventory, and Analysis of Fish and Wildlife Habitat.* FWS/OBS-78/76, U.S. Fish and Wildlife Service, Washington, D.C., pp. 461–489.

MABBERLEY, D. J. 1983. *Tropical Rain Forest Ecology.* Blackie, London.

MCALLEN, B. M., AND C. R. DICKMAN. 1986. The role of photoperiod in the timing of reproduction in the Dasyurid marsupial *Antechinus stuartii. Oecologica,* 68:259–264.

MCARDLE, R. E., W. H. MEYER, AND D. BRUCE. 1949. The yield of Douglas-fir in the Pacific Northwest. *U.S.D.A. Tech. Bull. No. 201* (rev.).

MACARTHUR, R. H. 1958. Population ecology of some warblers of northeastern coniferous forests. *Ecology,* 39:599–619.

———. 1960. On the relative abundance of species. *Am. Nat.,* 94:25–36.

———. 1972. *Geographical Ecology.* Harper & Row, New York.

MACARTHUR, R. H., AND R. LEVINS. 1967. The limiting similarity convergence and divergence of coexisting species. *Am. Nat.,* 101:377–385.

MACARTHUR, R. H., AND J. W. MACARTHUR. 1961. On bird species diversity. *Ecology,* 42:594–598.

MACARTHUR, R. H., AND E. R. PIANKA. 1966. On optimal use of a patchy environment. *Am. Nat.,* 100:603–609.

MACARTHUR, R. H., AND E. O. WILSON. 1963. An equilibrium theory of insular zoogeography. *Evolution,* 17:373–387.

———. 1967. *The Theory of Island Biogeography.* Princeton University Press, Princeton, N.J.

MCAULIFFE, J. R. 1984. Competition for space, disturbance,

and the structure of a benthic stream community. *Ecology,* 65:894–908.

MCBEE, R. H. 1971. Significance of intestinal microflora in herbivory. *Ann. Rev. Ecol. Syst.,* 2:165–176.

MCBRAYER, J. F. 1977. Contributions of cryptozoa to forest nutrient cycles. In W. J. Mattson (ed.), *The Role of Arthropods in Forest Ecosystems.* Springer-Verlag, New York, pp. 70–77.

MCBRIDE, G. I., P. PARE, AND F. FOENANDER. 1969. The social organization and behavior of the feral domestic fowl. *Anim. Behav. Monogr.,* 2:127–181.

MCCALLA, T. M. 1943. Microbiological studies of the effects of straw used as mulch. *Trans. Kansas Acad. Sci.,* 43:52–56.

MCCULLOUGH, D. R. 1981. Population dynamics of the Yellowstone grizzly. In C. W. Fowler and T. D. Smith (eds.), *Dynamics of Large Mammal Populations.* Wiley, New York, pp. 173–196.

———. 1982. Antler characteristics of George Reserve white-tailed deer. *J. Wildl. Manage.,* 46:823–826.

MCCUNE, B., AND G. COTTAM. 1985. The successional status of a southern Wisconsin woods. *Ecology,* 66:1270–1278.

MCDOWELL, D. M., AND R. J. MAIMAN. 1986. Structure and function of a benthic invertebrate stream community as influenced by beaver *(Castor canadensis). Oecologica,* 68:481–489.

MCGEE, C. E. 1984. Heavy mortality and succession in a virgin mixed mesophytic forest. *U.S.D.A. Southern For. Exp. Stat. Res. Paper SO-209.*

MCGINNES, W. G. 1972. North America. In C. M. McKell, J. P. Blaisdell, and J. R. Goodwin (eds.), *Wildland Shrubs: Their Biology and Utilization.* U.S.D.A. For. Serv. Gen. Tech. Rept. INT-1, pp. 55–66.

MCGRAW, J. B. 1989. Effects of age and size on life histories and population growth of *Rhododendron maximum* shoots. *Am. J. Bot.,* 76:113–123.

MCILROY, R. J. 1972. *An Introduction to Tropical Grassland Husbandry,* 2nd ed. Oxford, London.

MCINTOSH, R. P. 1980. The background of some current problems of theoretical ecology. *Synthese,* 43:195–255.

———. 1985. *The Background of Ecology: Concept and Theory.* Cambridge University Press, Cambridge, England.

———. 1987. Pluralism in ecology. *Ann. Rev. Ecol. Syst.,* 18:321–341.

MCKELL, C. M., J. P. BLAISDELL, AND J. R. GOODWIN (EDS.). 1972. *Wildland Shrubs: Their Biology and Utilization.* U.S.D.A. For. Serv. Gen. Tech. Rept. INT-1.

MCLAUGHLIN, S. B., T. J. BLASING, L. K. MANN, AND D. N. DUVICK. 1983. Effects of acid rain and gaseous pollutants on forest productivity: A regional scale approach. *J. Air Pollution Control Assoc.,* 33:1042–1045.

MACLEAN, S. F., JR. 1980. The detritus-based trophic system. In J. Brown, P. C. Miller, L. L. Tieszen, and F. L. Bunnell (eds.), *An Arctic Ecosystem: The Coastal Tundra at Barrow Alaska.* Dowden, Hutchinson, and Ross, Stroudsburg, Penn., pp. 411–457.

MCLELLAN, C. H., A. D. DOBSON, D. S. WILCOVE, AND J. F. LYNCH. 1986. Effects of forest fragmentation on new- and old-world bird communities: Empirical observations and theoretical implications. In J. Verner, M. L. Morrison, and C. J. Ralph (eds.), *Wildlife 2000: Modeling Habitat Relationships of Terrestrial Vertebrates.* University of Wisconsin Press, Madison, pp. 305–313.

MACLINTOCK, L., R. F. WHITCOMB, AND B. L. WHITCOMB. 1977. Island biogeography and the "habitat islands" of eastern forest. II. Evidence for the value of corridors and minimization of isolation in preservation of biotic diversity. *Am. Birds,* 31:6–12.

MACLULICH, D. A. 1937. Fluctuations in the numbers of varying hare *(Lepus americanus). Univ. Toronto Biol. Ser. No. 43.*

MACMAHON, J. A., AND F. H. WAGNER. 1985. The Mojave, Sonoran, and Chihuahuan deserts of North America. In M. Evenardi, I. Noy-meir, and D. W. Goodall (eds.), *Hot Deserts and Arid Shrublands A* (Ecosystems of the World 12). Elsevier, Amsterdam, pp. 105–202.

MACMILLAN, P. C. 1981. Log decomposition in Donaldson's woods, Spring Mill State Park, Indiana. *Am. Mid. Nat.,* 106:335–344.

MCNAB, B. K. 1963. Bioenergetics and the determination of home range size. *Am. Nat.,* 97:133–140.

———. 1973. Energetics and the distribution of vampires. *J. Mammal.,* 54:131–144.

———. 1978. The evolution of endothermy in the phylogeny of mammals. *Am. Nat.,* 112:1–21.

———. 1980. Food habits, energetics, and the population biology of mammals. *Am. Nat.,* 116:106–124.

———. 1982. Evolutionary alternatives in the physiological ecology of bats. In T. Kunz (ed.), *Ecology of Bats.* Plenum Press, New York, pp. 151–200.

MCNAUGHTON, S. J. 1975. r and K selection in *Typha. Am. Nat.,* 109:215–261.

———. 1979. Grazing as an optimization process: Grass–ungulate relationships in the Serengeti. *Am. Nat.,* 113:691–703.

———. 1983. Serengeti grassland ecology: The role of composite environmental factors and contingency in community organization. *Ecol. Monogr.,* 53:291–320.

———. 1984. Grazing lawns: Animals in herds, plant forms, and coevolution. *Am. Nat.,* 124:863–886.

———. 1985. Ecology of a grazing ecosystem: The Serengeti. *Ecol. Monogr.,* 55:259–294.

MCNAUGHTON, S. J., AND N. J. GEORGIADIS. 1988. Ecology of African grazing and browsing mammals. *Am. Rev. Ecol. Syst.,* 17:39–65.

MCNAUGHTON, S. J., R. W. RUESS, AND S. W. SEAGLE. 1988. Large mammals and process dynamics in African ecosystems. *Bioscience,* 38:794–800.

MCNAUGHTON, S. J., J. L. TARRANTS, M. M. MCNAUGHTON, AND T. H. DAVIS. 1985. Silica as a defense against herbivory and a growth promoter in African grasses. *Ecology,* 66:528–535.

MCNAUGHTON, S. J., AND L. L. WOLF. 1979. *General Ecology.* Holt, Rinehart, and Winston, New York.

MCPHERSON, J. K., AND C. H. MULLER. 1969. Allelopathic effects of *Adenostoma fasciculatum* "chamise" in the California Chaparral. *Ecol. Monogr.,* 39:177–179.

MADGWICK, H. A. I., AND J. D. OVINGTON. 1959. The chemical composition of precipitation in adjacent forest and open plots. *Forestry,* 32:14–22.

MAGUIRE, D. A., AND R. T. T. FORMAN. 1983. Herb cover effects on tree seedling patterns in a mature hemlock-hardwood forest. *Ecology,* 64:1367–1380.

MAHALL, B. E., AND R. B. PORK. 1976. The ecotone between *Spartina foliosa.* Trin. and *Salicornia virginica* L. in salt marshes of northern San Francisco Bay. 1. Biomass and production. *J. Ecol.,* 64:421–433.

MAIN, A. R. 1981. Fire tolerance of heathland animals. In R. Specht (ed.), *Heathlands and Related Shrublands* (Ecosystems of the World, 9B). Elsevier, Amsterdam, pp. 85–90.

MAIO, J. J. 1958 Predatory fungi. *Sci. Am.,* 199:67–72.

MAISUROW, D. K. 1941. The role of fire in the perpetuation of virgin forests of northern Wisconsin. *J. Fores.,* 39:201–207.

MALMER, N. 1975. Development of bog mires. In A. D. Hasler (ed.), *Coupling of Land and Water Systems.* Springer-Verlag, New York, pp. 85–92.

MALTHUS, T. R. 1798. *An Essay on Principles of Population.* Johnson, London (numerous reprints).

MALVIN, R. L., AND M. RAYNER. 1968. Renal function and blood chemistry in Cetacea. *Am. J. Physiol.,* 214:187–191.

MANABE, S., AND R. T. WETHERALD. 1986. Reduction in

summer soil wetness induced by an increase in atmospheric carbon dioxide. *Science,* 232:626–628.

MANUAT, J. 1983. The vegetation of African savannas. In F. Bourliere (ed.), *Tropical Savannas* (Ecosystems of the World, 13). Elsevier, Amsterdam, pp. 109–149.

MARGALEF, R. 1963. On certain unifying principles in ecology. *Am. Nat.,* 47:357–374.

———. 1968. *Perspectives in Ecological Theory.* University of Chicago Press, Chicago.

MARGULES, C., AND M. B. USHER. 1981. Criteria used in assessing wildlife conservation potential: A review. *Biol. Cons.,* 22:217–227.

MARKS, P. L. 1974. The role of pin cherry (*Prunus pensylvanica* L.) in the maintenance of stability in northern hardwood ecosystems. *Ecol. Monogr.,* 44:73–88.

MARKS, P. L., AND F. H. BORMANN. 1972. Revegetation following forest cutting: Mechanisms for return to steady-state nutrient cycling. *Science,* 176:914–915.

MARQUIS, D. A. 1974. The impact of deer browsing on Allegheny hardwood regeneration. *U.S.D.A. Forest Serv. Res. Paper NE-308.*

———. 1981. Effect of deer browsing on timber production in Allegheny hardwood forests of northwestern Pennsylvania. *U.S.D.A. For. Ser. Res. Paper NE-475.*

MARQUIS, D. A., AND T. J. GRISEZ. 1978. The effect of deer exclosures on the recovery of vegetation in failed clearcuts on the Allegheny plateau. *U.S.D.A. For. Res. Note, NE-270.*

MARTIN, C. W., R. S. PIERCE, G. E. LIKENS, AND F. H. BORMANN. 1986. Clearcutting affects stream chemistry in the White Mountains of New Hampshire. *U.S.D.A. For. Serv. Res. Paper NE-579.*

MARTIN, M. M. 1970. The biochemical basis of the fungus-attine ant symbiosis. *Science,* 169:16–20.

MARTIN, P. S. 1973. The discovery of America. *Science,* 179:969–974.

MARX, D. H. 1971. Ectomycorrhizae as biological deterrents to pathogenic root infections. In E. Hacskaylo (ed.), *Mycorrhizae. U.S.D.A. Misc. Pub. 1189,* pp. 81–96.

MASER, C., R. G. ANDERSON, K. CROMAC, J. T. WILLIAMS, AND R. E. MARTIN. 1979. Dead and down woody material. In J. W. Thomas (ed.), *Wildlife Habitats in Managed Forests: The Blue Mountains of Washington and Oregon. Agr. Handbook 533.* U.S. Department of Agriculture, Washington, D.C., pp. 78–95.

MASER, C., AND J. M. TRAPPE (EDS.). 1984. *The Seen and the Unseen World of the Fallen Tree.* U.S.D.A. For. Serv. Gen. Tech. Rept. PNW-164.

MASER, C., J. M. TRAPPE, AND R. A. NUSSBAUM. 1978. Fungal–Small mammal interrelationship with emphasis on Oregon coniferous forests. *Ecology,* 59:799–809.

MASSEY, A., AND J. D. VANDENBERG. 1980. Puberty delay by a urinary cue from female house mice in feral populations. *Science,* 209:821–822.

MATTHIESSEN, P. 1985. Contamination of wildlife with DDT insecticides in relation to tsetse fly control operations. *Env. Pollution (B),* 10:189–211.

MATTSON, W. J., AND N. D. ADDY. 1975. Phytophagous insects as regulators of forest primary productivity. *Science,* 190:515–521.

MATTSON, W. J., AND R. H. HOACH. 1987. The role of drought stress in provoking outbreaks of phytophagous insects. In P. Barbosa and J. Schultz (eds.), *Insect Outbreaks: Ecological and Evolutionary Perspectives.* Academic Press, Orlando, Fla.

———. 1987. The role of drought in outbreaks of plant-eating insects. *Bioscience,* 37:110–118.

MATVEYEVA, N. U., O. M. PARINKINA, AND Y. I. CHERNOV. 1975. Maria Pronchitsheva Bay, U.S.S.R. In T. Rosswall and O.

W. Heal (eds.), *Structure and Function of Tundra Ecosystems.* Swedish Natural Science Research Council, Stockholm. pp. 61–72.

MAY, M. L. 1976. Thermoregulation and adaptation to temperature in dragonflies (Odonata: Anisoptera). *Ecol. Monogr.,* 46:1–32.

———. 1979. Insect thermoregulation. *Ann. Rev. Entomol.,* 24:313–350.

MAY, R. 1973. *Stability and Complexity in Model Ecosystems.* Princeton University Press, Princeton, N.J.

———. 1981. Models for two interacting populations. In R. M. May (ed.), *Theoretical Ecology* 2nd ed. Sinauer Associates, Sunderland, Mass., pp. 78–104.

———. 1983. Parasitic infections as regulators of animal populations. *Am. Sci.,* 71:36–45.

MAY, R. (ED.). 1976. *Theoretical Ecology.* Saunders, Philadelphia.

MAY, R. M., AND R. M. ANDERSON. 1978. Regulation and stability of host-parasite population interactions. II. Destabilizing processes. *J. Anim. Ecol.,* 47:249–267.

———. 1979. Population biology of infectious diseases: Part II. *Nature,* 280:455–461.

———. 1983. Parasite–host coevolution. In D. J. Futuyma and M. Slatkin (eds.), *Coevolution.* Sinauer Associates, Sunderland, Mass., pp. 186–206.

MAY, R. M., AND D. I. RUBENSTEIN. 1985. Mammalian reproductive strategies. In C. R. Austin and R. V. Short (eds.), *Reproduction in Mammals, 4. Reproductive Fitness,* 2nd ed. Cambridge University Press, Cambridge, England, pp. 1–23.

MAYER, H. M. 1969. *The Spatial Expression of Urban Growth, Commission on College Geography Resource Paper No. 7.* Association of American Geographers, Washington, D.C.

MAYER, W. V. 1960. Histological changes during the hibernation cycle in the arctic ground squirrel. In C. P. Lyman and A. R. Dawe (eds.), *Mammalian Hibernation.* Museum of Comparative Zoology, Mass., Harvard University, Cambridge, Mass., pp. 131–148.

MAYFIELD, H. R. 1960. The Kirtland's warbler. *Cranbrook Inst. Sci. Bull. No. 40.*

MAYNARD SMITH, J. 1964. Group selection and kin selection. *Nature,* 201:1145–1147.

———. 1971. The origin and maintenance of sex. In G. C. Williams (ed.), *Group Selection.* Aldine, Chicago, pp. 163–175.

———. 1976. A comment on the Red Queen. *Am. Nat.,* 110:325–330.

MAYO, J. M., A. P. HARTGERING, D. E., DESPAIN, R. G. THOMPSON, E. M. B. VAN ZINDEREN BAKER, AND S. D. NELSON. 1977. Gas exchange studies of *Carex* and *Dryas* Truelove lowland. In L. C. Bliss (ed.), *Truelove Lowland, Devon Island Canada: A High Arctic Ecosystem.* University of Alberta Press, Edmonton, pp. 265–280.

MAYR, E. 1963. *Animal Species and Evolution.* Harvard University Press, Cambridge, Mass.

MECH, L. D. 1970. *The Wolf: The Ecology and Behavior of an Endangered Species.* Doubleday, Garden City, New York.

MECH, L. D., R. E. MCROBERTS, R. O. PETERSON, AND R. E. PAGE. 1987. Relationships of deer and moose populations to previous winter's snow. *J. Anim. Ecol.,* 56:615–627.

MEETINTEMEYER, V., E. O. BOX, AND R. THOMPSON. 1982. World patterns and amounts of terrestrial plant litter production. *Bioscience,* 32:108–113.

MEEUSE, B. J. 1975. Thermogenic respiration in aroids. *Ann. Rev. Plant Physiol.,* 25:117–126.

MEFFEE, G. K., AND M. L. CRUMP. 1987. Possible growth and reproductive benefits of cannibalism in the mosquitofish. *Am. Nat.,* 129:203–212.

MENDELSSHON, I. A. 1979. Nitrogen metabolism in the height form of *Spartina alterniflora* in North Carolina. *Ecology,* 60:514–584.

MENGEL, R. M. 1964. The probable history of species formation in some northern wood warblers (Parulidae). *Living Bird*, 3:9–43.

———. 1970. The North American Central Plains as an isolating agent in bird speciation. In W. Dort and J. K. Jones (eds.), *Pleistocene and Recent Environments of the Central Great Plains. Spec Pub No. 3.* Dept. Biology, University of Kansas, Lawrence, pp. 279–345.

MENZIE, C. M. 1969. Metabolism of pesticides. *U.S.D.I. Fish and Wildl. Serv. Sp. Sci. Rept. Wildl. No. 127.*

MERRIAM, C. H. 1898. Life zones and crop zones of the United States. *Bull. U.S. Bureau Biol. Survey*, 10:1–79.

MESLOW, E. C., AND L. B. KEITH. 1968. Demographic parameters of a snowshoe hare population. *J. Wildl. Manage.*, 32:812–835.

METTLER, L. E., AND T. G. GREGG. 1969. *Population Genetics and Evolution.* Prentice-Hall, Englewood Cliffs, N.J.

MICHOD, R. E. 1982. The theory of kin selection. *Ann. Rev. Ecol. Syst.*, 13:23–56.

MILLER, A. H. 1942. Habitat selection among higher vertebrates and its relation to intraspecific variation. *Am. Nat.*, 76:25–35.

MILLER, G. R. 1968. Evidence for selective feeding on fertilized plots by red grouse, hares and rabbits. *J. Wildl. Manage.*, 32:849–853.

MILLER, G. R., AND A. WATSON. 1978a. Territories and the food plants of individual red grouse: 1. Territory size, number of mates and brood size compared with the abundance, production, and diversity of heather. *J. Anim. Ecol.*, 47:293–305.

———. 1978b. Heather productivity and its relevance to the regulation of red grouse populations. In O. W. Heal and D. F. Perkins (eds.), *Production Ecology of British Moors and Grasslands.* Springer-Verlag, New York, pp. 277–285.

MILLER, W. R., AND F. E. EGLER. 1950. Vegetation of the Wequetequock-Pawcatuck tidal marshes, Connecticut. *Ecol. Monogr.* 20:141–172.

MILNE, A. 1957. Theories of natural control of insect populations. *Cold Spring Harbor Symp. Quant. Biol.*, 22:253–271.

MILTON, W. E. J. 1940. The effect of manuring, grazing and cutting on the yield; botanical and chemical composition of natural hill pastures: 1. Yield and botanical composition. *J. Ecol.*, 28:326–356.

MINSHALL, G. W. 1978. Autotrophy in stream ecosystems. *Bioscience*, 28:767–771.

MINSHALL, G. W., K. W. CUMMINS, R. C. PETERSEN, C. E. CUSHING, D. A. BRUNS, J. R. SEDELL, AND R. L. VANNOTE. 1985. Developments in stream ecology. *Can. J. Fish. Aquat. Sci.*, 42:1045–1055.

MINSHALL, G. W., R. C. PETERSEN, K. W. CUMMINS, T. L. BOTT, J. R. SEDELL, C. E. CUSHING, AND R. L. VANNOTE. 1983. Interbiome comparison of stream ecosystem dynamics. *Ecol. Monogr.*, 53:1–25.

MISHUSTIN, E. N., AND V. K. SHILNIKOVA. 1969. The biological fixation of atmospheric nitrogen by free-living bacteria. In *Soil Biology.* UNESCO, Paris, pp. 65–124.

MITSCH, W. J., AND J. G. GOSSELINK. 1986. *Wetlands.* Van Nostrand Reinhold, New York.

MOEN, A. M. 1968. The critical thermal environment. *Bioscience*, 18:1041–1043.

MOEN, A. M., AND C. W. SEVERINGHAUS. 1981. The annual weight cycle and survival of white-tailed deer in New York. *N.Y. Fish Game J.*, 28:162–177.

MONK, C. A. 1967. Tree species diversity in the eastern deciduous forest with particular reference to northcentral Florida. *Am. Nat.*, 101:173–187.

MONRO, J. 1967. The exploitation and conservation of resources by populations of insects. *J. Anim. Ecol.*, 36:531–547.

MONSI, M. 1968. Mathematical models of plant communities. In F. E. Eckardt (ed.), *Functioning of Terrestrial Ecosystems at the Primary Production Level.* UNESCO, Paris, pp. 131–149.

MONSON, R. K., G. E. EDWARDS, M. S. B. KU. 1984. C_3–C_4 intermediate photosynthesis in plants. *Bioscience*, 34:563–574. (3)

MONTEITH, J. L. 1962. Measurement and interpretation of carbon dioxide flues in the field. *Netherlands J. Agr. Sci.*, 10 (Sp. issue):334–346.

MONTEITH, J. L. (ED.). 1976. *Vegetation and the Atmosphere, Vol. 2, Case Studies.* Academic Press, London.

MOOK, L. J. 1963. Birds and spruce budworm. In R. Morris (ed.), *Entomol. Soc. Can. Mem. 31*, pp. 244–248.

MOONEY, H. A. (ED.). 1977. *Convergent Evolution in Chile and California Mediterranean Climate Ecosystems.* Academic Press, New York.

MOONEY, H. A. 1981. Primary production in Mediterranean-type shrubland. In F. di Castri, D. Goodall, and R. Spetch (eds.), *Mediterranean-type Shrublands.* Elsevier, Amsterdam, pp. 249–256.

MOONEY, H. A., AND W. D. BILLINGS. 1961. Comparative physiological ecology of arctic and alpine populations of *Oxyria digyna. Ecol. Monogr.*, 31:1–29

MOONEY, H. A., AND C. E. CONRAD. 1977. *Proc. Symp. Environmental Consequences of Fire and Fuel Management in Mediterranean Ecosystems.* U.S.D.A. For. Serv. Gen. Tech. Rept. WO-3. U.S. Department of Agriculture, Washington, D.C.

MOONEY, H. A., AND E. L. DUNN. 1970a. Convergent evolution of Mediterranean climate evergreen sclerophyll shrubs. *Evolution*, 24:292–303.

———. 1970b. Photosynthetic systems of Mediterranean climate shrubs and trees of California and Chile. *Am. Midl. Nat.*, 104:447–453.

MOONEY, H. A., AND D. J. PARSONS. 1973. Structure and function of the California chaparral—an example from San Dimas. In F. di Castri and H. A. Mooney, (eds.), *Mediterranean-type Ecosystems: Origin and Structure.* Springer-Verlag, New York, pp. 83–112.

MOONEY, H. A., J. EHLERINGER, AND O. BJORKMAN. 1977. The energy balance of leaves of the evergreen desert shrub *Atriplex hymenelytra. Oecologica*, 29:301–310.

MOONEY, H. A., D. J. PARSONS, AND J. KUMMEROW. 1974. Plant development in Mediterranean climates. In H. Leith (ed.), *Phenology and Seasonality Modeling.* Springer-Verlag, New York, pp. 225–267.

MOORE, H. B. 1958. *Marine Ecology.* Wiley, New York.

MOORE, J. A. 1949a. Geographic variation of adaptive characters in *Rana pipiens* Schreber. *Evolution*, 3:1–24.

MOORE, J. J., P. DOUDING, AND B. HEALY. 1975. Glenamoy, Ireland. In T. Rosswall and O. W. Heal (eds.), *Structure and Function of Tundra Ecosystems.* Swedish Natural Science Research Council, Stockholm, pp. 321–343.

MOORE, N. W., AND M. D. HOOPER. 1975. On the number of bird species in British woods. *Biol. Cons.*, 8:239–250.

MOORE, P. D., AND D. J. BELLAMY. 1974. *Peatlands.* Springer-Verlag, New York,

MORRIS, R. F. 1963. Predictive population equations based on key factors. *Entomol. Soc. Can. Mem. 32*, pp. 16–21.

MORRIS, R. F. (ED.). 1963. The dynamics of epidemic spruce budworm populations. *Entomol. Soc. Can. Mem. 31.*

MORRIS, R. F., W. F. CHESHIRE, C. A. MILLER, AND D. G. MOTT. 1958. The numerical response of avian and mammalian predators during a gradation of the spruce budworm. *Ecology*, 39:487–494.

MOORE, W. S., AND R. A. DOLBEER. 1989. The use of banding recovery data to estimate dispersal rates and gene flow in avian species: Case studies in the red-winged blackbird and common grackle. *Condor*, 91:242–253.

MOSS, R. 1972. Food selection by red grouse (*Lagopus*

lagopus scoticus Lath.) in relation to chemical composition. *J. Anim. Ecol.*, 41:411–428.

MOSS, R., G. R. MILLER, AND S. F. ALLEN. 1972. Selection of heather by captive red grouse in relation to the age of the plant. *J. Appl. Ecol.*, 9:771–781.

MOSSER, J. L., N. S. FISHER, T. TENG, AND C. F. WURSTER. 1972. Polychlorinated biphenyls toxicity to certain phytoplankton. *Science*, 175:191–192.

MOULDER, B. C., AND D. E. REICHLE. 1974. Significance of spider predation in the energy dynamics of forest floor arthropod communities. *Ecol. Monogr.*, 42:473–498.

MOULDER, B. C., D. E. REICHLE, AND S. I. AUERBACH. 1970. *Significance of Spider Predation in the Energy Dynamics of Forest Floor Arthropod Communities.* Oak Ridge National Laboratory Report ORNL 4452.

MOULTON, M. P., AND S. L. PIMM. 1986. The extent of competition in shaping an introduced avifauna. In J. Diamond and T. J. Case (eds.), *Community Ecology.* Harper & Row, New York, pp. 80–97.

MUELLER-DOMBOIS, D. 1987. Natural dieback in forests. *Bioscience*, 37:575–585.

MUELLER-DOMBOIS, D., AND H. ELLENBERG. 1974. *Aims and Methods of Vegetation Ecology.* Wiley, New York.

MULLER, C. H., R. B. HANAWALT, AND J. K. MCPHERSON. 1968. Allelopathic control of herb growth in the fire cycle of California chaparral. *Bull. Torrey Bot. Club*, 95:225–231.

MULROY, T. W., AND P. W. RUNDEL. 1977. Annual plants: Adaptations to desert environments. *Bioscience*, 27: 109–114.

MURDOCH, W. W. 1969. Switching in general predators: Experiments on predator specificity and stability of prey populations. *Ecol. Monogr.* 39:335–354.

———. 1973. The functional response of predators. *J. Appl. Ecol.*, 10:335–342.

MURDOCH, W. W., AND A. OATEN. 1975. Predation and population stability. *Adv. Ecol. Res.*, 9:1–131.

MURPHY, G. I. 1966. Population biology on the Pacific sardine. *Proc. Calif. Acad. Sci. 4th Series*, 34:1–84.

———. 1967. Vital statistics of the Pacific sardine and the population consequences. *Ecology*, 48:731–736.

MURPHY, P. W. 1952. Soil faunal investigations. In *Report on Forest Research for the Year Ending March, 1951.* Forest. Comm., London, pp. 130–134.

———. 1953. The biology of the forest soils with special reference to the mesofauna or meiofauna. *J. Soil Sci.*, 4:155–193.

MURRAY, B. G., JR. 1967. Dispersal in vertebrates. *Ecology*, 48:975–978.

MUSCATINE, L., AND J. W. PORTER. 1977. Reef corals: Mutualistic symbioses adapted to nutrient-poor environments. *Bioscience*, 27:454–460.

MUTCH, R. W. 1970. Wildland fires and ecosystems—a hypothesis. *Ecology*, 51:1046–1051.

MUUL, I. 1965. Daylength and food caches. *Nat. Hist.*, 74(3):22–27.

———. 1969. Photoperiod and reproduction flying squirrels, *Glaucomys volans. J. Mammal.*, 50:542–549.

MYERS, J. H., AND C. J. KREBS. 1971. Genetic, behavioral, and reproductive attributes of dispersing field voles *Microtus pennsylvanicus* and *Microtus ochrogaster. Ecol. Monogr.*, 41:53–78.

MYERS, K., AND W. E. POOLE. 1967. A study of the biology of the wild rabbit *Oryctolagus cuniculus* L. in confined populations: 4. The effects of rabbit grazing on sown pastures. *J. Ecol.*, 55:435–451.

MYERS, K., C. S. HALE, R. MYKYTOWYCZ, AND R. L. HUGHS. 1971. The effects of varying density and space on sociality and health in animals. In A. H. Esser, *Behavior and Environment: The Use of Space by Animals and Men.* Plenum, New York, pp. 148–187.

MYERS, N. 1983. *A Wealth of Wild Species.* Westview Press, Boulder Colo.

NACE, R. L. 1969. Human uses of ground water. In R. J. Chorley (ed.), *Water, Earth, and Man.* Methuen, London, pp. 285–294.

NADELHOFFER, K., J. D. ABER, AND J. M. MELILLO. 1985. Fine roots, net primary production, and soil nitrogen availability: A new hypothesis. *Ecology*, 66:1377–1390.

NAIMAN, R. J., J. M. MELILLIO, AND J. E. HOBBIE. 1986. Ecosystem alternation of boreal forest streams by beaver *(Castor Canadensis). Ecology*, 67:1254–1269.

NAPIER, J. R. 1966. Stratification and primate ecology. *J. Anim. Ecol.*, 35:411–412.

NATIONAL ACADEMY OF SCIENCES. 1974. *U.S. Participation in the International Biological Program, Rept. No. 6.* U.S. Committee for the International Biological Program, National Academy of Science, Washington, D.C.

———. 1975. *Productivity of World Ecosystems.* National Academy of Science, Washington, D.C.

NATIONAL RESEARCH COUNCIL. 1983. *Acid Deposition: Atmospheric Processes in Eastern North America.* National Academy Press, Washington, D.C.

NATIONAL RESEARCH COUNCIL OF CANADA. 1981. Acidification in the Canadian aquatic environment: Scientific criteria for assessing the effects of acid deposition on aquatic ecosystems, NRCC No. 18475. National Research Council of Canada, Ottawa.

NAVTH, Z. 1974. Effects of fire in the Mediterranean region. In T. F. Kozlowski and C. E. Ahlgren (eds.), *Fire and Ecosystems.* Academic Press, New York, pp. 401–434.

NEEL, J. K. 1951. Interrelations of certain physical and chemical features in headwater limestone streams. *Ecology*, 32:368–391.

NEGUS, N. C., P. J. BERGER, AND L. G. FORSLUND. 1977. Reproductive strategy of *Microtus montanus. J. Mammal.*, 58:347–353.

NELLIS, C. H., S. P. WETMORE, AND L. B. KIETH. 1972. Lynx–prey interaction in central Alberta. *J. Wildl. Manage.*, 36:320–328.

NELSON, D. J. 1962. Clams as indicators of strontium-90. *Science*, 138:38–39.

NELSON, D. J., ET AL. 1971. Hazards of mercury. *Envir. Res.* 4:3–69.

NELSON, D. J., AND F. E. EVANS (EDS.). 1967. *Symposium on Radioecology, Conf. 670503.* National Technical Information Service, Springfield, Va.

NELSON, D. J., AND C. C. SCOTT. 1962. Role of detritus in the productivity of a rock-outcrop community in a Piedmont stream. *Limnol. Oceanogr.*, 3:396–413.

NELSON, M. E., AND L. D. MECH. 1981. Deer social organization and wolf predation in northeastern Minnesota. *Wildl. Monogr.*, 77, 53pp.

NELSON, R. A. 1980. Protein and fat metabolism in hibernating bears. *Fed. Proc.*, 39:2955–2958.

NELSON, R. A., AND T. D. I. BECK. 1984. Hibernation adaptation in the black bear: Implications for management. *Proc. East. Workshop Black Bear Manage. Res.*, 7:48–53.

NELSON, R. A., T. D. I. BECK, AND D. L. STENGER. 1983. Ratio of serum urea to serum creatinine in wild black bears. *Science*, 226:841–842.

NELSON, R. A., G. E. FOLK, JR., E. W. PFEIFFER, J. J. CRAIGHEAD, C. J. JONKEL, AND D. L. STEIGER. 1983. Behavior, biochemistry, and hibernation in black, grizzly, and polar bears. *Int. Conf. Bear Res. Manage.*, 5:284–290.

NEUSOME, A. E., AND P. C. CATLING. 1979. Habitat preferences of mammals inhabiting heathlands of warm temperate coastal southeastern Queensland. In R. L. Spech (ed.), *Heathlands and Related Shrubland* (Ecosystems of the World 9A). Elsevier, Amsterdam, pp. 301–316.

NEUWIRTH, R. 1957. Some recent investigations into the chemistry of air and of precipitation and their significance for forestry. *Allg. Fost-v. Jagdztg.*, 128:147–150.

NEWBOLD, J. D., J. W. ELWOOD, R. V. O'NEILL, AND A. L. SHELDON. 1983. Phosphorous dynamics in a woodland stream: A study of nutrient spiraling. *Ecology,* 65:1249–1265.

———. 1984. Phosphorous dynamics in a woodland stream ecosystem. *Bioscience,* 34:43–44.

NEWBOLD, J. D., R. V. O'NEILL, J. W. ELWOOD, AND W. VAN WINKLE. 1982. Nutrient spiraling in streams: Implications for nutrient and invertebrate activity. *Am. Nat.,* 20:628–652.

NEWELL, R. C. 1965. The role of detritus on the nutrition of two marine deposit feeders, the prosobranch *Hydrobia ulvae* and the bivalva *Macoma balthica. Proc. Zool. Soc. London,* 144:25–45.

NEWELL, S. Y., R. D. FALLON, R. M. CAL RODRIGUEZ, AND L. C. GROENE. 1985. Influence of rain, tidal wetting, and relative humidity on release of carbon dioxide by standing-dead salt marsh plants. *Oecologica,* 68:73–79.

NICE, M. M. 1941. The role of territory in bird life. *Am. Midl. Nat.,* 26:441–487.

———. 1943. Studies in the life history of the song sparrow: 2. *Trans. Linn. Soc. New York,* 6:1–329.

———. 1962. Development of behavior in precocial birds. *Trans. Linn. Soc. New York, Vol. 8.*

NICHOLLS, A. G. 1933. On the biology of *Calanus finmarchius.* III. Vertical distribution and diurnal migration in the Clyde Sea area. *J. Marine Biol. Assoc. United Kingdom,* 19:139–164.

NICHOLSON, A. J. 1954. An outline of the dynamics of animal populations. *Aust. J. Zool.,* 2:9–65.

———. 1957. The self-adjustment of populations to change. *Cold Spring Harbor Symp. Quant. Biol.,* 22:153–173.

NICHOLSON, A. J., AND V. A. BAILEY. 1935. The balance of animal populations: Part 1. *Proc. Zool. Soc. London,* pp. 551–598.

NIELSEN, A. 1950. The torrential invertebrate fauna. *Oikos,* 2:176–196.

NIELSEN, C. O. 1961. Respiratory metabolism in some populations of enchytraeid worms and free-living nematodes. *Oikos,* 12:17–35.

NIERING, W. A., AND F. E. EGLER. 1955. A shrub community of *Viburnum lentago* stable for twenty-five years. *Ecology,* 36:356–360.

NIERING, W., AND R. GOODWIN. 1974. Creation of relatively stable shrublands with herbicides: Arresting "succession" on rights-of-way and pasture land. *Ecology,* 55:784–795.

NIXON, S. W., AND C. A. OVIATT. 1973. Ecology of a New England salt marsh. *Ecol. Monogr.,* 43:463–498.

NOBEL, P. S. 1978. Surface temperature of cacti: Influence of environmental and morphological factors. *Ecology,* 59:986–996.

NOBLE, J. C., A. D. BELL, AND J. L. HARPER. 1979. The population biology of plants with clonal growth. I. Morphology and structural demography of *Carex Arernaria. J. Ecol.,* 67:983–1008.

NOBLE, J. R., AND R. O. SLATYER. 1980. The use of vital attributes to predict successional changes in plant communities subject to recurrent disturbances. *Vegetatio,* 43:5–21.

NOMMIK, H. 1965. Ammonium fixation and other reactions involving nonenzymatic immobilization. In W. Bartholomew and F. Clark (eds.), *Soil Nitrogen.* American Society Agronomy, Madison, Wis., pp. 198–258.

NOY-MIER, I. 1973. Desert ecosystems: environment and producers. *Ann. Rev. Ecol. Syst.,* 4:25–51.

———. 1974. Desert ecosystems: Higher trophic levels. *Ann. Rev. Ecol. Syst.,* 5:195–214.

———. 1975. Stability of grazing systems: an application of predator-prey graphs. *J. Ecol.,* 63:459–481.

———. 1985. Desert ecosystem structure and function. In M. I. Evenardi, I. Noy-Mier, and D. W. Goodall (eds.),

Hot Deserts and Arid Shrublands A. Elsevier, Amsterdam, pp. 93–103.

NYBAKKEN, J. W. 1988. *Marine Biology: An Ecological: Approach,* 2nd ed. Harper & Row, New York.

OBERNDORFER, R. Y., J. V. MCARTHUR, J. R. BARNES, AND J. DIXON. 1984. The effects of invertebrate predators of leaf litter processing in an alpine stream. *Ecology,* 65:1325–1331.

O'BRIEN, S. J., D. E. WILDT, D. GOLDMAN, D. R. MERREL, AND M. BASH. 1983. The cheetah is depauperate in genetic variation. *Science,* 221:459–462.

O'DOWD, D. J., AND M. E. HAY. 1980. Mutualism between harvester ants and a desert ephemeral: Seed escape from rodents. *Ecology,* 61:531–540.

ODUM, E. P. 1964. The new ecology. *Bioscience,* 14:14–16.

———. 1969. The strategy of ecosystem development. *Science,* 164:262–270.

———. 1971. *Fundamentals of Ecology,* 3rd ed. Saunders, Philadelphia.

———. 1983. *Basic Ecology.* Saunders, Philadelphia.

ODUM, E. P., C. E. CONNELL, AND L. B. DAVENPORT. 1962. Population energy flow of three primary consumer components of old field ecosystems. *Ecology,* 43:88–96.

ODUM, E. P., AND E. J. KUENZLER. 1963. Experimental isolation of food chains in an old-field ecosystem with the use of phosphorus-32. In V. Schultz and A. Klement (eds.), 1983, pp. 113–120.

ODUM, H. T. 1956. Primary production in flowing water. *Limnol. Oceanogr.,* 1(2):102–117.

———. 1957a. Trophic structure and productivity of Silver Springs, Florida. *Ecol. Monogr.,* 27:55–112.

———. 1970. Summary: An emerging view of the ecological system at El Verde. In H. T. Odum and R. F. Pigeon (eds.), *A Tropical Rain Forest,* pp. I/191–I/218.

———. 1983. *Systems Ecology: An Introduction.* Wiley, New York.

ODUM, H. T., B. J. COPELAND, AND E. A. MCMAHAN. 1974. *Coastal Ecological Systems of the U.S.,* Vols. 1–4. Conservation Foundation, Washington, D.C.

ODUM, W. E., AND M. A. HEYWOOD. 1978. Decomposition of intertidal freshwater marsh plants. In R. E. Good, D. F. Whigham, and R. L. Simpson (eds.), *Freshwater Wetlands.* Academic Press, New York, pp. 89–97.

ODUM, W. E., T. J. SMITH, III, J. K. HOOVER, AND C. C. MCIVOR. 1984. *The Ecology of Tidal Freshwater Marshes of the United States East Coast: A Community Profile.* U.S. Fish and Wildlife Service FWS/OBS-87/17, Washington D.C., 177pp.

OECHEL, W. C., AND W. T. LAURENCE. 1985. Tiaga. In B. F. Chabot and H. A. Mooney (eds.), *Physiological Ecology of North American Plant Communities.* Chapman and Hall, New York, pp. 66–94.

OFFICER, C. B. 1983. Physics of estuarine circulation. In B. Ketchum (ed.), *Estuaries and Enclosed Seas* (Ecosystems of the World 26). Elsevier, Amsterdam, pp. 15–42.

OGREN, W. L., AND R. CHOLLET. 1982. Photorespiration. In Govindjee (ed.), *Photosynthesis Development, Carbon Metabolism, and Plant Productivity.* Academic Press, New York, pp. 191–230.

OHMART, C. P., G. G. STEWART, AND J. R. THOMAS. 1983. Leaf consumption by insects in three *Eucalyptus* forest types in southeastern Australia and their role in short-term nutrient cycling. *Oecologica,* 59:322–330.

OJASTI, J. 1983. Ungulates and large rodents of South America. In F. Bourliere (ed.), *Tropical Savannas* (Ecosystems of the World 9). Elsevier, Amsterdam, pp. 427–440.

OKE, T. R., AND C. EAST. 1971. The urban boundary layer in Montreal. *Boundary-layer Meteorol.,* 1:411.

OLSON, J. S. 1970. Carbon cycles and temperate woodlands. In D. E. Reichle (ed.), *Analysis of Temperate Forest Ecosystems.* Springer-Verlag, New York, pp. 226–241.

O'NEILL, R. V. 1976. Ecosystem persistence and heterotrophic regulations. *Ecology,* 57:1244–1253.

O'NEILL, R. V., W. F. HARRIS, B. S. AUSMUS, AND D. E. REICHLE. 1975. A theoretical basis for ecosystem analysis with particular reference to element cycling. In F. G. Howell, J. B. Gentry, and M. H. Smith (eds.), *Mineral Cycling in Southeastern Ecosystems.* ERDA Symposium Series, National Technical Information Service, U.S. Department of Commerce, pp. 28–40.

ORIANS, G. H. 1969. On the evolution of mating systems in birds and mammals. *Am. Nat.,* 103:589–603.

ORIANS, G. H., AND O. T. SOLBRIG. 1977. *Convergent Evolution in Warm Deserts,* Academic Press, New York.

ORING, L. W., AND M. L. KNUDSON. 1972. Monogamy and polygamy in the spotted sandpiper. *Living Bird,* 11:59–73.

ORITZ, C. L., B. J. LEBOEUF, AND D. P. COSTA. 1984. Milk intake of elephant seal pups: An index of parental investment. *Am. Nat.,* 124:416–422.

OSBORN, F. 1949. *Our Plundered Planet.* Little, Brown, Boston.

OSTBYE, E., ET AL. 1975. Hardangervidda, Norway. In T. Rosswall and O. W. Heal (eds.), 1975, pp. 225–264.

OVINGTON, J. D. 1961. Some aspects of energy flow in plantation of *Pinus sylvestris* L. *Ann. Bot., London, n.s.,* 25:12–20.

OWEN, D. F. 1980. How plants may benefit from animals that eat them. *Oikos* 35:230–235.

OWEN, D. F., AND R. G. WIEGERT. 1981. Mutualism between grasses and grazers: An evolutionary hypothesis. *Oikos,* 36:376–378.

PAINE, R. T. 1966. Food web complexity and species diversity. *Am. Nat.,* 100:65–75.

———. 1969. The *Pisaster-Tegula* interaction: Prey patches, predator food preference and intertidal community structure. *Ecology,* 50:950–961.

PALACA, S., AND J. ROUGHGARDEN. 1984. Control of arthropod abundance by *Anoles* lizards on St. Euatatius (Netl. Antilles). *Oecologica,* 64:160–162.

PALMER, H. E., W. C. HANSON, B. I. GRIFFIN, AND W. C. ROESCH. 1963. Cesium-137 in Alaskan Eskimos. *Science,* 142(3588):64–65.

PALMER, J. D. 1974. *Biological Clocks in Marine Organisms.* Wiley, New York.

———. 1976. *An Introduction to Biological Rhythms.* Academic Press, New York.

PALMGREN, P. 1949. Some remarks on the short-term fluctuations in the numbers of northern birds and mammals. *Oikos,* 1:114–121.

PALO, R. T., A. PEHRSON, AND P. KNUTSSON. 1983. Can birch phenolics be of importance in the defense against browsing vertebrates? *Finn. Game Res.,* 41:75–80.

PARIS, O. H. 1969. The function of soil fauna in grassland ecosystems. In R. L. Dix and R. G. Beidleman (eds.), The grassland ecosystem: A preliminary synthesis. *Range Sci. Dept. Sci Ser. No. 2,* Colorado State University, Fort Collins, pp. 331–360.

PARK, T. 1948. Experimental studies of interspecies competition: 1. Competition between populations of the flour beetles, *Trilobium confusum* Duval and *Trilobium castaneum* Herbst. *Ecol. Monogr.,* 18:265–308.

———. 1954. Experimental studies of interspecies competition: 2. Temperature, humidity and competition in two species of *Trilobium. Physiol. Zool.,* 27:177–238.

———. 1955. Experimental competition in beetles with some general implications. In J. B. Cragg and N. W. Pirie (eds.), *The Numbers of Man and Animals.* Oliver & Boyd, London.

PARKER, G. A., AND R. A. STUART. 1976. Animal behavior as a strategy optimizer: Evolution of resource assessment strata and optimal emigration thresholds. *Am. Nat.,* 110:1055–1076.

PASTOR, J., R. J. NAIMAN, B. DEWEY, AND P. MCINNES. 1988. Moose, microbes, and boreal forest. *Bioscience,* 88:770–777.

PATE, V. S. L. 1933. Studies on fish food in selected areas: A biological survey of Raquette Watershed, *N.Y. State Conserv. Dept. Biol. Survey No. 8,* pp. 136–157.

PATTERSON, D. T. 1975. Nutrient return in stemflow and throughfall of individual trees in the Piedmont deciduous forest. In F. G. Howell et al. (eds.), *Mineral Cycling in Southeastern Ecosystems.* National Technical Information Science, U.S. Dept. Commerce. pp. 800–812.

PATTON, D.R. 1975 A diversity index for quantifying habitat "edge." *Wildl. Soc. Bull.* 3:171–173

PAYETTE, S. 1988. Late-Holocene development of subarctic ombrotrophic peatlands: Allogenic and autogenic succession. *Ecology,* 62:516–531.

PAYNE, R. 1968. Among wild whales. *N.Y. Zool. Soc. Newsletter,* November 1968.

PEAKALL, D. B. 1970. Pesticides and the reproduction of birds. *Sci. Am.,* 222:72–78.

PEAKALL, D. B., AND J. L. LINGER. 1970. Polycholorinated biphenyls: Another long-life widespread chemical in the environment. *Bioscience,* 20:958–964.

PEARCE, R. B. 1967. Photosynthesis in plant communities as influenced by leaf angle. *Crop Sci.,* 7:321–326.

PEARCY, R. W. 1976. Temperature effects on growth and CO_2 exchange rates of *Atriplex leniformis. Oecologica,* 26:245–255.

———. 1977. Acclimation of photosynthetic and respiratory CO_2 to growth temperature in *Atriplex lentiformis* (Torr.) Wats. *Plant Physiol.,* 61:484–486.

PEARL, R. 1927. The growth of populations. *Quart. Rev. Biol.,* 2:532–548.

PEARL, R., AND L. J. REED. 1920. On the rate of growth of the population of the U.S. since 1790 and its mathematical representation. *Proc. Natur. Acad. Sci.,* 6:275–288.

PEARSON, D. L. 1971. Vertical stratification of birds in a tropical dry forest. *Condor,* 73:46–55.

PEASE, J. L., R. H. VOWLES, AND L. B. KEITH. 1979. Interaction of snowshoe hares and woody vegetation. *J. Wildl. Manage.,* 43:43–60.

PECARSKY, B. L. 1982. Aquatic insect predator–prey relations. *Bioscience,* 32:261–266.

PEET, R. K. 1974. The measurement of species diversity. *Ann. Rev. Ecol. Syst.,* 5:285–307.

PEET, R. K. 1981. Changes in biomass and production during secondary forest succession. In D. C. West, H. H. Shugart, and D. B. Botkin (eds.), *Forest Succession: Concepts and Applications.* Springer-Verlag, New York, pp. 324–338.

PEET, R. K., AND N. L. CHRISTENSEN. 1980. Succession: A population process. *Vegetatio,* 43:131–140.

PEFRANKA, J. W., AND A. SIH. 1986. Environmental instability, competition, and density-dependent growth and survivorship of a stream-dwelling salamander. *Ecology,* 67:729–736.

PENGELLEY, E. T., AND S. J. ASMUNDSON. 1974. Circannual rhythmicity in hibernating mammals. In E. T. Pengelley (ed.), *Circannual Clocks: Annual Biological Rhythms.* Academic Press, New York, pp. 95–106.

PENNAK, R. W., AND E. D. VAN GERPEN. 1947. Bottom fauna production and physical nature of a substrate in a northern Colorado trout stream. *Ecology,* 28:42–48.

PETERS, R. L., AND J. D. S. DARLING. 1985. The greenhouse effect and nature reserves. *Bioscience,* 3:707–717.

PETERS, R. P., AND L. D. MECH. 1975. Scent-marking in wolves. *Am. Sci.,* 63:628–637.

PETERSON, D. L., AND F. A. BAZZAZ. 1978. Life cycle characteristics of *Aster pilosus* in early successional habitats. *Ecology,* 59:1005–1013.

PETERSON, H., AND M. LUXTON. 1982. A comparative analysis

of soil fauna populations and their role in the decomposition process. *Oikos,* 39:287–388.

PETERSON, R. O. 1977. Wolf ecology and prey relationships in Isle Royale. *U.S. Nat. Park Serv. Sci. Monogr. Ser. 11,* 210 pp.

PETRINVICH, L., AND T. L. PATTERSON. 1982. The white-crowned sparrow: Stability, recruitment, and population structure in the Nuttall subspecies (1975–1980). *Auk,* 99:1–14.

PETROV, V. S. 1946. Aktevnaia reaktsiia pochvy pH kah faktor rasprpstaneniia dozhdevykh chorvei (Lumbricidae, Oligochaetae). *Zool. Zh.,* 25:107–110.

PETRUSEWICZ, K. (ED.). 1967. *Secondary Productivity of Terrestrial Ecosystems.* Polish Academy Sciences, Warsaw.

PFEIFFER, W. 1962. The fright reaction of fish. *Biol. Rev.,* 37:495–511.

PFISTER, R. D., B. L. KOVALCHIK, S. E. ARNO, AND P. C. PRESBY. 1977. *Forest Habitat Types of Montana.* U. S. D. A. For. Serv. Gen. Tech. Rept. INT-34. (31)

PHILLIPS, J. 1965. Fire—as master and servant: Its influence in the bioclimatic regions of trans-Saharan Africa. *Proc. 4th Tall Timbers Fire Ecol. Conf.,* pp. 7–109.

PHILPOT, C. W. 1977. Vegetation features as determinants of fire frequency and intensity. In H. A. Mooney and C. E. Conrad (eds.), *Environmental Consequences of Fire and Fuel Management in Mediterranean Ecosystems,* U.S.D.A. For. Serv. Gen. Tech. Rept. WO-26. Department of Agriculture, Washington, D.C., pp. 12–16.

PIANKA, E. R. 1966. Latitudinal gradients in species diversity: A review of concepts. *Am. Nat.,* 100:33–46.

———. 1967. On lizard species diversity, North American flatlands desert. *Ecology,* 48:333–351.

———. 1972. *r* and *K* selections or *b* and *d* selection? *Am. Nat.,* 100:65–75.

———. 1974 Niche overlap and diffuse competition. *Proc. Nat. Acad. Sci.,* 71:2141–2145.

———. 1975. Niche relations of desert lizards. In M. Clody and J. Diamond (eds.), *Ecology and Evolution of Communities.* Harvard University Press, Cambridge, Mass., pp. 292–314.

———. 1978 *Evolutionary Ecology,* 3rd ed. Harper & Row, New York.

———. 1980. On *r* and *K* selection. *Am. Nat.,* 102:592–597.

PIANKA, E., AND W. S. PARKER. 1975. Age specific reproductive tactics. *Am. Nat.,* 109:453–464.

PIEHLER, K. G. 1987. Habitat relationships of three grassland sparrow species on reclaimed surface mines in Pennsylvania. Unpublished MS thesis, West Virginia University.

PIELOU, E. C. 1972. Niche width and niche overlap: A method for measuring them. *Ecology,* 53:687–692.

———. 1974. *Population and Community Ecology,* Gordon and Breach, New York.

———. 1975. *Ecological Diversity.* Wiley, New York.

———. 1977. *Mathematical Ecology.* Wiley, New York.

———. 1981. The usefulness of ecological models: A stock-taking. *Quart. Rev. Biol.,* 56:1423–1437

PIERCE, B. A. 1985. Acid tolerance in amphibians. *Bioscience,* 35:239–243.

PIJL, L. VAN DER. 1969. *Principles of Dispersal in Higher Plants.* Springer-Verlag, New York.

PIJL, L. VAN DER, AND C. L. DOTSON. 1966. *Orchid Flowers: Their Pollination and Evolution.* University of Miami Press, Miami.

PILCHER, J. R., AND B. GRAY. 1982. The relationship between oak tree growth and climate in Butran. *J. Ecol.,* 70:297–304.

PIMENTEL, D. 1971a. *Ecological Effects of Pesticides on Non-target Species.* Executive Office of the President, Office of Science and Technology, Washington, D.C.

———. 1971b. Evolutionary and environmental impact of pesticides. *Bioscience,* 21:109.

PIMENTEL, D., J. E. DEWEY, AND H. H. SCHWARDT. 1951. An increase in the duration of the life cycle of DDT-resistant strains of the house fly. *J. Econ. Entomol.,* 44:477–481.

PIMENTEL, D., E. H. FEINBERG, P. W. WOOD, AND J. T. HAYES. 1965. Selection, spacial distribution, and the coexistence of competing fly species. *Am. Nat.,* 99:97–109.

PIMENTEL, D., W. P. NAGEL, AND J. L. MADDEN. 1963. Space-time structure of the environment and the survival of the parasite-host system. *Am. Nat.,* 97:141–167.

PIMM, S. L. 1980. Food web design and the effect of species delation. *Oikos,* 35:139–149.

———. 1982. *Food Webs.* Chapman & Hall, London.

PIMM, S. L., AND J. H. LAWTON. 1977. The number of trophic levels in ecological communities. *Nature,* 268:329–331.

PIMM, S. L., AND R. L. KITCHING. 1987. The determinants of food chain lengths. *Oikos,* 50:302–307.

PIMM, S. L., H. L. JONES, AND J. DIAMOND. 1988. On the risk of extinction. *Am. Nat.,* 132:757–785.

PIMM, S. L., AND J. W. PIMM. 1982. Resource use, competition, and resource availability in Hawaiian honeycreepers. *Ecology,* 63:1468–1486.

PISTOLE, D. H., AND J. A. CRANFORD. 1982. Photoperiodic effects of growth in *Microtus pennsylvanicus. J. Mammal.,* 63:547–553.

PITELKA, F. A. 1957a. Some aspects of population structure in the short term cycle of the brown lemming in northern Alaska. *Cold Spring Harbor Symp. Quant. Biol.,* 22:237–251.

———. 1957b. Some characteristics of microtine cycles in the Arctic. *Proc. 18th Biol. Coll.,* Oregon State College, Corvallis, pp. 73–88.

PITTENDRIDGH, C. S. 1966. The circadian oscillation in *Drosophila pseudoobscura* pupae: A model of the photoperiodic clock. *Z. Pfanzenphysiol,* 54:275–307.

PIVNICK, K. A., AND J. N. MCNEIL. 1986. Sexual differences in the thermoregulation of *Thymelicus lineola* adults (Lepidoptera: Hesperiidae). *Ecology,* 67:1024–1035

PLATTS, W. J., AND D. R. STRONG. 1989. Tree fall gaps and forest dynamics. *Ecology,* 70:535–576.

PLOWRIGHT, W. 1982. The effects of rinderpest and rinderpest control on wildlife in Africa. *Symp. Zool. Soc. Lond.,* 50:1–28.

POLICANSKY, D. 1982. Sex change in plants and animals. *Ann. Rev. Ecol. Syst.,* 13:471–495.

POLIS, G. 1981. The evolution of intraspecific predation. *Ann. Rev. Ecol. Syst.,* 12:225–251.

POLUNIN, N. 1955. Aspects of arctic botany. *Am. Sci.,* 43:307–322.

PODOLER, H., AND D. ROGERS. 1975. A new method for the identification of key factors from life table data. *J. Anim. Ecol.,* 44:85–114.

POMEROY, L. R. 1959. Algae productivity in salt marshes of Georgia. *Limnol. Oceanogr.,* 4:386–397.

POMEROY, L. R., AND E. J. KUENZLER. 1969. Phosphorous turnover by coral reef animals. In D. J. Nelson and F. E. Evans (eds.), *Symposium on Radioecology, Conf. 670503.* National Technical Information Services, Springfield, Va., pp. 478–483.

POMEROY, L. R., H. M. MATHEWS, AND H. SHIKMIN. 1963. Excretion of phosphate and soluble organic phosphorous compounds by zooplankton. *Limnol. Oceanogr.,* 4:50–55.

POMEROY, L. R., AND R. G. WIEGERT (EDS.). 1981. *The Ecology of a Salt Marsh.* Springer-Verlag, New York.

POOLE, R. W. 1974. *An Introduction to Quantitative Ecology.* McGraw-Hill, New York.

POORE, M. E. D. 1962. The method of successive approximation in descriptive ecology. *Adv. Ecol. Res.,* 2:35–68.

POORE, M. E. D. 1968 Studies in Malaysian rain forests. 1.

The forest on Triassic sediments in the Jenka forest reserve. *J. Ecol.*, 56:143–196.

POST, W. M., C. C. TRAVIS, AND D. L. DEANGELIS. 1980. In C. L. Cooke and S. Brisenberg (eds.), Evolution of mutualism between species. In *Differential Equations and Applications in Ecology, Epidemics, and Population Problems.* Academic Press, New York, pp. 183–201.

————. 1985. Mutualism, limited competition, and positive feedback. In D. H. Boucher (ed.), *The Biology of Mutualism.* Oxford University Press, New York, pp. 305–325.

POSTEL, S. 1984. Air pollution, acid rain, and the future of forests. *Worldwatch Paper*, 58:1–22, 44–49.

POTTS, G. R., S. C. TAPPE, AND P. J. HUDSON. 1984. Population fluctuations in red grouse: Analysis of bog records and a simulation model. *J. Anim. Ecol.*, 53:21–36.

POWELL, M. A., AND G. N. SOMERO. 1983. Blood components prevent sulfide poisoning of respiration of the hydrothermal vent tube worm *Riftia pachyptila*. *Science*, 219:297–299.

PREISTER, L. E. 1965. The accumulation in metabolism of DDT, parathion, and endrin by aquatic food chain organisms. Ph.D. thesis, Clemson University.

PRENTKI, R. T., T. D. GUFASON, AND M. S. ADAMS. 1978. Nutrient movements in lakeside marshes. In R. E. Good, D. F. Whigham, and R. L. Simpson (eds.), *Freshwater Wetlands: Ecological Processes and Management Potential.* Academic Press, New York. pp. 169–194.

PRESTON, F. W. 1948. The commonness and rarity of species. *Ecology*, 29:254–283.

PRESTON, F. W. 1960. Time and space and the variation of species. *Ecology*, 41:611–627.

————. 1962. The canonical distribution of commonness and rarity: Parts 1 and 2. *Ecology*, 43:185–215, 410–432.

PRICE, P. W., C. E. BOUTON, P. GROSS, B. A. MCPHERON, J. N. THOMPSON, AND E. E. WEIS. 1980. Interactions among three trophic levels: Influence of plants on interactions between insect herbivores and natural enemies. *Ann. Rev. Ecol. Syst.*, 11:41–65.

PRIMICK, R. B. 1979. Reproductive effort in annual and perennial species of *Plantago* (Plantaginaceae). *Am. Nat.*, 114:51–62.

PRITCHARD, D. W. 1952. Salinity distribution and circulation in the Chesapeake Bay estuarine system. *J. Marine Res.*, 11:106–123.

PROCTOR, J., AND S. R. J. WOODWELL. 1975. The ecology of serpentine soils. *Adv. Ecol. Res.*, 9:256–366.

PRUITT, W. O., JR. 1970 Some aspects of interrelationships of permafrost and tundra biotic communities. In *Productivity and Conservation in Northern Circumpolar Lands*, IUCN Publ. n.s. 10:33–41.

PUCKETT, L. J. 1982. Acid rain, air pollution, and tree growth in southeastern New York. *J. Environ. Qual.*, 11:376–381.

PUSEY, A. E., AND C. PACKER. 1986. The evolution of sex-biased dispersal in lions. *Behaviour*, 101:275–310.

PUTMAN, R. J. 1978a. Patterns of carbon dioxide evolution from decaying carrion: Decomposition of small mammal carrion in temperate systems. *Oikos*, 31:49–57.

————. 1978b. Flow of energy and organic matter from a carcass during decomposition: Decomposition of small mammal carrion in temperate systems. 2. *Oikos*, 31:58–68.

————. 1983. *Carrion and Dung: The Decomposition of Animal Wastes.* Edward Arnold, London.

PUTMAN, R. J., AND S. D. WRATTEN. 1984. *Principles of Ecology.* University of California Press, Berkeley.

PUTWAIN, P. D., AND J. L. HARPER. 1970. Studies of dynamics of plant populations: 3. The influence of associated species on populations of *Rumex acetosa* L. and *R. acetosella* L. in grassland. *J. Ecol.*, 58:251–264.

PUTWAIN, P. D., D. MACHIN, AND J. L. HARPER. 1968. Studies in the dynamics of plant population: 2. Components and

regulation of a natural population of *Rumex acetosella* L. *J. Ecol.*, 56:421–431.

RABATNOV, T. A. 1974. Differences between fluctuations and successions. In R. Knapp (ed.), *Vegetation Dynamics:* Part 8. *Handbook of Vegetation Science*, Junk. The Hague, Netherlands, pp. 21–24.

RALLS, K, P. H. HARVEY, AND M. A. LYLES. 1986. Inbreeding in natural populations of birds and mammals. In M. Soule (ed.), *Conservation Biology: The Science of Diversity.* Sinauer Associates, Sunderland, Mass., pp. 35–56.

RAMUS, J. 1983. A physiological test of the theory of complementary chromatic adaptation. II. Brown, green, and red seaweeds. *J. Phycol.*, 19:173–178.

RANDOLPH, S. E. 1975. Patterns of distribution of the tick *Ioxdes trianguliceps* Birula on its host. *J. Anim. Ecol.*

RANNEY, J. W. 1977. Forest island edges—their structure, development, and importance to regional forest ecosystem dynamics. *EDFB/IBP Cont. No. 77/1*, Oak Ridge National Laboratory, Oak Ridge, Tenn.

RANNEY, J. W., M. C. BRUNNER, AND J. B. LEVENSON. 1981. The importance of edge in the structure and dynamics of forest islands. In R. L. Burgess and D. M. Sharpe (eds.), *Forest Island Dynamics in Man-dominated Landscapes* (Ecological Studies No. 41). Springer-Verlag, New York, pp. 67–95.

RANWELL, D. S. 1961. *Spartina* salt marshes in southern England: 1. The effects of sheep grazing at the upper limits or *Spartina* marsh in Bridgwater Bay. *J. Ecol.*, 49:325–340.

RAO, S. S., A. A. JURKOVIC, AND O. NRIAGU. 1984. Bacterial activity in sediments of lakes receiving acid precipitation. *Environ. Pollution (A)*, 36:195–205.

RAPP, M., AND P. LOSSAINT. 1981. Some aspects of mineral cycling in the garrigue of southern France. In F. di Castri, D. Goodall, and R. L. Specht (eds.), *Mediterranean-type Shrublands* (Ecosystems of the World 14). Elsevier, Amsterdam, pp. 289–302.

RASMUSSEN, R. A., AND M. A. K. KAHLIL. 1986 Atmospheric trace gases: Trends and distribution over the last decade. *Science*, 232:1623–1624.

RATLIFF, R. D. 1982. A correction of Coles C_7 and Hurlburt's C_8 coefficients of interspecific association. *Ecology*, 63:1605–1606.

RATTI, J. T., AND K. P. REESE. 1988. Preliminary test of the ecological trap hypothesis. *J. Wildl. Manage.*, 52:484–491.

RAUNER, Y. L. 1972. *Heat Balance of the Vegetation Cover.* Gidrometeoizdat, Leningrad.

RAUNKIAER, C. 1934. *The Life Form of Plants and Statistical Plant Geography.* Clarendon Press, Oxford.

RAVEN, P. H. 1973. The evolution of Mediterranean flora. In F. di Castri and H. A. Mooney (eds.), *Mediterranean-type Ecosystems: Origin and Structure.* Springer-Verlag, New York, pp. 213–224.

RAWLINS, J. E. 1980. Thermoregulation by the black swallowtail butterfly *Papilo polypenes*. *Ecology*, 61:345–357.

RAYMONT, J. E. G. 1963. *Plankton and Productivity in Ocean.* Pergamon Press, Elmsford, N.Y.

RAYNOL, D. J. 1983. Atmospheric deposition and ionic input in Adirondack forests. *J. Air Pollution Control Assoc.*, 33:1032–1036.

REAL, L., AND T. CARACO. 1986. Risk and foraging in stochastic environments. *Ann. Rev. Ecol. Syst.*, 17:371–390.

REED, D. C., AND M. S. FOSTER. 1984. The effects of canopy shading on algal recruitment and growth in a giant kelp forest. *Ecology*, 65:937–948.

REEKIE, E. G., AND F. A. BAZZAZ. 1987. Reproductive efforts in plants. 3. Effect of reproduction on vegetative activity. *Am. Nat.*, 29:907–919.

REEM, C. H. 1976. Loon productivity, human disturbance, and pesticide residues in northern Minnesota. *Wilson Bull.*, 88:427–431.

REEMOLD, R. J. 1972. The movement of phosphorus through the salt marsh cord grass *Spartina alterniflora* Loisel. *Limnol. Oceanogr.*, 17:606–611.

REGEHR, D. L., AND F. A. BAZZAZ. 1976. Low temperature photosynthesis in successional winter annuals. *Ecology*, 57:1297–1303.

REGIER, H. A., AND K. H. LOFTUS. 1972. Effects of fisheries exploitation on salmonid communities in oligotrophic lakes. *J. Fish. Res. Board Can.*, 29:959–968.

REICHLE, D. E. 1971. Energy and nutrient metabolism of soil and litter invertebrates. In P. Duvigneaud (ed.), *Productivity of Forest Ecosystems.* UNESCO, Paris, pp. 465–477.

———. 1975. Advances in ecosystem analysis. *Bioscience*, 25:257–264.

REICHLE, D. E., (ED.). 1970. *Analysis of Temperate Forest Ecosystems, Ecological Studies 1.* Springer-Verlag, New York.

———. 1981. *Dynamic Properties of Forest Ecosystems.* Cambridge University Press, Cambridge, England.

REICHLE, D. E., ET AL. 1973. Carbon flow and storage in a forest ecosystem. In G. M. Woodwell and E. V. Pecan (eds.), *Carbon and the Biosphere, Conf. 72501.* National Technical Information Service, Springfield, Va., pp. 345–365.

REICHLE, D. E., AND D. A. CROSSLEY, JR. 1967. Investigation of heterotrophic productivity in forest insect communities. In K. Petrusewicz (ed.), *Secondary Productivity of Terrestrial Ecosystems.* Polish Academy of Sciences, Warsaw, pp. 563–587.

REICHLE, D. E., P. B. DUNAWAY, AND D. J. NELSON. 1970. Turnover and concentration of radionuclides in food chains. *Nuclear Safety*, 11:43–56.

REICHLE, D. E., R. A. GOLDSTEIN, R. I. VAN HOOK, AND G. J. DODSON. 1973. Analysis of insect consumption in a forest canopy. *Ecology*, 54:1076–1084.

REIFSNYDER, W. E., AND H. W. LULL. 1965. Radiant energy in relation to forests. *U.S.D.A. Tech. Bull. No. 1344.*

REIMOLD, R. J. 1972. Salt marsh ecology: The effects on marine food webs of direct harvest of marsh grass by man and the contribution of marsh grass to the food available to marine organisms. *Sea Grant Rept.,* University of Georgia, Athens.

REINERS, W. A. 1973. A summary of the world carbon cycle and recommendations for critical research. In G. M. Woodwell and E. Pecan (eds.), *Carbon and the Biosphere, Conf. 72501.* National Technical Information Service, Springfield, Va., pp. 368–382.

REITEMEIER, R. F. 1957. Soil potassium and fertility. *Yearbook of Agriculture.* U.S. Department of Agriculture, Washington, D.C., pp. 101–106.

REITER, R. J. 1981. The mammalian pineal gland: Structure and function. *Am. J. Anat.*, 162:287–313.

REX, M. A. 1981. Community structure in the deep-sea benthios. *Ann. Rev. Ecol. Syst.*, 12:331–354.

REY, J. R. 1981. Ecological biogeography of arthropods on *Spartina* islands in northwest Florida. *Ecol. Monogr.*, 51:237–265.

REYNOLDS, J. C. 1985. Details of the geographic replacement of the red squirrel *Sciurus vulgaris* by the gray squirrel *Sciurus carolinensis* in eastern England. *J. Anim. Ecol.*, 54:149.

REYNOLDS, D. N. 1984. Alpine annual plants: Phenology, germination, photosynthesis, and growth of three Rocky Mountain species. *Ecology*, 65:759–766.

RHOADES, D. F., AND R. G. CATES. 1976. A general theory of plant antiherbivore chemistry. *Recent Adv. Phytochemistry,* 10:168–213.

RICE, E. L. 1964. Inhibition of nitrogen-fixing and nitrifying bacteria by seed plants. *Ecology*, 45:824–837.

———. 1965. Inhibition of nitrogen-fixing and nitrifying bacteria by seed plants: 2. Characterization and identification of inhibitors. *Physiol. Plant*, 18:255–268.

———. 1972. Allelopathic effects of *Andropogon virginicus* and its persistence in old fields. *Am. J. Bot.*, 59:752–755.

RICE, E. L., AND R. L. PARENTI. 1967. Inhibition of nitrogen-fixing and nitrifying bacteria by seed plants. V. Inhibitors produced by *Bromus japonicus* Thunb. *Southwest Nat.*, 12:97–103.

RICH, P. H., AND R. G. WETZEL. 1978. Detritus in the lake ecosystem. *Am. Nat.*, 112:57–71.

RICHARDS, C. M. 1958. The inhibition of growth in crowded *Rana pipiens* tadpoles. *Physiol. Zool.*, 31:138–151.

———. 1962. The control of tadpole growth by algal-like cells. *Physiol. Zool.*, 35:285–296.

RICHARDS, P. W. 1952. *The Tropical Rain Forest.* Cambridge University Press, London.

RICKER, W. E. 1954. Stock and recruitment. *J. Fish. Res. Board Can.*, 11:559–623.

———. 1958a. Maximum sustained yields from fluctuating environments and mixed stocks. *J. Fish. Res. Board Can.*, 15:991–1006.

———. 1958b. Handbook of computations for biological statistics of fish populations. *Bull. 119, J. Fish. Res. Board Can.*, pp. 1–300.

RICKLEFS, R. 1979. *Ecology*, 2nd ed. Chiron Press, New York.

———. 1987. Community diversity: Relative roles of local and regional processes. *Science*, 235:167–171.

RIEBESELL, J. F. 1981. Photosynthetic adaptations in bog and alpine populations of *Ledum groenlandica. Ecology*, 62:579–586.

RIECHERT, S. E. 1981. The consequences of being territorial: Spiders, a case study. *Am. Nat.*, 117:871–892.

RIGBY, C., AND J. H. LAWTON. 1981. Species–area relationships of arthropods on host plants: Herbivores on bracken. *J. Biogeog.*, 8:125–133.

RIGLER, F. H. 1956. A tracer study of the phosphorus cycle in lake water. *Ecology*, 37:550–562.

———. 1964. The phosphorus fractions and turnover time of inorganic phosphorus in different types of lakes. *Limno. Oceanogr.*, 9:511–518.

———. 1973. A dynamic view of the phosphorus cycle in lakes. In E. J. Griffith et al. (eds.), *Environmental Phosphorus Handbook.* Wiley, New York, pp. 539–572.

RILEY, G. A. 1970. Particulate organic matter in sea water. *Adv. Marine Biol.*, 8:1–118.

RINGLER, N. 1979. Selective predation by drift-feeding brown trout *(Salmo trutta). J. Fish. Res. Bd. Can.*, 26:392–403.

RIPLEY, E. A., AND R. E. REDMANN. 1976. Grassland. In J. L. Monteith (ed.), *Vegetation and the Atmosphere, Vol. 2, Case Studies.* Academic Press, London, pp. 349–398.

RISEBROUGH, R. W., W. WALKER, T. T. SCHMIDT, B. W. DELAPPE, AND C. W. CONNERS. 1976. Transfer of chlorinated biphenyls to Antarctica. *Nature, Lond.*, 264:738–739.

RISSER, P. G., E. C. BIRNEY, H. D. BLOCKER, S. W. MAY, W. J. PARTON, AND J. A. WIENS. 1981. *The True Prairie Ecosystem* (US/IBP Synthesis Series 16). Hutchinson, Ross, Stroudsburg, Penn.

ROBINSON, M. H., AND B. ROBINSON. 1970. Prey caught by a sample population of the spider in Panama: A year's census data. *Zool. J. Linn. Soc.*, 49:345–358.

ROBINSON, S. K., AND R. T. HOLMES. 1984. Effects of plant species and foliage structure on the foraging behavior of forest birds. *Auk*, 101:672–684.

RODIN, L. E., AND N. I. BAZILEVIC. 1964. The biological productivity of the main vegetation types in the Northern Hemisphere of the Old World. *Forest. Abst.*, 27:369–372.

ROGERS, L. L. 1987. Effects of food supply and kinship on social behavior, movements, and population dynamics of black bears in northeastern Minnesota. *Wildl. Monogr.*, 97.

———. 1987. Factors influencing dispersal in black bears.

In B. D. Chepko-Sade and Z. T. Halpin (eds.), *Mammalian Dispersal Patterns.* University of Chicago Press, Chicago, pp. 75–84.

ROMESBURG, H. C. 1981. Wildlife science: Gaining reliable knowledge. *J. Wildl. Manage.,* 45:293–313.

ROMME, W. H., AND D. H. KNIGHT. 1982. Landscape diversity: The concept applied to Yellowstone Park. *Bioscience,* 32:664–670.

———. 1967. G. E. Fogg (ed.), *Production and Mineral Cycling in Terrestrial Vegetation* (transl. from Russian by Scripta Technica). Oliver & Boyd, London.

ROOT, R. B. 1967. The niche exploitation pattern of the blue gray gnatcatcher. *Ecol. Monogr.,* 37:317–350.

RORISON, I. H. (ED.). 1969. *Ecological Aspects of Mineral Nutrition of Plants.* Blackwell Scientific Publ., Oxford.

ROSEBERRY, J. L., AND W. D. KLIMSTRA. 1984. *Population Ecology of the Bobwhite.* Southern Illinois University Press, Carbondale.

ROSENZWEIG, M. L., AND R. H. MACARTHUR. 1963. Graphical representation and stability conditions of predator-prey interactions. *Am. Nat.,* 97:209–223.

ROSS, B. A., J. R BRAY, AND W. H. MARSHALL. 1970. Effects of a long-term deer exclusion on a *Pinus resinosa* forest in north-central Minnesota. *Ecology,* 51:1088–1093.

ROSS, H. H. 1970. The ecological history of the Great Plains—evidence from grassland insects. In W. Dort and J. K. Jones *Pleistocene and Recent Environments of the Central Great Plains. Spec. Publ. No. 3,* Dept. Biology, University of Kansas, Lawrence, pp. 225–240.

ROSS, J. 1982. Myxomatosis: The natural evolution of disease. *Symp. Zool. Soc. Lond.,* 50:77–95.

ROSSWALL, T., ET AL. 1975. Stordalen (Abisko) Sweden. In T. Rosswall and O. W. Heal (eds.), *Structure and Function of Tundra Ecosystems.* Swedish Natural Science Research Council, Stockholm, pp. 265–294.

ROSSWALL, T. AND U. GRANHALL. 1980. Nitrogen cycling in a subarctic ombrotrophic mire. In M. Sonesson (ed.), *Ecology of A Subarctic Mire Ecol. Bull. 30.* Swedish Natural Science Research Council, Stockholm.

ROUGHGARDEN, J. 1974. Species packing and the competition function with illustrations from coral reef fish. *Theor. Pop. Biol.,* 5:163–186.

———. 1986. A comparison of food-limited and space-limited animal competition communities. In J. Diamond and T. J. Case (eds.), *Community Ecology.* Harper & Row, New York, pp. 492–516.

ROWAN, W. R. 1925. Relation of flight to bird migration and developmental changes. *Nature,* 115:494–495.

———. 1929. Experiments in bird migration: 1. Manipulation of the reproductive cycle, seasonal histological changes in the gonads. *Proc. Boston Soc. Natur. Hist.,* 39:151–208.

ROWLAND, F. S. 1973. Mercury levels in swordfish and tuna. *Biol. Conser.,* 5:52–53.

ROYAMA, T. 1970. Factors governing the hunting behavior and selection of food by the great tit. *J. Anim. Ecol.,* 39:619–668.

RUDD, R. L. 1964. *Pesticides and the Living Landscape.* University of Wisconsin Press, Madison.

RUDD, R. L., AND R. E. GENELLY. 1956. Pesticides: Their use and toxicity in relation to wildlife. *Calif. Fish and Game Bull. No. 7.*

RUINEN, J. 1962. The phyllosphere: An ecologically neglected region. *Plant Soil,* 15:81–109.

RUNKLE, J. R. 1981. Gap regeneration in some old-growth forests of eastern United States. *Ecology,* 62:1041–1051.

———. 1984. Development of woody vegetation in treefall gaps in a beech-maple forest. *Holarctic Ecol.,* 7:157–164.

———. 1985. Disturbance regimes in temperate forests. In S. T. A. Pickett and P. S. White (eds.), *Ecology of Natural Disturbance and Patch Dynamics.* Academic Press, Orlando, Fla., pp. 17–34.

RUNKLE, J. R., AND T. C. YETTER. 1987. Treefalls revisited: Gap dynamics in the southern Appalachians. *Ecology,* 68:417–424.

RUNDEL, P. W. 1980. The ecological distribution of C_3 and C_4 grasses in the Hawaiian Islands. *Oecologica,* 45:354–359.

———. 1981. The mattoral zone of central Chile. In F. di Castri, D. Goodall, and R. Spetch (eds.), *Mediterranean-type Ecosystems.* Elsevier, Amsterdam, pp. 175–210.

RUSCH, D. H., E. C. MESLOW, P. D. DOERR, AND L. B. KEITH. 1972. Response of great horned owl populations to changing prey densities. *J. Wildl. Manage.,* 36:282–296.

RUTHERFORD, M. C. 1978. Primary production ecology in southern Africa. In M. J. Wegner (ed.), *Biogeography and Ecology of Southern Africa.* Junk, The Hague, Netherlands, pp. 621–659.

RYAN, D. F., AND F. H. BORMAN. 1982. Nutrient resorption in northern hardwood forests. *Bioscience,* 32:29–32.

RYTHER, J. H. 1969. Photosynthesis and fish production in the sea. *Science,* 166:72–75.

SABINE, W. S. 1959. The winter society of the Oregon junco: Intolerance, dominance, and the pecking order. *Condor,* 61:110–135.

SAFFO, M. B. 1987. New light on seaweeds. *Bioscience,* 37:654–664.

SAKAI, A. K., AND T. A. BURRIS. 1985. Growth in male and female aspen clones: A twenty-five year longitudinal study. *Ecology,* 66:1921–1927.

SALE, P. F. 1980. The ecology of fishes on coral reefs. *Oceanogr. Mar. Biol. Ann. Rev.,* 18:367–421.

SALISBURY, E. J. 1929. The biological equipment of species in relation to competition. *J. Ecol.,* 17:197–222.

SALT, G. W. 1957. An analysis of avifaunas in the Teton Mountains and Jackson Hole, Wyoming. *Condor,* 59:373–393.

SANDERS, H. L. 1968. Marine benthic diversity: A comparative study. *Am. Nat.,* 102:243–283.

SARGENT, R. C., AND M. R. GROSS. 1985. Parental investment decision rules and the Concorde fallacy. *Behav. Ecol. Sociobiol.,* 17:43–45.

SARGENT, R. C., P. D. TAYLOR, AND M. P. GROSS. 1987. Parental care and the evolution of egg size in fish. *Am. Nat.,* 129:32–46.

SARMIENTO, G. 1984. *The Ecology of Neotropical Savannas.* Harvard University Press, Cambridge, Mass.

SARMIENTO, G., AND M. MONASTERRIO. 1975. A critical consideration of environmental conditions associated with the occurrence of savanna ecosystems in tropical America. In F. B. Golly and E. Medina (eds.), *Tropical Ecological Systems.* Springer-Verlag, Berlin, pp. 223–250.

SARUKHAN, J. 1974. Studies on plant demography: *Ranunculus repens* L., *R. bulbosus* L., and *R. acris* L.: 2. Reproductive strategies and seed population dynamics. *J. Ecol.,* 62:151–177.

SARUKHAN, J., AND J. HARPER. 1974. Studies on plant demography: 1. Population flux and survivorship. *J. Ecol.,* 61:676–716.

SARUKHAN, J., M. MARTINEZ-RAMOS, AND D. PINERO. 1984. The analysis of demographic variability at the individual level and its population consequences. In R. Drizo and J. Sarukhan (eds.), *Perspectives in Plant Population Ecology.* Sinauer Associates, Sunderland, Mass., pp. 83–106.

SAUNDERS, D. S. 1982. *Insect Clocks,* 2nd ed. Pergamon Press, Oxford.

SCHAEFER, R. 1973. Microbial activity under seasonal conditions of drought in Mediterranean climates. In F. di Castri and H. A. Mooney (eds.), *Mediterranean-type Ecosystems; Origin and Structure.* Springer-Verlag, New York, pp. 191–198.

SCHAFFER, W. G. 1974. Selection for optimal life histories: The effects of age structure. *Ecology,* 55:291–303.

SCHAFFER, W. M. 1974. Optimal reproductive effort in fluctuating environments. *Am. Nat.*, 108:783–790.

———. 1981. Ecological abstraction: The consequences of reduced dimensionality in ecological models. *Ecol. Monogr.*, 51:383–401.

———. 1985. Order and chaos in ecological systems. *Ecology*, 66:93–106.

SCHAFFER, W. M., AND M. KOT. 1985. Differential systems in ecology and epidemiology. In A. V. Holder (ed.), *Chaos.* Manchester University Press, Manchester, England.

SCHALL, B. A. 1978. Age structure in *Liatris acidota*. *Oecologica*, 32:93–100.

SCHALLER, G. B. 1972. *Serengeti: A Kingdom of Predators.* Knopf, New York.

SCHARITZ, R. R., AND J. F. MCCORMICK. 1973. Population dynamics of two competing annual plant species. *Ecology*, 54:723–740.

SCHEFFER, V. C. 1951. The rise and fall of a reindeer herd. *Sci. Month.*, 73:356–362.

SCHELDERUP-EBBE, T. 1922. Beitrage zur Socialpsychologie des Haushuhns. *Zeitschr. Psychol.*, 88:225–252.

SCHEMISKE, D. W., AND T. BROKAW. 1981. Treefalls and the distribution of understory birds in a tropical forest. *Ecology*, 62:938–945.

SCHENKEL, R. 1948. Ausdrucksstudien an Wolfen. *Behaviour*, 1:81–130.

SCHLESINGER, W. H. 1977. Carbon balance in terrestrial detritus. *Ann. Rev. Ecol. Syst.*, 8:51–81.

SCHLICHTER, L. C. 1981. Low pH affects the fertilization and development of *Rana pipiens* eggs. *Can. J. Zool.*, 59:1693–1699.

SCHMID, A. 1982. Survival of frogs in low temperature. *Science*, 215:312–315.

SCHMIDT-NIELSEN, K. 1960. The salt secreting gland of marine birds. *Circulation*, 21:955–967.

———. 1964. *Desert Animals: Physiological Problems of Heat and Water.* Oxford University Press, London.

SCHODDEE, R. 1981. Bird communities of the Australian mallee: Composition, derivation, distribution, structure, seasonal cycles. In F. di Castri, D. Goodall, and R. Specht (eds.), *Mediterranean-type Shrublands* (Ecosystems of the World 11). Elsevier, Amsterdam, pp. 387–416.

SCHOENER, T. 1982. The controversy over interspecific competition. *Am. Sci.*, 70:586–595.

———. 1983. Simple models of optimal feeding-territory size: A reconciliation. *Am. Nat.*, 121:608–629.

———. 1983. Field experiments on interspecific competition. *Am. Nat.*, 122:240–285.

SCHOENER, T. W., AND D. A. SPILLER. 1987. Effect of lizards on spider populations: Manipulative reconstruction of a natural experiment. *Science*, 236:949–953.

SCHOENER, T. W., AND C. A. TOFT. 1983. Spider populations: Extraordinarily high densities on islands without top predators. *Science*, 21:1353–1355.

SCHOFIELD, C. L., AND J. R. TROJNAR. 1980. Aluminum toxicity to brook trout (*Salvelinus fontinalis*) in acidified waters. In T. Y. Toribara, M. W. Miller, and P. E. Morrow (eds.), *Polluted Rain.* Plenum Press, New York, pp. 341–365.

SCHULZE, E-D. 1989. Air pollution and forest decline in spruce *Picea abies* forest. *Science*, 244:776–783.

SCHOLANDER, P. F., R. HOCK, V. WALTERS, F. JOHNSON, AND L. IRVING. 1950. Heat regulation in some arctic and tropical birds and mammals. *Biol. Bull.*, 99:237–258.

SCHOLANDER, P. F., V. WALTERS, R. HOCK, L. IRVING, AND F. JOHNSON. 1950. Body insulation of some arctic and tropical mammals and birds. *Biol. Bull.*, 99:225–236.

SCHULTZ, V., AND A. W. KLEMENT (EDS.). 1963. *Radioecology.* Van Nostrand Rheinhold, New York.

SCHULZE, E. D., R. H. ROBICHAUX, J. GRACE, P. W. RUNDEL, AND J. R. EHLERINGER. 1987. Plant water balance. *Bioscience*, 37:30–37.

SCHWARTZ, S. E. 1989. Acid deposition: Unraveling a regional phenomenon. *Science*, 243:753–763.

SCIDENSLICKER, J. C., IV, M. G. HORNOCKER, W. V. WILES, AND J. P. MESSICK. 1973. Mountain lion social organization in the Idaho primitive area. *Wildl. Monogr.* 35:1–60.

SCLATER, P. L. 1858. On the general geographical distribution of the members of the class Aves. *J. Proc. Limnol. Soc. (Zool.)*, 2:130–145.

SCOTT, T. C. 1943. Some food coactions of the northern plains red fox. *Ecol. Monogr.*, 13:427–479.

———. 1955. An evaluation of the red fox. *Ill. Natur. Hist. Surv., Biol. Notes No. 35*, pp. 1–16.

SEARCY, W. A. 1979. Male characteristics and pairing success in red-winged blackbird. *Auk*, 96:353–363.

SEARCY, W. A., AND M. ANDERSSON. 1986. Sexual selection and the evolution of song. *Ann. Rev. Ecol. Syst.*, 17:507–534.

SEARS, P. B. 1935. *Deserts on the March.* University of Oklahoma Press, Norman.

SEGERSTRALE, S. G. 1947. New observations on the distribution and morphology of the amphipod *Gammarus zaddachi* Sexon, with notes on related species. *J. Marine Biol. Assoc. U.K.*, 27:219–244.

SELLECK, G. W. 1960. The climax concept. *Bot. Rev.*, 26:534–545.

SHAW, S. P., AND C. G. FREDINE. 1956. Wetlands of the United States. *U.S. Fish and Wildl. Circ. 39.*

SHEA, M. L., R. S. WARREN, AND W. A. NEIRING. 1975. Biochemical and transplantation studies of the growth form of *Spartina alterniflora* on Connecticut salt marshes. *Ecology*, 56:461–466.

SHELDON, W. G. 1967. *The Book of the American Woodcock.* University of Massachusetts Press, Amherst.

SHELFORD, V. E. 1913. Animal communities in temperate America. *Bull. Geog. Soc. Chicago*, 5:1–368.

SHELFORD, V. E., AND A. C. TWOMEY. 1941. Tundra animals in the vicinity of Churchill, Manitoba. *Ecology*, 22:47–69.

SHEPPARD, P. M. 1959. *Natural Selection and Heredity.* Hutchinson, London.

SHERMAN, P. W. 1977. Nepotism and the evolution of alarm calls. *Science*, 197:1246–1253.

———. 1981. Kinship, demography and Belding's ground squirrel nepotism. *Behav. Ecol. Sociobiol.*, 8:251–259.

SHIELDS, W. M. 1987. Dispersal and mating systems: Investigating their causal connections. In B. D. Chepho-Sade and Z. T. Halpin (eds.), *Mammalian Dispersal Patterns.* University of Chicago Press, Chicago, pp. 3–24.

SHUGART, H. H. 1984. *A Theory of Forest Dynamics: The Ecological Implications of Forest Succession.* Springer-Verlag, New York.

SHURE, D. J., AND H. S. RAGSDALE. 1977. Patterns of primary succession on granite outcrop surfaces. *Ecology*, 58:993–1006.

SIEBURTH, J. MCN., AND A. JENSEN. 1970. Production and transformation of extracellular organic matter from marine littoral algae: A resume. In D. E. Hood (ed.), *Organic Matter in Natural Waters.* University of Alaska, pp. 203–223.

SIEGENTHALER, U., AND H. OESCHGER. 1978. Predicting future atmospheric carbon dioxide levels. *Science*, 199:388–395.

SIMBERLOFF, D. S. 1974. Equilibrium theory of island biogeography and ecology. *Ann. Rev. Ecol. Sys.*, 5:161–182.

———. 1976. Species turnover and equilibrium island biogeography. *Science*, 194:572–578.

SIMBERLOFF, D. S., AND L. G. ABELE. 1976. Island biogeographic theory and conservation practice. *Science*, 191:285–286.

———. 1982. Refuge design and island biogeographic theory: Effects of fragmentation. *Am. Nat.*, 120:41–50.

———. 1984. Conservation and obfuscation: Subdivision of reserves. *Oikos*, 42:399–401.

SIMBERLOFF, D. S., AND E. O. WILSON. 1969. Experimental zoogeography of islands. The colonization of empty islands. *Ecology*, 50:278–296.

———. 1970. Experimental zoogeography of islands. A two-year record of colonization. *Ecology*, 50:278–296.

SIMPSON, G. G. 1964. Species density of North American recent mammals. *Syst. Zool.*, 13:57–73.

SIMS, P. L., AND J. S. SINGH. 1971. Herbage dynamics and net primary production in certain grazed and ungrazed grasslands in North America. In N. R. French (ed.), *Preliminary Analysis of Structure and Function in Grasslands*. Range Sci. Dept. Sci. Ser. No. 10. Colorado State University, Fort Collins, pp. 59–124.

SINCLAIR, A. R. E. 1977a. *The African Buffalo: A Study of Resource Limitation of Populations*. University of Chicago Press, Chicago.

SINCLAIR, A. R. E., AND J. N. M. SMITH. 1984. Do secondary compounds determine feeding preferences of snowshoe hares? *Oecologica*, 61:403–410.

SINGER, F. S. (ED.). 1970. *Global Effects of Environmental Pollution*. Springer-Verlag, New York.

SINGH, R. N. 1961. *Role of Blue-Green Algae in the Nitrogen Economy of Indian Agriculture*. Indian Council of Agricultural Research, New Delhi.

SKUTCH, A. 1986. *Helpers at Birds' Nests: A World-wide Survey of Cooperative Breeding and Related Behavior*. University of Iowa Press, Iowa City.

SLATKIN, M. 1987. Gene flow and the geographic structure of natural populations. *Science*, 236:787–792.

SLOBODKIN, L. B. 1962. *Growth and Regulation of Animal Population*. Holt, Rinehart, and Winston, New York.

SMALLEY, A. E. 1960. Energy flow of a salt marsh grasshopper population. *Ecology*, 41:672–677.

SMAYDA, T. J. 1983. The phytoplankton of estuaries. In B. Ketchum (ed.), *Estuaries and Enclosed Seas* (Ecosystems of the World 26). Elsevier, Amsterdam.

SMITH, A. D. M. 1985. A continuous time model: A deterministic model of temporal rabies. In P. J. Bacon (ed.), *Population Dynamics of Rabies in Wildlife*. Academic Press, London.

SMITH, B. D. 1974. Predator and prey relationships in southeastern Ozarks, A.D. 1300. *Human Ecol.*, 2:31–43.

SMITH, C. C., AND S. D. FRETWELL. 1974. The optimal balance between size and number of offspring. *Am. Nat.*, 108:499–506.

SMITH, J. N. M., AND H. P. A. SWEATMAN. 1974. Food searching behavior of titmice in patchy environments. *Ecology*, 55:1216–1232.

SMITH, M. 1974. Seasonality in mammals. In H. Leith (ed.), *Phenology and Seasonality Modeling*. Springer-Verlag, New York, pp. 149–162.

SMITH, M. H., R. K. CHESSER, E. C. COTHRAN, AND P. E. JOHNS. 1983. Genetic variability and antler growth in a natural population of white-tailed deer. In R. D. Brown (ed.), *Antler Development in Cervidae*. Caesar Kleberg Wildl. Inst. Kingsville, Tex.

SMITH, N. 1968. The advantage of being parasitized. *Nature*, 219:690–694.

SMITH, R. A. H., AND A. D. BRADSHAW. 1979. The use of heavy metal tolerant plant populations for the reclamation of metalliferous wastes. *J. Appl. Ecol.*, 16:595–612.

SMITH, R. E., AND B. A. HORWITZ. 1969. Brown fat and thermogenesis. *Physiol. Rev.*, 49:330–425.

SMITH, R. L. 1959a. Conifer plantations as wildlife habitat. *N.Y. Fish Game J.*, 5:101–132.

———. 1959b. The songs of the grasshopper sparrow. *Wilson Bull.*, 71:141–152.

———. 1962. Acorn consumption by white-footed mice (*Peromyscus leucopus*). *Bull. 482T, W. Va. Univ. Agr. Expt. Sta.*

———. 1963. Some ecological notes on the grasshopper sparrow. *Wilson Bull.*, 75:159–165.

———. 1966. Animals and the vegetation of West Virginia. In E. L. Core, *Vegetation of West Virginia*. McClain, Parsons, W. Va., pp. 17–24.

———. 1976. Socio-ecological evolution in the hill country of southwestern West Virginia. In J. Luchok et al. (eds.), *Hill Lands. Proc. International Symposium*, pp. 198–202.

———. 1977. Ecological genesis of endangered species: The philosophy of preservation. *Ann. Rev. Ecol. Syst.*, 7:33–55.

SMITH, S. M. 1978. The "underworld" in a territorial adaptive strategy for floaters. *Am. Nat.*, 112:570–582.

SMITH, T. M., AND P. GOODMAN. 1987. The effect of competition on the structure and dynamics of *Acacia* savannas in southern Africa. *J. Ecol.*, 75:1013–1044.

SMITH, T. M., AND K. GRANT. 1986. The role of competition in the spacing of trees in a *Burkea africana—Terminalia sericea* savanna. *Biotropica*, 18:219–223.

SMITH, T. M., AND H. H. SHUGART. 1987. Territory size variation in the ovenbird: The role of habitat structure. *Ecology*, 68:695–704.

SMITH, T. M., AND B. H. WALKER. 1983. The role of competition in the spacing of savanna trees. *Proc. Grassland Soc. S. Africa*, 18:159–164.

SMITH, W. 1976. Lead contamination of the roadside ecosystem. *J. Air Pollution Control Assoc.*, 26:753–766.

SMITH, W. H. 1981. *Air Pollution and Forests: Interactions Between Air Contamination and Forest Ecosystems*. Springer-Verlag, New York.

SMOCK, L. A., AND K. L. HARLOWE. 1983. Utilization and processing of freshwater wetland macrophytes by the detritivore *Asellus forbesi*. *Ecology*, 64:1156–1565.

SNOW, D. W. 1976. *The Web of Adaptation: Bird Studies in the American Tropics*. Quadrangle, New York.

SOLBRIG, O. 1970. *Principles and Methods of Plant Biosystematics*. Macmillan, New York.

SOLLINS, P. 1982. Input and decay of coarse woody debris in coniferous forest stands in western Oregon and Washington. *Can. J. For.*, 12:18–28.

SOLLINS, P., C. C. GRIER, F. M. MCCORSIN, K. CROMACK, JR., R. FOGEL, AND R. L. FREDRIKSEN. 1980. The internal element cycles of an old-growth Douglas-fir ecosystem in western Washington. *Ecol. Monogr.*, 50:275–282.

SOLOMON, A. M. 1986. Transient response of forests to CO_2 induced climate change: Simulation modeling experiments in eastern North America. *Oecologica*, 68:567–579.

SOLOMON, M. E. 1949. The natural control of animal populations. *J. Anim. Ecol.*, 18:1–32.

———. 1957. Dynamics of insect populations. *Ann. Rev. Entomol.*, 2:121–142.

SOLOMON, P. M., R. DE ZAFRA, A. PARRISH, AND J. W. BARRET. 1984. Diurnal variation of stratospheric chlorine monoxide: Critical test of chlorine chemistry in the ozone layer. *Science*, 224:1210–1214.

SOULÉ, M. E. 1986. *Conservation Biology: The Science of Scarcity and Diversity*. Sinauer Associates, Sunderland, Mass.

SOUSA, W. P. 1979. Disturbance in marine intertidal boulder fields: The nonequilibrium maintenance of species diversity. *Ecology*, 60:1225–1239.

———. 1984. Intertidal mosaics: Patch size, propagule availability, and spatially variable patterns of succession. *Ecology*, 65:1918–1935.

SOWLS, L. K. 1960. Results of a banding study of Gambel's quail in southern Arizona. *J. Wildl. Manage.*, 24:185–190.

SPAETH, J. N., AND C. H. DIEBOLD. 1938. Some interrelations between soil characteristics, water tables, soil temperature, and snow cover in the forest and adjacent open areas in south central New York. *Cornell Univ. Ag. Exp. Stat. Mem 213.*

SPECHT, R. L. 1973. Structure and functional response of ecosystems in the Mediterranean climate of Australia. In F. di Castri and H. A. Mooney (eds.), *Mediterranean Type Ecosystems, Origin and Structure.* Springer-Verlag, New York, pp. 113–120.

———. 1979. Heathlands and related shrublands of the world. In R. L. Specht (ed.), *Heathlands and Related Shrublands: Descriptive Studies* (Ecosystems of the World 9A). Elsevier, Amsterdam, pp. 1–18.

———. 1981. Mallee ecosystems in southern Australia. In F. di Castri, D. W. Goodall, and R. L. Specht (eds.), *Mediterranean-type Shrublands* (Ecosystems of the World 11). Elsevier, Amsterdam, pp. 203–231.

SPOONER, G. M. 1947. The distribution of *Gammarus* species in estuaries: Part 1. *J. Marine Biol. Assoc. U. K.,* 27:1–52.

SPRUGEL, D. G. 1976. Dynamic structure of wave generated *Abies balsamea* forests in north-eastern United States. *J. Ecol.,* 64:889–911.

SPURR, S. H. 1957. Local climate in the Harvard Forest. *Ecology,* 38:37–56.

STARK, N. 1972. Nutrient cycling pathways and litter fungi. *Bioscience,* 22:355–360.

———. 1973. *Nutrient Cycling in a Jeffey Pine Forest Ecosystem.* Montana Forest and Conservation Experiment Station, Missoula.

———. 1976. Fuel reduction and nutrient status and cycling relationships associated with understory burning in larch-Douglas-fir stands. In *Proc. Tall Timbers Fire Ecology Conf.,* 143:573–596.

STEARNS, S. C. 1976. Life history tactics: A review of ideas. *Quart. Rev. Biol.,* 51:3–47.

———. 1977. The evolution of life history traits. *Ann. Rev. Ecol. Syst.,* 8:145–171.

STECK, F. 1982. Rabies in wildlife. *Symp. Zool. Soc. Lond.,* 50:57–75.

1972. Evolution and diversity of arid-land shrubs. In C. McKell et al. (eds.), *Wildland Shrubs: Their Biology and Utilization.* U.S.D.A. Forest Serv. Gen. Tech. Rept. INT-1, pp. 111–116.

STEELE, J. H. 1974. *The Structure of Marine Ecosystems.* Harvard University Press, Cambridge, Mass.

STEGEMAN, L. C. 1960. A preliminary survey of earthworms of the Tully Forest in central New York. *Ecology,* 41:779–782.

STEINBERG, P. D. 1984. Algal defense against herbivores: Allocation of phenolic compounds in the kelp *Alaria marginala. Science,* 223:405–407.

STENGER, J., AND J. B. FALLS. 1959. The utilized territory of the ovenbird. *Wilson Bull.,* 71:125–140.

STENSETH, N. C. 1983. Causes and consequences of dispersal in small mammals. In I. R. Swingland and P. J. Greenwood (eds.), *The Ecology of Animal Movements.* Oxford University Press, Oxford, pp. 63–101.

STEPHENS, D. W. 1981. The logic of risk-sensitive foraging preferences. *Anim. Behav.,* 29:628–629

STEPHENSON, T. A., AND A. STEPHENSON. 1952. Life between tide-marks in North America: 2. North Florida and the Carolinas, *J. Ecol.,* 40:1–49

———. 1954. Life between the tide-marks in North America: 3A. Nova Scotia and Prince Edward Island: The geographical features of the region. *J. Ecol.,* 42:14–45, 46–70.

———. 1971. *Life Between the Tide-marks on Rocky Shores.* Freeman, San Francisco.

STEPONKUS, P. L. 1981. Responses to extreme temperatures: Cellular and subcellular bases. In O. Lange, P. S. Nobel, C. B. Osmund, and H. Zeigler (eds.), *Physiological Plant Ecology. I. Vol 12A. Encyclopedia of Plant Physiology,* pp. 371–402.

STERN, J. E., T. HOM, E. CASILLAS, A. FRIEDMAN, AND U.

VARANASI. 1987. Simultaneous exposure of English sole *(Parophrys vetulus)* to sediment-associated xenobiotics. Part 2. Chronic exposure to an urban estuarine sediment with added: ^{3}H-Benzo(a)pyrene and ^{14}C-polychlorinated biphenyls. *Marine Environ. Res.,* 22:123–151.

STERN, W. L., AND M. F. BUELL. 1951. Life-form spectra of New Jersey pine barren forest and Minnesota jack pine forest. *Bull. Torrey Bot. Club,* 78:61–65.

STEVENSON, J. R., AND E. F. STOERMER. 1982. Luxury consumption of phosphorus by benthic algae. *Bioscience,* 31:682–683.

STEWARD, G. A. 1970. High potential productivity of the tropics for cereal crops, grass forage crops, and beef. *J. Aust. Inst. Agr. Sci.,* 36:85.

STEWART, B. A., L. K. PORTER, F. G. VIETS. 1966a. Effects of sulfur content of straws on rates of decomposition and plant growth. *Soil Sci. Soc. Am Proc.,* 30:355–358.

———. 1966b. Sulfur requirements for decomposition of cellulose and glucose in soil. *Soil Sci. Soc. Am. Proc.,* 30:453–456.

STEWART, W. D. P. 1967. Nitrogen-fixing plants. *Science,* 158:1426–1432.

STILES, E. W. 1980. Patterns of fruit presentation and seed dispersal in bird disseminated woody plants in the eastern deciduous forest. *Am. Nat.,* 116:670–688.

———. 1982. Fruit flags: Two hypotheses. *Am. Nat.,* 120:500–509.

STILES, F. G. 1975. Ecology, flowering phenology, and hummingbird pollination of some Costa Rican *Heliconia* species. *Ecology,* 56:285–301.

STILES, F. G., AND L. L. WOLF. 1970. Hummingbird territoriality at a tropical flowering tree. *Auk,* 87:469–491.

STODDARD, H. 1932. *The Bobwhite Quail: Its Habits, Preservation and Increase.* Scribner, New York.

STOECKLER, J. H. 1962. Shelterbelt influence on Great Plains field environment and crops. *U.S.D.A. Prod. Res. Rept. No. 62.*

STOWE, L. G., AND J. A. TEERI. 1978. The geographic distribution of C_4 species of the Dicotyledonae in relation to relation to climate. *Am. Nat.,* 112:609–623.

STRAHLER, A. 1971. *The Earth Sciences.* Harper & Row, New York.

STRICKLAND, J. D. H. 1965a. Production of organic matter in the primary stages of the marine food chain. In J. P. Riley and G. Skirrow (eds.), *Chemical Oceanography, Vol. 1.* Academic Press, New York, pp. 477–610.

———. 1965b. Phytoplankton and marine primary production. *Ann. Rev. Microbiol.,* 19:127–162.

STRONG, D. R., J. H. LAWTON, AND R. SOUTHWARD. 1984. *Insects on Plants.* Harvard University Press, Cambridge, Mass.

STRONG, D. R., JR., AND J. R. REY. 1982. Testing for MacArthur-Wilson equilibrium with the arthropods of the miniature *Spartina* archipelago at Oyster Bay, Florida. *Am. Zool,* 22:350–360.

STRONG, D. R., D. SIMBERLOFF, L. G. ABELE, AND A. B. THISTLE (EDS.). 1986 *Ecological Communities: Conceptual Issues and the Evidence.* Princeton University Press, Princeton, N.J.

———. 1979. Atmospheric carbon dioxide in the nineteenth century. *Science,* 202:1109.

STRUMWASSER, F. 1960. Some physiological principles governing hibernation in *Citellus beecheyi.* In C. P. Lyman and A. R. Dawe (eds.), *Mammalian Hibernation.* Museum of Comparative Zoology, Harvard University, Cambridge, Mass., pp. 285–318.

STUDY OF CRITICAL ENVIRONMENTAL PROBLEMS. 1970. *Man's Impact on the Global Environment.* MIT Press, Cambridge, Mass.

STUDY OF MAN'S IMPACT ON CLIMATE. 1971. *Inadvertent Climate Modification.* MIT Press, Cambridge, Mass.

STUIVER, M. 1978. Atmospheric carbon dioxide and carbon reservoir changes. *Science,* 199:253–258.

B-37

SUBRAMANIAN, A. N., S. TANABE, H. TANEKA, H. HIDAKA, AND R. TATSUKAWA. 1987. Gain and loss rates and biological half-life of PCB's and DDE in the bodies of Adelie penguins. *Environ. Pollution*, 43:39–46.

SWANK, W. T., J. W. FITZGERALD, AND J. T. ASH. 1983. Microbial transformation of sulfate in forest soils. *Science*, 223:182–184.

SWANK, W. T., AND J. B. WAIDE. 1979. Interpretation of nutrient cycling research in a management context: Evaluating potential effects of alternative management strategies on site productivity. In R. Waring (ed.), *Forests: Fresh Perspectives from Ecosystem Analysis, Proc. Ann. Biol. Coll.* Oregon State University Press, Corvallis, pp. 137–158.

SWANK, W. T., J. B. WAIDE, D. A. CROSSLEY, AND R. L. TODD. 1981. Insect defoliation enhances nitrate export from forest ecosystems. *Oecologica*, 51:297–299.

SWIFT, M. J., O. W. HEAL, AND J. M. ANDERSON. 1979. *Decomposition in Terrestrial Ecosystems.* Blackwell, Oxford.

TABER, R. D., AND R. F. DASMANN. 1958. The black-tailed deer of the chaparral. *Calif. Dept. Fish Game, Game Bull. No. 8.*

TAHVANAINEN, J, E. HELLE, R. JULKUNEN-TITTO, AND A. LAVOLA. 1985. Phenolic compounds of willow bark as deterrents against feeding by mountain hare. *Oecologica*, 65:319–323.

TAIT, D. E. N. 1980. Abandonment as a reproductive tactic—the example of grizzly bears. *Am. Nat.*, 115:800–808.

TAIT, R. V. 1968. *Elements of Marine Ecology.* Plenum, New York.

TALLAMY, D. W. 1984. Insect parental care. *Bioscience*, 34:20–24.

TAMARIN, R. H. 1978. Dispersal, population regulation, and K-selection in field mice. *Am. Nat.*, 112:545–555.

TAMARIN, R. H., AND C. J. KREBS. 1969. Microtus population biology: 2. Genetic changes at the transferrin locus in fluctuating populations of two vole species. *Evolution*, 23:183–211.

TAMM, C. O. 1951. Removal of plant nutrients from tree crowns by rain. *Physiol. Plant.*, 4:184–188.

TANNER, J. T. 1975. The stability and intrinsic growth rates of prey and predator populations. *Ecology*, 56:855–867.

TANSLEY, A. G. 1935. The use and abuse of vegetational concepts and terms. *Ecology*, 16:284–307.

TANSLEY, A. G., AND R. S. ADAMSON. 1925. Studies on the vegetation of English chalk. III. The chalk grasslands of the Hampshire-Sussex border. *J. Ecol.*, 13:177–223.

TARRANT, R. F. 1971. Persistence of some chemicals in Pacific Northwest forest. In *Pesticides, Pest Control, and Safety on Forest Lands.* Continuing Education Books, Corvallis, Oregon, pp. 133–141.

TATUM, L. A. 1971. The southern corn leaf blight epidemic. *Science*, 171:1113–1116.

TAYLOR, C. R. 1969. The eland and the oryx, *Sci. Am.*, 220(1):88–95.

———. 1970a. Strategies of temperature regulation: Effect of evaporation on East African ungulates. *Am. J. Physiol.*, 219:1131–1135.

———. 1970b. Dehydration and heat: Effects on temperature regulation of East African ungulates. *Am. J. Physiol.*, 219:1136–1139.

TAYLOR, F. G., JR. 1974. Phenodynamics of production in a mesic deciduous forest. In H. Leith (ed.), *Phenology and Seasonality Modeling.* Springer-Verlag, New York, pp. 237–254.

TAYLOR, K. 1971. Biological flora of the British Isles. *Rubus chamaemorus* L. *J. Ecol.*, 59:293.

TAYLOR, R. J. 1984. *Predation.* Chapman & Hall, New York.

TAYLOR, W. R. 1961. Distribution in depth of marine algae in the Caribbean and adjacent seas. In *Recent Advances in Botany.* University of Toronto Press, Toronto, pp. 193–197.

TEAL, J. M. 1957. Community metabolism in a temperate cold spring. *Ecol. Monogr.*, 27:283–302.

———. 1962. Energy flow in the salt marsh ecosystem of Georgia. *Ecology*, 43:614–624.

TEAL, J. M., AND J. KANWISHER. 1961. Gas exchange in a Georgia salt marsh. *Limnol. Oceanogr.*, 6:388–399.

TEMPLE, S. A. 1977. The dodo and the tambalacoque tree. *Science*, 203:1364.

———. 1986. Predicting impacts of habitat fragmentation on forest birds: A comparison of two models. In J. Verner, M. L. Morrison, and C. T. Ralph (eds.), *Wildlife 2000: Modeling Habitat Relations of Terrestrial Vertebrates.* University of Wisconsin Press, Madison, pp. 301–304.

TERRI, J. A. 1979. The climatology of the C_4 photosynthetic pathway. In O. T. Solbrig, S. Jain, G. B. Johnson, P. H. Raven (eds.), *Topics in Plant Population Biology.* Columbia University Press, New York, pp. 356–374.

TERRI, J., AND L. STOWE. 1976. Climate patterns and distribution of C_4 grasses in North America. *Oecologia*, 23:1–12.

TESTER, J. R., AND W. H. MARSHALL. 1961. A study of certain plant and animal interrelations on a native prairie in Northwestern Minnesota. *Minn. Mus. Natur. Hist. Occasional Paper No. 8.*

THAYER, G. W., W. J. KENWORTHY, AND M. S. FONSECA. 1984. *The Ecology of Eelgrass Meadows of the Atlantic Coast.* U.S. Fish and Wildlife Service FWS/OBS-84/02.

THOMAS, J. W., R. G. ANDERSON, C. MASER, E. L. BULL. 1979. Snags. In J. W. Thomas (ed.), *Wildlife Habitats in Managed Forests* (The Blue Mountains of Oregon and Washington). U.S.D.A. Forest Serv. Ag. Handb. No. 553, pp. 60–77.

THOMAS, R. D. K., AND E. C. OLSON (EDS.). 1980. *A Cold Look at the Warm-blooded Dinosaurs.* AAAS Selected Symposia 28. Westview Press, Boulder, Colo.

THOMAS, J. W., C. MASER, AND J. E. RODICK. 1978. Edges—their interspersion, resulting diversity, and its measurement. In *Proceedings Workshop on Nongame Bird Habitat in Coniferous Forests of Western United States, U.S.D.A. Forest Ser, Gen. Tech. Rept. PNW-64,* pp. 91–100.

THOMPSON, H. V. 1953. The grazing behavior of the wild rabbit. *Br. J. Anim. Behav.*, 1:16–20.

———. 1954. The rabbit disease, myxomatosis. *Ann. Appl. Biol.*, 41:358–366.

THOMPSON, W. L. 1960. Agonistic behavior in the house finch: Part 1. Annual cycle and display patterns. *Condor*, 62:245–271.

THURSTON, J. M. 1969. The effect of liming and fertilizers on the botanical composition of permanent grassland and on the yield of hay. In I. H. Rorison (ed.), *Ecological Aspects of Mineral Nutrition of Plants.* Blackwell, Oxford, pp. 3–10.

TIEDEMANN, A. R., E. D. MCARTHUR, H. C. STUTZ, R. STEVENS, K. L. JOHNSON (COMPILERS). 1984. *Proceedings Symposia on the Biology of Atriplex and Related Chenopods.* U.S.D.A. For. Serv. Gen. Tech. Rep. INT-172.

TIESZEN, L. L. 1978. Photosynthesis in the principal Barrow, Alaska species: A summary of field and laboratory responses. In *Vegetation and Production Ecology of an Alaska Arctic Tundra.* Springer-Verlag, New York, pp. 241–268.

TIESZEN, L. L., P. C. MILLER, AND W. C. OECHEL. 1980. Photosynthesis. In J. Brown, P. C. Miller, L. L. Tieszen, and F. L. Bunnell (eds.), *The Arctic Ecosystem: The Coastal Tundra at Barrow, Alaska* (US/IBP Synthesis Series No. 12). Dowden, Hutchinson, and Ross, Stroudsburg, Penn., pp. 102–139.

TIESZEN, L. L., M. M. SENYIMBA, S. K. IMBAMBA, AND J. H. TROUGHTON. 1979. The distribution of C_3 and C_4 grasses and carbon isotope discrimination along an altitudinal and mositure gradient in Kenya. *Oecologica,* 37:337–350.

TILLMAN, D. 1980. Resources: A graphical-mechanistic approach to competition and predation. *Am. Nat.,* 116:362–393.

———. 1982. *Resource Competition and Community Structure.* Princeton University Press, Princeton, N.J.

———. 1985. The resource ratio hypothesis of succession. *Am. Nat.,* 125:827–852.

———. 1986. Evolution and differentiation in terrestrial plant communities: The importance of the soil resource-light gradient. In J. Diamond and T. Case (eds.), *Community Ecology.* Harper & Row, New York, pp. 359–380.

———. 1989. *Plant Strategies and the Dynamics and Structure of Plant Communities.* Princeton University Press, Princeton, N.J.

TILLMAN, D., M. MATTSON, AND S. LANGER. 1981. Competition and nutrient kinetics along a temperature gradient: An experimental test of a mechanistic approach to niche theory. *Limno. Oceanogr.,* 26:1020–1033.

TILLY, L. J. 1968. The structure and dynamics of Cone Spring. *Ecol. Monogr.,* 28:169–197.

TILZER, M. M., AND C. R. GOLDMAN. 1978. Importance of mixing, thermal stratification and light adaptation for phytoplankton productivity in Lake Tahoe (California, Nevada). *Ecology,* 59:810–821.

TINBERGEN, L. 1951. *The Study of Instinct.* Oxford University Press, New York.

———. 1960. The natural control of insects in pinewoods: 1. Factors influencing the intensity of predation by songbirds. *Arch. Neerl. Zool.,* 13:265–343.

TINKLE, D. W. 1969. The concept of reproductive effort and its relation to the evolution of life histories of lizards. *Am. Nat.,* 103:501–516.

TINKLE, D. W., AND R. E. BALLINGER. 1972. *Sceloporus undulatus,* a study of the intraspecific comparative demography of a lizard. *Ecology,* 53:570–585.

TINLEY, K. L. 1982. The influence of soil moisture balance on ecosystem patterns in southern Africa. In B. J. Huntley and B. H. Walker (eds.), *Ecology of Tropical Savannas.* Springer-Verlag, New York, pp. 175–192.

TOMANEK, G. W. 1969. Dynamics of mulch layer in a grassland ecosystem. In R. L. Dix and R. G. Beidleman (eds.), *The grassland ecosystem: A preliminary synthesis. Range Sci. Dept. Sci. Ser. No.* 2. Colorado State University, Fort Collins, pp. 225–240.

TOMASELLE, R. 1981. Main physiognomic types and geographic distribution of shrub systems related to Mediterranean climates. In F. di Castri, D. W. Goodall, and R. L. Specht (eds.), *Mediterranean-type Shrublands* (Ecosystems of the World 11). Elsevier, Amsterdam, pp. 95–106.

———. 1981b. Relations with other ecosystems: Temperate evergreen forests, coniferous forests, savannas, steppes, and desert shrubland. In F. di Castri, D. W. Goodall, and R. L. Specht (eds.), *Mediterranean-type Shrublands.* Elsevier, Amsterdam., pp. 123–136.

TOMBACH, D. F. 1982. Dispersal of whitebark pine seeds by Clark's nutcracker: A mutualism hypothesis. *J. Anim. Ecol.,* 51:451–467.

TOMLINSON, P. B. 1983. Structural elements of the rain forest. In F. B. Golley (ed.), *Tropical Rain Forest Ecosystems: Structure and Function* (Ecosystems of the World 14A). Elsevier, Amsterdam, pp. 9–28.

TORDOFF, H. B. 1954 Social organization and behaviour in a flock of captive, non-breeding red crossbills. *Condor,* 36:346–358.

TORNABENE, T. G., AND H. W. EDWARDS. 1972. Microbial uptake of lead. *Science,* 176:1334–1335.

TRABAUD, D. L. 1981. Man and fire: Impacts on Mediterranean vegetation. In F. di Castri, D. W. Goodall,

and R. L. Specht (eds.), *Mediterranean-type Shrublands* (Ecosystems of the World 11). Elsevier, Amsterdam, pp. 523–538.

TRACY, C. R. 1976. A model of the dynamic exchanges of water and energy between a terrestrial amphibian and its environment. *Ecol. Monogr.,* 46:293–326.

TRESHOW, M. 1970. *Environment and Plant Response.* McGraw-Hill, New York.

TRIMBLE, G. R., JR. 1973. The regeneration of central Appalachian hardwoods with emphasis on the effects of site quality and harvesting practice. *U.S.D.A. Forest Serv. Res. Paper NE-282.*

TRIMBLE, G. R., JR., ET AL. 1974. Some options for managing forest land in the central Appalachians. *U.S.D.A. Forest Serv. Gen. Tech. Rept. NE-12.*

TRIMBLE, G., JR., AND E. H. TYRON. 1966. Crown encroachment into openings cut into Appalachian hardwood stands. *J. For.,* 64:104–108.

TRISKA, F. J., AND K. CROMACH, JR. 1980. The role of wood debris in forest and streams. In R. H. Waring (ed.), *Forests: Fresh Perspectives from Ecosystem Analysis.* Oregon State University Press, Corvallis, pp. 171–190.

TRISKA, F. J., J. R. SEDELL, K. CROMACH, JR., S. V. GREGORY, AND F. M. MCCOUSON. 1984. Nitrogen budget for a small coniferous forest stream. *Ecol. Monogr.,* 54:119–140.

TRISKA, F. J., J. R. SEDELL, S. V. GREGORY. 1982. Coniferous forest streams. In R. L. Edmonds (ed.), *Analysis of Coniferous Forest Ecosystems in Western United States,* US/IBP Synthesis Ser. No. 14. Dowden, Hutchinson, and Ross, Stroudsburg, Penn., pp. 292–332.

TRUE, R. P., ET AL. 1960. Oak wilt in West Virginia. *West Va. Univ. Agr. Expt. Sta. Bull. 448T.*

TRYON, E. H., AND G. R. TRIMBLE, JR. 1969. Effect of distance from stand border on height of hardwood reproduction in openings. *W. Va. Acad. Sci. Proc.,* 41:125–132.

TULLER, B. F., JR. 1979. The management of foxes in New York State. *Conservationist,* 34(3):33–36.

TUREKIAN, K. K. (ED.) 1971. *The Late Cenozoic Glacial Ages.* Yale University Press, New Haven, Conn.

TURKINGTON, R. 1983. Leaf and flower demography of *Trifolium repens* L growth in a mixture with grasses. *New Phytol.,* 93:599–616.

TYRTIKOV, A. P. 1959. Perennially frozen ground. In *Principles of Geocryology,* Part I, *General Geocryology* (trans. from Russian by R. E. Brown). Nat. Res. Cun. Canada. Tech Trans., 1163(1964):399–421.

UDVARDY, M. D. F. 1958. Ecological and distributional analysis of North American birds. *Condor,* 60:50–66.

UHL, C., AND C. F. JORDAN. 1984. Succession and nutrient dynamics following forest cutting and burning in Amazonia. *Ecology,* 65:1467–1492.

UNDERWOOD, A. J. 1986. The analysis of competition by field experiments. In J. Kikkawa and D. J. Anderson (eds.), *Community Ecology: Pattern and Process.* Blackwell, Melbourne, pp. 240–268.

UNDERWOOD, A. J., AND E. J. DENLEY. 1984. Paradigms, explanations, and generalizations in models for the structure of intertidal communities on rocky shores. In D. R. Strong, Jr., D. Simberloff, L. G. Abele, and A. B. Thistle (eds.), *Ecological Communities: Conceptual Issues and the Evidence.* Princeton University Press, Princeton, N.J., pp. 151–180.

UNDERWOOD, A. J., E. J. DENLEY, AND M. J. MORAN. 1983. Experimental analyses of the structure and dynamics of mid-shore rocky intertidal communities in New South Wales. *Oecologica,* 56:202–219.

U.S.D.A. SOIL CONSERVATION SERVICE SOIL SURVEY STAFF. 1975. *Soil Taxonomy: A Basic System of Soil Classification for Making and Interpreting Soil Surveys.* Agr. Handbook 436. U.S. Department of Agriculture, Washington, D.C.

U.S. FISH AND WILDLIFE SERVICE. *See* L. M. COWARD, ET AL., 1979.

UYENOYAMA, M. K., AND M. W. FELDMAN. 1980. Theories of kin and group selection in large and small populations in fluctuating environments. *Theor. Popul. Biol.*, 17:380–414.

VADAS, R. L. 1977. Pereferential feeding: An optimization strategy in sea urchins. *Ecol. Monogr.*, 47:337–371.

VALIELA, I., AND J. M. TEAL. 1979. The nitrogen budget of a salt marsh ecosystem. *Nature*, 20:652–656.

VALIELA, I., J. M. TEAL, AND W. G. DENSER. 1978. The nature of growth forms in salt marsh grass *Spartina alterniflora*. *Am. Nat.*, 112:461–470.

VALLENTYNE, J. R. 1957. The principles of modern limnology. *Am. Sci.*, 45:218–244.

———. 1974. *The Algal Bowl—Lakes and Man*. Misc. Spec. Publ. 22. Department of Environment, Ottawa, Canada.

VANCE, B. D., AND W. DRUMMOND. 1969. Biological concentration of pesticides by algae. *J. Am. Water Works Assoc.*, 61:360–362.

VAN DER HAMMER, T., T. A. WIJMSTRA, AND W. H. ZAGWIGN. 1971. The floral record of late Cenozoic of Europe. In K. K. Turekian (ed.), *The Late Cenozoic Glacial Ages*. Yale University Press, New Haven, Conn., pp. 391–424.

VANDERMEER, J. H. 1972. Niche theory. *Am. Rev. Ecol. Syst.*, 3:107–132.

———. 1980. Indirect mutualism: Variations on a theme by Stephen Levine. *Am. Nat.*, 116:441–448.

VANDERMEER, J. H., AND D. H. BOUCHER. 1978. Varieties of mutualistic interaction in population models. *J. Theor. Biol.*, 74:594–558.

VAN DER VALK, A. G., AND C. B. DAVIS. 1978. The role of seed banks in the vegetation dynamics of prairie glacial marshes. *Ecology*, 59:322–335.

VANNOTE, R. L., G. W. MINSHALL, K. W. CUMMINS, J. R. SCHELL, AND C. E. CUSHING. 1980. The river continuum concept. *Can. J. Fish. Aq. Sci.*, 37:130–137.

VAN HOOKE, R. I. 1971. Energy and nutrient dynamics of spider and orthopteran populations in a grassland ecosystem. *Ecol. Monogr.*, 41:1–26.

VAN VALEN, L. 1965. Morphological variation and the width of the ecological niche. *Am. Nat.*, 99:377–389.

VASEK, F. C., AND L. J. LUND. 1980. Soil characteristics associated with a primary plant succession on a Mojave Desert dry lake. *Ecology*, 61:1013–1018.

VAUGHN, M. R., AND L. B. KEITH. 1981. Demographic response of experimental snowshoe hare populations to overwinter food shortage. *J. Wildl. Manage.*, 45:354–380.

VAUGHN, T. A. *Mammalogy*. Saunders, Philadelphia.

VEALE, P. T., AND H. L. WASCHER. 1956. Henderson County soils. *Illinois Univ. Agr. Expt. Stat. Soil Rept. No. 77*.

VEITH, G. D., D. W. KUEHL, F. A. PUGLISI, G. F. GLASS, AND J. G. EATON. 1977. Residues of PCB's and DDT in the western Lake Superior Ecosystem. *Arch. Environ. Contam. Toxicol.*, 5(4):487–499.

VERNER, J., M. L. MORRISON, AND C. J. RALPH. 1986. *Wildlife 2000: Modeling Habitat Relationships of Terrestrial Vertebrates*. University of Wisconsin Press, Madison.

VIESSMAN, W., JR. 1966. The hydrology of small impervious areas. *Water Resources Res.*, 2:405–412.

VIRO, P. J. 1953. Loss of nutrients and the natural nutrient balance of the soil in Finland. *Comm. Inst. Forest. Fenn.*, 42:1–50.

VITOUSAK, P. M., AND J. M. MELILLO. 1979. Nutrient losses from disturbed forests: Patterns and mechanisms. *For. Sci.*, 25:605–619.

VITOUSAK, P. M., AND R. L. SANFORD, JR. 1986. Nutrient cycling in moist tropical forests. *Ann. Rev. Ecol. Syst.*, 17:131–167.

VOGEL, S. 1969. Flowers offering fatty oil instead of nectar. *XI Proc. Intl. Bot. Congress*, Seattle, p. 229.

VOIGT, G. K. 1960. Distribution of rain under forest stands. *For. Sci.*, 6:2–10.

———. 1971. Mycorrhizae and nutrient mobilization. In E. Hacskaylo (ed.), *Mycorrhizae*, *U.S.D.A. Forest Serv. Misc. Pub. No. 1189*, pp. 122–131.

VOIGHT, W. 1948. *The Road to Survival*. William Sloan, New York.

VOLTERRA, V. 1926. Variation and fluctuations of the numbers of individuals in animal species living together. Reprinted in R. M. Chapman (1931), *Animal Ecology*. McGraw-Hill, New York, pp. 409–448.

WAGGONER, P. E. 1984. Agriculture and carbon dioxide. *Am. Sci.*, 72:179–184.

WAGNER, F. H., AND L. C. STODDART. 1972. Influence of coyote predation on black-tailed jack rabbit populations in Utah. *J. Wildl. Manage.*, 36:329–342.

WAIDE, J. B., W. H. CASHEY, R. I. TODD, AND L. R. BORING. 1988. Changes in soil nitrogen pools and transformations following forest clear-cutting. In W. T. Swank and D. A. Crossley (eds.), *Forest Hydrology and Ecology at Coweeta*. Springer-Verlag, New York.

WALBOT, V., AND C. A. CULLIS. 1985. Rapid genomic changes in higher plants. *Ann. Rev. Plant Physiol.*, 36:367–396.

WALDMAN, J. M., J. W. MUNGER, D. J. JACOB, R. C. FLAGAN, J. J. MORGAN, AND M. R. HOFFMAN. 1982. Chemical composition of acid fog. *Science*, 218:677–680.

WALES, B. A. 1972. Vegetation analysis of north and south edges in a mature oak-hickory forest. *Ecol. Monogr.*, 42:451–471.

WALKER, J. C. 1980. The oxygen cycle. In O. Hutzinger (ed.), *The Natural Environment and the Biogeochemical Cycles*. Springer-Verlag, New York, pp. 87–104.

———. 1984. How life affects the atmosphere. *Bioscience*, 43:486–491.

WALKER, L. R., J. C. ZASADA, F. S. CHAPIN, III. 1986. The role of life history processes in primary succession on Alaskan floodplain. *Ecology*, 67:1243–1253.

WALKER, P. C., AND R. T. HARTMAN. 1960. Forest sequence of the Hartstown bog area in western Pennsylvania. *Ecology*, 41:461–474.

WALKER, R. B. 1954. Ecology of serpentine soils. II. Factors affecting plant growth on serpentine soils. *Ecology*, 35:259–274.

WALKINGSHAW, L. 1983. *Kirtland's Warbler*. Cranbrook Institute of Science, Bloomfield Hills, Mich.

WALLACE, A. R. 1876. *The Geographical Distribution of Animals*, 2 vols. Macmillan, London.

WALLACE, B. 1968. *Topics in Population Genetics*. Norton, New York.

WALLACE, J. B., J. R. WEBSTER, AND T. F. COFFNEY. 1982. Stream detritus dynamics regulation by invertebrate consumers. *Oecologica*, 53:197–200.

WALLER, D. M. 1982. Jewelweed's sexual skills. *Nat. History*, 91(5):32–39.

WALTER, E. 1934. Grundlagen der allegmeinen fisherielichen Produktionslehre. *Handb. Binnenfisch. Mitteleur.*, 14:480–662.

WALTER, H. 1977. *Ecology of Tropical and Subtropical Vegetation*. Oliver and Boyd, Edinburgh.

WALTER, H. 1979. *Vegetation of the Earth and Ecological Systems of the Geosphere*, 2nd ed. Springer-Verlag, New York.

WARDLE, P. 1968. Englemann spruce (*Picea engelmannii Engel.*) at its upper limits on the Front Range, Colorado. *Ecology*, 49:483–495.

WARING, R. H., AND J. F. FRANKLIN. 1979. Evergreen coniferous forests of the Pacific Northwest. *Science*, 204:1380–1386.

WARNER, R. E. 1968. The role of introduced diseases in the extinction of the endemic Hawaiian avifauna. *Condor*, 70:101–120.

WASER, N. M., AND L. A. REAL. 1979. Effective mutualism between sequentially flowering plant species. *Nature*, 281:670–672.

WASER, P. M. 1985. Does competition drive dispersal? *Ecology*, 66:1170–1175.

WASSINK, E. C. 1968. Light energy conversion in photosynthesis and growth of plants. In F. E. Eckhardt (ed.), *Functioning of Terrestrial Ecosystems of the Primary Production Level*. UNESCO, Paris, pp. 53–66.

WATERHOUSE, F. L. 1955. Microclimatological profiles in grass cover in relation to biological problems. *Quart. J. Roy. Meteorol. Soc.*, 81:63–71.

WATERS, T. F. 1972. The drift of stream insects. *Ann. Rev. Entomol.*, 17:253–272.

WATKINSON, A. R. 1986. Plant population dynamics. In M. J. Crawley (ed.), *Plant Ecology*. Blackwell, Oxford, pp. 137–184.

WATSON, A., AND D. JENKINS 1968. Experiments on population control by territorial behaviour in red grouse. *J. Anim. Ecol.*, 37:595–614.

WATSON, A., AND R. MOSS. 1970. Dominance, spacing behaviour and aggression in relation to population limitation in vertebrates. In A. Watson (ed.), *Animal Populations in Relation to Their Food Resources*. Blackwell, Oxford, pp. 167–218.

———. 1971. Spacing as affected by territorial behavior, habitat, and nutrition in red grouse *(Lagopus l. scoticus)*. In A. E. Esser (ed.), *Behavior and Environment*. Plenum, New York, pp. 92–111.

———. 1972. A current model of population dynamics in red grouse. In *Proc. 15th Inter. Ornithol. Congr.*, pp. 134–149.

WATT, A. S. 1947. Pattern and process in the plant community. *J. Ecol.*, 35:1–22.

WATT, K. E. F. 1968. *Ecology and Resource Management: A Quantitative Approach*. McGraw-Hill, New York.

WEAVER, G. T. 1975. The quantity and distribution of four nutrient elements in high elevation forest ecosystems, Balsam Mountains, North Carolina. In F. G. Howell, J. B. Gentry, and M. H. Smith (eds.), *Mineral Cycling in Southeastern Ecosystems*. National Technical Information Service, U.S. Dept. Commerce, pp. 715–728.

WEAVER, J. A., AND F. C. CLEMENTS. 1938. *Plant Ecology*. McGraw-Hill, New York.

WEAVER, J. E. 1954. *North American Prairie*. Johnson, Lincoln, Neb.

WEAVER, J. E., AND F. W. ALBERTSON. 1956. *Grasslands of the Great Plains: Their Nature and Use*. Johnson, Lincoln, Neb.

WEAVER, J. E., AND N. W. ROWLAND. 1952. Effects of excessive natural mulch on development, yield and structure of native grassland. *Botan. Gaz.*, 114:1–19.

WEAVER, M., AND M. KELLMAN. 1981. The effects of forest fragmentation on woodlot tree biotas in Southern Ontario. *J. Biogeogr.*, 8:199–210.

WEBBER, P. J., P. C. MILLER, F. S. CHAPIN, III, AND B. H. MCCOWN. 1980. The vegetations: Pattern and succession. In J. Brown, P. C. Miller, L. L. Trieszen, F. L. Bunnell (eds.), *An Arctic Ecosystem: The Coastal Tundra at Barrow Alaska*. Dowden, Hutchinson, and Ross, Stroudsburg, Penn.

WEBER, C. A. 1907. Aufbau und Vegetation der Moore Norddeutschlands. *Bot. Jahrb.*, 40, *beibl.*, 90:19–34.

WEBSTER, J. 1956–1957. Succession of fungi on decaying cocksfoot culms: Parts 1 and 2. *J. Ecol.*, 44:517–544; 45:1–30.

WEBSTER, J. R., J. B. WAIDE, AND B. C. PATTEN. 1975. Nutrient recycling and the stability of ecosystems. In F. G. Howell, J. B. Gentry, and M. H. Smith (eds.), *Mineral Cycling in Southeastern Ecosystems*. National Technical Information Service, U.S. Dept. Commerce, pp. 1–27.

WECKER, S. C. 1963. The role of early experience in habitat selection by the prairie deer mouse, *Peromyscus maniculatus bairdi. Ecol. Monogr.*, 33:307–325.

WEEKS, H. P., JR., AND C. M. KIRKPATRICK. 1976. Adaptations of white-tailed deer to naturally occurring sodium deficiencies. *J. Wildl. Manage.*, 40:610–625.

———. 1978. Salt preferences and sodium drive phenology in fox squirrels and woodchuck. *J. Mamm.*, 59:531–542.

WEIR, B. J., AND I. W. ROWLANDS. 1973. Reproductive strategies in mammals. *Ann. Rev. Ecol. Syst.*, 4:139–163.

WEIR, J. S. 1972. Spatial distribution of elephants in an African national park in relation to environmental sodium. *Oikos*, 23:1–13.

WEISE, C. M. 1974. Seasonality in birds. In H. Leith (ed.), *Phenology and Seasonality Modeling*. Springer-Verlag, New York, pp. 139–147.

WEIS-FOGH, T. 1948. Ecological investigations of mites and collembola in the soil. *Nat. Jutland*, 1:135–270.

WELCH, P. S. 1952. *Limnology*. McGraw-Hill, New York.

WELLER, D. E. 1987a. A reevaluation of the −3/2 power rule of plant self-thinning. *Ecol. Monogr.*, 57:23–43.

———. 1987b. Self-thinning exponent correlated with allometric measures of plant geometry. *Ecology*, 68:813–821.

———. 1989. The interspecific size-density relationship among crowded plant stands and its implications for the −3/2 power rule of self-thinning. *Am. Nat.*, 133:20–41.

WELLER, M. W. 1959. Parasitic egg laying the redhead *(Aythya americana)* and other North American Anatidae. *Ecol. Monogr.*, 29:333–365.

———. 1981. *Freshwater Marshes*. University of Minnesota Press, Minneapolis.

WELLER, M. W., AND L. H. FREDRICKSON. 1974. Avian ecology of a managed glacial marsh. *The Living Bird*, 12:269–291.

WELLINGTON, G. W. 1982a. An experimental analysis of the effects of light and zooplankton on coral zonation. *Oecologia*, 52:311–320.

———. 1982b. Depth zonation of corals in the Gulf of Panama: Control and facilitation by resident reef fishes. *Ecol. Monogr.*, 52:223–241.

WELLS, H. G., J. S. HUXLEY, G. P. WELLS. 1939. *The Science of Life Book 6*, Part V. Garden City Publishing, Garden City, N.Y.

WELLS, H. W. 1961. The fauna of oyster beds, with special reference to the salinity factor. *Ecol. Monogr.*, 31:239–266.

WELLS, K. B. 1977. The social behavior of anuran amphibians. *Anim. Behav.*, 25:666–693.

WELLS, L. 1960. Seasonal abundance and vertical movements of planktonic crustaceans in Lake Michigan. *U.S.D.I. Fish. Bull.*, 60 (172):343–369.

WENDEL, G. 1987. Abundance and distribution of vegetation under four hardwood stands in north-central West Virginia. *U.S.D.A. Northeast For. Ext. Stat. Res. Paper NE-607*.

WENT, F. W., AND N. STARK. 1968. Mycorrhiza. *Bioscience*, 18(11):1035–1039.

WERNER, E. E., AND D. J. HALL. 1976. Niche shift in sunfishes: Experimental evidence and significance. *Science*, 191:404–406.

WESSEL, M. A., AND A. DOMINSKI. 1977. The children's daily lead. *Am. Sci.*, 65:294–298.

WEST, N. E. 1979. Formation, distribution, and function of plant litter in desert ecosystems. In D. W. Goodall and R. A. Perry (eds.), *Arid-Land Ecosystems: Structure, Functioning and Management*, Vol. 1. Cambridge University Press, London, pp. 647–659.

———. 1981. Nutrient cycling in desert ecosystems. In D. W. Goodall, R. A. Perry, K. M. W. Howes (eds.), *Arid-Land Ecosystems: Structure, Functioning and Management* II. Cambridge University Press, Cambridge, England, pp. 301–324.

WEST, N. E., AND J. J. SKUJINS (EDS.). 1978. *Nitrogen in Desert Ecosystems* (US/IBP Synthesis Series 9). Dowden, Hutchinson, and Ross, Stroudsburg, Penn.

WEST, N. E., AND P. T. TUELLER. 1972. Special approaches to studies of competition and succession in shrub communities. In C. M. McKell, J. P. Blaisdell, and J. R.

Goodin, *Wildland Shrubs: Their Biology and Utilization.* U.S.D.A. Forest Service Gen. Tech. Rept INT-1, pp. 172–181.

WESTMORELAND, D., AND L. B. BEST. 1987. What limits mourning dove nests to a clutch of two? *Condor,* 89:489–493.

WETZEL, R. G., AND H. L. ALLEN. 1970. Function and interactions of dissolved organic matter and the littoral zone in lake metabolism and eutrophication. In Z. Kabap and A. Hillbricht-Ilkowska (eds.), *Productivity Problems of Freshwater.* PWN Polish Sci. Publ., Warsaw, pp. 333–347.

WETZEL, R. G., P. H. RICH, M. C. MILLER, AND H. L. ALLEN. 1972. Metabolism of dissolved and particulate detrital carbon in a temperate hardwater lake. *Mem. 1st Ital. Idrobiol. 29 (Supp.),* pp. 185–243.

WHEELRIGHT, N. T., AND C. H. JANSON. 1985. Colors of fruit displays of bird-dispersed plants in two tropical forests. *Am. Nat.,* 126:777–799.

WHICKER, A. D., AND J. K. DETLING. 1988. Ecological consequences of prairie dog disturbances. *Bioscience,* 38:778–785.

WHIGHAM, D. F., J. MCCORMICK, R. E. GOOD, AND R. L. SIMPSON. 1978. Biomass and primary production of freshwater tidal marshes. In R. E. Good, D. F. Whigham, and R. L. Simpson (eds.), *Freshwater Wetlands: Ecological Processes and Management Potentials.* Academic Press, New York, pp. 243–257.

WHITCOMB, R. F., J. F. LYNCH, P. A. OPLER, AND C. S. ROBBINS. 1977. Long term turnover and effects of selective logging on the avifauna of forest fragments. *Am. Birds,* 31(1):17–23.

WHITCOMB, R. F., J. F. LYNCH, P. A. OPLER, AND C. S. ROBBINS. 1976. Island biogeography and conservation: Strategy and limitations. *Science,* 193:1030–1032.

WHITCOMB, R. F., C. F. ROBBINS, J. F. LYNCH, B. L. WHITCOMB, M. K. KLIMKIEWIEZ, AND D. BYSTRAK. 1981. Effects of forest fragmentation on avifauna of the eastern deciduous forest. In R. L. Burgess and D. M. Sharpe (eds.), *Forest Island Dynamics in Man-Dominated Landscapes.* Springer-Verlag, New York, pp. 125–206.

WHITE, J. 1979. The plant as a metapopulation. *Ann. Rev. Ecol. Syst.,* 10:109–145.

WHITE, J. 1980. Demographic factors in populations of plants. In O. T. Solbrig (ed.), *Demography and Evolution in Plant Populations.* Blackwell, Oxford, pp. 21–48.

———. 1984. Plant metamerism. In R. Divizo and J. Sarukhán (eds.), *Perspectives in Plant Population Ecology.* Sinauer Associates, Sunderland, Mass., pp. 15–47.

WHITEHEAD, D. R. 1967. Studies of full-glacial vegetation and climate in southeastern United States. In E. J. Cushing and H. E. Wright (eds.), *Quaternary Paleoecology.* Yale University Press, New Haven, Conn., pp. 237–248.

WHITHAM, T. G. 1980. The theory of habitat selection: Examined and extended using *Pemphigus* aphids. *Am. Nat.,* 115:449–466.

WHITMORE, T. C. 1984. *Tropical Rain Forest of the Far East,* 2nd ed. Oxford University Press, London.

WHITTAKER, R. H. 1952. A study of summer foliage insect communities in the Great Smoky Mountains. *Ecol. Monogr.,* 22:1–44.

———. 1953. A consideration of the climax theory: The climax as a population and pattern. *Ecol. Monogr.,* 23:41–78.

———. 1954. The ecology of serpentine soils. IV. The vegetational response to serpentine soils. *Ecology,* 35:275–288.

———. 1956. Vegetation of the Great Smoky Mountains. *Ecol. Monogr.,* 26:1–80.

———. 1960. Vegetation of the Siskiyou Mountains, Oregon and California. *Ecol. Monogr.,* 30:279–338.

———. 1961. Estimation of net primary production of forest and shrub communities. *Ecology,* 42:177–183.

———. 1962. Classification of natural communities. *Bot. Rev.,* 28:1–239.

———. 1963. Net production of heath balds and forest heaths in the Great Smoky Mountains. *Ecology,* 44:176–182.

———. 1965. Dominance and diversity in land plant communities. *Science,* 147:250–260.

———. 1967. Gradient analysis of vegetation. *Biol. Rev.,* 42:207–264.

———. 1970a. *Communities and Ecosystems,* Macmillan, New York.

———. 1970b. The biochemical ecology of higher plants. In E. Sondheimer and J. B. Simeone (eds.), *Chemical Ecology.* Academic Press, New York, pp. 43–70.

———. 1972. Evolution and the measurement of species diversity. *Taxon,* 21:213–251.

———. 1974. Climax concepts and recognition. In R. Knapp (ed.), *Vegetation Dynamics.* W. Junk, The Hague, pp. 137–154.

———. 1975. *Communities and Ecosystems,* 2nd ed. Macmillan, New York.

WHITTAKER, R. H., F. H. BORMANN, G. E. LIKENS, AND T. G. SICCAMA. 1974. The Hubbard Brook ecosystem study: Forest biomass and production. *Ecol. Monogr.,* 44:233–252.

———. 1975. The biosphere and man. In R. H. Whittaker and H. Leith (eds.), *Primary Productivity of the Biosphere.* Springer-Verlag, New York, pp. 305–328.

WHITTAKER, R. H., AND P. R. FEENEY. 1971. Allelochemics: Chemical interactions between species. *Science,* 171:757–770.

WHITTAKER, R. H., S. A. LEVIN, AND R. B. ROOT. 1973. Niche, habitat, and ecotope. *Am. Nat.,* 107:321–338.

WHITTAKER, R. H., AND G. E. LIKENS. 1973. Carbon in the biota. In G. M. Woodwell and E. V. Pecan (eds.), *Carbon and the Biosphere Conf. 72501.* National Technical Information Service Springfield, Va., pp. 281–300.

WHITTAKER, R. H., R. B. WALKER, AND A. R. KRUCKEBERG. 1954. The ecology of serpentine soils. *Ecology,* 35:258–288.

WHITTAKER, R. H., AND G. M. WOODWELL. 1969. Structure, production, and diversity of the oak-pine forest at Brookhaven, New York. *J. Ecol.,* 57:155–174.

WHYTE, R. O. 1968. *Grasslands of the Monsoon.* Praeger, New York.

WIEBES, J. T. 1979. Coevolution of figs and their insect pollenators. *Ann. Rev. Ecol. Syst.,* 10:1–12.

WIEGANDT, E. 1976. Past and present in the Swiss Alps. In J. Luchok, J. D. Cawthon, and M. J. Breslen (eds.), *Hill Lands Proc. Int. Symp.,* West Virginia University, Ofc. Publications: Morgantown, pp. 203–208.

WIEGERT, R. G. 1964. Population energetics of meadow spittlebugs as affected by migration and habitat. *Ecol. Monogr.,* 34:217–241.

———. 1965. Energy dynamics of the grasshopper populations in old field and alfalfa field ecosystems. *Oikos,* 16:161–176.

WIEGERT, R. G., AND D. F. OWEN. 1971. Trophic structure, available resources, and population density in terrestrial vs. aquatic ecosystems. *J. Theoret. Biol.,* 30:69–81.

WIELGOLASKI, F. E. 1975a. Productivity of tundra ecosystems. In National Academy of Science, *Fennoscandian Tundra Ecosystems Part 1, Plants and Microorganisms.* Springer-Verlag, New York, pp. 1–12.

———. 1975b. Primary productivity of alpine meadow communities. In F. E. Wielgolaski (ed.), *Fennoscandian Tundra Ecosystems Part 1, Plants and Microorganisms.* Springer-Verlag, New York, pp. 121–128.

WIELGOLASSI, F. E., AND S. KJELVIK. 1975. Energy content and use of solar radiation of Fennoscandian tundra plants. In F. E. Wielgolaski (ed.), *Fennoscandian Tundra*

Ecosystems: Part 1, Plants and Microorganisms, Springer-Verlag, New York, pp. 201–207.

WIELGOLASKI, F. E., S. KJELVIK, AND P. KALHIO. 1975. Mineral content of tundra and forest tundra plants in Fennoscandia. In F. E. Wielgolaski (ed.), *Fennoscandian Tundra Ecosystems: Part 1, Plants and Microorganisms,* Springer-Verlag, New York, pp. 316–332.

WIENS, J. A. 1972. *Ecosystem Structure and Function, Proc. 31st Ann. Biol. Colloq.* Oregon University Press, Salem.

———. 1973. Pattern and process in grassland bird communities. *Ecol. Monogr.,* 43:237–270.

———. 1975. Avian communities, energetics, and functions in coniferous forest habitats. In *Proc. Symp. Manage. Forest Range Habitats for Nongame Birds. U.S.D.A. Forest Serv. Gen. Tech. Rept. WO-1.*

———. 1976. Populations responses to patchy environments. *Ann. Rev. Ecol. Syst.,* 7:81–120.

———. 1977. On competition and variable environments. *Am. Sci.,* 65:590–597.

———. 1985. Habitat selection in variable environments: Shrub-steppe birds. In M. Cody (ed.), *Habitat Selection in Birds.* Academic Press, Orlando, Fla., pp. 227–251.

WIENS, J. A., J. F. ADDICOTT, T. J. CASE, J. DIAMOND. 1986. Overview of the importance of spatial and temporal scale in ecological investigations. In J. Diamond and T. Case (eds.), *Community Ecology.* Harper & Row, New York, pp. 229–239.

WIENS, J. A., AND J. T. ROTENBERRY. 1981. Habitat associations and community structure of birds in shrub-steppe environments. *Ecol. Managr.,* 51:21–41.

WIGGINS, I. L., AND J. H. THOMAS. 1962. A flora of the Alaskan arctic slope. *Publ. Arctic Inst. N. Amer. No. 4.* University of Toronto Press, Toronto.

WILCOX, B. A. 1980. Insular ecology and conservation. In M. E. Soulé and B. A. Wilcox (eds.), *Conservation Biology: An Evolutionary-Ecological Perspective.* Sinauer Associates, Sunderland, Mass.

WILCOVE, D. S. 1985. Nest predation in forest tracts and the decline of migratory songbirds. *Ecology,* 66:1211–1214.

WILCOVE, D. S., C. H. MCLELLAN, AND A. P. DOBSON. 1986. Habitat fragmentation in the temperate zone. In M. Soulé (ed.), *Conservation Biology.* Sinauer Associates, Sunderland, Mass., pp. 237–256.

WILEY, R. H. 1974. Evolution of social organization and life history patterns among grouse (Aves: Tetraonidae). *Quart. Rev. Biol.,* 49:207–227.

WILLIAMS, C. B. 1964. *Patterns in the Balance of Nature.* Academic Press, New York.

———. 1966. *Adaptation and Natural Selection.* Princeton University Press, Princeton, N.J.

WILLIAMS, C. E. 1965. Soil fertility and cottontail body weights: A reexamination. *J. Wildl. Manage.,* 28:329–337.

WILLIAMS, C. E., AND A. I. CASHEY. 1965. Soil fertility and cottontail fecundity in southeastern Missouri. *Am. Midl. Nat.,* 74:211–224.

WILLIAMS, G. C. 1966. *Adaptation and Natural Selection.* Princeton University Press, Princeton, N.J.

———. 1975. *Sex and Evolution* Princeton University Press, Princeton N.J.

WILLIAMS, K. S., AND L. E. GILBERT. 1981. Insects as selective agents on plant vegetative morphology: Egg mimicry reduces egg laying by butterflies. *Science,* 212:467–469.

WILLIAMSON, M. 1981. *Island Populations.* Oxford University Press, Oxford.

WILLIAMSON, P. 1978. Above-ground primary production of chalk grassland allowing for leaf death. *J. Ecology,* 64:1059–1075.

WILLIS, A. J. 1963. Braunton burrows: The effects on vegetation of the addition of mineral nutrients to the dune soils. *J. Ecol.,* 51:353–374.

WILLIS, E. O. 1984. Conservation, subdivision of reserves, and antidismemberment hypothesis. *Oikos,* 42:396–398.

WILLOUGHBY, L. C. 1974. Decomposition of litter in freshwater. In C. H. Dickinson and G. J. F. Pugh (eds.), *Biology of Plant Litter Decomposition,* Vol. II. Academic Press, London, pp. 659–661.

WILLSON, M. F. 1979. Sexual selection in plants. *Am. Nat.,* 113:777–790.

———. 1983. *Plant Reproduction Physiology.* Wiley, New York.

WILLSON, M. F., AND N. BURLEY. 1983. *Mate Choice in Plants: Tactics, Mechanisms, and Consequences.* Princeton University Press, Princeton, N.J.

WILSON, D. 1975. A theory of group selection. *Proc. Nat. Acad. Sci.,* 72:143–146.

———. 1977. Structured demes and the evolution of group-advantageous traits. *Am. Nat.,* 111:157–185.

———. 1979. Structured demes and trait-group variation. *Am. Nat.,* 113:606–610.

———. 1980. *The Natural Selection of Populations and Communities.* Benjamin Cummings, Menlo Park, Calif.

———. 1983. The group selection controversy: History and current status. *Ann. Rev. Ecol. Syst.,* 14:159–187.

WILSON, E. O. 1971. Competitive and aggressive behavior. In J. Eisenberg and W. Dillom (eds.), *Man and Beast: Comparative Social Behavior.* Smithsonian Institute Press, Washington, D.C., pp. 522–533.

———. 1975. *Sociobiology.* Harvard University Press, Cambridge, Mass.

WILSON, E. O., AND W. BOSSERT. 1977. *A Primer of Population Ecology.* Sinauer Associates, Sunderland, Mass.

WILSON, E. O., AND E. O. WILLIS. 1975. Applied biogeography. In M. L. Cody and J. Diamond (eds.), *Ecology and Evolution of Communities.* Cambridge University Press Cambridge, England, pp. 522–533.

WINSTON, F. W. 1956. The acorn microsere with special reference to arthropods. *Ecology,* 37:120–132.

WITHERSPOON, J. P., S. I. AVERBACH, AND J. S. OLSON. 1962. Cycling of Cesium-134 in white oak trees on sites of contrasting soil type and moisture. *Oak Ridge Natur. Lab.,* 3328:1–143.

WITKAMP, M. 1963. Microbial populations of leaf litter in relation to environmental conditions and decomposition. *Ecology,* 44:370–377.

WITKAMP, M., AND D. A. CROSSLEY. 1966. The role of arthropods and microflora on the breakdown of white oak litter. *Pedabiologia,* 6:293–303.

WITKAMP, M., AND J. S. OLSON. 1963. Breakdown of confined and nonconfined oak litter. *Oikos,* 14:138–147.

WOLF, D. D., AND D. SMITH. 1964. Yield and persistence of several legume-grass mixtures as affected by cutting frequency and nitrogen fertilization. *Agron. J.,* 56:130–133.

WOLF, H. W., AND R. G. CATES. 1987. Site and stand characteristics. In M. H. Brookes et al. (coordinators), *Western Spruce Budworm, U.S.D.A. For. Serv. Tech. Bull. No. 1694.* Cooperative State Research Service, Washington, D.C.

WOLF, L. L., F. G. STILES, AND F. R. HAINSWORTH. 1976. Ecological organization of a highland tropical hummingbird community. *J. Anim. Ecol.,* 45:349–379.

WOLFE, J. N., R. T. WAREHAM, AND H. T. SCOFIELD. 1949. Microclimates and macroclimates of Neotoma, a small valley in central Ohio. *Ohio Biol. Survey Bull. No. 41*

WOLFF, J. O. 1980. The role of habitat patchiness in the population dynamics of snowshoe hares. *Ecol. Monogr.,* 50:111–130.

WOLFSON, A. 1959. The role of light and darkness in the regulation of spring migration and reproductive cycles in birds. In R. B. Withrow (ed.), *Photoperiodism and Related Phenomena in Plants and Animals,* Pub. No. 55. American Association for the Advancement of Science, Washington, D.C., pp. 679–716.

B-43

WOOD, D. M., AND R. DELMORAL. 1987. Mechanisms of early primary succession in subalpine habitats on Mount St. Helens. *Ecology*, 68:780–790.

WOOD-GUSH, D. G. M. 1955. The behaviour of the domestic chicken: A review. *Br. J. Anim. Behav.*, 3:81–110.

WOODWARD, I. 1987. *Climate and Plant Distribution.* Cambridge University Press, Cambridge, England.

WOODWELL, G. M., AND D. B. BOTKIN. 1970. Metabolism of terrestrial ecosystems by gas exchange techniques: The Brookhaven approach. In D. E. Reichle (ed.), *Analysis of Temperate Forest Ecosystems, Ecological Studies 1.* Springer-Verlag, New York, pp. 73–85.

WOODWELL, G. M., P. P. CRAIG, AND H. A. JOHNSON. 1971. DDT in the biosphere: Where does it go? *Science*, 174:1101–1107.

WOODWELL, G. M., AND W. R. DYKEMAN. 1966. Respiration of a forest measured by carbon dioxide accumulations during temperature inversions. *Science*, 154:1031–1034.

WOODWELL, G. M., R. A. HOUGHTON, C. A. S. HALL, D. E. WHITNEY, R. A. MOLL, AND D. W. JUERS. 1979. The Flax Pond ecosystem study: The annual metabolism and nutrient budgets of a salt marsh. In R. L. Jeffries and A. J. Davies (eds.), *Ecological Processes in Coastal Environments.* Blackwell, Oxford, pp. 491–511.

WOODWELL, G. M., AND T. G. MARPLES. 1968. The influence of chronic gamma radiation on the production and decay of litter and humus in an oak-pine forest. *Ecology*, 49:456–465.

WOODWELL, G. M., AND E. V. PECAN (EDS.). 1973. *Carbon and the Biosphere, Conf. 72501.* National Technical Information Service, Springfield, Va.

WOODWELL, G. M., AND R. H. WHITTAKER. 1968. Primary productivity in terrestrial ecosystems. *Am. Zool.*, 8:19–30.

WOODWELL, G. M., R. H. WHITTAKER, W. A. REINERS, G. E. LIKENS, C. C. DELWICHE, AND D. B. BOTKIN. 1978. The biota and the world carbon cycle. *Science*, 199:141–145.

WOODWELL, G. M., C. F. WORSTER, JR., AND P. A. ISAACSON. 1967. DDT residues in an east coast estuary: A case of biological concentration of a persistent pesticide. *Science*, 156:821–823.

WOOLFENDEN, G. 1973. Nesting and survival in a population of Florida scrub jays. *Living Bird*, 12:25–49.

———. 1975. Florida scrub jay helpers at the nest. *Auk*, 92:1–15.

WOOLFENDEN, G. E., AND J. W. FITZPATRICK. 1984. *The Florida Scrub Jay: Demography of a Cooperatively Breeding Bird.* Princeton University Press, Princeton, N.J.

WOOLPY, J. H. 1968. The social organization of wolves. *Natur. Hist.*, 77(5):46–55.

WRIGHT, H. E., JR. 1970 Vegetational history of the Great Plains. In W. Dort and J. K. Jones (eds.), *Pleistocene and Recent Environments of the Central Great Plains, Spec. Pub. No. 3.* Dept. Biology, University of Kansas, Laurence; pp. 157–172.

WRIGHT, H. E., JR., AND D. G. FREY (EDS.). 1965. *The Quaternary of the U.S.* Princeton University Press, Princeton, N.J.

WRIGHT, R. F., ET AL. 1976. Impact of acid precipitation on freshwater systems in Norway. *Water, Air, Soil Pollution* 6(2,3,4,):483–499.

WRIGHT, R. T. 1970. Glycollic acid uptake by plankton bacteria. In D. Wood (ed.), *Organic Matter in Natural Waters, Occ. Pub. No. 1.* Institute Marine Science, University of Alaska.

WRIGHT, S. 1931a. Evolution in Mendelian populations. *Genetics*, 16:97–159.

———. 1931b. Statistical theory of evolution. *Am. Statist. J.*, March Suppl., pp. 201–208.

———. 1935. Evolution in population in approximate equilibrium. *J. Genet.*, 30:243–256.

WRIGHT, S. 1945. Tempo and mode in evolution: A critical review. *Ecology*, 26:415–419.

———. 1978. *Evolution and the Genetics of Populations. Vol 4. Variability Within and Among Populations.* University of Chicago Press, Chicago.

WURSTER, C. F., JR. 1968. DDT reduces photosynthesis by marine plankton. *Science*, 159:1477–1475.

———. 1969. Chlorinated hydrocarbon insecticides and the world ecosystem. *Biol. Conser.*, 1:123–129.

WYNNE-EDWARDS, V. C. 1962. *Animal Dispersion in Relation to Social Behavior.* Hafner, New York.

———. 1963. Intergroup selection in the evolution of social systems. *Nature*, 200:623–628.

———. 1965. Self-regulating system in populations of animals. *Science*, 147:1543–1548.

YAHNER, R. H. 1984. Effects of habitat patchiness created by a ruffed grouse management plan on breeding bird communities. *Am. Mid. Nat.*, 96:179–194.

———. 1986. Structure, seasonal dynamics, and habitat relationships of avian communities in small, even-aged forest stands. *Wilson Bull.*, 98:61–82.

YARIE, J. 1981. Forest fire cycles and life tables: A case study from interior Alaska. *Can. J. For. Res.*, 11:554–562.

YARRANTON, G. A., AND R. G. MORRISON. 1974. Spatial dynamics of a primary succession: Nucleation. *J. Ecol.*, 62:417–428.

YODA, K., T. KIRA, H. OGAWA, AND K. HOZUMI. 1963. Self-thinning in overcrowded pure stands under cultivated and natural conditions. *J. Biol. Osaka Univ.*, 14:107–129.

YODZIS, P. 1981a. The connectance of real ecosystems. *Nature*, 284:544–545.

———. 1981b. The structure of assembled communities. *J. Theor. Biol.*, 92:103–117.

———. 1982. The compartmentation of real and assembled communities. *Am. Nat.*, 120:551–570.

———. 1983. Community assembly, energy flow, and food web structure. In D. L. DeAngelis, W. M. Post, and G. Sugihara (eds.), *Current Trends in Food Web Theory.* ORNL 5983, Oak Ridge National Laboratory, Oak Ridge, Tenn., pp. 41–44.

———. 1988 *An Introduction to Theoretical Ecology,* Harper & Row, New York.

YONGE, C. M. 1949. *The Sea Shore.* Collins, London.

YONGE, C. M. 1963. The biology of coral reefs. *Adv. Mar. Biol.*, 1:209–260.

ZACH, R., AND J. B. FALLS. 1976a. Ovenbird hunting behavior in a patchy environment: An experimental study. *Can. J. Zool.*, 54:1863–1879.

———. 1976b. Foraging behavior, learning, and exploration by captive ovenbirds. *Can. J. Zool.*, 54:1880–1893.

———. 1976c. Do ovenbirds hunt by expectation? *Can. J. Zool.*, 54:1894–1903.

ZAK, B. 1964. Role of mycorrhizae in root disease. *Ann. Rev. Phytopathol.*, 2:377–392.

ZAMMUTO, R. M., AND J. S. MILLAR. 1985. Environmental predictability, variability, and *Spermophilus columbianus* life history over an environmental gradient. *Ecology*, 66:1784–1794.

———. 1985b. A consideration of bet hedging in *Spermophilus columbianus. J. Mamm.*, 66:652–660.

ZARET, T. M., AND R. T. PAINE. 1973. Species introduction in a tropical lake. *Science*, 182:449–455.

ZEDLER, P. H., C. R. GAUTIER, AND G. S. MCMASTER. 1983. Vegetation change in response to extreme events: The effect of a short interval between fires in California chaparral and coastal shrub. *Ecology*, 64:809–818.

ZEEVALKING, H. S., AND L. F. M. FRESCO. 1977. Rabbit

grazing and diversity in a dune area. *Vegatatio,* 35:193–196.

ZELLER, D. 1961. Certain mulch and soil characteristics of major range sites in western North Dakota as related to range conditions. MA thesis, North Dakota State University, Fargo.

ZIEMAN, J. C. 1982. The ecology of the seagrasses of South Florida: A community profile. U.S. Fish and Wildlife Service program FWS/OBS-82/25.

ZIMEN, E. 1978. *The Wolf: A Species in Danger.* 1981 translation. Dell, New York.

ZIMMERMAN, J. L. 1971. The territory and its density dependent effect in *Spiza americana. Auk,* 88:591–612.

CREDITS

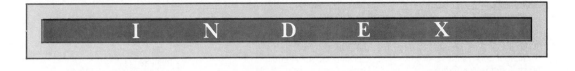

Note: Boldfaced numbers indicate pages with tables.

Biomass (*Continued*)
  and detritus, **254**
  grassland, 750
  plant, 379–383
    and net primary production, **217**
  pyramid of, 234–235, **236**
  of tropical forest, 807–808
  of tundra, 824–825
  vertical profile of, 212–214
Biomass yield, 536
Biomes, 718–720
Biophagy, 483
Biosphere, 26, **27**
Biotemperature, Holdridge's definition
    of, 722–723
Biotic provinces, 717–718
Birds
  biological clocks in, 135
  clutch size, 422–424
  of coniferous forest, 783–785
  and DDT, 280, 281
  density and fecundity, 377–378
  distraction displays, 528
  floating reserves, 412–414
  and habitat selection, 446–448
  interspecific competition among, 459
  maturity at hatching, 424–425
  migration in, 145–147, 335–337
  nesting, 147
  niche boundaries of, 475, **476**
  and niche overlap, 478, **479, 480**
  number of species, and island area,
    611–612
  peck order, 403, **404, 406**
  and resource partitioning, 468–469
  responses to light, 138–139
  seasonality in, 145–147
  sex ratios of, 360
  and size of forests, 618–620
  temperature regulation in, 111–112
  territoriality of, 406
  territory, quality of, 411–412
  of tundra, 818–819
  water balance of, 92–93
Black grass, 890
Black grub, **555**
Black spruce, 682
Black zone, 900
Blanket mire, 866, **873**
Blowflies, 229
  sheep, intraspecific competition in,
    374–376
Bogs, 870–871
Bollworm, diapause in, 138
Boron, 154
Botfly, 569–570
Bottleneck, of population, 316–317
Breeding dispersal, 396
British thermal unit (BTU), 209
Brown fat, 113
Browsing, 623
Buffalo, water balance of, 93
Buffer system, 204
Bumblebee, temperature regulation in,
    114–116
Bunch grass, 742–743
Bünning model, of biological clock,
    135, 135–136
Butterfly
  checkerspot, 384
  mimicry in, 526
  monarch, migration of, 335

Cactus, prickly pear, 508–509
Calcicoles, 156–157
Calcification, 184–186
Calcifuges, 156–157
Calcium, 154
Calcium carbonate, 205
Calvin–Benson cycle, 32, 34
Cannibalism, 483, 534–536
Capillary water, 79
Carbon, 152–153
  cycle of, 249–258
    global pattern, 251–255
    in lakes, 839–840, **844**
  immobilization of, 149
  –nutrient ratio, 43, **44**
  recycling of, in symbiosis, **576**
Carbonate ions (CO_3^{2+}),
    152–153
Carbon dioxide, 152
  cycle of, 249–258
    intrusions into, 255–257
  and decomposition, 149
  global compartments of, **253**
  levels, and hibernation, 117–118
  local patterns of, 251
  in water, 204–205
Caribou, 827
  migration of, **336**
Carnivores, 224
  of desert, 775
  of tundra, 818, 827
Carr, 866
Carrying capacity, 368
Catabolism, 38
Catena, 187
Cation exchange capacity, 178
Cesium-134, tracing nutrient cycle by,
    165–167
Cesium-137
  cycle of, **166,** 166–167
  in human food chain, **285**
Chamaephytes, 591, **592**
Chaparral, 762–763, 767–768
Characteristic species, 630
Chemical ecology, 8
Chernobyl accident, 282, 284
Cheetah, 316–317
Chistogamous flower, 438
Chlorinated hydrocarbons, **279,**
    279–281
Chlorine, 154
  in sea salts, 205–206
Chlorofluorocarbons, and ozone layer,
    249
Chromosomes, 295
Chronosequences, 189
Cicada, periodical, **293**
  and predator satiation, 528, **529**
Circadian rhythms, 132–135
  adaptive value of, 136–137
  biological clocks, 135–137
Circannual clocks, 140
Clam worm, and salinity, 883–884
Clark's nutcracker, 577
Classification
  floristic, 630
  systems of, 629–631
Clear-cutting, 673, **674,** 686
  biomass accumulation after, **655**
Cleistogamous flower, 438
Clements, F. E., 4–5, 9, 13,
    641–642

Climate
  changes in, and species dispersal,
    694–696
  of city compared to country, 68–72,
    72
  global, 48–55
  Mediterranean-type, 142–145
  microclimates, 64–72
  near ground, 64
  regional, 55–63
  and reproduction, 422–423
  and solar radiation, 48–49
Climatic stability theory, of diversity,
    608
Climax, 645–648
  in Clements' succession theory, 641
  urban settlement as, 662
Climax community, 635
Climax pattern theory, 646
Climbers, 803–804
Climograph, 56
  temperature-moisture, **59**
Clumped distribution, 328–330
Coarse-grained distribution, 330, **331,**
    497–498
Coarse particulate organic matter
    (CPOM), 852
Coastal renosterveld, 762
Cobalt, 154–155
Coefficient of attenuation, 196
Coefficient of extinction, 196
Coevolution, 572–573. *See also*
    Evolution; Predation
  of bees and orchids, 581
  diffuse, 573, 583
Coexistence, 464–465
Cohen, J. E., 241–242, 243
Cohort life table, 338–339
Cold desert, 770
Cold stress, in plants, 102–104
Colonization, 677, 678–679
Coloration
  cryptic, 526–527, 529–530
  warning, 525–526
Commensalism, 451
Community
  biological conditions of, 602–611
  defined, 590
  edges and ecotones, 596–602
  horizontal heterogeneity of, 595–596
  human, 661–663
  physical structure of, 591–602
  structure of
    and competition, 620–627
    and predation, 622–625
  vertical stratification of, 593–595
Community ordination, 630–631
Community pattern, 628–629
Compensation depth, 197–198
Compensation intensity, 126, 212
Competition
  allelopathy, 465–466
  classic theory of, 452–457
  in Clements' theory of succession,
    641
  and coexistence, 464–465
  contest, 376
  defined, 451
  exploitative, 375–376, 452
  field studies of, 458–464
  interaction types, 455–457
  interference, 376, 452

Gonochoristic animal, 436
Gopher, pocket, 821
Gouger, 855
Gram calorie (g-cal), 208
Grass, types of, 742–743
Grasslands, 741–756
  annual, 745
  belowground consumption, 753–754
  cultivation of, and nitrogen cycle, 263–264
  desert, 744–745
  development of, 699
  energy flow through, 230
  function, 749–756
  and grazing, 512, 751–754
  influence of topography and drainage of, **90**
  and light, 127–129
  mean standing crop, **750**
  and niche boundaries, 475–477
  primary production and energy flow, **228**
  production efficiency, 237–238, **238**
  productivity of, 214, 218
  profile of, **746**
  Rothamstead Experimental Farm (England), 156
  strata of, 594
  structure, 745–749
  successional, 749
  tropical, 745
  types of, 743–745
Grass shrimp, 894–895
Grazing, 622–624
  food chain, 225–227, 229–231
    and nuclear fallout, 283–284
  and grasslands, 751–754
  stimulation by, 511–512
Great tit, **391**
Greenhouse effect, **257**, 257–258
Gross primary production, 211
Gross production, 220
Ground beetles, 230
Ground fire, 669–670
Ground moss, 787–788
Ground water, 76
Group selection, 303–307
Groups, as defense, 528
Grouse, red
  and parasites, 626
  territoriality in, 413
Guild(s), 590
Gular fluttering, 112
Gyres, 53

Habitat(s)
  fragmentation of, 615–617
  hosts as, 550
  minimum size of, 618–620
  selection of, 446–448
  sink, 401, 402
Hadal zone, of ocean, 909
Halocline, 206
Halophytes, 85, 159
Haploid number, 295
Hardseededness, 680
Hardy–Weinberg equilibrium, 298–300, 308, 310
Hare
  Mountain, 523
  snowshoe, 18–19, **388**, 389
    ten-year cycle of, 522–523

-lynx cycle, 532–534
population density of, 495, **496**
Harvest
  of populations, 540–542
  rate of, 537–538, **538**
Harvest effort approach, 543
Heat budget, and water cycle, 81
Heat energy, 208
Heat increment, 219
Heat islands, 71
Heat stress, in plants, 101–102
Heath hen, 372–373
Heathlands, 764–766
Heavy metals, cycling of, 277–278
Heliothemism, 107, 108
Hemicryptophytes, 591, **592**
Hemiparasites, 553
Herbivore, 223–224. *See also* Food chain; Food web; Secondary production
  belowground, 230–231
  countermeasures to plant defenses, 519–520
  of desert, 773–774
  fitness of, 512
  and plant community, 622–624
  satiation of, 516–517
  of savanna, 759
  of tundra, 818, 827
Herbivory, 483
Herbs, 761
Hermaphroditic organism, 436, 437–438
Herring gull, 292–293
Heterotherms, 113–119
Heterotroph, 30. *See also* Food chain; Food web; Secondary production
Heterotrophic community, 591, 595
Heterozygosity, and inbreeding, **308,** 309
Heterozygous alleles, 295, 298
Hibernation, 116–119
Histogram, 293
Holarctic region, 715
Holdridge life zone system, 720–723
Holoparasites, 552–553
Home range, 416–417
Homoiohydric plants, 83
Homoiotherms, 105. *See also* Endotherms
  and body size, 110–111
Homozygosity, and inbreeding, **309,** 310
Homozygous allele, 295, 298
Horizons, of soil, 173–174
Horned lark, **332**
Host
  defenses against parasites, 558–560
  as habitat, 550
Hot desert, 770, **771**
Hourglass model, of biological clock, 136, **136**
Houseflies, resistance to DDT in, 300
Houston, D. B., 354
Hubbard Brook forest (New Hampshire), 794–799
Human(s)
  and animal habitat, 749
  and edge structure, **601**
  exploitation by, 536–546
  food chains, and radioactivity, 284

and grazing, 752
growth curve of, 369, **370**
impact on ecosystems, 671–673
intrusions into carbon cycle, 255–257
intrusions into nitrogen cycle, 263–265
intrusions into sulfur cycle, 267–269
and phosphorus cycle, 276–277
of Pleistocene Epoch, 711
sex ratios in, 357–360
succession in, 661–663
Human immunodeficiency virus (HIV), 559
Humboldt, F. H. A. von, 3–4, 7
Humidity, 81–83
  in temperate broadleaf forest, 793
Hummingbird, and plaintain, 582–583
Humus, 42, 251
  formation of, 180
  types of, 181–182
Hunting
  adaptations for, 530
  methods of, 528–529
  permit system, 543
Hybrid vigor, 311
Hybridization, 566–567
Hydrosphere, 26, **27**
Hydrothermal vents, 915, **916**
Hydroscopic water, 79
Hyperthermia, 91, 112
Hypertrophic lake, 847
Hypervolume, 474
Hyphomycetes, 735
Hypolimnion, 194
Hypothesis, 16
Hypothetico-deductive method, 19

Immigration, 317–318, 331–333
  and extinction, 371–372
Impala, feeding habits of, 514–518
Imprinting, 446
Inbreeding, 307–311
  coefficient of, 308–310
  consequences of, 310–311
Inbreeding depression, 310
Incipient lethal temperature, 106
Inclusive fitness, 306, 307
Increase, rate of, 537–538
Independent variable, 17
Indigo bunting, as edge species, **598,** 598
Indirect gradient analysis, 630–631
Indirect mutualism, 583
Individual(s)
  defining, 322–324
  energy flow through, 232–233, **233**
Individual distance, 315–316
Individual selection, 304–306
Induced edge, 597, **599**
Inductive method, 18
Infiltration, 78
Inflection point, of growth curve, 368
Infralittoral fringe, 901
Inherent edge, 596, **599**
Inhibitors, as defense, 518
Initial floristic composition, 642
Inland renosterveld, 762
Input, 28
  abiotic and biotic, 30
Insects
  as agents of disturbance, 670–671

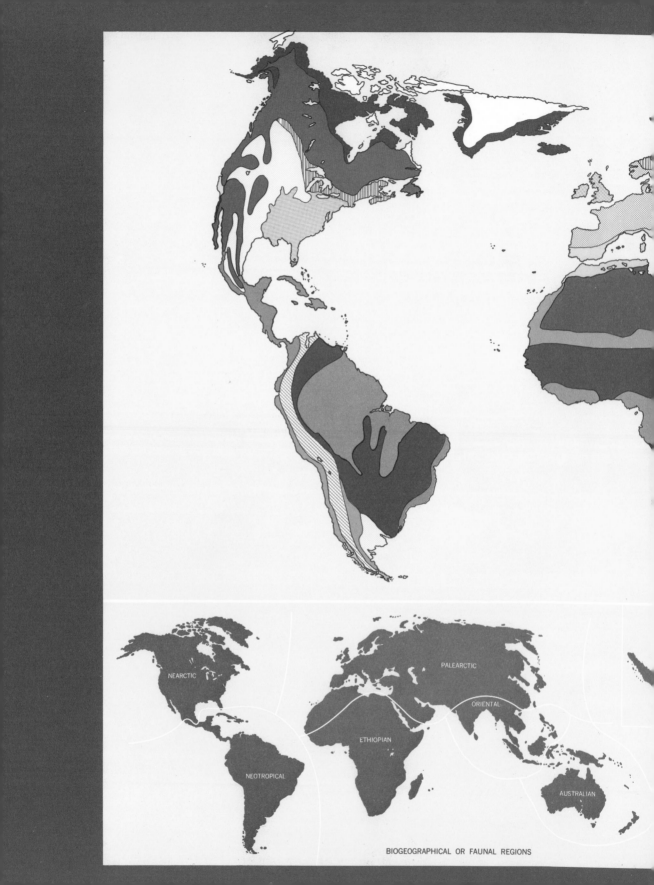

BIOGEOGRAPHICAL OR FAUNAL REGIONS

NEARCTIC

PALEARCTIC

ORIENTAL

ETHIOPIAN

NEOTROPICAL

AUSTRALIAN